EXERCISE PHYSIOLOGY
Human Bioenergetics and Its Applications

THIRD EDITION

George A. Brooks
University of California, Berkeley

Thomas D. Fahey
California State University, Chico

Timothy P. White
Oregon State University

Kenneth M. Baldwin
University of California, Irvine

Boston   Burr Ridge, IL   Dubuque, IA   Madison, WI   New York
San Francisco   St. Louis   Bangkok   Bogotá   Caracas   Kuala Lumpur
Lisbon   London   Madrid   Mexico City   Milan   Montreal   New Delhi
Santiago   Seoul   Singapore   Sydney   Taipei   Toronto

McGraw-Hill Higher Education

A Division of The **McGraw-Hill** Companies

EXERCISE PHYSIOLOGY, HUMAN BIOENERGETICS AND ITS APPLICATIONS, THIRD EDITION

Published by McGraw-Hill, a business unit of The McGraw-Hill Companies, Inc. 1221 Avenue of the Americas, New York, NY, 10020. Copyright © 2000, 1996 McGraw-Hill Companies, Inc. All rights reserved. No part of this publication may be reproduced or distributed in any form or by any means, or stored in a database or retrieval system, without the prior written consent of The McGraw-Hill Companies, Inc., including, but not limited to, in any network or other electronic storage or transmission, or broadcast for distance learning.

Some ancillaries, including electronic and print components, may not be available to customers outside the United States.

This book is printed on acid-free paper.

2 3 4 5 6 7 8 9 0 DOC/DOC 0 9 8 7 6 5 4 3 2 1

ISBN 0-7674-1024-6

Portions of this work were originally published under the titles, *Exercise Physiology: Human Bioenergetics and Its Applications,* copyright © 1984, and *Fundamentals of Human Performance,* copyright © 1987, by Macmillan Publishing Company.

Library of Congress Cataloging-in-Publication Data
Exercise physiology : human bioenergetics and its applications /
   George A. Brooks . . . [et al.] — 3rd ed.
       p.   cm.
   Second ed. gives Brooks as main entry.
   Includes bibliographical references and index.
   ISBN 0-7674-1024-6
   1. Exercise—Physiological aspects. 2. Energy metabolism.
3. Bioenergetics.   I. Brooks, George A. (George Austin).
QP301.B885   1999
612'.044—DC21   99-16015
                                                 CIP

Sponsoring editor, Michele Sordi; *production editor,* Carla White Kirschenbaum; *copyeditor,* Patricia Ohlenroth; *text designer,* Richard Kharibian; *cover designer,* Glenda King; *art manager,* Amy Folden and Jean Mailander; *illustrators,* Academy ArtWorks, Inc., Accurate Art, Inc., Joan Carol, Dartmouth Publishing, Inc., Judith Ogus; *photo researcher,* Brian Pecko; *manufacturing manager,* Randy Hurst. The text was set in 9/12 Palatino by G&S Typesetters, Inc. and printed on 45# Highland Plus by R. R. Donnelley & Sons Company.

.mhhe.com

To Ethan, Bruno, Phil, Lars, Roger, Ed, Bob, Mike and John for their inspiration, works and spirit.

CONTENTS

Chapter 10
Metabolic Response to Exercise: Lactate Metabolism During Exercise and Recovery, Excess Post-exercise O$_2$ Consumption (EPOC), O$_2$ Deficit, O$_2$ Debt, and the "Anaerobic Threshold"   197

In several major ways the first edition of *Exercise Physiology: Human Bioenergetics and Its Applications* was a departure from existing texts in the field. The text was unique in terms of focusing on human bioenergetics and attempting to describe muscle performance in terms of energy transduction at cellular levels. Our approach came out of the then (early 1980s) burgeoning field of exercise biochemistry. In the third edition of *Exercise Physiology* this theme has been retained, but the approach has become increasingly mechanistic due to many developments, including the use of molecular and cellular biology and isotope tracer technology in the field.

Such is the pace of current discovery that for the third edition it was necessary to add Dr. Kenneth M. Baldwin as a coauthor. Our thematic emphasis, our original scientific contributions using both the tools and techniques of contemporary biology and classical performance physiology, as well as our overall familiarity with the field, have allowed us to respond to the challenge of providing an up-to-date text for students of the field. The first two editions of *Exercise Physiology* found widespread use in institutions of higher learning in the United States and Canada. We hope that the current edition will also merit extended use, not only in North America, but elsewhere as well.

The ever-increasing pace of progress in biological sciences can be traced to the development and application of tools and strategies of cell and molecular biology. Many scientists and science administrators have written about these changes in the biological sciences. In his article titled "Research in sports medicine and exercise sciences in the 21st century,"
which appeared in the American College of Sports Medicine bulletin (vol. 34, #3, p. 9), Ken Baldwin wrote about 'the times that are a-changing.' To quote Ken, "within the next five to eight years the complete genomes for the mouse, rat, and human will be deciphered. These genetic maps will serve as the blueprints for evolving the science that will unfold; and with the new molecular tools coming aboard it will be possible to decipher clusters of genes involved in various disorders such as obesity, the metabolic syndrome, as well as ascertaining why our muscles waste away as one ages." In preparing the third edition of *Exercise Physiology* we have utilized the information available and anticipated "the exciting research initiatives ready to explode in the twenty-first century." Specifically, with regard to the rapidly developing influences of cell and molecular biology techniques and experimental approaches on exercise physiology, the reader is referred to sections on muscle membrane lactate transporters (Chapter 5) and influences of exercise overload on myosin expression (Chapter 17).

In writing or revising a textbook there are many challenges. Beyond dealing with the volume of ever-changing science, a far greater challenge lies in appropriately interpreting the results of classical as well as contemporary studies. At its essence, scientific discovery involves disputation and thus, in some areas, our judgments may be controversial. Notwithstanding, we believe we are scientifically justified in presenting concepts in particular areas, although we recognize that some scientists and educators will disagree with our emphases and interpretations. The student must realize, however, that in this respect our textbook is no different from any other. To minimize influences of our biases in composing the text, we have consciously included many original data sets. The intent in providing many figures and tables from the literature is to provide the impetus for students to recognize that the viability of the conclusions we reach is limited by the technologies available. Further, we have included original materials so that readers will be able to discuss the limitations in data and its interpretation. Anecdotally, with regard to the area of energy substrate utilization during exercise (Chapters 5–10), it is a great relief to know now that many of the conclusions reached on the basis of experiments with rodents in the 1970s and 1980s were correct for humans. Nonetheless, species differences are a real concern, and thus some of our interpretations (for example, the importance of lipid oxidation for human muscular exercise, Chapter 7) have needed to be modified. Fortunately, a burgeoning of recent literature employing stable, nonradioactive isotope technology has allowed the conclusions in that section of the book to be based on results of experiments on men and women.

As with the first edition, we remain extremely enthusiastic about our field. We believe that no greater interest can be generated for the study of physiology than that involved in analyzing human performance during motor activities, particularly when the student's object of study is himself or herself. Because so many of us are engaged in lifestyles that require high-energy outputs, it is important that we understand our physiological capacities for exercise, which largely determine our success and enjoyment in many areas. Studying exercise physiology not only allows students to understand the mechanisms governing their own performance but also enables them to be aware of performance in athletics, work, the performing arts, recreation, and preventive and rehabilitative medicine. Thus this book links the student of physiology to the reader's own self-interest.

In terms of its potential to enhance and promote the general health and well-being of the population, we are equally enthusiastic about potential applications of information from the field of exercise physiology. In our contemporary society, degenerative diseases have replaced infectious diseases as major causes of debilitation and death—most notably coronary heart disease (CHD) and Type II diabetes (NIDDM). The causes of CHD and NIDDM are complex, but lack of physical exercise is certainly involved. Recognition of the relationship between degenerative diseases and physical inactivity on the bases of epidemiological and intervention studies led to the Surgeon General's Report on Physical Activity and Health. Therefore, as described in Chapter 1 and throughout the text, not only is proper physical exercise essential for somatic development in the formative years, but it is also necessary for maintenance of physical capacity in adults. Today, exercise is used to diagnose coronary heart disease, to retard its development, and to treat it. Exercise conditioning and other forms of physical therapy are used to assist and improve recuperation from injury to, and surgery on, muscles and joints. It is our hope that the Surgeon General's advocacy is not too late—for the incidence of obesity, CHD, NIDDM, and related pathologies is ever-increasing, reaching epidemic proportions in contemporary Western societies.

Our considerable experience in the field has, we hope, engendered us with wisdom to interpret events shaping the field of exercise physiology and to place and portray those events within the broader discipline of biological science. Therefore, in the introductory chapter we have expanded the section on the history of exercise physiology, its events, and its personalities. In addition, we have written about the emergence of the American College of Sports Medicine and its coalition with other scholarly, clinical, and public health organizations that produced the Surgeon General's report. Also of note is the emergence of the European College of Sports Science and similar organizations around the world.

Many people have helped write this book, including our teachers, W. D. McArdle, J. A. Faulkner, R. E. Beyer, K. J. Hittelman, F. M. Henry, G. L. Rarick, C. M. Tipton, and J. H. Wilmore. We have been inspired by many contemporary and past researchers, whose work is referenced in our chapters. We also thank the authors of other texts, which we as students have used, including A. C. Guyton, A. L. Lehninger, P. O. Åstrand, V. R. Edgerton, E. Fox, M. Kleiber, A. J. Vander, J. S. Sherman, and D. S. Luciano. We also thank reviewers of our manuscript, including Sally E. Blank, Washington State University; Ronald E. De Meersman, Teacher's College, Columbia University; Dennis Dolny, University of Idaho; Casey M. Donovan, University of Southern California; Victor F. Froelicher, VA Medical Center, Stanford University; Mathew Hickey, Colorado State University; Ethan R. Nadel, John B. Pierce Laboratory, Yale University; Tim Rickabaugh, Central Michigan University; Mayra C. Santiago, Temple University; and Darlene A. Sedlock, Purdue University. So far as our research careers are concerned, those careers in science have served to lay the bases upon which we wrote this text. Therefore, we thank our many colleagues, students, and postdoctoral fellows in research with whom we have enjoyed working and who have helped our careers to prosper. They have supported us, inspired us, cheered us, and filled our lives with challenge and satisfaction. The ideas and efforts of all of these individuals are contained in our text.

As authors we would appreciate hearing your reviews and opinions of our text. We encourage you to write us with your criticisms and suggestions, which will be seriously considered for incorporation into subsequent printings and editions of this book.

CHAPTER 1

INTRODUCTION:
The Limits of Human Performance

What are the limits of physical performance? Can the discus be hurled 300 feet (Figure 1-1)? Is it possible to run the 100-m dash in 9 seconds (Figure 1-2)? Can the mile be run in less than 3 minutes (Figure 1-3)? Is it possible to swim 100 meters in less than 40 seconds (Figure 1-4)? Can exercise training be used to retard the advance of coronary heart disease and other illnesses, increasing vitality and extending the life span, and can the paralyzed limbs of paraplegics be exercised to prevent their further deterioration? Are these goals unrealistic, or do they underestimate the limits of human performance? These are important questions, of interest not only to physiologists, biochemists, molecular biologists, physicians, nurses, physical therapists, human factors engineers, physical anthropologists, and zoologists but also to physical educators and athletics coaches.

Based on the principles of physiology as applied to training, the answers to the questions just

Figure 1-1   Al Oerter has won four Olympic gold medals in the discus. Throwing the discus over 200 feet requires enormous power, coordination, and technique. It also requires use of the immediate (phosphagen) energy system. PHOTO: © Wayne Glusker.

Figure 1-2 Jesse Owens in action during the 1936 Olympics in Berlin. Owens was dubbed the "world's fastest human." Such tremendous physical performance requires use of the immediate (phosphagen) and non-oxidative (glycolytic) energy systems. PHOTO: AP/Wide World Photos.

Figure 1-3 Sir Roger Bannister was the first person to run the mile under 4 minutes. Athletic feats such as this possibly represent the supreme example of speed, power, and endurance in human endeavor. The immediate, non-oxidative and oxidative energy systems are all involved in competitive mile racing. PHOTO: © UPI/Corbis.

posed are somewhat predictable. A shot-putter, for example, who is able to propel a shot a distance of 80 feet has to generate the necessary force to propel the implement that distance. An athlete with insufficient muscle mass, less than optimal leverage, and inadequate metabolic capability will simply be incapable of this feat. Similarly, individuals with poor cardiovascular capacity, muscular strength, metabolism and coordination will lead physically restricted lives. The study of exercise physiology can lead to a better understanding of the physical capabilities and limitations of the human body, as well as of its underlying mechanisms.

■ The Scientific Basis of Exercise Physiology: Some Definitions

The scientific method involves the systematic solution of problems. The scientific approach to solving a problem involves the development and presentation of ideas, or hypotheses; the collection of information (data) relevant to those hypotheses; and the acceptance or rejection of the hypotheses based on evaluation of the data (conclusions). Although the

Figure 1-4   Competitors in the Olympic 50-meter free-style take to the water. In internationally recognized swim competition, there is no pure sprint event; athletes rely on muscular power and endurance, aerobic endurance, great coordination, and highly developed technique. PHOTO: © Reuters/Corbis.

Physiology is a branch of biological science concerned with the function of organisms and their parts. The study of physiology depends on and is intertwined with other disciplines, such as anatomy, biochemistry, molecular biology, and biophysics. This interdependence is based on the fact that the human body follows the natural laws of structure and function, which fall within the domain of these other disciplines.

Exercise physiology is a branch of physiology that deals with the functioning of the body during exercise. As we shall see, definite physiological responses to exercise depend on the intensity, duration, and frequency of the exercise and the environmental circumstances, diet, and physiological status of the individual.

■ Physiological Science in Sports Medicine, the Allied Health Professions, Physical Education, and Athletics

One might ask, Are physical educators, coaches, and other clinicians and practitioners scientists? The answer depends on their approach to problem solving. Teachers and coaches who systematically evaluate their selection and training of individuals can be considered scientists. The scientist-coach introduces an exercise stimulus and systematically evaluates the response. The non–scientist-coach, in contrast, administers the training problem according to whim—such as mimicking the techniques of successful athletes—or by conforming tenaciously to traditional practices. Although non–scientist-coaches are sometimes successful, they are rarely innovative and seldom help an individual perform optimally. Moreover, their accomplishment is episodic, and they rarely sustain success in training athletes. For progress to occur in any field, systematic innovation is absolutely essential.

One might also ask, Are physicians, physical therapists, nurses, and other health care professionals coaches? Again, the answer depends on their approach to improving the condition of their patients.

scientific method appears to be straightforward, the process of deriving appropriate hypotheses and systematically testing them can be complex. It is nevertheless evident that, in our increasingly technological society, those who systematically analyze their problems and take appropriate steps to solve them are most likely to acquire satisfactory answers to their questions. Individuals who make the best use of the scientific method are the most successful scientists, educators, coaches, and health professionals.

Are health care practitioners aware of the preventive and regenerative effects of exercise? Are they aware of principles of exercise training such as overload and specificity? Do they seek to motivate their patients personally and through design of a well-structured exercise and dietary program? Today, many successful physicians and other health care professionals evaluate their patients as successful coaches evaluate their athletes. Thus, in recent years there has been a tremendous convergence in the science and practices of preventive medicine and coaching. This convergence has centered around the science of exercise physiology.

Exercise and sport, particularly individual sports that demand extremes of strength, coordination, speed, or power, lend themselves to scientific analysis because the measures of success are easily quantified. If a weight-lifting coach attempts to improve a particular athlete's performance with a specific training technique, the results can be evaluated objectively: The athlete either improves or does not improve. Likewise, the distance a discus is thrown can be measured, the duration of a 100-m run can be timed, the number of baskets sunk in a game can be counted, and the distance an amputee can walk can be measured. The scientist-coach and health science professional observes and quantifies the factors affecting human performance and systematically varies them to achieve success. Admittedly, it is easier to predict performance in individual sports than in team sports or activities in which group interactions and extremes in environment operate to influence the outcome. It is not beyond the scope of our imagination, however, that, with continued development of exercise physiology and other branches of exercise science, it may someday be as possible to predict outcomes in team sports as in individual endeavors.

■ The Relevance of Physiology for the Allied Health Professions, Physical Education, and Athletics

Understanding the functioning of the body during exercise is a primary responsibility of each exercise physiologist, physical educator, coach, and health science professional. In sports, athletes and coaches who want to maximize performance require a knowledge of physiological processes. As competition in sports becomes more intense, continued improvement will be attained only by careful consideration of the most efficient means of attaining biological adaptation.

Exercise physiologists, physical educators, and health science professionals increasingly work with more and more diverse populations—new populations interested in assuming an active lifestyle and even in participating in competitive sports. Older adults flock to masters' sporting events; heart attack patients resume physical activity earlier than ever, geriatric populations weight train to maintain muscle mass and strength, and people with diseases such as asthma and diabetes use exercise to reduce the effects of their disabilities. These people all need guidance by trained professionals who understand responses to exercise in a variety of circumstances. Knowledge of exercise responses in these diverse populations requires a thorough understanding of both normal and abnormal physiology.

Exercise is also important to those with degenerative diseases such as coronary heart disease and osteoarthritis, which have replaced infectious diseases as primary health problems. Many of these degenerative disorders are amenable to change through modification of lifestyle, such as participation in regular exercise.

The importance of exercise in a program of preventive and rehabilitative medicine has reinforced the role of the physical educator as part of the interdisciplinary team concerned with health care and maintenance. The physical educator, exercise physiologist, physical therapist, team physician, and other health care professionals must speak the "language of the science" to become true professionals and interact with the other professionals on the health care team.

■ The Body as a Machine

In many ways the exercising human can be compared to a machine such as an automobile. The machine converts one form of energy into another in performing work; likewise, the human converts chemical energy to mechanical energy in the process of running, throwing, and jumping. Like a machine, a human can increase exercise intensity by increasing the rate at which energy is converted from one form to another. An athlete, for example, goes faster by increasing metabolic rate and speeding the breakdown of fuels, which provides more energy for muscular work.

At their roots, motor activities are based on principles of bioenergetics, which control and limit the performance of physical activities. In this sense, the body is a machine. When exercise starts, the mechanisms of performance are determined by physical and chemical factors. Understanding how to select and prepare the biological apparatus for exercise and how the exercise affects the machine over both the short and long term is important in exercise physiology and other fields.

This book emphasizes understanding the individual during exercise from the standpoint of the energetic systems that support the various activities. The discussion begins with energy and its importance to living organisms, emphasizing how we acquire, conserve, store, and release energy for everyday life. The functions of various physiological systems (ventilatory, circulatory, endocrine, and so on) are examined from the perspective of their function in supporting physical performance and their place in the process of energy conversion. Discussion of immediate as well as the long-term effects of exercise is integrated throughout the text. We begin with some background information on the general principles of physiological response and the field of exercise physiology.

■ The Rate-Limiting Factor

In a complex biological machine such as an exercising human, many physiological processes occur simultaneously. For example, when a person runs a mile, the heart's contractility and beating frequency increase, hormones are secreted, the metabolic rate increases, and body temperature is elevated. Despite the vast number of events occurring simultaneously, usually only a few control and limit the overall performance of the activity. Many scientists approach the understanding of physiological systems by studying the rate-limiting processes. Imagine an assembly line that manufactures a commodity such as an automobile. Although there are many steps in the manufacturing process, assume that one step—installing the engine—is the slowest. If we want to increase production, it will do us little good to increase the speed of the other steps, such as assembling the chassis. Rather, we should focus our attention on speeding up the process of installing the engine. We might hire extra people to do the task, or we might use more machinery, or we might remove some impediment to the process so workers can perform more rapidly. As we shall see, the body is controlled by and adjusts to exercise in a similar fashion.

In athletics, successful coaches are those who can identify the rate-limiting factor, sometimes called a weakness, and improve the individual's capacity to perform that process. Let us assume, for instance, we are coaching a novice wrestler who has been a successful competitive weight lifter. It makes no sense to emphasize strength training. Rather, we should emphasize technique development and other aspects of fitness, such as endurance. We would strive to maintain strength while concentrating on the performance-limiting factors. Similarly, we would be ill advised to have a 400-m runner do 100 miles of road running a week, because this type of fitness is of minimal use to this athlete and may even interfere with the enzymatic apparatus that facilitates the high rates of power output used in 400-m runs.

■ Maximal Oxygen Consumption ($\dot{V}_{O_2max}$) and Physical Fitness

The ability to supply energy for activities lasting more than 30 seconds depends on the consumption and use of oxygen (O_2). Because most physical activities in daily life, in athletics, and in physical medicine take more than 90 seconds, consumption of O_2 provides the energetic basis of our existence. The rate of consumption of a given volume of O_2 (abbreviated $\dot{V}_{O_2}$) increases as activities progress from rest to easy, to difficult, and finally to maxi-

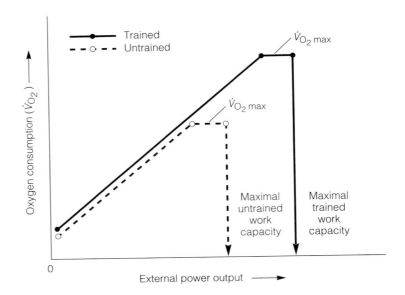

Figure 1-5   Relationship between oxygen consumption ($\dot{V}_{O_2}$) and external work rate (power output). In response to increments in power output, both trained and untrained individuals respond with an increase in $\dot{V}_{O_2}$. The greater ability of trained individuals to sustain a high power output is largely due to a greater maximal O_2 consumption ($\dot{V}_{O_2max}$).

mal work loads (Figure 1-5). The maximum rate at which an individual can consume oxygen ($\dot{V}_{O_2max}$) is an important determinant of the peak power output and the maximal sustained power output, or physical work capacity, of which an individual is capable. Moreover, an adequate good capacity to consume and utilize oxygen is essential for recovery from sprint (or burst) activity, as recovery is mainly an aerobic process.

As will be shown in later chapters, the capacity for $\dot{V}_{O_2max}$ depends on the capacity of the cardiovascular system. This realization that physical work capacity, $\dot{V}_{O_2max}$, and cardiovascular fitness are interrelated has resulted in a convergence of physical education (athletic performance) and medical (clinical) definitions of fitness. From the physical education–athletics perspective, cardiovascular function determines $\dot{V}_{O_2max}$, which in turn determines physical work capacity, or fitness (Figure 1-6). From the medicoclinical perspective, fitness involves, minimally, freedom from disease. Because cardiovascular disease is the greatest threat to the health of individuals in contemporary Western society, medical fitness is largely cardiovascular fitness. One of the major ways of determining cardiovascular fitness is measuring $\dot{V}_{O_2max}$. Therefore, $\dot{V}_{O_2}$ is not only an

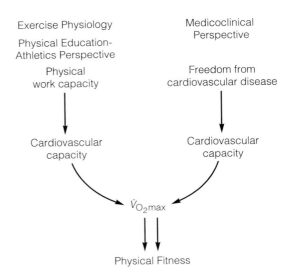

Figure 1-6   Because maximal oxygen consumption ($\dot{V}_{O_2max}$) depends on a high degree of cardiovascular health and because it is very important in aerobic (endurance) exercise, $\dot{V}_{O_2max}$ is becoming recognized as the most important index of physical fitness.

important parameter of metabolism but also a good measure of fitness for life in contemporary society. In fact, $\dot{V}_{O_2max}$ is so important from both the physical education–athletics and medicoclinical perspectives, it has emerged as the single most important criterion of physical fitness.

■ Factors Affecting the Performance of the Biological Machine

Although the body can be compared to a machine, it would be simplistic, and indeed dehumanizing, to leave it at that. Unlike a machine, the body can adapt to physical stresses and improve its function. Conversely, in the absence of appropriate stress, functional capacity deteriorates. In addition, whereas the functions of particular types of machines are set at the time of manufacture, the performances of human machines are quite variable. Performance capabilities change continuously throughout life according to several time-honored principles that account for many observed individual differences. These principles are examined next.

Stress and Response

Physiological systems respond to appropriate stimuli. Sometimes the stimulus is called "stress," and the response is called "strain." Repeated stresses on physical systems frequently lead to adaptations, resulting in an increase in functional capacity. Enlargement, or *hypertrophy*, in skeletal muscle occurs as a result of the stress of weight training. However, not all stresses are appropriate to enhance the functioning of physiological systems. For instance, although cigarette smoking is a stress, it does not improve lung function. Smoking is an example of an inappropriate stimulus.

Physiologically, the purpose of any training session is to stress the body so that adaptation results. Physical training is beneficial only as long as it forces the body to adapt to the stress of physical effort. If the stress is not sufficient to overload the body, then no adaptation occurs. If a stress is so great that it cannot be tolerated, then injury or overtraining result. The

greatest improvements in performance occur when appropriate exercise stresses are introduced into an individual's training program.

Dr. Hans Selye has made scientists aware of the phenomenon of the stress-response-adaptation process, which he called the *general adaptation syndrome (GAS).* Selye described three stages involved in response to a stressor: alarm reaction, resistance development, and exhaustion. Each of these stages should be familiar to every physical educator, athlete, coach, physician, nurse, physical therapist, and other health professionals who use exercise to improve physical capacity.

Alarm Reaction The alarm reaction, the initial response to the stressor, involves the mobilization of systems and processes within the organism. During exercise, for example, the stress of running is supported by the strain of increasing oxygen transport through an augmentation of cardiac output and a redistribution of blood flow to active muscle. The body has a limited capacity to adjust to various stressors; thus, it must adapt its capacity so that the stressor is less of a threat to its homeostasis in the future.

Resistance Development The body improves its capacity or builds its reserves during the resistance stage of GAS. This stage represents the goal of physical conditioning. Unfortunately, the attainment of optimal physiological resistance (or, *physical fitness,* the term more commonly used in athletics) does not occur in response to every random stressor. During physical training, for example, no training effect occurs if the stress is below a critical threshold. At the other extreme, if the stimulus cannot be tolerated, injury results.

The effectiveness of a stressor in creating an adaptive response is specific to an individual and relative to any given point in time and place. For example, running a 10-min mile may be exhausting to a sedentary 40-year-old man but would cause essentially no adaptive response in a world-class runner. Likewise, a training run that is easily tolerated one day may be completely inappropriate following a prolonged illness. Environment can also introduce intra-individual variability in performance.

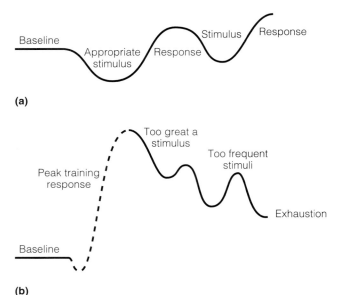

Figure 1-7 Illustration of the biological principle of stimulus and response, or the general adaptation syndrome, as applied to physical training. (a) Appropriate stimuli degrade the system slightly, but result in adaptive responses during recovery. Properly gauged and timed, training stimuli result in progressive improvements in physiological systems. (b) Application of too great a training stress and too frequent training can result in exhaustion of a physiological system.

An athlete will typically experience decreased performance in extreme heat and cold, at high altitudes, and in polluted air.

Exhaustion When stress becomes intolerable, the organism enters the third stage of GAS, exhaustion (or distress). The stress that results in exhaustion can be either acute or chronic. Examples of acute exhaustion include fractures, sprains, and strains. Chronic exhaustion (overtraining) is more subtle and includes stress fractures, emotional problems, and a variety of soft-tissue injuries. Again, the resistance development stage of GAS may require considerable time. It may therefore be inadvisable to elicit the alarm reaction frequently through severe training because the exhaustion stage of GAS may result. Periodization of training, involving training to a peak and then training at a reduced load before attempting a new peak, is discussed in Chapters 20 and 21 (Figure 1-7).

■ The Overload Principle

Application of an appropriate stressor is sometimes referred to as *overloading* the system. The principle of overload states that habitually overloading a system causes it to respond and adapt. Overload is a positive stressor that can be quantified according to load (intensity and duration), repetition, rest, and frequency.

Load refers to the intensity of the exercise stressor. In strength training, load refers to the amount of resistance; in running or swimming, it refers to speed. In general, the greater the load, the greater the fatigue and recovery time required.

Repetition refers to the number of times a load is administered. More favorable adaptation tends to occur (up to a point) when the load is administered more than once. In general, there is little agreement on the ideal number of repetitions in a given sport. The empirical maxims of sports training are in a constant state of flux as athletes become successful using overload combinations different from the norm. In middle-distance running and swimming, for example, interval training workouts have become extremely demanding as a result of the success of athletes who employed repetitions far in excess of those practiced only a few years ago.

Rest refers to the time interval between repetitions. Rest is vitally important for obtaining an adaptation and should be applied according to the nature of the desired physiological outcome. For example, a weight lifter who desires maximal strength is more concerned with load and less concerned with rest interval. However, too short a rest interval will impair the weight lifter's strength gain because inadequate recovery makes it impossible for him or her to exert maximal tension. Mountain climbers, by contrast, are more concerned with muscular endurance than peak strength, so they would use short rest intervals to maximize this fitness characteristic. Rest also refers to the interval between training sessions. Because some responses to stress are prolonged, adaptation to stress requires adequate rest and recovery. *Resting is a necessary part of training because adaptations occur during recovery.*

Frequency refers to the number of training sessions per week. In some sports, such as distance running, the tendency has been toward more frequent training sessions. Unfortunately, this often leads to increases in overuse injuries due to overtraining. Although more severe training regimens have resulted in improved performance in many sports, these workouts must be tempered with proper recovery periods or injury may result.

Specificity

It has repeatedly been observed that stressing a particular system or body part does little to affect other systems or body parts. For example, doing repeated biceps curls with the right arm may cause the right biceps to hypertrophy, but the right triceps or left biceps will be little affected. Any training program should reflect the desired adaptation. The closer the training routine is to the requirements of competition, the better the outcome will be.

Reversibility

In a way, the concept of reversibility is a restatement of the principle of overload and emphasizes that, whereas training may enhance performance, inactivity will lead to a performance decrement. For example, someone who built a robust circulatory system in college as a runner should expect little or no residual capacity at 40 years of age following 20 years of inactivity.

Individuality

We are all individuals and, whereas physiological responses to particular stimuli are generally predictable, one individual's precise response and adaptation to those stimuli are largely unpredictable and will vary from those of others. The same training regimen therefore may not benefit equally everyone who participates.

■ Development of the Field of Exercise Physiology

Many researchers have contributed the groundwork to the study of exercise physiology. Today, knowledge in this field is obtained from a wide range of studies. The testing of physical work capacities of athletes, laborers, and a variety of people to determine their metabolic responses to exercise remains an important area of interest. Questions relating to the caloric cost of exercise, the efficiency of exercise, and the fuels used to support the exercise are addressed. In addition to traditional methods of respiratory and blood metabolite determinations, newer techniques, including muscle sampling (biopsy), light and electron microscopy, enzymology, molecular biology, endocrinology, nuclear magnetic resonance (NMR), and radioactive and nonradioactive tracers have been developed that contribute to our understanding of the metabolic responses to exercise.

Along these lines, research has been carried out and is still under way to determine optimal training techniques for particular activities. The training regimens to improve such qualities as muscular strength and running endurance are vastly different.

When exercise tests are performed in a clinical setting, the term *stress test* is often used. Under controlled exercise conditions, cardiac and blood constituent responses are useful in determining the presence of underlying disease. Following an exercise stress test, an exercise prescription can be written to improve functional capacity.

Recent advances at Wright State University in Dayton, Ohio, in computerized control of paralyzed muscles now allow paraplegic and quadriplegic individuals to use their own muscle power to provide locomotion (See Figure 1-4.) As with non-paralyzed muscles, repeated overload of paralyzed muscles—in this case by computer-controlled electrical stimulation—results in significant improvements in strength and endurance. Consequently, some muscles paralyzed for years can be trained and restored to near-normal strength. Together with appropriate orthopedic devices, these strengthened muscles allow paralyzed patients a degree of independence.

The effects of environment on physical performance have long been a concern in exercise physiology. Science and medicine, for example, have long been associated with mountaineering. Environmental studies reached their peak of interest, however, during preparations for the high-altitude Olympics held in Mexico in 1968, and new information was gained during studies on Pikes Peak and in a simulated ascent of Mt. Everest in the U.S. Army Research Institute of Environmental Medicine Decompression chamber. Similarly, the Los Angeles Olympics stimulated research into the effects of heat and air pollution on human performance. At present, the physiological stress of work in the heat and in polluted environments remains a particularly active area of research.

■ Pioneers and Leaders in Exercise Physiology

Exercise physiology has at times been in the forefront of the advances made in basic science. The greatest respiratory physiologist of all time was the eighteenth-century Frenchman Antoine L. Lavoisier, who used exercise to study physiology. Lavoisier contributed more to the understanding of metabolism and respiration than anyone will ever again have the opportunity to do. During the late nineteenth century in Germany, great strides were made in studying metabolism and nutrition under conditions of rest and exercise. Nathan Zuntz and his associates (including Schumburg and Geppert) were particularly important. Tables developed by Zuntz and Schumburg (1901) relating metabolic rate to O_2 consumption, CO_2 production, and amount of carbohydrate and fat used are essentially the same as those frequently referred to today by respiratory physiologists, exercise physiologists, and nutritionists. Work in exercise (Arbeits) physiology was also carried out in Germany, centered at the Max-Planck Institut für Arbeits physiologie. The Nazi regime and the events of World War II resulted in the emigration of many of these scientists to the United States, among them Ernst Simonsen (University of Minnesota) and Bruno Balke (University of Wisconsin).

In the early twentieth century, F. G. Benedict and his associates, including E. P. Cathcart and Henry M. Smith at the Carnegie Nutrition Laboratory in Boston, performed detailed studies on metabolism on people at rest and during steady-rate exercise. The precision, thoroughness, and insightful interpretation of results by the Carnegie Nutrition group is seldom matched today. The works of Benedict and Cathcart (1913) on the efficiency of the body during cycling exercise and similar work by Smith (1922) on the efficiency of walking should be required reading for all graduate students specializing in exercise physiology.

The giant in the field of exercise physiology, however, is English physiologist and Nobel prize laureate Archibald Vivian (A. V.) Hill. Hill's tremendous understanding of physiology was coupled with a likewise tremendous stamina for work and the technical ability to develop experimental devices. Throughout his career, Hill performed detailed studies on the energetics of muscles isolated from small animals. Moreover, he sought to relate the results of those detailed studies to the functioning, intact human Hill and his associates (see Hill, Long, and Lupton, 1924) performed many studies on athletes and other individuals engaged in heavy exercise. In 1926, as a visiting lecturer at Cornell University, Hill studied acceleration in varsity sprinters. His influence was so great that he inspired a group of American scientists who later went on to help found the Harvard Fatigue Laboratory.

As recounted by David Bruce (D. B.) Dill (1967), the Harvard Fatigue Laboratory, established in the late 1920s at Harvard University to study exercise and environmental physiology, became a center for the study of applied physiology in the United States, attracting scientists from around the world. Included among visitors to the laboratory were Nobel laureate August Krogh of the University of Copenhagen (see below). In the field of exercise physiology, the laboratory is perhaps best known for the attempt by Margaria, Edwards, and Dill in the 1930s to understand metabolic responses to "non–steady-rate" exercise. As noted, earlier work in Germany and the United States had laid the foundation for understanding metabolic responses to continuous exercise of moderate intensity. The problems of understanding non–steady-rate metabolism, however, were—and remain—far more difficult. In tackling these problems, the Harvard group was carrying on in the tradition of A. V. Hill. The enormity of their task is revealed by the fact that these problems have still not been resolved (see Chapter 10).

When the laboratory was dissolved following World War II, its members dispersed, carrying with them their work and their commitment. Their dispersal led, perhaps, to a wider and more vigorous proliferation of work in exercise and environmental physiology than would have been possible had the laboratory remained the main site of research. Of particular note has been the work in the United States throughout the 1960s to 1980s, of Sid Robinson at the University of Illinois and Steve Horvath at the University of California, Santa Barbara, and, in Italy, of Rudolfo Margaria at the University of Milan.

During the early twentieth century in Copenhagen, August Krogh established a laboratory for the study of zoological physiology. Although Krogh's range of physiological interests was wide, he and his associates (J. L. Lindhard, Erik Hohwü-Christensen, Erling Asmussen, and Marius Neilsen) became known as exercise physiologists. When Krogh retired, Asmussen and Nielsen continued the work in Copenhagen; Hohwü-Christensen, in 1941, became professor at the Gymnastik-och Idrottshogkolan (GIH) in Stockholm. In 1960, Hohwü-Christensen was succeeded at the GIH by Per-Olaf (P.-O.) Åstrand.

The period from the late 1960s through the 1970s was an important time for exercise physiology. During this time, scientists began increasingly to apply the tools and techniques of biochemistry to the study of exercise physiology. In the United States, Phil Gollnick and John Holloszy, in large measure, invented the field of "exercise biochemistry." They and their associates developed the use of animal models to study basic metabolic and biochemical responses to exercise and exercise training. At about the same time in Sweden, Jonas Bergström and Eric Hultman first used the biopsy needle for studies of exercise physiology. This technical advance allowed lessons learned from animal experimentation to be extended to human subjects.

Also in the United States during this period, scientists at several Big-10 midwestern universities were playing essential roles in the development of the field. At the University of Wisconsin, Bruno Balke, whose contributions were many, was bringing to exercise physiology a European tradition, in which the disciplines and professions of medicine, exercise and environmental physiology, and physical education were all closely allied. At the University of Iowa, Charles M. Tipton's broad range of physiological interests were inspiring a generation of exercise physiologists, including Ken Baldwin, Jim Barnard, Frank Booth, and Ron Terjung. Today, Baldwin (a coauthor of this work) and Booth are among the leaders in applying the techniques of molecular biology to the study of the effects of exercise and other stresses on cardiac and skeletal muscle structure and function. At the University of Michigan, John A. Faulkner's work in exercise and environmental physiology was instrumental in gaining acceptance for these areas as legitimate parts of basic physiology. Today, John Faulkner remains a driving force in muscle physiology and the physiology of aging. Faulkner's former students and postdoctoral fellows—and their students and fellows—have had a wide impact on the field. Included are George Brooks and Tim White, two co-authors of this text.

Meanwhile, discoveries were being made by exercise physiologists on the effects of activity on nerves and muscles. At UCLA, R. J. (Jim) Barnard, who had studied with Tipton at the University of Iowa, and V. R. (Reggie) Edgerton began a series of investigations that have shaped our views of how muscle function is controlled during acute and chronic exercise. The basic groundwork for their investigations was laid by neurophysiologists such as Elwood Henneman.

In the area of pressure flow and other cardiovascular responses to exercise, progress was also being made. Significant contributions were made by many scientists, especially Jere Mitchell (Dallas), Peter Raven (Fort Worth), Alf Holmgren (Stockholm), Larry Rowell (Seattle), Peter Wagner (San Diego), and John T. (Jack) Reeves (Colorado). At the University of Wisconsin, Madison, and the University of California, Davis, Jerry Dempsey and Marc Kaufman are making major strides in understanding the control of breathing during exercise. Today in Italy, the tradition of Rudolfo Margaria in studying exercise energetics and basic environmental physiology is being carried on by Peter di Prampero (Udine) and Paolo Cerretelli (Milan and Geneva).

In Scandinavia today, the rich tradition of studies in exercise physiology is continuing. Krogh's laboratory in Copenhagen is now called the August Krogh Institute, and Erik Richter is now its director. Bengt Saltin, who now directs the Copenhagen Muscle Research Center, is best known for integrating diverse results from different aspects of physiology into an understanding of the human engaged in exercise. Also in Copenhagen, distinguished scientists such as Henrik Galbo, Bente Kiens, Michael Kjaer, and others are conducting important research in the field of fuel energy utilization in exercise. Together, they form the Copenhagen Muscle Research Center.

In Canada, which may have the strongest research program per capita, interest in exercise physiology is flourishing. Whether or not they consider themselves exercise physiologists, Canadian researchers continue to make major contributions to the field. From Quebec (with Claude Bouchard, Francois Peronnet, Jean-Aimé Simoneau, and Angelo Tremblay) to British Columbia (with Peter Hochachka), there exist remarkable research and educational institutions. The province of Ontario, in particular, is noted for its numerous exercise physiologists, including Roy Shephard and Mladen Vranic (University of Toronto), Enzo Cafarelli and Norman Gledhill (York), Arend Bonen and Howard Green (Waterloo), and Norman Jones and Duncan MacDougall (McMaster).

Australia, too, is experiencing a resurgence of interest in exercise physiology. After years on the faculty at McMaster University in Ontario, John Sutton returned to the University of Sydney and instituted a center for sports medicine and basic and applied exercise physiology. Today, Maria Firatone Singh holds the John Sutton Chair of Exercise and Sports Science at the University of Sydney. In the United Kingdom, Clyde Williams directs the very active and distinguished program at Loughborough. In Scotland, several scientists are noted for their accomplishments; among them are Ron Maughan (Aberdeen), Mike Rennie (Dundee), and Neil Spurway (Glasgow). In Cape Town, South Africa, Tim Noakes developed the Sports Sciences Institute, a research center that is making major contributions to both clinical sports medicine and the study of human energetics during exercise. In France, there is also a resurgence of effort; notable among French scientists are Jacques Mercier of Montpellier, who is known both as a scientist and a clinician.

Interest in exercise physiology is evident in Asia as well. Though long unknown in the Western world because of language differences, programs in the study of exercise physiology are thriving in Korea, as revealed by the Seoul Olympics. There, Sung Tae Chung (Seoul National University), Chang Kyu Kim (Kookmin University), and Sung Soo Kim (Korea University), are among leaders in the field. A similar tradition of effort also exists in Japan, where, for example, Mitsumasa Miyashita heads the sports medicine and exercise physiology group at the University of Tokyo. Hiroshi Nose is Professor of Sports Medicine at the Shinshu University School of Medicine, and Sadayoshi Taguchi is Professor of Human and Environmental Sciences at Kyoto University. In China, serious efforts are directed toward the application of exercise physiology to improving athletic performance, and serious efforts at more basic research are developing.

Interest in exercise physiology is truly international, and the interest is growing in scope and sophistication, as is interest in the Olympics.

■ The Ever-Changing Fields of Exercise Physiology and Exercise and Sports Science in the United States and Elsewhere

In the previous section, a historical perspective and report on status of the fields of exercise physiology and exercise and sports science were articulated; now, we attempt to address the future of these fields. As a biological science, exercise physiology is constantly changing as new tools and technologies arise, and as overall sophistication and standards in the field develop. For example, the application of stable isotope technology has resulted in major revisions to Chapters 5–7 in this third edition of *Exercise Physiology*. Similarly, tools of molecular biology applied to the study of skeletal and cardiac muscle have resulted in significant changes in Chapters 17–19. Consequently, the chapters on training, particularly Chapter 21, have also changed in major ways. Reflective of the changes in science are corresponding changes in the organization and administration of higher education. The march of science as well as the dynamics of change in institutions of higher education parallel change in the fields of exercise physiology and exercise and sports science. For students in the field, some of the current trends are noted here.

Analysis of the previous section reveals a widespread interest in exercise physiology that has neither institutional nor national boundaries. Researchers who do sophisticated work and make major contributions to exercise physiology are scattered throughout the biological sciences at many different colleges, universities, and medical schools. An example of two highly regarded scientists in the field

are Reggie Edgerton and Jim Barnard, who hold faculty positions at the Department of Physiological Sciences at UCLA. However, departments of physiology or biochemistry rarely house more than a few faculty doing work related to exercise physiology. Consequently, most such departments at major research universities, so-called research 1 (R1) institutions, do not advertise exercise and sports science as a focus of the department.

There are probably several reasons for the minimal emphasis placed on exercise physiology and sports science at R1 institutions. One reason is that those institutions have diverse missions that do not include a health-and-fitness-related purpose. Another reason is that R1 institutions gauge their standings against each other from the National Research Council (NRC) ranking of departments. Because the NRC does not rank exercise science, kinesiology, physical education or any similarly titled department, the presence of departments of exercise science at institutions such as Harvard, Stanford, and Berkeley has waned; those types of institutions are unwilling to allocate resources to departments that have no hope of attaining an NRC ranking. And even if distinguished R1 universities do assemble a critical mass of exercise physiologists under one roof, as at the Pierce Laboratory at Yale or the Noll Laboratory at Penn State, the laboratory members often hold faculty positions in diverse departments of biological science or medicine.

Fortunately, many R1 and other institutions do see the need to support departments and programs of exercise and sports science. Therefore, leaders at several institutions have increased the prominence of programs in the field of exercise physiology and exercise and sports science. Because there is no definitive study or accepted set of objective criteria to rank departments of exercise science, no such effort will be undertaken here. However, the following provides some information on trends.

At the time of this writing, the University of Colorado at Boulder is perhaps preeminent in the field of exercise physiology due to the leadership of Russ Moore, who heads a department that includes Bob Mazzeo, Roger Enoka, Doug Seals, and Bill Byrnes. In terms of the research productivity, the faculty at Boulder remains consistently at the top. It has also developed an exceptional curriculum, and the department is large and oversubscribed with students.

A rising star in the field is the Department of Kinesiology at Kansas State University. There, a very active and productive group including Tim Musch, David Poole, Tom Barstow, Craig Harms, and Richard McAllister has been assembled. The University of Texas at Austin also needs to be mentioned as a site of excellence simply because of the accomplishments of John Ivy and Ed Coyle. Priscilla Clarkson, president of the American College of sports medicine (ACSM) for the 2000–2001 year, Patty Freedson and Jane Kent-Braun make the University of Massachusetts at Amherst another site of excellence in the field, and one that is noted for the leadership of its women. Faculty in the department of Exercise Science and Physical Education at Arizona State University, where Wayne Willis conducts elegant studies in the field of exercise biochemistry, will rightfully argue that their department is preeminent due to the breadth of their faculty expertise, which includes the renowned sports psychologist Dan Landers.

Throughout the United States, the letters "USC" identify institutions of excellence in fields related to exercise physiology and kinesiology. At the University of Southern California, Casey Donovan and Lorraine Turcotte have made fundamental contributions to our understanding of the regulation of glucose and fatty acid metabolism. At the University of South Carolina, Russ Pate has not only advanced the field of exercise physiology, but he has also served as president of the American College of Sports Medicine. Pate's colleagues include Barbara Ainsworth, Mark Davis, and Larry Durstine. Their department is unique in that it is housed in the School of Public Health, a position that reflects the important relationship between physical activity and health. Some of their many contributions are contained in the following section on the Surgeon General's Report on Physical Activity and Health.

In conclusion, we reiterate the constancy of change in the fields of exercise physiology and exercise and sports science. Vision and leadership in these fields are important attributes, which com-

bined with other factors will result in future progress. Already, the programs at the University of Colorado at Boulder and Kansas State University have been mentioned as distinguished sites of exercise physiology research. In the coming years, these programs will be rivaled by departments and programs at Oregon State University, where Tim White, one of the coauthors of this text and a former president of the ACSM, is dean of the College of Health and Human Performance. White will build a program to rival any on the West Coast. Similarly, at Texas A & M, Bob Armstrong and Jack Wilmore are fashioning a faculty that may rival that at the University of Texas at Austin. Ultimately, there is every expectation that students beginning in the field today will emulate those mentioned here who have led the field to increasing levels of eminence.

■ Physical Activity and Health: A Report of the Surgeon General of the United States

The interim between publication of the second and third editions of *Exercise Physiology* was highlighted by publication of the Surgeon General's report titled "Physical Activity and Health" (http://www.cdc.gov/nccdphp/sgr/summary.htm). This report is a landmark in the history of muscle and exercise physiology and disciplines related to sports medicine, not only because it summarizes results of research from diverse fields, but also because the document serves as an articulation of public policy. The main message of this report is that Americans can substantially improve their health and quality of life by including moderate amounts of physical activity in their daily lives.

In 1994, the Office of the Surgeon General of the United States authorized the Centers for Disease Control and Prevention (CDC) to serve as the lead agency in preparing the first Surgeon General's report on physical activity and health. The CDC was joined in the effort by the President's Council on Physical Fitness and Sports (PCPFS), the Office of Public Health and Science, the Office of Disease Prevention at the National Institutes of Health (NIH), and several institutes from the NIH, including the National Heart, Lung, and Blood Institute; the National Institute of Child Health and Human Development; the National Institute of Diabetes and Digestive and Kidney Diseases; and the National Institute of Arthritis and Musculoskeletal and Skin Diseases. In addition, the CDC's efforts were buttressed by several nonfederal scholarly and professional organizations, including the American Alliance for Health, Physical Education, Recreation, and Dance (AAHPERD); the American College of Sports Medicine (ACSM); and the American Heart Association (AHA). Representatives of those organizations provided consultation throughout the development process.

The report is noteworthy in several respects. As previously stated, it recognized that physical activity is essential for the health and well-being of the general population, and it emphasized the importance of regular, moderate-intensity exercise as well as vigorous activity to achieve and maintain cardiorespiratory fitness. The report encourages people of all ages to include a minimum of 30 minutes of physical activity of moderate intensity (such as brisk walking) on most days. Further, the report was definite in recommending physical activity as a means to manage chronic diseases other than cardiovascular disease, such as diabetes, colon cancer, osteoarthritis, and osteoporosis. The Surgeon General's report noted also that in addition to promoting muscle strength and minimizing injury due to falls in the aged, regular exercise may be important in relieving symptoms of depression and anxiety, thereby improving mood and promoting a sense of well-being.

In addition, the report examined the literature and noted that life in the United States and other industrialized countries is becoming increasingly sedentary. Daily life contains less and less rigorous physical activity, a trend reflected in reduced rigor and frequency of activities offered in school physical education programs. Obesity is also on the rise in diverse segments of our population, particularly in the aged and among minorities and women.

Concluding the Surgeon General's report were the specific recommendations to increase physical activity in homes and in communities, at worksites, and in health-care settings and to establish intervention programs targeting physical education in elementary schools, thereby substantially increasing the amount of time students spend being physically active.

In short, like the Surgeon General's report on the health risks of cigarette smoking, the most recent report on physical activity represents a major departure from previous practices. It is testament to the growing recognition that physical activity is important in promoting physical and mental health throughout the life cycle. It is likely that the first Surgeon General's report will be succeeded by other reports on physical activity and health as the mortality and morbidity of degenerative diseases increase and associated health-care costs escalate.

■ Summary

Although the study of exercise physiology is in its infancy compared with other sciences such as chemistry and physics, it is bustling with activity. This area of research has a great deal of appeal because it concerns the limits of human potential. The results of these studies affect us all, whether we jog three times a week to improve our health, are concerned with the health of others, or are concerned with the training of world-class performers.

■ Selected Readings

Asmussen, E. Muscle metabolism during exercise in man: a historical survey. In Muscle Metabolism During Exercise, B. Pernow and B. Saltin (Eds.). New York: Plenum Press, 1971, pp. 1–12.

Åstrand, P.-O. Influence of Scandinavian scientists in exercise physiology. *Scand. J. Med. Sci. Sports* 1: 3–9, 1991.

Baldwin, K., G. Klinkerfuss, R. Terjung, P. A. Molé, and J. O. Holloszy. Respiratory capacity of white, red, and intermediate muscle: adaptive response to exercise. *Am. J. Physiol.* 22: 373–378, 1972.

Barnard, R., V. R. Edgerton, T. Furukawa, and J. B. Peter. Histochemical, biochemical and contractile properties of red, white, and intermediate fibers. *Am. J. Physiol.* 220: 41–414, 1971.

Barnard, R., V. R. Edgerton, and J. B. Peter. Effect of exercise on skeletal muscle I. Biochemical and histochemical properties I. *J. Appl. Physiol.* 28: 762–766, 1970.

Benedict, F. G., and E. P. Cathcart. Muscular Work. A Metabolic Study with Special Reference to the Efficiency of the Human Body as a Machine. (Publ. 187). Washington, D.C.: Carnegie Institute of Washington, 1913.

Brooks, G. A. (Ed.). Perspectives on the Academic Discipline of Physical Education. Champaign, Ill.: Human Kinetics, 1981.

Brooks, G. A., G. E. Butterfield, R. R. Wolfe, B. M. Groves, R. S. Mazzeo, J. R. Sutton, E. E. Wolfel, and J. T. Reeves. Increased dependence on blood glucose after acclimatization to 4,300 m. *J. Appl. Physiol.* 70: 919–927, 1991.

Brooks, G. A., G. E. Butterfield, R. R. Wolfe, B. M. Groves, R. S. Mazzeo, J. R. Sutton, E. E. Wolfel, and J. T. Reeves. Decreased reliance on lactate during exercise after acclimatization to 4,300 m. *J. Appl. Physiol.* 71: 333–341, 1991.

Brooks, G. A., E. E. Wolfel, B. M. Groves, P. R. Bender, G. E. Butterfield, A. Cymerman, R. S. Mazzeo, J. R. Sutton, R. R. Wolfe, and J. T. Reeves. Muscle accounts for glucose disposal but not lactate release during exercise after acclimatization to 4,300 m. *J. Appl. Physiol.* 72(6): 2435–2445, 1992.

Dill, D. B. The Harvard Fatigue Laboratory: its development, contributions and demise. *Circ. Resh.* S1: 161–170, 1967.

Fahey, T. Getting into Olympic Form. New York: Butterick, 1980.

Fenn, W. O. History of the American Physiological Society: The Third Quarter Century 1937–1962. Washington, D.C.: The American Physiological Society, 1963.

Glaser, R. M., J. S. Petrofsky, J. A. Gruner, and B. A. Green. Isometric strength and endurance of electrically stimulated leg muscles of quadriplegics. *Physiologist* 25: 253, 1982.

Grimaux, E. Lavoisier: 1743–1794. Ancienne Librairie Germer Bailliero et Cie, Paris, 1888.

Gunga, H.-C., and K. Kirsch. The life and work of Nathan Zuntz (1847–1920). In Hypoxia and Mountain Medicine, J. R. Sutton, G. Coates, and C. S. Houston (Eds.). Burlington, Vt.: Queen City Printers, 1992, pp. 279–291.

Henneman, E., and C. B. Olson. Relations between structure and function in the design of skeletal muscles. *J. Neuro. Physiol.* 28: 581–598, 1956.

Henry, F. M. Physical education: an academic discipline. *J. Health Phys. Ed. Recreation* 35: 32–33, 1964.

Hill, A. V., C. N. H. Long, and H. Lupton. Muscular exercise, lactic acid, and the supply and utilization of oxygen (Pt. I–III). Proc. Roy. Soc. (London) Ser. B. 96: 438–475, 1924.

Hill, A. V., C. N. H. Long, and H. Lupton. Muscular exercise, lactic acid, and the supply and utilization of oxygen (Pt. IV–VI). Proc. Roy. Soc. (London) Ser. B. 97: 84–138, 1924.

Hill, A. V., C. N. H. Long, and H. Lupton. Muscular exercise, lactic acid, and the supply and utilization of oxygen (Pt. VI–VIII). Proc. Roy. Soc. (London) Ser. B. 97: 155–176, 1924.

Lehninger, A. L. Biochemistry. New York: Worth, 1970.

Margaria, R., H. T. Edwards, and D. B. Dill. Possible mechanism of contracting and paying oxygen debt and the role of lactic acid in muscular contraction. *Am. J. Physiol.* 106: 689–715, 1933.

Mazzeo, R. S., G. A. Brooks, G. E. Butterfield, A. Cymerman, A. C. Roberts, M. Selland, E. E. Wolfel, and J. T. Reeves. b-Adrenergic blockade does not prevent lactate response to exercise after acclimatization to high altitude. *J. Appl. Physiol.* 76: 610–615, 1994.

Reeves, J. T., R. F. Grover, S. G. Blout, Jr., and F. G. Filley. The cardiac output response to standing and walking. *J. Appl. Physiol.* 6: 283–287, 1961.

Reeves, J. T., E. E. Wolfel, H. J. Green, R. S. Mazzeo, J. Young, J. R. Sutton, and G. A. Brooks. Oxygen transport during exercise at high altitude and the lactate paradox: lessons from Operation Everest II and Pikes Peak. Exercise and Sport Sciences Reviews, vol. 20. Baltimore: Williams and Wilkins, 1992, pp. 275–296.

Seyle, H. The Stress of Life. New York: McGraw-Hill, 1976.

Wolfel, E. E., P. R. Bender, G. A. Brooks, G. E. Butterfield, B. M. Groves, R. S. Mazzeo, J. R. Sutton, and J. T. Reeves. Oxygen transport during steady state, submaximal exercise in chronic hypoxia. *J. Appl. Physiol.* 70: 1129–1136, 1991.

Simonsen, E. Physiology and Work Capacity and Fatigue. Springfield, Ill.: C. C. Thomas, 1971.

Smith, H. M. Gaseous Exchange and Physiological Requirements for Level and Grade Walking. (Publ. 309). Washington, D.C.: Carnegie Institute of Washington, 1922.

Strange, S., N. H. Secher, J. A. Pawelczyk, J. Karpakka, N. J. Christensen, J. H. Mitchell, and B. Saltin. Neural control of cardiovascular responses and of ventilation during dynamic exercise in man. *J. Physiol.* (London) 470: 693–704, 1993.

Zuntz, N., and D. Schumburg. Studien zu einer Physiologie des Marches. Verlag von August Hirschwald, Berlin, 1901.

CHAPTER 2

BIOENERGETICS

Human locomotion and other physical activities in daily life and athletics are energetic events. Understanding what energy is and how the body acquires, converts, stores, and utilizes it is the key to understanding how the body performs in sports, recreational, and occupational activities. The science that studies the principles limiting energetic events is known by two names: *thermodynamics* and *energetics.* With some limitations, the same principles that govern energetic events in the physical world—for example, the explosion of dynamite—also govern events in the biological world—for example, a sprinter's first step out of the starting blocks. The science that involves studies of energetic events in the biological world is called *bioenergetics.* In describing energy in the body there are two things to keep in mind. First, energy is not created; rather, it is acquired in one form and converted to another. Second, the conversion process is fairly inefficient, and much of the energy released is in a nonusable form: heat. In this chapter, we discuss energy and begin to describe three energy systems that power and sustain muscular activities. They can function under different conditions, at different speeds, and for different durations. Together, these three energy systems determine our capacities for power events (e.g., shot putting), speed events (e.g., 100-m sprint running), and endurance events (e.g., marathon running).

Bob Bearman soars to a new world and Olympic record during the 1976 games in Mexico City. Long jumping is one of the classic events in human bioenergetics. SOURCE: © Corbis/Bettman.

To beginning students of biology, the study of energetics seems abstract, especially when they learn that thermodynamics does not tell them much about the specific steps in a process or the time it takes to complete a biological process. However, if one understands that the ability to do work, or exercise, depends on the conversion of one form of energy to another, then the importance of studying energetics becomes apparent.

Terminology

Before starting the formal presentation of bioenergetics, we need to define some important terms.

Energy   The capacity to do work.

Work   The product of a given force acting through a given distance.

Power   The rate of work production.

System   An organized, functional unit. The boundaries of systems vary from situation to situation and depend on the process under consideration. Systems can vary from sublight microscopic cellular organelles such as mitochondria, to muscles, to whole individuals.

Surroundings   Exchanges of energy and matter frequently occur between systems and their environments. Once we have defined a system, all else comprises the surroundings.

Universe   Together, the system and its surroundings make up the universe (Figure 2-1).

The definition of *work* is a Newtonian or mechanical definition. In actuality, cells more commonly perform chemical and electrical work than mechanical work. However, it is possible to exchange and convert energy from one form to another. The physical science dealing with energy exchange is called *thermodynamics.*

The study of thermodynamics began in the nineteenth century, with the desire to predict the work output of machines such as steam engines. It was so called because heat was the most common form of energy used. However, as there are six primary forms of energy (thermal, chemical, mechanical, electrical, radiant, and atomic), a more appropriate term—and one that is used more often today—is *energetics.* Again, the branch of science that deals with energy exchanges in living things is called *bioenergetics.* Although a few limitations

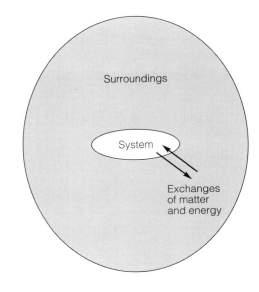

Universe

Figure 2-1   The universe consists of the system(s) plus the surroundings. At any one time, a reaction taking place in the system results in a decrease in the energy (free energy) available to do the work of the system. Consequently, over the long term, an input of energy from the surroundings is required to maintain the system. The energy content of the universe is always constant.

and properties involving heat and temperature are unique to biological systems, these systems and other types of machines follow the same general principles of energetics.

Heat, Temperature, and the Biological Apparatus

Steam and internal combustion engines are examples of machines that convert chemical energy (coal and gasoline) to heat. That heat energy is then converted to mechanical energy. Generally, the higher the heat, the more the power produced. Biological engines differ from mechanical engines with respect to their ability to use heat. Biological engines *cannot* convert heat energy to other forms, such as mechanical energy. In biological systems, heat is released as an essential but useless component of reactions in which other forms of work are accomplished. This is a fundamental difference between biological and mechanical engines.

Another characteristic of biological systems is their temperature dependence. Two factors merit consideration here. First, biological systems are sensitive to small increments in temperature. Above 45°C, and certainly above 60°C, tissue proteins become denatured, or degraded. Consequently, although a muscle might theoretically contract faster at 50°C than at 35°C, increasing the temperature that much would literally cook the muscle.

Second, the rates of biological or enzymatic reactions are also sensitive to temperature. Usually, the effect of temperature on reaction rates is studied by changing the temperature in multiples of 10°C. The result of such a change is called a "Q_{10} effect." Increasing the temperature 10°C, for example, doubles the rate of an enzymatic reaction, and the Q_{10} is 2. Figure 2-2 illustrates the Q_{10} effect. In that figure, note the large effect of temperature changes within the normal range of muscle temperature.

Changes in temperature can be advantageous as well as deleterious to individuals. For instance, even though a lizard cannot capture heat to do cellular work, it can sit in the sun to warm itself, and in this way it takes advantage of the Q_{10} effect. Some rodents can adapt biochemically when they are exposed to extreme cold; these adaptations allow them to consume extra oxygen and generate heat without shivering, which would endanger their survival. Also, many athletes have found that preliminary exercise ("warming up") makes for better performance. We shall see (in Chapter 21) that preliminary exercise has several beneficial effects, one of which is to warm up muscle, thus taking advantage of the local Q_{10} effect.

Excessively high as well as low temperatures can be harmful. Although warming up can increase the speed of particular enzymatic processes in muscles through the Q_{10} effect, temperatures of over 40°C have been observed to decrease the efficiency of oxygen use in muscle. This can negatively affect endurance. It has also been noted that cooling brain temperature by only a few degrees can affect think-

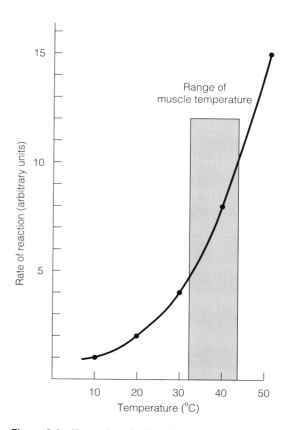

Figure 2-2   Illustration of a Q_{10} effect, where each 10°C increase in temperature doubles the rate of reaction. In the physiological range (shaded area) the curve is very steep.

ing and cause disorientation. This can be especially serious for campers, swimmers, or divers, whose survival in a hostile environment often depends on their behavior.

■ Laws of Thermodynamics

Repeated observations of events in the physical world reveal that two fundamental principles, or laws, always hold. The *first law of thermodynamics* states that energy can be neither created nor destroyed. Whenever there is an exchange of energy or matter between a system and its surroundings, the total energy content in the universe remains constant. The expression, "we get a burst of energy," is not quite correct. Rather, humans consume food (parcels of chemical energy) and degrade the foodstuffs, converting some of the chemical energy to heat and cell work, and releasing the remaining chemical energy unused in body excrements.

Equation 2-1, which describes weight loss or gain in healthy human bodies, is not a statement of the first law but does conform to the first law. According to this equation, the only way obese people can lose fat is to eat less and exercise more. If the energy input (food) is less than that expended as work and heat resulting from exercise, then the amount stored (as fat) will be reduced.

In addition to telling us that energy is not created or destroyed, but rather interconverted among forms, the first law tells us to be careful to account for all the energy.

The first law of thermodynamics implies that energy forms can be exchanged, but it does not tell us in what direction the exchanges will occur. The second law of thermodynamics tells us that processes always go in the direction of randomness, or disorder. A quantitative measure of this disorder is termed *entropy.* As the result of the second law, entropy always increases. In biology, the second law tells us that whenever energy is exchanged, the efficiency of the exchange will be imperfect and some energy will escape—usually in the form of heat—thus increasing entropy in the universe. Now consider a system that is increasing in organization, such as a child growing. Is this increasing organization of matter contrary to the second law? The answer is no. The process of growth requires a tremendous input of energy from the environment, and for a little growth—a little increase in order—a lot of heat (random energy) is released. In general, when one biological process moves some product toward a higher level of organization, it is driven by at least one "linked" or "coupled" entropic reaction.

There are many examples of concentrated units of energy being dispersed into random energy—the cooling of the sun, for example—and increasing entropy. What are the consequences of this increasing entropy for the biological and physical world? Ultimately, the universe will be complete randomness and disorder, which means that life is ultimately doomed. In the meantime, however, the energetic trend toward entropy allows us to capture energy in useful forms and perform biological reactions that require energy input. In this sense, then, for now, the trend toward entropy is necessary, and therefore, good.

■ Exergonic and Endergonic Reactions

An *exergonic* reaction is one that gives up energy. If heat is the form of energy given off, another term used is *exothermic. Spontaneous reaction* is another synonymous term. An *endergonic* reaction is one that absorbs energy from its surroundings. If heat is the energy form absorbed, the reaction is *endothermic.* In these terms, *erg* refers to work or energy, and *therm* refers to heat.

An example of a spontaneous reaction (A→B) is diagrammed in Figure 2-3. The energy content of the product B is less than that of the reactant A. The difference or change in energy is ΔE_1. In this example, ΔE is negative because the energy content of B is less than that of A.

Reaction (C→D) is an example of an endergonic, nonspontaneous, uphill reaction. Here, the energy level of the product is greater than that of the reactant. This reaction will not occur unless there is an energy input. Here, ΔE_2 is positive in sign because the energy content of D is greater than that of C. Many important reactions and processes in the physical and biological world are endergonic—that is, require energy inputs. In the biological world, endergonic reactions such as C→D are linked or coupled to, and driven by, exergonic reactions such as A→B. In the example given, the overall process is A→D. Note that the energy level of D, the final product, is less than that of A, the initial reactant.

Although chemical reactions and processes in biological systems are governed by laws of ener-

Energy in (food) = Energy out (work) + Energy out (heat) ± Energy stored (fat)   (2-1)

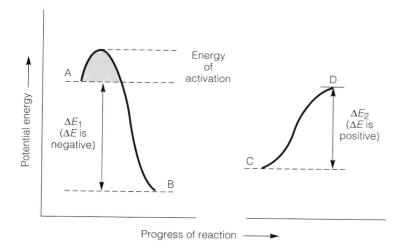

Figure 2-3 Examples of spontaneous (A→B) and nonspontaneous (C→D) reactions. Even though A→B is energetically downhill, the impediment imposed by the energy of activation may prohibit A→B unless energy is put in to overcome the energy of activation. Note that if C→D is driven by A→B, then the absolute value of ΔE_2 is less than ΔE_1.

getics, the linkages between exergonic (energy-yielding) and endergonic (energy-consuming) reactions is usually indirect. As we shall see in the following sections, part of the energy available from exergonic processes is captured in the form of a high-energy intermediate compound, adenosine triphosphate (ATP). In a complementary way, most endergonic reactions in mammalian systems are catalyzed by an enzyme that uses the energy of ATP.

Even though reaction A→B is downhill and "spontaneous," it is not likely to happen because an energy barrier, called the *energy of activation,* must first be overcome. In other words, even though the reaction A→B is classified as spontaneous, some energy has to be put in to activate the system and "prime the pump." As we shall see, several important biochemical pathways begin with such activating steps. We will also see that enzymes are important because they lower the energy of activation.

Although the energy of activation impedes some processes, the world as we know it depends on other processes having very high energies of activation. For example, oxides of nitrogen and other automobile exhaust emissions are currently very much in the news. Oxides of nitrogen are formed from nitrogen and oxygen according to the following reactions:

$$N_2 + 2\ O_2 \rightarrow 2\ NO_2 \qquad (2\text{-}2)$$
$$N + O_2 \rightarrow NO_2 \qquad (2\text{-}3)$$

These are spontaneous reactions with high energies of activation. In the automobile combustion chamber, extremes of temperature and pressure activate the reaction and produce noxious products. Without very high energies of activation, these reactions might well cause the atmosphere to catch fire.

In metabolic processes, initial reactants must first be activated. Because enzymes can lower the energy of activation, there is enzymatic control over these processes.

■ Enthalpy

In the example of a spontaneous reaction just cited (A→B in Figure 2-3), we saw that energy was released (ΔE_1) because the energy content of the products was less than that of the reactants. More specifically,

$$\Sigma(EA) = \Sigma(EB) + \Delta E_1 \qquad (2\text{-}4)$$

Here Σ means "sum of" or "content of." Accordingly, $\Sigma(EA)$ means the content of all the energy in A.

In natural events, however, it is difficult to determine the change in energy (ΔE) because some work is done in the atmosphere. *Enthalpy* (symbolized by H) takes into account this work done in the atmosphere by adjusting ΔE for changes in pressure and volume.

$$\Delta H = \Delta E + P\Delta V \qquad (2\text{-}5)$$

Fortunately, volume changes at constant pressures are extremely small, and for most biological reactions, $P\Delta V = 0$. Therefore,

$$\Delta E = \Delta H \qquad (2\text{-}6)$$

An example of this is the important reaction of carbohydrate oxidation:

$$C_6H_{12}O_6 + 6\ O_2 \rightarrow 6\ H_2O + 6\ CO_2 \qquad (2\text{-}7)$$

In the aqueous medium of a cell, the volume of O_2 consumed is equal to the volume of CO_2 formed, so the net volume change is zero.

Note: The trend of biologists to use the terms *energy* and *enthalpy* interchangeably is somewhat confusing. In addition, early theorists used the terms *enthalpy* and *entropy,* which sound alike but are very different. Be aware of this possible confusion in terminology and take care to avoid it.

■ Free Energy

Although the second law of thermodynamics states that the entropy of the universe always increases, and it is the drive toward the entropic condition that allows physical and chemical reactions to occur, the random energy of entropy is not directly useful in the performance of cell work. As implied by the first and second laws of thermodynamics, there exists another component to the total energy of a system. The term *free energy* is used because this energy is available, or free, to do work. In biological reactions, such as those involved in muscle contraction, free energy changes are of primary importance.

The following equations summarize what has been covered so far:

$$\begin{aligned}\text{Energy change} &= \text{Change in energy available} \\ \text{(enthalpy)} &\quad \text{to do work} + \\ &\quad \text{Change in unavailable energy}\end{aligned} \qquad (2\text{-}8)$$

For energy change let us substitute ΔH, and for unavailable energy let us substitute entropy. Therefore,

$$\begin{aligned}\Delta H &= \text{Change in available energy} \\ &\quad + \text{Change in entropy}\end{aligned} \qquad (2\text{-}9)$$

Recall that the energy available to do work was termed free energy. Therefore, we now have the following:

$$\begin{aligned}\Delta H &= \text{Change in free energy} \\ &\quad + \text{Change in entropy}\end{aligned} \qquad (2\text{-}10)$$

Formally, free energy is symbolized by G, after the scientist Willard Gibbs, and entropy is symbolized by S. With substitution, we then obtain the following:

$$\Delta H = \Delta G + T\Delta S \qquad (2\text{-}11)$$

where Δ means change and T is absolute temperature. The temperature term is necessary to get entropy in appropriate units. Because of the second law, and by convention, as any real process proceeds, the free energy carries a negative sign. Another way to remember that free energy (the useful component) carries a negative sign is to reason that a negative free energy term means that energy is given up by a reaction and is available to do work. Figure 2-4 summarizes the various components of chemical energy.

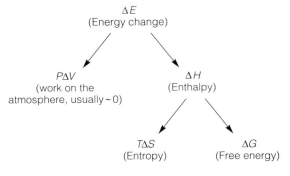

Figure 2-4 Enthalpy, free energy, and entropy. Only part of the energy change of a spontaneous reaction results in work. For a muscle, work may be only 25% of energy release.

■ **The Equilibrium Constant and Free Energy**

The second law of thermodynamics states that entropy is always increasing in the universe. Entropy was defined as a random, disordered condition. When a reaction has reached the end of the road and entropy is at a maximum, the reaction has reached equilibrium. At chemical equilibrium there is no longer any net change from reactant to product. At equilibrium, then, there is no further potential to do work. In other words, the further from equilibrium the reactants of a process are, the more potential use the reaction has.

The importance of the equilibrium constant in determining the free energy change for a reaction is illustrated in Figure 2-5, where A is the reactant, B is the product, and each of the reactions proceeds to equilibrium. In reaction (a), where very little or no product is formed, it is obvious that little has happened, so there is little opportunity for energy to be exchanged. At the other end, in reaction (c), almost all the reactant has become product. In this reaction, much has happened chemically, and there has been opportunity to capture some of the energy given off.

The equilibrium constant K_{eq} is used to denote the ratio of concentrations of products to reactants at equilibrium:

$$K_{eq} = \frac{[B]}{[A]} = \frac{[products]}{[reactants]} \qquad (2\text{-}12)$$

a. **A** $\rightleftharpoons$ B $K_{eq} = B/A$, a small fraction

b. A $\rightleftharpoons$ B

c. A $\rightleftharpoons$ **B** $K_{eq} = $ **B**/A, a large integer

Figure 2-5   Reactant and product levels as related to K_{eq}. At time zero, when the reaction begins, only the reactant A is present. Given enough time (a), A changes to B and reaches equilibrium with B. At equilibrium (b), the net amount of A→B equals the net amount of B→A. Useful reactions, therefore, are those in which a large fraction of A changes to B and the K_{eq} is large (c).

Here the brackets indicate "concentration of." The K_{eq} of a reaction is an immutable constant at specified conditions of temperature and pressure.

In examining K_{eq} and looking at the reactions in Figure 2-5, we can see that if the quotient is large, the reaction has potential use in driving a biological system.

Empirically, it has been determined that the change in free energy in a reaction is simply related to its equilibrium constant:

$$\Delta G^{\circ\prime} = -RT \ln K'_{eq} \qquad (2\text{-}13)$$

where R is the gas constant ($1.99 \text{ cal} \cdot \text{mol deg}^{-1}$), T is absolute temperature, and $\ln K'_{eq}$ is the natural logarithm of the equilibrium constant determined under standard laboratory conditions.

The symbol $\Delta G^{\circ\prime}$ thus refers to the standard free energy change determined in the laboratory when a reaction takes place at 25°C, at 1 atmosphere of pressure, and where the concentrations are maintained at 1 molal and in an aqueous medium at pH 7. In the notation the superscript ′ refers to pH 7; the ° refers to the other standard conditions. Although these "standard conditions" seem to be far removed from those in the body, categorizing reactions by this system allows us to compare free energy potentials of various reactions.

The relationship between K'_{eq} and $\Delta G^{\circ\prime}$ is further illustrated in Table 2-1.

TABLE 2-1		

Relationship Between Equilibrium Constant K'_{eq} and Free Energy Change $\Delta G^{\circ\prime}$ Determined Under Standard Conditions (25°C, or 298 K, and pH 7)

K'_{eq}	$\Delta G^{\circ\prime}$ (kcal · mol^{-1})	
0.001	+4.09	Endergonic reactions
0.010	+2.73	
0.100	+1.36	
1.000	0.00	
10.000	−1.36	Exergonic reactions
100.000	−2.73	
1000.000	−4.09	

The Actual Free Energy Change

In order to consider a more realistic evaluation of free energy change, let us consider the reaction

$$rR + sS \rightarrow pP + qQ \qquad (2\text{-}14)$$

By convention, for this reaction at equilibrium,

$$K'_{eq} = \frac{[P]^p\,[Q]^q}{[R]^r\,[S]^s} \qquad (2\text{-}15)$$

For the same reaction in a living cell, a determination of the actual, not equilibrium, concentrations gives the mass action ratio (MAR):

$$MAR = \frac{[P]^p\,[Q]^q}{[R]^r\,[S]^s} \qquad (2\text{-}16)$$

The formulas for the K'_{eq} and the MAR may look alike, but their values may be far apart. In fact, a key to having a flux of energy in living organisms is having MARs removed from the K'_{eq} in various reactions, or steps, of a process. It has been found that the actual (ΔG) and standard free energy changes ($\Delta G^{\circ\prime}$) are related by the equation

$$\Delta G = \Delta G^{\circ\prime} + RT \ln MAR \qquad (2\text{-}17)$$

Here we can clearly see what the effect of concentration change is on the actual free energy change of a reaction. Also, we can see that temperature has a direct effect on the free energy change. Although it is not directly shown, a change in pH would also affect the actual free energy. Some things to consider are the effects of exercise on muscle metabolite concentrations, pH, and temperature. These factors can affect exercise performance.

ATP: The Common Chemical Intermediate

The mechanisms of energy conversion are contained within each cell; therefore, cells require the presence of a substance that can receive energy input from energy-yielding reactions. Equally important, that substance must be able to yield energy to reactions requiring an energy input. In our cells,

that substance is almost always adenosine triphosphate (ATP).

In Chapter 4, we will discuss in some detail the processes of cell respiration and cell work. In the cell, respiration represents the conversion of the chemical energy of foodstuffs into useful chemical form. Cell work represents the conversion of that useful form to other forms of energy. In order to function, this coupled system of energy-yielding and energy-utilizing reactions depends on having a substance that can act both as an energy receiver and as an energy resource, or donor. In most cells of most organisms ATP is that substance. Because of its central role in metabolism, ATP is frequently referred to as the "common chemical intermediate." If most species had not evolved to use ATP, they would have evolved to use a related compound.

The history of the study of ATP is marked by several notable discoveries. In the 1920s, the work of Fletcher, Hopkins, and Hill led to the belief that the release of lactic acid in muscles was the stimulus for contraction. Subsequently, after Embden showed that rapid freezing of small isolated muscles after a contraction prevented the formation of lactic acid, it was concluded that lactate was not the cause of contraction. In the early 1930s, Lundsgaard showed that poisoning glucose metabolism in muscles with iodoacetic acid hastened fatigue and prevented lactate formation. Instead, there was a decrease in a phosphorylated compound (later shown to be creatine phosphate) and an increase in inorganic phosphate. In 1929, Fiske and Subbarow isolated and deduced the structure of ATP. Embden and Meyerhof independently found that ATP is formed by joining inorganic phosphate (P_i) to adenosine diphosphate (ADP) during the catabolism of glucose. Englehardt discovered that ATP was split to ADP by myosin, one of the two major contractile proteins of muscle. Szent Gyorgi found that injections of ATP into muscle fibers that had been soaked in glycerin to remove metabolites had two effects: An initial infusion of ATP caused the muscles to contract; a second injection frequently caused relaxation. Cori and Cori observed that ATP had a biosynthetic function and could stimulate the aggregation of glucose (a simple sugar) into glycogen (a polymeric form of glucose).

In 1940, Lipman deduced from these various bits of information that ATP was the substance that linked energy-yielding and energy-using functions in the cell. This hypothesis was confirmed and generally accepted when Cain and Davies used the poison dinitrofluorobenzene (DNFB) on small isolated muscles to show that contractions resulted in a utilization of ATP. In the normal muscle cell, little ATP is present. This rather low concentration of ATP is probably more advantageous than not. As will be shown, the energy-yielding processes of metabolism are finely tuned to comparative levels of ATP, ADP, P_i, and AMP. By keeping the normal level of ATP low, any small utilization immediately changes the level markedly and stimulates the processes that generate ATP.

More cellular energy is stored in the form of creatine phosphate (CP) than in the form of ATP. The metabolisms of ATP and CP are linked by the reaction governed by the enzyme creatine kinase.

$$ADP + CP \xrightarrow[\text{kinase}]{\text{Creatine}} ATP + C \qquad (2\text{-}18)$$

The poison DNFB blocks the action of creatine kinase.

$$ADP + CP \xrightarrow[\substack{\text{kinase} \\ \text{(reaction blocked)}}]{\text{Creatine \quad DNFB}} \qquad (2\text{-}19)$$

In most cells, such as muscle, there is enough CP and the action of creatine kinase is so rapid that it is difficult to determine an overall decrease in ATP level after a twitch or contraction. Therefore, although the muscle uses ATP as an immediate energy source for contraction, the ATP is replenished almost immediately by CP. By using DNFB, Cain and Davies demonstrated that muscle contractions are powered directly by ATP.

Structure of ATP

The structure of ATP is shown in Figure 2-6. One of the group of compounds called nucleotides, ATP

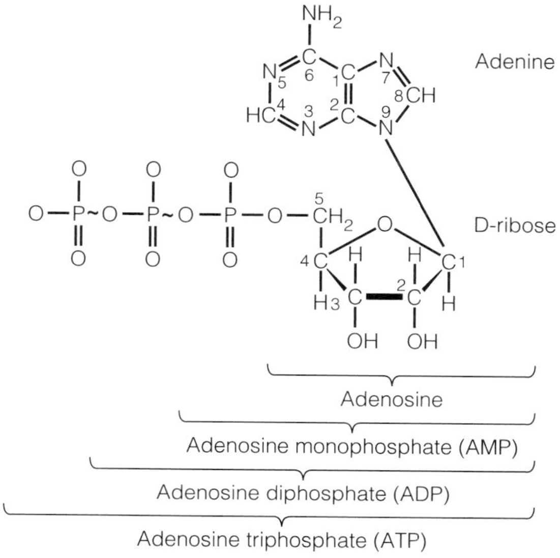

Figure 2-6 The structure of adenosine triphosphate (ATP). At pH 7 the molecule is ionized. Usually, the terminal phosphate (~bond) is hydrolyzed to provide energy.

contains a nitrogenous base (adenine), a five-carbon sugar (ribose), and three phosphates. Removal of the terminal phosphate results in adenosine diphosphate (ADP); cleaving two phosphates gives adenosine monophosphate (AMP). With no phosphates, the compound is simply the nucleotide, adenosine. In the cell, ATP is negatively charged, and the terminal phosphates of each ATP molecule often associate with a magnesium (Mg^{2+}). The Mg^{2+} is usually required for enzymatic activity.

The reactions in which ATP is split to ADP to liberate energy involve water. Accordingly these reactions are called *hydrolyses*, meaning "split by water."

$$ATP + H2O \xrightarrow[\text{enzyme}]{\text{ATPase}} ADP + P_i \qquad (2\text{-}20)$$

The standard free energy of ATP hydrolysis ($\Delta G°'$) is -7.3 kcal $\cdot$ mol^{-1}. As pointed out earlier, the factors of metabolite concentration, temperature, and pH affect the actual free energy of hydrolysis (ΔG) in the cell. The effect of pH on the $\Delta G°$

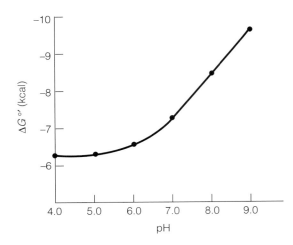

Figure 2-7   Effect of pH on $\Delta G°$ of ATP at 25°C (298 K).

of ATP is given in Figure 2-7. In the working muscle, ΔG for ATP is probably close to -11 kcal $\cdot$ mol^{-1}.

Three factors operate to give ATP a relatively high free energy of hydrolysis. First, the negative charges of the phosphates repel each other. Second, the products ADP and P_i form "resonance hybrids," which means that they can share electrons in ways

to reduce the energy state. Third, ADP and ATP have the proper configurations to be accepted by enzymes that regulate energy-yielding and energy-requiring reactions.

The hydrolysis of ATP almost always involves splitting the terminal phosphate group. Therefore, it is possible to write ATP as ADP~P, where the ~P is the "high-energy" phosphate. Similarly, creatine phosphate can be abbreviated as C~P. Exergonic reactions of metabolism that hydrolyze the second phosphate group of ATP are infrequent; the myokinase (adenylate kinase) reaction is a notable exception (Equation 3-4). Probably no cellular reactions of energy cleave the first phosphate from ATP.

ATP: The High-Energy Chemical Intermediate

The potential chemical energy of hydrolyzing the terminal phosphate of ATP is intermediate with regard to the hydrolysis of other phosphorylated biological compounds (Figure 2-8). The term *high energy,* when applied to ATP, more properly refers to the fact that the probability is high of hydrolysis transferring energy.

Figure 2-8 illustrates several aspects of the role

Figure 2-8   Standard free energies of hydrolysis of ATP compared to those of related physiological compounds. Because of its intermediate position, ATP can transfer phosphate group energy to "low-energy" phosphate compounds, whereas ADP can accept phosphate group energy from "high energy" compounds.

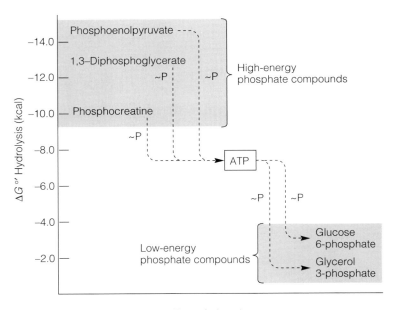

Flow of phosphate energy

of ATP in cellular energy transfer. First, the intermediate level of ATP in the scale of phosphate metabolites means that the phosphorylation of ADP to ATP can be driven by phosphorylated compounds of higher energy. These are energy-conserving reactions. The hydrolysis of ATP to ADP and P_i can transfer energy to other compounds or do cell work. These are energy-yielding reactions.

Second, it should be noted that many biological reactions of energy transfer involve phosphate exchange. The $ATP \leftrightarrows ADP + P_i$ system functions as a shuttle, or energy bridge, for exchanging phosphate energy.

■ Summary

Performance in muscular activity depends on energetics. The utilization of energy by machines, both mechanical and biological, can be precisely defined. Basically, two considerations govern physical and biological reactions involving energy exchanges: (1) Energy is not created but is acquired in one form and converted to another; and (2) the interconversion of energy among forms is inefficient, and a large fraction of the energy released appears in an unusable form, usually heat. In the cell, the transduction of chemical energy to other forms of energy depends on the metabolism of adenosine triphosphate (ATP). Elaborate enzymatic pathways exist to capture the energy content of foodstuffs in the form of ATP. Similarly, complex enzymatic pathways are evolved to utilize the potential energy in ATP to do cell work. Indeed, the overall processes sustaining working muscle during exercise can be considered as a balance between those processes that degrade ATP to its products ADP, AMP, and P_i, and those processes that restore and maintain the level of ATP. In coming chapters, we shall describe these processes in detail.

■ Selected Readings

Bray, H. G., and K. White. Kinetics and Thermodynamics in Biochemistry. New York: Academic Press, 1966.

Gaesser, G. A., and G. A. Brooks. Muscular efficiency during steady-rate exercise. *J. Appl. Physiol.* 38: 1132–1139, 1975.

Gibbs, C. L., and W. R. Gibson. Energy production of rat soleus muscle. *Am. J. Physiol.* 223: 864–871, 1972.

Helmholtz, H. Über die Erhaltung der Kraft, Berlin (1847). Reprinted in Ostwald's Klassiker, no. 1, Leipzig, 1902.

Hill, T. L. Free Energy Transductions in Biology. New York: Academic Press, 1977.

Kleiber, M. The Fire of Life. New York: Wiley, 1961, pp. 105–124.

Krebs, H. A., and H. L. Kornberg. Energy Transformations in Living Matter. Berlin: Springer-Verlag OHG, 1957.

Lehninger, A. L. Biochemistry. New York: Worth, 1970, pp. 289–312.

Lehninger, A. L. Bioenergetics. Menlo Park, Ca.: W. A. Benjamin, 1973, pp. 2–34.

McGilvery, R. W. Biochemical Concepts. Philadelphia: W. B. Saunders, 1975.

Merowitz, H. J. Entropy for Biologists. New York: Academic Press, 1970.

Mommaerts, W. F. H. M. Energetics of muscle contraction. *Physiol. Rev.* 49: 427–508, 1969.

Wendt, I. R., and C. L. Gibbs. Energy production of rat extensor digitorum longus muscle. *Am. J. Physiol.* 224: 1081–1086, 1973.

Wilkie, D. R. Heat work and phosphorylcreatine breakdown in muscle. *J. Physiol.* (London) 195: 157–183, 1968.

Wilkie, D. R. The efficiency of muscular contraction. *J. Mechanochem. Cell Motility* 2: 257–267, 1974.

CHAPTER 3

ENERGETICS AND HUMAN MOVEMENT

All forms of human movement, including athletic, occupational, and rehabilitation exercises can be described as energetic events, with the liberation and harnessing of energy central to maximal performance. Further, athletic activities can be classified as one of three groups: power, speed, and endurance events. (See Table 3-1.) Examples of these groups are the shot put, the 400-m sprint, and the marathon run, respectively. Success in each of these depends on energetics. Skeletal muscle has three energy systems, each of which is used in these three types of activities. In power events, where the activity lasts a few seconds or less, the muscle has several immediate energy sources (Figure 3-1). For rapid, forceful exercises lasting from a few seconds to approximately 1 minute, muscle depends mainly on nonoxidative, or glycolytic, energy sources, as well as on immediate sources. For activities lasting 2

TABLE 3-1

Energy Sources of Muscular Work for Different Types of Activities

	Power	Speed	Endurance
Duration of event	0 to 3 sec	4 to 50 sec	>2 min
Example of event	Shot put, discus, weight lifting	100- to 400-m run	≥ 1500-m run
Enzyme system	Single enzyme	One complex pathway	Several complex pathways
Enzyme location	Cytosol	Cytosol	Cytosol and mitochondria
Fuel storage site	Cytosol	Cytosol	Cytosol, blood, liver, adipose tissue
Rate of process	Immediate, very rapid	Rapid	Slower but prolonged
Storage form	ATP, creatine phosphate	Muscle glycogen and glucose	Muscle and liver glycogen, glucose; muscle, blood, and adipose tissue lipids; muscle, blood, and liver amino acids
Oxygen involved	No	No	Yes

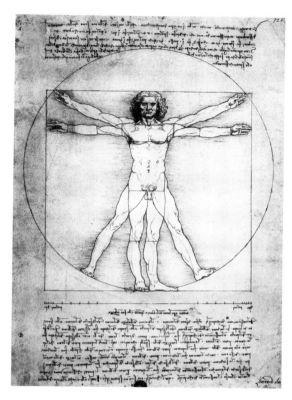

Leonardo's drawing of man's ability to move represents the fact that the laws of nature operate to control and limit human muscular performance. SOURCE: © Corbis/Bettman.

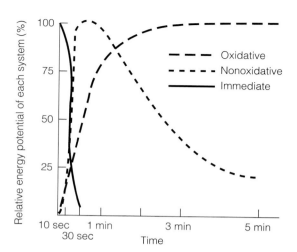

Figure 3-1 Energy sources for muscle as a function of activity duration. Schematic presentation showing how long each of the major energy systems can endure in supporting all-out work. SOURCE: D. W. Edington and V. R. Edgerton, 1976. Used with permission.

minutes or more, oxidative mechanisms become increasingly important. Before describing these three basic muscle energy sources, we need to describe the chemical-mechanical energy transduction of muscle contraction, as depicted in Equation 3-1.

In this reaction, actin and myosin are the two contractile proteins of muscle, and Ca^{2+} is the calcium ion whose presence triggers the combination of actin and myosin. Inorganic phosphate (P_i) is also produced by the reaction.

■ Immediate Energy Sources

Each of the three main energy sources in muscle is mediated by specific enzymes or enzyme systems, as described in Tables 3-1 and 3-2. In any muscle contraction, whether the activity is primarily one of power or endurance, the degradation of ATP supplies the chemical energy to power the contraction (Eq. 3-1).

The immediate energy source in muscle, as in most other cells, is composed of three components. First, there is ATP itself. ATP is degraded by enzymes that are generally called ATPases. Because the reaction involves combination with H_2O, the splitting of ATP is called hydrolysis (Eq. 3-2).

The chemical products of ATP hydrolysis are adenosine diphosphate (ADP) and inorganic phosphate (P_i). In the cyclic process of muscle contrac-

$$\text{ATP + Actin + Myosin} \xrightarrow{Ca^{2+}} \text{Actomyosin} + P_i + \text{ADP} + \text{Energy (heat + work)} \qquad (3\text{-}1)$$

$$\text{ATP} + H_2O \xrightarrow{\text{ATPase}} \text{ADP} + P_i \qquad (3\text{-}2)$$

TABLE 3-2

Estimation of the Energy Available in the Body Through Immediate (Phosphagen) Energy Sources

	ATP	CP	Myokinase ATP Equivalent	Total Phosphagen (ATP + CP)
Muscular concentration				
mmol · kg⁻¹ muscle[a]	6	28		34
mmol total muscle mass[b]	180	840	90	1110
Useful energy[c]				
kcal · kg⁻¹ muscle	0.06	0.28		0.34
kcal total muscle mass	1.8	8.4	0.9	11.1

[a] Based on data from Edwards et al., 1982.
[b] Assuming 30 kg of muscle in a 70-kg man.
[c] Assuming 10 kcal · mol⁻¹ ATP.

tion and recovery from contraction, ATP is continually being hydrolyzed to ADP, and ADP is continually being reenergized by phosphorylation back to ATP.

The standard free energy of hydrolysis ($\Delta G^{\circ\prime}$) of ATP measured under test tube conditions is -7.3 kcal · mol⁻¹. In the body, however, the actual free energy of ATP hydrolysis (i.e., the ΔG) is probably closer to -11 kcal · mol⁻¹ (Chapter 2). This is because conditions in the living cell are somewhat different from standard test tube conditions, and more energy is available from ATP splitting in our muscles than in test tubes under standard conditions.

The second cellular source of immediate energy is creatine phosphate (CP). This high-energy phosphorylated compound exists in five to six times greater concentration in resting muscle than does ATP. Creatine phosphate provides a reserve of phosphate energy to regenerate ATP, which is consumed as the result of muscle contraction. The interaction of CP and ADP (degraded ATP) is catalyzed by the enzyme creatine kinase.

$$CP + ADP \xrightarrow[\text{kinase}]{\text{Creatine}} ATP + C \qquad (3\text{-}3)$$

Thus, in the muscle, ATP that is hydrolyzed to ADP during muscle contraction is rephosphory-

lated by CP. Seen in this role, CP is particularly important as an intracellular energy shuttle. This is discussed more in Chapter 6. Here it is sufficient to note that hydrolysis of ATP by the contractile apparatus in muscle is rapidly compensated for by cytoplasmic creatine kinase and creatine phosphate stores. The resulting creatine is then rephosphorylated through the action of mitochondrial creatine kinase, which accesses mitochondrial ATP.

The third immediate energy source in muscle involves an enzyme called adenylate kinase, which in muscle is also referred to as myokinase. The enzyme has the ability to generate one ATP (and one AMP) from two ADPs.

$$ADP + ADP \xrightarrow[\text{kinase}]{\text{Adenylate}} ATP + AMP \qquad (3\text{-}4)$$

The three components of the immediate energy system in muscle are all H_2O-soluble. Therefore, they exist throughout the aqueous part of the cell, from near the cell's inner boundary to deep within it, surrounding the contractile elements, actin and myosin, and other important parts of the cell. The immediate energy sources are so named because they are immediately available to support muscle contraction.

Quantitatively, ATP and CP (which together are called *phosphagen*) make up a critical and important energy reserve. The amount of ATP on hand, however, cannot sustain maximal muscle contraction for more than a few seconds (Table 3-2). Even when ATP is augmented by both CP and the myokinase enzyme system, activities that must be sustained for more than a fraction of a minute (i.e., more than 5–15 seconds) require the assistance of other energy sources.

■ Nonoxidative (Glycolytic) Energy Sources

By convention, we use the terms *glycolytic* and *nonoxidative* interchangeably. Nonoxidative energy sources in muscle are the breakdowns of glucose (a simple sugar) and glycogen (stored carbohydrate made up of many glucose subunits). These processes are specifically termed *glycolysis* and *glycogenolysis*, respectively. Muscle tissue is densely packed with glycolytic and glycogenolytic enzymes; therefore, muscle is specialized in these processes and can break down glucose and glycogen rapidly, with the net formation of lactic acid. Glycolysis can be summarized as follows:

$$\text{Glucose} \xrightarrow[\text{Glycolysis}]{} 2 \text{ ATP} + 2 \text{ Lactate} \qquad (3\text{-}5)$$

In skeletal muscle, the concentration of free glucose is very low, so most of the potential energy available from nonoxidative energy sources comes from the breakdown of glycogen (Table 3-3).

Like the immediate energy system, the nonoxidative energy system is composed of elements that are H_2O-soluble and exist in the cell cytosol. The apparatus for nonoxidative energy metabolism therefore exists in immediate proximity to the contractile elements in muscle. Nonoxidative energy sources are called upon when muscle contraction lasts more than a few seconds.

Quantitatively, the energy available through nonoxidative metabolism (Table 3-3) is significantly greater than that available through immediate energy sources (Table 3-2). However, immediate and nonoxidative energy sources combined still provide only a small fraction of the energy available through

TABLE 3-3

Estimation of the Energy Availablein the Body Through Nonoxidative (Glycolytic) Available in Metabolism

	Per Kilogram Muscle	Total Muscle Mass[a]
Maximal lactic acid tolerance (g)	3.0	90
ATP formation (mmol)	50.0	1500
Useful energy (kcal)	0.5	15

[a] Assumes that all muscles were activated simultaneously.

SOURCE: Based on data from Karlsson, 1971.

oxidative metabolism. Therefore, intense muscular activities lasting longer than approximately 30 seconds cannot be sustained without the benefit of oxidative metabolism. (See Figure 3-1.)

Before leaving this topic, it is important to note some new and very important results concerning the pathways of glycolytic and oxidative metabolism and their linkage. As will be discussed in Chapter 6, glycolysis inevitably results in lactate formation, with 10 or more times more lactate than pyruvate present in resting muscle and other tissues. However, while resting muscle releases lactate on a net basis, and working muscle can actually consume lactate on a net basis, lactate formation occurs continuously as a function of the glycolytic flux rate. But, because most lactate enters mitochondria and is oxidized within muscle cells and tissues, lactate formation through glycolysis and removal through oxidation are in balance. Consequently, while glycolysis inevitably results in lactate production, the accumulation and net release of lactate from muscle and other tissues is much less than the amount of lactate produced in muscle and other tissues.

■ Oxidative Energy Sources

Potential oxidative energy sources for muscle include sugars, carbohydrates, fats, and amino acids.

As just noted, muscle tissue in healthy, fed individuals has significant reserves of glycogen. This fuel source can be supplemented by glucose supplied from the blood; liver glycogen, which can be broken down to glucose and delivered to muscle through the circulation; and fats and amino acids, which exist in muscle as well as in other depots around the body. Further, whereas the sugar glucose can be metabolized to an extent by glycolytic mechanisms (Eq. 3-5), oxidative mechanisms allow far more energy to be liberated from a glucose molecule.

$$\text{Glucose} + O_2 \xrightarrow[\text{metabolism}]{\text{Oxidative}} 36 \text{ ATP} + CO_2 + H_2O \quad (3\text{-}6)$$

This additional energy is liberated because the oxidative pathways of metabolism carry on the process of glucose catabolism to a far greater extent than the nonoxidative pathways. The oxidative breakdown of glucose is longer and more involved, so the opportunity for energy transduction and capture of glucose chemical energy in the form of ATP is greater. The details of oxidative metabolism will be discussed in more detail in Chapters 6, 7, and 8.

Fats can be catabolized by oxidative mechanisms only, but the energy yield is very large. For palmitate, an average-sized and commonly occurring fatty acid:

$$\text{Palmitate} + O_2 \xrightarrow[\text{metabolism}]{\text{Oxidative}} 129 \text{ ATP} + CO_2 + H_2O \quad (3\text{-}7)$$

Like fats, amino acids can be catabolized only by oxidative mechanisms. Before an amino acid can be oxidized, the nitrogen residue must be removed. This is generally done by switching the nitrogen to some other compound (a process called *transamina-*

TABLE 3-4

Estimation of Energy Available from Muscle and Liver Glycogen, Fat (Adipose Triglyceride), and Body Proteins

	Energy Equivalent (kcal)
Glycogen in muscle	480
Glycogen in liver	280
Fat (adipose triglyceride)	141,000
Body proteins	24,000

SOURCE: Based on data from Young and Scrimshaw, 1971.

tion) or by a unique process of nitrogen removal in the liver (oxidative deamination). Examples involving alanine, a three-carbon amino acid, are seen in Equations 3-8 and 3-9.

The significance of oxidative metabolism for energy production in the body is illustrated in Table 3-4. In comparison to the energy potential of the immediate energy system (Table 3-2) and the nonoxidative metabolism of glycogen (Table 3-3), the energy available from oxidative energy sources is far greater. Also, the energy available from the combustion of muscle glycogen is small compared with the much larger potential energy reserves in fat and body protein.

■ Aerobic and Anaerobic Metabolism

Of the body's three energy systems, two systems (the immediate and nonoxidative) do not require oxygen for their operation. Consequently, by convention these systems are referred to as *anaerobic*, meaning not dependent on O_2. The third, the oxidative energy system, is referred to as the *aerobic* en-

$$\text{Alanine} + \alpha\text{-Ketoglutarate} \xrightarrow[\text{pyruvate transaminase}]{\text{Glutamate}} \text{Pyruvate} + \text{Glutamate} \quad (3\text{-}8)$$

$$\text{Pyruvate} + O_2 \xrightarrow[\text{metabolism}]{\text{Oxidative}} 15 \text{ ATP} + CO_2 + H_2O \quad (3\text{-}9)$$

ergy system. Energy transduction in this system is dependent on the presence of O_2. The discovery of direct linkages between glycolytic and oxidative processes and its impact on use of the terms *aerobic* and *anaerobic* will be discussed in Chapter 6.

■ Power and Capacity of Muscle Energy Systems

Although the maximal energy capacity of immediate (Table 3-2) and nonoxidative (Table 3-3) energy systems is small compared with that of the oxidative energy system (Table 3-4), immediate and non-oxidative energy systems are important because they are activated very rapidly when muscles start to contract. By comparison, the oxidative energy system is activated more slowly and produces energy at a lower rate even when fully activated. In Table 3-5 the maximal rates (power) at which the various systems provide energy for muscle contraction are contrasted with the maximal capacities (total contribution available) for energy release. In this comparison, immediate and nonoxidative energy systems are revealed to have superior, though short-lived, power capacities. Thus, the three energy systems in muscle together provide a means to sustain short, intense bursts of activity as well as more sustained activities of lesser intensity.

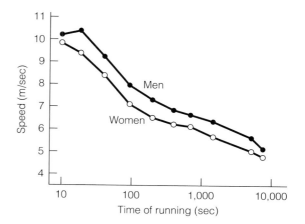

Figure 3-2 Plot of the average running speed maintained versus logarithm of time of event for men's and women's world running records. Note the presence of three curve components that suggest the presence of three energy systems. Data from the 1999 World list.

■ Energetics and Athletic Performance

Our interpretation that energy for the human engine comes from three sets of enzyme systems (Figure 3-1 and Table 3-5) is supported by an analysis of world running records (Figure 3-2). A plot of running speed versus time reveals three distinct curve components. Thus, it appears that the selection and training of athletes for athletic events requires knowledge of both the metabolic requirements of the activity and the metabolic characteristics of the individual. Because many athletic events and most of life's other activities take longer than 90 seconds, the primary importance of oxidative energy metabolism is obvious.

■ Enzymatic Regulation of Metabolism

The biochemical pathways that result in cellular energy transfer are discussed in the following chapters. Each of these pathways involves many steps, each of which is catalyzed by a specific enzyme. It is important to note here, therefore, some functions of enzymes. Although enzymes cannot change the

TABLE 3-5

Maximal Power and Capacity of the Three Energy Systems

System	Maximal Power (kcal · min⁻¹)	Maximal Capacity (Total kcal Available)
Immediate energy sources (ATP + CP)	36	11.1
Nonoxidative energy sources (anaerobic glycolysis)	16	15.0
Oxidative energy sources (from muscle glycogen only)	10	2000

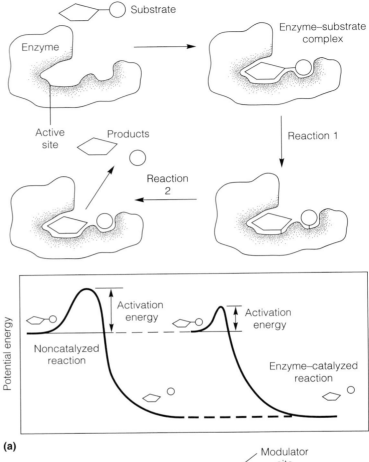

(a)

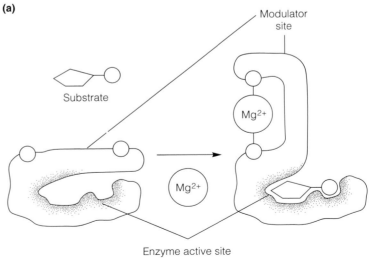

(b)

Figure 3-3 (a) Enzymes function, in part, by virtue of the substrate inducing a fit at the active site. By mechanisms still not completely understood, enzymes lower energies of activation, thereby increasing the probability of the reaction occurring. (b) Some enzymes have additional (modulatory) sites at which factors other than substrates bind. When binding occurs at these modulator sites, the three-dimensional shape of the enzyme is adjusted such that the probability of substrate binding at the active site is affected. Modulators can stimulate or inhibit catalytic function. Modified from A. J. Vander, J. H. Sherman, and D. S. Luciano, 1980. Used with permission.

equilibria of reactions, they can lower the energies of activation, thereby allowing spontaneous reactions to proceed. Also, by linking exergonic to endergonic reactions through the use of ATP or other high-energy intermediates, enzymes facilitate endergonic processes.

The precise mechanisms by which enzymes operate are not completely understood, but some details of enzyme action are known. As illustrated in Figure 3-3a, enzymes are usually large molecules, in most cases with only a single site at which a reactant, or *substrate*, attaches. This site is called the *active site*. With a few exceptions, only the appropriate substrate interacts with an enzyme to induce a fit at the enzyme's active site. The combined substrate and enzyme are referred to as the *enzyme–substrate complex*. After the enzyme catalyzes the reaction, the products are released. Foreign substances that have the capability of competing for and occupying enzymatic sites are frequently poisons. An example is the hallucinogen LSD, which competes with the chemical messenger serotonin in the brain. "Bad" and recurring "free" trips occur because LSD binds in such a way that serotonin has difficulty displacing it.

Note also in Figure 3-3b that certain enzymes have binding sites other than the active site. When appropriate substances bind to these other sites, they affect the *configuration*, or *conformation*, of the active site. Therefore, the binding of these other factors affects the interaction of the enzyme with its substrate, thereby changing the rate at which the enzyme can function. Consequently, these factors that change catalytic rates of particular enzymes are termed *modulators*.

Modulators can be classified into two groups: Those that increase catalytic rates of enzymes are termed *stimulators*, and those that slow enzymatic function are called *inhibitors*. In the control of energy metabolism, ATP is the classic example of an inhibitor; ADP and P_i are usually stimulators. In resting muscle, high levels of ATP inhibit carbohydrate, fat, and amino acid catabolism. However, when a muscle starts to twitch, ATP is degraded to ADP and P_i. When these are formed, in the locus of their presence, they stimulate the mechanisms of ATP resupply.

The effect of various modulators on particular enzymes is called *allosterism*, referring to the fact that modulators change the spatial orientation, or shapes, of the parts of the enzymes. When several modulators are capable of affecting the catalytic rate of a particular enzyme, that enzyme is said to be *multivalent*. In coming chapters, we shall encounter several allosteric modulators. Among the most potent are those that phosphorylate (add phosphate, P_i) or dephosphorylate enzymes. An example of a multivalent allosteric enzyme is phosphofructokinase (PFK), a key enzyme of carbohydrate catabolism (Chapter 5). An example of an enzyme whose activity depends on phosphorylation is phosphorylase, a key enzyme in the degradation of glycogen (Chapter 6).

Different enzymes have different properties that have great effects on metabolism. Among these properties is the rate at which the enzyme functions. Maximum velocity (V_{max}) is an important descriptive parameter of enzymes. The Michaelis–Menten constant (K_M) describes the interaction of substrate and enzyme. The K_M is defined as the substrate concentration that gives half of V_{max}. V_{max} and K_M are illustrated in Figure 3-4. An appreciation for V_{max} and K_M is necessary because these parameters of a reaction tell us much about how fast that reaction will proceed and how the reaction rate is affected by changes in substrate concentration. In the following example, as well as in following chapters, we shall encounter many enzymes, substrates, and modifiers of enzyme function. In evaluating this material we will be well served to remember that at its essence, the rate at which any process proceeds depends on the kinetic properties (i.e., V_{max} and K_M) of the responsible enzymes.

The enzymes glucokinase and hexokinase have the same catalytic function: to phosphorylate and activate sugars for further metabolism (Chapter 5). These enzymes, however, have different catalytic activities, as measured by different V_{max} and K_M values. These different catalytic activities also allow for different physiological functions. For example, at normal blood glucose concentration (5.5 mM), hexokinase in muscle would be maximally stimulated, whereas glucokinase in liver would not be

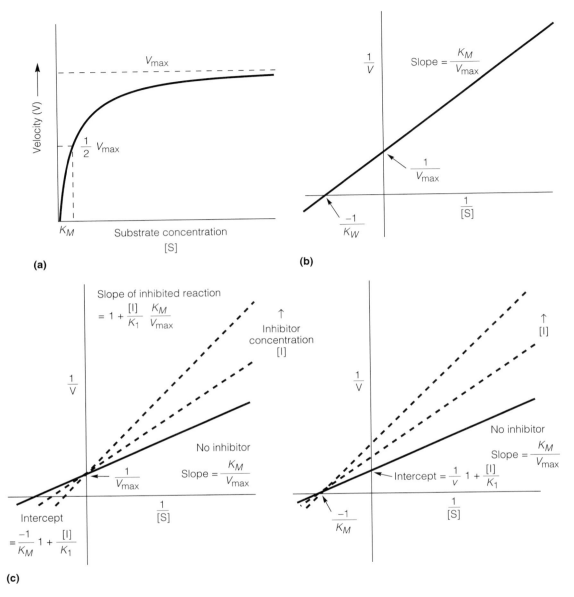

Figure 3-4 (a) Relationship between the maximum rate of catalysis for an enzyme and substrate concentration up to maximum (V_{max}) and its relation to the substrate (Michaelis–Menten constant, K_M). The K_M is the substrate concentration that gives 50% of V_{max}. Enzymes can vary in both V_{max} and K_M. (b) The Lineweaver–Burk double-reciprocal plot is a means for accurately predicting V_{max} and K_M, as well as for predicting the effect of inhibitors. (c) Lineweaver–Burk plots of competitive (left) and noncompetitive (right) inhibition. Different types of inhibitors have different effects on K_M and V_{max}.

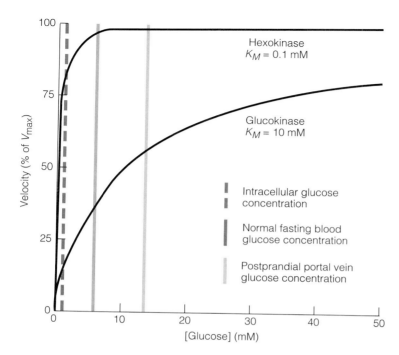

Figure 3-5 Illustration of the effects of K_M on physiological function. At levels of glucose in arterial blood and inside cells, hexokinase (a "low K_M enzyme") is fully activated. At the same concentrations, glucokinase (a "high K_M enzyme") is relatively inactive. Under postprandial (i.e., after eating) conditions, portal vein and hepatocyte glucose levels rise to where glucokinase becomes active. Thus, glucokinase is sensitive to changes in glucose availability, such as occurs after eating.

very active (Figure 3-5). However, intracellular free glucose concentration is usually low (<1.0 mM) compared to the concentration in blood. Therefore, normally hexokinase (K_M = 0.1 mM) is very active, whereas glucokinase is relatively inactive. However, after eating, blood glucose in the portal vein perfusing the liver from the GI tract can exceed 10 mM, or more than twice arterial. Under this circumstance, free glucose concentration in liver cells rises to the point where hexokinase becomes activated to half or more of V_{max}. Thus, in this example, a "high K_M enzyme" such as glucokinase is sensitive to physiological changes in glucose concentration. In contrast, "low K_M enzymes" such as hexokinase probably operate at close to V_{max} in vivo.

■ Summary

Processes of food and energy substrate catabolism in cells are usually linked to the process of ATP res-

titution. Thus, approximately 50% of the potential chemical energy released from foodstuffs is captured in the common chemical intermediate, ATP. ATP, together with its storage form, creatine phosphate (CP), then serves as the immediate cellular energy source on which endergonic processes depend. ATP and CP not only supply immediate cellular energy sources, but their relative levels also stimulate or inhibit processes of energy metabolism. At rest, normally high levels of ATP and CP inhibit energy metabolism. When exercise starts, however, the utilization and decreased levels of ATP and CP, and the increased levels of ADP and P_i, stimulate processes of energy metabolism. Enzymes interact with products of energy metabolism to regulate the rate at which specific processes proceed. Muscles utilize three different systems of energy release during exercise, each of which differs in mechanism, capacity, and endurance. Consequently, muscular capacity is limited by these three systems of energy release.

■ Selected Readings

Binzoni, T., G. Feretti, K. Schenker, F. Barbalan, E. Hiltbrand, and P. Cerretelli. Metabolic transients and NMR. *Int. J. Sports Med.* 13: S155–S157, 1992.

Brooks, G. A. Lactate: Glycolytic end product and oxidative substrate during sustained exercise in mammals—the "lactate shuttle." In Comparative Physiology and Biochemistry: Current Topics and Trends, R. Gilles (Ed.). Vol. A. Berlin: Springer-Verlag, 1984, pp. 208–218.

Brooks, G. A. Current concepts in lactate exchange. *Med. Sci. Sports Exerc.* 23: 895–906, 1991.

Brooks, G. A. Mammalian Fuel Utilization During Sustained Exercise. *Comp. Biochem. Physiol.* 120: 89–107, 1998.

Cain, D. F., and R. E. Davies. Breakdown of adenosine triphosphate during a single contraction of working muscle. *Biochem. Biophys. Res. Com.* 8: 361–366, 1962.

Cerretelli, P. Energy sources for muscle contraction. *Int. J. Sports Med.* 13: S106–S110, 1992.

Di Prampero, P. E. Energetics of muscle contraction. *Rev. Physiol. Biochem. Pharmacol.* 89: 143–152, 1981.

Duhaylongsod, F. G., J. A. Greibel, D. S. Bacon, W. G. Wolfe, and C. A. Piantadosi. Effects of muscle contraction on cytochrome a, a^3 redox state. *J. Appl. Physiol.* 75: 790–797, 1993.

Edwards, R. H. T., D. R. Wilkie, M. J. Dawson, R. E. Gordon, and D. Shaw. Clinical use of nuclear magnetic resonance in the investigation of myopathy. *Lancet* (March): 725–731, 1982.

Embden, G., and H. Lawaczeck. Über den zeitlichen Verlauf der Milchsäurebildung bei der Muskel Kontraktion. *Z. Physiol. Chem.* 170: 311–315, 1928.

Gaesser, G. A., and G. A. Brooks. Muscular efficiency during steady-rate exercise: effects of speed and work rate. *J. Appl. Physiol.* 38: 1132–1139, 1975.

Gaesser, G. A., and G. A. Brooks. Metabolic bases of excess post-exercise oxygen consumption: a review. *Med. Sci. Sports Exerc.* 16: 29–43, 1984.

Hill, A. V. The oxidative removal of lactic acid. *J. Physiol.* 48: x–xi, 1914.

Hill, A. V., and P. Kupalov. Anaerobic and aerobic activity in isolated muscle. *Proc. Roy. Soc.* (London) Ser. B. 105: 313–322, 1929.

Karlsson, J. Lactate and phosphagen concentrations in working muscle of man. *Acta Physiol. Scand.* 358 (Suppl.): 1–72, 1971.

Lehninger, A. L. Biochemistry. New York: Worth, 1970, pp. 313–327.

Lehninger, A. L. Bioenergetics. Menlo Park, Ca.: W. A. Benjamin, 1978, pp. 38–51.

Lipmann, F. Metabolic generation and utilization of phosphate bond energy. *Advan. Enzymol.* 1: 99–162, 1941.

Lundsgaard, E. Phosphagen—and Pyrophophatumsatz in Iodessigsäure-vegifteten Muskeln. *Biochem. Z.* 269: 308–328, 1934.

Meyerhof, O. Die Energie-wandlungen in Muskel. I. Über die Beziehungen der Milschsäure Wärmebildung und Arbeitstung des Muskels in der Anaerobiose. *Arch. Ges. Physiol.* 182: 232–282, 1920.

Meyerhof, O. Die Energie-wandlungen in Muskel. II. Das Schicksal der Milschsäure in der Erholungsperiode des Muskels. *Arch. Ges. Physiol.* 182: 284–317, 1920.

Molé, P. A., R. L. Coulson, J. R. Caton, B. G. Nichols, and T. J. Barstow. In vivo ^{31}PNMR in human muscle: transient patterns with exercise. *J. Appl. Physiol.* 59: 101–104, 1985.

Mommarts, W. F. H. M. Energetics of muscle contraction. *Physiol. Dev.* 49: 427–508, 1969.

Poole, D. C., G. A. Gaesser, M. C. Hogan, D. R. Knight, and P. D. Wagner. Pulmonary and leg V_{O_2} during submaximal exercise: implications for muscular efficiency. *J. Appl. Physiol.* 72: 805–810, 1992.

Szent-Györgi, A. Myosin and Muscular Contraction. Basel: Karger, 1942.

Szent-Györgi, A. Chemistry of Muscular Contraction. New York: Academic Press, 1951.

Wilkie, D. R. Thermodynamics and the interpretation of biological heat measurements. *Prog. Biophys. Chem.* 2: 260–298, 1961.

Wilkie, D. R. The efficiency of muscular contraction. *J. Mechanochem. Cell Motility.* 2: 257–267, 1982.

Young, V. R., and N. S. Scrimshaw. The physiology of starvation. *Sci. Amer.* 225: 14–22, 1971.

CHAPTER 4

BASICS OF METABOLISM

Metabolism can be defined as the sum total of processes occurring in a living organism. Because heat is produced by those processes, the *metabolic rate* is indicated by the rate of heat production. All processes of metabolism ultimately depend on biological oxidation, so measuring the rate of O_2 consumption yields a good estimate of the rate of heat production, or metabolic rate. The maximum capability of an individual to consume oxygen ($\dot{V}_{O_2max}$) is highly related to that individual's ability to perform hard work over prolonged periods. A high capacity to consume and utilize O_2 indicates a high metabolic capacity.

■ Energy Transductions in the Biosphere

Our lives depend on conversions of chemical energy to other forms of energy. These conversions, or *transductions,* of energy are limited by the two laws of thermodynamics, which apply to physical as well as biological energy transductions.

In the biological world (the biosphere), there are three major stages of energy transduction: photosynthesis, cell respiration, and cell work.

The photosynthesis of sugars is illustrated by Equation 4-1. In photosynthesis, the ΔG is positive in sign. Energy is put in.

Cell respiration can be illustrated by Equation 4-2. In cell respiration, the ΔG is negative in sign. Energy is given up and the process is associated with the production of the important high-energy intermediate compound, ATP.

There are many types of cellular work, including mechanical, synthetic, chemical, osmotic, and electrical rorms. Muscle contraction (a chemical-mechanical energy transduction) can be illustrated by Equation 4-3. Here actin and myosin are the contractile proteins and the release of Ca^{2+} within the muscle cell triggers the reaction.

Although it may appear that our functioning depends on only two of these three major energy transductions (respiration and cell work), in reality we are ultimately dependent on photosynthesis.

$$\text{Energy (sunlight)} + 6\ CO_2 + 6\ H_2O \rightarrow C_6H_{12}O_6 + 6\ O_2 \tag{4-1}$$

$$C_6H_{12}O_6 + 6\ O_2 \rightarrow 6\ CO_2 + 6\ H_2O + \text{Energy (heat + work)} \tag{4-2}$$

$$\text{ATP} + \text{Actin} + \text{Myosin} \xrightarrow{\ Ca^{2+}\ } \text{Actomyosin} + P_i + \text{ADP} + \text{Heat} + \text{Work} \tag{4-3}$$

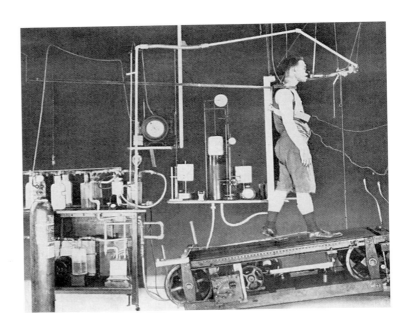

A photo from the classical study of human metabolic and cardioventilatory responses to exercise by H. M. Smith, 1922.

The products of photosynthesis give us the oxygen we breathe and the food we eat. Cell respiration is a reversal of photosynthesis. Have you thanked a green plant today?

■ Metabolism and Heat Production in Animals

One characteristic of living animals is that they give off heat. As illustrated in Figure 4-1, for a body at rest, life processes result in heat production.

Scientists have developed two definitions of metabolism. A functional definition is that metabolism is the sum of all transformations of energy and matter that occur within an organism. In other words, by this definition, metabolism is everything going on. It is not possible to measure that. Therefore, another operational definition has been developed, stating that metabolism is the rate of heat production. This definition takes advantage of the fact that all the cellular events result in heat. By determining the heat produced, one can obtain a measure of metabolism.

The basic unit of heat measurement is the calorie.

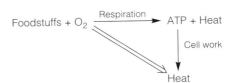

Figure 4-1 Metabolism and heat production. In a body at rest, all metabolic processes eventually result in heat production. Measuring heat production (calorimetry) gives the metabolic rate.

Simply defined, a *calorie* is the heat required to raise the temperature of 1 gram of water 1 degree Celsius. The calorie is a very small quantity, so the term *kilocalories* (kcal) is frequently used instead. A kilocalorie represents 1000 calories. Because heat must be measured to determine metabolic rate, this procedure is termed *calorimetry.* Several types of calorimetry are currently used. They are diagrammed in Figure 4-2.

Direct calorimetry, involving the direct measurement of heat, is technically very difficult. However, it has been determined that *indirect calorimetry,* the measurement of oxygen consumption, is also a valid and technically reliable procedure for measur-

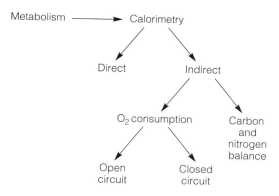

Figure 4-2 Relationship between metabolism and different methods of calorimetry. Because the processes of metabolism result in heat production, measuring heat production gives an estimate of the metabolic rate. Heat production can be measured directly (direct calorimetry), or it can be estimated from O_2 consumption or from the carbon and nitrogen excreted (indirect calorimetry).

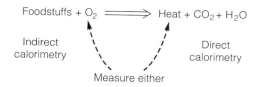

Figure 4-3 The principle of indirect calorimetry (measuring O_2 consumption) as a basis of estimating heat production. Instead of measuring the heat produced as the result of biological reactions, we measure the O_2 used to support biological oxidations.

Figure 4-4 Lavoisier's calorimeter of 1780. The animal's body heat melts the ice. Knowing that 80 kcal of heat melts 1000 grams of ice, we can measure the amount of water formed to estimate the heat produced. The ice water surrounding the calorimeter provides a perfect (adiabatic) insulation because it is at the same temperature as the ice in the inner jacket around the animal's chamber. The insulation will neither add heat to nor take heat from the calorimeter. Based on original sources and Kleiber, 1961. Used with permission.

ing metabolic rate. The principle of indirect calorimetry is illustrated in Figure 4-3. Another form of indirect calorimetry involves determining the carbon and nitrogen content of excreted materials.

■ Early Attempts at Calorimetry

To understand the relationship between heat production and O_2 consumption as alternative methods for determining metabolic rate, let us consider

some of the work of the eighteenth-century genius, French chemist Antoine Lavoisier.

Because of his interest in studying living creatures, Lavoisier came to recognize certain characteristics of living animals: They give off heat and they breathe. Dead animals do not give off heat and do not breathe. Lavoisier's calorimeter, diagrammed in Figure 4-4, is simple but beautiful in its design. By allowing the animal's warmth to melt the ice, and knowing the quantity of heat required to melt a given quantity of ice, Lavoisier could calculate the heat produced by the animal by measuring the volume of water produced. Such a device is called a *direct calorimeter* because it determines metabolism by measuring heat produced.

Lavoisier's respirometer (Figure 4-5) was another device that was novel for its time. With it Lavoisier could establish that something in the air (O_2) was consumed by the animal and that something else (CO_2) was produced in approximately equal amounts. Lavoisier also determined that matter gains weight when it burns. It has been thought previously that burning represented the loss of substance, sometimes called phlogiston.

With information obtained from his experiments, Lavoisier was able to interpret some earlier findings. For instance, Boyle had shown that air was necessary to have a flame, and Mayow had observed that a burning candle and an animal together in an airtight container expired at the same time. The fire of life and the fire of physical burning depended on the same substance in the air, which Lavoisier called *oxygène.*

The belief of Lavoisier and others that biological oxidation took place in the lungs has led to some confusion. Although it is true that breathing, or ventilation, takes place in the lungs and associated organs, respiration, or biological oxidation, takes place in most of the body's cells. Therefore, in this text, we shall use the term *respiration* to denote cellular oxidations and *ventilation* to denote pulmonary gas exchange.

Devices such as Lavoisier's respirometer are called *indirect calorimeters* because they estimate heat production by determining O_2 consumption or CO_2 production. Lavoisier's device is also referred to as a closed-circuit indirect calorimeter because the animal breathes gas within a sealed system.

Haldane's respirometer (Figure 4-6) is an example of an open-circuit indirect calorimeter. This system is open to the atmosphere, and the

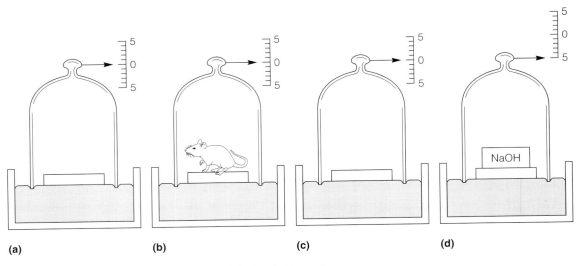

(a) (b) (c) (d)

Figure 4-5 Lavoisier's respirometer of 1784. (a) A glass bell jar rests on a bed of mercury. (b) An animal is placed in the jar from beneath the mercury seal and is left there for several hours. The apparent respirometer volume increases when the animal enters, but then the volume decreases very slowly. (c) The animal is removed, and the volume is observed to have decreased slightly. (d) Addition of NaOH (a CO_2 absorber) into the jar results in a decrease in the measured volume. From these volume changes, O_2 consumption ($\dot{V}_{O_2}$) *and carbon dioxide production ($\dot{V}_{CO_2}$) can be measured:* $V_{O_2} = V_a - V_d$; $\dot{V}_{CO_2} = V_c - V_d$. Based on original sources and Kleiber, 1961. Used with permission.

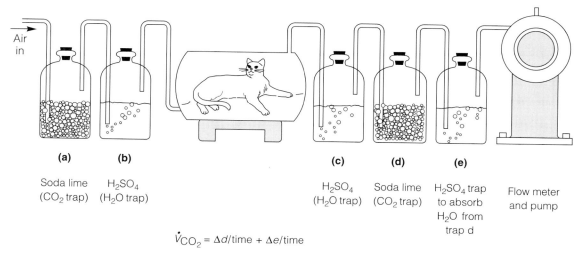

$$\dot{V}_{CO_2} = \Delta d/time + \Delta e/time$$

Figure 4-6 Haldane's respirometer. This device is an open-circuit indirect calorimeter, in which carbon dioxide and water vapor in air entering the system are removed by traps (a) and (b), respectively. Trap (c) removes the animal's expired H_2O vapor. Increase in weight of the soda lime CO_2 trap (d) gives the animal's CO_2 production. Based on original sources and Kleiber, 1961. Used with permission.

animal breathes air. Today, the type of calorimeters most frequently used are open-circuit indirect designs.

The Atwater and Rosa device (Figure 4-7) is an important apparatus for the study of metabolism. Large enough to accommodate a person, it has the capability of determining heat production, O_2 consumption, and CO_2 production simultaneously. Through this device, the relationship between direct and indirect calorimetry was established. Thus, it is now possible to predict metabolic rate (heat production) on the basis of determinations of O_2 consumption and CO_2 production in resting individuals.

The calorimeter illustrated in Figure 4-8 is called a *bomb calorimeter*. In this device, foodstuffs are ignited and burned in O_2 under pressure. Through this device, the heats of combustion (ΔH) of particular foods can be determined.

Table 4-1 presents the relationships among caloric equivalents for combustion of various foodstuffs as determined by indirect and direct calorimetry as well as by bomb calorimetry. Perhaps the

most interesting feature of this table is that, with a single exception, the caloric equivalents for the combustion of foodstuffs inside and outside the body are the same. Protein is the exception because nitrogen, an element unique to protein, is not oxidized within the body but is eliminated, chiefly in urine but also in sweat. Therefore, the caloric equivalent of protein metabolism is approximately 26% less than in a bomb calorimeter.

Table 4-1 also gives the caloric equivalents of foodstuffs in kilocalories per liter of O_2 consumed. Although fat, because of its relatively high carbon and hydrogen content (reduction), contains more potential chemical energy on a per-unit-weight basis, carbohydrates give more energy when combusted in a given volume of O_2.

In addition to providing an estimate of metabolic rate, indirect calorimetry provides a means of estimating the composition of the fuels oxidized. Similarly, determining the ratio of CO_2 produced ($\dot{V}_{CO_2}$) to O_2 consumed ($\dot{V}_{O_2}$), gives an indication of the type of foodstuff being combusted. This ratio ($\dot{V}_{CO_2}/\dot{V}_{O_2}$) is usually referred to as the *respiratory*

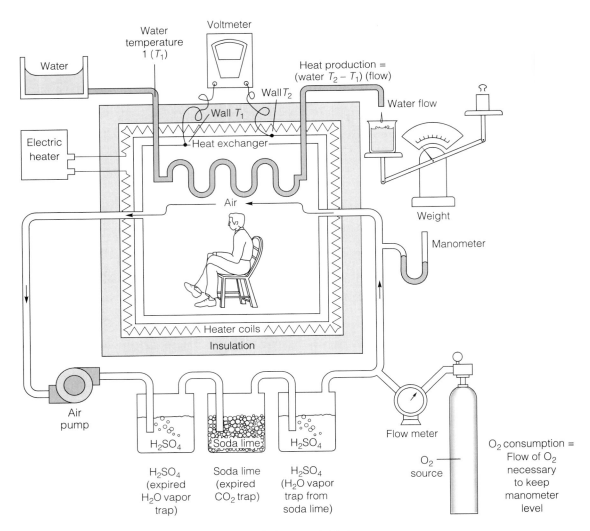

Figure 4-7 Atwater–Rosa calorimeter. A direct calorimeter suitable to accommodate a resting human and simultaneously determine that individual's O_2 consumption and CO_2 production. Through this device, direct and indirect calorimetry were correlated. O_2 consumption is equal to the volume of O_2 added to keep the internal (manometer) pressure constant. In the calorimeter, heat loss through the walls is prevented by heating the middle wall (wall T_2) to the temperature of the inner wall (wall T_1). Metabolic heat production is then picked up in the water heat exchanger. Based on original sources and Kleiber, 1961.

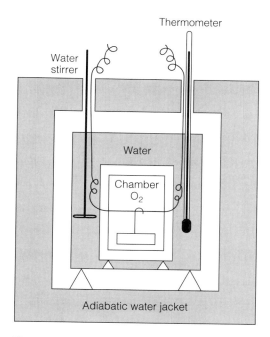

Thermometer

Water stirrer

Water

Chamber O_2

Adiabatic water jacket

Figure 4-8 Bomb calorimeter. A food substance is attached to the ignition wires and placed in the chamber under several atmospheres of O_2 pressure. The sample is then ignited and burns explosively. The stirrer distributes the heat of combustion uniformly throughout the water surrounding the chamber. The thermometer detects the heat released. Based on Kleiber, 1961.

quotient (RQ) and reflects cellular processes. Equation 4-4 shows why the RQ of glucose, a sugar carbohydrate, is unity:

$$C_6H_{12}O_6 + 6\ O_2 \rightarrow 6\ CO_2 + 6\ H_2O$$
$$6\ CO_2 \text{ produced } / 6\ O_2 \text{ consumed} \qquad (4\text{-}4)$$
$$= 1.0 = RQ$$

For the neutral fat trioleate, the RQ approximates 0.7:

$$C_{57}H_{104}O_6 + 80\ O_2 \rightarrow 57\ CO_2 + 52\ H_2O$$
$$RQ = \frac{57}{80} = 0.71 \qquad (4\text{-}5)$$

During hard exercise, an individual's respiratory gas exchange ratio [R, an estimate of RQ (see page 49)] approaches 1.0, whereas during prolonged exercise, the R may be somewhat lower, 0.9 or less. Figure 4-9 shows data on male subjects running a marathon on a treadmill where respiratory gas exchange could be measured. In one case, subjects ran at their race pace, whereas in the other case subjects ran more slowly. Note that subjects' race pace corresponds to an R of 0.95 to 1.0. Note also that in later stages, R declined, but then rose at the end. In the slower marathon runners, R was lower, indicating less carbohydrate and more lipid use than in the faster runners. The fuel mix was, however, still mostly carbohydrate.

TABLE 4-1

Caloric Equivalents of Foodstuffs Combusted Inside and Outside the Body

Food	kcal · liter O_2^{-1}	RQ ($\dot{V}_{CO_2}/\dot{V}_{O_2}$)	Inside Body (kcal · g^{-1})	Outside Body (kcal · g^{-1})
Carbohydrate	5.05	1.00	4.2	4.2
Fat	4.70	0.70	9.5	9.5
Protein	4.50	0.80	4.2	5.7[a]
Mixed diet	4.82	0.82		
Starving individual	4.70	0.70		

[a] The amount of protein combusted outside the body is greater than that combusted inside the body (see text):
$$\frac{5.7 - 4.2}{5.7} = 26\% \text{ difference}$$

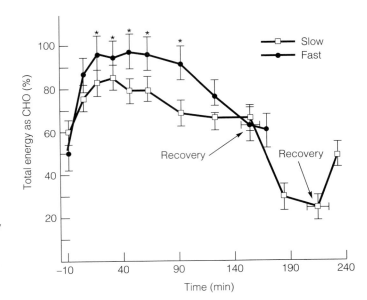

Figure 4-9 Calculated percentage of energy expenditure contributed by carbohydrates (CHO) before, during, and after a treadmill marathon in fast and slow groups. Values are means ± SEM; N = 6 per group. SOURCE: O'Brien et al., 1993. Used with permission.

Table 4-1 shows why it is an advantage for these changes in RQ to occur. During hard exercise, O_2 consumption can be limiting. Therefore, as shown in Equation 4-6, in oxidizing carbohydrate rather than fat the individual derives

$$\frac{5.0 - 4.7 \text{ kcal} \cdot \text{liter}^{-1} \text{ O}_2}{4.7 \text{ kcal} \cdot \text{liter}^{-1} \text{ O}_2} \qquad (4\text{-}6)$$

or 6.4% more energy per unit O_2 consumed. During prolonged exercise, however, it makes sense that RQ decreases, indicating that more fat is combusted. In prolonged work, glycogen supply rather than O_2 consumption can be limiting. Table 4-1 indicates that on a mass basis, fats provide about 9.5/4.2 kcal $\cdot$ g^{-1}, or 2.3 times as much energy as carbohydrate. Given this large difference, we can also see why endurance training improves the ability to use fat as a fuel during prolonged mild to moderate intensity exercise (i.e., 40–60% $\dot{V}_{\text{O}_2\text{max}}$).

The preceding type of discussion is sometimes referred to as a "teleological argument," meaning that the purpose of something is assumed to explain its operation. In actuality, as will be shown, the reason why relatively more carbohydrate is used in hard exercise is related to the quantity of activity and regulation of glycolytic enzymes. There are also enzymatic explanations for the preponderance of fat used in prolonged exercise.

In order to obtain a precise estimate of metabolic rate and fuel used by means of indirect calorimetry, we must know a few other details besides the quantities of O_2 consumed and CO_2 produced. These additional parameters include the food ingested and the nitrogen excreted. To provide a relatively simple example of the utility of indirect calorimetry, let us consider a starving man, in whom there is no food input to account for and no large excretion of urinary nitrogen (Table 4-2).

In exercise physiology, current estimates of fuels combusted are usually simplified by assuming there is no increase in the basal amino acid and protein degradation during exercise. The ventilatory exchange ratio R is then used to represent the nonprotein RQ. As we shall see later (Chapter 8), this assumption is not quite valid.

Although both RQ and R are given by the same formula ($\dot{V}_{\text{CO}_2} / \dot{V}_{\text{O}_2}$), over any short period of measurement of gas exchange at the lungs, changes in CO_2 storage may cause R not to equal RQ. Although RQ does not exceed 1.0, R can reach 1.5 or higher. For the present, let us consider RQ to be the ratio

	TABLE 4-2	
	Calculation of Nitrogen-Free RQ on a Resting Starving Man	

Given: (a) Protein is about 17% N by weight, or there is 1 g N $\cdot$ 5.9 g^{-1} protein (1/5.9 = 0.17).

          (b) For protein RQ = 4.9/5.9 = 0.83, or 4.9 liters CO_2 are derived from the catabolism of the protein associated with 1 g N, and 5.9 liters of O_2 are required to catabolize the protein.

The total O_2 consumption was 634 liters. The total CO_2 production was 461 liters, and urinary N losses were 14.7 g over 24 hours. We can use these data to calculate the nitrogen-free RQ.

Calculations	Total CO_2 (liters)	Total O_2 (liters)
	461	634
In the urine, there were 14.7 g N. The CO_2 produced by protein catabolism was (14.7) (4.9) = 72.0 liters CO_2.	72	
The O_2 consumed associated with protein catabolism was (14.7) (5.9) = 86.7 liters O_2.		86.7
	389	547.3

$$\text{Nonprotein RQ} = \frac{389}{547.3} = 0.71$$

Heat production
    From protein: (14.7 g N) (5.9 g protein $\cdot$ g^{-1} N) (4.2 kcal $\cdot$ g^{-1} protein) = 364.3 kcal

    From fat: The nonprotein RQ was 0.71, so fat comprised the remaining fuel. Therefore, (547.3 liters O_2) (4.7 kcal $\cdot$ liter^{-1} O_2) = 2572.3 kcal.

            Total heat production = 364.3 + 2572.3 kcal = 2936.6 kcal

$\dot{V}_{CO_2}/\dot{V}_{O_2}$ in the cell, where O_2 is consumed and CO_2 produced. Further, let us consider R to be the ratio $\dot{V}_{CO_2}/\dot{V}_{O_2}$ measured at the mouth. Over time, R must equal RQ, but during the onset and offset of exercise, as well as during hard exercise, $R \neq$ RQ because body CO_2 storage changes (see Figure 4-9).

■ Indirect Calorimetry

For individuals at rest, indirect calorimetric determinations on the effects of body size, growth, disease, gender, drugs, nutrition, age, and environment on metabolism are very useful. The resting metabolic rate per unit body mass is greater in males than in females, greater in children than in the aged, greater in small individuals than in large ones, and greater under extremes of heat and cold than under normal conditions.

■ The Utility of Indirect Calorimetry During Exercise

Physical exercise represents a special metabolic situation. As Figure 4-1 indicates, for a body at rest, all the energy liberated within appears as heat. If metabolism is constant, the quantity of heat produced within the body over a period of time will be the same as that leaving the body. However, during exercise, some of the energy liberated within the body appears as physical work outside the body. Therefore, devices to measure external work performed, such as bicycle ergometers (Figure 4-10) and treadmills (Figure 4-11) are used.

During exercise, direct calorimeters such as the Atwater–Rosa calorimeter (Figure 4-7) are of little use for several reasons. First, such devices are very expensive. Second, the heat generated by an ergometer, if it is electrically powered, may far exceed that of the subject. Third, body temperature increases during exercise because not all the heat produced is liberated from the body. Therefore, the sensors in the walls of the calorimeter do not pick up all the heat produced. Finally, the body sweats during exercise, which also affects the calorimeter and changes body mass. Changes in body mass and the unequal distribution of heat within the body make it very difficult to use direct calorimetry in exercise.

As with direct calorimetry during exercise, techniques of indirect calorimetry have certain limitations. These are summarized in Figure 4-12a. In order for determinations of $\dot{V}_{O_2}$ to reflect metabolism accurately, the situation in Figure 4-12a must hold. If another mechanism is used to supply energy, such as that shown in Figure 4-12b, then respiratory determinations do not completely reflect all metabolic processes. As will be seen, the body has the means to derive energy from the degradation of substances without the immediate use of O_2. These mechanisms include immediate sources and rapid glycogen (muscle carbohydrate) breakdown. Use of the $\dot{V}_{CO_2} / \dot{V}_{O_2}$ ratio is also limited during exercise. Although over time the O_2 consumed by and CO_2 liberated at the lungs (the respiratory, or ventilatory exchange ratio) equals the respiratory quotient, the

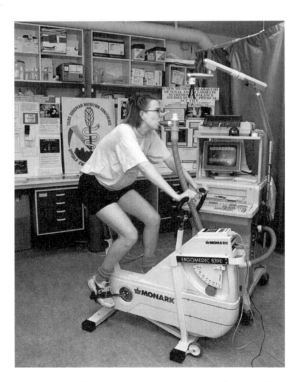

Figure 4-10 Bicycle ergometers are convenient, stationary laboratory devices to control the external work rate (power output) while physiological responses to standardized or experimental protocols are observed. Courtesy of Monark, Inc., Varberg, Sweden.

Figure 4-11 Treadmills are frequently used in the laboratory to apply exercise stress and record physiological responses on relatively stationary subjects during exercise. Compared to the bicycle ergometer (Figure 4-10), it is difficult to quantify external work on the treadmill. However, the treadmill does allow subjects to walk or run, which are perhaps more common modes of locomotion than is by-cycling. PHOTO: © David Madison.

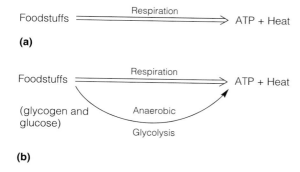

(a)

(b)

Figure 4-12 Respiration and ATP production. The validity of indirect calorimetric measurements depends on the O_2 consumption accurately representing the ATP formed. This is not always the case in exercise. In (a), the measurement is valid. In (b), it is invalid.

cellular events are not always immediately represented in expired air. This is because the cells are fluid systems, and they are surrounded by other fluid systems on both the arterial and venous sides. When exercise starts, CO_2 is frequently stored in cells. When exercise is very difficult, the blood bicarbonate buffer system buffers lactic acid, and extra nonrespiratory CO_2 is produced (Chapter 11).

Lactic acid (HLA) is a strong acid whose level in muscle and blood increases during heavy work. It is known as a strong acid in physiological systems because it can readily dissociate a proton (H^+). To lessen the effect of protons generated from lactic acid during hard exercise, the body has a system of chemicals that lessen, or buffer, the effects of the acid. In the blood, the bicarbonate (HCO_3^-)–carbonic acid (H_2CO_3) system is the main system by which the effects of lactic acid are buffered. In equations 4-7 to 4-9, HCO_3^- neutralizes the H^+, but CO_2 is produced. This is eliminated at the lungs and appears in the breath. Consequently, during hard exercise, $R \neq RQ$. After exercise, metabolic CO_2 may be stored in cells, blood, and other body compartments to make up for that lost during exercise.

$$HLA \rightarrow H^+ + LA^- \qquad (4\text{-}7)$$
$$H^+ + HCO_3^- \rightarrow H_2CO_3 \qquad (4\text{-}8)$$
$$H_2CO_3 \rightarrow H_2O + CO_2 \qquad (4\text{-}9)$$

Furthermore, during and immediately after exercise, urine production by the kidney is inhibited. Also, during exercise considerable nitrogen can be lost as urea in sweat. Therefore, it is difficult to determine the nitrogen excreted during exercise.

Determinations of indirect calorimetry are somewhat limited in their use by the fact that the respiratory gases give no specific information on the fuels used. If, for example, RQ is 1.0, then although we know that carbohydrate was the fuel catabolized, we do not know specifically which carbohydrate was involved. The possibilities could include, among others, glycogen, glucose, lactic acid, and pyruvic acid. However, radioactive and nonradioactive tracers to study metabolism at rest and during exercise have come into use in conjunction with indirect calorimetry to provide more detailed information on specific fuels.

Exercise is also a special situation in that the metabolic responses persist long after the exercise itself may have been completed. Consequently, physical activity results in an excess postexercise O_2 consumption (EPOC). This EPOC has sometimes been called the "O_2 debt" and has been used as a measure of anaerobic metabolism during exercise. A more detailed explanation of the O_2 debt is given later (Chapter 10); suffice it to say here that the mechanisms of the O_2 debt are complex and cannot be used to estimate anaerobic metabolism during exercise.

Whereas the body does present certain problems in determining metabolic rate during exercise, careful consideration of those various factors allows us to obtain important information about metabolic responses to exercise. Estimations of $\dot{V}_{O_2}$, for example, provide information on the cardioventilatory response to exercise. The caloric cost of various exercises can be estimated (Table 4-3), and information about the fuels used to support the exercise can be obtained.

Knowing that part of the energy liberated during exercise appears as external work is useful. By measuring the respiratory response to graded, submaximal exercise at specific external work rates, we can determine the fraction of the energy liberated within the human machine that appears as external

TABLE 4-3

Estimates of Caloric Expenditures of Sports Activities for a 70-kg Person

Activity	Caloric Expenditure (kcal · min⁻¹)	Activity	Caloric Expenditure (kcal · min⁻¹)
Archery	4.6	Resting	1.2
Badminton	6.4	Running	
Basketball	9.8	8 min · mi⁻¹	14.8
Canoeing	7.3	6 min · mi⁻¹	17.9
Cycling	12.0	Squash	15.1
Field hockey	9.5	Swimming	
Fishing	4.4	Backstroke	12.0
Football	9.4	Crawl	11.1
Golf	6.0	Tennis	7.7
Gymnastics	4.7	Volleyball	3.6
Judo	13.8	Walking, easy	5.7

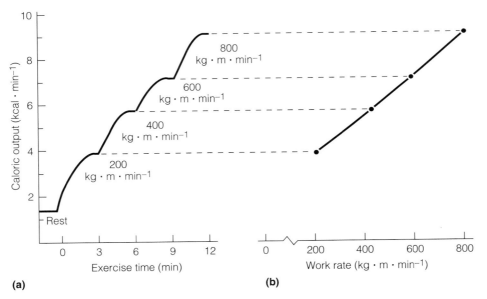

(a)   **(b)**

Figure 4-13   Respiratory response to graded submaximal bicycle ergometer work. Every 3 minutes the work rate is increased 200 kg · m · min⁻¹. The observed O_2 consumption $(\dot{V}_{O_2})$ is converted to kcal · min⁻¹. These values are then plotted as (a) a function of time and (b) a function of the steady-rate work load. Note that a plot of the caloric cost of exercise against work rate (b) yields a straight line, or one that bends upward slightly.

work. This fraction is frequently reported as a percentage and is called *efficiency*.

An example of how the efficiency of the human body is calculated during bicycle ergometer exercise is given in Equation 4-10. In Figure 4-13, we see that the O_2 consumption of an individual increases in direct response to increments in work load while pedaling at constant speed. In this case, efficiency can be calculated as in Equation 4-10.

The plateau steps in Figure 4-13a are referred to as *steady-rate* exercise. During the steady rate, the oxygen consumption ($\dot{V}_{O_2}$) is relatively constant and is directly proportional to the constant submaximal work load.

The calculation of body efficiency during bicycle exercise is given in Table 4-4. Here the calculated value of efficiency is 29.2%, which is close to a maximum value for bicycle ergometer work. Cycling at greater speeds and working at greater loads results in decrements in calculated efficiency. The efficiency of walking is slightly higher than that of cycling, but responds similarly to increments in speed and resistance. The reason it is usually easier to cycle from one place to another than to walk is that the rolling and wind resistance to cycling at a particular speed are far less than the work done in accelerating and decelerating the limbs during walking—that is, less work is done in cycling. Attempting to bicycle in soft sand will reveal that the work done in covering a given distance is far greater; yet measurements of the efficiency of movement would reveal no change or only a relatively small decrement.

In contrast to the bicycle ergometer, where the work done is the product of the pedaling speed and the resistance, the calculation of work done in walking is more involved. This is because the body walking on a level treadmill does no external work. Estimates of the work done in walking, therefore, depend on applying an external work load that can be measured, or by estimating the work done internally in the body as a result of accelerating and decelerating the limbs.

The most common way to apply external work during walking is to have a subject go up an incline. In Figure 4-14, the vertical external work performed is in lifting the body mass the distance $B-D$. The work done is calculated according to either of two formulas as seen in Equations 4-11 and 4-12.

TABLE 4-4

Calculation of Body Efficiency During Cycling Exercise

Given:

$\dot{V}_{O_2}$ at 200 kg $\cdot$ m $\cdot$ min^{-1} = 0.76 liter $\cdot$ min^{-1}
$\dot{V}_{O_2}$ at 400 kg $\cdot$ m $\cdot$ min^{-1} = 1.08 liters $\cdot$ min^{-1}
R = RQ = 1.0
When RQ = 1.0, 1 liter O_2 = 5 kcal
1 kg $\cdot$ m = 0.00234 kcal

$$\text{Efficiency} = \frac{\text{Change in work output}}{\text{Change in } \dot{V}_{O_2}}$$

$$\text{Efficiency}$$
$$= \frac{400 - 200 \text{ kg} \cdot \text{m} \cdot \text{min}^{-1}}{1.08 - 0.76 \text{ liter} \cdot \text{min}^{-1}}$$

$$= \frac{200 \text{ kg} \cdot \text{m} \cdot \text{min}^{-1} \times 0.00234 \text{ kcal} \cdot \text{kg}^{-1} \cdot \text{m}^{-1}}{0.32 \text{ liter} \cdot \text{min}^{-1} \times 5 \text{ kcal} \cdot \text{liter}^{-1} O_2}$$

$$= 0.292 = 29.2\%$$

$$\text{Efficiency} = \frac{\text{Caloric equivalent of change in external work}}{\text{Caloric equivalent of change in } O_2 \text{ consumption}} \tag{4-10}$$

$$\text{External work rate} = [\text{Body weight (kg)}] \, [\text{Speed (m} \cdot \text{min}^{-1})] \, [\% \text{ grade}/100] \tag{4-11}$$

$$\text{External work rate} = [\text{Body weight (kg)}] \, [\text{Speed (m} \cdot \text{min}^{-1})] \, [\sin \Theta] \tag{4-12}$$

where sin Θ is the angle $ACB = BD/CB$.

Recently, external work has been applied in studies of energetics by having subjects walk against a horizontal impeding force (Figure 4-15). The work done against the horizontal impeding force is calculated in Equation 4-13.

An example of how to calculate the efficiency of performing external work during incline walking is given in Table 4-5.

Another innovation for estimating the work involved in horizontal walking has been established by Ralston, Zarrugh, and other mechanical engineers at the University of California. They attached sensitive transducers to the joints so that their movements during walking could be recorded; these recordings, coupled with estimates of the masses of different body parts, made it possible to calculate on a computer the work done in moving the body parts

$$\text{External work} = [\text{Speed (m} \cdot \text{min}^{-1})] \, [\text{Weight pulled (kg)}] \qquad (4\text{-}13)$$

Figure 4-14 During horizontal treadmill walking, no external work is done; therefore, it is impossible to calculate a value for body efficiency. However, a way to determine external work is to measure the work done in lifting the body up a hill. Refer to Equations 4-11 and 4-12 in the text for details of work rate calculation.

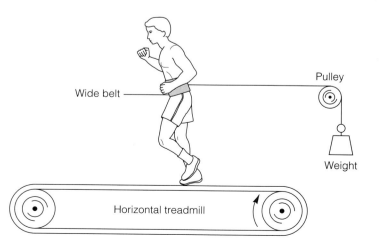

Figure 4-15 External work can be determined during horizontal treadmill walking by having subjects pull a training weight. Refer to Equation 4-13 in the text for work rate calculation.

TABLE 4-5

Estimation of the Whole-Body Efficiency of Doing Vertical Work During Steady-Rate Treadmill Walking at 3.0 km · hr^{-1}

Given: (a) Steady-rate caloric equivalent of $\dot{V}_{O_2}$ during horizontal, ungraded walking (i.e., zero vertical work) at 3.0 km · hr^{-1} = 5 kcal · min^{-1}

(b) Steady-rate caloric equivalent of $\dot{V}_{O_2}$ while performing 375 kg · m · min^{-1} of vertical work at 3.0 km · hr^{-1} = 7.9 kcal · min^{-1}

(c) 1.0 kg · m = 0.00234 kcal

$$\text{Efficiency} = \frac{\text{Caloric equivalent of change in vertical work}}{\text{Caloric equivalent of change in respiration}}$$

$$= \frac{(375 \text{ kg} \cdot \text{m} \cdot \text{min}^{-1} - 0 \text{ kg} \cdot \text{m} \cdot \text{min}^{-1})(0.00234 \text{ kcal} \cdot \text{kg}^{-1} \cdot \text{m}^{-1})}{7.9 - 5 \text{ kcal} \cdot \text{min}^{-1}}$$

$$= 0.30 \text{ or } 30\%$$

---

and the entire body. Because the various techniques of estimating work done in walking give similar results, it appears that the efficiencies with which the body does internal, horizontal, and lifting work during walking are similar.

Although the efficiency of the body during easy cycling and walking may be as high as 30%, it can only be surmised that the efficiency of running is somewhat lower. Evidence concerning the efficiency of running is lacking because running is not a true steady-rate situation. During running, the metabolic rate is so high that both situations a and b in Figure 4-12 occur. Because $\dot{V}_{O_2}$ does not account for all the ATP supplied during running, a proper estimation of efficiency during running awaits development of the technical ability to estimate nonoxidative ATP supply during exercise.

■ Summary

Metabolism can be estimated in two ways: by direct determinations of heat production and by determinations of O_2 consumption. Determinations of metabolic rate provide valuable information about the status of an individual. In resting individuals, both methods provide similar results. During exercise, direct calorimetry is not feasible; therefore, indirect calorimetry must be used. However, during hard and prolonged exercise, indirect calorimetry may not provide a precise estimate of metabolic rate. Under these conditions, determinations of O_2 consumption still provide important information about the cardioventilatory systems.

■ Selected Readings

Asmussen, E. Aerobic recovery after anaerobiosis in rest and work. *Acta Physiol. Scand.* 11: 197–210, 1946.

Atwater, W. O., and F. G. Benedict. Experiments on the metabolism of matter and energy in the human body. U.S. Dept. Agr. Off. Exp. Sta. Bull. 136: 1–357, 1903.

Atwater, W. O., and E. B. Rosa. Description of new respiration calorimeter and experiments on the conservation in the human body. U.S. Dept. Agr. Off. Exp. Sta. Bull. 63, 1899.

Benedict, F. G., and E. P. Cathcart. Muscular Work. A Metabolic Study with Special Reference to the Efficiency of the Human Body as a Machine. (Publ. 187). Washington, D.C.: Carnegie Institution of Washington, 1913.

Benedict, F. G., and H. Murchhauer. Energy Transformations During Horizontal Walking. (Publ. 231). Washington, D.C.: Carnegie Institution of Washington, 1945.

Brooks, G. A., C. M. Donovan, and T. P. White. Estimation of anaerobic energy production and efficiency in rats during exercise. *J. Appl. Physiol.: Respirat. Environ. Exercise Physiol.* 56: 520–525, 1984.

Dickensen, S. The efficiency of bicycle pedaling as affected by speed and load. *J. Physiol.* (London) 67: 242–255, 1929.

Donovan, C. M., and G. A. Brooks. Muscular efficiency during steady-rate exercise II: effects of walking speed on work rate. *J. Appl. Physiol.* 43: 431–439, 1977.

Gaesser, G. A., and G. A. Brooks. Muscular efficiency during steady-rate exercise: effects of speed and work rate. *J. Appl. Physiol.* 38: 1132, 1975.

Haldane, J. S. A new form of apparatus for measuring the respiratory exchange of animals. *J. Physiol.* (London) 13: 419–430, 1892.

Kleiber, M. Calorimetric measurements. In Biophysical Research Methods, F. Über (Ed.). New York: Interscience, 1950.

Kleiber, M. The Fire of Life: An Introduction to Animal Energetics. New York: Wiley, 1961, pp. 116–128, 291–311.

Krogh, A., and J. Lindhard. The relative value of fat and carbohydrate as sources of muscular energy. *Biochem. J.* 14: 290, 1920.

Lavoisier, A. L., and R. S. de La Place. Mémoire sur la Chaleur; Mémoires de l'Académie Royale (1789). Reprinted in Ostwald's Klassiker, no. 40, Leipzig, 1892.

Lloyd, B. B., and R. M. Zacks. The mechanical efficiency of treadmill running against a horizontal impeding force. *J. Physiol.* (London) 223: 355–363, 1972.

O'Brien, M. J., C. A. Viguie, R. S. Mazzeo, and G. A. Brooks. Carbohydrate dependence during marathon running. *Med. Sci. Sports Exer.* 25: 1009–1017, 1993.

Ralston, H. J. Energy-speed relation and optimal speed during level walking. *Intern. Z. Angew. Physiol.* 17: 277–282, 1958.

Smith, H. M. Gaseous Exchange and Physiological Requirements for Level Walking. (Publ. 309). Washington, D.C.: Carnegie Institution of Washington, 1922.

Wilkie, D. R. The efficiency of muscular contraction. *J. Mechanochem. Cell Motility* 2: 257–267, 1974.

Zarrugh, M. Y., F. M. Todd, and H. J. Ralston. Optimization of energy expenditure during level walking. *European J. Appl. Physiol. Occupational Physiol.* 33: 293–306, 1974.

Zinker, B. A., K. Britz, and G. A. Brooks. Effects of a 36-hour fast on human endurance and substrate utilization. *J. Appl. Physiol.* 69: 1849–1855, 1990.

CHAPTER 5

GLYCOGENOLYSIS AND GLYCOLYSIS IN MUSCLE:

The Cellular Degradation of Sugar and Carbohydrate to Pyruvate and Lactate

Of the three main foodstuffs (carbohydrates, fats, and proteins), only carbohydrates can be degraded without the direct participation of oxygen. The main product of dietary sugar and starch digestion is glucose, which is released into the blood of the systemic circulation. The simple sugar glucose enters cells, including muscle and liver cells, and is either used directly or stored for later use. The first stage of cellular glucose catabolism is called glycolysis. Glucose molecules not undergoing glycolysis can be linked together to form the carbohydrate storage form called glycogen. Glycogen stored in muscle is broken down in a process called glycogenolysis. These processes occur in all cells but are specialized in some muscle cells and red blood cells. Glycogenolysis and glycolysis provide the energy to sustain powerful muscular contractions for a period of a few seconds to about a minute. During prolonged exercise, glycogenolysis provides most of the fuel for muscle contractions. At the same time, glycogenolysis in liver provides glucose, which circulates via the bloodstream to working muscle. In Chapter 5, we emphasize the muscular use of carbohydrate. In Chapter 9, the role of gluconeogenic organs (liver and kidneys) will be emphasized.

Readers of this third edition of *Exercise Physi-*

Michael Johnson, one of the world's premier long sprinters, in action. Success in activities such as sprinting requires great muscular power and the ability of muscles to operate from chemical energy stores which do not immediately require O_2 for metabolism. PHOTO: © Steven E. Sutton/Duomo.

ology will find major changes in the current treatment of the relationship between glycolysis and oxidative metabolism. Indeed, the current treatment of the topic is not only different from that of the second edition, but it is also different from presentations of the topic found in most other texts on physiology and biochemistry. These differences are due to recent discoveries concerning the presence of lactate transporters, their distribution in cells and tissues of humans and other species, and the ability of mitochondria to take up and oxidize lactate directly. As will be described in the following section, since the 1920s, it has been believed that lactate is produced in muscle and other cells as the result of insufficient oxygen. However, it is now clear that lactate is continuously produced and removed. Hence, in resting muscle there is ten times more lactate than pyruvate. Even though glycolysis inevitably results in lactate formation, this formation is of little consequence so long as the lactate is removed. Lactate is removed mainly by oxidation. As such it is a major energy source.

Recognition that lactate moves between lactate-producing and lactate-consuming cells led to formulation of the *lactate shuttle concept.* Highly glycolytic (white, type IIb) muscle fibers were thought to be the primary sites of lactate formation. The classic, cell-cell lactate shuttle concept posited that highly oxidative cells (red, slow oxidative and cardiac muscle cells) were the sites of lactate removal. Recent discoveries indicating that mitochondria take up and oxidize lactate directly have led to formulation of the *"Intracellular Lactate Shuttle Concept."* According to this concept, glycolysis results in the formulation of lactate because of the abundance of the terminal glycolytic enzyme (lactate dehydrogenase, LDH), in which the equilibrium constant (K_{eq}) of LDH is $3.6 \times 10^4 \, \text{m}^{-1}$. Thus, glycolysis in the cytosol results in lactate production, most of which is consumed by mitochondria that have the enzymatic apparatus to take up and oxidize lactate. The cell-cell and intracellular lactate shuttles function because some cells, such as those found in red skeletal muscle and cardiac muscles and liver, contain high mitochondrial densities.

■ The Dietary Sources of Glucose

Glucose, a six-carbon sugar, is the primary product of photosynthesis (Figure 5-1). Glucose is produced by plants which use it as a fuel, much as we do, and store it by linking the molecules together to form starch. Plants such as potatoes that store large amounts of starch are very useful to us as foodstuffs. Plants also link glucose molecules together in a complex pattern to form cellulose for structural purposes, but humans lack the enzymes necessary to digest this glucose polymer.

There are numerous dietary sources of glucose, including starches, such as rice, pasta, and potatoes, and dietary sugars, such as granulated sugar and brown sugar. Most of these enter the bloodstream in the form of glucose and lactate. Large starch molecules are split fairly rapidly into disaccharides and glucose by the action of the pancreatic enzymes known as amylases. Sugars other than glucose are largely isomerized—that is, converted to glucose by the wall of the small intestine or by the liver. In Chapter 28, we shall see that the form of delivery influences the rate of glucose delivery to the systemic circulation, and so physiological responses to sugar and complex carbohydrate feeding can differ, even if the carbon content of meals is the same.

The uptake of glucose by cells depends on several factors, including the type of tissue, the levels

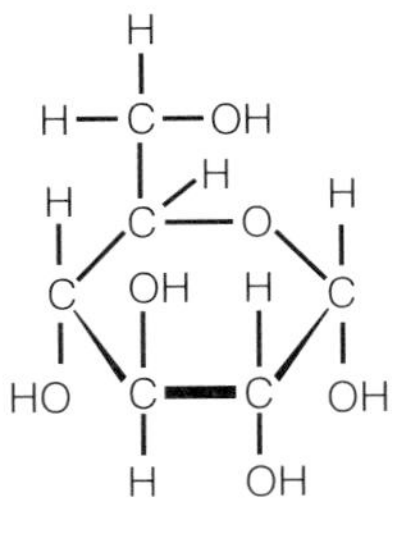

Figure 5-1 Structure of glucose, a simple sugar. Five carbons and an oxygen atom serve to create a hexagonal ring conformation. Shaded lines represent the three-dimensional platelike structure.

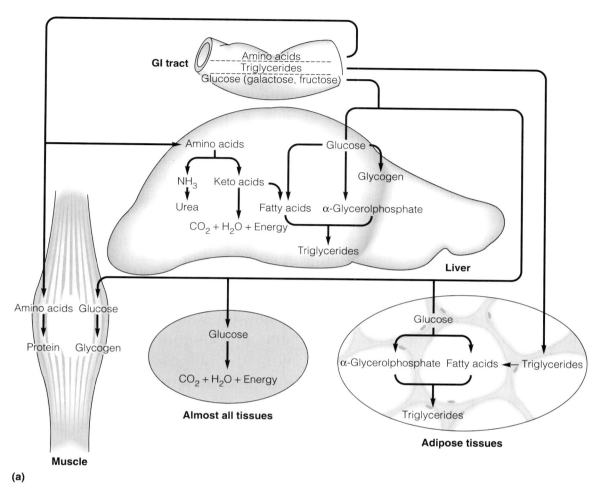

Figure 5-2   Flowchart of glucose and other metabolites in an individual (a) immediately after a meal and (b) during fasting. Modified from Vander, Sherman, and Luciano, 1980. Used with permission.

of glucose in the blood and tissue, the presence of the hormone insulin, and the physiological status of the tissue. Most tissues—with the notable exception of muscles during contraction—require insulin in order to take in glucose. The unique mechanism by which muscles take up glucose during exercise is currently being studied (Chapter 9). Nerve and brain tissues usually consume large amounts of glucose; the liver also usually takes up large amounts even if insulin is low. In the past, it was thought that the liver only stores and releases glucose but does not utilize it as a fuel. We now know, however, that under appropriate circumstances—that is, when the blood levels of both glucose and insulin are high—liver cells can use glucose. In adipose cells, the presence of glucose stimulates fat synthesis (Figure 5-2a).

The storage of dietary glucose under postprandial (after eating) conditions is illustrated in Figure 5-2a. Besides use as an energy substrate

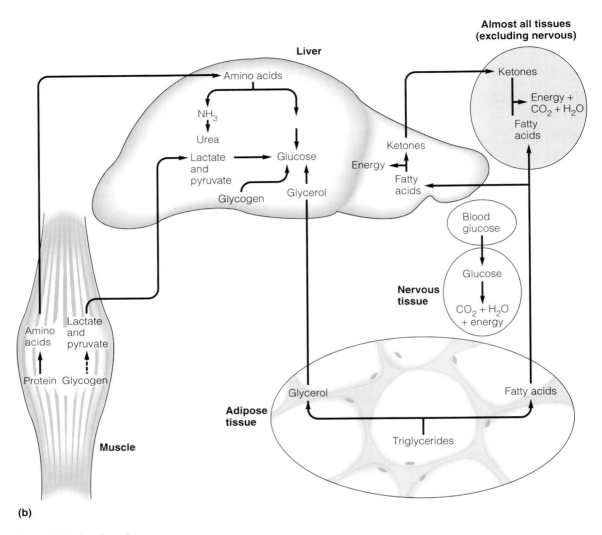

(b)

Figure 5-2 *(continued)*

(fuel), glucose is directed mainly to storage as glycogen in liver and muscle tissues and as triglyceride in adipose tissue. Figure 5-2b illustrates that the mobilization of substrates for energy and gluconeogenesis (the making of glucose) under postabsorptive (fasting) conditions is maintained by degradation of glycogen in the liver (i.e., glycogenolysis) and production of glucose in the liver and kidneys from precursors (mainly lactate) delivered in the circulation.

■ **Direct vs. Indirect Pathways of Liver Glycogen Synthesis: The "Glucose Paradox"**

Figure 5-3 diagrams the so-called new glucose to liver glycogen pathway. Of course, this pathway itself is not "new," as it likely has been inherited from our earliest ancestors; it is recognition of the pathway's importance that is "new." In contrast to Figure 5-2, Figure 5-3 shows that liver glycogen can

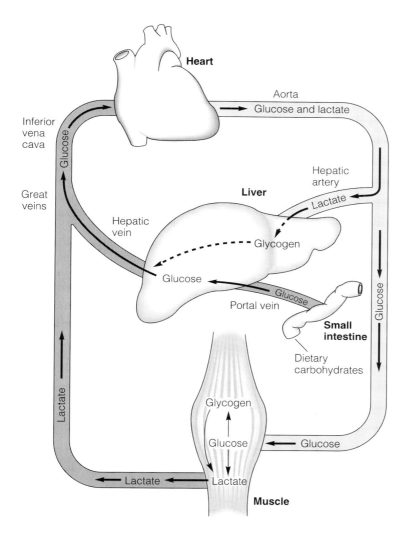

Figure 5-3 Diagram of the new glucose to hepatic glycogen pathway ("glucose paradox") by which the liver prefers to make glycogen from lactate as opposed to glucose. Glucose released into the blood from the digestion of dietary carbohydrate bypasses the liver and is taken up by skeletal muscle. The muscle can either synthesize glycogen or produce lactate. The lactate then recirculates to the liver and stimulates glucose and glycogen formation. See Foster (1984) for additional information.

be synthesized indirectly from three-carbon precursors, mainly lactate. When the pathway of liver glycogen synthesis from dietary carbohydrate involves hepatic uptake of glucose and addition of glucose to already existing glycogen molecules, the pathway of liver glycogen synthesis is said to be *direct*. However, when the pathway involves the degradation of glucose to lactate, and then conversion of lactate to glycogen, the pathway is said to be *indirect*. Because skeletal muscle is the largest tissue containing enzymes of glycolysis, much of the glucose-to-lactate conversion is thought to occur in muscle. Other cells

and tissues containing enzymes of glycolysis, including red blood cells (erythrocytes), adipose cells (adipocytes), and portions of the liver itself, are also likely to contribute to lactate formation in the indirect pathway. Because of the circuitous route of dietary carbohydrate carbon flow—that is, the use of lactate (an indirect precursor) rather than glucose (a direct precursor)—in liver glycogen synthesis, the phenomenon is sometimes also called the "glucose paradox." Current best estimates are that 60% of liver glycogen synthesis is by the direct pathway, whereas 40% is by the indirect, paradoxical pathway. However, conditions prior to and immediately after eating are likely to greatly affect the pathway of liver glycogen synthesis.

Recognition by biochemists that lactate is important in the synthesis of liver glycogen after eating comes as little surprise to exercise physiologists, who have long known that during and after exercise blood glucose supply is maintained by gluconeogenesis via the Cori cycle (discussed in more detail later in this chapter). However, it is important to note here that, like the Cori cycle, the concept of the indirect pathway is part of a revision in current thinking which indicates that the formation, distribution, and utilization of lactate is a central means by which carbohydrate metabolism in diverse tissues is coordinated. This overall conceptual scheme is termed the *lactate shuttle*, and is discussed more in a later section.

■ Blood Glucose Concentration During Rest and Exercise

The concentration of glucose in plasma is one of the most precisely regulated physiological variables (Chapter 9). Normal blood glucose in postabsorptive

humans is approximately 100 mg $\cdot$ dl^{-1} (100 mg %, or 5.55 mM). This level of glucose is required for function of the central nervous system (CNS) and other glucose-requiring systems, organs, and cells. During exercise, maintenance of blood glucose homeostasis becomes a major physiological challenge as skeletal muscle suddenly switches from a situation of little glucose uptake to a situation of greatly increased glucose uptake. Therefore, release of glucose by the liver (the liver being the main organ of glucose production) must rise from a value of approximately 1.8 mg $\cdot$ kg body weight^{-1} $\cdot$ min^{-1} to a much higher value. For instance, during exercise at 50% $\dot{V}_{O_2max}$, hepatic glucose production (HGP) rises to a value of approximately 3.5 mg $\cdot$ kg^{-1} $\cdot$ min^{-1}. Even greater values of HGP can be observed in maximal exercise.

A typical response of blood glucose in postabsorptive men is shown in Figure 5-4. From the normal value of about 100 mg/dl, glucose rises when exercise starts. This rise is due to a hormonally, and perhaps neurally, mediated feed-forward mechanism (Chapter 9). Then, over time, depending on liver glycogen reserves and other factors, glucose may fall, but remain within approximately 10% of the normative value until exercise stops.

Figure 5-4 Blood glucose concentration at rest, during exercise to exhaustion at 50% $\dot{V}_{O_2max}$, and 10-min postrecovery under postabsorptive (PA) and fasting (F) conditions. SE vertical bars are included for each sampling point, and horizontal bars are shown for time to exhaustion. *Significant difference between means (PA vs. F, $P < 0.05$). SOURCE: Zinker et al., 1990. Used with permission of the American Physiological Society.

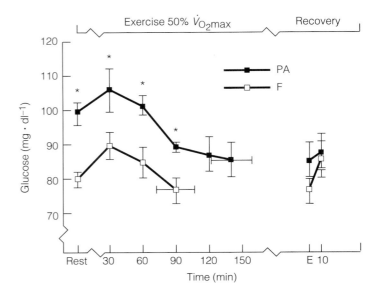

Figure 5-4 also shows the blood glucose response in the same subjects after a 36-hour fast. The concentration of glucose at rest is decreased approximately 20%, but this decrease is compensated for by doubling the levels of alternative fuels such as free fatty acids and ketones (Chapter 7). Even in the fasting state, glucose rises for a time during exercise but then falls, probably because liver glycogen reserves are depleted by the 36-hour fast, and gluconeogenesis cannot fully compensate. Low and falling blood glucose concentrations (hypoglycemia, glucose of ≤ 3.5 mM, 65 mg $\cdot$ dl^{-1}) are frequently associated with fatigue.

■ Glycolysis

The metabolic pathway of glucose breakdown in mammalian cells is called *glycolysis* (glyco / lysis), the dissolution of sugar. The process of glycolysis is frequently referred to as a metabolic pathway because the process occurs over a specific route, in either 11 or 12 specific steps, each of which is catalyzed and regulated by a specific enzyme. The pathway is sometimes called *anaerobic*, because oxygen is not directly involved.

The study and appreciation of glycolysis predates recorded history, thanks to the process of fermentation. Fermentation occurs when yeast carry glycolysis a few extra steps and produces ethyl alcohol, CO_2, and vinegar. There is some evidence that the earliest communities in ancient Egypt were organized around structures that served as bakeries and breweries. It was probably discovered quite by accident that grain left to soak fermented and became beer, and that leavening (yeast) added to dough fermented and yielded bread after baking. The practice of fermentation technology, however, has been extremely important in the development of many cultures. Louis Pasteur was famous even before he developed any vaccines to prevent diseases. Pasteur discovered a method for preventing ordinary table wine from fermenting to vinegar. Today, his process, called *pasteurization,* is more frequently used to preserve milk.

Glycolysis in Muscle

The process of glycolysis is very active in skeletal muscle, which is often termed a *glycolytic tissue.* In particular, pale or white skeletal muscles contain large quantities of glycolytic enzymes. As we will see later, highly oxidative red muscle fibers and heart cells are capable of rapid glycolysis, but glycolysis is the main energy source for white muscle fibers during exercise (Figure 5-5). Because it is the predominant metabolic pathway of energy conservation in amphibians and reptiles, glycolysis is frequently considered a "primitive pathway." The hearts and lungs of these animals are poorly developed and cannot utilize oxidative metabolism for bursts of activity; they must rely, therefore, on glycogenolysis and glycolysis.

Aerobic, Anaerobic, and "Nonrobic" Glycolysis in the Cytosol

There are two forms of glycolysis: aerobic ("with oxygen") and anaerobic ("without oxygen"). Historically, these terms were developed by scientists, such as Pasteur, who studied glycolysis in yeast and other unicellular organisms in test tubes where air was either present or was flushed out by gassing with substances like nitrogen. Pasteur noted that when oxygen was present (aerobic), glycolysis was slow and lactate did not accumulate. When oxygen was removed (anaerobic), Pasteur found that glycolysis was rapid. Today, we also know that in most cells glycolysis proceeds the additional step beyond pyruvic acid to form lactic acid even when oxygen is present in abundance (Figure 5-6).

The early experimentation and terminology has led to some confusion in contemporary physiology, because when pioneer scientists observed increased levels of lactic acid in muscle and blood as the result of exercise, they assumed that the tissue was anaerobic (without oxygen) during exercise. As we will see, however, there are several reasons for lactic acid formation; no oxygen or insufficient oxygen supply is seldom one of them. For instance, maximal exercise under the low oxygen condition of high altitude exposure results in little lactate accumulation. Further,

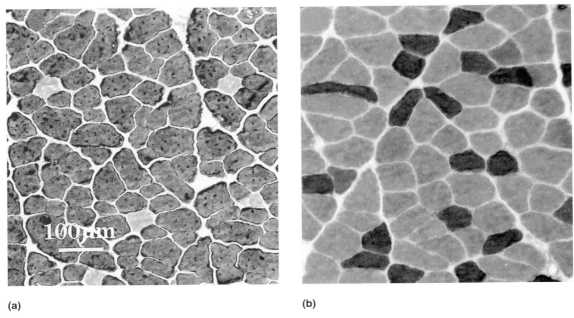

(a)   (b)

Figure 5-5 Transverse sections of skeletal muscle from a rodent, incubated for myofibrillar-ATPase at pH 10.3. Panel (a) is a section from the extensor digitorum longus (EDL) muscle, which is composed primarily of fibers that have high glycolytic capacity. Note the high proportion of dark-appearing fibers in the EDL muscle. In contrast, panel (b) is a section from the soleus (SOL) muscle, which is composed primarily of fibers low in glycolytic capacity (but high in oxidative type I fibers). Note the large number of light-appearing fibers. The details regarding the diversity of fiber types are developed in Chapter 18. SOURCE: Micrographs, courtesy of T. P. White.

a constant level of lactic acid in the blood does not mean that no lactic acid is formed, only that production and removal are equivalent. Also, there is good evidence that lactic acid is always produced, even at rest, as in the indirect pathway of liver glycogen synthesis after eating. Therefore, the terms *aerobic* and *anaerobic* as applied to mammalian systems are archaic. We can deal with these terms and use them as long as we keep in mind their actual meanings. Perhaps more descriptive terms such as "rapid" (for anaerobic) and "slow" (for aerobic) will eventually come into use. Because much of glycolysis has little to do with the presence of O_2, the process itself is essentially "nonrobic"; that is, it does not involve O_2. *Note:* Before closing this section it is important to mention that "nonrobic" is used for emphasis

and is *not* a generally appropriate or widely accepted and used term.

Glycolysis is a process that occurs mainly in the cytosol, where the enzymes are concentrated. However, some glycolytic enzymes, such as lactate dehydrogenase (LDH) exist in other cellular organelles, such as mitochondria and microsomes. The glycolysis pathway is depicted in two ways in Figures 5-7 and 5-8. Figure 5-7 is a detailed presentation that will serve as a reference. Figure 5-8 is designed to contribute to an understanding of how the pathway is controlled when the rate (flux) of glycolysis is slow (Figure 5-8a), and when flux is rapid (5-8b). In glycolysis, one six-carbon sugar is split into two three-carbon carboxylic acids. Pyruvic and lactic acid each possess a carboxyl group (Figure 5-9).

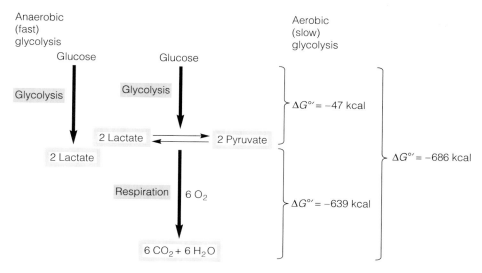

Figure 5-6 In anaerobic (fast) glycolysis, lactic acid is the product. In aerobic (slow) glycolysis, pyruvic acid is the main product. The terms *aerobic* (with O_2) and *anaerobic* (without O_2) refer to the test tube conditions used by early researchers to speed up or slow down glycolysis. In real life, pyruvate and lactate pools are in equilibrium, and the rapidity of glycolysis largely determines the product formed. Note the far greater energy released under aerobic conditions. Modified from Lehninger, 1973. Used with permission.

At physiological pH, these molecules dissociate a hydrogen ion (H^+) and therefore are acids. The terms *pyruvate* and *lactate,* properly meaning salts of the respective acids, are generally used interchangeably with pyruvic acid and lactic acid. Figure 5-9 illustrates why glycolysis inevitably results in lactate production. The standard free energy change for the reaction catalyzed by lactate dehydrogenase ($\Delta G^{\circ\prime}$) is very high (-6.0 kcal), as is the equilibrium constant (K_{eq}, 3.6×10^4 m^{-1}). Consequently, even considering K_M differences among the various LDH isoforms, in the cytosol, pyruvate not immediately entering mitochondria is reduced to lactate. For this reason, lactate accumulation is typical in red blood cells (erythrocytes), which do not have mitochondria, and type IIb muscle fibers with high glycolytic flux rates and low mitochondrial density.

The substance diagrammed in Figure 5-10a is called nicotinamide adenine dinucleotide (NAD^+). Because of its unique structure, NAD^+ can exist in two forms: NAD^+ (oxidized) and NADH (reduced,

Figure 5-10b). We shall encounter NAD^+ frequently, because it transfers hydrogen ions and electrons within cells and also because the cellular (NADH/ NAD^+) ratio (or redox) is important in the control of metabolism. Examination of its structure (Figure 5-10a) reveals that NAD^+ contains the nucleotides adenine and nicotinamide. We have already encountered adenine in the structure of ATP (Figure 2-6). Nicotinamide is the product of a B vitamin. The dietary deficiency disease pellagra is typified by poor energy metabolism. Flavine adenine dinucleotide (FAD) is a compound similar to NAD^+. FAD can be reduced to $FADH_2$ and, like NADH, functions to conserve and transport reducing equivalents within the cell.

The simplified flowchart for glycolysis (Figure 5-8) reveals that step 6 involves the reduction of NAD^+ (adding of hydrogen and electrons) to yield NADH. When glycolysis proceeds slowly (Figure 5-8a), NADH transports or "shuttles" the hydrogen and electron to mitochondria (those cellular

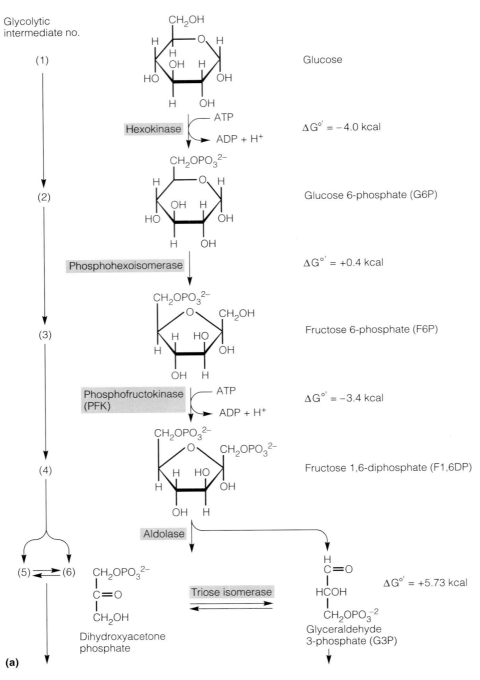

Figure 5-7   Detailed presentation of the glycolytic pathway. Glycolytic intermediates are identified by name to the right of each structure and by number to the left; catalyzing enzymes are noted on the left of each reaction step.

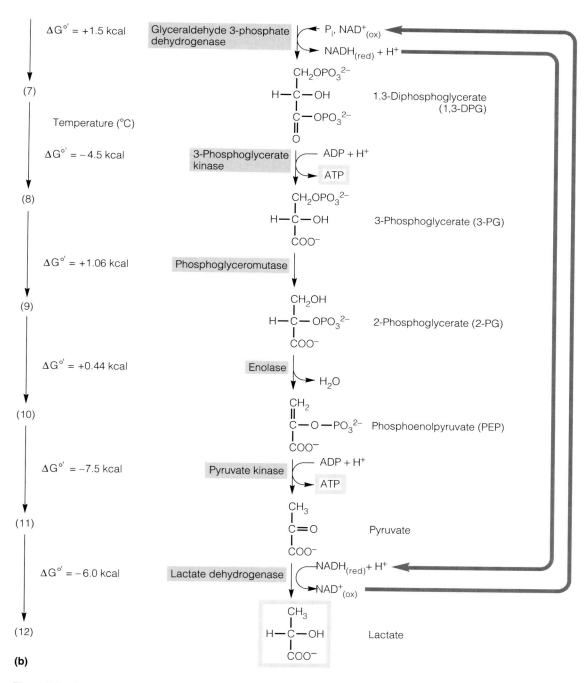

(b)

Figure 5-7 *(continued)*

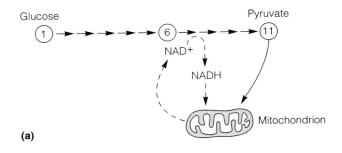

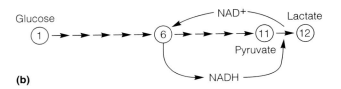

Figure 5-8 Simplified and conventional portrayal of glycolysis showing the beginning and ending substances for: (a) slow, and (b) rapid glycolysis.

Figure 5-9 Chemical structures of glucose, pyruvic acid, and lactic acid. Small numbers in parentheses identify the carbon atoms in the original glucose structure. At physiological pH, lactic and pyruvic acids dissociate a hydrogen ion. In glycolytic cells, such as erythrocytes and muscle cells, lactate is inevitably formed because the Keq of lactate dehydrogenase (LDH) is 36,000 and V_{max} is high.

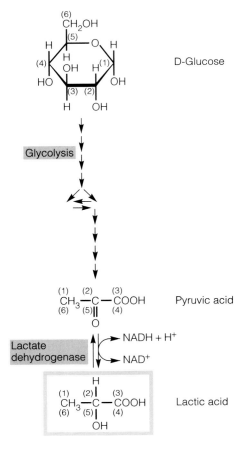

Adenine

D-Ribose

Nicotinamide

D-Ribose

(a)

$$NAD^+ + 2 H^\bullet \rightleftharpoons \quad \text{(structure)} \quad + H^+$$

(b)

R

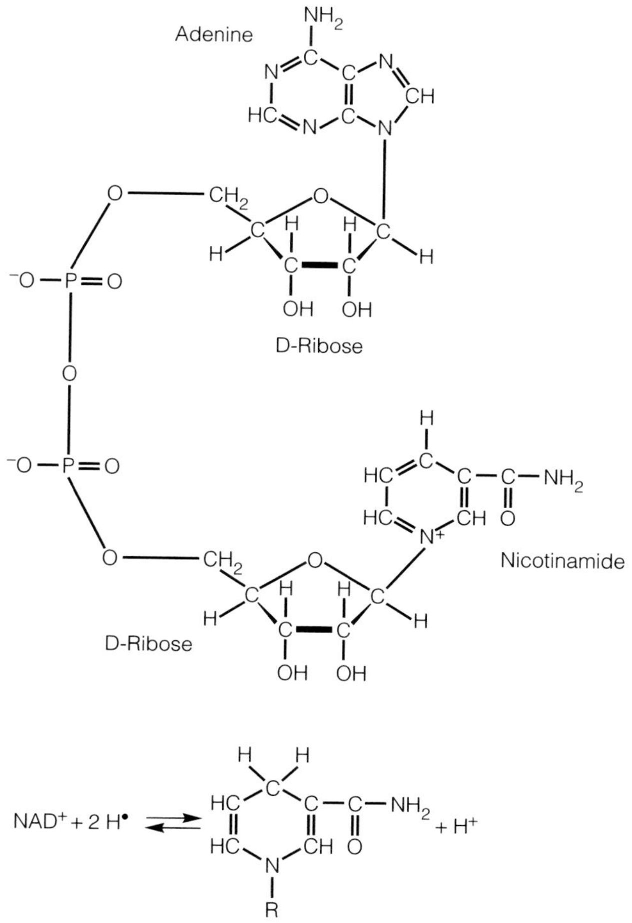

Figure 5-10 Structures of NAD$^+$ and NADH. NAD$^+$ can exist in two forms: (a) without added hydrogen and electrons—that is, oxidized—or (b) with added hydrogen and electrons—that is, reduced. NAD$^+$ serves to transfer these high-energy species within cells: NAD$^+$ + 2 H $\cdot$ $\rightleftharpoons$ NADH + H$^+$.

organelles where most oxygen is consumed). Under these circumstances, the end product of glycolysis, pyruvate, is also consumed by the mitochondria. However, if there is insufficient mitochondrial activity to accept the glycolytic flux, which can occur in type IIb fibers or type I fibers during maximal exercise or if the mitochondria are defective or poisoned (Figure 5-8b), then NADH is oxidized and pyruvate reduced to form lactate in the cytoplasm as the result of rapid glycolysis (Equation 5-1).

The net formation of lactate or pyruvate, then, depends on relative glycolytic and mitochondrial activities, and not on the presence of oxygen. As in the indirect pathway (Figure 5-3), which occurs during rest under fully aerobic conditions, glycolysis leading to lactate production occurs readily in fully aerobic contracting red skeletal muscle. Thus, in relation to mitochondrial respiration, which is tightly controlled and directly related to energy demand, glycolysis and glycogenolysis are less tightly controlled. Glycolytic flux in excess of mitochondrial demand results in lactate production simply because

$$\text{Pyruvate + NADH + H}^+ \xrightarrow[\text{dehydrogenase}]{\text{Lactate}} \text{Lactate + NAD}^+ \qquad (5\text{-}1)$$

LDH has the highest V_{max} of any glycolytic enzyme and because the K_{eq} and ΔG (Chapter 2) of pyruvate-to-lactate conversion favors product formation.

Malate-Aspartate and Glycerol-Phosphate Cytoplasmic-Mitochondrial Shuttle Systems

Scientists have elucidated the pathways and controls of the three glycolytic-mitochondrial shuttle systems. Because NADH and NAD⁺ poorly diffuse across the inner mitochondrial membrane, nature

utilizes three reduced compounds (malate, glycerol phosphate, and lactate) to transfer reducing equivalents from cytosol to the mitochondrial electron transport chain. The first two shuttles discovered are diagrammed in simple form in Figure 5-11. The malate–aspartate shuttle predominates in the heart, and the glycerol–phosphate shuttle predominates in skeletal muscle. In addition to differences in mechanism and tissue specificity, the shuttles differ in the energy level of the product that becomes available within mitochondria. As we will see, NADH

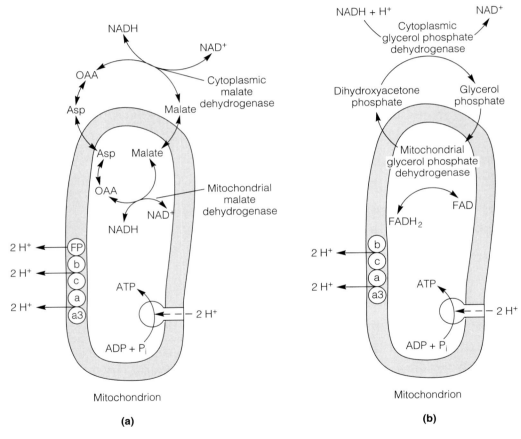

Figure 5-11   Shuttle systems for moving reducing equivalents generated from glycolysis in the cytoplasm to mitochondria. (a) The malate–aspartate shuttle is the main mechanism in the heart. (b) The glycerol–phosphate shuttle is the main mechanism in skeletal muscle. Each proton pair pumped out of mitochondria results in a sufficient chemical and osmotic energy gradient to form an ATP molecule. NADH ≈ 3 ATP; FADH ≈ 2 ATP. Modified from Lehninger, 1971. Used with permission.

TABLE 5-1

Control Enzymes of Glycolysis

Enzyme	Stimulators	Inhibitors
Phosphofructokinase	ADP, P_i, AMP, $\uparrow$pH, (NH_4^+)	ATP, CP, citrate
Pyruvate kinase		ATP, CP
Hexokinase		Glucose 6-phosphate
Lactic dehydrogenase		ATP

generates three ATP molecules within mitochondria for each atom of oxygen consumed (ATP : O = ADP : O = P_i : O = 3), and FADH generates two ATPs within mitochondria (P_i : O = 2).

The Intracellular Lactate Shuttle

Since the 1970s, electron microscope and biochemical evidence have indicated that mitochondria in skeletal muscles, liver, sperm, and other cells contain lactate dehydrogenase (LDH). Unfortunately, these results were either ignored or misinterpreted. However, recent observations that tissue homogenates and that mitochondria isolated from liver, heart, and skeletal muscle oxidize lactate at rates greater than pyruvate (Table 5-1) have rekindled interest in studying mitochondrial LDH and lactate-pyruvate (monocarboxylate) transporters. More importantly, the realization that mitochondria consume and oxidize lactate completely change our understanding of the relationship between glycolytic and oxidative metabolism. Rather than the conclusion that lactate is a dead-end metabolite formed as the result of O_2 insufficiency, we now realize that lactate, more than pyruvate, is the link between glycolytic (anaerobic) and oxidative (aerobic) metabolism.

Figure 5-12 is a transmission electron micrograph (TEM) of rat liver showing mitochondria, rough ER, and cytosol. The presence of the LDH-5 (M4) isoform of lactate dehydrogenase in mitochondria and cytosol is indicated by gold particles which appear as black dots. Similar results have been obtained on heart and skeletal muscle, and together these results have given rise to realization of an *intracellular lactate shuttle* (Figure 5-13). According to this model, when glycolysis is rapid and cytosolic

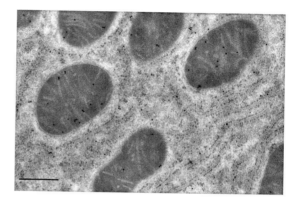

Figure 5-12: Transmission electron micrograph (TEM) of high pressure frozen rat liver showing mitochondria, rough ER, and cytosol. Immunolocalization of anti-LDH-5 (M4) antibodies is indicated by the 15 nm gold particles which appear as black dots. Magnification = 58,300 × and scale bar = 400 nm. Note presence of LDH-5 in mitochondria and surrounding matrix and organelles. From Brooks et al., 1999. Used with permission of the National Academy of Sciences.

lactate concentration rises, lactate enters mitochondria; lactate is oxidized to pyruvate by mitochondrial LDH with participation of the mitochondrial electron transport chain, and the resulting pyruvate is oxidized by the mitochondrial tricarboxylic acid cycle (Chapter 6).

In important ways, the intracellular lactate shuttle (Figure 5-14) is similar to the malate-aspartate and glycerol-phosphate shuttles (see Figure 5-11). Specifically, the shuttles are similar in that NADH-equivalent potential energy is transported into mitochondria by means of a compound for which there is a specific mitochondrial carrier protein. As already stated, this transport-carrier

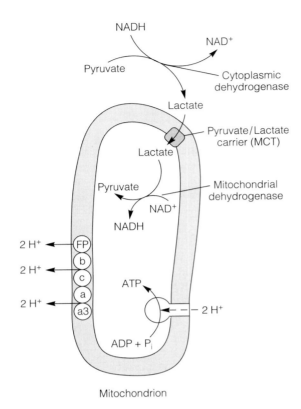

Mitochondrion

Figure 5-13 The *intracellular lactate shuttle* permits both reducing equivalents as well as oxidizable substrate generated from glycolysis in the cytosol to gain entry into mitochondria. The malate-aspartate and glycerol-phosphate shuttles (Figure 5-11) are believed to function when glycolysis is slow, whereas the lactate shuttle operates when glycolysis is rapid and lactate accumulates. MCT stands for monocarboxylate transporter.

mechanism is necessary because NADH and NAD^+ poorly diffuse across mitochondrial membranes. The shuttle systems differ in a major way in that while the malate-aspartate and glycerol-phosphate shuttles allow transfer of reducing equivalents from cytosol to mitochondria, there is no net transfer of substrate for oxidation. In contrast, the intracellular lactate shuttle transports from cytosol to mitochondria both NADH and a substrate for oxidation by the tricarboxylic acid cycle (TCA cycle). In this case, however, lactate, not malate or glycerol phosphate, is the carrier.

The intracellular lactate shuttle has a major advantage for working muscle in that it permits rapid *aerobic glycolysis* involving lactate formation to occur simultaneously with high rates of oxygen consumption. Thus, Figures 5-8a and 5-8b are combined in Figure 5-13 to illustrate how the intracellular lactate shuttle supports aerobic glycolysis.

ATP Yield by Glycolysis

At the beginning of glycolysis (Figure 5-7), there are two activating steps where ATP is consumed. However, following the cleavage of the six-carbon molecule into two three-carbon molecules, there are two steps where an ATP molecule is formed. Because these steps occur twice each for each six-carbon glucose that enters the pathway, the gross ATP yield is $2 \times 2 = 4$. However, if we then subtract the two ATP molecules used for activation, the net yield is 2.

Under "aerobic" conditions, the formation of two ATPs per glucose is complemented by the formation for mitochondrial consumption of two NADHs. Depending on the shuttle system used, these are equivalent to another four to six ATPs.

Figure 5-14 Diagram of the intracellular lactate shuttle by which lactate formed as the result of rapid glycolysis is shuttled into mitochondria and oxidized. By this means, glycolytic ("anaerobic") and oxidative ("aerobic") metabolism are linked. Comprehension of the intracellular lactate shuttle makes possible understanding of many observations of so-called *aerobic glycolysis* in which fully oxygenated cells consume glucose and produce lactate, or consume lactate and reduce glucose consumption.

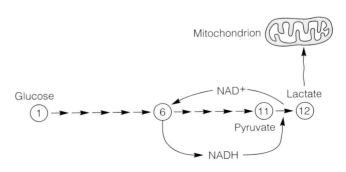

$$(2 \text{ NADH})(3 \text{ ATP/NADH}) = 6 \text{ ATP} \qquad (5\text{-}2)$$
$$(2 \text{ FADH})(2 \text{ ATP/FADH}) = 4 \text{ ATP} \qquad (5\text{-}3)$$

"Anaerobic" glycolysis is summarized in Equation 5-4 and "aerobic" glycolysis in Equation 5-5.

The Efficiency of Glycolysis

The formation of only two ATPs from each glucose in glycolysis seems small (Eqs. 5-1 and 5-2). For this reason, glycolysis is sometimes called an "inefficient" pathway. Fortunately, however, skeletal muscle (white and red glycolytic fibers) can break down glucose rapidly and can produce significant quantities of ATP for short periods during glycolysis.

Nevertheless, an appropriate energetic consideration reveals that, in fact, the efficiency of glycolysis is excellent. The energy change (ΔH) from glucose to lactate is $-47 \text{ kcal} \cdot \text{mol}^{-1}$.

If two ATPs are formed, and $\Delta G°'$ for ATP = $-7.3 \text{ kcal} \cdot \text{mol}^{-1}$, then

$$\text{Efficiency} = \frac{2(-7.3)}{-47} = 31\% \qquad (5\text{-}6)$$

If ΔG for ATP is $-11 \text{ kcal} \cdot \text{mol}^{-1}$,

$$\text{Efficiency} = \frac{2(-11)}{-47} = 46.8\% \qquad (5\text{-}7)$$

This latter result of almost 50% efficiency compares favorably with that of oxidative enzymatic processes. It is not correct, therefore, to assume that glycolysis, or "anaerobic metabolism," is inefficient—the enzymes actually conserve a good deal of the energy released.

The Control of Glycolysis

Glycolysis is a pathway controlled by many factors but, in general, two kinds of controls predominate: "feed-forward" and "feedback" controls. Feed-forward and feedback controls are illustrated by their various effects on the level and flux (movement of molecules) through the glucose 6-phosphate (G6P) pool (Figure 5-15).

In feed-forward control of glycolysis, factors that increase G6P levels tend to stimulate glycolysis. Feed-forward factors include stimulation of glycogenolysis (by epinephrine and contractions) and glucose uptake (by contractions and insulin). Thus, as seen in Figure 5-4 with exercise of moderate to high intensity, blood glucose rises. Feedback controls involve changes in levels of metabolites that result from glycolysis (e.g., citrate) or from muscle contraction (e.g., ADP). Also, a decline in blood glucose concentration, such as occurs at the end of the exercise (Figure 5-4), is probably the most important feedback control in normal, healthy subjects. Feedback control usually resides at the phosphofructokinase (PFK) step and can either speed (stimulate) or slow (inhibit) regulatory enzymes.

GLUT-4 and Glucose Transporter Translocation Recent advances indicate that glucose uptake into muscle and other cells occurs via glucose transport proteins. Transporter-mediated cell glucose uptake is discussed more fully in Chapter 9. (See Figure 9-5 and Table 9-1). Here, it needs to be emphasized that muscle and adipose cells possess both noninsulin (GLUT-1) and insulin-mediated (GLUT-4) glucose uptake transporters. In resting muscle, most glucose enters by the GLUT-1 carrier. However, when glucose and insulin levels are high, or during exercise, most glucose enters muscle cells by the insulin (and contraction) regulatable (GLUT-4) glucose transporter protein. Under resting conditions, insulin binding to the muscle cell surface receptor causes release of some messenger or messengers, which cause the GLUT-4 transporters to move (translocate) from intracellular locations to

$$\text{Glucose} + 2 \text{ P}_i + 2 \text{ ADP} \xrightarrow[\text{glycolysis}]{\text{"Anaerobic"}} 2 \text{ Lactate} + 2 \text{ ATP} + \text{H}_2\text{O} \qquad (5\text{-}4)$$

$$\text{Glucose} + 2 \text{ P}_i + 2 \text{ ADP} + 2 \text{ NAD}^+ \xrightarrow[\text{glycolysis}]{\text{"Aerobic"}} 2 \text{ Pyruvate} + 2 \text{ ATP} + 2 \text{ NADH} + 2 \text{ H}^+ + 2 \text{ H}_2\text{O} \qquad (5\text{-}5)$$

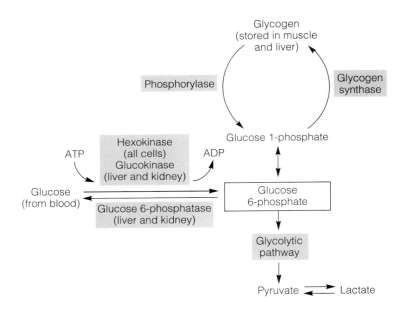

Figure 5-15 Illustration of the central role of glucose 6-phosphate in determining the direction of carbon flow in glycolysis (catabolism of glucose), glycogenolysis, glycogen synthesis, lactate and pyruvate production, and glucose release.

the cell surface. Translocation of GLUT-4 transporters to the muscle cell surface also occurs as the result of contractions. Recent results of Friedman, Dohm, and associates (1991) suggest that in the resting state GLUT-4 transporters reside near T-tubules and lateral cisternae of the sarcoplasmic reticulum. This finding offers the possibility that events associated with the ion fluxes of contraction (Na^+ or Ca^{2+}) may influence transporter translocation to T-tubules, which are extensions of the sarcolemma. Such a mechanism would explain the results of Richter and associates, which show that contracting muscle can take up glucose with "no need for insulin." Thus, with both insulin-dependent and insulin-independent mechanisms of glucose transport, working muscle can take up glucose even when insulin falls during exercise.

Phosphofructokinase Usually a dominant factor in the regulation of glycolysis is the activity of the enzyme phosphofructokinase (PFK). As we shall see in Chapter 9, two forms of PFK exist: one for glycolysis (PFK-1) and one for glyconeogenesis (PFK-2). In muscle, PFK-1 predominates. PFK-1, which catalyzes the third step of glycolysis, is a multivalent, allosteric enzyme. This means that sev-

eral metabolites bind to the enzyme and affect its catalytic capacity. Known modulators of PFK-1 are given in Table 5-1. PFK is probably the rate-limiting enzyme in muscle when glycolysis is rapid during exercise, but factors such as ATP, CP, and citrate affect the conformation of PFK to slow its activity during rest. When exercise starts, immediate changes in the relative concentrations of PFK modulators increase its activity. Pyruvate kinase, hexokinase, and lactic dehydrogenase are other important controlling enzymes whose activities are modulated.

The Cellular "Energy Charge" Earlier it was shown that the contractile actin–myosin system splits ATP to ADP and P_i, and that adenyl kinase maintains ATP levels by forming ATP and AMP from two ADPs. The contents of the ATP–ADP–AMP system therefore exist in one of three forms. The "energy charge" of the cell is defined as follows:

Adenine nucleotide energy charge

$$= \left(\frac{[ADP] + 2\,[ATP]}{[AMP] + [ADP] + [ATP]} \right) \frac{1}{2} \quad (5\text{-}8)$$

If all the adenine nucleotide is in the form of ATP, the energy charge is 1.0. If all the adenine nucleotide is in the form of AMP, the energy charge is

0.0. In the cell, the energy charge is usually around 0.8. Even small decrements below that will activate ATP-yielding systems.

Thermodynamic Control As indicated in Figures 5-6 and 5-7, significant free energy changes occur in several exergonic steps. These reactions are catalyzed by the enzymes hexokinase, phosphofructokinase, pyruvate kinase, and lactic dehydrogenase. The reactions keep glycolysis going in the direction of product; the other steps have small free energy changes and are freely reversible. The reverse of glycolysis is not possible without energy input and the intervention of specific enzymes. This reversal process is called *gluconeogenesis,* meaning "making new glucose." Gluconeogenesis is a function of tissues such as liver and kidney. It is a capability of muscle, but at present is not believed to be a primary function of muscle.

Control by Lactate Dehydrogenase (LDH)
The terminal enzyme of glycolysis, which results in the formation of lactic acid from pyruvic acid, is lactate dehydrogenase (LDH). As shown earlier (Figure 5-8a), when glycolysis is slow, LDH is in competition with mitochondria for pyruvate. However, LDH is an enzyme of significant content in muscle, especially white skeletal muscle. As already mentioned, the K'_{eq} and the $\Delta G^{\circ\prime}$ of LDH are large and the reaction proceeds actively to completion. Therefore, some lactate is always formed. As a consequence, resting muscle always produces and releases lactate on a net basis (Figure 10-4).

There are two basic types of LDH: muscle (M) and heart (H). These LDH types are found predominantly in white skeletal muscle and heart, respectively. The equilibria of the two types are identical, but they differ in their affinities for reactants (substrates) and products. The M type has a high affinity for the substrate pyruvate, and therefore has higher biological activity than the H type, which has a lower affinity for pyruvate.

Each molecule of LDH has four subunits. Considering the two basic types of LDH, there are then five possible arrangements: M_4, M_3H_1, M_2H_2, M_1H_3, and H_4. The population of these *isozymes* of the LDH varies among tissues, with M_4 being highest in white skeletal muscle and lowest in heart (Table 5-2).

TABLE 5-2

Distribution of lactate dehydrogenase (LDH) isoenzymes in various mammalian tissues.[a]

Tissue	M4 (LDH5)	M3H1 (LDH4)	M2H2 (LDH3)	M1H3 (LDH4)	H4 (LDH1)
Liver	94.0	4.0	1.0	0.8	0.2
Heart	2	3	5	30	60
Fast White Muscle (Tibialis Anterior)	80	14	3	2	1
Fast Red Muscle (Red Gastroc)	58	10	11	13	8
Slow Red Muscle (Soleus)	11	13	18	30	28
Kidney	6	11	21	34	28
Red Blood Cells	5	10	15	30	40
Lungs	21	23	38	18	10
Spleen	18	31	31	15	5

[a]Each enzyme molecule is comprised of four subunits, thus yielding five isoenzymes according to relative composition of muscle (M) and heart (H) subunits. Data are for whole tissues, not cytosolic and other cell compartments.

SOURCE: Lott and Wolfe (1986), McCullagh et al. (1996), York et al. (1975), Dubouchaud and Brooks (1999).

Because of the distribution of LDH isoenzymes in various cells and tissues, the biological activity of LDH depends to some extent on its concentration and isoenzyme type. An older idea was that lactate-consuming tissues contained predominantly LDH1 (H4) isoenzymes, whereas lactate-producing cells and tissues contained predominantly LDH5 (M4) isoenzymes. However, while the heart and red skeletal muscles can be net lactate consumers, liver and red skeletal muscle, which contain predominantly LDH5 (Table 5-2) can be net lactate consumers as well. Therefore, in determining whether a tissue is a net lactate consumer or producer, the presence of LDH in mitochondria, where lactate oxidation takes place by the intracellular lactate shuttle mechanism (Figures 5-13 and 5-14), is more important than the presence of LDH isoform in the cytosol of the tissue.

Control by Pyruvate Dehydrogenase   Although a mitochondrial and not a glycolytic enzyme, pyruvate dehydrogenase (PDH) is a key enzyme whose activity, which is regulated by phosphorylation (Chapter 9), can affect the rate of lactate production. As we will see in Chapter 6, when PDH is active, pyruvate can be diverted to the mitochondria for oxidation. By competing with LDH for pyruvate, PDH indirectly affects the $NADH/NAD^+$ ratio, and therefore, the rate of glycolysis.

Control by Cytoplasmic Redox (NADH/ NAD^+)   Glycolytic flux and the competition between cytoplasmic LDH and mitochondrial PDH for pyruvate affect the ratio of pyruvate to lactate as well as the ratio of NADH to NAD^+. The cytoplasmic $NADH/NAD^+$ ratio (or redox) affects the activity of glyceraldehyde 3-phosphate dehydrogenase, which requires NADH as a cofactor. In general, cytoplasmic reduction ($\uparrow NADH/NAD^+$) slows glycolysis, whereas oxidation ($\downarrow NADH/NAD^+$) speeds glycolysis.

Control by Glycogenolysis   To a certain extent, the rate at which glycolysis proceeds depends on the rate of glycogen breakdown. In resting muscle, little glycogen is broken down, so the rate of glycolysis is limited by muscle glucose uptake. However, during exercise, glycogen breakdown is greatly accelerated, and glycogen, not glucose, is the major precursor for glycolysis. For instance, during steady-rate exercise at 65% of $\dot{V}_{O_2max}$, glycogen breakdown can exceed glucose uptake by four to five times. Thus, glycolysis is said to be under "feed-forward" control.

■   Glycogenolysis

Skeletal muscle glycolysis is heavily dependent on the intramuscular storage form, glycogen. During heavy muscular exercise, glycogen may supply most of the immediate glucosyl residues for glycolysis. Because 80% or more of the carbon for glycolysis in muscle comes from the glycogen in muscle, and not from blood glucose, glycogen depletion results in fatigue (Chapter 33).

The structure of glycogen (Figure 5-16) consists mostly of end-to-end (C1–C4) linkages, with a few cross (C1–C6) linkages. The storage of glucose units as glycogen is dependent on the activity of the enzyme glycogen synthase (Figure 5-17). Breakdown of glycogen is dependent on the enzyme phosphorylase, which hydrolyzes the C1–C4 linkages. Another enzyme, called "debranching enzyme," hydrolyzes the C1–C6 branching, or side linkages.

The activity of phosphorylase appears to be controlled by two mechanisms. One system is hormonally mediated and depends on the extracellular action of epinephrine and the intracellular action of cyclic AMP (cAMP), the "intracellular hormone" (Figure 5-18). This mechanism is too slow to explain the rapid glycolysis during onset of heavy exercise. Therefore, faster mechanisms mediated by phosphate (Pi, or PO_4^{2-}) and calcium ion (Ca^{2+}) provide important control mechanisms for mobilizing glycogen in working skeletal muscle. These signals arise from the causes and consequences of muscle contraction and not extracellular, endocrine signaling.

Control by Phosphate

According to the work of J. O. Holloszy and associates, probably the control mechanism most impor-

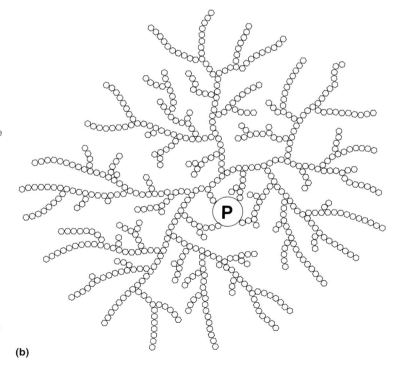

(a)

Figure 5-16 (a) The structure of glycogen is seen to be a polymer of glucose units. The linkages exist mainly end to end (C1–C4 linkages), but there is also some cross-bonding (C1–C6 linkages). (b) A pinwheel-like structure results from C1–C4 and C1–C6 linkages. The hexagons represent glucosyl units. Under high magnification in the electron microscope (e.g., Figure 5-5), the pinwheel-like structures of glycogen appear as dark granules. Glycogen forms around a "foundation" protein P-glycogenin.

(b)

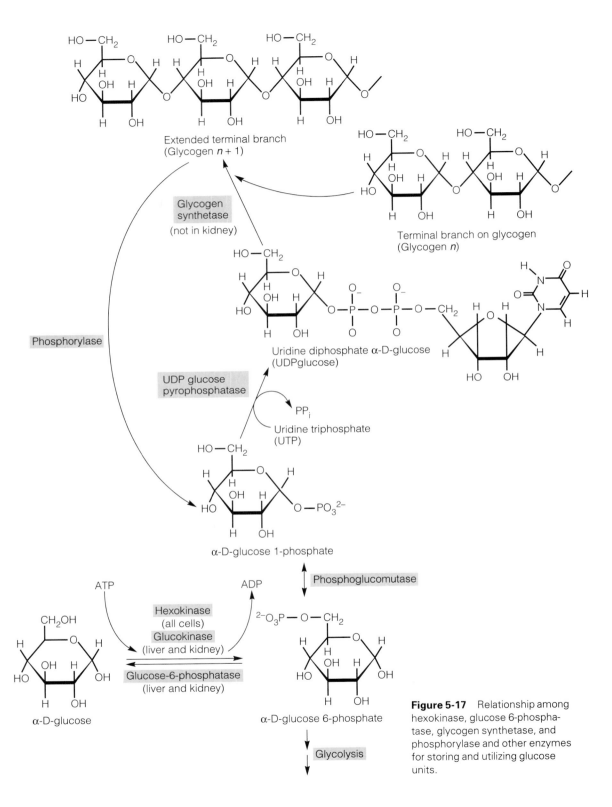

Extended terminal branch
(Glycogen $n + 1$)

Terminal branch on glycogen
(Glycogen n)

Glycogen synthetase
(not in kidney)

Phosphorylase

Uridine diphosphate α-D-glucose
(UDPglucose)

UDP glucose pyrophosphatase

PP$_i$

Uridine triphosphate
(UTP)

α-D-glucose 1-phosphate

Phosphoglucomutase

ATP

ADP

Hexokinase
(all cells)
Glucokinase
(liver and kidney)

Glucose-6-phosphatase
(liver and kidney)

α-D-glucose

α-D-glucose 6-phosphate

Glycolysis

Figure 5-17 Relationship among hexokinase, glucose 6-phosphatase, glycogen synthetase, and phosphorylase and other enzymes for storing and utilizing glucose units.

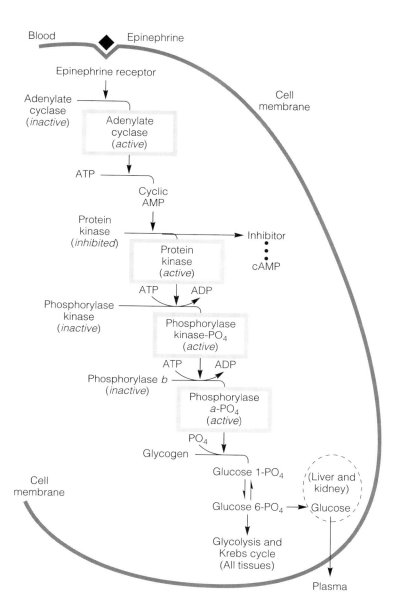

Blood — Epinephrine

Epinephrine receptor

Cell membrane

Adenylate cyclase (*inactive*)

Adenylate cyclase (*active*)

ATP — Cyclic AMP

Protein kinase (*inhibited*) ——→ Inhibitor

Protein kinase (*active*)

cAMP

ATP   ADP

Phosphorylase kinase (*inactive*)

Phosphorylase kinase-PO_4 (*active*)

ATP   ADP

Phosphorylase *b* (*inactive*)

Phosphorylase *a*-PO_4 (*active*)

PO_4

Glycogen

Glucose 1-PO_4

Cell membrane

(Liver and kidney)

Glucose 6-PO_4 ——→ Glucose

Glycolysis and Krebs cycle (All tissues)

Plasma

Figure 5-18 The breakdown of glycogen in muscle is heavily influenced by the enzyme phosphorylase. (a) Phosphorylase *b* (the inactive form) can be converted to phosphorylase *a* (the active form) through a series of events initiated by the hormone epinephrine. This mechanism involves cyclic AMP (cAMP), the intracellular hormone. During sudden bursts of activity, the cAMP mechanism is too slow to account for the observed glycogenolysis. Intracellular factors such as Ca^{2+} and AMP increase the catalytic activity of phosphorylase. Modified from Vander, Sherman, and Luciano, 1980. Used with permission.

tant for mobilizing glycogen is the free inorganic phosphate (Pi) that comes from the breakdown of ATP (Equation 3-2). Phosphate provides a potent stimulus for glycogen degradation in muscle because it is a substrate for phosphorylase *a* (Figure 5-18). The important role of phosphate in the regulation of muscle glycogenolysis is usually overlooked by biochemists, but is given attention by exercise physiologists, who know that epinephrine stimula-

tion that increases cyclic AMP (cAMP) in the absence of contraction does not result in appreciable glycogen breakdown because Pi level does not change without contraction.

Control by Calcium Ion

In addition, Ca^{2+}, which is released from the sarcoplasmic reticulum, constitutes another control

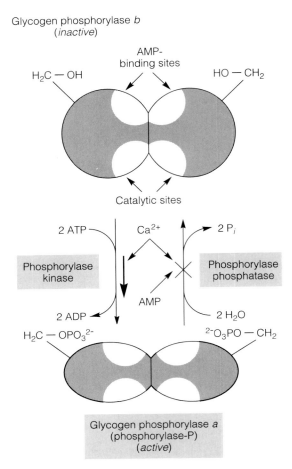

Figure 5-19   The conversion of phosphorylase *b* (the inactive form of the enzyme) to phosphorylase *a* (the active form) depends on the stimulation of phosphorylase kinase by Ca^{2+}. Calcium ions are released immediately when muscles contract, and this mechanism helps to link pathways of ATP supply with those of ATP utilization. During exercise, the levels of AMP increase; this helps to minimize the reconversion of phosphorylase *a* to *b* by inhibiting phosphorylase phosphatase. Modified from McGilvery, 1975.

mechanism (Figure 5-19) regulating glycogen catabolism.

The hormonally mediated, cAMP-dependent mechanism serves two purposes during exercise and recovery: (1) It amplifies the local, Pi^{-2}, and Ca^{2+}-mediated process in *active* muscle and (2) it mobilizes glycogen in *inactive* muscle to provide lactate as a fuel and as a glyconeogenic (Cori cycle) precursor (Chapter 9).

During rest, cellular uptake of glucose is usually sufficient to support glycogen synthesis and glycolysis. During maximal exercise, glycogenolysis and glucose uptake are sufficient to support rapid glycolysis. During prolonged exercise, the depletion of intramuscular glycogen and liver glycogen can result in the decreased capability of substrate for muscle glycolysis. Moreover, it is also likely that the ability to glycolyze limits the ability to utilize fat. Therefore, glycogen depletion during exercise may be doubly important. The interaction between carbohydrate and fat metabolism is discussed later (Chapter 7) in more detail.

■ The Cell-Cell Lactate Shuttle

Isotope tracer studies and arteriovenous difference measurements across muscles and other tissues have allowed precise estimation of the rates of lactate and glucose production and oxidation during sustained, submaximal exercise. The results indicate that lactate is actively oxidized in working muscle beds and may be a preferred fuel in heart and red skeletal muscle fibers. In the original (cell-cell) lactate shuttle concept, the idea was that within a muscle tissue during sustained exercise, lactate produced at some sites, such as type IIb (FG) fibers, diffuses or is transported into type I (SO) fibers (Figure 5-20). Some of the lactate produced in type IIb fibers shuttles directly to adjacent type I fibers. Alternatively, other lactate produced in type IIb fibers can reach type I fibers by recirculation through the blood. Thus, by this mechanism of shuttling lactate between cells, glycogenolysis in one cell can supply a fuel for oxidation to another cell. Skeletal muscle tissue then becomes not only the major site of lactate production, but also the major site of its removal. In addition, much of the lactate produced in a working muscle is consumed within the same tissue and never reaches the venous blood.

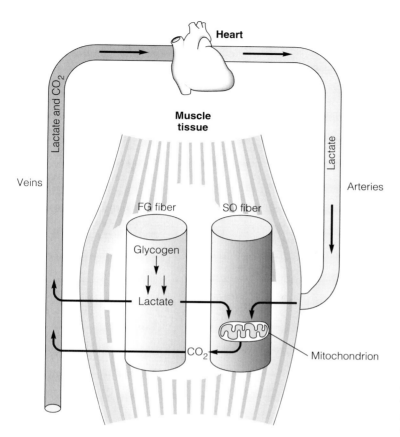

Figure 5-20 Diagram of the lactate shuttle. Lactate produced in some cells [e.g., fast glycolytic (FG, type IIb) muscle cells] can shuttle to other cells [e.g., slow oxidative (SO, type I) fibers] and be oxidized. Also, lactate released into the venous blood can recirculate to the active muscle tissue bed and be oxidized. During exercise, the lactate shuttle can provide significant amounts of fuel. Muscle cell membrane lactate transport proteins (MCT1 and MCT4) facilitate lactate release and uptake. (See Brooks et al. (1985 and 1999) for additional information.)

■ Monocarboxylate (Lactate/Pyruvate) Transport Proteins in Muscle Cell Membranes and Mitochondria

The original cell-cell lactate shuttle concept predicted that lactate exchange occurred between lactate-producing and -consuming muscle fibers during exercise and predicted presence of sarcolemmal lactate transport proteins to facilitate lactate exchange. The concept was well supported by results obtained from isotope tracer and limb metabolite balance studies. The concept has also been well supported by results of subsequent physiological, biochemical, and molecular studies. Some of these deserve mention here.

Working independently, Juel (1988) and Watt and associates (1988) provided convincing evidence of facilitated and proton-linked lactate transport into mammalian muscle tissue. Those observations were followed by work of Roth and Brooks (1990), who provided the first evidence of a lactate transporter in sarcolemmal membranes.

The field of study of cell membrane lactate transport proteins received a major boost when, in extending their earlier work on the mevalonate transporter (Mev) in Chinese hamster ovary cells, Garcia and associates cloned and sequenced a monocarboxylate transporter, which they termed MCT1, or monocarboxylate transporter 1. MCT1 differed from Mev by only one amino acid substitution (Cys for

Phe), and the amino acid structures of MCT1 and Mev predicted a membrane-bound protein with 12 membrane-spanning regions. Transfection of a breast cancer cell line lacking MCT1 with a plasmid containing cDNA encoding for MCT1 conferred properties reported for the erythrocyte transporter, including increased pyruvate uptake, proton symport, transstimulation, partial inhibition by other monocarboxylates (including lactate), and sensitivity to the known inhibitor of lactate cinnamate (CIN). Garcia and associates also raised rabbit polyclonal antibodies against the C-terminus of MCT1, and conducted immunofluorescence studies to locate MCT1 and succinic dehydrogenase (SDH, a mitochondrial marker) on several hamster tissues.

MCT1 was found to be abundant in erythrocytes and heart and basolateral intestinal epithelium. In skeletal muscle, MCT1 was detectable only in oxidative muscle fiber types and not at all in liver. With an interest in describing a role for MCT isoforms in the Cori cycle (see below), Garcia and associates subsequently described isolation of a second isoform (MCT2) by screening a Syrian hamster liver library; and MCT2 was found in liver and testes.

Independent of Garcia and associates, V. N. Jackson, A. P. Halestrap, and associates cloned and sequenced MCT1 and MCT2 isoforms from rabbit and rat tissues, respectively. Subsequently, a unique isoform (MCT3) was found in the chicken eye. Candidates for seven putative cell membrane mono-

Figure 5-21 Two human multitissue Northern blots showing distribution of RNAs encoding for MCT isoforms in human tissues. Sizes of transcripts were estimated by comparison with the position of RNA standards marked on the blot. The break in identification of RNAs encoding for MCT isoforms is due to the discovery of MCT3, a unique isoform so far found only in chicken eye. To date, MCT1, -2, -3, and -4 are lactate/pyruvate transporters, and it is unknown which monocarboxylic acids are transported by other isoforms. MCT1 is abundant in numerous cell types, including erythrocytes, red skeletal muscle cells, and heart cells. In muscle and heart cells, MCT1 is present in mitochondria as well as cell membranes. MCT4 is a sarcolemmal lactate transporter and is more abundant in white, glycolytic striated muscle. From Price and associates (1998), courtesy of A. Halestrap and Portland Press.

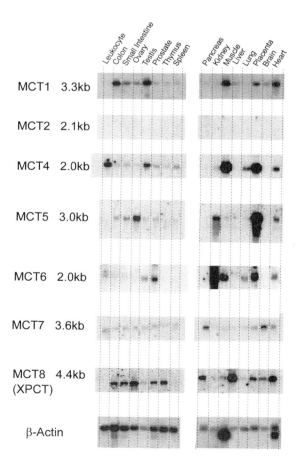

carboxylate transporters (MCT4-MCT8) were later cloned and sequenced by Price, Jackson, and Halestrap (1998). The distribution of RNAs for seven of eight known MCTs is shown in Figure 5-21. This number is likely to change significantly in the near future, and it is likely to be discovered that *not* all the MCTs are lactate- or pyruvate-specific transporters. For the present, it is certain that muscle cell membrane lactate and other monocarboxylate transporters are highly abundant in diverse tissues. Again, MCT1 appears more abundant in oxidative striated muscle fibers, whereas Wilson and associates (1998) have shown MCT4 to be more abundant in white, glycolytic fibers.

The Muscle Mitochondrial Lactate/Pyruvate Transporter Is MCT1

The original cell-cell lactate shuttle concept successfully predicted results obtained from isotope tracer and limb metabolite balance studies as well as results of subsequent biochemical and molecular studies. However, while the original concept worked for exercise, it was less satisfactory for rest, during which type IIb fibers were (obviously) not recruited; glycolysis was slow and oxygen was present in abundance. Therefore, G. A. Brooks and associates turned their attention to a model that would permit glycolysis leading to lactate formation and mitochondrial lactate oxidation to occur simultaneously in noncontracting, fully oxygenated muscle. Subsequently, they showed that isolated mitochondria readily oxidize lactate and that mitochondrial lactate oxidation is blocked by known LDH and MCT inhibitors. In addition to showing that mitochondria from heart and skeletal muscle contain LDH (Figure 5-12), Brooks and associates (1999) reported that MCT1 is abundant in mitochondria of skeletal and cardiac muscle. Further, Dubouchaud and Brooks (1999) have found that training increases both the amounts of mitochondrial and sarcolemmal MCT1 in human muscle. Thus, not only are there variations in the distribution of MCTs, but tissues can express multiple isoforms that have different domains (locations) in the cells.

Gluconeogenesis

Although red mammalian skeletal muscle itself has only a vestigial enzymatic capacity to make glucose or glycogen, mammalian muscle does indirectly participate in gluconeogenesis, the making of new glucose.

Cori and Cori were among the first to recognize that the lactate and pyruvate produced by skeletal muscle could circulate to the liver and be made into glucose. The glucose so produced could then recirculate to muscle. This cycle of carbon flow is called the *Cori cycle* (Figure 5-22). Rapid glycolysis in skeletal muscle inevitably results in lactate production because of the activity and K'_{eq} of LDH, gluconeogenesis is an efficient way to reutilize the products of glycolysis, thereby providing for the maintenance of blood glucose and prolonged muscle glycolysis. The Cori cycle is complemented by the glucose-alanine cycle, which is discussed in Chapter 8.

Because glycolysis is an exergonic process, gluconeogenesis must be endergonic. Gluconeogenesis is under endocrine control, and gluconeogenic tissues such as liver and kidney possess enzymes that bypass the controlling exergonic reactions of glycolysis. Gluconeogenesis is discussed more in Chapter 9.

Mammalian skeletal muscle has poor capacity to convert pyruvate or lactate to glycogen, but recently Gleeson and associates (1993) showed that reptilian and probably amphibian muscles are capable of glycogen synthesis (glyconeogenesis) from lactate *in situ*. In white reptilian muscle fibers, contractions produce extremely high lactate levels, but circulation is poor, so lactate needs to be removed within the tissue bed of formation. During recovery, lactate moves from white to red fibers by facilitated diffusion. In reptilian red fibers, lactate undergoes glyconeogenesis, as opposed to oxidation in mammalian fibers. (See Figure 5-20 and Chapter 10). Thus the reptilian lactate shuttle supports glyconeogenesis mainly, and lactate oxidation to a lesser extent.

In mammalian white, not red, fibers, some of the

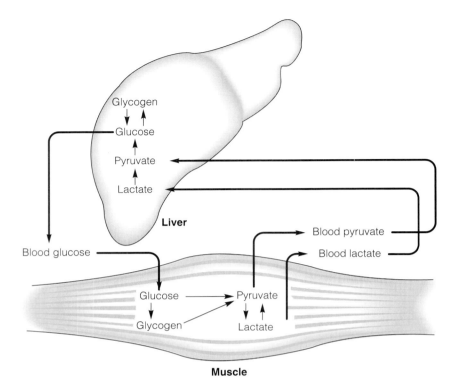

Figure 5-22 The Cori cycle, showing that pyruvate and lactate formed in muscle can circulate to liver and kidney. There carboxylic acids can be synthesized to glucose. The glucose thus formed can then reenter the circulation.

lactate resulting from maximal exercise can be converted to glycogen *in situ*. However, in contrast to reptilian muscle, which is poorly perfused with blood, lactate present in mammalian muscle is either removed by oxidation within the tissue bed, or the lactate circulates to the heart or active red muscle, where it is oxidized, or to the liver, where it is converted to glucose. Some of the newly formed glucose can recirculate to recovering muscle and be synthesized to glycogen. However, restoration of muscle glycogen to preexercise levels in mammalian muscle depends on carbohydrate feeding (Chapters 10 and 28). Thus, in mammalian muscle, the lactate shuttle favors direct muscle lactate oxidation and, to a minor extent, gluconeogenesis and glycogen synthesis by a version of the indirect pathway (i.e., muscle glycogen → muscle lactate → blood lactate → liver glucose → blood glucose → muscle glycogen).

Glycogen Synthesis

Recent work by Wehlan and associates (Lomako et al., 1990) indicates that glycogen is synthesized around a core protein called *glycogenin.* Glycogenin is autocatalyzed, or self-activated, by addition of glucose molecules.

Studies of the role of glycogenin in regulating glycogen synthesis in resting muscle and during recovery from exercise are in their infancy, with the initial work conducted by Terry Graham of the University of Guelph (Adamo et al., 1998, 1998). The analyses of human muscle biopsies taken after exercise by Graham, Adamo, and others indicates that large molecular glycogen (i.e., macroglycogen) is degraded to smaller glycogen particles (i.e., proglycogen), but that exercise does not unmask glycogenin itself. Hence, total muscle glycogen restoration mainly involves building pro- to macroglycogen.

■ Effects of Training on Glycolysis

Studies on the effects of training on the glycolytic capability of skeletal muscle have mainly utilized the technique of catalysis. In this experimental approach, a muscle sample of the experimental individual (animal or human) is taken. Then, either an attempt is made to isolate the enzyme or, more usually, the muscle sample is homogenized and the activity of the enzyme is studied. This can be done by observing the disappearance of a substrate or the appearance of a product.

Compared to its considerable effect on oxidative enzymes of muscle (Chapter 6), endurance training appears to have relatively insignificant effects on catalytic activities of glycolytic enzymes. This may be because skeletal muscle, especially fast-twitch muscle (see Figure 5-5 and Chapter 17), has an intrinsically high glycolytic capability.

Endurance training apparently has little effect on most glycolytic enzymes. Reports of the effects of endurance training on PFK vary, so we may conclude for the present that endurance training has no significant effect on PFK activity.

Hexokinase activity increases significantly in endurance-trained animals. This adaptation, which is thought to facilitate the entry of sugars from the blood into the glycolytic pathway of muscle, would be of benefit in prolonged exercise, where the liver serves to supply muscle with glucose.

Endurance training has been observed to *decrease* total LDH activity in fast glycolytic muscle and to influence the LDH isozymes in muscle to include more of the heart type. However, by increasing muscle mitochondrial mass (Chapter 6), training actually increases mitochondrial LDH content. These adaptations serve to decrease lactate accumulation per unit glycolytic carbon flow in endurance-trained muscle during exercise, and to allow some muscle fibers to take up and oxidize lactate produced in other fibers.

Studies of the effects of speed and power training on the activities of glycolytic enzymes are not as numerous as studies on the effects of endurance training. However, as with endurance training, the effects of speed and power training on the specific activities of glycolytic enzymes are not great. If the muscle hypertrophies as the result of speed and power training, then the total catalytic activity of the muscle probably improves.

■ Quantitative and Relative Uses of Glucose, Glycogen, and Other Substrates During Exercise

With the advent of stable (nonradioactive) isotope tracer technology, and with simultaneous application of tracer and classical arterial-venous difference (a-v) measurements of metabolites across resting and exercising limbs, much recent progress has been made in measuring use of glucose and other substrates during exercise. Data in Figure 5-23 were obtained in the Berkeley Exercise Physiology Laboratory (Friedlander et al., 1998). These results show several important things about the effects of exercise and training on glucose flux during exercise among young women in the mid-follicular menstrual phase using [6,6-^{2}H]glucose. The results show that glucose-use rate (Rd, or rate of disposal) increases as an exponential function of relative power output. This is also illustrated in Figure 5-24, which contains data on both men and women. Training decreased glucose Rd for a given specific (absolute) power output, but when normalized to %$\dot{V}_{O_2max}$, an exponential relationship for both men and women is clear. Thus, at high relative intensity power outputs (in this case, 65% of $\dot{V}_{O_2max}$), glucose use is high, even after training.

Despite the exponential rise in glucose use as a function of exercise power output, in reality, the rise compared to $\dot{V}_{O_2}$ rise is relatively small. For instance, in the studies referenced, $\dot{V}_{O2}$ rose more than sixfold in the transition from rest to exercise at 65% of $\dot{V}_{O_2max}$ after training; however, glucose disposal and oxidation rates little more than doubled. As shown in Figure 5-25, use of other carbohydrate energy sources, mainly glycogen and lactate, increase far more during exercise than does glucose.

That glucose supplies only 10–25% of the carbohydrate energy used for exercise, and that working muscle relies on endogenous glycogen and the

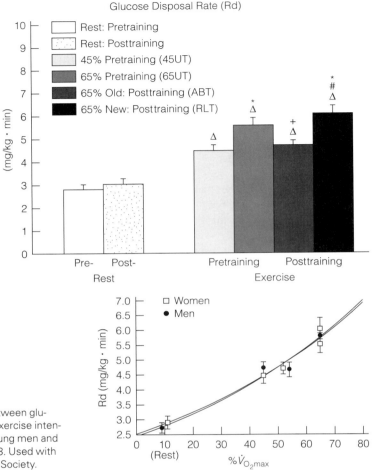

Figure 5-23   The effect of exercise intensity and training on the plasma glucose rate of disappearance (Rd) measured with [6,6-²H]glucose tracer. Values are mean ± SEM of the last 15 and 30 minutes for rest and exercise, respectively; n = 17; Δ significantly different from rest: * significantly different from 45UT; + significantly different from 65UT; # significantly different from ABT (p < 0.05). Data on young women from Friedlander et al., 1998. Used with permission of the American Physiological Society.

Figure 5-24   Exponential relationship between glucose use (disposal rate, Rd) and relative exercise intensity ($\dot{V}_{O_2max}$) before and after training in young men and women. Data from Friedlander et al., 1998. Used with permission of the American Physiological Society.

lactate shuttle probably indicates a protective mechanism preventing the fall in blood glucose during exercise. Our capacity for hepatic glucose production from liver glycogen breakdown and gluconeogenesis is limited to a maximum fourfold increase. Therefore, should glucose demand by muscle increase too much during exercise, the supply of glucose in the blood could be drained in a matter of minutes during exercise, thus leaving the brain starved for substrate and the person hypoglycemic and unable to think or function physically.

Figure 5-26 illustrates absolute and relative (fractional) glucose consumption rates in resting and working limbs. These results were determined from the product of the arterial-venous concentration difference for glucose and limb blood flow. The results are also from the Berkeley Exercise Physiology Laboratory and are comparable with those obtained in Figures 5-23 and 5-24 because experimental protocols were identical. Note in Figure 5-26b that fractional glucose extraction is restricted to the range of 2–8%. In other words, most (92–98%) glucose courses through working tissue beds without being taken up. Again, this "pseudo-glucose re-

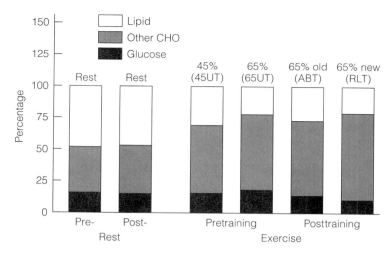

Figure 5-25 Relative percentages of substrate supply by percentage of energy supplied in resting and exercising men, before and after ten weeks of endurance training. Subjects studied at 45 and 65% of $\dot{V}_{O_2max}$ before training, and after training at the same power output that elicited 65% of $\dot{V}_{O_2max}$ before training (then 54% $\dot{V}_{O_2max}$), and 65% of the posttraining $\dot{V}_{O_2max}$. Even though use of glucose increases in relation to relative exercise intensity (Figure 5-24), glycogen and lactate are more important as substrates. Lipid use is discussed in Chapter 7. RLT is the relative exercise intensity; ABT is the absolute power output. Values are mean ± SEM of the last 15 and 30 minutes for rest and exercise, respectively; n = 17; Δ significantly different from rest; * significantly different from 45UT; + significantly different from 65UT; # significantly different from ABT (p < 0.05). Data on young women from Friedlander et al., 1998. Used with permission of the American Physiological Society.

sistance" by working muscle is possible only because of the use of intramuscular glycogen and other energy sources.

◾ Effects of Exercise and Endurance Training on Blood Lactate Concentration, Appearance, and Clearance Rates

Using a combination of stable (nonradioactive) [3-^{13}C]lactate tracer technology, and simultaneous arterial and arterial-venous difference measurements of lactate across resting and exercising limbs, Brooks and colleagues have provided long needed data on the effects of training on the blood lactate response to exercise. The results in Figure 5-27 reproduce what has long been known: that exercise increases the blood lactate concentration, but that endurance training decreases blood lactate concentration whether measured during given absolute or relative exercise intensities.

Consistent with results of previous studies on laboratory rats by Donovan and Brooks using radioactive tracers (Chapter 10), results in Figure 5-27b show that blood lactate appearance (entry) rate is exponentially related to the metabolic rate during exercise. Importantly, there is no training effect on blood lactate appearance during rest or exercise.

How then are the seemingly disparate results in Figures 5-27a and 5-27b to be understood? The answers are in Figure 5-27c. Blood lactate appearance

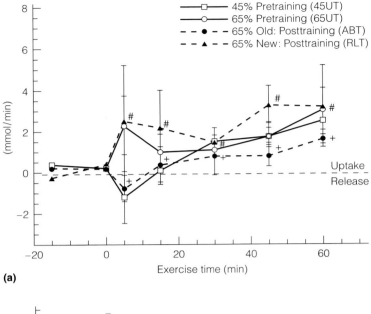

(a)

Figure 5-26 Absolute (a) and fractional leg glucose uptake (b) in resting and exercising men, before and after ten weeks of endurance training. Subjects studied at 45% and 65% of $\dot{V}_{O_2max}$ before training, and after training at the same absolute (ABT) power output that elicited 65% of $\dot{V}_{O_2max}$ before training (then 54% $\dot{V}_{O_2max}$), and 65% of the posttraining $\dot{V}_{O_2max}$, i.e., the same relative exercise intensity (RLT). Even though the increase in blood flow and glucose delivery to working muscle increases during exercise and after endurance training, only a few percent of the glucose delivered is taken up and used. Values are mean ± SEM of the last 15 and 30 minutes for rest and exercise, respectively; n = 17; Δ significantly different from rest; * significantly different from 45UT; + significantly different from 65UT; # significantly different from ABT (p < 0.05). Data from Bergman et al., 1999. Used with permission of the American Physiological Society.

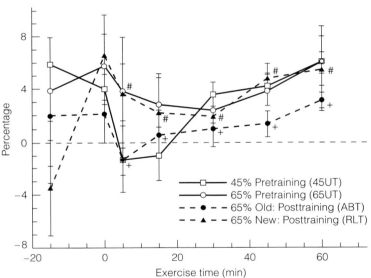

(b)

+ significantly different from pretraining (65%) at p<0.05.
# significantly different from posttraining (65% old) at p<0.05.

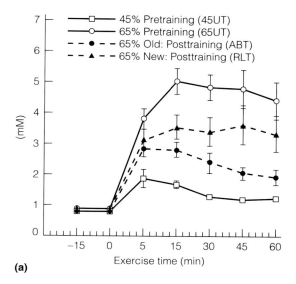

(a)

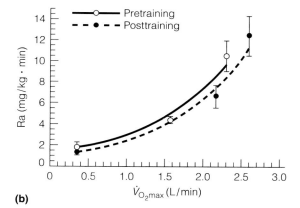

(b)

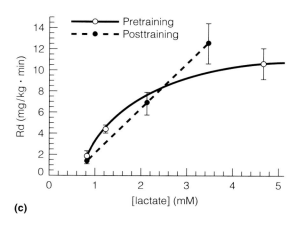

(c)

Figure 5-27 Arterial blood lactate concentration and blood lactate kinetics in resting and exercising men, before and after ten weeks of endurance training. Subjects studied at 45 and 65% of $\dot{V}_{O_2max}$ before training and after training at the same absolute power output (ABT) that elicited 65% of $\dot{V}_{O_2max}$ before training (then 54% $\dot{V}_{O_2max}$), and 65% of the posttraining $\dot{V}_{O_2max}$ (i.e., the same relative exercise intensity (RLT)). In (a), the blood lactate response during exercise is given; training reduces blood lactate concentration at given relative and absolute power outputs after training. In (b), the appearance rate of lactate as determined by infusion of [3-^{13}C]lactate is given as a function of metabolic rate ($\dot{V}_{O_2}$) during rest and exercise; note the absence of a training effect. In (c), the metabolic clearance rate is increased after training. Values are mean ± SEM of the last 15 and 30 minutes for rest and exercise, respectively; n = 17; Δ significantly different from rest; * significantly different from 45UT; + significantly different from 65UT; # significantly different from ABT (p < 0.05). Data from Bergman et al., 1999. Used with permission of the American Physiological Society.

(Ra) rises during even mild exercise (Figure 5-27b) with little effect on blood lactate concentration (Figure 5-27a) because blood lactate disposal rate (Rd) rises to compensate for the increased Ra. After training, the slope of the Rd/[Lactate] curve in Figure 5-27c is greater than before training, indicating improved lactate clearance. Further, before training the lactate-clearance curve shows a limitation (saturation), whereas after training the clearance curve rose continuously over the range of power outputs studied. In other words, after training blood lactate has to rise less to achieve a given removal rate. The results in Figure 5-27c are consistent with training increasing the number of sarcolemmal and mitochondrial lactate transport proteins (MCTs), thereby facilitating function of the cell-cell and intracellular lactate shuttles (Figures 5-13 and 5-20).

■ Summary

Glycolysis and glycogenolysis are specialized functions in skeletal muscle. Large quantities of glycolytic and glycogenolytic enzymes in muscle can support powerful contractions for brief periods. In resting muscle, little glycogenolysis occurs, so the muscle depends primarily on exogenously supplied glucose and free fatty acids (Chapter 7). However, when exercise starts, muscle glycogen provides most (80–90%) of the carbohydrate requirement for exercise. Working muscle glucose uptake rises exponentially as a function of the relative exercise intensity, but the scope of the rise (two- to fourfold) is small in comparison to the increment in overall energy supply. Working muscle takes up an average of only 4% of the glucose circulated to it. The "pseudo-glucose resistance" of working muscle is considered a protective mechanism, allowing glucose to be available for the brain, which depends on it.

Recent discoveries that mitochondria in muscle, liver, and other tissues consume and oxidize lactate, along with documentation of the existence of mitochondrial LDH and MCT, allow a new view of how glycolytic (nonoxidative or anaerobic) metabolism is linked to oxidative (or aerobic) metabolism. Lactate is always produced in muscle and other tissues because of the abundance, activity, and characteristics of cytoplasmic lactic dehydrogenase. The lactate to pyruvate ratio (L/P) is approximately 10 in resting muscle, and the ratio can rise many times during exercise even when there is no limitation in oxygen supply. Because of the *intracellular lactate shuttle mechanism,* lactate produced as the result of glycolysis in the cytosol is balanced by oxidation in mitochondria of the same cell. Hence, while resting muscle produces and releases lactate on a net basis, the release is small compared to the rate of production. In the steady state, working muscle releases little lactate so long as mitochondrial respiration is adequate to keep pace with cytosolic lactate production.

At the start of exercise, large amounts of lactate are released from muscle until the rate of oxygen consumption can rise and balance lactate production and removal. Thereafter, working muscle stops releasing lactate and, like the heart, becomes a net lactate consumer (Chapter 10). During exercise, the intracellular lactate shuttle is assisted in removing lactate by the cell-cell lactate shuttle. By this mechanism, lactate moves through the interstitium and vasculature from sites of production and net release (e.g., type IIb fibers) to highly oxidative (type I and cardiac fibers) as well as liver where removal is via gluconeogenesis as opposed to oxidation.

■ Selected Readings

Adamo, K. B., and T. E. Graham. Comparison of traditional measurements with macroglycogen and proglycogen analysis of muscle glycogen. *J. Appl. Physiol.* 84: 908–913, 1998.

Adamo, K. B., M. A. Tarnopolsky, and T. E. Graham. Dietary carbohydrate and postexercise synthesis of proglycogen and macroglycogen in human skeletal muscle. *Am. J. Physiol.* 275: E229–234, 1998.

Ahlborg, G. Mechanism of glycogenolysis in nonexercising human muscle during and after exercise. *Am. J. Physiol.* 248: E540–E545, 1985.

Baba, N. and H. M. Sharma. Histochemistry of lactic dehydrogenase in heart and pectoralis muscles of rat. *J. Cell. Biol.* 51: 621–635, 1971.

Baldwin, K. M., A. M. Hooker, and R. E. Herrick. Lactate oxidative capacity in different types of muscle. *Biochem. Biophys. Res. Com.* 83: 151–157, 1978.

Baldwin, K. M., W. W. Winder, R. L. Terjung, and J. O. Holfozy. Glycolytic enzyme in different types of skeletal muscle: adaptation to exercise. *Am. J. Physiol.* 225: 962–966, 1973.

Barnard, R. J., V. R. Edgerton, T. Furukawa, and J. B. Peter. Histochemical, biochemical and contractile properties of red, white and intermediate fibers. *Am. J. Physiol.* 220: 410–414, 1971.

Barnard, R. J., and J. B. Peter. Effect of training and exhaustion on hexokinase activity of skeletal muscle. *J. Appl. Physiol.* 27: 691–695, 1969.

Bergman, B. C., and G. A. Brooks. Respiratory gas exchange ratios during graded exercise in fed and fasted trained and untrained men. *J. Appl. Physiol.* 86: 479–487, 1999.

Bergman, B. C., G. E. Butterfield, E. E. Wolfel, G. A. Casazza, and G. A. Brooks. An evaluation of exercise and training on muscle lipid metabolism. *Am. J. Physiol.* 276 (*Endocrinol. Metab.* 39): E106–E117, 1999.

Bergman, B. C., G. E. Butterfield, E. E. Wolfel, G. Lopaschuk, G. A. Casazza, M. A. Horning, and G. A. Brooks. Net glucose uptake and glucose kinetics after endurance training in men. *A. J. Physiol.* 277 (*Endocrinol. Metab.* 40): E81–E92, 1999.

Bolli, R., K. A. Nalecz, and A. Azzi. Monocarboxylate and α-ketoglutarate carriers in bovine heart mitochondria. Purification by affinity chromatography on immobilized 2-cyano-4-hydroxycinnamate. *J. Biol. Chem.* 264: 18024–18030, 1989.

Brandt, R. B., J. E. Laux, S. E. Spainhour, and E. S. Kline. Lactate dehydrogenase in mitochondria. *Arch. Biochem. Biophys.* 259: 412–422, 1987.

Brooks, G. A. Lactate: glycolytic end product and oxidative substrate during sustained exercise in mammals—the "lactate shuttle." In Circulation, Respiration, and Metabolism: Current Comparative Approaches, vol. A, Respiration, Metabolism, Circulation, R. Gillis (Ed.). Berlin: Springer-Verlag, 1985, pp. 208–218.

Brooks, G. A. Mammalian fuel utilization during sustained exercise. *Comp. Biochem. Physiol.* 120: 89–107, 1998.

Brooks, G. A., G. E. Butterfield, R. R. Wolfe, B. M. Groves, R. S. Mazzeo, J. R. Sutton, E. E. Wolfel, and J. T. Reeves. Increased dependence on blood glucose after acclimatization to 4300 m. *J. Appl. Physiol.* 70: 919–927, 1991.

Brooks, G. A., G. E. Butterfield, R. R. Wolfe, B. M. Groves, R. S. Mazzeo, J. R. Sutton, E. E. Wolfel, and J. T. Reeves. Decreased reliance on lactate during exercise after acclimatization to 4300 m. *J. Appl. Physiol.* 71: 333–341, 1991.

Brooks, G. A., H. Dubouchaud, M. Brown, J. P. Sicurello, and C. E. Butz. Role of mitochondrial lactic dehydrogenase and lactate oxidation in the 'intra-cellular lactate shuttle.' *Proc. Natl. Acad. Sci.* 96: 1129–1134, 1999.

Brooks, G. A., and G. A. Gaesser. End points of lactate and glucose metabolism after exhausting exercise. *J. Appl. Physiol.: Respirat. Environ. Exercise Physiol.* 49: 1057–1069, 1980.

Brooks, G. A., and J. Mercier. The balance of carbohydrate and lipid utilization during exercise: the "crossover" concept. *J. Appl. Physiol.* 76: 2253–2261, 1994.

Brooks, G. A., E. E. Wolfel, B. M. Groves, P. R. Bender, G. E. Butterfield, A. Cymerman, R. S. Mazzeo, J. R. Sutton, R. R. Wolfe, and J. T. Reeves. Muscle accounts for glucose disposal but not blood lactate appearance during exercise after acclimatization to 4300 m. *J. Appl. Physiol.* 72: 2435–2445, 1992.

Connett, R. J., C. R. Honig, T. E. J. Gayeski, and G. A. Brooks. Defining hypoxia: a systems view of $\dot{V}O_2$, glycolysis, energetics and intracellular PO_2. *J. Appl. Physiol.* 68: 833–842, 1990.

Cori, C. F. Mammalian carbohydrate metabolism. *Physiol. Rev.* 11: 143–275, 1931.

Crabtree, B., and E. A. Newsholme. The activities of phosphorylase, hexokinase, phosphofructokinase, lactic dehydrogenase and glycerol 3-phosphate dehydrogenases in muscle of vertebrates and invertebrates. *Biochem. J.* 126: 49–58, 1972.

Davies, K. J. A., L. Packer, and G. A. Brooks. Exercise bioenergetics following sprint training. *Arch. Biochem. Biophys.* 215: 260–265, 1982.

Depocas, F., J. Minaire, and J. Chattonet. Rates of formation of lactic acid in dogs at rest and during moderate exercise. *Can. J. Physiol. Pharmacol.* 47: 603–610, 1964.

Donovan, C. M., and G. A. Brooks. Endurance training affects lactate clearance, not lactate production. *Am. J. Physiol.* 244 (*Endocrinol. Metab.*): E83–E92, 1983.

Eldridge, F. L. Relationship between turnover rate and blood concentration of lactate in exercising dogs. *J. Appl. Physiol.* 39: 231–234, 1975.

Embden, G., E. Lehnartz, and H. Hentschel. Der zeitliche Verlauf der Milchsäurebildung bei der Muskelkontraktion. *Mitteilung. z. Physiol. Chem.* 176: 231–248, 1928.

Foster, D. W. From glycogen to ketones and back. *Diabetes* 33: 1188–1199, 1984.

Friedlander, A. L., G. A. Casazza, M. Huie, M. A. Horning, and G. A. Brooks. Endurance training alters glucose kinetics in response to the same absolute, but not the same relative workload. *J. Appl. Physiol.* 82: 1360–1369, 1997.

Friedlander, A. L. G. A. Casazza, M. A. Horning, M. J. Huie, M.-F. Piacentini, J. K. Trimmer, and G. A. Brooks. Training-induced alterations of glucose flux in young women: Gender differences in carbohydrate oxidation. *J. Appl. Physiol.* 85: 1175–1186, 1998.

Friedlander, A. L., G. A. Casazza, M. A. Horning, T. F. Budinger, and G. A. Brooks. Effects of exercise intensity and training on lipid metabolism in young women. *Am. J. Physiol.* 275 (*Endocrinol. Metab.* 38) E853–863, 1998.

Friedman, J. E., R. W. Dudek, D. S. Whitehead, D. L. Downes, W. R. Frisell, J. F. Caro, and G. L. Dohm. Immunolocalization of glucose transporter GLUT4 within human skeletal muscle. *Diabetes* 40: 150–154, 1991.

Garcia, C. K., J. L. Goldstein, R. K. Pathak, R. G. Anderson, and M. S. Brown. Molecular characterization of a membrane transporter for lactate, pyruvate, and other monocarboxylates: implications for the Cori cycle. *Cell.* 76: 865–873, 1994.

Garcia, C. K., X. Lie, J. Ulna, and U. France. cDNA cloning of the human monocarboxylate transporter 1 and chromosomal localization of the SLC16A1 locus to 1p13.2-p12. *Genomics* 23: 500–503, 1994.

Gladden, L. B. Lactate transport and exchange. American Physiological Society Handbook of Physiology-Exercise: Regulation and Integration of Multiple Systems, Oxford University Press, 1996, 614–648.

Gladden, L. B, R. E. Crawford, and M. J. Webster. Effect of lactate concentration and metabolic rate on net lactate uptake by canine skeletal muscle. *Am. J. Physiol.* 266: R1095–1101, 1994.

Gleeson, T. T., P. M. Dalessio, J. A. Carr, S. J. Wickler, and R. S. Mazzeo. Plasma catecholamine and corticosterone and their in vitro effects on lizard skeletal muscle lactate metabolism. *Am. J. Physiol.* (*Regulatory Integrative Comp. Physiol.* 34): R632–R639, 1993.

Gollnick, P. D., R. Armstrong, C. Saubert, W. Sembrowich, R. Shepherd, and B. Saltin. Glycogen depletion patterns in human skeletal muscle fibers during prolonged work. *Pflügers Arch.* 344: 1–12, 1973.

Gollnick, P. D., R. B. Armstrong, W. L. Sembrowich, and B. Saltin. Glycogen depletion pattern in human skeletal muscle fibers after heavy exercise. *J. Appl. Physiol.* 34: 615–618, 1973.

Gollnick, P. D., and L. Hermansen. Biochemical adaptations to exercise: anaerobic metabolism. In Exercise and Sport Sciences Reviews, vol. 1, J. H. Wilmore (Ed.). New York: Academic Press, 1973, pp. 1–43.

Gollnick, P. D., and D. W. King. Energy release in the muscle cell. *Med. Sci. Sports* 1: 23–31, 1969.

Gollnick, P. D., K. Piehl, and B. Saltin. Selective glycogen depletion pattern in human muscle fibers after exercise of varying intensity and various pedaling rates. *J. Physiol.* (London) 241: 45–57, 1974.

Gollnick, P. D., P. J. Struck, and T. P. Bogyo. Lactic dehydrogenase activities of rat heart and skeletal muscle after exercise and training. *J. Appl. Physiol.* 22: 623–627, 1967.

Halestrap, A. P. The mitochondrial pyruvate carrier. Kinetics and specificity for substrates and inhibitors. *Biochem. J.* 148: 85–96, 1975.

Halestrap, A. P., and R. M. Denton. Specific inhibition of pyruvate transport in rat liver mitochondria and human erythrocytes by alpha-cyano-4-hydroxycinnamate. *Biochem. J.* 138: 313–316, 1974.

Hermansen, L. Anaerobic energy release. *Med. Sci. Sports.* 1: 32–38, 1969.

Hickson, R. C., W. W. Heusner, and W. D. Van Huss. Skeletal muscle enzyme alterations after sprint and endurance training. *J. Appl. Physiol.* 40: 868–872, 1976.

Houmard, J. A., P. C. Egan, P. D. Neufer, J. E. Friedman, W. S. Wheeler, R. G. Israel, and G. L. Dohm. Elevated skeletal muscle glucose transporter levels in exercise-trained middle-aged men. *Am. J. Physiol.* 261 (*Endocrinol. Metab.* 24): E437–E443, 1991.

Issekutz, B., W. A. S. Shaw, and A. C. Issekutz. Lactate metabolism in resting and exercising dogs. *J. Appl. Physiol.* 40: 312–319, 1976.

Jost, J. P., and H. V. Rickenberg. Cyclic AMP. *Ann. Rev. Biochem.* 40: 741, 1971.

Juel, C. Intracellular pH recovery and lactate efflux in mouse soleus muscles stimulated in vitro: the involvement of sodium/proton exchange and a lactate carrier. *Acta Physiol. Scand.* 132: 363–371, 1988.

Katz, J., J. Rostami, and A. Dunn. Evaluation of glucose turnover, body mass, and recycling with reversible and irreversible tracers. *Biochem. J.* 194: 513–524, 1981.

Kiens, B., B. Essen-Gustavsson, N. J. Christensen, and B. Saltin. Skeletal muscle substrate utilization during submaximal exercise in man: effect of endurance training. *J. Physiol.* 469: 459–478, 1993.

Kjaer, M., K. Engfred, A. Fernandes, N. H. Secher, and H. Galbo. Regulation of hepatic glucose production during exercise in humans: role of sympathoadrenergic activity. *Am. J. Physiol.* 265 (*Endocrinol. Metab.* 28): E275–E283, 1993.

Kjaer, M., P. A. Farrell, N. J. Chistensen, and H. Galbo. Increase epinephrine response and inaccurate glucoregulation in exercising athletes. *J. Appl. Physiol.* 61: 1693–1700, 1986.

Kjaer, M., B. Kiens, M. Hargreaves, and E. A. Richter. Influence of active muscle mass on glucose homeostasis during exercise. *J. Appl. Physiol.* 71: 552–557, 1991.

Kline, E. S., R. B. Brandt, J. E. Laux, S. E. Spainhour, E. S. Higgins, K. S. Rogers, S. B. Tinsley, and M. G. Waters. Localization of L-lactate dehydrogenase in mitochondria. *Arch. Biochem. Biophys.* 246: 673–680, 1986.

Koshland, D. E., and K. E. Neet. The catalytic and regulatory properties of enzymes. *Ann. Rev. Biochem.* 37: 359, 1968.

Krebs, H. A., and M. Woodford. Fructose 1,6-diphosphatase in striated muscle. *Biochem. J.* 94: 436–445, 1965.

Lamb, D. R., J. B. Peter, R. N. Jeffress, and H. A. Wallace. Glycogen, hexokinase, and glycogen synthetase adaptations to exercise. *Am. J. Physiol.* 217: 1628–1632, 1969.

Lehninger, A. L. Biochemistry. New York: Worth, 1971, pp. 313–335.

Lehninger, A. L. Bioenergetics. New York: W. A. Benjamin, 1973, pp. 53–72.

Lomako, J., W. M. Lomako, and W. J. Whelan. The biogenesis of glycogen: nature of the carbohydrate in the protein primer. *Biochem. Intl.* 21: 251–260, 1990.

Lott, J. A., and P. L. Wolfe. Clinical Enzymology. Field and Rich/Yearbook Publ., New York, 1986.

MacRae, H. S. H., S. C. Dennis, A. N. Bosch, and T. D. Noakes. Effects of training on lactate production and removal during progressive exercise. *J. Appl. Physiol.* 72: 1649–1657, 1992.

Mansour, T. E. Phosphofructokinase. *Curr. Topics Cell Regul.* 5:1, 1972.

Mazzeo, R. S., G. A. Brooks, D. A. Schoeller, and T. F. Budinger. Disposal of [1-^{13}]lactate during rest and exercise. *J. Appl. Physiol.* 60: 232–241, 1986.

McCullagh, K. J., R. C. Poole, A. P. Halestrap, M. O'Brien, and A. Bonen. Role of the lactate transporter (MCT1) in skeletal muscles. *Am. J. Physiol.* 271: E143–150, 1996.

McGilvery, R. W. Biochemical Concepts. Philadelphia: W. B. Saunders, 1975, pp. 230–266.

Meyerhof, O. Die Energie-wandlungen in Muskel. III. Kohlenhydrat und Milschsäureumsatz in Froschmuskel. *Arch. Ges. Physiol.* 185: 11–25, 1920.

Molé, P. A., P. A. VanHandel, and W. R. Sandel. Extra O_2 consumption attributable to NADH2 during maximum lactate oxidation in the heart. *Biochem. Biophys. Resh. Comm.* 85: 1143–1149, 1978.

Newsholme, E. A. The regulation of phosphofructokinase in muscle. *Cardiology.* 56: 22, 1971.

Opie, L. H., and E. A. Newsholme. The activities of fructose 1,6-diphosphatase, phosphofructokinase and phosphenolpyruvate carboxykinase in white and red muscle. *Biochem. J.* 103: 391–399, 1967.

Paradies, G., and S. Papa. The transport of monocarboxylic oxoacids in rat liver mitochondria. *Febs Letts.* 52(1): 149–152, 1975.

Price, N. T., V. N. Jackson, and A. P. Halestrap. Cloning and sequencing of four new mammalian monocarboxylate transporter (MCT) homologues confirms the existence of a transporter family with an ancient past. *Biochem. J.,* 329: 321–328, 1998.

Richter, E. A., B. Kienes, B. Saltin, N. J. Christensen, and G. Savard. Skeletal muscle glucose uptake during dynamic exercise in humans: role of muscle mass. *Am. J. Physiol.* 254: E555–E561, 1988.

Richter, E. A., T. Plough, and H. Galbo. Increased muscle glucose uptake after exercise. No need for insulin during exercise. *Diabetes* 34: 1041–1048, 1985.

Richter, E. A., N. B. Ruderman, H. Gavras, E. R. Belur, and H. Galbo. Muscle glycogenolysis during exercise: dual control by epinephrine and contractions. *Am. J. Physiol.* 242 (*Endocrinol. Metab.* 5): E25–E32, 1982.

Roth, D. A., and G. A. Brooks. Lactate transport is mediated by a membrane-borne carrier in rat skeletal muscle sarcolemmal vesicles. *Arch. Biochem. Biophys.* 279: 377–385, 1990.

Roth, D. A., and G. A. Brooks. Lactate and pyruvate transport is dominated using a pH gradient-sensitive carrier in rat skeletal muscle sarcolemmal vesicles. *Arch. Biochem. Biophys.* 279: 386–394, 1990.

Scrutton, M. C., and M. F. Utter. The regulation of glycolysis and gluconeogenesis in animal tissues. *Ann. Rev. Biochem.* 37: 269–302, 1968.

Stanley, W. C., E. W. Gertz, J. A. Wisneski, D. L. Morris, R. Neese, and G. A. Brooks. Lactate metabolism in exercising human skeletal muscle: Evidence for lactate extraction during net lactate release. *J. Appl. Physiol.* 60: 1116–1120, 1986.

Stanley, W. C., J. A. Wisneski, E. W. Gertz, R. A. Neese, and G. A. Brooks. Glucose and lactate interrelations during moderate intensity exercise in man. *Metabolism* 37: 850–858, 1988.

Taylor, A. W., J. Stothart, R. Thayer, M. Booth, and S. Rao. Human skeletal muscle debranching enzyme activities with exercise and training. *Europ. J. Appl. Physiol.* 33: 327–330, 1974.

Taylor, A. W., R. Thayor, and S. Rao. Human skeletal muscle glycogen synthetase activities with exercise and training. *Can. J. Physiol. Pharmacol.* 50: 411–415, 1972.

Vander, A. J., J. H. Sherman, and D. S. Luciano. Human Physiology. 3d Ed. New York: McGraw-Hill, 1980.

Watt, P. W., P. A. MacLennan, H. S. Hundal, C. M. Kuret, and M. J. Rennie. L(+)-lactate transport in perfused rat skeletal muscle: kinetic characteristics and sensitivity to pH and transport inhibitors. *Biochem. Biophys. Acta.* 944: 213–222, 1988.

Wilson, M. C., V. N. Jackson, C. Heddle, N. T. Price, H. Pilegaard, C. Juel, A. Bonen, I. Montgomery, O. F. Hutter, and A. P. Halestrap. Lactic acid efflux from white skeletal muscle is catalyzed by the monocarboxylate transporter isoform MCT3. *J. Biol. Chem.* 273: 15920–15926, 1998.

Winder, W. W., S. R. Fisher, S. P. Gygi, J. A. Mitchell, E. Ojuka, and D. A. Weidman. Divergence of muscle and liver fructose 2,6-diphosphate in fasted exercised rats. *Am. J. Physiol.* 260 (*Endocrinol. Metab.* 23): E756–E761, 1991.

Winder, W. W., H. T. Yang, A. W. Jaussi, and C. R. Hopkins. Epinephrine, glucose, and lactate infusion in exercising adrenodemedullated rats. *J. Appl. Physiol.* 62: 1442–1447, 1987.

York, J., L. B. Oscai, and D. G. Penny. Alterations in skeletal muscle lactic dehydrogenase isozymes following exercise training. *Biochem. Biophys. Res. Commun.* 61: 1387–1393, 1974.

York, J. W., D. G. Penney, and L. B. Oscai. Effects of physical training on several glycolytic enzymes in rat heart. *Biochem. Biophys. Acta.* 381: 22–27, 1975.

CHAPTER 6

CELLULAR OXIDATION OF PYRUVATE AND LACTATE

Physical activities lasting a minute or more absolutely require the presence and use of oxygen in active muscle. Moreover, recovery from all-out, fatiguing exercise is essentially an aerobic process. Far more energy can be realized from a substrate through oxidation than through glycolytic processes. Within muscle cells are specialized structures called mitochondria which link the breakdown of foodstuffs, the consumption of oxygen, and the maintenance of ATP and CP levels. As opposed to glycolysis, which involves carbohydrate materials exclusively, cellular oxidative mechanisms allow for the continued metabolism of carbohydrates as well as for the breakdown of derivatives of fats and proteins. Even though the processes of cellular oxidation are far removed from the anatomical sites of breathing and the pumping of blood, it is the processes of cellular oxidation that breathing and beating of the heart serve. Seen in their proper perspective, the two most familiar physiological processes (breathing and the heartbeat) play a key role in energy transduction.

Hicham El Guerrouj of Morocco sets a new World record in the mile run (3:43.13). Performances such as this one require the highest sustained metabolic power output humanly possible. Such performances depend on the ability to consume and utilize oxygen as well as the ability to access immediate and non-oxidative energy sources.
SOURCE: © AP/Wide World Photos.

■ Mitochondrial Structure

Cellular oxidation takes place in cellular organelles called *mitochondria* (mitochondrion, singular). Pyruvate and lactate (products of glycolysis) as well as products of lipid and amino acid metabolism are metabolized within mitochondria. These organelles have been called "the powerhouses of the cell." It is within mitochondria that almost all oxygen is consumed and ADP is phosphorylated to ATP. Consequently, those activities that last more than a minute are powered by mitochondria. As suggested in Figure 6-1, mitochondria appear in two places in skeletal muscle. A significant population of mitochondria is located immediately beneath the cell membrane (sarcolemma); these are subsarcolemmal *mitochondria* and are in a position to receive O_2 provided by the arterial circulation. Subsarcolemmal mitochondria are believed to provide the energy required to maintain the integrity of the sarcolemma. Energy-requiring exchanges of ions and metabolites

across the sarcolemma are most probably supported by subsarcolemmal mitochondria activity. Deeper within muscle cells are the *intermyofibrillar* mitochondria. As the name implies, these mitochondria exist among the contractile elements of the muscle. Intermyofibrillar mitochondria probably have a higher activity per unit mass (specific activity) than subsarcolemmal mitochondria, and probably play a major role in maintaining the ATP supply for energy transduction during contraction.

Because of their appearance in cross-sectional electron micrographs and in micrographs of isolated mitochondria, it has long been believed that the mitochondria exist as discrete capsule-shaped organelles. However, seminal work on diaphragm muscle by the Russian scientists Bakeeva and Skulachev indicates that mitochondria are interconnected in a network (the mitochondrial reticulum), much like the sarcoplasmic reticulum. Instead of thousands of mitochondria, a muscle cell may actually contain relatively few subsarcolemmal and intermyofibrillar mitochondria, but each mitochondrion has thousands of branches.

Figure 6-2 is the result of collaborative efforts by researchers in Berkeley, California, and Babraham, England. From serial cross sections through rat soleus muscle (Figure 6-2a), a particular "mitochondrion" can be identified and traced, and a model constructed, based on knowledge of the magnification and section thickness. The same "mitochondrion" can continue to be modeled in subsequent slices through the muscle (i.e., serial cross sections). These models of individual sections can then be stacked up to reveal the appearance of a mitochondrial reticulum (Figure 6-2b).

Red-pigmented muscle fibers obtain their color, in part, from their number of mitochondria, which are red. By comparison, pale muscle fibers contain few mitochondria. (See Figure 5-5.) More detailed studies on animal muscle that suggest the presence of a mitochondrial reticulum are supported by cross-sectional analyses of human muscle biopsy specimens. Figure 6-3 from Hoppeler (1986) in Bern, Switzerland, shows muscle ultrastructure at three magnifications and illustrates the anatomical relationship between capillary O_2 delivery and the mitochondrial reticulum.

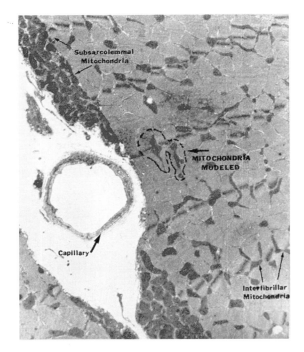

Figure 6-1 Mitochondria are found to be distributed within two areas of a muscle cell. Subsarcolemmal mitochondria are found immediately beneath the cell membrane (i.e., the sarcolemma). Intermyofibrillar mitochondria are found among the muscle cell's contractile elements. Micrograph courtesy of E. E. Munn, C. Greenwood, S. P. Kirkwood, L. Packer, and G. A. Brooks.

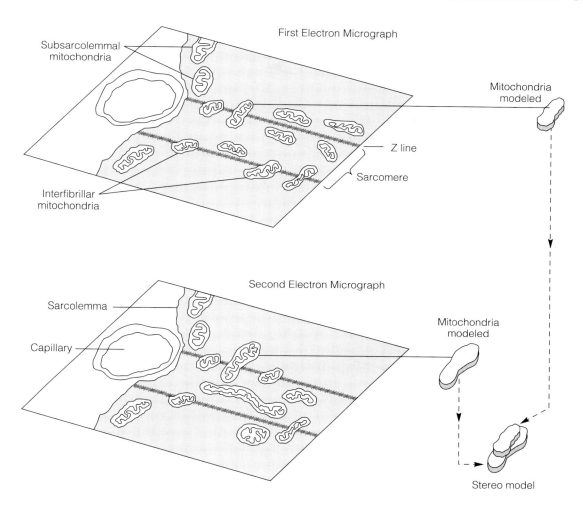

First Electron Micrograph

Subsarcolemmal mitochondria

Mitochondria modeled

Z line

Sarcomere

Interfibrillar mitochondria

Second Electron Micrograph

Sarcolemma

Capillary

Mitochondria modeled

Stereo model

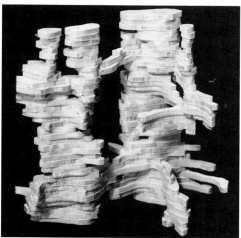

Figure 6-2 In all probability, "mitochondria" do not exist as separate, individual entities within muscle cells but rather as parts of a network, or "reticulum." Evidence for the mitochondrial reticulum in limb skeletal muscles was obtained by modeling the mitochondria seen in electron micrographs of serial cross sections through rat soleus muscles. The process of modeling is illustrated in (a), and the resulting model is pictured in (b). [Model (part b) courtesy of E. A. Munn and G. Greenwood, Babraham, England.]

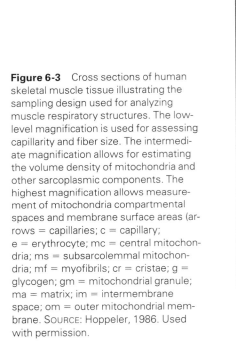
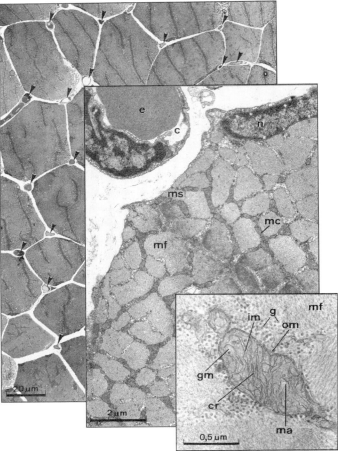

Figure 6-3 Cross sections of human skeletal muscle tissue illustrating the sampling design used for analyzing muscle respiratory structures. The low-level magnification is used for assessing capillarity and fiber size. The intermediate magnification allows for estimating the volume density of mitochondria and other sarcoplasmic components. The highest magnification allows measurement of mitochondria compartmental spaces and membrane surface areas (arrows = capillaries; c = capillary; e = erythrocyte; mc = central mitochondria; ms = subsarcolemmal mitochondria; mf = myofibrils; cr = cristae; g = glycogen; gm = mitochondrial granule; ma = matrix; im = intermembrane space; om = outer mitochondrial membrane. SOURCE: Hoppeler, 1986. Used with permission.

As noted by Hoppeler and Weibel and associates, the essentially cylindrical shape of muscle fibers means that 50% of the cell volume is in the outer 25% of cross-sectional diameter. This means that subsarcolemmal and interfibrillar components of the mitochondrial reticulum occupy approximately equal portions of muscle cell volume.

■ The Mitochondrial Reticulum: Mitochondria Are Interconnected Tubes, Not Individual Spheres

Recognition of the existence of a muscle mitochondrial reticulum has brought about the development of new concepts of the distribution of oxygen, energy metabolites, and energy. Bakeeva and Skulachev (Bakeeva et al., 1978) have pointed out that oxygen is more soluble in the phospholipid environment of mitochondrial membranes than in the cytosol. Most importantly, presence of the mitochondrial reticulum (Figures 6-3 and 6-4) creates a chemiosmotic (energy) gradient whereby energy can be transmitted from the cell surface, where oxygen is in high concentration, through the mitochondrial reticulum to deep within the cell, where oxygen delivery is less and oxygen concentration is lower. The mitochondrial chemiosmotic gradient is discussed more later (see Figure 6-10) but here it is important to note that oxygen and all mito-

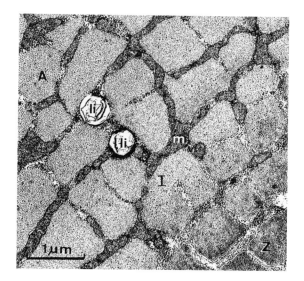

Figure 6-4 Cross section of a portion of a human muscle fiber exposing the A- and I-band and the Z-line regions. Lipid droplets (li) are seen in contact with mitochondria (m). It is evident that the mitochondria in this muscle fiber form an extensively branched tubular network, or reticulum. SOURCE: Hoppeler, 1986. Used with permission.

chondrial constituents are necessary to maintain the chemiosmotic gradient over the distance from the sarcolemmal surface to deep within the fiber.

■ Mitochondrial Structure and Function

Figure 6-3 is an electron micrograph of a muscle mitochondrion, whereas Figure 6-5 provides a schematic representation of a mitochondrion. Area 1, the outer membrane, functions as a barrier to maintain important internal constituents (e.g., NADH) and to exclude exterior factors. The outer membrane contains specific transport mechanisms to regulate the influx and efflux of various materials. Area 2, the intermembrane space, also contains enzymes for exchange and transport. Area 3, the inner membrane, can be divided into two surfaces. Area 3a functions, in part, along with the outer and inner membrane constituents in a transport capacity. For example, the enzyme carnitine transferase is found on the inner wall of fragmented mitochondria. Carnitine transferase is involved in moving lipids into mitochondria. The mitochondrial inner membrane is also relatively impermeable to protons, which helps preserve the chemiosmotic gradient. Area 3b of the inner membrane is referred to as the cristae membrane, so called because it is made up of many folds, or *cristae* (*crista,* singular). The cristae membrane is the main mitochondrial site where oxidative phosphorylation takes place. Because area 3a is

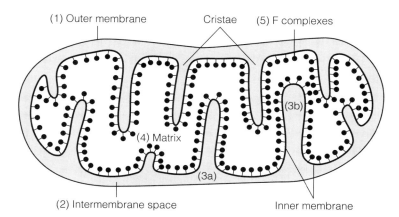

(1) Outer membrane   Cristae   (5) F complexes

(3b)

(4) Matrix

(3a)

(2) Intermembrane space   Inner membrane

Figure 6-5 Schematic representation of a mitochondrion (reticulum fragment). Areas indicated are (1) the outer membrane, (2) the intermembrane space, (3) inner membrane, and (4) the matrix. The inner membrane is diagrammed as existing in either of two orientations: juxtaposed to the outer membrane (3a) or protruding into the matrix (3b). F complexes (5) protrude from the inner membrane, which is a mobile phospholipid structure. Inward folds of the inner membrane form the cristae. See the text for a description of the functions of each area.

confluent with area 3b, and the chemical compositions are similar, area 3a, the part of the inner membrane that is juxtaposed to the outer membrane, can phosphorylate as well.

Keep in mind that the membrane constituents of mitochondria are not rigidly locked into position the way a leg is attached to the trunk of a body. Rather, the inner membranes change in conformation (shape) as mitochondria function, and the membrane constituents themselves display a high degree of mobility such that various components can move both along and within the membranes.

The actual mitochondrial site of phosphorylation is the F complex, which looks like a ball on a stalk on the mitochondrial inner membrane. The F complex is alternatively termed the *elementary particle,* and is made up of two subunits: the stalk (F_0 complex) and ball (F_1 complex).

Area 4, the matrix, is not simply a space but contains nearly half the mitochondrial protein. In the matrix are located LDH and the Krebs cycle enzymes. We describe those in detail next.

■ The Krebs Cycle

Pyruvate (and lactate) gain entry to the mitochondrial matrix via a carrier protein (MCT) located in the inner membrane (see Figure 5-13). Additionally, there are possibilities for lactate and pyruvate transporters to exist in the outer membrane. However, the mitochondrial outer membrane is permeable to low molecular weight substances, so a transporter is unnecessary. The sequence of events of pyruvate metabolism, catalyzed first by the enzyme complex pyruvate dehydrogenase (PDH) and then by the Krebs cycle enzymes, is illustrated in Figures 6-6 and 6-7.

The series of enzymes depicted in Figures 6-6 and 6-7 is called the *Krebs cycle* after Sir Hans Krebs, who did much of the work elaborating the pathway. This series of enzymes is also often referred to as the *citric acid cycle* (the first constituent is citric acid) and the *tricarboxylic acid cycle* (the initial constituents have three carboxyl groups, the TCA cycle).

Although the TCA cycle is generally referred to as a cycle, it is important to realize that it is imperfect. Various substances can leave the cycle, and others can gain entry to it. This will be discussed later in detail. Figure 6-6 is intended to show the key role of pyruvate dehydrogenase (PDH) in regulating flux into the TCA cycle. The purposes of PDH and the TCA cycle are revealed to be decarboxylation (CO_2 formation), ATP production and most importantly, NADH production. The figure shows that there are four places where NAD^+ is reduced to NADH, and one place each where FADH and ATP are formed. Recalling that each NADH is equivalent to three ATPs, and that each FADH is equivalent to two ATPs, the purpose of the TCA cycle is revealed: to continue the metabolism of pyruvate from glucose (as well as from the intermediate products of lipid and protein catabolism) and to trap part of the energy released in the forms of ATP and the high-energy reduced compounds NADH and FADH.

The enzyme pyruvate dehydrogenase is a complex that has three parts and functions in several steps to convert pyruvate to acetyl-CoA, and one additional step to reduce NAD^+ to NADH. Both thiamine pyrophosphate (TPP) and flavine adenine dinucleotide (FAD) serve as cofactors for the complex. In addition, PDH is an enzyme controlled by phosphorylation state (Figure 6-8). The PDH enzyme component of PDH complex is inhibited when phosphorylated by the action of a specific kinase that uses ATP. Thus, high ATP/ADP, acetyl-CoA/CoA, and $NADH/NAD^+$ act to reduce glycolytic flux to the TCA by inactivation of PDH. Conversely, dephosphorylation of PDH by a specific phosphatase serves to activate the enzyme. Dephosphorylation is promoted by high levels of pyruvate and Ca^{2+} as well as by decreases in ATP/ADP, acetyl-CoA/CoA, and $NADH/NAD^+$. Also important, insulin binding to the cell surface promotes dephosphorylation (activation) of PDH by some, as yet unknown, second messenger. The chemical agent dichloroacetate (DCA) serves to dephosphorylate and to activate PDH, thus increasing flux from glycolysis to mitochondria. For this reason, DCA is sometimes used to reduce lactic acidosis.

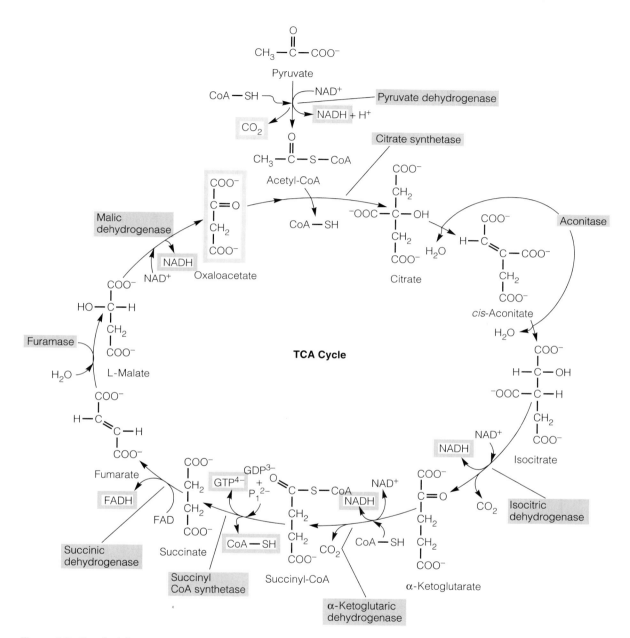

Figure 6-6   Detailed diagram of the Krebs cycle intermediates and catalyzing enzymes. As the result of one acetyl-CoA entering and traversing the cycle, one GTP (guanosine triphosphate, energetically equivalent to an ATP) and several high-energy reducing equivalent compounds (NADH and $FADH_2$) are formed. These substances result in significant mitochondrial electron transport and ATP production. Modified from McGilvery, 1975.

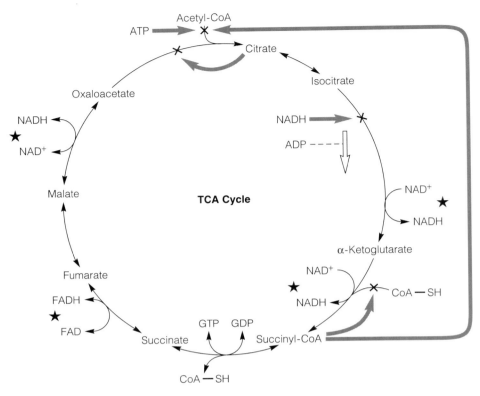

Figure 6-7 Regulation of the citric acid cycle. The starred reactions require oxidized coenzymes; the ratio of oxidized to reduced coenzymes is governed by the availability of ADP and P_i for oxidative phosphorylation, thereby regulating the cycle to demand. However, the nearly irreversible isocitric and α-ketoglutaric dehydrogenase reactions can still proceed at low NAD^+ levels. They are controlled through inhibition by NADH and succinyl-CoA, respectively. Also, isocitric dehydrogenase is activated by ADP and the dehydrogenase is activated by Ca^{2+}, which enters mitochondria during muscle contraction. Finally, when the high-energy phosphate supply is high and GTP accumulates, the accumulating succinyl-CoA inhibits the initial citrate synthetase reaction in competition with oxaloacetate. This effect also tends to balance the initial part of the cycle, which consumes oxaloacetate, against the final part, which produces oxaloacetate. With the exception of succinic dehydrogenase, which bonds to the inner membrane, TCA cycle enzymes exist in the mitochondrial matrix. SOURCE: McGilvery, 1975.

As implied by the name, pyruvate dehydrogenase, in the conversion of pyruvate to acetyl-CoA, an NAD^+ molecule is reduced to NADH. NADH is termed the "universal hydrogen carrier." In addition, pyruvate dehydrogenase functions in concert with coenzyme A (CoA), which is referred to as a coenzyme or a cofactor. A cofactor is a low-molecular-weight substance whose presence is required to allow an enzyme to work. CoA is sometimes referred to as coenzyme 1, as it was the first such factor discovered; it is also sometimes referred to as the "universal acetate carrier." In Figure 6-9,

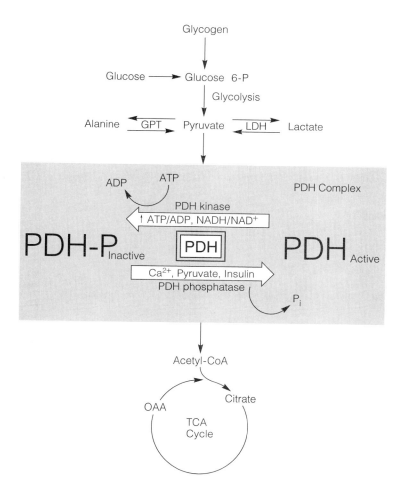

Figure 6-8 The enzyme pyruvate dehydrogenase (PDH) is one component of the "PDH complex," which regulates glycolytic flux (flow) to the TCA cycle or, alternatively, to lactate and alanine. Phosphorylation inhibits PDH, whereas dephosphorylation activates the enzyme and enzyme complex.

the terminal sulfur bond of CoA is shown to be the site at which acetate binds. As the result of pyruvate dehydrogenase, pyruvate is decarboxylated, releasing a CO_2 molecule, and the remaining two-carbon unit (acetate) is combined with CoA to give acetyl-CoA. In summary, pyruvate dehydrogenase is an important enzyme, not only because of its complex function, but also because it is a rate-limiting enzyme. The activity of PDH plays a major role in determining the rates of glycolysis lactate production and carbohydrate supply for mitochondrial oxidation.

Acetyl-CoA is the entry substance into the TCA cycle. Acetyl-CoA can be formed from pyruvate as

well as from fatty and amino acids. Under the influence of the enzyme citrate synthetase, acetyl-CoA and oxaloacetic acid (OAA) condense to give citric acid. As we shall see, the presence of OAA may be a regulating factor controlling the rate of the TCA cycle.

Several steps into the TCA cycle is the enzyme isocitric dehydrogenase (IDH). This is the rate-limiting enzyme of the TCA cycle, much as PFK is the rate-limiting enzyme in glycolysis. Like PFK, IDH is an allosteric enzyme stimulated by ADP. IDH, together with the other dehydrogenases of the TCA cycle, as well as pyruvate dehydrogenase, are sensitive to the redox (reduction-oxidation) poten-

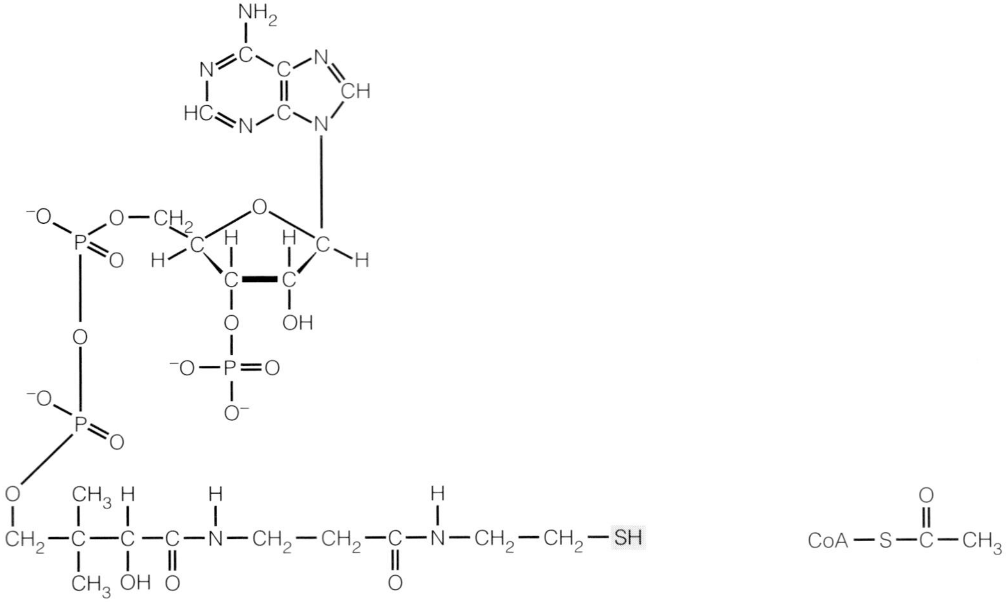

Figure 6-9 (a) Structure of coenzyme A (CoA) showing the binding site (-SH) for acetate. (b) Acetyl-CoA is formed from the union of CoA and acetate.

tial in the cell. Simply defined, the redox potential is the $NADH/NAD^+$ ratio. The dehydrogenases are inhibited by a high redox potential and are stimulated by a decline in redox potential.

■ The Electron Transport Chain

The *electron transport chain* (*ETC*) is located on the mitochondrial inner membrane, probably in both areas 3a and 3b. (See Figure 6-5.) The ETC constituents are arranged in the sequence indicated in Figure 6-10. We owe much of our present understanding of how mitochondria function to Peter Mitchell (1965) and his chemiosmotic theory of oxidative phosphorylation. The term *oxidative phosphorylation* refers to two separate processes that usually function together. Oxidation is a spontaneous process that is linked or coupled to the phosphorylation, the union of P_i with ADP to make ATP. Phosphorylation is an endergonic process driven by oxidation. It is

important to note that although the linkage in oxidative phosphorylation is tight, certain situations— for example, heat buildup in muscles from prolonged work—can cause the linkage to be loosened or uncoupled.

The function of the ETC can be simply described as follows. Reducing equivalents containing a high-energy hydrogen and electron pair gain entry to the beginning of the chain. The hydrogen and electron then move from areas of electronegativity (NAD^+) toward areas of electropositivity (atomic oxygen). Along the ETC, the electron is stripped from the hydrogen, which continues along the chain; the resulting proton (H^+) is pumped outside the mitochondria (see Figure 6-10). For each NADH entering, three pairs, or a total of six protons, are pumped out. For each FADH entering, two pairs of protons are pumped out. Outside the mitochondrion, a region of decreased pH and positive charge is created. This chemical and osmotic potential ultimately supplies the energy to phosphorylate ADP.

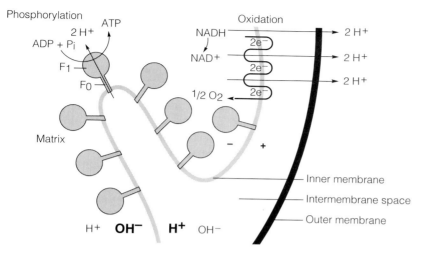

Figure 6-10 Orientation of components of the electron transport chain (ETC) within the mitochondrial inner membrane. Entry of hydride ions (H) from high-energy reduced compounds (e.g., NADH) into the ETC results in the oxidation of those compounds, electron transport, and the expulsion of protons (H^+). This creates a chemical and electrical gradient across the inner membrane. The entry of protons through specific portals into the F_0–F_1 complex (an elementary particle like a stalk and ball) provides the energy for the phosphorylation of ADP to ATP.

It is theorized that portals exist through the outer wall of the mitochondrion, which apparently can channel the potential energy of the external proton field to the ATPase on the inner membrane.

According to Mitchell, pairs of protons enter the mitochondrion and are directed to the stalk (or F_0) part of the elementary particle. These protons attack the oxygen of phosphate groups previously bound to the ball (or F_1) part of the elementary particle. The combination of phosphate and atomic oxygen results in the formation of water and an energized phosphate ion. This phosphate ion is then in position to unite with ADP, which has also previously been bound to F_1. This forms ATP. Phosphorylation may thus be written as two partial reactions:

$$2\,H^+ + PO_4^{3-} \rightarrow H_2O + PO_3^{-5} \qquad (6\text{-}1)$$
$$PO_3^{-5} + ADP^{3-} \rightarrow ATP^{4-} \qquad (6\text{-}2)$$

The formation of ATP according to Mitchell's chemiosmotic theory represents a reversal of the hydrolysis of ATP. Recalling Equation 3-2, the hydrolysis of ATP can be written as:

$$\begin{array}{c} \xrightarrow{\text{Hydrolysis}} \\ ATP + HOH \quad \underset{\xleftarrow{\text{Phosphorylation}}}{ATPase} \quad ADP + P_i \quad (6\text{-}3) \end{array}$$

In the mitochondrion, the phosphorylation of ADP is made possible by linking ATP production to the formation of water. With energy input, the mitochondrial ATPase is driven against its equilibrium, toward ATP production.

Function of the ETC

The electron transport chain functions based on three factors. First, each constituent can exist in reduced (higher energy, with electron) and oxidized (lower energy, without electron) forms. Second, the ETC constituents are sequentially arranged in close proximity on the inner membrane, forming a "respiratory assembly." Again, these respiratory assemblies are linked through the mitochondrial reticulum and run from the cell surface to deep within the fiber. Distance, therefore, does not impede electron movement. Third, and most important, the ETC constituents are arranged such that the redox potential of each constituent is greater (i.e., more positive) than that of the previous constituent. As a result, electrons can move from NADH (electronegative redox potential) to atomic oxygen (electropositive). Oxygen is called "the final electron acceptor."

Control of the ETC

We have seen that the adenine nucleotide charge regulates both glycolysis and the Krebs cycle. It should therefore not be surprising that ADP and ATP, respectively, stimulate and inhibit the ETC. The control of muscle metabolism is elegantly simple. As soon as muscle contracts, ATP is split and ADP is formed. The change in the relative amounts of these substances then sets in motion biochemical events to re-form the spent ATP. When exercise stops, the cellular respiratory mechanisms soon reestablish normal levels of ATP, ADP, and AMP. Consequently, whole-body O_2 consumption rapidly declines toward resting levels after exercise. The mechanism by which ATPase activity of the contractile elements is buffered by cytoplasmic creatine phosphate, with the ultimate result being phosphorylation of mitochondrial ADP to ATP is termed the *creatine phosphate shuttle* (Figure 6-11).

The Number of ATP from a Glucose Molecule

The breakdown of glucose and glucose 6-phosphate in muscle is extremely important because it can occur rapidly and therefore can supply energy rapidly. Further, glycolysis can occur in the presence or absence of oxygen. This means that energy requirements in muscle can be supplied by glycolysis for finite periods of time under anaerobic conditions.

Under aerobic conditions, for each glucose unit a net of two ATPs are formed in the cytoplasm, and the energy equivalents of two cytoplasmic NADH can be shuttled into the mitochondria. Recalling that the reducing equivalent energy of cytoplasmic NADH can give rise to either NADH or FADH within mitochondria, and that the P:O of NADH is 3 and the P:O of FADH is 2, then, under aerobic conditions, glycolysis yields four to six ATPs in mitochondria in addition to the two cytoplasmic ATPs. Together with the ATP formed in the TCA cycle and the ATP produced as the result of reducing equivalents donated from the TCA cycle to the ETC, the ATP yield per glucose is 36 to 38 molecules of ATP per molecule of glucose.

As we noted earlier, when glycolysis is anaerobic, a net of two ATPs is formed per glucose.

When the substrate for glycolysis originates from glycogen—that is, when glucose 6-phosphate is supplied by glycogenolysis—it is more difficult to estimate the ATP yield per glucosyl unit cleaved from glycogen. When a molecule of glucose enters the glycolytic pathway through the action of the enzyme hexokinase, a molecule of ATP is required to energize glucose to the level of glucose 6-phosphate. In glycogenolysis, this energizing step is bypassed, and theoretically we could consider that the energy of one phosphorylation is saved. Thus, the ATP yield for a glucosyl unit derived from glycogen under anaerobic conditions may be 3. However, it may be incorrect to assume that glycogen yields relatively more ATP than does glucose. This is because it takes energy to synthesize glycogen from glucose or glucose 6-phosphate. Although glycogen synthesis may have preceded glycogenolysis in time by hours or days, the energy requirement for the process should be taken into account. Further, ATP was required to activate the phosphorylase, the enzyme catalyzing glycogenolysis. In starting glycolysis from glucose, each molecule has to be phosphorylated. In glycogenolysis, the enzyme rather than the substrate is phosphorylated.

The activity of phosphorylase is much higher in muscle than is the activity of hexokinase. Consequently, entry of glucosyl units into glycolysis during exercise is more rapid from glycogen than from glucose. It follows, therefore, that the ATP yield for anaerobic glycolysis during heavy exercise reflects

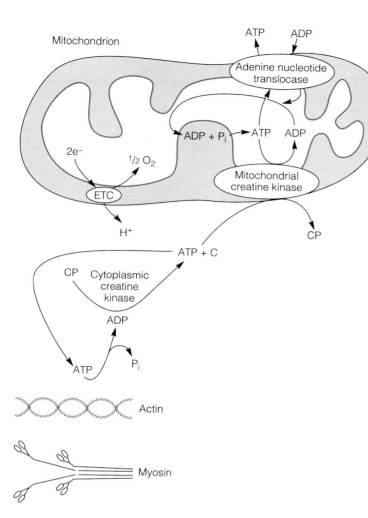

Figure 6-11 A model of the control of cellular respiration by creatine phosphate and ADP (the creatine phosphate shuttle). The model begins at the lower left, where the contractile proteins actin and myosin hydrolyze ATP. The resulting ADP is phosphorylated by cytoplasmic creatine kinase, with CP serving as the phosphate donor. The resulting cytoplasmic creatine is rephosphorylated by mitochondrial creatine kinase. Thus, ATP hydrolysis in the cytoplasm results in ADP formation in mitochondria. The rates of electron transport and O_2 consumption in mitochondria respond to the presence of ADP, phosphorylating it to ATP.

the dominant role of glycogenolysis and is closer to 3 than to 2.

■ Effects of Training on Skeletal Muscle Mitochondria

Beginning in the late 1960s and early 1970s, John Holloszy in St. Louis, Missouri, and Philip Gollnick in Pullman, Washington, ushered in a new era of study on what has been called *exercise biochemistry*. In their initial papers and in subsequent reports with associates, Gollnick (1969) and Hollo-szy (1967, 1975; Baldwin et al., 1972; Holloszy and Booth, 1976; Holloszy et al., 1971) have identified a number of specific effects of endurance training on skeletal muscle mitochondria and respiratory capacity. Researchers in other laboratories have obtained similar results. Table 6-1 summarizes some of the results of these studies. In response to the endurance-training procedure utilized, several enzymes of the TCA cycle and constituents of the ETC have been observed to double in activity.

A major question in exercise biochemistry has been: How is the increase in muscle mitochondrial capacity (Table 6-1) accomplished? Do the mitochon-

TABLE 6-1

Effect of Endurance Training on Respiratory Capacity of Rat Skeletal Muscle Mitochondria[a]

Muscle	Group	Muscle Citrate Synthase (μmol $\cdot$ g^{-1} $\cdot$ min^{-1})	Muscle Carnitine Palmityl-transferase (μmol $\cdot$ g^{-1} $\cdot$ min^{-1})	Muscle Cytochrome c (nmol $\cdot$ g^{-1})
Fast-twitch white	Exercised	18 ± 1	0.20 ± 0.02	6.3 ± 0.7
	Sedentary	10 ± 1	0.11 ± 0.01	3.2 ± 0.3
Fast-twitch red	Exercised	70 ± 4	1.20 ± 0.09	28.4 ± 2.1
	Sedentary	36 ± 3	0.72 ± 0.06	16.5 ± 1.6
Slow-twitch red	Exercised	41 ± 3	1.20 ± 0.05	—
	Sedentary	24 ± 2	0.63 ± 0.07	—
Heart	Exercised	160 ± 4	—	46.6 ± 0.9
	Sedentary	158 ± 1	—	47.1 ± 1.0

[a] In response to endurance training, mitochondrial components in different types of rat skeletal muscles double in concentration. The heart does not change.

SOURCE: Baldwin et al. (1972), Oscai et al. (1971), and Holloszy (1967, 1975; Holloszy and Booth, 1976; Holloszy et al., 1971).

dria adapt in specific ways, such as by an increase in the number or density of enzymes on the mitochondrial cristae, or are there simply more mitochondria (a larger mitochondrial reticulum)? The answer to this question has been investigated by scientists at the University of California, Berkeley. Their results clearly indicate that mitochondria do not increase in specific activity—that is, in enzymatic activity per unit of mitochondrial protein (Table 6-2, top). Rather, there are more mitochondria or there is a more elaborate reticulum (Table 6-2, bottom). With a few minor exceptions, such as the labile, nonstructural enzyme α-glycerol phosphate dehydrogenase, most mitochondrial constituents increase in direct proportion to the amount of mitochondrial material. Further, the Berkeley group showed no change in protein/lipid ratio with training. Training does affect the mechanisms of mitochondrial replication and destruction, however. As a result of endurance training, skeletal muscle contains more mitochondrial material, but its activity is the same as that of untrained individuals.

Examinations of muscle from untrained and trained animals under the electron microscope (Figure 6-12) provide results consistent with the biochemical analyses (Tables 6-1 and 6-2). Mitochon-

dria in tissues of trained rats may appear to be more numerous. We may say that the mitochondrial reticulum is more elaborate in response to training.

As dramatic as the effects of training on the respiratory capacity of muscle are, at least two problems need to be resolved. First, what specific aspects of endurance training result in these changes? Why does progressive resistance (weight) training not improve mitochondrial capacity? Obviously, endurance training presents some stimulus to the muscle that weight training does not. The identities of these training stimuli that cause mitochondrial protein to increase in response to endurance training and contractile protein to increase in response to heavy resistance training are under investigation.

A second major question concerns the significance of the doubling of mitochondrial activity. Table 6-1 indicates large increments in the respiratory capacities of mitochondria in all three types of skeletal muscle fibers. In humans and in laboratory animals, training may increase by severalfold the length of time a submaximal work load can be endured. However, in both humans and laboratory animals, maximal O_2 consumption capacity increases only 10 to 15% in response to endurance training. In results obtained by the U.C. group,

TABLE 6-2

Specific Activity (Top) and Mitochondrial Content (Bottom) in Skeletal Muscle from Endurance-Trained Rats and Sedentary Controls[a-c]

Parameter	Mitochondrial Protein Specific Activity (nmol · mg^{-1})	
	Control Group ($n = 10$)	Endurance Group ($n = 10$)
Cytochrome a	0.37 ± 0.02	0.35 ± 0.08
Cytochrome b	0.32 ± 0.02	0.36 ± 0.02
Cytochrome c ($+c_1$)	0.91 ± 0.05	0.82 ± 0.03
Flavoprotein	1.13 ± 0.08	1.04 ± 0.05
b/a	0.86	1.03
c ($+c_1$)$/a$	2.46	2.34
Flavoprotein$/a$	3.05	2.97

Parameter	Mitochondrial Content of Muscle (mg · g^{-1})		
	Control Group ($n = 9$)	Endurance Group ($n = 9$)[b]	Percentage Difference of Control from Endurance Trained
Pyruvate–malate oxidase	15.5 ± 0.7	25.1 ± 1.2	+ 62
Succinate oxidase	20.1 ± 1.0	43.6 ± 2.1	+ 117
Palmitoyl carnitine oxidase	21.1 ± 2.6	50.4 ± 4.7	+ 138
Cytochrome oxidase	19.2 ± 0.8	38.3 ± 1.8	+ 99
Succinate dehydrogenase	14.9 ± 0.8	30.9 ± 1.8	+ 108
NADH dehydrogenase	17.5 ± 0.8	27.9 ± 1.7	+ 59
Choline dehydrogenase	25.9 ± 1.7	55.7 ± 3.8	+ 115
Cytochrome c ($+ c_1$)	16.2 ± 0.6	31.6 ± 1.2	+ 95
Cytochrome a	18.2 ± 0.7	36.5 ± 1.6	+ 101
Average	18.7 ± 1.2	37.8 ± 3.7	+ 99

[a] Values are mg mitochondrial protein · g wet muscle^{-1} (means $\pm$ SE). Mitochondrial content of muscle was calculated as muscle activity/mitochondrial specific activity, for the relevant oxidases, dehydrogenases, and cytochromes.
[b] All endurance-trained values were significantly higher than controls, $P < 0.01$ (one-tailed t test).
[c] In terms of specific activity (units of activity or component content per unit of mitochondrial protein), no difference is seen between endurance-trained and sedentary control groups (top table), whereas the muscle mitochondrial content doubles (bottom table).

SOURCE: Davies et al. (1981, 1982).

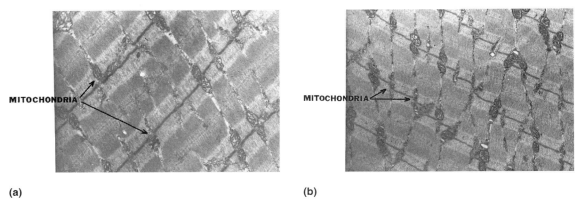

(a)   **(b)**

Figure 6-12   Electron micrographs of mitochondria in tissues of (a) untrained and (b) trained rats. Mitochondria in trained animals appear to be more numerous, probably because there is a more elaborate mitochondrial reticulum. Micrographs courtesy of P. D. Gollnick.

TABLE 6-3

Correlation Matrix for Muscle Oxidases, $\dot{V}_{O_2max}$, and Maximal Endurance in Rats on a Treadmill[a,b]

	Pyruvate–malate Oxidase	Palmitoyl Carnitine Oxidase	$\dot{V}_{O_2max}$ (weight normalized)	Maximal Endurance
Cytochrome oxidase	0.95	0.93	0.74	0.92
Pyruvate–malate oxidase	—	0.89	0.68	0.89
Palmitoyl carnitine oxidase	—	—	0.71	0.91
$\dot{V}_{O_2max}$ (weight normalized)	—	—	—	0.70

[a] All correlations reported were statistically significant ($P < 0.01$).
[b] Running endurance of rats as measured on a standardized treadmill test correlates significantly better with skeletal muscle mitochondrial activity (i.e., cytochrome oxidase activity) than with maximal O_2 consumption ($\dot{V}_{O_2max}$).

SOURCE: Davies et al. (1981, 1982).

muscle mitochondrial cytochrome oxidase activity correlated 0.92 with running endurance but only 0.70 with $\dot{V}_{O_2max}$ (Table 6-3). Consequently, it appears that mitochondrial capacity may be better related to endurance than $\dot{V}_{O_2max}$. The reason for this needs to be elaborated, but most likely increased mitochondrial mass increases the sensitivity of respiratory control. This training effect is illustrated in Figure 6-13. A greater mitochondrial mass means that a given rate of oxygen consumption can be achieved with a higher ATP/ADP ratio. Superior

respiratory control due to increased mitochondrial mass is thought to feed back on regulatory enzymes of carbohydrate degradation, thereby allowing more lipid and less carbohydrate to be consumed at a given $\dot{V}_{O_2}$ after training. The increase in mitochondrial mass due to endurance training may also be the means for increasing fat utilization as a fuel during exercise (Chapter 7). It may be that training has a greater effect on subsarcolemmal mitochondria, thereby improving the ability to maintain the integrity of the cell membrane and

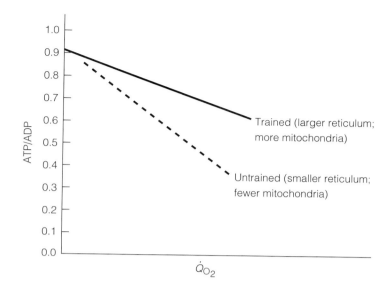

Figure 6-13 Increasing the mitochondrial mass in a muscle, or any cell, allows a given rate of mitochondrial oxygen consumption ($\dot{Q}_{O_2}$) to be accomplished at a higher ATP/ADP ratio. This increased sensitivity of respiratory control is thought to down regulate glycolysis, thus allowing for greater lipid oxidation at a given $\dot{Q}_{O_2}$ and $\dot{V}_{O_2}$.

thus improving endurance during heavy exercise. Training may have a lesser effect on the interfibrillar mitochondria—those supportive of muscular contraction. Alternatively, it is likely that $\dot{V}_{O_2max}$ is limited by blood flow (cardiac output) but that endurance is dependent on mitochondrial function. From this perspective, it may be that exercise exhausts the functional capability of mitochondria and that having a large, elaborate reticulum, or many mitochondria, retards the eventual fatigue point. This question is being investigated.

Doubling the mitochondrial content in muscle due to training (as determined by biochemical procedures) is consistent with an elaboration of the mitochondrial reticulum due to training. This elaboration of the mitochondrial reticulum due to training, however, has yet to be demonstrated by appropriate morphometric studies. Similarly, it has not yet been established whether subsarcolemmal mitochondria are arranged in a reticulum, or whether the subsarcolemmal reticulum (if it exists) is continuous with the intrafibrillar reticulum. However, on the basis of electron microscopic information now available (Figures 6-2, 6-3, 6-4, 6-12, and 6-14), it appears that the mitochondrial reticulum is continuous from the subsarcolemmal area to deep regions within a muscle cell.

■ The Training Adaptation and Coordination of Mitochondrial and Nuclear Genes

Despite remarkable discoveries in the areas of mitochondrial energetics and biogenesis, we do not yet know the signal for the proliferation of the mitochondrial reticulum in response to endurance training. At present, we do know that the muscle mitochondrial mass increases in response to chronic electrical stimulation, altitude hypoxia, and intermittent interruption of blood flow as well as endurance training. Therefore, it is suspected that things such as oxygen free radicals, hypoxia, ADP, and inorganic phosphate affect the balance of synthesis and degradation of mitochondrial constituents. We hope that in the not-too-distant future, the signals and mechanisms of action of mitochondrial genesis will be understood, as such knowledge will have wide clinical application.

In the recent past, investigators such as Booth, Williams and Essig, and their associates (Booth and Thompson, 1991; Williams, 1986, and Williams et al., 1986; and Essig and McNabney, 1991) have made progress in explaining the molecular biology of mitochondrial biogenesis. Progress in this field has been complicated by the fact that the several

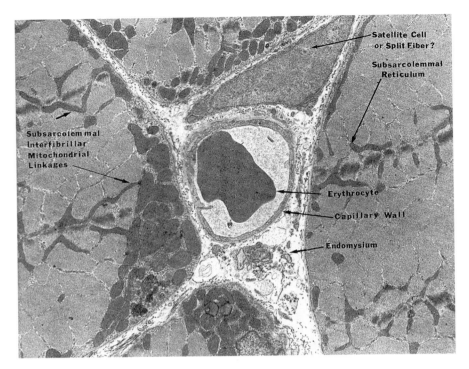

Figure 6-14 Electron micrograph of rat deep (red) vastus thigh muscle showing linkages between subsarcolemmal and interfibrillar mitochondria. Micrograph courtesy of S. P. Kirkwood, E. A. Munn, L. Packer, and G. A. Brooks.

hundred nuclear mitochondrial genes have to be coordinated with the 13 genes of mitochondrial DNA (mtDNA). However, more recently Ordway, Williams and their associates (1993) studied the expression of a small RNA encoded within the nucleus, which is an essential subunit of mitochondrial RNA-processing endoribonuclease. The nuclear mitochondrial RNA-processing RNA (MRP-RNA), is exported to mitochondria, where it generates primers for mtDNA replication. Ordway and associates found that the expression of MRP-RNA in rabbit skeletal muscle is directly related to the oxidative capacity of that muscle. Further, they showed that chronic electrical stimulation, a profound stimulator of muscle mitochondrial replication, was associated with the increased expression of MRP-RNA. This most recent information on the coordination of nuclear and mitochondrial genes may lead to an improved understanding of the mechanism of mitochondrial biogenesis in different muscle fiber types and following endurance exercise training.

■ What is the Mitochondrial Oxygen Partial Pressure During Exercise?

For years, it has been known that the critical mitochondrial oxygen tension is very low, on the order of 1 mm Hg (Figure 6-15). The critical oxygen tension is the partial pressure below which there is insufficient oxygen for mitochondria to achieve maximal rates of respiration. A major technical limitation in measuring critical oxygen tension has been the difficulty of measuring oxygen tension in working skeletal muscle as well as the difficulty of understanding the relationship between working muscle P_{O_2} and lactate formation. With dog muscle preparations, first Jöbsis and Stainsby (1968), and then

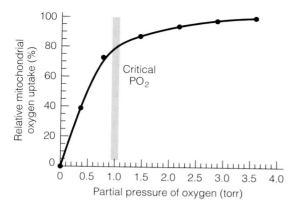

Figure 6-15   Relationship between oxygen consumption and partial pressure of oxygen in mitochondria isolated from heart. The critical mitochondrial P_{O_2} is the pressure below which maximal respiratory rate (V_{max}) cannot be maintained. Modified from Rumsey and Wilson et al. (1990). As discussed by Gladden (1996), the critical mitochondrial P_{O_2} may even be lower ($\cong 0.5$ torr). Used with permission of the American Physiological Society.

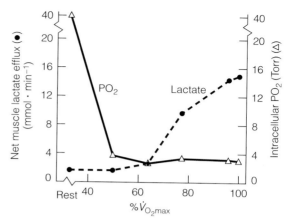

Figure 6-16   Muscle intracellular P_{O_2} (determined from NMR myoglobin spectroscopy) and net lactate release (determined from arterial-venous difference measurements) in resting and exercising men. Note that when exercise starts, muscle P_{O_2} falls but remains well above the critical mitochondrial O_2 tension of 1 torr. In contrast, resting muscle releases a small amount of lactate, and muscle net lactate release does not change until a power output eliciting 65% of $\dot{V}_{O_2max}$ is achieved. Lactate is formed under fully aerobic conditions, during both rest and exercise. Modified from Richardson et al. (1998) with inclusion of resting values from parallel experiments. Used with permission.

Connett and associates (1984) provided data that working muscles maintain a P_{O_2} above the critical oxygen tension while lactate is released. In other words, lactate was formed under fully aerobic conditions (see Figure 5-14). Unfortunately, the techniques used on isolated muscles in anesthetized dogs were not readily transferable to human exercise studies.

Most recently, Richardson and associates (1998) utilized classical (a-v) lactate balance measurements for measuring lactate release along with NMR spectroscopy to measure myoglobin saturation in the working human quadriceps muscle. Using the myoglobin spectra from muscle, and knowing the shape of the myoglobin dissociation curve, Richardson and colleagues were able to calculate the intramuscular P_{O_2}. Their results for intracellular P_{O_2} and net lactate release during progressive exercise to maximum are portrayed in Figure 6-16. At rest, for healthy subjects breathing normal air muscle P_{O_2} is

quite high, very close to that in the venous effluent from the limb, approximately 40 torr. However, when exercise starts, muscle P_{O_2} falls dramatically to approximately 4 torr, a value well above the critical mitochondrial P_{O_2}. Moreover, the intracellular P_{O_2} (triangles in Figure 6-16) was well maintained above the critical mitochondrial P_{O_2} as power output increased.

In contrast to intramuscular P_{O_2}, which fell rapidly at exercise onset, as has been seen many times before, muscle lactate release (circles in Figure 5-16) changed little at exercise onset. However, at approximately a power output corresponding to 65% of $\dot{V}_{O_2max}$, coinciding with a rise in arterial epinephrine, muscle lactate release began a steep rise, again an observation made previously. Thus, as predicted from the model of an intramuscular lactate shuttle, during both rest and exercise lactate is formed and released under fully aerobic conditions. During easy to moderate intensity exercise (i.e., between 50% and 65% $\dot{V}_{O_2max}$), muscles do not release lactate, probably because clearance by oxidation in the muscle balances production (see Figure 5-26). However, when the glycolytic flux and lactate production exceeds that which can be cleared by mitochondria, then the excess glycolytic flux is released as lactate.

Before closing this section it needs to be acknowledged that at the present time investigators are pushing the limits of technology in applying NMR spectroscopy to contracting human skeletal muscle. In their efforts at the University of California at Davis, Thomas Jue and Paul Molé have also used NMR and myoglobin saturation measurements to measure P_{O_2} in working human muscle. Their data show a *less* pronounced decline in muscle P_{O_2} at exercise onset than evident in Figure 6-16. However, the values during hard and maximal exercise ($\cong$ 4 torr) are similar. With continued efforts investigators will soon agree on the shape of the curve relating muscle P_{O_2} and power output. For the present, however, it is probable that the results showing lactate production in muscle during submaximal exercise when P_{O_2} is well above the critical mitochondrial P_{O_2} will be confirmed.

Students interested in reading more about this exciting contemporary research are encouraged to read Bruce Gladden's review on the subject that appeared in the *Handbook of Physiology,* Section 12, 1996.

■ Summary

Oxygen supplied to active muscle by the lungs, heart, and blood supports cellular production of ATP. Through the process of cell respiration, derivatives of carbohydrates as well as fats and proteins can yield substantial sources of potential energy for phosphorylating ADP to ATP. At present, we know that red skeletal muscle fibers are more abundant in mitochondria than are pale muscle fibers. Further, we know that the oxidative capacity of muscle improves greatly in response to endurance training because of an elaboration of the mitochondrial reticulum. Thus, *endurance training makes skeletal muscle more like the heart* in terms of its oxidative capacity.

The training-induced increase in muscle mitochondrial mass and other related training adaptations, such as increased capillarity, significantly increase the ability of working muscle to extract oxygen from the blood coursing through it. Thus, improving the muscle mitochondrial mass through training serves to increase $\dot{V}_{O_2max}$. However, the major result of increasing the muscle mitochondrial content through training is that respiratory control in skeletal muscle is improved. Improved respiratory control allows for muscle glycogen sparing, increased intramuscular lactate clearance, and increased lipid metabolism to occur during submaximal exercise. At present, our knowledge of why and how muscle cells increase their mitochondrial content in response to endurance exercise is incomplete, but important progress is being made.

■ Selected Readings

Annex, B. H., W. E. Kraus, G. L. Dohm, and R. S. Williams. Mitochondrial biogenesis in striated muscles: rapid induction of citrate synthase mRNA by nerve stimulation. *Am. J. Physiol.* 260 (*Cell Physiol.* 29): C266–C270, 1991.

Bakeeva, L. E., Y. S. Chentsov, and V. P. Skulachev. Mitochondrial framework (reticulum mitochondriale) in rat diaphragm muscle. *Biochem. Biophys. Acta.* 501: 349–369, 1978.

Baldwin, K., G. Klinkerfuss, R. Terjung, P. A. Molé, and J. O. Holloszy. Respiratory capacity of white, red, and intermediate muscle: adaptive response to exercise. *Am. J. Physiol.* 22: 373–378, 1972.

Barnard, R., V. R. Edgerton, T. Furukawa, and J. B. Peter. Histochemical, biochemical and contractile properties of red, white, and intermediate fibers. *Am. J. Physiol.* 220: 410–414, 1971.

Barnard R., V. R. Edgerton, and J. B. Peter. Effect of exercise on skeletal muscle. I. Biochemical and histochemical properties. *Appl. Physiol.* 28: 762–766, 1970.

Barnard, R. J., and J. B. Peter. Effect of exercise on skeletal muscle. III. Cytochrome changes. *Am. J. Physiol.* 31: 904–908, 1971.

Booth, F. W., and D. B. Thompson. Molecular and cellular adaptation of muscle in response to exercise: perspectives of various models. *Physiol. Reviews* 71: 541–585, 1991.

Brooks, G. A. Mammalian Fuel Utilization During Sustained Exercise. *Comp. Biochem. Physiol.* 120: 89–107, 1998.

Brooks, G. A., H. Dubouchaud, M. Brown, J. P. Sicurello, and C. E. Butz. Role of mitochondrial lactic dehydrogenase and lactate oxidation in the intra-cellular lactate shuttle. *Proc. Natl. Acad. Sci.* 96: 1129–1134, 1999.

Burke, F., F. Cerny, D. Costill, and D. Fink. Characteristics of skeletal muscle in competitive cyclists. *Med. Sci. Sports* 9: 109–112, 1977.

Chance, B., and C. R. Williams. The respiratory chain and oxidative phosphorylation. In Advances in Enzymology, vol. 17. New York: Interscience, 1956, pp. 65–134.

Connett, R. J., T. E. J. Gayeski, and C. R. Honig. Lactate accumulation in fully aerobic, working, dog gracilis muscle. *Am. J. Physiol.* 246: H120–128, 1984.

Connett, R. J., C. R. Honig, T. E. J. Gayeski, and G. A. Brooks. Defining hypoxia: a systems view of $\dot{V}_{O_2}$, glycolysis, energetics and intracellular P_{O_2}. *J. Appl. Physiol.* 68: 833–842, 1990.

Costill, D. L., J. Daniels, W. Evans, W. Fink, G. Krahenbuhl, and B. Saltin. Skeletal muscle enzymes and fiber composition in male and female track athletes. *J. Appl. Physiol.* 40: 149–154, 1976.

Costill, D. L., W. J. Fink, and M. L. Pollock. Muscle fiber composition and enzyme activities of elite distance runners. *Med. Sci. Sports* 8: 96–100, 1976.

Davies, K. J. A., J. J. Maguire, G. A. Brooks, P. R. Dallman, and L. Packer. Muscle mitochondrial bioenergetics, oxygen supply, and work capacity during dietary iron deficiency and repletion. *Am. J. Physiol.* 242 (*Endocrinol. Metab.* 5): E418–E427, 1982.

Davies, K. J. A., L. Packer, and G. A. Brooks. Biochemical adaptation of mitochondria, muscle, and whole-animal respiration to endurance training. *Arch. Biochem. Biophys.* 209: 538–553, 1981.

Davies, K. J. A., L. Packer, and G. A. Brooks. Exercise bioenergetics following sprint training. *Arch. Biochem. Biophys.* 215: 260–265, 1982.

Dohm, G. L., R. L. Huston, E. W. Askew, and H. L. Fleshood. Effect of exercise, training and diet on muscle citric acid cycle enzyme activity. *Can. J. Biochem.* 51: 849–854, 1973.

Ernster, L., and Z. Drahota (Eds.). Mitochondria Structure and Function. New York: American Press, 1969.

Essig, D. A., and A. McNabney. Muscle-specific regulation of the heme biosynthetic enzyme 5'-aminolevulinate synthase. *Am. J. Physiol.* 261 (*Cell Physiol.* 30): C691–C698, 1991.

Gale, J. B. Skeletal muscle mitochondrial swelling with exhaustive exercise. *Med. Sci. Sports* 6: 182–187, 1974.

Gladden, L. B. Lactate transport and exchange during exercise. In Handbook of Physiology, Section 12, Exercise: Regulation and Integration of Multiple Systems, L. B. Rowell and J. T. Shepherd (Eds.), Oxford University Press, New York, 1996, pp. 614–648.

Gollnick, P. D., and D. W. King. Effect of exercise and training on mitochondria of rat skeletal muscle. *Am. J. Physiol.* 216: 1502–1509, 1969.

Holloszy, J. O. Effects of exercise on mitochondrial oxygen uptake and respiratory enzyme activity in skeletal muscle. *J. Biol. Chem.* 242: 2278–2282, 1967.

Holloszy, J. O. Adaptation of skeletal muscle to endurance exercise. *Med. Sci. Sports* 7: 155–164, 1975.

Holloszy, J. O., and F. W. Booth. Biochemical adaptation to endurance exercise in muscle. *Ann. Rev. Physiol.* 38: 273–291, 1976.

Holloszy, J. O., L. B. Oscai, P. A. Molé, and I. J. Don. Biochemical adaptations to endurance exercise in skeletal muscle. In Muscle Metabolism During Exercise, B. Pernow and B. Saltin (Eds.). New York: Plenum Press, 1971.

Hood, D. A., A. R. Zak, and D. Pette. Chronic stimulation of rat skeletal muscle induces coordinate increases in mitochondrial and nuclear mRNAs of cytochrome-c-oxidase subunits. *Eur. J. Biochem.* 179: 275–280, 1991.

Hoppeler, H. Exercise-induced changes in skeletal muscle. *Int. J. Sports Med.* 7: 187–204, 1986.

Jöbsis, F. F., and W. N. Stainsby. Oxidation of NADH during contractions of circulated skeletal muscle. *Resp. Physiol.* 4: 292–300, 1968.

Keilin, D. The History of Cell Respiration and Cytochromes. New York: Cambridge University Press, 1966.

King, D. W., and P. Gollnick. Ultrastructure of rat heart and liver after exhaustive exercise. *Am. J. Physiol.* 218: 1150–1155, 1970.

Kirkwood, S. P., E. A. Munn, L. Packer, and G. A. Brooks. Mitochondrial reticulum in limb skeletal muscle. *Am. J. Physiol.* 251 (*Cell Physiology* 20): C395–C402, 1986.

Kirkwood, S. P., E. A. Munn, L. Packer, and G. A. Brooks. Effects of endurance training on mitochondrial reticulum in limb skeletal muscle. *Arch. Biochem. Biophys.* 255: 80–88, 1986.

Krebs, H. A., and H. L. Kornberg. Energy Transformation in Living Matter. Berlin: Springer-Verlag OHG, 1957.

Lehninger, A. L. Biochemistry. New York: Worth, 1970, pp. 337–416.

Lehninger, A. L. Bioenergetics. Menlo Park, Ca.: W. A. Benjamin, 1973, pp. 75–98.

Lowenstein, J. M. The tricarboxylic acid cycle. In Metabolic Pathways, vol. 1, D. M. Greenberg (Ed.). New York: Academic Press, 1967, pp. 146–270.

McGilvery, R. W. Biochemical Concepts. Philadelphia: W. B. Saunders, 1975, pp. 157–249.

Mitchell, P. Chemiosmotic coupling in oxidative and photosynthetic phosphorylation. *Biol. Rev.* 41: 455, 1965.

Mitchell, P., and J. Moyle. Estimation of membrane potential and pH difference across the cristae membrane of rat liver mitochondria. *Eur. J. Biochem.* 7: 471–484, 1969.

Molé, P., K. Baldwin, R. Terjung, and J. O. Holloszy. Enzymatic pathways of pyruvate metabolism in skeletal muscle: adaptations to exercise. *Am. J. Physiol.* 224: 50–54, 1973.

Munn, E. A. The Structure of Mitochondria. London: Academic Press, 1974.

Ordway, G. A., K. Li, G. A. Hand, and R. S. Williams. RNA subunit of mitochondrial RNA-processing enzyme is induced by contractile activity in striated muscle. *Am. J. Physiol.* 265 (*Cell Physiol.* 34): C1511–1516, 1993.

Oscai, L. B., P. A. Molé, B. Brei, and J. O. Holloszy. Cardiac growth and respiratory enzyme levels in male rat subjected to a running program. *Am. J. Physiol.* 220: 1238–1241, 1971.

Oscai, L. B., P. A. Molé, and J. O. Holloszy. Effects of exercise on cardiac weight and mitochondria in male and female rats. *Am. J. Physiol.* 220: 1944–1948, 1971.

Richardson, R. S., E. A. Noyszewski, K. F. Kendrick, J. S. Leigh, and P. D. Wagner. Myoglobin O_2 desaturation during exercise. Evidence of limited O_2 transport. *J. Clin. Invest.* 96: 1916–1926, 1995.

Richardson, R. S., E. A. Noyszewski, J. S. Leigh, and P. D. Wagner. Lactate efflux from exercising human skeletal muscle: role of intracellular P_{O_2}. *J. Appl. Physiol.* 85: 627–634, 1998.

Rumsey, W. L., C. Schlosser, E. M. Nuutinen, M. Robiolio, and D. F. Wilson. Cellular energetics and the oxygen dependence of respiration in cardiac myocytes isolated from adult rat. *J. Biol. Chem.* 265: 15392–15402, 1990.

Terjung, R. L. Muscle fiber involvement during training of various intensities. *Am. J. Physiol.* 230: 946–950, 1976.

Terjung, R. L., K. M. Baldwin, W. W. Winder, and J. O. Holloszy. Glycogen repletion in different types of muscle and in liver after exhaustive exercise. *Am. J. Physiol.* 226: 1387–1391, 1974.

Terjung, R. L., G. H. Kinkerfuss, K. M. Baldwin, W. W. Winder, and J. O. Holloszy. Effect of exhausting exercise on rat heart mitochondria. *Am. J. Physiol.* 225: 300–305, 1973.

Tzagoloff, A. Mitochondria. New York: Plenum Press, 1982.

Vander, A. J., J. H. Sherman, and D. S. Luciano. Human Physiology. 3d Ed. New York: McGraw-Hill, 1980.

Williams, R. S. Mitochondrial gene expression in mammalian striated muscle. *J. Biol. Chem.* 261: 12390–12394, 1986.

Williams, R. S., S. Salmons, E. A. Newsholme, R. E. Kaufman, and J. Mellor. Regulation of nuclear and mitochondrial gene expression by contractile activity. *J. Biol. Chem.* 262: 2764–2767, 1986.

CHAPTER 7

LIPID METABOLISM

Dietary sources contain a variety of different kind of lipids (fats), but quantitatively triglycerides are the most important. The high energy content of triglycerides and their vast, efficient storage reserve in adipose tissue, liver, and skeletal muscle provide an almost inexhaustible fuel supply for muscular exercise. However, despite the large quantity of lipid available as fuel, the processes of lipid utilization are slow to be activated and proceed at rates significantly slower than the processes controlling sugar and carbohydrate catabolism. Nevertheless, fats are an important segment of the fuel used during prolonged work, and even small increases in the ability to use fats as fuel during exercise can effectively slow sugar and carbohydrate metabolism. As opposed to carbohydrates, which can yield energy through glycolysis without the use of oxygen, fat catabolism is purely

Greta Waitz and Rosa Mora lead the pack in the Seoul Olympics. Success in endurance activities such as the marathon depend critically upon the ability to oxidize all substrates, including fats, as energy sources. In humans, the ability to utilize fats is only indirectly related to the amount of stored body fat. PHOTO: © Claus Andersen.

an aerobic process, which is best developed in heart and red skeletal muscle fibers. The sparing of glucose and glycogen metabolism by fats during prolonged exercise in highly trained individuals slows the depletion of these essential nutrients. Thus, increased fat oxidation capability due to training and genetic endowment greatly enhances endurance.

■ Lipid Defined

A *lipid* can be simply defined on the basis of its physical solubility characteristics: It is a substance that is soluble in organic solvents but not in water. The solubility characteristics of lipids are determined by their nonpolar physical characteristics. Figure 7-1

gives structural formulas for several common lipids of physiological importance. Because of its broad definition, many different substances with different structures and functions are included in the lipid category. Fatty acids such as stearate (Figure 7-1a) are long-chain carboxylic acids. Fatty acids can combine with glycerol (Figure 7-1b) at its hydroxyl groups to form mono-, di-, or triglycerides. In Figure 7-1c, fatty acids have been substituted at each of the three hydroxyl groups of glycerol to yield a triglyceride. Triglycerides constitute the vast majority of the lipid, and calories, taken into and stored within the body.

Naturally occurring fatty acids vary in size, but they commonly contain an even number (14 to 24) of carbon atoms in a straight chain. Because fatty

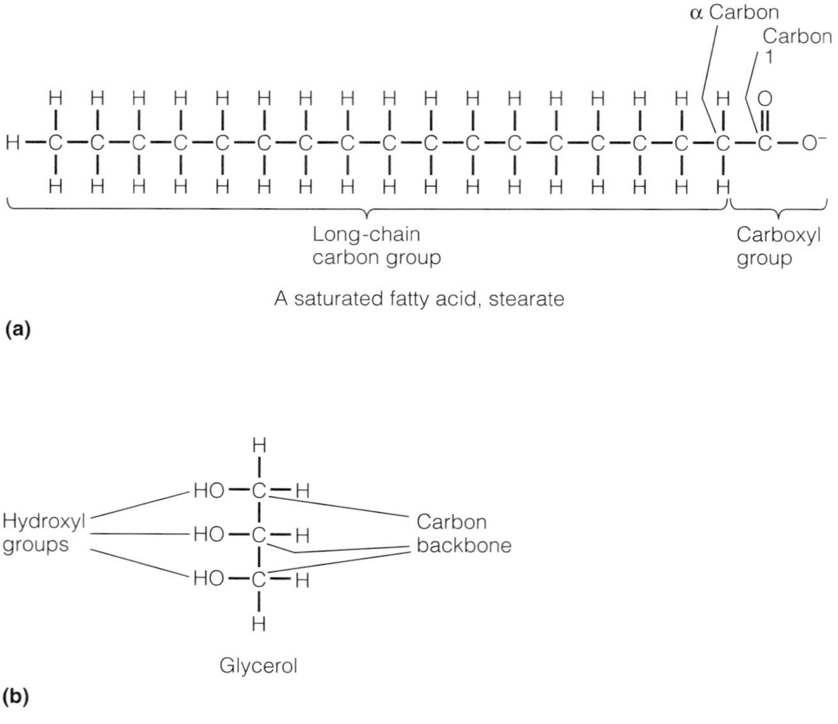

Long-chain carbon group

Carboxyl group

A saturated fatty acid, stearate

(a)

Hydroxyl groups

Carbon backbone

Glycerol

(b)

Figure 7-1   Typical lipids and important metabolites in lipid metabolism. (a) A saturated fatty acid, stearate; (b) Glycerol; (c) A triglyceride, stearyl-palmitoyl-linoleal-glycerol; (d) A polyunsaturated fatty acid, linoleate; (e) A common phospholipid, lecithin, (f) Cholesterol, (g) α-Glycerolphosphate, or glycerol 3-phosphate, (h) Acetoacetate, a ketone.

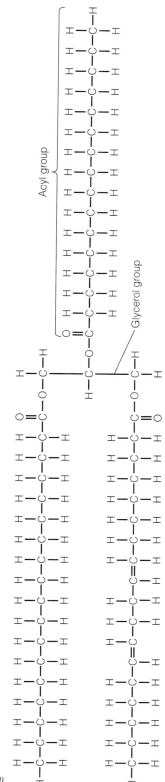

Figure 7-1 *(continued)*

Acyl group

Glycerol group

A triglyceride, stearyl-palmitoyl-linoleal-glycerol

(c)

Carboxyl group

Unsaturations (double bonds)

A polyunsaturated fatty acid, linoleate

(d)

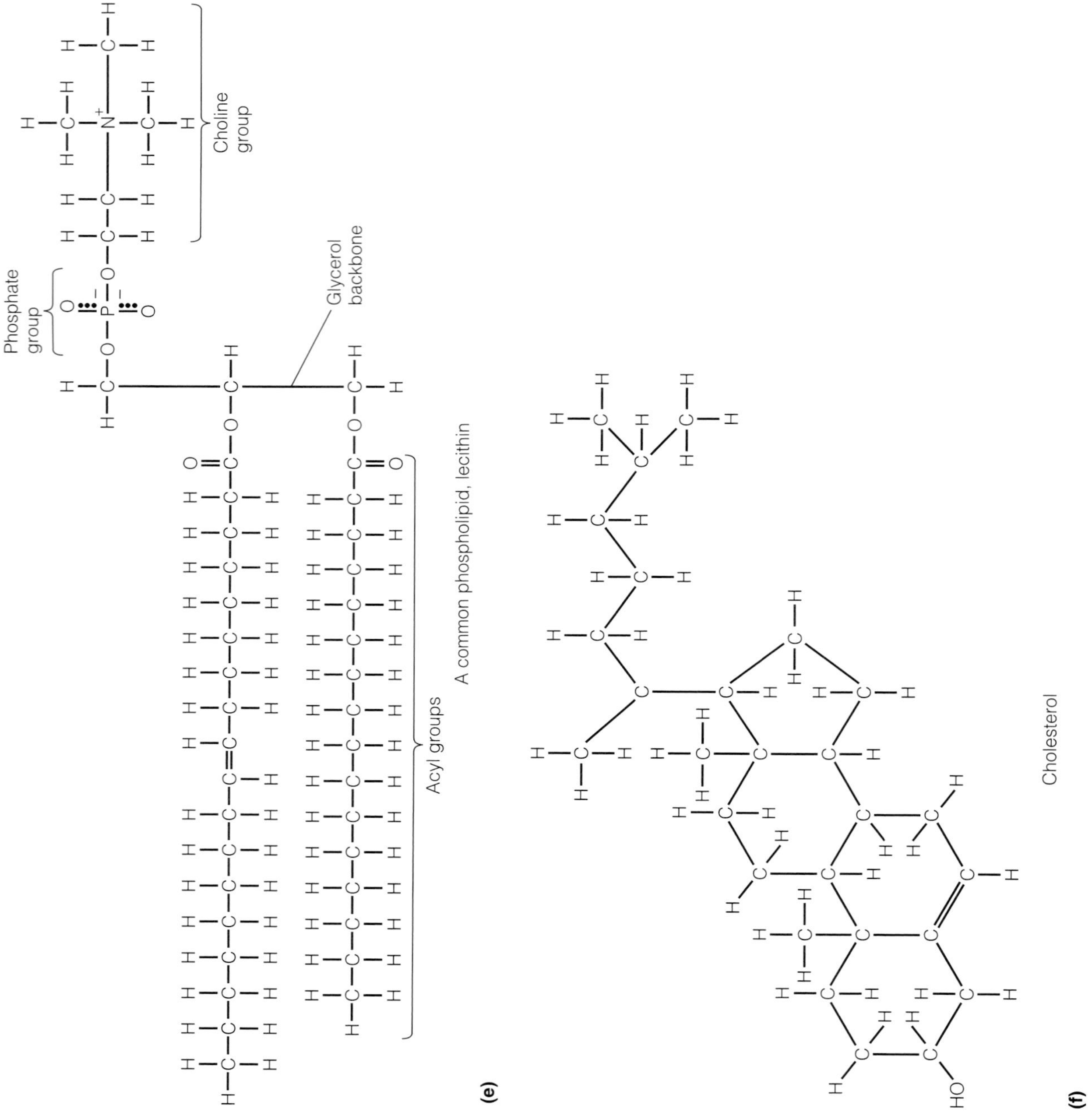

A common phospholipid, lecithin

Phosphate group

Choline group

Glycerol backbone

Acyl groups

(e)

Cholesterol

(f)

Figure 7-1 *(continued)*

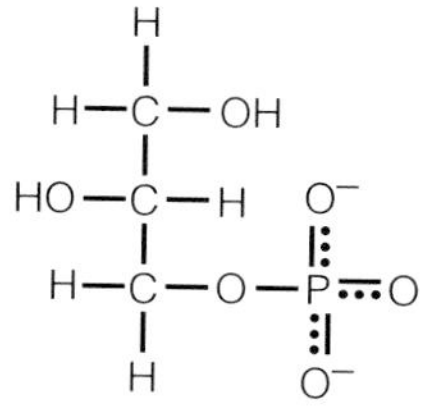

α-Glycerol phosphate,
or glycerol-3-phosphate

(g)

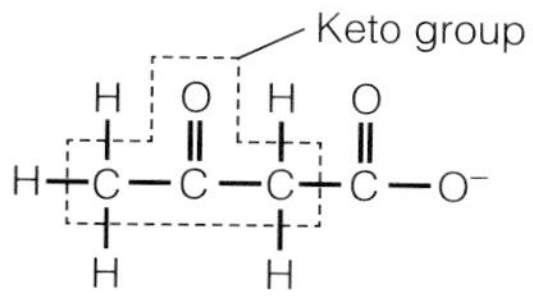

Acetoacetate, a ketone

(h)

Figure 7-1 *(continued)*

acids are synthesized from acetyl-CoA (two-carbon units) and are catabolyzed in two-carbon units, the even number of carbon atoms in the fatty acid chain is no accident.

Stearate (Figure 7-1a) contains 18 carbon atoms and is one of the most abundant saturated fatty acids in the diet. Palmitate, a 16-carbon (16 C) saturated fatty acid, is another physiologically abundant and highly used fatty acid. Unsaturation means that not all the valence bonds in the carbon chain are filled (saturated) with hydrogen; this also means that a carbon-to-carbon double bond exists.

Figure 7-1d illustrates linoleate, a polyunsaturated fatty acid. Like stearate, linoleate contains 18 carbon atoms, but linoleate contains two double bonds. The bonds usually exist at three-carbon intervals (e.g., between carbons 9 and 10 and between 12 and 13), which is a frequent arrangement of double bonds in fatty acids. Because mammals have lost the ability to synthesize unsaturated fatty acids with double bonds beyond the ninth carbon, fatty acids such as linoleate and oleate, which has a single unsaturation between C9 and C10, are "essential" and must be in the diet.

The presence of polyunsaturated fatty acids (two or more double bonds) in the diet has, in the past, been thought to lower levels of cholesterol in the blood. Elevated blood cholesterol has been implicated as a risk factor predisposing people to coronary heart disease. Consequently, substitution of polyunsaturated dietary fats for saturated fats has been recommended to protect against coronary artery disease. This is discussed more in Chapter 28.

Other common lipids include phospholipids, such as lecithin (Figure 7-1e), and cholesterol (Figure 7-1f). Phospholipids are crucial constituents of membranes; cholesterol is necessary for the formation of steroid hormones and the bile salts used in lipid digestion.

■ Esterification and Hydrolysis

The making of triglycerides is called *esterification* because it involves attachment of a fatty acid to glycerol by means of an oxygen atom. *Lipolysis* is the process of triglyceride hydrolysis. Esterification and lipolysis are essentially reversals of each other. Triglycerides synthesized in plants or animals are consumed by humans and then digested in a process involving lipolysis or stored through the reesterification of fatty acids delivered to adipose tissue. The mobilization of triglycerides stored in adipose tissue involves another lipolysis. Synthesis, digestion, storage, and mobilization can be simply diagrammed as a series of reversal reactions (Equation 7-1).

Recent advances in understanding factors that regulate the balance of carbohydrate and lipid utilization during exercise (see the section on the "crossover concept" later in the chapter) require that increased attention be paid to the tissue sites and intracellular signals that regulate lipid mobili-

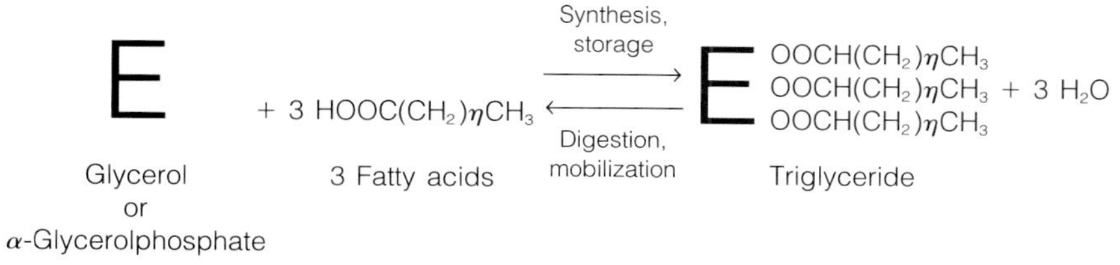

Glycerol
or
α-Glycerolphosphate

3 Fatty acids

Synthesis,
storage

Digestion,
mobilization

Triglyceride

(7-1)

zation and reesterification. In skeletal muscle, adipose tissue, and probably heart α-glycerolphosphate (not glycerol), is the backbone substrate for triglyceride synthesis. However, because cells in these tissues lack the enzyme glycerol kinase, the enzyme responsible for phosphorylation of glycerol to α-glycerolphosphate, reesterification of glycerol to fatty acids to form triglyceride is not thought to be possible in myocytes, cardiocytes, or adipocytes. Therefore, reesterification is a process relegated to the liver.

Skeletal muscle can readily oxidize fatty acids delivered from arterial blood as well as fatty acids provided from lipolysis within muscle itself. However, because muscle can neither oxidize glycerol nor convert it to α-glycerolphosphate for reesterification, glycerol release from muscle is a better indicator of intramuscular lipolysis than is fatty acid release. For the same reason, the circulating level of glycerol is usually a better marker for lipolysis than is the circulating level of fatty acids. These facts have been reviewed recently by Brooks and Mercier (1994) and are discussed more later.

■ Fats in the Diet

Triglycerides constitute the greatest part by far of lipid mass consumed in the diet. Dietary triglycerides, along with ingested cholesterol and phospholipids, undergo no significant digestion in the stomach. Rather, fat digestion begins in the small intestine through the emulsification (dispersion into a water medium) of fats by the action of bile salts from the liver via the gallbladder. Bile salts reduce the surface (interfacial) tension of the fat globules so the agitation action of the small intestine can disperse them. In this emulsified form, fat globules are then hydrolyzed into monoglycerides, free fatty acids (FFA), and glycerol by the action of pancreatic lipase, a digestive enzyme. Following lipolysis, digestion of fats continues, again through the action of bile salts, which now allow the products to aggregate into very small fatty particles called *micelles.* When micelles come into contact with epithelial cells lining the small intestine, they allow the diffusion of lipid digestion products into the wall of the intestine.

After entry into the epithelial cell, monoglycerides are further hydrolyzed by an intracellular lipase. Then the fatty acids are reconstituted by the epithelial cell endoplasmic reticulum through α-glycerolphosphate (Figure 7-1g), an activated form of glycerol, into triglycerides. These new triglycerides, along with absorbed cholesterol and phospholipids, are encased by the endoplasmic reticulum into globules with a protein coat. These particles, termed *chylomicrons,* are then extruded from the epithelial cells and released into the lymph. From there, chylomicrons eventually enter the great veins in the neck through the thoracic duct. Thus, the route through the lymphatics constitutes the major route by which dietary lipids gain entry into the central circulation system. However, a small quantity of fatty acids, particularly short-chain fatty acids, are absorbed directly into the portal circulation system and reach the liver directly.

Chylomicrons and Lipoproteins

Dietary fats appearing in the blood after a meal (see Figure 5-2a) are largely cleared (removed) from the plasma within several hours. Two major mechanisms exist for this.

One mechanism for the removal of chylomicron lipid from the blood is the transport of chylomicrons into liver cells. The liver is highly adapted to use lipid as a fuel and to convert lipids from chylomicrons into a variety of lipid-containing products including lipoproteins. Between meals, and during fasting and starvation, the liver is particularly active in producing lipoproteins to maintain circulating levels of lipid. Lipoproteins from the liver are mixtures of triglycerides, phospholipids, cholesterol, and proteins, as are the chylomicrons from the small intestine. Also as with chylomicrons, protein encapsulates the lipid of lipoproteins to facilitate their solubilization and transport in the blood.

Lipoproteins may be classified by means of their density. Increased lipid content reduces the density

TABLE 7-1

Major Constituents of Lipoproteins (%)

Type of Lipoprotein	Triglycerides	Phospholipids	Cholesterol	Protein
Chylomicrons	85	7	7	1
Very-low-density lipoproteins	50	18	23	9
Low-density lipoproteins	10	22	47	21
High-density lipoproteins	8	29	30	33

of lipoproteins. The constituents of lipoproteins and chylomicrons are given in Table 7-1. In general, the low-density (LDL) and very-low-density lipoproteins (VLDL) transport triglycerides from liver to adipose tissue. High-density lipoproteins (HDL) are thought to be involved in cholesterol catabolism, and a high circulating level of HDL in relation to LDL (HDL/LDL) is thought to be beneficial in terms of protection from coronary heart disease (Chapter 24).

Lipoprotein lipase (LPL) is an enzyme located in the capillary walls of most tissues in the body. Relatively large quantities of the enzyme are found particularly in adipose tissue and in heart and skeletal muscle. Most chylomicrons, as well as LDL and VLDL, are cleared from the blood by the action of LPL. After a meal, when blood glucose and insulin levels are elevated and glucagon is low (i.e., the insulin/glucagon ratio is high), capillary wall LPL is activated, and the hydrolysis rate of the triglycerides in lipoproteins arriving at adipose tissue is accelerated. Free fatty acids are thereby made available for storage in adipose tissue.

In addition to capillary LPL, muscle cells contain a second lipoprotein lipase that is essentially opposite in function to that in the capillary endothelium. Muscle cell LPL is hormone-sensitive and operates much like the adipose hormone-sensitive lipase (discussed later in this chapter). This form of muscle LPL is termed Type L-HSL (lipoprotein hormone-sensitive lipase). During exercise, when insulin falls and glucagon rises (i.e., when the insulin/glucagon ratio falls), muscle L-HSL is activated and adipose lipoprotein lipase is inhibited.

An elevated insulin level following a meal promotes fat storage in other ways as well. Insulin allows the entry of glucose into cells, including adipose cells (adipocytes). Again, a by-product of glucose metabolism, α-glycerolphosphate, provides the glycerol backbone for the synthesis of adipose triglyceride.

■ Free Fatty Acid Levels in Blood During Rest and Exercise

Figure 7-2 shows that free fatty acid (FFA) levels in the blood of postabsorptive (PA) men are less than 1 mM. During prolonged exercise at 50% $\dot{V}_{O_2max}$, the blood FFA levels of these subjects rise continuously. Also shown is the effect of a 36-hour fast (F) on blood FFA levels. Clearly, fasting increases the mobilization of triglyceride to FFA and glycerol such that circulating levels of FFA and glycerol increase to levels more than double those in postabsorptive or postprandial conditions.

Although Figure 7-2a shows that blood FFA levels increase during prolonged exercise, a rise in blood FFA concentration does not always occur. Increased FFA levels occur during mild to moderate intensity exercises of 50% $\dot{V}_{O_2max}$ or less. During more severe intensity exercises, when blood lactate concentration rises and pH declines, lipolysis is inhibited and the esterification of free fatty acids to triglyceride is promoted. For this reason, during hard exercise of about 65% $\dot{V}_{O_2max}$ or greater intensity, blood FFA levels decline, limiting the availability of FFAs as fuel sources.

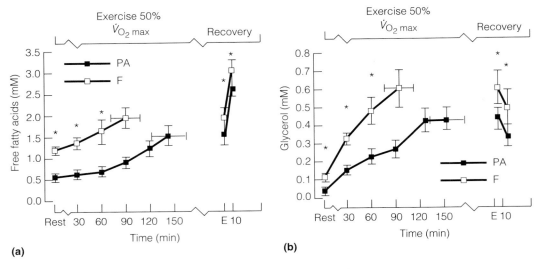

Figure 7-2 (a) Plasma free fatty acid (FFA); (b) blood glycerol concentrations at rest and during exercise to exhaustion at 50% $\dot{V}_{O_2max}$, and at 10-min post-recovery under postabsorptive (PA) and fasted (F) conditions. SE vertical bars are included for each sampling point, and horizontal bars are shown for time to exhaustion. * = Significant difference between means (PA vs. F, $P < 0.05$). SOURCE: Zinker et al., 1990. Used with permission of the American Physiological Society.

■ The Utilization of Lipids During Exercise

As noted earlier, lipids are an important energy source for prolonged mild to moderate intensity exercise. Also, lipids are primary fuels following prolonged exercise leading to glycogen depletion. Lipid utilization is a complicated process that usually begins at one site (adipose tissue) and ends at another (skeletal muscle mitochondria). The metabolism at each of these sites and at intervening sites is well controlled, so the ultimate process, lipid oxidation, is a highly integrated process.

The processes of lipid metabolism during exercise can be summarized as follows:

1. **Mobilization**   the breakdown of adipose and intramuscular triglyceride

2. **Circulation**   the transport of free fatty acids (FFAs) from adipose to muscle

3. **Uptake**   the entry of FFAs into muscles from blood

4. **Activation**   raising the energy level of fatty acids preparatory to catabolism

5. **Translocation**   the entry of activated fatty acids into mitochondria

6. **β Oxidation**   the production of acetyl-CoA from activated fatty acids and the production of reducing equivalents (NADH and FADH)

7. **Mitochondrial oxidation**   Krebs cycle and electron transport chain activity

Mobilization from Adipose

Figure 7-3 illustrates the processes of lipid mobilization, circulation, and uptake. In addition to the lipoprotein lipase in adipose tissue, adipocytes contain a second lipase enzyme called *hormone-sensitive lipase* (HSL). The control and action of these two

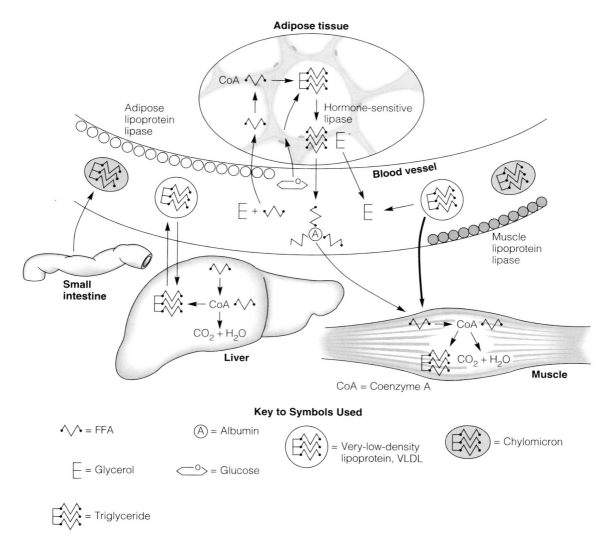

Adipose tissue

Adipose lipoprotein lipase

Hormone-sensitive lipase

Blood vessel

Muscle lipoprotein lipase

Small intestine

Liver

$CO_2 + H_2O$

$CO_2 + H_2O$

Muscle

CoA = Coenzyme A

Key to Symbols Used

= FFA

(A) = Albumin

= Very-low-density lipoprotein, VLDL

= Chylomicron

= Glycerol

= Glucose

= Triglyceride

Figure 7-3 Mobilization, circulation, and uptake. Most of the lipid oxidized in muscle during exercise is delivered by the circulating blood. The blood contains free fatty acids (FFAs) that are released from adipose tissue after the mobilization of triglyceride stores by the activation of hormone-sensitive lipase (HSL). The blood also contains lipoproteins (both chylomicrons from the gastrointestinal tract and VLDL from the liver) that deliver a much smaller quantity of lipid to muscle during exercise. Thus, the ability to use fat as a fuel during exercise depends on the arterial blood level (a function of mobilization) and muscle blood flow (a function of the circulation). Modified from Havel, 1970. Used with permission.

lipases are essentially reversed. Whereas the adipose capillary wall LPL is stimulated by insulin and glucose and promotes fat storage, the HSL stimulates fat breakdown, is inhibited by insulin, and is stimulated by other hormones, including the catecholamines (epinephrine and norepinephrine) and growth hormone (Chapter 9).

The activity of HSL is directly controlled by the presence of cyclic AMP, which, in turn, is regulated by the adenylate cyclase system. Thus, the activation of fat breakdown is much like the initiation of glucose breakdown (see Figures 5-15 and 5-17). Two activators of the HSL system (epinephrine and growth hormone) reach adipose tissue via circulation, whereas norepinephrine is released locally by sympathetic nerve endings within adipose tissue. Compared to the release of growth hormone, which is slow, the release of the catecholamines is relatively rapid. The catecholamines are therefore responsible for the initiation of lipolysis at exercise onset. It takes 10 to 15 minutes for blood levels of growth hormone to increase during exercise, and the activation of adipocyte lipase activity by growth hormone requires protein synthesis. Growth hormone helps to maintain lipolysis during prolonged

exercise. The intracellular actions of these hormones is discussed in Chapter 9.

An important question in current exercise physiology research concerns the source of blood glycerol during exercise (Figure 7-2b). To date, no investigator has been able to document presence of the enzyme glycerol kinase in either adipocytes (fat cells) or myocytes (muscle cells). Glycerol kinase exists mainly in the liver where, in presence of ATP, glycerol is phosphorylated to α-glycerolphosphate. Therefore, the glycerol produced as the result of lipolysis in adipose or skeletal muscle cannot be reutilized in those cells and tissues but must be released into the circulation. Recent results of G. A. Brooks and associates obtained by simultaneous measurements of glycerol in femoral artery and vein indicate that small amounts of glycerol are released from resting muscle. Thus, lipolysis is indicated. However, in neither trained nor untrained subjects is there significant net glycerol release from working legs during exercise (Figure 7-4). Consequently, the rise in arterial glycerol observed during exercise as seen in Figure 7-2b is a sign of lipolysis in adipose tissue, not working skeletal muscle.

Figure 7-4: Net glycerol release in resting and exercising men, before and after ten weeks of endurance training. Subjects studied at 45% and 65% of $\dot{V}_{O_2max}$ before training and after training at the same absolute (ABT) power output that elicited 65% of $\dot{V}_{O_2max}$ before training (then 54% $\dot{V}_{O_2max}$), and 65% of $\dot{V}_{O_2max}$ after training; i.e., same relative exercise intensity (RLT). Even though there is an increase in blood flow during exercise, essentially no glycerol is released. Because muscle (and adipose) in the leg lack glycerol kinase, results indicate no hydrolysis of muscle triglyceride. Note: # exercise mean for 65%pre different from 45%pre and 65%old at p < 0.05. Data from Bergman et al., 1999. Used with permission of the American Physiological Society.

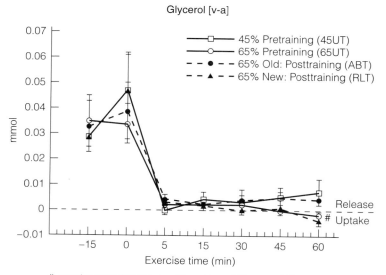

Glycerol [v-a]

— □ — 45% Pretraining (45UT)
— ○ — 65% Pretraining (65UT)
– – ● – – 65% Old: Posttraining (ABT)
– – ▲ – – 65% New: Posttraining (RLT)

exercise mean for 65% pre different from 45% pre and 65% old at p<0.05

TABLE 7-2

Blood Lipid Levels in Normal Resting Adults

Lipid Type	Blood Level (mg $\cdot$ dl^{-1})
Free fatty acids	8–20
Triglycerides	50–150
Cholesterol	150–250
Phospholipid	150–250

In liver, glycerol serves a minor role as a gluconeogenic precursor during exercise. Alternatively, after conversion to α-glycerolphosphate and dihydroxyacetone phosphate in liver, the carbon skeleton from glycerol can be used in glycolysis. Because glycerol release by adipose tissue and muscle is greater than the hepatic uptake of glycerol during exercise, glycerol levels rise sooner and relatively higher than do levels of FFA concentration (Figure 7-2a). Therefore, the elevation of blood glycerol during exercise provides a rough index of the rate of adipose lipolysis.

Because they are insoluble in aqueous media, fatty acids must be carried in the blood. The blood protein albumin carries almost all the FFAs and therefore a majority of the total lipids transported in the blood. Even though the quantity of lipids existing as FFAs in the blood at any one time is only a small part of the total blood lipid content (Table 7-2), the turnover (entry and exit) of blood FFAs is very rapid. Therefore, the contribution of FFAs to the fuel supply during rest and exercise far exceeds the contribution of other blood lipids such as triglycerides.

Circulation and Uptake

Several physiological as well as biochemical factors affect lipid metabolism during exercise. Because the uptake of FFAs in lipid delivery depends to a large extent on the arterial fatty acid content, the rate of adipose lipolysis directly affects FFA uptake by muscle. That is, a key to lipid oxidation during exercise is the arterial level of FFAs. As will

be noted in Chapter 29 on ergogenic aids, any factor (e.g., caffeine) that stimulates adipose lipolysis and raises blood FFA levels could promote exercise endurance.

The rate of blood flow through muscle is a major determinant of FFA uptake and utilization during exercise. Although this factor is usually overlooked in a discussion of the control of fat metabolism, the enhancement of cardiac output and muscle blood flow by endurance training is a major factor determining the ability to oxidize lipids during exercise. The greater the muscle blood flow, the greater the delivery, uptake, and utilization of FFAs by muscle during exercise.

The uptake of fatty acids into muscle from blood albumin is accomplished via a specific receptor site on the sarcolemma. According to Turcotte, the receptor is now called *sarcolemmal fatty acid binding protein* (S-FABP). The sarcolemmal FABP is one of a family of FABPs that exist throughout the muscle cell matrix. Together, the muscle cell FABPs function to transport fatty acids throughout the cell. As a result of there being a variable number of receptor (FABP) binding sites in sarcolemmal membranes, the entry of fatty acids from blood into muscle cells is more rapid in the heart than in red or white skeletal muscle. However, red muscle has more FABPs than white, and endurance training increases the number of sarcolemmal FABPs. While there appears to be long-term adaptation and regulation of the number of S-FABPs, there is no evidence of acute regulation, such as that noted with GLUT-4 translocation (Chapter 4; Figure 9-5). Therefore, acute regulation of FFA oxidation in heart and muscle is downstream of the FABPs, such as in the mitochondria.

Activation and Translocation

The first step in intracellular metabolism of lipids resembles the first step in glycolysis; this process, the activation of the fatty acid, is diagrammed in Figure 7-5. The activation process raises the fatty acids to a higher energy level and involves ATP. However, the process differs from the activation in glycolysis in that the fatty acid is attached to

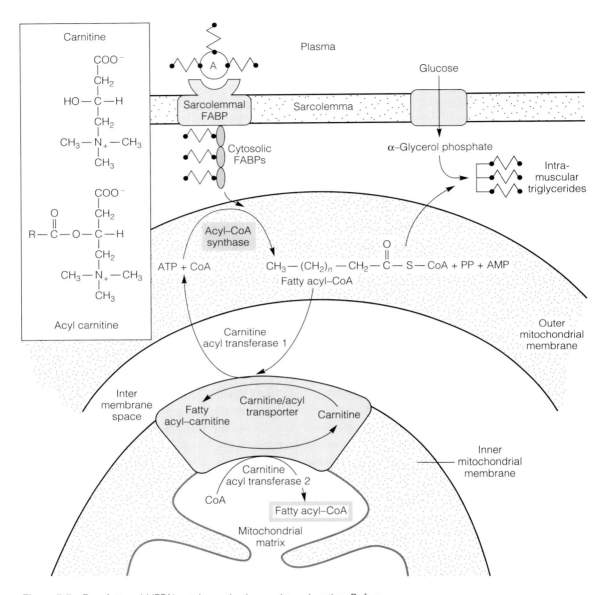

Figure 7-5 Free fatty acid (FFA) uptake, activation, and translocation. Before being either oxidized or stored as triglyceride in muscle, fatty acids delivered by the circulation must be activated; this involves ATP and coenzyme A (CoA). To gain entry into the mitochondrial matrix for oxidation, the activated fatty acid must combine with a carrier substance, carnitine, which exists in the mitochondrial inner membrane. An enzyme, carnitine-acyl transferase (CAT) enzymes 1 & 2, facilitates both the formation of the fatty acyl–carnitine complex and the dissociation of the complex and release of the activated fatty acid into the matrix. The translocation of activated fatty acids from the cytosol into mitochondria via CAT1 is rate-limiting step in lipid metabolism.

coenzyme A (CoA), with the formation of a CoA derivative, termed fatty acyl-CoA.

The site of fatty acyl-CoA formation is the inner mitochondrial membrane. However, the site of fatty acid oxidation is the mitochondrial matrix. Like a wide variety of substances, activated fatty acids gain entry into or exit from mitochondria by a transport mechanism. For fatty acids, the mechanism involves a carrier, carnitine, and the carnitine acyltransferase enzymes. This mechanism (Figure 7-5) involves the stripping off of CoA and its return to the cytosol, and the acceptance of the fatty acid by carnitine, with the formation of fatty acyl–carnitine. This process is catalyzed by a family of enzymes collectively called *carnitine acyl transferase* 1 (CAT1). For each long chain fatty acid, there is a specific CAT, e.g., for palmitate, the enzyme is carnitine palmitoyl transferase 1 (CPT1). The fatty acyl–carnitine complex is free to move across the mitochondrial membrane, where, on the inner side, carnitine transferase 2 catalyzes the reverse reaction, leaving carnitine within the membrane, and releasing fatty acyl-CoA into the mitochondrial matrix.

Malonyl-CoA and Fatty Acid Translocation

In general, and particularly during exercise, the catalytic activity of carnitine translocase is extremely important in the overall control of fat oxidation. Translocation of fatty acid derivatives into the mitochondrial organelles (where the fats can be oxidized) is directly affected by the translocation process. Although no oxygen is directly involved, translocation is a very aerobic process because it depends on the mitochondrial mass present. It has been known for some time that in liver, carnitine transferase-1 is inhibited by malonyl-CoA, an intermediate in fatty acid synthesis. Winder and associates (1993) have shown that rat skeletal muscle also contains malonyl-CoA, and that its concentration declines during moderate intensity exercise.

While the work of Lopaschuk and others clearly shows that the level of malonyl-CoA is an important regulator of FFA oxidation in the heart, at present it is less certain that a parallel mechanism operates in skeletal muscle. As already indicated, Winder and associates have shown that malonyl-CoA declines in

rat muscle during exercise. However, while the amount of malonyl-CoA fell during exercise, the level was many times higher than the concentration of malonyl-CoA sufficient to inhibit CAT1 by 50% (i.e., the I_{50}). Similarly, in their experiments on exercising men, L. M. Odland, G. J. Heigenhauser, and associates failed to see significant and sustained decrements in malonyl-CoA, even when subjects were studied at the mild power output of 35% $\dot{V}_{O_2max}$ (Figure 7-6). These results give rise to the idea that there must be compartments in which the carnitine-acyl transferases are sequestered in compartments isolated from the prevailing malonyl-CoA concentration. Alternatively, as will be described below, the data of Odland and Heigenhauser may be correct for working muscle. The lipid metabolism determined at the whole-body level in exercising people may take place in nonworking muscle and other tissues.

Both genetic endowment, with red skeletal muscle fiber types rich in mitochondria, and endurance training to increase the mitochondrial mass, increase the amount of carnitine-acyl transferase enzymes that are present in constant amounts within mitochondria. One reason a lean, endurance-trained athlete is more successful than an obese,

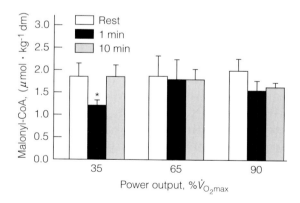

Figure 7-6 Muscle malonyl-CoA content in human skeletal muscle at rest and during exercise at varying power outputs. The predicted fall in human muscle malonyl-CoA during exercise has not yet been observed. Values are means ± SE. * Significantly different from rest and 10 min at same power output. dm = dry muscle. From Odland et al., 1998. Used with permission of the authors and the American Physiological Society.

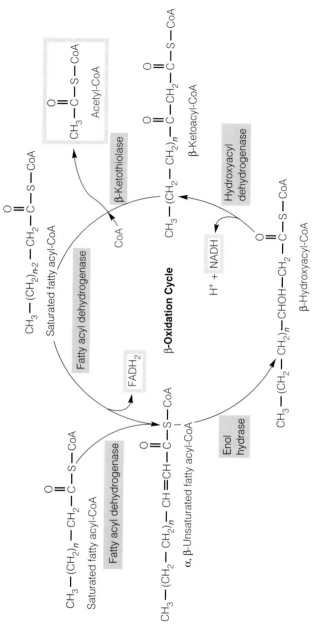

Figure 7-7 In the β-oxidation cycle, activated long-chain saturated fatty acids that have been translocated into the mitochondrial matrix are processed with the formation of three products (usually) for each cycle of the metabolic apparatus: acetyl-CoA, FADH, and NADH. These products will result in significant mitochondrial ATP production. The β-oxidation cycle is controlled by product inhibition. A reduced state of mitochondrial cofactors (i.e., NADH and FADH$_2$) will inhibit the dehydrogenases, and acetyl-CoA will inhibit the thiolase. Conversely, oxidized cofactors (NAD$^+$ and FAD) and low levels of acetyl-CoA will stimulate the β-oxidation cycle.

sedentary counterpart in using fats as a fuel is that the athlete has a larger number of mitochondria, which control the amount of activated fatty acids to be oxidized.

β-Oxidation

The β-oxidation cycle, located in the mitochondrial matrix (Figure 7-7), serves several purposes. First it degrades the fatty acyl-CoA to acetyl-CoA by cleaving the carbon atoms two at a time. Starting from the carboxyl end of a fatty acid (see Figure 7-1a), the first carbon in is the α carbon and the second is the β carbon. In the β-oxidation pathway, cleavage occurs between the α and β carbons. The acetyl-CoA formed as the result of β oxidation can then enter the tricarboxylic acid (TCA) cycle, each acetyl-CoA resulting in the formation of 12 ATPs. For each cycle of the β-oxidation pathway, two carbons are lost from fatty acyl-CoA. For palmitate, which contains 16 carbons, 8 acetyl-CoA molecules are formed, but only $7[(n/2) - 1]$ cycles of the β-oxidation pathway will occur because the last unit formed will be acetyl-CoA itself, which will enter the TCA cycle rather than traversing the β-oxidation pathway.

The rate-limiting step in β-oxidation is probably the last step, which is catalyzed by the enzyme β-ketothiolase. This enzyme is inhibited by its product, acetyl-CoA. Thus, when the acetyl-CoA level is elevated, such as after a meal rich in carbohydrates, fat catabolism is slowed. When the acetyl-CoA level is depleted, such as from glycogen depletion following exhausting exercise, fat utilization is promoted.

Like the other metabolic enzyme systems we have discussed, the β-oxidation pathway is also controlled by mitochondrial redox (i.e., the NADH/ NAD^+ ratio). Reduction (i.e., high NADH/NAD^+) inhibits the dehydrogenases, whereas oxidation activates them. Because of its importance in lipid oxidation, the enzyme hydroxyacyl dehydrogenase (HOAD) is frequently used by scientists as a marker for a cell's ability to use fat.

As seen in Figure 7-7, a second function of the β-oxidation pathway is to produce the high-energy reducing equivalents NADH and FADH. For each

cycle of the β-oxidation pathway, one each of NADH and FADH is formed. These are collectively worth $3 + 2 = 5$ ATP.

In calculating the number of reducing equivalents formed from a fatty acid in β oxidation, remember that the number of cycles is $[(n/2) - 1]$. For palmitate (16 carbons), there are 7 cycles, and 7 NADH and 7 FADH are formed, plus one acetyl-CoA is formed after the seventh cycle. Therefore, from a single palmitate, 129 ATPs are formed. It may be helpful to make this calculation, remembering the ATP used in activation that produces AMP.

Mitochondrial Oxidation

After fatty acids are converted to acetyl-CoA, the metabolism of their residues is the same as that of the residues from sugar and carbohydrate. The formation of citrate in the Krebs cycle represents a common entry point for the metabolism of acetyl-CoA derived from the various fuel sources. (See Figure 6-6.)

■ Intramuscular Triglycerides and Lipoproteins as Fuel Sources

Although fatty acids provide most of the lipid material for oxidation, triglyceride reserves in muscle (Figure 7-8) also provide a significant reserve of oxidizable substrate. As already mentioned, muscle tissue contains two lipolytic enzymes which, unfortunately, have the same name, *lipoprotein lipase* (LPL). Lipoprotein lipase in the muscle capillary wall hydrolyzes triglycerides in blood lipoproteins and makes the resulting fatty acids available in muscle. The form of lipoprotein lipase (Type L-HSL) that exists within muscle cells is responsible for hydrolyzing triglycerides in circulating lipoproteins as well as for hydrolyzing the triglyceride stores within muscle cells. At present, there is a major controversy about the role of intramuscular triglyceride during exercise.

According to the work of Larry Oscai and Warren Palmer, both LPLs in muscle are hormonally regulated. Type L-HSL is inhibited by high levels of

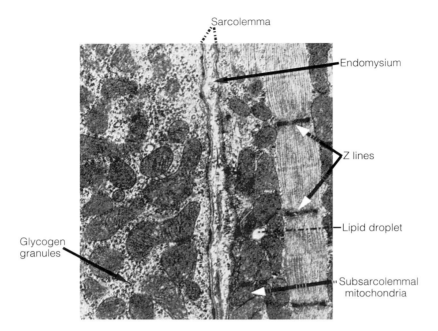

Figure 7-8 Electron micrograph of red skeletal muscle fiber showing mitochondria, lipid droplets near the mitochondria, and glycogen granules. Micrograph courtesy of J. B. Gale.

insulin and stimulated by glucagon. Thus, regulation of muscle L-HSL is similar to that of adipose HSL but opposite that of capillary bed LPL. This antagonistic regulation of triglyceride lipases within the body allows for triglyceride synthesis after eating and for triglyceride mobilization during exercise and starvation. In addition, growth hormone release activates L-HSL during prolonged exercise and helps to supply fat for combustion in muscle.

Endurance training has the effect of increasing muscle L-HSL activity. For this reason, lipid reserves within muscle cells are more available as fuel sources in the muscle of trained individuals.

As shown in Figures 6-4 and 7-8, skeletal muscle tissue contains several fat storage depots. In addition to receiving bloodborne lipids, muscle tissue contains extracellular (interstitial) fat tissue among muscle fascicles and fibers, as well as the intracellular lipid stores. Also, subcutaneous adipose can

contain large amounts of triglyceride. Because of the presence of these multiple intramuscular lipid storage sites, and because neither biopsy nor arteriovenous difference measures are capable of discriminating FFA mobilization among the sites, some of the major unresolved questions of lipid metabolism at present involve understanding how important each compartment is in supplying energy for the muscle and other parts of body during exercise and during recovery from exercise.

Despite the large potential energy source for mitochondrial oxidation represented by intramuscular triglycerides, it is unclear that they are used during the usual activities performed in daily life or in most sports activities. Initial reports from muscle biopsy studies showing a decrease in intramuscular triglyceride during exercise have not been replicated. Further, as shown in Figure 7-4, little glycerol is released from working human muscle, and biop-

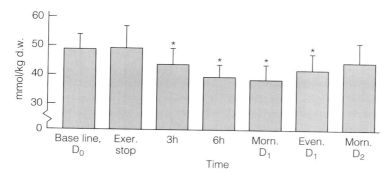

Figure 7-9 Concentrations of triglyceride in vastus lateralis muscle (TG_m) at rest (D_0), immediately after exercise (Exer. Stop), and during postexercise recovery. Results show disappearance of intramuscular triglycerides (IMTG) after, not during, exercise. Morn. (Morning) D_1 = hour 18 of recovery; even. (evening) D_1 = hour 30 of recovery; and morn. D_2 = hour 42 of recovery. Values are means ± SE of 8 observations. dw = dry muscle weight. * $P < 0.05$ vs. resting values. From Kiens and Richter, 1998. Used with permission of the authors and the American Physiological Society.

sies taken in the same study showed unchanged intramuscular triglycerides during exercise. In contrast, results from the same study showed a small but significant effect of training increasing working muscle plasma FFA uptake. However, the amount of fatty acids taken up was minor in comparison to the glucose taken up or glycogen catabolized.

If not to provide energy sources for routine exercises, when are intramuscular triglycerides used? There are two possibilities. Muscle biopsies obtained from athletes show disappearance of glycogen and intramuscular lipids after ultra-marathon running. These energy reserves can be seen as lipid droplets in the electron micrograph of a resting muscle sample that is shown in Figure 7-8. Thus, it is certain that intramuscular triglycerides are used during prolonged, submaximal exercise leading to glycogen depletion. Additionally, using muscle biopsy technique on human subjects, Kiens and Richter (Figure 7-9) have shown that intramuscular triglyceride declines after, not during, exercise.

Thus, for the present, we are left with the following conclusions about lipid oxidation in working human muscle:

1. The uptake of free fatty acids from working muscle is small, but training increases FFA up-take and oxidation, thereby sparing carbohydrate oxidation to a small but significant extent.

2. Intramuscular triglycerides are not mobilized during most activities, but probably are recruited after glycogen depletion.

3. Intramuscular triglycerides are mobilized during recovery from fatiguing exercise leading to glycogen depletion.

■ Tissue Specificity in Lipid Utilization

Various tissues in the body, such as the heart and liver, are highly adapted for lipid utilization. Other cells, such as brain and red blood cells, rely almost exclusively on glycolysis for energy supplies. Among skeletal muscle cells, the ability to utilize fats as an energy source varies greatly. White, fast-contracting fibers, with less than optimal blood supply, low mitochondrial density, and low numbers of sarcolemmal FABPs are limited in their ability to utilize fat. These fibers (sometimes called "fast glycolytic," or type IIb, fibers; see Figure 5-5) depend on glycogenolysis and glycolysis for energy supplies. In contrast, red skeletal muscle fibers, with a rich blood supply, many capillaries, and high myoglobin as well as mitochondrial contents, are well adapted to utilize fats as a fuel. The ability of muscle tissue to sustain prolonged activity (muscle tissue being made up of a heterogeneous mixture of muscle cells) by fat metabolism varies greatly. Endurance training has been demonstrated to double a cell's mitochondrial capacity to utilize fat. However, red skeletal muscles, on a unit weight basis, can have 8 to 10 times the ability to utilize fats as white muscle cells (Table 6-1). Therefore, both genetics and training affect the capacity for fat utilization during exercise.

Mitochondrial Adaptation to Enhance Fat Oxidation

Several groups of researchers have estimated that skeletal muscle in an untrained individual has a greater oxidative capacity than can be supplied by O_2 delivery through the circulation. Why, then, does muscle mitochondrial mass increase 100% after training but maximal cardiac output only increases 10% to 20%? The answer may be related to an effect on the muscle's ability to utilize FFAs as fuels. Although the mitochondrial enzymes specific to fatty acid utilization (e.g., carnitine translocase) constitute only a small fraction of the mitochondrial mass, the fact that mitochondria are assembled as complete units requires the whole organelle to be synthesized in order to enhance the capacity for fat utilization.

In the muscles of trained individuals during exercise, an increased ability to generate ATP and citrate from β oxidation would inhibit phosphofructokinase (PFK) and pyruvate dehydrogenase (PDH), thereby slowing the rate of glycolysis and the catabolism of glucose and glycogen.

The effect of mitochondrial proliferation due to training has been described by Gollnick and Saltin (1983) as having the effect of raising the apparent V_{max} of fat oxidation (Figure 7-10). Thus, in trained individuals, the absolute utilization of fat is greater at any given FFA concentration. For reasons related to the kinetics of enzymes of fat utilization, then, we can explain the effect of endurance training on muscle mitochondrial mass.

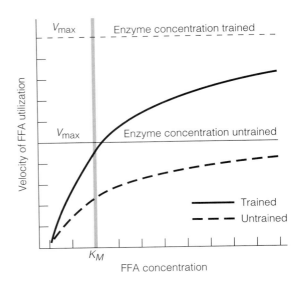

Figure 7-10 Effect of doubling enzyme concentrations on the utilization of high K_M substrates, such as fatty acids. At any given fatty acid concentration, the rate of utilization is greatly increased if the enzyme content is raised. Refer to Figure 3-4. Modified from Gollnick and Saltin, 1983.

■ Carbohydrate–Lipid Interactions: The Glucose–Fatty Acid Cycle

Based on observations of the effect of elevated FFA levels on suppressing glucose utilization by isolated hearts, the British biochemist Sir Philip Randle and associates (1964) conceived the *glucose–fatty acid cycle* (Figure 7-11). Conceptually, the glucose–fatty acid cycle is important because it explains how the availability of one fuel source (in this case, fatty acids) can influence the use of an alternative (in this case, glucose). Simply described, when citrate is abundant from β oxidation, citrate inhibits PFK, thereby slowing glucose and glycogen catabolism. Though conceived for cardiac metabolism, the glucose–fatty acid cycle has been essential to understanding how the mitochondrial adaptations resulting from endurance training help to promote lipid oxidation and spare carbohydrate utilization in skeletal muscle during mild exercise.

A major problem in the field of exercise bio-

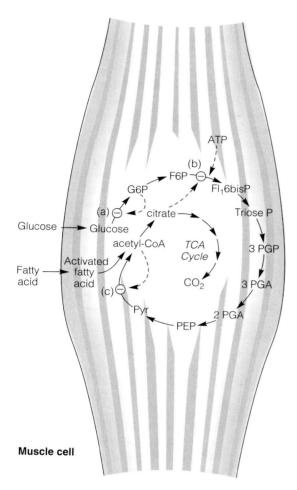

Muscle cell

Figure 7-11 Operation of the glucose–fatty acid cycle: (a) Hexokinase is inhibited by its product, glucose 6-phosphate; (b) PFK-1 is inhibited by ATP and citrate; (c) PDH is inhibited by acetyl-CoA. Modified from Randle et al., 1964.

however, require another conceptual basis, which is termed the *crossover concept*.

■ Substrate Utilization During Exercise: The "Crossover Concept"

That power output is the most important factor in determining the fuels used during exercise was recognized by Jacques Mercier of Montpellier, France, and George Brooks of Berkeley, California. Other factors, such as diet, training status, gender, and age, are of secondary importance in determining the fuels used by working muscles and the remainder of the body during exercise. Simply put, the crossover concept recognizes that in the postabsorptive person, resting muscles and the remainder of the body utilize lipids predominantly as fuels. However, as exercise starts and intensity progresses from mild through moderate to hard intensities, the fuel mix switches ("crosses over") from lipid to carbohydrate. The evidence supporting the concept follows.

Results of classic indirect calorimetry studies (Chapter 4) indicate that in a resting postabsorptive person, most energy ($\approx 60\%$) is from lipid oxidation, most of the remainder is from carbohydrate oxidation ($\approx 35\%$), and approximately 5% is from proteins. However, as soon as exercise starts, even though the same amount or more of lipid is used, the relative contribution of lipid declines, and that of carbohydrate rises (Figure 7-12). Then, as exercise power output increases from mild to moderate and then to hard intensity, the relative contribution of lipids declines while that of carbohydrate increases. Thus, hard intensity exercise is accomplished by "crossover" to dependence on carbohydrate oxidation and utilization.

When contractions start, GLUT-4 is translocated to the sarcolemma and phosphorylase is activated by elevated cytosolic Ca^{2+}. These events, in combination with high V_{max} rates and an abundance of glycolytic enzymes in the muscle, give rise to rapid glycolysis leading to pyruvate and lactate formation. Arterial FFA concentration falls, due in part to the inhibition of lipolysis by lactic acidosis, and muscle L-HSL may be inhibited by the same mecha-

chemistry has been lack of appreciation for how the glucose–fatty acid cycle operates in an intact, functioning person. In the past, the bias has been that because endurance training increases the capacity to use fat as opposed to carbohydrate and because the glucose–fatty acid cycle explains substrate interactions, the glucose–fatty acid cycle must be important for explaining fuel use during exercise. Recent evidence, and a reevaluation of older data,

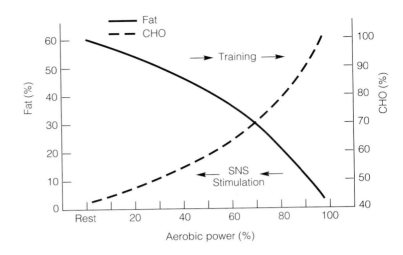

Figure 7-12 The balance of carbohydrate and lipid use during exercise is explained by the *crossover concept*. At low to moderate intensity exercise, carbohydrates and lipids both play major roles as energy substrates. However, when relative aerobic power output reaches 60–65%, then carbohydrates (CHO) become increasingly important and lipids become less important. Because of the crossover phenomenon, in most athletic activities, glycogen stores provide the greatest fuel for exercise. Lipids become important energy sources during recovery. SNS refers to the sympathetic nervous system and the metabolic effects of epinephrine and norepinephrine (see Chapter 9). Adapted from Brooks and Mercier, 1994, used with permission of the American Physiological Society.

nism. Thus, in hard intensity exercise, carbohydrate is the predominant energy source, regardless of training state.

Endurance training has the effect of displacing to a higher absolute level the point where crossover occurs. By suppressing epinephrine secretion (Chapter 9), by increasing capacity for lactate clearance, and by increasing sensitivity of respiratory control through an increase in mitochondrial mass (Chapter 6), training shifts the crossover point to higher absolute and relative power outputs. However, because athletes usually train and compete at hard or greater relative intensities, they invariably cross over to the region where most energy is from carbohydrate.

Even though athletes utilize relatively more energy from carbohydrate than lipid, lipid utilization in the athlete remains an important means of suppressing muscle glycogenolysis, thereby prolonging the time to muscle glycogen depletion and fatigue. Moreover, during recovery from hard training and competition, utilization of lipid is essential to allow normal functioning of the individual and repletion of muscle glycogen. Thus, under the larger umbrella of the "crossover concept," the glucose–fatty acid cycle is understood to function during mild to moderate intensity exercise, especially in highly trained persons, but mainly to function after exercise that depletes muscle glycogen.

Recent evidence for the crossover concept has been provided by G. A. Brooks and associates. They used stable, nonradioactive tracers (Friedlander et al. 1997, 1998, 1998, 1999), and classical mass-balance (arterial-venous) difference and muscle biopsy measurements (Bergman et al. 1999, 1999) to study glucose, glycogen, glycerol, fatty acid, and intramuscular triglyceride (IMTG) metabolism during exercise. Results not only support the crossover concept, but they also show that working muscles and the rest of the body coordinate their use of substrates by shifting their substrate utilization patterns during exercise.

Figure 7-13 shows blood glycerol and free fatty acid concentrations in subjects during rest and moderate and hard exercise, both before and after 10 weeks of endurance training. When exercise starts, blood glycerol concentration rises as a function of relative power output. Thus, in untrained subjects the rise in glycerol concentration is greater during exercise at 65% of $\dot{V}_{O_2max}$ than at 45%. Training does increase glycerol somewhat for a given absolute, but not relative power output. Com-

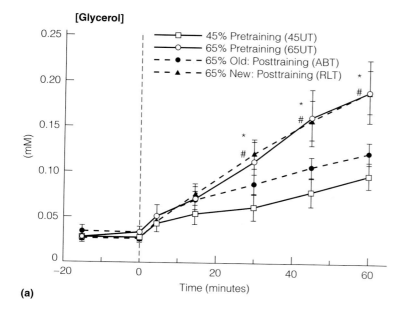

(a)

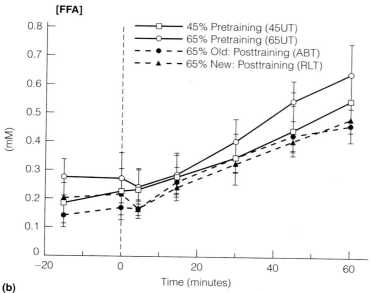

(b)

Figure 7-13 Blood glycerol (a) and free fatty acid [FFA] concentrations (b) in female subjects, during rest and exercise before and after training. Subjects were studied at 45 and 65% of $\dot{V}_{O_2max}$ before training and after training at the same absolute power output (ABT) that elicited 65% of $\dot{V}_{O_2max}$ before training and 65% of the post-training $\dot{V}_{O_2max}$ (i.e., the same relative intensity (RLT)). Typically, glycerol rises during exercise, whereas FFAs fall and then rise. Values are mean $\pm$ SEM of the last 15 and 30 minutes for rest and exercise, respectively; n = 17; Δ significantly different from rest; * significantly different from 45UT; + significantly different from 65UT; # significantly different from ABT (p < 0.05). Data on young women from Freidlander et al., 1998. Used with permission of the American Physiological Society.

parison of Figures 7-13a and 7-14a indicate that blood glycerol concentration is a good marker of its appearance (production), and therefore, the rate of lipolysis in adipose tissue. This is because glycerol can not be reesterified back to triglyceride in peripheral tissues, such as adipose and striated muscle, and because glycerol is a poor gluconeogenic precursor that is removed slowly by the liver for that purpose. Therefore, when glycerol, a water-soluble metabolite, is released from adipose, glycerol accumulates in the blood.

In contrast to that for glycerol, the pattern of

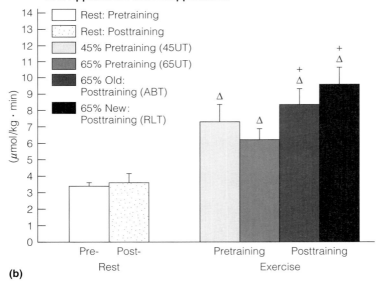

Figure 7-14 The effect of exercise intensity and training on the plasma FFA rate of appearance (a) and disappearance (b). Subjects studied at 45 and 65% of $\dot{V}_{O_2max}$ before training, and after training at the same absolute power output (ABT) that elicited 65% of the pretraining $\dot{V}_{O_2max}$, and 65% of the posttraining $\dot{V}_{O_2max}$—i.e., at the same relative exercise intensity (RLT). Values are mean ± SEM of the last 15 and 30 minutes for rest and exercise, respectively; n = 8; Δ significantly different from rest; * significantly differently from 45UT; # significantly different from ABT (p < 0.05). From Friedlander et al., 1998. Used with permission of the American Physiological Society.

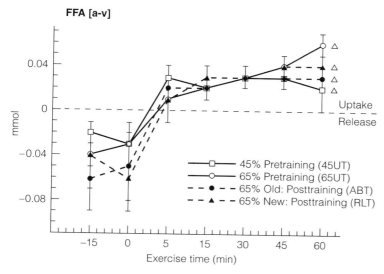

Figure 7-15 Arterial–venous differences of FFA measured by femoral catheterizations of men during rest and exercise, both before and after training. Values are mean ± SEM of the last 15 and 30 minutes for rest and exercise, respectively; n = 17; Δ significantly different from rest; (p < 0.05). From Bergman et al., 1999. Used with permission of the American Physiological Society.

FFA concentration response (Figure 7-13) and use (Figure 7-14b) shows some similarities, as well as differences. These reflect mainly the greater uptake of FFA by working muscle, but also the limited ability of plasma albumin to access FFAs released in adipose. Typically, when exercise starts, plasma FFA concentration falls because uptake by working muscles suddenly rises. Thereafter, concentration of FFAs rise in the plasma, but the rise depends on variations in the relationship between release from adipose depots and uptake by working muscles and other tissues. Consequently, plasma FFA concentration is not a good indicator of either lipolysis or FFA disposal.

Results in Figure 7-14b show several things about the effects of training on FFA mobilization and use. Prior to training, FFA rate of appearance in the blood is higher at low power output (45% $\dot{V}_{O_2max}$) than high power output (65% $\dot{V}_{O_2max}$). However, training increases whole body free fatty acid Ra and Rd at given absolute and relative power outputs. Thus, in Figure 7-14b, the highest FFA Ra is seen at the relatively high power output of 65%

$\dot{V}_{O_2max}$. However, even with the training-induced rise in FFA use, the amount of fat use by working human muscle is small.

Figure 7-15 comes from a study in which working muscle FFA use was assessed by measurements of arterial-venous (a-v) concentration differences measured across legs of men. The experimental design was the same as the studies using stable isotopes (Figures 7-13 and 7-14) and the results obtained by different methods lead to the same conclusions. Training increases slightly whole body (tracer-measured) FFA disposal (Figure 7-14b) and working muscle FFA uptake (Figure 15). In both cases, however, the change due to training is small.

■ The Substrate Shunt During Exercise

If all the data are considered (Figures 5-23 through 5-26, Figures 7-13 through 7-15 and data of J. A. Romijn et al.) then a new and exquisite view of how different body parts support energy distribution and use during exercise becomes evident. For the data obtained by various techniques by scientists in different laboratories actually fit if one recognizes that active and inactive tissue beds reciprocally alter their demands for carbohydrates during exercise.

Estimates of total body carbohydrate and fat oxidation during exercise from indirect calorimetry are reconciled with results of tracers, biopsies, and (a–v) difference measurements when it is realized that even though working muscles utilize mainly carbohydrate-derived fuels, the rest of the body is free to utilize lipids. Training does decrease muscle glucose uptake for a given submaximal power output, and this adaptation decreases the demand for hepatic glucose production. However, during hard exercise, inactive tissues become glucose resistant, thus leaving glucose to be shunted to and taken up by active muscles. Seen from this perspective, it is reasonable for breath measurements to show that some fat oxidation occurs somewhere in the body, whereas (a–v) difference measurements across working limbs and isotope tracers indicate little lipid oxidation in working muscles.

■ Ketones as Fuels

The pathways described in the preceding sections constitute the major ways in which mammals utilize fats. However, another, somewhat indirect, mechanism exists whereby fats can be used as fuels. Under conditions of carbohydrate starvation, such as during fasting, prolonged exercise, and diabetes, ketones can be used as fuels. During this process, diagrammed in Figure 7-16, the liver catabolizes fat to acetyl-CoA, and then converts acetyl-CoA to ke-

Figure 7-16 Acetoacetate and β-hydroxybutyrate are formed in the liver and kidneys from fatty acid catabolism. These ''ketones'' are released into the circulation and are taken up by the brain (e.g., during starvation) and red skeletal muscle (e.g., during prolonged exercise). Modified from McGilvery, 1975.

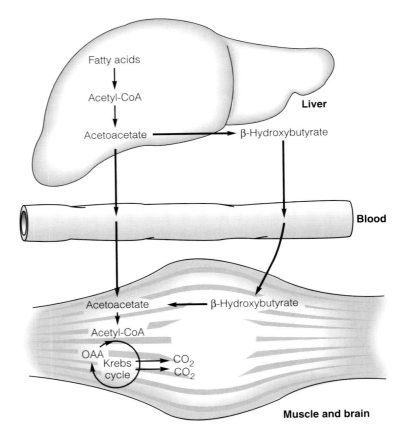

tones. The ketones circulate and can enter cells such as muscle, nerves, and brain. During dietary starvation, the ability to sustain life depends on the body's ability to form and utilize ketones. Within a day of fasting, the liver's glycogen supply is exhausted in the attempt to maintain blood glucose levels. The brain, and to some extent the kidneys, depend heavily on glucose as a fuel. Because the entry of fatty acids into the brain is very limited, in the first days of starvation, skeletal muscle is catabolized rapidly to provide material for glucose synthesis. This rapid loss of lean tissue cannot be maintained very long without severely affecting the individual. However, after several days of starvation, ketone levels in the blood elevate to the point where they can provide an alternative energy source for the brain and other nervous tissues. During untreated diabetes, blood glucose levels can be extremely high, but many cells can be starving for glucose because glucose cannot enter without insulin, which is lacking. Very high levels of ketones results, which frequently constitutes another problem because ketones are acidic and can affect an individual's physiology and biochemistry.

A *ketone* is a compound with the essential structure given in Figure 7-17. Starting from acetyl-CoA,

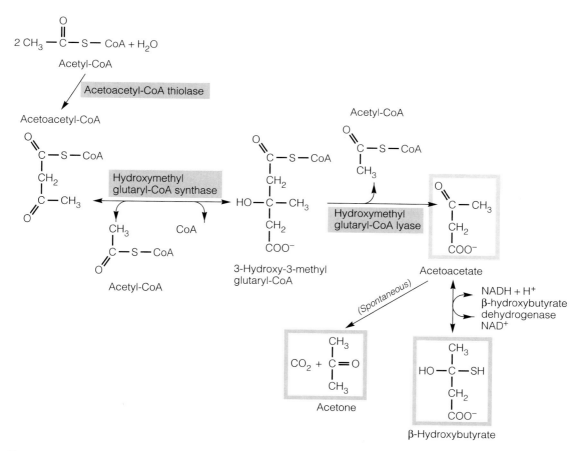

Figure 7-17 In the liver and kidneys, acetoacetyl coenzyme A is formed from acetyl-CoA produced in fatty acid catabolism. Acetoacetyl-CoA is then converted to acetoacetate and β-hydroxybutyrate. A small amount of acetone is also formed from the spontaneous decarboxylation of acetoacetate.

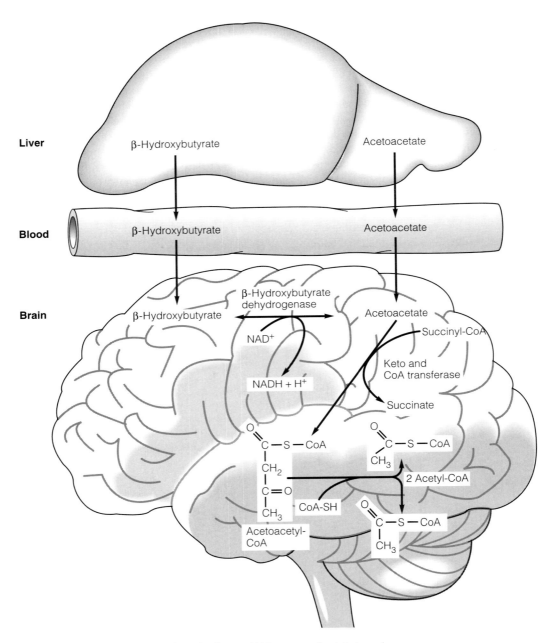

Figure 7-18 Ketones released from the liver and kidneys can circulate to and enter muscle and brain. In these tissues, β-hydroxybutyrate is converted to acetoacetate, which is converted to two molecules of acetyl-CoA. Each acetyl-CoA is equivalent to 12 ATPs after metabolism in the Krebs cycle and mitochondrial oxidative phosphorylation.

the basic chemistry of ketone formation is given in Figure 7-17. From acetyl-CoA, there are two main products (acetoacetate and β-hydroxybutyrate) and one minor product (acetone). Acetone is the most obvious product, as it is extremely volatile and can be smelled on someone's breath. Acetoacetate can be formed in both the liver and kidneys. Much of the acetoacetate formed in the liver is converted to β-hydroxybutyrate. β-hydroxybutyrate formed in the liver is a more reduced compound than acetoacetate. Therefore, when β-hydroxybutyrate reaches the muscles and brain, it can cause the formation of an extra NADH. As diagrammed in Figure 7-18, both acetoacetate and β-hydroxybutyrate result in acetyl-CoA formation in muscle and brain.

It has been demonstrated that endurance training increases the muscle enzyme content of enzymes that allow ketone utilization. In prolonged exercise, blood ketone levels are usually elevated. However, when trained individuals exercise, the increase in blood ketones is not as dramatic as it is in untrained individuals because of the uptake and utilization of ketones in the former by working skeletal muscle.

■ Summary

During muscular activities of prolonged duration and mild to moderate intensity, fuels must be provided at a rate sufficient to generate aerobically the ATP for muscular work and also to maintain an adequate level of blood glucose. Unfortunately, blood glucose and liver glycogen stores are inadequate over protracted periods to both fuel exercise and maintain blood glucose and muscle glycogen levels. Therefore, the ability to use fats as fuels is essential. The abundant supply of fats and their high energy content per unit weight provides, even in healthy lean people, a vast potential energy reserve (see Table 3-3). Moreover, the utilization of fats as fuels spares glucose and glycogen, which have essential roles in sustaining high rates of energy metabolism. In comparison to the use of carbohydrates and sugars as fuels, fats cannot be used except by oxidative processes, and thus the rate of fat catabolism is less

than the rate of sugar and glycogen utilization. The capacity for fat utilization is specialized in red (oxidative) tissues such as liver and heart. Red skeletal muscle is likewise rich in mitochondrial content and can utilize fats at significant rates. Endurance training enhances the ability to utilize fats by improving the circulation and mitochondrial content in skeletal muscle. Increased muscle mitochondrial capacity due to training, or genetic endowment with a high percentage of red high-oxidative muscle fibers, allows for superior endurance capacity, in part because of enhanced ability to oxidize bloodborne free fatty acids. However, even after training, at high power outputs, muscle is dependent on carbohydrates, not lipids, for energy. The carbohydrates of primary use are glycogen, lactate, and glucose. The effect of increasing exercise intensity on the balance of lipid and carbohydrate utilization is explained by the crossover concept, in which the balance of fuel utilization changes (crosses over) from lipid to carbohydrate in the transition from mild to high-intensity exercise. Intramuscular triglycerides (fats) are probably used most during recovery from exercises that result in muscle glycogen depletion.

■ Selected Readings

Ahlborg, G., L. Hagenfeldt, and J. Wahren. Influence of lactate infusion on glucose during prolonged exercise in man. *J. Clin. Invest.* 53: 1080–1090, 1974.

Ahlborg, G., L. Hagenfeldt, and J. Wahren. Influence of lactate infusion on glucose and FFA metabolism in man. *Scand. J. Clin. Lab Invest.* 36: 193–201, 1976.

Bergman, B. C., and G. A. Brooks. Respiratory gas exchange ratios during graded exercise in fed and fasted trained and untrained men. *J. Appl. Physiol.* 86: 479–487, 1999.

Bergman, B. C., G. E. Butterfield, E. E. Wolfel, G. A. Casazza, and G. A. Brooks. An evaluation of exercise and training on muscle lipid metabolism. *Am. J. Physiol.* 276 (*Endocrinol. Metab.* 39) E106–E117, 1999.

Bergman, B. C., G. E. Butterfield, E. E. Wolfel, G. Lopaschuk, G. A. Casazza, M. A. Horning, and G. A. Brooks. Net glucose uptake and glucose kinetics after endurance training in men. *Am. J. Physiol.* (*Endocrinol. Metab.*) [In Press, July 1999].

Brooks, G. A., and J. Mercier. The balance of carbohydrate and lipid utilization during exercise: the "crossover" concept. *J. Appl. Physiol.* 76: 2253–2261, 1994.

Brooks, G. A. Mammalian fuel utilization during sustained exercise. *Comp. Biochem. Physiol.* 120: 89–107, 1998.

Carlson, L., L. Ekelund, and S. Froberg. Concentration of triglycerides, phospholipids, and glycogen in skeletal muscle and FFA and beta hydroxybutric acid in blood in man in response to exercise. *Eur. J. Clin. Invest.* 1: 248–254, 1976.

Costill, D., E. Coyle, G. Dalsky, W. Evans, W. Fink, and D. Hoopes. Effect of elevated FFA and insulin on muscle glycogen usage during exercise. *J. Appl. Physiol.* 43: 695–699, 1977.

Dagenais, G., R. Tancredi, and K. Zierler. Free fatty acid oxidation by forearm muscle at rest, and evidence for an intramuscular lipid pool in human forearm. *J. Clin. Invest.* 58: 421–431, 1976.

Davies, K. J. A., L. Packer, and G. A. Brooks. Biochemical adaptation of mitochondria, muscle, and whole-animal respiration to endurance training. *Arch. Biochem. Biophys.* 209: 538–553, 1981.

Fitts, R. H., F. W. Booth, W. W. Winder, and J. O. Holloszy. Skeletal muscle respiratory capacity, endurance, and glycogen utilization. *Am. J. Physiol.* 228: 1029–1033, 1975.

Frieberg, S., R. Klein, D. Trout, M. Bodgonoff, and E. Estes. The characteristics of peripheral transport of ^{14}C-labeled palmitic acid. *J. Clin. Invest.* 9: 1511–1515, 1960.

Friedlander, A. L., G. A. Casazza, M. A. Horning, and G. A. Brooks. Plasma free fatty acid rate of appearance is increased in men following endurance training. *J. Appl. Physiol.* 86: 2097-2105, 1999.

Friedlander, A. L., G. A. Casazza, M. A. Horning, T. F. Budinger, and G. A. Brooks. Effects of exercise intensity and training on lipid metabolism in young women. *Am. J. Physiol.* 275 (*Endocrinol. Metab.* 38) E853–E863, 1998.

Friedlander, A. L., G. A. Casazza, M. A. Horning, M. J. Huie, M.-F. Piacentini, J. K. Trimmer, and G. A. Brooks. Training-induced alterations of glucose flux in young women: gender differences in carbohydrate oxidation. *J. Appl. Physiol.* 85: 1175–1186, 1998.

Friedlander, A. L., G. A. Casazza, M. Huie, M. A. Horning, and G. A. Brooks. Endurance training alters glucose kinetics in response to the same absolute, but not the same relative workload. *J. Appl. Physiol.* 82: 1360–1369, 1997.

Gollnick, P. D. Free fatty acid turnover and the availability of substrates as a limiting factor in prolonged exercise. *Ann. N.Y. Acad. Sci.* 301: 64–71, 1977.

Gollnick, P. D. Metabolism of substrates: energy substrate metabolism during exercise and as modified by training. *Federation Proc.* 44: 353–357, 1985.

Gollnick, P. D., and B. Saltin. Hypothesis: significance of skeletal muscle oxidative enzyme enhancement with endurance training. *Clin. Physiol.* 2: 1–12, 1983.

Guyton, A. C. Textbook of Medical Physiology. Philadelphia: W. B. Saunders, 1981, pp. 787–825.

Gyntelberg, F. M., M. J. Rennie, R. C. Hickson, and J. O. Holloszy. Effect of training on the response of plasma glucagon to exercise. *J. Appl. Physiol.* 43: 302–305, 1977.

Hall, S. E. H., J. T. Bratten, T. Bulton, M. Vranic, and I. Thoden. Substrate utilization during normal and loading diet treadmill marathons. In Biochemistry of Exercise, H. G. Knuttgen, J. A. Vogel, and J. Poortmans (Eds.). Champaign, Ill.: Human Kinetics, 1983, pp. 536–542.

Havel, R. J. Lipid as an energy source. In Physiology and Biochemistry of Muscle as a Food, E. J. Briskey (Ed.). Madison: University of Wisconsin Press, 1970, pp. 109–622.

Havel, R. J., G. Carlson, L. Ekelund, and A. Holmgren. Turnover rate and oxidation of different free fatty acids in man during exercise. *J. Appl. Physiol.* 19: 613–618, 1964.

Havel, R. J., A. Naimark, and C. F. Borchgrerink. Turnover rate and oxidation of free fatty acids of blood plasma in man during exercise: studies during continuous infusion of palmitate-1-^{14}C. *J. Clin. Invest.* 42: 1054–1063, 1963.

Havel, R. J., A. Naimark, and C. F. Borchgrerink. Turnover rate and oxidation of free fatty acids of blood plasma in man during exercise. *J. Appl. Physiol.* 236: 90–99, 1967.

Hickson, R., M. Rennie, K. R. Conlee, W. Winder, and J. Holloszy. Effects of increased plasma FFA on glycogen utilization and endurance. *J. Appl. Physiol.* 43: 829–833, 1977.

Hultman, E. A. Physiological role of muscle glycogen in man. In Physiology of Muscular Exercise. Monograph 15. New York: American Heart Association, 1967, pp. I99–I112.

Issekutz, B., and H. I. Miller. Plasma free fatty acids during exercise and the effect of lactic acid. *Proc. Soc. Exp. Biol. Med.* 110: 237–245, 1962.

Issekutz, B., H. I. Miller, P. Paul, and K. Rodahl. Source of fat oxidation in exercising dogs. *Am. J. Physiol.* 207: 583–589, 1964.

Issekutz, B., H. Miller, P. Paul, and K. Rodahl. Aerobic work capacity and plasma FFA turnover. *J. Appl. Physiol.* 20: 293–296, 1965.

Jones, N., G. Heigenhauser, A. Kuksis, C. Matsos, J. Sutton, and C. Toews. Fat metabolism in heavy exercise. *Clin. Sci.* 59: 469–478, 1980.

Karlsson, J., and B. Saltin. Muscle glycogen utilization during work of different intensities. In Muscle Metabolism During Exercise, B. Pernow and B. Saltin (Eds.). New York: Plenum Press, 1971, pp. 289–299.

Kiens, B., B. Essen-Gustavsson, N. J. Christensen, and B. Saltin. Skeletal muscle substrate utilization during submaximal exercise in man: effect of endurance training. *J. Physiol.* 469, 1993.

Kiens, B., and E. A. Richter. Utilization of skeletal muscle triacylglycerol during postexercise recovery in humans. *Am. J. Physiol.* 275(2): E332–E337, 1998.

Kozlowski, S., L. Budohoski, E. Pohoska, and K. Nazar. Lipoprotein lipase activity in the skeletal muscle during physical exercise in dogs. *Pflügers Arch.* 322: 105–107, 1979.

Lehninger, A. L. Biochemistry. New York: Worth Publishing, 1970, pp. 417–432.

Lehninger, A. L. Bioenergetics. Menlo Park, Ca.: W. A. Benjamin, 1973, pp. 75–98.

Masoro, E. J. Physiological Chemistry of Lipids in Mammals. Philadelphia: W. B. Saunders, 1968.

McGilvery, R. W. Biochemical Concepts. Philadelphia: W. B. Saunders, 1975, pp. 345–358, 463–476.

Molé, P., L. Oscai, and J. Holloszy. Adaptations in muscle to exercise. *J. Clin. Invest.* 50: 2323–2330, 1971.

Newsholme, E. Carbohydrate metabolism in vivo: regulation of the blood glucose level. *Clin. Endocrin. Metab.* 5: 543–578, 1976.

Newsholme, E. The control of fuel utilization by muscle during exercise and starvation. *Diabetes* 28 (Suppl.): 1–7, 1979.

Oscai, L. B., R. A. Caruso, and A. C. Wergeles. Lipoprotein lipase hydrolyzes endogenous triacylglycerols in muscle of exercised rats. *J. Appl. Physiol.* 52: 1059–1063, 1982.

Oscai, L. B., D. A. Essig, and W. K. Palmer. Lipase regulation of muscle triglyceride hydrolysis. *J. Appl. Physiol.* 69: 1571–1577, 1990.

Paul, P. FFA metabolism in normal dogs during steady-state exercise at different work loads. *J. Appl. Physiol.* 28: 127–132, 1970.

Phinney, S. D., B. R. Bistrian, W. J. Evans, E. Gervino, and G. L. Blackman. The human metabolic response to ketosis without caloric restriction: preservation of submaximal exercise capacity with reduced carbohydrate oxidation. *Metabolism* 32: 769–776, 1983.

Pruett, E. FFA mobilization during and after prolonged severe muscular work in men. *J. Appl. Physiol.* 29: 809–815, 1970.

Randle, P. J., E. A. Newsholme, and P. B. Garland. Effects of fatty acids, ketone bodies and pyruvate and of alloxan-diabetes and starvation on the uptake and metabolic fate of glucose in rat heart and diaphragm muscles. *Biochem. J.* 93: 652–665, 1964.

Richter, E. A., B. Kienes, B. Saltin, N. J. Christensen, and G. Savard. Skeletal muscle glucose uptake during dynamic exercise in humans: role of muscle mass. *Am. J. Physiol.* 254: E555–E561, 1988.

Rodbell, M. Modulation of lipolysis in adipose tissue by fatty acid concentration in the fat cell. *Ann. N.Y. Acad. Sci.* 131: 302–314, 1965.

Romijn, J. A., E. F. Coyle, L. S. Sidossis, A. Gastaldelli, J. F. Horowitz, E. Endert, and R. R. Wolfe. Regulation of endogenous fat and carbohydrate metabolism in relation to exercise intensity and duration. *Am. J. Physiol.* 265 (Endocrinol. Metab. 28): E380–E391, 1993.

Rowell, L. B., B. Saltin, B. Kiens, and N. J. Christensen. Is peak quadriceps blood flow in humans even higher during exercise with hypoxemia? *Am. J. Physiol.* 251: H1038–H1044, 1986.

Terblanche, S. E., R. D. Fell, A. C. Juhlin-Dannfelt, B. W. Craig, and J. O. Holloszy. Effects of glycerol feeding before and after exercise. *J. Appl. Physiol.: Respirat. Environ. Exercise Physiol.* 50: 94–101, 1981.

Winder, W. W., J. Argyasami, R. J. Barton, I. M. Elayan, and P. R. Vahrs. Muscle malonyl-CoA decreases during exercise. *J. Appl. Physiol.* 67: 2230–2233, 1989.

Winder, W. W., K. Baldwin, and J. Holloszy. Exercise-induced increase in the capacity of rat skeletal muscle to oxidize ketones. *Can. J. Physiol. Pharmacol.* 53: 86–91, 1974.

Winder, W. W., R. W. Braiden, D. C. Cartmill, C. A. Hutber, and J. P. Jones. Effect of adrenodemedulation on the decline in muscle malonyl-CoA during exercise. *J. Appl. Physiol.* 74: 2548–2551, 1993.

Wolfe, R. R., S. Klein, F. Carraro, and J. M. Weber. Role of triglyceride-fatty acid cycle in controlling fat metabolism in humans during and after exercise. *Am. J. Physiol.* 258 (Endocrinol. Metab. 21): E382–E389, 1990.

Zierler, K. Fatty acids as substrate for heart and skeletal muscle. *Circl. Res.* 38: 459–463, 1976.

Zinker, B. A., K. Britz, and G. A. Brooks. Effects of a 36-hr fast upon human endurance and substrate utilization. *J. Appl. Physiol.* 69: 1849–1855, 1990.

CHAPTER 8

METABOLISM OF PROTEINS AND AMINO ACIDS

Of the three major categories of foodstuffs, only proteins have the characteristics necessary to form body structures and enzymes. Proteins are assemblages of individual *amino acid* units. Amino acids contain peptide ($-NH_3^+$) groups that can be chemically linked to the carboxyl groups of other amino acids. These linkages are called *peptide bonds* and are the basis of protein structures. Because of the difficulties of studying amino acid and protein metabolism, and because of their essential structural roles, it was thought for a long time that amino acids do not serve as fuels to sustain muscular

One of the world's premier decathlon athletes, Dan O'Brien, at the start of a 100-m dash. Performances such as these depend upon having a sizable muscle mass as well as the ability to draw upon body reserves of carbohydrates, fats, amino acids, and proteins as fuel sources. PHOTO: © Reuters/Corbis.

work. However, it is now apparent that particular amino acids such as alanine, leucine, and glutamine are, in fact, integrated into the flow of substrates powering the body during prolonged exercise. Furthermore, amino acids may be important not only because they are fuel sources, but also because they give rise to glucose (through the process gluconeogenesis) as well as support the utilization of other fuels, such as fats.

TABLE 8-1

Nonessential (a) and essential (b) free amino acid concentrations in mM for human plasma, and mM intracellular H₂O for skeletal muscle. [a]

(a) Nonessential			Intracellular/ Plasma
Amino Acid	Plasma	Muscle	Gradient
Alanine	0.33	2.34	7.3
Arginine	0.08	0.51	6.4
Asparagine	0.05	0.47	9.5
Citrulline	0.3	0.40	1.6
Cysteine	0.11	0.18	1.6
Glutamate	0.06	4.38	73.2
Glutamine	0.57	19.45	33.8
Glycine	0.21	1.33	6.5
Ornithine	0.06	0.30	5.1
Proline	0.17	0.83	4.9
Serine	0.12	0.98	6.9
Taurine	0.07	15.44	220.4

(b) Essential			Intracellular/ Plasma
Amino Acid	Plasma	Muscle	Gradient
Histidine	0.08	0.37	4.6
Isoleucine	0.06	0.11	1.8
Leucine	0.12	0.15	1.2
Lysine	0.18	1.15	6.4
Phenylalanine	0.05	0.07	1.3
Methionine	0.02	0.11	5.6
Threonine	0.15	1.03	6.8
Tyrosine	0.05	0.10	2.00
Valine	0.22	0.26	1.2

[a] Data on 18 young healthy volunteers (16 males and 2 females) after an overnight fast.
[b] Essential in children.

SOURCE: Bergstrom et al., 1974, nonessential, essential categorization modified Rev. Food and Nutrition Board, NAS, 1980.

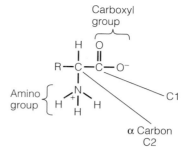

Figure 8-1 General structure of an amino acid, an α-amino carboxylic acid. At physiological pH, most amino acids carry a negative (−) charge, and usually a positive (+) charge as well. *Note:* R groups are side chains.

■ Structure of Amino Acids and Proteins

There are over 20 amino acids in the body (Table 8-1). A generalized structure of an amino acid is given in Figure 8-1. Some of the body's amino acids can be synthesized internally from existing amino acids and other substances; amino acids that the body can synthesize are termed *nonessential.* Ten other amino acids cannot be synthesized in the body and are *essential* constituents in the diet. Generally, these essential amino acids have structures that are quite different from structures in intermediary metabolism. Some essential amino acids contain ring structures, branched end chains, or sulfur.

By convention, the structures of amino acids and other organic acids are described by considering the carboxyl carbon as carbon 1 in the structure. By convention also, the carboxyl end of an acid is usually placed at the right when drawing it (see Figure 8-1). The carbon next to carbon 1 (i.e., carbon 2) is termed the α carbon. When a carboxylic acid has an amino group (—NH₃⁺) on the α carbon, it can be classified as an amino acid. Amino acids are sometimes referred to as α-amino carboxylic acids.

Amino groups on amino acids provide the structure necessary for attachment to carboxyl groups of other amino acids. When a bond is formed between the α amine of one amino acid and the carboxyl group of another, a peptide bond is formed; this is diagrammed in Figure 8-2. Given

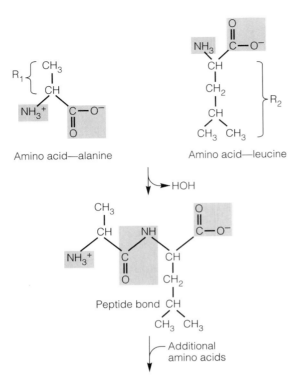

Figure 8-2 Proteins are formed by the linkage of the carboxyl group of one amino acid to the amino group of another. Such linkages are called peptide bonds and are enzymatically formed and broken. Modified from Vander, Sherman, and Luciano, 1980. Used with permission.

that there are more than 20 body amino acids, it is possible to assemble these in an almost infinite number of ways. Hence, it is possible for the body to synthesize a large number of very different protein structures as well as smaller amino acid structures and enzymes.

■ Proteins in the Diet

Meats, pasta, beans, and vegetables contribute the bulk of dietary proteins (Chapter 28). Digestion of proteins begins in the stomach under the influence of the digestive enzyme pepsin, which is very active in acidic media; thus, the stomach also secretes HCl.

There is almost no absorption of amino acids or proteins in the stomach. Digestion continues in the small intestine.

When the contents of a protein meal leave the stomach, only about 15% leave as amino acids. In the small intestine, the larger by-products are attacked by the pancreatic enzymes trypsin, chymotrypsin, and carboxypolypeptidase. Together, these proteolytic enzymes reduce most of the proteins in the meal contents to amino acids and slightly larger units, polypeptides. Epithelial cells of the small intestine contain additional amino polypeptidase and dipeptidase enzymes, which are responsible for hydrolyzing the remaining peptide bonds.

Absorption of amino acids occurs through the mucosal cells lining the small intestine. The absorption process is relatively more rapid than the digestive processes in the stomach and small intestine. Consequently, during the two to three hours it takes to digest and absorb protein, only low levels of amino acids exist freely in the digestive organs.

Amino acids are absorbed by specific carrier mechanisms. There are four such carrier systems, one each for neutral, basic, and acidic amino acids as well as one for proline and hydroxyproline. The transport of amino acids is an active process (requiring ATP) and is linked to the transport of sodium ions. Apparently, the carriers for amino acid transport, like those for glucose and lactate, exist on the brush border of intestinal villi and have sites for binding both sodium and the amino acid. The carrier mechanism is thought to operate because of the sodium gradient across the brush border; sodium movement pulls the carrier and the attached amino acid into the mucosal cells. From there, the amino acids diffuse into the portal circulation.

■ The Amino Acid Pool

When amino acids from the diet enter the circulation (see Figure 5-2a), they are entering into one of the important compartments comprising the body's amino acid pool (Figure 8-3). Liver and skeletal muscle are other major compartments comprising

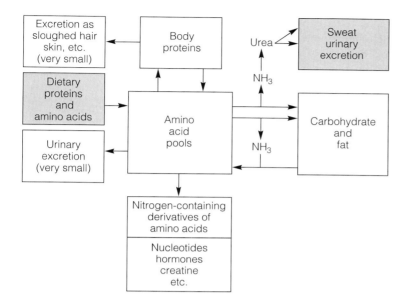

Figure 8-3   Amino acids and proteins enter the body through one route (foods) and leave through one major route (urea). The body's proteins and amino acids are not bound up in static compounds and structures, but the amino acids exist as part of various pools, some of which turn over slowly and some of which turn over more rapidly. When the total amount of dietary protein nitrogen entering a person equals that leaving, the person is said to be in nitrogen balance. Modified from Vander, Sherman, and Luciano, 1980. Used with permission.

the total pool. Amino acids in these compartments are in direct equilibrium with those in the blood. Therefore, amino acid metabolism in one compartment affects amino acids not only in that compartment, but also in other compartments. There are relatively few amino acids in the blood compartment compared to the quantity of amino acids and proteins in the other compartments. However, movement (flux) through the blood is very rapid, and the blood provides communication and exchange with the other compartments.

The continuous elimination of nitrogenous products from the body (Figure 8-3) is due to the ongoing catabolism of proteins and amino acids in the body. Different proteins turn over at different rates, but the half-life (time for removal of half the protein existing at any one time) ranges from a few days to a few months. Therefore, a constant input of new amino acids into the central pool is required to compensate for the loss and the constant rearrangement of material within the body. The fact that a central pool of amino acids exists that is in balance with other compartments is of great advantage to the organism. It ensures that the amino acid requirements in any one compartment can be met for a pe-

riod of time even though the diet does not contain the proper amount or combination of amino acids required by that compartment. Ultimately, however, the central blood pool and its allied compartments will be depleted if dietary protein input is inadequate.

Nitrogen Balance

When the dietary input (in terms of grams of nitrogen per day) is equal to that excreted, the individual is in *nitrogen balance.* Most young, healthy adults on an adequate, balanced diet are in nitrogen balance. Children given an adequate diet are in *positive* nitrogen balance, indicating that amino acids are being stored and lean tissue synthesized. Adults who engage in weight lifting, eat large amounts of protein, and/or take anabolic steroid drugs can also enter into positive nitrogen balance. This is discussed further in Chapter 28. People who are sick or injured as the result of burns or other trauma are frequently in *negative* nitrogen balance. In mature adults, nitrogen balance can usually be maintained on 0.57 g of good quality dietary protein per kilogram of body weight per day. Allowing an almost 50–100% error for individual differences and fluctuations in daily requirements, having 0.8–1.0 g of protein per kilogram of body weight per day is a usual daily recommended protein consumption (RDA, or recommended dietary allowance). The normal, balanced North American or European diet usually contains at least this quantity. In the belief that they require extra amounts of dietary protein, athletes frequently consume several times the recommended quantity (i.e., > 1.0 g · kg^{-1} · day^{-1}). The protein requirements of athletes are discussed in Chapter 28. In reality, the information available indicates that endurance athletes may be more in need of dietary protein than resistance, speed, or power athletes.

As emphasized by Butterfield and others (1984), nitrogen balance depends on having both adequate protein and energy in the diet. A diet deficient in energy (calories) will be unable to maintain the body in nitrogen balance, no matter the protein content. Similarly, providing energy in the diet from nonprotein foods can help maintain nitrogen balance, even on low protein contents, *as long as the calories taken in are adequate to maintain the energy balance.*

Figure 8-4 from Meredith and associates (1989) shows the effect of increasing protein intake in endurance-trained men in which dietary energy was sufficient to meet need. As predicted, nitrogen balance was achieved at the RDA, but in these active men a dietary protein intake of 1.2–1.3 g · kg^{-1} · day^{-1} was recommended to supply the 50–100% safety factor.

The Removal of Nitrogen from Amino Acids— The Role of Glutamate

Before amino acids can be used as fuels, the nitrogen-containing amine group (or groups) must be removed. Removal of nitrogen is especially important for carnivores and for humans on high-protein diets. Consequently, the use of such diets to sustain endurance activities, where large supplies of sugars and carbohydrates are utilized, is both expensive (in terms of the dollar cost of the protein foods) and inefficient (in terms of the extra metabolism involved).

Nitrogen is removed from amino acids by two major mechanisms: (1) oxidative deamination and (2) transamination. Although there are diverse routes by which amino acids can be deaminated or transaminated, a basic strategy in nitrogen removal is to convert amino acids to glutamate. The pathways of glutamate metabolism are well developed, and the funneling of amino acids into glutamate is a common method for dealing with amino acid nitrogen (Figure 8-5).

Oxidative Deamination *Oxidative deamination* is a process that occurs in the mitochondrial matrix of liver and involves NAD$^+$ as the oxidizing agent. The oxidative deamination of glutamate can be seen in Equation 8.1.

This reaction is catalyzed by glutamate dehydrogenase, an enzyme that is freely reversible and can function in either direction, depending on the conditions. When sufficient quantities of substrates

Figure 8-4 Relationship between protein intake (g · kg^{-1} · day^{-1}) and nitrogen balances (mg · kg^{-1} · day^{-1}) in young and middle-aged endurance-trained men. Each data point represents an individual's average for the final 5 days of each 10-day diet period. Source: Meredith et al., 1989. Used with permission.

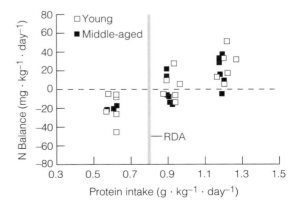

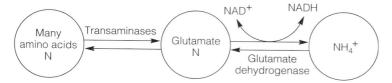

Figure 8-5 The most common means of removing nitrogen from amino acids is to transaminate the amino acid with α-ketoglutarate to form glutamate. The glutamate is then oxidized by NAD^+ through the action of glutamic dehydrogenase, with the formation of ammonium ion (NH_4^+). This is a specialized enzymatic process in the liver. Modified from McGilvery, 1975.

are available to provide material for the tricarboxylic acid (TCA) cycle and to reduce NAD^+ to NADH, the reaction can form glutamate. When there is a shortage of substrates, then glutamate can be broken down to two useful products for energy metabolism: α-ketoglutarate (α-KG) and NADH. α-Ketoglutarate is a TCA cycle intermediate, and NADH can yield several ATP molecules as the result of mitochondrial oxidative phosphorylation.

Transamination *Transamination* is by far the most common route for exchange of amino acid nitrogen in most tissues, including muscle. As implied in the name, transamination involves the transfer of an amine group. More specifically, transaminations involve the transfer of an amine from an amino acid to a keto analog of the original amino

acid. Like amino acid dehydrogenases, transaminases are enzymes that can function in either direction depending on the circumstances. As with deamination, major transaminases involve the amino acid glutamate.

A very common and important transaminase is glutamate-pyruvate transaminase (GPT). This is a major route by which alanine is utilized. (See Equation 8-2.) A second common transaminase is glutamate–oxaloacetate transaminase (GOT). (See Equation 8-3.) This reaction yields the TCA cycle intermediate (α-ketoglutarate) as well as the amino acid aspartate. Passage through aspartate is the major route by which most nitrogen is excreted from the body. Of course, pyruvate can be used as a fuel or, in liver, a gluconeogenic precursor.

Most transaminase reactions, such as the GPT

$$H_2O + NH_3^+ - \underset{\underset{\underset{COO^-}{|}}{\underset{CH_2}{|}}{\overset{\overset{COO^-}{|}}{\underset{|}{C}}} - H + NAD^+ \underset{dehydrogenase}{\overset{Glutamate}{\rightleftharpoons}} \underset{\underset{\underset{COO^-}{|}}{\underset{CH_2}{|}}{\overset{\overset{COO^-}{|}}{C}}=O + NH_4^+ + NADH$$

$$\underset{\text{Glutamate}}{} \qquad \qquad \underset{\alpha\text{-Ketoglutarate}}{} \tag{8-1}$$

$$\text{Pyruvate} + \text{Glutamate} \underset{transaminase}{\overset{Glutamate–pyruvate}{\rightleftharpoons}} \text{Alanine} + \alpha\text{-Ketoglutarate} \tag{8-2}$$

$$\text{Oxaloacetate} + \text{Glutamate} \underset{transaminase}{\overset{Glutamate–oxaloacetate}{\rightleftharpoons}} \text{Aspartate} + \alpha\text{-Ketoglutarate} \tag{8-3}$$

reaction (Eq. 8-2), are freely reversible, and function depending on substrate availability. For example, pyruvate and glutamate have high concentrations, in muscle, so the enzyme tends to operate left to right, with the formation and release of alanine (see the right side of Figure 8-9). Conversely, GPT operates from right to left in the liver, using alanine to form pyruvate, which can be made into glucose (see the left side of Figure 8-9).

In addition to the above-mentioned reactions involving glutamate, another reaction that has recently been recognized for its importance is that involving glutamine synthase (Equation 8-7, below). Until recently, most amino acid analyzers did not measure glutamine, an amino acid that has been found to be the most abundant free amino acid in human plasma and skeletal muscle. Because it contains not one, but two amino groups, glutamine is a key vehicle for interorgan nitrogen transport and excretion. This interorgan transport function also means that glutamine is an important gluconeogenic precursor, carrying carbon as well as nitrogen between skeletal muscles and the kidneys. Glutamine is also an important fuel source for cells of the gastrointestinal tract and for the immune system.

The Excretion of Nitrogenous Wastes

Although small quantities of nitrogen are excreted as ammonia and other compounds (see Figure 8-3), most nitrogen by far is excreted in the form of urea. The urea cycle (Figure 8-6) is a process centralized in the liver. Sometimes this process of urea synthesis is termed the "other Krebs cycle," for it was elaborated in large part by the efforts of Sir Hans Krebs. As in the TCA cycle, in the urea cycle there is a union of two compounds—one entering the cycle (in this case, carbamoyl phosphate) and the other being the last constituent in the cycle (in this case, ornithine).

The formation of carbamoyl phosphate from ammonia and CO_2 is an energy-requiring and irreversible process. Normally, carbamoyl phosphate synthetase activity in liver is sufficient to "cleanse" liver and blood of almost all ammonium ions. In a subsequent step of the urea cycle, aspartate, which is derived from the transamination of glutamate, enters the cycle, bringing its nitrogen with it. Consequently, the product of the cycle (urea) contains two nitrogens. The urea synthesized in the liver is released into the blood. It is then the function of the kidneys to remove the circulating urea and secrete it into the urine. During exercise, sweat glands also excrete some urea. Therefore, studies of nitrogen balance on athletes frequently require that sweat be collected.

Also of note in the urea cycle (Figure 8-6) is the formation of fumarate; this TCA cycle intermediate is a useful compound for gluconeogenesis.

Sites of Amino Acid and Protein Degradation

Skeletal muscle proteins are the major storehouse of amino acids in the body. Other significant though lesser amino acid sources are the proteins found in the liver, blood, and the intestinal wall. The role of skeletal muscle in the destruction of amino acids is limited, but it is capable of catabolizing considerable amounts of its protein content into amino acids. Skeletal muscle also contains large quantities of transaminase enzymes and can exchange amine groups among amino acids and keto acids. With the exception of the branched-chain amino acids, the capability of skeletal muscle for net degradation of amino acids is, as noted above, limited. This is because transaminases can change the type of amino acids present but not the molal quantity of amino acids. Oxidative deamination and the urea cycle are processes of the liver; thus the liver is the major site of amino acid degradation.

The Fate of Amino Acid Carbon Skeletons

Whereas urea is the major end product of nitrogen from degraded amino acids, there are two general routes by which carbon skeletons of amino acids are degraded: (1) by converting the carbon atoms to glucose and (2) by converting the carbon to the ketone acetoacetate or to acetyl-CoA. The former process is said to be glucogenic; the latter process is ketogenic (Figure 8-7). After removal of their nitrogen groups, most amino acid residues appear as pyruvate or as TCA cycle intermediates.

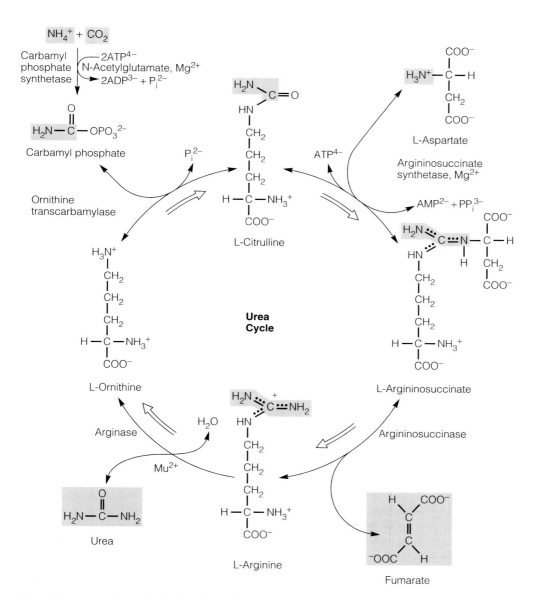

Figure 8-6   Urea is formed in the liver; the carbon and the nitrogen of urea come from CO_2 and NH_4^+ (top left), and the other nitrogen comes from aspartate (top right). Urea is released into the blood from the liver and is removed by the kidneys and sweat glands. Three ATP molecules are utilized and one fumarate (TCA cycle intermediate) is formed as the result of each urea molecule synthesized. SOURCE: McGilvery, 1975.

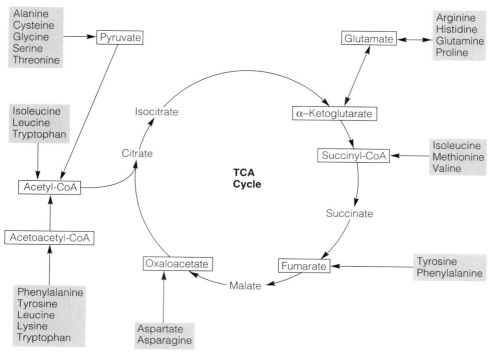

Figure 8-7 Carbon skeletons of the various amino acids gain access to the TCA cycle by various pathways. Leucine carbon enters only as acetyl-CoA. SOURCE: Lehninger, 1970. Used with permission.

■ Gluconeogenic Amino Acids

Amino acids can sometimes supply most of the carbon used in gluconeogenesis. During fasting and starvation, the catabolism of proteins to amino acids and the conversion of amino acids are very important processes in maintaining the levels of glucose essential for brain and kidney function. In addition, gluconeogenesis from amino acid occurs early each morning. Even after a large supper, the liver cannot store sufficient glycogen to maintain blood glucose levels until morning, when the nutrients in breakfast become available. In the very early morning, then, as the potential for glycogenolysis falls, the liver increases the rate of gluconeogenesis.

In the last several years researchers have realized that prolonged exercise creates in the body a

situation analogous to fasting. In both instances, caloric demands are in excess of supply, and glucose is required. It has become evident that, during these situations, some of the same mechanisms come into play to maintain substrate supply.

Amino acids that give rise to pyruvate, oxaloacetate, and malate can become precursors to phosphoenolpyruvate (PEP), which can be converted to glucose. As noted earlier, this is an expensive process for several reasons. Amino acids are frequently the most costly component in the diet. If the amino acids are derived from body proteins, then their cost is even more expensive. Also, from the standpoint of energetics, gluconeogenesis from amino acids through PEP is expensive because the process requires a significant energy input. When glucose supply is adequate, pyruvate, malate, and oxaloacetate can be converted to fatty acids.

Pathways of Phosphoenolypyruvate Formation

The formation of PEP can occur in several ways. The process usually begins in the mitochondria of the liver and kidneys, and mitochondrial processes are often complemented by cytoplasmic processes. The formation of PEP is an important step in the overall process of gluconeogenesis from most three-carbon precursors. Gluconeogenesis is discussed in more detail in Chapter 9.

Pyruvate can be converted to PEP in a two-step process. Pyruvate is first converted to oxaloacetate by the enzyme pyruvate carboxylase, which is stimulated by the presence of acetyl-coenzyme A (acetyl-CoA). (See Equation 8-4.) Oxaloacetate is then converted to phosphoenolpyruvate. (See Equation 8-5.)

Malate can probably be converted directly to PEP by malic enzyme. Recent evidence suggests that malic enzyme may exist in muscle as well as in the gluconeogenic organs (the liver and kidneys). (See Equation 8-6.)

Amino Acids in Anaplerotic and Cataplerotic Reactions

The terms *anaplerotic* and *cataplerotic* refer to the filling up (addition to) and emptying (loss from), respectively, of TCA cycle intermediates. The concept is important because there are relatively small amounts of each TCA cycle intermediate, but the TCA cycle can undergo high rates of turnover, especially during exercise. Therefore, any cataplerotic loss of TCA cycle material can result in diminished or halted TCA cycle capacity. One example of cataplerotic loss from the TCA cycle involves glutamine release from muscle during exercise (see Figure 8-7). The source of glutamine is glutamate, which is, in turn, derived from α-ketoglutarate. Although glutamine formation serves beneficially as an ammonia-scavenging mechanism (see the section on ammonia scavengers later in the chapter), glutamine formation also places a strain on TCA cycle function. To compensate for cataplerotic losses, such as through α-ketoglutarate, there must be compensating

$$\begin{array}{c} CH_3 \\ | \\ C{=}O + CO_2 + ATP \\ | \\ COO^- \end{array} \xrightarrow[\text{carboxylase}]{\text{Pyruvate}} \begin{array}{c} COO^- \\ | \\ CH_2 + ADP + P_i \\ | \\ C{=}O \\ | \\ COO^- \end{array} \qquad (8\text{-}4)$$

Pyruvate   Oxaloacetate

$$\begin{array}{c} COO^- \\ | \\ C{=}O + GTP \\ | \\ CH_2 \\ | \\ COO^- \end{array} \xrightarrow[\text{carboxykinase}]{\text{Phosphoenolpyruvate}} \begin{array}{c} COO^- \\ | \\ C{-}OPO_3^- + CO_2 + GDP \\ | \\ CH_3 \end{array} \qquad (8\text{-}5)$$

Oxaloacetate   Phosphoenolpyruvate

$$\begin{array}{c} COO^- \\ | \\ HO{-}C{-}H \\ | \\ CH_2 + ATP \\ | \\ COO^- \end{array} \xrightarrow[\text{enzyme}]{\text{Malic}} \begin{array}{c} COO^- \\ | \\ C{-}OPO_3^- + ADP + CO_2 \\ | \\ CH_3 \end{array} \qquad (8\text{-}6)$$

Malate   PEP

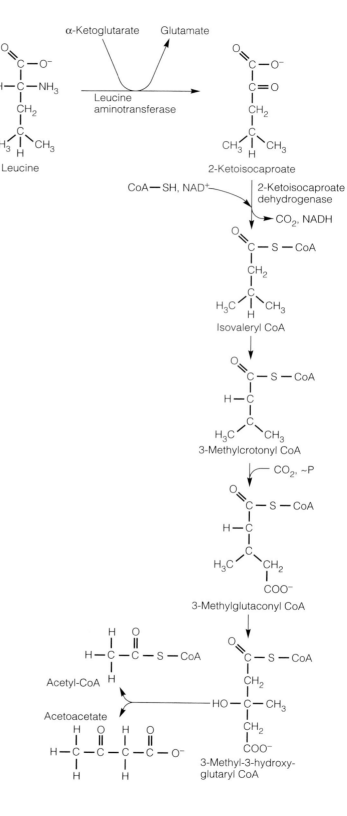

Figure 8-8 The catabolism of leucine begins with a transamination followed by a dehydrogenation that is also a decarboxylation. The transamination product glutamate can react to form alanine. The carbon skeleton of leucine is ketogenic.

anaplerotic additions to the TCA cycle, such as by aspartate at OAA, or at fumarate, by the action of the purine nucleotide cycle (discussed later in the chapter). The sites of these additions, and the roles played by particular amino acids, are illustrated in Figure 8-7.

■ Branched-Chain Amino Acids

The branched-chain amino acids (leucine, isoleucine, and valine) are essential amino acids. They are unusual in that they are catabolized mainly in skeletal muscle—where carbon skeletons provide an oxidizable source of substrate and where their nitrogen residues participate in alanine formation. The metabolism of branched-chain amino acids (Figure 8-8) begins with a transamination and the formation of glutamate and a keto acid. The glutamate so formed can then donate a nitrogen to pyruvate and form alanine (Eq. 8-2). The second step in leucine catabolism is the dehydrogenase step (Figure 8-8), which is also a decarboxylase. The remaining carbon atoms of leucine are then converted either to acetyl-CoA or to acetoacetate. The fate of these products is oxidation. Leucine is therefore purely ketogenic. The catabolism of the other branched-chain amino acids proceeds somewhat differently. Isoleucine forms both acetyl-CoA and succinyl-CoA and is therefore both ketogenic and glucogenic. Valine produces succinyl-CoA and is glucogenic. According to the results of work by Kasperek (1989), the dehydrogenase step is rate-limiting in the catabolism of branched-chain amino acids during exercise and recovery.

■ The Glucose–Alanine Cycle

Basic Studies

For some time it has been known that the body's reserves of amino acids provide precursors for gluconeogenesis during fasting and starvation. One mechanism by which blood glucose homeostasis is maintained is the *glucose–alanine cycle* (Figure 8-9).

During fasting, proteins are degraded rapidly to provide 100 g (400 kcal) of glucose per day, which is used almost exclusively by the brain, nerves, and kidney. Amino acids from degraded muscle proteins circulate to the liver, where deamination, transamination, and gluconeogenesis take place. Of the amino acids reaching the liver, alanine is by far the most important; half or more of the amino acids taken up by the liver are in the form of alanine. During fasting and starvation, the arterial level of alanine largely determines the rate of gluconeogenesis.

The alanine formed in muscle and released into the circulation most likely does not represent the catabolism of a protein rich in alanine content (i.e., a polyalanine). Rather, the alanine is newly (*de novo*) synthesized in muscle. At present, considerable controversy exists over the sources of both the carbon and nitrogen in alanine synthesis. Goldberg and Odessey (1972) have provided results to indicate that branched-chain amino acids released from liver and muscle protein may provide the nitrogen precursor. The carbon skeletons provided in such a process would yield a source of oxidizable substrate in muscle. Other scientists have provided data to suggest that other amino acids also provide nitrogen for alanine synthesis.

The source of the carbon atoms for alanine synthesis has provided even greater scientific controversy. Goldberg and associates (1972, 1978) have obtained results to indicate that the carbon source is glycolytic. Other scientists, including Goldstein and Newsholme (1976), believe that the carbon skeletons for alanine synthesis are derived from other amino acids. The mechanism of Goldstein and Newsholme is shown in Figure 8-10.

Which scientific group is correct with respect to the precursors of alanine synthesis remains to be determined. Perhaps both mechanisms operate with one or the other predominating at different times. For example, the model displayed in Figure 8-9 may operate when fasting starts, but the quantity of glucose and glycogen present in muscle and liver would be inadequate to sustain the cycle for very long unless other substances donated carbon atoms. Therefore, in prolonged starvation or exercise, the mechanism in Figure 8-10 may predominate.

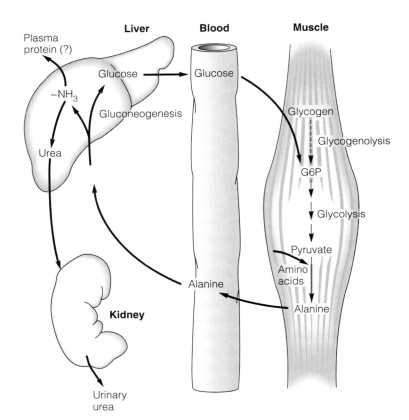

Figure 8-9 The glucose–alanine cycle as proposed by Cahill complements the Cori cycle and lactate shuttle and provides a means for transporting carbon atoms from skeletal muscle to the liver for gluconeogenesis. According to Odessey and Goldberg (1972), the branched-chain amino acids (mainly leucine) provide the nitrogen for amino acid formation in muscle. According to them, the carbon source for alanine formation is glycolytic.

Figure 8-10 The scheme of Goldstein and Newsholme, whereby muscle amino acids could provide *both* nitrogen and carbon precursors to alanine in situations such as starvation and prolonged exercise, when blood glucose is low and glycogen is depleted. SOURCE: Goldstein and Newsholme, 1976.

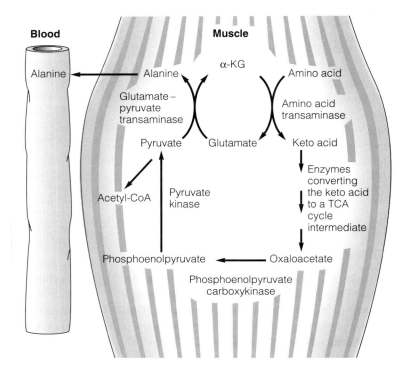

Results of Experiments on Exercising Animals

Because the glucose–alanine cycle contributes to glucose homeostasis during fasting and starvation, it was logical for researchers to hypothesize that the cycle also operates during prolonged exhaustive exercise. This hypothesis was put forth by Felig and associates (1971), including Wahren and Ahlborg (1973). Their results, based on quantitative measurements of muscle alanine release, liver uptake, and liver glucose release, indicate greatly elevated glucose–alanine cycle activity during exercise as compared to rest. According to their data, the glucose–alanine cycle may provide 5% of the total fuels used during exercise.

Also, White and Brooks (1981), by injecting [^{14}C] leucine and [^{14}C]alanine into animals during rest and exercise, demonstrated that oxidation of the substances increases during exercise in proportion to the rate of oxygen consumption. Because several amino acids contribute to maintaining glucose homeostasis and substrate supply during exercise, perhaps the total contribution to the fuel supply by all amino acids is closer to 10 than 5%. This percentage may not seem significant, but when it is recalled that record-breaking performances in athletics never supersede the existing standard by 2%, the 5 to 10% of the fuel supply contributed by amino acids may be critical. In addition, amino acids, like glycolytic substances, may have anaplerotic and cataplerotic roles in sustaining fatty acid metabolism during prolonged work.

Muscle Proteolysis During Exercise

White and Brooks (1981) have demonstrated that circulating levels of alanine and leucine increase in animals during prolonged hard exercise. Because alanine can be synthesized in muscle *de novo,* the significance of this result is difficult to interpret. However, leucine is an essential amino acid, and the results indicating an increase in its level must indicate elevated proteolysis in liver and muscle. During similar types of experiments, Dohm and associates (1977, 1978, 1980, 1982) have demonstrated that the activity of particular enzymes that catabolize sarcoplasmic proteins are elevated after prolonged exhausting exercise.

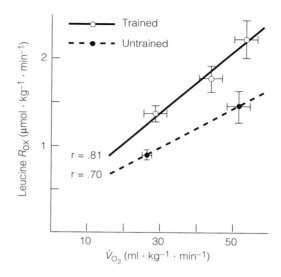

Figure 8-11 Effects of metabolic rate on leucine oxidation in trained and untrained control rats during rest and exercise. Data given in absolute terms. Data from Henderson et al., 1983.

Effects of Endurance Training on Amino Acid Metabolism

It has long been known that muscle contains high levels of transaminase enzymes. Molé and associates (1971) have demonstrated that endurance training can double the levels of important transaminases such as GPT. Using radioactive tracers, Henderson, Black, and Brooks (1983) demonstrated that endurance-trained animals have an increased ability to utilize the branched-chain amino acid leucine as a fuel during exercise (Figure 8-11). They also demonstrated that, even when not exercising, trained animals use amino acids more efficiently than do untrained animals. Therefore, although trained animals use relatively greater amounts of amino acids during exercise, their nutritional protein needs are similar to those of sedentary animals.

■ Results of Human Experimentation

The results of studies on humans during rest and exercise have confirmed and, in some cases, extended the results of animal experimentation. Rennie, Millward, and associates (1981; Figure 8-12)

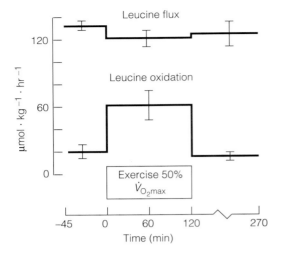

Figure 8-12 Changes in leucine oxidation and flux during a 2-hr exercise bout at 50% $\dot{V}_{O_2max}$ in the fasted state. SOURCE: Millward et al., 1982.

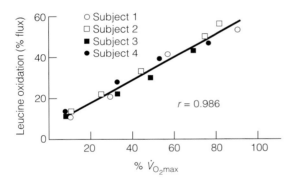

Figure 8-13 Relationship between work rate expressed as % $\dot{V}_{O_2max}$ and leucine oxidation rate during 0.5-hr period of exercise on a bicycle ergometer for the fasted state. Leucine oxidation was measured by means of a primed constant infusion of [1-^{13}C]leucine. Because the ^{13}C enrichment of expired CO_2 was the same at 15 and 30 minutes of each exercise period, it was assumed that a new isotopic steady state had been achieved during that exercise period. SOURCE: Millward et al., 1982.

dation rate is directly related to the rate of oxygen consumption (Figure 8-13).

The suspected interrelationships between leucine catabolism and alanine formation were demonstrated by Wolfe and associates (1982, 1984) using stable isotopes (both ^{13}C- and ^{15}N). In Figure 8-14, Wolfe and associates show that the rate of appearance of alanine in blood (R_a) increases 100%, an event directly related to the incorporation of nitrogen from leucine into alanine. Thus, the transamination of leucine to glutamate (see Figure 8-8), and the subsequent reaction of glutamate with pyruvate leads to alanine formation (Eq. 8-2).

The effect of alternative fuel utilization on amino acid use in humans during exercise has been shown. Rennie, Davies, and associates (1981; Figure 8-15) showed that giving glucose could decrease

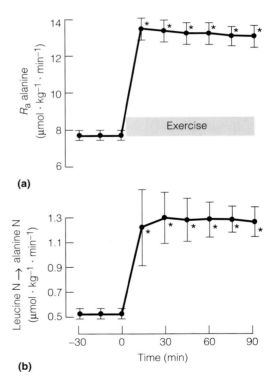

Figure 8-14 Rate of appearance (R_a) of alanine at rest and during exercise in four human subjects as determined by the primed-constant infusion of [2,3-^{13}C]alanine (a) and rate of leucine-nitrogen transfer to alanine (b). Data from Wolfe et al., 1982, 1984.

showed, through the use of stable isotopes (^{13}C-leucine), that during exercise the flux (overall use) of leucine does not change. However, a greater percentage of that flux is directed to oxidation. Further, Rennie and associates showed that the leucine oxi-

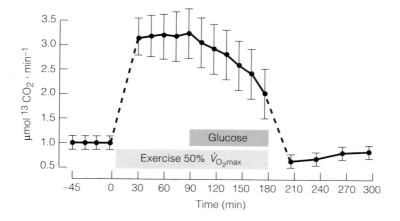

Figure 8-15 Production of $^{13}CO_2$ at rest and during treadmill exercise at 50% of $\dot{V}_{O_2max}$ by four men infused with L-[^{13}C]leucine. Ingestion of glucose at the rate of 15 g per 15 minutes had the effect of decreasing leucine oxidation. Data from Davies et al., 1982.

leucine oxidation during prolonged submaximal exercise. Further, Lemon and Mullin (1980) showed that compared to a low, carbohydrate-depletion diet, a high carbohydrate diet reduced urea excretion in sweat.

Glutamate and Glutamine as Ammonia Scavengers

Basic Studies

It has long been known that muscle liberates ammonia (NH_4^+) when it contracts. If creatine phosphate stores are maintained by intermediary metabolism in contracting muscle, then the ammonia must be provided by a source other than creatine phosphate. That source is thought to be the *purine nucleotide cycle,* described by Lowenstein (1972; Figure 8-16). The purine nucleotide cycle is very active in contracting muscle; the cycle produces IMP (ino-sine monophosphate) from AMP, a TCA cycle intermediate fumarate, and ammonia. The AMP is potentially useful in regulating metabolism, and the fumarate provides material that will eventually form oxaloacetate and combine with acetyl-CoA in operation of the TCA cycle. Ammonia is one of the stimulators of a key glycolytic enzyme (phosphofructokinase), but accumulation of ammonia can be toxic to the tissue. This accumulation is minimized by the formation of glutamate and glutamine (Eq. 8-7). Studies by Meyer and Terjung (1980) indicate that PNC activity is particularly high in fast skeletal muscle during exercise. (See Equation 8-7.)

Ruderman (1974, 1975) has observed that under some conditions, contracting muscle releases glutamine in amounts comparable to those of alanine. Brooks and coworkers (1987) have produced radiochromatograms from the blood and other tissues of exercised animals showing the incorporation of isotopic label from glycolytic metabolites into glu-

COO⁻ — C=O — CH₂ — CH₂ — COO⁻   (α-KG)
$\xrightarrow{NH_4^+}$ Glutamate dehydrogenase (NADPH or NADH → NADP⁺ or NAD⁺)
COO⁻ — NH₃⁺—C—H — CH₂ — CH₂ — COO⁻ (Glutamate)
$\xleftarrow{NH_4^+}$ Glutamine synthetase (ATP → ADP + P_i)
COO⁻ — NH₃⁺—C—H — CH₂ — CH₂ — C(=O)NH₂ (Glutamine)

(8-7)

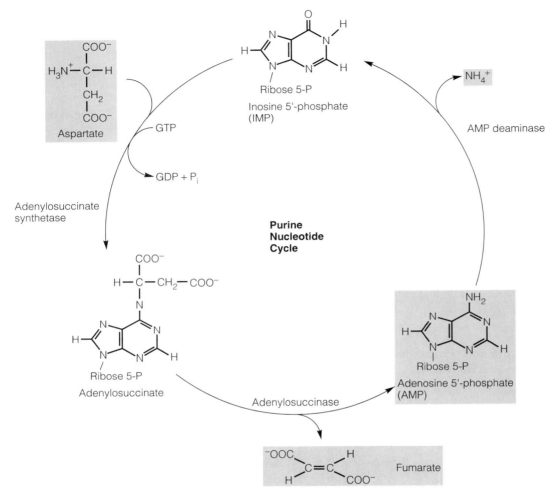

Figure 8-16 The purine nucleotide cycle (PNC) of Lowenstein, whereby the deamination of AMP (lower right) to IMP results in the deamination of aspartate to fumarate and the formation of NH_4^+. During muscular contraction, the PNC may result in purine nucleotide (AMP) consumption but the anaplerotic formation of fumarate. The starting point of the cycle is at AMP (lower right). SOURCE: Lowenstein, 1972.

tamine and alanine (see Chapter 10 and Figure 10-6). Because amino acid analyzers used in the initial studies of amino acid metabolism were incapable of detecting it, glutamate was not known to be released from working muscle. Therefore, glutamine's role as an ammonia scavenger and gluconeogenic precursor were slow to be recognized. Glutamine is particu-

larly well adapted for preventing ammonia toxicity during exercise, as it carries two amine groups to the liver and kidneys for disposal. During recovery, accumulated glutamine and glutamate function in the synthesis of amino acids and proteins.

Results of recent experiments by Gerich and associates (Perriello 1995, 1997; Kreider, 1997)

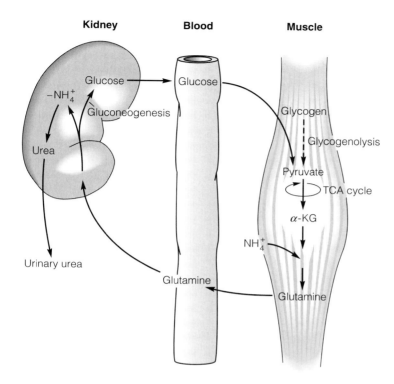

Figure 8-17 The glucose-glutamine cycle as proposed by Gerich and colleagues compliments the Cori and glucose-alanine cycles and provides a means for transporting carbon and nitrogen atoms from skeletal muscle to the kidneys for nitrogen (urea) excretion and gluconeogenesis. The importance of this cycle for exercise has not yet been determined.

have shown that a glucose-glutamine cycle exists in the body as a major means for transporting carbon and nitrogen atoms from skeletal muscle to the kidneys for nitrogen (urea) excretion and gluconeogenesis (Figure 8-17). With important differences, this cycle is similar to the glucose-alanine cycle (Figure 8-9). The differences are that in the glucose-glutamine cycle, glutamine (not alanine) is the transport mechanism and the kidneys (not the liver) are sites of transamination and gluconeogenesis (Stumvoll, 1998). In resting postabsorptive individuals, the two glucose-amino acid cycles provide similar rates of gluconeogenesis, but unfortunately the quantitative role of glutamine as a gluconeogenic precursor in exercising humans has not yet been determined. Because far more quantities of lactate are produced during exercise than alanine or glutamine, it is likely that the Cori cycle (with lactate as the gluconeogenic precursor) will prove to be the more important mechanism for maintenance of glycemia during exercise than

either the glucose-alanine or glucose-glutamine cycles.

■ Stimulation of Anabolic Processes Immediately After Exercise

Because during exercise energy expenditure exceeds energy input, exercise is a catabolic situation during which both intra- and extramuscular energy reserves are mobilized. However, immediately upon entering the period of recovery, the body is primed to reverse the catabolic processes of exercise and to replenish and restore body energy reserves and structures. In Chapter 28, nutritional practices for exercise and healthful living are discussed. However, because of their work on protein metabolism it is appropriate to note here the classical results of Butterfield and Calloway (1984), as well as the more recent work of Tarnopolsky and associates (Roy et al., 1997, 1998; Tarnopolsky et al.,

1997). In recovery, restoration of both glycogen and protein reserves is stimulated mainly by the input of carbohydrate energy. In some experiments, investigators have tried to accentuate protein synthesis in recovery by supplying dietary amino acids; however, in comparison to providing carbohydrates, which are most rapidly assimilated and which stimulate insulin and other anabolic hormonal responses, the addition of amino acids does little to protein levels. In practice, athletes have discovered this effect and many pack PowerBars in their training kits for consumption immediately after training and competition.

Summary

Although it was long (inappropriately) believed that amino acids and proteins play no significant role in supplying fuels for muscular exercise, the results of recent studies show that amino acids and proteins are, in fact, integrally involved in the metabolic adjustment to exercise. Amino acids and proteins appear to be involved in at least three important ways. During prolonged exercise, the exchanges of energy and matter are quite complex. In particular, the Krebs (TCA) cycle supports a number of functions, some of which deplete metabolic intermediates from the cycle. For TCA cycle activity to continue functioning, the levels of TCA cycle intermediates must be maintained by balancing cataplerotic losses with anaplerotic additions. Amino acids can contribute to this important anaplerotic function.

During prolonged exercise, the amino acid alanine is an important gluconeogenic precursor. The glucose–alanine cycle, whereby alanine formed in skeletal muscle is released into the circulation and reaches the liver where conversion to glucose takes place, is an important adjunct to the Cori cycle for maintaining blood glucose homeostasis during exercise.

Recently, it has been discovered that amino acids, including even the essential amino acid leucine, are oxidized as fuels to support muscular exercise. The contribution of amino acids to the substrate supply is relatively small (5–10%), but because of their large tissue masses, amino acids and proteins represent a potentially significant fuel supply to support prolonged exercise bouts. Because of the release of the proteolytic hormone cortisol during prolonged exercise, endurance athletes require high dietary protein contents so that, in recovery, the catabolic effect of training can be reversed.

When amino acid- and protein-containing foods are consumed, nitrogenous products are excreted. Nitrogenous wastes are also excreted when body proteins are catabolized. Between the consumption of amino acids and proteins and the excretion of urea, important metabolic processes occur. These processes occur during exercise as well as during rest.

Selected Readings

Adibi, S. A., E. L. Morse, and P. M. Amin. Amino acid levels in plasma, liver, and skeletal muscle during protein deprivation. *Am. J. Physiol.* 225: 408–414, 1973.

Ahlborg, G., P. Felig, L. Hagenfeldt, R. Hendler, and J. Wahren. Substrate turnover during prolonged exercise in man: splanchnic and leg metabolism of glucose free fatty acids, and amino acids. *J. Clin. Invest.* 53: 1080, 1974.

Aragon, J. J., and J. M. Lowenstein. Purine-nucleotide cycle: comparison of the levels of citric acid cycle intermediates with operation of the purine nucleotide cycle in rat skeletal muscle during exercise and recovery from exercise. *Eur. J. Biochem.* 110: 371–377, 1980.

Bergström, J., P. Fürst, L.-O. Norée, and E. Vinnars. Intracellular free amino acid concentration in human muscle tissue. *J. Appl. Physiol.* 36: 693–697, 1974.

Brooks, G. A. Amino acid and protein metabolism during exercise and recovery. *Med. Sci. Sports Exer.* 19 (Suppl.): S150–S156, 1987.

Buse, M. G., J. Biggers, C. Drier, and J. Buse. The effect of epinephrine, glucagon, and the nutritional state on the oxidation of branched chain amino acids and pyruvate by isolated hearts and diaphragms of the rat. *J. Biol. Chem.* 248: 697–706, 1973.

Butterfield, G. E., and D. Calloway. Physical activity improves protein utilization in young men. *Br. J. Nutr.* 51: 171–184, 1984.

Calloway, D. H., A. C. F. Odell, and S. Margen. Sweat and miscellaneous nitrogen losses in human balance studies. *J. Nutr.* 101: 775–786, 1971.

Calloway, D. H., and H. Spector. Nitrogen balance as related to caloric and protein intake in active young men. *Am. J. Clin. Nutr.* 2: 405–411, 1954.

Cathcart, E. P. The Physiology of Protein Metabolism. London: Longmans, Green, 1921, pp. 129–132.

Cathcart, E. P. The influence of muscle work on protein metabolism. *Physiol. Rev.* 5: 225–243, 1925.

Celejowa, I., and M. Homa. Food intake, nitrogen and energy balance in Polish weight lifters during training camp. *Nutr. and Metab.* 12: 259–274, 1970.

Consolazio, C. F., H. L. Johnson, R. A. Nelson, J. G. Dramize, and J. H. Skala. Nitrogen metabolism during intensive physical training in the young adult. *Am. J. Clin. Nutr.* 28: 29–35, 1975.

Cuthbertson, D. P., J. L. McGirr, and H. N. Munro. A study of the effect of overfeeding on protein metabolism in man. IV. The effect of muscular work at different levels of energy intake with particular reference to the timing of the work in relation to the taking of food. *Biochem. J.* 31: 2293–2305, 1937.

Davies, C. T. M., D. Halliday, D. J. Millward, M. J. Rennie, and J. R. Sutton. Glucose inhibits CO_2 production from leucine during whole-body exercise in man. *J. Physiol.* 332: 40–41, 1982.

Dohm, G. L., A. L. Heckler, W. E. Brown, G. J. Klain, F. R. Puente, E. W. Askew, and G. R. Beecher. Adaptation of protein metabolism to endurance training: increased amino acid oxidation in response to training. *Biochem. J.* 164: 705–708, 1977.

Dohm, G. L., G. J. Kasperek, E. B. Tapscott, and G. R. Beecher. Effect of exercise on synthesis and degradation of muscle protein. *Biochem. J.* 188: 255–262, 1980.

Dohm, G. L., F. R. Puente, C. P. Smith, and A. Edge. Changes in tissue protein levels as a result of endurance exercise. *Life Sci.* 23: 845–850, 1978.

Dohm, G. L., R. T. Williams, G. J. Kasperek, and A. M. van Rij. Increased excretion of urea and N^+-methylhistidine in rats and humans after a bout of exercise. *J. Appl. Physiol.* 52: 458–466, 1982.

Felig, P., and J. Wahren. Amino acid metabolism in exercising man. *J. Clin. Invest.* 50: 2703–2714, 1971.

Goldberg, A. L., and T. W. Chang. Regulation and significance of amino acid metabolism in skeletal muscle. *Fed. Proc.* 37: 2301–2307, 1978.

Goldberg, A. L., and R. Odessey. Oxidation of amino acids by diaphragms. *Am. J. Physiol.* 223: 1384–1391, 1972.

Goldstein, L., and E. A. Newsholme. The formation of alanine from amino acids in diaphragm muscle of the rat. *Biochem. J.* 154: 555–558, 1976.

Gontzea, I., P. Sutzescu, and S. Dumitriache. The influence of muscular activity on nitrogen balance and on the need of man for proteins. *Nutr. Reports Inter.* 10: 35–43, 1974.

Gontzea, I., P. Sutzescu, and S. Dumitriache. The influence of adaptation to physical effort on nitrogen balance in man. *Nutr. Reports Inter.* 11: 231–233, 1975.

Guyton, A. C. Textbook of Medical Physiology. Philadelphia: W. B. Saunders, 1981, pp. 816–825, 861–867.

Hagg, S. A., E. L. Morse, and S. A. Adibi. Effect of exercise on rates of oxidation, turnover, and plasma clearance of leucine in human subjects. *Am. J. Physiol.* 242: E407–E410, 1982.

Henderson, S. A., A. L. Black, and G. A. Brooks. Leucine turnover and oxidation on trained and untrained rats during rest and exercise. *Med. Sci. Sports Exer.* 15: 98, 1983.

Iyengar, A., and B. S. Narasinga Roa. Effect of varying energy and protein intake on nitrogen balance in adults engaged in heavy manual labor. *Br. J. Nutr.* 41: 19–25, 1979.

Kasperek, G. J. Regulation of branched-chain 2-oxo acid dehydrogenase activity during exercise. *Am. J. Physiol.* 256: (*Endocrinol. Metab.* 19): E186–E190, 1989.

Kreider, M. E., M. Stumvoll, C. Meyer, D. Overkamp, S. Welle, and J. Gerich. Steady-state and non-steady-state measurements of plasma glutamine turnover in humans. *Am. J. Physiol.* 272: E621–E627, 1997.

Lehninger, A. L. Biochemistry. New York: Worth, 1970, pp. 433–454.

Lemon, P. W., and J. P. Mullin. Effect of initial muscle glycogen levels on protein catabolism during exercise. *J. Appl. Physiol.: Respirat. Environ. Exercise Physiol.* 48: 624–629, 1980.

Lowenstein, J. M. Ammonia production in muscle and other tissues. The purine nucleotide cycle. *Physiol. Rev.* 52: 382–414, 1972.

Manchester, K. L. Oxidation of amino acids by isolated rat diaphragm and the influence of insulin. *Biochim. Biophys. Acta.* 100: 295–298, 1965.

Manchester, K. L. Control by insulin of amino acid accumulation in muscle. *Biochem. J.* 117: 457–466, 1970.

McGilvery, R. W. Biochemical Concepts. Philadelphia: W. B. Saunders, 1975, pp. 359–384.

Meredith, C. N., M. J. Zackin, W. R. Frontera, and W. J. Evans. Dietary protein requirements and body protein metabolism in endurance-trained men. *J. Appl. Physiol.* 66: 2850–2856, 1989.

Meyer, R. A., and R. L. Terjung. Differences in ammonia and adenylate metabolism in contracting fast and slow muscle. *Am. J. Physiol.* 239: C32–C38, 1980.

Millward, D. J., C. T. M. Davies, D. Halliday, S. L. Wolman, D. Matthews, and M. Rennie. Effect of exercise on protein and metabolism in humans as explored with stable isotopes. *Federation Proc.* 41: 2686–2691, 1982.

Molé, R. A., and R. E. Johnson. Disclosure by dietary modification of an exercise induced protein catabolism in man. *J. Appl. Physiol.* 31: 185–190, 1971.

Odessey, R., and A. L. Goldberg. Oxidation of leucine by rat skeletal muscle. *Am. J. Physiol.* 223: 1376–1383, 1972.

Perriello, G., R. Jorde, N. Nurjhan, M. Stumvoll, G. Dailey, T. Jenssen, D. M. Bier, and J. E. Gerich. Estimation of glucose-alanine-lactate-glutamine cycles in postabsorptive humans: role of skeletal muscle. *Am. J. Physiol.* 269: E443–E450, 1995.

Perriello, G., N. Nurjhan, M. Stumvoll, A. Bucci, S. Welle, G. Dailey, D. M. Bier, I. Toft, T. G. Jenssen, and J. E. Gerich. Regulation of gluconeogenesis by glutamine in normal, postabsorptive humans. *Am. J. Physiol.* 272: E437–E445, 1997.

Poortmans, J. R. Protein turnover and amino acid oxidation during and after exercise. *Medicine and Sport Sci.* 17: 130–147, 1984.

Rennie, M. J., R. H. T. Edwards, D. Halliday, C. T. M. Davies, E. E. Mathews, and D. J. Millward. Protein metabolism during exercise. In Nitrogen Metabolism in Man, J. C. Warterlow and J. M. L. Stephensen (Eds.). London: Applied Science, 1981, pp. 509–523.

Roy, B. D., and M. A. Tarnopolsky. Influence of differing macronutrient intakes on muscle glycogen resynthesis after resistance exercise. *J. Appl. Physiol.* 84: 890–896, 1998.

Roy, B. D., M. A. Tarnopolsky, J. D. MacDougall, J. Fowles, and K. E. Yarasheski. Effect of glucose supplement timing on protein metabolism after resistance training. *J. Appl. Physiol.* 82: 1882–1888, 1997.

Ruderman, N. B. Amino acid metabolism and gluconeogenesis. *Ann. Rev. Med.* 26: 245–258, 1975.

Ruderman, N. B., and M. Berger. The formation of glutamine and alanine in skeletal muscle. *J. Biol. Chem.* 249: 5500–5506, 1974.

Stumvoll, M., C. Meyer, G. Perriello, M. Kreider, S. Welle, and J. Gerich. Human kidney and liver gluconeogenesis: evidence for organ substrate selectivity. *Am. J. Physiol.* 274: E817–E826, 1998.

Tarnopolsky, M. A., M. Bosman, J. R. Macdonald, D. Vandeputte, J. Martin, and R. D. Roy. Postexercise protein-carbohydrate and carbohydrate supplements increase muscle glycogen in men and women. *J. Appl. Physiol.* 83: 1877–1883, 1997.

Tischler, M. E., and A. F. Goldberg. Amino acid degradation and effect of leucine on pyruvate oxidation in rat atrial muscle. *Am. J. Physiol.* 238 (*Endocrinol. Metab.*): E480–E486, 1980.

Vander, A. J., J. H. Sherman, and D. S. Luciano. Human Physiology. New York: McGraw-Hill, 1980, pp. 402–478.

Wahren, J., P. Felig, R. Hendler, and G. Ahlborg. Glucose and alanine metabolism during recovery from exercise. *J. Appl. Physiol.* 34: 838–845, 1973.

White, T. P., and G. A. Brooks. [U-14C]glucose, -alanine, and -leucine oxidation in rats at rest and two intensities of running. *Am. J. Physiol.* 240 (*Endocrinol. Metab.* 3): E155–E165, 1981.

Wolfe, R. R., R. D. Goodenough, M. H. Wolfe, G. T. Royle, and E. R. Nadel. Isotopic analysis of leucine and urea metabolism in exercising humans. *J. Appl. Physiol.* 52: 458–466, 1982.

Wolfe, R. R., M. H. Wolfe, E. R. Nadel, and J. H. F. Shaw. Isotopic determination of amino acid-urea interactions in exercise in humans. *J. Appl. Physiol.* 56: 221–229, 1984.

CHAPTER 9

NEURAL-ENDOCRINE CONTROL OF METABOLISM
Blood Glucose Homeostasis During Exercise

A primary consideration during the stress of exercise is the maintenance of nearly "normal resting" levels of blood and cellular metabolites. In particular, the maintenance of blood glucose levels in the range of 4–5 mM (90–100 mg · dl⁻¹) is critical. Exercise causes increased glucose uptake from the blood. During exercise, blood glucose level can be maintained or increased by augmented release into blood of glucose from the liver and kidneys, as well as by the mobilization of other fuels that may serve as alternatives. The coordinated physiological response to maintain blood glucose homeostasis during exercise is governed by two related body systems: the autonomic nervous system (ANS) and the endocrine (hormonal) system. Specifically with regard to the control of metabolism during exercise, the sympathetic part of the ANS (i.e., the SNS) is most important. Chemical mediators are released by both the ANS and hormonal system to help maintain an adequate blood glucose level. When blood glucose falls during prolonged hard exercise, powerful counterregulatory feedback controls come into play to increase glucose production and maintain circulating glucose concentration. However, during moderate or greater intensity exercise, the liver is under "feed-forward" control to maintain or raise blood glucose concentration. By feed-forward, we mean neurally or hormonally mediated stimulation of hepatic glucose production (HGP) in

Miguel Indurain leads the pack during a mountain stage in the Tour de France. In this event, all the body's energy reserves are called upon. The effects of hormones in regulating metabolism are always important, especially in human endeavors such as the Tour de France. PHOTO: © Reuters/Corbis.

advance of muscular uptake. Feed-forward mechanisms cause arterial glucose concentration to rise at the start of moderate and greater intensity exercises.

Hormones are chemical substances that are secreted into body fluids, usually by endocrine glands. The target tissues of hormones can be anatomically close to or quite far removed from the glands of secretion, and the targets can be one or several tissues. At the targets of their action, hormones have powerful effect on metabolism. In the resting person, metabolism is largely controlled by hormones. During exercise, both hormonal and intracellular factors control metabolism.

■ Glucose Homeostasis: Hepatic Glucose Production and Glucose Shunting During Exercise

The maintenance of nearly "normal resting" blood glucose levels is always of primary importance, including during prolonged hard exercise. Glucose from the blood and its storage form (glycogen) in muscle are necessary for continued muscular activity. Glucose and glycogen are important fuel sources, and they have anaplerotic effects in allowing fat utilization during activity. Glucose is also usually the only fuel acceptable to the brain and other central nervous system (CNS) tissues, so maintaining a reasonable blood glucose level during prolonged work to supply fuel for the brain may be even more important than supplying substrate to muscle, which has recourse to alternative fuels. Several hormones act specifically to maintain blood glucose levels; these hormones are said to be *glucoregulatory.*

During exercise, the draw on blood glucose reserves is accelerated. To compensate for this draw (or uptake), glucose production is increased in two ways. One is via the increased release into the blood of glucose from the gut, liver, and kidneys. In addition to release from the gut of digested contents from a previous meal, the liver can release glucosyl units previously stored as glycogen, and both the liver and kidneys can attempt to make new glucose from precursor molecules (i.e., gluconeogenesis). Those precursors to glucose (a six-carbon molecule) are mainly three-carbon molecules (lactate, pyruvate, glycerol, alanine); of these, lactate is by far the most important. In a recently fed individual, digestion products of the meal raise the blood glucose level (Figure 5-2a). The elevated glucose concentration causes insulin to be secreted. Insulin promotes uptake and utilization of glucose by most tissues, glycogen synthesis in muscle and liver, and triglyceride synthesis in adipose tissue. When a meal has not been eaten for several hours or exercise has intervened, glucose is taken up from the blood, lowering its level there. A lowered blood glucose level inhibits insulin secretion, which in turn decreases glucose uptake by nonactive tissues, leaving or sparing the existing supply for muscle, brain, and nerve.

Several hormones are said to be insulin *antagonists;* that is, their actions oppose those of insulin. Alternatively, the insulin antagonists are termed *counter-regulatory hormones.* This term is effective in conveying the meaning that blood glucose homeostasis is achieved by means of a balance of hormonal actions. Whereas insulin stimulates glycogen storage, epinephrine and glucagon stimulate glycogen breakdown (glycogenolysis in muscle and liver, respectively). Similarly, whereas insulin stimulates protein synthesis, cortisol promotes protein catabolism and the release of amino acids from muscle into venous blood. Several of the amino acids in muscle are glycogenic when metabolized in the liver and kidneys.

The body's second approach to the problem of maintaining blood glucose during exercise is to raise the circulating levels of alternative substances to glucose and deliver these to active tissues. When those substances (fatty acids, triglycerides, lactate, and the amino acid leucine) are utilized by contracting muscle, glucose is spared. Insulin antagonists such as epinephrine and growth hormone mobilize fatty acids and triglycerides during exercise. Use of these fuels slows glucose and glycogen utilization, thereby postponing the time when the fall in blood glucose level becomes critical.

Before describing the various factors responsible for blood glucose regulation during exercise, we first illustrate the precision of control in mild to moderate intensity exercise. We also illustrate the response to hard exercise, and the restoration of control upon return to moderate intensity exercise. Then, after describing the controls of endocrine secretion and the resulting cellular responses, we address training and glucoregulation during hard intensity exercise.

Feed-Forward Control of Glycemia During Exercise

Several groups of investigators (Ahlborg et al., 1974; Kjaer et al., 1991; Brooks et al., 1992; and Wahren et al., 1971) have attempted to account for glucose production and utilization during exercise. The quality of regulation of blood glucose homeostasis during exercise is illustrated using isotopic tracers to estimate glucose production (see Chapter 10 for an explanation of how tracers can be used) and direct [Net uptake = $(a-v)$Glucose · Limb blood flow] measurements of glucose uptake across active tissue beds.

Figure 9-1 (a composite from Kjaer et al., 1991) shows that in resting, postabsorptive people, the legs take up little glucose; this leaves most of the hepatic glucose production for other tissues. However, when leg exercise starts, glucose uptake by the legs increases dramatically. As a result, glucose up-

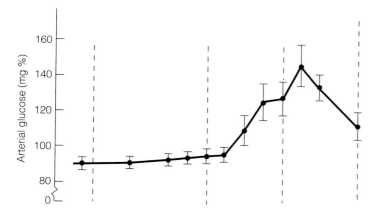

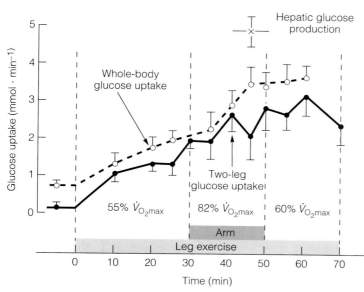

Figure 9-1 Bottom: Glucose uptake across the exercising leg (solid line) measured directly as the $(a-v)$ difference for glucose times blood flow, and whole-body glucose uptake (dotted line) measured, with tracer glucose, at rest and during exercise, with legs only or legs plus arms. The hepatic glucose production (X) is shown to be greater than leg or whole-body uptake. Consequently, blood glucose concentration rises (top). Modified from Kjaer et al., 1991. Used with permission.

take by the whole body (measured by tracers) increases also, with continued but relatively constant uptake by the rest of the body indicated. Similarly, the results of Brooks and associates (1992) show that during submaximal moderate intensity exercise most or all of the increased glucose taken up during exercise is accounted for by the active tissue beds. Thus, like blood flow distribution during exercise, glucose is now recognized to be shunted to active muscle during exercise, with other organs and tissues accounting for a constant, but relatively lesser, portion of the total glucose used.

Figure 9-1 shows also that increasing active muscle mass by about 50% through the addition of arm exercise results in a further rise in whole-body glucose uptake while leg uptake remains essentially constant. When arm exercise is added to increase the total effort, blood catecholamines (norepinephrine and epinephrine) rise and insulin falls (Figure 9-2). This is because a feed-forward, catecholamine-mediated stimulation of hepatic glucose production causes HGP to rise and exceed use (uptake). During the onset of exercise, then, or during moderate or greater intensity exercises, SNS and adrenal responses provide glucose for working muscles and the rest of the body as well. Thus, largely through the efforts of a remarkable group of Danish and Swedish scientists (listed alphabetically: Ahlborg, Galbo, Kiens, Kjaer, Richter, Sonne, Vissing, Wahren) and some others, our understanding of glucoregulation during vigorous exercise has changed dramatically: *Rather than a feedback, it is a feed-forward controlled system.*

■ Characteristics of Hormones

Having seen in Figure 5-4 the effects of diet and prolonged submaximal exercise, and now observing the effects of graded exercise, we return to a discussion of the endocrine factors that control glycemia. A hormone is a chemical messenger that is produced and stored in glandular tissues. Because endocrine glands are ductless, hormones are released (secreted) into body fluids such as blood or lymph. In this way, hormones circulate throughout the body and affect a variety of particular (target)

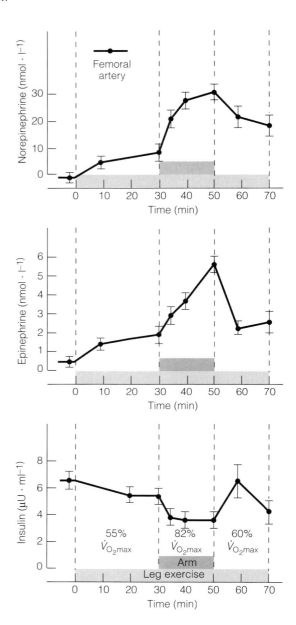

Figure 9-2 Catecholamine and insulin concentrations in femoral arterial plasma before and during exercise with legs or arms plus legs in seven healthy young subjects. Values are means ± SE. Modified from Kjaer et al., 1991. Used with permission.

tissues. Most hormones are generalized in their action and circulate widely. Other hormones, such as acetylcholine, are local hormones and are released by parasympathetic and skeletal muscle nerve endings. Local hormones have specific local effects, and they are usually metabolized within a limited area. At least one hormone, norepinephrine, is both local and general. Norepinephrine released by sympathetic nerve endings has local effects; a significant amount of this sympathetic release of norepinephrine also reaches the circulation. This circulating norepinephrine, along with norepinephrine released into the circulation from the adrenal medulla, has general effects as well.

Chemically there are two basic types of hormones: (1) steroids and (2) large polypeptides or small proteins. *Steroids* are produced from cholesterol by the adrenal cortex and gonads. *Polypeptide hormones* are derived from amino acids in the other endocrine glands.

Feedback and the Control of Hormonal Secretion

Hormones have powerful effects on metabolism. They exert their influence in concentrations of only nanograms per milliliter (ng · ml^{-1}), where *nano* means 10^{-9}. Precise regulation of hormonal secretion is, therefore, essential for normal functioning as well as for adjusting to a variety of stressful situations, such as exercise or high altitude (or even exercise at high altitude). Hormonal secretion (production in and release from endocrine glands) is regulated by feedback mechanisms, by which the secretion of a hormone is inhibited (turned off) if a particular end result of the hormonal action is achieved. In other words, a positive result has a negative effect on hormonal secretion, a situation termed *negative feedback*. Should hormonal secretion not have the desired effect, then, by negative feedback, the hormonal secretion is stimulated.

■ Mechanisms of Hormonal Action

The wide variety of hormonal actions appears to be accomplished by only a relatively few basic mechanisms. The specific effects of hormones on target tissues are accomplished by the binding of the hormones to stereospecific binding sites on membranes of the target tissues. Polypeptide hormones interact with receptors on the cell's surface; steroid hormones have mobility through the cell membrane and interact with the nucleus. Binding of the hormone may then (1) affect the permeability of the target cell membrane to a metabolite or ion, (2) activate an enzyme or enzyme system, or (3) activate the genetic apparatus to manufacture intracellular proteins or other substances. For instance, the binding of insulin to most types of cells increases permeability of those cells to glucose; the binding of epinephrine to the muscle cell membrane (sarcolemma) causes glycogenolysis in muscle by activation of the cAMP cascade (Chapter 5), and growth hormone stimulates protein synthesis in most cells.

■ Cyclic AMP (cAMP)— The Intracellular Hormone

For many hormones, their binding to the cell membranes of target tissues causes the formation of cyclic 3′, 5′-adenosine monophosphate (cyclic AMP, or cAMP) in the cell (Figure 9-3). Depending on the hormone and the target cell, the cAMP formed can then activate a variety of enzyme systems. We saw in Chapter 5 that epinephrine causes glycogenolysis in muscle and that glucagon causes glycogenolysis in liver. Both of these hormones act through cAMP. Epinephrine also activates lipolysis in adipose tissue through a cAMP-dependent mechanism.

■ Insulin and Glucagon—The Immediate Control of Blood Glucose Level

Insulin is secreted from the β cells of the pancreatic islets of Langerhans (Figure 9-4). Insulin is a polypeptide hormone that stimulates glucose uptake by many cells, of which muscle and adipose tissue are quantitatively the most important. Brain cells and erythrocytes depend on glucose for fuel but not on insulin for glucose uptake. Increased glucose

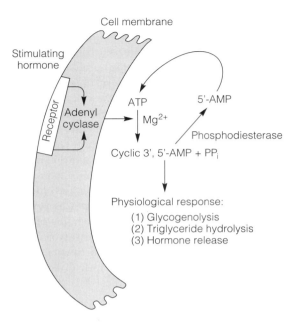

Figure 9-3 Many hormones exert their cellular action through a cyclic AMP (cAMP) mechanism. Through this mechanism, the circulating hormone binds to a specific receptor on a target cell's surface. As a result, the cell's level of cAMP, the intracellular hormone, is elevated. Depending on the target cell, specific physiological responses occur. Modified from Guyton, 1976. Used with permission.

uptake usually stimulates glycogen synthesis in muscle and fat synthesis in adipose tissue. The effect of this glucose uptake from blood is lowered blood glucose levels. The effect of insulin on glucoregulation is so profound that traditionally the topic of glucoregulation has been presented as a discussion of insulin and, secondarily, its antagonists. Now, with the recognition of feed-forward regulatory mechanisms, the roles of insulin antagonists need to be emphasized as well.

Facilitated Glucose Transport— The "Translocation Hypothesis" as a Mechanism of Insulin Action

A major mechanism of insulin action is facilitating the transport of glucose through cell membranes. As mentioned briefly in Chapter 5, glucose gains access to cells by means of a glucose transporter, or carrier protein. Different cells have different isoforms of the glucose transporter; the distribution of these glucose transporters is listed in Table 9-1. Cells, such as muscle, contain more than one transporter isoform (in this case, GLUT-1 and GLUT-4), and these transporters vary in their response to insulin and their response to glucose concentration (K_M). GLUT-4 in muscle and adipose tissue is

Figure 9-4 The pancreatic acini secrete digestive juices into the small intestine, whereas the islets of Langerhans secrete insulin and amylin (β cells), glucagon (α cells), and somatostatin (δ cells) into blood. Modified from Guyton, 1976. Used with permission.

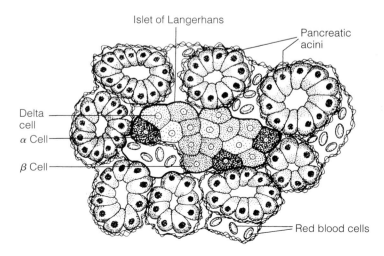

TABLE 9-1

Characteristics of Human Glucose Transport Proteins

Isoform	Tissue Distribution
GLUT-1	Erythrocytes, brain, microvessels, kidneys, placenta, muscle[a]
SGLUT-1	Sodium–glucose transporter of the small intestine
GLUT-2	Liver, β cells of pancreas, kidneys, small intestine
GLUT-3	Brain, placenta, kidneys
GLUT-4 (Insulin regulatable)	Skeletal muscle, fat, heart
GLUT-5	Small intestine

[a] At present, there is some controversy over whether the GLUT-1 in muscle tissue is in the small blood vessels, or in the vessel walls and in the muscle cell (sarcolemmal) membranes.

known as the *insulin-regulatable transporter;* other transporters are apparently unaffected by insulin.

Our understanding of GLUT-4 comes from the work of Cushman and Wardzala (1980) and Suzuki and Kono (1980), and is termed the *translocation hypothesis.* Simply described, the binding of insulin to its receptor on the cell surface gives rise to a series of events involving second messengers within the cell. These second messengers precipitate a cascade of events leading to movement (translocation) of GLUT-4 proteins from inside the cell to the cell surface. At present, our understanding of the location of the cell surface sites is incomplete although the work of Friedman, Dohm, and associates (1991) suggest that, because muscle contraction possesses "insulin-like effects," the site is the T-tubule. Figure 9-5 illustrates this hypothesis that the translocation site is the T-tubule, a major structure involved in excitation-contraction coupling in muscle. Also, because contraction causes an insulin-like increase in muscle

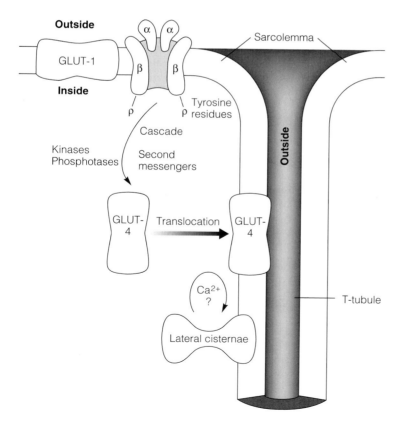

Figure 9-5 Schematic of how insulin and contractions stimulate glucose transport in skeletal muscle. The sarcolemma contains GLUT-1, which probably facilitates continuous glucose uptake, regardless of insulin or contractions. Insulin action begins with its binding to the α subunit of the cell membrane receptor. Insulin binding causes autophosphorylation of tyrosine residues of the β subunit and a cascade of events leading to the translocation of GLUT-4 proteins to the cell surface. Independent of insulin, contractions also stimulate GLUT-4 translocation, through either a separate or joined mechanism calcium ion (Ca^{2+}) is thought to be involved in the contraction-induced increase in glucose uptake.

glucose uptake, it has been speculated that calcium ion (Ca^{2+}) is one of the "second messengers" in the translocation of GLUT-4 carriers to the muscle cell surface.

Currently, the action of insulin is perceived to operate as follows. Insulin binding to the outer α subunit of the insulin receptor causes the autophosphorylation of tyrosine residues of the inner β subunit. The phosphorylated receptor also has tyrosine kinase activity, which phosphorylates several other intracellular proteins, which then gives rise to a cascade (series) of events leading to GLUT-4 translocation. The identity of the proteins, kinases, and phosphatases regulating the system is now being studied.

Facilitated Glucose Transport—The Mechanism of Muscle Contractions

Richter and associates (1982) have shown that contracting muscle can achieve maximal glucose uptake without insulin. The effect of exercise on increasing insulin action has long been described. However, recognition that contractions induce GLUT-4 translocation through a separate, insulin-independent mechanism is important in our understanding of the physiology of glucose in healthy people and diabetics during exercise. As shown in Figure 9-5, the insulin- and contraction-signaling pathways merge and cause translocation. However, in fact, there may be two separate mechanisms by which insulin and contractions stimulate glucose transport. As already indicated by the work of Goodyear and associates (1992), contraction does not increase tyrosine kinase activity or stimulate the cascade of events in the insulin-signaling pathway.

Role of the Liver in Stabilizing Blood Glucose Level

The actions of the liver are crucial to the overall process of glucose homeostasis. When blood glucose levels are high, the liver stores glucose as glycogen to release later when blood glucose is low. When blood glucose levels fall, the liver releases glucose into the blood. Both glycogenolysis and gluconeogenesis assist in the process. Because of the liver's central role in producing glucose, hepatic (liver) glucose production (HGP) is considered to represent most glucose production.

Liver contains GLUT-2, a high K_M (20 mM) and insulin-insensitive glucose transporter, so glucose uptake or release by the liver depends on concentration gradients. Thus, after a meal, when portal vein glucose is very high, the liver can take up and store glucose as glycogen or triglyceride. Then, during fasting or prolonged exercise, when blood glucose is low, GLUT-2 in the liver facilitates the release of glucose into the hepatic vein and hence the systemic circulation.

As we shall see, the metabolic effects of insulin and its antagonist, glucagon, have a powerful effect on hepatic metabolism that is independent of phenomena related to glucose transport proteins. One effect of insulin that is independent of glucose transport is to stimulate the synthesis of glucokinase. This enzyme (Chapter 5) phosphorylates glucose-6-phosphate (G6P) and causes the uptake of glucose by the liver. The phosphate group of G6P prevents efflux from the liver while G6P serves as a substrate for glycogen synthesis.

The activity of glucokinase (sometimes called high-K_M hexokinase; Chapter 5) is very sensitive to the circulating level of glucose. When blood glucose level falls, the activity of glucokinase decreases, whereas the activity of the bypass enzyme for glucokinase (glucose-6-phosphatase) is activated. Glucose-6-phosphatase forms glucose from G6P; the hepatic glucose so formed then follows the concentration gradient and moves into the circulation, thereby replenishing depleted blood glucose (Chapter 5).

While increasing glycogen storage in the liver through its effect on glucokinase, insulin has other important effects. One of these is to inhibit hepatic glucose production from gluconeogenesis. This role of insulin is antagonized by glucagon and epinephrine. These effects are discussed in the following sections.

Insulin and Hepatic Fat Metabolism

High levels of circulating glucose and insulin inhibit HGP and promote glycogen synthesis in the liver. However, the liver can store at most 5 to 6% of its net weight ($5\ g \cdot 100\ g^{-1}$) as glycogen. Thereafter, the excess G6P stimulates glycolysis and leads to acetyl-CoA and fatty acid formation. These, together with α-glycerolphosphate, promote triglyceride synthesis in liver. Triglycerides produced in liver are combined into very-low-density lipoproteins (VLDLs) and circulate with chylomicrons from the gastrointestinal (GI) tract to adipose tissue, where most fat is stored. Conversely, when insulin levels are low, triglycerides are hydrolyzed in adipose tissue (Chapter 7), and free fatty acids and glycerol are released into the blood. These then circulate to supply fuels that are alternative to glucose.

Low levels of glucose and insulin in the blood cause fatty acids to become greatly elevated. Through the process called the *glucose–fatty acid cycle* (Chapter 7), very low insulin and high fatty acid levels greatly reduce sugar and carbohydrate catabolism in resting persons. The K_M of fatty acids is higher than normal physiological levels. Consequently, an elevation in the circulating level of free fatty acids (FFA) greatly promotes the catabolism of FFA to acetyl-CoA. Acetyl-CoA from FFA inhibits pyruvate dehydrogenase, thus slowing the entry of pyruvate into the TCA cycle. High acetyl-CoA levels also elevate mitochondrial and cytoplasmic citrate and ATP levels. These inhibit phosphofructokinase (the rate-limiting reaction in glycolysis). This limitation in turn causes G6P to accumulate, which further slows cellular glucose uptake and utilization.

Excess acetyl-CoA formation from FFA can be condensed into acetoacetic acid (Chapter 7). This ketone can be converted into β-hydroxybutyrate and acetone, two other ketones. Therefore, insulin lack can lead to ketone formation. During prolonged exercise, blood insulin levels fall and blood ketones inevitably rise. Fortunately, the rise is counterbalanced by the fact that muscle and brain can utilize the ketones as fuels. Endurance training promotes ketone oxidation in muscle. However, in starving or diabetic individuals, elevated keto acids can result in an acidotic state that can cause severe discomfort, coma, and death.

The Insulin Response to Exercise

The requirements for glucose in muscle during even moderate intensity exercise tend to cause a decline in blood glucose (Figure 5-4). This decline can, for a time, be compensated for by the release of glucose, mainly from the liver but also from the kidneys to some extent. For a time during exercise, blood glucose level may actually rise as a result of this accelerated release associated with catecholamine secretion (Figures 9-1 and 9-2). Eventually, however, even if glucose rises during exercise, insulin falls (Figure 9-2). The decline in insulin during exercise is likely due to epinephrine, which suppresses insulin secretion. By suppressing secretion of insulin, epinephrine not only operates directly but also indirectly to increase hepatic glucose production. The indirect effect involves relieving the inhibitory effects of insulin on glucose production from gluconeogenesis. The direct effect involves stimulating glycogenolysis (see Table 9-2). The decline in blood insulin levels during exercise helps to minimize glucose uptake by nonactive tissues, therefore spar-

TABLE 9-2

Adrenergic Receptors and Their Functions

Effect	Receptor
Vasoconstriction	α
Vasodilatation	β_2
Cardiac acceleration	β_1
Increased myocardial contractility	β_1
Bronchodilatation	β_2
Calorigenesis	β_2
Glycogenolysis	β_2
Lipolysis	β_1
Intestinal relaxation	α
Pilomotor erection	α
Bladder sphincter contraction	α

ing blood glucose for active muscle and brain. During prolonged exercise, glucose and insulin both decline, which helps to spare blood glucose and muscle glycogen by enhancing lipolysis and making FFA available in the circulation for active and nonactive tissues alike.

Training and Insulin Release in Exercise

The general effect that training has on hormonal secretion is to reduce the hormonal response during submaximal exercises of given intensities. Glucoregulatory hormones that are released during exercise (e.g., glucagon and catecholamines) are released to lesser extents in trained individuals. In trained individuals during exercise, insulin does not fall as far as in the untrained (Figure 9-6). This lesser decrement in circulating insulin may be associated with more normal (higher) blood glucose levels in the trained during exercise. In trained individuals engaged in mild to moderate intensity exercise, increased FFA utilization and gluconeogenesis result in better control of blood glucose levels.

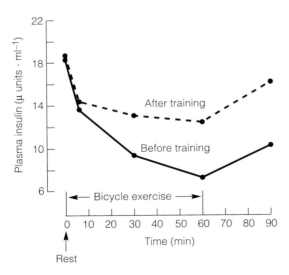

Figure 9-6 During prolonged exercise, as blood glucose level falls, so does the level of insulin. After training, the decrease in insulin is not as pronounced during exercise. Modified from Gyntelberg et al., 1977. Used with permission.

Training and Muscle GLUT-4

Kern, Dohm, and associates (1990) have shown that red-slow skeletal muscle has more GLUT-4 and greater glucose uptake capacity than white-fast muscle. Thus, oxidative muscle is adept at taking up and utilizing glucose as a fuel during exercise, even when insulin declines. Houmard, Dohm, and associates (1991, 1993) have shown that endurance training increases the amount of GLUT-4 in human muscle. Though current assays are imprecise, it is also likely that as the result of translocation, trained muscle—at least temporarily—has more of its GLUT-4 near the cell surface than does untrained muscle. By these effects, endurance training is thought to improve insulin action during exercise and recovery, even if insulin itself is not responsible for the observed effects.

Glucagon—The Insulin Antagonist

The α cells of the pancreas (see Figure 9-4) secrete the protein structure hormone glucagon. Whereas insulin is secreted when blood glucose levels are high, promoting removal of glucose from the blood, glucagon is secreted when blood glucose levels are low, acting to raise those levels. Glucagon has two effects on hepatic metabolism: (1) It enhances glycogenolysis and (2) it increases gluconeogenesis.

Glucagon activates the adenylate cyclase cascade mechanism (Chapter 5) in liver. Glucagon has a much smaller role in muscle glycogenolysis, which is activated by epinephrine.

Glucagon level in blood responds not only to glucose but also follows the blood alanine level. When alanine and other amino acids are released from muscle as a result of the actions of the catabolic steroid hormone cortisol and proteolytic enzymes, glucagon promotes hepatic amino acid uptake and gluconeogenesis from amino acids (Chapter 8). During prolonged exercise (Figure 9-7), the blood glucagon level rises as glucose and insulin levels fall. In this way, both insulin and glucagon responses help maintain blood glucose homeostasis. Gluconeogenesis is accelerated not only during exercise but also in fasting. In the average adult, about 100 g (400 kcal)

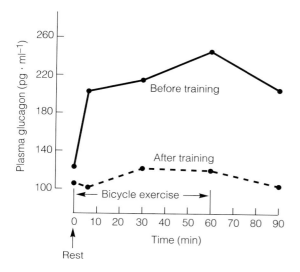

Figure 9-7 As part of the neuroendocrine response to exercise, blood glucagon rises to help maintain blood glucose levels. In trained individuals, the rise in glucagon is much less pronounced. Modified from Gyntelberg et al., 1977. Used with permission.

of glucose can be produced each day. This production approximates the obligatory needs of the CNS for glucose.

As with blood insulin level (Figure 9-6), after training the glucagon response to exercise is dampened (Figure 9-7).

■ The Autonomic Nervous System and the Adrenal Medulla

The autonomic nervous system (ANS) is composed of the sympathetic and parasympathetic nervous systems. The *parasympathetic nervous system* (Figure 9-8) controls resting functions. Parasympathetic function is dominated by the vagus, or tenth cranial nerve (X in the figure). The nerve processes of the parasympathetic nervous system are composed of two neurons, each of which releases acetylcholine (ACH). Therefore, it is the ACH released from the ending of the second parasympathetic neuron that affects the target tissue. In keeping with its general function of controlling resting metabolism, para-

sympathetic activity has effects such as slowing the heart rate and stimulating digestion.

The *sympathetic nervous system* (Figure 9-9) controls fight-or-flight responses. Like the parasympathetic system, the sympathetic nerve processes are composed of two neurons. The first releases ACH (as in the parasympathetics), but the second usually releases norepinephrine (also known as noradrenaline).

As with the endocrine system, chemical mediators released from the autonomic nerve endings bind to receptors in the cell membranes of target tissues. The chemical mediators frequently change postsynaptic target tissue membrane permeability to ions. For instance, in the heart, ACH tends to promote entry of Cl^-, whereas norepinephrine tends to increase entries of Na^+ and Ca^{2+}; ACH slows the heart by lowering the resting membrane potential whereas norepinephrine speeds the heart by stimulating cation influx.

Sympathetic activity stimulates secretion from the adrenal medulla of norepinephrine and also epinephrine. Together, epinephrine and norepinephrine are called the blood *catecholamines*; they function together to effect powerful physiological responses. These responses are brought about not only by changes in ion permeability (norepinephrine and epinephrine), but also by activating enzymes such as adenylate cyclase (epinephrine). The ratio of epinephrine to norepinephrine in adrenal secretions is 4 to 1. However, the circulating level of norepinephrine exceeds that of epinephrine fivefold. Therefore, much of the circulating norepinephrine originates from sympathetic release or "spill."

The catecholamines interact with two receptors, referred to as α and β receptors. Norepinephrine affects mainly the α receptors, whereas epinephrine affects both α and β receptors. The β receptors can be further subdivided into two groups (β_1 and β_2), depending on the actions of sympathomimetic drugs. In simplified form, the β_1 receptors affect heart function and β_2 receptors affect tissue metabolism. The actions of these receptors are outlined in Table 9-2. It may be noted here that the term *adrenergic* means *activated or transmitted by epinephrine* (adrenaline).

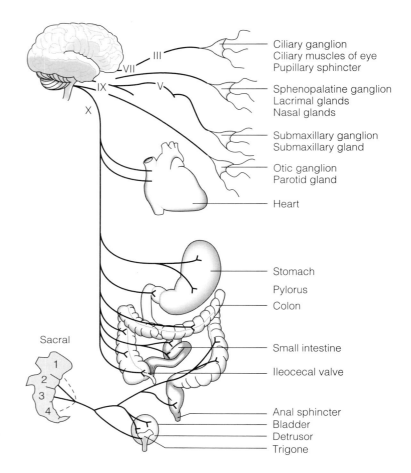

Ciliary ganglion
Ciliary muscles of eye
Pupillary sphincter

Sphenopalatine ganglion
Lacrimal glands
Nasal glands

Submaxillary ganglion
Submaxillary gland

Otic ganglion
Parotid gland

Heart

Stomach

Pylorus

Colon

Sacral

Small intestine

Ileocecal valve

Anal sphincter
Bladder
Detrusor
Trigone

Figure 9-8 The parasympathetic nervous system. Modified from Guyton, 1976. Used with permission.

Effects of Exercise Intensity and Training on Catecholamine Responses

Unless it is prolonged and results in blood glucose levels falling, moderate exercise has no or minimal effect on circulating catecholamine levels. However, as the level of exercise increases to an intensity of 50 to 60% of $\dot{V}_{O_2max}$, blood catecholamine levels increase dramatically (Figure 9-10). Because during hard exercise blood levels of catecholamines rise before blood levels of glucose fall, and because norepinephrine levels continue to exceed epinephrine levels by approximately four- to fivefold, catecholamine release is mediated by the sympathetic nervous system. Where a person is to perform in an athletic competition or hard training, sympathetic

activity may cause a rise in catecholamine levels prior to exercise.

Mild to Moderate Intensity Exercise

As is the case with insulin (see Figure 9-6) and glucagon (see Figure 9-7), the release of catecholamines is minimized by endurance training. After training, exercise bouts of a given absolute intensity present less of a stress, and this is reflected in the lesser catecholamine response to exercise (Figure 9-11).

Hard to Maximal Intensity Exercise

As suggested in Figures 9-2 and 9-10, exercise bouts of moderate or greater intensities (i.e., 60% $\dot{V}_{O_2max}$

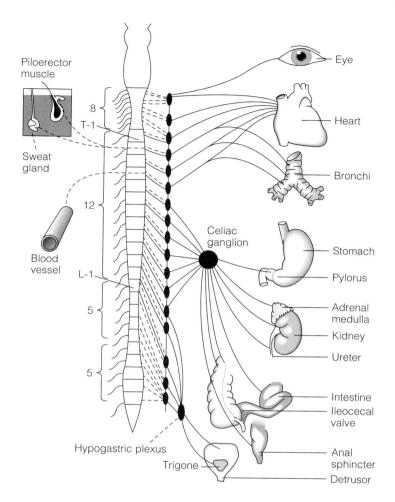

Piloerector muscle

Sweat gland

Blood vessel

8

T-1

12

L-1

5

5

Celiac ganglion

Hypogastric plexus

Trigone

Eye

Heart

Bronchi

Stomach

Pylorus

Adrenal medulla

Kidney

Ureter

Intestine

Ileocecal valve

Anal sphincter

Detrusor

Figure 9-9 The adrenal medulla is innervated by the sympathetic nervous system. Modified from Guyton, 1976. Used with permission.

or greater) are accompanied by large increases in circulating catecholamine levels. In contrast to the situation in mild to moderate intensity exercise, in which training lowers catecholamine response, during hard to maximal intensity exercise in trained individuals, catecholamine release is exaggerated over that in untrained individuals. The effects of hard exercise and prior training on blood glucose regulation are illustrated in Figure 9-12, a composite from Kjaer and associates (1986). In these experiments, subjects cycled at power outputs that elicited 60% and 100% $\dot{V}_{O_2max}$. They followed this with a supramaximal sprint (termed 110%). Exercise at 60% $\dot{V}_{O_2max}$ caused slight rises in catecholamines

and blood glucose appearance (R_a, a measure of HGP). Blood glucose concentration was the same or a bit higher than it was at rest.

Exercise at 100% $\dot{V}_{O_2max}$ (Figure 9-12) caused catecholamines to rise, which precipitated such a large rise in glucose appearance (HGP) that glucose concentration rose greatly. The sprint then caused even greater catecholamine responses, leading to hyperglycemia. Of special note are the exaggerated responses in trained subjects, who can generate hyperglycemia during maximal exercise due to their exaggerated and powerful feed-forward mechanisms.

Therefore, although training dampens hor-

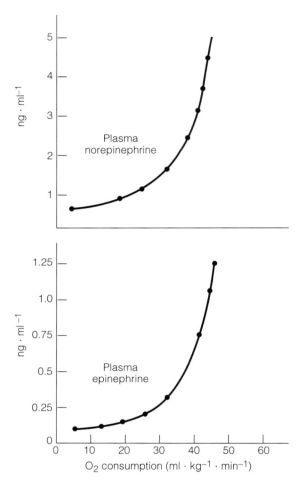

Figure 9-10 Circulating levels of catecholamines (epinephrine and norepinephrine) depend on the relative intensity of exercise. Moderate exercise intensities result in almost no increase in catecholamine levels. Beyond 50 to 70% of $\dot{V}_{O_2max}$, however, catecholamine levels rise disproportionately. Modified from Keul et al., 1981.

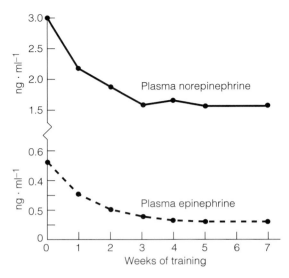

Figure 9-11 Blood catecholamine levels decrease during exercise bouts of given intensity as the result of endurance training. Modified from Winder et al., 1978.

Catecholamines and Blood Glucose Homeostasis

Epinephrine and norepinephrine secretion have powerful effects on blood glucose and carbohydrate metabolism during exercise. In muscle, through β-receptor action, epinephrine activates the adenylate cyclase mechanism (Chapter 5). That mechanism, along with changes in intramuscular free Pi and Ca^{2+}, serves to stimulate muscle glycogenolysis. By stimulating glycogenolysis and lactate production in muscle, epinephrine supports the Cori cycle and gluconeogenesis by raising the precursor (lactate) supply.

Catecholamine secretion also has the effect of stimulating glycogenolysis in the liver. The sensitivity of the mechanisms of liver glycogenolysis to epinephrine is much less than their sensitivity to glucagon. However, the rapid release of large amounts of catecholamines during high-intensity exercise (see Figures 9-10 and 9-12) is sufficient to elevate blood glucose significantly by stimulating hepatic glycogenolysis. The strong association

monal responses to exercise, these dampened responses occur during submaximal, not hard and maximal exercises; during hard and maximal exercises, endocrine responses in trained people can be the same as or greater than those in untrained individuals.

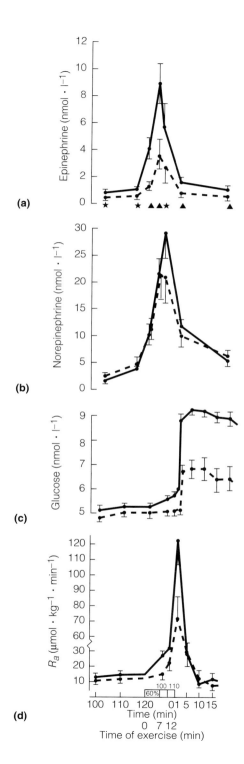

Figure 9-12   Epinephrine (a) and norepinephrine (b) concentrations in plasma at rest and during exercise (running and bicycling) in eight trained (solid line) and eight untrained (dashed lines) subjects. (c) Effect of graded exercise on glucose concentration in plasma, and (d) glucose kinetics during and after treadmill running in trained and untrained subjects. R_a, rate of glucose appearance. For all figures: Values are means $\pm$ SE. Stars and triangles denote differences ($P < 0.05$ and $P < 0.01$, respectively) between groups (trained and untrained). Modified from Kjaer et al., 1986. Used with permission.

Figure 9-13 Glucose Ra and Arterial [Norepi]. Correlations between arterial glucose rate of appearance (Ra), a measure of hepatic glucose production (HGP), and arterial norepinephrine concentration in seven men studied at rest and during exercise at sea level upon acute exposure to 4300-m altitude on Pikes Peak and after chronic (3-week) altitude exposure. The strong association between norepinephrine and HGP under all conditions is seen. In addition, the effects of acute and chronic altitude exposure are shown (see Chapter 23). Data from Brooks et al., 1991. Used with permission of the American Physiological Society.

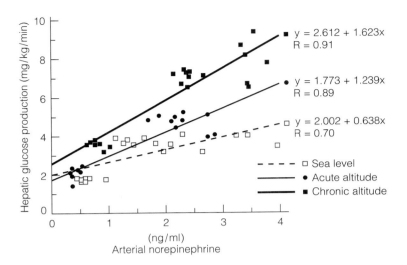

between circulating norepinephrine concentration and glucose appearance rate (a measure of HGP) is seen in Figure 9-13, which is based on studies of men exposed to high altitude on Pikes Peak (4300 m).

Norepinephrine reaches the liver by direct release from the sympathetic hepatic nerves. Also, during exercise, sympathetic spill from numerous sites results in a rise in plasma norepinephrine, which also reaches and affects liver function. As part of the feed-forward mechanism of glycemia regulation, the rapid norepinephrine rise during difficult exercise (see Figures 9-2 and 9-12) is thought to support HGP, mainly through gluconeogenesis.

In addition to affecting glucose and glycogen metabolism directly, epinephrine has an equally important indirect effect on blood glucose homeostasis during exercise. This is through the effect of a fatty acid mobilization from adipose tissue (Chapter 7). By stimulating the hormone-sensitive lipase (HSL) in adipose tissue, epinephrine acts at exercise onset to raise arterial fatty acid levels. This rapid effect of epinephrine in stimulating lipolysis is followed by a slower but more prolonged effect of growth hormone in maintaining lipolysis.

The Endocrine Control of Hepatic Glucose Production

Previously, researchers (e.g., Brooks and Fahey, 1984) described gluconeogenesis (the making of sugar) as a reversal of glycolysis and presented glycolysis and gluconeogenesis together. However, as in nature, we have now segregated the two, discussing glycolysis in Chapter 5 and addressing gluconeogenesis here. As alluded to in Chapter 5, in mammals gluconeogenesis is a specialized function in the liver and kidneys. Only white-fast muscle is capable of gluconeogenesis, and then only transiently when lactate levels are extremely high.

Gluconeogenesis in liver is depicted in Figure 9-14. The process, from pyruvate, lactate, and alanine, involves three bypass enzymes [pyruvate carboxylase (PC), phosphoenolpyruvate carboxykinase (PEPCK), and fructose 1,6-bisphosphatase (1,6-diphosphatase)]. In addition, the degradation of glucose 6-phosphate (G6P) and the release of glucose involves another enzyme, glucose-6-phosphatase. These four enzymes are in low supply or are nonexistent in mammalian muscle, but are found in large supply in the liver and kidneys. Therefore, these gut organs are termed *gluconeogenic,* as opposed to *glycolytic,* like muscle.

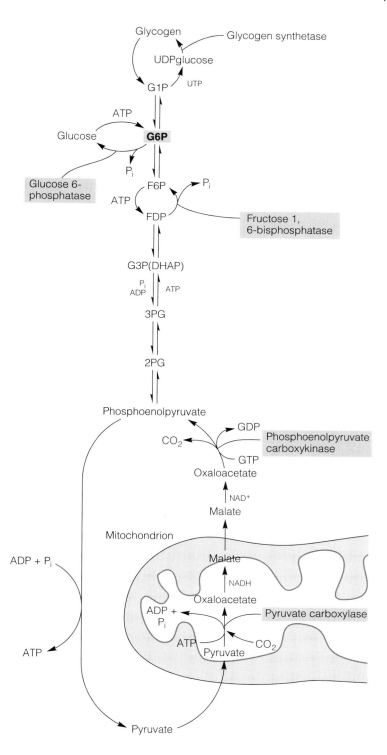

Figure 9-14 Gluconeogenesis is the process of making new glucose. This process occurs mainly in the liver and to some extent in the kidneys. Gluconeogenesis depends on the activities of specialized enzymes that can bypass the exergonic steps in glycolysis by catalyzing energy inputs from ATP and NADH into endergonic steps. In this scheme, note the central role played by glucose 6-phosphate (G6P). Modified from Lehninger, 1970. Used with permission.

Recent advances indicate that cAMP and Ca^{2+} are extremely important in regulating gluconeogenesis in liver. Thus, many of the antagonistic effects of insulin, glucagon, and catecholamines on liver function are now understood at an intracellular level.

In muscle, pyruvate kinase (PK) is very active and has a high K_{eq}; consequently, it strongly promotes glycolysis. In muscle, the process catalyzed by PK is said to be thermodynamically irreversible. However, liver has another form of pyruvate kinase called liver pyruvate kinase or Type L-PK. In liver, pyruvate kinase can be phosphorylated and inhibited by cAMP and a Ca^{2+}-dependent protein kinase. Thus, in liver, L-PK phosphorylation promotes gluconeogenesis by reducing glycolysis (Figure 9-15).

The discovery that in liver the metabolite fructose 2,6-bisphosphate (Fru $2,6-P_2$) has powerful effects as an activator of glycolytic phosphokinase (PFK-1) and as an inhibitor of gluconeogenic fructose 1,6-bisphosphatase (Fru-1,6 P_2ase) has led to new understanding of how glycolysis and gluconeogenesis are regulated in liver. Today, we appreciate that conditions such as eating, which increase Fru $2,6-P_2$, favor glycolysis, whereas a fall in Fru $2,6-P_2$, as in starvation, favors gluconeogenesis.

Recognition of the importance of Fru $2,6-P_2$ led to the discovery that its level is controlled by a special form of phosphofructokinase, simply termed PFK-2. PFK-2 is a unique bifunctional enzyme that can operate either as a kinase or phosphatase (see Figure 9-15). The enzyme is also sometimes called

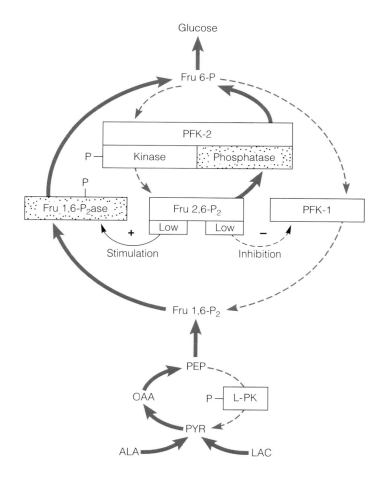

Figure 9-15 Hepatic gluconeogenesis from pyruvate (PYR), lactate (LAC), and alanine (ALA) is stimulated by a cAMP-dependent protein kinase, which results in phosphorylation (inhibition) of the kinase and dephosphorylation (activation) of the bisphosphatase of PFK-2. Activation of the phosphatase and inactivation of the kinase of PFK-2 *decreases fructose 2,6-bisphosphate* (Fru $2,6-P_2$) levels, stimulating fructose 1,6-bisphosphatase (Fru $1,6-P_2$ase), while at the same time dephosphorylating (inhibiting) PFK-1. Thus, cAMP stimulates gluconeogenesis, as indicated by the bold arrows, and inhibits glycolysis, as indicated by the dashed lines. Stippled boxes indicate activated enzymes and open boxes indicate inhibited enzymes. Phosphorylation is indicated by -P. Adapted from Pilkis and Claus, 1991, and Benkovic and deMaine, 1983.

TABLE 9-3

Summary of Hormonal Regulation of Gluconeogenesis, L-PK, PFK-2 (6PF-2-K/Fru-2,6-P$_2$ase), and Fru 2,6-P$_2$ levels

Hormone	Gluco-neogenesis	Fru 2,6-P$_2$	Pyruvate Kinase	6PF-2-K/ FRU-2,6-P$_2$ase	Mechanism
Glucagon	Stimulates	↓	Inhibits	Inhibits/activates	↑ cAMP/↑ cAMP-protein kinase
Insulin	Inhibits	↑	Activates	Activates/inhibits	↓ cAMP/↓ cAMP-protein kinase and ↑ protein phosphatase
Catecholamines					
Epinephrine (β-agonist)	Stimulates	↓	Inhibits	Inhibits/activates	↑ cAMP/↑ cAMP-protein kinase
Norepinephrine (α-agonist)	Stimulates	↑	Inhibits	— —	↑ Ca^{2+}/↑ Ca^{2+}/Calmodulin-protein kinase

SOURCE: Modified from Pilkis and Claus, 1991.

PFK2/FRU-P$_2$ase to indicate its dual function. Depending on conditions, this enzyme can synthesize Fru 2,6-P$_2$:

$$\text{Fru 6-P} + \text{ATP} \xrightarrow{\text{Kinase}} \text{Fru 2,6-P}_2 + \text{ADP} \quad (9\text{-}1)$$

or degrade Fru 2,6-P$_2$:

$$\text{Fru 2,6-P}_2 \xrightarrow{\text{Phosphatase}} \text{Fru 6-P} + \text{P}_i \quad (9\text{-}2)$$

The phosphorylation of PFK-2 is by cAMP-dependent protein kinase, whereas protein phosphatase 2A is primarily responsible for dephosphorylation.

The discovery of Fru 2,6-P$_2$ and PFK-2 has allowed us to understand how different segments of the gluconeogenesis pathway are coordinated and controlled by hormones. As illustrated in Figure 9-15, high Fru 1,6-P$_2$ from active PFK-1 promotes dephosphorylation of L-PK, whereas low Fru 1,6-P$_2$ promotes phosphorylation and inhibition of L-PK. Thus, the inhibition of pyruvate kinase and the activation of fructose- 1,6-bisphosphatase allows coordination of both ends of the gluconeogenic pathway.

The actions of insulin on inhibiting gluconeo-genesis and the antagonistic effects of glucagon and the catecholamines are understood in terms of their effects on the activities of L-PK and PFK-2 and the level of Fru 2,6-P$_2$ (Table 9-3). Thus, insulin lowers blood glucose *both* by increasing peripheral uptake (in muscle and adipose tissue) and by decreasing hepatic glucose production (HGP). In contrast, norepinephrine released from sympathetic nerves to the liver and delivered from the circulation raises free Ca^{2+} and increases Ca^{2+}-protein kinase activity. By this means norepinephrine stimulates HGP. In addition, glucagon delivered from the portal and systemic circulations, and epinephrine delivered from the systemic circulation, raise cAMP and cAMP-protein kinase activities. These counter-regulatory hormones increase HGP through gluco-neogenesis as well as glycogenolysis.

Insulin, Glucagon, Norepinephrine, and Epinephrine: Redundant Controls of Glycemia

Considering the importance of maintaining blood glucose concentration at a reasonable level, it is not surprising that nature has yielded a system with redundant controls. By *redundancy,* we mean that if one aspect of a system fails, other aspects can compensate to accomplish the same task. To the scientist, the complexity of control makes understanding the regulation of glycemia extremely difficult. However, to the athlete on variable training and competitive schedules, engaged in both short-term and prolonged exercise, as well as having a daily life, the beauty of the control mechanisms is that they allow good regulation over a wide range of circumstances. Nevertheless, if pressed, we may be able to develop a list of the hierarchy of hormones.

Of the glucoregulatory hormones in resting people, insulin is most important because without it, Type I diabetes results, or if we lack the capability to respond to it, Type II diabetes is the consequence. In terms of the antagonistic effects of insulin and glucagon, possibly the best way to conceive of the differing effects is to focus on the insulin to glucagon ratio (i.e., the I/G). A high I/G tends to lower blood glucose, whereas a low I/G tends to raise it. Again, glucagon is especially important as it stimulates hepatic glycogenolysis as well as gluconeogenesis. From the work of Alan Cherrington, David Wasserman, and associates at Vanderbilt (1989) on dogs, we know that the portal I/G has profound effects on HGP.

At exercise onset, feed-forward regulation involves norepinephrine and epinephrine (see Figure 9-2). Distinguishing between the effects of epinephrine and norepinephrine is difficult because they tend to rise in a coordinated manner. However, in studies of men exercising on Pikes Peak in Colorado, Brooks and associates (1991, 1991, 1992) observed that altitude and altitude acclimatization affected epinephrine and norepinephrine responses differently. In their experiments, blood glucose appearance rate (R_a, an estimate of HGP) corresponded better to norepinephrine than to epinephrine. Subsequently, in studies on dogs, Miles, Vranic, and associates (1992) observed catecholamines and found the catecholamine to insulin ratio to be an important regulatory factor.

Due to control redundancy, our placement of norepinephrine high in the hierarchy of glucoregulatory control is complicated by results of experiments on rats and dogs involving the cutting of hepatic nerves, as well as by the recent experiments of Kjaer and associates (1987, 1991) on men with pharmacological blockade of the abdominal nerves. Despite the blocking of hepatic nerve traffic, very good control of blood glucose concentration during exercise was observed. Again, these experiments point to a role for circulating norepinephrine as well as to redundancy of control.

When norepinephrine rises, epinephrine inevitably rises also. Besides stimulating glycogenolysis to a limited extent, epinephrine also suppresses insulin secretion, and therefore during exercise insulin falls (see Figure 9-2), even if glucose rises. The fall in insulin increases the effects of glucagon by changing the I/G, even if glucagon remains constant.

In extremely elegant studies on rats, in which the adrenal glands were surgically removed, Winder and associates (1988) observed good control of blood glucose concentration during exercise. Although these adrenaldemedulation studies show that epinephrine is not essential to glucoregulation during exercise, the results do not preclude the importance of epinephrine as part of a redundant system of controls.

Amylin—The Missing Ingredient?

Part of the difficulty in unraveling the various roles of insulin and the recognized counterregulatory hormones is that there may be another important hormone that has gone unrecognized. One candidate for the role of missing glucoregulatory hormone is amylin.

Amylin is a small, 37–amino acid peptide hormone that, like insulin, is secreted by the β cells of the pancreas. At present, little is known about amylin function in humans during exercise, but Young and associates (1991, 1993) believe that amylin complements insulin by antagonizing aspects of insulin action. Whereas insulin stimulates muscle glucose uptake, amylin stimulates glycogenolysis and lactate production in muscle. Thus, in parallel with epinephrine, amylin may support the Cori cycle and hepatic gluconeogenesis. Researchers are hopeful that work in progress will soon clarify the role, if any, of amylin.

Studies on the role of amylin in regulating substrate supply during exercise are in their infancy, but pioneering work has been conducted by John Ivy of the University of Texas at Austin and associates (Castle et al., 1998, 1998) on rat muscle preparations. Their work indicates that amylin inhibits muscle glucose uptake while at the same time stimulating glycogenolysis and lactate release. To date, it has not been possible to demonstrate in exercising humans relationships between the amylin level, blood glucose and lactate appearance rates, lactate conversion to glucose, or glucose recycling rate (a measure of Cori cycle activity) (Colberg et al., 1994; Huie et al., 1996).

■ Can Muscle Make Glucose?

As already mentioned, results of recent studies indicate that white mammalian skeletal muscle can convert lactate to glycogen. The proper term for this process is *glyconeogenesis,* meaning the synthesis of new glycogen. The means by which muscle can reverse some of the so-called irreversible steps in glycolysis (e.g., the conversion of phosphoenolpyruvate to pyruvate by the enzyme pyruvate kinase, Figure 5-7) without the bypass enzyme pyruvate carboxylase (see Figure 9-14) is now a subject of intense investigation. (Can malic enzyme convert pyruvate to malate in white muscle, perhaps?) However, regardless of whether muscle can reverse glycolysis, the process is not likely to occur during exercise, but rather early in recovery, when intramuscular lactate levels are high. Moreover, muscle lacks the enzyme glucose 6-phosphatase (see Figure 9-14), and so, although the reversal of glycolysis can potentially synthesize glycogen, muscle is unable to produce free glucose for export as the result of gluconeogenesis.

Although muscle is not supposed to make free glucose, and although glucose production as the result of gluconeogenesis is unlikely during exercise, there are reports, based on arteriovenous difference $[(a-v)]$ measurements, of net glucose release from muscle. For this reason, François Peronnet and other researchers have attempted to determine whether some other mechanism, such as epinephrine stimu-

lation of muscle glycogenolysis, which occurs in maximal exercise, can yield free glucose in quantities sufficient to spill into the circulation. Sometimes, muscle does appear to release small amounts of glucose at the start of exercise (see negative values in Figure 5-26a), but such observations are inconsistent.

In Figure 5-16, we illustrated that although glycogen is composed mostly of 1–4 linkages, 1–6 linkages also exist, and these latter form the branches in the glycogen structure. Hydrolysis of these 1–6 linkages is catalyzed by debranching enzyme, which is apparently controlled much like phosphorylase, which splits the 1–4 linkages. Some side branches in the glycogen molecule are short, so debranching enzyme could potentially yield free glucose. Whether muscle glycogenolysis can raise blood glucose concentration during maximal exercise has yet to be determined, but data such as those in Figure 9-12 require further examination. Because the rise in circulating glucose concentration may exceed the assumed capability of the liver, the rapid doubling of arterial glucose concentration may be due in part to epinephrine-stimulated net glucose release, and in part to hepatic glucose release.

■ Growth Hormone Response to Continuous and Intermittent Exercise

The integration of the neural and endocrine systems is best illustrated by describing the release of growth hormone from the anterior pituitary gland. The neural-endocrine integration is illustrated in Figure 9-16. The hypothalamus receives neural inputs and is sensitive to blood metabolite levels (e.g., glucose levels). In response to neural and blood-borne input, the hypothalamus synthesizes chemical factors that either inhibit or stimulate the synthesis and release of anterior pituitary hormones. One of the anterior pituitary hormones is growth hormone (GH).

Growth hormone, a polypeptide molecule released from the anterior pituitary, stimulates protein synthesis, especially in the young. In young as well as older individuals, GH is one of the major

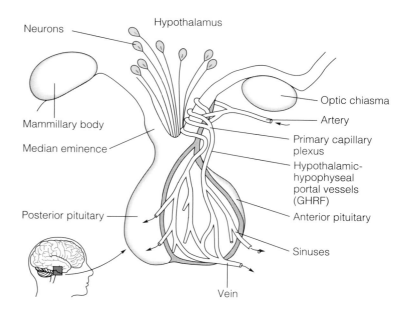

Figure 9-16 The hypothalamus of the brain and the pituitary hypophysis are connected by a series of neurons and a blood (portal) system. Modified from Guyton, 1976. Used with permission.

lipolytic hormones. Therefore, GH directly stimulates fat metabolism and indirectly suppresses carbohydrate metabolism. GH stimulates lipolysis in adipose tissue through synthesis of enzymatic or protein factors that stimulate lipase activity.

In a fasting individual, low blood glucose levels stimulate the release of growth hormone–releasing factor (GHRF) from the hypothalamus. The factors responsible for GH release during exercise are less well understood. At different times, low blood glucose, high blood lactate, low blood pH, and elevated body temperature have been thought to be responsible for GH secretion during exercise. Through various experimental manipulations, however, scientists have gradually eliminated these factors in the direct regulation of GH release during exercise.

In Figure 9-17, the GH response to continuous exercise at about 50% of $\dot{V}_{O_2max}$ is contrasted with the GH response to intermittent exercise (1 min exercise, 1 min rest), where the exercise power output was twice as great as in continuous exercise. In Figure 9-17a, a lag of approximately 15 minutes is indicated between exercise onset and accelerated GH release. Also, because the activity of GH on adipose lipolysis

is indirect, GH at best has a delayed effect on the release of FFAs from adipose tissue during exercise. In Figure 9-17b, the blood lactate responses to continuous and intermittent exercise bouts are graphed. Although blood lactate was several times higher in intermittent exercise, GH levels were not significantly affected. Further, although intermittent exercise resulted in great differences in blood alanine and pyruvate levels, these also did not have a significant effect on GH release. Blood glucose level was higher during intermittent (Figure 9-17c) than during continuous exercise, so glucose as well could not have been an important regulatory factor for GH release during exercise. Furthermore, body temperature did not correlate with GH level. Therefore, at present, neural factors are implicated as exerting primary control over GH secretion during exercise.

■ Cortisol and the Pituitary— Adrenal Axis

The steroid hormone cortisol assists in maintaining blood glucose homeostasis by stimulating amino acid release from muscle, by stimulating hepatic

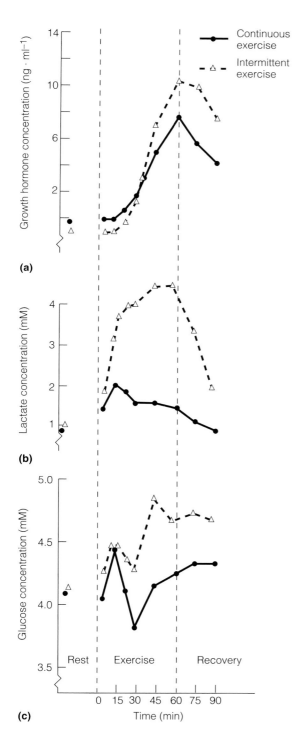

(a)

(b)

(c)

Figure 9-17 After about a 15-min lag, human growth hormone is released in response to both continuous and intermittent exercise bouts (a). In these experiments, growth hormone levels do not correlate with levels of blood lactate (b) or blood glucose (c). Modified from Karagiorgos et al., 1979. Used with permission.

gluconeogenesis from amino acids, and by helping mobilize FFAs from adipose tissue.

The mechanism of cortisol secretion is illustrated in Figure 9-18. Stress (either physical or emotional) or declining blood glucose levels stimulate the hypothalamus to secrete corticotropin-releasing factor (CRF). In turn, CRF stimulates the anterior pituitary to release adrenocorticotropin (ACTH), which causes the adrenal cortex to release cortisol into the circulation. Cortisol has a number of effects, including providing negative feedback on its own secretion.

Of the four corticosteroids released by the adre-

nal gland, two (cortisol and cortisone) stimulate glucose formation and are termed *glucocorticoids.* Two other corticosteroids (aldosterone and deoxycorticosterone) are important in electrolyte metabolism and are termed *mineralcorticoids.* Aldosterone is of primary importance for sodium resorption in the kidneys and ultimately for fluid and electrolyte balance. This will be discussed in Chapter 22. Cortisol and aldosterone are the major corticosteroids; cortisone and deoxycorticosterone are released in lesser amounts.

The action of ACTH is to stimulate adenylate cyclase activity in adrenocortical cells. The resulting

Figure 9-18 In response to stress, cortisol secretion is stimulated by the adrenal cortex through a series of events initiated in the hypothalamus. Modified from Guyton, 1976. Used with permission.

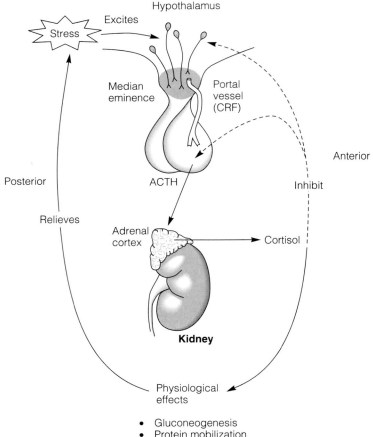

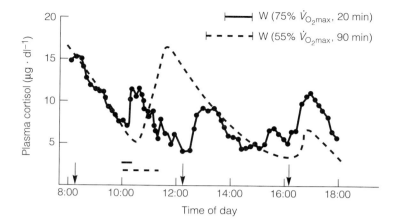

Figure 9-19 Blood cortisol level follows a daily pattern that is affected by circadian rhythms, eating, and exercise. Cortisol levels are high during the early morning but fall after breakfast (first arrow). After a lag when exercise starts, cortisol levels increase at a rate proportional to the exercise intensity, but reach a final level dependent on the duration of exercise. Later in the day, cortisol levels are again low, but they rise again after lunch and supper (second and third arrows, respectively). W = Work. Modified from Brandenberger and Follenius, 1975.

formation of cAMP stimulates the secretion of cortisol as well as of other corticosteroids. The particular steroid secreted appears to depend on the particular action of the adrenocortical cell stimulated.

Role of Cortisol in Prolonged Exercise and During Recovery from Exhausting Exercise

During starvation or during prolonged, hard exercise, ACTH is secreted in response to the level of stress and to falling blood glucose levels. In turn, ACTH stimulates cortisol release, which stimulates proteolysis in muscle (Figure 9-19). Some of the resulting amino acids formed in muscle are released directly into venous blood, and some are channeled through alanine and released into venous blood. During the period of recovery from exhausting exercise, the return to normal blood glucose levels is a primary physiological concern. Gaesser and Brooks (1980) and Fell and coworkers (1980) have observed that normalization of blood glucose levels and restitution of cardiac glycogen is possible in exercise-exhausted animals that were subsequently starved. The source of the glucosyl units is thought to be muscle protein. Cortisol (through the mobilization of muscle proteins) and glucagon (through the stimulation of gluconeogenesis from amino acids) are thought to affect glucose and glycogen replenishment in recovery.

■ The Permissive Action of Thyroid Hormone

The thyroid gland secretes the amino acid–iodine-bound hormones thyroxine (T_4) and triiodothyronine (T_3). Of these, thyroxine is released in the greatest amount, while triiodothyronine has the greatest activity. Under the influence of thyroid-stimulating hormone (TSH) released from the pituitary, the thyroid secretes and releases thyroxine. Most of this is bound to plasma proteins; the bound T_4 is in equilibrium with free T_4, which is available to interact with target tissues.

Most cells in the body are targets for T_4. Target cells apparently metabolize T_4 to T_3, which is the active form of the hormone. The thyroid itself releases some T_3, but the circulating levels of T_3 cannot be accounted for by thyroid secretion alone. Thyroxine and triiodothyronine generally stimulate metabolism and increase the rate of such processes as oxygen consumption, protein synthesis, glycogenolysis, and lipolysis. Thyroid hormones appear to act indirectly to promote the general increase in metabolic rate by enhancing the effects of other hormones. This enhancement, or potentiating, of other hormonal actions is termed *permissiveness*. Thyroid hormones have the effect of raising cellular cAMP levels, and therefore their action (via T_3) may be to stimulate adenylate cyclase. In this manner, the

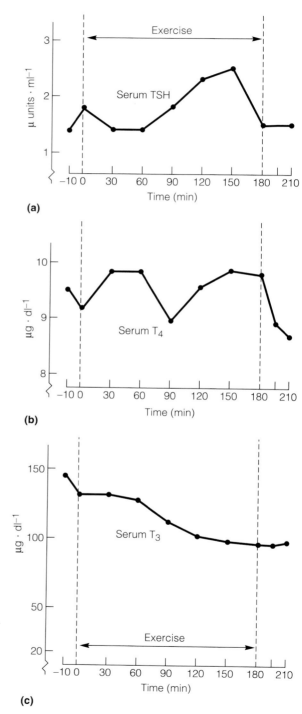

Figure 9-20 During prolonged submaximal exercise, changes in circulating levels of pituitary–thyroid hormones are small in comparison to the changes in glucoregulatory hormones (e.g., insulin and glucagon). The level of TSH tends to increase, whereas T_4 remains constant but then falls during recovery. The level of T_3 falls continuously during exercise. Increased T_4 and T_3 utilization during exercise is matched by thyroid hormone release during exercise. Modified from Berchtold et al., 1978.

effects of hormones that act through cAMP would be amplified.

Compared to the glycoregulatory hormones (insulin, glucagon, epinephrine, and cortisol), the levels of thyroid hormones and ACTH do not change much during exercise or as the result of training. This is because the secretion of these hormones is linked to their utilization during exercise (Figure 9-20). In the case of the thyroid hormones, measurements of circulating levels do not give a clear indication of their increased utilization in active individuals. Winder and associates (1988) have clearly shown the turnover of thyroxine to be increased as a result of physical activity.

■ Posterior Pituitary and ADH Secretion

Antidiuretic hormone (ADH) functions together with aldosterone to maintain fluid and electrolyte balance. ADH is an amino acid protein hormone released from the posterior pituitary. (See Figure 9-16.) Two types of stimuli appear to stimulate ADH secretion: osmolality and arterial pressure.

Particular hypothalamic nuclei are sensitive to osmolality (electrolyte concentration) in arterial blood. When severe sweating causes an increase in blood osmolality (sweat is more dilute than plasma, so sweating results in a concentration of blood constituents), the hypothalamic supraoptic nuclei transmit action potentials to the posterior pituitary (neural hypophysis). These signals result in ADH secretion. In response to a fluid overload and hemodilution, the supraoptic nuclei decrease in their activity, and ADH release is suppressed. In the kidneys, ADH allows the reabsorption of H_2O after glomerular filtration. Consequently, sweating stimulates ADH secretion, which stimulates renal H_2O retention and a decrease in urine volume. In response to a fluid overload, ADH secretion is suppressed and urine volume increases.

The second mechanism of ADH secretion involves pressure receptors in the left atrium and other vascular baroreceptors. When dehydration results in a fall in blood pressure, the vascular baroreceptors develop action potentials that are trans-

mitted to the hypothalamus via afferent neurons. The end result is increased stimulation of the posterior pituitary and ADH secretion. Thus, a fall in blood pressure results in a decrease in urinary output, which helps to minimize further dehydration.

During exercise where significant sweating is involved and hemoconcentration is a usual result, levels of ADH and aldosterone increase considerably. In the case of exercise, increased plasma osmolality is probably the major factor stimulating ADH secretion.

■ Summary

The maintenance of "normal" blood glucose levels and the provision of a supply of fuels that are alternatives to glucose depend on the integrated functioning of several systems. This integration is affected by transmitter substances released from the sympathetic nervous system and the endocrine hormonal system. Hormones have powerful effects on metabolism even if they are present in only minute amounts. Table 9-4 summarizes the effects of the various hormones.

During prolonged submaximal exercise, the increased demand for glucose by contracting muscle tends to cause a fall in blood glucose level. This fall is prevented, or compensated for, by increases in catecholamines and glucagon and a decrease in the circulating insulin level. Approximately constant rates of glucose uptake by nonworking muscle allow most of the hepatic glucose production to be shunted to working muscle and other tissues with an essential requirement for glucose (e.g., brain).

Glucose homeostasis is also maintained by the stimulation of liver glycogen breakdown (glycogenolysis) and glucose release, and by the stimulation of the synthesis of glucose from other substances (gluconeogenesis).

Both feed-forward and feedback mechanisms allow glycemia to be controlled during exercise. At the start of exercise, and during moderate to hard intensity exercise, feed-forward mechanisms operate to maintain or increase circulating blood glucose. However, if blood glucose concentration falls

TABLE 9-4

Summary of Metabolic Effects of Hormones

Metabolic Effect	Hormone(s)
Cellular glucose uptake	Insulin
Glycolysis and glycogen synthesis	Insulin
Triglyceride synthesis	Insulin
Decrease in blood glucose level	Insulin
Liver glycogenolysis	Epinephrine (nonspecific) Glucagon (specific) Norepinephrine (?)[a]
Liver gluconeogenesis	Glucagon
Muscle glycogenolysis	Epinephrine Norepinephrine (?)
Lipolysis	Cortisol Epinephrine Growth hormone
Protein synthesis	Growth hormone Insulin
Protein catabolism	Cortisol
Increase in blood glucose level	Epinephrine
Direct effect	Glucagon
Indirect effect	Cortisol Epinephrine Glucagon
Increased metabolic rate	Epinephrine Norepinephrine Thyroxine

[a] Question mark indicates that specificity of effect is unknown.

which stimulates proteolysis in muscle and the release from muscle of amino acids (e.g., alanine) that serve as gluconeogenic precursors.

By stimulating the release of fatty acids from adipose tissue into the circulation, epinephrine and growth hormone provide alternative fuels to blood glucose and muscle glycogen. Because glucose and glycogen are necessary for continued functioning during prolonged, hard exercise, the "sparing" of glucose and glycogen use by increasing FFA use postpones the time when glucose and glycogen stores become critical. Epinephrine has a rapid effect after exercise onset, stimulating lipolysis in adipose tissue, whereas growth hormone becomes increasingly more important for maintaining lipid mobilization as exercise duration progresses beyond 15 minutes.

Just as the hormonal response is geared to maintain a "normal" level of blood glucose during exercise, training is also geared to preserve glucose homeostasis during exercise. Given exercise bouts of absolute or relative intensity, trained individuals experience less metabolic stress and have dampened endocrine responses. In trained individuals, blood insulin level does not fall as much, and catecholamines, growth hormone, and glucagon do not rise as much as in untrained individuals. However, during maximal exercise the hormonal (e.g., catecholamine) responses in trained individuals are the same or greater than in untrained individuals.

during prolonged submaximal exercise, then powerful feedback responses are elicited to maintain blood glucose. The control of glycemia during exercise is a multicomponent and highly redundant system, with several hormones (catecholamines, glucagon, cortisol, and possibly amylin) antagonizing the actions of insulin.

The hormone glucagon (which is synthesized in the pancreas, as are insulin and amylin) is important in the control of hepatic glycogenolysis and gluconeogenesis. The gluconeogenic function of glucagon is helped by the steroid hormone cortisol,

■ Selected Readings

Ahlborg, G., P. Felig, L. Hagenfeldt, and J. Wahren. Substrate turnover during prolonged exercise in man. *J. Clin. Invest.* 53: 1080–1090, 1974.

Benkovic, S. J., and M. M. deMaine. Hepatic fructose 1,6-bisphosphatases. *Adv. Enzymol. Relat. Areas Mol. Biol.* 53: 45, 1983.

Berchtold, P., M. Berger, H. J. Cüppers, J. Herrmann, E. Nieschlag, K. Rudorff, H. Zimmerman, and H. L. Kruskemper. Non-gluco regulatory hormones (T4, T3, vT3, TSH, and testosterone) during physical exercise in juvenile type diabetics. *Horm. Metab. Res.* 10: 269–273, 1978.

Bergman, B. C., G. E. Butterfield, E. E. Wolfel, G. Lopaschuk, G. A. Casazza, M. A. Horning, and G. A. Brooks. Net glucose uptake and glucose kinetics after endurance training in men. *Am. J. Physiol.* 277 (*Endocrinol. Metab.* 40): E81–E92, 1999.

Bradford, M. M. A rapid and sensitive method for the quantitation of mg quantities of protein utilizing the principle of protein dye binding. *Anal. Biochem.* 72: 248–254, 1976.

Brandenberger, G., and M. Follenius. Influence of timing and intensity of muscular exercise on temporal patterns of plasma cortisol levels. *I. Clin. Endocrinol. Metab.* 40: 845–849, 1975.

Brooks, G. A. Lactate: glycolytic end product and oxidative substrate during sustained exercise in mammals—the "lactate shuttle." In Proceedings of the First International Congress of Comparative Physiology and Biochemistry. Berlin: Springer-Verlag, 1985, pp. 208–218.

Brooks, G. A. Lactate production under fully aerobic conditions: the Lactate Shuttle during rest and exercise. *Fed. Proc.* 45: 2924–2929, 1986.

Brooks, G. A., G. E. Butterfield, R. R. Wolfe, B. M. Groves, R. S. Mazzeo, J. R. Sutton, E. E. Wolfel, and J. T. Reeves. Increased dependence on blood glucose after acclimatization to 4300 m. *J. Appl. Physiol.* 70: 919–927, 1991.

Brooks, G. A., G. E. Butterfield, R. R. Wolfe, B. M. Groves, R. S. Mazzeo, J. R. Sutton, E. E. Wolfel, and J. T. Reeves. Decreased reliance on lactate during exercise after acclimatization to 4300 m. *J. Appl. Physiol.* 71: 333–341, 1991.

Brooks, G. A., and T. D. Fahey. Exercise Physiology: Human Bioenergetics and Its Applications, New York: Wiley, 1984, pp. 90–92.

Brooks, G. A., and G. A. Gaesser. End points of lactate and glucose metabolism after exhausting exercise. *J. Appl. Physiol.: Respirat. Environ. Exercise Physiol.* 49: 1057–1069, 1980.

Brooks, G. A., E. E. Wolfel, B. M. Groves, P. R. Bender, G. E. Butterfield, A. Cymerman, R. S. Mazzeo, J. R. Sutton, R. R. Wolfe, and J. T. Reeves. Muscle accounts for glucose disposal but not blood lactate appearance during exercise after acclimatization to 4300 m. *J. Appl. Physiol.* 72: 2435–2445, 1992.

Butterfield, G. E., J. Gates, G. A. Brooks, J. R. Sutton, and J. T. Reeves. Energy balance in men during three weeks at 4300 m. *J. Appl. Physiol.* 72: 1741–1748, 1992.

Cain, D. F., and R. E. Davies. Breakdown of adenosine triphosphate during a single contraction of working muscle. *Biochem. Biophys. Res. Com.* 8: 361–366, 1962.

Castle, A. L., C. H. Kuo, D. H. Han, and J. L. Ivy. Amylin-mediataed inhibition of insulin-stimulated glucose transport in skeletal muscle. *Am. J. Physiol.* 275: E531–E536, 1998.

Castle, A. L., C. H. Kuo, and J. L. Ivy. Amylin influences insulin-stimulated glucose metabolism by two independent mechanisms. *Am. J. Physiol.* 274: E6–E12, 1998.

Colberg, S. R., G. A. Casazza, M. A. Horning, and G. A. Brooks. Increased dependence on blood glucose in smokers during rest and exercise. *J. Appl. Physiol.* 76: 26–32, 1994.

Cushman, S. W., and L. J. Wardzala. Potential mechanism of insulin action on glucose transport in the isolated adipose cell. Apparent translocation of intracellular transport systems to the plasma membrane. *J. Biol. Chem.* 255: 4558–4762, 1980.

Davies, K. J. A., L. Packer, and G. A. Brooks. Biochemical adaptation of mitochondria, muscle and whole-animal respiration to endurance training. *Arch. Biochem. Biophys.* 209: 539–554, 1981.

Deuticke, B. Monocarboxylate transport in erythrocytes. *J. Membr. Biol.* 70: 89–103, 1982.

Donovan, C. M., and G. A. Brooks. Endurance training affects lactate clearance, not lactate production. *Am. J. Physiol.* 244 (*Endocrinol. Metab.* 7): E83–E92, 1983.

Donovan, C. M., and M. J. Pagliassotti. Endurance training enhances lactate clearance during hyperlactatemia. *Am. J. Physiol.* 257: E782–E789, 1989.

Donovan, C. M., and K. D. Sumida. Training improves glucose homeostasis in rats during exercise via glucose production. *Am. J. Physiol.* 258 (*Regulatory Integrative Comp. Physiol.* 27): R770–R776, 1990.

Dudley, G. A., P. C. Tullson, and R. C. Terjung. Influence of mitochondrial content on the sensitivity of respiratory control. *J. Biol. Chem.* 262: 9109–9114, 1987.

Embden, G., and H. Lawaczeck. Über den zeitlichen Verlauf der Milchsäurebildung bei der Muskel Kontraktion. *Z. Physiol. Chem.* 170: 311–315, 1928.

Fell, R., J. McLane, W. Winder, and J. Holloszy. Preferential resynthesis of muscle glycogen in fasting rats after exhausting exercise. *Am. J. Physiol.* 238 (*Regulatory Integrative Comp. Physiol.* 7): R328–R332, 1980.

Friedman, J. E., R. W. Dudek, D. S. Whitehead, D. L. Downes, W. R. Frisell, J. F. Caro, and G. L. Dohm. Immunolocalization of glucose transporter GLUT4 within human skeletal muscle. *Diabetes* 40: 150–154, 1991.

Friedlander, A. L., G. A. Casazza, M. A. Horning, M. J. Huie, M.-F. Piacentini, J. K. Trimmer, and G. A. Brooks. Training-induced alterations of glucose flux in young women: Gender differences in carbohydrate oxidation. *J. Appl. Physiol.* 85: 1175–1186, 1998.

Friedlander, A. L., G. A. Casazza, M. A. Horning, T. F. Budinger, and G. A. Brooks. Effects of exercise intensity and training on lipid metabolism in young women. *Am. J. Physiol.* 275 (*Endocrinol. Metab.* 38) E853–E863, 1998.

Gaesser, G. A., and G. A. Brooks. Glycogen repletion following continuous and intermittent exercise to exhaustion. *J. Appl. Physiol.: Respirat. Environ. Exercise Physiol.* 49: 722–728, 1980.

Gertz, E. W., J. A. Wisneski, W. C. Stanley, and R. A. Neese. Myocardial substrate utilization during exercise in humans: dual carbon-labeled carbohydrate isotope experiments. *J. Clin. Invest.* 82: 2017–2025, 1988.

Gleeson, T. T., and K. M. Baldwin. Cardiovascular response to treadmill exercise in untrained rats. *J. Appl. Physiol.: Resp. Environ. Exercise Physiol.* 50: 1206–1211, 1981.

Goodyear, L. J., M. F. Hirshman, P. M. Valyou, and E. S. Horton. Glucose transporter number, function, and subcellular distribution in rat skeletal muscle after exercise training. *Diabetes* 41: 1091–1099, 1992.

Guyton, A. C. Textbook of Medical Physiology. Philadelphia: W. B. Saunders, 1976.

Gyntelberg, F., M. J. Rennie, R. C. Hickson, and J. O. Holloszy. Effect of training on the response of plasma glucagon to exercise. *J. Appl. Physiol.: Respirat. Environ. Exercise Physiol.* 43: 302–305, 1977.

Hall, S. E. H., J. T. Bratten, T. Bulton, M. Vranic, and I. Thoden. Substrate utilization during normal and loading diet treadmill marathons. In Biochemistry of Exercise, H. G. Knuttgen, J. A. Vogel, and J. R. Poortmans (Eds.). Champaign, Ill.: Human Kinetics, 1983, pp. 536–542.

Hill, A. V. The oxidative removal of lactic acid. *J. Physiol.* 48: x–xi, 1914.

Hill, A. V., and P. Kupalov. Anaerobic and aerobic activity in isolated muscle. *Proc. Roy. Soc.* (London). Ser. B. 105: 313–322, 1929.

Houmard, J. A., P. C. Egan, P. D. Neufer, J. E. Friedman, W. S. Wheeler, R. G. Israel, and G. L. Dohm. Elevated skeletal muscle glucose transporter levels in exercise-trained middle-aged men. *Am. J. Physiol.* 261 (*Endocrinol. Metab.* 24): E437–E443, 1991.

Houmard, J. A., M. H. Shinebarger, P. L. Dolan, N. Leggett-Frazier, R. K. Bruner, M. R. McCammon, R. G. Israel, and G. L. Dohm. Exercise training increases GLUT-4 protein concentration in previously sedentary middle-aged men. *Am. J. Physiol.* 264 (*Endocrinol. Metab.* 27): E896–E901, 1993.

Huie, M. J., G. A. Casazza, M. A. Horning, and G. A. Brooks. Smoking increases conversion of lactate to glucose. *J. Appl. Physiol.* 80: 1554–1559, 1996.

John-Alder, H. B., R. M. McAlister, and R. L. Terjung. Reduced running endurance in gluconeogenesis-inhibited rats. *Am. J. Physiol.* 251: R137–R142, 1986.

Jones, W. L., G. J. F. Heigenhauser, A. Kuksis, C. G. Matos, J. R. Sutton, and C. J. Toews. Fat metabolism in heavy exercise. *Clinical Science* 59: 469–478, 1980.

Joost, H. G., and T. M. Weber. The regulation of glucose transport in insulin-sensitive cells. *Diabetologia* 32: 831–838, 1989.

Juel, C. Intracellular pH recovery and lactate efflux in mouse soleus muscle stimulated in vitro: the involvement of sodium/protein exchange and a lactate carrier. *Acta. Physiol. Scand.* 132: 363–371, 1988.

Karagiorgos, A., J. F. Garcia, and G. A. Brooks. Growth hormone response to continuous and intermittent exercise. *Med. Sci. Sports* 11: 302–307, 1979.

Kern, M., J. A. Wells, J. M. Stephens, C. W. Elton, J. E. Friedman, E. B. Tapscott, P. H. Pekala, and G. L. Dohm. Insulin responsiveness in skeletal muscle is determined by glucose transporter (GLUT4) protein level. *Biochem. J.* 270: 397–400, 1990.

Keul, V. J., M. Lehmann, and K. Wybitul. Zur siung von burnitrolol auf hertzfrequenz, metabolishe Grössen bei Korperarbeit und Leistungsverhalten. *Arzneim.-Gorsch/Drug Res.* 31: 1–16, 1981.

Kjaer, M., P. A. Farrell, N. J. Christensen, and H. Galbo. Increased epinephrine response and inaccurate glucoregulation in exercising athletes. *J. Appl. Physiol.* 61(5): 1693–1700, 1986.

Kjaer, M., and H. Galbo. Effect of physical training on the capacity to secrete epinephrine. *J. Appl. Physiol.* 64(1): 11–16, 1988.

Kjaer, M., B. Kiens, M. Hargreaves, and E. A. Richter. Influence of active muscle mass on glucose homeostasis during exercise in humans. *J. Appl. Physiol.* 71(2): 552–557, 1991.

Kjaer, M., N. H. Secher, F. W. Bach, and H. Galbo. Role of motor center activity for hormonal changes and substrate mobilization in humans. *Am. J. Physiol.* 253 (*Regulatory Integrative Comp. Physiol.* 22): R687–R695, 1987.

Kjaer, M., N. H. Secher, F. W. Bach, H. Galbo, C. R. Reeves, Jr., and J. H. Mitchell. Hormonal, metabolic, and cardiovascular responses to static exercise in humans: influence of epidural anesthesia. *Am. J. Physiol.* 261 (*Endocrinol. Metab.* 24): E214–E220, 1991.

Lavoie, J. M., J. Bongbélé, S. Cardin, M. Bélisle, J. Terrettaz, and G. Van de Werve. Increased insulin suppression of plasma free fatty acid concentration in exercise-trained rats. *J. Appl. Phys.* 74: 293–296, 1993.

Lavoie, J. M., M. Lord, and A. Paulin. Effect of selective hepatic vagotomy on plasma FFA levels in resting and exercising rats. *Am. J. Physiol.* 254 *(Regulatory Integrative Comp. Physiol.* 23): R602–R606, 1988.

Lehninger, A. L. Biochemistry. New York: Worth, 1970, pp. 313–327, 488.

Lehninger, A. L. Bioenergetics. Menlo Park, Ca.: W. A. Benjamin, 1978, pp. 38–51.

Lipmann, F. Metabolic generation and utilization of phosphate bond energy. *Advan. Enzymol.* 1: 99–162, 1941.

Lundsgaard, E. Phosphagen– and PyrophoPhatumsatz in Iodessigsäure-vegifteten Muskeln. *Biochem. Z.* 269: 308–328, 1934.

MacRae, H. S-H., S. C. Dennis, A. N. Bosch, and T. D. Noakes. Effects of training on lactate production and removal during progressive exercise in humans. *J. Appl. Physiol.* 72: 1649–1656, 1992.

Merrill, G. F., E. J. Zambraski, and S. M. Grassl. Effect of dichloroacetate on plasma lactic acid in exercising dogs. *J. Appl. Physiol.* 48: 427–431, 1980.

Meyerhof, O. Die Energie-wandlungen in Muskel. I. Über die Beziehungen der Milchsäure Wärmebildung und Arbeitstung des Muskels in der Anaerobiose. *Arch. Ges. Physiol.* 182: 232–282, 1920.

Meyerhof, O. Die Energie-wandlungen in Muskel. II. Das Schicksal der Milchsäure in der Erholungs-periode des Muskels. *Arch. Ges. Physiol.* 182: 284–317, 1920.

Miles, P. D. G., D. T. Finegood, H. Lavina, A. Lickley, and M. Vranic. Regulation of glucose turnover at the onset of exercise in the dog. *J. Appl. Physiol.* 72(6): 2487–2494, 1992.

Mommarts, W. F. H. M. Energetics of muscle contraction. *Physiol. Dev.* 49: 427–508, 1969.

Pilkis, S. J., and T. H. Claus. Hepatic gluconeogenesis/glycolysis: regulation and structure/function relationships of substrate cycle enzymes. *Ann. Rev. Nutr.* 11: 465–515, 1991.

Richter, E. A., N. B. Ruderman, H. Gavras, E. R. Belur, and H. Galbo. Muscle glycogenolysis during exercise: dual control by epinephrine and contractions. *Am. J. Physiol.* 242 *(Endocrinol. Metab.* 5): E25–E32, 1982.

Roth, D. A., and G. A. Brooks. Lactate transport is mediated by a membrane-borne carrier in rat skeletal muscle sarcolemmal vesicles. *Arch. of Biochem. Biophys.* 279: 377–385, 1990.

Searle, G. L., K. R. Feingold, F. S. F. Hsu, O. H. Clark, E. W. Gertz, and W. C. Stanley. Inhibition of endogenous lactate turnover with lactate infusion in humans. *Metabolism* 38: 1120–1123, 1989.

Spriet, L., J. M. Ren, and E. Hultman. Epinephrine infusion enhances muscle glycogenolysis during prolonged electrical stimulation. *J. Appl. Physiol.* 64(4): 1439–1444, 1988.

Stanley, W. C., E. W. Gertz, J. A. Wisneski, D. L. Morris, R. Neese, and G. A. Brooks. Lactate metabolism in exercising human skeletal muscle: evidence for lactate extraction during net lactate release. *J. Appl. Physiol.* 60: 1116–1120, 1986.

Stanley, W. C., J. A. Wisneski, E. W. Gertz, R. A. Neese, and G. A. Brooks. Glucose and lactate interrelations during moderate-intensity exercise in humans. *Metabolism* 37(9): 850–858, 1988.

Suzuki, K., and T. Kono. Evidence that insulin causes translocation of glucose transport activity to the plasma membrane from an intracellular storage site. *Proc. Natl. Acad. Sci. USA.* 77: 2542–2545, 1980.

Szent-Györgi, A. Myosin and Muscular Contraction. Basel: Karger, 1942.

Szent-Györgi, A. Chemistry of Muscular Contraction. New York: Academic Press, 1951.

Turcotte, L. P., and G. A. Brooks. Effects of training on glucose metabolism of gluconeogenesis-inhibited short-term-fasted rats. *J. Appl. Physiol.* 68(3): 944–954, 1990.

Turcotte, L. P., A. S. Rovner, R. R. Roark, and G. A. Brooks. Glucose kinetics in gluconeogenesis-inhibited rats during rest and exercise. *Am. J. Physiol.* 258 *(Endocrinol. Metab.* 21): E203–E211, 1990.

Wahren, J., P. Felig, G. Ahlborg, and L. Jorfeldt. Glucose metabolism during leg exercise in man. *J. Clin. Invest.* 50: 2715–2725, 1971.

Wasserman, D. H., J. S. Spalding, D. B. Lacy, C. A. Colburn, R. E. Goldstein, and A. D. Cherrington. Glucagon is a primary controller of the increments in hepatic glycogenolysis and gluconeogenesis during muscular work. *Am. J. Physiol.* 257 *(Endocrinol. Metab.* 20): E108–E117, 1989.

Wasserman, D. H., P. E. Williams, D. Brooks Lacy, D. Bracy, and A. D. Cherington. Hepatic nerves are not essential to the increase in hepatic glucose production during muscular work. *Am. J. Physiol.* 259 *(Endocrinol. Metab.* 22): E195–E203, 1990.

Williams, C., J. Brewer, and A. Patton. The metabolic challenge of the marathon. *Brit. J. Sports Med.* 18: 245–252, 1984.

Winder, W. W. Role of cyclic AMP in regulation of hepatic glucose production during exercise. *Med. Sci. Sports Exer.* 20: 551–560, 1988.

Winder, W. W., J. M. Hagberg, R. C. Hickson, A. A. Ehsani, and J. A. McLane. Time course of sympathoadrenal adaptation to endurance exercise training in man. *J. Appl. Physiol.* 45: 370–374, 1978.

Winder, W. W., H. T. Yang, and J. Arogyasami. Liver fructose 2,6-bisphosphate in rats running at different treadmill speeds. *Am. J. Physiol.* 255 (*Regulatory Integrative Comp. Physiol.* 24): R38–R41, 1988.

Young, A. A., G. J. S. Cooper, P. Carlo, T. J. Rink, and M.-W. Wang. Response to intravenous injections of amylin and glucagon in fasted, fed, and hypoglycemic rats. *Am. J. Physiol.* 264: E943–E950, 1993.

Young, A. A., M.-W. Wang, and G. J. S. Cooper. Amylin injection causes elevated plasma lactate and glucose in the rat. *FEBS Lett.* 291: 101–104, 1991.

Zinker, B. A., D. Brooks Lacy, D. P. Bracy, and D. H. Wasserman. Role of glucose and insulin loads to the exercising limb in increasing glucose uptake and metabolism. *J. Appl. Physiol.* 74(6): 2915–2921, 1993.

METABOLIC RESPONSE TO EXERCISE:

Lactate Metabolism During Exercise and Recovery, Excess Post-exercise O₂ Consumption (EPOC), O₂ Deficit, O₂ Debt, and the "Anaerobic Threshold"

It has long been known that respiratory ($\dot{V}_{O_2}$) responses to exercise are at times inadequate to provide the ATP necessary to sustain work. Although aerobic metabolism has been well understood, the role of glycolytic (anaerobic) metabolism has not. In this chapter, therefore, we discuss three misconceptions concerning "anaerobic" metabolism and exercise. These are (1) that anaerobic metabolism during exercise results in an "O₂ debt" to be repaid after exercise, (2) that lactic acid is a "dead-end metabolite" that is only formed and not removed during exercise, and (3) that the elevation of blood lactic acid level during exercise represents anaerobiosis (O₂ insufficiency) in muscle. We need to discuss these three misconceptions because of their pandemic scope, and also to discuss contemporary explanations of lactic acid metabolism during exercise and recovery. Prior to reading Chapter 10, it is assumed that the reader is familiar with the contents of Chapter 5, particularly the concepts of cell-cell and intracellular lactate shuttles, the presence of mitochondrial LDH and MCT1, and factors governing the formation and oxidation of lactate in muscle.

■ Why Measure the Metabolic Response to Exercise?

Knowing the metabolic response to exercise is often the most important means of evaluating the effects of exercise—immediate as well as long term—on the body. Exercise itself is often described in terms of the metabolic response it elicits. Comparing metabolic response during a particular exercise bout with resting metabolic rate, or with the maximal response, gives us the absolute as well as relative intensity of the exercise bout. Exercise bouts are therefore described in terms of the absolute $\dot{V}_{O_2}$, multiples of the resting metabolic rate (MET), or as a percentage of $\dot{V}_{O_2max}$. Determining metabolic response to exercise also allows us to estimate the energy (calorie) cost of the exercise. Knowing this cost can be important if the nutritional requirements of the exercise bout need to be provided for or if the efficiency of the body during the performance of the exercise is to be calculated.

The ravages of time. (a) A young George Brooks (*left*), running the 400 meters at the Intercollegiate Championships, shows amazing lactate clearance capacity and power while easily defeating his competitors. (b) An aging Tom Fahey, throwing the discus in the US Master's National Championships, searches for the Fountion of Middle Age.

■ Validity of Indirect Calorimetry in Measuring Exercise Response

Metabolic responses to exercise are studied almost exclusively using indirect calorimetry (i.e., measurements of O_2 consumption). In the main, attempts to validate indirect calorimetry as a measure of metabolic rate have been made on resting individuals. These and the more limited number of attempts to validate indirect calorimetry on exercising individuals yield excellent results. However, we must be sure to ask the following question when using O_2 consumption measures to estimate metabolic rate during exercise: Is indirect calorimetry valid for these circumstances? For exercise bouts that are not overly intense or too long in duration, the $\dot{V}_{O_2}$ can be counted on to provide an excellent measure of metabolic rate.

In order for indirect calorimetry to be valid, all the ATP used must come from respiration (O_2 consumption). Figure 10-1 is similar to Figure 4-13, except that Figure 10-1b includes the blood lactate level, which is seen to increase nonlinearly after a work intensity of approximately 60% of $\dot{V}_{O_2max}$ during a progressive intensity exercise test. In addition to being confident that glycolytic lactate production is managed by cell respiration, two other assumptions have to be made in using indirect calorimetry to estimate metabolic rate during exercise. One is that ATP and CP stores are maintained. This, like the assumption of insignificant anaerobic glycolysis, addresses the issue of whether all the ATP utilized for exercise is provided by respiration.

The second assumption is that amino acid or protein catabolism during exercise is insignificant. This assumption is most likely invalid (see Chapter 8), but it is a compromise we accept for convenience and necessity. The assumption allows us to

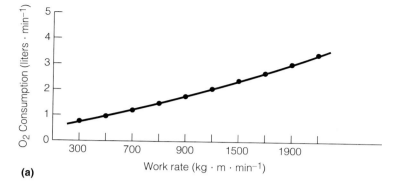

(a)

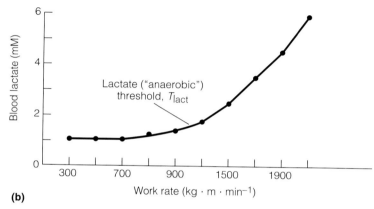

(b)

Figure 10-1 O_2 consumption (a) and blood lactate (b) responses to a continuous, progressive cycle ergometer test in healthy young subjects. $\dot{V}_{O_2}$ responds in almost a linear fashion (a), whereas lactate level does not change at first but then begins to rise nonlinearly (b). The lactate inflection point, or lactate threshold, T_{lact}, represents the point at which lactate entry into the blood exceeds its removal. Modified from Hughes, Turner, and Brooks, 1982.

use the respiratory exchange ratio (R) as the non-protein RQ (convenience), and it eliminates the need to collect nitrogenous excretions and to match N_2 excretion through sweat and urine in time with exercise (necessity). The error incurred by this assumption is usually small.

As seen in Chapter 4, and as illustrated in Figure 10-1a, the steady-rate O_2 consumption responses can be used to represent the caloric, or O_2, cost of exercise—provided, of course, that all the ATP required for exercise is supplied by respiration. When, during exercise, $\dot{V}_{O_2}$ is submaximal and constant, and when blood lactate concentration is constant, a "steady-state" condition of O_2 consumption is considered to exist. More precisely, a "steady rate" of O_2 consumption exists.

The robust nature of $\dot{V}_{O_2}$ as a measure of metabolic rate is indicated from results of isotopic tracer studies on animals and people. Figure 10-2 from

Stanley and associates (1985) shows blood lactate appearance rate (R_a) during exercise of progressively increasing intensity. In resting, postabsorptive people, blood lactate appearance is about half of glucose appearance and disappearance (production and use). However, at an exercise intensity eliciting approximately 40–50% of $\dot{V}_{O_2max}$, blood lactate utilization increases to approximate that of glucose, even though glucose use has increased in the transition from rest to exercise (Figure 10-3). Therefore, even in mild to moderate intensity exercise, a lot of lactate is formed. Moreover, in hard intensity exercise, even more lactate is produced and appears in the blood (Figure 10-2). The question then arises, Does lactate production during rest and during mild to hard intensity exercise represent a situation where significant anaerobic production leads to a violation of the steady-rate assumptions? Fortunately, because most (about 75%) lactate pro-

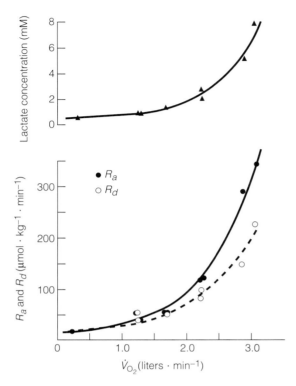

Figure 10-2 Arterial lactate concentration and rates of lactate appearance (R_a) and disappearance (R_d) plotted as a function of oxygen consumption ($\dot{V}_{O_2}$) in one subject. SOURCE: Stanley et al., 1985.

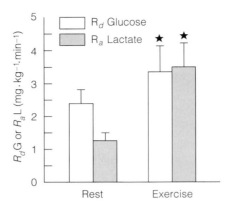

Figure 10-3 The mean values (±SD) for glucose disappearance and lactate appearance at rest and at 25 to 30 minutes of exercise ($n = 10$). The star is significantly different from rest ($P < 0.005$). SOURCE: Stanley et al., 1988.

duced as the result of anaerobic glycolysis is removed by oxidation, mainly in working muscle, the $\dot{V}_{O_2}$ still provides a valid measure of the energy flux. In this situation, lactate formed in some sites (e.g., in contracting fast-glycolytic fibers) will be removed at either site (e.g., in heart and oxidative fibers; see Figure 5-20) or in mitochondria of the lactate-producing cell (see Figure 5-14), but lactate simply represents the vehicle for shuttling chemical potential energy from one site in the body to another. In this case, the accumulation—not production—of lactate represents anaerobic energy production, which is related on a stoichiometric basis (i.e., 1 mM ATP produced/mM lactate accumulated). In terms of net energy production, the anaerobic component from accumulated lactate is small (only a few percent) compared to that from oxidative metabolism.

Muscle as a Consumer of Lactate During Exercise

Although it has long been believed that during exercise working muscle produces lactate, this belief is not always justified. Figure 10-4a shows the mean arterial blood lactate concentration in six men during leg cycling exercise at 50% $\dot{V}_{O_2max}$ at sea level. Figure 10-4b shows the average net lactate release from the leg. Here, net release equals the venous-arterial lactate concentration difference times leg blood flow. In Figure 10-4b, we see that a resting leg releases lactate slowly. Further, we see that when exercise starts, net lactate release increases dramatically. However, as exercise continues, lactate release declines, and after 30 minutes the standard error overlaps with zero, indicating insignificant net release. An examination of individual subjects confirms this, showing that lactate production and net release are associated with the onset of exercise, but that as exercise continues, lactate release declines, and in some cases, as in subject 7, the working leg becomes a net lactate consumer (Figure 10-4c). In addition, isotope tracer studies show that muscle tissue continuously consumes and releases lactate, with the greatest consumption occurring during exercise, when blood lactate level is highest. Thus, the

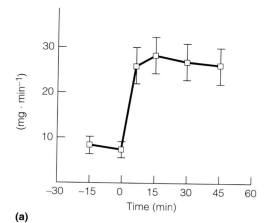

(a)

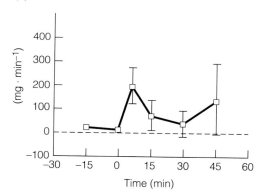

(b)

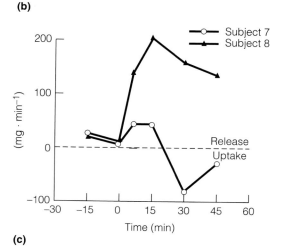

(c)

Figure 10-4   (a) Mean arterial blood lactate concentration ($\pm$ SEM) in six men during rest and cycle ergometer exercise at 50% $\dot{V}_{O_2max}$. (b) Net leg lactate release [$\dot{L} = (v - a)_{lactate} \cdot$ Leg blood flow] in the same subjects. Note that after 15 minutes the legs stop releasing lactate even though exercise continues and arterial level is constant. (c) Leg net lactate exchange in two subjects. When exercise starts, legs increase lactate release, but transiently. Further, individual subjects show great variability, with subject 7 showing a switch from release to consumption. Data from Brooks et al., 1991.

net release of lactate (Figure 10-4b) underestimates the total lactate formed in muscle.

■ The Excess Postexercise Oxygen Consumption (EPOC), or the "O₂ Debt"

Estimating the metabolic cost of hard or maximal exercise that results in large lactate accumulation has long fascinated physiologists. This problem is far more complex than just understanding steady-rate exercise. Because the $\dot{V}_{O_2}$ obviously does not represent the metabolic rate during extreme exercise, investigators have tried to take advantage of the extra O_2 consumed by the body during recovery to estimate the total "O_2 cost" of exercise. For a long time, it was thought that if O_2 consumption during exercise was inadequate to meet energy demands— that is, if there was a "deficit" in the O_2 consumption (Figure 10-5)—then the body borrowed on its energy reserves (or credits). After exercise, then, the body was thought to pay back those credits, plus some interest. The extra O_2 consumed, above a resting baseline during recovery, was referred to as the "O_2 debt" and was used as a measure of anaerobic metabolism during exercise. As may be surmised, however, this explanation of the relationship between the O_2 deficit and O_2 debt is too simplistic. *Measuring the O_2 deficit or debt may have a purpose in contemporary research, but measuring the O_2 debt after exercise is inadequate for estimating anaerobic metabolism during exercise.*

After exercise, O_2 consumption does not return to resting levels immediately but does so rather in a curvilinear fashion (Figure 10-6a). Plotting the postexercise O_2 consumption on semilogarithmic coordinates reveals that the curve is composed of one exponential component after mild exercise (Figure 10-6b) or two exponential components after moderate to maximal exercise (Figure 10-6c). The extra O_2 consumed during recovery above a resting baseline has been referred to by Brooks and associates (1971, 1980) and Gaesser and Brooks (1984) as the *excess postexercise oxygen consumption* (EPOC). This phenomenon of extra O_2 consumption during recovery has long been observed and was termed

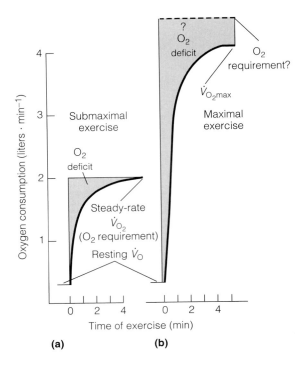

Figure 10-5 Oxygen consumption before and during continuous bouts of steady-rate submaximal and maximal exercise. In submaximal exercise (a), the O_2 deficit can be estimated as the difference between the steady-rate $\dot{V}_{O_2}$ and the actual $\dot{V}_{O_2}$ prior to attainment of the steady-rate $\dot{V}_{O_2}$. In maximal exercise (b), the O_2 deficit cannot be estimated with certainty for lack of a precise value of the O_2 required. For this reason, scientists have attempted to utilize the "O_2 debt" (Figure 10-6) as a measure of anaerobic metabolism during exercise. SOURCE: Brooks, 1981.

the "O_2 debt" by A. V. Hill and coworkers (1923, 1924). Because so much confusion exists at present about the mechanism of postexercise O_2 consumption, we now address the topic in some detail.

■ Classical O₂ Debt Theory: The Early Twentieth Century

The exponential decline in O_2 consumption after exercise was first reported by August Krogh in Denmark. That report was apparently noted by A. V. Hill in England, who was at that time per-

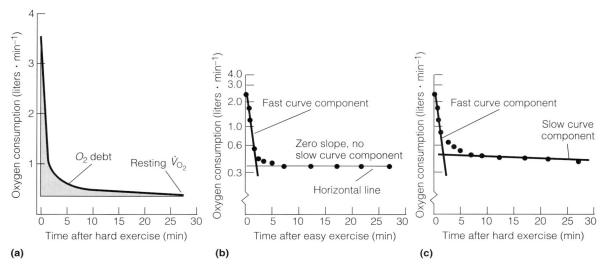

Figure 10-6 O_2 consumption after a period of exercise requiring a constant work output. (a) Plotted on linear coordinates, the O_2 debt is the area under the postexercise $\dot{V}_{O_2}$ curve above the resting $\dot{V}_{O_2}$ baseline. (b) and (c) $\dot{V}_{O_2}$ after easy and hard exercise bouts, plotted on semilogarithmic coordinates. Note one postexercise $\dot{V}_{O_2}$ curve component after easy exercise (b) and two components after hard exercise (c). SOURCE: Brooks, 1981.

forming detailed studies on the energetics of small muscles isolated from frogs. From the earlier work of Fletcher and Hopkins (1907) in England, it was known that muscles stimulated to contract produced lactic acid that needed to be removed, and that doing so required that O_2 be present. Hill observed that a muscle recovering from a contraction displayed a second burst of heat release (a latent heat), which approximated in magnitude the initial heat associated with contraction. However, the latent, or recovery, heat appeared only if O_2 was present. According to Hill's calculations, the recovery heat represented the heat that would be released if one-fifth of the lactic acid formed during the contraction was combusted. That is, Hill thought the combustion coefficient of lactic acid (i.e., the fraction of lactate oxidized in recovery) was 1/5.

In the 1920s, Otto Meyerhof in Germany discovered that the precursor of lactic acid was glycogen. When frog muscle was made to contract, glycogen broke down (disappeared), and similar amounts of lactic acid appeared. During recovery, lactic acid

disappeared and glycogen was re-formed in only slightly smaller amounts. According to Meyerhof (1920), the combustion coefficient of lactate was 1/3.

Shortly thereafter, A. V. Hill and his associates in England experimented on humans recovering from exercise. Those experiments were a brilliant attempt to unify the biochemical and energetic results obtained on isolated muscles with an understanding of human physiology. The researchers knew that during contraction, frog muscles broke down glycogen and formed lactic acid. They also knew that during recovery those isolated muscles combusted a small portion (one-fifth to one-third) of the lactate formed and that most of the lactate (two-thirds to four-fifths) was reconverted to glycogen. Although we now know this to be correct for amphibia and reptiles but incorrect for mammals, it was logical at the time for Hill to assume that a similar phenomenon occurred in humans during exercise. The Hill-Meyerhof theory of the "O₂ debt" (as it came to be known) was that during recovery, one-fifth of the lactate was oxidized and four-fifths was reconverted to glycogen. In 1922, Hill and Meyerhof shared a Nobel Prize awarded, in part, for the work described here, in which detailed biochemical and biophysical measurements on isolated muscles were linked to the intact functioning human body.

Subsequent to Hill and associates, Margaria, Edwards, and Dill (1933) at the Harvard Fatigue Laboratory directed their efforts to studying recovery metabolism. Following some of the exercise protocols used by these investigators (approximately three to eight minutes of difficult exercise), blood lactate did not decline immediately; rather, there was a delay in the decline of blood lactate (Figure 10-7). Therefore, Margaria and associates surmised that the first, fast phase of the postexercise O_2 consumption curve, which was not temporally associated with a change in blood lactate, had nothing to do with lactate metabolism. This phase (Figure 10-7) they termed *alactacid*, meaning not associated with lactate metabolism. They proposed that the slow postexercise O_2 consumption curve, which temporarily coincided with the decline in blood lactate, was due to the reconversion of lactate to glycogen. They termed the slow phase the *lactacid* phase. In this, Margaria and coworkers departed

from the work of Hill and coworkers and interpreted the O_2 debt as being due to these two phases. Margaria and coworkers were incorrect in their interpretation; they could not have known that, in fact, lactate was rapidly entering and leaving the blood immediately after exercise and that oxidation was the major fate of lactate during exercise and recovery. Nevertheless, the world came to know and accept the hypothesis of lactic and alactic O_2 debts.

Since its formulation in the 1920s, and subsequent modification in the 1930s, the O_2 debt theory has been the target of several serious challenges. The challenges have gone unanswered in the scientific literature, however, and the O_2 debt theory has persisted in both textbooks and popular literature. This is an anomaly in science.

One of the first challenges to the O_2 debt theory was by Ole Bang (1936), who showed by using exercises of varied intensities and durations that the results of Margaria and colleagues were fortuitous results of the duration of their experiments. With prolonged exercise, Bang showed, blood lactate level reaches a maximum after about 10 minutes of exercise and then declines whether exercise ceases or not (Figure 10-8b). In some cases, basal lactate

Figure 10-7 Oxygen consumption and blood lactate levels in a subject after a hard 3-min treadmill run. Initially, blood lactate did not appear to change, whereas $\dot{V}_{O_2}$ declined rapidly. Later, $\dot{V}_{O_2}$ declined slowly and blood lactate declined. On this basis, Margaria, Edwards, and Dill separated the "O_2 debt" into fast (alactacid) and slow (lactacid) components. Contemporary research reveals these terms to be inappropriate. See the text for details. Modified from Margaria et al., 1933.

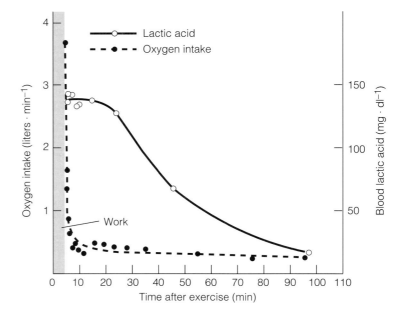

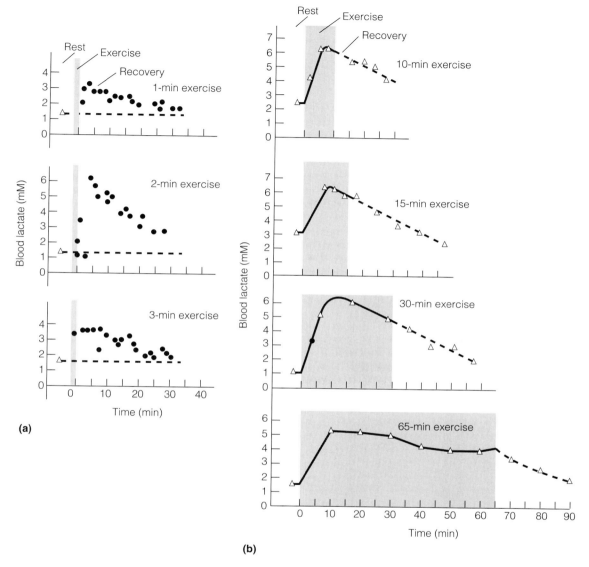

Figure 10-8 Blood lactate levels in subjects after bicycle ergometer exercise of the same intensity but different durations. In series (a), involving short exercise bouts of from 1 to 3 minutes, blood lactate clearly increases even though exercise has stopped. In series (b), involving longer periods of exercise, lactate is clearly decreasing even though exercise continues at a steady rate. Although blood lactate varies greatly, the kinetics of the postexercise O₂ consumption curve (O₂ debt) are unaffected. Modified from Bang, 1936.

levels can be achieved during exercise itself. After exercise, however, there is always an O_2 debt, with predictable kinetics to be "paid." In contrast, by using exercise bouts lasting only a few minutes, Bang found that the concentration of lactic acid in the blood reached a maximum after exercise had ended, and depending on the intensity of the exercise, remained elevated long after the oxygen intake had returned to preexercise levels (Figure 10-8a). These results cannot be reconciled with the idea that lactic acid determines oxygen consumption after exercise.

■ The Metabolic Fate of Lactic Acid After Exercise

What happens to lactic acid after exercise (i.e., the metabolic fate of lactic acid) has been studied using radioisotopes in animals by Brooks, Brauner, and Cassens (1973), and by Brooks and Gaesser (1980).

In the over 60-year history of interest in this subject, there have been remarkably few attempts to determine directly what happens to lactic acid after exercise in intact mammals. In their study, Brooks and Gaesser injected [U-^{14}C]lactate into rats at the point of fatigue from hard exercise. At various times during recovery, blood, liver, kidney, heart, and muscle tissues were sampled. Metabolites were separated and quantified using two-dimensional radiochromatography (illustrated in Figure 10-9). From these radiochromatograms, the pathways of lactate metabolism were traced. The metabolic endpoints reached are displayed in Figure 10-10. The results of these experiments reveal that the metabolic pathways taken by lactate during recovery are diverse. Lactate is mainly oxidized after exercise, but it does participate in a number of other processes. As in amphibia and reptiles, in white mammalian fibers a small amount of lactate can be converted to glycogen *in situ*. However, in contrast to

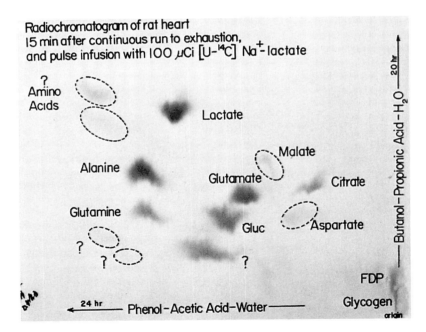

Figure 10-9 Two-dimensional radiochromatogram from the heart of a rat, 15 minutes after exercise to exhaustion and injection with [U-^{14}C]lactate. Note the incorporation of the carbon label from tracer lactate into a wide variety of compounds. SOURCE: Brooks and Gaesser, 1980. Reprinted with permission of the American Journal of Physiology.

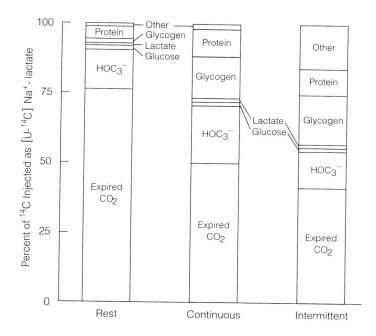

Figure 10-10   Histograms representing the quantitative recovery of tracer carbon injected as [U-^{14}C]lactate four hours after injection under three conditions. Note that oxidation is the major pathway of removal, and that after exercise, reconversion to glycogen represents less than 20% of the total. SOURCE: Brooks and Gaesser, 1980. Reprinted with permission.

the assumptions of Hill and Meyerhof, in mammals most lactate formed during exercise is either removed during exercise or combusted during recovery. Further, in contrast to the case in amphibia, reptiles, and some fish, which have poorly circulated muscles, in humans and other mammals after exercise significant amounts of lactate efflux (leave) muscle for removal at other sites, including the liver.

Lactate as a Carbon Reservoir During Recovery

As indicated in Figure 10-9, lactate can traverse a number of different metabolic pathways. Lactate, in effect, sits close to a metabolic crossroads. Because of the proximity of lactate to the TCA cycle, its entry into that cycle and subsequent oxidation constitute a major pathway of metabolism. In addition to serving as an oxidizable substrate, lactate can also serve as a gluconeogenic precursor or be incorporated into amino acids and proteins. The pathways of lactic acid metabolism after exercise appear to depend to some extent on the internal metabolic conditions when exercise stops. High levels of lactate and near-normal concentrations of other substrates such as

liver glycogen and blood glucose appear to favor lactate oxidation. However, the effects of prolonged exercise leading to exhaustion—such as glycogen depletion and hypoglycemia (low blood sugar)—may favor lesser oxidation and greater conversion of lactate to glucose (gluconeogenesis).

Lactic Acid Does Not Cause the O₂ Debt

Even though most of the lactate present when recovery begins may be oxidized, lactic acid cannot be said to cause the O_2 debt. Because lactate merely supplies fuel to power the recovery processes, the combustion of lactic acid does not result in extra O_2 consumption. Lactate is converted to pyruvate, and this pyruvate in effect substitutes for the pyruvate that would have been supplied by glucose or glycogen. Levels of these compounds may be low after prolonged or difficult exercise.

In contrast to the lactic acid theory of oxygen debt, contemporary understanding of lactic acid metabolism involves the concept of the lactate shuttle, in which lactate serves as a fuel source and gluconeogenic precursor that can be exchanged through the interstitium and vasculature during rest, exercise, and recovery from exercise. Please refer to Chapter 5 and Figure 5-20.

■ Exercise-Related Disturbances to Mitochondrial Function

If lactic acid accumulation does not by itself cause the extra O_2 consumption after exercise, what does? The answer may well lie in the mitochondria and the cellular sites at which O_2 is consumed.

Temperature

The major metabolic waste product during exercise is heat. Exercise heat production elevates temperature in active muscle as well as other tissues. Using mitochondria isolated from muscle and liver, Brooks and coworkers (1971) demonstrated that elevated tissue temperatures produced by exercise cause an increase in the rate of mitochondrial O_2 consumption and an eventual decline in energy-trapping efficiency. A mitochondrion and a mechanical model for oxidative phosphorylation are diagrammed in Figure 10-11. It must be emphasized that the process of oxidation (O_2 consumption) and phosphorylation (ATP production) are separate (Chance and Williams, 1956). Ordinarily, the two processes are linked in that oxidation provides the energy for phosphorylation. In Figure 10-11, the linkage is simplistically diagrammed as a mechanical gear linkage, a situation analogous to that in an automobile. The engine can run full speed, consume oxygen, and utilize fuel, but unless the transmission and differential are in good operating order, nothing meaningful will happen in terms of rotation of the driving wheels. Elevated temperatures of the magnitude experienced in heavy exercise have the effect of loosening the coupling or linkage between oxidation and phosphorylation. Since the beginning of the twentieth century, when Dubois (1921) thoroughly studied basal metabolic rate (BMR), it has been known that the BMR increase in humans during fever is about 13% per degree Celsius increase in body temperature. Elevated tissue temperatures in humans after exercise have been demonstrated to be associated with EPOC volume (Claremont, Nagle, Reddan, and Brooks, 1975).

Figure 10-11 Diagram of a mitochondrion with the coupling mechanism between oxidation and phosphorylation depicted as a mechanical linkage. Factors suspected of affecting the linkage between oxidation and phosphorylation during the postexercise period are indicated. In addition, ion pumping and ATP utilization by the sarcolemma are promoted by muscle contraction and by thyroid hormones and glucocorticoids. Modified from Brooks, 1981.

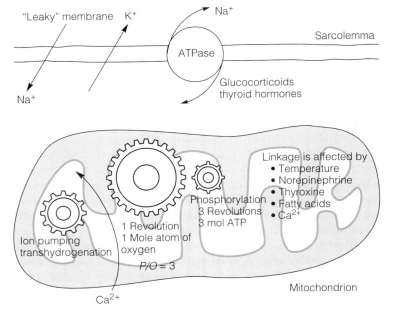

Fatty Acids and Ions

One very interesting physiological phenomenon, not generally familiar to students interested in exercise, is the phenomenon of nonshivering thermogenesis in certain animals. This well-studied phenomenon provides an interesting model for comparison to the individual recovering from exercise. When placed in cold environments, most mammals will shiver to develop adequate body heat. After a week or 10 days, some species, such as the rat, will stop shivering yet continue to generate adequate heat to maintain internal temperature. The animal's basal metabolic rate will have increased two- to threefold. There are several theories about how the rat accomplishes this. One theory is that oxidation is uncoupled from phosphorylation (Smith and Hoijer, 1962). The sites of this uncoupling may be skeletal muscle as well as brown fat deposits in the body core. Brown fat deposits in hibernating animals have a similar capability and, in this instance, the brown fat functions to maintain an adequate temperature around the hibernating animal's vital organs. Thus, the specific agent that uncouples oxidation from phosphorylation may be a fatty acid (Hittelman, Lindberg, and Cannon, 1969). It is possible that the lipolysis and release of fatty acids that occur during exercise have a similar effect on O_2 consumption after exercise.

Another theory to explain the mechanism of nonshivering thermogenesis is that the hormone norepinephrine directly or indirectly causes cell membranes to become more permeable to sodium ion (Na^+) and potassium ion (K^+) (Horwitz, 1979). In order to compensate for the ion leaks, the cell membrane Na^+–K^+ pump activity is increased. This pump requires an energy input in the form of ATP. Increased ATP demands for the pump are met by mitochondrial oxidative phosphorylation. Consequently, O_2 consumption is increased. It has also been discovered that thyroid hormone and glucocorticoids can contribute to Na^+–K^+ pump activity (Horwitz, 1979; see Figure 10-11). Because exercise affects both Na^+ and K^+ balance, as well as hormone levels, these factors probably serve to elevate O_2 consumption after exercise.

Calcium Ion

Carafoli and Lehninger (1971) have shown that Ca^{2+} has similar effects on mitochondria isolated from all tissues on all animals studied. The effect of calcium ion is to increase oxygen consumption in two ways. First, mitochondria have a great affinity for calcium ions and, given the chance, will sequester Ca^{2+} at a high rate. This process requires energy, which is reflected in an increased rate of O_2 consumption that is not associated with ATP generation. Second, increased amounts of Ca^{2+} within mitochondria ultimately affect the mechanism linking oxidation and phosphorylation. Again, the result is an increased rate of O_2 consumption.

During each muscle contraction in exercise, Ca^{2+} is liberated from the sarcoplasmic reticulum. In heart as well as skeletal muscle, some of this Ca^{2+} is probably taken up by mitochondria. The mechanism by which mitochondria, during recovery, dispose of the Ca^{2+} accumulated during exercise is at present not understood. However, it is likely that Ca^{2+} accumulated by mitochondria during exercise affects mitochondrial respiration during exercise as well as during recovery.

Sympathetic Stimulation

Increased stimulation of the sympathetic nervous system (SNS) is known to increase the metabolic rate of many tissues and systems by both alpha (α)- and beta (β)-mediated mechanisms (Chapter 9). Because catecholamines are elevated during hard exercise as well as for a time during recovery, it has been believed that part of the EPOC is due to the persistent effects of catecholamines. The effects of β-adrenoceptor on EPOC have been evaluated by Elisabet Børsheim in Norway (1998). In separate experiments, she gave both the β_1-specific blocker (atenolol) and the nonspecific ($\beta_1 + \beta_2$) β-adrenergic blocker (propranolol) to subjects after exercise. In comparison to control (unblocked) trials, the β-blockers had no effects on EPOC volume. Because of compensatory actions of the α- and β-adrenergic arms of the SNS, it is clear that β-adrenergic stimulation of metabolism does not occur after exercise during the EPOC period. However, at present factors other than epinephrine are better candidates for causing elevations in O_2 consummption during recovery from exercise.

■ Use of the Term "O_2 Debt"

Throughout this discussion, the descriptive term *excess (or elevated) postexercise oxygen consumption* (EPOC) has been used, as has the classical term, O_2 *debt*. Because it is evident that at present no complete explanation of postexercise metabolic phenomena exists, it may be advantageous to substitute another term for O_2 debt. Other investigators agree that a name change is due, but there is some disagreement as to what that should be. An alternative suggested by Harris (1969) was *phlogiston debt.* This whimsical suggestion, which projects O_2 debt theory back to the days of alchemy, was obviously meant to draw attention to the problem of terminology. An alternative suggested by Stainsby and Barclay (1970) was the term *recovery O_2.* A disadvantage of this term is that it does not completely escape the implication of a mechanism or explanation. It is now considered probable that two factors responsible for significant portions of the elevated postexercise O_2 consumption (i.e., the calorigenic effects of catecholamines and temperature) occur during recovery from exercise but are not obligatory and necessary for recovery. Therefore, the purely descriptive term, *excess postexercise oxygen consumption* (EPOC), is probably most appropriate.

■ Continued Utility of the O_2 Deficit Measure

Despite recent advances in the study of exercise physiology, including tools and technologies such as isotopic tracers (discussed later in this chapter) and nuclear magnetic resonance (NMR, Chapter 33), estimating the anaerobic component to the energy flux sustaining exercise remains difficult. Because the more recent technologies require steady-state, or

near steady-state, conditions for interpreting data, researchers have recently again begun employing the classic approach of measuring the O_2 deficit (see Figure 10-6) to address the problem. However, because the same (unsatisfactory) assumptions apply to contemporary as well as classic data utilizing the O_2 deficit method, uncertainty remains about the validity of interpretations of data on muscle and whole-body energetics during all-out, non–steady-state exercise. Although, for the present, the O_2 deficit method of estimating the energy cost of non–steady-state exercise remains the main tool available, researchers hope the future will yield another method that will possess independent criteria of validity—and possibly even support continued use of the O_2 deficit method to estimate anaerobic metabolism during exercise.

■ Lactic Acid Turnover During Exercise (Production, Effects of Training on Blood, Removal, and Clearance)

As already shown (Figures 10-2 through 10-4), detailed studies using both radioactive and nonradioactive tracers on animals and humans have shown that lactic acid is a dynamic metabolite both at rest and during exercise. At rest and during easy exercise, lactic acid is produced and removed at equal rates. This balance of production and removal is called *turnover.* Even though a metabolite such as lactic acid turns over very rapidly, its concentration in the blood may not change so long as the removal (disappearance from the blood) keeps pace with the production (entry or appearance in the blood). Tracer studies have clearly shown that for a given blood lactate level, the turnover of lactic acid during exercise is several times greater than it is at rest. Therefore, if blood lactic acid level remains at resting levels during exercise (Figure 10-1b, left-hand portion of curve), it is erroneous to conclude that no lactic acid is being produced.

At heavier exercise intensities, blood lactate level increases as compared to that at rest (Figures 10-1b and 10-8). If the exercise is maintained long enough, the lactate level can remain elevated, but by a constant amount. If the lactate level is constant, then lactate production and removal rates are equal. Tracer studies indicate that the metabolic clearance rate (i.e., turnover rate divided by the blood lactate level) is increased in heavy exercise. This means that lactate removal is concentration-dependent; lactic acid needs to be raised to a higher level to force its removal.

In other exercise situations, blood lactate falls toward resting levels after an initial rise (Figures 10-4b and 10-8b), or it continues to increase throughout the exercise (Figure 10-1b). In the former case of declining blood lactate, removal exceeds production once circulation supplies the O_2 necessary for muscle respiration and lactate in the blood is delivered to sites of removal. In the latter case of continuously rising lactate levels, production exceeds removal.

■ Principles of Tracer Methodology

To understand the dynamics of any blood metabolite, such as glucose or lactate, it is helpful to become familiar with the use of isotopic tracers. A *tracer* is a molecule in which a normally occurring atom or atoms provided by nature is substituted for by a less frequently occurring isotope of that same elemental atom or atoms. Isotopes of an element have the same atomic number but different weight. They can be either radioactive or nonradioactive. Radioactive isotopes are usually far easier to detect and follow in the body or expired air, but because of the hazards of radioactivity, these isotopes are today used infrequently in human experimentation. Carbon, with molecular weight 12 (^{12}C), is the most abundant isotope of carbon, and it is not radioactive. Carbon-14 (^{14}C) is radioactive and occurs naturally but infrequently in nature. Carbon-14 has a half-life of over 5000 years. This stability allows it to be produced in a cyclotron and to be chemically incorporated into a variety of compounds. These compounds can be stored for a time and then used to follow metabolic pathways in organisms. Carbon-11 (^{11}C) is another radioactive isotope of carbon, with a half-life of about 20 minutes. This

isotope is being used more and more in studies on humans because it breaks down so rapidly, dissipating its potential for radiation exposure. To use this isotope, scientists must have a cyclotron close at hand to produce the material. Carbon-13 (^{13}C) is a nonradioactive isotope of carbon that is naturally abundant in nature. About 1% of all carbon is this heavier isotope, ^{13}C. This isotope is more difficult to utilize and trace in metabolism because it is not radioactive, but, for the same reason, it is becoming the isotope of preference for human experimentation. The presence of carbon-13 is detected on the basis of its mass.

In addition to carbon, isotopes of hydrogen are frequently used to study metabolism. Hydrogen-2 (deuterium, ^2H) is not radioactive, whereas hydrogen-3 (tritium, ^3H) is. Simultaneous use of the isotopes of carbon and hydrogen (e.g., ^{14}C and ^3H) has been helpful in determining how metabolites are split apart and joined together in metabolism.

There are two basic strategies for using isotopes to study metabolism: (1) pulse injection and (2) continuous infusion. In the first approach, all of the isotope is injected into a rapidly mixing pool (e.g., the blood) in a single rapid (bolus) injection. In the second, the isotope is added at a continuous, set rate. A third approach is the primed continuous-infusion technique, which is a combination of the other two. The process of using tracers can be illustrated by the analogy of a rain barrel. Figure 10-12 illustrates the pulse injection strategy. Observing the unchanging water level in the barrel (Figure 10-12a), one might conclude that nothing is happening to the H_2O. Hidden from view, however, are the facts of entry into and removal from the barrel. Suppose the observer decides to perform an experiment by adding a dye, all at once, to the water. The intensity of color developed in the water depends on the amount of dye used and the volume of H_2O; the greater the concentration of dye, the darker the color (Figure 10-12b). If there is no movement (flux) of water through the barrel, the intensity of color developed remains constant. If, however, new clear water enters and the original colored water is removed, then the color of the water in the barrel will fade. The speed with which the color fades indicates the speed of the

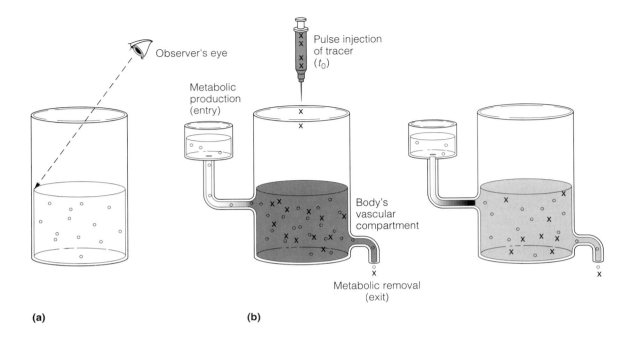

(a)

(b)

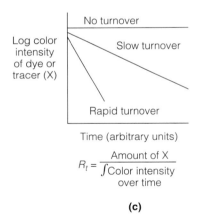

(c)

$$R_t = \frac{\text{Amount of X}}{\int \text{Color intensity over time}}$$

Figure 10-12 Principle of the pulse (bolus) injection tracer technique illustrated with a water barrel. (a) Noting no change in water level, one is tempted to conclude that there is no water movement through the barrel. (b) The observer decides to perform an experiment and adds a dye that colors the H_2O. This is time zero (t_0) of the experiment. At various times after t_0 (i.e., t_1, $t_2, \ldots, t_i$) the observer records the color intensity in the barrel (c). The more rapid the H_2O flux through the barrel, the more rapidly the color changes. In real-life experiments, the dye represents an isotopic tracer and the barrel represents the vascular compartment of the body.

water's movement (flux) through the barrel (Figure 10-12c). This is analogous to the pulse injection technique, where the isotopic tracer is represented by the dye.

A second way of using dye (representing isotopic tracer in our analogy) to study the movement of water through the rain barrel is to add the dye at a continuous, constant rate by placing it in a sy-

ringe and pushing it into the water at a constant rate. After the infusion of dye begins (Figure 10-13a), the clear water gradually becomes darker (Figure 10-13b) until an equilibrium point is reached (Figure 10-13c). At that point, the entry and removal of the dye are proportional to the entry and removal of water into and from the barrel. The darker the water becomes, the slower the movement of water

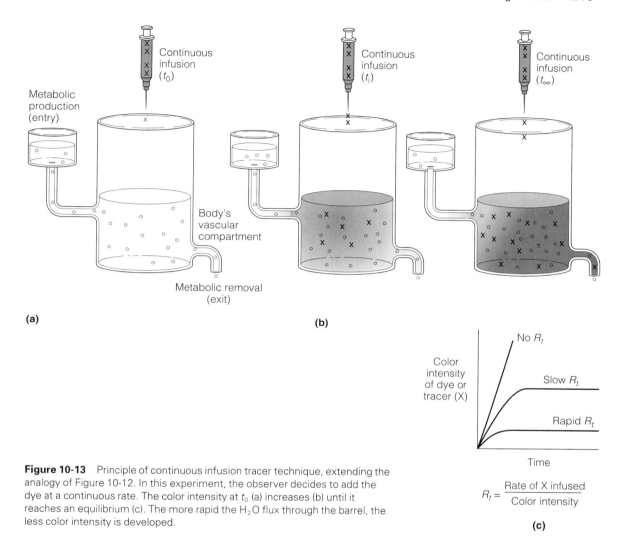

Figure 10-13   Principle of continuous infusion tracer technique, extending the analogy of Figure 10-12. In this experiment, the observer decides to add the dye at a continuous rate. The color intensity at t_0 (a) increases (b) until it reaches an equilibrium (c). The more rapid the H_2O flux through the barrel, the less color intensity is developed.

$$R_t = \frac{\text{Rate of X infused}}{\text{Color intensity}}$$

(c)

through the barrel; the clearer the water remains, the more rapid the movement of water. This is analogous to the continuous-infusion technique of studying metabolite flux in organisms.

In the third type of tracer approach, the primed continuous-infusion method, the tracer is added in a bolus, after which a continuous infusion is begun. This approach essentially relies on the same calculations as the continuous-infusion method, but the initial bolus is intended to decrease the time to equilibrium.

Effect of Endurance Training on Lactate Metabolism During Exercise

At the University of California, Donovan and Brooks (1983) studied the effects of endurance training on glucose and lactate turnover during rest and exercise using the primed continuous-infusion technique and both ^{14}C and ^{3}H tracers. In those studies, trained animals had lower lactate levels during both easy and hard exercises than did untrained animals. However, lactate turnover rates in trained animals during exercise were the same as in untrained animals. Lower blood lactate levels in trained animals concealed the fact that lactate production was the same in trained and untrained animals. The difference due to training was the greater lactate clearance rate from the blood in trained animals (Figure 10-14). In subsequent studies on human subjects, Mazzeo et al. (1982), Stanley et al. (1985, 1986, 1988),

and MacRae et al. (1992) have demonstrated that training reduces blood lactate during given levels of submaximal exercise by improving the body's capacity for lactate clearance. Figure 10-15 shows blood lactate kinetics in two young healthy men, one of whom (subject 8) is a trained bicyclist. When

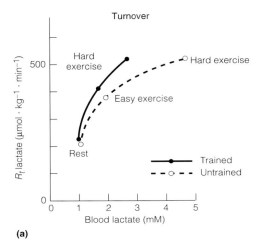

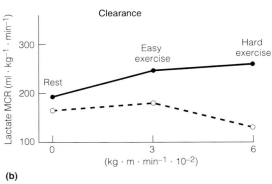

(a)

(b)

Figure 10-14 The effects of endurance training on lactate metabolism during steady-rate, submaximal exercise. (a) The relationship between lactate turnover (as measured with [U-14C]lactate) and blood lactate concentration. For a given lactate production (turnover, R_t), lactate concentration is much lower in trained rats. (b) The relationship between the metabolic clearance rate of lactate (MCR = R_t/concentration) and external work rate in trained and untrained rats. Modified from Donovan and Brooks, 1983.

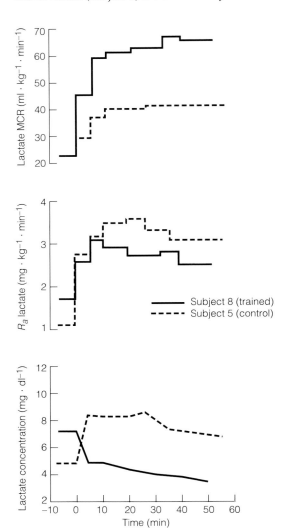

Figure 10-15 Lactate concentration, rate of appearance of lactate in the blood (R_aL) and lactate metabolic clearance rate (MCR) plotted as a function of time for two men, subjects 5 and 8. Subject 8 is a trained endurance cyclist and subject 5 is a control. SOURCE: Stanley et al., 1988.

the cycling exercise starts, subject 5 displays a typical response: Blood lactate appearance and clearance rises, but the rise in clearance is insufficient to prevent a rise in arterial lactate concentration. In contrast to subject 5—but sometimes characteristic of highly trained athletes engaged in mild exercise (40% $\dot{V}_{O_2max}$)—when the trained cyclist (subject 8) starts cycling, his blood lactate level declines despite a rise in appearance (R_a), because clearance (MCR) rises relatively more. For the effects of training on blood lactate kinetics, see Figure 5-27.

■ The "Anaerobic Threshold": A Misnomer

As exercise intensity increases, $\dot{V}_{O_2}$ increases linearly, but blood lactate level does not change until about 60% of $\dot{V}_{O_2max}$ has been reached. Therefore, blood lactate level increases nonlinearly (Figure 10-1b). The inflection point in the blood lactate curve has mistakenly been termed the *anaerobic threshold*. However, the blood lactate inflection point per se gives no information about anaerobic metabolism; rather, it reflects the balance between lactate entry into and removal from the blood.

As described by Jones and Ehrsam (1982), the anaerobic threshold model is a long-standing paradigm that was advanced by Wasserman and coworkers (1973; Figure 10-16). The model, which proposed causal linkages among muscle O_2 insufficiency (anaerobiosis), lactate production, and changes in pulmonary ventilation, was particularly attractive because it offered the possibility of detecting muscle anaerobiosis through changes in breathing. Unfortunately, it was found that various factors affect anaerobic threshold determination, including nutritional status, body mass, mode of exercise, and speed of movement. Perhaps most revealing of the fallacy of the anaerobic threshold model were studies on McArdle's syndrome patients by Hagberg and others (1982) at Washington University in St. Louis. McArdle's syndrome is a genetically linked disease in which victims lack the enzyme phosphorylase (see Figure 5-19), which renders them incapable of catabolizing glycogen and forming lactic

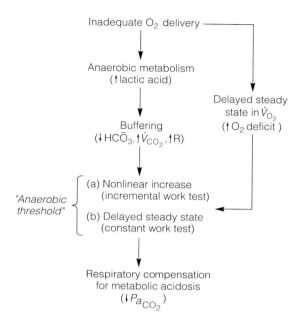

Figure 10-16 The model of the "anaerobic threshold" of Wasserman and coworkers is a permutation of the O_2 deficit–O_2 debt theory. According to this model, tissue anaerobiosis results in lactate production and entry into the venous blood. This entry of fixed (nonvolatile) acid displaces the volatile acid CO_2 from the blood through the bicarbonate buffer system (Chapter 11). The change in CO_2 flux to the lungs is sensed and ventilation is stimulated. Although this model usually provides a method for determining the point at which blood lactate level starts to increase, under some circumstances it fails, which is understandable on theoretical grounds. See the text for additional details. Modified from Wasserman et al., 1973. Used with permission of the Journal of Applied Physiology.

acid. Nevertheless, McArdle's syndrome patients demonstrate ventilatory, or anaerobic, thresholds at the usual place on the lactate–work load curve, even though there is no change in blood lactate (Figure 10-17).

Beyond studies on McArdle's syndrome patients, which represent a special population, perhaps a more telling blow to the anaerobic threshold theory were the results of Hughes, Turner, and Brooks (1982) and Heigenhauser, Sutton, and Jones (1983) on healthy young men studied in normally

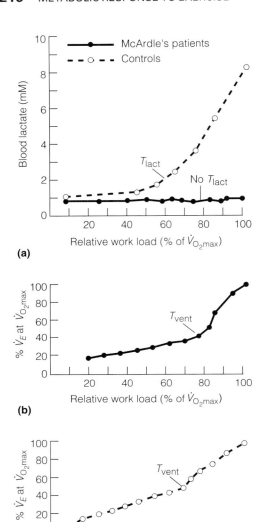

(a)

(b)

(c)

Figure 10-17 Blood lactate and ventilatory ($\dot{V}_E$) responses in McArdle's syndrome patients and normal controls during continuous, progressive exercise tests. (a) Controls reach a lactate threshold, T_{lact}, at about 70% of $\dot{V}_{O_2max}$, whereas McArdle's patients show no lactate inflection. (b) McArdle's patients display a ventilatory threshold, T_{vent}, despite the fact that there is no T_{lact}. (c) The usual ventilatory change is observed in controls. These results show that T_{vent} is not solely due to T_{lact}. Modified from Hagberg et al., 1982. Used with permission.

fed and glycogen-depleted states. After glycogen depletion, the ventilatory threshold (T_{vent}) comes at a lower exercise power output than normal, but the blood lactate threshold (T_{lact}) comes at a higher power output. Similarly, Gaesser and Poole (1988) have observed a dissociation of the (T_{vent}) and (T_{lact}) in young men after endurance training.

■ Causes of the Lactate Inflection Point

If the inflection point in blood lactic acid concentration is not necessarily due to muscle anoxia, then (1) why is there a lactate inflection point, (2) why is there a ventilatory inflection point, and (3) why do the two inflection points sometimes coincide?

As exercise intensity increases, several factors may operate to change the blood lactate level. Given that blood lactate concentration depends on the entry into and exit from the blood, any factor that affects entry and/or removal changes the blood concentration. Fast-contracting white skeletal muscle fibers produce lactic acid when they are recruited to contract, whether O_2 is present or not. This is a result of the high activity of muscle-type (M-type) lactic dehydrogenase (LDHase) (Chapter 5) in the cycotosol and a result of low mitochondrial density. Moritani and colleagues (1984) have provided electromyographic evidence that more white-fast lactate-producing fibers are recruited as exercise intensity increases. This change in recruitment pattern appears to coincide with the change in blood lactate level.

There may also be a neuroendocrine component to the explanation of lactate inflection point. As exercise difficulty increases, afferent (nervous) signals to the CNS increase in frequency and amplitude. This activates the "fight-or-flight" response mechanism and results in sympathetic autonomic nervous output to the various circulatory beds (including those in muscle, liver, kidney, and adipose tissue) and to the adrenal gland. The neurotransmitter released by the sympathetic nerve endings, norepinephrine, causes local vasoconstriction in arterioles, which increases resistance to blood flow in those areas. In contracting muscle itself, local factors override the constrictor effect and cause arterioles and precapillary sphincters to dilate, allowing relatively more of the cardiac output to flow to active areas.

Sympathetic, autonomic stimulation of the adrenal gland—specifically its inner segment, the adrenal medulla—results in the secretion of epinephrine (adrenaline) into the circulation. Sympathetic stimulation of the α cells in the pancreas causes release of glucagon. Epinephrine and glucagon cause glycogenolysis in muscle and liver, respectively (see Chapters 5 and 9). In muscle, glycogen is broken down rapidly to glucose 6-phosphate (G6P) by activated phosphorylase. Glucose released from liver and picked up in muscle is converted to G6P by hexokinase. Elevated levels of G6P increase glycolytic rate and the formation of pyruvate. Because in this case the rate of glycolysis is not necessarily matched to the activity of the TCA cycle, the "excess" pyruvate produced will be converted to lactate. In other words, difficult exercise results in the secretion of hormones that cause glycogenolysis and glycolysis to be greatly accelerated, resulting in pyruvate production in excess of TCA cycle needs. The excess pyruvate is converted to lactate by LDHase. This lactate formed in muscle appears in the blood. This is the so-called *mass action effect.*

The autonomic effect of vasoconstriction has another effect on lactate level. Blood flow is shunted toward active muscle and away from other tissues, such as liver and kidney (areas of increased resistance). These organs usually remove lactate and produce glucose. As blood flow to these gluconeogenic tissues decreases, they can compensate somewhat by extracting a greater fraction of the lactate passing through them in the systemic circulation. This ability to increase the fractional extraction, however, is limited, and blood lactate level increases because of limited capacity of gluconeogenic tissues to remove the lactate.

Although sympathetic nervous activity and epinephrine secretion during exercise augment muscle lactate production, the process of muscle excitation-contraction coupling involving release of Ca^{2+} from the sarcoplasmic reticulum (Chapter 18) causes glycogenolysis by activation of phosphorylase kinase (Chapter 5). Thus, even if epinephrine is not secreted, or if its action is blocked by a drug, contraction still results in glycogenolysis, glycolysis, and lactate production, at least transiently.

The relative effects of epinephrine stimulation and exercise on blood lactate appearance are illustrated in Figure 10-18 based on studies made on Pikes Peak in Colorado. Upon acute altitude exposure, epinephrine is high, an event that corresponds to very high lactate appearance. After acclimatization to altitude, epinephrine declines, and so does blood lactate appearance. However, the slope of the relationship between blood lactate R_a and epinephrine concentration is far greater for exercise than for rest. Therefore, both exercise (muscle contractions)

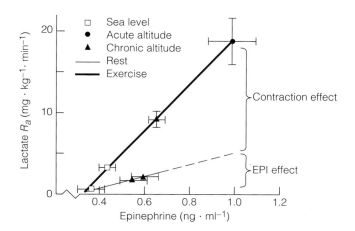

Figure 10-18 Mean blood lactate rate of appearance (R_a) vs. mean epinephrine concentration in men at sea level, on acute exposure to 4300-m altitude, and after 3 weeks of acclimatization to 4300 m. The total and individual effects of epinephrine stimulation and muscle contraction on lactate production (Ra) are indicated. Values are means ± SE. SOURCE: Brooks et al., 1991.

and epinephrine affect lactate production, with the smaller effect of epinephrine amplifying the larger exercise-related increase in lactate production rate.

The control of breathing during exercise is described in detail in Chapter 12. Various mechanisms ensure that ventilation is matched to the metabolic rate. The motor cortex affects the breathing centers in the brain to adjust the frequency and depth of breathing. These neural mechanisms operate rapidly and account for most of the breathing control during exercise. The fine-tuning of breathing to match exercise intensity depends on neural signals, and metabolite levels in blood, reaching the brain. The carotid and aortic bodies (see Figure 12-17) are sensitive to the partial pressure of O_2 and to the concentration of hydrogen ion (H^+). Carbon dioxide tension in blood affects the cerebrospinal fluid, which in turn affects respiratory centers in the brain. Receptors in joints and muscles are sensitive to movement and transmit signals to the brain.

Of the peripheral controls of breathing, two factors appear to be most important: (1) the pH of arterial blood, which affects the arterial chemoreceptors, and (2) to a lesser extent, the pH of the cerebrospinal fluid, which directly affects the respiratory center. Arterial pH is influenced by the blood lactate level and the venous delivery (flux) of CO_2 to the lungs. When CO_2 is dissolved in water, carbonic acid is formed. This acid dissociates a proton, as does lactic acid.

Activity level, metabolism, CO_2 production, and lactate production are usually closely matched. Therefore, CO_2 flux to the lungs is highly correlated with pulmonary minute ventilation. This effect, however, does not amount to a verification that muscle lacked sufficient oxygen. The effect simply means that there was acidosis. If there is a disturbance in blood pH, such as reduced glycolysis and lactate production due to glycogen depletion or McArdle's syndrome, or if acidifying or alkalinizing agents are ingested, CO_2 flux to the lungs and arterial pH are affected. The usual relationships among pulmonary ventilation, CO_2 flux to the lungs, and arterial pH are disturbed. Ventilation continues to be highly related to the metabolic rate because the neural controls of ventilation are of primary importance; the peripheral factors are of secondary importance.

■ Summary

Aerobic responses dominate the metabolic adjustments to exercise that lasts more than a minute. So important are aerobic responses to exercise that anaerobic responses have often been classified in terms of oxygen equivalency. This tendency has resulted in the evolution of three terms: O_2 debt, O_2 deficit, and anaerobic threshold. These terms originally provided theoretical bases for observed phenomena, but the theories involved have been shown to be incorrect. Because these terms have been used so widely, however—not only in exercise physiology and sports, but also in many other situations—it is necessary for the reader to understand their uses and limitations.

After exercise, O_2 consumption remains elevated for a while. This extra O_2 consumption has in the past been called the O_2 debt, to represent a repayment for the nonoxidative (anaerobic) metabolism performed during exercise. The extra O_2 consumed after exercise was thought to support the oxidation of a minor fraction (one-fifth) of the lactate formed during exercise, thereby providing the energy to convert the remainder (four-fifths) to glycogen. In reality, however, the cause of excess postexercise O_2 consumption (EPOC) is the general disturbance to homeostasis brought on by exercise. Exercise causes a rise in tissue temperatures, changes in intra- and extracellular ion concentrations, and changes in metabolite and hormone levels. Because these physiological changes persist into recovery, their effects serve to elevate O_2 consumption immediately after exercise.

Immediately after exercise, lactic acid provides a readily metabolizable reservoir of substrate. Depending on the conditions of recovery, this lactate reservoir may be mostly oxidized, or it may also contribute to the reestablishment of normal blood glucose levels. Lactic acid oxidation does not in itself result in elevated O_2 consumption because the lactate substitutes for other substrates. Rather, the energy necessary to return the organism to the preexercise condition results in EPOC, with lactate merely supplying the fuel.

Lactic acid is produced constantly, not just during hard exercise. In fact, lactic acid may be the most dynamic metabolite produced during exercise; lactate turnover exceeds that of any other metabolite yet studied. The most recent data indicate that mitochondria consume and oxidize lactate directly (Chapter 5). The constancy of the blood lactic acid level, during rest or exercise, means that the entry into and removal of lactate from the blood are in balance. Increasing blood lactate levels indicate that entry exceeds removal; declining levels indicate the opposite. Radio tracer studies have shown the turnover of lactic acid during exercise to be several times greater for a given blood lactate level during exercise than during rest. Similarly, for a given blood lactate level, blood lactate removal is several times greater in trained than in untrained individuals.

During exercise tests of increasing intensity, at about 60% of $\dot{V}_{O_2max}$, blood lactate levels begin increasing dramatically. This inflection point in blood lactic acid level has mistakenly been called the anaerobic threshold, meaning the point at which muscle becomes O_2 deficient. This is an archaic concept akin to the beliefs that muscle develops an O_2 debt during exercise and that lactic acid is a dead-end metabolite during exercise. Several factors appear to be responsible for the lactate inflection point during graded exercise: Contraction itself stimulates glycogenolysis and lactate production; in addition, hormone-mediated accelerations in glycogenolysis and glycolysis, recruitment of fast-glycolytic muscle fibers, and a redistribution of blood flow from lactate-removing, gluconeogenic tissues to lactate-producing, glycolytic tissues cause blood lactate to rise during exercise protocols that call for continually increasing power output. Simply, the rise in blood lactate during exercise means that removal mechanisms are inadequate to accommodate the rising tide of lactate production in muscle and other tissues. During exercise that demands continuously increasing power output, lactate production and lactate release into blood rise faster than clearance mechanisms can accommodate. However, during continuous, submaximal exercise, tissues other than muscle produce lactate for consumption by working muscle and blood lactate level can stabilize or fall as clearance mechanisms accommodate the increased production.

■ Selected Readings

Abramson, H. A., M. G. Eggleton, and P. Eggleton. The utilization of intravenous sodium r-lactate. III. Glycogen synthesis by the liver. Blood sugar. Oxygen consumption. *J. Biol. Chem.* 75: 763–778, 1927.

Ahlborg, G., and P. Felig. Lactate and glucose exchange across the forearm, legs, and splanchnic bed during and after prolonged leg exercise. *J. Clin. Invest.* 69: 45–54, 1982.

Alpert, N. R. Lactate production and removal and the regulation of metabolism. *Ann. N.Y. Acad. Sci.* 119: 995–1012, 1965.

Alpert, N. R., and W. S. Root. Relationship between excess respiratory metabolism and utilization of intravenously infused sodium racemic lactate and sodium L-(+)-lactate. *Am. J. Physiol.* 177: 455–562, 1954.

Bang, O. The lactate content of the blood during and after muscular exercise in man. *Skand. Arch. Physiol.* 74 (Suppl 10): 49–82, 1936.

Barnard, R. J., and M. L. Foss. Oxygen debt: effect of beta-adrenergic blockade on the lactacid and alactacid components. *J. Appl. Physiol.* 27: 813–816, 1969.

Barnard, R. J., M. L. Foss, and C. M. Tipton. Oxygen debt: involvement of the Cori cycle. *Int. Z. Angew. Physiol.* 28: 105–119, 1970.

Bendall, J. R., and A. A. Taylor. The Meyerhof quotient and the synthesis of glycogen from lactate in frog and rabbit muscle. *Biochem. J.* 118: 887–893, 1970.

Benedict, F. G., and E. P. Cathcart. *Muscular Work.* (Publ. 187). Washington, D. C.: Carnegie Institution of Washington, 1913.

Bennett, A. F., and P. Licht. Relative contributions of anaerobic and aerobic energy production during activity in Amphibia. *J. Comp. Physiol.* 87: 351–360, 1973.

Børsheim, E., R. Bahr, and S. Knardahl. Effect of beta-adrenoceptor stimulation on oxygen consumption and triglyceride/fatty acid cycling after exercise. *Acta Physiol. Scand.* 164: 157–166, 1998.

Børsheim, E., R. Bahr, A. T. Hostmark, and E. S. Knardahl. Effect of beta-adrenoceptor blockade on postexercise oxygen consumption and triglyceride/fatty acid cycling. *Metabolism.* 47: 439–448, 1998.

Børsheim, E. S. Knardahl, A. T. Hostmark, and R. Bahr. Adrenergic control of post-exercise metabolism. *Acta Physiol. Scand.* 162: 313–323, 1998.

Brooks, G. A. Lactate: glycolytic end product and oxidative substrate during sustained exercise in mammals—the "lactate shuttle." In Circulation, Respiration, and Metabolism: Current Comparative Approaches, R. Gilles (Ed.). Berlin: Springer-Verlag, 1985, pp. 208–218.

Brooks, G. A. Training improves lactate clearance. In Membranes and Muscle, M. C. Berman, W. Gevers, and L. H. Opie (Eds.). Oxford: RL Press, 1985, pp. 257–275.

Brooks, G. A. Lactate production under fully aerobic conditions: the lactate shuttle during rest and exercise. *Fed. Proc.* 45: 2924–2929, 1986.

Brooks, G. A. The lactate shuttle during exercise and recovery. *Med. Sci. Sports Exer.* 18: 360–368, 1986.

Brooks, G. A. (Ed.). Perspectives on the Academic Discipline of Physical Education. Champaign, Ill.: Human Kinetics, 1981, pp. 97–120.

Brooks, G. A., K. E. Brauner, and R. G. Cassens. Glycogen synthesis and metabolism of lactic acid after exercise. *Am. J. Physiol.* 224: 1162–1166, 1973.

Brooks, G. A., G. E. Butterfield, R. R. Wolfe, B. M. Groves, R. S. Mazzeo, J. R. Sutton, E. E. Wolfel, and J. T. Reeves. Increased dependence on blood glucose after acclimatization to 4,300 m. *J. Appl. Physiol.* 70: 919–927, 1991.

Brooks, G. A., G. E. Butterfield, R. R. Wolfe, B. M. Groves, R. S. Mazzeo, J. R. Sutton, E. E. Wolfel, and J. T. Reeves. Decreased reliance on lactate during exercise after acclimatization to 4,300 m. *J. Appl. Physiol.* 71: 333–341, 1991.

Brooks, G. A., and C. M. Donovan. Effect of endurance training on glucose kinetics during exercise. *Am. J. Physiol. (Endocrinol. Metab.* 7). (In press).

Brooks, G. A., and G. A. Gaesser. End points of lactate and glucose metabolism after exhausting exercise. *J. Appl. Physiol.: Respirat. Environ. Exercise Physiol.* 49: 1057–1069, 1980.

Brooks, G. A., K. J. Hittelman, J. A. Faulkner, and R. E. Beyer. Temperature, liver mitochondrial respiratory functions, and oxygen debt. *Med. Sci. Sports* 2: 72–74, 1971.

Brooks, G. A., K. J. Hittelman, J. A. Faulkner, and R. E. Beyer. Temperature, skeletal muscle mitochondrial functions, and oxygen debt. *Am. J. Physiol.* 220: 1053–1059, 1971.

Brooks, G. A., K. J. Hittelman, J. A. Faulkner, and R. E. Beyer. Tissue temperatures and whole-animal oxygen consumption after exercise. *Am. J. Physiol.* 221: 427–431, 1971.

Cain, S. M. Exercise O_2 debts of dogs at ground level and at altitude with and without β-block. *J. Appl. Physiol.* 30: 838–843, 1971.

Cain, S. M., and C. K. Chapler. Effects of norepinephrine and α-block on O_2 uptake and blood flow in dog hindlimb. *J. Appl. Physiol.: Respirat. Environ. Exercise Physiol.* 51: 1245–1250, 1981.

Carafoli, E., and A. L. Lehninger. A survey of the interaction of calcium ions with mitochondria from different tissues and species. *Biochem. J.* 122: 681–690, 1971.

Chance, B., and C. R. Williams. The respiratory chain and oxidative phosphorylation. In Advances in Enzymology, vol. 17. New York: Interscience, 1956, pp. 65–134.

Chapler, C. K., W. M. Stainsby, and L. B. Gladden. Effect of changes in blood flow, norepinephrine, and pH on oxygen uptake by resting skeletal muscle. *Can. J. Physiol. Pharmacol.* 58: 91–96, 1980.

Claremont, A. D., F. Nagle, W. D. Reddan, and G. A. Brooks. Comparison of metabolic, temperature, heart rate and ventilatory responses to exercise at extreme ambient temperatures (0°C and 35°C). *Med. Sci. Sports* 7: 150–154, 1975.

Cori, C. F. Mammalian carbohydrate metabolism. *Physiol. Rev.* 11: 143–275, 1931.

Depocas, F., Y. Minaire, and J. Chatonnet. Rate of formation and oxidation of lactic acid in dogs at rest and during moderate exercise. *Can. J. Physiol. Pharmacol.* 47: 603–610, 1969.

Donovan, C. M., and G. A. Brooks. Endurance training affects lactate clearance, not lactate production. *Am. J. Physiol.* 244 (*Endocrinol. Metab.* 7): E83–E92, 1983.

Drury, D. R., and A. N. Wick. Metabolism of lactic acid in the intact rabbit. *Am. J. Physiol.* 184: 304–308, 1956.

Drury, D. R., and A. N. Wick. Chemistry and metabolism of L(+) and D(−) lactic acids. *Ann. N.Y. Acad. Sci.* 119: 1061–1069, 1965.

Dubois, E. F. The basal metabolism in fever. *J.A.M.A.* 77: 352–355, 1921.

Eggleton, M. G., and C. L. Evans. The lactic acid content of the blood after muscular contraction under experimental conditions. *J. Physiol.* (London) 70: 269–293, 1930.

Eldridge, F. L. Relationships between lactate turnover rate and blood concentration in hemorrhagic shock. *J. Appl. Physiol.* 37: 321–323, 1974.

Eldridge, F. L. Relationship between turnover rate and blood concentration of lactate in exercising dogs. *J. Appl. Physiol.* 39: 231–234, 1975.

Eldridge, F. L., L. T'so, and H. Chang. Relationship between turnover rate and blood concentration of lactate in normal dogs. *J. Appl. Physiol.* 37: 316–320, 1974.

Fell, R., J. McLane, W. Winder, and J. Holloszy. Preferential resynthesis of muscle glycogen in fasting rats after exhausting exercise. *Am. J. Physiol.* 238 (*Regulatory Integrative Comp. Physiol.* 7): R328–R332, 1980.

Fletcher, W. M., and F. G. Hopkins. Lactic acid in amphibian muscle. *J. Physiol.* 35: 247–309, 1907.

Freminet, A., E. Bursaux, and C. F. Poyart. Effect of elevated lactataemia on the rates of lactate turnover and oxidation in rats. *Pflugers Arch.* 346: 75–86, 1974.

Gaesser, G. A., and G. A. Brooks. Glycogen repletion following continuous and intermittent exercise to exhaustion. *J. Appl. Physiol.: Respirat. Environ. Exercise Physiol.* 49: 722–728, 1980.

Gaesser, G. A., and G. A. Brooks. Metabolic bases of excess post-exercise oxygen consumption: a review. *Med. Sci. Sports Exer.* 16: 29–43, 1984.

Gaesser, G. A., and D. C. Poole. Blood lactate during exercise: time course of training adaptations in humans. *Int. J. Sports Med.* 9: 284–288, 1988.

Gladden, L., W. Stainsby, and B. MacIntosh. Norepinephrine increase canine, skeletal muscle $\dot{V}_{O_2}$ during recovery. *Med. Sci. Sports Exer.* 14: 471–476, 1982.

Gladden, L. B., and J. W. Yates. Lactate infusion in dogs: effects of varying infusate pH. *J. Appl. Physiol.* 54: 1254–1260, 1983.

Gleeson, T. T. Metabolic recovery from exhaustive activity by a large lizard. *J. Appl. Physiol.: Respirat. Environ. Exercise Physiol.* 48: 689–694, 1980.

Hagberg, J. M., E. F. Coyle, J. E. Carroll, J. M. Miller, W. H. Martin, and M. H. Brooke. Exercise hyperventilation in patients with McArdle's disease. *J. Appl. Physiol.: Respirat. Environ. Exercise Physiol.* 52: 991–994, 1982.

Hagberg, J. M., J. P. Mullin, and F. J. Nagle. Effect of work intensity and duration on recovery O_2. *J. Appl. Physiol.: Respirat. Environ. Exercise Physiol.* 48: 540–544, 1980.

Harris, P. Lactic acid and the phlogiston debt. *Cardiov. Res.* 3: 381–390, 1969.

Harris, P., M. Bateman, T. J. Kayley, K. W. Donald, J. Gloster, and T. Whitehead. Observations on the course of the metabolic events accompanying mild exercise. *Quart. J. Exp. Physiol.* 53: 43–64, 1968.

Harris, R., R. Edwards, E. Hultman, L.-O. Nordesjö, B. Nyland, and K. Sahlin. The time course of phosphorylcreatine resynthesis during recovery of the quadriceps muscle in man. *Pflügers Arch.* 367: 137–142, 1976.

Hartree, W., and A. V. Hill. The recovery heat-production in muscle. *J. Physiol.* 56: 367–381, 1922.

Heigenhauser, G. J. F., J. R. Sutton, and N. L. Jones. Effect of glycogen depletion on ventilatory response to exercise. *J. Appl. Physiol.* 54: 470–474, 1983.

Henry, F. M., and J. DeMoor. Lactic and alactic oxygen consumption in moderate exercise of graded intensity. *J. Appl. Physiol.* 8: 608–614, 1956.

Hermansen, L., and Il Stensvold. Production and removal of lactate during exercise in man. *Acta Physiol. Scand.* 86: 191–201, 1972.

Hill, A. V. The energy degraded in the recovery processes of stimulated muscles. *J. Physiol.* 46: 28–80, 1913.

Hill, A. V. The oxidative removal of lactic acid. *J. Physiol.* 48 (*Proc. Physiol. Soc.*) x–xi, 1914.

Hill, A. V., C. N. H. Long, and H. Lupton. Muscular exercise, lactic acid and the supply and utilization of oxygen. Pt. I–III *Proc. Roy. Soc. B.* 96: 438–475, 1924.

Hill, A. V., C. N. H. Long, and H. Lupton. Muscular exercise, lactic acid and the supply and utilization of oxygen. Pt. IV–VI. *Proc. Roy. Soc. B.* 97: 84–138, 1924.

Hill, A. V., C. N. H. Long, and H. Lupton. Muscular exercise, lactic acid and the supply and utilization of oxygen. Pt. VII–IX. *Proc. Roy. Soc. B.* 97: 155–176, 1924.

Hill, A. V., and H. Lupton. Muscular exercise, lactic acid, and the supply and utilization of oxygen. *Quart. J. Med.* 16: 135–171, 1923.

Hittelman, K. J., O. Lindberg, and B. Cannon. Oxidative phosphorylation and compartmentation of fatty acid metabolism in brown fat mitochondria. *Europ. J. Biochem.* 11: 183–192, 1969.

Horwitz, H. A. Metabolic aspects of thermogenesis: neuronal and hormonal control. *Fed. Proc.* 38: 2147–2149, 1979.

Hubbard, J. L. The effect of exercise on lactate metabolism. *J. Physiol.* 231: 1–18, 1973.

Hughes, E. F., S. C. Turner, and G. A. Brooks. Effects of glycogen depletion and pedaling speed on the "anaerobic threshold." *J. Appl. Physiol: Respirat. Environ. Exercise Physiol.* 52: 1598–1607, 1982.

Issekutz, B., W. A. S. Shaw, and A. C. Issekutz. Lactate metabolism in resting and exercising dogs. *J. Appl. Physiol.* 40: 312–319, 1976.

Ivy, J. L., D. L. Costill, P. J. Van Handel, D. A. Essig, and R. W. Lower. Alteration in the lactate threshold with changes in substrate availability. *Int. J. Sports Med.* 2: 139–142, 1981.

Jöbsis, F. F., and W. N. Stainsby. Oxidation of NADH during contractions of circulated mammalian skeletal muscle. *Resp. Physiol.* 4: 292–300, 1968.

Jones, N., and R. Ehrsam. The anaerobic threshold. In Exercise and Sports Sciences Reviews, R. Terjung (Ed.). 10: 49–83, 1982.

Jorfeldt, L. Metabolism of L(+)-lactate in human skeletal muscle during exercise. *Acta. Physiol. Scand.* 338 (Suppl.), 1970.

Jorfeldt, L., A. Juhlin-Dannfeldt, and J. Karlsson. Lactate release in relation to tissue lactate in human skeletal muscle during exercise. *J. Appl. Physiol.: Respirat. Environ. Exercise Physiol.* 44: 350–352, 1978.

Kayne, H. L., and N. R. Alpert. Oxygen consumption following exercise in the anesthetized dog. *Am. J. Physiol.* 206: 51–56, 1964.

Keul, J., E. Doll, and D. Keppler. The substrate supply of the human skeletal muscle at rest, during and after work. *Experientia.* 23: 974–979, 1967.

Knuttgen, H. Oxygen debt, lactate, pyruvate, and excess lactate after muscular work. *J. Appl. Physiol.* 17: 639–644, 1962.

Knuttgen, H. Oxygen debt after submaximal exercise. *J. Appl. Physiol.* 29: 651–657, 1970.

Knuttgen, H. Lactate and oxygen debt: an introduction. In Muscle Metabolism During Exercise, B. Pernow, and B. Saltin (Eds.). New York: Plenum Press, 1971, pp. 361–369.

Krebs, H. A., R. Hems, M. J. Weidmann, and R. N. Speake. The fate of isotopic carbon in kidney cortex synthesizing glucose and lactate. *Biochem. J.* 101: 242–249, 1966.

Krebs, H. A., and M. Woodford. Fructose 1,6-diphosphatase in striated muscle. *Biochem. J.* 94: 436–445, 1965.

Lee, S.-H., and E. J. Davis. Carboxylation and decarboxylation reactions. *J. Biol. Chem.* 254: 420–430, 1979.

Lundsgaard, E. Untersuchungen über Muskel-kontraktionen ohne Milchsäurebildung. *Biochem. Z.* 217: 162–177, 1930.

MacDougall, J., G. Ward, D. Sale, and J. Sutton. Muscle glycogen repletion after high intensity intermittent exercise. *J. Appl. Physiol.: Respirat. Environ. Exercise Physiol.* 42: 129–132, 1977.

MacRae, H. S.-H., S. C. Dennis, A. N. Bosch, and T. D. Noakes. Effects of training on lactate production and removal during progressive exercise in humans. *J. Appl. Physiol.* 72: 1649–1656, 1992.

Maehlum, S., P. Felig, and J. Wahren. Splanchnic glucose and muscle glycogen metabolism after glucose feeding during postexercise recovery. *Am. J. Physiol.* 235 (*Endocrinol. Metab. Gastrointest. Physiol.* 4): E255–E260, 1978.

Maehlum, S., and L. Hermansen. Muscle glycogen concentration during recovery after prolonged severe exercise in fasting subjects. *Scand. J. Clin. Lab. Invest.* 38: 557–560, 1978.

Margaria, R., H. T. Edwards, and D. B. Dill. The possible mechanisms of contracting and paying the oxygen debt and the role of lactic acid in muscular contraction. *Am. J. Physiol.* 106: 689–715, 1933.

Mazzeo, R. S., G. A. Brooks, D. A. Schoeller, and T. F. Budinger. Pulse injection ^{13}C-tracer studies of lactate metabolism in humans during rest and two levels of exercise. *Biomed. Mass. Spect.* 9: 310–314, 1982.

McLane, J. A., and J. O. Holloszy. Glycogen synthesis from lactate in the three types of skeletal muscle. *J. Biol. Chem.* 254: 6548–6553, 1979.

Meyerhof, O. Die Energie-wandlungen im Muskel. I. Über die Beziehungen der Milschsäure zur Wärmebildung und Arbeitstung des Muskels in der Anaerobiose. *Arch. Ges. Physiol.* 182: 232–282, 1920.

Meyerhof, O. Über die Energieumwandlungen im Muskel. II. Das Schicksal der Milchsäure in der Erholungsperiode des Muskels. *Pflügers Arch. Ges. Physiol.* 182: 284–317, 1920.

Meyerhof, O. Die Energieuwandlungen im Muskel. III. Kohlenhydrat und Milchsaureumsatz in Froschmuskel. *Arch. Ges. Physiol.* 185: 11–25, 1920.

Moorthy, K. A., and M. K. Gould. Synthesis of glycogen from glucose and lactate in isolated rat soleus muscle. *Arch. Biochem. Biophys.* 130: 399–407, 1969.

Moritani, T., H. Tanaka, T. Yoshida, C. Ishi, T. Yoshida, and M. Shindo. Relationship between myoelectric signals and blood lactate during incremental forearm exercise. *Am. J. Phys. Med.* 63: 122–132, 1984.

Newsholme, E. Substrate cycles: their metabolic, energetic and thermic consequences in man. *Biochem. Soc. Symp.* 43: 183–205, 1978.

Opie, L. H., and E. A. Newsholme. The activities of fructose 1,6-diphosphatase, phosphofructokinase and phosphoenolypyruvate carboxykinase in white muscle and red muscle. *Biochem. J.* 103: 391–399, 1967.

Piiper, J., and P. Spiller. Repayment of the O_2 debt and resynthesis of high energy phosphates in gastrocnemius muscle of the dog. *J. Appl. Physiol.* 28: 657–662, 1970.

Richter, E., N. Ruderman, H. Gavras, E. Belur, and H. Galbo. Muscle glycogenolysis during exercise: dual control by epinephrine and contractions. *Am. J. Physiol.* 242 (*Endocrinol. Metab.* 5): E25–E32, 1982.

Ross, R. D., R. Hems, and H. A. Krebs. The rate of gluconeogenesis from various precursors in the perfused rat liver. *Biochem. J.* 102: 942–951, 1967.

Rowell, L. B., K. K. Kraning II, T. O. Evans, J. W. Kennedy, J. R. Blackmon, and F. Kusumi. Splanchnic removal of lactate and pyruvate during prolonged exercise in man. *J. Appl. Physiol.* 21: 1773–1783, 1966.

Rowell, L. B., B. Saltin, B. Kiens, and N. J. Christensen. Is peak quadriceps blood glow in humans even higher during exercise with hypoxemia? *Am. J. Physiol.* 251: H1038–H1044, 1986.

Ryan, W., J. Sutton, C. Toews, and N. Jones. Metabolism of infuse L(+)-lactate during exercise. *Clin. Sci.* 56: 139–146, 1979.

Sacks, J., and W. Sacks. Carbohydrate changes during recovery from muscular contraction. *Am. J. Physiol.* 112: 565–572, 1935.

Saltin, B., K. Nazar, D. L. Costill, E. Stein, E. Jansson, B. Essen, and P. D. Gollnick. The nature of the training response: peripheral and central adaptations to one-legged exercise. *Acta Physiol. Scand.* 96: 289–305, 1976.

Scheen, A., J. Juchmes, and A. Cession-Fossion. Critical analysis of the "anaerobic threshold" during exercise at constant workloads. *Eur. J. Appl. Physiol.* 46: 367–377, 1981.

Scrutton, M. C., and M. F. Utter. The regulation of glycolysis and gluconeogenesis in animal tissues. *Ann. Rev. Biochem.* 37: 250–302, 1968.

Segal, S. S., and G. A. Brooks. Effects of glycogen depletion and workload on postexercise O_2 consumption and blood lactate. *J. Appl. Physiol.: Respirat. Environ. Exercise Physiol.* 47: 514–521, 1979.

Smith, R. E., and D. J. Hoijer. Metabolism and cellular function in cold acclimation. *Physiol. Rev.* 42: 50–142, 1962.

Stainsby, W. N., and J. K. Barclay. Exercise metabolism: O_2 deficit, steady level O_2 uptake and O_2 uptake for recovery. *Med. Sci. Sports* 2: 177–186, 1970.

Stanley, W. C., E. W. Gertz, J. A. Wisneski, D. L. Morris, R. Neese, and G. A. Brooks. Systemic lactate turnover during graded exercise in man. *Am. J. Physiol. (Endocrinol. Metab.* 12) 249: E595–E602, 1985.

Stanley, W. C., E. W. Gertz, J. A. Wisneski, D. L. Morris, R. Neese, and G. A. Brooks. Lactate metabolism in exercising human skeletal muscle: evidence for lactate extraction during net lactate release. *J. Appl. Physiol.* 60: 1116–1120, 1986.

Stanley, W. C., J. A. Wisneski, E. W. Gertz, R. A. Neese, and G. A. Brooks. Glucose and lactate interrelations during moderate intensity exercise in man. *Metabolism* 37: 850–858, 1988.

Warnock, L. G., R. E. Keoppe, N. F. Inciardi, and W. E. Wilson. L(+) and D(−) lactate as precursors of muscle glycogen. *Ann. N.Y. Acad. Sci.* 119: 1048–1060, 1965.

Wasserman, K., B. J. Whipp, S. N. Koyal, and W. L. Beaver. Anaerobic threshold and respiratory gas exchange during exercise. *J. Appl. Physiol.* 35: 236–243, 1973.

Welch, H. G., J. A. Faulkner, J. K. Barclay, and G. A. Brooks. Ventilatory response during recovery from muscular work and its relation with O_2 debt. *Med. Sci. Sports* 2: 15–19, 1970.

Welch, H. G., and W. N. Stainsby. Oxygen debt in contracting dog skeletal muscle *in situ. Resp. Physiol.* 3: 229–242, 1967.

CHAPTER 11

THE WHY OF PULMONARY VENTILATION

The movement of air into and out of the pulmonary system is called *breathing,* or *ventilation.* Four specific purposes are accomplished by breathing: (1) the exchange of O_2, (2) the exchange of CO_2, (3) the control of blood acidity, and (4) oral communication. The rapidity and depth of breathing affect the amount of O_2 and CO_2 exchanged between body and atmosphere. In general, breathing is essential to the cellular bioenergetic processes of life. Gases such as O_2 and CO_2 move from areas of high concentration (or pressure) to areas of lower concentration or pressure. The pressure exerted by a particular type of gas is called *partial pressure.*

Oxygen and CO_2 are exchanged between the atmosphere and the blood perfusing the alveoli (air sacs) of the lungs. This process of ventilation results in a relatively higher partial pressure of O_2 in the lungs' alveoli than in the metabolizing tissues or the venous blood draining those tissues. By keeping the alveolar partial pressure of O_2 at about 105 mmHg, then, ventilation creates a positive pressure gradient whereby O_2 moves from the alveoli into the blood, which then circulates around the body to deliver O_2 to metabolizing tissues. Also through the action of ventilation, the partial pressure of the metabolic waste, CO_2, is kept relatively low in the alveoli. This creates a negative pressure gradient for CO_2 to move from tissues, through blood and alveoli, to the atmosphere.

Metabolic CO_2 dissolved in the body fluid forms carbonic acid, which then dissociates a proton or H^+ ($CO_2 + H_2O \rightleftharpoons H_2CO_3 \rightleftharpoons H^+ + HCO_3^-$). Because ventilation affects CO_2 exchange and storage, and because CO_2 affects H^+ concentration (or pH), ventilation affects blood acid–base balance (i.e., the balance of acids and bases in the blood).

■ Breathing, Ventilation, and Respiration

During the eighteenth century, the great French scientist Lavoisier and others believed that biological oxidation occurred in the lungs. For this reason, ventilation or breathing came to be known as respiration. Today, by convention in the study of physiology, the three terms *ventilation, breathing,* and *respiration* are used synonymously. Properly speaking, however, ventilation is the breathing of air into and out of the pulmonary system (nose, mouth, trachea, lungs), whereas respiration is the cellular utilization of O_2. Ventilation is but one step leading to respiration. Respiration is the major cellular mechanism of energy conversion (Chapters 2–6).

Rhythmicity in Ventilation

We breathe continually by means of a complex neural control mechanism. Sometimes, such as dur-

Climbers on Mt. Everest. © Art Wolfe

ing heavy exercise, the vigor of ventilation makes breathing noticeable; usually, however, breathing goes on unnoticed. Breathing is usually controlled at an unconscious level, but sometimes we can volitionally modify our pattern of breathing. The trumpeter and the opera singer must breathe continually, but they coordinate ventilation with the muscle requirements of their tasks. Similarly, the swimmer must coordinate breathing into complex stroke mechanics. With practice, complex conscious breathing patterns can be learned and precisely integrated into motor performance. Breathing then becomes, again, automatic and nearly unconscious.

The movement of air into the pulmonary system (inhalation) and the movement out of previously inhaled air (expiration) is regulated in two basic ways: by frequency and by amount or volume. Within seconds after an inhalation, the partial pressure of O_2 in alveoli falls below its optimum, and the partial pressure of CO_2 becomes greater than optimal. Very shortly after beginning to hold a breath, the sensation to breathe again becomes very strong. This sen-

sation is delayed only slightly if we take a large breath and hold it, as opposed to taking a small breath. It is a good thing that we ventilate so frequently, because the O_2 content of our lungs, inflated to maximum, can sustain us for only a few minutes, even at rest.

Pulmonary Minute Volume

The rate of pulmonary ventilation is usually expressed in terms of volume (liters) per minute. The abbreviation for pulmonary ventilation per minute (pulmonary minute volume) is ($\dot{V}$), where the V and the dot refer to volume and per minute, respectively. Pulmonary ventilation can be measured either as volume expired per minute ($\dot{V}_E$) or volume inspired per minute ($\dot{V}_I$). Because the respiratory exchange ratio (R) is not necessarily 1.0, because H_2O vapor is added to inspired air, and because inspired air is warmed during ventilation, $\dot{V}_I$ and $\dot{V}_E$ are not necessarily equal.

Pulmonary minute volume ($\dot{V}$) is equal to the product of the frequency of breathing during a minute (f) and the average volume of air moved on each ventilatory excursion ($\bar{V}_T$, or average tidal volume):

$$\dot{V}_E = (f)(\bar{V}_T) \qquad (11\text{-}1)$$

■ Environmental Influences on Pulmonary Gas Volumes

Environmental conditions have a significant effect on pulmonary gas volumes. [Refer to Appendix I for a review of pulmonary (respiratory) symbols and terminology.] Pulmonary gas volumes can be defined by three sets of conditions, or standards. The first of these is the STPD volume, where ST = standard temperature = 0°C, P = standard pressure = 760 mmHg = 1 atmosphere (atm), and D = dry = 0.0 mmHg H_2O vapor pressure. Clearly, the STPD condition is nonphysiological, but it is a norm by which results obtained at different times and places can be compared.

The second pulmonary gas condition is the BTPS volume, where BT = body temperature,

P = ambient pressure, S = saturated with H_2O. At approximately 1 atm pressure, the P_{H_2O} depends on temperature; for body temperature (37°C), the P_{H_2O} = 47 mmHg pressure. The BTPS volume is the volume that a subject actually exhales.

The third pulmonary gas condition of measurement is the ATPS volume, where AT = the ambient temperature in degrees Celsius, P = the ambient pressure, and S = the ambient P_{H_2O}. The ATPS volume is the volume that a subject actually inhales.

In expressing pulmonary gas volumes, $\dot{V}_{O_2}$ is usually given in liters $\cdot$ min^{-1} (STPD), $\dot{V}_E$ is usually given in liters $\cdot$ min^{-1} (BTPS), and $\dot{V}_I$ is usually given in liters $\cdot$ min^{-1} (ATPS).

■ Entry of O_2 into Blood

We ventilate to keep the partial pressure of O_2 in the alveoli at about 105 mmHg. From the alveoli, the diffusion distance for O_2 into erythrocytes (red blood cells) in the blood perfusing the alveolar walls is relatively short (Figure 11-1). The short distance is necessary because the solubility of O_2 in body water at 37°C is low; only 0.3 ml $O_2 \cdot$ dl^{-1} blood is physically dissolved. Fortunately, erythrocytes contain the heme-iron compound hemoglobin, which can bind O_2 according to its partial pressure (Figure 11-2). At an O_2 partial pressure of 100 mmHg, which exists in alveolar capillaries at sea level, hemoglobin is nearly 100% saturated with O_2. Because a very small percentage of blood passing through the lungs passes through alveoli that are not ventilated, the saturation (S) of blood with oxygen returning to the left heart from the lungs is about 96–98%. This impressive figure is maintained not only in the individual resting at a sea-level altitude, but also during maximal exercise (Figure 11-3).

Quantitatively, normal hemoglobin can bind 1.34 ml of $O_2 \cdot$ g^{-1}. In the average male, blood hemoglobin is about 15 g $\cdot$ dl^{-1} blood. With an arterial saturation (S_{aO_2}) of close to 100%, the arterial O_2 content (C_{aO_2}) is then equal to the sum of the dissolved O_2 plus that combined with hemoglobin as seen in Equation 11-2. By convention, this figure can also be referred to as 20.4 vol %, or 20.4 ml $O_2 \cdot$ 100 ml^{-1}; note that 1 dl = 100 ml, and vol % = vol $\cdot$ 100 ml^{-1}. In females, where the blood hemoglobin concentration is less than that in males (about 13 g $\cdot$ dl^{-1} in females), the C_{aO_2} = 17.7 vol %.

To summarize the relationships among pulmonary ventilation, P_{AO_2}, P_{aO_2}, S_{aO_2}, and C_{aO_2}, we state the following for healthy young people under sea-level conditions. Appropriate ventilation of the alveoli allows the alveolar partial pressure of oxygen (i.e., the P_{AO_2}) to remain at resting levels (about 100 mmHg) or to increase during exercise. The partial pressure of oxygen in the pulmonary vein (and, hence, in arterial blood, P_{aO_2}) is thus about 100 mmHg. Assuming normal hemoglobin and blood hemoglobin content, a P_{aO_2} of 100 mmHg results in an arterial saturation (S_{aO_2}) of close to 100%.

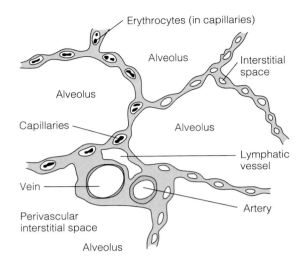

Figure 11-1 Schematic of a cross-sectional view through alveolar walls showing the relationship between the "alveolar air" and blood supply. Modified from Guyton, 1976, p. 538. Used with permission.

Arterial O_2 content = O_2 physically dissolved + O_2 in combination with hemoglobin

$$C_{aO_2} = 0.3 \text{ ml } O_2 \cdot dl^{-1} \text{ blood} + (1.34 \text{ ml } O_2 \cdot g^{-1} \text{ Hb})(15 \text{ g Hb} \cdot dl^{-1} \text{ blood}) \quad (11\text{-}2)$$
$$= 20.4 \text{ ml } O_2 \cdot dl^{-1} \text{ blood}$$

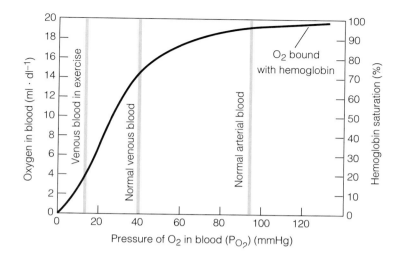

Figure 11-2 According to the oxygen–hemoglobin dissociation curve, the O_2 carried in blood in combination with hemoglobin depends mainly on the partial pressure of O_2. At normal arterial pressure of O_2 (P_{O_2} approximately 95 mmHg), hemoglobin is 95 to 98% saturated with O_2.

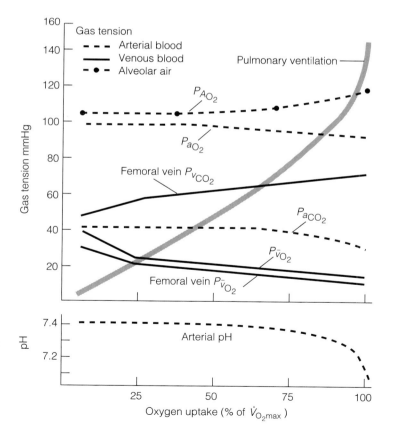

Figure 11-3 Oxygen and carbon dioxide partial pressures in alveolar air (P_A), arterial blood (P_a), and mixed venous blood ($P_{\bar{v}}$) during rest and graded exercise. Note that as relative effort increases, the partial pressure of O_2 in arterial blood (P_{aO_2}) remains constant or falls only slightly. Due to the shape of the oxyhemoglobin dissociation curve (Figure 11-2), arterial O_2 content remains close to resting levels of approximately 95 to 98% saturation. Modified from Åstrand and Rodahl, 1970. Used with permission.

At close to complete saturation, arterial blood has an oxygen content ($C_{a_{O_2}}$) of about 20 ml $\cdot$ 100 dl^{-1}, or 200 ml $\cdot$ liter^{-1}. Because $P_{a_{O_2}}$ is maintained at about 100 mmHg up to maximum effort, it is generally believed that in healthy young people ventilation does not limit oxygen transport during exercise at sea-level altitudes.

Blood Hemoglobin and Hematocrit

In Table 11-1, normal values for blood hemoglobin and hematocrit (i.e., % of red blood cells in blood), and arterial O_2 content are shown in young men before and after endurance training. Note that plasma and total blood volume increase as much or more than total hemoglobin content as a result of training. This plasma and blood volume expansion is associated with albumin synthesis and the necessity to transport heat and resist dehydration during exercise (Chapter 22). Hematocrit values in some heat acclimatized endurance athletes have been found to be significantly lower than in sedentary controls even though total hemoglobin content is higher in athletes than in non-athletes. Consequently, a mis-

diagnosis of "athlete anemia" can be made in some endurance athletes if plasma volume is not measured. As we shall also see in Chapter 23, acute altitude exposure can result in a diuresis and loss of plasma volume, especially if dietary energy intake is inadequate. Acute loss of plasma volume can increase blood hemoglobin concentration, thus partially compensating for the decrease in $P_{a_{O_2}}$ due to the low ambient barometric pressure at high altitude.

■ Pulmonary Diffusion

Solubility and Diffusion of Gases in Liquids

The diffusion of a gas through a liquid depends on several factors. Of these, perhaps the most important is pressure difference. As already indicated, gas diffuses from areas of high to low relative pressure. Moreover, breathing helps regulate pulmonary partial gas pressures.

In addition to pressure difference, other factors also effect the net rate of diffusion of gases in fluids. The greater the solubility of a gas, the greater the amount that can diffuse for a given pressure difference. *Solubility coefficients* for respiratory gases are given in Table 11-2. Some types of gas molecules (e.g., CO_2) are attracted to water and are, therefore, more soluble than gas molecules that are repelled. For this reason, CO_2 is over 20 times more soluble than O_2.

Diffusion is facilitated by increasing the liquid cross-sectional area and decreasing the distance through which gases must diffuse. Therefore, diffusion rate (D) is:

$$D \propto \frac{\Delta P \; A \; S}{d\sqrt{MW}} \qquad (11\text{-}3)$$

where: D is the diffusion rate, ΔP is the pressure difference between the two ends of the diffusion pathway, A is the cross-sectional area of the pathway, S is the solubility of the gas, d is the diffusion distance, and $\sqrt{MW}$ is the square root of the molecular weight of the gas.

The above formula can be simplified somewhat because both the solubility and molecular weights

TABLE 11-1

Representative values for blood hemoglobin concentration, hematocrit, and arterial oxygen content in young men before and after endurance training.

Variable	Before Training	After Training	Change (%)
Hemoglobin (g/dl)	15.3	15.1	−1
Blood volume (l)	5.25	6.58	25
Total hemoglobin (g)	803	994	24
Arterial O_2 content (ml/dl)	20.8	20.5	
Hematocrit (%)	42	41	−2

SOURCE: Kjellberg et al., 1949; Wolfel et al., 1991, 1998; and other sources.

TABLE 11-2

Solubility and relative diffusion characteristics of respiratory gases in humans.

Gas	Solubility Coefficient	Relative Diffusion Coefficient	Diffusing Capacity Rest (ml/min/mmHg)	Diffusing Capacity Exercise (ml/min/mmHg)
Oxygen	0.024	1.0	20	65
Carbon dioxide	0.57	20.3	400	1300
Nitrogen	0.012	0.53		
Helium	0.008	0.95		

of individual gases are unique characteristics of those gases. The diffusion coefficient of a gas is proportional to $S/\sqrt{MW}$. Respiratory physiologists compute relative diffusion coefficients of gases in aqueous solution with oxygen as the standard.

Diffusion of Gases Through Respiratory Membranes (Pulmonary Diffusing Capacity)

The same general principles that govern diffusion of gases through solutions govern movement of gases through membranes. Thus, pressure difference across the membrane, thickness of the membrane, membrane surface area, and the diffusion coefficient of the gas in the membrane all affect movement of gases in the lungs and peripheral tissues. This is why diseases such as fibrosis, which thickens pulmonary membranes, and emphysema, which causes loss of alveolar membranes, are so devastating to oxygen transport. In quantitative terms, *diffusing capacity is the volume of a gas that diffuses through a membrane each minute per mmHG pressure gradient.*

In resting people, the diffusing capacity for oxygen approximates 22–25 ml O_2/min/mmHg. At rest, the pressure difference for O_2 across pulmonary membranes approximates 11 torr. Thus, for a resting person with a $\dot{V}_{O_2}$ of 250 ml/min, the diffusing capacity is $250/11 = 22.7$ ml/min. During exercise, diffusing capacity increases in proportion to the metabolic rate, and values in the range of 65 ml/min/mmHg pressure can be measured. Pulmonary diffusing capacity for O_2 increases several times

during exercise because alveoli in the upper lungs are ventilated and because circulation (perfusion) and ventilation of various parts of the lung are better matched during exercise than at rest. In comparison to that for O_2, the diffusing capacity of CO_2 is approximately 20 times greater.

Oxygen Transport

As we have just seen, if we know the partial pressure of O_2 in arterial blood and the concentration of hemoglobin in the blood, then, barring any unusual circumstances, we can calculate the arterial O_2 content. Further, if we know the cardiac output ($\dot{Q}$, the volume of blood ejected from the left ventricle each minute), we can calculate the O_2 transport capacity ($\dot{T}_{O_2}$) from heart (or lungs) to the rest of the body. For example, during rest,

$$\dot{T}_{O_2} = (C_{aO_2})(\dot{Q})$$
$$O_2 \text{ transport} = (20 \text{ ml } O_2 \cdot dl^{-1} \text{ blood}) \quad (11\text{-}4)$$
$$(50 \text{ dl blood} \cdot min^{-1})$$
$$= 1000 \text{ ml } O_2 \cdot min^{-1}$$

During maximal exercise in a young, fit individual with a 30 l/min (300 dl/min) cardiac output, O_2 transport can be much greater—for example,

$$= (20 \text{ ml } O_2 \cdot dl^{-1} \text{ blood})$$
$$(300 \text{ dl blood} \cdot min^{-1}) \quad (11\text{-}5)$$
$$= 6000 \text{ ml } O_2 \cdot min^{-1}$$

During rest, actual whole-body O_2 consumption is much less than O_2 transport capacity because

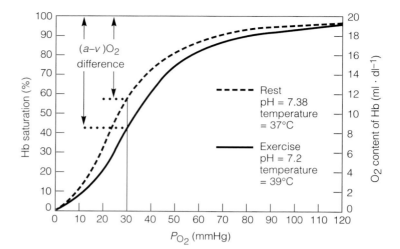

Figure 11-4 Decreases in blood pH and increases in temperature, such as experienced during exercise, cause the oxyhemoglobin curve to shift down and to the right. This Bohr effect facilitates the unloading of O_2 from hemoglobin in contracting muscles.

mixed venous blood returning to the right heart contains substantial amounts of O_2. During maximal exercise, as just illustrated for a very fit male, the actual O_2 consumption approaches the limits of the O_2 transport capacity. This is because during exercise, as compared to rest, most of the O_2 present in arterial blood is removed during each circulatory passage.

Effects of CO_2 and H^+ on O_2 Transport

In addition to the partial pressure of O_2 (see Figure 11-2), other factors can affect the combination of O_2 with hemoglobin. These factors include temperature, pH, and concentration of 2, 3-diphosphoglycerate (2, 3-DPG).

The effects of elevated temperature and H^+ (lower pH) on the O_2 dissociation curve are given in Figure 11-4. This shifting of the dissociation curve down and to the right during exercise is termed the *Bohr effect,* and it facilitates the unloading of O_2 from the hemoglobin of blood passing through active muscle beds.

Erythrocytes are unique in that they are specialized for nonoxidative metabolism. In addition to the usual glycolytic enzymes, erythrocytes also have the enzyme diphosphoglycerate mutase, which catalyzes the formation of 2, 3-DPG from 1, 3-DPG. Levels of 2, 3-DPG are elevated during exercise,

particularly during exercise at high altitude. The binding of 2, 3-DPG to hemoglobin occurs at a site that negatively affects the binding of O_2. Consequently, increased levels of 2, 3-DPG cause a rightward shift in the oxyhemoglobin dissociation curve; 2, 3-DPG is hydrolyzed to 3-phosphoglycerate (a common glycolytic intermediate) by diphosphoglycerate phosphatase. Low blood P_{O_2} levels, such as those which occur at high altitudes (Chapter 23), stimulate erythrocyte glycolysis and the formation of 2, 3-DPG.

Because of the sigmoidal (S) shape of the O_2 dissociation curve (Figure 11-4), the Bohr effect during exercise has only a minimal impact on the combination of O_2 with hemoglobin in the lungs. In active tissues, however, where the P_{O_2} is below 40 mmHg, the binding of O_2 to hemoglobin can be reduced to 10 to 15% as a result of the Bohr effect. This difference represents additional O_2 available in tissues to support metabolism.

■ The Red Blood Cells and Hemoglobin in CO_2 Transport

In almost everyone's mind, the erythrocyte (red blood cell) and hemoglobin are synonymous with oxygen transport. As described earlier, O_2 transport is a paramount role for the erythrocyte. In addition, however, the erythrocyte, which contains hemoglobin and the enzyme carbonic anhydrase, is crucial for CO_2 transport in the blood.

Carbon dioxide is a by-product of cellular respiration. Because the respiratory quotient of metabolism ($RQ = \dot{V}_{CO_2}/\dot{V}_{O_2}$) approximates unity, quantitatively the problem of CO_2 transport from cells to the lungs is as great as that of O_2 transport

from the lungs to body cells. The cellular formation and accumulation of CO_2 results in diffusion out of the cell, mostly in gaseous form. Once CO_2 reaches the capillary blood, the reactions diagrammed in Figure 11-5 occur rapidly. Relative to O_2, CO_2 is more soluble in the aqueous phase of the blood; however, only about 5 to 7% of CO_2 is carried in the dissolved form.

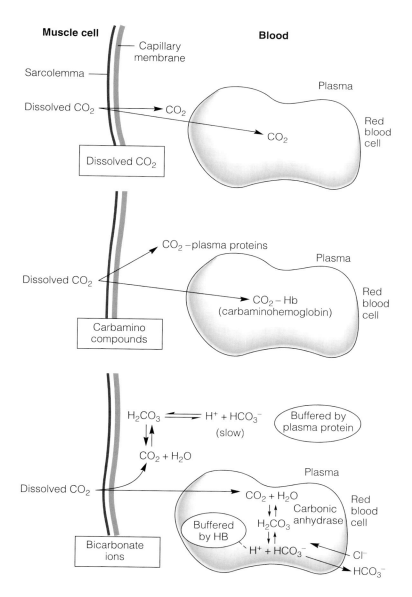

Figure 11-5   Carbon dioxide is transported in the blood by three mechanisms: (a) dissolved CO_2, (b) carbamino compounds, and (c) as bicarbonate ion. Hemoglobin and carbonic anhydrase enzyme in red blood cells are essential in the transport of CO_2 from sites of tissue formation to elimination in the lung.

The CO_2 that diffuses from plasma into the red blood cells reacts with water in the erythrocyte to form carbonic acid. In the red blood cell, this reaction is catalyzed by the enzyme carbonic anhydrase.

$$CO_2 + H_2O \xrightleftharpoons[\text{anhydrase}]{\text{Carbonic}} H_2CO_3 \qquad (11\text{-}6)$$

Carbonic anhydrase, the enzyme that wets CO_2, is not present in plasma, so although carbonic acid (H_2CO_3) is also formed there from the physical interaction of CO_2 and H_2O, it is formed at a rate several thousand times slower than that in the erythrocytes. Although the formation of carbonic acid is slow and enzyme-limited, the dissociation of carbonic acid to bicarbonate ion (HCO_3^-) and a proton is not enzyme-catalyzed and proceeds spontaneously.

$$H_2CO_3 \rightleftharpoons HCO_3^- + H^+ \qquad (11\text{-}7)$$

The HCO_3^- so formed leaves the erythrocyte for the plasma by exchange with chloride ion (Cl^-), which moves from plasma into the erythrocyte (see Figure 11-5). Although this chloride shift was thought to occur via passive diffusion to maintain the transmembrane chemical-electrical gradient, it is now clear that a specific carrier protein in the erythrocyte cell membrane, the Cl^--HCO_3^- anion exchanger, facilitates this Cl^- for HCO_3^- exchange.

The Cl^--HCO_3^- – exchange protein is but one of a group of transport proteins (transporters or carriers) involved in erythrocyte and muscle cell function. Others of note include the Na^+-H^+ exchanger, and the H^+-lactate$^-$ cotransporter or MCT1 (see Chapter 5 for a discussion of cell membrane lactate transporters). These carriers maintain intracellular pH by facilitating, in muscle tissue, the efflux of hydrogen ions (H^+) and lactate anions (LA^-) from contracting and lactate-producing cells into plasma and erythrocytes. The carriers also maintain pH by facilitating, in areas of lactate removal (e.g., liver, heart, highly oxidative muscle), the flux of H^+ and LA^- from erythrocytes and plasma into cell sites of lactic acid removal. About 70% of all CO_2 trans-

ported from tissues to the lungs is carried in the form of bicarbonate ion (HCO_3^-).

The H^+ so formed as the result of carbonic acid dissociation reacts rapidly with reduced hemoglobin (i.e., hemoglobin that has dissociated its O_2 molecule). The stoichiometry of hemoglobin O_2 and H^+ binding is interesting in that there is almost a one-to-one exchange of H^+ for O_2. In fact, reduced hemoglobin is such a strong buffer that it can take up almost all the H^+ formed as the result of CO_2 transport.

$$Hb^- + H^+ \rightleftharpoons HHb \qquad (11\text{-}8)$$

The reaction goes to the right as presented here when the P_{O_2} is low (hemoglobin dissociates O_2) and when the P_{CO_2} and $[H^+]$ are high. In fact, elevated CO_2 and H^+ concentration help to dissociate the oxyhemoglobin complex as part of the Bohr effect. In the lung, where P_{CO_2} is low and P_{O_2} is high, the reactions proceed from right to left. We shall discuss this shortly as part of the Haldane effect.

In the past, controversy existed over whether muscle contains the enzyme carbonic anhydrase. Now it is generally accepted that muscle and other cells *do contain* carbonic anhydrase. An advantage of having carbonic anhydrase in myocytes is that the cytosol can participate in the plasma acid–base buffering mechanism. However, because charged ions move through membranes in conjunction with a protein carrier (transporter), under non–steady-state conditions, such as heavy exercise, H^+ and HCO_3^- do not appear in or disappear from the blood at identical rates. These differences in rates of appearance of HCO_3^- and H^+ during non–steady-state conditions are due to the separate movements of HCO_3^- in exchange for Cl^-, and H^+ in exchange with Na^+, or H^+ cotransported with lactate anion.

Figure 11-5 also illustrates that CO_2 is carried in a third way in the blood. Some of the CO_2 entering the erythrocyte reacts directly with hemoglobin to form carbaminohemoglobin. Carbon dioxide reacts more readily with reduced hemoglobin than with oxyhemoglobin, although both reactions are possible. Some CO_2 also combines with plasma proteins. Together, the carbamino compounds account for approximately 25% of the CO_2 transported.

CO_2 Content of Blood Depends on the P_{CO_2}

Just as the combination of O_2 with hemoglobin can be expressed as a function of the P_{O_2} (Figure 11-2), the CO_2 content of blood can be described as a function of the P_{CO_2} (Figure 11-6). The normal resting $P_{a_{CO_2}}$ of 40 mmHg corresponds to an arterial CO_2 content ($C_{a_{CO_2}}$) of about 48 vol %. During rest, the mixed venous P_{CO_2} (i.e., $P_{\bar{v}_{CO_2}}$) rises to about 45 mmHg, and the mixed venous content of CO_2 ($C_{\bar{v}_{CO_2}}$) rises to about 52 vol %.

During rest, the $C_{a_{O_2}}$ is about 20 vol %, and mixed venous O_2 content is about 16 vol %. Therefore, the $(a - v)_{O_2}$ is about 4 vol %. During maximal exercise, the $C_{a_{O_2}}$ remains about constant, but the $C_{v_{O_2}}$ falls to about 4 vol %; the $(a - v)_{O_2}$, therefore, rises to about 16 vol %. During heavy exercise, the respiratory exchange ratio (RQ = $\dot{V}_{CO_2} / \dot{V}_{O_2}$) reaches values of 1.0 or greater. This means that the $(v - a)_{CO_2}$ reaches ≥ 16 vol %. As we will see, the addition of acid (e.g., a metabolic acid such as lactic

acid) to blood causes the $P_{a_{CO_2}}$ to decrease. During hard exercise, the entry of CO_2 into venous blood causes the $P_{\bar{v}_{CO_2}}$ to increase (see Figure 11-3). The physiological range of approximately 40 mmHg corresponds to a $(v–a)_{CO_2}$ of approximately 20 vol % (Figure 11-6).

Before leaving this section, one additional comment is appropriate. Although the beginning student of physiology may sometimes find the cryptic nature of symbols confusing, with experience the system can be informative as well as convenient. For instance, consider the abbreviations $P_{a_{O_2}}$, $P_{a_{CO_2}}$, $P_{\bar{v}_{O_2}}$, and $P_{v_{O_2}}$, $P_{\bar{v}_{CO_2}}$, and $P_{v_{CO_2}}$.

In this system, where we discriminate between $P_{\bar{v}_{O_2}}$ and $P_{v_{O_2}}$, and between $P_{\bar{v}_{CO_2}}$ and $P_{v_{CO_2}}$, we acknowledge that blood draining different tissues under diverse, or the same, circumstances may contain different amounts of oxygen (and carbon dioxide). Thus, when venous return reaches the right atrium, the blood will be from different tissue sites (legs, head, etc.), and due to the often laminar nature of blood flow, a sample of blood from a great vein, right atrium, or even right ventricle will not necessarily be a representative, mixed sample giving the average concentration of oxygen, carbon dioxide, or

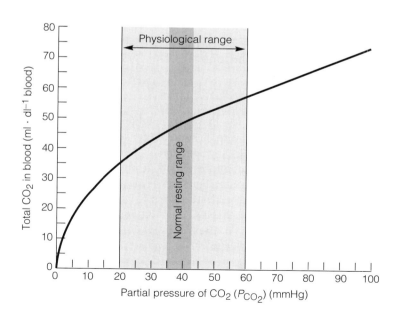

Figure 11-6 The CO_2 carried in the blood depends mainly on the partial pressure of CO_2. The carbon dioxide dissociation curve has a positive, almost linear, slope within the physiological range.

any single metabolite. However, it has been determined that once venous return reaches the pulmonary artery, the blood has been mixed. Hence, in the pulmonary artery, $P_{\bar{v}_{O_2}} = P_{v_{O_2}}$ and $P_{\bar{v}_{CO_2}} = P_{v_{CO_2}}$.

Because the pulmonary circulatory system acts as a mixing manifold for blood, and because oxygen and carbon dioxide are exchanged in the lungs, blood leaving the lungs via the pulmonary vein and then left heart will be mixed. Hence, arterial blood sampled at any site (aorta, wrist, groin, etc.) will be uniform in terms of P_{O_2} and P_{CO_2}. This fact of uniform contents of oxygen and carbon dioxide in the arterial circulation obviates use of the bar (average) symbol over a in the terms $P_{a_{O_2}}$, $P_{a_{CO_2}}$. For this reason, the seemingly cryptic use of symbols $P_{\bar{v}_{O_2}}$ and $P_{v_{O_2}}$, $P_{\bar{v}_{CO_2}}$ and $P_{v_{CO_2}}$, and $P_{a_{O_2}}$ and $P_{a_{CO_2}}$ actually tells us a lot about physiology.

Effects of O_2 on Hemoglobin and CO_2 Transport

As described earlier for the Bohr effect, the pressure of CO_2 and H^+ can loosen the binding of O_2 with hemoglobin. Conversely, oxygen can also act to displace CO_2 and H^+ from hemoglobin. This displacement of CO_2 from the blood by O_2 (a reversal of the Bohr effect) is termed the *Haldane effect*.

The Haldane effect results from the fact that O_2 causes hemoglobin to dissociate hydrogen ions. Hemoglobin in effect becomes a stronger acid. The increase in acidity in erythrocytes and the surrounding plasma brought about by a large increase in the P_{O_2} in the alveolar capillary causes a reversal of all the reactions diagrammed in Figure 11-5.

Figure 11-7 is an exploded version of Figure 11-6, modified to show the effects of tissue and alveolar O_2 partial pressures on the CO_2 dissociation curve. In actuality, there is not one CO_2 dissociation curve but rather a family of curves depending on the P_{O_2}. As mixed venous blood enters the alveolar capillary, the P_{CO_2} falls and the P_{O_2} rises. Therefore, CO_2 does not dissociate from the blood in a way described solely by the dashed curve in Figure 11-7; rather, the dissociation is actually described by the drop from point A on the dashed curve to point B on the solid curve. This shift from one curve to another due to the increased P_{O_2}, as well as to the effect

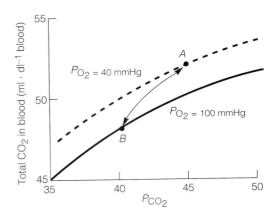

Figure 11-7 The effect of oxygenated hemoglobin on CO_2 transport in the blood (Haldane effect). As blood moves from tissues (where CO_2 tension is relatively high and O_2 is low) to the lungs (where O_2 tension is high and CO_2 is lower), the CO_2 dissociation curve in effect shifts from A to B. This Haldane effect causes CO_2 to be unloaded at the lungs. Modified from Guyton, 1976, p. 555. Used with permission.

of the reduced P_{CO_2}, causes a doubling of CO_2 dissociation from blood. Thus, the Haldane effect has a relatively larger impact on CO_2 transport than does the Bohr effect on O_2 transport.

■ The Buffering of Metabolic Acids by the Bicarbonate Buffer System

The fact that most CO_2 is transported in forms other than as CO_2 means that the dissolved CO_2 in venous blood is less than what might be expected. This is a highly desirable situation for several reasons. First, cellular CO_2 production is greater than the amount that can be transported in physical solution, and, just as hemoglobin acts as a carrier for O_2, so there needs to be a carrier for CO_2. Important also is the fact that dissolution of CO_2 in H_2O causes the formation of carbonic acid (H_2CO_3). However, by converting CO_2 to HCO_3^- and H^+ in the erythrocyte, by buffering the H^+ in hemoglobin, and by shifting the HCO_3^- into plasma, a mechanism is set up whereby the effects of metabolic acids, or bases, on plasma pH can be minimized (buffered).

Figure 11-8 describes the relationship between change in pH and added acid (or base) to the bicarbonate buffer system that exists in plasma. By definition, the system is half dissociated at the pK. Therefore, when the pH is at the pK, the pH changes least when acid or base is added to the system. The pK of the bicarbonate system is 6.1. At another pH, such as at the normal arterial pH of 7.4, added acid or base will have a major effect on pH unless something else is done.

A buffer system consists of a mixture of an acid and its dissociated components. For the bicarbonate (carbonic acid) buffer system we have:

$$H_2CO_3 \overset{K'}{\rightleftharpoons} H^+ + HCO_3^- \qquad (11\text{-}9)$$

where K' is the dissociation constant for the system.

The rate of dissociation of carbonic acid depends on the ratio of products to reactants, and is given by the dissociation constant:

$$K' = \frac{[H^+][HCO_3^-]}{[H_2CO_3]} \qquad (11\text{-}10)$$

Therefore, the carbonic acid dissociation constant is given by the ratio of the product of hydrogen ion and bicarbonate concentration to the concentration of carbonic acid. Unfortunately, whereas it is practical to measure H^+ and HCO_3^-, it is very diffi-

cult to measure the H_2CO_3 concentration. However, scientists have determined that the H_2CO_3 concentration is directly proportional to the P_{CO_2}. Therefore, because P_{CO_2} is readily measurable, H_2CO_3 concentration can be accurately estimated, or the P_{CO_2} may be used instead.

$$P_{CO_2}(\text{mmHg}) \times 0.03 = [H_2CO_3](\text{mM}) \quad (11\text{-}11)$$

Because in Equation 11-10 it is the P_{CO_2} that is measured, Equation 11-11 can be rewritten as follows:

$$K = \frac{[H^+][HCO_3^-]}{[CO_2]} \qquad (11\text{-}12)$$

Further, because acidity is usually the variable of interest, Equation 11-12 can be rearranged:

$$H^+ = K\frac{[CO_2]}{[HCO_3^-]} \qquad (11\text{-}13)$$

Then, taking the logarithm of each side, we get:

$$\log[H^+] = \log K + \log \left(\frac{[CO_2]}{[HCO_3^-]}\right) \quad (11\text{-}14)$$

By convention, the negative logarithm of the hydrogen ion concentration (i.e., $-\log [H^+]$) is termed the pH. Similarly, the $-\log K$ is termed the pK.

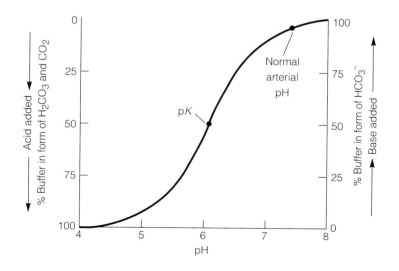

Figure 11-8 At its pK, a buffer system has the greatest resistance to additions of base or acid. Despite the fact that the pK of the bicarbonate buffer system (6.1) is removed from the normal arterial pH (7.4), the bicarbonate system works effectively to minimize the effects of acid or base added to the blood. This is because the blood CO_2 content can be varied through ventilation.

Therefore, by substituting pH and pK into Equation 11-14, by changing signs (inverting the log ratio) and knowing that the pK of the bicarbonate buffer system is 6.1, we get:

$$pH = 6.1 + \log \left(\frac{[HCO_3^-]}{[CO_2]} \right) \qquad (11\text{-}15)$$

This is called the Henderson–Hasselbalch equation; it describes how plasma pH is regulated by the relative, *not* absolute, quantities of bicarbonate ion and CO_2.

The Control of Blood pH by Ventilation (and Vice Versa)

Inspection of Figure 11-9 reveals that the pK of the bicarbonate–carbonic acid buffer system (6.1) is quite removed from the normal arterial pH (7.4). At pH 7.4, the strength of the buffer system is weak, and small additions of acid or base could markedly affect blood pH. However, despite this deficiency, the bicarbonate buffer system is an effective physiological buffer system because the concentrations of the elements of the system can be regulated.

The main way that blood pH is regulated during metabolic transient periods such as during exercise is by ventilation. As noted at the outset of this chapter, ventilation is important for several reasons, one of them being the regulation of blood pH. Ventilation affects pH by changing the CO_2 content of the blood. Arterial CO_2 content is inversely related to the pulmonary minute volume: Too little breathing causes CO_2 build up; too much breathing causes CO_2 to be eliminated from the blood. For example, if a person volitionally hyperventilates at rest, the alveolar partial pressure of CO_2 ($P_{A_{CO_2}}$) will fall and the $P_{A_{O_2}}$ will rise. These changes will cause CO_2 to move from blood to alveoli and be expired.

The loss of CO_2 by increased breathing (hyperventilation) affects pH in two ways. First, CO_2 forms an acid in H_2O (i.e., H_2CO_3). Thus, eliminating CO_2 in effect eliminates an acid from the blood. In addition, according to the Henderson–Hasselbalch equation (Eq. 11-15), the pH depends on a constant plus the logarithm of a ratio; CO_2 is in the denominator of that ratio. Therefore, if the denominator de-

creases, the ratio increases: Eliminating CO_2 from the blood causes pH to increase. In actuality, this description is a bit simplistic because the concentrations of H^+, HCO_3^-, and CO_2 are interrelated. If the P_{CO_2} falls as the result of hyperventilation, then HCO_3^- will ultimately also fall.

In the long term, blood pH is determined by those metabolic processes that form or remove metabolic acids, as well as by the kidneys, which can regulate blood concentrations of H^+, HCO_3^-, and NH_4^+.

The Buffering of Metabolic Acids

During heavy exercise, production of several metabolic acids is increased, lactic acid in particular. The entry of lactic acid into the blood would cause a large drop in pH if it were not for the bicarbonate buffer system and its ventilatory regulation. The action of the bicarbonate system in buffering lactic acid is illustrated by the following equations:

$$\underset{\text{(Lactic acid)}}{HLA} \rightarrow \underset{\text{(Hydrogen ion)}}{H^+} + \underset{\text{(Lactate ion)}}{LA^-} \qquad (11\text{-}16)$$

$$H^+ + HCO_3^- \rightarrow H_2CO_3 \qquad (11\text{-}17)$$

$$H_2CO_3 \rightarrow H_2O + CO_2 \qquad (11\text{-}18)$$

In summary, lactic acid gives rise to CO_2. This CO_2 is termed *nonmetabolic CO_2*, as it does not arise from the immediate combustion of a substrate. The nonmetabolic CO_2, or its volume equivalent of CO_2, can be eliminated from the blood at the lungs. In other words, the effects of adding a strong acid to blood are lessened by forming a weaker acid (H_2CO_3) and then by eliminating the weaker acid as CO_2. In this way, the R ($R = \dot{V}_{CO_2} / \dot{V}_{O_2}$) can exceed 1.0 during hard exercise when lactic acid enters the blood. Conversely, during recovery from hard exercise, retention of CO_2 and bicarbonate to make up for stores lost during exercise can result in $R < RQ$, and Rs of less than 0.7 can be observed.

■ Breathing for Talking

In addition to the exchanges of O_2 and CO_2, and the control of blood pH, another major purpose of ventilation is for vocal communication. The details of the control of speech by breathing are not germane to us at present. It is critical to note, however, that the control of speech (a voluntary activity) and breathing (a necessary activity) are closely integrated. In fact, the control mechanism must be one and the same. This control mechanism will become apparent in Chapter 12.

■ Summary

Life depends on a constant flow of energy to the body's cells and a controlled conversion of that energy into useful forms. The main cellular mechanism of bioenergetics is respiration. *Respiration* is the proper term for biological oxidation. Therefore, life also depends on a continuous flow of O_2 to the body's cells. Entry of O_2 into the body begins by breathing O_2 into the lungs. Ventilating the lungs has the effect of raising the partial pressure of O_2 in the alveoli (i.e., the $P_{A_{O_2}}$). Oxygen then diffuses from alveoli into blood, where it combines with hemoglobin in a way determined mostly by the arterial P_{O_2} (i.e., the $P_{a_{O_2}}$).

The gaseous by-products of respiration are H_2O and CO_2. It is important to eliminate most CO_2 because it is toxic even in relatively low concentrations. During ventilation of the lungs, the $P_{A_{CO_2}}$ is lowered relative to the tissue and mixed venous P_{CO_2}. Therefore, CO_2 tends to diffuse toward the lungs, where it can be expired. In the blood, CO_2 is transported in three ways: in simple solution, as bicarbonate ion, and in union with hemoglobin (carbaminohemoglobin).

In addition to CO_2, other by-products of cellular metabolism include strong organic acids such as lactic acid. The effects of metabolic acids on blood pH can be lessened (buffered) by increasing the ventilatory rate, causing diffusion of CO_2 out of the blood. Because CO_2 forms carbonic acid in H_2O, the exit of CO_2 in effect makes room for another acid, such as lactic acid.

Breathing, then, has three critically important metabolic functions for exercise: the consumption of O_2, the elimination of CO_2, and the buffering of metabolic acids. A fourth function of breathing, communication, requires extremely complex control of breathing frequency and volume, especially when talking and exercise are performed simultaneously.

■ Selected Readings

American College of Sports Medicine. Symposium on ventilatory control during exercise. *Med. Sci. Sports* 11: 190–226, 1979.

Åstrand, P.-O., and K. Rodahl. Textbook of Work Physiology. New York: McGraw-Hill, 1970, pp. 185–254.

Cherniak, R. M., and L. Cherniak. Respiration in Health and Disease. Philadelphia: W. B. Saunders, 1961.

Comroe, J. Physiology of Respiration. Chicago: Year Book Medical Publishers, 1974.

Dejours, P. Control of respiration in muscular exercise. In Handbook of Physiology, section 3, Respiration, vol. 1, W. O. Fenn and H. Rahn (Eds.). Washington, D.C.: American Physiological Society, 1964.

Dejours, P. Respiration. New York: Oxford University Press, 1966.

Dempsey, J. A., and C. E. Reed (Eds.). Muscular Exercise and the Lung. Madison: University of Wisconsin Press, 1977.

Fenn, W. O., and H. Rahn (Eds.). Handbook of Physiology, section 3, Respiration, vols. I and II. Washington, D.C.: American Physiological Society, 1964.

Guyton, A. C. Textbook of Medical Physiology. Philadelphia: W. B. Saunders, 1976, pp. 516–529.

Kjellberg, S., U. Rudhe, and T. Sjöstrand. Increase of the amount of hemoglobin and blood volume in conjunction with physical training. *Acta Physiol. Scand.* 19: 136–145, 1949.

Krogh, A. The Comparative Physiology of Respiratory Mechanisms. Philadelphia: University of Pennsylvania Press, 1941.

Lavoisier, A. L., and R. S. de La Place. Mémoire sur la Chaleur; Mémoires de l'Academie Royal (1780). Reprinted in Ostwald's Klassiker, no. 40, Leipzig, 1892.

McClintic, J. R. Physiology of the Human Body. New York: Wiley, 1975, pp. 208–215.

Miller, W. S. The Lung. Springfield, Ill.: Charles Thomas, 1947.

Otis, A. The work of breathing. In Handbook of Physiology, section 3, Respiration, vol. 1, W. O. Fenn and H. Rahn (Eds.). Washington, D.C.: American Physiological Society, 1964.

Pappenheimer, J. R. Standardization of definitions and symbols in respiratory physiology. *Fed. Proc.* 9: 602–605, 1950.

Riley, R. Pulmonary function in relation to exercise. In Science and Medicine of Exercise and Sports, W. Johnson (Ed.). New York: Harper and Brothers, 1960, pp. 162–177.

Vander, A. J., J. H. Sherman, and D. S. Luciano. Human Physiology. 3d Ed. New York: McGraw-Hill, 1980, pp. 327–365.

West, J. B. Respiratory Physiology—The Essentials. Baltimore: Williams and Wilkins, 1974.

Wolfel, E. E., P. R. Bender, G. A. Brooks, G. E. Butterfield, B. M. Groves, R. S. Mazzeo, J. R. Sutton, and J. T. Reeves. Oxygen transport during steady state, submaximal exercise in chronic hypoxia. *J. Appl. Physiol.* 70: 1129–1136, 1991.

Wolfel, E. E., M. A. Selland, A. Cymerman, G. A. Brooks, G. E. Butterfield, R. S. Mazzeo, R. F. Grover, and J. T. Reeves. O_2 extraction maintains O_2 uptake during exercise with β-adrenergic blockade at 4300 m. *J. Appl. Physiol.* 85: 1092–1102, 1998.

CHAPTER **12**

THE HOW OF VENTILATION

Our lungs represent a protected space wherein O_2 from the atmosphere can gain entry into the blood, and CO_2 in the blood can move to the atmosphere. By continually moving air in and out of the lungs from the atmosphere, we can render the partial pressures of O_2 and CO_2 in the alveolar air spaces of the lungs more like those in the atmosphere than like those in the systemic venous circulation reaching the lungs from the tissues. Because gases move from areas of higher concentration, or partial pressure, to areas of lower partial pressure, the exchange of O_2 and CO_2 between "alveolar air" and the blood becomes possible. The process of ventilation, then, becomes a means to set up a diffusion gradient between the respiring cells in various tissues around the body and the atmosphere.

In addition to creating pressure gradients for the movement of respiratory gases (O_2 and CO_2), the ventilatory system and its allied cardiovascular system affect O_2 and CO_2 transport in other ways. One way is to minimize diffusion distances between alveoli and surrounding capillaries, and between capillaries and extraalveolar cells. (The small distance between an alveolus and erythrocytes passing through is illustrated in Figure 12-2.) Another way is to optimize O_2 and CO_2 transport by having carrier mechanisms whereby larger quantities of both O_2 and CO_2 can be transported in the blood than would be possible on the basis of physical solution.

Athletes such as 5 and 10k world record holder Haile Gebreselassie of Ethiopia must control ventilatory movements in such a way that they are integrated with other body movements. The controls of breathing are both conscious and unconscious. PHOTO: © Corbis/AFP.

239

Ventilation rate is under the control of the respiratory center in the medulla of the brain. This respiratory center receives several different types of input that affect the center and the resulting pulmonary minute ventilation. Some inputs are central and come from synaptic connections within the brain itself; these are referred to as *neural* factors. Other neural inputs are peripheral and come from receptors and nerves in the lungs and skeletal muscles. Still other inputs are bloodborne and are called *humoral* factors. Humoral factors generally operate peripherally, and therefore involve receptors with nerve connections to the brain, but they can also operate centrally.

When exercise starts, pulmonary ventilation accelerates first rapidly and then slowly. When exercise stops, ventilation decelerates first rapidly, then slowly. The rapid control of ventilation is by neural mechanisms; the slower control is by humoral mechanisms. These mechanisms operate to cause rhythmical ventilatory movements. These movements result in the pulmonary exchange of O_2 and CO_2, so that for the sea-level resident the partial pressure of O_2 in arterial blood ($P_{a_{O_2}}$) remains at about 100 mmHg even during maximal exercise, and the resting $P_{a_{CO_2}}$ of 40 mmHg actually decreases during exercise.

■ Pulmonary Anatomy

Gross anatomy of the pulmonary system is diagrammed in Figure 12-1a, and the terminal anatomy is diagrammed in Figure 12-1b. Although it is only in the alveolar air sacs that O_2 and CO_2 are exchanged, the remaining upper respiratory tract (nose, mouth, trachea) performs important functions, including adding H_2O vapor to inspired air, warming (usually, or sometimes cooling) it to body temperature, and trapping particulate material (e.g., dust, yeast, and bacteria) as well as noxious fumes (e.g., smoke, ozone). The upper respiratory tract is extremely efficient in performing these tasks and is adequate to prevent all but the most overwhelming invasions from reaching the delicate alveolar membranes.

As suggested in Figure 12-1, the lower end of

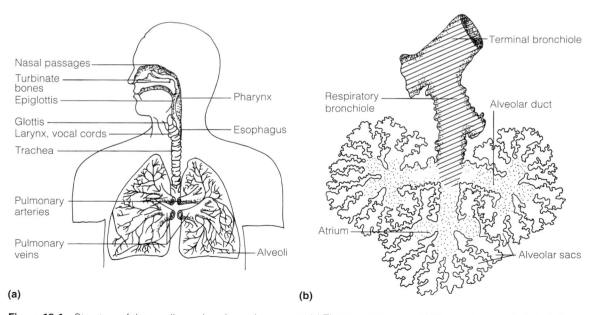

(a)   **(b)**

Figure 12-1 Structure of the ventilatory (respiratory) passages. (a) The gross anatomy. (b) The respiratory lobule including the alveolar sites of gas exchange. SOURCE: (a) Guyton, 1976, p. 526. Used with permission. (b) Miller, 1947, p. 42. Used with permission.

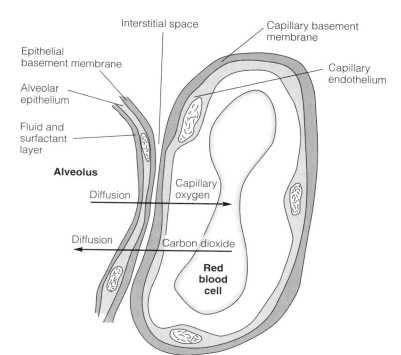

Figure 12-2 The ultrastructure of the alveolar capillary provides a minimum distance and mass of tissue between alveolar air and red blood cells in the pulmonary circulation. Modified from Guyton, 1976, p. 539. Used with permission.

the pulmonary tree anatomy provides a large surface area for respiratory gas exchange between air (alveoli) and blood (pulmonary capillaries). The exchange diffusion of O_2 and CO_2 between an alveolus and one of its surrounding capillaries is illustrated in Figure 12-2. Because the cross-sectional area of the capillary is barely adequate to allow for passage of red blood cells (erythrocytes), and because no other structure intervenes between the outer walls of the alveolus and the capillary, the diffusion distance is held to a minimum. This short distance facilitates the exchange of O_2 and CO_2. The exchange of respiratory gases, particularly O_2, is further facilitated by the large lipid content of the alveolar and capillary membrane walls, in which O_2 has a greater solubility than it does in H_2O.

■ Mechanics of Ventilation

Movement of air into and out of the lungs is caused by changes in thoracic volume, which result in intrapulmonary pressure changes. The structures re-

sponsible for this bellows-like action are diagrammed in Figure 12-3. During rest, an inspiration begins with the contraction of the diaphragm and external intercostal muscles. These actions lower the floor of the thorax and lift the ribs up and out. The volume of the thorax increases and the intrapulmonary pressure momentarily decreases. Atmospheric air moves into the pulmonary system to equilibrate the pressure gradient between lung and atmosphere.

During rest, expiration is a passive action wherein the diaphragm and external intercostals relax. The diaphragm recoils, moving up and "raising the floor" of the thorax, and the lung and ribs recoil to their original positions. The movements decrease the volume of the thorax, transiently increasing intrapulmonary pressure and forcing pulmonary air out.

During exercise, inspiratory movements are assisted by *accessory inspiratory muscles,* which include the sternocleidomastoid, scalene, and trapezius muscles. These muscles act to lift the ribs and clavicles vertically and transversely, allowing for large increases in tidal volume (V_T) during exercise.

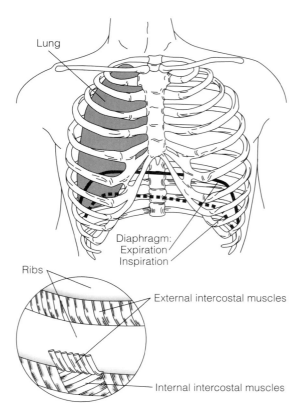

Lung

Diaphragm:
Expiration
Inspiration

Ribs

External intercostal muscles

Internal intercostal muscles

Figure 12-3 The major ventilatory muscles for breathing at rest include the diaphragm and external intercostal muscles. During the elevated breathing accompanying exercise (hyperpnea), other thoracic muscles including the sternocleidomastoid, scalene, and trapezius assist in ventilatory movements.

Also during exercise, expiration becomes an active (forced) movement. Contractions of the internal intercostals pull the ribs down and in, and contractions of the abdominal muscles increase abdominal pressure, forcing the diaphragm up into the thorax. The rapid and forceful movements of accessory ventilatory muscles during exercise greatly increase the maximal rate of ventilatory airflow. Consequently, pulmonary minute flow ($\dot{V}_E$) can increase tremendously without a dramatic increase in breathing frequency (f). Typical changes in pulmonary minute volume, tidal volume, and breathing frequency during the transition from rest to exercise are given in

---

	TABLE 12-1		
	Pulmonary Ventilation at Rest and during Maximal Exercise in a Large, Healthy, Fit Male (Values in BTPS)		
Condition	$\dot{V}_E$ (liters · min^{-1})	V_T (liters · breath^{-1})	f (breath · min^{-1})
Rest	6	0.5	12
Maximal exercise	192	4.00	48
Relative increase from rest to exercise	32 × Rest	8 × Rest	4 × Rest

Table 12-1. Note that by a relatively greater expansion in tidal volume (V_T) than in respiratory frequency (f), sufficient time is allowed for efficient gas exchange in the alveoli, and ventilation of respiratory dead space is minimized.

■ Dead Space and Alveolar Ventilation

Due to the anatomy of the pulmonary system (see Figure 12-1), not all the inspired air reaches the alveoli where O_2 and CO_2 are exchanged. Therefore, the alveolar minute ventilation ($\dot{V}_A$) is less than the pulmonary minute ventilation ($\dot{V}_E$). The part of each breath that remains in the upper respiratory tract does not exchange and is called anatomical dead space (D). Air in the dead space is warmed and humidified but remains at the relative concentrations of O_2 and CO_2 as they exist in atmospheric air. Conversely, as indicated in Figure 12-4, the last air leaving the mouth during an expiration most closely reflects alveolar composition. In general, in a tidal volume, the sequence is, last air in, first out—first in, last out. As the figure shows, the fractions of O_2 and CO_2 expired from the mouth during a breath change continuously until the alveolar air streams out. The last air out during a tidal volume (i.e., the end-tidal air) has the highest P_{CO_2} and the lowest P_{O_2} of gas moved during a breath. This air reflects the alveolar composition.

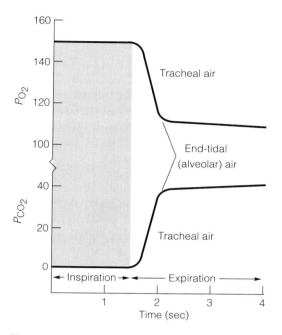

Figure 12-4 Partial pressures of O_2 and CO_2 measured in the mouth during a ventilatory cycle (inspiration and expiration) in a resting subject. The last air out during an expiration (i.e., the end-tidal air) represents the alveolar air. The precise values and timing of changes depend on exercise and environmental conditions.

Recall from Chapter 11 (Eq. 11-1) that $\dot{V}_E = (f)$ (V_T). However, as just described, not all of the tidal volume represents air entering the alveoli. The difference between tidal volume and alveolar volume ventilated is the dead space (D) (Figure 12-5).

$$\dot{V}_A = (\bar{V}_T - D)(f) \qquad (12\text{-}1)$$

The anatomical dead space is not a fixed volume but does increase slightly during exercise as tidal volume increases. Bronchiolar dilation and greater distances for air to flow within the lungs before air reaches the alveoli increase the D volume during exercise. As opposed to the rather small effect on D of increasing V_T, the effect of increasing f on ventilation is relatively greater. This effect of dead space ventilation on alveolar ventilation is illustrated in Table 12-2 for a resting individual. Clearly the effects of rapid, shallow breathing (panting) is to cause $\dot{V}_A$ to be much less than $\dot{V}_E$.

In addition to anatomical dead space, where respiratory gas exchange is not possible, there exists some "physiological dead space." This space is represented by alveoli that receive blood flow inadequate to effect an equilibration of alveolar and pulmonary capillary blood. In the seated resting

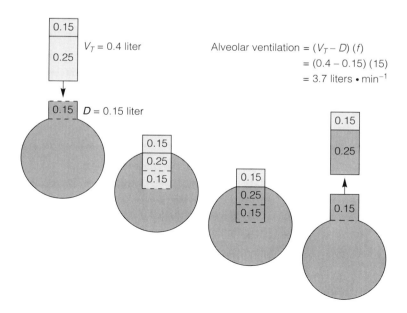

Alveolar ventilation = $(V_T - D)$ (f)
= $(0.4 - 0.15)$ (15)
= 3.7 liters • min^{-1}

Figure 12-5 During each breath (tidal volume, V_T), inspired air enters structures in which no exchange of ventilatory gas is possible. This dead space volume (D) comprises part of each tidal volume. Consequently, the alveolar minute ventilation equals the pulmonary minute ventilation less the dead-space minute ventilation.

TABLE 12-2

Effect of Panting on Alveolar Ventilation
($\dot{V}_A$) in a Resting Individual
(Ventilatory Volumes in BTPS)

Value	Breathing Normally	Panting
$\dot{V}_E$	6 liters · min^{-1}	6 liters · min^{-1}
$\dot{V}_T$	0.4 liter · min^{-1}	0.2 liter · min^{-1}
f	15 breaths · min^{-1}	30 breaths · min^{-1}
$\dot{V}_A$	3.7 liters · min^{-1}	1.5 liters · min^{-1}

person, not all the alveoli are open—particularly those at the top of the lungs. Similarly, because blood reaching the lungs tends to flow to the bottom of the lungs under the influence of gravity, the lower lungs tend to be better perfused than the top. Consequently, alveoli can exist that are ventilated but not circulated. Such areas comprise the physiological dead space.

The pulmonary volumes measured during exercise (i.e., $\dot{V}_E$ and V_T) are called *dynamic lung volumes*. These are opposed to the *static lung volumes,* which describe the pulmonary dimensions in resting individuals.

■ Static Lung Volumes—Physical Dimensions of the Lungs

Static lung volumes (Figure 12-6a) are measured with a device called a spirometer (Figure 12-6b). To use the displacement spirometer, a person breathes into and out of a chamber that consists of a counterbalanced cylinder suspended over a water seal. On exhalation, the person's breath raises the cylinder. On inhalation, the cylinder falls. A written record of ventilatory movements and volumes is obtained by means of a pen attached to the cylinder. The displacement spirometer illustrated in the figure is gradually being replaced with an electronic device called a pneumotachometer.

In general, lung volumes are correlated with body size. Lung volumes tend to be larger in tall people than in short people and larger in males than in females. Figure 12-6a illustrates that the normal

resting tidal volume can be increased or decreased in size by expiring a bit more. This has the effect of increasing or decreasing the *inspiratory reserve volume* (IRV) or the *expiratory reserve volume* (ERV). When tidal volume is maximal, it is termed *vital capacity* (VC). Vital capacity is a commonly measured pulmonary parameter, as is the 1-sec timed vital capacity. In some types of pulmonary problems (e.g., emphysema), distensibility of the lungs is reduced, but the total volume is unaffected. Therefore, VC is normal, but the timed VC, or the percentage of VC achieved in one second, is greatly reduced.

Even after a maximal expiration, some part of the air in the lungs cannot be forced out. This volume is termed the *residual volume* (RV). Together, the VC and the RV make up the *total lung capacity* (TLC).

By means of a much larger spirometer than that illustrated in Figure 12-6, which has been filled with a mixture of 95% O_2 and 5% CO_2, the maximum amount of air a subject can ventilate during a minute can be measured. This volume is termed *maximum voluntary ventilation* (MVV). To perform the MVV test, a subject wears a special breathing valve or mask in order to inspire from the spirometer and expire into the atmosphere. The MVV is measured by emptying the spirometer. The high (95%) O_2 in the spirometer during the MVV test supplies O_2 for respiration, and the high (5%) CO_2 prevents dizziness from a decrease in the $P_{a_{CO_2}}$. In most individuals, the MVV is much greater than the $\dot{V}_E$ observed during maximal exercise at sea-level altitudes. In other words, even during maximal exercise, a ventilatory reserve exists. This ventilatory reserve allows people to exercise intensely and for prolonged periods, to inhabit very high altitudes (14,000 ft), occasionally to sojourn at altitudes exceeding 20,000 ft, and even to abuse their pulmonary systems through habits like smoking without immediate effects.

■ Physics of Ventilatory Gases

Partial Pressures

The purpose of ventilating the pulmonary alveoli is to provide a place where the partial pressure of O_2 can be kept relatively high and the partial pressure

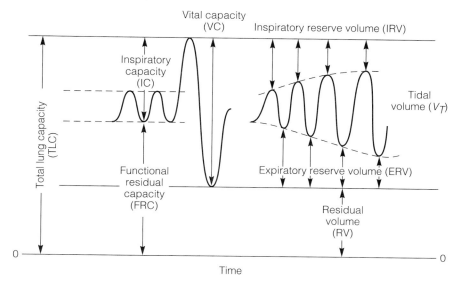

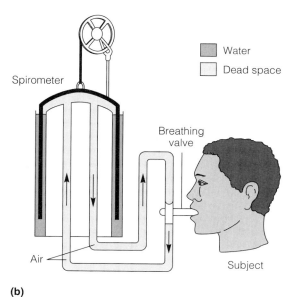

Figure 12-6 Lung volumes and capacities (a) can be measured during rest and exercise on devices called spirometers (b). Both tidal volume and ventilatory frequency can change from rest to exercise.

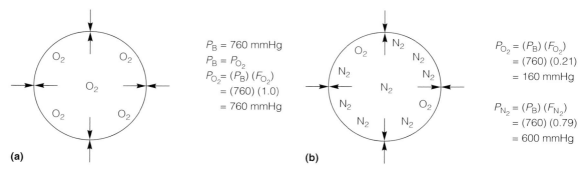

Figure 12-7 In a gas mixture, the total pressure is equal to the sum of the partial pressures exerted by the individual gases. For dry environments, the partial pressure exerted by a gas equals the total pressure times the fractional concentration of that gas. In environment (a), the partial pressure of O_2 equals the total pressure. In environment (b), the total pressure equals the sum of the partial pressures exerted by O_2 and N_2.

of CO_2 can be kept low, so that these respiratory gases exchange with the blood. We have previously defined the partial pressure of a gas as the pressure exerted by that species of gas. The partial pressure of a dry gas depends on the total barometric pressure (P_B) and the fractional composition (F_{O_2}) of that gas. (See Equation 12-2.)

For oxygen,

$$P_{O_2} = (P_B \text{ mmHg}) (F_{O_2})$$
$$= 760 \text{ mmHg} (0.21)$$
$$= 159.6 \text{ mmHg or torr}$$

This equation is illustrated in Figure 12-7. In a dry container at 1 atm pressure (760 mmHg) containing 100% O_2, the P_{O_2} is 760 mmHg (Figure 12-7a). In a container containing 21% O_2 and 79% N_2, the partial pressures of O_2 and N_2 are approximately 160 and 600 mmHg, respectively (Figure 12-7b).

Water Vapor

In studying the movement of gas around the body, it is important to realize that the body is not a dry environment; water is always present, and respira-

tory gases exist either dissolved or in a gaseous environment saturated with H_2O vapor.

The partial pressure exerted by water is somewhat different from the partial pressures exerted by other ventilatory gases, in that the P_{H_2O} depends on environmental temperature. At body temperature (37°C), the P_{H_2O} is 47 mmHg. Therefore, as illustrated in Figure 12-8, if gas from a tank containing dry 100% O_2 is allowed to flush through and fill a wet container warmed to 37°C, then 47 mmHg of the 760 mmHg total pressure will be occupied by H_2O vapor. The diluting effect of water vapor reduces the partial pressures of other gases present. Therefore, Equation 12-2 must be rewritten to include H_2O vapor as in Equation 12-3.

The effect of H_2O vapor on the P_{O_2} of inspired tracheal air reaching the alveoli is

$$P_{O_2} = (P_B - 47 \text{ mmHg}) (F_{I_{O_2}})$$
$$= (760 - 47 \text{ mmHg}) (0.21) \quad (12\text{-}4)$$
$$= 150 \text{ mmHg or 150 torr}$$

Similarly, for CO_2, which is 0.03% of the inspired air, the tracheal P_{CO_2} will be

Partial pressure of gas X in a dry environment = $(P_B \text{ mmHg})$(fractional composition of gas X) $\quad$ (12-2)

Partial pressure of gas X in a wet environment = $(P_B - P_{H_2O})$(fractional composition of gas X) $\quad$ (12-3)

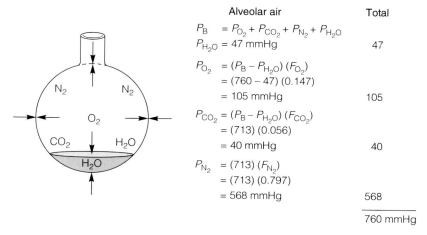

Alveolar air

		Total
P_B	$= P_{O_2} + P_{CO_2} + P_{N_2} + P_{H_2O}$	
P_{H_2O}	$= 47$ mmHg	47
P_{O_2}	$= (P_B - P_{H_2O})\,(F_{O_2})$	
	$= (760 - 47)\,(0.147)$	
	$= 105$ mmHg	105
P_{CO_2}	$= (P_B - P_{H_2O})\,(F_{CO_2})$	
	$= (713)\,(0.056)$	
	$= 40$ mmHg	40
P_{N_2}	$= (713)\,(F_{N_2})$	
	$= (713)\,(0.797)$	
	$= 568$ mmHg	568
		_____
		760 mmHg

Figure 12-8 In a warm, moist environment (such as the alveolus), H_2O vapor saturates the environment, contributing significantly to the total pressure and diluting the relative fractional concentrations of O_2 and the other ventilatory gases. At a body temperature of 37°C, H_2O exerts a pressure of 47 mmHg. Carbon dioxide delivered to the alveolus from the venous circulation also dilutes the concentration of O_2.

$$P_{CO_2} = (760 - 47 \text{ mmHg})\,(0.0003) \qquad (12\text{-}5)$$
$$= 0.2 \text{ mmHg}$$

In the alveolus itself, the partial pressure of O_2 decreases as a result of the ongoing consumption of O_2 and the diluting effect of CO_2, and the partial pressure of CO_2 becomes higher because of diffusion into the alveolus from the blood. In healthy resting individuals at sea-level altitudes, the alveolar partial pressure of O_2 ($P_{A_{O_2}}$) is about 105 mmHg, and the $P_{A_{CO_2}}$ is about 40 mmHg (Figure 12-8). As we shall see, ventilation is controlled to maintain these partial pressures.

In the pulmonary system, not only does H_2O move into the gas phase, but the respiratory gases O_2 and CO_2 also dissolve in the aqueous phases of fluids lining the alveoli and in the plasma (see Figure 12-2). For the movement of respiratory gases, it is important that the partial pressures of O_2 and CO_2 in the gas (alveolar) phase equilibrate with pressures in the aqueous (blood) phase. As illustrated in Figure 12-9, the molecules of a gas continuously move in all directions. Some of the molecules of a gas in contact with water penetrate the surface

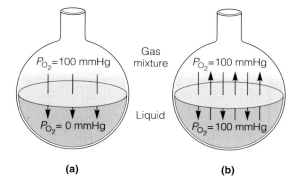

Figure 12-9 Given sufficient time, a gas in a closed environment will equilibrate across the gas–liquid interface. The partial pressure of the gas will then be equal in gaseous and liquid phases. O_2 admitted to a closed environment (a) will equilibrate in the liquid phase (b). By analogy, the sphere represents an alveolus, the O_2 is from inspired air, and the water represents body fluids. Modified from Guyton, 1976, p. 531. Used with permission.

and occupy spaces between the molecules as well as (in the case of CO_2) react chemically with the water. These gas molecules entering the fluid phase dissolve in the fluid. Gas molecules dissolved in a fluid are not trapped there, but can move around within the fluid or leave it. Therefore the gas exerts a partial pressure both within and around the fluid. Given enough time at a particular temperature and pressure, the gas molecules in liquid and gas phases will come to equilibrium. At that point, the same number of molecules will be entering and leaving each phase for the other. At equilibrium, then, the partial pressures of the gas in liquid and gas phases will be equal. By this means, O_2 in alveolar gas can move through the body fluids to cellular mitochon-

dria, and the reverse pathway is possible for CO_2 (Figure 12-10).

In the pulmonary capillaries, the equilibration of O_2 and CO_2 with alveolar air depends on the pressure gradient and the time the blood is in the capillary. During rest, blood (erythrocytes) is in the pulmonary capillary an average of 0.75 second. This time period is referred to as the *capillary transit time* (Figure 12-11). This is an adequate amount of time for O_2 and CO_2 to equilibrate between the pulmonary capillary and the alveoli. During maximal exercise, capillary transit time has been estimated to decrease to 0.4 to 0.5 second. At sea-level altitudes this is still more than adequate time for equilibration of CO_2, and marginally adequate time for

Figure 12-10 As CO_2 follows its pressure gradient (high to low), it flows from sites of cellular production to excretion in alveoli. Along the way, CO_2 passes through a number of spatial, fluid, and membrane barriers. Additionally, CO_2 flux through the body involves hemoglobin and the enzyme carbonic anhydrase (see Figure 11-5). Modified from Vander, Sherman, and Luciano, 1975, p. 305. Used with permission.

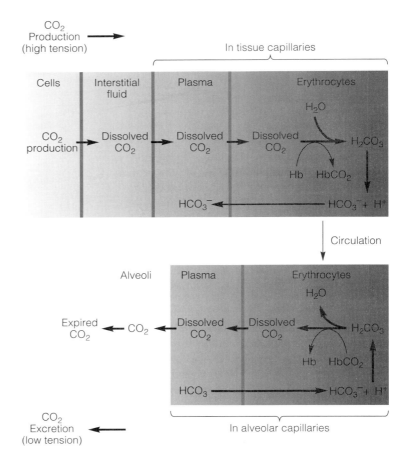

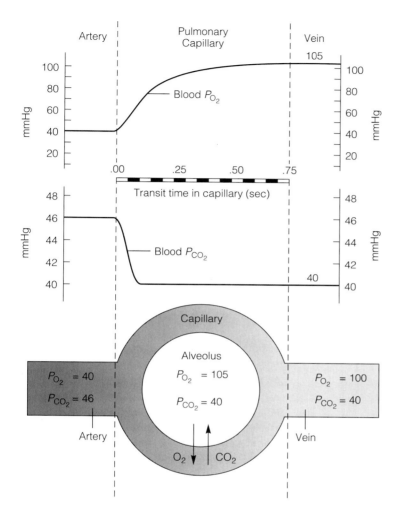

Figure 12-11 The exchange of O_2 and CO_2 in alveolar capillaries depends on time as well as partial pressure. Note in the figure that CO_2 equilibrates faster than O_2. During hard exercise at sea-level altitudes, there is usually sufficient time (0.5 sec) to complete the exchange. Source: Åstrand and Rodahl, 1970. Used with permission.

equilibration of O_2. At higher altitudes, where the P_B decreases and with it also the $P_{I_{O_2}}$ and $P_{A_{O_2}}$, the $P_{a_{O_2}}$ necessarily decreases in comparison to sea-level values of about 100 mmHg.

The amount of O_2 (in milliliters of O_2 and STPD) diffusing across the pulmonary membranes per minute per mmHg pressure difference between alveolar air and pulmonary capillary blood is defined as the *lung diffusing capacity* (D_L). Diffusing capacity increases from rest to exercise and as a result of endurance training. However, altitude hypobaria (low pressure) decreases D_L.

■ Respiration, Circulation, and Ventilation

Because life depends on energy transduction, and because most of the body's cells depend on O_2 for energy transduction, a constant delivery of O_2 is necessary to sustain life. During exercise, energy (ATP) demands are increased and cell respiration must be accelerated. As depicted in Figure 12-12, continuous muscular activity depends on a flux of O_2 from the alveolus to the muscle. Conversely, CO_2 formed in muscle must flow around to the alveolus.

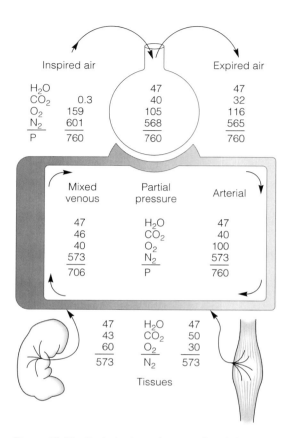

Figure 12-12 Typical values of gas tensions in inspired air, alveolar air, the blood, and expired air in a resting individual. Increased pulmonary ventilation during exercise compared to rest allows arterial oxygen pressure ($P_{a_{O_2}}$) to be maintained. Standard conditions are assumed for inspired air. Modified from Åstrand and Rodahl, 1970. Used with permission.

Therefore, the circulatory system plays a critical role in sustaining cellular respiration by keeping arterial oxygen content constant in the transition from rest to exercise. The pulmonary system provides an equally important role. For more about the integration of reflexes controlling circulatory and ventilatory response to exercise, see the review on the subject by Kaufman and Forster in the American Physiological Society handbook on exercise (1996).

■ The Control of Alveolar Minute Ventilation

If a person is forewarned and can anticipate the beginning of exercise, ventilation starts to increase before the exercise starts (Figure 12-13). When exercise does begin, ventilation increases very rapidly with a half-response time of 20 to 30 seconds. After approximately 2 minutes, ventilation stops increasing rapidly, but continues to increase at a slower rate. During submaximal exercise, $\dot{V}_E$ plateaus (stops rising) over 4 to 5 minutes. During maximal exercise, ventilation continues until either MVV or the point of fatigue is reached.

The pulmonary minute volume reached during exercise depends on a number of factors, including the work rate, the state of training, and the muscle group used. Typical ventilatory volumes reached during leg exercise for trained and untrained individuals are illustrated in Figure 12-14. The relationship between $\dot{V}_E$ and $\dot{V}_{O_2}$ (or work rate) has two components. These are a linear rise followed by a curvilinear, accelerated increase in response to exercise work rates eliciting more than 65 to 75% of $\dot{V}_{O_2max}$. In general, ventilation is higher in untrained than in trained subjects for given absolute and relative work loads. Ventilation is also higher when small muscles (e.g., arms) perform a given amount of work in comparison to the same work performed by larger muscles (e.g., legs).

■ Control of Tidal Volume and Breathing Frequency

Pulmonary minute ventilation ($\dot{V}_E$) increases during exercise because both tidal volume (V_T) and frequency of breathing (f) increase (Eq. 11-1). In the transition from rest to maximal exercise, both V_T and f are capable of increasing about threefold (Figure 12-15). At relatively low exercise intensities, both breathing frequency and tidal volume increase proportionally. However, at high relative exercise intensities V_T plateaus (Figure 12-15A) and further increases in $\dot{V}_E$ are due to increases in f (Figure 12-15B).

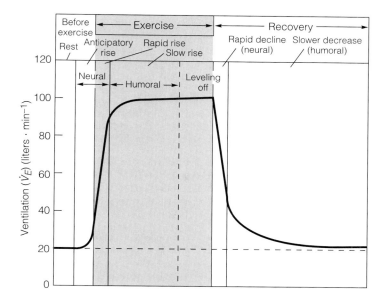

Figure 12-13 The neural-humoral control of ventilation during exercise. The elevated pulmonary ventilation during exercise (exercise hyperpnea) is controlled by at least two sets of mechanisms. One set acts very rapidly when exercise starts and stops. This mechanism has a neural basis and even causes ventilation to increase before exercise. The other set of factors that affect ventilation acts slowly and results from the effects of bloodborne factors on the ventilatory (respiratory) center.

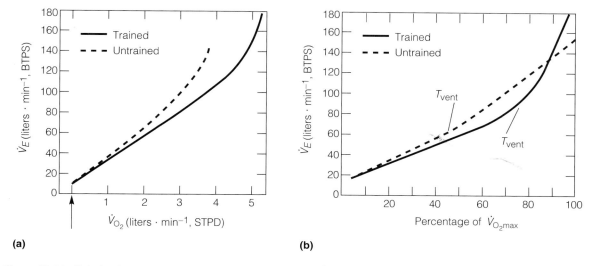

Figure 12-14 Relationship between pulmonary minute ventilation ($\dot{V}_E$) and metabolic rate ($\dot{V}_{O_2}$) during rest and exercise. In both untrained (broken line) and trained (solid line) subjects, $\dot{V}_E$ increases linearly with $\dot{V}_{O_2}$ up to about 50% to 65% of $\dot{V}_{O_2max}$. Thereafter, $\dot{V}_E$ increases at a rate disproportionately greater than the change in $\dot{V}_{O_2}$. Note that an effect of endurance training is to delay the ventilatory inflection point (T_{vent}).

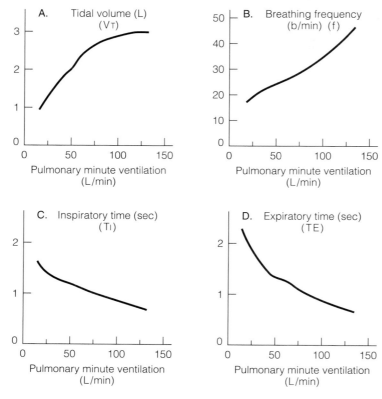

Figure 12-15 Comparison of the components determining pulmonary minute ventilation ($\dot{V}_E$) during progressive exercise up to maximum. When exercise starts, both (a) tidal volume (V_T) and (b) breathing frequency (f) increase proportionally. However, V_T plateaus; therefore, high ventilatory rates during hard exercise are due to incremental increases in f. Because of the increased breathing frequency, (c) inspiratory and (d) expiratory times decrease during progressive exercise. Consequently, peak expiratory flow rate increases more than peak inspiratory flow rate (see Figure 13-3). Modified from Dempsey et al., 1996.

When exercise starts, the increase in V_T is due to decreases in both IRV and ERV (Figure 12-6). However, as exercise intensity and V_E increase, ERV stabilizes, and further increases in V_T are due to decreases in IRV.

As breathing frequency increases during exercise (Figure 12-15B), both inspiratory (T_I) and expiratory (T_E) times must fall (Figures 12-15 c and d). However, T_E falls relatively more than T_I, and so

peak expiratory flow rates and pressures must rise more than the corresponding inspiratory flow rates and pressure during progressive exercise to maximum. The relationships among ventilatory gas flows and pressures during progressive intensity exercise are illustrated in Figure 13-3. Further discussion of these relationships can be found in a review by Dempsey and associates in the American Physiological Society handbook on exercise (1996).

In the following section, we describe the factors that control ventilation during exercise. Our understanding of ventilatory control is still incomplete—in particular, finding an explanation for elevated ventilation (hyperpnea) during exercise has been one of the major challenges in modern physiology. Nevertheless, significant progress has been made, and we look forward to continued research on the regulation of exercise hyperpnea.

■ Control of Ventilation: An Integrated, Redundant Neural-Humoral Mechanism

The neural center that controls ventilation (i.e., the respiratory center) is located in the lower brain, below the thalamus (Figure 12-16). The hypothalamic respiratory center is designed to alternate inspiration and expiration rhythmically. The rate and amplitude of these ventilatory movements is

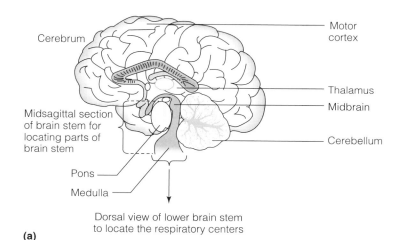

(a)

Figure 12-16   (a) Schematic of a sagittal (cross) section through the brain showing parts of the respiratory center in relation to the rest of the brain. (b) Schematic of the brain stem (pons and medulla) depicting the respiratory centers.

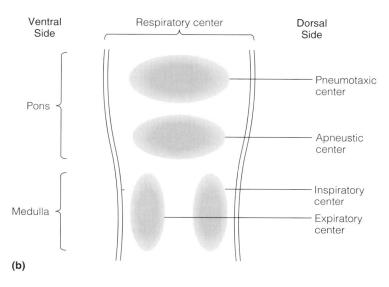

(b)

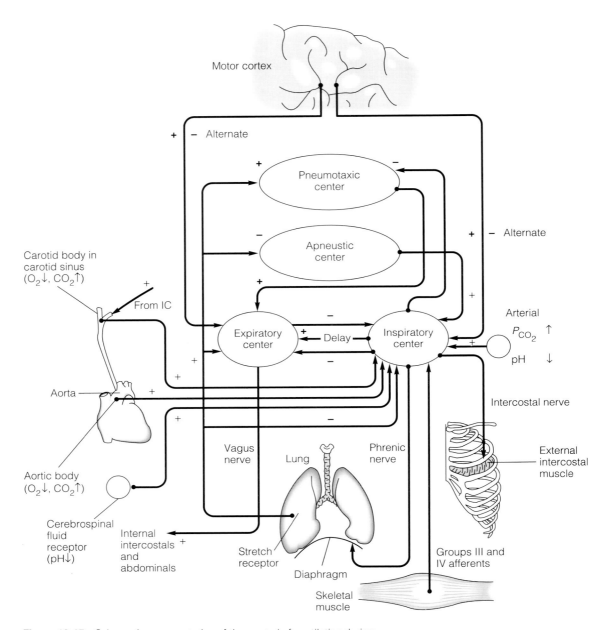

Figure 12-17 Schematic representation of the control of ventilation during rest and exercise in various environmental conditions. See the text for a description.

under the direct control of the respiratory center. Impinging on the center are a large number of neural and chemical (humoral) inputs. These inputs operate singularly and in concert to set the frequency and amplitude of output from the respiratory center. Not only are the various inputs into the respiratory center integrated there to produce a given output, but also the center appears to have redundant mechanisms (Figure 12-17). Following nature's example, human engineers designing sophisticated machines for which there is sometimes little opportunity for repair (such as airliners and the NASA space shuttle) build in redundant control systems. In this way, if one system fails there will be at least one other—and probably several other—systems to take over. The human respiratory center includes a number of redundancies, and these, as well as the anatomical location of the center, have made it a very difficult system to study. Scientists usually study a system by varying inputs into the system and seeing what happens to the output. Because the various inputs into the respiratory center alter the relationships among the different inputs, and because the center can compensate for the loss of a usual input, the output from the center will be different from the sum total of the actions of the individual inputs. This system of operation is most beneficial for us, who must breathe continually under a wide range of circumstances. However, the complexity of the system makes the respiratory physiologist's job a real challenge.

■ The Respiratory (Ventilatory) Center

The respiratory center is divided into four different areas: (1) the medullary expiratory center, (2) the medullary inspiratory center, (3) the apneustic center, and (4) the pneumotaxic center. (See Figure 12-16.) Both the apneustic and the pneumotaxic centers are located in the pons. The basic rhythmicity of the center is set in the medulla. If the brain of an anesthetized experimental animal is cut (transected) above and below the medullary area, breathing will be irregular or spasmodic, but inspirations will continue to alternate with expirations. The apneustic center facilitates (stimulates) the inspiratory center. Transection between the apneustic and pneumotaxic centers (thereby eliminating the pneumotaxic center) results in emphasized (prolonged) inspiration, and short expiration. Conversely, the pneumotaxic center facilitates expiration. Obliteration of the apneustic center allows dominance of the pneumotaxic center. The resulting breathing pattern is characterized by prolonged expiration.

There are two populations of neurons in the medullary rhythmicity center—the inspiratory and expiratory neurons—and they tend to be localized in individual areas. The *inspiratory center* is located in the nucleus of the *tractus solitarius* of the dorsal medulla. In contrast, the *expiratory center* is located in the ventral side of the medulla in the *nucleus ambiguus* and the *nucleus retroambiguus*. However, the inspiratory and expiratory neurons intermingle as well as have interconnections with the apneustic and pneumotaxic centers. Rhythmicity of the respiratory center is set in the medullary neurons, which spontaneously generate action potentials.

In describing the functioning of any system that acts rhythmically, such as the ventilatory system, it is difficult to select a point at which to begin describing events within that system. However, the inspiratory center appears to predominate, and so we shall begin by describing events leading to an inspiration. Basically, the system works as follows:

Due to facilitatory inputs from other inspiratory neurons, from neurons in the apneustic center, and from central and peripheral sources, inspiratory neuronal activity increases and results in increased phrenic and intercostal nerve activity and contraction of the muscles of inspiration (Figure 12-3). Because the inspiratory and expiratory neurons of the medullary center are reciprocally innervated, inspiratory center activity temporarily inhibits expiratory center activity. However, as inspiratory activity begins to wane, the reciprocal inhibition of the expiratory neurons decreases also, and because of their own intrinsic rhythmicity, the activity of the expiratory neurons increases. This activity is facilitated by synaptic interconnections from other expiratory neurons and from neuronal loops from the inspiratory center. Again, because of the reciprocal inhibitory interconnections between inspiratory and expiratory areas, neuronal discharge in the expiratory area leads to an inhibition of activity in the inspiratory areas. During rest, when expiration is passive, inhibition of the inspiratory area is sufficient to result in relaxation of the muscles of inspiration. During exercise, when expiration is forced, accessory expiratory muscles are contracted. The ventilatory cycle begins again as expiratory neuronal activity decreases and inspiratory neuronal activity increases.

Central Inputs to the Inspiratory Center

Neural Input—Central Command from the Motor Cortex

The motor cortex is located above the pons and is, therefore, said to exert *suprapontine* control over ventilation. The motor cortex is primarily responsible for stimulating the respiratory center to achieve the elevated ventilatory rates seen in exercise. It is responsible as well for the voluntary control of breathing that allows an individual speaking, singing, or playing a wind instrument to do so successfully. The motor cortex also allows movement and breathing patterns to be integrated during exercise. For instance, a swimmer must take large inspirations when the mouth and nose are out of the water, and expire when they are under the water. A wide range of breathing patterns is learned by swimmers who ventilate at high rates in a variety of stroke events (freestyle, butterfly, backstroke, breaststroke, and so on). During some sports activities, such as sprint running and weight lifting, ventilation must be inhibited temporarily in order to stabilize the rib cage. Because of the predominance of signals from the motor cortex and other higher brain centers in governing breathing during exercise, investigators such as Mitchell and Eldridge have emphasized the concept that during exercise breathing is under *"central command."* For a more in-depth description of the central neural control of breathing, see a review on the subject by Waldorp, Eldridge,

Iwamoto, and Mitchell in the American Physiological Society handbook on exercise.

The rapid increase in pulmonary ventilation with exercise, which is in fact preceded by an anticipatory rise (see Figure 12-13), is the key to the long-standing hypothesis of cortical control over the respiratory center during exercise. Input from the motor cortex into the respiratory center during exercise is an example of a feed-forward mechanism in which both ventilation and work rate are proportional to the output from the cortical controller. Admittedly, few direct data exist on the output of the cortical area to the respiratory center during exercise in humans. The invasive procedures necessary to make such measurements preclude such study. However, considerable experimental evidence exists, obtained on anesthetized animal preparations, that suprapontine areas have major facilitatory effects on the respiratory center. In the near future, the development of several noninvasive techniques may reach the stage where cortical influences on the respiratory center can be documented on humans during exercise.

In addition to the motor cortex, other areas that facilitate the medullary respiratory center include the hypothalamus, the cerebellum, and the reticular formation. The hypothalamus is particularly sensitive to increases in body temperature (Chapter 22). Increases in body temperature during exercise have been correlated with hypothalamic activity and ventilatory patterns. In some species that cool themselves by panting (e.g., the dog), hypothalamic effects on ventilation are pronounced. In humans, the effects of the hypothalamic mechanisms are apparently not well developed and are probably not observable except during recovery from prolonged hard exercise, which elevates body temperature.

Whereas the cortex exerts feed-forward control over the respiratory center, the cerebellum provides feedback control. Afferents from contracting muscle reach the cerebellum, which then provides facilitatory input to the respiratory center.

The reticular formation is a diffuse subcortical, suprapontine area. The reticular formulation is responsible for the general state of central nervous system (CNS) arousal. Exercise is associated with heightened activity in the reticular formation. Facilitatory input to the respiratory center from the reticular formation is thought to amplify the ventilatory response to exercise.

Humoral Input—The Medullary Extracellular Fluid

Any substance that circulates in the blood and that has general effects at a site or sites removed from the location of the secretion of the substance is termed a *humor*. These bloodborne substances are thought to be responsible for the slow phase of ventilatory adjustment to exercise (see Figure 12-13).

Most ventilatory responses in resting individuals at sea-level altitudes are mediated by central chemoreceptors. Specialized cells on the ventral surface of the medulla, which are distinct from inspiratory and expiratory neurons, are sensitive to changes in hydrogen ion (H^+) concentration. Some of these chemosensitive cells are influenced by the pH of the medullary interstitial fluid, whereas other cells are sensitive to the pH of the cerebrospinal fluid (CSF).

Increases in medullary H^+ concentration (decreases in pH) stimulate ventilation. Increases in arterial H^+ concentration decrease medullary interstitial pH; this pH change, in turn, activates the chemosensitive cells that monitor interstitial pH.

Whereas medullary chemosensitive cells respond to cerebrospinal fluid (CSF) pH, the pH perturbations to CSF come, not from changes in arterial pH, but rather from changes in the $P_{a_{CO_2}}$. Hydrogen ions diffuse slowly through the blood–brain barrier separating CSF and arterial blood. This barrier is rather more permeable to arterial CO_2. Because the buffer capacity of CSF is rather low, CO_2 penetrating the blood–brain barrier results in the formation of carbonic acid, which lowers the pH.

It should be emphasized that the central chemoreceptors sensitive to H^+ and CO_2 act to control ventilation during rest when the $P_{a_{O_2}}$ is around 100 mmHg. At other times, such as during exercise or acute exposure to high altitude, there is actually an alkalotic response in the medullary extracellular fluid. An increase in pH acts to restrain ventilatory response during exercise or acute hypobaric hypoxia.

■ Peripheral Inputs to the Respiratory System

Neural inputs to the medullary respiratory center that originate outside the brain are said to be peripheral to it. Peripheral inputs to the respiratory center include those from chemoreceptors and mechanical receptors.

Peripheral Chemoreceptors

The carotid bodies are located at the bifurcation of the common carotid artery into the internal and external carotids (see Figure 12-17). Afferent signals from the carotid bodies reach the medullary respiratory center via the carotid sinus and glossopharyngeal nerves. The carotid bodies are stimulated by decreased $P_{a_{O_2}}$, and the response is heightened by increased $P_{a_{CO_2}}$ and decreased pH. The carotid bodies are in an ideal position to detect a decrease in $P_{a_{O_2}}$ if, for some reason, the partial pressure of inspired O_2 ($P_{I_{O_2}}$) decreases or pulmonary ventilation becomes limited. The $P_{I_{O_2}}$ of sea-level residents could fall suddenly if they breathed air in which the O_2 content was lowered by the presence of a contaminating gas, or if they went up in altitude. The mechanical requirements of many sports activities such as swimming frequently impair pulmonary ventilation.

Although a fall in $P_{a_{O_2}}$ rapidly stimulates the carotid bodies, the contribution (if any) of the carotid bodies to stimulating ventilation during exercise has remained a puzzle. Breathing is so regulated during exercise that mean $P_{a_{O_2}}$ hardly changes (see Figure 11-3). During exercise there are a range of normal arterial CO_2 responses, but one frequent response is for $P_{a_{CO_2}}$ to decrease. Therefore, except during transient situations, such as at the start of exercise, the carotid bodies would not see an increase in the $P_{a_{CO_2}}$ or a decrease in the $P_{a_{O_2}}$.

During exercise, arterial pH does decrease, and the $P_{a_{O_2}}$ does fluctuate subtly with the arterial pulse wave. It has been suggested that these factors stimulate the carotid bodies during exercise. It is known that hypoxia exaggerates the CO_2 response of the carotid bodies in resting individuals. Therefore, it is possible that during exercise, pulsatile variations in $P_{a_{O_2}}$ increase the sensitivity of the carotid bodies to CO_2 and H^+. In addition, under the influence of the motor cortex, the medullary respiratory center may be more influenced by input from the carotid bodies during exercise than during rest.

In addition to the carotid bodies, other arterial chemoreceptors exist in the aorta and brachiocephalic arteries. These *aortic bodies* are sensitive to the $P_{a_{O_2}}$, $P_{a_{CO_2}}$, and pH; they send afferent signals to the medulla via the vagus nerve.

Other Peripheral Chemoreceptors

In addition to the arterial chemoreceptors, several other peripheral chemoreceptors have been postulated to exist. Because $\dot{V}_E$ and $\dot{V}_{CO_2}$ are so closely related whether at rest or during exercise, it has long been hypothesized that increased CO_2 flux to the lungs is sensed by some receptor in the right heart, pulmonary artery, or the lung itself. However, concerted efforts to detect the existence of such a receptor have failed. In the absence of such direct evidence, the high correlation between $\dot{V}_E$ and $\dot{V}_{CO_2}$, in which the end-tidal CO_2 is maintained as a constant during exercise, is taken as evidence that ventilation is well coordinated to metabolism. The high correlation between $\dot{V}_E$ and $\dot{V}_{CO_2}$ during exercise may in fact give a misleading impression that a pulmonary CO_2 receptor exists. The recent development of devices to measure changes in $\dot{V}_E$, $\dot{V}_{CO_2}$, and $\dot{V}_{O_2}$ breath by breath have allowed investigators such as Hildebrandt and colleagues (1979) to study in detail the role of CO_2 flux to the lungs in controlling $\dot{V}_E$ during exercise. In the experiments of Hildebrandt and colleagues, subjects performed bicycle ergometer exercises with pressure cuffs causing a venous occlusion of the legs. Following release of the cuffs, it took 5 to 10 seconds for the CO_2 trapped in the legs to reach the lungs. That event was marked by an increase in the partial pressure of CO_2 in end-tidal air

($P_{ET_{CO_2}}$). This increase in $P_{ET_{CO_2}}$ was followed 10 to 18 seconds later by an increase in $\dot{V}_E$. This time delay was sufficient for the pH effects of increased CO_2 load to reach and stimulate the arterial chemoreceptors. Had there been a pulmonary CO_2 receptor, $\dot{V}_E$ would have increased immediately as the $P_{ET_{CO_2}}$ rose.

Peripheral Neural Inputs to the Inspiratory Center

On the basis of an extensive body of evidence produced by Mitchell and Kaufman and associates in Dallas, TX, and Davis, CA, it is now apparent that neural input from muscles to the respiratory centers can provide important information for the control of ventilatory and cardiovascular responses to exercise. Initial work involved electrically induced static contractions in anesthetized laboratory animals (e.g., Mitchell et al., 1977), but now experiments have been performed on conscious, but spinally paralyzed and electrically stimulated humans (e.g., Strange et al., 1993). According to Kaufman and associates (1983), lactic acid accumulation and other results of contraction stimulate muscle receptors, which transmit signals to the brain by Group III (fast-conducting, myelinated) and Group IV (slow-conducting, nonmyelinated) afferent nerve fibers. Inputs from Group III and IV afferents are part of the redundant mechanism of ventilatory control, but they are likely more important for control of cardiovascular responses (blood pressure, heart rate, and cardiac output) than for breathing.

Peripheral Mechanoreceptors

hen the lungs and chest wall expand rapidly, mechanical receptors there are stimulated; these mechanoreceptors transmit to the respiratory center signals that have the effect of inhibiting inspiration. This action is known as the Hering–Bruer reflex. Because mechanoreceptors are known to exist and because of the difficulty in identifying a main peripheral receptor that regulates ventilation during exercise, scientists have looked for the presence of mechanical receptors that could be activated during exercise. The muscle spindles, Golgi tendon bodies, and skeletal joint receptors are known to send afferent signals to the sensory cortex, which relays information to the respiratory center. Evidence for peripheral mechanoreceptors was provided by Dejours, who used passive limb movement on awake as well as on lightly anesthetized human subjects to demonstrate an increase in $\dot{V}_E$ without an increase in $\dot{V}_{O_2}$. The lack of change in $\dot{V}_{O_2}$ was taken as evidence of no change in muscle metabolism. Furthermore, because the ventilatory response to passive limb movement was not blocked by venous occlusion, a potential role of chemoreceptors was eliminated.

Unfortunately, whereas peripheral mechanoreceptor stimulation has been observed to increase $\dot{V}_E$ in humans, the increase is small compared to the large and abrupt changes seen during exercise. Further, although the results of some experiments on animal preparations support a role of peripheral mechanoreceptors in controlling ventilation during muscular work, other experiments do not.

■ Control of Exercise Hyperpnea

The abrupt and very large change in ventilation that occurs as exercise starts and continues (see Figure 12-13) has been and remains a difficult phenomenon for scientists to explain. For the present, the neuro-humoral theory of Dejours remains a reasonable explanation, but Eldridge and associates have emphasized neural control or central command as being primarily important. Basic ventilatory rhythmicity is set by the medullary respiratory center. When exercise starts, the output of the respiratory center is increased tremendously by neural inputs from the motor cortex, muscles, and joints. Once the broad range of ventilatory responses is set by these neural components, it is fine-tuned by humoral factors that affect the peripheral and central chemoreceptors. When exercise stops, the exact opposite happens. Cortical and other neural inputs to the respiratory center cease, and ventilation slows dramatically.

Then, as humoral disturbances wane, ventilation slowly returns to resting values.

■ Summary

Most of the body's cells require O_2 to sustain life. Cellular respiration produces CO_2 as a by-product. Because CO_2 is toxic in high concentrations, it must be removed. The atmosphere is not only the source of O_2, it is also the dump for metabolic CO_2. The sites of exchange of O_2 and CO_2 between the body and atmosphere are the alveoli in the lungs. Through the process of moving air into and out of the lungs, the partial pressure of O_2 is kept relatively high in the alveoli, and the partial pressure of CO_2 is kept low in comparison to the pressures that exist in most tissues. The anatomy of the pulmonary system, its allied circulatory system, and the peripheral tissue anatomy are designed to optimize the opportunity for transport and exchange of respiratory gases.

Ventilation of the lungs is under the control of a respiratory center in the medulla of the brain. The respiratory center receives a variety of neural and humoral (chemical) inputs both directly (internally) and indirectly (externally) from receptors in the periphery. These inputs are integrated within the respiratory center to produce an appropriate ventilatory response; nevertheless, the control circuits within the respiratory center's system of operation contain redundancies. This means that the respiratory center can usually produce an appropriate ventilatory response even if one or several inputs are cut off, or if the inputs are contradictory.

Physical exercise results in rapid and large ventilatory adjustments, noted by both initial (rapid) and secondary (slow) response characteristics. When exercise stops, ventilation first slows abruptly and then declines slowly to resting levels. This fast-slow ventilatory response to the initiation and cessation of exercise is best explained by the neural-humoral theory of ventilatory control. During exercise, neural inputs to the respiratory center come from the motor cortex and from peripheral mechanoreceptors and chemoreceptors. Of these, the cortical inputs are likely the most important. As exercise continues, humoral (chemical) factors such as fluctuating P_{O_2} and increased P_{CO_2} and [H$^+$] are sensed in the peripheral chemoreceptors. Of these, the arterial chemoreceptors have been identified as being the most important for controlling the ventilatory response to exercise. Chemoreceptors sensitive mainly to CO_2 have long been hypothesized to exist in muscles and in the right heart or lung. Now it is apparent that lactate and other ion accumulation can affect both ventilatory and, more significantly, cardiovascular function. The muscle metaboreceptors probably operate through Group III and IV afferents. The role of Group III and IV afferents in the regulation of blood pressure during exercise is discussed more in Chapter 15.

■ Selected Readings

Andreani, C. M., J. M. Hill, and M. P. Kaufman. Responses of group III and IV afferents to dynamic exercise. *J Appl. Physiol.* 82: 1811–1817, 1997.

Andreani, C. M., and M. P. Kaufman. Effect of arterial occlusion on responses of group III and IV afferents to dynamic exercise. *J. Appl. Physiol.* 84: 1827–1833, 1998.

Asmussen, E., E. H. Christensen, and M. Nielsen. Humoral or nervous control of respiration during muscular work? *Acta Physiol. Scand.* 6: 160–167, 1943.

Asmussen, E., S. H. Johansen, M. Jorgensen, and M. Nielsen. On the nervous factors controlling respiration and circulation during exercise. Experiments with curarization. *Acta Physiol. Scand.* 63: 343–350, 1965.

Asmussen, E., and M. Nielsen. Experiments on nervous factors controlling respiration and circulation during exercise employing blocking of the blood flow. *Acta Physiol. Scand.* 60: 103–111, 1964.

Åstrand, P.-O., and K. Rodahl. Textbook of Work Physiology. New York: McGraw-Hill, 1970, pp. 187–254.

Barcroft, H., V. Basnyake, O. Celander, A. F. Cobbold, D. J. C. Cunningham, M. G. M. Jukes, and I. M. Young. The effects of carbon dioxide in the respiratory response to noradrenaline in man. *J. Physiol.* (London) 137: 365–373, 1957.

Band, D. M., I. R. Cameron, and S. J. G. Semple. Oscillations in arterial pH with breathing in the cat. *J. Appl. Physiol.* 26: 261–267, 1969.

Band, D. M., I. R. Cameron, and S. J. G. Semple. Effect of different methods of CO_2 administration on oscillations of arterial pH in cat. *J. Appl. Physiol.* 26: 268–273, 1969.

Barman, J. M., M. F. Moreira, and F. Consolazio. Metabolic effects of local ischemia during muscular exercise. *Am. J. Physiol.* 138: 20–26, 1942.

Barman, J. M., M. F. Moreira, and F. Consolazio. Effective stimulus for increased pulmonary ventilation during muscular exertion. *J. Clin. Invest.* 22: 53–56, 1943.

Beaver, W. L., K. Wasserman, and B. J. Whipp. On-line computer analysis and breathing by-breath display of exercise function tests. *J. Appl. Physiol.* 34: 123–132, 1973.

Belmonte, C., and C. Eyzaguirre. Efferent influences on carotid body chemoreceptors. *J. Neurophysiol.* 37: 1131–1143, 1974.

Biscoe, T. J. Carotid body: structure and function. *Physiol. Rev.* 51: 427–495, 1971.

Bisgard, G. E., H. V. Forster, B. Byrnes, K. Stark, J. Klein, and M. Manohor. Cerebrospinal fluid acid–base balance during muscular exercise. *J. Appl. Physiol.* 45: 94–101, 1978.

Byrne-Quinn, E., J. V. Weil, I. E. Sodal, G. F. Filley, and R. F. Grover. Ventilatory control in the athlete. *J. Appl. Physiol.* 30: 91–98, 1971.

Casaburi, R., J. Daly, J. E. Hansen, and R. M. Effros. Abrupt changes in mixed venous blood gas composition after the onset of exercise. *J. Appl. Physiol.* 67: 1106–1112, 1989.

Casaburi, R., B. J. Whipp, K. Wasserman, W. L. Beaver, and S. N. Koyal. Ventilatory and gas exchange dynamics in response to sinusoidal work. *J. Appl. Physiol.: Respirat. Environ. Exercise Physiol.* 42: 300–311, 1977.

Chermak, R. M., and N. S. Cherniack. Respiration in Health and Disease. Philadelphia: W. B. Saunders, 1961.

Coles, D. R., F. Duff, W. H. T. Shepherd, and R. F. Whelan. The effect on respiration of infusions of adrenalin and noradrenaline into the carotid and vertebral arteries in man. *Brit. J. Pharmacol.* 11: 346–350, 1956.

Comroe, J. H. Physiology of Respiration. Chicago: Year Book Medical Publishers, 1974, p. 234.

Cropp, G. J. A., and J. H. Comroe, Jr. Role of mixed venous blood P_{CO_2} in respiratory control. *J. Appl. Physiol.* 16: 1029–1033, 1961.

Cunningham, D. J. C., E. N. Hey, and B. B. Lloyd. The effect of intravenous infusion of noradrenaline on the respiratory response of carbon dioxide in man. *Q. J. Exp. Physiol.* 43: 394–399, 1958.

Cunningham, D. J. C., E. N. Hey, J. M. Patrick, and B. B. Lloyd. The effect of noradrenaline infusion on the relation between pulmonary ventilation and alveolar P_{O_2} and P_{CO_2} in man. *Ann. N.Y. Acad. Sci.* 109: 756–770, 1963.

Cunningham, D. J. C., M. G. Howson, and S. B. Pearson. The respiratory effects in man of altering the time profile of alveolar carbon dioxide and oxygen within each respiratory cycle. *J. Physiol.* (London) 234: 1–28, 1973.

Dejours, P. Control of respiration in muscular exercise. In Handbook of Physiology, section 3, Respiration, vol. 1, W. O. Fenn and H. Rahn (Eds.). Washington, D.C.: American Physiological Society, 1964.

Dejours, P., J. C. Mithoefer, and J. Raynaud. Evidence against the existence of specific ventilatory chemoreceptors in the legs. *J. Appl. Physiol.* 10: 367–371, 1957.

Dejours, P., J. C. Mithoefer, and A. Teillac. Essai de mise en evidence de chemorecepteurs veineux de ventilation. *J. Physiol.* (Paris) 47: 160–163, 1955.

Dempsey, J. A., L. Adams, D. M. Ainsworth, R. F. Fregosi, C. S. Gallagher, A. Guz, B. D. Johnson, and S. K. Powers. Airway, lung, and respiratory muscle function during exercise. American Physiological Society Handbook of Physiology-Exercise: Regulation and Integration of Multiple Systems. Oxford University Press, 1996, pp. 449–514.

Dempsey, J. A., N. Gledhill, W. G. Reddan, H. V. Forster, P. G. Hanson, and A. D. Claremont. Pulmonary adaptation to exercise: effects of exercise type and duration, chronic hypoxia, and physical training. *Ann. N.Y. Sci.* 301: 243–261, 1977.

Dempsey, J. A., D. A. Pelligrino, D. Aggarwal, and E. B. Olson. The brain's role in exercise hyperpnea. *Med. Sci. Sports* 11: 213–220, 1979.

Dempsey, J. A., E. H. Vidruk, and S. M. Mastenbrook. Pulmonary control systems in exercise. *Fed. Proc.* 39: 1498–1505, 1980.

Dempsey, J. A., E. H. Vidruk, and G. S. Mitchell. Pulmonary control systems in exercise: update. *Fed. Proc.* 44: 2260–2270, 1985.

Dutton, R. E., and S. Permutt. Ventilatory responses to transient changes in carbon dioxide. In Arterial Chemoreceptors, R. W. Torrance (Ed.). Oxford, England: Blackwell, 1968, pp. 373–386.

Eccles, J. C. Analysis of electrical potentials evoked in the cerebullar anterior lobe by stimulation of hind and forelimb nerves. *Exper. Brain Res.* 6: 171–194, 1968.

Eldridge, F. L. Central neural respiratory stimulatory effect of active respiration. *J. Appl. Physiol.* 37: 723–735, 1974.

Eldridge, F. L. Maintenance of respiration by central neural feedback mechanisms. *Fed. Proc.* 36: 2400–2404, 1974.

Eldridge, F. L. Relationship between respiratory nerve and muscle activity and muscle force output. *J. Appl. Physiol.* 39: 567–574, 1975.

Eldridge, F. L. Central neural stimulation of respiration in unanesthetized decerebrate cats. *J. Appl. Physiol.* 40: 23–28, 1976.

Eldridge, F. L., D. E. Millhorn, J. P. Kiley, and T. G. Waldrop. Stimulation by central command of locomotion, respiration and circulation during exercise. *Resp. Physiol.* 59: 313–337, 1985.

Eldridge, F. L., D. E. Millhorn, and T. G. Waldrop. Exercise hyperpnea and locomotion: parallel activation from the hypothalamus. *Science* 211: 844–846, 1981.

Flandrois, R., R. Fravier, and J. M. Pequignot. Role of adrenaline in gas exchange and respiration in the dog at rest and exercise. *Resp. Physiol.* 30: 291–303, 1971.

Fordyce, W. E., F. M. Bennett, S. K. Edelman, and F. S. Grodins. Evidence in man for a fast neural mechanism during the early phase of exercise hyperpnea. *Resp. Physiol.* 48: 27–43, 1982.

Freund, P. R., S. F. Hoggs, and L. B. Rowell. Cardiovascular responses to muscle ischemia in man—dependency on muscle mass. *J. Appl. Physiol.: Respirat. Environ. Exercise Physiol.* 45: 762–767, 1978.

Ganong, W. R. Review of Medical Physiology. Los Altos, Ca.: Lange Medical Publications, 1979, pp. 517–524.

Gonzalez, F., Jr., W. E. Fordyce, and F. S. Grodins. Mechanism of respiratory responses to intravenous $NaHCO_3$, HCl, and KCN. *J. Appl. Physiol.: Respirat. Environ. Exercise Physiol.* 43: 1075–1079, 1977.

Guyton, A. C. Textbook of Medical Physiology. Philadelphia: W. B. Saunders, 1976, pp. 516–528.

Hagberg, J. M., E. F. Carroll, J. M. Miller, W. H. Martin, and M. H. Brooke. Exercise hyperventilation in patients with McArdle's disease. *J. Appl. Physiol.: Respirat. Environ. Exercise Physiol.* 52: 991–994, 1982.

Heistad, D. D., R. C. Wheeler, A. L. Mark, P. G. Schid, and F. M. Abboud. Effects of adrenergic stimulation on ventilation in man. *J. Clin. Invest.* 51: 1469–1475, 1972.

Henry, J. P., and C. R. Bainton. Human core temperature increase as a stimulus to breathing during moderate exercise. *Respir. Physiol.* 21: 183–191, 1974.

Henry, J. P., and W. V. Whitehorn. Effect of cooling of orbital cortex on exercise hyperpnea in the dog. *J. Appl. Physiol.* 14: 241–244, 1959.

Hildebrandt, J. R., R. K. Winn, and J. Hildebrandt. Cardiorespiratory response to sudden releases of circulatory occlusion during exercise. *Resp. Physiol.* 38: 83–92, 1979.

Hughes, R. L., M. Clode, R. H. T. Edwards, T. J. Goodwin, and N. L. Jones. Effect of inspired O_2 on cardiopulmonary and metabolic responses to exercise in man. *J. Appl. Physiol.* 24: 336–347, 1968.

Innes, J. A., I. Solarte, A. Huszczuk, E. Yeh, B. J. Whipp, and K. Wasserman. Respiration during recovery from exercise: effects of trapping and release of femoral blood flow. *J. Appl. Physiol.* 67: 2608–2613, 1989.

Joels, N., and H. White. The contribution of the arterial chemoreceptors to the stimulation of respiration by adrenaline and noradrenaline in the cat. *J. Physiol.* (London) 197: 1–24, 1968.

Jones, N. L., D. G. Robertson, and J. W. Kane. Difference between end-tidal and arterial PCO_2 in exercise. *J. Appl. Physiol.: Environ. Exercise Physiol.* 47: 954–960, 1979.

Kao, F. F., C. C. Michel, S. S. Mei, and W. K. Li. Somatic afferent influence on respiration. *Ann. N.Y. Acad. Sci.* 109: 696–708, 1963.

Kaufman, M. P., and H. V. Forster. Reflexes controlling circulatory, ventilatory and airway responses to exercise. American Physiological Society Handbook of Physiology-Exercise: Regulation and Integration of Multiple Systems. Oxford University Press, 1996, pp. 381–448.

Kaufman, M. P., J. C. Longhurst, K. J. Rybicki, J. H. Wallach, and J. H. Mitchell. Effects of static muscular contraction on impulse activity of group III and IV afferents in cats. *J. Appl. Physiol.* 55: 105–112, 1983.

Krogh, A., and J. Linhard. The regulation of respiration and circulation during the initial stages of muscular work. *J. Physiol.* (London) 47: 112–136, 1913.

Krogh, A., and J. Linhard. A comparison between voluntary and electrically induced muscular work in man. *J. Physiol.* (London) 51: 182–201, 1917.

Lewis, S. M. Awake baboon's ventilatory response to venous and inhaled CO_2 loading. *J. Appl. Physiol.* 39: 417–422, 1975.

Linton, R. A. F., R. Miller, and I. R. Cameron. Ventilatory response to CO_2 inhalation and intravenous infusion of hypercapnic blood. *Respir. Physiol.* 26: 383–394, 1976.

Majcherczyk, S., J. C. G. Coleridge, H. M. Coleridge, M. P. Kaufman, and D. G. Baker. Carotid sinus nerve efferents: properties and physiological significance. *Fed. Proc.* 39: 2662–2667, 1980.

Majcherczyk, S., and P. Willshaw. Inhibition of peripheral chemoreceptor activity during superfusion with an alkaline c.s.f. of the ventral brainstem surface of the cat. *J. Physiol.* (London) 231: 26P–27P, 1973.

Majcherczyk, S., and P. Willshaw. The influence of hyperventilation of efferent control of peripheral chemoreceptors. *Brain Res.* 124: 561–564, 1977.

McClintic, J. F. Physiology of the Human Body. New York: Wiley, 1975.

McCloskey, D. I., and J. H. Mitchell. Reflex cardiovascular and respiratory responses originating in exercising muscle. *J. Physiol.* (London) 224: 173–186, 1972.

Miller, W. S. The Lung. Springfield, Ill.: C. C. Thomas, 1947.

Mitchell, J. H., W. C. Reardon, and D. I. McCloskey. Reflex effects on circulation and respiration from contracting skeletal muscle. *Am. J. Physiol.* 233 (*Heart Circ. Physiol.* 2): H374–H378, 1977.

Mitchell, J. H., and A. Berger. State of the art: review of neural regulation of respiration. *Am. Rev. Respir. Dis.* 111: 206, 1975.

Miyamoto, Y. Neural and humoral factors affecting ventilatory response during exercise. *Jap. J. Physiol.* 39: 199–214, 1989.

Nice, L. B., J. L. Rock, and R. O. Courtight. The influence of adrenaline on respiration. *Am. J. Physiol.* 34: 326–331, 1914.

Noguchi, H., Y. Ogushi, I. Yoshiya, N. Itakura, and H. Yambayashi. Breath-by-breath VCO_2 and VO_2 require compensation for transport delay and dynamic response. *J. Appl. Physiol.* 52: 79–84, 1982.

Pearce, D. H., H. T. Milhorn, G. H. Holloman, and W. J. Reynolds. Computer-based system for analysis of respiratory responses to exercise. *J. Appl. Physiol.: Respirat. Environ. Exercise Physiol.* 42: 968–975, 1977.

Ponte, J., and M. J. Purves. Frequency response of carotid body chemoreceptors in the cat to changes of $PaCO_2$, and pH_a. *J. Appl. Physiol.* 37: 635–647, 1974.

Purves, M. J. The effect of eliminating fluctuations of gas tensions in arterial blood on carotid chemoreceptor activity and respiration. *J. Physiol.* (London) 186: 63P, 1975.

Richard, C. A., T. G. Waldrop, R. M. Bauer, J. H. Mitchell, and R. W. Stremel. The nucleus reticularis gigantocellularis modulates the cardiopulmonary responses to central and peripheral drives related to exercise. *Brain Res.* 482: 49–56, 1989.

Rotto, D. M., and M. P. Kaufman. Effect of metabolic products of muscular contraction on discharge of group III and IV afferents. *J. Appl. Physiol.* 64: 2306–2313, 1988.

Rowell, L. B., L. Hermansen, and J. R. Blackmon. Human cardiovascular and respiratory responses to graded muscle ischemia. *J. Appl. Physiol.* 41: 693–701, 1976.

Salmoiraghi, C. G. Functional organization of brain stem respiratory neurons. *Ann. N.Y. Acad. Sci.* 109: 571–585, 1963.

Sargeant, A. J., M. Y. Rouleau, J. R. Sutton, and N. L. Jones. Ventilation in exercise studied with circulatory occlusion. *J. Appl. Physiol.* 50: 718–723, 1981.

Segal, S. S., and G. A. Brooks. Effects of glycogen depletion and work load on postexercise O_2 consumption and blood lactate. *J. Appl. Physiol.: Respirat. Environ. Exercise Physiol.* 47: 514–521, 1979.

Senapati, J. The role of the cerebellum in the hyperpnea produced by muscle afferents. In Respiratory Adaptation, Capillary Exchange and Reflex Mechanisms, A. S. Paintal and P. Gill-Kumar (Eds.). New Delhi: Navchetan Press (Private) Limited, 1977, pp. 115–120.

Strange, S., N. H. Secher, J. A. Pawelczyk, J. Karpakka, N. J. Christensen, J. H. Mitchell, and B. Saltin. Neural control of cardiovascular responses and of ventilation during dynamic exercise in man. *J. Physiol.* (London) 470: 693–704, 1993.

Strange-Peterson, E., B. J. Whipp, D. B. Drysdale, and D. J. C. Cunningham. Carotid arterial blood gas oscillations and the phase of the respiratory cycle during exercise in man: testing a model. *Adv. Ex. Med. Biol.* 99: 335–342, 1978.

Sutton, J. R., R. L. Hughson, R. McDonald, A. C. P. Powles, N. L. Jones, and J. D. Fitzgerald. Oral and intravenous propranolol during exercise. *Clin. Pharmacol. Thera.* 21: 700–705, 1975.

Sylvester, J. T., B. J. Whipp, and K. Wasserman. Ventilatory control during brief infusions of CO_2-laden blood in the awake dog. *J. Appl. Physiol.* 35: 178–186, 1973.

Thimm, F., and U. Tibes. Effect of K^+, osmolality, lactic acid, orthophosphate and epinephrine on muscular receptors with group I, III, and IV afferents. *J. Physiol.* (London) 284: 182P–183P, 1978.

Tibes, U., B. Hemmer, and D. Boning. Heart rate and ventilation in relation to venous K^+, osmolality, pH, PCO_2, PO_2, orthophosphate, and lactate at transition from rest to exercise in athletes and nonathletes. *Europ. J. Appl. Physiol.* 36: 127–140, 1977.

Vander, A. J., J. H. Sherman, and D. S. Luciano. Human Physiology. New York: McGraw-Hill, 1975.

Waldorp, T. G., F. L. Eldridge, G. A. Iwamoto, and J. H. Mitchell. Central neural control of respiration and circulation during exercise. American Physiological Society Handbook of Physiology-Exercise: Regulation and Integration of Multiple Systems. Oxford University Press, 1996, pp. 334–380.

Wasserman, D. H., and B. J. Whipp. Coupling of ventilation in pulmonary gas exchange during nonsteady-steady work in man. *J. Appl. Physiol.: Respirat. Environ. Exercise Physiol.* 54: 587–593, 1983.

Wasserman, K., B. J. Whipp, R. Casaburi, D. J. Huntsman, J. Castagna, and R. Lugliani. Regulation of PCO_2 during intravenous CO_2 loading. *J. Appl. Physiol.* 36: 651–656, 1975.

Wasserman, K., B. J. Whipp, and J. Castagna. Cardiodynamic hyperpnea secondary to cardiac output increase. *J. Appl. Physiol.* 36: 457–464, 1974.

Wasserman, K., B. J. Whipp, S. N. Koyal, and M. G. Cleary. Effect of carotid body resection of ventilatory and acid–base control during exercise. *J. Appl. Physiol.* 39: 354–358, 1975.

Weissman, M. L., B. J. Whipp, D. J. Huntsman, and K. Wasserman. Role of neural afferents from working limbs in exercise hyperpnea. *J. Appl. Physiol.: Environ. Exercise Physiol.* 49: 239–248, 1980.

Whelan, R. F., and I. M. Young. The effect of adrenaline infusion on respiration in man. *Br. J. Pharmacol.* 8: 98–102, 1953.

Whipp, B. J., and J. A. Davis. Peripheral chemoreceptors and exercise hyperpnea. *Med. Sci. Sports.* 11: 204–212, 1979.

Whipp, B. J., D. J. Huntsman, and K. Wasserman. Evidence for a CO_2-mediated pulmonary chemoreflex on dog. (Abstract.) *Physiologist* 18: 447, 1975.

Whipp, B. J., and K. Wasserman. Carotid bodies and ventilatory control dynamics in man. *Fed. Proc.* 39: 2668–2673, 1980.

Willshaw, P. Mechanism of inhibition of chemoreceptor activity by sinus nerve efferents. In *Chemoreception in the Carotid Body*, H. Acker, S. Fidone, E. Pallot, C. Eyzaguirre, and D. W. Lubbers (Eds.). New York: Springer-Verlag, 1977, pp. 168–172.

Willshaw, P., and S. Majcherczyk. The effects of changes in arterial pressure on sinus nerve efferent activity. *Adv. Exp. Med. Biol.* 99: 275–280, 1978.

Yamamoto, W. S. Mathematical analysis of the time course of alveolar CO_2. *J. Appl. Physiol.* 15: 215–219, 1960.

Yamamoto, W. S., and M. W. Edwards, Jr. Homeostasis of carbon dioxide during intravenous infusion of carbon dioxide. *J. Appl. Physiol.* 15: 807–818, 1960.

Young, M. I. Some observations on the mechanism of adrenaline hyperpnea. *J. Physiol.* (London) 137: 374–395, 1957.

Zuntz, N., and J. Geppert. Ueber die Natur der normalen Atomreize um den Ort ihrer Wirkung. *Arch. Ges. Physiol.* 38: 337–338, 1886.

CHAPTER 13

VENTILATION AS A LIMITING FACTOR IN AEROBIC PERFORMANCE AT SEA LEVEL

In healthy young people who are not endurance athletes, ventilation is not usually considered the factor limiting aerobic performance at altitudes close to sea level. The term *aerobic performance* is sometimes defined as the ability to endure hard and prolonged tasks of submaximal intensity. Sometimes, *aerobic performance* is also defined as $\dot{V}_{O_2max}$. However, whether it is defined in terms of submaximal endurance or short-time capacity to achieve $\dot{V}_{O_2max}$, the process of ventilation is sufficiently robust to continue at high rates for prolonged periods, oxygenating the blood passing through the lungs.

The capacity to increase ventilation during exercise is relatively much greater than the body's capacity to increase cardiac output or oxygen consumption. Alveolar surface area is large compared to pulmonary blood volume; the alveolar partial pressure of O_2 ($P_{A_{O_2}}$) increases during exercise, pulmonary capillary transit time for erythrocytes is sufficient for oxygenation, and the arterial partial pressure of O_2 ($P_{a_{O_2}}$) is maintained close to resting levels even during exercise eliciting $\dot{V}_{O_2max}$ at sea level. The alveolar-arterial (*A-a*) oxygen gradient widens slightly during maximal effort owing to a rise in $P_{A_{O_2}}$. Therefore, considerable ventilatory reserve exists to oxygenate blood passing through the lungs, and this reserve allows us to perform effectively at altitudes significantly above sea level.

■ Ventilatory Perfusion Ratio ($\dot{V}_E/\dot{Q}$) During Rest and Exercise

When scientists attempt to understand the limitations in a system such as the O_2 transport system, they frequently attempt to identify the factor or step that limits the system (Chapter 1). Finding the rate-limiting step in a series of linked reactions is analogous to finding the weak link in a chain that must support a great weight. In the overall scheme of O_2 transport from the atmosphere to tissues, the process of pulmonary ventilation is not generally considered to be limiting at normally inhabited altitudes near sea level.

As exercise work rate increases from resting levels up through easy to moderate intensities, pulmonary minute ventilation ($\dot{V}_E$) increases linearly (see Figure 12-14). However, as exercise intensity becomes more severe, a ventilatory threshold exists beyond which further increments in exercise work rate or $\dot{V}_{O_2}$ result in exaggerated, nonlinear increments in $\dot{V}_E$ (Chapter 10). Resting values of pulmonary minute ventilatory volume (5 liters · min^{-1}, BTPS) can increase to values of around 190 liters · min^{-1} in a healthy young adult male during exercise. This represents a 35-fold increase. As exercise intensity increases, the volume of blood flowing through (perfusing) the pulmonary vessels also in-

In most healthy young people arterial oxygen transport is not limited by breathing when performing at maximum at altitudes close to sea level. However, in Olympic champions such as Fermin Cacho, winner of the 1500-m run at Barcelona, breathing may limit the capacity for oxygen transport. PHOTO: © Claus Andersen.

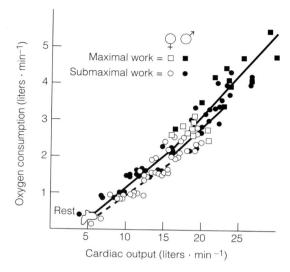

Figure 13-1 Relationship between oxygen consumption and cardiac output during rest and exercise. Given physical limits for arterial O_2 content and arteriovenous O_2 difference [$(a - v)_{O_2}$], maximal O_2 consumption ($\dot{V}_{O_2max}$) is largely a function of the ability to increase cardiac output. Modified from Holmgren et al. (1960) and Åstrand and Rodahl, 1970. Used with permission.

creases. However, this increase in cardiac output is essentially linear from resting levels up through those achieved during maximal exercise (Figure 13-1). In absolute terms, resting cardiac output (5 liters · min⁻¹) can increase five or six times (25 to 30 liters · min⁻¹) during exercise in a young, healthy, and fit adult male.

The ratio of pulmonary minute ventilation to cardiac output (i.e., the ventilation/perfusion ratio, $\dot{V}_E / \dot{Q}$) approximates unity during rest in most individuals. In the robust individual just described, the $\dot{V}_E / \dot{Q}$ may increase five- to sixfold during the transition from rest to maximal exercise. Because the capacity for expanding $\dot{V}_E$ (range 5 to 190 liters · min⁻¹)

is far greater than the capacity for expanding $\dot{Q}$ (range 5 to 30 liters · min⁻¹), the $\dot{V}_E / \dot{Q}$ ratio increases during exercise. The increase in $\dot{V}_E / \dot{Q}$ seen during exercise is usually presented as one reason why pulmonary ventilation is not thought to limit aerobic performances. In individuals less fit than the one just described, the absolute increments in $\dot{V}_E$ and $\dot{Q}$ achieved during exercise will be less, but the $\dot{V}_E / \dot{Q}$ ratio will expand to a similar extent. A new and emerging concept developed by Tim Noakes of South Africa describes how in healthy young people the heart protects itself against oxygen lack during exercise through a "self-limiting" mechanism (Chapter 16).

■ The Ventilatory Equivalent of O₂ During Exercise

An argument similar to that on the $\dot{V}_E / \dot{Q}$ is used with the ventilatory equivalent of O₂ (i.e., $\dot{V}_E / \dot{V}_{O_2}$) to exclude pulmonary ventilation as a factor limiting aerobic performance. During rest, $\dot{V}_E$ approximates 5 liters $\cdot$ min^{-1}, whereas $\dot{V}_{O_2}$ can be as low as 0.25 liter $\cdot$ min^{-1}; the $\dot{V}_E / \dot{V}_{O_2}$ is 20. During maximal exercise, for the individual we are describing, $\dot{V}_{E_{max}}$ may expand to 190 liters $\cdot$ min^{-1}, and $\dot{V}_{O_2max}$ might be 5 liters $\cdot$ min^{-1}; the $\dot{V}_E / \dot{V}_{O_2}$ increases to 35. Therefore, the ability to expand ventilation is relatively greater than the ability to expand oxidative metabolism.

■ $\dot{V}_{Emax}$ Versus MVV During Exercise

Pulmonary minute ventilation can increase tremendously from rest to maximal exercise (see Figure 12-14). The greatest $\dot{V}_E$ observed in an individual during exercise, however, is usually less than the maximum voluntary ventilatory (MVV, Chapter 12) capacity of that individual. When a young recreational athlete turns into the home stretch in finishing a track race, he or she will be breathing very heavily and will probably feel a degree of breathlessness. However, if that person were to volitionally attempt to increase $\dot{V}_E$ still further, he or she could probably do it. The fact that in most young healthy individuals the maximal $\dot{V}_E$ observed during exercise is less than the MVV capacity is another reason ventilation is not thought to limit aerobic performance at altitudes near sea level. However, as we will see later, in highly trained endurance athletes mechanical limitations to breathing may occur.

■ Partial Pressures of Alveolar ($P_{A_{O_2}}$) and Arterial Oxygen ($P_{a_{O_2}}$) During Exercise

The real test of the adequacy of pulmonary ventilation rests with the partial pressures of O_2 in alveoli ($P_{A_{O_2}}$) and arterial blood ($P_{a_{O_2}}$) during exercise. As described in Chapters 11 and 12, O_2 transport around the body is accomplished only because O_2 (like other gases) moves from areas of high concentration (or partial pressure) to areas of lower partial pressure. Respiratory gas exchange is accomplished through ventilation by maintaining a high alveolar partial pressure for O_2 and a low alveolar partial pressure for CO_2. The adequacy of pulmonary ventilation in maintaining the partial pressure of O_2 in alveoli during exercise up to $\dot{V}_{O_2max}$ is illustrated in Figure 11-3. If anything, increased ventilation during exercise raises the $P_{A_{O_2}}$ in most people.

The partial pressure of O_2 ($P_{a_{O_2}}$) in systemic arterial blood (i.e., blood that has circulated through the lungs and heart, and into the aorta) is also well maintained during exercise (see Figure 11-3). This apparently results from several major factors. First, the $P_{A_{O_2}}$ is maintained or increased. Second, the erythrocytes passing through the pulmonary capillaries remain there sufficiently long for equilibration with alveolar O_2 (see Figure 12-11). Third, the sigmoid shape of the O_2 dissociation curve (see Figure 11-2) is such that when the P_{O_2} is about 100 mm Hg, the saturation of hemoglobin with O_2 is maintained, even if the P_{O_2} falls off several millimeters of mercury.

■ Alveolar Surface Area for Exchange

As noted earlier, the alveolar surface area estimated to exist in the average-sized individual is 50 m^2, or 35 times the surface area of the person. This is the same area represented by one-half of a single tennis court. Keeping in mind that the average blood volume is 5 liters, picture the following scene. You are trying to spread the liquid contents of a 5-liter container over one side of a tennis court. How far do you think you could get in spreading the liquid? Could you cover the one side? Now consider that in reality about 4% (200 ml) of the blood volume (5 liters) is in the pulmonary system at any one instant during maximal exercise. This means that the ratio of alveolar surface area to pulmonary capillary blood volume is enormous. This disproportionality ensures a large capacity for the exchange of respiratory gases between blood and alveolar air during ventilation.

■ Fatigue of Ventilatory Muscles and Other Limitations in Ventilation

Any muscle, including the diaphragm and accessory ventilatory muscles, can be made to fatigue.

When diaphragm muscle preparations isolated from experimental animals are electrically stimulated to contract at high rates, the muscles demonstrate distinct fatigue characteristics. In addition, during a MVV test, human subjects usually demonstrate a degree of fatigue and are not able to ventilate at as high a rate at the end of the test as at the beginning of the test. Maximum voluntary ventilation tests repeated in rapid succession also usually produce decreasing values. The central question regarding ventilatory muscle fatigue during exercise, however, is whether ventilatory fatigue precedes or coincides with and results in decrements in body exercise performance. The answer to this question is usually no. As noted previously, $\dot{V}_{E\,max}$ during exercise is usually less than MVV tests, even MVV tests determined after exercise. Further, most athletes can volitionally raise their ventilatory flow at the end of exercise.

■ Pulmonary Limitations in Highly Trained Athletes

The issue of adequacy of the ventilatory system during maximal exercise in very fit individuals has been raised by Dempsey and colleagues at the University of Wisconsin (1977, 1981, 1984, 1986). They observed decreases in $P_{a_{O_2}}$ (i.e., hypoxemia, low blood oxygen) when some very fit subjects were stressed to $\dot{V}_{O_2 max}$ (Figure 13-2). Because the $P_{a_{O_2}}$ decreased, the pulmonary system did not adequately oxygenate blood passing through it. The authors suspected that compliance in the ventilatory system prevented $\dot{V}_E$ from rising sufficiently to maintain arterial O_2 concentration. *Compliance* is a measure of the expandability of the lungs and thorax. These structures provide a resistance to ventilatory movements and ventilatory flow; that is, ventilatory flow is limited by the internal resistance of the pulmonary tissues. The University of Wisconsin scientists have apparently observed that in some very fit individuals, $\dot{V}_E$ during exercise approaches a limit imposed by their ability to generate expiratory and inspiratory flows and pressures sufficient to prevent hypoxemia. Figure 13-3 presents flow–volume and pressure–volume loop diagrams for two very fit,

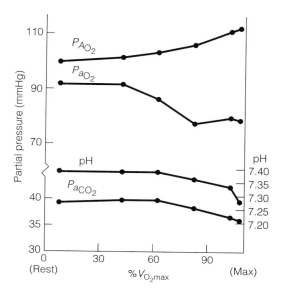

Figure 13-2 Average blood gases, acid–base status, and alveolar-to-arterial (*A-a*) O_2 difference during progressive exercise in eight male athletes. Note that in contrast to data on healthy, young, but not highly trained subjects (Figure 11-3), in these athletes $P_{a_{O_2}}$ falls and the (*A-a*) widens both because $P_{A_{O_2}}$ increases and $P_{a_{O_2}}$ decreases. SOURCE: Johnson, Dempsey, and associates, 1992. Used with permission.

young male athletes. In the diagrams, the maximal volitional flows are given by the dotted and dashed lines in the left-hand figures; the maximal effective pressures developed in expiration ($P_{E\,max}$) and inspiration ($P_{I\,max}$) are shown on the right. In the diagrams, the loops refer to flows or pressures generated during graded intensity exercises, with flows in easy exercise at the inner core, and flows at high levels of effort at the perimeters.

Results for subject S. J. indicates that he achieved maximal ventilatory flow rates and inspiratory and expiratory pressures during exercise. In contrast, subject T. H. still had a substantial reserve for inspiratory pressure development even in maximal exercise. However, T. H. became more hypoxemic at $\dot{V}_{O_2 max}$ than S. J. and his alveolar-arterial (*A-a*) oxygen gradient was greater.

Besides illustrating individual differences in the extent of arterial hypoxemia during maximal sea-

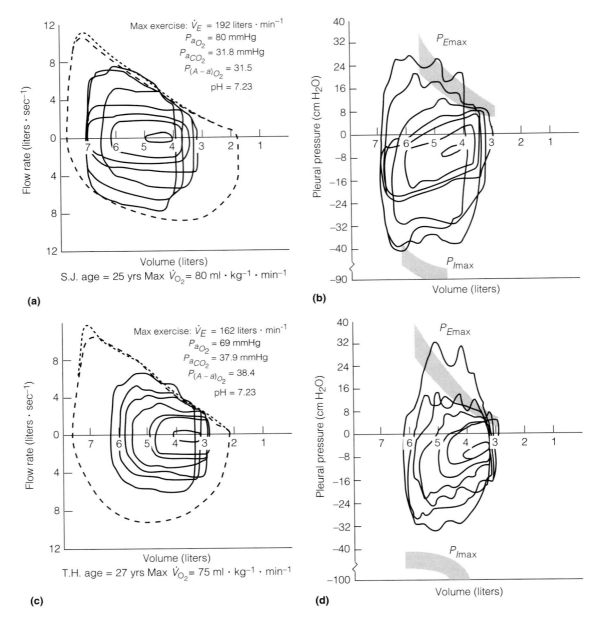

(a)

Max exercise: $\dot{V}_E$ = 192 liters · min^{-1}
$P_{a_{O_2}}$ = 80 mmHg
$P_{a_{CO_2}}$ = 31.8 mmHg
$P_{(A-a)_{O_2}}$ = 31.5
pH = 7.23

S.J. age = 25 yrs Max $\dot{V}_{O_2}$ = 80 ml · kg^{-1} · min^{-1}

(b)

(c)

Max exercise: $\dot{V}_E$ = 162 liters · min^{-1}
$P_{a_{O_2}}$ = 69 mmHg
$P_{a_{CO_2}}$ = 37.9 mmHg
$P_{(A-a)_{O_2}}$ = 38.4
pH = 7.23

T.H. age = 27 yrs Max $\dot{V}_{O_2}$ = 75 ml · kg^{-1} · min^{-1}

(d)

Figure 13-3 Flow–volume (left) and pressure–volume loop diagrams on two male athletes during progressive intensity exercise. The outer limits of maximal voluntary flow rate are given by the dotted and dashed lines in the left panels. The maximum values for expiratory ($P_{E_{max}}$) and inspiratory ($P_{I_{max}}$) pressures are given on the right. Easy exercise loops are at the core, and maximal exercise values are at the outer loops. Both subjects demonstrate limitations in expiratory flow rate and pressure, and subject S. J. (top) also reaches maximal inspiratory flow and pressure during exercise. Subject T. H., however, does not fully utilize inspiratory flow or pressure, but suffers a greater fall in $P_{a_{O_2}}$ (hypoxemia) than subject S. J. SOURCE: Johnson, Dempsey, and associates, 1992. Used with permission.

level exercise by trained endurance athletes, Figure 13-3 illustrates another interesting result. This is that T. H.'s ventilatory response to hypoxemia was apparently blunted because he could have exerted greater effort on inspiration in an attempt to compensate for the hypoxemia. The investigators suspected that athletes such as T. H. do not attempt to ventilate more during maximal exercise due to an apparent economy. Because at high pulmonary flows the oxygen cost of breathing equals the rise in $\dot{V}_{O_2}$ achieved by breathing more, athletes may learn to tolerate hypoxemia in order to lessen the overall effort associated with maximal exertion.

How general the presence of arterial hypoxemia is among top aerobic athletes needs to be determined, for it may be that Olympic champions do not experience hypoxemia even at very high rates of oxygen consumption. Certainly Dempsey and associates have demonstrated large *A-a* gradients, pulmonary limitations, and apparently also insufficient pulmonary capillary transit times in some highly trained athletes. However, for the present, we must consider it unlikely that the $\dot{V}_{O_2max}$ of normal asymptomatic individuals is limited by pulmonary ventilation.

■ Success at High Altitude

We have addressed ventilation as limiting performance at sea-level altitude. In addition, we have addressed the matter of ventilation as limiting the $\dot{V}_{O_2max}$ of elite athletes performing at sea level. Although in Chapter 23, we discuss the stresses imposed by high altitude, we need to mention here also important aspects of the relationships between high-altitude exercise performance and pulmonary minute ventilation.

In many respects, as shown by Olez and associates (1986), mountain climbers such as Reinhold Messner and Peter Habeler, who first climbed Mt. Everest without benefit of supplemental oxygen, are similar to other good, though not exceptional, sea-level athletes. In contrast to sea-level endurance athletes who may desaturate and hypoventilate during exertion at sea level (see Figures 13-2

and 13-3), Schoene (1982) has shown that elite climbers are noted for a strong hypoxic ventilatory drive. Thus, those endowed by physiology and experience for mountain climbing breathe more and maintain higher arterial oxygen saturation when climbing at high altitudes. In fact, Lawler and associates (1988) have shown that highly trained endurance athletes are probably more susceptible to gas exchange impairments than less fit individuals during severe exertion in hypoxic conditions.

■ Summary

Mechanisms for the control of ventilation are redundant in their operation, thus ensuring adequate responses under a variety of conditions (Chapter 12). Similarly, the robust capacity of the pulmonary system usually ensures that at sea level, O_2 transport from the atmosphere to metabolizing tissues is not limited by ventilation. The ventilatory perfusion ratio, $\dot{V}_E/\dot{Q}$, increases several times during exercise, as does the ventilatory equivalent of O_2, $\dot{V}_E/\dot{V}O_2$. The alveolar partial pressure of O_2 rises during exertion, and oxygenation of systemic arterial blood ($P_{a_{O_2}}$) does not decrease appreciably. Consequently, arterial oxygen saturation ($S_{a_{O_2}}$) and content ($C_{a_{O_2}}$) are maintained as well.

This robust, "overbuilt" ventilatory capacity allows us not only to sustain high rates of oxidative metabolism at altitudes near sea level, but also to withstand additional stresses imposed by unique situations. For instance, the limitations of stroke mechanics may limit ventilatory frequency during swimming; yet swimmers are perhaps the premier aerobic athletes. The ventilatory reserve allows athletes such as soccer and basketball players to converse even during very heavy exercise. Although most of the earth's surface is covered by water, and much of the remaining land surface is not much above sea level, significant land masses do exist at relatively high altitudes. The ventilatory reserve that we humans have allows us to inhabit areas as high as 14,000-ft altitude and to sojourn at even higher altitudes. Mt. Everest has recently been climbed without benefit of auxiliary O_2 supplies.

In subsequent chapters, we describe other systems and factors that limit O_2 transport during exercise.

■ **Selected Readings**

Anderson, P., and J. Henriksson. Capillary supply to the quadriceps femoris muscle of man: adaptive response to exercise. *J. Physiol.* (London) 270: 677–690, 1977.

Åstrand, P.-O., T. E. Cuddy, B. Saltin, and J. Stenberg. Cardiac output during submaximal and maximal work. *J. Appl. Physiol.* 19: 268–273, 1964.

Åstrand, P.-O., and K. Rodahl. Textbook of Work Physiology. New York: McGraw-Hill, 1970, pp. 154–178, 187–254.

Åstrand, P.-O., and B. Saltin. Maximal oxygen uptake and heart rate in various types of muscular activity. *J. Appl. Physiol.* 16: 977–981, 1961.

Bannister, R. G., and C. J. C. Cunningham. The effects on the respiration and performance during exercise of adding oxygen to the inspired air. *J. Physiol.* (London) 125: 118–120, 1954.

Barclay, J. K., and W. N. Stainsby. The role of blood flow in limiting maximal metabolic rate in muscle. *Med. Sci. Sports* 7: 116–119, 1975.

Bergh, U., I.-L. Kaustrap, and B. Ekbloom. Maximal oxygen uptake during exercise with various combinations of arm and leg work. *J. Appl. Physiol.* 41: 191–196, 1976.

Bevegård, S., A. Holmgren, and B. Jonsson. Circulatory studies in well trained athletes at rest and during heavy exercise, with special reference to stroke volume and the influence of body position. *Acta Physiol. Scand.* S7: 26, 1963.

Bishop, J. M., K. W. Harold, S. W. Taylor, and P. N. Wormald. Changes in arterial-hepatic venous oxygen content difference during and after supine leg exercise. *J. Physiol.* (London) 137: 309–314, 1957.

Booth, F. W., and K. A. Narahara. Vastus lateralis cytochrome oxidase activity and its relationship to maximal oxygen consumption in man. *Pflügers Archiv.* 369: 319–324, 1974.

Buick, F. J., N. Gledhill, A. B. Frosese, L. Spriet, and E. C. Meyeis. Effect of induced erythrocythemia on aerobic work capacity. *J. Appl. Physiol.: Respirat. Environ. Exercise Physiol.* 48: 636–642, 1980.

Byrne-Quinn, E., J. V. Weil, I. Sodal, G. F. Fillez, and R. F. Grover. Ventilatory control in the athlete. *J. Appl. Physiol.* 30: 91–98, 1971.

Davies, C. T. M., and A. J. Sargent. Physiological response to one- and two-leg exercise breathing air and 45% oxygen. *J. Appl. Physiol.* 36: 142–148, 1974.

Davies, K. J. A., J. L. Maguire, G. A. Brooks, P. R. Dallman, and L. Packer. Muscle mitochondrial bioenergetics, oxygen supply, and work capacity during dietary iron deficiency and repletion. *Am. J. Physiol.* 242 (*Endocrinol. Metabol.* 5): E418–E427, 1982.

Dempsey, J. A. Is the lung built for exercise? *Med. Sci. Sports Exerc.* 18: 143–155, 1986.

Dempsey, J. A., N. Gledhill, W. G. Reddan, H. V. Forester, P. G. Hanson, and A. D. Claremont. Pulmonary adaptation to exercise: effects of exercise type and duration, chronic hypoxia and physical training. *Ann. N.Y. Acad. Sci.* 301: 242–261, 1977.

Dempsey, J. A., P. G. Hanson, and K. S. Henderson. Exercise-induced arterial hypoxaemia in healthy human subjects at sea level. *J. Physiol.* (London) 355: 161–175, 1984.

Dempsey, J. A., P. E. Hanson, and S. M. Mastenbrook. Arterial hypoxemia during heavy exercise in highly trained runners. *Fed. Proc.* 40: 932, 1981.

Ekbloom, B., A. N. Goldbarg, and B. Bullbring. Response to exercise after blood loss and reinfusion. *J. Appl. Physiol.* 33: 175–189, 1972.

Ekelund, L. G., and A. Holmbren. Circulatory and respiratory adaptation during long-term, non-steady state exercise in the sitting position. *Acta Physiol. Scand.* 62: 240, 1964.

Ekelund, L. G., and A. Holmbren. Central hemodynamics during exercise. In Physiology of Muscular Exercise, C. B. Chapman (Ed.). Monograph 15. New York: American Heart Association, 1967, pp. I33–I43.

Faulkner, J. A., D. E. Roberts, R. L. Elk, and J. Conway. Cardiovascular responses to submaximum and maximum effort cycling and running. *J. Appl. Physiol.* 30: 457–461, 1971.

Gardner, G. W., V. R. Edgerton, B. Senesirathe, R. J. Barnard, and Y. Ohira. Physical work capacity and metabolic stress in subjects with iron deficiency anaemia. *Am. J. Clin. Nutr.* 30: 910–917, 1977.

Gleser, M. A., D. H. Horstman, and R. P. Mello. The effects of $\dot{V}_{O_2max}$ of adding arm work to maximal leg work. *Med. Sci. Sports* 6: 104–107, 1974.

Grimby, G., E. Haggendal, and B. Saltin. Local xenon 133 clearance from the quadriceps muscle during exercise in man. *J. Appl. Physiol.* 22: 305–310, 1967.

Hanson, P., A. Claremont, J. Dempsey, and W. Reddan. Determinants and consequences of ventilatory responses to competitive endurance running. *J. Appl. Physiol.: Respirat. Environ. Exercise Physiol.* 52: 615–623, 1982.

Henriksson, J., and J. S. Reitman. Time course of changes in human skeletal muscle succinate dehydrogenase and cytochrome oxidase activities and maximal oxygen uptake with physical activity and inactivity. *Acta Physiol. Scand.* 99: 91–97, 1977.

Holmgren, A., and P.-O. Åstrand. Pulmonary diffusing capacity and the dimensions and functional capacities of the oxygen transport system in humans. *J. Appl. Physiol.* 21: 1463–1467, 1966.

Holmgren, A., B. Jonsson, and T. Sjöstrand. Circulatory data in normal subjects at rest and during exercise in the recumbent position with special reference to the stroke volume at different intensities. *Acta Physiol. Scand.* 49: 343, 1960.

Holmgren, A., and M. B. McIlroy. Effect of temperature on arterial blood gas tensions and pH during exercise. *J. Appl. Physiol.* 19: 243, 1964.

Hughes, R. L., M. Clode, R. H. Edwards, T. J. Goodwin, and N. Jones. Effect of inspired O_2 on cardiopulmonary and metabolic responses to exercise in man. *J. Appl. Physiol.* 24: 336–347, 1968.

Johnson, B. D., K. W. Saupe, and J. S. Dempsey. Mechanical constraints on exercise hyperpnea in endurance athletes. *J. Appl. Physiol.* 73: 874–886, 1992.

Lawler, J., S. K. Powers, and D. Thompson. Linear relationship between $\dot{V}_{O_2max}$ and $\dot{V}_{O_2max}$ decrement during exposure to acute hypoxia. *J. Appl. Physiol.* 64: 1486–1492, 1988.

Leith, D. E., and M. E. Bradley. Ventilatory muscle strength and endurance training. *J. Appl. Physiol.* 41: 508–516, 1976.

Martin, B. J., K. E. Sparks, C. W. Z. Willich, and J. V. Weil. Low exercise ventilation in endurance athletes. *Med. Sci. Sports* 11: 181–185, 1979.

Mitchell, J. H., B. J. Sproule, and C. B. Chapman. Physiological meaning of the maximal oxygen uptake test. *J. Clin. Invest.* 37: 538, 1958.

Olez, O., H. Howald, P. E. Di Prampero, H. Hoppeler, H. Classen, R. Jenni, A. Buhlmann, G. Ferretti, J. C. Bruckner, A. Veiscteinas, M. Gussoni, and P. Cerre-

telli. Physiological profile of world-class climbers. *J. Appl. Physiol.* 60: 1734–1742, 1986.

Pirnay, F., M. Lamy, J. Dujarding, R. Deroanne, and J. M. Petit. Analysis of removal venous blood during maximum muscular exercise. *J. Appl. Physiol.* 33: 289–292, 1972.

Pirnay, F., R. Marechal, R. Radermecker, and J. M. Petit. Muscle blood flow during submaximum and maximum exercise on a bicycle ergometer. *J. Appl. Physiol.* 32: 210–212, 1972.

Reeves, J. T., R. F. Grover, S. G. Blout, Jr., and F. G. Filley. The circulatory changes in man during mild supine exercise. *J. Appl. Physiol.* 6: 279–282, 1961.

Reeves, J. T., R. F. Grover, S. G. Blout, Jr., and F. G. Filley. The cardiac output response to standing and walking. *J. Appl. Physiol.* 6: 283–287, 1961.

Reeves, J. T., R. F. Grover, F. G. Filley, and S. G. Blout, Jr. The cardiac output in normal resting man. *J. Appl. Physiol.* 16: 276–278, 1961.

Saltin, B., R. F. Grover, C. G. Blomquist, L. H. Hartley, and R. J. Johnson. Maximal oxygen uptake and cardiac output after 2 weeks at 4300 m. *J. Appl. Physiol.* 25: 406–409, 1968.

Schoene, R. B. Control of ventilation in climbers. *J. Appl. Physiol.* 53: 886–890, 1982.

Secher, N. H., N. Ruberg-Larsen, R. Binkhorst, and F. Bonde-Peterson. Maximal oxygen uptake during arm cranking and combined arm plus leg exercise. *J. Appl. Physiol.* 36: 515–518, 1974.

Sutton, J. R., and W. L. Jones. Control of pulmonary ventilation during exercise and mediators in the blood: CO_2 and hydrogen ion. *Med. Sci. Sports* 11: 198–203, 1979.

Taylor, H. E., E. R. Buskirk, and A. Henschel. Maximal oxygen uptake as an objective measure of cardiorespiratory performance. *J. Appl. Physiol.* 8: 73–84, 1955.

CHAPTER 14

THE HEART

The body, whether resting or exercising, depends on the cardiovascular system (CVS) for most of its transport needs. The CVS helps to maintain a constant internal environment by delivering important substances to tissues and eliminating metabolic end products. It is responsible for delivering oxygen and fuels to tissues, and it helps maintain pH and core temperature. The system also delivers hormones, which are often produced away from their sites of action.

Control of the CVS is directed toward maintaining blood pressure and meeting the tissues' metabolic requirements. During exercise, the need to deliver oxygen and fuels to working muscles is greater than it is when the body is at rest. The system increases its output to deliver more blood. Cardiac output and muscle blood flow are increased. Blood flow to less active tissues is decreased, while that to critical areas such as the brain and heart is either maintained or increased.

The heart and circulation work as a unit, with the heart pushing blood into the blood vessels. The ejected blood stretches the arteries, which then recoil to help propel blood through the circulation. Blood vessels are very compliant. This means they change their shape when stretched. They are also elastic, which means they tend to return to their resting shape after being stretched. The compliance and elasticity of blood vessels helps maintain blood

A biathlon athlete in action at the Norway Olympics. Aerobic capacities of cross-country skiers compare favorably with those of other world-class athletes. Aerobic capacity (as measured by $\dot{V}_{O_2max}$) depends largely upon cardiovascular capacity. PHOTO: © Claus Andersen.

pressure at rest and during exercise. Blood pressure is maintained during exercise by an increase in cardiac output and regulation of blood flow to tissues.

Maintaining blood pressure and meeting tissue demands requires coordination between the heart and circulation. Cardiovascular control involves neural, mechanical, and hormonal regulatory mechanisms. This chapter examines the structure and function of the heart during rest and exercise. Chapter 15 will discuss the circulation and cardiovascular control mechanisms. Chapter 16 will discuss oxygen transport dynamics. Although we review basic cardiovascular anatomy and physiology in these chapters, readers should refer to basic texts for more in-depth coverage.

■ The Structure of the Heart

The heart is a hollow four-chambered muscular organ. It pumps blood to the lungs and general circulation. It rests in the chest cavity on the diaphragm (Figure 14-1), behind the lower parts of the lungs and below the middle and left of the sternum. The apex of the heart descends to about the fifth rib.

The heart comprises three layers: The outer layer, the *pericardium,* is composed of fibrous tissue interspersed with adipose tissue; the middle layer, the *myocardium,* is composed of cardiac muscle; and the inner layer, the *endocardium,* is composed of squamous endothelial cells. This layer is continuous with the inner lining of the arteries.

The heart is divided into the right and left atria (single, *atrium*) and ventricles. The chambers are distinct, separated by walls called *septa.* The *atria* are thin-walled, low-pressure chambers that serve as reservoirs for the *ventricles.* The smaller right ventricle pumps blood into the pulmonary artery to the lungs. The larger, more muscular left ventricle pumps blood into the aorta to the general circulation.

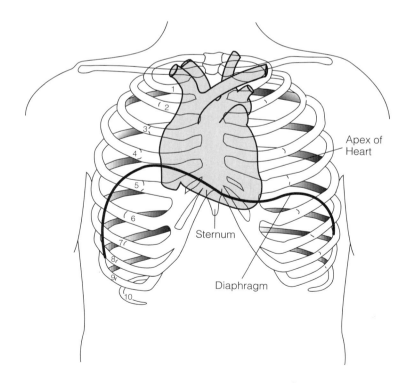

Figure 14-1 Location of the heart in the chest cavity.

Apex of Heart

Sternum

Diaphragm

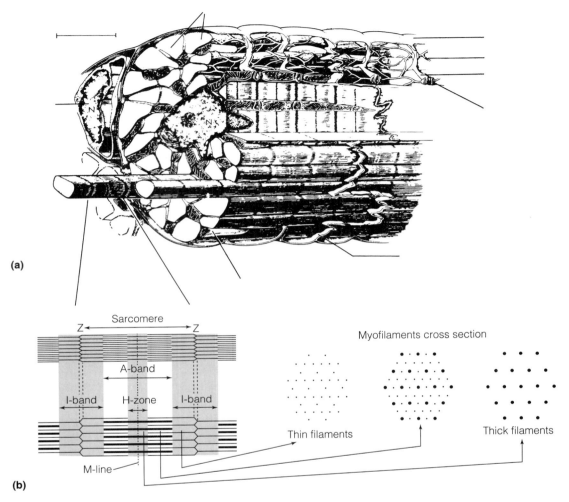

Figure 14-2 Schematic diagram of cardiac muscle as seen under the electron microscope. (a) Myocardial cell, showing arrangement of multiple parallel fibrils. (b) Individual sarcomere from a myofibril. A representation of the arrangement of myofilaments that make up the sarcomere is shown in bottom portion. At right are cross sections of the sarcomere, showing the specific lattice arrangement of myofilaments. Reproduced, with permission, from Braunwald, Ross, Jr., and Sonnenblick, 1967.

The Myocardium

All cells in the myocardium have an inherent rhythmicity. This means they contract at regular intervals. Myocardial cells are also highly conductive. They can propagate impulses that move rapidly from one cell to the next. All myocardial cells can also contract in response to an impulse. From a practical standpoint, cardiac cells are either contractile or electrical. Contractile cells make up the bulk of the myocardium, causing the pumping action of the heart. Electrical cells consist of specialized excitatory and conductive fibers that are responsible for the rapid conduction of impulses throughout the heart (Figure 14-2).

Cardiac and skeletal muscles are similar in many ways (Chapter 19). Both are striated and have myofibrils containing actin and myosin filaments that participate in contraction. Both fibers are surrounded by a membrane called the *sarcolemma* and depolarize before contracting.

Cardiac and skeletal muscles are also different in several ways. Unlike multinucleated skeletal muscle cells, cardiac cells have only one or two centrally located nuclei. Each cardiac muscle cell is surrounded by a delicate sheath of connective tissue and an extensive capillary network that is more developed than the capillary network in skeletal muscle. Mitochondrial mass is also considerably greater in cardiac cells than it is in skeletal muscle cells. Mitochondria occupy 40% of the cytoplasm volume of cardiac cells compared to only 2-6% of skeletal muscle cells. This reflects the need for continuous aerobic metabolism in heart muscle. Cardiac muscle is also involuntary. Unlike skeletal fibers, cardiac muscle cells are connected in tight series, and the stimulation of one cell results in the stimulation of all the cells. In addition, cardiac muscle cells are shorter than skeletal muscle cells. They are joined together by *intercalated* disks, which serve as boundaries between the cardiac muscle cells. The electrical resistance between the cells is minimal, so all the cardiac muscles in the atria or ventricles can contract synchronously. Although this "all-or-none" principle applies to an entire muscle area in the heart, it applies to only a single motor unit in skeletal muscle. The cardiac contractile unit is called a *syncytium*.

All cardiac muscle cells have intrinsic rhythmicity. The cardiac tissue with the highest intrinsic rhythmicity determines the heart rate. The most important concentration of autorhythmic cells is found in the sinoatrial node (SA node), which is located in the right atrium. Normally, the SA node acts as the pacemaker for the heart. It starts an impulse at a rate of approximately 72 beats a minute. Many other pacemakers in the heart can become operative or dominant under certain circumstances.

A special conduction system exists in the heart (Figure 14-3). A communication network connects the SA node with the left atrium. The neuromuscular conduction system of the ventricles consists of the atrioventricular node (AV node, AV bundle, or Bundle of His), left and right bundle branches,

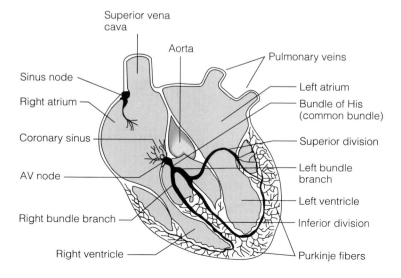

Figure 14-3 The electrical conduction system of the heart.

and Purkinje fibers. This system allows impulses to travel much faster and more uniformly than if they traveled directly across cardiac muscle.

Structure and the Cardiac Cycle

The heart receives blood from the veins and pumps it to the lungs and circulation. This is accomplished by a rhythmic contraction-relaxation process called the *cardiac cycle.* The cardiac cycle consists of an active phase (cardiac contraction) called *systole* and a relaxation phase called *diastole.* During diastole, the ventricles relax and the chambers fill with blood in preparation for the next systole. At rest, the duration of atrial systole is approximately 0.15 sec, while

ventricular systole is about 0.3 sec. Atrial diastole lasts about 0.65 sec, while ventricular diastole lasts about 0.5 sec. At a resting heart rate of 75 beats · min^{-1}, a cardiac cycle (systole and diastole) takes about 0.8 sec. During exercise, the cardiac cycle accelerates. In abnormal heart rhythms, such as ventricular tachycardia, heart rate can increase so high that it compromises filling and stroke volume.

Low-oxygenated blood returns to the heart from the circulation via the inferior and superior vena cavae to the right atrium (Figure 14-4). It then passes through the tricuspid valve into the right ventricle. The blood is then propelled from the right ventricle through the pulmonary valve into the pulmonary arteries to the lungs, where gas exchange occurs.

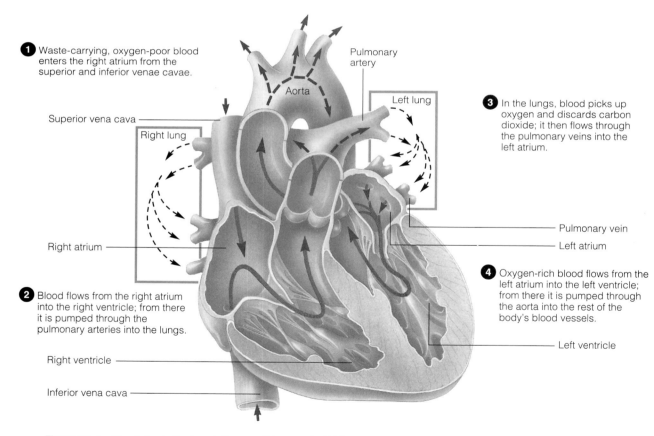

1. Waste-carrying, oxygen-poor blood enters the right atrium from the superior and inferior venae cavae.

Pulmonary artery

Aorta

Left lung

3. In the lungs, blood picks up oxygen and discards carbon dioxide; it then flows through the pulmonary veins into the left atrium.

Superior vena cava

Right lung

Pulmonary vein

Left atrium

Right atrium

4. Oxygen-rich blood flows from the left atrium into the left ventricle; from there it is pumped through the aorta into the rest of the body's blood vessels.

2. Blood flows from the right atrium into the right ventricle; from there it is pumped through the pulmonary arteries into the lungs.

Left ventricle

Right ventricle

Inferior vena cava

Figure 14-4 Circulation in the heart. SOURCE: Fahey et al., 1999.

The blood returns to the left atrium via the pulmonary veins and passes through the mitral valve into the left ventricle. It is finally propelled through the aortic valve into the aorta to the systemic circulation.

The atria act as reservoirs for the ventricles. At rest, the right and left atria play a small role in helping the heart maintain necessary cardiac output. During exercise, they help with the large increase in blood returning from the veins. The atria act as primer pumps, increasing the output of the ventricles.

The ventricles pump blood to the lungs and circulation, both ventricles pumping essentially the same amount of blood. The walls of the left ventricle are thicker than those of the right because of the greater resistance provided by the peripheral circulation. The left ventricle acts as a pressure pump and decreases its transverse diameter during systole. The right ventricle is more of a volume pump. Its large surface-to-volume ratio produces a bellows action when it contracts.

The amount of blood ejected from each ventricle during systole is called the *stroke volume.* The stroke volume is determined by the difference between ventricular filling and ventricular emptying. The normal resting stroke volume is 82 ± 20 ml. The volume at the peak of ventricular filling is called *end-diastolic volume* (EDV). The normal resting EDV is 125 ± 31 ml. The volume of blood remaining in the heart after contraction is called *end-systolic volume* (ESV). The normal resting ESV is 42 ± 17 ml. The percentage of the end-diastolic volume pumped from the ventricles is called the *ejection fraction* (EF). The normal resting EF is 0.67 for the left heart and 0.54 for the right heart.

The amount of blood pumped by the heart per unit of time is called *cardiac output.* Cardiac output ($\dot{Q}$) is a product of heart rate (f_H or HR) and stroke volume (V_S) (Eq. 14-1). At rest, the cardiac output is approximately 5 to 6 liters $\cdot$ min^{-1}. During exercise, it increases to about 20 liters $\cdot$ min^{-1}. In some world-class endurance athletes, it can exceed 35 liters $\cdot$ min^{-1}.

$$\dot{Q} = (V_S) \cdot (f_H) \qquad (14\text{-}1)$$

■ The Electrical Activity of the Heart and the Electrocardiogram

Resting heart muscle is polarized. Cells are negatively charged on the inside and positively charged on the surface. This electrical potential difference is due to different concentrations of ions on the inside and outside of the cell. The ions include sodium, potassium, calcium, chloride, and nondiffusible anions. Sodium and potassium are the most important ions determining this difference in potential across the cell membrane. There are approximately 130 meq $\cdot$ liter^{-1} of sodium on the outside of the cell and 10 meq $\cdot$ liter^{-1} on the inside. For potassium, the concentration gradient is reversed. Potassium concentrations are about 130 meq $\cdot$ liter^{-1} inside the cell and 5 meq $\cdot$ liter^{-1} outside the cell.

Heart muscle must depolarize before it can contract. The inside of the cell becomes positively charged and the outside becomes negatively charged. Depolarization is caused by an increased conductance of sodium across the membrane. As sodium moves into the cell, it is accompanied by calcium, which activates the contractile proteins in the heart responsible for the contraction of heart muscle. During repolarization, the cell's permeability for potassium increases and the movement of sodium and calcium into the cell slows. This returns the cell to its resting action potential. The cardiac cycle involves a series of such depolarizations, contractions, and repolarizations.

As we mentioned, contraction is stimulated by the release of calcium ion. Calcium release occurs when the impulse spreads from the sarcolemma to calcium storage sites in the terminal cisternae of the sarcoplasmic reticulum. The *sarcoplasmic reticulum* is a series of membrane tubes located close to the *sarcomeres,* which are the structures that cause muscle contraction. The impulse travels from the sarcolemma to the sarcoplasmic reticulum by way of transverse tubules (T-tubules). *T-tubules* are channels filled with interstitial fluid that connect the sarcolemma to the interior of the cell. These structures are illustrated in Figure 14-2. After contraction, the muscle repolarizes back to its resting state, and calcium is pumped back into its storage sites.

The Electrocardiogram (ECG or EKG)

Differences in membrane potentials on the heart's surface can be measured by determining the differences in potentials between two electrodes on the body's surface. This technique, called *electrocardiography,* has been used extensively as a noninvasive tool for detecting heart disease and assessing cardiac function. The *electrocardiogram, or ECG,* is an important tool in medicine and physiology.

The elements of the ECG precede contraction of the heart (Figure 14-5). Changing polarity or depolarization stimulates the myocardium to contract. In a normal heart, recall, all the muscle fibers in a given chamber contract together; this occurs because the depolarization occurs rapidly.

The SA node begins the impulse that starts the cardiac cycle. The impulse causes a wave of depolarization across both atria called a *P wave.* The P wave immediately precedes atrial contraction.

Figure 14-5   The elements of an electrocardiogram and the cardiac cycle. SOURCE: Crouch and McClintic.

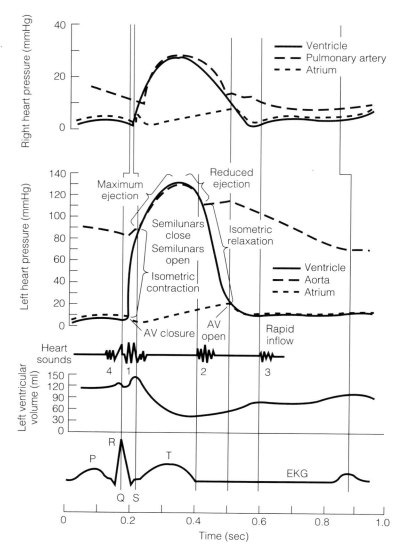

Conduction tracts from the SA node to the left atrium allow simultaneous contraction of the right and left atria.

A slight delay occurs when the depolarization wave reaches the AV node. This delay allows the blood to pass through the AV valve into the ventricles. After the delay, the AV node is stimulated, which causes depolarization of the ventricles and produces a QRS complex on the ECG. The impulse then enters a special neural conduction system consisting of the AV bundle, left and right bundle branches, and Purkinje fibers. This system allows for rapid stimulation of ventricular cardiac muscle. The QRS complex occurs just before the ventricles contract. The repolarization of the atria is usually masked by the QRS complex, so it cannot be seen on the ECG recording.

The *T wave* is caused by the repolarization of the ventricles. It happens just after ventricular contraction. The period between the S and T waves is called the *ST segment*. The ST segment is important in exercise stress testing because it is used to detect deficiencies in coronary artery blood flow.

A refractory period, an interval during which the muscle is incapable of full contraction, occurs after the contraction of cardiac muscle. An attempt at depolarization during this period reduces the force of cardiac contraction.

Training and the ECG

Training leads to changes in the ECG that can be confused with symptoms of heart disease. The most common ECG change is the appearance of sinus bradycardia (low heart rate). Studies of endurance athletes, football players, basketball players, and even weight lifters have shown reduced resting heart rates. Resting heart rates in elite distance runners have been reported as low as 28 beats per minute. The probable cause of reduced heart rate is a reduction in stimulation from the sympathetic nervous system and an increased stimulation from the parasympathetic system. Lower intrinsic heart rate may also contribute to the bradycardia.

A variety of ECG rhythm disturbances has also been reported in athletes. These disturbances include atrioventricular conduction delays, wandering atrial pacemaker, ST segment elevation, T-wave inversion, and substitution of the AV node as the primary pacemaker. These findings are common in people with ischemic heart disease. In athletes, however, most of these changes are due to increased parasympathetic influences from the vagus nerve. During exercise, it is common for these ECG variations to disappear. As exercise begins, parasympathetic influences decrease and sympathetic stimulation increases. Nevertheless, ECG abnormalities in athletes should not be automatically dismissed as a normal effect of exercise training. They should also not be used as a reason for excluding athletes from participation without further evaluation.

■ Cardiac Performance at Rest and During Exercise

Cardiac output is determined by heart rate and stroke volume. The heart is essentially a slave to central control mechanisms that aim to maintain blood pressure. Increased cardiac output is largely dictated by these control systems. When metabolic rate increases during exercise, the control mechanisms increase cardiac output. They also regulate blood flow so blood pressure is maintained. Circulatory control is described in Chapter 15.

Endurance training improves cardiac performance. Changes include increased stroke volume, decreased heart rate, increased ventricular compliance, improved calcium release and transport, and increased ventricular mass and volume. These changes are summarized in Table 14-1.

■ Factors Determining Cardiac Performance

Cardiac performance is determined by preload, afterload, contractility, and heart rate (Figure 14-6). *Preload,* or end-diastolic pressure, is the extent to which the heart chambers are stretched when they fill with blood. *Afterload* is the resistance the heart

TABLE 14-1

Changes in the Heart with Endurance Training

Change in Heart	Mechanism
• Decreased resting and submaximal exercise heart rate (bradycardia)	Increased parasympathetic influence Decreased sympathetic influence Lower intrinsic heart rate
• Increased stroke volume	Decreased heart rate (rest and submaximal exercise) Increased blood volume Increased heart size and volume Increased cardiac contractility Increased ventricular compliance Increased ventricular filling pressure
• Altered electrocardiogram: AV blocks, wandering atrial pacemaker, AV nodal pacemaker rhythm, ST segment elevation, PVCs, T-wave inversion	Increased parasympathetic influence
• Increased prevalence of third and fourth heart sounds	More rapid filling of ventricles Prolonged P–R interval on ECG Thinner chest walls in athletes
• Improved calcium release and transport	Increased strength of contraction

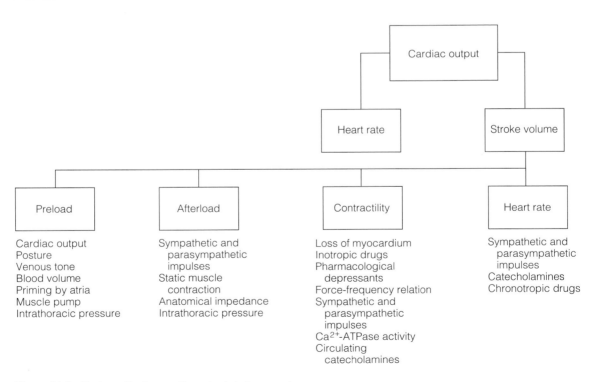

Figure 14-6 Factors affecting cardiac output during exercise.

meets as it attempts to pump blood into the circulation. *Contractility* is the strength of cardiac contraction. *Heart rate* is the speed at which the heart is beating. Each factor helps determine cardiac output.

Preload

Frank–Starling Mechanism The relationship between myocardial stretch and stroke volume is known as the *Frank–Starling mechanism* or the *Frank–Starling Law of the Heart* (Figure 14-7). Increased myocardial stretch (preload) causes increased pressure in the ventricles, which leads to an increase in stroke volume and stroke work. Stroke work is the work the heart does to eject blood from each ventricle against pressure caused by afterload. The heart is more efficient at performing *flow work* (work due to preload) than *pressure work* (work due to afterload). Exercises such as running tend to affect flow work, and those such as weight lifting tend to affect pressure work. Myocardial oxygen consumption increases more with an increase in pressure work than with an increase in flow work.

Further, an increase in myocardial oxygen consumption correlates to an increase in stress on the heart. As a result, exercises such as walking and running tend to be less stressful on the heart than lifting weights or shoveling snow.

A good measure of myocardial oxygen consumption is the *rate-pressure product* or *double produce* (heart rate x systolic blood pressure). This measure takes into consideration the effects of the four factors that load the heart. It is a particularly valuable measure during exercise and is used extensively in cardiac rehabilitation programs. The ventricles are stretched when they receive more blood, which causes them to contract more forcefully. Any sudden increase in output in the right ventricle is automatically matched by the left. During exercise, more blood is returned to the heart. The Frank–Starling mechanism helps to increase stroke volume and cardiac output.

This mechanism is limited in its ability to increase output by the structure of cardiac muscle. The strength of a cardiac contraction depends on the number of actin and myosin crossbridges

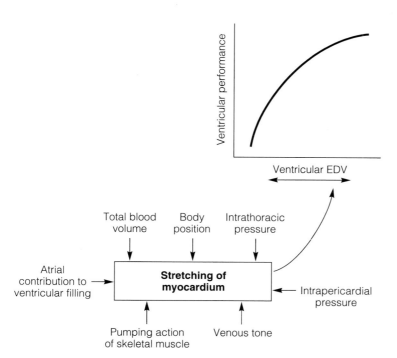

Figure 14-7 The Frank–Starling curve and the factors affecting end-diastolic volume (EDV). SOURCE: Braunwald, Ross, Jr., and Sonnenblick, 1967.

formed. To a point, the capacity for cardiac muscle tension increases as the fibers are lengthened. Individual fibers are capable of producing maximum tension when a maximum number of myosin filaments have access to actin receptor sites. When individual sarcomeres are stretched excessively, then the capacity for tension development decreases. Other factors increasing preload include mitral or aortic valve insufficiency, increased blood volume, and drugs that cause vasoconstriction; and other factors decreasing preload include mitral stenosis, decreased blood volume, and drugs that cause vasodilation.

Law of Laplace The extent to which the chambers of the heart are stretched also affects energy requirements. Any increase in the size of the heart geometrically increases the energy required for contraction. If the linear dimensions of the heart are doubled, for example, the ventricular muscles must produce a tension four times greater to secure the same systolic pressure. This is expressed in the law of Laplace, which states:

$$\text{Tension} = \frac{\text{Pressure} \cdot \text{Radius}}{2} \qquad (14\text{-}2)$$

During exercise, most of the increase in cardiac output is due to increased heart rate rather than increased stroke volume. There is a greater mechanical advantage in contracting the smaller ventricular volume.

Factors Affecting Preload During Exercise

A number of factors affect preload during exercise (Figure 14-7), including cardiac output, blood volume, the pumping action of respiration and skeletal muscles, body position or posture, heart size, and venous tone. Endurance activities such as distance running significantly contribute to preload.

Cardiac Output Preload is determined by the return of blood from the veins to the heart, which, in turn, is directly affected by cardiac output. Over the course of a few beats, cardiac output equals venous return.

Blood Volume Blood volume has an important effect on preload. The circulation is a closed-loop system, so increasing blood volume increases preload by making more blood available for delivery back to the heart, and reduced blood volume decreases preload. For example, exercise in the heat without adequate fluid replacement can lead to dehydration and thermal injury (Chapter 22). In exercise-related heat exhaustion, the heart cannot satisfy the demand for blood by the muscles and by the skin for cooling. Lower blood volume decreases the venous return of blood to the heart and impairs cardiac output. Endurance training increases blood volume.

Muscle and Thoracic Pump The pumping actions of muscles and ventilation increase the venous return of blood during exercise. Veins have one-way valves that prevent blood from moving away from the heart. (See Chapter 15.) Blood enters the capillary beds of muscles when they are relaxed and is pushed toward the heart when the muscles contract. During breathing, the alternating increase and decrease in intra-abdominal and -thoracic pressures also help to return blood to the heart.

Posture Posture has an effect on preload, particularly during brief bursts of exertion. At rest, end-diastolic volume is highest when the body is in the reclining position. It decreases progressively as the body shifts into sitting and standing postures. During exercise in the supine position, end-diastolic volume stays the same. Changes in preload play little role in increasing stroke volume in this type of exercise. During exercise in the erect posture, end-diastolic volume increases at intensities less than 50% of maximum capacity.

At maximum exercise intensities in the upright posture, end-diastolic volume decreases in some subjects (Figure 14-8). Studies of endurance athletes using two-dimensional echocardiography have shown that end-diastolic and stroke volumes decrease at maximum exercise. This decrease occurs because high heart rates compromise ventricular filling and limit cardiac output during maximal exercise.

The Frank–Starling mechanism is most important in increasing end-diastolic volume during low intensity exercise in athletes. During maximal exercise, the Frank–Starling mechanism can enhance

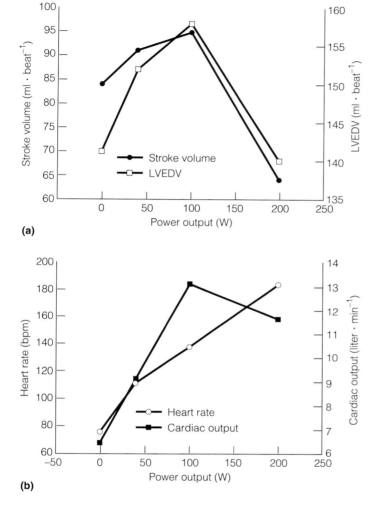

Figure 14-8 Changes in left ventricular end-diastolic volume (LVEDV), stroke volume, heart rate, and cardiac output in a sedentary 20-year-old male. Exercise intensity progressed to approximately 90% of maximum. Adapted from Concu, 1995.

cardiac output if heart rate is reduced. For example, administering β-blocking drugs, which decrease heart rate and cardiac contractility, to healthy subjects running on a treadmill decreases maximal exercise by 40%. However, stroke volume increases by the Frank–Starling mechanism, so cardiac outputs remain the same.

Prolonged upright exercise at a constant exercise intensity places an increasing load on the heart. Although the metabolic requirement of the exercise does not change, there is a progressive decrease in venous return of blood to the heart. This leads to a

decreased stroke volume and a progressive rise in heart rate, a phenomenon called *cardiovascular drift*. Cardiovascular drift is probably caused by a breakdown in sympathetic blood flow control mechanisms, increased distribution of blood to the skin for cooling, or both.

The effects of gravity on venous return are particularly acute in static postures such as standing. Soldiers standing at attention for prolonged periods have been known to faint. The cause is orthostatic intolerance—the inability to maintain central blood volume. In this case, venous return and cardiac output become inadequate. Because the soldier is standing still, the muscle pump mechanism is not operating to help return blood to the heart, and gradually blood collects in the veins. This occurs as compensatory mechanisms by the sympathetic nervous system become less effective. Orthostatic intolerance is more pronounced in warmer, more humid weather because blood is increasingly shunted to the periphery to cool the body.

The tilt table test is often used to measure the capacity of autonomic reflexes to adjust to changes in posture. Subjects are placed on the tilt table in a stationary upright posture. Heart rate and blood pressure measurements are taken during the test. Gradually, heart rate increases and systolic blood pressure decreases until the subject faints.

The tilt table test is frequently used in space and aviation physiology. In space flight, chronic lack of hydrostatic pressure produced by weightlessness results in lower venous tone. This reduces the capacity of the cardiovascular system to adjust to the effects of gravity.

Heart Size Heart size is highly related to stroke volume, as well as to $\dot{V}_{O_2max}$. Maximum stroke volume in a sedentary college-age male is approximately 90–100 ml. In elite endurance athletes, such as cross-country skiers, maximal stroke volume may exceed 170 ml. Much of this is due to a larger heart. Large heart size accounts for the large difference in maximal cardiac output capacity between endurance athletes and sedentary people.

Heart size increases in response to stretch or tension. In the endurance-trained heart, the left ventricle enlarges through the addition of sarcomeres. In the strength-trained heart, the size of the heart wall increases through the addition of sarcomeres in parallel. The walls thicken instead of lengthen.

The relative importance of genetics and training for increased heart volume is not completely understood. Studies of cardiac dimensions in trained and untrained school children show no evidence of congenitally enlarged hearts in children judged high in potential for success in endurance sports. Studies of heart size in monozygotic and dizygotic twins, a technique for determining the relative contribution of genetics and environment on heart size, suggest that training is much more important than genetics in determining heart size.

The two most important tools for determining heart size are echocardiography and magnetic resonance imaging (MRI). Other research tools used to study heart size have included the ECG and X-ray. *Echocardiography* measures heart wall size and volumes by determining the rate that sound waves travel through the tissue. *Magnetic resonance imaging (MRI)* produces a tissue image by surrounding the body with a magnetic field. Molecules within the tissue resonate, which allows an accurate image to be made using special sensors in the instrument. These studies consistently show that endurance and strength athletes increase left ventricular diameter and wall thicknesses. Endurance athletes increase more in left ventricular volume, and strength athletes increase more in wall size.

Venous Tone Because most blood volume resides in the veins, controlling venous blood volume is vital to ensure the return of blood to the heart. Veins are sympathetically innervated, and, as discussed in Chapter 15, during exercise, sympathetic stimulation from the central nervous system causes venoconstriction, which helps propel blood toward the heart.

Other Factors Other factors besides size may contribute to larger left ventricular volumes in trained athletes. Blood volume has already been mentioned. Athletes may have a higher left ventricular filling pressure, as well as more compliant ventricles. This means that the walls stretch more when filled with blood. These three factors, combined with larger heart size, give athletes higher left ventricular volumes, both at rest and during exercise.

Afterload

Afterload is impedance or resistance to ventricular emptying. Increased afterload has a negative influence on cardiac performance because it creates an increased work load for the heart. Increasing afterload decreases the shortening velocity of the cardiac muscle. The heart compensates by increasing its contractility so that cardiac output can be maintained. During exercise, such as repetitive weight lifting, increased afterload causes an immediate decrease in stroke volume, which leads to an increase in end-diastolic volume in the next cardiac cycle. The increased EDV stimulates the Frank–Starling mechanism, which increases stoke volume at a higher pressure than normal.

The extent of afterload on the heart depends on the absolute load on the muscle. Males tend to experience more afterload stress than females when both are doing identical high-tension exercise. During isometric dead lifts, for example, both sexes increase cardiac contractility in response to afterload stress but males experience greater afterload stress. This is due to greater muscle mass and their capability of exerting greater force.

Exercise involving high muscle tension, such as weight lifting, can increase afterload. Strength exercises often involve a *Valsalva maneuver*—expiration against a closed glottis. Valsalva increases intrathoracic pressure, which is partially responsible for increased afterload. High-tension upper-body exercises use groups of smaller muscles than those used in lower-body exercises. Blood flow through these small muscles is slowed. Brachial arterial pressures of 400 mmHg have been measured in athletes performing heavyweight lifts.

Repeated static exercise can result in prolonged elevation of heart rate during a workout. This has led some people to believe that weight lifting can produce a significant endurance training effect. However, weight lifting places a different load on the heart than endurance exercise does. Endurance exercise places a volume load on the heart; strength exercise causes an increased afterload.

Consequences of Chronic Afterload Stress

In hypertension, high peripheral resistance to blood flow causes increased afterload. This can cause the heart to hypertrophy and sometimes fail. Larger heart mass increases myocardial oxygen consumption. Hypertension is often accompanied by coronary artery disease. The combination of cardiac hypertrophy and impaired blood flow in the coronary arteries places an extraordinary load on the heart.

The destructive effect of chronic afterload stress of hypertension lies partly in the nature of hypertrophy. Increased heart size is due mostly to an increase in the size and number of connective tissue cells rather than cardiac muscle cells. Increased size results in a marked increase in the load of the heart muscle, as well as an increase in metabolic requirements. There is very little increase in the heart's functional capacity.

Static exercises involving a dynamic motion put much more stress on the heart than do purely isometric exercises. An example of static-dynamic exercise is shoveling snow. This type of exercise increases intrathoracic pressure and impedes blood flow with near maximal contractions in relatively large muscle groups. This causes a large afterload on the heart. Static-dynamic exercise results in many deaths each year in people with heart disease, who are not accustomed to such severe effort.

An overloaded myocardium hypertrophies in an attempt to restore normal wall stress. Strength training results in a greater left ventricular mass and wall thickness; endurance exercise increases the left ventricular dimensions, with less ventricular hypertrophy. The nature of cardiac hypertrophy differs in response to volume and pressure overloads. In volume overload, which occurs during endurance exercise, the heart responds with eccentric hypertrophy in which myofibrils are added in series (i.e., the heart volume gets larger). Pressure overloads, which occur during weight lifting, for example, lead to eccentric hypertrophy, where myofibrils are added in parallel (i.e., the heart muscle gets thicker).

A number of hormones and growth factors influence cardiac hypertrophy. These include transforming fibroblast growth factor beta (TGF-β), fibroblast growth factor (FGF), angiotensin II, atrial natriuretic peptide (ANP), thyroid hormone (T3), and catecholamines. TGF-β, FGF, and angiotensin II help modulate cell proliferation and differentiation in the heart muscle. ANP, which increases sodium removal, is synthesized in the heart and is secreted in an effort to reduce blood pressure, and thus the load on the left ventricle. T3 helps convert myosin isoforms (subtypes of the myosin filament—these can be fast or slow) from slow to fast. Catecholamines increase cardiac contractility. In response to overload, β-adrenergic receptor density decreases, which reduces cardiac contractility.

The pressure load hypertrophy caused by weight lifting is similar to that resulting from hypertension. In hypertension, the heart eventually enters a failure phase. Here, the ventricular walls become thinner and left ventricular dilation occurs. During this phase, the collagen content of the walls increases, which decreases their functional capacity. The changes due to weight training are probably not harmful, however. No long-term effects on the heart have been noted, even in athletes who have lifted weights for more than 20 years. In fact, most elite weight-trained athletes have better developed cardiovascular systems than the average population.

Contractility

Contractility refers to the quality of ventricular performance at constant conditions of loading and heart rate. The strength of cardiac contractility is determined by the rate of reaction at the contractile sites in the heart. Higher contraction strength increases stroke volume at a given end-diastolic volume. Less strength decreases stroke volume (Figure 14-9). Changes in cardiac contractility which take place independently of changes in preload or afterload, affect myocardial force development and the velocity and extent of myocardial shortening. Cardiac contractility depends on the availability of intracellular Ca^{2+}.

Myocardial contractility has a large effect on the Frank–Starling mechanism and stroke volume. In

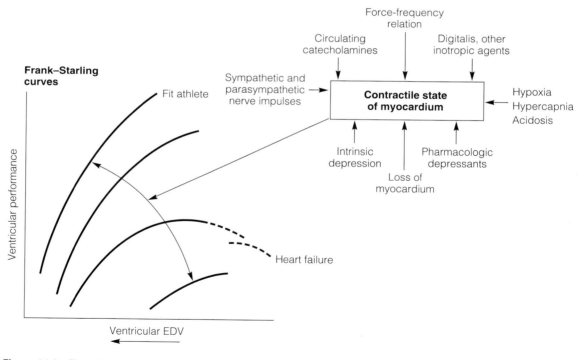

Figure 14-9 The effects of changes in myocardial contractility on the Frank–Starling curve, and the major factors affecting contractility. Each curve represents different levels of cardiac contractility. EDV = End-diastolic volume. SOURCE: Braunwald, Ross, Jr., and Sonnenblick, 1967.

heart failure, the contractile capacity of the heart is reduced. Increased preload is much less effective in increasing stroke volume. In addition, the Frank–Starling curve shifts to the right and stroke volume is decreased. In an athlete, the heart has greater contractile strength, and a given level of ventricular filling results in a greater stroke volume than normal.

Factors Affecting Cardiac Contractiiity Sympathetic stimulation, circulating catecholamines, and inotropic agents such as digitalis improve cardiac performance by increasing cardiac contractility and shift the Frank–Starling curve to the left. Factors decreasing cardiac contractility include decreased sympathetic stimulation, pharmacologic depressants, hypoxemia, hypercapnia, acidosis, and death of a portion of the heart muscle due to heart attack.

Cardiac muscle depends on increasing the strength of contraction in order to produce more force. Skeletal muscles, in contrast, rely on recruiting more motor units in order to increase force of contraction. The most important factor dictating inotropic capacity of the heart is the availability of calcium ion (Ca^{2+}), its rate of release, and its rate of uptake. Calcium ion–stimulated myosin–ATPase activity in the heart helps determine the strength of heart contractions.

Biochemical Changes Due to Training That Affect Cardiac Contractility Biochemical changes due to training that increase contractility are not completely understood. Training appears to increase calcium release and transport by the sarcoplasmic reticulum. Training also affects myosin–ATPase isoenzyme characteristics. Myosin ATPase partially determines the strength of cardiac contraction. High concentrations of the V1 isoform result in greater contraction capacity of the heart. In heart disease that causes cardiac hypertrophy, there is a shift in myosin–ATPase characteristics from the high activity V1 isoform to the lower activity V2 and V3 isoforms. Exercise training partially reverses the shift from V1 to V2 and V3 isoforms that occur after myocardial infarction. Endurance exercise increases the V1 isoform.

Increased stroke volume plays a greater role in increasing cardiac output during submaximal exercise intensities in trained individuals than in untrained subjects. Increased contractility plays an important role in the increased stroke volume that occurs with training. This adaptation can increase maximum cardiac output by 15 to 20%.

Blood flow through the coronary arteries also has a large effect on cardiac contractility. Impaired coronary circulation can lead to decreased contractility and, eventually, to destruction of cardiac muscle. Loss of cardiac muscle further affects myocardial contractility and exercise performance.

Heart Rate

Heart rate is the major determinant of cardiac output, particularly during moderate to maximal exercise. It is the most important factor affecting oxygen consumption in the heart. Increased heart rate quickly increases blood flow and oxygen transport during exercise. Even anticipation of exercise results in centrally mediated inhibition of the cardiac parasympathetics and stimulation of the sympathetics that stimulate the SA node.

Heart rate increases with exercise intensity, leveling off at maximum exercise. Training increases impulses to the heart from the vagus nerve, which slows heart rate at rest and during submaximal heart rate. Cardiac output is maintained by increasing stroke volume. As discussed, endurance athletes depend on the Frank–Starling mechanism to increase cardiac output at rest and low intensities of exercise. As exercise intensity exceeds approximately 50% of maximum, cardiac output increases by increasing heart rate.

Maximal heart rate is fairly consistent under a variety of circumstances. It changes very little with training, but tends to decrease with age. An estimate of maximal heart rate can be obtained by subtracting age from 220. However, this method is subject to considerable error. Whenever possible, it is better to measure maximal heart rate directly. This can be done accurately by measuring the highest heart rate reached during a maximum treadmill test.

Coronary artery blood flow is highly related to heart rate. In healthy people, coronary blood flow is adequate at all intensities of exercise, but in people with coronary artery disease, coronary blood flow may not be adequate during exercise. This can sometimes be detected in the ECG as a depression of the ST segment. Coronary ischemia can depress cardiac performance by decreasing heart rate and the strength of cardiac contraction.

■ Summary

The heart is a critical initiator of fluid transport in the body. It works with the elastic blood vessels to maintain a relatively constant blood pressure in the arteries. It is a hollow, four-chambered organ that pumps blood to the lungs and general circulation.

The myocardium or heart muscle is involved in cardiac contraction and the propagation of nervous impulses. Cardiac muscle is similar to skeletal muscle in many ways. However, it does not recruit motor units for increased strength of contraction. Rather, contraction is regulated by the amount of calcium released and transported by the sarcoplasmic reticulum.

The heart has a special conduction system for rapidly moving impulses through the heart. This conduction system can be monitored by an ECG. Electrical conduction is closely coupled to mechanical contraction, as the chambers of the heart must depolarize before they can contract. Training affects the relationship between electrical stimulation and mechanical contraction. Vagus nerve influences in the trained heart increase while sympathetic influences decrease, as reflected by a decreased resting heart rate and a variety of abnormalities on an ECG. These ECG abnormalities are thought to be benign because they usually disappear during exercise as vagus nerve influence on the heart decreases.

Cardiac performance is determined by preload, afterload, contractility, and heart rate. Preload is the extent to which the heart chambers are stretched when they fill with blood. Afterload is the resistance the heart meets as it attempts to pump blood into the circulation. Contractility is the strength of cardiac contraction. Heart rate is the speed the heart is beating. Endurance training improves heart performance by increasing left ventricular volume and

increasing cardiac contractility. Cardiac contractility is improved by enhancing calcium release and transport from the sarcoplasmic reticulum. It is also increased by increasing the portion of the V1 isoenzyme of myosin ATPase. Myosin ATPase is an important enzyme determining the strength of cardiac contraction.

■ Selected Readings

Adams, T. D., F. G. Yanowitz, A. G. Fisher, J. D. Ridges, A. G. Nelson, A. D. Hagen, R. R. Williams, and S. C. Hunt. Heritability of cardiac size: an echocardiographic study of monozygotic and dizygotic twins. *Circulation* 71: 39–44, 1985.

Bevilacqua, M., S. Savonitto, E. Bosisio, E. Chebat, P. L. Bertora, M. Sardina, and G. Norbiato. Role of the Frank–Starling mechanism in maintaining cardiac output during increasing levels of treadmill exercise in beta-blocked normal men. *Am. J. Cardiol.* 63: 853–857, 1989.

Bielen, E., R. Fagard, and A. Amery. Inheritance of heart structure and physical exercise capacity: a study of left ventricular structure and exercise capacity in 7-year-old twins. *Europ. Heart J.* 11: 7–16, 1990.

Braunwald, E., J. Ross, Jr., and E. H. Sonnenblick. Mechanisms of contraction of the normal and failing heart. *N. Engl. J. Med.* 277: 794–1022, 1967.

Bugaisky, L. B., M. Gupta, M. P. Gupta, and R. Zak. Cellular and molecular mechanisms of cardiac hypertrophy. In The Heart and Cardiovascular System, 2d Ed., H. A. Fozzard (Ed.). New York: Raven Press, 1992, pp. 1621–1640.

Chien, K. R., K. U. Knowlton, H. Zhu, and S. Chien. Regulation of cardiac gene expression and hypertrophy: molecular studies of an adaptive physiologic response. *FASEB J.* 5: 3037–3046, 1991.

Colan, S. D. Mechanics of left ventricular systolic and diastolic function in physiologic hypertrophy of the athlete's heart. *Cardiol. Clin.* 15: 355–372, 1997.

Concu, A., and C. Marcello. Stroke volume response to progressive exercise in athletes engaged in different types of training. *Eur. J. Appl. Physiol.* 66: 11–17, 1993.

Fahey, T. D., P. Insel, and W. Roth. Fit and Well. 3d Ed. Mountain View, Ca.: Mayfield Publishing Co., 1999.

George, K. P., L. A. Wolfe, and G. W. Burggraf. The athletic heart syndrome: a critical review. *Sports Med.* 11: 300–331, 1991.

Ginzton, L. E., R. Conant, M. Brizendine, and M. M. Laks. Effect of long-term high intensity aerobic training on left ventricular volume during maximal upright exercise. *J. Am. Coll. Cardiol.* 14: 364–371, 1989.

Haskin-Popp, C., D. Nazareno, J. Wegner, B. A. Franklin, J. Schafer, S. Gordon, and G. C. Timmis. Aerobic and myocardial demands of lawn mowing in patients with coronary artery disease. *Am. J. Cardiol.* 81: 1243–1245, 1998.

Katz, A. Angiotensin II: hemodynamic regulator or growth factor? *J. Mol. Cell Cardiol.* 22: 739–747, 1990.

Komuro, I., and Y. Yazaki. Control of cardiac gene expression by mechanical stress. *Annu. Rev. Physiology.* 55: 55–75, 1993.

Lankford, E. B., D. H. Korzick, B. M. Palmer, B. L. Stauffer, J. Y. Cheung, and R. L. Moore. Endurance exercise alters the contractile responsiveness of rat heart to extracellular Na^+ and Ca^{2+} *Med. Sci. Sports Exerc.* 30: 1502–1509, 1998.

Libonati, J. R., J. P. Gaughan, C. A. Hefner, A. Gow, A. M. Paolone, and S. R. Houser. Reduced ischemia and reperfusion injury following exercise training. *Med. Sci. Sports Exerc.* 29: 509–516, 1997.

Mair, J., W. Schobersberger, A. Koller, P. Bialk, B. Villiger, W. Frey, and B. Puschendorf. Risk for exercise-induced myocardial injury for athletes performing prolonged strenuous endurance exercise. *Am. J. Cardiol.* 80: 543–544, 1997.

Marwick, T. H. The viable myocardium: epidemiology, detection, and clinical implications. *Lancet* 351: 815–819, 1998.

Morgan, H. E., and K. E. Baker. Cardiac hypertrophy: mechanical, neural, and endocrine dependence. *Circulation* 83: 13–25, 1991.

Morris, G. S., R. R. Roy, T. P. Martin, and K. M. Baldwin. The effect of weight lifting exercise on cardiac myosin isoenzyme distribution. *Med. Sci. Sports Exerc.* 2: S29, 1990.

Musch, T. I., R. L. Moore, P. G. Smaldone, M. Reidt, and R. Zelis. Cardiac adaptations to endurance training in rats with a chronic myocardial infarction. *J. Appl. Physiol.* 66: 712–719, 1989.

Nadal-Ginard, B., and V. Mahdavi. Molecular basis of cardiac performance: plasticity of the myocardium generated through protein isoform switches. *J. Clin. Invest.* 84: 1693–1700, 1989.

Nagai, R., A. Zarain-Herzberg, C. J. Brandl, J. Fujii, M. Tada, D. H. MacLennan, N. R. Alpert, and M. Periasamy. Regulation of myocardial Ca2+-ATPase and phospholamban mRNA expression in response to overload and thyroid hormone. *Proc. Natl. Acad. Sci. USA* 86: 2966–2970, 1989.

Powers, S. K., H. A. Demirel, H. K. Vincent, J. S. Coombes, H. Naito, K. L. Hamilton, R. A. Shanely, and J. Jessup. Exercise training improves myocardial tolerance to *in vivo* ischemia-reperfusion in the rat. *Am. J. Physiol.* 275: R1468–R1477, 1998.

Rost, R. The athlete's heart. What we did learn from Henschen, what Henschen could have learned from us! *J. Sports Med. Phys. Fit.* 30: 339–346, 1990.

Sagiv, M., R. Metrany, N. Fisher, E. Z. Fisman, and J. J. Kellermann. Comparison of hemodynamic and left ventricular responses to increased after-load in healthy males and females. *Int. J. Sports Med.* 12: 41–45, 1991.

Sagiv, M., A. Sagiv, D. Ben-Sira, S. Ben-Gal, and M. Soudry. Effects of chronic overload training and aging on left ventricular systolic function. *Gerontology* 43: 307–315, 1997.

Schartz, K., K. R. Boheler, D. de la Bastie, A. M. Lompre, and J. J. Mercadier. Switches in cardiac muscle gene expression as a result of pressure and volume overload. *Am. J. Physiol.* 262: R364–R369, 1992.

Seymour, R. S., A. R. Hargens, and T. J. Pedley. The heart works against gravity. *Am. J. Physiol.* 265: R715–R720, 1993.

Shapiro, L. M. The morphologic consequences of systemic training. *Cardiol. Clin.* 15: 373–379, 1997.

Siegel, A. J., M. Sholar, J. Yang, E. Dhanak, and K. B. Lewandrowski. Elevated serum cardiac markers in asymptomatic marathon runners after competition: is the myocardium stunned? *Cardiology* 88: 487–491, 1997.

Smith, M. L., D. L. Hudson, H. M. Graitzer, and P. B. Raven. Exercise training bradycardia: the role of autonomic balance. *Med. Sci. Sports Exerc.* 21: 40–44, 1989.

Spina, R. J., M. J. Turner, and A. A. Ehsani. Beta-adrenergic-mediated improvement in left ventricular function by exercise training in older men. *Am. J. Physiol.* 274: H397–H404, 1998.

Spina, R. J., M. J. Turner, and A. A. Ehsani. Exercise training enhances cardiac function in response to an afterload stress in older men. *Am. J. Physiol.* 272: H995–H1000, 1997.

Sugihara, H., K. Shiga, K. Terada, N. Kinoshita, Y. Taniguchi, K. Ito, Y. Adachi, Y. Ushijima, M. Nakagawa, and T. Maeda. Effect of exercise-induced activation of sympathetic nerve activity on clearance of 123I-MIBG from the myocardium. *Ann. Nucl. Med.* 12: 175–178, 1998.

Voipio-Pulkki, L. M. Myocardial adaptation to exercise training: a view through positron emission tomography. *Ann. Med.* 1 (Suppl.): 54–60, 1998.

Zhang, X. Q., Y. C. Ng, T. I. Musch, R. L. Moore, R. Zelis, and J. V. Cheung. Sprint training attenuates myocyte hypertrophy and improves Ca^{2+} homeostasis in postinfarction myocytes. *J. Appl. Physiol.* 84: 544–552, 1998.

CHAPTER 15

CIRCULATION AND ITS CONTROL

Control of blood flow is critical during exercise. Blood must be rapidly directed to working muscles in order to meet their demands for oxygen and fuels. Blood pressure must be maintained as well. In addition, uninterrupted blood flow to the heart is needed to ensure blood supply to critical areas such as the central nervous system.

At rest, a large portion of the cardiac output is directed to the spleen, liver, kidneys, brain, and heart. Muscles comprise over 40% of the body's tissue. Yet, they receive only about 20% of total blood flow. During exercise, however, the muscles can receive more than 85% of cardiac output. Circulatory control mechanisms carefully balance their central role of maintaining blood pressure with the increased metabolic demands of exercising muscle.

Although the average blood volume is about 5 liters, the circulation has the capacity to hold over 25 liters. Because peripheral resistance decreases during exercise as a result of increasing blood flow to active muscles, blood vessels in less active tissues are constricted and cardiac output is increased to maintain blood pressure. In other words, circulatory controls make it possible to maintain blood pressure, supply blood to important areas, and satisfy the metabolic needs of working muscles. The heart and circulation are mainly controlled by higher brain centers (central command) and car-

Road cycling competitions, like other endurance activities, require high rates of oxygen consumption and utilization. These, in turn, require a high cardiac output and the ability to circulate blood to active muscles and other metabolically active areas. Photo: © David Madison 1991.

diovascular control areas in the brain stem. Control is also affected by baroreceptors, chemoreceptors, muscle afferents, local tissue metabolism, and circulating hormones.

■ The Circulation

The blood vessels consist of arteries, arterioles, capillaries, venules, and veins. *Arteries* transport blood under pressure to the tissues. *Arterioles,* metarterioles, and precapillary sphincters act as valves to control blood flow to the various tissues. *Capillaries* act as an exchange medium between blood and the interstitial spaces. *Venules* collect blood from the capillaries and *veins* return blood to the heart. The *lymphatics,* a related system, acts as an exchange medium among all the capillaries, cells, and circulation.

The circulation is subdivided into the systemic and pulmonary circulations. Blood is pumped from the right ventricle into the lungs and from the left ventricle into the aorta. The many branches of the aorta deliver blood to most tissues of the body (Figure 15-1).

Systemic arteries are elastic and muscular, which

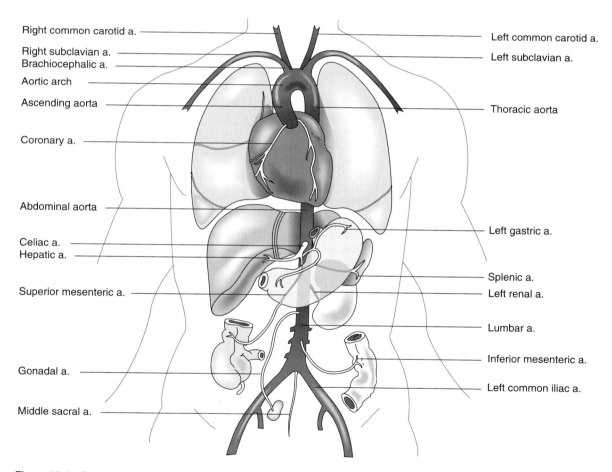

Right common carotid a.
Right subclavian a.
Brachiocephalic a.
Aortic arch
Ascending aorta
Coronary a.
Abdominal aorta
Celiac a.
Hepatic a.
Superior mesenteric a.
Gonadal a.
Middle sacral a.

Left common carotid a.
Left subclavian a.
Thoracic aorta
Left gastric a.
Splenic a.
Left renal a.
Lumbar a.
Inferior mesenteric a.
Left common iliac a.

Figure 15-1   Principal branches of the aorta.

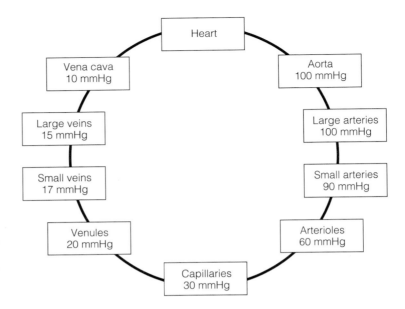

Figure 15-2 Mean internal pressure in the circulation. Blood pressure decreases as blood encounters progressively less resistance throughout the circulation.

makes them capable of withstanding high pressures. Blood pressure in the arteries is relatively high but diminishes throughout the circulation, approaching 0 mmHg when it reaches the right atrium. The pressure in the circulation diminishes with decreased resistance to flow, unless the resistance is offset by increased cardiac output (Figure 15-2).

Blood passes from the arteries into the arterioles, metarterioles, and precapillary sphincters. The largest resistance to flow occurs in these areas, which are important for maintaining blood pressure and ensuring tissue blood supply, particularly during exercise. This regulation of blood flow is directed by sympathetic neural control mechanisms and local responses to metabolism and stretch.

At rest, the arterioles in the inactive muscle remain mostly vasoconstricted, thus allowing blood to be directed to other areas. During exercise, when more blood is needed by the working muscles, the arterioles supplying blood to the muscles vasodilate, allowing transport of the oxygen and nutrients to the tissue. As the intensity of exercise progresses, even vascular beds in active muscles begin to vasoconstrict to maintain blood pressure.

Blood passes through the arterioles into the capillaries, where gases, substrates, fluids, and metabolites are exchanged. Capillary blood flow is controlled by the circulatory driving pressure provided by the heart. It is also affected, however, by the resistance provided by arterioles, precapillary sphincters, and veins. The capillaries contain no smooth muscle, so they play no role in regulating their own blood flow.

By far the largest portion of the blood volume resides in the veins, which are sometimes called the *venous capacitance vessels* because of their storage capacity. The veins can be constricted by sympathetic stimulation and by mechanical compression—both important factors during exercise. Every time muscles are contracted, the veins are compressed and blood is propelled toward the heart. Valves in the veins prevent the blood from flowing away from the heart (Figure 15-3).

Vascular Smooth Muscle

The arrangement of vascular smooth muscle supplying skeletal muscle is ideal for regulating blood

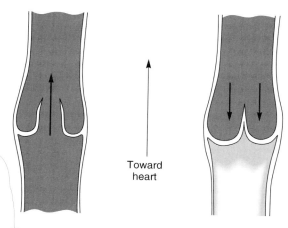

Figure 15-3 Venous valves allow blood to flow toward the heart but prevent it from moving away from the heart.

Toward heart

flow to active tissue and maintaining systemic blood pressure. Smooth muscle differs from skeletal muscles in that the fibers are smaller, they are not striated, they contract more slowly but develop more force, and they contract over a wider range of length. They also contract using a latch mechanism; that is, after they contract, they can maintain full tension while reducing the amount of nervous activation. This allows vascular tone to be maintained with very little energy. During exercise, vascular smooth muscles are controlled by combinations of metabolic, myogenic (intrinsic to the muscle), and neural factors.

Two types of smooth muscle are found in blood vessels:

1. **Unitary smooth muscle.** This tissue is not directly innervated and has a high degree of myogenic activity. It is highly responsive to stretch and chemical stimulators and is the most predominant smooth muscle type in the precapillary sphincter. Because it is responsible for local control of blood flow, unitary smooth muscle is important in directing blood flow to the muscles during exercise.

2. **Multiunit smooth muscle.** This tissue is sympathetically innervated and is the predominant

smooth muscle type in the outer portion of the precapillary sphincters, large arteries, veins, and arteriovenous anastomoses (AVA) in the skin. AVA blood vessels shunt blood directly from the arteries to the veins without crossing capillaries, and thus are important for cooling.

During intense exercise, multiunit smooth muscle appears to override the action of unitary smooth muscle to maintain blood pressure. This results in vasoconstriction in active muscle, which limits maximum muscle blood flow.

■ Determinants of Blood Flow

The rate at which blood flows through the circulation is determined by blood pressure (Eq. 15-1). Blood pressure (BP) is the product of cardiac output ($\dot{Q}$) and total peripheral resistance (TPR). Cardiac output is the amount of blood pumped from the heart per minute. Total peripheral resistance is the resistance to blood flow provided by the circulation. At a constant cardiac output, tissue blood flow is regulated by increasing or decreasing the size or resistance in the local blood vessels.

$$BP = \dot{Q} \cdot TPR \qquad (15\text{-}1)$$

Cardiac output depends on adequate venous return of blood to the heart. The most important factor affecting venous return is the pumping pressure of the heart that pushes blood through the circulation system from the arteries to the veins. Other mechanisms that affect venous return include heart "suction," negative thoracic pressure, and the respiratory and skeletal muscle pumps. Circulation is also greatly affected by basic biophysical principles, such as Poiseuille's equation and the siphon effect.

The contraction and relaxation of the heart creates a rhythmic positive and negative pressure that helps propel blood through the circulation. During diastole (relaxation phase of the cardiac cycle), relaxation of the ventricles expands ventricular volume and lowers pressure, which aids filling. Likewise, during systole (contractile phase of the cardiac cycle) ventricular contractions pull the atria downward. This expands them and helps return blood to the heart. These conditions create a negative thoracic pressure, averaging 4 mmHg; that enhances venous return.

Breathing also causes a pumping action that helps draw blood to the heart. Ventilation oscillates intrathoracic pressure from -7 to -3 mmHg, which causes the central veins to expand with each inspiration.

During exercise, the skeletal muscle pump is critical for boosting cardiac output. It results from muscular compression of the veins and one-way venous valves (Figure 15-2). During intense exercise, the muscle pump mechanism may provide almost one-third of total pumping power, while the heart provides the rest. The muscle pump boosts cardiac output by stimulating the Frank–Starling mechanism. Through this mechanism, as discussed in Chapter 14, an increased return of blood to the heart stretches the ventricles, which then recoil and eject an increased volume from the heart. The muscle pump also reduces total venous volume, shifts blood to arteries, and increases arterial pressure.

Poiseuille's Equation

The control of blood flow by pressure and resistance is expressed in Poiseuille's equation (see Equation 15-2), which states that flow through a tube varies directly with the differences in pressure. It varies to the fourth power of the radius of the tube and is inversely proportional to the length of the tube and the viscosity of the fluid.

Of the four factors in this equation, length is the only one that does not vary in the physiological system.

$$F = \frac{(P_1 - P_2)\pi R^4}{8LN} \qquad (15\text{-}2)$$

where F is flow rate, $(P_1 - P_2)$ is the drop in pressure, R is the radius of the tube, L is the length of the tube, and N is the viscosity of the fluid.

Poiseuille's equation assumes that the flow of fluid is *laminar* (i.e., the fluid travels in layers). While blood flow tends to be laminar, it is sometimes turbulent. With turbulence, blood tumbles through the vessels rather than traveling in layers. Turbulence usually occurs in the aorta, where the vessel radius and blood velocity is greatest, and it may be detected as murmurs.

Pressure Pressure is very important in determining blood flow. Blood flow between two points is directly proportional to the driving force between them. During exercise, blood flow to active muscle can be increased as much as 25 times. This is accomplished by increasing cardiac output and by directing a greater fraction of that output to active muscle.

Blood Vessel Radius The distribution of blood is mainly controlled by variations in the size of the arterioles. These controls are extremely sensitive. Blood flow varies to the fourth power of the radius. For example, if a blood vessel decreases its radius by half, blood flow will decrease to one-sixteenth of its original value. A very small increase in a vessel's radius, therefore, results in a significant change in its blood flow (Figure 15-4).

Circulatory control mechanisms help to ensure a constant vascular volume by regulating the size of the blood vessels. Thus, blood flow regulation de-

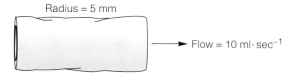

Radius = 5 mm
Flow = 10 ml·sec⁻¹

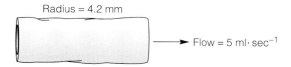

Radius = 4.2 mm
Flow = 5 ml·sec⁻¹

Figure 15-4   Effect of decreasing the radius of a tube 16% on flow rate at a constant pressure.

pends on integrating control of cardiac output and resistance in the blood vessels.

Viscosity Viscosity is the "slipperiness" of the layers of a fluid. Blood viscosity is a function of hematocrit (percent of cells in the blood). Normally, viscosity does not affect blood flow very much. However, when the hematocrit becomes abnormally high, such as in polycythemia vera, the increased viscosity decreases blood flow. Extreme cold, such as occurs in frostbite, can increase blood viscosity enough to severely impede blood flow. Erythropoietin (EPO, a substance that stimulates the production of red blood cells) and blood doping, used as ergogenic aids by some endurance athletes, also increases viscosity. During exercise, increased body temperature decreases blood viscosity, which has a small effect in increasing blood flow to the active muscles.

Principle of the Siphon

Blood moves from the left ventricle back to the right atrium very easily, in spite of the heart's having to pump blood against gravity in the upright posture. It is able to do so mainly because of the pumping pressure of the heart and the principle of the siphon, which states that the flow of fluid between two points depends on the differences in pressure between them.

The siphon principle is at work when a person empties a water bed from the second story of an apartment house. The water flows downward because there is a pressure difference between the two ends of the hose. This occurs even though the hose must rise above the level of the bed so it can pass through the window. Flow is not affected by the level of the tube between the points of high and low pressure (Figure 15-5a).

In the circulation, blood is siphoned from the high pressure of the left ventricle, which is approximately 100 mmHg, through the circulation to the low pressure of the right atrium, which approaches 0 mmHg (Figure 15-5a). Systemic blood flow is maintained even though blood must be pumped to

tissues with widely differing pressure located below and above the heart. In the upright posture, for example, hydrostatic pressure is much higher in the lower than upper extremities (Figure 15-5b).

The circulation is not a rigid system of inflexible tubes. When the volume of the circulation changes, the siphon effect is momentarily interrupted. This occurs, for example, when blood vessels in active muscles vasodilate to increase muscle blood flow. Interruption is only temporary, however, until the system adjusts to the new volume. Baroreceptors monitor such changes in blood pressure, stimulating the cardiovascular control centers to maintain blood pressure by vasoconstricting blood vessels and increasing heart rate.

Limitations of Poiseuille's Equation and the Siphon *In Vivo*

Seymour and associates (1993) questioned the effects of the siphon and Poiseuille's equation on circulatory performance in vertebrate animals. They argue that the siphon effect does not help blood flow in blood vessels that travel against gravity (i.e., blood vessels that travel toward the head) if any of the blood vessels that travel in the same direction as gravity (i.e., travel toward the feet) are collapsible. They define a tube as collapsible if it cannot sustain an internal pressure less than the pressure of its surroundings. Outside of rigid cavities, such as the brain, the arterioles and venules are collapsible. Gravity directly affects the collapsibility of these vessels—gravity directly affects the work of the heart. If central arterial blood pressure is insufficient to support a blood column between the heart and the head, blood flow ceases because of vascular collapse. Therefore, circulation is maintained because of the dynamic action of the heart.

Their work also suggests that the application of the Poiseuille equation in collapsible vessels is also limited. Their data suggest that the pressure developed by the heart to establish a given flow rate is independent of the resistance occurring in the partially collapsed vessels. The pressure depends only on the height of the blood column and the resistance

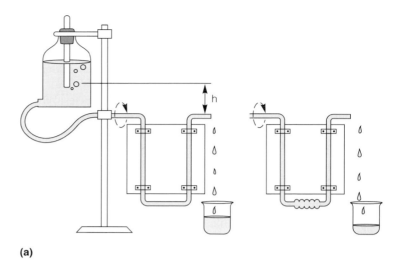

(a)

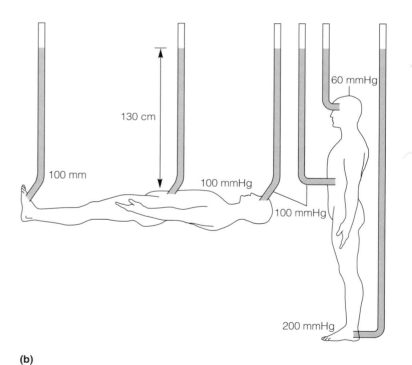

Figure 15-5 (a) An apparatus that demonstrates the principle of the siphon in nondistensible (left) and distensible (right) tubes. Flow will continue as long as the output is at a lower pressure than the source—regardless of the position of the tube in the middle of the siphon. (b) Hydrostatic pressure in operation on arterial pressure in the supine and erect positions. SOURCE: Ruch and Patton, 1965.

(b)

in the noncollapsed parts of the system. Both positions of these authors remain controversial.

■ Cardiovascular Regulation and Control

Regulation of blood pressure and tissue circulation is affected by cardiovascular neural control centers, local tissue metabolism and intrinsic regulation, and muscle afferents. Extrinsic neural control is directed toward maintaining blood pressure. Local regulation of blood flow is directed toward meeting tissue metabolic requirements. During exercise, the careful regulation of cardiac output and muscle blood flow is critical in maintaining blood pressure.

During exercise, various circulatory control mechanisms act to increase or maintain blood pressure. They also supply blood to tissues and cool the body. Unfortunately, the maximum demands of each requirement cannot be met simultaneously. This means that during intense exercise, a limitation is placed on maximum blood flow to active muscle and the skin so blood pressure can be maintained. Neural and hormonal control mechanisms help balance the demand for blood to active tissues with this critical need to maintain blood pressure. The cardiovascular control mechanisms (muscle afferents, central command, baroreceptors, chemoreceptors, hypothalamus, hormonal) are redundant rather than additive. If one mechanism is blocked, others will produce a response.

Neural Control of the Heart

Neural control of the heart is integrated into regulation of the cardiovascular system. Mechanisms that regulate arterial blood pressure, blood volume, and the intake of fluid and electrolytes affect the heart and circulation.

Circulatory requirements during exercise easily exceed the ability of the Frank–Starling mechanism (see Chapter 14) to increase cardiac output. Fortunately, the heart is supplied with sympathetic and parasympathetic nerves from the autonomic nervous system (Figure 15-6). These nerves affect the contractile strength of the heart and the heart rate.

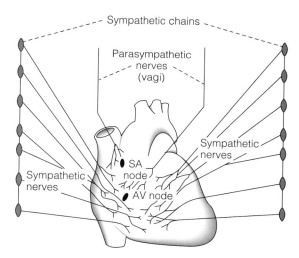

Figure 15-6 The cardiac nerves. SOURCE: Guyton, 1986, p. 171.

The sympathetic nerves stimulate the heart, and the parasympathetics suppress it. Parasympathetic neurons innervate the SA node, the atrial myocardium, the AV node, and the ventricular myocardium. Sympathetic nerves not only stimulate heart rate but also control the force of cardiac contraction. They increase blood pressure by peripheral vasoconstriction of blood vessels, and allow more blood to pass through the coronary arteries.

Neural Control of the Cardiovascular System

Neural control of circulation involves the cerebral cortex (central command), cardiovascular control center (CVC), and peripheral afferents (Figure 15-7). Peripheral afferents include the hypothalamus, baroreceptors, chemoreceptors, and muscle afferents. Neural control mechanisms match vascular conductance (blood flow) with cardiac output. The heart, arteries, arterioles, venules, and veins are sympathetically innervated and are partially regulated by peripheral cardiovascular afferents (Figure 15-8).

Central Circulatory Command Neural control of circulation arises from a central area of

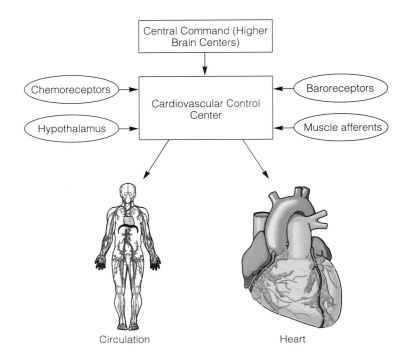

Figure 15-7 Neural control of circulation.

the brain called the subthalamic locomotor region. This area has also been associated with motor unit recruitment and ventilatory control. Central command is responsible for controlling heart rate, arterial blood pressure, and left ventricular contractility. It does this by controlling the level of efferent activity of the sympathetic (cardiovascular control center) and parasympathetic (vagus nerve) nervous systems. Through its stimulation of the cardioinhibitory center in the medulla (see the next section), the central command is involved in the withdrawal of parasympathetic influences during exercise.

Central command starts sympathetic activity by directly influencing the cardiovascular control areas, which are located in the reticular formations. During exercise, stimulation of the CVC by central command results in circulatory vasoconstriction, increased heart rate, increased strength of contraction by the myocardium, and increased impulse conduction velocity in the heart.

Cardiovascular Control Center (CVC) The CVC is a loose connection of nerve cells in the re-

ticular formation of the brain stem, centered mainly in the pons and medulla. Four areas in the medulla, sometimes called the vasomotor center, have been identified as being particularly important in cardiovascular control:

1. **Pressor area** Increases blood pressure by vasoconstricting blood vessels

2. **Depressor area** Decreases blood pressure by inhibiting nerves that cause vasoconstriction. Blood pressure falls because vasoconstrictor tone and heart rate decrease.

3. **Cardioacceleration center** Increases heart rate. It is activated when the pressor area is stimulated.

4. **Cardioinhibitory center** Depresses cardiac activity. This center is associated with the vagus nerve.

During exercise, the CVC receives impulses from central command, the hypothalamus, baroreceptors, chemoreceptors, and muscle afferents that

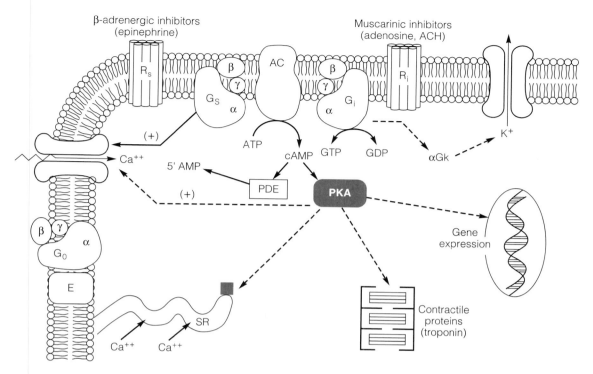

Figure 15-8 Diagram of the receptor-G protein-adenylyl cyclase transmembrane signaling system. The binding of catecholamines and ACH to their specific receptors serve to stimulate or inhibit cardiac function. Their signals are mediated through cAMP and cAMP-dependent protein kinase. See text for explanation. Modified from Feldman, 1993.
R_S = stimulatory receptor
R_i = inhibitory receptor
AC = adenylyl cyclase
G_S = guanine nucleotide binding regulatory protein that stimulates AC
G_i = guanine nucleotide binding regulatory protein that inhibits AC
PDE = phosphodiesterase
PKA = cAMP-dependent protein kinase
GTP = guanosine triphosphate
SR = sarcoplasmic reticulum
Ca^{2+} = calcium ion
G_O = other guanine nucleotide binding regulatory protein of unknown function
E = effector enzyme

determine the level of cardiac output and peripheral resistance. The CVC is also directly affected by the levels of O_2 and CO_2 in blood. Decreased P_{CO_2}, which can result during hyperventilation, causes a decrease in blood pressure. Low P_{O_2} stimulates the CVC.

Hypothalamus The hypothalamus lies in the upper part of the brain stem, above the pons, and is involved in many physiological processes. These include neuroendocrine control, hunger and satiety, body water control, temperature regulation, and circulatory control. The hypothalamus is particularly active in influencing circulation during temperature challenges from the environment or exercise. The anterior hypothalamus influences the cardiovascular control areas to vasodilate skin vessels in response to increased body temperature. The posterior hypothalamus induces vasoconstriction in response to falling body temperature.

Baroreceptors Baroreceptors are stretch receptors located in the heart, major arteries (particularly in the carotid sinus and aortic bodies), and pulmonary vessels. They influence the CVC, affecting heart rate, cardiac contractility, and vascular resistance and compliance. They are involved in the chronic as well as acute regulation of blood pressure. Atrial baroreceptors stimulate the release of vasopressin from the pituitary.

Baroreceptors work by establishing a blood pressure set-point. If pressure increases above the set-point, the baroreceptors send impulses to the CVC, which reduces the level of sympathetic activity and causes blood pressure to decrease. When pressure again reaches the set-point, the activity of the baroreceptors decreases. The reverse is true when blood pressure is below the set-point. Impulses from the baroreceptors to the CVC decrease, which causes blood pressure to increase.

At the onset of exercise, the baroreceptors' set-point increases, causing an immediate increase in heart rate and blood pressure. This baroreflex may be critical in the rapid cardiovascular response that occurs during exercise, and has been found to be higher in fit subjects. Baroreceptors have an important effect on active skeletal muscle circulation. By inhibiting the CVC pressor center, the baroreceptors stimulate muscle circulation to vasodilate when blood pressure increases and to vasoconstrict when blood pressure begins to fall.

Chemoreceptors Chemoreceptors are located in the aortic and carotid bodies. They respond to decreased P_{O_2}, decreased pH, and increased P_{CO_2} by sending impulses to the pressor area of the CVC. As with baroreceptors, they establish a set-point and work by negative feedback: They increase impulses to the CVC when P_{O_2} or the pH is below and P_{CO_2}

is above the set-point, and they increase impulses when the values move in the other direction.

Muscle Afferents Blood flow to skeletal muscles during exercise is determined by a balance between sympathetic vasoconstriction and metabolic vasodilation. Muscle afferents send signals to the central nervous system regarding the metabolic status of the muscle. They help balance the muscles' need for blood with the systemic need for blood pressure control.

Muscle afferents are classified as Types I, II, III, and IV. Types I and II are highly myelinated nerve fibers that serve as endings for the muscle spindles and Golgi tendon organs. They have no effect on the cardiovascular function. Type III and IV afferents, however, are excited by mechanical, thermal, and chemical stimuli. When stimulated, they send signals to the CVC that increase blood pressure by accelerating heart rate. The signals also increase the strength of cardiac contraction and vasoconstrict blood vessels.

Type III afferents are particularly sensitive to stretch and mechanical deformation. They have been called ergoreceptors because they are activated by muscle contraction. During exercise (particularly isometric exercise), they respond more quickly than Type IV fibers. Type IV fibers are more responsive to chemical stimuli, including potassium, decreased pH, and prostaglandins.

Muscle afferents are most active during intense exercise, when increases in muscle metabolism tend to cause local vasodilation through circulatory autoregulation (see the discussion of autoregulation later in this chapter). If muscles were allowed to maximally vasodilate, blood pressure would fall. However, feedback from muscle afferents limits this local muscle vasodilation so blood pressure is maintained.

Hormonal Control Mechanisms

The release of the catecholamines epinephrine and norepinephrine from the adrenal medulla affects the circulation as well as the heart. In the heart, both hormones enhance the effect of specific sympathetic stimulation. However, they have dissimilar effects on circulatory vasoconstriction. Both hormones stimulate the heart by increasing its rate, force and speed of contraction, speed of conduction, and its irritability. However, although norepinephrine and epinephrine both stimulate the α-adrenergic receptors to stimulate vasoconstriction, epinephrine stimulates the β-adrenergic receptors to induce vasodilation.

Catecholamine release increases during exercise. However, trained people tend to have lower blood catecholamine levels than sedentary people during submaximal exercise. This response is important because it allows more blood to go to the liver, kidneys, gut, and skin.

During maximal exercise, catecholamine levels are higher in trained people. This adaptation maximizes blood pressure control in trained people during heavy exercise by increasing cardiac contractility and maximizing blood flow to working muscles. Blood vessels in type I muscle fibers are much less influenced by sympathetic adrenergic stimulation than are type II fibers. Delp (1998) has hypothesized that the increased blood flow to the high oxidative muscles may result from:

- Increased recruitment of high-oxidative motor units.

- Increased local release of metabolic vasodilator substances.

- Qualitative changes in the metabolic substances released (see section on metabolic regulation of blood flow).

- Decreased muscle sympathetic nervous activity.

- Enhanced endothelium-mediated dilation in the resistance vasculature.

- Increased effectiveness of the skeletal muscle pump.

Adrenergic Receptors Catecholamines (e.g., epinephrine and norepinephrine) interact with specific receptors in vascular smooth muscle membranes and endothelial cells to cause vasoconstriction or vasodilation (see Chapter 9). In muscle vasculature, α-adrenergic receptors cause vasoconstriction when stimulated. β_2 receptors, the most predominant adrenergic receptor type in muscle, cause vasodilation when stimulated.

Adrenergic Influences on Myocardial Activity As discussed in Chapter 9, a single hormone can have many effects on the body (see Table 9-2). For example, catecholamines, such as norepinephrine, speed glycogen breakdown and increase cardiac contractility and heart rate. A common property of many hormones is that cyclic AMP 3′, 5′-cyclic adenosine monophosphate—cAMP) actually triggers their effects within the cells.

The catecholamines epinephrine and norepinephrine increase heart rate and cardiac contractility, while acetylcholine (ACH) and adenosine protect the heart by decreasing heart rate and contractility. Together, these stimulatory and inhibitory substances regulate the sympathetic-adrenergic signaling pathway that coordinates chemical signals from outside the heart with factors within the heart (e.g., the Frank–Starling mechanism). This extracellular signaling allows heart muscle to respond to neural and hormonal inputs to regulate the rate and force of muscular contractions. In skeletal muscle, similar signaling pathways exist, but they regulate muscle force rather than contractile frequency.

Catecholamines stimulate heart function by binding to highly specific β-adrenergic receptors (βARs) on the heart cell (sarcolemmal) membranes (see Figure 15-8). This triggers the interaction of three sarcolemmal proteins: (1) multiple coupled stimulatory guanosine triphosphate (GTP), (2) binding regulatory proteins (G-proteins, or G_s), (3) catalytic subunits of adenyl cyclase (AC). The interaction of these three proteins increases the production of cAMP, which in turn stimulates the enzyme cAMP-dependent protein kinase (PKA). PKA has several important effects on cardiac function.

PKA increases cardiac output by enhancing myocardial relaxation and diastolic filling and by increasing cardiac contractility. Myocardial relaxation and diastolic filling increase through the phosphorylation of troponin I and stimulation of the Ca^{2+}-ATPase pump in the sarcoplasmic reticulum.

PKA increases cardiac contractility by enhancing Ca^{2+} release from the sarcoplasmic reticulum and by closing sarcolemmal Na^+ channels during membrane depolarization.

Acetylcholine and adenosine inhibit heart function by a mechanism that works opposite to that described for catecholamines. They bind to inhibitory receptors, which triggers inhibitory G-proteins to inhibit adenyl cyclase and thus decrease cAMP production. Decreased cAMP production leads to lower levels of PKA activity, which decreases cardiac contractility and diastolic filling.

Cyclic AMP production, stimulated by catecholamines and inhibited by ACH and adenosine, determine the relative activation of the heart at rest and during exercise. The intracellular concentration of cAMP in heart muscle cells is regulated quickly via a rapid balance between cAMP production and removal. Thus, cyclic AMP dynamics in cardiac muscle cells plays a pivotal role in cardiac performance in health and disease.

Vasopressin, Renin, and Angiotensin II Vasopressin (sometimes called arginine-vasopressin), a hormone formed in the hypothalamus and secreted by the posterior pituitary, is a potent vasoconstrictor that is released during prolonged exercise. It helps, during exercise, to combat cardiovascular drift, the tendency for blood pressure to drop and heart rate to increase during endurance exercise.

The renin–angiotensin mechanism is another hormone control system that exerts a relatively acute effect on blood pressure and blood flow. When blood pressure falls, the enzyme renin is released from the kidneys. The production of renin leads to the production of angiotensin II, a potent vasoconstrictor of both arteries and veins. The resulting vasoconstriction causes an increase in peripheral resistance and venous return that raises blood pressure. The system works in only one direction, raising pressure when it falls too low; it has no effect when pressure is too high. There is some evidence that this mechanism plays a role in increasing blood pressure during exercise.

The renin–angiotensin system also causes kidneys to retain fluid and salt and thus increases blood volume. This may be its most potent effect on blood pressure. In addition, angiotensin has an acute (24-hour), stimulating effect on aldosterone secretion. Aldosterone also increases water and salt retention. Aberrations in this control mechanism are suspected to be related to the development of some types of hypertension.

Vasopressin and angiotensin increase as a function of exercise intensity. However, their release is less in endurance-trained people during submaximal exercise. This effect contributes to higher blood flow to the kidneys and gut in trained individuals during submaximal exercise.

Several other substances are involved in vasoconstriction of kidney, liver, gut, and inactive muscle blood flow during exercise. They include dopamine, neuropeptide Y, and endothelin-1 (a vasoconstricting substance released by the endothelial cells).

Metabolic Regulation of Blood Flow

Metabolic, or local, control of blood flow is critical for redirecting blood to active muscles during exercise. Metabolic regulation allows local tissues to vasodilate, while less active tissues are vasoconstricting to maintain blood pressure. At rest, metabolic control of blood flow is directed toward maintaining tissue perfusion at a relatively constant rate. During exercise, it is directed toward increasing blood flow to working muscles. Local metabolic control of blood flow occurs in response to tissue demands for oxygen and fuels and responses to CO_2, hydrogen ion, and temperature.

Skeletal muscle blood flow is determined by the balance between metabolic requirements of the muscle and the need to maintain blood pressure. When blood flow is inadequate, vasodilator metabolites accumulate, which stimulates blood flow. Factors that stimulate vasodilation of the active blood vessels include adenosine, low P_{O_2}, high P_{CO_2}, low pH, and lactic acid. Of these, adenosine, a breakdown product of adenosine triphosphate (ATP), is thought to be most important.

Vascular smooth muscles contract and relax also in response to chemical substances released by the endothelium (cells comprising the inner lining of the vessels). Relaxing factors released by the endothelium include nitric oxide, prostacyclin, and endothelium-derived hyperpolarizing factor. Contracting factors include endothelin and vasoconstrictor prostaglandins. During exercise, circulating catecholamines are particularly important in stimulating the release of nitric oxide (NO), the most potent smooth muscle relaxing factor. Aerobic training appears to increase the capacity of the endothelium to release vasodilatory substances.

Stamler and associates (1997) showed that hemoglobin also distributes nitric oxide and thus contributes to muscle blood flow during exercise. Hemoglobin may act as a biosensor that adjusts blood flow to muscle and other tissues and organs.

Nitric oxide released from endothelial cells in large "feeder" arteries (large arteries supplying blood to muscle beds) appear to be critical in supplying the necessary blood during exercise. Skeletal muscles have no pre-capillary sphincters, so muscle blood flow is regulated by the terminal arterioles and large muscular arteries branching from main arteries, such as the femoral and brachial arteries. Nitric oxide appears to stimulate vasodilation in both the terminal arterioles and the large feeder arteries. It is released by the endothelium in response to pulsatile blood flow and blood vessel wall stress, both of which increase during exercise. The vasodilatory effects of nitric oxide during exercise increase the metabolically stimulated vasodilation caused by substances such as adenosine.

Muscle blood flow increases exponentially with metabolism. As the oxygen requirements of muscles increase, the arterioles vasodilate to allow more blood flow. At lower levels of metabolism, the pre-capillary sphincters open and then close. At higher intensities, the vessels tend to stay open. However, as discussed earlier, there is a limit to maximal muscle blood flow.

Sympathetic nervous activity to most tissues, including active muscle, increases with the intensity of the exercise in order to maintain blood pressure. At high exercise intensities, for example, skeletal muscle becomes a primary site of vasoconstriction because high muscle blood flow becomes a threat to maintaining blood pressure. With vasoconstriction, although blood flow remains relatively high (60–70 ml $\cdot$ 100 g^{-1} muscle $\cdot$ min^{-1}), it is still well short of the maximum flow capacity of the muscle.

When sympathetic stimulation causes the veins to constrict, there is also an increased return of blood to the heart. Venous return of blood can also be affected by such factors as increased blood volume, increased tone of large blood vessels, and dilation of small blood vessels.

The sympathetic nervous system has other actions that are important during exercise. For example, it can independently affect blood flow to the skin to cause either vasodilation or vasoconstriction. This means that while blood vessels in other inactive areas are vasoconstricted during exercise, blood vessels in the skin can be vasodilated to facilitate body cooling. A summary of the effects of the autonomic nervous system on cardiovascular function is shown in Table 15-1.

TABLE 15-1

The Autonomic Nervous System and Cardiovascular Function

Sympathetics	Parasympathetics
↑ Heart rate	↓ Heart rate
↑ Strength of contraction	↓ Strength of atrial contraction
Vasodilation of coronary arteries	Vasoconstriction of coronary arteries
Mild vasoconstriction of pulmonary vessels	Dilation of skin blood vessels
Vasoconstriction in abdomen, muscle (adrenergic), skin (adrenergic), and kidneys	
Vasodilation of muscle (cholinergic) and skin (cholinergic)	

Reactive Hyperemia Muscle blood flow can be blocked during isometric or heavy isotonic exercise because of the compression of blood vessels. When free flow is restored, blood flow to the area can increase many times above normal, resulting in a phenomenon known as *reactive hyperemia*. Reactive hyperemia is a local tissue response to increased metabolites and low oxygen that is similar to what normally occurs during exercise. A good example of reactive hyperemia is the increase in muscle size (called "the muscle pump") that follows weight-lifting exercise. Many novice weight lifters interpret the increased girth as evidence that the muscle grew before their eyes. However, the growth is only the result of reactive hyperemia; the muscles will revert to their normal size in a short time.

Autoregulation

Vascular smooth muscle is sensitive to stretch (or lack of stretch). When pressure increases in the circulation (i.e., during exercise), the vascular smooth muscle contracts in response to stretch. This is important for maintaining a constant blood flow to the tissues. When pressure is low, conversely, the smooth muscle relaxes in response to the lack of stretch.

Summary of Cardiovascular Control During Exercise

Cardiovascular control is aimed at maintaining blood pressure. Local circulatory control can override central command (to a point), so local tissue requirements are met. The following control mechanisms occur during exercise to meet the simultaneous demands of maintaining blood pressure, satisfying local tissue demands, and cooling the body:

- **When exercise begins, central command stimulates the cardiovascular control center (CVC)** to begin cardiovascular responses. These responses include increased heart rate, increased strength of cardiac contraction, and vasoconstriction. The CVC works in coordination with CNS centers, which control motor unit recruitment and breathing.

- **Vagus tone decreases.** This removes the parasympathetic inhibition on the heart, which, in turn, increases blood pressure.

- **The muscle pump facilitates venous return of blood to the heart.** Large muscles activated during exercise, as well as cyclical changes in intrathoracic pressure caused by increased breathing, help push blood toward the heart.

- **The baroreceptor set-point is raised.** This results in stimulation of the CVC to begin cardiovascular activity, including increased heart rate, increased strength of myocardial contraction, and circulatory vasoconstriction.

- **The hypothalamus is stimulated by increased temperature,** which results in stimulation of the CVC.

- **Type III muscle afferents are stimulated by muscle contraction and stretch** (particularly during isometric exercise). This stimulates the CVC to increase cardiovascular activity.

- **The release of catecholamines from the adrenal glands** results in stimulation of the heart and vasodilation in muscle blood vessels. It also has widespread effects on metabolism.

- **Sympathetic stimulation results in vasoconstriction** in tissues such as the spleen, kidneys, gastrointestinal tract, and inactive muscles. Vasoconstriction in the veins is particularly important for maintaining venous return of blood to the heart. Other important causes of vasoconstriction during exercise include the release of vasopressin and angiotensin II.

- **Local changes in metabolism stimulate muscle blood flow.** This results in vasodilation in active muscle. Adenosine and nitric oxide are the two most important local factors causing increased muscle blood flow.

- **As the intensity of exercise increases, the activity of Type III and IV muscle afferents increases.** This causes vasoconstriction in active muscle. During exercise using many large muscle groups, maximum blood flow is limited so blood pressure can be maintained. Even so, during maximal exercise, muscle blood flow increases to 85%.

- **As exercise progresses, signals from central command, the CVC, hypothalamus, baroreceptors, chemoreceptors, and muscle afferents are balanced.** Cardiac output and circulatory conductance are balanced to maintain blood pressure.

Regional Circulations

Muscle blood flow is balanced by the requirement to maintain blood pressure and by the circulatory requirements of other tissues. Circulation to areas such as the brain, spleen, liver, gastrointestinal tract, heart, and kidneys have important implications for exercise performance and well-being. Blood flow to major organs at rest and during exercise is shown in Table 15-2.

Brain Blood Flow Constant blood flow to the brain is critical. Cessation for only a few seconds leads to loss of consciousness. Although there is some sympathetic cerebral innervation, local regulation of blood flow in response to metabolism is the most important agent for controlling blood flow. A particularly potent metabolic stimulator of cerebral blood flow is CO_2. Vasodilation occurs when CO_2 increases; vasoconstriction occurs when CO_2 decreases.

Until recently, it was thought that brain blood flow did not change from rest to exercise. However, studies using the ^{133}Xe clearance technique have shown that cerebral blood flow increases over 25% during dynamic leg exercise to over 75 ml. This technique measures the rate at which the radioactive tracer ^{133}Xe clears the brain circulation. During static exercise (e.g., hand-grip), there is no increase in cerebral blood flow above the normal resting value of approximately $55-60$ ml $\cdot 100$ g$^{-1} \cdot$ min^{-1}.

Coronary Artery Blood Flow Blood is supplied to the heart by the right and left coronary arteries (Figure 15-9). The left ventricle is mainly supplied by the left coronary artery. The right ventricle and some of the left ventricle is supplied by the right coronary artery. These arteries originate just past the aortic valve. Thus, although large quantities of blood pass through the chambers of the heart, the myocardium requires a circulatory supply system similar to other tissues in the body. Coronary arterial blood flow is regulated by adjustments in the diameter of arteries throughout the coronary arterial tree. These arteries are of obvious importance: If they fail to perform adequately, loss of function or death will ensue.

The elevated stroke volume and heart rate during exercise increase the oxygen consumption of the heart. Thus, enhanced cardiac performance must be

TABLE 15-2

Tissue Blood Flow at Rest and During Dynamic Exercise

Tissue	Rest		Max Exercise	
	Blood flow (ml $\cdot$ min^{-1})	Flow rate (ml $\cdot 100$ g$^{-1} \cdot$ min^{-1})	Blood flow (ml $\cdot$ min^{-1})	Flow rate (ml $\cdot 100$ g$^{-1} \cdot$ min^{-1})
CNS	825	55	1125	75
Heart	260	87	900	300
Muscle	1200	25	18000	60–100
Viscera	2400	65	500	14
Skin	500	24	500	24

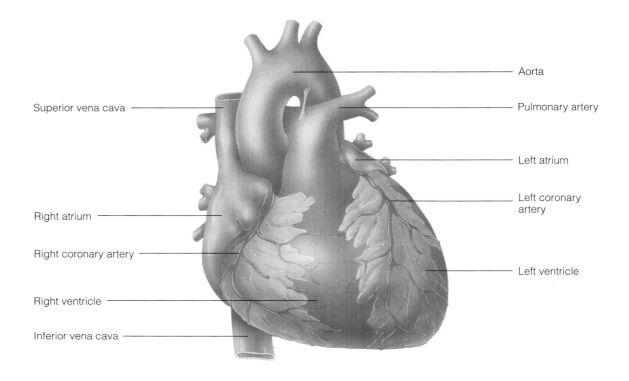

Figure 15-9   Blood supply to the heart. Blood is supplied to the heart from the right and left coronary arteries, which branch off the aorta. If a coronary artery becomes blocked by plaque buildup and a blood clot, a heart attack occurs; part of the heart muscle may die due to lack of oxygen. SOURCE: Fahey et al., 1999.

met by increased coronary artery blood flow. Resting coronary blood flow is approximately 260 ml · min^{-1}. During maximal exercise, it rises to about 900 ml · min^{-1}. Endurance training has little or no effect on maximal coronary flow.

Arterial pressure is the most important factor directing blood flow through the coronary arteries. There is some neurogenic control of blood flow in the coronaries, but metabolic regulation and autoregulation are the principle mechanisms increasing blood flow during exercise. Another important mechanism causing vasodilation in the coronary arteries is the release of nitric oxide by the endothelial cells. This may account for 15–20% of the dilation that occurs in the coronary arteries during exercise.

Because there is a limited ability to increase oxygen extraction with increased levels of myocardial metabolism, blood flow must increase with increasing work loads. If the coronaries are limited by disease, for example, in their ability to vasodilate, then the heart can become hypoxemic when its demand for oxygen is high. This can lead to diminished contractile capacity and heart pain called *angina pectoris*.

Blood flow through the coronary capillaries of the left ventricle is affected by the cardiac cycle. During systole, these capillaries are compressed by the contracting cardiac muscles. Blood flow is restored during diastole, when the ventricle is relaxed. This is an important consideration in exercise physiology and exercise stress testing. If diastolic pressure rises during exercise, then blood flow may be impeded when myocardial oxygen demand is greater. If diastolic pressure increases 20 mmHg during exercise, there is about an 80% chance that a person has a serious narrowing of at least one coronary artery. The coronary arteries and disease will be further discussed in Chapters 24 and 25.

Training improves calcium regulation in the coronary arteries by opening more sarcoplasmic reticulum release channels in the vascular smooth muscle. Training may also increase the release of endothelial relaxing factors, which aids coronary artery dilation during exercise. This increase is thought to occur by a rise in the production of nitric oxide synthase or an increased activation of the nitric oxide synthase pathway. Training may also stimulate an increase in coronary arterial diameter.

Skeletal Muscle Blood Flow In healthy adults, the typical maximal exercise cardiac output of 21 liters $\cdot$ min^{-1} results in a skeletal muscle blood flow rate of approximately 60 ml $\cdot$ 100 g^{-1} $\cdot$ min^{-1}. However, peak flow rates of greater than 100 ml $\cdot$ 100 g^{-1} $\cdot$ min^{-1} have been measured in specific muscles during exercise. In experiments where isolated muscles performed endurance exercise, peak muscle blood flow has been measured at almost

300 ml $\cdot$ 100 g^{-1} $\cdot$ min^{-1}. The heart is not capable of supplying blood at such a high rate to a large muscle mass. Therefore, during exercise, blood flow to active muscle is limited so blood pressure can be maintained.

Figure 15-10 compares leg blood flow during one- and two-leg exercise. Muscle blood flow during maximal exercise is determined by the amount of active muscle. When an isolated muscle group is exercised, muscle blood flow and oxygen consumption are very high. As the mass of active muscle increases, however, vasoconstriction increases muscle blood flow so blood pressure can be maintained.

Muscle blood flow increases even before exercise begins. Interestingly, the extent of pre-exercise muscle blood flow increase is specific to the type of exercise. If the exercise is prolonged, blood flow increases to slow-twitch fibers, while it increases to fast-twitch fibers if the exercise is intense.

Submaximal exercise muscle blood flow is either unaffected or decreased after a period of training. Some studies suggest that submaximal exercise blood flow decreases to fast-twitch fibers but increases to fast-twitch fibers.

Maximal exercise blood flow following training varies with the increase in cardiac output. Other

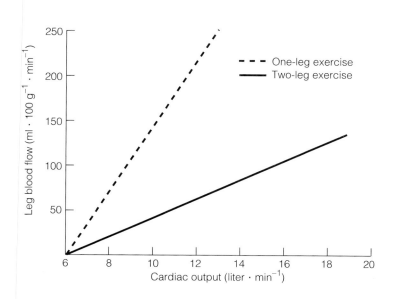

- - - One-leg exercise
——— Two-leg exercise

Figure 15-10 Leg muscle blood flow during one- and two-leg exercise. When an isolated muscle is exercised (e.g., one-leg exercise), muscle blood flow is high. As the mass of exercising muscle increases (e.g., during two-leg exercise), however, cardiac output capacity limits the blood flow available to working muscles. In order to maintain blood pressure, increased sympathetic activity causes vasoconstriction in exercising muscles, overriding the chemical signals within the muscles that normally cause vasodilation of muscle blood vessels. Adapted from Saltin, 1988.

Leg blood flow (ml $\cdot$ 100 g^{-1} $\cdot$ min^{-1})

Cardiac output (liter $\cdot$ min^{-1})

factors increasing blood flow during intense exercise include increased blood volume and increased capacity of the muscle pump mechanism due to increased mass and muscular vascularization. Recruiting more motor units will also enhance muscle blood flow during maximal exercise.

Kidney and Splanchnic Blood Flow Kidney (renal) function has important acute and chronic effects on exercising individuals. Kidneys are vital for maintaining acid-base balance in the blood, regulating water and sodium in the body, eliminating nitrogen waste products, secreting erthropoietin (EPO), and synthesizing such substances as bradykinin, prostaglandins, nitrogen oxide, and dopamine.

The spleen plays an important role in enhancing venous return of blood to the heart during exercise. It serves as an important reservoir of blood that can be released by sympathetically induced vasoconstriction.

At the onset of exercise, renal and splanchnic blood flows decrease quickly. While the renal and splanchnic circulations receive about 50% of the cardiac output at rest, they receive only 5% during maximal-intensity exercise. This decreased blood flow is due in part to increases in sympathetic nervous activity in the kidneys and the spleen at the onset of exercise. The sympathetic nervous activity in these tissues is intensity dependent—the more intense the exercise, the greater the sympathetic nerve stimulation to the tissues—and it can increase by as much as 400% during exercise. As discussed, vasopressin and angiotensin II also increase with exercise intensity, an effect that contributes to decreased blood flow to the kidneys and the spleen. The splanchnic circulation is also especially sensitive to stimulation from baroreceptors and chemoreceptors. Stimulation from these receptors leads to the release of blood from splanchnic circulation.

Glomerular filtration rate remains normal during submaximal exercise, but it is compromised as intensity increases. The glomerular filtration rate is the rate that blood is filtered by the glomeruli. During maximal exercise, the rate is only 60% of resting values.

Such decreases in kidney and splanchnic blood flow during exercise are extremely important modifications to circulation. They both help boost cardiac output distribution to muscles and help maintain blood pressure. Blood-flow redistribution from these areas increases blood volume to the muscles by 2 liters per minute during exercise.

Endurance training leads to an opposite effect on blood flow to the kidneys and spleen during submaximal exercise and actually increases renal and splanchnic blood flow. This increase probably occurs because endurance training causes lower activity of the sympathetic nervous system, decreased α-adrenergic receptor density in the kidney and spleen, and a decreased secretion of vasopressin and angiotensin II during exercise. Such improved blood flow has several benefits (McAllister, 1988):

- **Greater blood flow during exercise** places less stress on renal tissue. Severe kidney damage due to dehydration and loss of plasma volume has been reported in athletes during endurance events, such as marathons and ultramarathons. However, intense exercise seldom has long-term effects on renal function. Regular exercise enhances kidney function.

- **Enhanced gluconeogenesis:** Blood is delivered to the liver via the splanchnic vasculature—the hepatic artery and portal vein. Increased blood flow during exercise following a period of endurance training increases the flow of gluconeogenic substrates and enhances liver glucose production during exercise.

- **Improved absorption of glucose:** Carbohydrate feeding as part of a fluid replacement beverage improves exercise performance in a variety of circumstances. Improved splanchnic blood flow might enhance the absorption of glucose from the gut.

The Kidneys and Control of Blood and Fluid Volumes

Total blood volume is related to endurance fitness. Athletes have a higher blood volume, plasma volume, total hemoglobin content, and erythrocyte volume. Increased blood volume facilitates venous return of blood to the heart, which enhances the Frank–Starling mechanism of increasing cardiac output. As discussed in Chapter 14, the Frank–Starling mechanism is important in increasing stroke volume during submaximal exercise.

The control of blood volume is an important consideration in cardiovascular performance at rest and during exercise. Cardiac output capacity depends on venous return of blood to the heart. If blood volume is too low, the capacity of the heart is diminished because it has less blood to pump. If blood volume is high, cardiac output and arterial blood pressure increase.

Blood volume control is part of an integrated system regulating extracellular fluid, intracellular fluid, and fluid intake. Fluid volumes are controlled by (1) the renal pressure diuresis mechanism, (2) vasopressin stimulation of water reabsorption by the kidneys, (3) control of water reabsorption by thirst, and (4) stimulation of sodium reabsorption and potassium secretion in the renal tubules by aldosterone.

The kidney regulates the rate of fluid excretion by determining the rate of urine formation and electrolyte reabsorption. The hormone vasopressin, which is formed in the hypothalamus and secreted by the posterior pituitary gland, controls water output in the kidney. Sodium and potassium are controlled by the hormones aldosterone, vasopressin, and angiotensin.

Fluid intake is extremely important for the maintenance of blood volume. Usually, the thirst mechanism stays abreast of the body's fluid requirements (for important exceptions, see Chapter 22). Fluid intake is an important consideration in exercise physiology. When fluids are ingested, both blood volume and interstitial fluid volume increase.

The capillaries also play an important role in the control of fluid volumes. This function is important in regulating blood volume and maintaining cellular homeostasis. Fluid movement at the capillaries is regulated by hydrostatic pressure, interstitial pressure, plasma protein oncotic pressure (the tendency of protein to cause an osmotic pressure that

draws water toward it), interstitial oncotic pressure, and the filtration constant of the capillary membrane, as seen in the following equation:

$$\text{Fluid movement} = K[(P_c + O_i) - (P_i + O_p)] \quad (15\text{-}3)$$

where P_c is capillary hydrostatic pressure, P_i is interstitial fluid pressure, O_i is interstitial fluid oncotic pressure, O_p is plasma protein oncotic pressure, and K is the filtration constant of the membrane.

Hydrostatic and interstitial pressures are the forces exerted against the capillary walls by these fluids. Plasma protein and interstitial oncotic pressures refer to the osmolality of those respective fluid compartments. *Osmolality* is the ratio of water to dissolved substances and the tendency of fluids to diffuse into the compartment. The filtration constant of the membrane refers to the ability of the membrane to allow passage of the fluid. Other factors that affect the movement of fluid in the capillaries are capillary wall area, distance across the wall, and fluid viscosity.

An increase in capillary pressure tends to cause fluid to move from the circulation to the interstitium. A decrease in pressure results in the osmosis of fluid in the opposite direction. The lymphatics are involved in draining fluid from the interstitium back into the circulation, but only about 3 liters of fluid are returned.

The lymphatic system plays an important role in maintaining blood volume. Protein leaks continuously from the circulation into the tissue, and the lymphatics supply the only means of returning that protein to the circulation. Without this action, blood volume would decrease and tissues would swell with fluid.

Blood volume undergoes dynamic changes during exercise, largely because of increases in hydrostatic pressure. Plasma volume decreases at least 10% during prolonged exercise, which places an increasing load on the circulation as the physical activity continues.

Fortunately, blood volume increases with exercise training—and these adaptations occur rapidly. After a few days of endurance training, plasma volume increases by almost 400 ml, even in relatively fit people. This adaptation is important for increasing stroke volume.

Two mechanisms increase blood volume after endurance training. The initial increase is due to large increases in plasma renin and vasopressin levels during exercise, which result in increased retention of sodium and water by the kidneys. Second, chronic training leads to an increase in plasma protein, mainly albumin, which increases the osmolality of the blood, allowing it to hold more fluid.

Gastrointestinal Tract Blood Flow Redistribution of blood from the gastrointestinal (GI) tract during exercise facilitates muscle blood flow. Compared to rest, blood flow to the GI tract may decrease by as much as 80% during exercise. This process can sometimes cause GI distress and result in side pains ("runner's stitch"), acid reflux, diarrhea, bleeding, and gas. During exercise, dehydration and increased body temperature can also exacerbate GI ischemia.

Gravity and vigorous body movements contribute to GI problems during exercise. Tissue trauma is an important cause of "runner's stitch." When running, each stride jostles the internal organs and may traumatize the stomach, intestines, and liver. Side pains are less common in such sports as biking or swimming, in which there is less impact.

Exercise also impairs gastric motility and nutrient absorption into the blood stream. Normally, the GI tract contracts rhythmically after a meal to propel food through the system and allow the absorption of nutrients. Exercise interferes with this process, causing symptoms such as heartburn, vomiting, cramps, diarrhea, and the urge to defecate. It may also impair carbohydrate absorption, causing bacterial degradation of carbohydrates, which can produce painful gas and diarrhea.

Decreased GI blood flow can also increase the risk of gastroesophageal reflux by causing a relaxation of the esophageal valve. This valve prevents the stomach contents from purging into the esophagus. Also, exercise increases thorax pressure, which tends to put pressure on the stomach. The combination of valvular incompetence and pressure on the stomach force (reflux) stomach acid into the esophagus. In the stomach, acid causes no problems. However, in the esophagus, stomach acid causes pain and discomfort, symptoms of the condition known as heartburn. Over time, chronic heartburn can lead to ulcers and even cancer.

GI distress during exercise may be prevented by improving fitness. Fit people divert less blood from the GI tract during submaximal exercise. Other techniques to reduce GI distress include improving running techniques to decrease impact; avoiding high fat, high fiber, gas-producing meals before exercise; preventing dehydration; avoiding caffeine and alcohol; and possibly using antacids or H-2 blockers (e.g., Zantac, Pepcid, and Axid) to help manage stomach acid. People should seek medical attention for persistent problems. Heartburn and exercise side pains may mask serious medical conditions, such as coronary heart disease.

Skin Blood Flow Skin circulation is largely controlled by the cardiovascular control center and the anterior hypothalamus and is integrated with other temperature-regulating functions. Changes in skin blood flow occur through input from skin temperature receptors, sympathetic arousal from exercise and stress, and baroreceptors. These inputs can be modified by factors that include circadian rhythms, the menstrual cycle, heat acclimatization, and exercise training. CNS reflexes can also modify skin blood flow. They can cause decreases in skin blood flow via an adrenergic vasoconstrictor pathway or increases via a nonadrenergic vasodilator pathway. The vasoconstrictor pathway is triggered by norepinephrine. The triggering mechanism of the vasodilator pathway is unknown.

Skin blood flow increases during increasing intensities of exercise but decreases again as exercise intensity approaches maximum. Blood flow exceeds its metabolic demands, obviously because of its role in temperature regulation.

Skin blood flow helps regulate body temperature through heat exchange with the environment. During exercise, as body temperature increases, blood flow to the skin also increases. Heat is carried

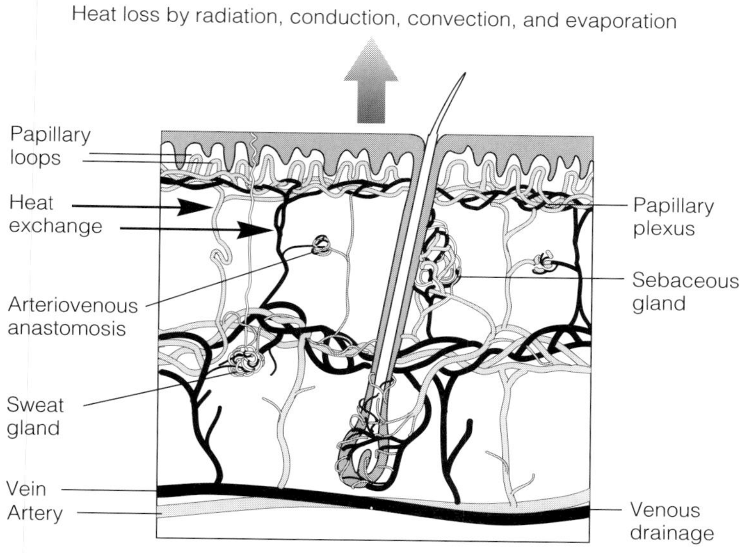

Heat loss by radiation, conduction, convection, and evaporation

Papillary loops

Heat exchange

Arteriovenous anastomosis

Sweat gland

Vein
Artery

Papillary plexus

Sebaceous gland

Venous drainage

Figure 15-11   Skin circulation. The rich vascular supply of the skin facilitates heat exchange. Arteriovenous anastomoses enhance heat exchange by transporting blood directly from arteries to veins.

toward the surface. There it is lost through conduction, convection, radiation, or evaporation (see Chapter 22).

The general structure of the skin provides a large surface area for heat exchange (Figure 15-11). Specific structures such as arteriovenous anastomoses and capillary plexi allow maximum exposure of blood flow to favorable thermal gradients, which allows heat from blood to be lost through heat dissipation mechanisms. The arteries and veins in the skin move toward the surface in parallel. This arrangement allows countercurrent heat exchange. In cold weather, cold blood from the surface can be warmed by warm blood from adjacent arteries.

The arteriovenous anastomoses shunt blood directly from the arteries to the veins and are particularly active when skin temperature is high. Some exercise scientists have attempted to obtain arterialized venous blood from hands that were heated to increase blood flow through the arteriovenous anastomoses. (Arterial blood is preferable to venous blood for many exercise studies because it is less influenced by regional changes in metabolism.) This practice is controversial.

Improved fitness increases skin blood flow and

sweat rate (see Chapter 22). This is particularly true when people train in the heat. The fit person also has the capacity to dissipate heat better than the sedentary person.

■ Vascularization and Exercise Training

Endurance training and long-term exposure to altitude will lead to an increased number of blood vessels in skeletal muscle. This adaptation occurs most readily in type I (slow-twitch) muscle fibers. In general, the degree of adaptation reflects the chronic metabolic demands placed on the tissues. In addition, this vascular response is stronger in the young than the old. The increase in blood vessels reverses itself if the metabolic demand diminishes.

Coronary Collateral Circulation   Coronary collateral vessels link one portion of a coronary artery to another or link two coronary arteries together. They are formed in response to the narrowing of a coronary vessel. Coronary collaterals aid in delivering blood to heart tissue and can protect the heart muscle from ischemia and myocardial infarction. Exercise training may enhance the development of coronary collaterals in the ischemic heart.

There is no evidence that collateral circulation develops in a healthy heart in response to exercise training. The coronaries have a large reserve, and the ischemic threshold necessary to produce coronary collaterals is not reached in healthy humans. Studies of coronary collaterals in healthy and diseased hearts are contradictory and controversial. There appear to be responders and nonresponders (i.e., people who develop coronary collaterals and people who don't). New investigative techniques will doubtlessly improve our knowledge in the years to come.

■ Summary

During exercise, blood pressure must be maintained and blood must be rapidly directed to working muscles to meet their demands for oxygen and substrates. The mechanisms that control blood flow make it possible to maintain or increase blood pressure, to continue supplying blood to the tissues, and to satisfy the metabolic requirements of working muscles.

The blood vessels consist of arteries, arterioles, capillaries, venules, and veins. Blood pressure is high in the arteries but diminishes throughout the circulation until it reaches the right atrium, where it approaches 0 mmHg. The factors controlling blood flow in individual vessels are described in Poiseuille's equation, which states that flow through a tube varies directly with the difference in pressure, varies to the fourth power of the radius of the tube, and varies inversely with the length of tube and viscosity of the fluid.

Metabolic regulation is the principal factor channeling blood to muscles during exercise. Local metabolic regulation of blood flow occurs in response to tissue demands for oxygen and fuels and in response to adenosine, nitric oxide, carbon dioxide, hydrogen ion, and temperature. Muscle blood flow increases exponentially with metabolism. At maximum exercise, some vasoconstriction occurs in active muscle in order to maintain blood pressure. Maximum muscle blood flow is largely determined by the amount of active muscle. The larger the portion of active muscle mass, the less total muscle blood flow.

Neural-hormonal control mechanisms are important in maintaining a balance between cardiac output and systemic profusion. They are important in redistributing blood to active muscles and causing vasoconstriction in the viscera and inactive muscles. Circulatory control centers are located in the cerebral cortex, hypothalamus, cardiovascular control center (CVC) in the medulla, baroreceptors, chemoreceptors, and muscle afferents. Circulating catecholamines and the renin–angiotensin system are involved in the acute control of blood pressure and flow. Changes in vascularization and fluid volumes are more important long-term control mechanisms.

■ Selected Readings

Adreani, C. M., and M. P. Kaufman. Effect of arterial occlusion on responses of group III and IV afferents to dynamic exercise. *J. Appl. Physiol.* 84: 1827–1833, 1988.

Barney, J. A., T. R. J. Ebert, L. Groban, P. A. Farrell, C. V. Hughes, and J. J. Smith. Carotid baroreflex responsiveness in high-fit and sedentary young men. *J. Appl. Physiol.* 65: 2190–2194, 1988.

Bloor, C. M., F. C. White, and T. M. Sanders. Effects of exercise on collateral development in myocardial ischemia in pigs. *J. Appl. Physiol.: Respirat. Environ. Exercise Physiol.* 56: 656–665, 1984.

Brock, R. W., M. E. Tschakovsky, J. K. Shoemaker, J. R. Halliwill, M. J. Joyner, and R. L. Hughson. Effects of acetylcholine and nitric oxide on forearm blood flow at rest and after a single muscle contraction. *J. Appl. Physiol.* 85: 2249–2254, 1998.

Brooks, G. A., E. E. Wolfel, G. E. Butterfield, A. Cymerman, A. C. Roberts, R. S. Mazzeo, and J. T. Reeves. Poor relationship between arterial [lactate] and leg net release during exercise at 4,300 m altitude. *Am. J. Physiol.* 275: R1192–1201, 1998.

Bryan, P. T., and J. M. Marshall. Cellular mechanisms by which adenosine induces vasodilation in rat skeletal muscle: significance for systemic hypoxia. *J Physiol.* (London) 514: 163–175, 1999.

Buckwalter, J. B., P. J. Mueller, and P. S. Clifford. Alpha1-adrenergic-receptor responsiveness in skeletal muscle during dynamic exercise. *J. Appl. Physiol.* 85: 2277–2283, 1998.

Buckwalter, J. B., P. J. Mueller, and P. S. Clifford. Autonomic control of skeletal muscle vasodilation during exercise. *J. Appl. Physiol.* 83: 2037–2042, 1997.

Buckwalter, J. B., S. B. Ruble, P. J. Mueller, and P. S. Clifford. Skeletal muscle vasodilation at the onset of exercise. *J. Appl. Physiol.* 85: 1649–1654, 1998.

Caru, B., E. Colombo, F. Santoro, A. Laporta, and F. Maslowsky. Regional flow responses to exercise. *Chest* 101 (Suppl.): 223S–225S, 1992.

Convertino, V. A., G. W. Mack, and E. R. Nadel. Elevated central venous pressure: a consequence of exercise training-induced hypervolemia? *Am. J. Physiol.* 29 (*Reg. Integr. Comp. Physiol.*): R273–R277, 1991.

Crandall, C. G., D. P. Stephens, and J. M. Johnson. Muscle metaboreceptor modulation of cutaneous active vasodilation. *Med. Sci. Sports Exerc.* 30: 490–496, 1998.

Delp, M. D. Differential effects of training on the control of skeletal muscle perfusion. *Med. Sci. Sports Exerc.* 30: 361–374, 1998.

Delp, M. D., and M. H. Laughlin. Regulation of skeletal muscle perfusion during exercise. *Acta Physiol. Scand.* 162: 411–419, 1998.

DiCarlo, S. E., and C.-Y. Chen. Vascular smooth muscle and exercise. In Encyclopedia of Sports Medicine and Exercise Physiology, T. D. Fahey (Ed.). New York: Garland, unpublished.

Egginton, S., O. Hudlicka, M. D. Brown, H. Walter, J. B. Weiss, and A. Bate. Capillary growth in relation to blood flow and performance in overloaded rat skeletal muscle. *J. Appl. Physiol.* 85: 2025–2032, 1998.

Fahey, T. D., P. Insel, and W. Roth. Fit and Well. 3d. Ed. Mountain View, Ca.: Mayfield Publishing Company, 1999.

Feldman, A. M. Modulation of adrenergic receptors and G-transduction proteins in failing human ventricular myocardium. *Circulation* 87 (Suppl. IV): 27–34, 1993.

Fleming, J. W., and A. M. Watanabe. Muscarinic-cholinergic receptor stimulation of specific GTP hydrolysis related to adenylate cyclase activity in canine cardiac sarcolemma. *Circ. Res.* 64: 340–350, 1988.

Fleming, J. W., P. L. Wisler, A. M. Watanabe. Signal transduction by G-proteins in cardiac tissues. *Circulation* 85: 420–433, 1992.

Fuglevand, A. J., and S. S. Segal. Simulation of motor unit recruitment and microvascular unit perfusion: spatial considerations. *J. Appl. Physiol.* 83: 1223–1234, 1997.

Gonzalez-Alonso, J., J. A. Calbet, and B. Nielsen. Muscle blood flow is reduced with dehydration during prolonged exercise in humans. *J. Physiol.* (London) 513: 895–905, 1998.

Green, D. J., G. O'Driscoll, B. A. Blanksby, and R. R. Taylor. Control of skeletal muscle blood flow during dynamic exercise: contribution of endothelium-derived nitric oxide. *Sports Med.* 21: 119–146, 1996.

Guyton, A. C. Textbook of Medical Physiology. Philadelphia: W. B. Saunders, 1986, pp. 206–217.

Hammond, H. K., D. A. Roth, M. D. McKirnan, and P. Ping. Regional myocardial down-regulation of the inhibitory GTP-binding protein ($G_i\alpha2$) and β-adrenergic receptors in a porcine model for chronic episodic myocardial ischemia. *J. Clin. Invest.* 92: 2644–2652, 1993.

Hammond, H. K., D. A. Roth, F. C. White, C. E. Ford, P. A. Insel, and C. M. Bloor. Myocardial β-adrenergic receptor expression and signal transduction after chronic volume overload hypertrophy and circulatory congestion in pigs. *Circulation* 85: 269–280, 1992.

Harms, C. A., M. A. Babcock, S. R. McClaran, D. F. Pegelow, G. A. Nickele, W. B. Nelson, and J. A. Dempsey. Respiratory muscle work compromises leg blood flow during maximal exercise. *J. Appl. Physiol.* 82: 1573–1583, 1997.

Hartshorne, D. J., and T. Kawamura. Regulation of contraction-relaxation in smooth muscle. *News in Physiol. Sci.* 7: 59–64, 1992.

Hayashi, N., A. Tanaka, M. Ishihara, and T. Yoshida. Delayed vagal withdrawal slows circulatory but not oxygen uptake responses at work increase. *Am. J. Physiol.* 274: R1268–1273, 1998.

Hepple, R. T., S. L. Mackinnon, J. M. Goodman, S. G. Thomas, M. J. Plyley. Resistance and aerobic training in older men: effects on $\dot{V}_{O_2}$ peak and the capillary supply to skeletal muscle. *J. Appl. Physiol.* 81: 1305–1310, 1997.

Hickner, R. C., J. S. Fisher, A. A. Ehsani, W. M. Kohrt. Role of nitric oxide in skeletal muscle blood flow at rest and during dynamic exercise in humans. *Am. J. Physiol.* 273: H405–410, 1997.

Howard, M. G., S. E. DiCarlo, and J. N. Stallone. Acute exercise attenuates phenylephrine-induced contraction of rabbit isolated aortic rings. *Med. Sci. Sports Exerc.* 24: 1102–1107, 1992.

Imms, F. J., F. Russo, V. I. Iyawe, and M. B. Segal. Cerebral blood flow velocity during and after sustained isometric skeletal muscle contractions in man. *Clin. Sci.* 94: 353–358, 1998.

Ingjer, F. Maximal aerobic power related to the capillary supply of the quadriceps femoris muscle in man. *Acta Physiol. Scand.* 104: 238–240, 1978.

Insel, P. A., and L. A. Ransnas. G-proteins and cardiovascular disease. *Circulation* 78: 1511–1513, 1988.

Insel, P. A., K. Urasawa, D. Leiber, D. A. Roth, and H. K. Hammond. Regulation of G_s protein in the heart. In *New Aspects of the Treatment of Failing Heart.* Springer-Verlag, Tokyo, Japan, 1993.

Jardine, D. L., H. Ikram, C. M. Frampton, R. Frethey, S. I. Bennett, and I. G. Crozier. Autonomic control of vasovagal syncope. *Am. J. Physiol.* 274: H2110–2115, 1998.

Johnson, J. M. Physical training and the control of skin blood flow. *Med. Sci. Sports Exerc.* 30: 382–386, 1998.

Joyner, M. J., and N. M. Dietz. Nitric oxide and vasodilation in human limbs. *J. Appl. Physiol.* 83: 1785–1796, 1997.

Kingwell, B. A., B. Sherrard, G. L. Jennings, and A. M. Dart. Four weeks of cycle training increases basal production of nitric oxide from the forearm. *Am. J. Physiol.* 272: H1070–1077, 1997.

Kjaer, M. Adrenal medulla and exercise training. *Eur. J. Appl. Physiol.* 77: 195–199, 1998.

Kohzuki, H., H. Ishidate, T. Kishi, Y. Ohga, S. Sakata, and M. Takaki. Unloaded skeletal muscle O_2 uptake decreases with decreased venous PO_2 at high-frequency stimulation. *Jpn. J. Physiol.* 48: 347–354, 1998.

Koller, A., G. Dornyei, and G. Kaley. Flow-induced responses in skeletal muscle venules: modulation by nitric oxide and prostaglandins. *Am. J. Physiol.* 275: H831–836, 1998.

Koller-Strametz, J., B. Matulla, M. Wolzt, M. Muller, J. Entlicher, H. G. Eichler, and L. Schmetterer. Role of nitric oxide in exercise-induced vasodilation in man. *Life Sci.* 62: 1035–1042, 1998.

Krajcar, M., and G. Heusch. Local and neurohumoral control of coronary blood flow. *Basic Res. Cardiol.* 88 (Suppl.): 25–42, 1993.

Kuo, L., M. J. Davis, and W. M. Chiliam. Endothelial modulation of arteriolar tone. *News in Physiol. Sci.* 7: 5–9, 1992.

Lash, J. M. Training-induced alterations in contractile function and excitation-contraction coupling in vascular smooth muscle. *Med. Sci. Sports Exerc.* 30: 60–66, 1998.

Laughlin, M. H., C. C. Hale, L. Novela, D. Gute, N. Hamilton, and C. D. Ianuzzo. Biochemical characterization of exercise-trained porcine myocardium. *J. Appl. Physiol.* 71: 229–235, 1991.

Laughlin, M. H., and R. M. McAllister. Exercise training induced coronary vascular adaptation. *J. Appl. Physiol.* 73(6): 2209–2225, 1992.

Laughlin, M. H., C. L. Oltman, and D. K. Bowles. Exercise training-induced adaptations in the coronary circulation. *Med. Sci. Sports Exerc.* 30: 352–360, 1998.

MacLean, D. A., L. I. Sinoway, and U. Leuenberger. Systemic hypoxia elevates skeletal muscle interstitial adenosine levels in humans. *Circulation* 98: 1990–1992, 1998.

Madsen, P. L. Blood flow and oxygen uptake in the human brain during various states of sleep and wakefulness. *Acta Neurol. Scand.* 148 (Suppl.): 3–27, 1993.

Maeda, S., T. Miyauchi, M. Sakane, M. Saito, S. Maki, K. Goto, and M. Matsuda. Does endothelin-1 participate in the exercise-induced changes of blood flow distribution of muscles in humans? *J. Appl. Physiol.* 82: 1107–1111, 1997.

McAllister, R. M. Adaptations in control of blood flow with training: splanchnic and renal blood flows. *Med. Sci. Sports Exerc.* 30: 375–381, 1998.

Mellander, S., and B. Johansson. Control of resistance, exchange, and capacitance vessels in the peripheral circulation. *Pharmacol. Rev.* 20: 117–196, 1968.

Minson, C. T., S. L. Wladkowski, J. A. Pawelczyk, and W. L. Kenney. Age, splanchnic vasoconstriction, and heat stress during tilting. *Am. J. Physiol.* 276: R203–R212, 1999.

Mitchell, J. H., and R. F. Schmidt. Cardiovascular reflex control by afferent fibers from skeletal muscle receptors. In Handbook of Physiology. The Cardiovascular System. Peripheral Circulation and Organ Blood Flow. Bethesda, Md.: American Physiological Society, 1983, pp. 623–658.

Mitchell, J. H., and R. G. Victor. Neural control of the cardiovascular system: insights from muscle sympathetic nerve recordings in humans. *Med. Sci. Sports Exerc.* 28 (Suppl.): S60–69, 1996.

Moore, R. L., and D. H. Korzick. Cellular adaptations of the myocardium to chronic exercise training. *Prog. Cardiovasc. Dis.* 37(6): 1–26, 1995.

Musch, T. I., G. H. Haidet, G. A. Ordway, J. C. Longhurst, and J. H. Mitchell. Training effects on regional blood flow response to maximal exercise in foxhounds. *J. Appl. Physiol.* 62: 1724–1732, 1987.

Naeije, R. Pulmonary circulation in hypoxia. *Int. J. Sports Med.* 13 (Suppl. 1): S27–S30, 1992.

Neer, E. J., and D. E. Clapham. Signal transduction through G-proteins in the cardiac myocyte. *Trends Cardiovasc. Med.* 2: 6–11, 1992.

Nioka, S., D. Moser, G. Lech, M. Evengelisti, T. Verde, B. Chance, and S. Kuno. Muscle deoxygenation in aerobic and anaerobic exercise. *Adv. Exp. Med. Biol.* 454: 63–70, 1998.

Noakes, T. D. 1996 J. B. Wolffe Memorial Lecture. Challenging beliefs: ex Africa semper aliquid novi. *Med. Sci. Sports Exerc.* 29: 571–590, 1997.

O'Leary, D. S., L. B. Rowell, and A. M. Scher. Baroreflex-induced vasoconstriction in active skeletal muscle of conscious dogs. *Am. J. Physiol.* 260: H37–H41, 1991.

Poortmans, J. R. Exercise and renal function. *Sports Med.* 1: 125–153, 1984.

Poortmans, J. R., and J. Vanderstraeten. Kidney function during exercise in healthy and diseased humans—an update. *Sports Med.* 18: 419–437, 1994.

Radegran, G., and B. Saltin. Muscle blood flow at onset of dynamic exercise in humans. *Am. J. Physiol.* 274: H314–322, 1998.

Ray, C. A., and K. M. Hume. Sympathetic neural adaptations to exercise training in humans: insights from microneurography. *Med. Sci. Sports Exerc.* 30: 387–391, 1998.

Richardson, R. S. Oxygen transport: air to muscle cell. *Med. Sci. Sports Exerc.* 30: 53–59, 1998.

Richardson, R. S., E. A. Noyszewski, J. S. Leigh, and P. D. Wagner. Lactate efflux from exercising human skeletal muscle: role of intracellular PO_2. *J. Appl. Physiol.* 85: 627–634, 1998.

Rogers, H. B., T. Schroeder, N. H. Secher, and J. H. Mitchell. Cerebral blood flow during static exercise in humans. *J. Appl. Physiol.* 68: 2358–2361, 1990.

Roth, D. A., K. Urasawa, G. A. Helmer, and H. K. Hammond. Down-regulation of cardiac GTP-binding proteins in right atrium and left ventricle in pacing-induced congestive heart failure. *J. Clin. Invest.* 91: 939–949, 1993.

Roth, D. A., C. D. White, C. D. Hamilton, J. L. Hall, and W. C. Stanley. Adrenergic desensitization in left ventricle from streptozotocin diabetic swine. *J. Mol. Cell. Cardiol.* 27 (10): 2315–2325, 1995.

Rowell, L. B. Human Circulatory Regulation During Physical Stress. London: Oxford University Press, 1986.

Rowell, L. B. Blood pressure regulation during exercise. *Ann. Med.* 23: 329–333, 1991.

Rowell, L. B. General principles of vascular control. In Human Circulation Regulation During Physical Stress, L. B. Rowell (Ed.). New York: Oxford University Press, 1986, pp. 8–43.

Rowell, L. B. Reflex control of the circulation during exercise. *Int. J. Sports Med.* (Suppl. 1): S25–S27, 1992.

Rowell, L. B., and D. S. O'Leary. Reflex control of the circulation during exercise: chemoreflexes and mechanoreflexes. *J. Appl. Physiol.* 69: 407–418, 1990.

Rowell, L. B., M. V. Savage, J. Chambers, and J. R. Blackmon. Cardiovascular responses to graded reductions in leg perfusion in exercising humans. *Am. J. Physiol.* 261: H1545–H1553, 1991.

Ruch, T., and H. Patton. Physiology and Biophysics. Philadelphia: W. B. Saunders, 1965, pp. 523–542.

Saltin, B. Capacity of blood flow delivery to exercising skeletal muscle in humans. *Amer. J. Cardiol.* 62: 30E–35E, 1988.

Saltin, B., G. Radegran, M. D. Koskolou, and R. C. Roach. Skeletal muscle blood flow in humans and its regulation during exercise. *Acta Physiol. Scand.* 162: 421–436, 1998.

Segal, S. S. Communication among endothelial and smooth muscle cells coordinates blood flow during exercise. *News in Physiol. Sci.* 7: 152–156, 1992.

Seymour, R. S., A. R. Hargens, and T. J. Pedley. The heart works against gravity. *Am. J. Physiol.* 265: R715–R720, 1993.

Shamsuzzaman, A. S. M., Y. Sugiyama, A. Kamiya, Q. Fu, and T. Mano. Head-up suspension in humans: effects on sympathetic vasomotor activity and cardiovascular responses. *J. Appl. Physiol.* 84: 1513–1519, 1998.

Sheriff, D. D., R. A. Augustyniak, and D. S. O'Leary. Muscle chemoreflex induced increases in right atrial pressure. *Am. J. Physiol.* 275: H767–775, 1998.

Shoemaker, J. K., J. R. Halliwill, R. L. Hughson, and M. J. Joyner. Contributions of acetylcholine and nitric oxide to forearm blood flow at exercise onset and recovery. *Am. J. Physiol.* 273: H2388–2395, 1997.

Shoemaker, J. K., M. E. Tschakovsky, and R. L. Hughson. Vasodilation contributes to the rapid hyperemia with rhythmic contractions in humans. *Can. J. Physiol. Pharmacol.* 76: 418–427, 1998.

Sinoway, L., S. Prophet, I. Gorman, T. Mosher, J. Shenberger, M. Dolecki, R. Briggs, and R. Zelis. Muscle acidosis during static exercise is associated with calf vasoconstriction. *J. Appl. Physiol.* 66: 429–436, 1989.

Stamler, J. S., L. Jia, J. P. Eu, T. J. McMahon, I. T. Demchenko, J. Bonaventura, K. Gernert, and C. A. Piantadosi. Blood flow regulation by S-nitrosohemoglobin in the physiological oxygen gradient. *Science* 276: 2034–2037, 1997.

Stehno-Bittel, L., and M. H. Laughlin. Exercise training depletes sarcoplasmic reticulum calcium in coronary smooth muscle. *J. Appl. Physiol.* 71: 1764–1773, 1991.

Strange, S. Cardiovascular control during concomitant dynamic leg exercise and static arm exercise in humans. *J. Physiol.* (London) 514: 283–291, 1999.

Strange, S., N. H. Secher, J. A. Pawelczyk, J. Karpakka, N. J. Christensen, J. H. Mitchell, and B. Saltin. Neural control of cardiovascular responses and of ventilation during dynamic exercise in man. *J. Physiol.* (London) 470: 693–704, 1993.

Sun, D., A. Huang, A. Koller, and G. Kaley. Adaptation of flow-induced dilation of arterioles to daily exercise. *Microvasc. Res.* 56: 54–61, 1998.

Thomas, S. N., T. Schroeder, N. H. Secher, and J. H. Mitchell. Cerebral blood flow during submaximal and maximal dynamic exercise in man. *J. Appl. Physiol.* 67: 744–748, 1989.

Tuominen, J. A., J. E. Peltonen, and V. A. Koivisto. Blood flow, lipid oxidation, and muscle glycogen synthesis after glycogen depletion by strenuous exercise. *Med. Sci. Sports Exerc.* 29: 874–881, 1997.

Vatner, S. F., and L. Hittinger. Myocardial perfusion dependent and independent mechanisms of regional myocardial dysfunction in hypertrophy. *Basic Res. Cardiol.* 88 (Suppl. 1): 81–95, 1993.

Vicini, P., R. C. Bonadonna, M. Lehtovirta, L. C. Groop, and C. Cobelli. Estimation of blood flow heterogeneity in human skeletal muscle using intravascular tracer data: importance for modeling transcapillary exchange. *Ann. Biomed. Eng.* 26: 764–774, 1998.

Viru, M., E. Jansson, A. Viru, and C. J. Sundberg. Effect of restricted blood flow on exercise-induced hormone changes in healthy men. *Eur. J. Appl. Physiol.* 77: 517–522, 1998.

Walgenbach-Telford, S. Arterial baroreflex and cardiopulmonary mechanoreflex function during exercise. In Reflex Control of the Circulation, J. P. Gilmore and I. H. Zucker (Eds.). Caldwell, N.J.: Telford, 1990.

Yang, H. T., R. W. Ogilvie, and R. L. Terjung. Exercise training enhances basic fibroblast growth factor-induced collateral blood flow. *Am. J. Physiol.* 274: H2053–2061, 1998.

CHAPTER 16

CARDIOVASCULAR DYNAMICS DURING EXERCISE

At rest, the cardiovascular system in healthy people has little difficulty supplying oxygen and fuels to the tissues. The system removes waste products easily and helps maintain a stable cellular environment. During exercise, however, demands on the system increase considerably. The amount of oxygen delivered to the working muscles increases with the intensity of the exercise. Cardiac output increases because of a higher heart rate and stroke volume. Blood is directed to working skeletal and respiratory muscles, and blood flow is maintained or increased to the heart and brain. At the same time, blood is directed away from less active tissues, such as the viscera and inactive muscles, and directed toward the skin to help dissipate heat produced by the increased metabolism of the exercise.

Successful endurance athletes have a high cardiorespiratory capacity. The benchmark of the cardiorespiratory system's capacity is maximal oxygen consumption ($\dot{V}_{O_2max}$), the maximum rate at which oxygen can be taken up and used by the body during exercise. $\dot{V}_{O_2max}$ predicts performance over short distances—1-2 miles—but beyond that, fatigue resistance becomes a more significant factor in performance. Many people with a high $\dot{V}_{O_2max}$ perform poorly because they lack resistance to fatigue.

The limits of oxygen transport capacity have been the subject of intense controversy (see Basset

Lynn Jennings and others in World Championship competition. Performances such as these are dependent upon a high cardiac output and peripheral metabolic adaptations.
PHOTO: © Claus Andersen.

317

and Howley, 1997; and Noakes, 1997 and 1998). For almost 100 years, since the pioneering work of researchers such as A. V. Hill, the pumping capacity of the heart (i.e., cardiac output) was thought to limit oxygen transport capacity. During exercise, the heart is responsible for supplying the circulatory needs of tissue metabolism, breathing, and thermoregulation, as well as maintaining blood supply to the brain and the heart itself. During many types of exercise, fatigue was thought to occur when cardiac output reached its maximum capacity, causing the muscles to contract without an adequate oxygen supply (anaerobically). However, inconsistencies in the relationship between maximum cardiac output capacity, $\dot{V}_{O_2max}$, and fatigue have led researchers (most notably Tim Noakes of South Africa) to question whether the muscles can ever become anaerobic without the heart first developing anaerobic conditions itself. He has proposed that whereas maximum cardiac output may indeed be the limiting factor for oxygen transport capacity, maximal exercise must terminate through a regulatory process before either the heart or the skeletal muscles develop a limiting oxygen consumption. Noakes has proposed further that the blood supply to the heart is likely to be an important factor in determining when exercise must terminate.

During exercise, the cardiorespiratory system and its control mechanisms must carefully balance the critical need for blood flow in the coronary artery and brain with the metabolic requirements of active and inactive tissues. The system appears to be self-limiting—that is, as the heart, lungs, and blood vessels reach their capacity to satisfy the metabolic requirements of the tissues and simultaneously deliver blood to the heart and central nervous system, the system protects itself from coronary and central nervous system ischemia by limiting the activity of the exercising muscles so that a maximal cardiac output is achieved but without the production of anaerobic conditions in either the heart or the active muscles. This is necessary because the heart's pumping capacity depends on the blood that it can deliver to its own coronary arteries. Maintenance of blood pressure is the central regulating factor in cardiorespiratory physiology. The capacity to regulate blood flow to working tissues

while maintaining central blood volume is critical to exercise performance.

This chapter discusses the responses of the cardiovascular system to exercise.

■ Cardiovascular Responses to Exercise

The heart and circulation respond to the requirements of metabolism during exercise by increasing blood flow to active areas and decreasing it to less critical areas. The principal cardiovascular responses to exercise include:

- **Increased cardiac output.** This occurs by increasing heart rate and stroke volume. It enhances oxygen and substrate (fuel) delivery to active skeletal muscle and the heart and speeds the removal of CO_2 and metabolites.

- **Increased skin blood flow** helps remove heat.

- **Decreased blood flow to the kidneys,** resulting in diminished urinary output and maintenance of blood volume.

- **Decreased visceral blood flow,** resulting in reduced activity in the gastrointestinal tract.

- **Vasoconstriction in the spleen,** which increases blood volume.

- **Maintenance or slight increase in brain blood flow.**

- **Increased blood flow to the coronary arteries of the heart.**

- **Increased muscle blood flow.** Maximum muscle blood flow, however, is limited by the need to maintain blood pressure. Active muscles vasoconstrict at high rates of blood flow if blood pressure cannot be maintained.

To summarize: Cardiovascular regulation is directed toward maintaining blood pressure. Increased metabolism in skeletal muscle and the heart can stimulate more blood flow to these tissues. Cardiovascular regulation during exercise balances the need for more blood to active tissue with the need to maintain blood pressure and thus coronary artery and brain blood flow. Cardiovascular changes with exercise and with training are shown in Table 16-1.

/ lutagan

TABLE 16-1

Changes in Cardiovascular and Pulmonary Function as a Result of Endurance-Type Physical Training[a]

Measurement	Resting Pre		Resting Post	Upright Submaximal "Steady-State" Exercise Pre		Post	Upright Maximal Exercise Pre		Post
Heart rate (beats · min^{-1})	70	$\bar{0}$	63	150	−	130	185	$\bar{0}$	182
Stroke volume (ml · beat^{-1})	72	+	80	90	+	102	90	+	105
Cardiac output (liters · min^{-1})	5.0	$\bar{0}$	5.0	13.5	$\bar{0}$	13.2	16.6	+	19.1
$(a - v)_{O_2}$ (arteriovenous oxygen difference) (vol %)	5.6	$\overset{+}{0}$	5.6	11.0	+	11.3	16.2	$\overset{+}{0}$	16.5
O$_2$ uptake (liters · min)	0.280	0	0.280	1.485	$\bar{0}$	1.485	2.685	+	3.150
(ml · kg^{-1} · min^{-1})	3.7	0	3.7	19.8	$\bar{0}$	19.8	35.8	+	42.0
(METs)[b]	1.0	0	1.0	5.7	0	5.7	10.2	+	12.0
Work load (kg · kg^{-1} · min^{-1})	—		—	600	0	600	1050	+	1500
Blood pressure (mmHg)									
Systemic arterial systolic BP	120	$\bar{0}$	114	156	$\bar{0}$	140	200	$\bar{0}$	200
Systemic arterial diastolic BP	75	$\bar{0}$	70	80	$\bar{0}$	75	85	$\bar{0}$	75
Systemic arterial mean	90	$\bar{0}$	88	126	$\bar{0}$	118	155	$\bar{0}$	152
Total peripheral resistance (dyne sec · cm^{-5})	1250	0	1250	750	$\bar{0}$	750	450	$\bar{0}$	390
Blood flow (ml · min^{-1})									
Coronary	260	−	250	600	−	560	900	$\overset{+}{0}$	940
Brain	750	0	740	740	0	740	740	0	740
Viscera	2400	0	2500	900	$\overset{+}{0}$	1000	500	0	500
Inactive muscle	600	±	555	500	0	500	300	0	300
Active muscle	600	−	555	10,360	$\bar{0}$	10,000	13,760	+	16,220
Skin	400	0	400	400	0	400	400	0	400
Total	5000		5000	13,500		13,200	16,600		19,100
Blood volume (liters)	5.1	$\overset{+}{0}$	5.3						
Plasma volume (liters)	2.8	$\overset{+}{0}$	3.0						
Red cell mass (liters)	2.3	$\overset{+}{0}$	2.3						
Heart volume (ml)	730	$\overset{+}{0}$	785						
Pulmonary ventilation (liters · min^{-1})	10.2	0	10.3	44.8	$\bar{0}$	38.2	129	+	145
Respiratory rate (breaths · min^{-1})	12	$\bar{0}$	12	30	−	24	43	+	52
Tidal volume	850	$\overset{+}{0}$	855	1.5	$\overset{+}{0}$	1.6	3.0	$\bar{0}$	2.8
Lung diffusing capacity (D_L) (ml at STPD)[c]	34.1	$\overset{+}{0}$	35.2	40.6	$\overset{+}{0}$	42.8	48.2	+	50.6
Pulmonary capillary blood volume (ml)	90.1	$\overset{+}{0}$	97.2	129.3	+	141.2	124.5	+	220.0
Vital capacity (liters)	5.1	$\overset{+}{0}$	5.2						
Blood lactic acid (mM)	0.7	0	0.7	3.9	−	3.0	11.0	$\overset{+}{0}$	12.4
Blood pH	7.43	0	7.43	7.41	$\overset{+}{0}$	7.43	7.33	$\bar{0}$	7.29
Recovery rate					+			+	

[a] Estimated for a healthy man, age 45, weighing 75 kg. Pre = pretraining; post = posttraining; minus (−) sign usually means a decrease in value with training; plus (+) sign usually means an increase in value with training; zero (0) sign usually means no change in value with training.

[b] A MET is equal to the O$_2$ cost at rest. One MET is generally equal to 3.5 ml · kg^{-1} of body weight per minute of O$_2$ uptake or 1.2 cal · min^{-1}.

[c] STPD is standard temperature (0°C), pressure 760 mmHg, and dry.

SOURCE: Data courtesy of W. Haskell.

Cardiovascular responses to physical activity depend on the type and intensity of exercise. Dynamic exercise requiring a large muscle mass (i.e., large muscle groups performing rhythmical contractions) causes the largest response from the cardiovascular system. This kind of exercise leads to large increases in cardiac output, heart rate, and systolic blood pressure, but there is little change in diastolic blood pressure. Strength exercises cause marked increases in systolic, diastolic, and mean blood pressures, with more moderate increases in heart rate and cardiac output.

Oxygen Consumption

Oxygen consumption is proportional to the intensity of exercise (Figure 1-5): As intensity increases, so does oxygen consumption. It is determined by the rate at which oxygen is transported to the tissues, the oxygen-carrying capacity of blood, and the amount of oxygen extracted from the blood (Eq. 16-1):

$$\dot{V}_{O_2} = (f_H)(V_s)(a-v)_{O_2}$$
$$\dot{V}_{O_2} = \dot{Q}(a-v)_{O_2}$$

(16-1)

where:

f_H = = frequency of the heart or heart rate (i.e.,
HR   the number of heart beats per minute)

V_s = = stroke volume (the amount of blood
SV   ejected from the left ventricle per cardiac contraction; expressed in milliliters)

$(a-v)_{O_2}$ = arteriovenous oxygen difference (the difference in oxygen content between the arteries and veins; expressed in milliliters of oxygen per 100 ml of blood or volumes percent (vol %); a measure of oxygen extraction by the tissues.

$\dot{Q}$ = cardiac output (the amount of blood pumped by the left ventricle of the heart; expressed in liters · min^{-1})

During exercise in which the intensity gradually increases from rest to maximal intensity, stroke volume increases during the early phase of exercise. Heart rate and $(a-v)_{O_2}$ increase almost linearly with exercise intensity (Figures 16-1, 16-2, 16-3). Cardiac output and $(a-v)_{O_2}$ each account for about 50% of the increase in oxygen consumption during submaximal exercise. Cardiac output (mainly heart rate) plays a more important role in increasing oxygen consumption as the intensity of exercise approaches maximum. At maximum exercise, it accounts for approximately 75% of the increased oxygen uptake above rest.

The arteriovenous oxygen difference and cardiac output are related. The former depends not only on the capacity of the mitochondria to use oxygen, but also on the rate of diffusion of oxygen from

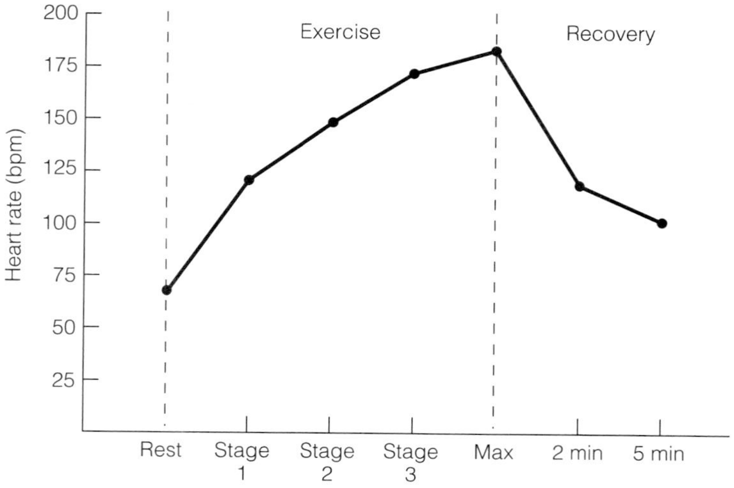

Figure 16-1 Heart rate response of a 30-year-old man during and after a Bruce treadmill test (for a description of this test, see Chapter 27). Maximal heart rate is relatively independent of functional capacity. Resting, submaximal, and recovery heart rates tend to decrease following an extended period of endurance training.

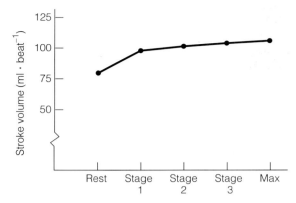

Figure 16-2 Stroke volume of a 30-year-old man during a Bruce treadmill test. Stroke volume increases during the early stages of exercise. Heart rate contributes relatively more to the total cardiac output as the intensity of exercise approaches maximum. Some studies suggest that in some subjects, stroke volume declines during exercise above about 75% of max.

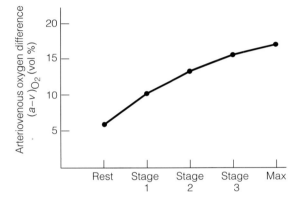

Figure 16-3 Arteriovenous oxygen difference [$(a-v)_{O_2}$] of a 30-year-old man during a Bruce treadmill test. This measure changes very little with training.

the blood into the cell. Factors such as blood flow, the oxyhemoglobin dissociation curve, hemoglobin, and myoglobin also affect the $(a-v)_{O_2}$ difference.

The oxygen-carrying capacity of blood is approximately 18 to 20 ml $O_2 \cdot 100$ ml^{-1} in most healthy people living at sea level. Because of their higher hemoglobin content, males have a higher oxygen-carrying capacity than females. Oxygen-carrying capacity is also higher in high-altitude natives (as high as 30 vol %) and in people with artificially induced increases in red blood cells. This latter group includes people who have taken the drug erythropoietin (EPO) or practiced induced erythrocythemia (blood doping). In blood doping, blood is withdrawn and stored. After the person's blood volume and red blood cells (RBCs, erythrocytes) rise to normal, the RBCs are reinfused, thus increasing the blood's oxygen-carrying capacity (see Chapter 29). EPO increases red blood cell production. In most

people, oxygen-carrying capacity is stable and is a constant in the oxygen consumption equation.

In some subjects, oxygen consumption plateaus near maximum effort. The highest achievable level of oxygen consumption is symbolized by $\dot{V}_{O_2max}$. (*Note*: The classic definition of $\dot{V}_{O_2max}$ is the point at which no further increases in $\dot{V}_{O_2}$ with increasing exercise intensity are possible.) In most people, the relationship between $\dot{V}_{O_2}$ and exercise intensity is linear—that is, the highest exercise intensity coincides with $\dot{V}_{O_2max}$. Further discussion of $\dot{V}_{O_2max}$ occurs later in this chapter and again in Chapter 33.

There is some variability between people in oxygen consumption during exercise done at the same power output or intensity. Some people are more efficient than others. If a world-class cyclist and a sedentary person rode a bicycle ergometer at the same power output, the oxygen consumption of the trained cyclist would probably be lower.

Also, the power output would represent a lower percentage of maximum for the cyclist.

Heart Rate

Heart rate is the most important factor increasing cardiac output during exercise. In dynamic exercise, heart rate increases with exercise intensity and oxygen consumption. It levels off at $\dot{V}_{O_2max}$. Typically, heart rate increases from approximately 70 beats · min^{-1} at rest to 180–200 beats · min^{-1} at maximal exercise. As discussed in Chapters 14 and 15, heart rate increases during exercise due to withdrawal of vagus nerve tone and sympathetic stimulation of the heart. Exercise and maximum heart rates are affected by fitness, age, and sex. In adults, maximum heart rate can be estimated by $HR_{max} = 220 - age$ ($\pm$ 12 bpm).

During a constant level of submaximal exercise, heart rate increases and then levels off as the oxygen requirements of the activity have been satisfied. As exercise intensity increases, it takes longer for heart rate to level off; at high exercise intensities, heart rate may not level off. During prolonged exercise, heart rate increases steadily at the same work rate. Recall from Chapter 15 that this phenomenon, which results from decreased stroke volume, is called cardiovascular drift. Heart rate must increase in order to maintain cardiac output and blood pressure at the same level. Cardiovascular drift is caused by a diminished capacity of the circulation to return blood to the heart (i.e., decreased venous return). Decreased venous return may be due, in turn, to decreased plasma volume caused by filtration of fluid from the blood or by sweating. It may also be due to decreased sympathetic tone.

At rest and during low intensity exercise, heart rate may be elevated above normal. Reasons for this include anxiety, dehydration, high ambient temperature, altitude, or digestion. Resting heart rates of over 90–130 beats · min^{-1} are not unusual before anxiety-producing events such as treadmill testing and sports competitions.

Heart rates are lower during strength exercises such as weight lifting than during endurance exercise such as running. During strength exercises, the heart rate increases in proportion to the muscle mass used. It also increases according to the percentage of maximum voluntary contraction. A dynamic lift, such as a clean and jerk using heavy weights, will increase the heart rate more than will a one-arm biceps curl using minimum resistance.

At the same power output, heart rate is higher during upper-body exercise than lower-body exercise. Upper-body exercise also results in higher $\dot{V}_{O_2}$, mean arterial pressure, and total peripheral resistance. The higher circulatory load in upper-body exercise results from the use of a smaller muscle mass, increased intrathoracic pressure, and a less effective muscle pump. Near maximal contraction using smaller muscle mass restricts blood flow. All three factors decrease venous return of blood to the heart.

Heart rate can be valuable in writing an exercise prescription. By measuring the heart rate during or immediately after exercise, the metabolic cost of the activity can be estimated (see Chapter 33). Exercise heart rate can provide a good estimate of cardiac load. Heart rate multiplied by systolic blood pressure gives the *rate–pressure product (RPP),* a rough index for coronary blood flow myocardial oxygen consumption.

Stroke Volume

Stroke volume increases during exercise in the upright posture (see Figure 16-2). Stroke volume increases steadily until about 25–50% of maximum, then tends to level off. As discussed in Chapter 14, echocardiography measurements suggest that stroke volume decreases as exercise intensity increases toward maximum in some people. This finding is controversial, however, and has not been confirmed by all investigators. If it proves to be true, the probable causes are that high heart rates decrease ventricular filling time and that peripheral shunting of blood to active skeletal muscles decreases central blood volume, which is necessary for maintaining ventricular volume. Heart rate is responsible for increasing cardiac output at higher intensities of exercise. Athletes have a higher exercise cardiac output because they have a higher stroke volume.

The Effects of Upright and Supine Postures

Stroke volume increases during upright exercise. When the subject is supine, however, stroke volume does not change from rest to exercise. At rest, stroke volume is higher in the supine than in the upright posture. Studies using radionuclide angiography have shown that left ventricular end-diastolic volume (EDV) increases during upright exercise but remains unchanged during supine exercise. EDV is the amount of blood in the left ventricle during the resting or diastolic portion of the cardiac cycle. Ejection fraction, the percent of EDV pumped from the heart with each cardiac contraction, increases during upright exercise because of an increased cardiac contractility. When exercise intensity is the same, stroke volume is equal in the supine and erect postures.

Stroke volume is perhaps the most important factor determining individual differences in V_{O_2max}. This is readily apparent when comparing the components of cardiac output of a sedentary man with those of a champion cross-country skier. Both men have maximum heart rates of 185 beats · min⁻¹. Yet, the maximum cardiac output of the untrained man is 16.6 liters · min⁻¹ and of the skier, 32 liters · min⁻¹. The maximum stroke volumes of the skier and the sedentary man are 173 and 90 ml, respectively.

Arteriovenous Oxygen Difference

Arteriovenous oxygen difference increases with exercise intensity (see Figure 16-3). The resting value of about 5.6 vol % (ml O_2 · 100 ml⁻¹) is increased to about 16 vol % at maximal exercise. Some oxygenated blood is always returning to the heart, even at exhaustive levels of exercise. This is because some blood continues to flow through metabolically less active tissues, which do not fully extract the oxygen from the blood. However, oxygen extraction approaches 100% when $(a-v)_{O_2}$ is measured across a maximally exercising muscle.

Blood Pressure

It is very important for blood pressure to increase during exercise. Blood flow must be maintained to critical areas such as the heart and brain, while, at the same time, the requirements of working muscles and skin must also be met. Blood pressure is a function of cardiac output and peripheral resistance (Eq. 16-2).

$$\text{Blood pressure (mmHg)} = \dot{Q} \cdot \text{TPR} \quad (16\text{-}2)$$

where $\dot{Q}$ = Cardiac output (liters · min⁻¹) and TPR = Total peripheral resistance (dyne sec · cm⁻⁵).

Peripheral resistance decreases during exercise because of the tremendous increase in blood flow to working skeletal muscles. However, although vasoconstriction in nonexercising tissue is not enough to compensate for the vasodilation in active muscles, blood pressure does not fall during exercise—it increases. And it does so because cardiac output increases greatly during exercise, more than compensating for the fall in peripheral resistance. For example, in a fit 20-year-old male, cardiac output will increase from about 5 liters at rest to approximately 20 liters during maximal exercise. Even though peripheral resistance may fall to ⅓ of resting during exercise, systolic blood pressure increases.

Systolic blood pressure rises steadily during exercise. It increases from about 120 mmHg at rest to 180 mmHg or more during maximal exercise (Figure 16-4). It follows the same general trend as heart rate. The increase in exercise blood pressure varies between people. Maximum systolic pressure may be as little as 150 to more than 250 mmHg in a normal person. The mean arterial pressure (MAP), described in Equation 16-3, increases from about 90 mmHg at rest to about 155 mmHg at maximal exercise.

$$\text{MAP} = \tfrac{1}{3}(\text{Systolic blood pressure} - \text{Diastolic blood pressure}) + \text{Diastolic blood pressure} \quad (16\text{-}3)$$

Failure of the systolic and mean arterial blood pressures to increase during exercise suggests heart failure. A fall in pressure near the end of an exercise test is particularly dangerous. Falling blood pressure during exercise is an absolute indication for stopping an exercise tolerance test (see Chapter 27).

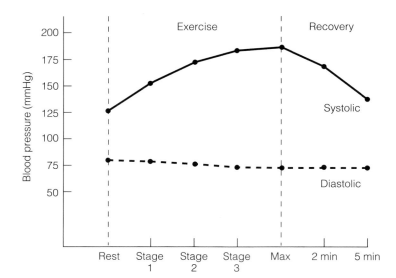

Figure 16-4 Blood pressure response of a 30-year-old man during and after a Bruce treadmill test. Falling systolic blood pressure or excessive increases in systolic or diastolic blood pressure are absolute indications for stopping an exercise test.

There is debate about the maximum safe systolic pressure during exercise. Some experts are comfortable with systolic blood pressures even higher than 250 mmHg; others suggest terminating exercise tests when the systolic blood pressure exceeds 220 mmHg. In this debate, the characteristics of the subject are important and must be considered. A high systolic pressure is very significant in a hypertensive patient but is probably meaningless in a world-class endurance athlete. Endurance athletes have extremely high cardiac output capacities, and because systolic blood pressure increases with cardiac output, their systolic blood pressures will naturally be higher. Systolic blood pressures of greater than 450 mmHg have been reported in weight lifters during exercise, with no ill effects resulting.

High systolic pressure during exercise in untrained people is a matter of concern. The combination of a high heart rate and high systolic blood pressure suggest a high oxygen consumption by the heart. In fact, the rate–pressure product is an excellent predictor of myocardial load. As discussed earlier, rate–pressure product is systolic blood pressure multiplied by heart rate. If a person has heart disease, extreme levels of systolic blood pressure could easily result in myocardial hypoxemia, or insufficient oxygen to the heart.

Diastolic pressure changes little during exercise in normal people. Typically, there is either no change or a slight decrease of less than 10 mmHg during exercise. There is also a small decrease during recovery of less than 4 mmHg. A significant increase in diastolic pressure (>15 mmHg or above 110 mmHg, see Chapters 24 and 27) is associated with a greater prevalence of coronary artery disease.

Blood Flow and "Cardiovascular Triage"

During exercise, blood is redistributed from inactive to active tissues. Critical areas such as the brain and heart are spared the vasoconstriction that occurs in other areas. Sympathetic vasomotor activity increases progressively with the increasing severity of exercise. Because maintenance of blood pressure takes precedence over delivering maximum blood flow to active muscle, even active skeletal muscle is not spared some vasoconstriction during intense exercise.

There appear to be protective mechanisms, which could be called "cardiovascular triage," associated with circulatory regulation that prevent coronary and central nervous system ischemia and maintain central blood volume. During exercise, these cardiovascular control mechanisms limit blood flow to muscles when they cannot simultaneously meet the needs of the heart, central nervous system, working muscles, lungs, and thermoregulatory system. Such limits on blood flow to muscles effectively limit exercise intensity so that a maximal cardiac output is achieved but without the heart and muscles resorting to anaerobic metabolism to sustain exercise.

A number of studies have examined redistribution of blood flow during exercise. Some focus on the circulatory load posed by the work of breathing. Several studies have shown that decreasing the work of breathing during exercise increases muscle blood flow. Two studies by Harms and associates (1977, 1998) showed that reducing the work of breathing during maximal exercise using a proportional-assist ventilator increased leg blood flow while increasing the work of breathing decreased leg blood flow. They concluded that up to a range of 14–16% of the cardiac output during exercise is directed to the respiratory muscles, and that local reflex vasoconstriction significantly compromises blood flow to leg locomotor muscles. This local reflex vasoconstriction is part of the cardiovascular triage. In another study with chronic obstructive pulmonary disease (COPD) patients, administration of morphine reduced the work of breathing during exercise and increased exercise capacity (Santiago, 1979). The findings of these studies are consistent with the concept that critical organs, such as the heart and brain, protect themselves first and allow increased blood flow to tissues when circumstances permit.

Perhaps the best example of cardiovascular triage during exercise is the cardiac output response at high altitude. In the Everest II study (Sutton et al., 1988), a laboratory experiment that simulated a climb of Mount Everest, ventilation increased and peak cardiac output, heart rate, and power output during maximal exercise decreased with increasing altitude (Figure 16-5). Thus, though the oxygen demand of skeletal muscle is a significant factor during exercise at sea level, it is not the most important factor in cardiovascular regulation during exercise at high altitude. Instead, the circulatory system appears to protect the heart by reducing blood flow to the muscles (i.e., limiting exercise) and reducing the work of the heart.

Resting blood flow to the spleen and kidneys is about 2.8 liters · min^{-1}. This amount is reduced to about 500 ml during maximal exercise. The marked reduction in blood flow to these areas is caused by sympathetic stimulation and circulating catecholamines. Sympathetic innervation is particularly strong in these tissues. In the kidneys, the reduction in blood flow during exercise is accompanied by postexercise proteinuria, hemoglobinuria, and myoglobinuria. These terms mean increased protein, hemoglobin, and myoglobin in the urine, respectively. Urine changes are probably due to decreased plasma volume during exercise. Other

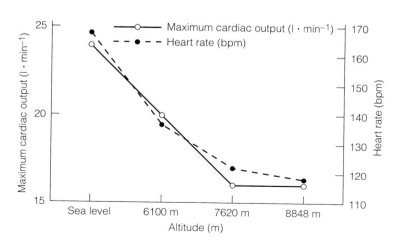

Figure 16-5 Cardiac output and heart rate during maximum exercise decrease at high altitude. The extent that cardiac output increases during exercise is a regulated process in which maximal exercise terminates before either the heart or the skeletal muscles develop a limiting oxygen consumption, becoming "anaerobic." Blood supply to the heart is likely to be an important factor determining when exercise must terminate. SOURCE: Sutton, et al., 1988.

causes include increased permeability of the glomeruli and partial inhibition of reabsorption in the kidney tubules.

Skin blood flow increases during submaximal exercise but decreases to resting values during maximal exercise, when muscle blood flow is highest. Relative skin blood flow, as a percentage of cardiac output, changes very little during exercise. Peripheral perfusion of skin blood vessels during exercise contributes to cardiovascular drift and to fatigue during endurance exercise. This is particularly true when a person exercises in the heat.

The coronary arteries have a large capacity for increasing blood flow. Coronary blood flow increases with intensity during exercise, going from about 260 ml $\cdot$ min^{-1} at rest to 900 ml $\cdot$ min^{-1} at maximal exercise. This large coronary reserve is extremely important, providing adequate blood flow even in the face of significant coronary artery disease. Coronary blood flow is not thought to limit oxygen transport capacity in people without coronary artery disease. As discussed in Chapter 15, coronary blood flow increases occur mainly by metabolic regulation and mainly during diastole. Severe coronary artery disease that impedes blood flow will interfere with control of blood flow and may cause coronary ischemia.

Warming up before endurance exercise is important in facilitating an increase in coronary blood flow during the early stages of exercise. Electrocardiographic changes have occurred in healthy people subjected to sudden strenuous exercise. These changes include ST segment depression, which may indicate coronary ischemia (inadequate coronary bloodflow). Ischemic changes with sudden exercise are usually benign in healthy people, but they could be dangerous in people with heart disease.

■ The Limits of Cardiovascular Performance

Maximal oxygen consumption ($\dot{V}_{O_2max}$) has long been considered the best measure of the capacity of the cardiovascular system and of aerobic exercise performance, provided there is no pulmonary disease. $\dot{V}_{O_2max}$ is the product of maximum cardiac output and maximum arteriovenous oxygen difference (Eq. 16-4):

$$\dot{V}_{O_2max} = \dot{Q}_{max}(a-v)_{O_2max} \qquad (16\text{-}4)$$

where

$\dot{V}_{O_2max}$ = Maximal oxygen uptake, expressed in ml $\cdot$ min^{-1}

$\dot{Q}_{max}$ = Maximal cardiac output, expressed in ml $\cdot$ min^{-1}

$(a-v)_{O_2max}$ = Maximum arteriovenous oxygen difference, expressed in ml $O_2 \cdot$ dl^{-1}

According to theories first developed by A. V. Hill in the 1920s and further advanced by many other scientists, $\dot{V}_{O_2max}$ is the point at which oxygen consumption fails to rise despite an increased exercise intensity or power output. After reaching $\dot{V}_{O_2max}$, a person is able to exercise at a higher intensity using nonoxidative metabolism (Figure 16-6). According to this theory, as the heart reaches its maximum capacity to pump blood (i.e., maximum cardiac output), skeletal muscles must resort to anaerobic metabolism to continue exercising. Maximal cardiac output and muscle anaerobiosis limit $\dot{V}_{O_2max}$. The theory further states that oxygen transport capacity, as measured by $\dot{V}_{O_2max}$ and maximum cardiac output, determines fitness in most kinds of exercise and is the best predictor of performance in endurance sports.

These basic concepts (referred to here as the $\dot{V}_{O_2max}$/anaerobic hypothesis) have been the cornerstones of exercise physiology for nearly 80 years. However, recently they have come under criticism, particularly by Tim Noakes from the University of Cape Town in South Africa (see Noakes 1988, 1991, 1997, 1998). Noakes analyzed the data of the classic studies by Hill and Lupton (1923), Wyndham and associates (1959), Åstrand (1952), and Taylor and associates (1955) that established $\dot{V}_{O_2max}$ as the laboratory benchmark for cardiovascular performance. He found that in most subjects in these studies, $\dot{V}_{O_2}$ did not pleateau (level off) during maximum exercise. Most subjects never satisfied the criteria of be-

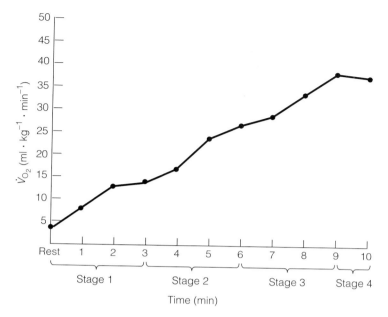

Figure 16-6 $\dot{V}_{O_2}$ (in ml · kg^{-1} · min^{-1}) during a Bruce treadmill test in a 25-year-old woman. It is likely that maximal exercise terminates before either the heart or skeletal muscles develop a limiting oxygen consumption, becoming "anaerobic."

ing able to do more work after oxygen consumption had leveled off.

Why is the presence or absence of a plateau in oxygen consumption and cardiac output during maximal exercise important? Cardiac output capacity is both determined by and dependent on blood flow through the coronary arteries. Contraction of heart muscle, like skeletal muscle, depends on oxygen, which the heart muscle extracts from the coronary blood flow. A maximum in cardiac output implies cardiac fatigue caused by coronary artery ischemia. If the heart typically became ischemic during maximal exercise, then most people would develop angina pectoris as they fatigued. But angina seldom occurs during maximum exercise in people without coronary artery disease.

There are several other problems and inconsistencies associated with the $\dot{V}_{O_2max}$/anaerobic hypothesis:

- Blood transfusions and oxygen breathing have each been shown by some studies to increase exercise performance. These techniques increase oxygen delivery to the tissues, thus having the same effect that increased cardiac

output would have. Some scientists concluded that if oxygen delivery is improved with these techniques, then cardiac output must limit performance. However, none of these studies showed that subjects reached a plateau in $\dot{V}_{O_2}$ during normal exercise. There was no evidence of an oxygen transport limitation before the subjects were given oxygen or blood.

- The hypothesis cannot explain why exercise performance in the heat decreases, even though muscle oxygen and blood supplies and metabolism are normal.

- In blood doping studies, there is a dissociation between changes in $\dot{V}_{O_2max}$ and performance. Performance changes last only a few days while changes in $\dot{V}_{O_2max}$ last longer.

- Dichloroacetate (DCA), a pyruvate dehydrogenase stimulator, increases $\dot{V}_{O_2max}$, maximum power output, and muscle creatine phosphate levels, and reduces exercise blood lactate concentrations. DCA alters performance and metabolism without affecting muscle oxygenation.

- There is a large discrepancy between $\dot{V}_{O_2max}$ and running performance in many elite athletes.

- Exercise at extreme altitudes is not limited by high blood lactate levels or by indications of limitations in cardiac or respiratory function. As discussed, during maximal exercise at increasingly higher altitudes, cardiac output actually decreases. This is strong evidence for protective mechanisms that prevent cardiac, central nervous system, and skeletal muscle damage. From the standpoint of the organism and its physiological control mechanisms, fatigue is preferable to death.

- Exhaustion during maximal exercise occurs at a lower oxygen consumption during cycling than during running in the same subjects. If cardiac output was limiting, then the $\dot{V}_{O_2}$ values should have been equal.

- Blood lactate levels at exhaustion during progressive treadmill exercise testing are lowest in elite athletes.

- Changes in running performance with training occur without equivalent changes in $\dot{V}_{O_2max}$. Changes in $\dot{V}_{O_2max}$ typically occur during the first six months of training. However, running performance can improve for many years after the initiation of training. In elite runners, career-best performances tend to occur after college competition, long after the peak in $\dot{V}_{O_2max}$ was achieved.

- Local muscle factors often appear to be more closely related to fatigue than a limitation in cardiac output. Muscle weakness that occurs during marathon running can persist for a week or more after the event. Decreased contractile capacity in these athletes is accompanied by alterations in the electrocardiogram.

Protection of the Heart and Muscles During Exercise

Noakes (1998) presented an alternative hypothesis that is consistent with the cardiovascular control mechanisms described in Chapter 15. Cardiovascular regulation and muscle recruitment are regulated by neural and chemical control mechanisms that prevent damage to the heart, central nervous system, muscles, and other tissues. In skeletal muscle, severe anaerobiosis would quickly lead to rigor and necrosis. As discussed in Chapter 14, anaerobiosis does not just impair the function of the heart, it destroys it. During maximal exercise, control mechanisms prevent myocardial ischemia that would lead to decreased cardiac output and skeletal muscle anaerobiosis. This is accomplished by regulating force and power output in muscle and controlling tissue blood flow.

Noakes's data suggest that a good predictor of endurance performance is peak treadmill velocity. He hypothesized that maximum treadmill speed may be related to the muscles' capacity for high crossbridge cycling (rapid binding and unbinding of actin and myosin) and respiratory adaptations. Respiratory adaptations may make it possible to prevent the onset of exercise induced dyspnea.

Biochemical factors, such as oxidative enzyme activity and mitochondrial volume, are also better predictors of endurance capacity than $\dot{V}_{O_2max}$. Endurance is the ability to sustain a particular level of physical effort. In a running race, such as the 5000-m run, performance is inversely correlated to $\dot{V}_{O_2max}$. Biochemical factors that allow the maintenance of a particular level of oxygen consumption are important as well.

Using an improper physiological paradigm for exercise training can have negative consequences in training athletes or patients. For example, in the 1970s, some swim coaches suggested that swimmers should run as part of their training because running was better than swimming for increasing maximum cardiac output capacity and $\dot{V}_{O_2max}$. Based on the $\dot{V}_{O_2max}$/anaerobic hypothesis, this made a lot of sense. However, the technique didn't work. Increasing cardiac output through running had little or no transfer to swimming, largely because cardiac output was probably not the limiting factor of swim performance.

The Noakes hypothesis that oxygen transport and muscle blood flow is limited to avoid possible tissue and organ injury during exercise has important practical implications. The practical basis of the Noakes hypothesis is that:

- The primary regulatory mechanism of the cardiorespiratory and neuromuscular systems facilitate intense exercise until it perceives a risk of ischemic injury to the heart, central nervous system, muscles, and other tissues and organs. The body anticipates organ damage and prevents skeletal muscles from overworking.

- Fitness should be attained through programs and techniques that specifically improve muscle power output capacity, substrate utilization (fuel use), and thermoregulatory capacity and that decrease the work of breathing. These adaptations lessen load on the heart and tissues so that the circulatory system allows more intense or prolonged exercise before instigating protective mechanisms that result in fatigue.

- Cardiovascular adaptations occur with training and are beneficial. Improved muscle and respiratory fitness, power output capacity, and improved thermoregulation are accompanied by increased cardiovascular capacity (including maximal cardiac output). The cardiovascular system develops at the same time that other adaptations occur from training.

Dynamics of Oxygen Transport Capacity

Regardless of whether $\dot{V}_{O_2max}$ and cardiac output capacity limit exercise performance, oxygen is critical to tissue function and life itself, and cardiac output and oxygen consumption increase with exercise intensity. Thus, athletes with high exercise capacities also have well-developed oxygen transport systems, and training helps develop these systems.

Cardiac output is the most important factor determining the increase in oxygen consumption during exercise. Cardiac output can increase 20% through endurance training. Elite endurance athletes have been known to have maximum cardiac outputs of more than 35 liters per min—almost

twice that of the average sedentary person. However, maximum arteriovenous oxygen difference, a measure of oxygen extraction, changes very little with training. Also, there is little difference in arteriovenous oxygen difference between endurance athletes and sedentary people.

The oxygen consumption capacity of a muscle varies according to the fiber type. The ability of the mitochondria to extract oxygen from blood is approximately three to five times greater in slow-twitch red fibers than in fast-twitch white fibers. However, training can double the mitochondrial mass, so it is possible for elite endurance athletes to have 10 times the oxygen-extracting capacity in their trained muscles as sedentary people have in their muscles. There is a high correlation (r = 0.80) between $\dot{V}_{O_2max}$ and leg muscle mitochondrial activity.

Criteria for Identifying $\dot{V}_{O_2max}$

While $\dot{V}_{O_2max}$ has its limitations for predicting exercise performance and varies under a wide variety of circumstances (e.g., mode of exercise and altitude), it is still useful for describing the function of the cardiovascular system during exercise. For this reason, it is important to review criteria for identifying $\dot{V}_{O_2max}$. The measurement of maximal oxygen consumption must satisfy several objective criteria:

- The exercise must use at least 50% of the total muscle mass. The exercise must be continuous and rhythmical and must be done for a prolonged period.

- The results must be independent of motivation or skill.

- The subject must reach maximum capacity. Traditionally, this has meant that oxygen consumption must plateau while exercise intensity increases. As discussed, this concept is theoretically problematic. Guidelines for judging if a subject has exercised to maximum capacity are discussed below.

- The measurement must be made under standard experimental conditions, avoiding stressful environments that expose the subject to excessive heat, humidity, air pollution, or altitude.

The mode of exercise is important for measuring $\dot{V}_{O_2max}$. It cannot be measured during upper-body exercise on untrained people. An untrained person fatigues rapidly during this type exercise. Typically, $\dot{V}_{O_2max}$ on an arm ergometer will be 70% of the $\dot{V}_{O_2max}$ measured on a treadmill. However, in a trained rower or canoeist, the difference between upper- and lower-body exercise measurements may be as small as a few percent.

$\dot{V}_{O_2max}$ is often measured on a bicycle ergometer. Most studies show that $\dot{V}_{O_2max}$ measured on a bicycle ergometer is 10 to 15% less than that measured on a treadmill. Cycling skills, muscle power capacity, and body weight affect the results.

There are several practical guidelines for judging if a subject has exercised to maximum capacity. Perceived exertion is one excellent means of determining relative exercise intensity. Perceived exertion is a rating scale of relative fatigue (see Chapter 27). A high correlation exists between the perception of exhaustion and $\dot{V}_{O_2max}$. There are also objective physiological measurements. These include a respiratory exchange ratio ($R = \dot{V}_{O_2}/\dot{V}_{CO}$ of at least 1.1, a blood lactate level greater than 8 mM, a peak heart rate similar to an age-predicted maximum, or progressively diminishing differences between successive oxygen consumption measurements.

Performance Efficiency

Oxygen consumption increases with exercise intensity. Individual differences in performance efficiency can be responsible for the difference between winning and losing in athletic competition. The measurement of efficiency was discussed in detail in Chapter 3. Efficiency is decreased by energy lost as heat, wasted movement, and mechanical factors, including wind resistance, friction, and drag. The efficiency of walking and cycle ergometry is slightly less than 30%. The efficiencies of running, cycling swimming, and cross-country skiing at competitive exercise intensities are probably less than that.

High-intensity exercise is not performed at a steady rate. $\dot{V}_{O_2}$ does not account for all of the ATP supplied during exercise; a portion is also supplied through anaerobic glycolysis. Consequently, efficiency cannot be accurately calculated even when power output can be measured. In fact, during high intensity exercise the oxygen requirement (Chapter 10) is often underestimated.

The relative change in performance efficiency can be estimated by measuring changes in oxygen consumption under different conditions—such as under different wind resistance, with various mechanical aids (e.g., toe clips in cycling and wax in cross-country skiing), and with different techniques. A fundamental problem in measuring relative change is determining how much of the efficiency is due to mechanical factors (i.e., technique and equipment) and how much is due to physiological factors (i.e., mitochondrial respiratory capacity). For example, one runner looks more efficient than another—but it is difficult to identify whether the greater efficiency is due to a more efficient running style or to a superior mitochondrial respiratory capacity.

Running efficiency has not been shown to be a good predictor of performance. At present, the effect of running efficiency on performance is not well understood. The most promising methods for improving running efficiency may be manipulating ventilation, changing body composition, improving training status, increasing power output, and improving running style.

Technique is as important as metabolic capacity for performance. In swimming, athletes should develop good hydrodynamics. They should use strokes that employ efficient propulsive force and minimize drag. Doing so may contribute almost as much to success as improving the physiological aspects of endurance. Likewise, the frequent use of skating in cross-country skiing has revolutionized the sport. Efficient runners are thought to have a lower vertical component in their technique. Efficient cyclists pedal smoothly at relatively high revolutions per minute without engaging muscle groups that do not contribute to pedaling speed. Wind resistance is also a factor in running and cycling. It is reduced by wearing clothing that enhances aerodynamics.

$\dot{V}_{O_2max}$ as Predictor of Endurance Performance

If $\dot{V}_{O_2max}$ were the only predictor of endurance performance, then endurance contests could be decided in the laboratory. Research scientists could administer treadmill tests, and the person with the highest $\dot{V}_{O_2max}$ would be the winner. This might be easier and more precise than conducting athletic contests on the track, road, or swimming pool. However, $\dot{V}_{O_2max}$ is a poor predictor of success in endurance events. Indeed, as discussed, a high $\dot{V}_{O_2max}$ may coincide with high capacity muscle characteristics that actually limit performance.

In a sample of people with a wide range of fitness levels, those with a high $\dot{V}_{O_2max}$ tended to run faster in such endurance events as the marathon. This relationship does not exist though when the sample is homogeneous (i.e., the runners are of the same fitness level). For example, Grete Waitz and Derek Clayton, two former elite marathon runners, had $\dot{V}_{O_2max}$ values of 73 and 69 ml · kg^{-1} · min^{-1}, respectively. These values were measured shortly after they set world records for the women's and men's marathons. Yet, Clayton's time was over 15 minutes faster than Waitz's. Thus, in addition to a high $\dot{V}_{O_2max}$, other factors important for success in certain endurance events include speed, the ability to continue exercising at a high percentage of $\dot{V}_{O_2max}$, lactic acid clearance capacity, and performance economy.

In general, a high $\dot{V}_{O_2max}$ is regarded as a prerequisite to performing at elite levels in endurance events. The minimum values for elite female and male endurance athletes are approximately 65 and 70 ml · kg^{-1} · min^{-1}, respectively, for runners and cross-country skiers. Appropriate values for swimmers are 55 to 60 ml · kg^{-1} · min^{-1} for women and 65 to 70 ml · kg^{-1} · min^{-1} for men. Circumstantial evidence indicates that there is also a minimum aerobic capacity requirement for elite performance in endurance events. What these high $\dot{V}_{O_2max}$ values really reflect is the ability of elite athletes to exercise at high intensities before the system limits itself; and such high oxygen transport capacities are matched

by other adaptations in the cardiovascular system of elite athletes.

■ Changes in Cardiovascular Parameters with Training

Endurance training results in adaptive changes in many aspects of cardiovascular function. These changes are summarized in Tables 16-1 and 16-2. The heart improves its ability to pump blood, mainly by increasing its stroke volume, which occurs because of an increase in end-diastolic volume and a small increase in left ventricular mass. These changes are induced by the increased volume load placed on the heart during endurance exercise. In contrast, strength exercises, such as weight lifting, subject the heart to a pressure load. This results in larger increases in left ventricular mass. There is little or no change in ventricular volume. Endurance exercise also decreases the metabolic load on the heart at rest and at any submaximal exercise intensity. It does so by increasing stroke volume and decreasing heart rate. The result is a more efficient pressure–time relationship.

Endurance (aerobic) exercise is best for improving the capacity of the cardiovascular system. These exercises require the use of at least 50% of the body's muscle mass in rhythmical exercise. Exercise sessions should last for at least 15 to 20 minutes and should take place 3 to 5 days a week. Intensity should be above 50 to 60% of $\dot{V}_{O_2max}$. Obviously, athletes must train harder than this for high-level performance.

Adaptation to endurance training is specific. Swimming, for example, will improve cardiovascular performance in swimming. It will do little to improve endurance in running. This principle is very important in multi-endurance events, such as triathlons. Cycling, running, and swimming must all be practiced if one is to be a successful triathlete.

Interval training is repeated bouts of short- to moderate-duration exercise. It is used to improve speed and cardiovascular conditioning. This mode of training manipulates distance, speed, repetition, and rest. Interval training allows training at a higher

TABLE 16-2

Cardiovascular Adaptations Resulting from Endurance Training[a]

Factor	Rest	Submaximal Exercise	Maximal Exercise
Heart rate	↓	↓	0
Stroke volume	↑	↑	↑
$(a–v)_{O_2}$ (arteriovenous oxygen difference)	0↑	↑	↑
Cardiac output	0↓	0↓	↑
$\dot{V}_{O_2}$	0	0	↑
Work capacity	———	———	↑
Systolic blood pressure	0↓	0↓	0
Diastolic blood pressure	0↓	0↓	0
Mean arterial blood pressure	0↓	0↓	0
Total peripheral resistance	0	0↓	0
Coronary blood flow	↓	↓	↑
Brain blood flow	0	0	0
Visceral blood flow	0	↑	0
Inactive muscle blood flow	0	0	0
Active muscle blood flow	↓0	↓0	↑
Skin blood flow	0	0↑	0
Blood volume	↑	———	———
Plasma volume	↑	———	———
Red cell mass	0↑	———	———
Heart volume	↑	———	———

[a] Symbols:
↑ increase
↓ decrease
0 no change
——— not applicable

intensity than that typically used during competition. It acts as an overload. However, although interval training is excellent for developing cardiovascular capacity, it is less effective in eliciting the biochemical changes critical for optimal endurance performance. To improve at the fastest rate, the endurance athlete must practice interval and over-distance training.

Oxygen Consumption

Improvement in maximal oxygen consumption with training depends on current fitness, type of training, and age. Intense endurance training results in a maximum increase in $\dot{V}_{O_2max}$ of approximately 20%. However, greater increases are possible if the initial physical fitness of the subject is low. Only certain types of exercise promote the cardiac alterations necessary for increased $\dot{V}_{O_2max}$. Maximal stroke volume can be increased in response to a volume overload in sports such as running, cycling, and swimming. In pressure-overload sports, such as weight lifting, left ventricular wall thickness increases, with no increase in left ventricular volume.

Changes in $\dot{V}_{O_2max}$ and endurance capacity are not the same. Endurance performance can be improved by much more than 20%. This is possible by improving mitochondrial density, speed, running economy, and body composition.

An athlete must be endowed with a relatively high capacity oxygen transport system to be successful in endurance events. This is not to say that training is not also extremely important. Rather, there is only a limited opportunity to improve aerobic capacity. Factors other than $\dot{V}_{O_2max}$ are important for success in endurance events. These include speed, mitochondrial density, body composition, and resistance to fatigue (Figure 16-7).

Heart Rate

Endurance training reduces resting and submaximal exercise heart rate. Training increases parasympathetic tone to the SA node from the vagus nerve. Training may also decrease the intrinsic rhythmicity of the heart. Resting heart rates of less than 40 beats · min^{-1} are not unusual in elite endurance athletes. There is probably a strong genetic influence in some low heart rates seen in endurance athletes, but aerobic training is responsible for most of it.

A low resting heart rate is not always a sign of physical fitness. On the contrary, bradycardia (low heart rate) is sometimes a sign of disease. For example, bradycardia is characteristic of sick sinus syndrome. The important sign of fitness is the reduction in resting heart rate with training rather than the low heart rate itself.

Training decreases the submaximal exercise heart rate although the metabolic requirement of the exercise remains the same. The decline in heart rate does not reduce cardiac output because stroke volume increases.

Training has only a small effect on maximal heart rate. It usually decreases it by about 3 beats · min^{-1}. Maximum heart rate decreases slowly with age. Because maximum heart rate is stable, it is a useful reference point for determining the relative intensity of exercise. For example, a person has a

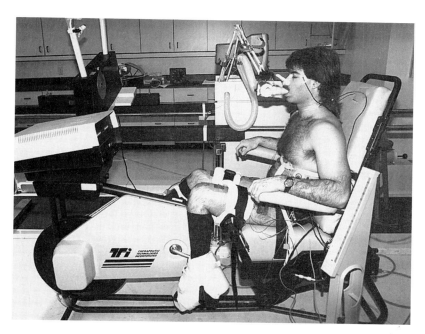

Figure 16-7 Computerized functional electrical stimulation (CFES) leg cycle ergometer exercise being performed by a spinal cord injured person. Physiological monitoring is being conducted to determine metabolic and cardiopulmonary responses. Spinal cord injured athletes often have low values of maximal heart rate in spite of impressive performance adaptations. PHOTO: Courtesy Roger M. Glaser, Ph.D., Institute for Rehabilitation Research and Medicine at Wright State University.

heart rate of 140 beats · min^{-1} while running a 10-min mile. The person returns after 6 months of endurance training and runs another 10-min mile. This time the heart rate is only 130 beats · min^{-1}. The exercise intensity no longer represents the same percentage of maximum heart rate (which remains the same). Because heart rate decreases at an identical submaximal exercise intensity, running speed has to be increased to achieve the same relative intensity of exercise.

Stroke Volume

Endurance training increases stroke volume at rest and during submaximal and maximal exercise. Most studies have found that training can increase stroke volume by no more than about 20%. This is the same percentage change as seen in $\dot{V}_{O_2max}$. The mechanism behind the very high stroke volumes in elite endurance athletes is not known.

Increased stroke volume is due to increased heart volume and contractility. There is a high relationship between heart size and stroke volume. Decreased heart rate increases stroke volume. Diastole is longer when heart rate is slower. Consequently, the heart has longer to fill up between contractions. The heart ejects the extra blood due to the Frank–Starling mechanism (see Chapter 14). End-diastolic volume also increases with training due to increased left ventricular volume, blood volume, and left ventricular compliance. Increased blood volume causes more blood flow to the heart. Increased left ventricular compliance allows the heart walls to stretch more in response to increased venous return of blood.

Myocardial contractility also increases with training. The probable mechanism here is an increase in the release and transport of calcium from the sarcoplasmic reticulum. There is also an increase in the more contractile isoform of myosin ATPase. End-systolic volume is similar in trained and untrained people. End-diastolic volume is greater in trained people, so the ejection fraction is also greater. This results in a greater stroke volume in endurance-trained people.

Arteriovenous Oxygen Difference

Arteriovenous oxygen difference, $(a-v)_{O_2}$, increases slightly with training. Possible mechanisms for the change include a rightward shift of the oxyhemoglobin dissociation curve, mitochondrial adaptations, elevated hemoglobin and myoglobin concentrations, and increased muscle capillary density. The rightward shift of the oxyhemoglobin dissociation curve loosens the binding to oxygen and hemoglobin. This makes it easier for red blood cells to release oxygen to the cell. Mitochondrial adaptations speed reactions involving oxygen. Elevated hemoglobin and myoglobin speed transport of oxygen to the mitochondria. Increased capillary density results in a shorter diffusion distance between the circulation and muscle mitochondria.

Circulatory and biochemical changes with training are linked. The number of capillaries around each fiber is a good predictor of aerobic power. It is related to the mitochondrial content of the fiber. Increased capillarization around muscle fibers is thought to facilitate diffusion during exercise. The adaptation in $(a-v)_{O_2}$ does not occur to the same extent in older people.

Blood Pressure

Endurance training reduces resting and submaximal exercise systolic, diastolic, and mean arterial blood pressures. Diastolic and mean arterial blood pressures are reduced at maximal exercise. Training has no effect on maximum systolic blood pressure. However, weight loss, which may occur with endurance training, tends to reduce exercise blood pressure. The mechanism of reduced blood pressure at rest is not known. Exercise training is important in reducing the risk of coronary heart disease. However, in some patients with hypertension (high blood pressure), exercise training may increase resting blood pressure.

Blood Flow

With training, coronary blood flow decreases slightly at rest and during submaximal exercise. The increased stroke volume and decreased heart rate reduce myocardial oxygen consumption, which decreases the heart's requirements for blood. Coronary blood flow increases at maximal exercise with training. Increased blood flow supports the metabolic requirement of a higher cardiac output. Heart blood flow increases with the metabolic load. Training does not increase myocardial vascularity in the normal heart.

Skeletal muscle vascularity (blood vessels in skeletal muscles) increases with endurance training. This facilitates diffusion of oxygen, substrates, and metabolites in exercising muscles. Increased vascularity also decreases peripheral resistance (resistance to blood flow in the circulation). Decreased peripheral resistance is one of the mechanisms that increases cardiac output with endurance training.

Training decreases muscle blood flow during submaximal exercise. The trained muscle has an increased oxygen extraction capacity due to improved diffusion capability and muscle respiratory capacity. Decreased muscle blood flow allows more blood to be directed to the viscera and skin. During maximal exercise, muscle blood flow increases by about 10%. Maximal muscle blood flow increases due to greater cardiac output and muscle vascularity.

The volume of skin blood flow is unaffected by endurance training. The onset of vasodilation skin vessels occurs earlier. As discussed in Chapter 22, training also causes an earlier onset of sweating and an increased sweat rate. These changes aid temperature regulation, enhancing cooling during endurance exercise.

■ Summary

During the transition from rest to exercise, cardiovascular activity accelerates. Cardiac output and blood flow to active muscles and the heart increases. Blood flow to the brain is maintained or increased slightly, and blood vessels in the viscera and inactive muscles vasoconstrict. These changes maintain blood pressure while delivering blood to active areas and the skin.

Successful endurance athletes have a high cardiorespiratory capacity. The benchmark of the cardiorespiratory system's capacity is maximal oxygen consumption ($\dot{V}_{O_2max}$), the maximum rate at which oxygen can be taken up and used by the body during exercise. $\dot{V}_{O_2max}$ predicts performance over short distances (e.g., 1–2 miles), but beyond that mechanisms that help the body resist fatigue begin to play a more important role. Many people with a high $\dot{V}_{O_2max}$ perform poorly because they lack fatigue resistance.

While maximum cardiac output may be the limiting factor for oxygen transport capacity, this must be a regulated process in which maximal exercise terminates before either the heart or the skeletal muscles develop a limiting oxygen consumption, becoming "anaerobic." The blood supply to the heart is likely to be an important factor in determining when exercise must terminate.

Endurance training is best for improving the capacity of the cardiovascular system. Training results in increased maximum cardiac output, stroke volume, and blood volume. Resting and submaximal exercise heart rates decrease with endurance training.

■ Selected Readings

Abe, D., K. Yanagawa, K. Yamanobe, and K. Tamura. Assessment of middle-distance running performance in sub-elite young runners using energy cost of running. *Eur. J. Appl. Physiol.* 77: 320–325, 1998.

Ameredes, B. T., W. F. Brechue, and W. N. Stainsby. Mechanical and metabolic determination of $\dot{V}_{O_2}$ and fatigue during repetitive isometric contractions in situ. *J. Appl. Physiol.* 84: 1909–1916, 1998.

Åstrand, P.-O. Experimental Studies of Physical Work Capacity in Relation to Sex and Age. Copenhagen: Munksgaard, 1952.

Åstrand, P. O., U. Bergh, and A. Kilbom. A 33-yr follow-up of peak oxygen uptake and related variables of former physical education students. *J. Appl. Physiol.* 82: 1844–1852, 1997.

Bailey, S. P., and R. R. Pate. Feasibility of improving running economy. *Sports Med.* 12: 228–236, 1991.

Bassett, D. R., Jr., and E. T. Howley. Maximal oxygen uptake: "classical" versus "contemporary" viewpoints. *Med. Sci. Sports Exerc.* 29: 591–603, 1997.

Behncke, H. Optimization models for the force and energy in competitive running. *J. Math. Biol.* 35: 375–390, 1997.

Bhambhani, Y., S. Buckley, and T. Susaki. Muscle oxygenation trends during constant work rate cycle exercise in men and women. *Med. Sci. Sports Exerc.* 31: 90–98, 1999.

Billat, V. L., R. Richard, V. M. Binsse, J. P. Koralsztein, and P. Haouzi. The $\dot{V}(O_2)$ slow component for severe exercise depends on type of exercise and is not correlated with time to fatigue. *J. Appl. Physiol.* 85: 2118–2124, 1998.

Blomstrand, E., G. Radegran, and B. Saltin. Maximum rate of oxygen uptake by human skeletal muscle in relation to maximal activities of enzymes in the Krebs cycle. *J. Physiol.* (London) 501: 455–460, 1997.

Bogdanis, G. C., M. E. Nevill, H. K. Lakomy, and L. H. Boobis. Power output and muscle metabolism during and following recovery from 10 and 20 s of maximal sprint exercise in humans. *Acta Physiol. Scand.* 163: 261–272, 1998.

Boning, D. Altitude and hypoxia training—a short review. *Int. J. Sports Med.* 18: 565–570, 1997.

Booth, F. W., and K. A. Narahara. Vastus lateralis cytochrome oxidase activity and its relationship to maximal oxygen consumption in man. *Pflügers Arch.* 349: 319–326, 1974.

Bouchard, C., F. T. Dionne, J. A. Simoneau, and M. R. Boulay. Genetics of aerobic and anaerobic performances. In Exercise and Sport Sciences Reviews, J. O. Holloszy (Ed.). Baltimore: Williams and Wilkins, 1992, pp. 27–58.

Capelli, C., D. R. Pendergast, and B. Termin. Energetics of swimming at maximal speeds in humans. *Eur. J. Appl. Physiol.* 78: 385–393, 1998.

Carter, R., B. Nicotra, and G. Huber. Differing effects of airway obstruction on physical work capacity and ventilation in men and women with COPD. *Chest* 106: 1730–1739, 1994.

Connent, R. J., and C. R. Honig. Regulation of $\dot{V}_{O_2max}$ in red muscle: do current biochemical hypotheses fit in vivo data? *Amer. J. Physiol.* 256: R898–R906, 1989.

Cooke, R., and E. Pate. The inhibition of muscle contraction by the products of ATP hydrolysis. In Biochemistry of Exercise, A. W. Taylor et al. (Eds.). Champaign, Ill.: Human Kinetics, 1990, pp. 59–72.

Daniels, J. T. A physiologist's view of running economy. *Med. Sci. Sports Exerc.* 17: 332–338, 1985.

Dempsey, J. A., P. Hanson, and K. Henderson. Exercise-induced arterial hypoxemia in healthy humans at sea level. *J. Physiol.* (London) 355: 161–175, 1984.

Dempsey, J. A., P. E. Hanson, and S. M. Mastenbrook. Arterial hypoxemia during heavy exercise in highly trained runners. *Fed. Proc.* 40: 93, 1981.

Dempsey, J. A., C. A. Harms, and D. M. Ainsworth. Respiratory muscle perfusion and energetics during exercise. *Med. Sci. Sports Exerc.* 28: 1123–1128, 1996.

Evertsen, F., J. I. Medbo, E. Jebens, and K. Nicolaysen. Hard training for 5 mo increases Na^+-K^+ pump concentration in skeletal muscle of cross-country skiers. *Am. J. Physiol.* 272: R1417–R1424, 1997.

Gayeski, T. E. J., R. J. Connett, and C. R. Honig. Minimum intracellular PO_2 for maximum cytochrome turnover in red muscle in situ. *Advanced Exper. Med. Biol.* 200: 487–494, 1987.

Goldsmith, R. L., J. T. Bigger, Jr., D. M. Bloomfield, and R. C. Steinman. Physical fitness as a determinant of vagal modulation. *Med. Sci. Sports Exerc.* 29: 812–817, 1997.

Green, J. J., and A. E. Patla. Maximal aerobic power: neuromuscular and metabolic considerations. *Med. Sci. Sports Exerc.* 24: 38–46, 1992.

Grucza, R., J. Smorawiński, G. Cybulski, W. Niewiadomski, J-F. Kahn, B. Kapitaniak, and H. Monod. Cardiovascular response to static handgrip in trained and untrained men. *Eur. J. Appl. Physiol.* 62: 337–341, 1991.

Harms, C. A., M. A. Babcock, S. R. McClaran, D. F. Pegelow, G. A. Nickele, W. B. Nelson, and J. A. Dempsey. Respiratory muscle work compromises leg blood flow during maximal exercise. *J. Appl. Physiol.* 82: 1573–1583, 1997.

Harms, C. A., T. J. Wetter, S. R. McClaran, D. F. Pegelow, G. A. Nickele, W. B. Nelson, P. Hanson, and J. A. Dempsey. Effects of respiratory muscle work on cardiac output and its distribution during maximal exercise. *J. Appl. Physiol.* 85: 609–618, 1998.

Hauber, C., R. L. Sharp, W. D. Franke. Heart rate response to submaximal and maximal workloads during running and swimming. *Int. J. Sports Med.* 18: 347–353, 1997.

Haußwirth, C., A. X. Bigard, and C. Y. Guezennec. Relationships between running mechanics and energy cost of running at the end of a triathlon and a marathon. *Int. J. Sports Med.* 18: 330–339, 1997.

Helge, J. W., and B. Kiens. Muscle enzyme activity in humans: role of substrate availability and training. *Am. J. Physiol.* 272: R1620–R1624, 1997.

Hill, A. V., and H. Lupton. Muscular exercise, lactic acid, and the supply and utilization of oxygen. *Q. J. Med.* 16: 135–171, 1923.

Hogan, M. C., J. Roca, J. B. West, and P. D. Wagner. Dissociation of maximal O_2 uptake from O_2 delivery in canine gastrocnemius in situ. *J. Appl. Physiol.* 66: 1219–1226, 1989.

Honig, C. R., R. J. Connett, and T. E. J. Gayeski. O_2 transport and its interaction with metabolism: a systems view of aerobic capacity. *Med. Sci. Sports Exerc.* 24: 47–53, 1992.

Howley, E. T., D. R. Bassett, Jr., and H. G. Welch. Criteria for maximal oxygen uptake: review and commentary. *Med. Sci. Sports Exerc.* 27: 1292–1301, 1995.

Ivy, J. L., D. L. Costill, and B. D. Maxwell. Skeletal muscle determinants of maximal aerobic power in man. *Eur. J. Appl. Physiol.* 44: 1, 1980.

Jensen, J., S. T. Jacobsen, S. Hetland, and P. Tveit. Effect of combined endurance, strength and sprint training on maximal oxygen uptake, isometric strength and sprint performance in female elite handball players during a season. *Int. J. Sports Med.* 18: 354–358, 1997.

Jones, A. M. A five-year physiological case study of an Olympic runner. *Br. J. Sports Med.* 32: 39–43, 1998.

Jones, D. P., F. G. Kennedy, and T. Yee. Intracellular O_2 gradients. In Hypoxia: The Tolerable Limits, J. Sutton (Ed.). Indianapolis: Benchmark Press, 1988, pp. 59–75.

Koskolou, M. D., J. A. Calbet, G. Radegran, and R. C. Roach. Hypoxia and the cardiovascular response to dynamic knee-extensor exercise. *Am. J. Physiol.* 272: H2655–H2663, 1997.

Loudon, J. K., P. E. Cagle, S. F. Figoni, K. L. Nau, and R. M. Klein. A submaximal all-extremity exercise test to predict maximal oxygen consumption. *Med. Sci. Sports Exerc.* 30: 1299–1303, 1998.

Maxwell, A. J., E. Schauble, D. Bernstein, and J. P. Cooke. Limb blood flow during exercise is dependent on nitric oxide. *Circulation* 98: 369–374, 1998.

McClaran, S. R., C. A. Harms, D. F. Pegelow, and J. A. Dempsey. Smaller lungs in women affect exercise hyperpnea. *J. Appl. Physiol.* 84: 1872–1881, 1998.

McConell, T. R. Practical considerations in the testing of $\dot{V}_{O_2max}$ in runners. *Sports Med.* 5: 57–68, 1988.

McKenna, M. J., G. J. Heigenhauser, R. S. McKelvie, G. Obminski, J. D. MacDougall, and N. L. Jones. Enhanced pulmonary and active skeletal muscle gas exchange during intense exercise after sprint training in men. *J. Physiol.* (London) 501: 703–716, 1997.

McMahon, S., and H. A. Wenger. The relationship between aerobic fitness and both power output and subsequent recovery during maximal intermittent exercise. *J. Sci. Med. Sport* 1: 219–227, 1998.

Myers, J., D. Walsh, M. Sullivan, and V. Froelicher. Effect of sampling on variability and plateau in oxygen uptake. *J. Appl. Physiol.* 68: 404–410, 1990.

Nielsen, H. B., P. Madsen, L. B. Svendsen, R. C. Roach, and N. H. Secher. The influence of PaO_2, pH and SaO_2 on maximal oxygen uptake. *Acta Physiol. Scand.* 164: 89–97, 1998.

Noakes, T. D. Challenging beliefs: ex Africa semper aliquid novi. *Med. Sci. Sports Exerc.* 29: 571–590, 1997.

Noakes, T. D. Implications of exercise testing for prediction of athletic performance: a contemporary perspective. *Med. Sci. Sports Exerc.* 20: 319–330, 1988.

Noakes, T. D. Lore of Running, 3d Ed. Champaign, Ill.: Leisure Press, 1991.

Noakes, T. D. Maximal oxygen uptake: "classical" versus "contemporary" viewpoints: a rebuttal. *Med. Sci. Sports Exerc.* 30: 1381–1398, 1998.

Oelberg, D. A., R. M. Kacmarek, P. P. Pappagianopoulos, L. C. Ginns, and D. M. Systrom. Ventilatory and cardiovascular responses to inspired He-O_2 during exercise in chronic obstructive pulmonary disease. *Am. J. Respir. Crit. Care Med.* 158: 1876–1882, 1998.

Oshima, Y., T. Miyamoto, S. Tanaka, T. Wadazumi, N. Kurihara, and S. Fujimoto. Relationship between isocapnic buffering and maximal aerobic capacity in athletes. *Eur. J. Appl. Physiol.* 76: 409–414, 1997.

Paavolainen, L. M., A. T. Nummela, and H. K. Rusko. Neuromusclar characteristics and muscle power as determinants of 5-km running performance. *Med. Sci. Sports Exerc.* 31: 124–130, 1999.

Paterson, D. H., and D. A. Cunningham. The gas transporting systems: limits and modifications with age and training. *Can. J. Appl. Physiol.* 24: 28–40, 1999.

Pollock, M. L., L. J. Mengelkoch, J. E. Graves, D. T. Lowenthal, M. C. Limacher, C. Foster, and J. H. Wilmore. Twenty-year follow-up of aerobic power and body composition of older track athletes. *J. Appl. Physiol.* 82: 1508–1516, 1997.

Poole, D. C., and R. S. Richardson. Determinants of oxygen uptake. Implications for exercise testing. *Sports Med.* 24: 308–320, 1997.

Poole, P. J., A. G. Veale, and P. N. Black. The effect of sustained-release morphine on breathlessness and quality of life in severe chronic obstructive pulmonary disease. *Am. J. Respir. Crit. Care Med.* 157: 1877–1880, 1998.

Proctor, D. N., and M. J. Joyner. Skeletal muscle mass and the reduction of $\dot{V}_{O_2max}$ in trained older subjects. *J. Appl. Physiol.* 82: 1411–1415, 1997.

Richardson, R. S. Oxygen transport: air to muscle cell. *Med. Sci. Sports Exerc.* 30: 53–59, 1998.

Richardson, R. S., K. Tagore, L. J. Haseler, M. Jordan, and P. D. Wagner. Increased $\dot{V}_{O_2max}$ with right-shifted $Hb-O_2$ dissociation curve at a constant O_2 delivery in dog muscle in situ. *J. Appl. Physiol.* 84: 995–1002, 1998.

Rodas, G., M. Calvo, A. Estruch, E. Garrido, G. Ercilla, A. Arcas, R. Segura, and J. L. Ventura. Heritability of running economy: a study made on twin brothers. *Eur. J. Appl. Physiol.* 77: 511–516, 1998.

Rowell, L. B., B. Saltin, B. Kiens, and N. J. Christensen. Is peak quadriceps blood flow in humans even higher during exercise with hypoxemia? *Amer. J. Physiol.* 251: H1038–H1044, 1986.

Rowland, T. W., E. L. Melanson, B. E. Popowski, and L. C. Ferrone. Test-retest reproducibility of maximum cardiac output by Doppler echocardiography. *Am. J. Cardiol.* 81: 1228–1230, 1998.

Saltin, B. Cardiovascular and pulmonary adaptation to physical activity. In Exercise, Fitness, and Health, C. Bouchar, R. J. Shephard, T. Stephens, J. R. Sutton, and B. D. McPherson (Eds.). Champaign, Ill.: Human Kinetics, 1990, pp. 187–203.

Saltin, B. Hemodynamic adaptations to exercise. *Amer. J. Cardiol.* 55: 42D–47D, 1985.

Saltin, B., and S. Strange. Maximal oxygen uptake: "old" and "new" arguments for a cardiovascular limitation. *Med. Sci. Sports Exerc.* 24: 30–37, 1992.

Santiago, T. V. and N. H. Edelman. Opioids and breathing. *J. Appl. Physiol.* 59: 1675–1685, 1985.

Santiago, T. V., J. Johnson, D. J. Riley, and N. H. Edelman. Effects of morphine on ventilatory response to exercise. *J. Appl. Physiol.* 47: 112–118, 1979.

Shephard, R. J. Tests of maximum oxygen intake: a critical review. *Sports Med.* 1: 99–124, 1984.

Stainsby, W. N., W. F. Brechue, D. M. O'Drobinak, and J. K. Barclay. Oxidation/reduction state of cytochrome oxidase during repetitive contractions. *J. Appl. Physiol.* 67: 2158–2162, 1989.

Sutton, J. R., J. T. Reeves, P. D. Wagner, B. M. Groves, A. Cymerman, M. K. Malconian, P. B. Rock, P. M. Young, S. D. Walter, and C. S. Houston. Operation Everest II: oxygen transport during exercise at extreme simulated altitude. *J. Appl. Physiol.* 64: 1309–1321, 1988.

Tabata, I., K. Irisawa, M. Kouzaki, K. Nishimura, F. Ogita, and M. Miyachi. Metabolic profile of high intensity intermittent exercises. *Med. Sci. Sports Exerc.* 390–395, 1997.

Uusitalo, A. L., A. J. Uusitalo, and H. K. Rusko. Endurance training, overtraining and baroreflex sensitivity in female athletes. *Clin. Physiol.* 18: 510–520, 1998.

Uusitalo, A. L., A. J. Uusitalo, and H. K. Rusko. Exhaustive endurance training for 6–9 weeks did not induce changes in intrinsic heart rate and cardiac autonomic modulation in female athletes. *Int. J. Sports Med.* 19: 532–540, 1998.

Vandewalle, H., J. F. Vautier, M. Kachouri, J. M. Lechevalier, and H. Monod. Work-exhaustion time relationships and the critical power concept: a critical review. *J. Sports Med. Phys. Fitness* 37: 89–102, 1997.

Verstappen, F. T., J. W. Veldhuizen, M. Twellaar, M. R. Drost, and H. Kuipers. Isokinetic aerobic power output testing of the quadriceps muscle. *Int. J. Sports Med.* 19: 485–489, 1998.

Wagner, P. D. Central and peripheral aspects of oxygen transport and adaptations with exercise. *Sports Med.* 11: 133–142, 1991.

Wagner, P. D. Insensitivity of $\dot{V}_{O_2max}$ to hemoglobin-P50 at sea level and altitude. *Respir. Physiol.* 107: 205–212, 1997.

Warren, G. L., and K. J. Cureton. Modeling the effect of alterations in hemoglobin concentration on $\dot{V}_{O_2max}$. *Med. Sci. Sports Exerc.* 21: 526–531, 1989.

Wolfel, E. E., M. A. Selland, A. Cymerman, G. A. Brooks, G. E. Butterfield, R. S. Mazzeo, R. F. Grover, and J. T. Reeves. O_2 extraction maintains O_2 uptake during submaximal exercise with beta-adrenergic blockade at 4,300 m. *J. Appl. Physiol.* 85: 1092–1102, 1998.

Wyndham, C. H., N. B. Strydom, J. S. Maritz, J. F. Morrison, J. Peter, and Z. U. Potgieter. Maximal oxygen intake and maximum heart rate during strenuous work. *J. Appl. Physiol.* 14: 927–936, 1959.

CHAPTER 17

SKELETAL MUSCLE STRUCTURE AND CONTRACTILE PROPERTIES

There are 660 skeletal muscles in the adult human being. These muscles constitute approximately 45% of body weight. Based simply on its large relative mass, it is clear that skeletal muscle is not only the largest organ system in the body, but it is also an important tissue for bioenergetic homeostasis during rest and exercise. Skeletal muscle is both the major site of energy transduction and a major site of energy storage. Moreover, skeletal muscle is the end organ for the primary support systems involved in exercise, such as cardiovascular and pulmonary. The bioenergetic and support system requirements for skeletal muscle are described elsewhere in this book. The purpose of this chapter is to describe muscle from the perspective of its primary purpose as a motor.

Purposeful human movement is the defining characteristic in arts, sports, occupational and domestic tasks, and recreational activities. Appropriate skeletal muscle behavior is the central thesis of human biodynamics, as it provides for self-propulsion, the ventilation of lungs, the ability to mate and produce offspring, the capacity to interact socially, and, in the animal kingdom, to capture prey, avoid predators, and migrate. The physical activity patterns of a given individual arise principally from the properties of skeletal muscle fibers, the joints through which fibers act, and the coordination and

Canadian gymnast Lori Strong in World Cup competition. Performances such as these are dependent upon the integrated function of nerves, muscles, spinal cord, brain, and a variety of sensory receptors. PHOTO: Claus Andersen.

integration of recruitment patterns of motor units in the involved musculature.

With an overall organization similar to that of chapters on the heart and cardiovascular system (Chapters 14–16), chapter 17 discusses the details of skeletal muscle structure and the fundamental contractile properties of muscle. The basic information presented in this chapter allows the student to more fully understand how muscle functions in partnership with the nervous system, as developed in Chapter 18. Chapter 19 builds on this material, by presenting the principles by which muscles adapt to different types of training, including variations due to age and gender.

■ Skeletal Muscle Functional Anatomy

Organizational Hierarchy

The structure of skeletal muscle can be considered from a hierarchical perspective (Figure 17-1). A typical skeletal *muscle* is composed of muscle *fascicles*, which in turn are composed of muscle *fibers*. Because a fiber is a single multinucleated cell, the terms *cell* and *fiber* are used interchangeably. Each fiber is composed of *myofibrils* arranged in a three dimensional mosaic pattern (Figure 17-1). Within this structure, the sarcomeres form the basic contractile units of the fiber; they are composed of interdigitating thick and thin *myofilaments*.

This hierarchical organization of muscle is achieved with several connective tissue membranes. The connective tissue that surrounds an entire muscle is called the *epimysium* (i.e., on top of muscle); the membrane that binds groups of fibers into fascicles is called the *perimysium* (i.e., around muscle). Two separate membranes surround the interior milieu of individual muscle fibers (Figure 17-2). The outermost membrane is most frequently referred to as the *basement membrane*. The basement membrane is not a membrane in the usual sense. Rather than having the normal structure of a lipid bilayer, the basement membrane is a loose collection of glycoproteins and collagen network. It is freely permeable to proteins, solutes, and other metabolites. An additional thin elastic membrane is found just beneath the basement membrane and is termed the *plasma membrane* or *sarcolemma*. The plasma membrane is, in fact, the true cell boundary.

At a muscle fiber's resting length, the plasma membrane has the morphological characteristic of small indentations termed *caveolae*. The caveolae provide additional length during fiber stretching. Muscle fiber lengths extend approximately 10 to 15% during normal physical activity. The caveolae allow this lengthening to occur without damaging the plasma membrane. Compared to the basement membrane, the plasma membrane is much more selective to ions, solutes, and substrates crossing it. An intact plasma membrane is of critical importance to cell function. This membrane maintains the proper acid–base balance of the fiber, allowing it to contract repeatedly during exercise. The membrane is involved in propagating an action potential that will lead to muscle contraction. The membrane also transports metabolites from the blood in the capillaries to the center cytosol of the muscle fiber. Many metabolites that can be used as a fuel for exercise, such as lactic acid and glucose, are transported by specific transporter mechanisms. Also, the plasma membrane has, at the neuromuscular junction, an even more elaborate region of functional folds than elsewhere along the fiber. This elaboration helps the transmission of an action potential from the nerve to the muscle fiber.

Wedged between the basement membrane and plasma membrane are cells known as satellite cells, which are discussed further below. Interior to the plasma membrane is the *cytoplasm* (i.e., cytosol), which is rich in soluble proteins, myofilaments, and true myonuclei, as well as stored high-energy intermediates (ATP, PC), substrates (glycogen and lipids), enzymes of metabolism, mitochondrial protein, ribosomes for protein synthesis, and so on.

True myonuclei and *satellite cells* are of fundamental importance for the growth and development of muscle, for the adaptive capacity of skeletal muscle to various forms of training or disuse, and for the recovery from exercise-induced or traumatic injury. When muscle fibers are viewed longitudinally with a light microscope, the nuclear material appears to be located along the peripheral edge of

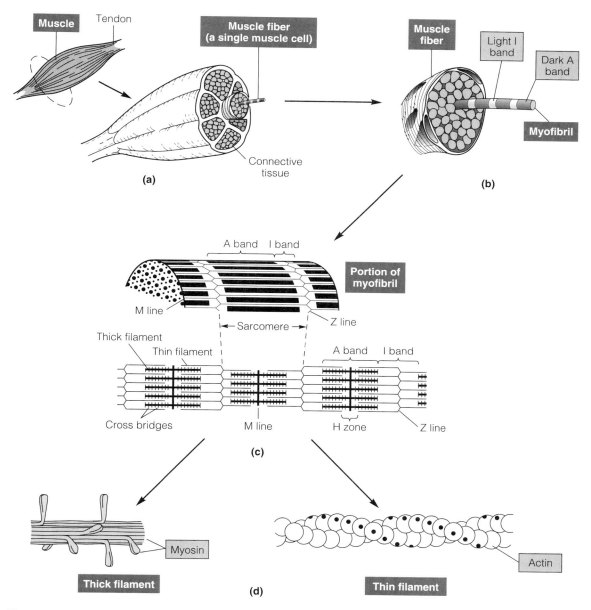

Figure 17-1   Levels of organization of a skeletal muscle. (a) Enlargement of a cross section of a whole muscle. (b) Enlargement of a myofibril within a muscle fiber. (c) Cytoskeletal component of a myofibril. (d) Protein components of a thick and thin filament. Modified from Sherwood, *Human Physiology*, 1997, p. 223. Used with permission.

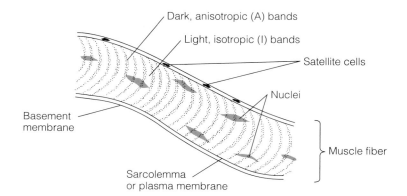

Figure 17-2 Skeletal muscle fibers have a striated (striped) appearance due to the banding of thick and thin myofilaments.

the fibers (see Figure 17-2). However, with an electron microscope one can see that, depending on the age and muscle type, 85 to 95% of the nuclear material is true myonuclei (i.e., located inside the plasma membrane) and the balance (5 to 15%) is satellite cells (i.e., located between the basal lamina and plasma membrane). There are approximately 200–300 nuclei per millimeter of fiber length. This is in contrast to many other cells in the human body, which have a single nucleus. Muscle fiber lengths range from a few millimeters in intraocular muscles of the eye to over 45 cm in the sartorius muscle. Thus, there is tremendous potential for altered gene expression in response to exercise, neuroendocrines, nutrition, and other factors. This potential for gene alteration leads, in turn, to alterations in structural proteins, regulatory proteins, and transport and metabolic proteins, and has an impact on the muscle's functional capacity. This adaptive process is described in Chapter 19 in the section on the principle of myoplasticity.

Alexander Mauro discovered the satellite cell and named it based on its peripheral location (Figure 17-3). From a functional point of view, these cells are adult myoblasts. Satellite cells are normally dormant, but under conditions of mechanical stress or injury they are essential for the compensatory growth of existing fibers and for the growth of regenerating fibers. Satellite cells have chemotaxic properties, which means they migrate from one location to another area of higher need within a muscle fiber, and then undergo the normal pro-

cess of developing a new muscle fiber. Under certain conditions, muscles have the capacity to generate new fibers. This process of new fiber formation begins with satellite cells entering a mitotic phase to produce additional satellite cells. These cells then migrate across the plasma membrane into the cytosol, where they recognize each other, align, and fuse into a myotube. A myotube is an immature form of a muscle fiber. The multinucleated myotube will then differentiate into a mature fiber. In case of serious injury or neuromuscular disease, satellite cells are, as noted, critical for the regenerative growth of new muscle fibers and entire new muscles. In the case of exercise training, where there is hypertrophy of existing cells, satellite cells play an important role. Recent research has demonstrated that satellite cells are activated when muscle fibers hypertrophy in experimental models of functional loading (stretch). It remains to be determined how much this occurs when muscle fibers hypertrophy to resistance training, but it seems very likely that satellite cell activation is one important mechanism of exercise-induced muscle growth.

Myotendinous Junction

The ends of muscle fibers connect to tendons at the *myotendinous junction,* thereby allowing the force generated by muscle fibers to be applied through the tendons to the bones to produce the desired movement. These junctions have a complex specialized structure, as seen in Figure 17-4. Jim Tidball

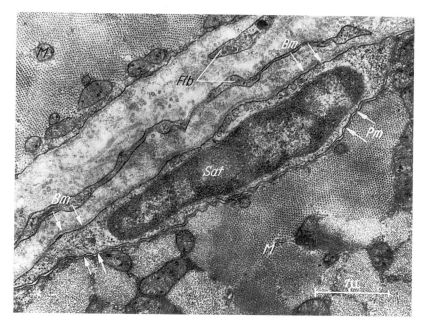

Figure 17-3 An electron micrograph of a mature skeletal muscle from a ten-year-old boy. A satellite cell (*Sat*) is visible between the muscle (*M*) fiber's plasma membrane (*Pm*) and the basement membrane (*Bm*). Fibroblasts (*Fib*) are visible in the extracellular space. Bar = 7 μm. SOURCE: H. Ishikawa, 1966. Used with permission.

(1991) has identified that in some musculoskeletal injuries, such as those associated with exercise, lesions occur at the myotendinous junction, perhaps more so than at other areas of the muscle fiber–tendon complex.

Cell Dimensions

Individual myofibrils are approximately 1 μm in diameter and comprise approximately 80% of the volume of a whole muscle. The number of myofibrils is a regulated variable during the hypertrophy of muscle fibers associated with growth. The number of myofibrils ranges from approximately 50 per muscle fiber in the muscles of a fetus to approximately 2000 per fiber in the muscles of an untrained adult. The hypertrophy and atrophy of adult mammalian muscle associated with certain types of training and disuse, respectively, also results from the regulation of the number of myofi-

brils per fiber. However, training and disuse have negligible effects on the number of fibers that are expressed in the muscles of mammals.

In adult human beings, the cross-sectional area of an individual muscle fiber ranges from approximately 2000 μm^2 to 7500 μm^2 with the mean and median in the 3000–4000 μm^2 range. Muscle fiber length and muscle length itself vary considerably in human beings. For example, the length of the medial gastrocnemius muscle is approximately 250 mm, with fiber lengths of 35 mm. On the other hand, the sartorius muscle is approximately 500 mm long, with fiber lengths of 450 mm. The number of fibers can range from several hundred in small muscles to over a million in large muscles, such as those involved in hip flexion and knee extension. The task of integrating the coordinated contraction and relaxation of this many individual cells during standing, running, walking, and other activities underscores the complexity of this remarkable system.

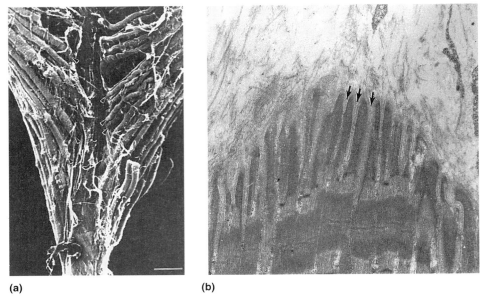

(a) (b)

Figure 17-4 Panel (a) is a scanning electron micrograph of a myotendinous junction (MTJ) of frog semitendinosus muscle. The epimysium has been dissected, allowing for visualization of the MTJs. The tendon lies vertically in the center of the micrograph and extends downward. Long cylindrical muscle cells insert on the tendon from either side. Several of the MTJs are indicated by arrowheads, and the scale bar = 250 μm. In panel (b), a transmission electron micrograph of a MTJ is shown that is magnified approximately 130-fold compared to panel (a) (for scale, note the A-bands are 1.6 μm wide). Visible at the lower portion of the micrograph are terminal sarcomeres, whose terminal Z-discs are bunched as they extend upward to the folded cell membrane. The MTJ of the muscle cell is folded and is visible in the mid-portion of the micrograph, where three folds are identified by arrows. The folded muscle cell appears to interdigitate with tendon, which is visible in the upper portion of this panel. SOURCES: (a) from Tidball, 1991; (b) courtesy of James Tidball, UCLA.

The radius of muscle cells, typically from 25 to 50 μm, is an important variable for sustained muscular performance, as it affects the diffusion distance from the capillary network (which is exterior to the muscle cell) to the cell's interior. As the radius of muscle cells increases, the distance through which gasses, such as oxygen, must travel to diffuse from the capillary blood to the center of the muscle cell increases. This can be a problem, limiting the muscle's ability to sustain endurance exercise, for sufficient oxygen delivery is needed for the mitochondria, where the majority of energy for muscle contraction is produced. Of equal importance to the delivery of oxygen is the delivery of metabolic fuels

like glucose, as well as the removal of waste products such as carbon dioxide gas. These events also occur by diffusion. Thus, for a variety of reasons, large diffusion distances are much less compatible with sustained performance than the smaller diffusion distances typical of smaller cells.

Sarcomeres

When muscle is viewed with a light microscope, one can see the cross striations of the myofibrils. These striations are called the birefringent properties of the myofibrils and are visible with a microscope when light is shined through the muscle cell (see

Figure 17-2). The birefringent properties provide insight into how sarcomeres contract and gives rise to the reason skeletal muscles are often termed striated muscle. In the mid-portion of the sarcomere, the area that appears dark is termed anisotropic and hence is known as the *A-band* (Figure 17-5). Areas at the outer ends of each sarcomere appear much lighter and are known as *I-bands* because they are isotropic with respect to their birefringent properties. The *H-zone* (from *helle,* German for light) is in the central region of the A-band, where there is no thick and thin filament overlap. The H-zone is bisected by the *M line,* which is composed of proteins that keep the sarcomere in proper spatial orientation as it lengthens and shortens. At the ends of each sarcomere are the *Z-disks* (from *zwischen*—German for *between;* also termed Z lines or Z-baskets). The I-band and H-zone are less dense than other areas as there is no overlap of thick and thin filaments; this allows for greater penetration of light when examined with a microscope. The sarcomere length is the distance from one Z-disk to the next; optimal sarcomere length in mammalian muscle is 2.4 to 2.5 μm. The length of a sarcomere relative to its optimal length is of fundamental importance to the capacity for force generation, as presented in greater detail later in this chapter.

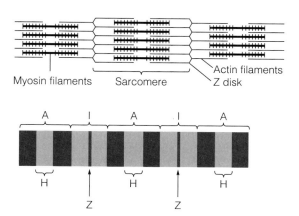

Figure 17-5 Dark (anisotropic, A) bands in striated muscle are due to the overlap of thick (myosin) and thin (actin) myofilaments. Light (isotropic, I) bands are due to a lesser density and greater penetration of light.

Contractile and Regulatory Proteins

Overall Protein Composition In mammals, the overall composition of skeletal muscle is approximately 75% water; 5% inorganic salts, pigments, and substrates; and 20% a mixture of proteins. Of the 20% mixed proteins, 12% is myofibrillar proteins, and the other 8% consists of enzymes, membrane proteins, transport channels, and other proteins. The myofibrillar protein fraction of skeletal muscle is primarily composed of proteins involved directly in contraction (i.e., contractile proteins) and others involved in the regulation of contraction (i.e., regulatory proteins). Other important structural proteins are found in much smaller proportion than the contractile and regulatory proteins.

The myofibrillar fraction of skeletal muscle is made up of several specific types of protein. The most prominent is *myosin,* which constitutes approximately one-half of the total myofibrillar protein. The other major contractile protein, *actin,* comprises about one-fifth of the myofibrillar protein fraction. Other myofibrillar proteins include the regulatory proteins *tropomyosin* and *troponin.* The way in which the contractile and regulatory proteins interact together to produce muscle contraction is described later in this chapter in the section on the steric block model.

In addition to the contractile and regulatory proteins, several other essential structural proteins are present in very small amounts (Figure 17-6). The most important include: *C protein* (part of the thick filament, it is involved in holding the tails of myosin in a correct spatial agreement); *titin* (links the end of the thick filament to the Z-disk); *M-line protein* (i.e., myomesin; functions to keep the thick and thin filaments in their correct spatial arrangement); *α-actinin* (attaches actin filaments together at the Z-disk); *desmin* (links Z-disks of adjacent myofibrils together); and *spectrin* and *dystrophin* (have structural and perhaps functional roles as sarcolemmal membrane proteins).

Thick Filament. The thick filament is made principally of the protein myosin (with the remainder being C protein). It is important to understand

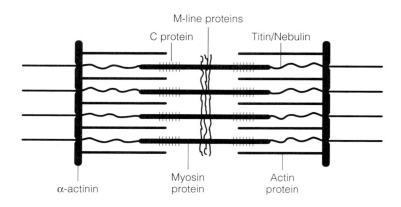

Figure 17-6 Structural proteins in the sarcomere. C protein is found exclusively in the thick filament, where it holds the myosin tails in proper orientation. M-line protein holds the thick filament in the proper horizontal arrangement. Titin and nebulin attach at the Z-disk and at either end of the thick filament to keep it centered in the sarcomere. From Jones and Round, 1990, p. 6.

some of the details of myosin, as this protein is of key importance for the development of muscular force and velocity of contraction. Myosin is a hexameric (six-component) molecule consisting of one pair of heavy chains (MHCs) of high molecular weight (200 kilodaltons) and two pairs of light chains (LCs) of low molecular weight (16–28 kilodaltons). The MHC component exists as a multigene family in which there are several different genes specifically encoded for distinct MHC protein molecules called *isoforms.* These MHC isoforms have slight variations in their amino acid composition, particularly in the head region of the molecule (S-1 subfragment), where energy transduction processes occur during the crossbridge cycle (Figure 17-7a and b). In most mammalian limb muscles, at least four adult MHC isoforms have been identified, and they are designated as slow type I (also called β in the heart), fast type IIa, fast type IIx (IId), and fast type IIb, in order of their increasing ATPase activity. Available evidence suggests that these MHC isoforms provide different energy transduction kinetics and crossbridge turnover rates during contraction.

In most muscles, two pairs of LCs (Figure 17-7c) are associated with the MHCs. One pair, LC1 and LC3, is called the *essential LC.* The other pair, two chains of LC_2, is called the regulatory LC. These LCs are encoded by specific LC genes; and in combination with alternate splicing techniques several LC isoforms can be expressed. The essential and regulatory LCs in combination with the MHC are

thought to modulate the contractile response by enhancing myosin-actin interactions during the contraction process.

Muscle fibers are organized into functional units called *motor units* (e.g., a motor neuron innervating hundreds of fibers sharing common structural and functional properties). As discussed in more detail in Chapter 19, muscle fibers comprising the so-called type I or slow-motor unit express chiefly the type I MHC; and these fibers are designed for both antigravity function and low-intensity repetitive activities. The fast IIa motor unit appears to be comprised of chiefly fibers expressing either IIa or IIx MHC (including hybrid combinations of the two) and these units are recruited during higher intensity movement activities that can be sustained for long durations. Finally, motor units that are used exclusively for brief explosive activities (such as sprinting and jumping) are designated as the IIb motor unit and these fibers express both the IIb and IIx MHC in varying proportions to one another.

There are approximately 300 molecules of myosin in one thick filament. Approximately one-half of the MHCs are aligned at one end of the thick filament; the other half has their HMM toward the opposite end of the thick filament—a tail-to-tail arrangement (Figure 17-7). When molecules are properly aligned, they are rotated 60° relative to one another, and they are offset slightly in the longitudinal plane. As a consequence of these three-dimensional structural factors, a thick filament has a character-

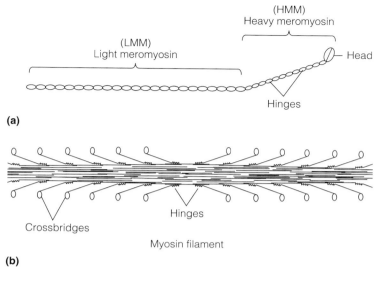

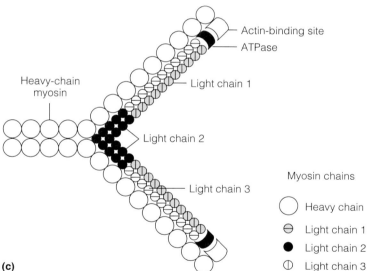

Figure 17-7   Myosin heavy and light chains. (a) Myosin heavy chain (MHC) is composed of light meromyosin (LMM) and heavy meromyosin (HMM). LMM provides a connecting link to other LMM filaments to form the thick filament. (b) The HMM fragments stick out at regular intervals along the thick filament. The head of heavy meromyosin contains binding sites for actin and ATP. Myosin in fast-contracting muscle has a higher ATPase activity than does myosin in slow muscle. (c) These differences in ATPase activity are due to differences in the light- and heavy-chain composition in the myosin head.

istic "bottlebrush" appearance, with HMM projecting out along most of the thick filament.

Thin Filament   A thin filament is made up of approximately 350 monomers of the contractile protein actin and 50 molecules each of the regulatory proteins tropomyosin and troponin (Figures 17-8 and 17-9). The actin monomers are termed G-actin because they are *g*lobular and have molecular weights of approximately 42 kD. G-actin is normally polymerized to F-actin (i.e., *f*ilamentous actin), which is arranged in a double helix. Think of a necklace: Each G-actin is an individual pearl, and when several G-actins polymerize into F-actin, a string of pearls is formed. The polymerization from G- to F-actin involves the hydrolysis of ATP and the

Figure 17-8 Filamentous (F) actin is a helix analogous to a string of pearls. F-actin is composed of individual globular (G) actin monomers. Each G-actin monomer, analogous to an individual pearl, contains a myosin head binding site. Tropomyosin and troponin (not shown), together with F-actin, form the thin filament (see Figure 17-9).

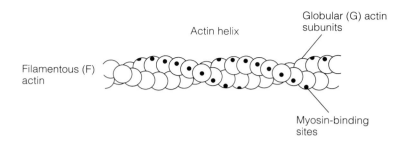

Figure 17-9 The binding of Ca²⁺ to troponin causes a conformational (shape) change in that protein and a movement of the tropomyosin helix. The tropomyosin shift causes the myosin-binding sites on actin to become unmasked. Thus, Ca²⁺ binding to troponin is essential for muscle contraction and indirectly allows for actin and myosin to interact.

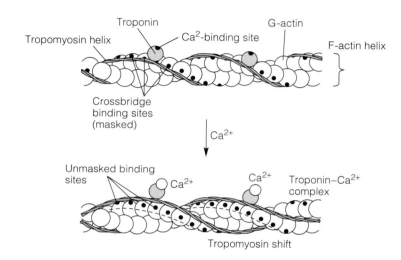

binding of ADP to actin. Indeed, 90% of ADP in skeletal muscle is bound to actin.

The actin protein has a binding site that, when exposed, attaches to the myosin crossbridge. The subsequent cycling of crossbridges causes the development of muscular force. Filamentous actin is arranged in an α-helical arrangement, thereby creating a groove along the thin filament's length. In this groove lies the regulatory protein *tropomyosin* (T_m), which, under resting conditions, blocks the binding site on actin for myosin (Figure 17-9). A T_m molecule is itself a long, ropelike protein that extends over the entire length of F-actin. Also found at intervals along the thin filament, spaced at every seventh G-actin, is the regulatory protein *troponin*. The interaction of the contractile and regulatory proteins is described in the following section on the steric block model.

One end of each actin filament is attached at the interface between two sarcomeres in the area of the Z-disk (Figure 17-10). Alpha-actinin is the protein that holds the actin filaments in the three-dimensional array that forms the Z line. Because four α-actinins are associated with each actin fiber, the structure of the Z-disk appears woven like a basket (Figure 17-11).

Steric Block Model Troponin is made up of three subunits, each having a specialized function (Figure 17-12). The subunits work together to move tropomyosin away from the active binding sites on actin. This process consists of a conformational change that permits actin and the myosin crossbridge to interact, thus allowing for the generation of force. The leading conceptual theory on conformational change is the steric block model (see Figure 17-9). In resting muscle, tropomyosin covers the myosin-binding site on actin. *Troponin-T* ($T_n T$,

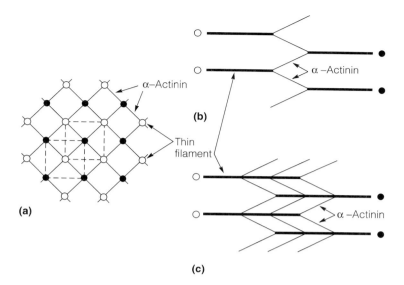

Figure 17-10 Actin filaments joining to form a Z line. (a) View of the Z line along the axis of the thin filaments, showing the square array. Solid circles indicate thin filaments coming out of the page (or Z line); open circles indicate thin filaments going into the page. (b) View from the side; note that only the α-actinin connections in the plane of the page are shown; for every thin filament there will be another two connections in a plane at right angles to the page. View (b) represents a simple Z-line structure found in fast muscle. View (c) represents a more complex Z line found in slow skeletal and cardiac muscle. SOURCE: Jones and Round, 1990.

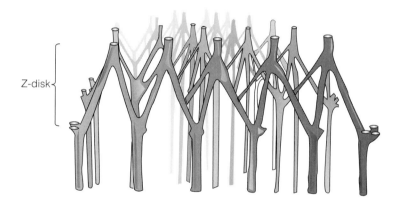

Figure 17-11 Diagram showing the interconnection of thin filaments at the Z-disk. From Carlson and Wilkie, 1979.

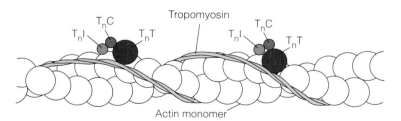

Figure 17-12 Part of an actin filament, together with tropomyosin and the troponin subunits. The thin filament containing F-actin, tropomyosin, and the three troponin subunits. Troponin T (T_nT) binds tropomyosin, troponin C (T_nC) binds Ca^{2+} ions, and troponin I (T_nI) is the inhibitory subunit. The entire troponin complex is found every seven G-actin monomers on the thin filament. SOURCE: Jones and Round, 1990.

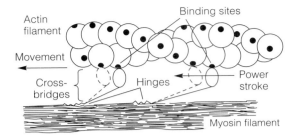

Figure 17-13 Muscle contraction (shortening) occurs when crossbridges extend from myosin to actin and a conformational change occurs in the crossbridge. Modified from H. E. Huxley, 1958.

the tropomyosin-binding subunit) binds loosely onto tropomyosin to prevent it from moving off of actin. *Troponin-I* (T_nI, the inhibitory subunit) also positions tropomyosin on the binding site. When the intracellular calcium concentration rises to a critical level, four molecules of calcium bind to *troponin-C* (T_nC, the calcium-binding subunit), thereby causing the entire three-subunit configuration to change. T_nT then binds tightly to tropomyosin, and the entire troponin protein physically moves the tropomyosin to expose the myosin-binding sites on actin. Once the myosin-binding site is exposed, the S-1 subfragment of myosin is permitted to insert and this creates the crossbridge attachment (Figure 17-13).

◼ Sliding Filament Theory and the Crossbridge Cycle

In 1954, H. E. Huxley and A. F. Huxley, working independently, were instrumental in establishing the sliding filament theory of muscle contraction. As elaborated below, this theory explains how the thick and thin filaments, which themselves are of fixed length, move in relation to each other, resulting in a change in sarcomere and, therefore, muscle length. A. F. Huxley further proposed, in 1957, a theory of crossbridge behavior. His work was very insightful, and since that time it has been elaborated upon and refined by many scientists.

During a shortening contraction, the overlapping and interdigitating thick and thin filaments in each sarcomere move past each other, propelled by the crossbridge cycle. A crossbridge is the HMM portion of myosin heavy chain (i.e., the S-1 and S-2 portions). This portion of the thick filament projects out from the myosin tail and attaches to an actin monomer in the thin filament. The crossbridges then function as ratchets, forcing the thin filaments to slide toward the M line and causing a small amount of sarcomere shortening. This process repeats many times along the length of the fiber, as long as the force-generating mechanism is activated, thereby producing a large overall movement due to the shortening of each sarcomere arranged in series. The process is conceptually similar to the oars of a rowboat. Connecting at one end with the water, the oars are pulled repetitively, thereby causing the boat to move relative to the water.

A crossbridge cycle is the sequence of events from the time a crossbridge first attaches to actin to the time when it binds again and repeats the process. Each crossbridge undergoes four steps, and it does so independent of other crossbridges in that sarcomere. Thus, at any given moment approximately one-half of the total crossbridges are attached to actin and producing force, although the fraction varies depending on functional needs.

The biochemical and physical events that occur in each step of the crossbridge cycle are depicted in Figure 17-14. At the conclusion of the previous crossbridge cycle, the ATP that was bound to myosin (M) is hydrolyzed to an intermediate state (ADP·Pi), thereby energizing the myosin (M*). The resultant ADP and inorganic phosphate (P_i) continue to be bound to the energized myosin. When the muscle fiber is next activated for contraction by the increase in intracellular calcium, the energized myosin attaches to an actin (A) molecule in the thin filament (step 1):

$$A + M^* \cdot ADP \cdot P_i \xrightarrow{\text{Actin binding}} A \cdot M^* \cdot ADP \cdot P_i$$

The attachment of the energized myosin to actin releases the stored energy and results in movement of the bound crossbridge. The ADP and P_i are re-

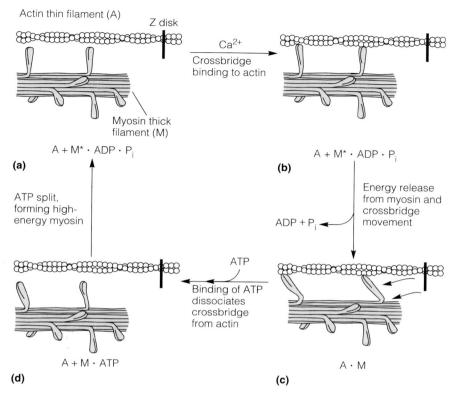

Figure 17-14 The cyclic process of muscle contraction and relaxation. In the resting state (a), actin (A) and energized myosin (M* · ADP · P$_i$) cannot interact because of the effect of tropomyosin. Upon the release of Ca^{2+} from the sarcoplasmic reticulum, M* · ADP · P$_i$ binds to actin (b). Tension is developed, and movement occurs with the release of ADP and P$_i$ (c). Dissociation of actin and myosin requires the presence of ATP to bind to myosin and to pump Ca^{2+} into the SR (d). Myosin is energized upon return to the resting state (a). From Vander, Sherman, and Luciano, 1980, p. 218. Used with permission.

leased from myosin during the crossbridge movement (step 2):

$$A \cdot M^* \cdot ADP \cdot P_i \xrightarrow[\text{movement}]{\text{Crossbridge}} A \cdot M + ADP + P_i$$

Myosin is attached very strongly to actin during the crossbridge movement. This attachment must be broken in order to allow the crossbridge to reattach to a new actin monomer so it can repeat the cycle. The linkage between actin and myosin is broken when ATP binds to myosin (step 3):

$$A \cdot M + ATP \xrightarrow[\substack{\text{Crossbridge dissociation} \\ \text{from actin}}]{} A + M \cdot ATP$$

Following the dissociation of actin and myosin, the ATP bound to myosin is hydrolyzed. The free energy from ATP hydrolysis is bound to the myosin, thereby re-forming an energized myosin that is poised to repeat the crossbridge cycle (step 4):

$$A + M \cdot ATP \xrightarrow[\text{ATP hydrolysis}]{} A + M^* \cdot ADP \cdot P_i$$

The crossbridge cycle continues by repeating steps 1 through 4 as long as calcium remains bound to troponin, thereby maintaining exposed attachment sites on actin for myosin.

■ Capillarity and the Microvascular Unit

In most physical activities, a degree of muscular endurance is required for proper performance. This is the case for life's daily activities, such as walking or riding a bicycle to your exercise physiology course. It is also the case in occupational circumstances, such as for assembly-line workers in an automobile plant and in the forearms and shoulders of concert pianists. It is also the case, obviously, in most individual and team sports and dance.

Important metabolic, bioenergetic, and cardiovascular factors, discussed in greater detail in earlier chapters of this book, contribute to muscular endurance. However, also important to sustained muscular performance are the concepts of *capillarity* and the *microvascular unit*. Often overlooked, these concepts are, nevertheless, of vital importance to exercise physiology. The perfusion of skeletal muscle fibers by blood circulating in the capillaries allows for the delivery and removal of gasses and nutrients that are essential for muscular endurance.

Capillarity

As presented in Chapter 15 on the circulation, the systemic circulation branches from large arteries to the smallest vessels, the one-cell-thick capillaries. These capillaries intertwine with individual skeletal muscle fibers, as shown in Figure 17-15. In the microcirculatory system, arterial blood enters the beginning of the capillary. As the blood slowly travels through the capillary, gasses and metabolites diffuse from the blood into the muscle cell, and vice versa. Thus, by the time blood reaches the end of a given capillary, it is considered to be venous blood, as it has given up much of its oxygen and substrates. The venous blood has also picked up carbon dioxide and other products from the muscle cell. Because the exchange of material from capillary blood

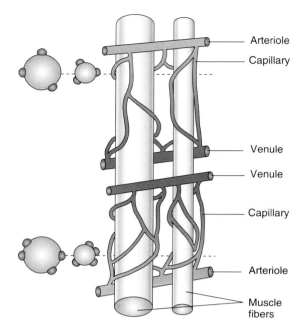

Figure 17-15 Drawing of the microcirculation of two muscle fibers. The fibers are positioned vertically. To the left is the view taken in cross section. Notice that in the upper microcirculatory bed, three capillaries are associated with each fiber; in the lower bed, there are four.

to the muscle cell occurs by the process of diffusion, the greater the difference in concentration— of gas, for example—the greater will be the actual exchange of gas. Also, the diffusion of material between capillary blood and the fiber's cytosol is improved when blood remains in the capillary for longer, rather than shorter, periods of time, that is, longer "transit times" enhance the exchange of nutrients and gas between blood and muscle cell.

Most studies of capillarity examine muscle in a cross-sectional plane. This approach is very common, but it limits interpretation somewhat. To make these measurements, one obtains muscle tissue from a needle biopsy or, in the case of experimental animals, an open biopsy. The muscle sample is then cut perpendicular to the long axis of its fibers—that is, a transverse section is cut. Very thin cross sections of muscle are then incubated for an enzyme reaction in the capillary endothelium, typically capillary membrane phosphatase. The incubation reaction makes the enzyme, which is normally invisible to the eye, visible under a light microscope by producing a dark color where the enzyme is located. This technique allows the number of capillaries in the muscle to be seen and counted, using a light microscope. The technique of measuring capillary membrane phosphatase activity reveals the structural presence of capillaries, whether or not their lumen is open at the time the muscle section is taken. As a result, this technique is generally considered superior to other techniques, which rely on perfusion by blood or other substances, such as a vital dye. These latter approaches allow the capillary to be seen only if it is perfused at the time the section is taken. Because the blood flow in muscle can change over twofold, depending on the body's physiological state, perfusion-based measures of structure are subject to errors in interpretation.

As noted in Figure 17-16, there are three common measures of muscle capillarity: (1) number of adjacent capillaries per fiber, (2) capillaries per square millimeter of cross-sectional fiber area, and (3) the capillary-to-muscle-fiber ratio. An insightful student of exercise physiology will realize that a shortcoming of the first measure is that a given single capillary will be counted two or three times (as it lies adjacent to two or three muscle fibers), thereby providing a misleading high total number of capillaries. The second measure, *capillary density,* is affected directly by the diameter of the fibers in the muscle, which can lead to errors in interpreting experiments. For example, with training-induced cellular hypertrophy, the ratio of capillary per square millimeter decreases, and with cellular atrophy, the ratio increases. With fiber hypertrophy or atrophy, capillarity per se may seem to be affected, but in fact it is not; rather, the measured ratio is altered only because of changes in cell diameter. The third common index of capillarity, the *capillary-to-muscle-fiber ratio,* is least sensitive to erroneous interpretations because it is unaffected by fiber diameter and the sharing of capillaries among adjacent fibers.

Now that we have presented some of the issues involved in interpreting capillarity data, we can go on to ask what are the normal values for cap-

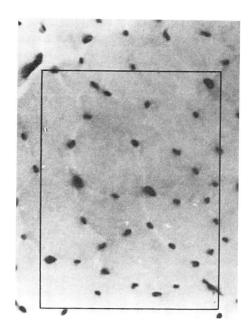

Figure 17-16   Photomicrograph of a cross-section of a cat soleus muscle. The section has been incubated for capillary endothelial phosphatase activity. The box illustrates a sample site, in which different measurements of capillarity can be made: (1) Number of adjacent capillaries per fiber; (2) capillaries per mm^2; and (3) capillary to fiber ratio (i.e., capillaries per mm^2 divided by fibers per mm^2). Magnification × 165. SOURCE: Maxwell, et al., 1980. Used with permission.

illarity, can they change with certain types of exercise training, and what functional importance might training-induced adaptations in capillarity have?

In an analysis of muscles from many members of the mammalian kingdom (including human, monkey, dog, cat, guinea pig, and rat), the number of adjacent capillaries averages four per muscle fiber. Although there is a range of values around the average, there is remarkable similarity within muscles of a given species and among the species, with dogs having slightly higher indices of capillarity than other species. Some species have more capillaries adjacent to high-oxidative cells than low-oxidative cells; in other species, this is not the case. Interpreting data that assign fiber-type-specific capillarity is complicated by the fact that any single capillary is adjacent to several muscle fibers. As the fibers are often of a different histochemical type, it is not possible to decide whether the capillary's location is most related to the characteristic of one fiber type or to that of another. The average number of capillaries per square millimeter typically ranges from 500 to 900. These values are related inversely to the cross-sectional area of the muscle fibers, so larger muscle fibers have fewer capillaries per square millimeter. In the example above, the value of 500 capillaries per square millimeter is associated with fiber areas of 4000 μm^2, while 900 capillaries per square millimeter is associated with fiber areas of approximately 1250 μm^2. The average capillary-to-muscle fiber ratio of various muscles across the animal kingdom is approximately 2.0 capillaries per fiber. In limb muscles of cats and untrained humans, values average approximately 1.5 capillaries per fiber, and in limb muscles of dogs, over 3 capillaries per fiber are common.

The fact that limb skeletal muscles of dogs have a higher capillarity than those of cats (and untrained humans), and that dogs have a greater endurance capacity than the other species mentioned, raises the interesting question of whether capillarity can be increased in humans by certain types of exercise training. The capillary-to-fiber ratio has been shown to increase in humans by 5 to 20% in muscles subjected to 8 to 12 weeks of endurance training. In contrast, a more dramatic activity-induced increase of the capillary-to-fiber ratio, approximately 55%, is produced by experimental chronic low-frequency electrical stimulation of a targeted fast muscle. This relatively large increase is only observed in muscles that are initially low in capillarity, such as those of rabbits, but not in muscles that are initially high in capillarity, such as those of dogs. An investigator must therefore be mindful of the initial values before making a definitive interpretation. Comparing elite endurance-trained athletes to untrained individuals can reveal differences of up to 40% in measures of capillarity, but it is unclear in these experiments if the difference is due to genetic predisposition, to training, or to a combination of both factors.

Small increases in the capillarity with training of humans are unlikely to have a major impact on maximum blood flow during exercise, for several reasons. The most significant is that very large changes can be produced in muscle blood flow by a process of local blood flow control. This process is described in a later section on exercise blood flow. The process uses the normal complement of capillaries via vasodilation of larger "feed" vessels that are "upstream" of the capillaries. However, the major physiological benefit of increased capillarity with endurance training in humans may very well be an increase in transit time for red blood cells in their excursion through a given capillary. The time the blood stays in any single capillary increases when a given volume of blood exits an arteriole and passes through a larger number of capillaries. An increase in transit time facilitates the passive exchange of nutrients and gasses, which is essential for prolonged physical performance.

Microvascular Unit

Capillarity studies are often done on skeletal muscle cross sections. However, as mentioned previously, this approach can be limited in scope because capillaries exist in rather tortuous patterns that are essentially parallel with the muscle fibers. The structure of a capillary is best seen in experimental preparations of whole intact skeletal muscle in which the microvasculature is first perfused with a synthetic material that fills each vessel, no matter how small, and then solidifies. The muscle is then removed from the animal and treated to dissolve the muscle fibers and other tissues, leaving a cast of the microvasculature. An example of this cast— often called a vascular corrosion cast—is shown in Figure 17-17. This preparation provides a clear view of the complexity of the microcirculatory pathway in skeletal muscle.

Brian Duling and his coworkers, among other groups, have advanced our understanding of the complexity of the structure–function relationship of the microvasculature. Duling's group has been instrumental in identifying the existence of functional *microvascular units* (Figure 17-18). Capillarity organization cannot be functionally understood by studying capillaries in cross section or as isolated longitudinal vessels because in many tissues, includ-

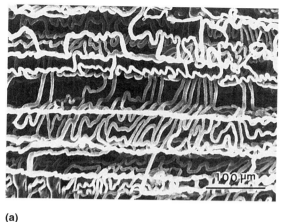

(a)

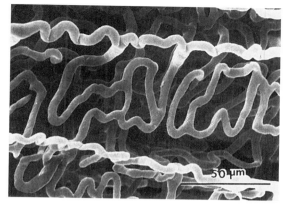

(b)

Figure 17-17   Vascular corrosion cast of mouse soleus muscle. (a) Low-power scanning electron micrograph (SEM). (b) High-power SEM. Capillary networks show a ladderlike pattern in this contracted state of muscle and are arranged in layers surrounding individual muscle fibers, which are dissolved away with all other tissue components. Note that there are few occurrences of broken ends in the capillary casts. SOURCE: Ishikawa, *Handbook of Physiology,* 1983.

ing skeletal muscle, several capillaries are fed by a common arteriole, thereby forming a microvascular unit. Although unequivocal evidence that microvascular units are the basic building blocks for the regulation of peripheral circulation still needs to be found, what is currently known is highly consistent with this premise. For example, in the tibialis anterior muscle of hamsters, a microvascular unit consists of 15 capillaries supplied by a common arteriole and drained by a common venule. The length from arteriole to venule averages 850 μm, and the average capillary length is 800 μm. Blood within the capillaries of a given unit flows in the same direction, and there is little interchange of blood among capillaries of adjacent microvascular units. The microvascular unit as a functional entity is analogous to—but separate from—the motor unit.

Exercise Blood Flow

The capillary beds are one critical component of the microcirculation. The other elements of the microcirculation, as introduced in Chapter 15 on the circulation, include the arterioles and venules, which

are also found within skeletal muscle intermingling with the muscle fibers. An increase in muscle blood flow during exercise is termed active or exercise hyperemia. Exercise hyperemia, which is mediated in large part by local mechanisms, is one of the important functional features of the microcirculation. In other words, the local control of blood flow is by mechanisms other than nerves or hormones. Often called *autoregulation,* this mechanism is a means by which muscle can regulate the resistance in its own arterioles, thereby regulating the flow of blood in the capillaries.

Exercise hyperemia occurs in essentially direct proportion to the increased metabolic activity of the muscle. The increased activity causes a local release or production of specific chemicals, as well as cell-to-cell communication, that cause the smooth muscle of arterioles to relax, thereby increasing blood flow. This increase in blood flow is due to a decrease in resistance in the arterioles. Several chemical elements have been identified that can explain many of the aspects of the blood flow response to exercise. However, which of several mechanisms predominate in controlling blood flow depends on

M. tenuissimus

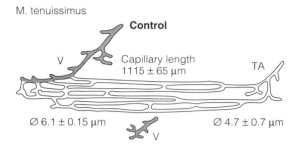

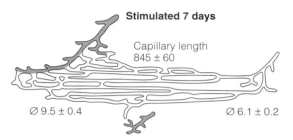

Figure 17-18 Capillary growth in skeletal muscle. Schematic representation of the capillary network illustrating the branching pattern and capillary dimensions in rabbit tenuissimus muscle based on observations in a control muscle (top) and after seven days of stimulation (bottom) *in vivo*. TA = terminal arteriole; V = two collecting venules. Note the circle close to the collecting venule and the sprouts that occur after stimulation. At the bottom of the picture, a tortuous sprouting capillary is about to be connected to a "new" collecting venule. SOURCE: Myrhage and Hudlicka, 1978.

the intensity and duration of the activity. Chemical changes, such as decreased oxygen concentration, or increased concentrations of adenosine, carbon dioxide, and hydrogen ion have been identified. Increased potassium concentration and increased osmolality, as a result of the breakdown of metabolites, have also been identified. Eicosanoids, a family of local chemical messengers, also increase in exercise, and seem to have an important regulatory role with smooth muscle in arterioles (along with many other tissues). Recent experiments by Tom Balon and Jerry Nadler (1994), and by Dong Sun and coworkers (1994), have identified one such eicosanoid—nitric oxide—that increases its release

in active skeletal muscle. Steve Segal (1994) has suggested that blood flow through the capillaries can stimulate release of nitric oxide from the endothelial cells of the capillary, thereby providing a very effective means of local regulation of blood flow. Indeed, there is experimental evidence that nitric oxide functions as an important mediator, not only of vascular delivery of blood, but also of oxygen utilization within muscle fibers.

■ Architectural Factors

The influence of muscle design on the production of force and the velocity of contractions is quite profound, and is independent of other biochemical and physiological factors. In general, as the number of sarcomeres that are arranged in series increases in a muscle fiber, so will the overall velocity of shortening of the fiber. Alternatively, the more sarcomeres that are arranged parallel to each other, the greater the capacity for maximum force production. The angle of pinnation also influences contractile properties, as Carl Gans (1982) has well described. The angle of pinnation is the angle at which individual fibers are oriented in relation to the overall muscle's longitudinal axis, which is the direction of force generated by the muscle when contracting (Figure 17-19).

In a parallel-fiber muscle, the muscle fibers are arranged essentially in parallel with the longitudinal axis of the muscle itself. Common examples of this design include the sartorious muscle in the leg, the biceps brachii muscle in the arm, and the sternohyoid muscle in the neck. In pinnate muscles, the fibers are aligned at an angle to the muscle's longitudinal axis. These designs can be fairly simple unipinnate structures, such as the soleus muscle of the leg, or be more complex bipinnate and multipinnate designs. Examples of the latter two include the rectus femoris muscle in the leg and the deltoid muscle in the shoulder, respectively.

Each sarcomere has a given maximum velocity of contraction, based primarily on its myosin ATPase activity. When sarcomeres are arranged in series in a fiber, the velocities become additive and

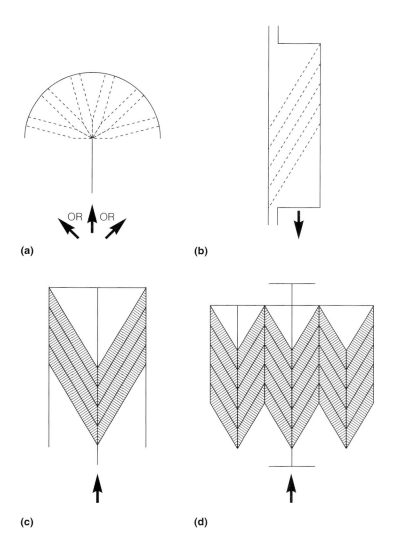

Figure 17-19 Permutations of muscle fibers that are arranged at an angle to the line of force generation. The hatched side is assumed to be the fixed "origin"; the open arrow shows the movement of the "insertion" site. It is assumed that the individual fibers do not interfere with each other and that all fibers extend from origin to insertion. Arrangement (a) is radial and (b)–(d) are pinnate. (a) If all fibers are stimulated simultaneously, the tendon is lifted. However, asymmetric action will shift the position of the tendon off the center line and may even stretch some of the fibers on the opposite side. (b) Singly pinnate muscle. The fibers lie at constant angles between the surfaces of origin and insertion, and the shortening of the fibers consequently causes the surfaces to slide past each other. (c) Doubly pinnate muscle. In this arrangement, the fibers contact one of the medial tendons from both sides, and their symmetrical contraction shifts it centrally. (d) Multipinnate muscle. In this arrangement, the fibers run between interleaved tendinous sheets, with the fiber directions alternating. In (b), (c), and (d), the force will be the resultant of that of all the fibers stimulated, and tendon movement will be greater than the absolute shortening of the fibers. SOURCE: Gans, 1982, pp. 176–177.

the overall velocity of shortening measured at the tendons increases (Figure 17-20). The functional advantage of this design is to increase the overall velocity of shortening of the intact muscle while maintaining the length of individual sarcomeres near their optimal length–tension relationship. The length of individual sarcomeres need to change relatively little with this architectural design, which allows for optimal displacement of the tendon.

Each sarcomere is capable of generating a fixed amount of force. The term *specific force* defines this

intrinsic property of muscle, which is approximately 22 to 28 newtons per square centimeter of muscle fiber cross-sectional area. Thus, the more sarcomeres that are arranged in parallel, the greater the absolute force production (see Figure 17-20).

When muscles are designed with angles of pinnation, which is by far the most common architecture, more sarcomeres can be packed in parallel between the origin and insertion of the muscle. By packing more sarcomeres in a muscle, more force can be developed (Figure 17-21). As the angle of

Figure 17-20 Force generated by sarcomeres in series and in parallel. (a) Sarcomeres in series. The forces F1 and F2 are opposed, leaving only F3 to exert force at the ends of the muscle. (b) The same number of actin and myosin filaments arranged in parallel to give four times the isometric force of (a) SOURCE: Jones and Round, 1990.

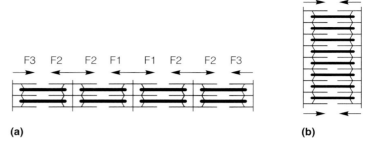

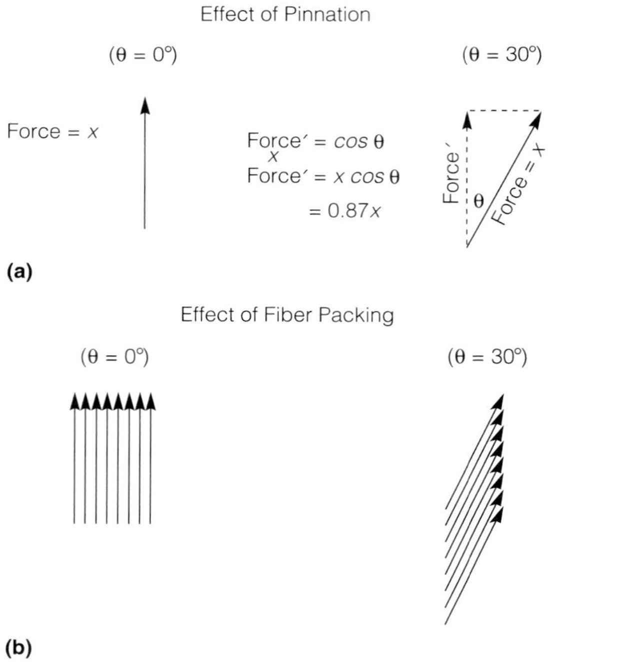

(a)

(b)

Figure 17-21 Schematic illustration of the effect of pinnation. (a) Muscle fibers oriented parallel to the axis of force generation transmit all of their force to the tendon. However, muscle fibers oriented at a 30° angle relative to the force-generating axis transmit only a portion of their force (cos 30° = 0.87, or 87%) to the tendon. (b) Although only about 90% of the muscle fiber force is transmitted to the tendon due to pinnation, pinnation itself permits packing of a large number of fibers into a smaller cross-sectional area. SOURCE: Lieber, 1992, p. 41.

Effect of Pinnation

$(\theta = 0°)$

Force = x

$(\theta = 30°)$

$$\text{Force}' = \cos \theta$$
$$x$$
$$\text{Force}' = x \cos \theta$$
$$= 0.87x$$

(a)

Effect of Fiber Packing

$(\theta = 0°)$

$(\theta = 30°)$

(b)

pinnation increases, however, an increasing portion of the force developed by sarcomeres is displaced away from the tendons (and toward the sides of the muscle). However, as long as the angle of pinnation is less than 30°, the force lost due to the angle of pinnation is more than compensated for by the increased packing of sarcomeres in parallel, producing an overall benefit to the force-producing capacity of muscle.

As depicted in Figure 17-22, muscles with long fibers that are arranged essentially in parallel have an advantage in velocity of shortening and in the amount of shortening or tendon displacement (i.e., excursion) that can occur during contraction. Examples in the leg include muscles of the hamstring group (i.e., knee flexor and hip extensor). In contrast, when fibers are relatively short and at a significant angle of pinnation, the functional advantage is in force development. Examples of this design shown in Figure 17-22 are muscles of the quadricep group (i.e., knee extensor and hip flexor) and ankle dorsiflexors (e.g., anterior tibialis muscle).

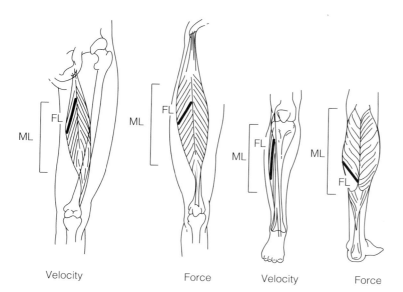

Figure 17-22 Schematic illustration of muscle architectural properties in the lower limb. Functionally, quadriceps and plantar flexors are designed for force production due to their low fiber length (FL)–muscle length (ML) ratios and large cross-sectional areas. Conversely, in general, hamstrings and dorsiflexors are designed for high excursions and velocity by nature of their high fiber length–muscle length ratios and long muscle fibers. SOURCE: Lieber, 1992, p. 43.

■ Contractile Properties

Types of Contraction

The term *contraction* is used here to mean activation of the force-generating capacity of the actomyosin complex within fibers and initiation of the four-step crossbridge cycle. *Contraction* does not imply that the muscle "contracts" or shortens. The muscle either shortens, remains at the same length, or lengthens during contraction, depending on the external load placed on the muscle relative to the amount of force developed by the fibers. When the muscle develops force, if the external load applied to the muscle is equal to the amount of force being developed, there is no change in muscle length. This type of contraction is termed *isometric* or *fixed-end.* If the external force is less than the force the muscle is generating, there is a *shortening* contraction, also called a *concentric* or *miometric* contraction. If the external force is greater than the force the muscle is developing, there is a *lengthening* contraction, which is also called a *stretching* or *pliometric* contraction (oftentimes referred to as an eccentric contraction). During physical activities, most limb muscles will be involved in equivalent amounts of pliometric and miometric contractions.

Experimental Models of Muscle Contraction

An important distinction exists between experimental studies of muscle contraction in a research laboratory and the voluntary contractions of human beings during the performance of routine movements. In the laboratory, it is common to study isolated muscles of experimental animals. Skeletal muscle can be fully excised from an animal, such as a rodent or frog, and studied in an artificial solution in a glass tissue bath as an *in vitro* (i.e., in solution) preparation. Alternatively, an *in situ* (i.e., in site) preparation requires that a muscle be surgically exposed in the anesthetized animal, and typically the distal tendon is attached to force and displacement transducers. In both of these preparations, the muscle is usually stimulated electrically, which results in the simultaneous recruitment of all the fibers (i.e., all of the motor units) in the muscle. In these type of experiments, an investigator can precisely control the frequency of recruitment, muscle length, and load characteristics, and thus precisely measure the intrinsic contractile properties of skeletal muscle.

When skeletal muscle is studied during normal physical activity, it is termed an *in vivo* (i.e., in the living being) preparation. In contrast to preparations *in situ* and *in vitro,* the amount of force developed *in vivo* can be controlled intrinsically by the nervous system in two ways. One way is to recruit a varying number of muscle fibers (which are functionally organized into motor units). The number of motor units active at any moment relates directly to the number of myofibrils in parallel that are active, which in turn directly affects the amount of force produced. A second physiological mechanism to control force output *in vivo* is to regulate the neural frequency by which recruited motor units are activated. This control mechanism is termed *rate-coding.* In this section, we review the intrinsic contractile properties of skeletal muscle, as they provide insights into the expression of human power in the intact system.

Length–Tension Relationship

The amount of force or tension that can be developed by a muscle fiber is critically dependent on the length of the fiber relative to its optimal length. Optimal length (i.e., L_o) is defined as the sarcomere length that provides for optimal overlap of the thick and thin filaments. Thus, when a sarcomere is at L_o, the greatest potential exists for force production upon activation of the crossbridge cycle. As seen in Figure 17-23, when a muscle is shorter than L_o, maximum force development is impaired. In con-

trast, when a muscle is extended beyond L_o, tension does not drop appreciably until the length is extended by 10–15%.

Isometric Contraction

When an intact muscle (or a single fiber) is set to L_o *in vitro* or *in situ* and held at that length during stimulation, an isometric contraction occurs because the ends of the muscle are fixed. As seen in Figure 17-24, a force recording can be made, allowing for the characterization of the twitch response. Commonly used descriptors of the twitch include the twitch tension (P_t), the time to peak tension (TPT), and the time it takes to relax from peak tension to 50% of peak tension (i.e., the half-relaxation time, $\frac{1}{2}$RT). An investigator can then compare the twitch characteristics, for example, between muscles expressing different types of myosin, or between untrained and trained muscles, and so on. A muscle that has a preponderance of fast myosin heavy chain (MHC) and an abundance of sarcoplasmic reticulum (SR) will have short TPTs and $\frac{1}{2}$RTs, approximately 12 to 15 msec, compared to values of 50–70 msec for a muscle with slow MHC and a less developed SR. As a consequence, the former muscle is termed *fast-twitch,* whereas the latter is *slow-twitch.*

In the isometric condition, progressive stimulation frequencies increase the amount of force devel-

Figure 17-23 The length–tension relationship for skeletal muscle. At full resting muscle length (L_o), the probability of actin–myosin interaction (*B* and *C*) is maximal, and the greatest isometric tension is recorded. At muscle lengths significantly less than L_o (*A*), or greater than L_o (*D*), isometric tension declines.

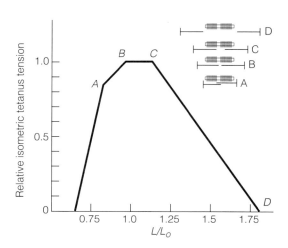

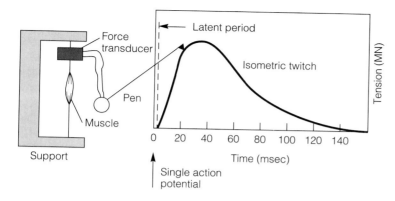

Figure 17-24 Schematics of apparatus to study the contractile properties of isolated muscles. When a muscle's length is fixed, isometric twitch characteristics can be recorded. Adapted from Vander, Sherman, and Luciano, 1980, p. 226. Used with permission.

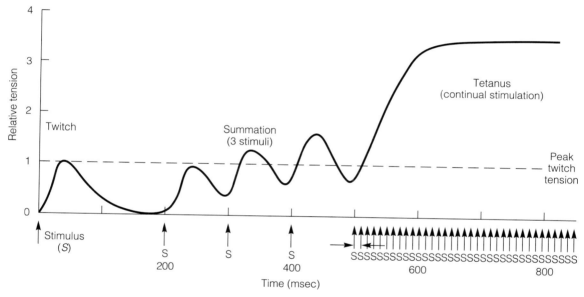

Figure 17-25 Repeated stimuli, each of a given strength (*S*) can produce a tension that sums to greater than the twitch tension. Continual stimulation results in a tetanic contraction three to five times stronger than twitch tension. Adapted from Vander, Sherman, and Luciano, 1980, p. 228. Used with permission.

oped because the muscle cannot completely relax from the previous stimulus before the next stimulus arrives. This phenomenon is called force *summation*. As seen in Figure 17-25, at low frequencies the force oscillates in a pattern that is termed *unfused tetanus*. At higher frequencies, the overall force increases but the oscillations, although lower in amplitude, are still visible. Once the frequency of stimulation is high enough, full summation is achieved, and the force curve is smooth and is termed *fused tetanus*. This value is also commonly called the maximum isometric force, and abbreviated P_o. The P_o is approximately four- to fivefold greater than P_t, the twitch force. For fast-twitch muscles, a fused tetanus occurs at frequencies of 150 to 200 Hz; for slow muscle, fusion typically occurs in the 80–100 Hz range.

From the type of *in vitro* protocol described in the preceding paragraph, a frequency–force curve can be drawn (Figure 17-26). This curve is plotted from the peak force developed at each frequency of stimulation (typically an investigator will study five or six frequencies ranging from 10 to 200 Hz). In normal human physical activity, controlling the frequency of motor unit recruitment (termed rate-coding) is one of two principal ways in which muscular force is regulated. The frequency–force curve is very instructive, for it demonstrates the profound influence of modulating the frequency of recruitment on force output. This curve is also useful in identifying phenotypic differences between muscles—a fast-twitch muscle has a frequency–force curve that is shifted to the right of that of a slow-twitch muscle. This phenomena is due chiefly to the fact that fast muscles contain more sarcoplasmic reticulum, the organelle system responsible for calcium sequestering, thereby enabling the muscle to relax faster. Consequently, a higher stimulation frequency is required to increase intracellular calcium to activate the crossbridge machinery.

Force–Velocity Relationship During Shortening and Lengthening Contractions

When a muscle is set to L_o, *in vitro* or *in situ*, and the length is not fixed during stimulation, an isotonic (concentric) contraction ensues (Figure 17-27). The muscle shortens at different velocities depending on the load placed on the muscle during this contraction. As load is increased, the velocity decreases (Figure 17-28). When the load exceeds the maximum force capable of being developed by the muscle, a lengthening contraction ensues. Investigators will record both the velocity of contraction and the force developed during the contraction at different load conditions. As seen in Figure 17-29, the force developed during a shortening contraction is less than P_o (i.e., the maximum isometric force). Indeed, the force developed at P_o is greater than during any shortening contraction. However, the force

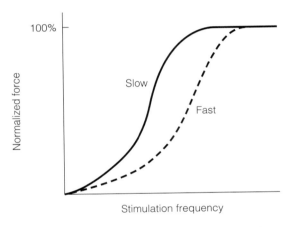

Figure 17-26 A force–frequency curve for slow- and fast-contracting muscle. Note the shift up and to the left for the slow muscle. At any given submaximal stimulation frequency, the slow muscle produces a greater relative amount of its maximal force.

Figure 17-27 Schematics of apparatus to study the contractile properties of isolated muscles. When a loaded muscle shortens against a load, the shortening record produces a linear, isotonic response. Adapted from Vander, Sherman, and Luciano, 1980, p. 226. Used with permission.

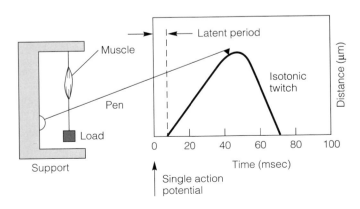

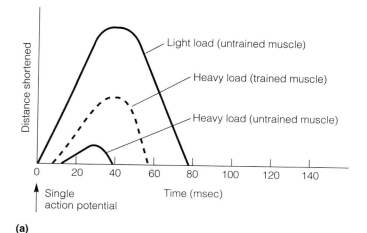

(a)

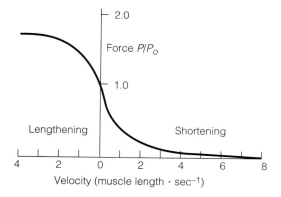

(b)

Figure 17-28 Characteristics of isotonic contractions. As compared to lifting light loads, isotonic responses to given stimuli when lifting heavy loads results in a greater latent period, slower movement, and less movement. (a) The effect of strength training is to make the load appear lighter. The force–velocity relationship (b) is hyperbolic in nature. Greater loads produce slower speeds but greater tension. The effect of strength training is to increase P_o (maximum isometric tension) but not V_o (maximum unloaded velocity). However, a stronger trained muscle can move a given isotonic load (X) at a greater velocity ($V_t > V_{ut}$).

Figure 17-29 Force–velocity curve of mouse soleus muscle. Note that the right of the figure represents shortening contractions while the left of the figure represents lengthening contractions. Force continues to rise as the muscle is lengthened. When the curve crosses the y-axis, the velocity is zero; this is by definition P_o (i.e., maximum isometric tetanic tension).

developed during a lengthening contraction exceeds P_o by 50 to 100% because of the greater amount of force that must be applied to the muscle to detach the crossbridges during the lengthening of the sarcomere. These very high forces have been shown to be the primary reason for exercise-induced muscle injury, which is much more prevalent following exercise that has a bias toward lengthening contractions (such as running downhill) than toward isometric or shortening contractions.

A plot of the force–velocity relationship during shortening and lengthening contractions is depicted in Figure 17-29. During the shortening contraction, the curve demonstrates a rectangular hyperbolic relationship. Muscle physiologists often use a linearized form of the relationship to discover basic mechanistic factors of muscle contraction.

During typical human physical activity, "pure" isometric or isotonic contractions rarely occur. Rather, muscular activity is a mixture of these simple components. Indeed, the notion that normal muscle function is in fact a "lengthening–shortening cycle" is generally accepted. Lengthening contractions occur normally as the body's limbs, for example, are subjected to impact forces associated with walking and running, or because of an external force such as gravity. The lengthening contraction is then followed by a shortening contraction.

The product of the force and velocity at which muscle shortens or lengthens determines the power output or power absorption, respectively. A number of major factors combine to influence the power developed by skeletal muscle fibers, including the frequency of stimulation (force); the number and size of motor units and therefore the number of crossbridges cycling in parallel (force); the position of fiber length relative to the optimal length–tension relationship (force); the length of muscle fibers and therefore sarcomeres in series (velocity); and the myosin ATPase activity (velocity).

The intrinsic property of velocity of shortening is a function of the myosin phenotype. The phenotype of myosin is the outward expression of the myosin protein, which reflects the underlying genotype. This attribute can be assessed by measuring the velocity of shortening, measuring both the heavy and light chains of myosin, or by assaying myosin or myofibrillar ATPase activity. Usually the time to peak twitch force (TPT) inversely correlates with the maximum velocity of shortening. This means that a shorter TPT for a muscle during an isometric contraction is associated with a faster velocity measurement made on the same fiber during an isotonic contraction. The TPT is often used to classify motor units and muscle fibers as either slow or fast. It is important to note that under circumstances where developmental or aging changes—or experimental interventions such as training—have differential effects on different proteins, erroneous interpretations may arise from comparisons within and among the classification systems.

Appropriate training can influence the structure and functional attributes of muscle. For example, appropriate resistance training will lead to muscle fiber hypertrophy, thereby increasing the number of sarcomeres arranged in parallel. This adaptive response will lead to an increase in the force-generating capacity of muscle (see Chapters 19, 20, and 21).

■ **Summary**

In this chapter, we have learned about skeletal muscle from the perspective of its primary structural and functional properties. The physical activity patterns of a given individual arise principally from the properties of skeletal muscle fibers, the levers and joints through which fibers act, and the coordination and integration of recruitment patterns of motor units in the involved musculature. The organizational hierarchy of skeletal muscle provides a useful perspective from which to gain an understanding of the most important components of muscle, including the diversity of contractile, regulatory, and structural proteins. The hierarchy of organization within muscle is achieved with several connective tissue membranes.

Muscle contraction is defined as the activation of the force-generating capacity of the actomyosin complex within fibers and the initiation of the crossbridge cycle. The muscle will either shorten, remain at the same length, or lengthen during contraction, depending on the external load placed on the muscle relative to the amount of force developed by the fibers. During physical activities, most limb muscles will be involved in equal amounts of lengthening and shortening contractions depending on the movement pattern.

The intricacies of capillary microvasculature, and its importance to sustained contractile activity, is now understood. The architectural features of muscle can, in and of themselves, confer significant impact on the force and velocity characteristics of muscle.

An important distinction exists between experimental studies of muscle contraction in a research laboratory and the voluntary contractions of human beings. In the laboratory, it is common to study isolated muscles of experimental animals, and force, velocity, and power characteristics can be precisely controlled and measured. An understanding of these basic mechanics is critically important for gaining insight into the more complex issues that are involved when muscles are studied during complex movements.

■ Selected Readings

Baldwin, K. M. Muscle development: neonatal to adult. In Exercise and Sports Sciences Review, R. L. Terjung (Ed.). Vol. 12. Lexington, Mass.: Collamore Press. 1984, pp. 1–19.

Baldwin, K. M., G. H. Klinkerfuss, R. L. Terjung, P. A. Molé, and J. O. Holloszy. Respiratory capacity of white, red and intermediate muscle, adaptive response to exercise. Am. J. Physiol. 222: 373–378, 1972.

Balon, T. W., and J. L. Nadler. Nitric oxide release is present from incubated skeletal muscle preparations. J. Appl. Physiol. 77 (6): 2519–2521, 1994.

Briskey, E. J., R. G. Cassens, and B. B. Marsh (Eds.). The Physiology and Biochemistry of Muscle as a Food, II. Madison: University of Wisconsin Press, 1970.

Buchtal, F., and H. Schmalbruch. Contraction times and fiber types in intact muscle. Acta Physiol. Scand. 79: 435–452, 1970.

Burke, R. E., and V. R. Edgerton. Motor unit properties and selective involvement in movement. In Exercise and Sports Sciences Reviews, J. Wilmore and J. Keogh (Eds.). New York: Academic Press, 1975, pp. 31–83.

Carlson, F. D., and D. R. Wilkie. Muscle Physiology. Englewood Cliffs, N.J.: Prentice-Hall, 1974.

Close, R. The relation between intrinsic speed of shortening and duration of the active state of muscle. J. Physiol. (London) 180: 542–559, 1965.

Ebashi, S. Muscle contraction and pharmacology. Trends Pharmacol. Sci. 1: 29–31, 1979.

Fenn, W. O. The relation between the work performed and the energy liberated in muscle contraction. J. Physiol. (London) 85: 343–345, 1923.

Fuchs, F. Striated muscle. Ann. Rev. Physiol. 36: 461–502, 1974.

Gans, C. Fiber architecture and muscle function. In Exercise and Sports Sciences Reviews, R. L. Terjung (Ed.). Vol. 10. Philadelphia: Franklin, 1982, pp. 160–208.

Goldman, Y. E. Special topic: molecular mechanism of muscle contraction. Ann. Rev. Physiol. 49: 629–636, 1987.

Henneman, E. Skeletal muscle. The servant of the nervous system. In Medical Physiology, 14th ed., V. B. Mountcastle (Ed.). Vol. 1. St. Louis: Mosby, 1980, pp. 674–702.

Hill, A. V. The revolution in muscle physiology. Physiol. Rev. 12: 56–67, 1932.

Hill, A. V. The heat of shortening and the dynamic constants of muscle. Proc. Roy. Soc. (London) Ser. B. 126: 136–195, 1938.

Hill, A. V. The heat of activation and the heat of shortening in a muscle twitch. Proc. Roy. Soc. (London) Ser. B. 136: 195–211, 1949.

Hill, A. V. Chemical change and the mechanical response in stimulated muscle. Proc. Roy. Soc. (London) Ser. B. 141: 314–321, 1953.

Hill, A. V. Production and absorption of work by muscle. Science 131: 897–903, 1960.

Hill, A. V. The effect of tension in prolonging the active state in a twitch. Proc. Roy. Soc. (London) Ser. B. 159: 589–595, 1964.

Hochachka, P. W. Muscles as Molecular and Metabolic Machines. Ann Arbor: CRC Press, 1994.

Hudlicka, O., M. Brown, and S. Egginton. Angiogenesis in skeletal and cardiac muscle. Physiol. Rev. 72: 369–400, 1992.

Huxley, A. F. The activation of striated muscle and its mechanical response. Proc. Roy. Soc. (London) Ser. B. 178: 1–27, 1971.

Huxley, A. F. A note suggesting that the cross-bridge attachment during muscle contraction may take place in two stages. Proc. Roy. Soc. 183: 83–86, 1973.

Huxley, H. E. The mechanism of muscular contraction. Science. 164: 1356–1366, 1969.

Huxley, H. E. Structural aspects of energy conversion in muscle. Ann. N.Y. Acad. Sci. 227: 500–503, 1974.

Ishikawa, H. Electron microscopic observations of satellite cells with special reference to the development of mammalian skeletal muscles. *Zeitschrift fur Anatomie und Entwicklungsgeschicte,* 125(1):43–63, 1966.

Ishikawa, H., H. Sawada, and E. Yamada. Surface and internal morphology of skeletal muscle. In Handbook of Physiology, Section 10: Skeletal muscle, L. D. Peachey, R. H. Adrian, and S. R. Geiger (Eds.). Bethesda, Md.: American Physiological Society, 1983, pp. 1–22.

Jones, D. A., and J. M. Round. Skeletal muscle in health and disease. New York: Manchester University Press, 1990.

Kurebayashi, N., A. B. Harkins, and S. M. Baylor. Use of fura red as an intracellular calcium indicator in frog skeletal muscle fibers. *Biophy. J.* 64: 1934–1960, 1993.

Laughlin, M. H., and J. Ripperger. Vascular transport capacity of hindlimb muscles of exercise-trained rats. *J. Appl. Physiol.* 62: 438–443, 1987.

Lehman, W., and A. G. Szent-Gyorgi. Regulation of muscular contraction. Distribution of actin control and myosin control in the animal kingdom. *J. Gen. Physiol.* 66: 1–30, 1975.

Lieber, R. L. Skeletal Muscle Structure and Function. Baltimore: Williams and Wilkins, 1992.

Mauro, A. Satellite cell of skeletal muscle fibers. *J. Biophys. Biochem. Cytol.* 9: 493–495, 1961.

Maxwell, L. C., T. P. White, and J. A. Faulkner. Oxidative capacity, blood flow, and capillarity of skeletal muscles. *J. Appl. Physiol.* 49: 627–633, 1980.

Merton, P. A. Voluntary strength and fatigue. *J. Physiol.* (London) 123: 553–564, 1954.

Miller, J. B. Myoblast diversity in skeletal myogenesis: how much and to what end? *Cell* 69: 1–3, 1992.

Mommaerts, W. F. H. B. Energetics of muscle contraction. *Physiol. Rev.* 49: 427–508, 1969.

Moss, F. P., and C. P. LeBlond. Satellite cells as the source of nuclei in muscles of growing rats. *Anat. Rec.* 170: 421–436, 1971.

Murray, J. M., and A. Weber. The cooperative action of muscle proteins. *Sci. Amer.* 230: 58–71, 1972.

Myrhage, R., and O. Hudlicka. Capillary growth in chronically stimulated adult skeletal muscle as studied by intravital microscopy and histological methods in rabbits and rats. *Microv. Res.* 16: 73–90, 1978.

Peachey, L. D., R. H. Adrian, and S. R. Geiger (Eds.). Handbook of Physiology. Section 10: Skeletal Muscle. Baltimore: American Physiological Society, 1983.

Peachey, L. D., and C. Franzini-Armstrong. Structure and function of membrane systems of skeletal muscle cells. In Handbook of Physiology, Section 10: Skeletal Muscle, L. D. Peachey, R. H. Adrian, and S. R. Geiger (Eds.). Baltimore: American Physiological Society, 1983, pp. 23–72.

Pette, D., and R. S. Staron. Cellular and molecular diversities of mammalian skeletal muscle fibers. *Rev. Physiol. Biochem. Pharmocol.* 116: 1–76, 1990.

Rayment, I., H. M. Halden, M. Whittaker, C. B. Yohn, M. Lorenz, K. C. Holmes, and R. A. Milligan. Structure of the actin-myosin complex and its implications for muscle contraction. *Science* 261: 58–65, 1993.

Rayment, I., W. R. Rypniewski, K. Schmidt–Bäse, R. Smith, D. R. Tomchick, M. M. Benning, D. A. Winkelmann, G. Wesenberg, and H. M. Holden. Three-dimensional structure of myosin subfragment-1: a molecular motor. *Science* 261: 50–58, 1993.

Schultz, E. Satellite cell behavior during skeletal muscle growth and regeneration. *Med. Sci. Sports Exerc.* 21: S181–S187, 1989.

Segal, S. Invited editorial on "Nitric oxide release is present from incubated skeletal muscle preparations." *J. Appl. Physiol.* 77 (6): 2517–2518, 1994.

Segal, S. S., D. N. Damon, and B. R. Duling. Propagation of vasomotor responses coordinates arteriolar resistance. *Am. J. Physiol.* H832-H837, 1989.

Segal, S. S., and B. R. Duling. Conduction of vasomotor responses in arterioles: a role for cell-to-cell coupling? *Am. J. Physiol.* H838-H845, 1989.

Squire, J. M. Molecular Mechanisms in Muscular Contraction. Boca Raton, Fla.: CRC Press, 1990.

Sun, D., A. Huang, A. Koller, and G. Kaley. Short-term daily exercise enhances endothelial NO synthesis in skeletal muscle arterioles of rats. *J. Appl. Physiol.* 76 (5): 2241–2247, 1994.

Szeng-Gyorgi, A. Chemistry of Muscular Contraction. New York: Academic Press, 1951.

Tidball, J. G. Myotendious junction injury in relation to junction structure and molecular composition. In Exercise and Sports Sciences Reviews, J. O. Holloszy (Ed.). Vol. 19. Baltimore: Williams and Wilkins, 1991, pp. 419–446.

Vander, A. J., J. H. Sherman, and D. S. Luciano. Human Physiology. 3d ed. New York: McGraw-Hill, 1980, pp. 144–190, 211–252.

Vander, A. J., J. Sherman, and D. S. Luciano. Human Physiology. 5th ed. New York: McGraw-Hill, 1990.

Weber, A., and J. Murray. Molecular control mechanisms in muscle contraction. *Physiol. Rev.* 53: 612–673, 1973.

White, T. P., and K. A. Esser. Satellite cell and growth factor involvement in skeletal muscle growth. *Med. Sci. Sports and Exer.* 21: S158–S163, 1989.

Woledge, R. C., N. A. Curtin, and E. Homsher. Energetic aspects of muscle contraction. Monograph of the Physiological Society, No. 41. London: Academic Press, 1985.

CHAPTER 18

NEURONS, MOTOR UNIT RECRUITMENT, AND INTEGRATIVE CONTROL OF MOVEMENT

In Chapter 17, we learned of the structure and function of skeletal muscle. In this chapter, we focus on the important role of the excitable membranes in muscle and nerve, for it is this property that allows us to voluntarily, and involuntarily, use our muscles. Indeed, the inherent excitability of nerve and muscle cells is vital for human movement and for survival of a species.

The ability to create and propagate electrical signals, which are called action potentials, is the defining characteristic that allows nerve cells, termed *neurons,* and muscle cells to be excitable and to communicate with each other. This characteristic of excitability depends on three membrane properties: (1) a sodium–potassium ion exchange pump; (2) a greater membrane permeability to potassium than

Olympic and World Cup champion Alberto Tomba in action at the Calgary Winter Olympics. Athletes such as Tomba must generate extremely high metabolic rates and display precise neuromuscular control while competing in freezing environments. PHOTO: © Reuters/The Bettmann Archive.

to sodium; and (3) the impermeability of the membrane to negatively charged organic molecules (e.g., proteins). These characteristics allow neurons and muscle cells to become polarized in the resting state, which means there is a negative charge on the inside of the cell relative to the outside. Membrane depolarization, whereby the polarity is partially or fully reversed, usually gives rise to an action potential that propagates over the cell membrane surface.

Terminal ends of neurons release chemical transmitter substances that affect the adjacent neurons or muscle cells. A *synapse* is formed when the terminal end of a neuron is in close contact with another neuron, whereas the general term *neuroeffector junction* is used when the postsynaptic cell is a muscle or gland. Specifically in the case of muscle, we use the term *neuromuscular junction*. When sufficient neurotransmitter substance is released from the presynaptic neuron to depolarize the postsynaptic neuron or muscle cell, then an action potential is transferred from one cell to the next.

Many synapses exist between neurons in the central nervous system (CNS—the brain and spinal cord). At the end of the peripheral nervous system (PNS), neurons can communicate with skeletal muscle cells at the *neuromuscular junction.* Major motor nerve cells and their associated muscle cells constitute a *motor unit.* Motor units are the basic functional units of the body's motor control system.

The spinal cord contains the appropriate neural circuits to produce voluntary and reflexive limb movement. The spinal cord also contains circuits that can produce some repetitive movement patterns, such as locomotion. However, for a person to respond to the environment around them, such as in sports, higher brain centers must influence the circuits in the spinal cord. This complex system requires integration of multiple and redundant feedback and feed-forward control mechanisms.

■ Excitability

Muscular movement is under the direct control of nerves. Not only do neurons send the action potential signals that cause muscles to contract, but a variety of sensory receptors provide constant feedback to the central nervous system (CNS) about movement status. This information is constantly transmitted to and utilized in the CNS to modify the ongoing movement as well as to initiate new movements. The ability of nerve cells to develop action potentials and to signal or communicate with other nerve or muscle cells depends on the presence of a voltage difference across the membrane of excitable tissues at rest.

Resting Membrane Potential

Under conditions of rest, nerve and muscle cells have an electrical potential difference across their membranes. The inside of the cell is negatively charged compared to the outside. This potential is called the *resting membrane potential.* In this section, we will examine how this resting membrane potential is established and maintained.

The resting membrane potential results from a greater number of negative ions (or negatively charged molecules) being inside the cell and an excess of positive ions being outside. The net excess of the positive and negative ions accumulates on either side of the cell's plasma membrane, creating a potential difference that essentially straddles the membrane. Although many ions and negatively charged proteins are involved in establishing the membrane potential, sodium, potassium, and chloride are three of the most prominent factors due to their inherent diffusion capabilities. Indeed, in a neuron, the extracellular concentration of sodium (150 mM) and chloride (110 mM) are greater than their intracellular concentrations (15 and 10 mM, respectively). On the other hand, potassium concentration is much less in the extracellular space (5 mM) than intracellular (150 mM). These differences result in a resting membrane potential of about -70 mV in neurons. The concentration difference for sodium and potassium is caused and maintained by two unique characteristics: (1) a sodium–potassium ion (Na^+–K^+) exchange pump driven by ATP hydrolysis that expels sodium from the cell and pumps potassium into the cell and (2) greater membrane permeability to potassium than to sodium.

In describing development of the resting membrane potential, also known as the transmembrane voltage, two characteristics of the current flow must

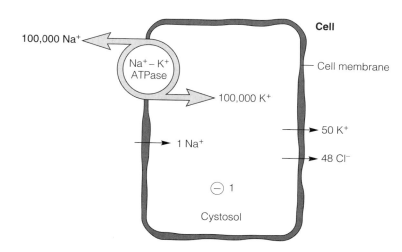

Figure 18-1 Two features of the excitable cell membrane allow for the development of a resting membrane electrical potential: (1) a (Na^+–K^+) exchange pump and (2) a greater permeability for K^+ than for Na^+. As shown in the figure and explained in the text, the pumping of relatively large amounts of ions eventually results in charge separation across the cell membrane.

be noted. The first is that the charge developed is of an electrochemical nature, depending mainly on separation of Na^+, K^+, and Cl^- ions. Electrical current depends on the movement of these ions. This is different from the usual form of electricity used by electronic devices, which depend on the flow of electrons, not ions. The second characteristic is that current always flows in complete loops. The movement of charge from one area into another will be accompanied by a flow of charge into the area from which the ion flow began. The following example illustrates how these two unique characteristics result in the generation of a transmembrane voltage.

Figure 18-1 depicts the hypothetical starting condition in a nerve cell for generation of a transmembrane voltage. At the outset in our model, the cell has equal intra- and extracellular ion concentrations and is uncharged. The cellular Na^+–K^+ exchange pumps activate on the first pump cycle, expelling 100,000 Na^+ ions and bringing in an equal number of K^+ ions. No charge difference develops between the inside and outside of the cell because of this electrically neutral ion exchange. However, the concentration of K^+ inside the cell is 30-fold greater than outside and this ion will diffuse down its concentration gradient, which means it will diffuse from a region of high concentration to a region of lower concentration. Outside the cell, the opposite occurs: Na^+ ions tend to diffuse down their concentration gradient by entering the cell. Because the

permeability of K^+ is about 50 times greater than that of Na^+, 50 K^+ ions will diffuse out of the cell, whereas only 1 Na^+ ion will reenter. Thus, the exit of K^+ [carrying a net 49 positive ($+$) electron charge units] leaves the inside of the cell with a slight negative ($-$) charge. Contributing to the net negative charge inside the cell are negatively charged proteins that are not permeable to the cell membrane. As a result of this developing electronegativity within the cell, chloride ions (Cl^-) are electrically repelled and exit the cell along with the K^+. In sum, then, because of the activation of the Na^+–K^+ pump, the relatively greater membrane permeability to K^+ than to Na^+, the impermeability of the cell to negatively charged proteins, and the efflux of Cl^- ions, the inside of the cell carries a negative charge relative to the outside.

Action Potential

Nerves and muscles communicate by developing and propagating action potentials. A nerve or muscle cell has a resting membrane potential, as described in the preceding subsection. An action potential is a rapid alteration, lasting perhaps only 1 millisecond, in the membrane potential, whereby the polarity across the cell becomes reversed, i.e., the inside of the cell becomes charged positive relative to the outside. During this very brief time, the membrane potential may depolarize from a resting value of −70 mV to +40 mV, and then repolarize to its resting value again. This wave of depolarization and repolarization is propagated along the cell membrane in a way analogous to "the wave," which has been popular in many sport stadiums and arenas. The concepts described earlier for the origin of resting membrane potentials also readily explain the events associated with the development of an action potential.

An action potential is a wave of depolarization that moves along the surface of a nerve or muscle cell (Figure 18-2). An action potential occurs because of a sudden transient increase in the permeability of Na$^+$ and the cascade of events that ensue, as just described. Figure 18-3 shows a glass electrode system used to measure voltage differences across excitable membranes. Figure 18-4 shows the electrical and chemical events that occur at the site of an action potential. Minor stimulation of an excitable membrane and perturbation of the resting

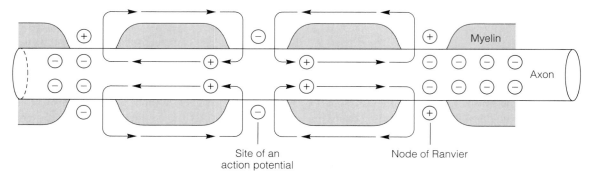

Figure 18-2 The movements of charged ions across excitable cell membranes causes the resting membrane voltage to reverse. Because ion current always flows in a complete circuit, and because current always flows at the point of least resistance, in myelinated nerves such as exist in mammals the conduction velocity of action potentials is very rapid. Breaks in the myelin are called nodes of Ranvier.

Figure 18-3 Microelectrodes are made from glass capillary tubes that are heated and then rapidly stretched to become microscopically thin. The ends remain open and can then be filled with a conducting saline solution and inserted into a cell. A voltmeter then records electrical potential differences across the cell membrane. Modified from Vander, Sherman, and Luciano, 1980, p. 158. Used with permission.

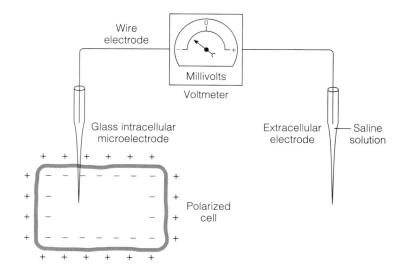

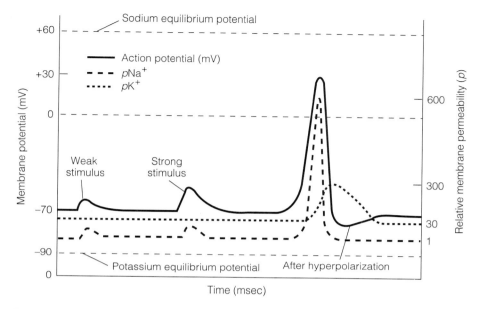

Figure 18-4 Weak stimuli can perturb the resting membrane voltage in the locus of the stimulus. Together, several weak stimuli can add up to produce a greater local effect. When the membrane voltage is disturbed to a threshold value (approximately 10 to 15 mV), then an action potential (a change in polarity all along the membrane) is propagated (see text and Figure 18-7). Modified from Vander, Sherman, and Luciano, 1980, p. 147.

membrane voltage does not necessarily result in an action potential. The subtle changes in the permeability of sodium ion (pNa^+), however, do result in a small increase in resting membrane potential (i.e., slight depolarization). When sufficient perturbation of the membrane potential occurs, a threshold voltage is reached at which the influx of Na^+ increases dramatically as sodium channels open. Indeed, during the time that the influx of Na^+ is high, the pNa^+ increases approximately 600 times, and the membrane potential approaches the equilibrium voltage of Na^+ (approximately +60mV).

The influx of sodium ions is short-lived, and the membrane potential rapidly changes back to resting negative levels. The influx of Na^+ is followed in time by an efflux of K^+. During the time K^+ is leaving the cell (thereby taking away positive charge), the membrane potential actually becomes more negative than usual. This is called *hyperpolarization*. During this time of hyperpolarization, the membrane

is in a refractory period. It is either impossible to stimulate the membrane (absolute refractory period) or more difficult than usual to stimulate the cell (relative refractory period) while the efflux of K^+ is taking place.

Although there are dramatic changes in the permeability of Na^+ and K^+ during the course of an action potential, the number of ions that move is relatively small compared to the total number of these ions in the intracellular and extracellular compartments. Consequently, the resting membrane potential recovers rapidly following an action potential (as mentioned, the entire event occurs in about 1 millisecond). During high levels of exercise, including those that are of long duration, in which action potentials are continuously being developed, the neurons do not show significant signs of fatigue. Thus, with prolonged exercise it is generally accepted that the site of fatigue does not lie in the nerves themselves. See Chapter 33 for a full treatment of fatigue.

Most excitable cells need an extrinsic stimulus to develop an action potential. However, other important regulatory cells spontaneously and regularly depolarize themselves. This spontaneity is due to the presence of a cell membrane that has a relatively high permeability to Na^+. At regular intervals, the membrane potential spontaneously reaches threshold and an action potential results. Examples include the specialized cardiac muscle cells comprising the SA and AV nodes in the heart (Chapter 14) and neurons in the medullar respiratory center (Chapter 12), which control heart and breathing rates, respectively, by their spontaneous excitability. As we will see, the frequency of depolarization can be affected by a variety of factors, but the characteristic of rhythmicity is intrinsic to these specialized cells.

Neuron Anatomy

The anatomy of an α motoneuron is shown in Figure 18-5. The α motoneuron consists of a cell body (or soma), short projections from the soma (dendrites), a long projection (axon), and terminal endings on the axon that contain and can release transmitter substances. The transmitter substance released by neurons varies from organ to organ. In the case of an α motoneuron, the transmitter substance is acetylcholine (ACh). In the brain, norepinephrine, serotonin, and γ-aminobutyric acid (GABA), as well as other neurotransmitters, are released in addition to ACh.

Characteristic of mammalian axons is the presence of *myelin* (a white substance high in lipid content). The myelin sheath covering axons is intermittently broken up, and the break points are called *nodes of Ranvier.* The electrical properties of myelin are such that it increases the capacitance (i.e., the charge or ion density) of the sections of the axon that are sheathed in it. This property increases the probability that ions will flow first at the unmyelinated nodes of Ranvier during the propagation of action potentials. Consequently, in myelinated axons the action potential conduction velocity is increased as a result of the jumping of action potentials from one node of Ranvier to the next. This is called *salti-*

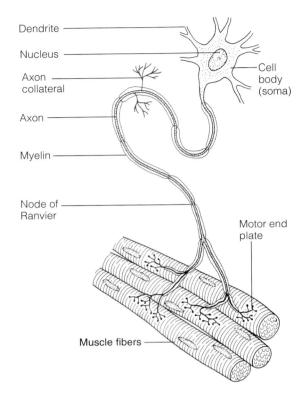

Figure 18-5 A motor unit, consisting of a cell body, the outgrowing α motoneuron, and all of the muscle fibers it innervates. In this drawing, only three fibers are shown; in reality the number of muscle cells in a single motor unit ranges from several hundred to several thousand.

tory conduction and is illustrated in Figure 18-2. In general, the thicker the myelin sheath around a nerve axon, the faster the conduction velocity.

Facilitation and Inhibition of Action Potentials

When there is an action potential in an α motoneuron, there is always an action potential in each of the postsynaptic muscle cells (see Figure 18-5). However, within the central nervous system (CNS), where one nerve cell synapses with another nerve cell, such as the soma of an α motoneuron, presynaptic action potentials do not always result in postsynaptic action potentials. In some instances, just the opposite happens. Figure 18-6 illustrates that

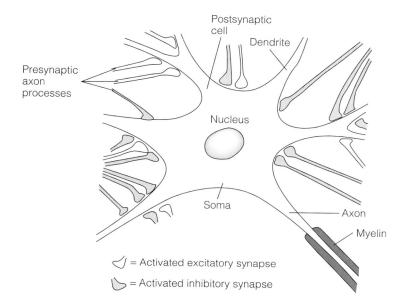

Figure 18-6 The α-motoneuron cell body (soma), located in the gray matter of the spinal cord, has many excitatory and inhibitory synapses. Activity from these presynaptic cell endings greatly affects membrane voltage in the soma.

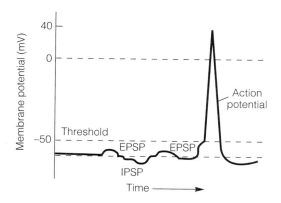

Figure 18-7 Recording from inside an α-motoneuron soma reveals the effects of excitatory presynaptic activity and inhibitory presynaptic activity. When the postsynaptic potential reaches threshold, an action potential results. Modified from Vander, Sherman, and Luciano, 1980, p. 172. Used with permission.

presynaptic terminal branches on neuron cell bodies can have either excitatory or inhibitory effects on the postsynaptic cell. The release of one neurotransmitter substance, such as acetylcholine, increases the postsynaptic permeability of sodium so that the membrane potential rises toward, but does not necessarily reach, a depolarizing threshold. This localized depolarization as the synaptic site is called an *excitatory postsynaptic potential* (EPSP). The release of another neurotransmitter into the synaptic cleft from another presynaptic terminal ending can result in a decrease in postsynaptic sodium perme-

ability or an increase in chloride permeability. These events result in a decrease in postsynaptic membrane potential and reduce the chance of the membrane potential reaching threshold. The localized, nonpropagating decrease in membrane potential illustrated in Figure 18-7 is called an *inhibitory postsynaptic potential* (IPSP).

The activity, or lack of it, in a neuron within the CNS depends on the sum total of EPSPs and IPSPs occurring within a narrow range of time and space. EPSPs can summate to give rise to an action potential in two basic ways. Excitatory presynaptic nerve endings distributed around a postsynaptic soma can release sufficient neurotransmitter to evoke a postsynaptic action potential. This is referred to as *spatial summation*. Alternatively, one or a few presynaptic endings can be repeatedly active within a short period of time. This summation to threshold is termed *temporal summation*.

■ The Neuromuscular Junction

Electrical signals that arise in the CNS are propagated as action potentials to skeletal muscles by *motoneurons* (i.e., motor neurons). These somatic efferent nerves have their cell bodies either in the brainstem or spinal cord. The axons of motoneurons are large in diameter and myelinated, and are thus able to propagate the action potential to the muscle fibers at high velocity. The motoneuron divides into several branches when it reaches a muscle, and each branch forms a junction with a single muscle fiber. A neuromuscular junction is seen clearly in Figure 18-8. In this figure, a single muscle fiber is seen in a left-to-right orientation; the striations seen are sarcomeres. A single axon is seen branching out from a nerve trunk that is composed of several axons. As the axon nears a muscle fiber, the myelin sheath ends and the axon divides into a small number of terminal branches that lie on the surface of the fiber. The region of the fiber's plasma membrane, which lies underneath this array of terminal branches, is the *motor end plate* (MEP). The MEP is seen in Figure 18-8 as a diffuse gray area that is located underneath the axon's terminal branches. The muscle section seen in this figure has been incubated for the enzyme acetylcholinesterase. During this incubation procedure, a pigment that is visible with a light microscope is deposited on the muscle membrane where the acetylcholinesterase enzyme is located, thereby marking the location of the MEP. Each muscle fiber has one neuromuscular junction, and yet each motoneuron innervates many muscle fibers within a given motor unit.

As shown in Figure 18-9, vesicles containing the neurotransmitter acetylcholine (ACh) are in the terminal axon. When the nerve's action potential

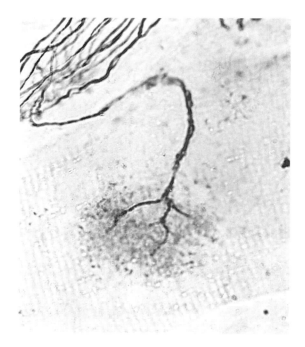

Figure 18-8 A neuromuscular junction (NMJ) for a single muscle fiber. The fiber is oriented left to right, and the sarcomere striations are visible. A single arm branches out from the bundle of α motoneurons, and then branches into several terminal axons at the fiber's motor end plate.

reaches the terminal axon, it depolarizes the plasma membrane of the nerve and opens voltage-gated calcium channels. Calcium diffuses into the terminal axon from the extracellular space. Before the calcium is pumped out by calcium–ATPase pumps in the nerve membrane, it triggers the fusion of the synaptic vesicles with the presynaptic membrane of the terminal axon. The vesicles then release their ACh by a process called *exocytosis*. Acetylcholine is released into the synaptic cleft that separates the terminal axon and the motor end plate. The ACh diffuses across the cleft and binds to its receptor in the muscle fiber membrane. Once ACh is bound to the receptor, sodium and potassium will pass through these channels at different rates, producing a depolarization of the motor end plate. This is termed an *end plate potential* (EPP). Normally, the EPP will, in turn, depolarize the plasma membrane of the fiber on either side of the end plate, resulting in an action

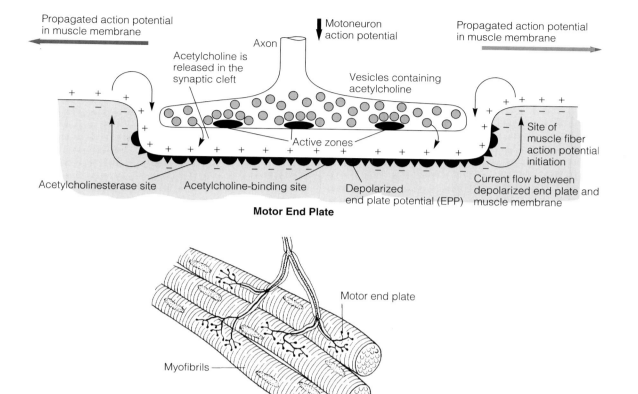

Propagated action potential in muscle membrane

Motoneuron action potential

Propagated action potential in muscle membrane

Axon

Acetylcholine is released in the synaptic cleft

Vesicles containing acetylcholine

Active zones

Site of muscle fiber action potential initiation

Acetylcholinesterase site

Acetylcholine-binding site

Depolarized end plate potential (EPP)

Current flow between depolarized end plate and muscle membrane

Motor End Plate

Motor end plate

Myofibrils

Figure 18-9 The motor end plate is the site where α motoneurons communicate with muscle cells. At the motor end plate (myoneural junction), terminal axon branches contain acetylcholine (ACh), which can be released into the synaptic cleft and bind to receptor sites on the muscle cell surface. Enzymatic degradation of ACh by acetylcholinesterase means that continual α motoneuron activity is required for continued muscle cell contraction. From Vander, Sherman, and Luciano, 1980, p. 224. Used with permission.

potential being propagated toward each end of the fiber and into the interior of the cell via the T-tubules (Figure 18-10). (The structure and function of T-tubules are developed later in this chapter.) There is a one-to-one correspondence between an action potential that reaches a terminal axon and an action potential in the fiber that it innervates.

As mentioned, the motor end plate also contains the enzyme acetylcholinesterase, which degrades ACh into acetate and choline. The reuptake of choline is into the terminal axon area of the nerve. Ace-

tate is a local waste product and is removed by the microcirculation. Acetylcholine that is bound to its receptors is in equilibrium with the free ACh in the cleft between nerve and muscle. Thus, as free ACh drops because of acetylcholinesterase activity, so does the amount of ACh bound to the receptors. This, in turn, leads to closing of the ion channels, and the depolarized end plate returns to its resting repolarized state. It is then ready to respond to a subsequent nerve action potential.

Figure 18-10 The T system and sarcoplasmic reticulum (SR) of frog twitch fiber. The longitudinal axis of the fiber is vertical, as drawn. Slightly more than 1 sarcomere's length of fiber is shown. The SR forms a three-dimensional network of cisternae and tubules around myofibrils, with specific structural differentiation of membrane forms adjacent to specific bands of myofibrillar striations. Two levels of T-tubule networks are shown: (1) adjacent to the Z lines within I-bands of myofibrils and (2) forming central elements of three-part structures (triads), which also include two SR terminal cisternae (TC). The TC of SR connect to SR longitudinal tubules (L) either directly or through SR intermediate cisternae (IC). Near the center of the A-band, longitudinal tubules join the fenestrated collar (FC) of SR. From Peachey et al., 1983, p. 35.

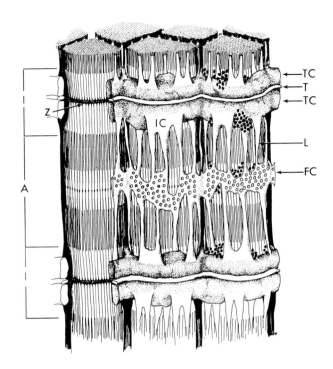

Excitation-Contraction Coupling

Voluntary muscle contraction typically originates in the motor cortex, and an action potential is propagated along the motoneuron to the motor end plate region of a muscle fiber. By virtue of a neurotransmitter (e.g., acetylcholine) at the motor end plate, a new action potential is initiated on the muscle fiber plasma membrane. The process by which the muscle membrane action potential leads to the release of intracellular calcium from the sarcoplasmic reticulum, thereby causing muscle contraction, is termed *excitation-contraction coupling.*

Muscle Membranes Involved in Excitation

In previous sections of this chapter and in Chapter 17, we discussed how the muscle sarcolemmal membrane propagates an action potential. Because this membrane is a surface membrane, other specialized membranes are needed to take the action potential from the surface down into the center of the cell

and, ultimately, allow the contractile proteins to interact. These specialized membranes are the transverse tubules and sarcoplasmic reticulum. *Transverse tubules,* commonly referred to as *T-tubules,* are invaginations of the surface membranes, each approximately 0.04 μm in diameter (see Figure 18-10). The tubules protrude transversely into the fibers, and, as a consequence, the lumen of the T-tubule remain as part of the extracellular space. Conceptually, the structure of the T-tubule is analogous to what occurs when one pokes a finger into a balloon. The T-tubule transmits the excitation signal (i.e., an action potential), which is propagating along the muscle fiber sarcolemma, to deep within the muscle fiber. At this point, the T-tubule is positioned closely to a separate membrane, the sarcoplasmic reticulum.

The *sarcoplasmic reticulum* (SR) is a complex, membranous bag that envelops each myofibril (see Figure 18-10). The muscle cell's cytoplasm is separate from the interior of the SR. The SR functions to store, release, and reuptake calcium and is thus integrally involved in fiber contraction. Portions of the SR surround the A-band and the I-band. Two enlarged regions of the SR known as *lateral sacs* (or *terminal cisternae*) are located at the end of the reticulum, connected by a network of small tubules. The lateral sacs are also in close proximity to the T-tubule network, giving the SR a key intermediate role between neural excitation and fiber contraction. The geometric relationship between the T-tubules and the SR is complex and three-dimensional. However, when viewed in a two-dimensional microscopic section, we see one T-tubule surrounded by a pair of SR lateral sacs; this arrangement is referred to as a *triad* (Figure 18-11).

Coupling of Excitation to Contraction

In a resting muscle fiber, the cytosolic concentration of free calcium is very low (i.e., 10^{-7} M). At low con-

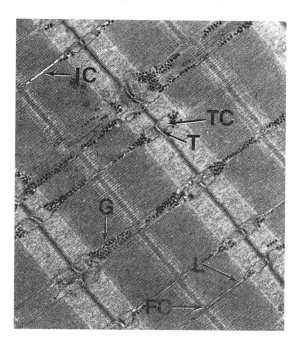

Figure 18-11 Longitudinal section of frog sartorius muscle fiber, seen in thin section in an electron microscope. T = tubules; TC = terminal cisternae; IC = intermediate cisternae; L = longitudinal tubules; FC = fenestrated collar. Glycogen granules (G) are intermingled with SR elements between the myofibrils. (× 23,000.) SOURCE: Peachey et al., 1983, p. 36.

centrations, few calcium-binding sites on troponin are occupied, and thus tropomyosin prevents the attachment of myosin crossbridges to actin. Following an action potential that is propagated to the cell's interior by a T-tubule, the cytosolic calcium concentration increases rapidly by up to 100-fold (i.e., 10^{-5} M). Calcium binds to troponin, tropomyosin is moved off the attachment sites on actin, and the crossbridge cycle is initiated (Figure 18-12).

The regulation of muscle force by calcium can be readily observed in a "skinned" muscle fiber preparation. In this type of experiment, the membranes surrounding an individual muscle fiber are disrupted, by either mechanical or chemical means, such as a detergent. This procedure then creates an experimental preparation in which calcium diffuses freely between the inside of the cell and the solution in which the cell is studied. Thus, by changing the calcium concentration in the solution that surrounds the skinned fiber, one actually is changing the intracellular concentration as well. In this preparation, the force generated by the fiber can be measured. As the calcium concentration is increased, the force that is developed also increases, as shown in Figure 18-13.

The precise mechanism by which the action potential in one membrane, the T-tubule, opens the calcium channels in another membrane, the sarcoplasmic reticulum, resulting in the diffusion of calcium into the cytosol from the sarcoplasmic reticulum, remains to be discovered. One likely mechanism involves a specialized tissue known as "electron-dense feet," or bridges that link the T-tubule and the lateral sacs of the SR. These "feet" (see Figure 18-12) are thought to serve as a communication link between certain "voltage-sensing" receptors (proteins called dihydropyradine, or DHP) in the T-tubule and the calcium-releasing channel (or ryanodine receptor) of the lateral sacs of the SR. Although the mechanism for this communication in skeletal muscle is not yet known, it appears that in cardiac muscles such communication between the receptors is essential for excitation-contraction processes (called calcium-induced–calcium-regulated contraction processes).

The membranes of the lateral sacs have active

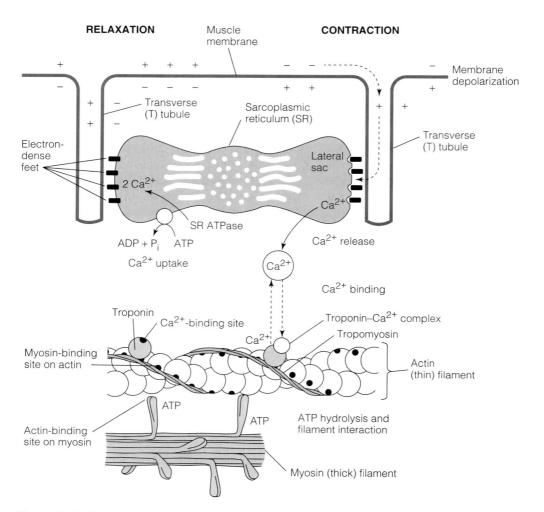

Figure 18-12   Excitation-contraction coupling is regulated in several steps. Muscle cell membrane depolarization causes the release of Ca^{2+} from the terminal cisternae of the sarcoplasmic reticulum. Calcium ion released into sarcoplasm finds binding sites on troponin, forming a troponin–Ca^{2+} complex. Binding of Ca^{2+} to troponin causes a direct conformational change in troponin and an indirect shift of tropomyosin. The shift in tropomyosin relieves its inhibition between actin and myosin, so that crossbridges can be formed and contraction can occur. Relaxation depends on pumping Ca^{2+} back into the sarcoplasmic reticulum. Modified from Vander, Sherman, and Luciano, 1980, p. 222. Used with permission.

transport proteins (i.e., calcium pumps) that pump calcium ions from the cytosol back into the lumen of the sarcoplasmic reticulum, thereby lowering the free calcium concentration in the cytosol and causing the muscle to relax. As the release of calcium is much more rapid than its reuptake, the crossbridge cycle continues for a period of time after each action potential. Thus, the process of calcium cycling is predicated on the coupling of two processes: a) the release of calcium from the sarcoplasmic reticulum

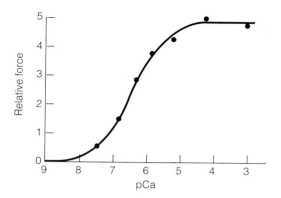

Figure 18-13 A force–pCa curve. pCa is analogous to pH in that the lower the pCa, the higher the concentration of Ca^{2+} ion in the solution. The closed symbols represent the amount of force produced by a muscle bathed in solutions with increasingly higher concentrations of Ca^{2+}. Force increases rapidly between pCa 7.5 and pCa 5.5.

by the ryanodine machinery, and b) the sequestering of calcium by the calcium pumps.

■ Motor Units, Fiber Types, and Recruitment

The Motor Unit

A *motor unit* is the functional unit of neural control for muscular activity. Motor units consist of a cell body, an alpha (α) motoneuron, and all the muscle fibers innervated by the α motoneuron (see Figure 18-5). The fibers of a given motor unit are distributed in regions of a single muscle in such a way that the fibers adjacent to each other are from different motor units. When an action potential is propagated in a single α motoneuron, all the fibers in that motor unit are stimulated and contract. The number of fibers in a given motor unit varies. In muscles that require fine motor control, such as those in the intraocular muscles of the eye, there are relatively few fibers per motor unit—they will number in the 10s to 100s. In large muscles of the leg, such as the vastus lateralis, fine control is not required, and the number of fibers per motor unit is in the 1000s.

Muscle Fiber Types

Within a given motor unit, all the muscle fibers normally have nearly identical biochemical and physiological properties. This means that fibers within a single motor unit are relatively homogeneous with respect to the concentrations and activities of metabolic, regulatory, and contractile proteins. As a consequence, the contractile properties of fibers within a motor unit are similar. Conversely, the concentrations and activities of proteins in fibers of different motor units and their resultant contractile properties may show as many as a threefold difference.

Fiber types have been determined by two general approaches. One approach has been to study small slices of muscle, termed "sections," with histochemical and immunocytochemical techniques. The second approach has been to study single fibers, motor units, and whole muscles with physiological techniques. These two general approaches are described further in the following sections.

Histochemical and Immunocytochemical Classification Muscle fiber types of motor units are typically identified by histochemical or immunocytochemical techniques done on muscle samples taken during biopsy. In histochemistry, various identifying characteristics are measured following incubations with substrates or by stains. Immunocytochemistry involves the reaction between specific protein isoforms with antibodies to the isoform. The fiber characteristics are then identified with the assistance of a light microscope and other imaging instruments. One common procedure is to characterize fibers on the basis of their immunoreactivity to antibodies that specifically bind to different myosin heavy chain isoforms (see below).

Histochemically, the "speed" of contraction of a fiber is estimated by the myofibrillar-ATPase (M-ATPase) reaction. This estimate is based on the work of R. Close (1965) and Michael Barany (1967), which demonstrated that the maximum velocity of shortening of fast and slow muscle fibers correlates positively with the presence of myosin isozymes that have high and low ATPase activity, respectively. This means that a fiber with high M-ATPase activity would normally have a high maximum velocity of shortening, and is referred to as "fast." Also, when contracting under isometric conditions, these fast fibers require a short time to reach peak tension (TPT) when stimulated in a twitch contraction. Hence, the term "fast-twitch" is often used synonymously with fast fibers. The converse is usually true for "slow" and "slow-twitch" fibers, in which low M-ATPase activity is associated with slow velocity of shortening and long TPTs. However, in some cases, such as in developing or immature muscle and muscle that is regenerating following injury, the relationship between M-ATPase activity and the speed of shortening is altered. Thus, we must be cautious in interpreting "speed" from histochemical measurements. Indeed, in response to this concern many scientists use type I and type II to indicate fibers with low and high ATPase activity, respectively. This terminology doesn't imply knowledge of the physiological speeds.

Fast-twitch fibers react dark with M-ATPase when preincubated under alkaline conditions (i.e., high pH). The reaction is reversed with acid preincubation, which then results in fast fibers appearing light. This technique, termed "acid-reversal," results in muscle sections that take on a mosaic appearance when viewed in cross-section, as in Figure 18-14.

In addition to measuring the myosin characteristics histochemically, the oxidative metabolic characteristics of a muscle fiber are often identified with a substrate that reacts with an oxidative enzyme or enzyme complex in the fiber. A reaction that identifies succinic dehydrogenase (SDHase) is frequently used. If a fiber reacts strongly for SDHase, it is interpreted to be an oxidative fiber. If a fiber reacts weakly with SDHase, it is usually assumed to have primarily a nonoxidative or glycolytic metabolism. When oxidative criteria are applied to the same fibers that have been differentiated into types I and II based on M-ATPase activity, we see that type I fibers are relatively high in oxidative capacity. Type II fibers, however, can either be high or low. This has led to the use of type IIa and IIb to indicate those that are high and low in oxidative capacity, respectively.

The glycolytic characteristics of a fiber are some-

times determined by incubating specifically for a glycolytic enzyme such as α-glycerophosphate dehydrogenase (α-GPDase). It should be noted, however, that because M-ATPase activity correlates highly with glycolytic capacity in practice, high M-ATPase activity is also often taken as evidence for a high glycolytic capacity. Thus, α-GPDase reactions are not typically done with M-ATPase incubations.

On the basis of histochemically determined characteristics, three prominent skeletal muscle fiber types have been identified: (1) type I, the slow oxidative (SO) fiber; (2) type IIa, the fast-oxidative-glycolytic (FOG) fiber; and (3) type IIb, the fast-glycolytic (FG) fiber. To the naked eye, type IIb fibers are pale or white in color, whereas type I is pink-red and IIa is red.

Recently, some powerful tools of molecular biology have been used to study the diversity of muscle fiber types. With the techniques of immunocytochemistry, it is now apparent that at least four distinct fiber types exist, at least in rat skeletal muscle. Stefano Schiaffino, among others, has shown types I, IIa, IIx, and IIb fibers (Schiaffino and Reggiani, 1994). Because IIx fibers are not distinguishable from IIa and IIb fibers when measured by traditional M-ATPase based histochemistry, some confusion exists in the scientific literature on the number of "types." However, IIx fibers in rat are numerous in most leg muscles and, in particular, in the diaphragm muscle. These fibers in rat are rich in oxidative enzymes, resist fatigue, and are thought to be positioned between IIa and IIb with respect to maximum power capacity. In humans, the gene for IIx fibers has been identified, and it may turn out that in humans there is no IIb, just IIx, and that traditional approaches have, in the past, mistaken IIx for IIb.

Physiological Classification Robert Burke and coworkers (1973, 1975, 1976) classified the motor units in cat skeletal muscles based on physiologically determined contraction time, a "sag" in the force record during unfused tetanus, and a fatigue test. These contractile events are demonstrated in the traces shown in the midregion of Figure 18-15. These classic physiological experiments resulted in a classification of motor units as

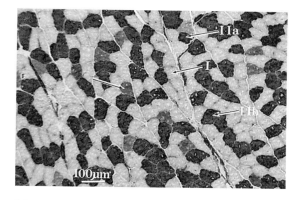

(a)

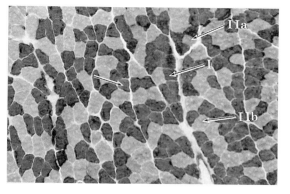

(b)

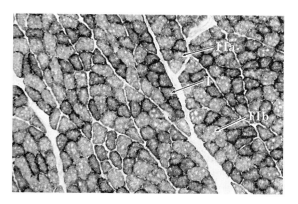

(c)

Figure 18-14 Serial cross-sections of medial gastrocnemius skeletal muscle fibers of a rodent incubated with traditional histochemical procedures. Panel (a), the section was incubated for M-ATPase at pH 10.3. Fibers that appear dark are high in ATPase values—and are often termed type II or fast twitch. Fibers low in M-ATPase are type I or slow twitch and appear to be light. In panel (b), the very same muscle fibers (the next serial section) are incubated for M-ATPase under acid conditions (i.e., pH 4.3). The response of the fibers is reversed from that of alkaline conditions. Thus, the high ATPase fast twitch fibers now appear light, while the slow type I fibers appear dark. Panel (c) depicts the same fibers from the next serial section, which has been incubated for the oxidative enzyme succinate dehydrogenase (SDH). In panel (c), high SDH is visualized in the darker areas of fibers, whereas low SDH is seen as a light fiber. Type I fibers are of high SDH, whereas type II fibers can be either high (type IIa) or low (type IIb). Some fibers do not reverse the intensity of M-ATPase when pH is changed from alkaline to acid conditions. One such fiber is identified in panels (a) and (b) by an arrow. Historically, fibers that do not demonstrate "acid reversal" have been designated as "type IIc" fibers. With our evolving understanding of myosin chemistry, it is possible that "IIc" fibers are actually composed of IIx myosin.

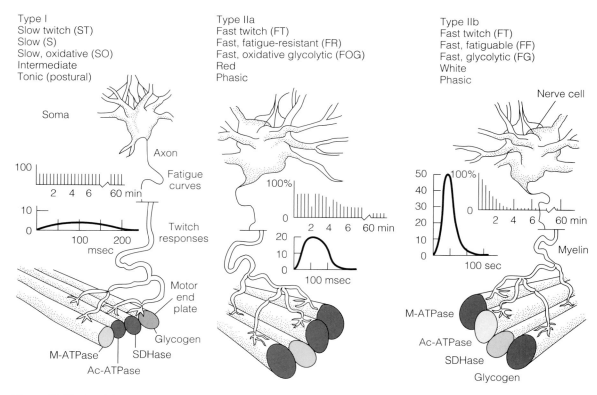

Type I
Slow twitch (ST)
Slow (S)
Slow, oxidative (SO)
Intermediate
Tonic (postural)

Type IIa
Fast twitch (FT)
Fast, fatigue-resistant (FR)
Fast, oxidative glycolytic (FOG)
Red
Phasic

Type IIb
Fast twitch (FT)
Fast, fatiguable (FF)
Fast, glycolytic (FG)
White
Phasic

Figure 18-15 In adult mammalian muscles, microscopic examination after transverse (cross) sectioning and histological staining reveals three fiber types. Fibers are classified as either type I (slow) or type II (fast), depending on intensity of staining with alkaline myofibrillar ATPase. Note that type I and II fibers demonstrate an "acid reverse" when stained for myofibrillar ATPase at low pH (Ac-ATPase). With the oxidative marker succinic dehydrogenase (SDH), type I and type IIa fibers stain dark, whereas type IIb fibers stain lightly. Glycogen content is revealed by the periodic acid Schiff (PAS) stain. The three fiber types differ in ease of recruitment, metabolism, twitch characteristics, and rate of fatigue. Alternative terminologies used to classify skeletal muscle fiber types are indicated at the top. Modified from Edington and Edgerton, 1976, p. 53. Used with permission of the authors.

slow (S), fast fatigue-resistant (FR), and fast fatigable (FF).

Comparison of Classification Schemes Classification of fibers by the different histochemical and physiological techniques provide similar results, but the similarity is not sufficient to allow unambiguous interchange among these different classification schemes. For normal limb muscles, a reasonable correlation has been established between the motor units classified by the histochemical techniques: (a) type I, IIa, and IIb and (2) slow (S), fast-oxidative-glycolytic (FOG), and fast-glycolytic (FG); and those classified by contractile properties (S, FR, FF). Figure 18-15 illustrates the three major fiber types that are commonly classified and indicates the various nomenclatures used to describe them.

Keep in mind that the technique of histochemistry, although useful, is qualitative and not quantitative. Therefore, interpreting results from histochemistry generally leads to the creation of a false dichotomy (i.e., high vs. low) for characteristics that are, in fact, on a continuum. For example, myosin ATPase activity is known to range biochemically from high to low values. Furthermore, when myosin heavy chain (MHC) is identified from muscle fibers by the techniques of electrophoresis and immunoblotting, a large number of separate isoforms can be identified. Indeed, current understanding is that, considering the number of possible combinations of MHCs and myosin light chains (MLCs), there are innumerable "fiber types."

Using contemporary biological techniques, like electrophoresis, immunoblotting, and in situ hybridization, it appears that muscle fibers from rodents and humans can express MHC types I, IIa, IIx, and IIb. These terms, which are presented in more detail in Chapter 17 for the myosin protein, correspond well with the similar terminology used in histochemistry, as described earlier in this section. Indeed, when one compares the ATPase hydrolysis rate during contraction, we see type I myosin to be low, with incremental increases in IIa, IIx, and IIb (Figure 18-16). Furthermore, some individual fibers express two different MHCs at a given moment in time, and the genes appear to be present to change the protein, given an appropriate stimulus. Clearly, as contemporary techniques in biology, and in particular molecular biology, become more powerful and discriminating, we will gain a clearer understanding of the complexity of fiber types.

Recently, investigators in the Baldwin laboratory at the University of California at Irvine developed techniques to quantitatively separate the myosin heavy chains (MHCs) from single fibers in adult and neonatal rodent skeletal muscle so that the relative content of neonatal, embryonic, type I, IIa, IIx, and IIb MHCs could be determined in different muscle fibers. They found that with the exception of the antigravity slow soleus muscle, which primarily contains fibers that express only the slow, type I MHC, most of the fibers comprising so-called "mixed fiber-type" muscles appear to normally express multiple combinations of both fast and slow

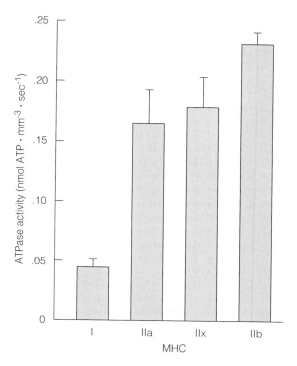

Figure 18-16   The relationship between M-ATPase activity and myosin heavy chain (MHC) isoforms. Four sets of fibers were studied, each containing a single MHC isoform: I, IIa, IIx, and IIb. The ATPase hydrolysis rate was measured during isometric contractions. Type I MHC is low in ATPase activity; this increases incrementally for IIa, IIx, and IIb. Adapted from Schiaffino, 1994.

MHCs (i.e., type I/IIa; type IIx/IIb; type IIaIIx/IIb and even type I/IIa/IIx/IIb). Thus, most fibers in mammalian muscles appear to be polymorphic with regard to the expression of the motor protein genes. The functional significance of such polymorphic expression of contractile proteins remains to be determined. For another example of current research in this area that is attempting to unravel the complexity, consider work in Stefano Schiaffino's lab. Vika Smerdu and coworkers (1994) demonstrated the existence of two human skeletal muscle genes as fast IIa and fast IIx MHC. This observation, along with earlier discoveries of slow MHC, has led to the observation in human muscle that there are three distinct fiber types (the slow type I and fast types IIa and IIx) that have only a single gene product. Two other hybrid fiber types were identified that contain I with IIa, and IIa with IIx.

The Size Principle of Motor Unit Recruitment

According to the size principle proposed by E. Henneman (1974), the frequency of motor unit use, called *recruitment,* is directly related to the size and ease of triggering an action potential in the soma. Bob Burke and Reggie Edgerton (1975) have been instrumental in enhancing our understanding of these principles in physical activity. In general, the smaller the soma (neuron cell body), the easier the motor unit can be recruited. According to the size principle, those motor units with the smaller cell bodies (e.g., slow motor units) will be used first and, overall, most frequently. Those motor units with larger cell bodies (e.g., fast-fatigable and fast fatigue-resistant motor units) will be used last during a recruitment and, overall, least frequently. Recruitment patterns are additive, not sequential. This means, for example, that during slow walking, the movement is supported primarily by slow motor units. When one breaks into a sprint, the slow units are still recruited, but, in addition, the fast units are recruited. Despite the terminology used, fiber recruitment is usually determined not by the speed of a movement but rather by the force or power necessary to perform that movement. For instance, slow motor units may be exclusively recruited while lifting a very light weight. However, in lifting a very heavy weight very slowly, many fast motor units will be recruited.

Electromyography

Electromyography (EMG) is a measurement tool that is used to identify the recruitment of motor units in skeletal muscle. Using recording electrodes that are placed over the skin or, more precisely, using fine wire electrodes that are inserted through the skin into skeletal muscle, an investigator can measure muscle electrical activity. Both the frequency and amplitude of the electrical activity can be measured, and this provides insight into actual recruitment patterns and the type and number of motor units being used during physical activity. In addition to measuring EMG, measurements of muscle force can be made, thus allowing an analysis of the relationship between force and EMG.

By studying muscle preparations in the laboratory, we have learned that the force of the overall muscle is controlled by two mechanisms during voluntary contractions. One mechanism is to alter the frequency by which a given motor unit is recruited, a concept that is called *rate-coding.* If you refer back to the frequency–force relationship illustrated in Figures 17-25 and 17-26, it will remind you that as the frequency of stimulation increases, so does the force output. This is true for a single muscle fiber, a single motor unit, and for an intact muscle. The second mechanism to increase overall force output during voluntary contractions is to recruit more motor units. This concept is clarified in Figure 18-17, in which the force recording from one, and then two, motor units is presented. It is clear from this figure that the forces are additive to each other. Figure 18-18 shows the incremental increases in overall force production of a muscle as more and larger motor units are recruited.

During normal physical activity, it is possible to record EMG and force using sophisticated implanted devices. These experiments are quite interesting, as they allow one to extrapolate information from the tightly controlled laboratory setting and

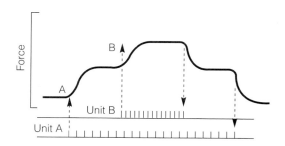

Figure 18-17   Recruitment of two motor units at different forces. The upper line is the record of force and the lower two lines the activity recorded by two separate microelectrodes. Unit A has a low force threshold, while Unit B is recruited at a higher force. Adapted from Jones and Round, 1990.

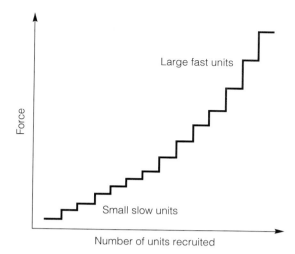

Figure 18-18   Regulation of force by recruitment. Small motor units are recruited first with the larger units coming in at the higher forces. From Jones and Round, 1990.

give insight into the relationship that occurs during normal exercise and participation in sport. An example of this relationship during walking is shown in Figure 18-19. Although there is generally a positive relationship between EMG and muscle force, many conditions including age, fatigue, velocity of movement, and so on can alter the relationship.

■ Integrative Control of Movement

Reflexes represent the simplest movements of which we are capable. They involve the integrated function of receptors, receptor afferent neurons, central nervous system (CNS), α motoneurons, and skeletal muscle. Reflexes are involuntary and automatically follow the activation of receptors. Reflexes can be very simple in function (such as withdrawal from pain) or a bit more complex, involving movements of a limb on the opposite side of the body. Also, there are certain reflex processes that are highly coordinated. For example, it is known that the motor processes that govern locomotor patterns (walking) in both humans and animals can be regulated at the spinal level because individuals with complete spinal lesions can still demonstrate normal step patterns.

More complicated and volitional movements involve some part of the brain. Most learned movement patterns, especially those movements that require sensory input (like hitting a baseball), require the participation of one or several areas of the brain. In particular, the motor cortex is in direct control of learned complex movement patterns. The coordination and modification of various movement patterns depend on the participation of various subcortical areas, including the reticular formation, cerebellum, and thalamus. The learning of intricate motor tasks involves the programming of the motor cortex. The initiation of motor tasks involves the playing back of stored programs in which motor units in active and supporting muscles are recruited in a precise pattern of time, space, frequency, and amplitude.

A few very essential movement patterns appear to be built into the CNS through the process of evolution. These primitive movements involve locomotion (e.g., swimming, walking, and running) and appear to be located in the spinal cord, as mentioned above. This segmentation hierarchy of control is advantageous, in that it allows more than one function to be performed simultaneously. These primitive spinal cord movement patterns are activated under the control of higher cortical centers.

Human movement during sports, like movement during other human endeavors, is under the ultimate control of the brain and central nervous system. Therefore, understanding the neural control of movement is of primary importance. Before describing how muscular activities are controlled by the nervous system, we first need to review its gross anatomy.

Anatomical Considerations

The central nervous system (CNS) consists of the brain and spinal cord (Figure 18-20). The peripheral nervous system (PNS) consists of all the nerves extending from the brain or spinal cord. Nerves consist of bundles of myelinated neuronal axons that may carry either sensory information to the CNS (affer-

TREADMILL LOCOMOTION, 1 mph

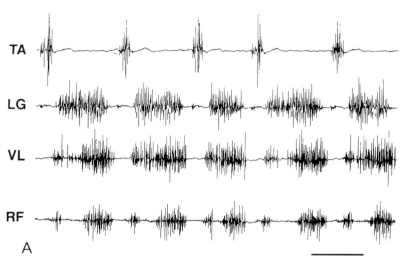

Figure 18-19 (a) Raw electromyograms obtained from rat leg muscles at a slow-walking speed. Note that muscles fire cyclically and relatively regularly during these cycles. Calibration bar = 0.25 sec. (b) Data from A rectified and averaged over the cycles. This processed EMG permits description of the typical muscle activation signal. TA = tibialis anterior; LG = lateral gastrocnemius; VL = vastus lateralis; RF = rectus femoris. (Experimental records courtesy of Dr. Sue Bodine-Fowler.) From Lieber, 1992, p. 107.

ent) or motor information from the CNS (efferent). In connection with the nervous system, the terms *afferent* (toward) and *efferent* (away from) are used to describe the direction of action potential flow.

Brain The brain consists of six major areas: the cerebrum, diencephalon, cerebellum, midbrain, pons, and medulla (Figure 18-21). The cerebrum

and diencephalon together constitute the *forebrain,* whereas the midbrain, pons, and medulla together form the *brainstem.* The outer portion of the cerebrum is termed the *cerebral cortex* (from the Greek meaning "tree bark"), and it covers most of the brain's surface (Figure 18-21a). The cortex is responsible for those functions of intellect and motor control that set human beings apart from other mam-

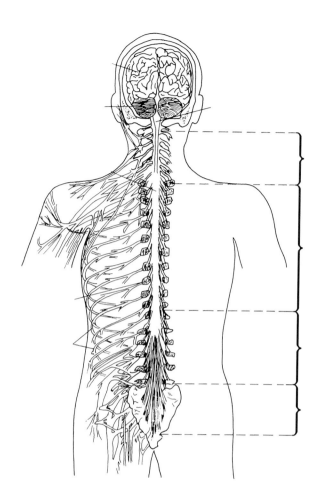

Figure 18-20 Dorsal (back) view of the nervous system. The central nervous system (CNS) consists of the brain and spinal cord. The peripheral nervous system (PNS) consists of nerves (bundles of axons) extending from the brain and spinal cord. Adapted from Vander, Sherman, and Luciano, 1980, p. 180, and from Woodburne, 1973.

malian species. The four sections (lobes) of each hemisphere of the cortex are associated with specific functions. The *frontal* lobe is responsible for high intellect and motor control. The *motor cortex* is at the rear of the frontal lobe and is anatomically and functionally associated with the red nucleus (a relay center for motor pathways located in the brain stem). The *parietal* lobe is associated with sensation and interpretation of sensory information. The *sensory cortex* is at the front of the parietal lobe. The *temporal* lobe is associated with auditory sensation and interpretation, and the *occipital* lobe is associated with visual sensation and interpretation.

The large core of the brain, the *diencephalon,* is made up of two areas: the thalamus and hypothala-

mus. The *thalamus* is an important integration center through which most sensory inputs pass. From the thalamus, neuronal signals arise for input to the cortex and cerebellum. In addition, descending signals from the cortex can be integrated with incoming sensory information by synapses in the thalamus. The *hypothalamus* is an important area where neural and hormonal functions effect a constancy of the internal body environment. It is in the hypothalamus that appetite is governed and body temperature is regulated.

The *cerebellum* is an area especially important for motor control. The cerebellum consists of an outer layer of cells, termed the *cerebellar cortex,* and deeper cells termed the *cerebellar nuclei.* The cere-

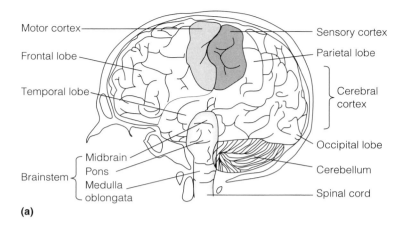

(a)

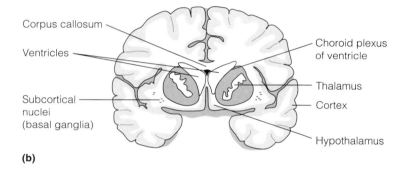

Figure 18-21 The human brain. The side view (a) shows major divisions of the brain, and the cross-sectional view (b) reveals the interior structures. From Vander, Sherman, and Luciano, 1980, pp. 181–183. Used with permission.

(b)

bellum is connected to the underlying brainstem by cerebellar peduncles. This area of the brain is mainly involved with skeletal muscle function, particularly with respect to coordinating and learning movement patterns. The cerebellum is important for posture and balance, and injuries to this portion of the brain often have catastrophic consequences for purposeful movement patterns.

All afferent and efferent signals pass through the brainstem, which, as mentioned, consist of the *midbrain, pons,* and *medulla.* The *brainstem* is an area that sets rhythmicity of breathing and controls the rate and force of breathing movements and heartbeat. Also, within the brainstem is an area called the *reticular formation,* which allows us to focus on specific sensory inputs and influences arousal and wakefulness. For instance, in baseball, the reticular

formation allows a batter to focus on the pitch rather than the taunts of opposing players and spectators.

Spinal Cord Seen in cross section, the spinal cord is divided into white and gray areas (Figure 18-22). The central gray matter consists of neuronal cell bodies with associated dendrites, short interneurons that do not leave the spinal cord, and terminal axon processes from neurons whose cell bodies are located elsewhere. Synapses occur and integration takes place in the gray area. Some areas in the gray matter can appear especially dark because of the high concentration of cell bodies. Such an area, where cell bodies having regulated functions cluster together, is called a *nucleus.*

The gray matter of the spinal cord is surrounded by white matter consisting of bundles of myelinated

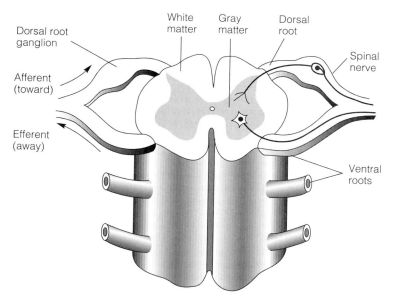

Figure 18-22 Cross-sectional view through the spinal cord (SC) and joining nerves. Gray matter contains cell bodies for motor nerves, which send axon processes out of the ventral root, as well as cell bodies of interneurons. Cell bodies for sensory nerves exist in the dorsal root ganglion. Sensory neurons also synapse in the gray matter. The SC white matter contains tracts for ascending (sensory) and descending (motor) signals.

neuronal axons running in parallel. These neuronal bundles are called tracts or pathways and are analogous to nerves in the PNS. Nerve tracts run both up (ascending) and down (descending) the spinal cord and carry sensory and motor signals. No synapses are possible in white matter, but axons may enter the gray matter to receive or transmit information (action potentials).

Groups of afferent fibers carrying sensory information enter the spinal cord on the dorsal (back) side of the body. The cell bodies of these sensory nerves are located immediately outside the spinal cord, in the dorsal root ganglion (see Figure 18-22). Groups of efferent fibers carrying motor signals leave the spinal cord on the ventral (belly) side.

Neural Control of Reflexes

Reflexes, as noted, are the simplest type of movement of which we are capable. The knee jerk reflex (Figure 18-23) involves at least four components: (1) the receptor (in this case a muscle stretch receptor, the *spindle*); (2) the gamma (γ)-afferent neurons, which synapse in the gray matter of the spinal cord (e.g., at A in Figure 18-23); (3) the α motoneurons with a cell body in the spinal cord; and (4) the

muscle fibers within the motor unit. Contrary to immediate appearance, the knee jerk reflex and other reflexes are not all-or-none responses.

Although the knee jerk response is involuntary, control is experienced at several levels. Stretching does depolarize the spindle receptor, for example, but an action potential in the γ afferent does not always result. To generate an afferent action potential or a train (series) of potentials, there must be a summation of receptor potentials. At the level of synapse with the α motoneuron, γ-afferent activity results in excitatory postsynaptic potentials. These must summate to result in an action potential in the α motoneuron. Because the α motoneuron resting membrane potential is also subject to inhibitory postsynaptic potentials, a degree of voluntary inhibition is also possible.

The simplest reflexes are monosynaptic. As illustrated in Figure 18-23, synapses with other neurons such as interneurons are also possible. At point B in the figure, the interneuron stimulated will release an inhibitory neurotransmitter on the soma of an antagonistic motoneuron—in this case, one that innervates a motor unit in one muscle of the hamstring muscle group. Thus, the antagonist hamstring muscle will stay relaxed while an agonist muscle in

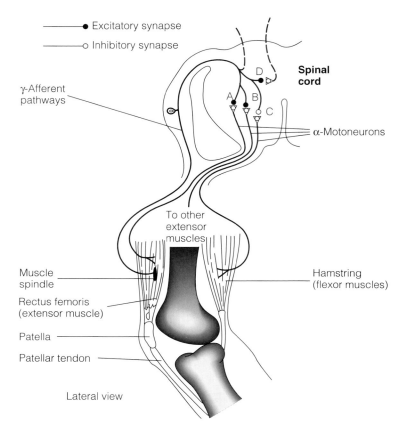

Figure 18-23   The knee jerk reflex involves a minimum of four components: the muscle spindle stretch receptor, the γ-afferent neuron, the α motoneuron for the rectus femoris extensor muscle, and the rectus femoris. Sufficient rapid stretching of the spindle results in reflex contraction of the rectus femoris (A) and other extensors in the quadriceps (B), inhibition of the soma of motor neurons innervating the antagonist hamstring flexors (C), and ascending signals (D), perhaps to motor units in arms and back. From Vander, Sherman, and Luciano, 1980, p. 594. Used with permission.

the quadricep group—for example, the rectus femoris muscle—contracts to perform the actual knee jerk. In this terminology, the *agonist* is the muscle primarily responsible for movement, and the *antagonist,* the muscle that retards movement.

Other synapses possible during a knee jerk reflex are those that activate motor units in other muscles that function synergistic to the knee jerk, such as the vastus lateralis muscle (at point C in Figure 18-23), or that send ascending signals to the brain or to motor units innervating a contralateral limb (at point D).

Muscle Spindles and the Gamma Loop

Other organelles of importance to coordinated physical activity are *muscle spindles,* connective tissue capsules in the shape of footballs that are filled

with lymph and specialized fibers. Spindles are receptor organelles that have both afferent and efferent innervation and are implanted between muscle fibers. The fibers within spindles are termed *intrafusal* fibers (in distinction to muscle fibers, which are *extrafusal* fibers). Intrafusal fibers lie parallel to the extrafusal fibers, and when the muscle is stretched, the intrafusal fibers are also stretched and the receptor endings are activated. The most important functional attribute of spindles is that they provide information to integrating centers in the spinal cord and brain regarding the absolute length of muscle fibers as well as the rate of change of fiber length (i.e., the change in length over time).

Muscle spindles (Figure 18-24) respond to the amount and rapidity of stretch. The rapid stretching (lengthening) of spindles in the quadriceps that results when the patellar tendon is struck is a

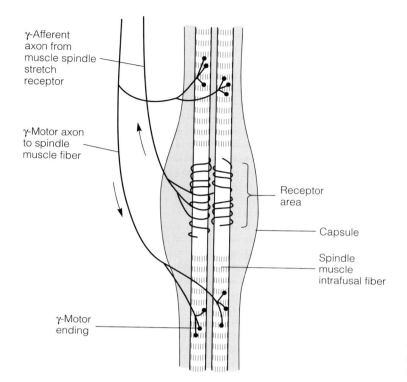

γ-Afferent axon from muscle spindle stretch receptor

γ-Motor axon to spindle muscle fiber

Receptor area

Capsule

Spindle muscle intrafusal fiber

γ-Motor ending

Figure 18-24 Muscle spindles exist in parallel with normal extrafusal skeletal muscle fibers. Rapid stretching of the muscle also stretches the sensitive receptor area of the spindle, and action potentials are sent to the spinal cord via the γ-afferent neuron.

rather extreme example of the spindle's range of responses. In real life, we use the full jerk response only in emergency situations, such as when a foot slips and it is necessary to maintain posture to keep from falling. The spindle, however, works continuously to monitor muscular movement. Muscle spindles exist in parallel with normal extrafusal muscle fibers (Figure 18-25). When a volitional muscular activity is initiated by activation of α motoneurons, there is also activation (coactivation) of smaller, γ motoneurons. These efferent γ neurons cause contraction of muscle fibers within the muscle spindle (see Figure 18-24). Contraction of the fibers within the muscle spindle capsules (intrafusal fibers) in effect takes up slack within the spindle capsule during contraction of the larger extrafusal fibers. In this way, the central receptor portion of the spindle stays at relatively the same length. The receptor can then respond to sudden changes in length throughout almost the full range of limb movement. The system through which muscle spindles partici-

pate in monitoring muscular activity is the γ-loop (Figure 18-26).

Golgi Tendon Organs

Golgi tendon organs (GTOs) are receptors that respond to tension rather than to length (as do muscle spindles). Golgi tendon organs are high-threshold receptors that exert inhibitory effects on agonist muscles and facilitatory effects on antagonist muscles. When the forces of muscle contraction and the forces resulting from external factors sum to the point where injury to the muscle tendon or bone becomes possible, then the GTOs cause inhibitory postsynaptic potentials (PSPs) on the cell body of the agonistic motor units. Similarly, when shortening during muscle contraction progresses to the point where continued shortening could damage the joint because of hyperextension or hyperflexion, the GTOs act to shut off the agonist (with ISPSs) and stimulate the antagonist [with excitatory

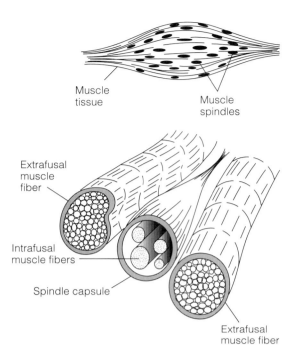

Muscle tissue

Muscle spindles

Extrafusal muscle fiber

Intrafusal muscle fibers

Spindle capsule

Extrafusal muscle fiber

Figure 18-25 Muscle spindles exist in parallel with extrafusal muscle fibers. Stretch of the muscle results in stretch of the spindle capsule and its contents. Modified from Edington and Edgerton, 1976. Used with permission.

postsynaptic potentials (ESPS)]. In this way, GTOs bring about smooth retardation of muscular contractions.

The GTO mechanism is not, however, a fail-safe mechanism. Because the GTOs influence motoneuron cell bodies with inhibitory postsynaptic potentials, their effects can be counterbalanced by additional excitatory postsynaptic potentials from higher centers. The process of minimizing the influence of GTOs is referred to as *disinhibition*. Indeed, practicing disinhibition appears to be part of athletic training, the purpose of which is to push performance to the limits of tissue capacity. Indeed, in the sport of wrist wrestling, ruptured muscles or tendons and broken bones occasionally occur. In highly motivated and disinhibited individuals, the combination of active muscle contraction plus

tension exerted by the opponent can exceed the strength of tissues.

Corticospinal Tract

The corticospinal tract (Figure 18-27) consists of bundles of neurons whose cell bodies exist in the motor cortex, whose axons extend through the spinal cord, and whose terminal processes communicate with α motoneurons in the gray matter of the spinal cord. Because portions of the corticospinal tract run from the top of the head to the lower back, the tract can be quite long. The cell bodies from which the corticospinal tract originates are pyramidal in shape; therefore, the tract is alternatively referred to as the *pyramidal tract, pathway,* or *system.*

The corticospinal pathway is the major effector of complicated and rapid volitional movements. The late Dr. Franklin Henry compared the function of the motor cortex to the running of a computer program. Because at the time that Henry described his hypothesis, computer programs were stored on magnetic drums (rather than on the more contemporary diskettes and CDs), the theory is now known as Henry's memory drum theory. According to this theory, intricate movements—which depend on the precise contraction of motor units in different muscles that are ordered specifically in terms of space, time, intensity, and duration—are "programmed" by an incredibly precise order of depolarization of the pyramidal cells in the motor cortex. At present, the precise mechanism by which memory is accomplished is not understood. Learning may well involve the synthesis of proteins or neurotransmitters in specific neurons or areas of neurons. The precise imprinting mechanism involved in learning rapid, precise performance of an intricate motor task depends on a computer-program-like sequencing of events in the motor cortex.

Multineuronal, Extrapyramidal Pathways

As shown in Figure 18-28, in addition to the pyramidal pathway, descending motor pathways also

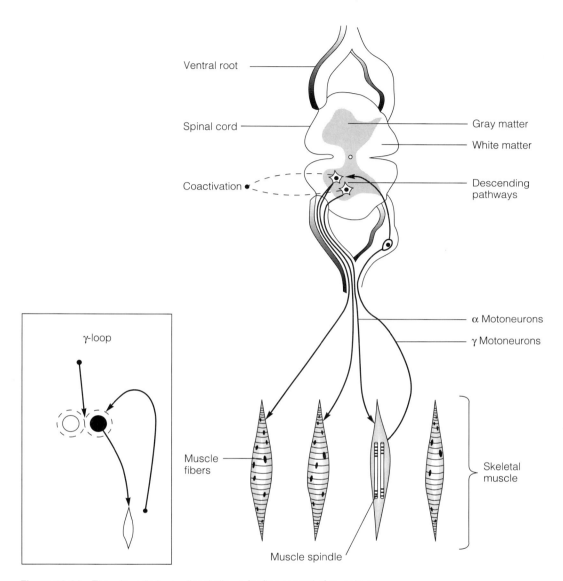

Ventral root

Spinal cord

Coactivation

Gray matter

White matter

Descending pathways

α Motoneurons

γ Motoneurons

γ-loop

Muscle fibers

Skeletal muscle

Muscle spindle

Figure 18-26 The γ-loop (in heavy lines) allows for fine control of muscle position in complex movement patterns. Descending signals result in activation of both α and γ motoneurons (coactivation). Stretch on the spindle receptors provides feedback on muscle position via the γ afferents. Input of the γ-loop back to the α motoneuron cell body affects its firing.

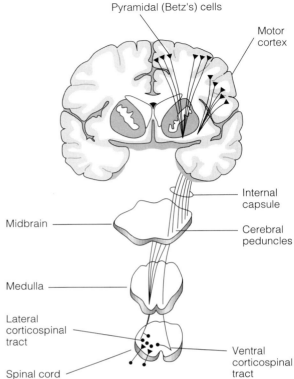

Figure 18-27 The pyramidal (corticospinal) tract originates from the pyramid-shaped Betz's cells in the cerebral cortex. Axons from the motor cortex cross over in the medulla and run down in the ventral and lateral white matter of the spinal cord. Bundles of descending axons make up a tract. Synapses occur with motor neurons in the gray area of the spinal cord at appropriate levels. Adapted from Vander, Sherman, and Luciano, 1980, p. 598. Used with permission.

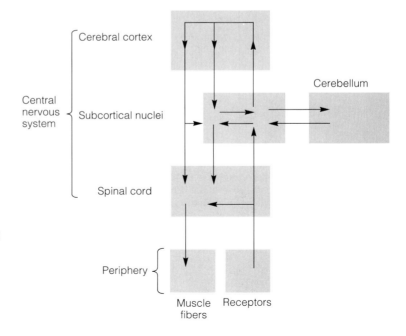

Figure 18-28 General scheme of motor control. The cerebral motor cortex initiates movement by sending descending signals to muscle fibers and to subcortical areas, which relay signals to the cerebellum. Various receptors provide sensory feedback, and the cerebellum allows a comparison (integration) of actual and intended movements. Subsequent activity initiated by the motor cortex is varied depending on the initial results. From Vander, Sherman, and Luciano, 1980. Used with permission.

exist. These extrapyramidal pathways are multineuronal in structure. Extra synaptic connections slow conduction velocity, but they allow the integration of motor programs with sensory inputs, including progress on the result of programs originating in pyramidal cells. In addition, the extrapyramidal system allows the cerebellum to become involved in the coordination of motor activities.

Spinal Movements

The colloquial expression, "running like a chicken with its head cut off," carries for us some useful information concerning the spinal control of locomotion. Although swimming, walking, and running are usually considered to be learned activities controlled by the pyramidal system, many anesthetized species will sprint when suddenly decapitated. Some of the victims of the guillotine during the French Revolution are reported to have run away from the block, but only after the blade had fallen. In residential areas where swimming pools are common, infants too young to walk are frequently taught to swim.

On the basis of these observations, we must conclude that motor patterns either can be learned by or are built into the spinal cord. These movement patterns are not usually apparent but are probably under the ultimate control of the motor cortex. For these movement patterns, the cortex does not necessarily determine each action potential that will occur in each motor unit involved; rather, the cortex may function to activate a spinal locomotory pattern and to set the frequency and amplitude of the movement pattern. Thus, it is not surprising that when the spinal cord is suddenly cut, a procedure called spinal transection, the mechanical stimulation of descending tracts leads to activation of a whole series of responses, one of which occasionally is to activate running motions.

V. Reggie Edgerton and colleagues at UCLA have been able to induce cats with complete spinal transections to walk. They can do this by providing appropriate sensory stimuli to the spinal cord in animals that cannot walk under their own control. John Hodgson and coworkers in Reggie Edgerton's lab (1994) have found that when the spinal cord is transected at the level of thoracic vertebrae 12–13, an immediate loss of locomotor function occurs in the hindlimbs. After several months, limited locomotor function returns. In their experiments, the stimuli are provided by placing the cat that has a spinal transection in a body sling so it can be suspended carefully over a slowly moving treadmill. As the feet come in contact with the treadmill belt, the animals emulate a stepping pattern that is quite similar in outward appearance to that of a normal cat during locomotion. In contrast to the several months it takes for locomotory function to return in the absence of therapy, when transected cats are trained to step on the treadmill, stepping ability develops much sooner. Indeed, when stepping therapy is initiated early following cord transection, the results were the most beneficial. In comparing cats with and without training for stepping and for standing, these investigators conclude that the spinal cord is capable of learning specific motor tasks. This research holds great promise not only for advancing our understanding of motor control, but also for the rehabilitation of para- and quadriplegics. Presently, these techniques are being extended to humans with spinal injuries.

Removing the immediate control of locomotion from the cortex and allowing a lower level in the CNS to control movement has several theoretical advantages. It may allow for more direct and immediate (faster) control, and it frees higher centers to perform other tasks. When chasing and catching a fly ball, an outfielder in baseball does not have to think about how each muscle functions during running; similarly, the relay runner can concentrate on timing his or her start and speed while focusing on receiving the baton.

■ Motor Control and Learning

Volitional and Learned Movements

The general scheme by which the brain controls motor activities is illustrated in Figure 18-28. According to this scheme, movement is initiated in the motor cortex, which sends commands (action potentials) directly via a spinal axon to α motoneurons whose cell bodies are located in the spinal cord. Simultaneously, or perhaps with a slight delay, the cortex signals subcortical nuclei, which receive input from sensory receptors. Meanwhile, the cerebellum constantly compares the intended with actual movements and integrates each sequential movement component into a coordinated effort.

Sensory Input During Movement

During motor activities, an individual uses, at various times, most of the sensory information available. The contributions of various forms of sensory information in leading to the successful completion of an activity will vary depending on the activity, but afferent inputs from the eyes, inner ear, muscle spindles, joints, and skin are very important. Different individuals may, in fact, perform identical motor tasks while relying on different sensory inputs to various degrees. Depending on the activity and sense involved, some individuals can adapt and successfully complete motor tasks even though deprived of a usually important form of sensory input. Sometimes also in the performance of motor tasks, receipt of too much information confuses individuals, distracting them from relevant inputs. Concentration and focusing are tasks governed by the frontal cortex and the reticular formation.

A detailed treatment of motor control is beyond the scope of this book and most courses in exercise physiology. But we would be remiss to not mention the provocative work of James C. Houk and coworkers (Keifer and Houk, 1994), in which they linked the known anatomical and physiological features of the cerebellorubrospinal circuit and proposed a model for function.

The cerebellorubrospinal tract is a major crossed pathway arising from neurons in the red nucleus in the motor cortex area of the brain. These neurons project to the contralateral brain stem and to specific targets on the contralateral spinal cord (Figure 18-29). These targets are organized on a somatotopic or topographic basis, meaning they influence different limbs, for example, at different targets

along the spinal cord. Innumerable feedback connections exist in the cerebellorubrospinal pathway.

A significant amount of sensory input to motor output occurs in the cerebellorubrospinal system. A significant amount of sensory information regarding the motion of a limb, for example, is transmitted to the cerebellum. On the other hand, signals in the cerebellorubrospinal circuit on the descending output side correspond to movement patterns and have little sensory activity. Also, there is good evidence of sensory inhibition during specific types of movements, which leads to the conclusion that motor commands are not generated by continuous sensory feedback from the periphery, although this is certainly part of it. But, in addition, Houk and coworkers identify the cerebellorubrospinal system as also containing a significant open-loop feed-forward component, as shown in Figure 18-30. This model postulates how internal signals for movement could take the place of sensory signals.

Motor Learning

Earlier in this chapter, in the subsection on the corticospinal tract, we discussed the concept of Henry's memory drum theory with regard to the learning of motor skills. Although that theoretical concept is useful, the location and nature of learning by the CNS remains largely unknown. Jonathan Wolpaw (1994) suggests that with the expansion of knowledge in the neurosciences, it is becoming increasingly clear that learning can occur, not only in the cerebellum and hippocampus, but at many other places throughout the CNS. In other words, the plasticity, which constitutes motor learning and improved performance, is widely distributed throughout the CNS.

Consider one example, in which human and nonhuman primates can slowly increase or decrease the size of the tendon jerk response (more properly termed the spinal stretch reflex) when rewards are given based on change. This is true when all supraspinal influence is removed, implying that the learning has occurred actually in the spinal cord or its branches. Current work is focusing attention on group Ia afferent synapses on the α motoneuron

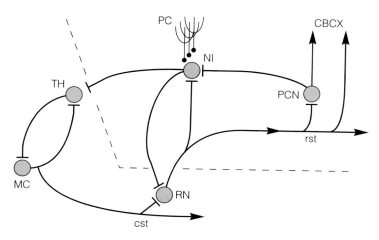

Figure 18-29 Schematic diagram of the cerebellorubral circuit. The red nucleus (RN) projects to all levels of the contralateral spinal cord and sends collateral axon branches to the precerebellar nuclei (PCN) and the cerebellar cortex (CBCX). Rubrospinal tract (rst) axons also form a minor pathway to nucleus interpositus (NI). The major input to the red nucleus is from the nucleus interpositus in the intermediate cerebellum. Inputs to interpositus arise from axon collaterals of precerebellar nuclei. Another major input to the red nucleus is from the motor cortex (MC). This pathway is restricted to mammals. The red nucleus receives signals from the intermediate cerebellum by way of the ventral thalamus (TH) and its reciprocal connections with the motor cortex. Thus, the cerebellorubral circuit is characterized by numerous feedback loops that are postulated to provide a neuroanatomic basis for a positive feedback circuit. SOURCE: Keifer and Houk, 1994, p. 514.

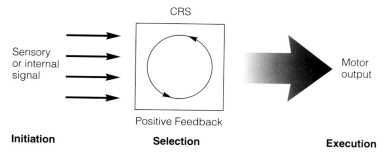

Figure 18-30 The theory of motor command generation in the cerebellorubral circuit. Detailed sensory information about limbs is highly represented in inputs to the circuit. These inputs may serve to initiate movement commands, or movement can be triggered by internal signals. Excitation diverges to other clusters of neurons in the circuit not originally activated by the sensory cue. Output signals of the cerebellorubral circuit are characterized by representing movement parameters and have little sensory content. The execution of a motor command results in coordinated movements of the whole limb with distal extremities such as the hand. Hence, function of the cerebellorubral circuit operates under a general plan of open-loop feed-forward control, while using positive feedback to sustain activity underlying the formation of motor commands. SOURCE: Keifer and Houk, 1994, p. 532.

and the motoneuron itself, as these are the most probable sites for change. This work has implication for reflex actions, like the tendon jerk, but also for more complicated movements involved in sports and dance.

■ Summary

The ability of nerve and muscle cells to respond to chemical (neurotransmitter) stimuli ultimately governs all human motor behavior. The ability to think, to breathe, to have a heartbeat, and to move reflexively or volitionally all depend on excitability in nerve and muscle cells. The characteristic of excitability depends on two cell membrane phenomena: (1) a $Na^+–K^+$ ATPase exchange pump and (2) a greater permeability to potassium (outward) than to sodium (inward). In response to appropriate stimuli, the resting membrane voltage of excitable cells is perturbed sufficiently to generate action potentials. Action potentials represent the basic message unit of communication among neurons and between neurons and muscle fibers. The transmitter substance released from the terminal ends of neurons as the result of action potentials, the frequency of action potentials, their spatial distribution, and related factors determine the recruitment of motor units.

The motor unit, which comprises the basic functional unit of the muscular system, also represents the final means by which the nervous system originates and controls muscular activity. All muscle fibers innervated by an α motoneuron in a motor unit have the same physiological characteristics. On the basis of metabolic and contractile characteristics, three basic muscle fiber types have been identified. This is known, however, to be a simplification of a very complex state of underlying attributes of muscle myosin and metabolic parameters. The amount of force developed by a muscle *in vivo* is controlled by two means. One way is to recruit a varying number of motor units, and the other is to regulate the neural frequency by which recruited motor units are activated.

The biological control of muscle movement depends on the central nervous system (CNS). Even the simplest form of movement involves at least two neurons and a synapse in the CNS. As motor activities become more complex, the center of neuromuscular control tends to move up the CNS and include the cerebral cortex of the brain. Learning intricate motor tasks involves programming the area of the brain's frontal lobe, called the motor cortex. Once the motor cortex is imprinted with a program, the program can be played back very rapidly and accurately.

Perhaps the most difficult motor tasks include those in which locomotion and a fine eye–hand or eye–foot skill are performed simultaneously. For instance, consider running to and hitting a tennis ball. In this situation, programs learned at various levels of the CNS are integrated with sensory input to produce a coordinated response. The CNS represents a very sophisticated computer that can contain large quantities of information, can simultaneously make decisions and affect actions of differing nature, and can constantly monitor and correct the results obtained.

■ Selected Readings

Alford, E., R. R. Roy, J. A. Hodgson, and V. R. Edgerton. Electromyography of rat soleus, gastrocnemius, and tibialis anterior during hind limb suspension. *Exp. Neurol.* 96: 635–649, 1987.

Arsharsky, Yu I., M. B. Berkinblit, I. M. Gelfand, G. M. Orlovsky, and O. I. Fukson. Activity of the neurons of the ventral spinocerebellar tract during locomotion. *Biophysics* 17: 926–935, 1972.

Barany, M. ATPase activity of myosin correlated with speed of muscle shortening. *J. Gen. Physiol.* 50: 197–218, 1967.

Brooke, M. H., and K. K. Kaiser. Muscle fiber types: how many and what kind? *Arc. Neurol.* 23: 369–379, 1970.

Burke, R. E., and V. R. Edgerton. Motor unit properties and selective involvement in movement. In Exercise and Sports Sciences Reviews, J. Wilmore and J. Keogh (Eds.). New York: Academic Press, 1975, pp. 31–83.

Burke, R. E., D. N. Levine, F. E. Zajac III, P. Tsairis, and W. K. Engel. Physiological types and histochemical profiles in motor units of the cat gastrocnemius. *J. Physiol.* (London) 234: 723–748, 1973.

Burke, R. E., P. Rodomin, and F. E. Zajac III. The effect of activation history or tension produced by individual muscle units. *Brain Res.* 109: 515–529, 1976.

Caiozzo, U. S., M. J. Baker, and K. M. Baldwin. Novel transitions in MHC: isoforms: separate and combined effects of thyroid hormone and mechanical unloading. *J. Appl. Physiol.* 85: 2237–2248, 1998.

Carew, T. J. Descending control of spinal circuits. In Principles of Neural Science, E. R. Kandel and J. H. Schwartz (Eds.). New York: Elsevier/North-Holland, 1981, pp. 312–322.

Carew, T. J. Spinal cord I: muscles and muscle receptors. In Principles of Neural Science, E. R. Kandel and J. H. Schwartz (Eds.). New York: Elsevier/North-Holland, 1981, pp. 284–292.

Carew, T. J. Spinal cord II: reflex action. In Principles of Neural Science, E. R. Kandel and J. H. Schwartz (Eds.). New York: Elsevier/North-Holland, 1981, pp. 293–304.

Close, R. The relation between intrinsic speed of shortening and duration of the active state of muscle. *J. Physiol.* (London) 180: 542–559, 1965.

Edgerton, V. R., C. P. de Guzman, R. J. Gregor, R. R. Roy, J. A. Hodgson, and R. G. Lovely. Trainability of the spinal cord to generate hindlimb stepping patterns in adult spinalized cats. In Neurophysiological Bases of Human Locomotion, M. Shimamura, S. Grillner, and V. R. Edgerton (Eds.). Berlin: Springer-Verlag, 1990.

Edington, D. W., and V. R. Edgerton. The Biology of Physical Activity. Boston: Houghton Mifflin, 1976, pp. 51–72.

Evarts, E. V., and J. Tanji. Reflex and intended responses in motor cortex pyramidal tract neurons of monkey. *J. Neurophysiol.* 39: 1069–1080, 1976.

Forssberg, H., S. Grillner, and S. Rossignol. Phase dependent reflex reversal during walking in chronic spinal cats. *Brain Res.* 85: 103–107, 1975.

Ghez, C. Cortical control of voluntary movement. In Principles of Neural Science, E. R. Kandel and J. H. Schwartz (Eds.). New York: Elsevier/North-Holland, 1981, pp. 324–333.

Grillner, S., and P. Zangger. How detailed is the central pattern generation for locomotion? *Brain Res.* 88: 367–371, 1975.

Henneman, E. Organization of the spinal cord and its reflexes. In Medical Physiology, 14th ed., V. B. Mountcastle (Ed.). Vol. 1. St. Louis: Mosby, 1980, pp. 762–786.

Henneman, E., H. P. Clamann, J. D. Gillies, and R. D. Skinner. Rank order of motoneurons within a pool, law of combination. *J. Neurophysiol.* 37: 1338–1349, 1974.

Henry, F. M. Increased response latency of complicated movements and a "memory drum" theory of neuromotor reaction. *Res. Quart.* 31: 448–458, 1960.

Henry, F. M. Influence of motor and sensory sets on reaction latency and speed of discrete movements. *Res. Quart.* 31: 459–468, 1960.

Henry, F. M. The evolution of the memory drum theory of neuromotor reaction. In Perspectives on the Academic Discipline of Physical Education, G. A. Brooks (Ed.). Champaign, Ill.: Human Kinetics, 1981, pp. 301–322.

Hodgson, J. A., R. R. Roy, R. deLeon, B. Dobkin, and V. R. Edgerton. Can the mammalian lumbar spinal cord learn a motor task? *Med. Sci. Sports Exer.* 26(12): 1491–1497, 1994.

Houk, J. C. Motor control processes: new data concerning motoservo mechanisms and a tentative model for stimulus-response processing. In Posture and Movement, R. E. Talbot and D. R. Humphrey (Eds.). New York: Raven Press, 1979, pp. 231–241.

Houk, J., and E. Henneman. Responses of Golgi tendon organs to active contractions of the soleus muscle of the cat. *J. Neurophysiol.* 30: 466–481, 1967.

Johnson, E. J., A. Smith, E. Eldred, and V. R. Edgerton. Exercise-induced changes of biochemical, histochemical, and contractile properties of muscle in cordotomized kittens. *Exp. Neurol.* 76: 414–427, 1982.

Jones, D. A., and J. M. Round. Skeletal muscle in health and disease. New York: Manchester University Press, 1990.

Kandel, E. R., and J. H. Schwartz (Eds.). Principles of Neural Science. New York: Elsevier/North-Holland, 1981.

Katz, B. Nerve, Muscle and Synapse. New York: McGraw-Hill, 1966.

Keifer, J., and J. C. Houk. Motor function of the cerebellorubrospinal system. *Physiol. Rev.* 74: 509–533, 1994.

Kuypers, H. G. J. M. The anatomical organization of the descending pathways and their contributions to motor control especially in primates. In New Developments in Electromyography and Clinical Neurophysiology, J. E. Desmedt (Ed.). Vol. 3. Basel: Karger, 1973, pp. 38–68.

Liddell, E. G. T., and C. Sherrington. Reflex in response to stretch (myotatic reflexes). *Proc. Roy. Soc.* (London) Ser. B. Biol. Sci. 96: 212–242, 1924.

Lieber, R. L. Skeletal Muscle Structure and Function. Baltimore: Williams and Wilkins, 1992.

Lundberg, A. Integration in the propriospinal motor center controlling the forelimb in the cat. In Integration in the Nervous System, H. Asanuma and V. J. Wilson (Eds.). Tokyo: Igaku-Shoin, 1979, pp. 47–64.

Mathews, B. H. C. Nerve endings in mammalian muscle. *J. Physiol.* (London) 78: 1–53, 1933.

Mathews, P. B. C. Muscle spindles and their motor control. *J. Physiol. Rev.* 44: 219–288, 1964.

McMahon, T. A. Muscles, Reflexes and Locomotion. Princeton, N.J.: Princeton University Press, 1984.

Merton, P. A. Voluntary strength and fatigue. *J. Physiol.* (London) 123: 553–564, 1954.

Merton, P. A. How we control the contraction of our muscles. *Sci. Amer.* 226: 30–37, 1972.

Nemeth, P., and D. Pette. Succinate dehydrogenase activity in fibers classified by myosin ATPase in hind limb muscles of rat. *J. Physiol.* (London) 320: 73–80, 1981.

Peter, J. B., R. J. Barnard, V. R. Edgerton, C. A. Gillespie, and K. E. Stemple. Metabolic profiles of three fiber types of skeletal muscle in guinea pigs and rabbits. *Biochem.* 14: 2627–2633, 1972.

Pette, D., and R. S. Staron. Cellular and molecular diversities of mammalian skeletal muscle fibers. *Rev. Physiol. Biochem. Pharmacol.* 116: 1–76, 1990.

Phillips, C. G., and R. Porter. Cortical Spinal Neurons: Their Role in Movement. London: Academic Press, 1977.

Reback, P. A., A. B. Scheibel, and J. L. Smith. Development and maintenance of dendrite bundles after cordotomy in exercised and non-exercised cats. *Exp. Neurol.* 76: 428–440, 1982.

Schiaffino, S., and C. Reggiani. Myosin isoforms in mammalian skeletal muscle. *J. Appl. Physiol.* 77 (2): 493–501, 1994.

Sherrington, C. The Integrative Action of the Nervous System. 2d ed. New Haven: Yale University Press, 1947.

Smerdu, V., I. Karsch-Mizraehi, M. Campione, L. Leniward, and S. Schiaffino. Type IIx myosin heavy chain transcripts are expressed in type IIb fiber of human skeletal muscle. *Am. J. Physiol.* 267: C1723–C1728, 1994.

Stein, R. B. Peripheral control of movement. *Physiol. Rev.* 54: 215–242, 1974.

Towe, A. L., and E. S. Luschei. Motor coordination. In Handbook of Behavioral Neurobiology, J. Schoff (Ed.). Vol. 4. Biological Rhythms. New York: Plenum Press, 1970.

Vallbo, A. B. Muscle spindle response at the onset of isometric voluntary contractions in man. Time difference between fusimotor and skeletomotor effects. *J. Physiol.* (London) 218: 405–431, 1971.

Vander, A. J., J. H. Sherman, and D. S. Luciano. Human Physiology. 3d ed. New York: McGraw-Hill, 1980, pp. 144–190.

Vander, A. J., J. Sherman, and D. S. Luciano. Human Physiology. 5th ed. New York: McGraw-Hill, 1990.

Walmsley, B. J., J. A. Hodgson, and R. E. Burke. Forces produced by medial gastrocnemius and soleus muscles during locomotion in freely moving cats. *J. Neurophysiol.* 41: 1202–1216, 1978.

Wolpaw, J. R. Acquisition and maintenance of the simplest motor skill: investigation of CNS mechanisms. *Med. Sci. Sports Exer.* 26(12): 1475–1479, 1994.

Woodburne, R. T. Essentials of Human Anatomy. 5th ed. New York: Oxford University Press, 1973.

CHAPTER 19

PRINCIPLES OF SKELETAL MUSCLE ADAPTATIONS

In Chapters 17 and 18, we learned about how skeletal muscle functions to produce force and movement and about how the nervous system interacts with skeletal muscle, resulting in coordinated patterns of physical activity. Having established these basic parameters, we now examine how repetitive use of skeletal muscle, which we typically call physical training, can cause adaptive changes in the structure and functional properties of muscle.

This chapter builds on the earlier material by presenting the principles by which muscle structure and function adapt to increases and decreases in habitual levels and types of physical activity. It also links the reader to Chapters 20 and 21, which focus on the practical applications of the principles of muscle adaptation developed here.

The elegance of a skilled movement reflects the contributions of heredity and the adaptations of training. Specific structural and functional attributes of skeletal muscles are inherited. This genetic inheritance sets unknown limits—a performance envelope if you will—within which attributes of

A tensing pose by Albert Treloar. Bodybuilders have developed muscle size and definition to an amazing degree. This adaptive response results from a combination of a rigorous training program, proper nutrition, and a genetic predisposition for a large muscle mass. The principle underlying muscle adaptation to training is developed in this chapter. PHOTO: Gollnick, et al. 1972. Used with permission.

skeletal muscle fibers are able to adapt to habitual patterns of use and disuse. Physical training, then, is a means by which individuals can push to the limits of their own individual performance envelope and express their potential for a given physical activity. One of the inevitable consequences of an active lifestyle is injury to skeletal muscle fibers, to which muscle also demonstrates a remarkable adaptive response. Thus, we have included in this chapter on adaptations resulting from exercise training a description of the regenerative process.

In this context of adaptive change, we also present important details on the intrinsic adaptations of skeletal muscle to the aging process, as well as the intrinsic differences in skeletal muscle between the genders. Skeletal muscle, perhaps more than any other system, defines aging and gender differences with respect to the potential for human performance in exercise and sports. Chapters 30 and 32 focus more broadly on differences in performance and in many of the physiological systems due to gender and aging, respectively.

■ Principle of Myoplasticity

There is tremendous potential to alter the gene expression of skeletal muscle. When gene expression is altered, it results in an increase and/or decrease in the amount of specific muscle proteins. This capacity for adaptive change is termed *plasticity*. The term *myoplasticity* refers specifically to the capacity of skeletal muscle for adaptive change. Altered gene expression is the molecular basis for adaptations that occur due to exercise training in skeletal muscle proteins.

Approximately 20% of skeletal muscle is composed of proteins, with the balance being water, salts, and other organic molecules. All types of proteins can be regulated by alterations in gene expression. The cascade of regulatory events impacting expression of any given gene is depicted in Figure 19-1. In the context of gene expression, *myoplasticity* is predicated on the ability of the muscle fiber's genetic machinery to change either the quantity (amount) or quality (specific type) of protein it

expresses. For example, a particular muscle fiber (a fast IIb fiber) may undergo hypertrophy (enlargement) in response to resistance training by increasing the net amount of all the protein normally comprising the fiber. On the other hand, another IIB fiber may not only enlarge its fiber size, but may also repress the gene encoded for expression of the fast IIb myosin heavy chain (MHC) while turning on expression of the fast IIa MHC. In the latter case, the fiber has not only enlarged, but it has also changed its contractile phenotype, thereby becoming a larger but slower contracting fiber; whereas in the former case, the fiber merely becomes a larger fast IIb fiber.

The above example also illustrates an important process called *protein turnover.* In this process, every protein in the cell is transcribed (DNA to mRNA), translated (mRNA to protein), and degraded (protein breakdown to amino acids), and each protein has a distinct time frame or half life for its existence. Thus, the level of any protein in a muscle cell is governed by its rate of synthesis relative to its rate of degradation, often referred to as the *synthesis/degradation ratio.* When synthesis exceeds degradation, protein content increases within the fiber and vice versa. In this way, by controlling either the activity of the gene in creating (transcribing) blueprints (mRNA) for a given protein, its synthesis rate (transcription), and/or its degradation rate, the expression of any protein can be precisely regulated. Thus, by changing the relative content or amount of any given protein, both the structural and functional properties of the muscle can be modified over time, implying the plasticity of the muscle.

The principle of myoplasticity also applies not only to alterations in the structural, contractile, and regulatory proteins described in some detail in Chapter 17, but also to the proteins involved in the metabolism of fat, carbohydrate, and amino acids. In Chapters 5 through 8, you learned of many metabolic factors that regulate exercise. In each of these chapters, the details of the metabolic adaptations to specific types of training were presented. It is important to recognize that the principle of myoplasticity governs the exercise-induced adaptive change that is described in the earlier chapters, as well

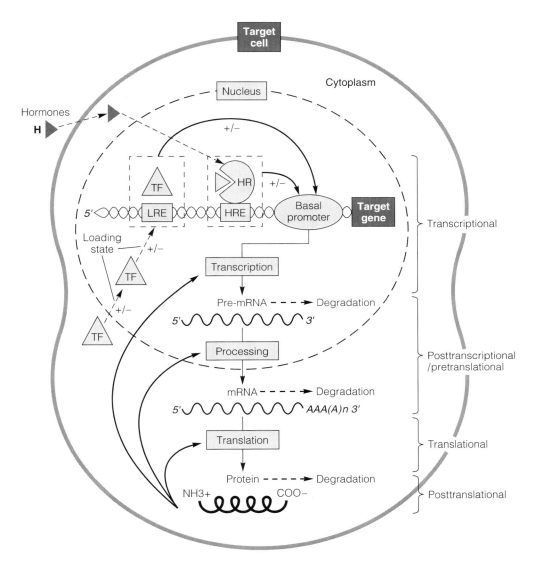

Figure 19-1 Muscle gene expression is affected by both the loading state (the chronic level of work performed by the muscle) and circulating hormones (e.g., thyroid hormone). Protein expression can be regulated at different levels—at the transcriptional level (i.e., the genetic level at which mRNA is made, the genetic blueprint for a protein), the posttranscriptional level (the level at which mRNA is maintained) and the translational and posttranslational level (at which proteins are synthesized and maintained). There are specific DNA sequences located upstream of the target gene (in an area on the gene that is designated the 5′ flanking region), which when bound by specific nuclear proteins (transcription factors) control the rate of transcription, e.g., the rate at which the gene synthesizes its mRNA. Transcriptional regulation involves complex interaction of these nuclear proteins with their designated DNA sequences that in turn affect the activity of the basal promoter (i.e., the basic transcription machinery) that drives the transcription process. Hormonal action results from a given hormone interacting with its high affinity nuclear receptor (HR), which is normally strongly bound to a sequence of DNA called a hormone response element (HRE) upstream of the target gene. When the HRE is activated by the hormone receptor complex, the transcriptional activity of the gene is altered. Alteration in the amount of force/work that the muscle produces also can change the level of certain transcription factors (TF), which when bound to their responsive elements (depicted in figure as LRE, loading or force sensitive-responsive element) change the level of transcriptional activity.

as the morphological and functional changes described in later sections of this chapter.

Adaptations in muscle due to exercise training, along with adaptations to nutritional and endocrine factors, are governed by the *principle of myoplasticity.* Skeletal muscle adaptations are characterized by modifications of morphological, biochemical, and molecular variables that alter the functional attributes of fibers in specific motor units. Adaptations range from a diminished capacity to generate or maintain power in response to reduced physical activity, to an enhanced capacity to maintain power for long periods of time following endurance training, and to an enhanced capacity to develop maximum strength and power following resistance training. Adaptations are readily reversible when the stimulus for adaptation is diminished or eliminated.

The principle of myoplasticity (Figure 19-2) indicates that several factors can influence the microenvironment of a skeletal muscle fiber and hence the regulation of its gene pool. The microenvironment is defined as the intracellular milieu and the fiber's immediate extracellular space. Alterations in the microenvironment that occur by a sufficient amount of stimuli spanning a sufficient period of time lead to changes in the relative expression of specific proteins.

What are the signals that initiate the cascade of events that we have identified as the principle of myoplasticity? One signal is insufficient energy balance, where caloric intake does not meet demand. The nutritional state can also influence specific endocrines, such as insulin, which in turn influence the microenvironment. Other endocrines independent of those influenced by nutritional state, such as thyroid hormone and insulin-like growth factor (IGF-1), also are powerful modulators of the cell's microenvironment. (For an excellent review on IGF-1 and thyroid hormone, see G. R. Adams and V. J. Caiozzo and F. Haddad, respectively.) Finally, the power developed by a motor unit, which results from the recruitment of fibers in motor units and the load

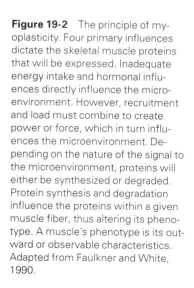

Figure 19-2 The principle of myoplasticity. Four primary influences dictate the skeletal muscle proteins that will be expressed. Inadequate energy intake and hormonal influences directly influence the microenvironment. However, recruitment and load must combine to create power or force, which in turn influences the microenvironment. Depending on the nature of the signal to the microenvironment, proteins will either be synthesized or degraded. Protein synthesis and degradation influence the proteins within a given muscle fiber, thus altering its phenotype. A muscle's phenotype is its outward or observable characteristics. Adapted from Faulkner and White, 1990.

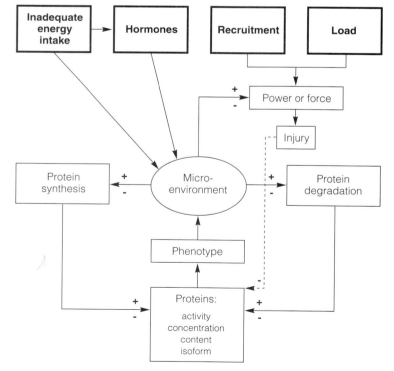

against which the fibers contract, also affect the cell's microenvironment. Motor unit recruitment principles are described in Chapter 18. Reduced power (i.e., inactivity), sustained power (i.e., endurance training), or high power (i.e., resistance training) result in acute changes in the cellular environment.

As seen in Figure 19-2, acute changes in the microenvironment lead to altered rates of protein synthesis and degradation. This in turn leads to changes in the concentrations and/or activities of specific proteins. When the protein structure of muscle is altered in this way, we say that the muscle's *phenotype* has been changed. Phenotype is the outward or observable characteristics of muscle; it reflects the underlying genes (i.e., genotype) and their regulation by several factors, including exercise training. The altered phenotype chronically, but not permanently, affects the cellular environment. The adaptations may increase or decrease the capability of fibers in a motor unit to develop and maintain power. The adaptive responses to many

physical activities are known, but the receptors, integrating centers, signal transduction factors, and effectors involved have not been fully identified. Under certain circumstances, muscle fibers are injured during contractions, and the concentration of myofibrillar proteins is decreased directly. When this occurs, however, the cell typically enters into a repair cycle to restore its integrity.

■ Muscle Fiber Types in Elite Athletes

Outstanding performers at the far end of the athletic spectrum have been shown to have specialized muscle fiber type characteristics. For instance, sprint runners have been found to have predominantly fast-twitch glycolytic fiber types, whereas slow-twitch, high-oxidative fiber types have been found to predominate in distance runners. Figure 19-3 is from work by the late Phillip Gollnick and associates. In that figure, panels (a) and (b) are derived

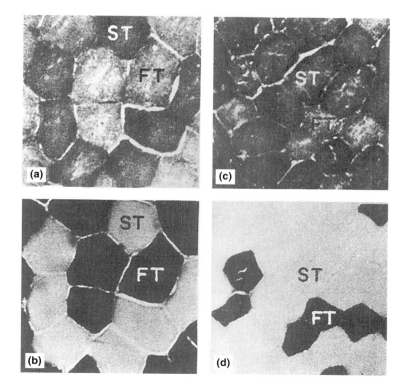

Figure 19-3 Serial sections of quadriceps muscle from two different athletes, stained with two different stains: (a) and (b) from an outstanding sprinter, (c) and (d) from an outstanding distance runner. Sections (a) and (c) are stained for succinic dehydrogenase (SDHase); (b) and (d) stained for alkaline myofibrillar-ATPase stain (M-ATPase). Note that fast fibers, which stain dark with M-ATPase, often are pale and stain weakly with SDHase. FT = fast-twitch fibers; ST = slow-twitch fibers. Note the two dark FT fibers in (d), which also stain dark for SDHase in (c). These are FOG fibers (see Figure 18-15). Source: Gollnick, et al., 1972. Used with permission.

from one subject, a sprinter, and panels (c) and (d) are from a distance runner. Sections (a) and (c) are reacted for SDHase, whereas (b) and (d) are reacted for M-ATPase. Note the pale appearance in (a), the oxidative marker of the sprinter. In the sprinter, however, note the dark appearance of fibers when reacted for the contractile protein marker, M-ATPase (b). Results obtained on the distance runner are the inverse: high oxidative capacity is indicated in (c), but low contractile protein enzyme activity is suggested in (d).

It should again be pointed out that results such as those in Figure 19-3 typify histochemical characteristics in athletes at the far end of the athletic spectrum. These results are consistent with other observations made in the animal kingdom, where species noted for speed (e.g., cats) have fast-twitch fiber characteristics, and species noted for endurance (e.g., dogs) have slow-twitch fiber type characteristics. With humans, Dave Costill and associates (1976) have observed a significant range of fiber type characteristics in athletes successful in middle-distance competition requiring both speed and endurance. Clearly, muscle fiber type is but one factor affecting human motor performance.

Classic observations of muscle fiber types that differ so strikingly among specific sports populations fueled speculation that these differences resulted from training. However, cross-sectional studies of elite athletes do not exclude the possibility that athletes who have high percentages of certain fiber types select and excel in specific activities. Furthermore, there is a strong genetic influence on fiber typing, as is observed in studies of identical twins.

A number of experiments on rats and humans have been designed to determine if exercise training can modify the type of myosin isozymes expressed in skeletal muscle fibers. Physical training can significantly affect muscle biochemistry and therefore the histochemical appearance of muscle cells. However, muscle fiber type, as determined by myosin isoforms, is genetically determined and influenced little by volitional training. Therefore, it is not possible for the muscle cells of the sprint athlete displayed in Figure 19-3a and b to assume, by training, the histochemical appearance of those of the endur-

ance athlete displayed in Figure 19-3c and d. Similarly, the endurance athlete cannot by training become like the sprint athlete. Training cannot exceed a muscle's intrinsic, genetically determined qualities. However, in spite of these generalities, there is mounting evidence to suggest that intermediate transitions can occur in MHC gene expression in response to different training paradigms, for example, transformations in IIx to IIz and vice versa that are not detected by the conventional histochemical staining techniques presented in Figure 19-3.

Clearly, though, endurance training can increase the oxidative capacity of each of the three fiber types. If the intensity and duration of the endurance training program is great enough, a percentage of the FG, or IIb, fibers may increase oxidative capacity sufficiently to change their histochemically determined fiber classification to FOG, or IIa.

With stimuli for adaptation other than voluntary activity, there can be an almost complete change from fast to slow or slow to fast fiber types. This has been shown with cross-innervation and cross-transplantation experiments. Also, muscles exposed to chronic low frequency stimulation demonstrate an increase in the percentage of type I fibers and a decrease in the percentage of type II. These experiments have established that under certain experimental and therapeutic circumstances, the expression of myosin proteins can be changed from fast to slow and vice versa.

■ Adaptations in Muscle Structure to Endurance Training

Endurance training of skeletal muscle occurs when there is a large increase in recruitment frequency of motor units and a more modest increase in the load that the motor units contract against. Typical physical activities that elicit this stimulus to skeletal muscle include jogging or running, swimming, Nordic skiing, and cycling.

Endurance exercise training has minimal impact on the cross-sectional area of muscle and muscle fibers. However, significant adaptations to the metabolic features of muscle occur with endur-

TABLE 19-1

Succinate Dehydrogenase Activity of Thigh Muscle Fiber Types in Response to Conditioning and Deconditioning

Condition	Range for Maximal Oxygen Uptake	Muscle Fiber Types			Mixed Muscle
		ST	FT$_a$	FT$_b$	
	ml $\cdot$ kg^{-1} $\cdot$ min^{-1}	μmol $\cdot$ g^{-1} $\cdot$ min^{-1}			
Deconditioning	30–40	5.0	4.0	3.5	4.0
Sedentary	40–50	9.2	5.8	4.9	7.0
Conditioning (months)	45–55	12.1	10.2	5.5	11.0
Endurance athletes	>70	23.2	22.1	22.0	22.5

ST = slow twitch. FT$_a$ and FT$_b$ = fast-twitch fibers of the thigh muscles. Approximate values for maximal oxygen uptake are included.

(Adapted from Saltin & Gollnick, 1983, p. 594.)

ance training. The details of these adaptations were developed in Chapters 5 through 8. The adaptations are primarily expressed as an increase in mitochondrial proteins and in many of the glycolytic enzymes. This, in turn, reflects an approximate twofold increase for oxidative metabolism in skeletal muscle. The amount of adaptive response for a given individual depends in part on his or her pre-training values; those with low values have a greater potential for improvement. The intensity and duration of training also has an impact on the amount of adaptive response. As seen in Table 19-1, a comparison of muscle oxidative capacity of thigh muscles among untrained and trained groups shows a range of responses. Furthermore, the response varies among the motor unit types, depending on the degree of involvement of the motor unit in the person's training. For example, in fibers identified in the table as fast-twitch b (FT$_b$; more properly identified as IIb), moderately trained individuals who trained for months did not demonstrate an adaptive response to SDH compared to sedentary control persons; endurance athletes showed a fourfold increase in this fiber type.

The functional significance of these changes is seen primarily during sustained exercise in which there will be a delay in the onset of metabolic acidosis, an increase in the capacity to oxidize free fatty acids and other fuels, and a conservation of carbohydrate. There is evidence that muscle blood flow is not altered appreciably, or even decreased, during submaximal exercise, whereas maximum blood flow increases following endurance training in the active motor units. Furthermore, capillarity seems to increase in human studies by 5 to 10%, which increases the transit time for red blood cells as they pass through the muscle's capillary network. This increased transit time provides a greater opportunity for gasses and metabolites to diffuse between the cell cytosol and the red blood cells and plasma in the capillary. Increased transit time then yields a greater extraction of material from arterial blood and enhances the arteriovenous difference locally in active muscle.

■ Adaptations in Muscle Structure to Resistance Training

Resistance exercise causes an increase in the recruitment frequency of motor units and a significant increase in the load against which the recruited motor units contract. Typical physical activities that elicit this stimulus to skeletal muscle include lifting free weights or using a resistance machine.

Contribution of Cellular Hypertrophy and Hyperplasia

The major adaptation that occurs to resistance training is an increase in the cross-sectional area of muscle, which is termed *hypertrophy*. Muscle hypertrophy leads to an increase in maximum force-generating capacity. There has been interest in whether training-induced hypertrophy of the muscle at the organ level is due to hypertrophy of the existing muscle cells, due to an increase in the number of cells (a phenomenon termed *hyperplasia*), or due to a combination of these factors.

Overall muscle hypertrophy is primarily due to an increase in cellular cross-sectional area. The number of muscle fibers is minimally affected by resistance training, at least in studies using mammals. In earlier work by Gonyea (1980), up to a 5% increase in the number of fibers (i.e., hyperplasia) was reported for resistance-trained limb muscles in cats. In a different model (hypertrophy compensatory to the removal of synergistic muscles), Gollnick found muscle hypertrophy to result solely from cellular enlargement, as cell number was unaltered. However, in experimental studies of birds in which muscle hypertrophy is induced by stretch, there is credible evidence of hyperplasia. The relevance of the hyperplastic observations for the human circumstance remains to be resolved. The principal mechanism for muscle hypertrophy in adult mammals is cellular hypertrophy and not hyperplasia.

The functional significance of the morphological change is primarily a greater capacity for strength and power development. Recall that Figure 17-28b shows that the force–velocity relationship for skeletal muscle is altered in such a way that a trained (hypertrophied) muscle is capable of moving a given submaximal load at a much higher velocity of shortening. This translates directly into an enhanced capacity for power.

Fiber-Type Specific Adaptations

High-resistance training results in an increase in cross-sectional area of both type I and type II fibers. J. D. MacDougall and coworkers at McMaster University (1980) provided much of our insight into human training studies. In one longitudinal study of the tricep brachii muscle, these investigators compared fiber cross-sectional area before and after five to six months of resistance training. As seen in Figure 19-4, on average, type II fibers hypertrophied by 33%, and type I fibers by 27%. Does this reflect preferential hypertrophy? Probably not, and one should not overinterpret the difference between fiber types as an inherently different capacity to adapt. Indeed, the different response may merely reflect chance or an increased stimulus for adaptation in the type II

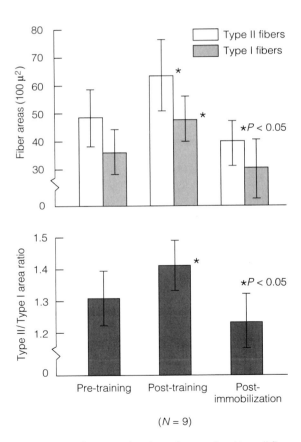

Figure 19-4 Cross-sectional area for type I and type II fibers in the control condition and following training and immobilization. The lower figure illustrates this data as type II/type I area ratios. Values are means ±1 SD. SOURCE: MacDougall, 1986.

fibers in this study by virtue of a greater involvement of type II fibers, compared to type I, in the training protocols. It is interesting to note that others have reported studies, for example, in which type I fibers hypertrophied slightly more than type II (39% vs. 31%). In addition to fiber areas, the angle in fiber pinnation can also adapt to training. The importance of muscle architecture was described in Chapter 17. In a comparison of bodybuilders to untrained men, a recent study from Japan has indicated that the angle of pinnation is greater in those persons with the thickest muscles.

Resistance training studies involving both humans and rodents have shown fiber-typing changes whereby the fastest MHCs in each species are repressed at the expense of increased expression of intermediate fast MHC isoforms (e.g., type IIx to IIa in humans and type IIb to IIx in rodents). These findings suggest that when a muscle is chronically stressed with high loading requirements (e.g., a heavy resistance training regimen), it shifts contractile protein phenotype to a more economical crossbridge cycling system, i.e., the crossbridges cycle at a slower rate in maintaining the force of a contraction.

Mitochondrial volume density and capillary density actually decrease with a program of high-resistance training. This is a direct result of a dilution of these elements due to an increase in cell volume with training. Figure 19-5 provides evidence of a 25% decrease in mitochondrial protein. Figure 19-6 shows that when capillary density is expressed relative to fiber cross-sectional area, there is a 13% decrease. From a practical perspective, individuals who train solely with resistance exercises risk decreasing endurance capacity. Fortunately, significant endurance training can be added to a workout plan to counteract this. Indeed, Hickson (Hickson et al., 1980) demonstrated that the combination of resistance training and endurance training can increase strength and endurance concurrently. Others have shown that previously hypertrophied muscle can subsequently adapt its metabolic properties with endurance training. Thus, from a practical perspective, when training stimuli are combined, a decrease in metabolic capacity is not inevitable.

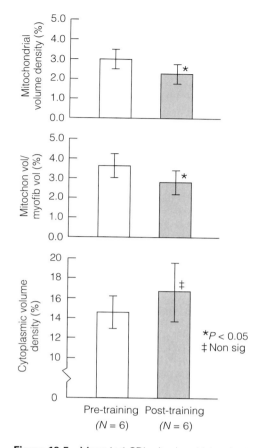

Figure 19-5   Mean (± 1 SD) mitochondrial and cytoplasmic volume density and mitochondrial-to-myofibrillar volume ratio before and after training. SOURCE: MacDougall, 1986.

Alterations in Specific Force with Hypertrophy

In studies of humans, it is common to find increases in strength of 30–40%, which exceed the amount of muscle hypertrophy. Also, it is common to see rapid strength gains at the beginning of a training schedule, particularly in previously untrained and inexperienced persons. These increases have been ascribed to the optimization of recruitment patterns, so-called neurological training. Digby Sale (1988) of McMaster University has been a leader in identifying this phenomena. During voluntary maximal contractions of large muscle masses, a person be-

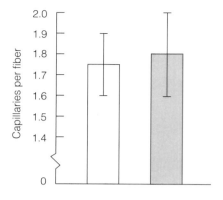

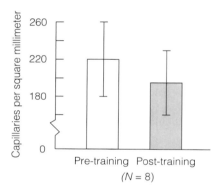

Figure 19-6 Capillary supply to biceps brachii before and after training. Values are expressed as a capillary-to-fiber ratio as well as capillary density (capillaries per mm²). Values are means (±1 SD). SOURCE: MacDougall, 1986.

comes, with training, more adept in effectively using the muscle mass he or she already has. This is independent of any improvements that will ultimately occur with time as muscle cross-sectional area increases.

The true maximum force of skeletal muscle can be measured more precisely in experimental studies, as detailed in Chapter 17. The concept of specific force involves maximum force being normalized to the physiological cross-sectional area of the muscle. The measurement of specific force allows for a comparison of the intrinsic capacity to develop force, regardless of the size of the muscle. Figure 19-7 shows that the increase in maximum isometric force (P_O, N) with compensatory hypertrophy is not as great as one would expect based on the amount of cellular hypertrophy. This results in a decrease in specific force (P_O, $N \cdot cm^{-2}$), also seen in the figure. The decrease in specific force has been seen in muscles following space flight, transplantation, and hypertrophy. A mechanism for this interesting observation has not yet been identified.

Consider for a moment, from a practical point of view, the points raised in the previous two paragraphs. During the onset of training, functional improvements exceed what you expect from the structural changes—you get more than you see! With continued training, however, functional improvements lag behind the structural—you get less than you see! This, more or less, provides a good summary.

Figure 19-7 Muscle cross-sectional area (CSA), maximum isometric force (P_o), and specific P_o (N/cm²) for hypertrophied soleus muscles are plotted as percent change compared with age-matched control value. Data are from days 1 to 30 after hypertrophy was induced. SOURCE: Kandarian and White, 1989.

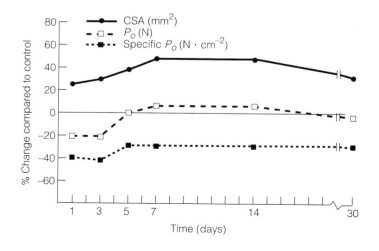

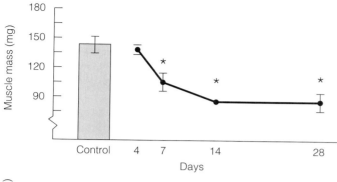

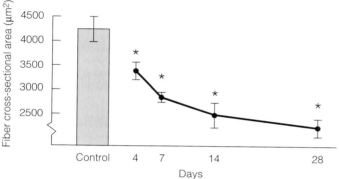

Figure 19-8   Body mass, muscle mass, and overall fiber cross-sectional area of soleus muscles from control animals and from experimental animals during 28 days of hindlimb suspension. Values are means ± 0SE. * Values differ from control value; ($P \leq 0.05$). From Kasper et al., 1990. Used with permission.

■ Adaptations in Muscle Structure to Decreased Physical Activity

Decreased physical activity occurs when there is a large reduction in the recruitment frequency of motor units and/or the load that the motor units contract against. Reduced physical activity is found, from a skeletal muscle perspective, in prolonged bed rest, in muscle immobilized by a cast, and in muscles that are nonweightbearing due to space flight or suspension from the floor.

The major adaptations that occur when physical activity is decreased are a reduction in muscle and muscle fiber cross-sectional area and a decrease in metabolic proteins that support endurance performance. Figure 19-8 shows the rapid decrease in muscle mass and fiber cross-sectional area that occurs when the muscle is chronically unweighted. Although the experiment represented in the figure involves rats in which plantarflexor muscles are unweighted because the hindlimbs are suspended off the cage floor, the rate of relative decay is almost

identical to that of rats that have been involved in space flight, or to that of humans whose legs are suspended and not contacting the floor. There is great interest in developing exercise countermeasures for human beings, so that the aspiration of long-term space flight can be realized without untoward effects on the muscles of the men and women who embark on such adventures. Baldwin and associates, as well as Booth and coworkers, have been instrumental in using atrophy models to study the molecular mechanisms of muscle adaptation.

■ Muscle Adaptation: Injury and Regeneration

Skeletal muscle fiber degeneration and subsequent regeneration follows from widespread damage to the fiber induced by a variety of insults, including free grafting operations, other mechanical and chemical trauma, ischemia, exposure to extreme heat and cold, and some diseases. Moreover, subtle and focal areas of degeneration–regeneration in skeletal muscle can result from excessive stretch, specific types and durations of exercise (particularly those with a bias toward lengthening contractions), and denervation or mild compression. The continuous presence of a population of regenerating skeletal muscle fibers is an inevitable and normal consequence of an active lifestyle.

Muscle injury will occur when the amount of force generated is high relative to capacity. This is particularly the case during lengthening contractions, which occur when the external load on muscle exceeds the tension that has been developed during activation. Bob Armstrong and coworkers with rats (1990) and David Jones and coworkers with humans (Jones and Rutherford, 1987), have demonstrated that downhill exercise, which has a bias to lengthening contractions, causes significantly more injury than other types of exercise. McCully and Faulkner demonstrated in an in situ preparation of mouse dorsiflexor muscle that lengthening contractions induced significant injury, whereas isometric and shortening contractions did not (McCully and Faulkner, 1986). The injury occurs essentially in two phases. The first is immediate and mechanical. The second phase reaches a peak several days later and is mediated through biochemical processes associated with calcium and oxygen free radicals. The injury causes significant muscle cell death, and the degenerative phase is followed by regeneration.

To explore facets of muscle fiber degeneration and regeneration, a graft-ischemia model has been employed in many laboratories. Carlson, Faulkner, and White have done much of the work in this area.

The grafting of skeletal muscle and the ensuing degeneration–regeneration of muscle fibers is a useful model for the study of many regulatory aspects of skeletal muscle development and maturation. Unlike ontogenetic development studied in utero or in ovo, in the graft model a host animal of sufficient size and age is employed; this allows researchers to experimentally manipulate variables such as the degree of reinnervation, the components of physical activity, and the level of circulating endocrines. With some liberties of literal interpretation, the skeletal muscle graft approach is a model of tissue culture in vivo.

Following skeletal muscle grafting, many structural and functional characteristics change with time until they reach a stable level. Successful regeneration requires revascularization, cellular infiltration, phagocytosis of necrotic muscle fibers, proliferation and fusion of muscle precursor cells (i.e., satellite cells or adult myoblasts; see Chapter 17), reinnervation, and recruitment and loading. The time taken to reach stable values varies among different structural and functional variables, and many reach stable values that are less than those of control skeletal muscle.

This line of research has applied clinical significance in the postgrafting management of patients who have had skeletal muscles grafted to repair sites of muscle impairment. Skeletal muscle graft operations are useful procedures for reconstructing sites of muscle dysfunction, such as partial facial paralysis, anal or urinary incontinence, forelimb muscle ablation, and so on. When these surgical procedures are done with large muscle masses, it is most common today to repair the vasculature, thereby allowing the majority of muscle cells to survive the graft operation. However, muscle fibers will degenerate and regenerate in cases where vascular repair of small or large muscle grafts is not possible or fully successful, or if the clinical decision is made not to repair the vasculature.

A change in physical activity during skeletal muscle regeneration can alter several attributes of the graft phenotype. It has been shown that proper recruitment and force development by grafts are essential in regulating the development and maturation of muscle grafts (White and Devor, 1993). Morphological and physiological attributes of grafts adapt to changes in the habitual level of physical activity in a qualitatively similar fashion to control muscle.

■ Gender Differences in Skeletal Muscle

Skeletal muscles of men typically differ from those of women in having a greater mass and muscle cross-sectional area. Thus, muscles from men are, on average, capable of generating a greater absolute amount of maximum force and maximum power. However, in human and animal studies, the intrinsic capacity of muscle to generate force and power—that is, when normalized for muscle cross-sectional area—does not differ by gender.

Although there is clearly some degree of overlap between men and women, the genders constitute two populations with respect to skeletal muscle attributes. The notion of dichotomous populations results principally from endocrine differences between the genders. The difference may also have a cultural component in cases where women are denied the same level of physical activity as men.

An interesting study published by Frontera and coworkers (1991) measured the muscle strength and mass of adult men and women. One notable feature of this study was that the 200 subjects studied were healthy and ranged in age from 45 to 78 years. Their overall height, weight, and an estimate of their muscle mass is found in Table 19-2. Strength was measured on a isokinetic dynamometer for knee and elbow extension and flexion; we discuss only knee extension here, as it is representative of all the other measurements made. Using the dynamometer, each person did maximum knee extensions at different speeds that were controlled by the dynamometer. The knee extensor muscles were studied at a slow speed, in which the joint angle moved at 60° per second, and at a fast speed of 240° per second.

Inspection of the absolute strength values, expressed in newton-meters (N · m) (Table 19-3) reveals that the strength of women was approximately 60% of the value for men of similar ages. However, when strength was corrected for the estimated differences in the amount of muscle mass (MM) in the men and women, the strength differences were no longer evident (see column where strength is expressed in $N \cdot m \cdot MM^{-1}$). One can conclude from this study that the profound gender-related differences in muscle strength are accounted for almost exclusively by the differences in body and muscle mass between the genders.

TABLE 19-2

General Characteristics of Subjects

	n	Age (yr)	Weight (kg)	Height (cm)	MM (kg)*
			Men		
	24	50.5 ± 2.8	80.8 ± 13.7	176.3 ± 7.4	27.9 ± 4.3
	28	60.1 ± 3.0	76.4 ± 10.1	177.0 ± 6.7	26.6 ± 3.2
	34	68.5 ± 2.8	79.0 ± 9.1	174.4 ± 7.5	25.2 ± 3.9
			Women		
	28	50.2 ± 2.6	61.6 ± 10.0	163.8 ± 5.9	18.3 ± 2.2
	52	60.1 ± 2.8	66.6 ± 11.2	162.0 ± 5.9	16.8 ± 2.9
	34	69.0 ± 3.8	63.3 ± 10.6	159.4 ± 5.0	14.8 ± 2.1
			P		
Age		<0.001	0.821	<0.055	0.002
Gender		0.865	<0.001	<0.001	<0.001

Values are mean ± SD; n = number of subjects in each group; *MM = muscle mass estimated from urinary creatinine excretion.

(Adapted from Frontera et al., 1991.)

TABLE 19-3

Isokinetic Muscle Strength of the Knee Extensors

Age Range (yr)	$60° \cdot s^{-1}$		$240° \cdot s^{-1}$	
	$N \cdot m$	$N \cdot m \cdot MM^{-1}$	$N \cdot m$	$N \cdot m \cdot MM^{-1}$
	Men			
45–54	180 ± 35	6.7 ± 1.0	101 ± 20	3.6 ± 0.5
55–64	163 ± 30	6.0 ± 0.9	92 ± 24	3.4 ± 0.9
65–78	144 ± 30	5.7 ± 1.5	78 ± 22	3.0 ± 0.9
	Women			
45–54	108 ± 22	6.1 ± 0.9	60 ± 14	3.4 ± 0.7
55–64	98 ± 20	5.9 ± 1.2	54 ± 13	3.2 ± 0.7
65–78	89 ± 15	5.8 ± 1.1	46 ± 12	3.0 ± 1.0
	P			
Age	<0.001	0.06	<0.001	0.03
Gender	<0.001	0.35	<0.001	0.29

Values are means $\pm$ SD.

(Adapted from Frontera et al., 1991.)

■ Age-Associated Changes in Skeletal Muscle

There is an inevitable reduction with age in the ability to produce and sustain muscular power. This age-associated phenomenon is of profound importance for the maintenance of posture, locomotion, and the ability to perform life's daily activities in the home or workplace. The age-associated decline in muscular strength and endurance also affects the performance capacity of recreational and elite performers in sport and dance. Neuromuscular impairments are also likely to contribute to an increased incidence of falls, muscle injury, and a reduced regenerative capacity of muscle.

In some circumstances and for some variables, physical inactivity may contribute to the age-associated decline in muscle structure and function, but this is by no means the predominant explanation. Inevitable age-associated changes are intrinsic to skeletal muscle and to its component molecules, cells, and organelles. The cause or causes of these intrinsic changes are unknown. The intrinsic aging factors predominate over those that can be ascribed to physical inactivity.

Age-Associated Muscle Atrophy

By 65 years of age, the muscle mass in humans decreases by approximately 25 to 30% from peak values observed at ages 25–30 years. Indeed, inspection of Table 19-2 indicates a significant decline in the estimated muscle mass of both men and women studied from the average age of 50 to 69 years. The possible explanations for the decline in muscle mass include muscle fiber atrophy, loss of muscle fibers (i.e., hypoplasia), or a combination of these factors.

Fiber Diameter In humans, a large number of studies indicate that the diameter of muscle fibers in limb muscles remains relatively constant or shows only modest decreases, at least through age 70 years. Thereafter, significant decreases are more common, and certainly in very old experimental rats, decline in muscle fiber diameter is on the order of 20–30%.

Fiber Number There are widely differing claims regarding the magnitude of fiber loss associated with aging. Much of the controversy exists because studies on humans are based on indirect approaches and employ cross-sectional designs, which can easily lead to imprecise results. In humans, one report suggested an approximate 25% drop in fiber number from adult through old age. These estimates were based on measuring fiber cross-sectional area from percutaneous needle biopsy samples and then calculating fiber number based on measures of muscle cross-sectional area by computer tomography scans. Other workers have studied whole vastus lateralis muscle obtained at autopsy in persons ranging from 15 to 83 years. Total fiber number was estimated from the product of total muscle cross-sectional area and the number of fibers in samples of a known area (i.e., fiber density). The results indicate that muscle atrophy begins in the third decade of life and is explained primarily by a 48% decrease in fiber number over the age range.

In a study of adult and old rats, the soleus and extensor digitorum longus (EDL) muscles were excised and weighed, and fiber number was quantified following nitric acid digestion. The mass of soleus and EDL muscles of old rats was 83% and 70%, respectively, of the adult values. Fiber number for soleus muscle declined with age by 6%, and for EDL muscles, by 4%. The evidence in this study indicates that with aging in rats, hypoplasia can account at most for approximately 25% of the observed skeletal muscle atrophy, the balance being ascribed to cell atrophy.

The conclusion from human studies regarding the underlying causes of muscle atrophy is opposite to that for rats, inasmuch as hypoplasia appears in humans to be a major mechanism of muscle atrophy. However, keep in mind that there are numerous technical problems in obtaining credible human data on this variable. Nevertheless, there seems to be significant hypoplasia in humans with age. The explanation may in part be that the old persons studied had some type of neurological disease that led to fiber loss. The reason for such a discrepancy between human and rodent models remains for further study.

Motor Unit Remodeling Age-associated motor unit remodeling is a consequence of alterations in the normal turnover of synaptic junctions and likely contributes in part to muscle atrophy. The turnover results from a cycle of denervation, axonal sprouting, and reinnervation. In young adults, turnover occurs without any alteration in the type or amount of innervation reaching fibers. With age, however, it is common to observe an aggregation of type I fibers. This age-associated change reflects some denervated type II fibers becoming reinnervated by axonal sprouting from adjacent innervated type I fibers. There is also evidence that some loss of α motoneurons occurs at the ventral root. Although incomplete, some degree of reinnervation of surviving motoneurons subsequently leads to rearrangement of the motor units in their preexisting anatomical location. Any loss in fibers presumably occurs because a population of the type II fibers is not reinnervated and thus atrophies, or because both type I and II fibers denervate with similar propensity but that type I motoneurons are more effective in reinnervating fibers.

Motor unit remodeling in humans was studied by Doherty and colleagues (1993). These workers examined the relationship between strength and motor unit number in arm muscles. They studied biceps brachii and brachialis muscles in men and women of two age groups: 24 persons ranged from 22 to 38 years and 20 persons ranged from 60 to 81 years. The quotient of the amplitude of the maximum compound muscle action potential to the single motor unit action potential (S-MUAP) yielded an estimate of the number of motor units. The number of motor units declined 47% with age (Figure 19-9). The older subjects had 189 ± 77 motor units compared with 357 ± 97 in younger subjects (mean ± SD), with comparable decreases in both genders. The size of the S-MUAPs increased 23% in the older persons, and twitch force as well as maximum voluntary contraction force declined by 33%.

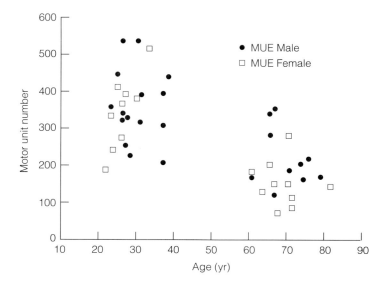

Figure 19-9 Relationship between number of motor units (MUs) and age in young and older men and women. There was a significant reduction in numbers of MUs with age ($P < 0.001$). Adapted from Doherty et al., 1993.

Age-Associated Change in Muscle Function

Muscle Group Strength Arm, leg, and back strength decline at an overall rate of 8% per decade, starting in the third decade of life. The rate of decline is not linear, and is slightly lower than this value early in the decline, and greater late in life. Referring back to Table 19-3, declines in absolute strength of knee extensors occurred in men and women from the average age of 50 to 69 years. It is interesting to note that when strength was corrected for declines in muscle mass, the age-associated decrease in strength was much less. Thus, a large portion of the decreased strength in old age is due to muscle atrophy, but it also appears that at least some degree of weakness goes beyond what you would expect from muscle atrophy per se. Results similar to those found for muscle groups of humans are found in isolated muscles of aging rodents. Decreases in absolute values for maximum isometric tetanic force of 20–35% occur in fast and slow limb muscles of old mice and rats.

Power and Endurance Brooks and Faulkner published an elegant paper on maximum and sustained power (i.e., endurance) in muscles of young, adult, and old mice (Brooks and Faulkner, 1991). Al-though the study was done with mice, it is thought to be quite indicative of what occurs in humans, in which it is not possible to make such definitive muscle measurements. The maximum muscle power in old mice was about 30% less than for adults. Approximately one-third of this decline was accounted for by atrophy, while the balance reflected intrinsic change in the muscle fibers. These investigators established a progressive exercise test with muscle in situ. The test consisted of repeated shortening contractions with progressive increases in duty cycles. A duty cycle is the fraction of time that a muscle is active during repeated bouts of contraction. In their study, the maximum absolute (in watts) and the normalized (in $W \cdot kg^{-1}$) power during single contractions were 20 to 30% less for young and old mice, compared to the adult mice (Figure 19-10). The ability to sustain power at any given duty cycle (in both absolute and normalized terms) was greatest for the muscles of young mice, intermediate for adults, and lowest for the old mice (Figure 19-11). Furthermore, the muscles of the young were able to tolerate higher duty cycles than the adult, who, in turn, were able to tolerate higher duty cycles than the old. A prolonged relaxation time with age is typically, but not always, reported for rodent muscle. This observation is likely mecha-

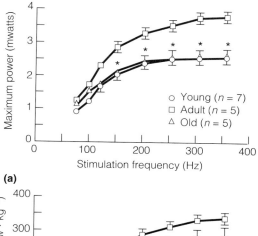

(a)

(b)

Figure 19-10 Relationship of stimulation frequency (Hz) and (a) absolute maximum power (mwatts) and (b) normalized maximum power (W · kg⁻¹) developed by EDL muscles in young, adult, and old mice. Error bars are shown whenever the SE is larger than the symbol for the mean. The asterisks indicate values for muscles in young and old mice that are different from the values for muscles in adult mice ($P < .05$). From Brooks and Faulkner, 1991. Used with permission.

A good understanding of the principle of myoplasticity will lead you to the conclusion that resistance training is required.

We now have some insight into the inherent potential for adaptation of skeletal muscle to changes in habitual activity levels associated with aging. When training is of high intensity, the old adapt in a manner comparable to the young. It is clear that adaptability of muscle to appropriate exercise remains robust even in very old age.

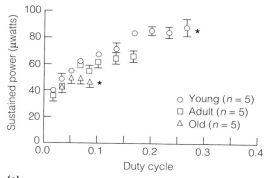

(a)

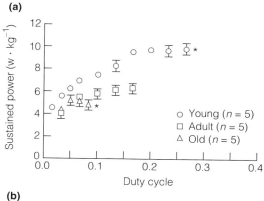

(b)

Figure 19-11 Relationship of the duty cycle and (a) absolute sustained power (μwatts) and (b) normalized sustained power (W · kg⁻¹) for EDL muscles in young, adult, and old mice. Error bars are shown whenever the SE is larger than the symbol for the mean. The asterisks indicate that the maximum values for muscles in young and old mice are different from the maximum value for muscles in adult mice, and the crosses indicate that the maximum value for young mice is also greater than that of old mice ($P < .05$). From Brooks and Faulkner, 1991. Used with permission.

nistically related to the reduced sustainable duty cycles in the old.

Structure-Function Responses to Resistance Training

We are gradually coming to understand the role of chronic exercise training in older people and animals. Because the key variable that declines with age is muscle mass, it is appropriate to focus primarily on training to delay or reverse this decline.

Frontera and associates published a study of resistance training of the legs of older men (1988). This study demonstrated convincingly that muscle hypertrophy, muscle fiber hypertrophy, and performance improvements can occur together with training. In this study, men 60–72 years of age demon- strated a 110% increase in force development for a one-repetition-maximum contraction of the knee ex- tensors following a 12-week resistance training pro- gram (Figure 19-12). The increase in performance was accompanied by a 9% increase in quadriceps area (Figure 19-13), a 34% increase in the cross-

Figure 19-12 Weekly measure- ments of dynamic muscle strength (1-repetition maximum) of left knee extensors and flexors. Results are means ±SE. From Frontera et al., 1988. Used with permission.

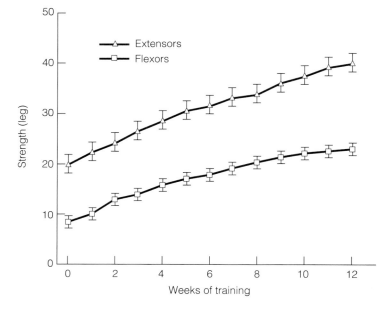

Figure 19-13 Changes in quadriceps muscle cross- sectional area of the right and left legs from planimetric analysis of computerized tomography scans. Results are means ±SE. * Different from pretraining measurements ($P < 0.05$). SOURCE: Frontera et al., 1988. Used with permission.

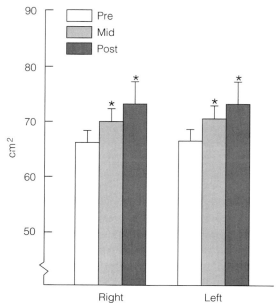

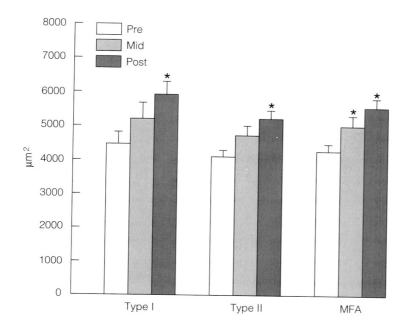

Figure 19-14 Effects of strength training on the area of type I and type II fibers of vastus lateralis muscle of the left leg. Results are means ±SE. * Different from pretraining measurements ($P < 0.05$). SOURCE: Frontera et al., 1988. Used with permission.

sectional area of type I, and a 28% increase in type II fibers in vastus lateralis muscle (Figure 19-14).

■ Summary

Within genetically set limits, appropriate training stimuli can change the structure and functional properties of skeletal muscle. An understanding of the underlying principle of myoplasticity will help you understand the application of training principles described elsewhere in the book. The components of physical activity, namely recruitment and load, are key stimuli that can initiate a chain of events that leads to an altered phenotype for many muscle proteins. Other factors, including nutrition and endocrines, also are key stimuli for adaptive change.

Individuals with exceptional muscular speed and power are characterized by a predominance of fast-twitch muscle fibers. Individuals noted for their endurance capability have a preponderance of slow-twitch but highly oxidative muscle fibers. Athletes whose competitive roles require aspects of both speed and endurance are successful with a wide range of muscle fiber types. Endurance training principally affects the metabolic features of skeletal muscle associated with substrate utilization and ATP turnover. Resistance training principally results in myofiber hypertrophy, whereas reductions in activity cause the opposite effect as well as decreases in metabolic capacity. Skeletal muscle fiber degeneration and subsequent regeneration follows from widespread damage to the fiber induced by a variety of insults including free grafting operations. Moreover, subtle and focal areas of degeneration–regeneration in skeletal muscle can result from excessive stretch and specific types and durations of exercise (particularly those with a bias toward lengthening contractions). The continuous presence of a population of regenerating skeletal muscle fibers is an inevitable and normal consequence of an active lifestyle.

Skeletal muscles of men typically differ from those of women by having a greater mass and muscle cross-sectional area. Thus, muscles from men are, on average, capable of generating a greater absolute amount of maximum force and maximum power. However, in human and animal studies, the intrinsic capacity of muscle to generate force and power does not differ due to gender.

There is an inevitable reduction with age in the

ability to produce and sustain muscular power. In some circumstances and for some variables, physical inactivity may contribute to the age-associated decline in muscle structure and function, but this is by no means the predominant explanation. Physical training, of the appropriate intensity, duration and frequency, can induce adaptations in the skeletal muscle of old mammals as well as in humans. In many cases, the magnitude of adaptive response is similar to that observed in younger subjects. However, it is not possible to fully reverse or permanently prolong the age-associated decline in muscular function—if you live a long life, neuromuscular declines are inevitable.

■ **Selected Readings**

Adams, G. R. Role of insulin-like growth factor-1 in the regulation of skeletal muscle adaptation to increased loading. In Exercise and Sport Sciences Reviews, J. O. Holloszy (Ed.). Vol. 26. Baltimore: Williams and Wilkins, 1998, pp. 31-60.

Adams, G. R., B. M. Hather, K. M. Baldwin, and G. A. Dudley. Skeletal muscle myosin heavy chain composition and resistance training. *J. Appl. Physiol.* 74: 911–915, 1993.

Alford, E., R. R. Roy, J. A. Hodgson, and V. R. Edgerton. Electromyography of rat soleus, gastrocnemius, and tibialis anterior during hind limb suspension. *Exp. Neurol.* 96: 635–649, 1987.

Armstrong, R. B. Initial events in exercise-induced muscular injury. *Med. Sci. Sports Exer.* 22: 429–435, 1990.

Armstrong, R. B., and M. H. Laughlin. Exercise blood flow patterns within and among rat muscles after training. *Am. J. Physiol.* 246: H59–H68, 1984.

Asmussen, E. Aging and exercise. In Environmental Physiology: Aging, Heat and Altitude (Sec. 3), S. M. Horvath and M. K. Yousef (Eds.). New York: Elsevier, North-Holland, 1980, pp. 419–428.

Babij, P., and F. W. Booth. Alpha-actin and cytochrome C mRNAs in atrophied adult rat skeletal muscle. *Am. J. Physiol.* 254: C651–656, 1988.

Baldwin, K. M., G. H. Klinkerfuss, R. L. Terjung, P. A. Molé, and J. O. Holloszy. Respiratory capacity of white, red and intermediate muscle, adaptive response to exercise. *Am. J. Physiol.* 222: 373–378, 1972.

Baldwin, K. M., V. Valdez, R. E. Herrick, A. M. MacIntosch, and R. R. Roy. Biochemical properties of overloaded

fast-twitch skeletal muscle. *J. Appl. Physiol.* 52: 467–472, 1982.

Baldwin, K. M., W. W. Winder, R. L. Terjung, and J. O. Holloszy. Glycolytic enzymes in different types of skeletal muscle: adaptation to exercise. *Am. J. Physiol.* 225: 962–966, 1973.

Berg, H. E., G. A. Dudley, T. Häggmark, H. Ohlsén, and P. A. Tesch. Effects of lower limb unloading on skeletal muscle mass and function in humans. *J. Appl. Physiol.* 70: 1882–1885, 1991.

Bischoff, R. Analysis of muscle regeneration using single myofibers in culture. *Med. Sci. Sports Exer.* 21: 164–172, 1989.

Booth, F. W., and D. B. Thomason. Molecular and cellular adaptation of muscle in response to exercise: perspectives of various models. *Physiol. Rev.* 71: 541–585, 1991.

Brooks, S. V., and J. A. Faulkner. Recovery from contraction-induced injury to skeletal muscles in young and old mice. *Am. J. Physiol.* 258 (*Cell Physiol.* 27): C436–C442, 1990.

Brooks, S. V., and J. A. Faulkner. Maximum and sustained power of extensor digitorum longus muscles from young, adult, and old mice. *J. Gerontol.: Biol. Sci.* 46: B28–B33, 1991.

Brown, A. B., N. McCartney, and D. G. Sale. Positive adaptations to weight-lifting training in the elderly. *J. Appl. Physiol.* 69: 1725–1733, 1990.

Burton, H. W., T. R. Stevenson, T. P. White, J. Hartman, and J. A. Faulkner. Force deficit of vascularized skeletal muscle grafts in rabbits. *J. Appl. Physiol.* 66: 675–679, 1989.

Caiozzo, V. J., and F. Haddad. Thyroid hormone: Modulation of muscle structure, function, and adaptive changes to mechanical loading. In Exercise and Sports Sciences Reviews, J. O. Holloszy (Ed.). Vol. 24. Baltimore: Williams and Wilkins, 1996, pp. 321–361.

Caiozzo, V. J., F. Haddad, M. J. Baker, and K. M. Baldwin. The influence of mechanical loading myosin heavy chain protein and mRNA isoform expression. *J. Appl. Physiol.* 80: 1503–1512, 1996.

Carlson, B. M. Regeneration of entire skeletal muscles. *Fed. Proc.* 45: 1456–1480, 1986.

Carlson, B. M., and J. A. Faulkner. The regeneration of skeletal muscle fibers following injury: a review. *Med. Sci. Sports Exer.* 15: 187–198, 1983.

Carlson, B. M., F. M. Hansen-Smith, and D. K. Magon. The life history of a free muscle graft. In Muscle Regeneration, A. Mauro (Ed.). New York: Raven Press, 1979, pp. 493–507.

Cartee, G. D. Aging skeletal muscle: response to exercise. In Exercise and Sports Sciences Reviews, J. O. Holloszy (Ed.). Vol. 22. Baltimore: Williams and Wilkins, 1994, pp. 91–120.

Clark, K. I., P. G. Morales, and T. P. White. Mass and fiber cross-sectional area of soleus muscle grafts following training. *Med. Sci. Sports Exer.* 21: 432–436, 1989.

Clark, K. I., and T. P. White. Morphology of stable muscle grafts of rats: effects of gender and muscle type. *Muscle Nerve* 8: 99–104, 1985.

Clark, K. I., and T. P. White. Neuromuscular adaptations to cross-reinnervation in 12- and 29-month Fischer 344 rats. *Am. J. Physiol.* 260 (*Cell Physiol.* 29): C96–C103, 1991.

Costill, D. L., J. Daniels, W. Evans, W. Fink, G. Krahenbuhl, and B. Saltin. Skeletal muscle enzymes and fiber composition in male and female athletes. *J. Appl. Physiol.* 40: 149–154, 1976.

Coyle, E. F., W. M. Martin III, D. R. Sinacore, M. J. Joyner, J. M. Hagberg, and J. O. Holloszy. Time course of loss of adaptations after stopping prolonged intense endurance training. *J. Appl. Physiol.* 57: 1857–1864, 1984.

Daw, C. K., J. W. Starnes, and T. P. White. Muscle atrophy and hypoplasia with aging: impact of training and food restriction. *J. Appl. Physiol.* 64: 2428–2432, 1988.

Diffee, G. M., V. J. Ciaozzo, R. E. Herrick, and K. M. Baldwin. Contractile and biochemical properties of rat soleus and plantaris after hindlimb suspension. *Am J. Physiol.* 260: C528–C534, 1991.

Doherty, T. J., A. A. Vandervoort, A. W. Taylor, and W. F. Brown. Effects of motor unit losses on strength in older men and women. *J. Appl. Physiol.* 74: 868–874, 1993.

Donovan, C. M., and J. A. Faulkner. Plasticity of skeletal muscle: regenerating fibers adapt more rapidly than surviving fibers. *J. Appl. Physiol.* 62: 2507–2511, 1987.

Dudley, G. A., W. A. Abraham, and R. L. Terjung. Influence of exercise intensity and duration on biochemical adaptations in skeletal muscle. *J. Appl. Physiol.* 53: 844–850, 1982.

Esser, K. A., and T. P. White. Mechanical load affects growth and maturation of skeletal muscle grafts. *J. Appl. Physiol.* 78: 30–37, 1995.

Faulkner, J. A., and C. Coté. Functional deficits in skeletal muscle grafts. *Fed. Proc.* 45: 1466–1469, 1986.

Faulkner, J. A., H. J. Green, and T. P. White. Response and adaptation of skeletal muscle to changes in physical activity. In Physical Activity, Fitness and Health, C. Bouchard et al. (Eds.). Champaign, Ill.: Human Kinetics, 1994, pp. 343–357.

Faulkner, J. A., and T. P. White. Adaptations of skeletal muscle to physical activity. In Exercise, Fitness, and Health, C. Bouchard, R. J. Shephard, J. R. Stephens,

and B. D. McPherson (Eds.). Champaign, Ill.: Human Kinetics, 1990, pp. 265–278.

Fitzsimons, D. P., G. M. Diffee, R. E. Herrick, and K. M. Baldwin. Effects of endurance exercise on isomyosin patterns in fast- and slow-twitch skeletal muscles. *J. Appl. Physiol.* 68: 1950–1955, 1990.

Frontera, W. R., V. A. Hughes, K. A. Lutz, and W. J. Evans. A cross-sectional study of muscle strength and mass in 45- to 78-yr-old men and women. *J. Appl. Physiol.* 71: 644–650, 1991.

Frontera, W. R., C. N. Meredith, K. P. O'Reilly, H. Knuttgen, and W. J. Evans. Strength conditioning in older men: skeletal muscle hypertrophy and improved function. *J. Appl. Physiol.* 64: 1038–1044, 1988.

Gollnick, P. D., R. B. Armstrong, C. W. Saubert IV, K. Piehl, and B. Saltin. Enzyme activity and fiber composition in skeletal muscle of untrained and trained men. *J. Appl. Physiol.* 33: 312–319, 1972.

Gollnick, P. D., L. A. Bertorci, T. B. Kelso, E. H. Witt, and D. R. Hodgson. The effect of high intensity exercise on the respiratory capacity of skeletal muscle. *Pflügers. Arch.* 415: 405–413, 1990.

Gollnick, P. D., B. F. Timson, R. L. Moore, and M. Riedy. Muscular enlargement and number of fibers in skeletal muscles of rats. *J. Appl. Physiol.* 50: 936–943, 1981.

Gonyea, W. J. Role of exercise in inducing increases in skeletal muscle fiber number. *J. Appl. Physiol.* 48: 421–426, 1980.

Green, H. J., R. Helyar, M. Ball-Burnett, N. Kowalchuk, S. Symon, and B. Farrance. Metabolic adaptations to training precede changes in muscle mitochondrial capacity. *J. Appl. Physiol.* 72: 484–491, 1992.

Grimby, G., A. Aniansson, M. Hedberg, G.-B. Henning, U. Grangard, and H. Kvist. Training can improve muscle strength and endurance in 78- to 84-yr-old men. *J. Appl. Physiol.* 78: 2517–2523, 1992.

Grimby, G., and B. Saltin. The aging muscle. *Clin. Physiol.* 3: 209–218, 1983.

Grounds, M. D. Towards understanding skeletal muscle regeneration. *Path. Res. Pract.* 187: 1–22, 1991.

Gunning, P., and E. Hardeman. Multiple mechanisms regulate muscle fiber diversity. *FASEB J.* 5: 3064–3070, 1991.

Hickson, R. C., Interference of strength development by simultaneously training for strength and endurance. *European J. of Appl. Physiol.* 45: 255–263, 1980.

Holloszy, J. O. Biochemical adaptations in muscle. Effects of exercise on mitochondrial oxygen uptake and respiratory enzyme activity in skeletal muscle. *J. Biol. Chem.* 242: 2278–2282, 1967.

Holloszy, J. O., and F. W. Booth. Biochemical adaptations to endurance exercise in muscle. *Annu. Rev. Physiol.* 38: 273–291, 1976.

Jones, D. A., and O. M. Rutherford. Human muscle strength training: the effects of three different regimes and the nature of the resultant changes. *J. Physiol.* 391: 1–11, 1987.

Kandarian, S. C., and T. P. White. Force deficit during the onset of muscle hypertrophy. *J. Appl. Physiol.* 67: 2600–2607, 1989.

Kandarian, S. C., and T. P. White. Mechanical deficit persists during long-term muscle hypertrophy. *J. Appl. Physiol.* 69: 861–867, 1990.

Kasper, C. E., T. P. White, and L. C. Maxwell. Running during recovery from hindlimb suspension induces transient muscle injury. *J. Appl. Physiol.* 68: 533–539, 1990.

Kawakami, Y., T. Abe, and T. Fukunga. Muscle fiber pennation angles are greater in hypertrophied than in normal muscles. *J. Appl. Physiol.* 74: 2740–2744, 1993.

Kennedy, J. M., B. R. Eisenberg, S. K. Reid, L. J. Sweeney, and R. Zak. Nascent muscle fiber appearance in overloaded chicken slow-tonic muscle. *Am. J. Anat.* 81: 203–215, 1988.

Larsson, L. Histochemical characteristics of human skeletal muscle during aging. *Acta Physiologica Scand.* 117: 469–471, 1983.

Larsson, L., and G. Salviati. Effects of age on calcium transport activity of sarcoplasmic reticulum in fast- and slow-twitch rat muscle fibres. *J. Physiol.* 419: 253–264, 1989.

Laughlin, M. H., and J. Ripperger. Vascular transport capacity of hindlimb muscles of exercise-trained rats. *J. Appl. Physiol.* 62: 438–443, 1987.

Lexell, J., C. Taylor, and M. Sjostrom. What is the cause of the aging atrophy? Total number, size and proportion of different fibre types studied in whole vastus lateralis muscle from 15- to 83-year-old men. *J. Neurolog. Sci.* 84: 275–294, 1988.

MacDougall, J. D. Morphological changes in human skeletal muscle following strength training and immobilization. In Human Muscle Power, N. L. Jones et al. (Eds.). Champaign, Ill.: Human Kinetics, 1986, pp. 269–288.

MacDougall, J. D., G. C. B. Elder, D. G. Sale, J. R. Moroz, and J. R. Sutton. Effects of strength training and immobilization on human muscle fibers. *Eur. J. Appl. Physiol.* 43: 25–34, 1980.

Markley, J. M., J. A. Faulkner, J. H. Niemeyer, and T. P. White. Functional properties of palmaris longus muscles of rhesus monkeys transplanted as index finger flexors. *Plast. Reconstr. Surg.* 76: 574–577, 1985.

Mauro, A. Satellite cell of skeletal muscle fibers. *J. Biophys. Biochem. Cytol.* 9: 493–495, 1961.

McComas, A. J. Human neuromuscular adaptations that accompany changes in activity. *Med. Sci. Sports Exer.* 26 (12): 1498–1509, 1994.

McCully, K. K., and J. A. Faulkner. Injury to skeletal muscle fibers of mice following lengthening contractions. *J. Appl. Physiol.* 59: 119–126, 1985.

McCully, K. K., and J. A. Faulkner. Characteristics of lengthening contractions associated with injury to skeletal muscle fibers. *J. Appl. Physiol.* 61: 293–299, 1986.

Miller, J. B. Myoblast diversity in skeletal myogenesis: how much and to what end? *Cell* 69: 1–3, 1992.

Musch, T. I., G. C. Haidet, G. A. Ordway, J. C. Longhurst, and J. H. Mitchell. Training effects on regional blood flow response to maximal exercise in fox hounds. *J. Appl. Physiol.* 62: 1724–1732, 1987.

Pette, D. (Ed.). Plasticity of Muscle. Berlin: Walter de Gruyter, 1980.

Pette, D., and R. S. Staron. Cellular and molecular diversities of mammalian skeletal muscle fibers. *Rev. Physiol. Biochem. Pharmocol.* 116: 1–76, 1990.

Phillips, S. K., S. A. Bruce, D. Newton, and R. C. Woledge. The weakness of old age is not due to failure of muscle activation. *J. Gerontol.: Med. Sci.* 47: M45–M49, 1992.

Phillips, S. K., S. A. Bruce, and R. C. Woledge. In mice, the muscle weakness due to age is absent during stretching. *J. Physiol.* (London) 437: 63–70, 1991.

Riedy, M., R. L. Moore, and P. D. Gollnick. Adaptive response of hypertrophied skeletal muscle to endurance training. *J. Appl. Physiol.* 59: 127–131, 1985.

Roy, R. R., K. M. Baldwin, and V. R. Edgerton. The plasticity of skeletal muscle: effects of neuromuscular activity. In Exercise and Sports Sciences Reviews, J. O. Holloszy (Ed.). Vol. 19. Baltimore: Williams and Wilkins, 1991, pp. 269–312.

Russell, B., D. J. Dix, D. L. Haller, and J. Jacobs-El. Repair of injured skeletal muscle: a molecular approach. *Med. Sci. Sports Exer.* 24: 189–196, 1992.

Sale, D. G. Neural adaptation to resistance training. *Med. Sci. Sports Exer.* 20: S135–S145, 1988.

Salmons, S., and J. Hendrickson. The adaptive response of skeletal muscle to increased use. *Muscle Nerve* 4: 94–105, 1981.

Saltin, B., and P. D. Gollnick. Skeletal muscle adaptability: significance for metabolism and performance. In Handbook of Physiology, Section 10: Skeletal Muscle, L. D. Peachey, R. H. Adrain, and S. R. Geiger (Eds.). Bethesda, Md.: American Physiological Society, 1983, pp. 555–631.

Schultz, E. Satellite cell behavior during skeletal muscle growth and regeneration. *Med. Sci. Sports Exer.* 21: S181–S187, 1989.

Segal, S. S., T. P. White, and J. A. Faulkner. Architecture, composition and contractile properties of rat soleus muscle grafts. *Am. J. Physiol.* 250 (*Cell Physiol.* 19): C474–C479, 1986.

Smith, D. O., and J. L. Rosenheimer. Decreased sprouting and degeneration of nerve terminals of active muscles in aged rats. *J. Neurophysiol.* 48: 100–109, 1982.

Sreter, F. A., K. Pinter, F. Jolesz, and K. Mabuchi. Fast to slow transformation of fast muscles in response to long-term phasic stimulation. *Exp. Neurol.* 75: 95–102, 1982.

Staron, R. S., M. J. Leonardi, D. L. Karapondo, E. S. Malicky, J. E. Falkel, F. C. Hagerman, and R. S. Hikida. Strength and skeletal muscle adaptations in heavy-resistance-trained women after detraining and retraining. *J. Appl. Physiol.* 70: 631–640, 1991.

Tesch, P. A. Skeletal muscle adaptations consequent to long-term heavy resistance exercise. *Med. Sci. Sports Exer.* 20: S132–S134, 1988.

Thomason, D. B., and F. W. Booth. Atrophy of the soleus muscle by hindlimb unweighting. *J. Appl. Physiol.* 68: 1–12, 1990.

Thomason, D. B., R. E. Herrick, D. Surdyka, and K. M. Baldwin. Time course of soleus muscle myosin expression during hindlimb suspension and recovery. *J. Appl. Physiol.* 63: 130–137, 1987.

Tsika, R. W., R. E. Herrick, and K. M. Baldwin. Interaction of compensatory overload and hindlimb suspension on myosin isoform expression. *J. Appl. Physiol.* 62: 2180–2186, 1987.

Vandenburgh, H. H., P. Karlisch, J. Shansky, and R. Feldstein. Insulin and IGF-I induce pronounced hypertrophy of skeletal myofibers in tissue culture. *Am. J. Physiol.* 260: C475–C484, 1991a.

Vandenburgh, H. H., S. Swasdison, and P. Karlisch. Computer-aided mechanogenesis of skeletal muscle organs from single cells in vitro. *FASEB J.* 5: 2860–2867, 1991b.

White, T. P. Adaptations of skeletal muscle grafts to chronic changes of physical activity. *Fed. Proc.* 45: 1470–1473, 1986.

White, T. P., G. J. Alderink, K. A. Esser, and K. I. Clark. Postoperative physical activity alters growth of skeletal muscle grafts. In Third Vienna Muscle Symposium, G. Freilinger and M. Deutinger (Eds.). Wien: Blackwell-MZV, 1992, pp. 53–60.

White, T. P., and S. T. Devor. Skeletal muscle regeneration and plasticity of grafts. In Exercise and Sports Sciences Reviews, J. O. Holloszy (Ed.). Vol. 21. 1993, pp. 263–296.

White, T. P., and K. A. Esser. Satellite cell and growth factor involvement in skeletal muscle growth. *Med. Sci. Sports and Exer.* 21: S158–S163, 1989.

White, T. P., J. A. Faulkner, J. M. Markley, Jr., and L. C. Maxwell. Translocation of the temporalis muscle for treatment of facial paralysis. *Muscle Nerve* 5: 500–504, 1982.

White, T. P., L. C. Maxwell, D. M. Sosin, and J. A. Faulkner. Capillarity and blood flow of transplanted skeletal muscles of cats. *Am. J. Physiol.* 241 (*Heart Circ. Physiol.* 10): H630–H636, 1981.

White, T. P., J. F. Villanacci, P. G. Morales, S. S. Segal, and D. A. Essig. Exercise-induced adaptations of rat soleus muscle grafts. *J. Appl. Physiol.* 56: 1325–1334, 1984.

Zerba, E., T. E. Komorowski, and J. A. Faulkner. The role of free radicals in skeletal muscle injury in young, adult, and old mice. *Am. J. Physiol.* 258: C429–C435, 1990.

Zhang, Y. L., and S. G. Kelsen. Effects of aging on diaphragm contractile function in golden hamsters. *Am. Rev. Respir. Dis.*, 142: 1396–1401, 1990.

CHAPTER 20

MUSCLE STRENGTH, POWER, AND FLEXIBILITY

Muscle strength, power, endurance, and flexibility are fitness components that are important for health and physical performance. Muscular strength is the amount of force that a muscle can produce with a single maximum effort. Power, a related fitness component, can be defined as the ability to exert force rapidly (power=Work/Time). Muscle endurance is the ability to sustain a given level of muscle tension—that is, to hold a muscle contraction for a long period of time or to contract a muscle over and over again for a long period of time. Flexibility is the ability to move joints through their full range of motion. In the 1970s and 1980s, exercise recommendations by professional organizations, such as the American College of Sports Medicine and American Heart Association, emphasized aerobic (or endurance) exercise as the central component of a fitness program. They scarcely mentioned muscular strength, power, and flexibility. However, recent recommendations place muscle fitness—strength, power, and flexibility—on equal footing with endurance exercises, such as walking, running, and swimming.

Strong, powerful muscles are important for the smooth and easy performance of everyday activities, such as carrying groceries, lifting boxes, and climbing stairs, as well as for emergency situations. They help keep the skeleton in proper alignment, preventing back and leg pain and providing the support necessary for good posture. Muscular strength and

power have obvious importance in sports. Powerful athletes can hit a tennis ball harder, throw a discus farther, and ride a bicycle uphill more easily. Athletes develop strength and power through intense resistance training as well as through techniques such as plyometrics—exercises involving sudden lengthening followed by immediate shortening of the muscles—and specific skill practices (e.g., throwing the discus to increase strength for that activity).

Muscular strength and power are related. However, as will be discussed, an incremental increase in strength does not necessarily produce the same incremental increase in power. In most activities, power is much more important than strength. Powerful motor performance in movements such as hitting a golf ball, throwing a baseball, or recovering rapidly after slipping in a tub, depend on the rate and sequence of motor-unit activation. Optimal training techniques for power, particularly in complex motor skills, is poorly understood and relies largely on empirical observations of coaches, athletes, and therapists. Exercise scientists are only beginning to understand the complex relationship between strength, power, and skill development.

Muscle tissue is an important element of overall body composition. Greater muscle mass means a higher rate of metabolism and faster energy use. Maintaining strength and muscle mass is also vital for healthy aging. Older people tend to lose muscle

Professional athletes such as Mark McGuire have utilized scientific techniques to result in fantasitic feats of human muscular power. PHOTO: AP/ Wide World Photos.

cells, and many of their remaining muscle cells become nonfunctional because the motor units become denervated (see Chapter 19). Strength training helps maintain muscle mass and function in older people and possibly helps decrease the risk of osteoporosis. Such benefits can greatly enhance quality of life and help prevent life-threatening injuries.

Muscular endurance is important for good posture and for injury prevention. For example, if abdominal and back muscles cannot hold the spine correctly, the chances of low-back pain and back injury are increased. Muscular endurance also helps people cope with the physical demands of everyday life and enhances performance in sports and work. In addition, it is important for most leisure and fitness activities. Recent studies by Scandinavian researchers have shown that muscle power and the capacity to sustain it are critical predictors of success in such endurance sports as distance running and cycling.

Although range of motion in the joints is not a significant factor in everyday activities for most people, inactivity causes the joints to become stiffer with age. Stiffness often causes older people to assume unnatural body postures that can stress joints and muscles. Stretching exercises can help ensure a normal range of motion for all the major joints.

Skeletal muscle is highly adaptable. The biologi-

cal principles underlying muscle adaptation were developed in Chapter 19. If muscle is overloaded— that is, subjected to a greater load than usual—it improves its function. This chapter explores the adaptation of skeletal muscle to the stresses of progressive resistance training. Along with Chapter 21, it also examines factors involved in muscle adaptation, the physiological effects of training, and examples of athletes' training programs. For a discussion of the adaptation to endurance training, see Chapter 21.

■ Progressive Resistance Training

Progressive resistance training has a long history, beginning with the Olympic champion Milo of Crotona, who lived in Greece in the sixth century B.C. Milo is said to have hoisted a baby bull on his shoulders every day to improve his strength. As the bull grew heavier with age, Milo's strength also grew. Since that time, progressive resistance training has been an important part of the training programs of many athletes, from football players and track and field participants to swimmers and figure skaters.

Nevertheless, progressive resistance training is a controversial area of exercise physiology. For one thing, it has been the subject of many poorly conducted studies in which the subjects were untrained and the investigations were carried out over only a short period of time. In addition, extrapolating the results of animal studies to humans and the results of studies of untrained subjects to athletes requires caution and insight. Further, the rampant commercialism connected with various types of progressive resistance training equipment makes it difficult to separate fact from exaggeration. It is also often difficult to distinguish the effects of motor learning from strength gains.

Some studies did not allow athletes to adequately recover from a training program before evaluating the effectiveness of the program. Every accomplished athlete knows that rest is necessary for peak performance following a period of heavy training. Yet, strength was often evaluated in the midst of the training program. Consequently, to assess properly the knowledge in this area, we must look at the results of both animal and human studies and evaluate the countless empirical observations that have been made.

■ Classification of Strength Exercises

Strength exercises can be divided into two categories: *isometric* (static) and *isotonic.* Isometric exercise involves the application of force without movement; isotonic exercise involves force with movement.

Isometric Exercise

Hettinger and Mueller generated much interest in 1953 when they found that six seconds of isometric exercise at 75% effort increased strength. Although later research showed that isometrics have limited applications in the training programs of athletes, this type of exercise received considerable attention in the 1950s and 1960s. Today, it is seldom practiced unless it is included with other techniques in the training program.

Isometric exercise usually does not increase strength throughout the range of motion of a joint because it is specific to the joint angle of training. For this reason, athletes sometimes use isometrics to help them overcome "sticking points" in the range of motion of an exercise. For example, an athlete who has difficulty pushing a weight past a certain point in the squat may perform the exercise isometrically at that point (Figure 20-1). Under some circumstances isometric training can increase strength 20° on either side of the training angle. Ways to increase joint-angle strength include doing more repetitions of an isometric exercise or holding a contraction longer. Another limitation of isometric training is that it does not improve (and, indeed, may hamper) the ability to exert force rapidly.

Isometric training caused a 14–44% increase in isometric force in studies lasting as long as 16 weeks. It has been shown that the total length of isometric contractions in workouts determines strength improvements. Optimal strength gains require a minimum of 15–20 maximum isometric contractions (contractions held at highest level) sustained for 3–5 seconds.

Most of the benefits of isometrics occur early in training, and maximal contraction is essential for optimal effect. The duration of contraction should be long enough to recruit as many fibers in a muscle group as possible. The largest gains in strength occur when isometrics are practiced several times a day. However, as with other progressive resistance training techniques, excessive training eventually leads to a deterioration in performance (overtraining).

Maximum isometric force is an important measure of strength and power. It is related to muscle cross-sectional area. However, curiously, changes in strength from isometric strength training are poorly related to changes in muscle cross-sectional area (Garfinkel and Cafarelli, 1992). Strength increases from isometrics are caused by increasing muscle size and improved neural activation (i.e., the capacity of the nervous system to recruit motor units). Isometrics train both type I and type II fibers.

Electrical Muscle Stimulation (EMS)   Electrical muscle stimulation causes an isometric contraction. EMS is valuable in the rehabilitation of muscles and joints after injury because joint mobilization is difficult at those times. Gains from EMS are small compared to traditional methods of progressive resistance training. However, several reports suggest that it may be valuable as an adjunct training method. Some reports suggest that EMS results in preferential recruitment of fast-twitch motor units. Also, EMS makes it possible to activate muscle at intensities above maximal voluntary isometric force. The effectiveness of EMS may depend on factors such as stimulation frequency and rest intervals between contractions. Matheson and coworkers (1997) showed that stimulating the muscle for 10 sec with a 10 sec rest decreased force output and caused fatigue, creatine phosphate depletion, and intramuscular acidosis to a much greater extent than a protocol that stimulated the muscle for 10 sec followed by 50 sec of rest. Stimulation frequencies of 60–100 Hz are necessary to produce maximum force. The effect of EMS on possible changes in strength and hypertrophy is unknown.

Isotonic Exercise

Isotonic exercise involves the application of force resulting in movement. It is the progressive resistance training technique most familiar to athletes and coaches. Isotonic loading methods include constant, variable, eccentric, plyometric, speed, proprioceptive neuromuscular facilitation, and isokinetic.

Constant-resistance exercises use a constant load, such as a barbell or dumbbell. The difficulty in overcoming the resistance to the load varies with the angle of the joint. An example is the "free-weight" bench press. It is easier to move the weight when it is at the end of the range of motion than when it is on the chest. Barbells and dumbbells are good examples of constant resistance exercise devices. Free-weights continue to be most popular with athletes in strength and power sports (Figure 20-2).

Variable-resistance exercises are done on specially designed resistance machines. They impose an increasing load throughout the range of motion, and so place a more constant stress on the muscles (Figure 20-3). This stress is accomplished by the changing relationship of the fulcrum and lever arm in the weight machine as the exercise progresses. Placing a muscle group under a relatively uniform stress throughout the range of motion, however, has not been shown to be superior to more traditional forms of progressive resistance training, in spite of the popularity of exercise machines using this technique. At least four studies have shown that freeweights are just as effective as variable-resistance training. The chief advantages of these exercise machines may be their ease of use and safety.

Figure 20-1   Isometric squat on the power rack. PHOTO: Wayne Glusker.

Figure 20-2   Constant resistance exercise using a fixed resistance: the power snatch. PHOTO: Wayne Glusker.

Figure 20-3   Variable resistance exercise: pullover machine. PHOTO: Wayne Glusker.

Muscle contractions resulting in shortening and lengthening are correctly termed *myometric* and *plyometric*, respectively. In sports, myometric muscle contractions are commonly termed concentric and seldom occur as isolated movements. Typically, a concentric contraction is preceded by a plyometric contraction (commonly called eccentric). Energy is stored in the muscle during the eccentric phase. This stored energy can be used during the concentric phase (called a stretch-shortening cycle). An example of an exercise in which this occurs is the bench press. The pectoralis major muscle works eccentrically controlling the bar while the weight is lowered to the chest. The muscle is elastically loaded as the bar reaches the chest. If there is a minimal delay, the stored elastic energy can be used to assist with the pushing motion. Eccentric, elastic, and concentric phases of the movement are thus combined. An athlete can push much more weight with this type of movement than he or she could during a purely concentric bench press (i.e., a lift starting from the chest).

Eccentric loading is tension exerted during the lengthening of a muscle. In a bench press, for example, the muscles work eccentrically by resisting the movement of the bar as it approaches the chest. Several studies have shown that this is an effective means of gaining strength, although it is not superior to other isotonic techniques. Maximal eccentric loading has been shown to recruit preferentially faster motor units.

One drawback of eccentric training is that it creates more muscle soreness than other methods. Soreness occurs because muscles can generate more force eccentrically than concentrically, which results in greater tension-induced damage to the muscle during eccentric contraction (see Chapter 19). Eccentrics are not widely practiced by strength athletes except as an adjunct to other training methods. Of course, eccentric contractions are a component of most progressive resistance training exercises and thus contribute to the strength gains made by most of the training methods described here.

Studies by researchers such as Higbie and associates (1996) show that both eccentric and concentric training improve concentric and eccentric muscle strength. However, concentric training improves concentric strength most, while eccentric training is better for improving eccentric strength. This simple and obvious finding has important implications for training and rehabilitation. For example, hamstring strains in sprinters often occur when the hamstrings are contracting eccentrically. It is likely that optimal rehabilitation or preventive strengthening of the hamstrings is better accomplished through eccentric than concentric exercise.

Plyometric loading uses a stretch-shortening cycle. It involves a sudden eccentric stress, muscle stretch, followed by a rapid concentric contraction. An eccentric stress occurs when the muscle exerts force while it lengthens. The muscles are loaded suddenly and forced to stretch before they can concentrically contract and elicit movement. An example of plyometrics is box jumping. The athlete jumps down from a box, then immediately jumps up onto another box. This type of training is particularly useful for training the neural and elastic components of strength (discussed below).

Plyometrics are becoming accepted in many sports. Because of their dynamic nature, they pose a higher risk of injury than do standard weight-training techniques. If practiced progressively and within individual tolerance, they contribute an added dimension to the training program.

Speed loading involves moving the resistance as rapidly as possible. Most studies have found that speed loading is less effective than constant resistance isotonic exercise for gaining strength because speed loading may not allow sufficient tension to elicit a training effect to the contractile aspects of the muscle cell. In the 1920s, Hill established that tension diminishes as the speed of contraction increases—and adequate tension is necessary for muscle hypertrophy. However, in most sports power is more important than strength.

Speed loading is effective for improving movement time. It may also enhance the rate of force development although it may have little effect on absolute force. The rate of force development is important in skills such as sprinting, jumping, throwing, and punching. Speed loading is often practiced by strength athletes in their training schedule, particularly during the competitive period, when maximum power is desired.

Training at or slightly above peak power output (weight · reps/time) seems to be necessary to achieve the full benefits of speed loading. Data collected by Mastropaolo (1992) suggest that maximum power is an important stimulus for gaining strength. His studies show that training at approximately 10% above maximum power output increases maximum power and strength. This type of training may be much more effective in transferring strength gains from the weight room to the playing field.

Proprioceptive neuromuscular facilitation is manual exercise involving a combination of muscle stretching and isotonic and isometric loading (Figure 20-4). This technique is widely used by physical therapists and athletic trainers in the treatment and prevention of athletic injuries. Although it is usually used to increase flexibility, it can also improve muscle strength. Unfortunately, there are few data available comparing this technique with traditional loading techniques. However, it is a promising method of resistance training during rehabilitation from injury.

Isokinetic exercise controls the rate of muscle shortening. It is sometimes called "accommodating resistance" because the exerted force is resisted by an equal force from the isokinetic dynamometer. As with variable-resistance isotonic exercise, isokinetics require a specially designed machine to produce the isokinetic loading. Isokinetics have become extremely popular with athletic trainers and physical therapists because they allow the training of injured joints with a lower risk of further injury. In addition, isokinetic dynamometers provide a speed-specific indication of the absolute strength of a muscle group.

The most effective strength gains from isokinetic exercise have come from a slower training speed (60 degrees per second or less). Training at fast speeds of motion has been found to increase the ability to exert force rapidly, but no more so than traditional isotonic techniques. Although this is a promising training method, more research is needed to establish its role in progressive resistance training and to determine the ideal isokinetic training protocol.

Isokinetic dynamometry is subject to several technical limitations. Because the torque registered by the instrument is due to muscle and gravitational forces, error is introduced during certain portions of the range of motion. Error is also introduced because of the effects of inertia. Some of the measured force reflects the momentum of the moving body part rather than muscular force. Various investigators have suggested corrections for these limitations.

Peak torque at specific movement velocities is the most common isokinetic measurement. Other commonly used measurements include torque at specific joint angles, torque ratios between antagonistic muscle groups, and decrease in torque output with repeated contractions. Because these devices are often interfaced to microprocessors, sophisticated measurements such as time to peak torque and integrated torque (total torque) are possible. Modern isokinetic dynamometers also allow measurement of concentric and eccentric strength.

■ Factors Involved in Muscle Adaptation to Progressive Resistance Exercise

To achieve maximum effectiveness in the strength training program, we must consider factors involved in the adaptation of muscle to stress and conditioning. Building upon the principle of myoplasticity (Chapter 19), the factors include overload, specificity, reversibility, and individual differences.

An interesting aspect of skeletal muscle is its adaptability. If a muscle is stressed (within tolerable limits), it adapts and improves its function. For example, weight lifters exercise their arms and shoulders, so their muscles hypertrophy and gain strength. Larger muscles allow them to accommodate an increased load. Likewise, if a muscle receives less stress than it's used to, it atrophies. For example, the muscles of a leg in a cast atrophy in response to disuse.

The purpose of physical training is to stress the body systematically so it improves its capacity to exercise. Physical training is beneficial only as long as it forces the body to adapt to the stress of physical effort. If the stress isn't sufficient to overload the body, then no adaptation occurs. If a stress can't be tolerated, then injury or overtraining results. When

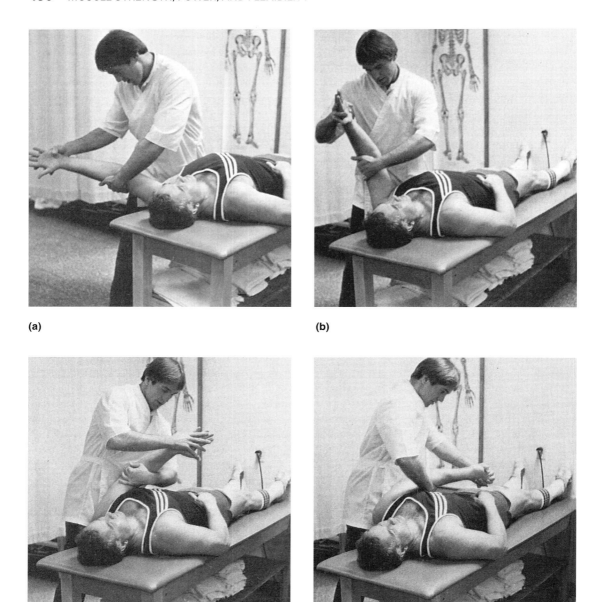

(a)

(b)

(c)

(d)

Figure 20-4 Proprioceptive neuromuscular facilitation. PHOTO: Wayne Glusker.

appropriate exercise stresses are introduced into an athlete's training program, the athlete undergoes significant improvements in performance.

Physical fitness is largely a reflection of the level of training. When an athlete is working hard, fitness is high. However, when heavy training ceases, fitness begins to deteriorate. For a better understanding of stress and adaptation in physical training, see the discussion of the general adaptation syndrome in Chapter 1.

Countless exercise devices and training programs have been hailed as the best methods for gaining strength. In most instances, however, as long as a threshold tension is developed, increases in strength will occur. It is the type of strength that develops that is the important consideration in exercise and sports. Long-distance running up steep hills, for example, develops a certain amount of muscle strength. These resulting muscle adaptations differ from those produced by high-resistance, low-repetition squats. The distance runner develops mainly sarcoplasmic protein (oxidative enzymes, mitochondrial mass, etc.), while the weight lifter develops mainly contractile protein. When designing a training program, the nature of the adaptive response must always be considered. As noted earlier, factors that determine the rate and type of strength gains include overload, specificity, reversibility, and individual differences.

Overload

Muscles increase their strength and size when they are forced to contract at tensions close to their maximum. Muscles must be overloaded to hypertrophy and improve strength. The amount of overload necessary for strength gains can be generalized from experimental and empirical observations.

Muscle protein accumulation occurs when the rate of protein synthesis increases, the rate of protein degradation decreases, or both. The rate of protein synthesis in a muscle is directly related to the rate of entry of amino acids into the cells. Amino acid transport into muscle is influenced directly by the intensity and duration of muscle tension. This was determined by experiments with isolated muscle. The muscles were bathed in a solution containing labeled amino acids, such as ^{14}C labeled α-amino isobutyric acid. It was found that amino acid uptake was highest when muscles were contracted. Uptake was greater as tension and the duration of tension increased.

Weight-training studies and empirical observations of athletes have reinforced the importance of generating muscle tension. For optimal development of strength, muscle tension must be developed at an adequate intensity and duration. Most studies have found that the ideal number of repetitions is between four and eight (repetitions maximum, 4–8 RM). Repetitions should be done in multiple sets (three or more). Less of a strength gain is made when either fewer or greater numbers of repetitions are performed. These findings are consistent with the progressive resistance training practices of athletes.

Athletes involved in speed-strength sports practice low repetition, high-intensity exercise during or immediately preceding the competitive season. These athletes include shot putters and discus throwers. Such training improves explosive strength, while allowing sufficient energy reserves for practicing motor skills. However, the effectiveness of this practice has not been established experimentally.

Strength-speed athletes must specifically overload the neuromuscular system with plyometric and speed-building exercises to stimulate increases in power and speed of contraction. As discussed in Chapter 19, muscle hypertrophy from resistance training is accompanied by a conversion of type IIb to type IIa muscle fibers (type IIx to IIa in humans). Caiozzo and coworkers (1996) found that high-resistance training produced a rapid (i.e., after two training sessions) elevation in the fast type IIx myosin heavy chain (MHC) mRNA isoform and a corresponding repression of the fast type IIb MHC mRNA isoform. This change slows the contractile velocity of the muscle by slowing the crossbridge cycle between actin and myosin.

Power and speed-building exercises may attenuate the effects of heavy resistance exercise on the muscles' contractile apparatus and increase the speed that the nervous system can activate the muscles. This concept has been supported by recent studies. For example, Delecluse and coworkers (1995, 1997) found that a combination of speed and resistance exercise markedly decreased sprint times, while sprinting or resistance exercise alone had no effect on the sprint. Harridge and associates (1998) found that sprint training on a cycle ergometer did not affect maximum shortening velocity in single fibers or in the relative distribution of myosin heavy chain isoforms. Yet the subjects improved their speed by 7% and leg muscle strength by 7–16%.

Training for power and speed sports requires the development of muscle hypertrophy to improve the force generating capacity of the muscle. This training to develop muscle hypertrophy has mixed effects. While myosin cycling rates slow slightly, the force the muscle exerts at any particular velocity increases. In order to maximize power for sport, athletes must improve the speed that the nervous system can "turn-on" the muscle fibers in the motor units. Fortunately, exercises that cause muscle hypertrophy also stimulate neural activation capacity to a certain extent. However, high-power exercises, such as plyometrics and speed exercises, are much more effective for enhancing neural activation than are intense progressive overload exercises. Thus, training for strength-speed sports requires a balance between intense progressive overload and power exercises.

Bodybuilders usually do more sets and repetitions of exercises, as well as more exercises per body part, than weight lifters or strength athletes. Their goal is to build large, defined, symmetrical muscles. It is not known if this typical bodybuilding training method is the most effective way to achieve this goal. Numerous studies have shown that high-resistance, low-repetition exercises are more effective than low-resistance, high-repetition exercises in promoting muscle hypertrophy.

How much overload is necessary for the novice or nonathlete who practices resistance exercise to enhance health and well-being? For years, most weight-training experts have recommended that people do 3 sets of 10 repetitions for 8–10 exercises. However, in the late 1990s, the American College of Sports Medicine (ACSM) recommended that people

do a minimum of 1 set of 10 reps for 8–10 exercises (ACSM 1998). This recommendation was based on studies that found that beginning weight trainers gain about the same muscle mass and strength doing one set or three sets of an exercise. Yet according to a conflicting report by Fleck and Kraemer (1997), there is ample evidence that doing multiple sets provide additional benefits, particularly in more experienced weight trainers.

Proper rest intervals are important for maximizing tension, between both exercises and training sessions. Insufficient rest results in inadequate recovery and a diminished capacity of the muscle to exert full force. Unfortunately, the ideal rest interval between exercises has not been determined. Most athletes strength-train three to four days per week, doing large-muscle exercises, such as the squat and bench press, no more than twice a week. This practice has been empirically derived and allows adequate recovery between training sessions.

The overload must be progressively increased if consistent strength gains are to occur. However, because of the danger of overtraining in strength-building exercises, constantly increasing the resistance is sometimes counterproductive. A relatively new practice among strength-trained athletes is periodization of training. This practice varies the volume and intensity of exercises so that the nature of the exercise stress frequently changes. Many athletes believe that this practice produces a faster rate of adaptation. Periodization of training is discussed further in the section on the progressive resistance training programs of athletes.

Specificity

Muscles adapt specifically according to the nature of the exercise stress, and thus progressive resistance training programs should stress the muscles that are to perform. This means that the muscle exercised is the muscle that adapts to training. If you train the leg muscles, they—and not the shoulder muscles, for example—are the ones that hypertrophy.

Specific recruitment of motor units occurs within a muscle depending on the requirements of the contraction. As discussed in Chapters 17 and 18, the different muscle fiber types have characteristic contractile properties. Slow-twitch fibers are relatively fatigue-resistant, but have a lower tension capacity than the fast-twitch fibers. Fast-twitch fibers can contract more rapidly and forcefully, but they also fatigue rapidly.

The use of a motor unit depends on the threshold levels of its α motoneuron. The low-threshold, slow-twitch fibers are recruited for low-intensity activities such as jogging (and for that matter, most tasks of human motion). However, for high-speed or high-intensity activities such as weight lifting, the fast-twitch motor units are recruited. Figure 20-5 shows the effects of running speed on the selective use of the gastrocnemius and the soleus. The gastrocnemius is largely a fast-twitch muscle; the soleus is mainly a slow-twitch muscle. Note that as the velocity increased, reliance on the fast-twitch gastrocnemius also increased.

The amount of training that occurs in a muscle fiber is determined by the extent to which it is recruited. High-repetition, low-intensity exercise, such as distance running, uses mainly slow-twitch fibers. Endurance training improves the fibers' oxidative capacity. Low-repetition, high-intensity activity, such as weight training, causes hypertrophy of fast-twitch fibers, with some changes to the lower threshold slow-twitch fibers. Again, the training program should be structured to produce the desired training effect.

Increases in strength are very specific to the type of exercise, even when the same muscle groups are used. For example, as shown in Figure 20-6, subjects performed squats for eight weeks and made impressive improvements in strength. However, strength gains over the same period in the leg press were only half as much, and gains in knee extension strength were negligible—even though the same muscle groups were used in all three activities. Specific motor units are recruited for specific tasks. If a person is weight training to improve strength for another activity, the exercises should be as close as possible to the desired movements of that activity. Likewise, when attempting to increase strength after an injury or surgery, rehabilitation should include muscle movements as close as possible to normal activities.

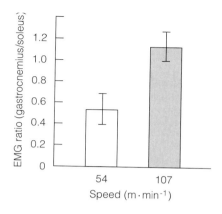

Figure 20-5 Electromyogram (EMG) ratio of gastrocnemius to soleus in a cat during treadmill runs of 54 and 107 m · min⁻¹. Fibers from the predominantly fast-twitch gastrocnemius are recruited more frequently as the intensity of exercise increases. Adapted from Edgerton, 1976.

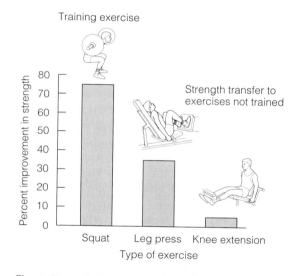

Figure 20-6 The importance of specificity during strength training. Subjects performed squats for eight weeks and made impressive strength gains. On different exercises that used the same muscles, strength gains were much less. SOURCE: Sale, 1988.

Weight-training exercises transfer power to motor skills best when the neurological activation pattern of the muscles most closely resembles the skill. Closed kinetic-chain exercises, such as squats or power cleans, transfer power to motor skills (e.g., jumping), better than open kinetic-chain exercises, such as knee extensions or leg curls. In a closed kinetic-chain exercise, movement begins at the segment most free to move and ends at the segment that is fixed. Movements typically involve several joints moving simultaneously or in sequence, much the way they do during most motor tasks. In open kinetic-chain exercises, movement begins at the fixed segment and involves typically only one joint. Exercises done on many popular exercise machines are open kinetic-chain, while free-weight exercises, such as squats, cleans, and snatches, are closed kinetic-chain.

The research of Blackburn and Morrissey (1998) supports the concept of neural specificity and motor performance. They found that closed kinetic-chain strength (squats) showed a higher correlation with jumping performance (vertical and standing long-jump performance) than open kinetic-chain strength (knee, hip, or ankle extension strength). Likewise, Wilson and coworkers (1996) demonstrated the importance of performing strength-training exercises in postures specific to the target motor skill. Their subjects were tested in a variety of motor skills and weight trained doing exercises in erect and supine postures. They found that subjects improved most in those motor skills that used the same posture used during weight training. These studies reinforce the concept that weight-training exercises should be as close as possible to the target motor skill.

Another aspect of specificity concerns muscle fiber type. Muscle fiber type appears to play an important role in determining success in some sports. Successful distance runners often have a high proportion of slow-twitch muscles (percent slow-twitch fibers is highly related to $\dot{V}_{O_2max}$). Sprinters often have a predominance of fast-twitch muscles. Several studies have shown that a high content of fast-twitch fibers is an element for success in progressive resistance training.

However, all sports do not require prerequisite fiber characteristics. For example, in world-class shot putters there is a surprisingly diverse muscle fiber composition. In those athletes, larger muscle fibers rather than percent fiber type accounted for their performance. There are differences in the rela-

tive percentage of fast-twitch fibers in explosive-strength athletes. Having a high percentage of fast-twitch fibers is not necessarily critical for success. Many strength athletes have a higher fast-to-slow twitch fiber area ratio than do sedentary subjects and endurance athletes. Individual differences in training intensity and technique can make up for deficiencies in the relative percentage of fast-twitch fibers in these athletes. It is interesting to speculate about the performance of a shot putter with a high percentage of fast-twitch fibers. What would performance be like in an athlete who developed good strength and technique? The high percentage of fast-twitch fibers would probably be a decided advantage.

Participation in a training program designed to stimulate both strength and endurance has been found to interfere with gains in strength (Figure 20-7). Hickson (1980) showed that there was more than a 20% difference in strength gain in subjects who participated in a strength-endurance program compared to subjects who trained only for strength. He was unable to demonstrate that strength training

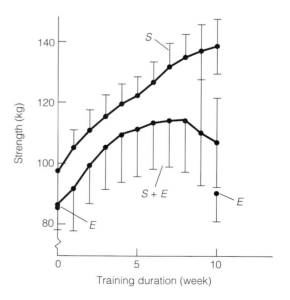

Figure 20-7 Simultaneously training for strength (*S*) and endurance (*E*) interferes with strength development. SOURCE: Hickson, 1980. Used with permission.

had a deleterious effect on endurance, suggesting that strength athletes may inhibit their ability to gain strength by participating in vigorous endurance activities. He also hypothesized that muscles may be unable to adapt optimally to both forms of exercise.

Hickson's paper generated a great deal of controversy. Many studies have supported his findings, while others have attributed his results to overtraining. McCarthy and coworkers (1995), using a training program more typical of recreational exercisers, found that strength and endurance can improve simultaneously. However, recent data showing that endurance training increases slower myosin heavy chain isoforms (Stone et al., 1996) provides a physiological basis for Hickson's findings.

Reversibility

Muscles atrophy as a result of disuse, immobilization, and starvation. Conversely, they adapt to increasing levels of stress by increasing their function. Disuse leads to decreasing strength and muscle mass. Atrophy results in a decrease in both contractile and sarcoplasmic protein.

Muscle fiber types do not atrophy at the same rate. Joint immobilization results in a faster rate of atrophy for the slow-twitch muscle. This has important implications for rehabilitation. Following immobilization, increasing strength is often a major goal. However, endurance should also be stressed because of this relatively greater loss of slow-twitch muscle capacity.

Immobilization affects muscle length. If a muscle is fixed in a lengthened position, sarcomeres are added; if the muscle is immobilized in a shortened position, sarcomeres are lost. Immobilization also leads to a variety of biochemical changes, including decreased amounts of glycogen, adenosine diphosphate (ADP), creatine phosphate (CP), and creatine. All of these factors can affect muscle performance when immobilization ends.

Individual Differences

As with other forms of exercise, people vary in the rate at which they gain strength. Some of this varia-

tion can be attributed to the relative predominance of fast- and slow-twitch motor units in muscles. Usually, endurance athletes have more slow-twitch fibers in their active muscles; strength athletes have more fast-twitch fibers, which tend to be stronger than other fiber types. Compared to slow motor units, fast motor units have more fibers of larger diameter and thus are stronger.

Several studies have shown that fiber composition is genetically determined. Thorstensson (1976), Komi (1984, 1992), and coworkers studied monozygotic and dizygotic twins (identical and non-identical twins). They found that fiber distribution and muscle enzyme activity in each pair of subjects were almost identical. Genetics, however, is not the sole determinant of individual strength differences. Many studies have shown that successful speed-strength athletes do not have a predominance of fast-twitch motor units in critical muscles. Further, fiber composition in athletes is only marginally related to the time for which subjects can maintain isometric force. Genetics influences the capacity to gain strength from a training program. Thomis and coworkers (1998) showed that gains in strength from a 10-week resistance-exercise training program were more similar in monozygotic twins (r = 0.49) than among dizygotic twins (r = 0.22). Their results showed that genetics plays a role in the amount of strength gained in a weight-training program. However, strength training results between people were highly variable. While genetics is important for gaining strength, a good training program can allow you to reach your capacity within a genetically determined "performance envelope."

■ Components of Muscle Strength: Neural-Motor, Contractile, Elastic

Progressive resistance training programs must also consider the neural-motor, contractile, and elastic components of muscle contraction and movement. (See Chapters 17–19 for related structural and functional considerations.) Many athletes and coaches focus on contractile tissue adaptations when designing training programs. However, maximum strength and power requires a coordination of neural activation, storage, and release of muscle elastic energy and contraction of muscle.

Neural-Motor Component of Strength

As discussed in Chapters 17 and 18, strength depends on muscle size and neural activation. Progressive resistance training causes the nervous system to more fully activate important muscles required in specific movements and to better coordinate their actions. These neural changes result in the ability to exert more force.

The neural component of strength is studied primarily with electromyography. Electromyography (EMG) measures changes in electrical potential in muscle. The EMG instrument contains an integrator that quantifies the electrical activity of the muscle over time; this is shown in an integrated electromyogram (IEMG). Greater electrical activity in the muscle reflects greater motor unit recruitment or an increase in motor unit firing rate.

Electromyography employs either surface or needle electrodes. Surface electrodes reflect the gross electrical activity of the entire muscle, while needle electrodes measure the electrical activity in a specific region of the muscle. Both techniques are valuable because they provide information about the rate, extent, and coordination of motor unit recruitment.

Table 20-1 summarizes some of the effects of training on the neural component of strength. Neural adaptations to progressive resistance training include:

- Increased electrical activity of the muscle (increased IEMG activity)
- Increased rate of motor unit stimulation
- Enhanced recruitment of high-threshold motor units (see the discussion of size principle in Chapter 18)
- Increased time that high-threshold motor units can be activated
- Nerve-reflex muscle contraction (motor units turned on when the muscle is suddenly stretched)
- Improved coordination of antagonistic muscle groups (co-contraction of antagonists during gross motor movements)
- Enhanced motor unit synchronization
- Cross-training effects (for example, training one limb results in a training effect in the limb on the other side)

TABLE 20-1	
Neural Adaptations from Strength Training	
Physiological Effect	**Significance**
↑ IEMG activity	↑ Force
↑ Rate of motor unit activation	↑ Rate of force development
↑ Reflex motor unit facilitation of contraction	↑ Rate of force development (assists elastic components to increase force)
Activity of the Golgi tendon organs	Disinhibition of maximal muscle contraction
↑ Coordination of antagonistic muscle groups	↑ Effectiveness of force application
↑ Cross training between muscles on left and right sides of body	↑ Balance between left and right sides of the body
↑ Motor unit synchronization	↑ Force; ↑ efficiency in application of force
↑ Recruitment of high-threshold motor units	↑ Rate of force development; ↑ ability to train high-threshold motor units
↑ Time high-threshold motor units can be activated	↑ Time maximum force can be maintained

↑ = increase.

- Decreased neural disinhibition, allowing more forceful muscle contractions

Increased IEMG   Electrical activity of the muscle increases early in the strength-training process. This is reflected by increased integrated electromyographic (IEMG) activity. Novice weight trainers typically experience significant increases in strength during the first few weeks of training, with only minimal changes in the cross-sectional areas of the muscles (Figure 20-8). This is attributable to an enhanced ability to recruit motor units. Age and sex differences may play a part in the early relative con-

tributions of neural and contractile contributions to strength increases. Several studies have suggested that neural increases play a greater relative role in increasing strength during the early phase of training in older people and women than in young men. IEMG changes from training are also greater for eccentric than concentric exercise. Hortobagyi and coworkers (1996) found that eccentric training increased IEMG activity seven times more during eccentric testing (86% increase) than concentric training increased IEMG activity during concentric testing (12% increase). Higbie and associates (1996) found similar results.

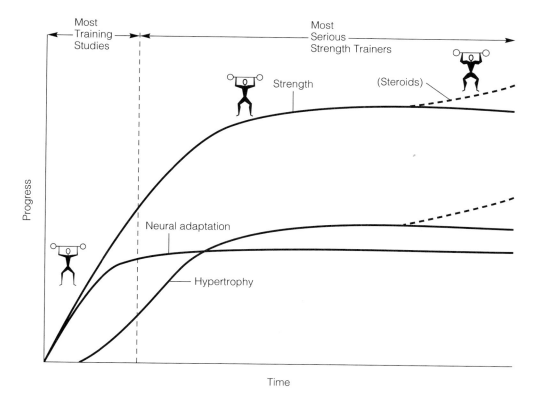

Figure 20-8   The contribution of neural adaptation and hypertrophy to muscle strength. Early changes in strength are due mainly to neural adaptation. Later, increasing the cross-sectional area of the muscle by hypertrophy becomes more important. After training for many years, rapid training gains are no longer possible, unless substances such as anabolic steroids are used. Source: Sale, 1988.

Increased Motor Unit Firing Rate Increasing the firing rate of a motor unit increases the strength and duration of a muscle contraction (Figure 20-9). Firing rate is controlled by the central nervous system. Larger, stronger motor units have a higher firing rate than smaller, weaker motor units. Enoka (1995) estimated that the force that a single motor unit can exert will increase 300–1500% when the firing rate is increased from a minimum to a maximum.

Motor unit firing rate affects the rate of force development. Force rate is important in sports requiring quick, powerful motions. Athletes such as ski jumpers can generate force more rapidly than other athletes. Their maximum strength may not be unusual. Motor unit firing rate can be enhanced by training explosively—that is, attempting to move the resistance rapidly.

Recruitment of High-Threshold Motor Units Motor units differ in ease of recruitment. As discussed in Chapter 18, fibers are recruited by the size principle, with larger, stronger motor units recruited last. To exert maximal force, as many mo-

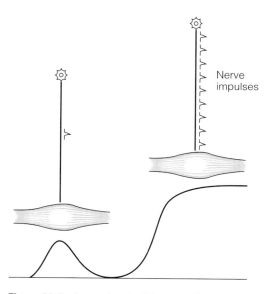

Figure 20-9 Increasing the firing rate of a motor unit increases the strength and duration of a muscle contraction. Adapted from Sale, 1988.

tor units as possible must be recruited. Untrained people cannot recruit high-threshold motor units to the same extent as trained athletes.

Two related neural adaptations—enhanced recruitment of high-threshold motor units and increased time that high-threshold motor units can be activated—have important implications for overall strength development. *A motor unit is trained in direct proportion to its recruitment.* This means that high-threshold motor units will not be trained unless they are recruited during training. In other words, neural adaptations that result in activating high-threshold motor units must occur before these motor units can be trained. For sports requiring explosive power, such as Olympic weight lifting, training programs should emphasize developing neural activation capacity, along with contractile and elastic adaptation.

Motor Unit Coordination Many neural adaptations increase strength and improve coordination. It is commonly observed that elite athletes make powerful movements look almost effortless and graceful. Such movements are possible because the athletes have coordinated neural control of movement, with increased contractile capacity and synchronized release of elastic energy within the muscles. These neural adaptations include reflex motor unit facilitation of contraction, improved coordination of antagonistic muscle groups, enhanced motor unit synchronization, cross-training effects, and inhibition of the Golgi tendon organs.

Neural Facilitation of Movement Training methods such as plyometrics may cause the nervous system to develop reflexes to sudden high-stretch loads. This may be helpful in producing more power in movements such as jumping. For example, the leg muscles of a trained high jumper will be more rapidly and fully activated before a jump than the leg muscles of an untrained person. This phenomenon is illustrated very well in trained versus untrained people performing a plyometric box jump (Figure 20-10). In this movement, the person jumps from a box to the ground, then jumps into the air as soon as possible after landing. There is immediate activation of the quadriceps muscle (as

Nerve impulses

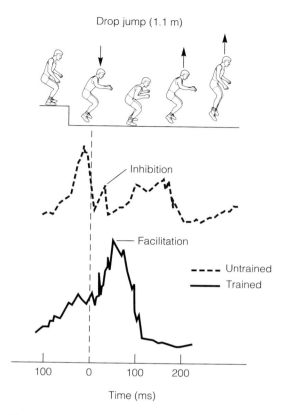

Drop jump (1.1 m)

Inhibition

Facilitation

- - - - Untrained
—— Trained

100   0   100   200

Time (ms)

Figure 20-10 Neural facilitation of movement. When box jumping (a plyometric exercise), the trained person has increased EMG activity in the legs compared to the untrained person. Upon landing, the untrained person is neurally inhibited (decreased DMB activity), while the trained person is neurally facilitated (increased EMG activity). SOURCE: Sale, 1988.

suggested by the IEMG) in the trained person, while the untrained person shows an inhibited response.

Co-contraction of Antagonists Training also enhances control of flexors and extensors during rapid movements. For example, sprinters exhibit greater activation of the hamstring muscles during forceful knee extension than do distance runners. This added control helps in controlling the rapid cycling of extensor and flexor muscles during high-intensity sprinting and in protecting joints from excessive torque.

Synchronization of Motor Unit Firing Rates Synchronized motor unit firing rates lead to smooth acceleration. Such synchronization also causes greater power and increases the time during which high muscle tension can be maintained. As the muscle lengthens suddenly, neural activation of motor units is very important for optimizing its elastic loading. Elastic loading results in a greater application of force when the muscles contract concentrically.

Cross-Training Although popular use of the term *cross-training* describes a training program incorporating running, cycling, and swimming, in this context it refers to training a muscle group on one side of the body and having it result in strength increases in the inactive complementary muscle group on the other side of the body. The inactive body part is called the *contralateral limb.* Although the increase in strength in the contralateral limb is less than that in the active muscle group, the strength gains can be impressive nonetheless. Moritani and deVries (1980) trained elbow flexors in one arm for eight weeks using isometric exercise. They showed a 36.4% increase in strength in the active arm and a 24.7% increase in the inactive arm. In more than 20 studies reported in the literature, contralateral limb strength increases (strength increases in the inactive limb) have varied from 0 to 40%.

Cross-training appears to play an important role in integrating strength gains into improved motor performance. It has been shown to affect interlimb timing and muscle stretch-shortening reflexes in the contralateral limb. It also increases muscle enzyme content. Training one limb increases in the other enzymes important in contraction. This phenomenon is useful in rehabilitation of injuries. For example, following knee surgery, some benefit could be secured for the injured knee by training the healthy leg. Cross-training effects are highest when the contractions used in training are near maximal.

Neural Disinhibition Progressive resistance training also decreases neural disinhibition. As discussed in Chapter 18, Golgi tendon organs act as controls on the muscle to prevent full contraction. Progressive resistance training is thought to inhibit these muscle proprioceptors and allow more forceful muscle contractions. Much of the evidence for this is circumstantial, however. In studies using hypnosis, untrained people can increase strength when told they are stronger, while trained people cannot. The mechanism of this phenomenon is unknown but may be related to Golgi tendon organ inhibition in the trained subjects. Amazing feats of strength performed by untrained people at times of great stress may be due to neural disinhibition.

After injury or surgery, muscle weakness often occurs due to reflex inhibition. Pain or joint swelling induces this reflex, which decreases the ability to exert force. Pain prevents people from exerting full effort during strength exercises, while swelling causes increased joint pressure, thus stimulating the Golgi tendon organs to inhibit force development. Reflex inhibition, then, delays healing and restoration of normal strength. Before normal progress can be achieved during a rehabilitation program, pain control and decreased joint swelling must be achieved.

In summary, neural activation is an important component of strength. Neural adaptation must take place before changes can occur in muscle size. Neu-ral changes are also critical for integrating the application of force into coordinated motor movements.

Contractile Component of Strength

Progressive resistance training tends to increase body weight, lean body mass, and the cross-sectional area of muscle. (See Table 20-2.) Progressive resistance training is often associated with an increase in muscle size. Indeed, there is a high relationship between a muscle's cross-sectional area and its strength. Muscles enlarge (hypertrophy) when the number of myofibrils within the muscle cell increases. This leads to an increase in fiber size rather than to an increase in the number of fibers (see Chapter 19 and the next section for a discussion of hypertrophy and hyperplasia). Muscle hypertrophy is often accompanied by an increase in fiber pinnation angles (Kawakami et al., 1993). Also, there is a greater increase in the area of fast-twitch fibers compared to slow-twitch fibers. Further, as discussed, strength increases during the early phases of a training program are due more to

---

TABLE 20-2

Adaptation of Muscle Tissue to Strength Training

Physiological Effect	Significance
↑ Muscle mass	↑ Muscle strength
↑ Cross-sectional area of muscle (↑ number of myofibrils causing fiber and muscle hypertrophy)	↑ Contractile capacity
↑ Myosin heavy chain IIx and IIa isoforms	Slows myosin cycling rate
↑ Angle of pinnation	↑ Muscle size
↑ Type I and II fiber area	↑ Strength (reflects selective recruitment)
↓ Capillary density (weight lifters); ↑ capillary density (bodybuilders)	↓ Diffusion capacity (weight lifters); Ø diffusion capacity (bodybuilders)
↓ Mitochondrial density per fiber; Ø mitochondrial volume	↓ Oxidative capacity
↑ Intracellular lipids	↑ Capacity to use lipids as fuel
↑ Intracellular glycogen	↑ Glycolytic capacity
↑ Intramuscular high-energy phosphate pool, ATP utilization rate, phosphokinase, myokinase	Improved phosphagen metabolism; ↑ capacity for maximum muscle contractions
Ø Glycolytic enzymes	Capacity apparently adequate
↑ Androgen receptor sites	↑ Effectiveness of androgens in promoting muscle hypertrophy

↑ = increase, ↓ = decrease, Ø = no change, ? = controversial.

improved neural activation than to cellular hypertrophy.

While the cross-sectional area of the muscle is an important determinant of muscle strength, the amount of myosin heavy chain II isoforms (MHC II) in the muscle cell also determines force capacity. In physically active subjects, Aagaard and Anderson (1998) showed moderate to high correlations (r = 0.61–0.93) between isokinetic knee strength and amount of type IIb myosin isoforms in the muscle. As discussed, training increases the amount of type IIx myosin isoforms. In another study, Caiozzo and coworkers (1996) found that eight resistance-exercise-training sessions (16 days) produced a substantial increase in type IIx myosin isoform content. They also found that the program produced a rapid (i.e., after two training sessions) elevation in the fast type IIx MHC messenger RNA (mRNA) isoform and a corresponding repression of the fast type IIB MHC mRNAisoform. Messenger RNA is a substance involved in protein synthesis. The Caiozzo study showed that biochemical processes necessary for changing the speed at which myosin binds to actin occurs very early in training.

Intensity is a key factor determining the extent of cellular hypertrophy. Type II muscle fiber area is over 20% greater in powerlifters than in bodybuilders. Powerlifters typically practice high-intensity, low-repetition (usually <6 repetitions per set) exercise, while bodybuilders often use lower-intensity, high-repetition (>8 repetitions per set) exercise, and perform more sets. In competitive weight lifters, fast-twitch fibers are almost two times larger in diameter than the slow-twitch fibers in the same muscle.

High-intensity progressive resistance training also apparently affects the aerobic capacity of muscle fiber. In weight lifters, as the size of the fast-twitch fibers increase, mitochondrial density decreases but mitochondrial volume remains unaffected. There is a decrease in capillary density, but the capillary-to-fiber ratio is unchanged. High-repetition progressive resistance training (bodybuilding) increases the number of capillaries per muscle fiber. It also increases strength-endurance capacity—the capacity to maintain a level of force for prolonged periods.

For substance delivery, capillaries depend on diffusion. Because weight lifting increases the distance between capillaries, O_2 uptake, substrate delivery, and the removal of CO_2 and other metabolites should be hampered. Decreased endurance capacity could theoretically result. However, an impaired aerobic capacity in weight-trained athletes has not been demonstrated. In fact, weight-trained athletes, who often include endurance exercise in their programs, have normal or above normal aerobic capacity for their age.

Several biochemical alterations accompany hypertrophy during progressive resistance training. High-intensity, slow-speed training using isokinetic loading is associated with increases in the amounts of muscle glycogen, CP, ATP, ADP, creatine, phosphorylase, phosphofructokinase (PFK), and Krebs cycle enzyme activity. Training at faster speeds does not induce these changes. Weight-training studies have replicated the results of this slow-speed, isokinetic training.

Few changes in enzyme concentration or activity occur with progressive resistance training. Most studies have shown no change, or even a decrease in glycolytic enzymes. However, intense weight training increases ATP production and the intramuscular high-energy phosphate pool. It also enhances the activities of creatine phosphokinase and myokinase (enzymes important in the metabolism of high-energy phosphates). So, the immediate energy system is affected by progressive resistance training, but the glycolytic system is not. Oxidative enzymes are unaffected by progressive resistance training, but intracellular lipid stores are increased.

Hypertrophy or Hyperplasia? The cross-sectional area (CSA) of a muscle organ increases with progressive resistance training. A muscle organ can hypertrophy, at least theoretically, by increasing the CSA of individual fibers, by increasing the number of fibers, or by a combination of both adaptive responses. An increase in fiber area is termed *cell hypertrophy,* and an increase in the number of fibers is termed *hyperplasia.* Although the relative contribution of cell hypertrophy and hyperplasia to organ hypertrophy has been controversial in the past, there is now scientific consensus that hypertrophy is the predominant mechanism in mammalian muscle. Cell hypertrophy typically accounts for 95–100% of the organ hypertrophy. The balance of organ hypertrophy is explained by hyperplasia and, to a lesser degree, increases in the angle of fiber pinnation.

In scientific studies of experimental models in which credible evidence is provided for a small degree of hyperplasia, it appears that either further branching of existing fibers occurs, or that there is growth of some new cells from satellite cells, which are the precursor muscle cells. In contrast to results observed in mammals, when hypertrophy is induced by stretch in muscles of birds, such as the Japanese quail, there is strong evidence of significant hyperplasia. The applied significance of the results from birds for human training remains to be established.

Several studies have shown that athletes with well-developed muscles, such as bodybuilders and kayakers, apparently had more muscle fibers than untrained subjects (Larsson and Tesch, 1986). These observations have not been supported by other investigators (MacDougall et al., 1984). It may be that some athletes have more muscle fibers due to genetic selection or that muscle fiber hyperplasia takes many years to occur. It may also be a common occurrence only in elite athletes who have trained for many years.

A meta-analysis conducted by Kelley (1996) suggested that muscle overload can increase muscle fiber number in several animal species. However, a carefully conducted study by McCall and coworkers (1996) failed to find such changes in humans following an intense weight-training program.

In summary, hypertrophy is the major mechanism involved in enlarging muscle in response to overload stress.

Muscle Elastic Component of Strength

Muscles contain elastic structures that can be stretched, absorb energy, and enhance the development of force. The elastic properties of muscle have been divided into series elastic and parallel elastic components (Figure 20-11). These components work in concert with the contractile component of muscle to produce force. The contractile components include the various structures of the sarcomere. The series elastic component includes the tendons and crossbridges of the muscle fibers. This is the most important component for translating stretch into force. The parallel elastic component includes collagenous structures, such as the fasicular membranes. It provides stability and protection to the muscle. The contractile and elastic components have elements in common, such as the crossbridges, which affect contraction and give elasticity to the muscle.

Figure 20-11 Elastic model of muscle. The contractile component includes the contractile elements of the sarcomere. The series elastic component is the most important in assisting force development. The parallel elastic component gives structure and protection to the muscle. Adapted from Hill, 1970.

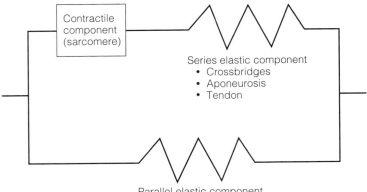

Elastic muscle energy enhances muscle force development in a process called the stretch-shortening cycle. During sudden movements, like jumping or throwing, elastic structures are stretched. The stretch represents potential energy that can enhance the force produced by the contractile component. Immediately after the muscle is stretched, it actively contracts, with the recoil of the elastic elements assisting in force development. The storage and reuse of elastic energy during the stretch-shortening cycle are influenced by fiber type and jumping techniques. Avela and Komi (1998) demonstrated that muscle fatigue is accompanied by a decrease in the performance of the stretch-shortening cycle. Thus, the muscles' capacity for increasing force by using elastic recoil may be a factor in decreased performance as athletes tire.

Many types of human motion incorporate stretch-shortening cycles. When jumping in basketball or the high jump, leg muscles are stretched immediately prior to the jump. The elastic loading of the leg enhances the force development during the jump. Hitting a baseball or throwing a discus also includes a stretch-shortening cycle. Movement of the legs and hips precedes movement of the upper body and elastically loads the torso, resulting in a whiplike motion. More power is generated by this motion than would have been possible using the upper body muscles alone.

■ Coordinating Neural-Motor, Contractile, and Elastic Components of Strength

Maximizing force development requires coordination of the three components of strength. Some evidence suggests that muscle stiffness and elasticity can be altered slightly with training. However, elastic energy is most useful if it can be coordinated into motor movements. Neural changes occur early in the progressive resistance training program. Contractile changes occur later and are probably the main limiting factor of strength (see Figure 20-8). However, improvements in skill can occur throughout life and can thus contribute to maximum force development.

■ Muscle Soreness

Delayed muscle soreness is an overuse injury common in people trying to develop muscle strength. It usually appears 24 to 48 hours after strenuous exercise. Several causes of muscle soreness have been proposed over the years. These include lactic acid buildup, torn tissue, muscle spasm, and connective tissue damage.

Available evidence suggests that delayed muscle soreness results from tissue injury caused by excessive mechanical forces on the muscle and connective tissue. Direct examination by electron microscopy of sore muscles that were previously eccentrically loaded has shown extensive tissue damage, principally in the area of the Z-disk. Soreness was induced by running downstairs. Similar findings have been observed in well-trained bodybuilders.

Biochemical damage appears to follow the initial mechanical damage to the muscle. While the exact nature of this secondary damage is not fully understood, the following sequence of events in the development of delayed-onset muscle soreness (DOMS) has been suggested by various researchers (Armstrong et al., 1991; Macintyre et al., 1995; and Clarkson and Newham, 1995); (Figure 20-12).

- Excessive mechanical forces (particularly during eccentric exercise) disrupt structural proteins in muscle fibers, connective tissue, extrasarcomeric cytoskeleton (titin, nebulin, desmin, synemin, talin, integrin, veniculin, and α-actinin), and sarcolemma.

- Tissue injury initiates inflammatory reaction in the damaged muscle. Elements of this inflammatory process include increased blood flow and tissue permeability—triggered by agents such as histamine, serotonin, prostaglandins (PGE_2), and nitric oxide—vasodilation, edema, increased concentrations of leukocytes, monocytes, macrophages, and neutrophils, and increased muscle temperature—triggered by cytokines, such as interleukin-1, tumor necrosis factor, and α-interferon). The physiological purpose of the inflammatory process is to rid the cells of damaged tissue and prepare the tissue for repair.

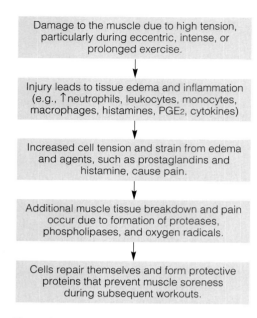

Damage to the muscle due to high tension, particularly during eccentric, intense, or prolonged exercise.

↓

Injury leads to tissue edema and inflammation (e.g., ↑neutrophils, leukocytes, monocytes, macrophages, histamines, PGE_2, cytokines)

↓

Increased cell tension and strain from edema and agents, such as prostaglandins and histamine, cause pain.

↓

Additional muscle tissue breakdown and pain occur due to formation of proteases, phospholipases, and oxygen radicals.

↓

Cells repair themselves and form protective proteins that prevent muscle soreness during subsequent workouts.

Figure 20-12 Possible mechanism for delayed onset muscle soreness.

- Edema and chemical substances (particularly PGE_2) stimulate type III and IV muscle afferents and increase sensitivity of pain receptors. The muscle damage and reaction to the edema and PGE_2 appear to cause the soreness experienced the day after the DOMS-producing workout.

- The inflammatory reaction causes secondary chemical damage through the formation of oxygen radicals, proteases and phospholipases, lysosomal enzymes, and nitric oxide. While this secondary chemical damage is initiated early in the injury process, its full effects manifest themselves 1–3 days following the initiation of DOMS.

- Inflammation is followed by a healing phase and the formation of protective proteins. During tissue healing, there are increases in growth factors, collagen and fibronectin fragments, enzyme inhibitors, oxygen scavengers, and remodeling collagenase. This process heals the tissue and prevents further incidence of DOMS during subsequent exercise sessions.

Several recent studies (including Lecomte et al., 1998, and Croisier et al., 1996) have described the structural and biological changes occurring with DOMS. These studies have sought to clarify the role of the inflammatory response that may be responsible for initiating, amplifying, or resolving skeletal muscle injury. Understanding the process of DOMS can help attenuate its effects. Unfortunately, treatments such as nonsteroidal anti-inflammatants (e.g., ibuprofen and naproxen), transcutaneous nerve stimulation, and post-exercise stretching have rendered mixed results. But in a recent study by Fahey and Pearl (1998) phophatidlyserine, a phospholipid found in all cells, supplementation was shown to attenuate muscle soreness in weight-trained athletes who had been subjected to overtraining.

■ Progressive Resistance Training Programs of Athletes

Athletes are faced with many methods and programs to help them develop strength and power. No doubt all of them will increase strength to a certain extent. It is beyond the scope of this text to provide in-depth analysis of the many programs available. Instead, this discussion deals strictly with techniques that work for athletes noted for high levels of strength.

Although there are undoubtedly exceptions, most strength athletes train with free-weights. Athletes tend to use techniques that improve performance, and although they may gravitate to the fad of the day, their defection will only be temporary if the technique is not effective. Athletes are extremely pragmatic and usually go with the techniques that produce good results.

The progressive resistance training programs of athletes involved in speed-strength sports employ presses, pulls, and squats. Examples of presses include the bench press, incline press, jerk, seated press, and behind the neck press. These lifts are important for developing the muscles of the shoulders, chest, and arms. Pulling exercises include the clean, snatch, high pull, and dead lift. These lifts develop the muscles of the legs, hips, back, and arms. Squats include the squat and leg press. These exercises develop the legs and back (leg press develops very little back strength). These three types of lifts are the most important part of the program. They are usually supplemented by auxiliary exercises, including biceps curls, sit-ups, pullovers, and bar dips (exercises designed to develop strength for specific sports).

The importance of neural development has become increasingly evident. Exercises using stretch-shortening cycles, such as plyometric box jumping, are much more effective than weight training in developing these neural components of strength. Plyometrics are particularly effective for developing the rate and extent of motor unit recruitment (Figure 20-13). The neural component of strength

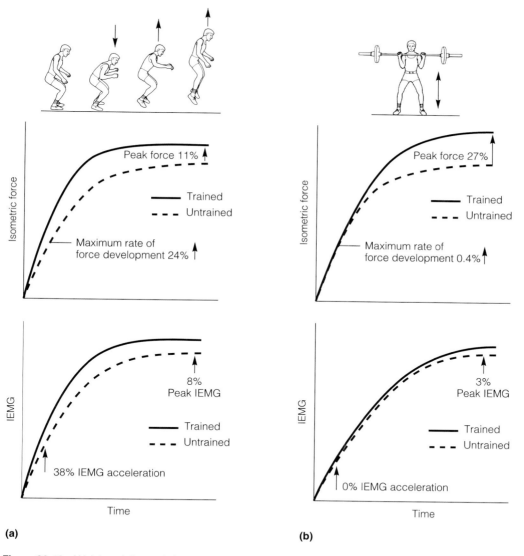

(a)   **(b)**

Figure 20-13 Weight training and plyometrics develop different aspects of strength. Plyometric training results in increases in motor unit recruitment. Weight training increases isometric force capacity and the cross-sectional area of the muscle. Source: Sale, 1988.

cannot be ignored by speed-strength athletes. Appropriate exercises should be integrated into the program.

Successful strength athletes usually train three or four days per week during strength-building periods and one to three days per week during competitive periods. However, the major lifts (press, pulls, and squats) are seldom practiced more than two days per week. If these lifts are practiced too frequently and too intensely, overtraining usually results.

Overtraining results in an increased risk of injury and a decrease in performance, probably due to the inability to train heavily during training sessions. Constant, severe training prevents recovery and maximal training. The training program should be structured so that athletes have recovered enough to train maximally when appropriate. The intensity and duration of tension are the most important factors eliciting strength increases.

Periodization of training was made popular by Soviet weight lifters and track and field athletes. This method may prevent overtraining and the problems that go with it. Periodization employs several types of training cycles throughout the year, including load cycles, recovery cycles, peak cycles, and conditioning cycles. The *load cycle* is the building portion of the program. It is employed during the off-season and in the precompetitive period. The *recovery cycle* is a transition period of active rest that separates the building and competitive periods. The *peak cycle* is the period for developing maximum strength, while affording time to work on motor skills. The *conditioning cycle* is also a period of active rest; it is usually employed during the few months following the competitive season. Examples of load and peak cycles are shown in Tables 20-3 and 20-4.

During the load cycle, athletes perform relatively high numbers of sets (5 to 7) and a moderate number of repetitions (4 to 7). They use moderate amounts of weight (approximately 80% of 1-repetition maximum). The intensity of the load is not large because the concentrated volume results in high-intensity training. Within each load cycle are microcycles lasting approximately two weeks. Notice in Table 20-3 that the highest levels of resistance for a particular exercise are practiced only once during a microcycle. Microcycles are day-to-day variations in exercise. A typical intensity progression of a microcycle is as follows: heavy, light, moderate, light. The cycle is repeated using more weight.

The progression of the load cycle enables the athlete to exert maximum effort during a heavy workout. Microcycles provide a certain amount of rest between heavy workouts. However, load cycles are exhausting. They cannot be kept up for more than two to three months without the risk of overtraining. Functional indicators of strength (i.e., 1 repetition maximum) begin to decrease toward the end of the load cycle because of fatigue. They quickly rebound and exceed the initial level during the recovery phase. Verkhoshansky (1979) has called this phenomenon the "delayed training effect," and it is analogous to glycogen supracompensation.

The recovery cycle is designed to give the athlete rest. It is a period of active rest that prepares the athlete for the high-intensity, low-volume workouts of the peak cycle. This cycle involves low to moderate volume at light to moderate intensity. Recovery cycles typically last approximately two to three weeks.

The peak cycle is designed to produce maximum strength and employs low-volume, high-intensity exercise. Peak cycles are employed during, or immediately preceding, the competitive period. As with the load cycle, the peak cycle is composed of a microcycle lasting two to three weeks. Again, maximum resistance for each exercise is employed only once during each microcycle.

The load cycle prepares the athlete for maximum performance during the peak cycle. However, the athlete must be aware of factors that interfere with maximum strength gains during this cycle. The peak cycle must be continued long enough for the athlete to experience maximum strength gains. The duration of the cycle should equal that of the load cycle. The athlete must not introduce high-volume workouts during the peak period or maximum strength gains will be compromised.

The conditioning cycle is a period of active rest that typically follows the competitive season. The

TABLE 20-3		
Four Weeks of an Eight-Week Load Cycle of a World-Class Discus Thrower		
Week	Monday	Wednesday	Friday
1	Power clean 2s 4r 100 kg Squat 2s 8r 170 kg 2s 6r 185 kg Good morning ex 3s 10r 90 kg Behind neck press 3s 5r 95 kg	Snatch high pull 3s 8r 90 kg Squat 1s 3r 200 kg Bench press 3s 8r 145 kg Dumbbell bench press 4s 8r 45 kg	Power clean 1s 5r 135 kg 1s 5r 142.5 kg Hack squat 3s 10r 100 kg Squats 3s 10r 130 kg Behind neck press 2s 4r 95 kg
2	Snatch high pull 1s 4r 120 kg 1s 4r 135 kg 1s 4r 150 kg Squat 4s 6r 165 kg Bench press 1s 4r 150 kg 1s 4r 155 kg 1s 4r 160 kg	Squat 2s 10r 100 kg Good morning ex 4s 10r 90 kg	Bench press 6s 6r 175 kg Hack squat 3s 10r 100 kg Squat 3s 10r 148 kg Dumbbell bench pr. 4s 8r 45 kg
3	Power clean 1s 4r 120 kg Squat 1s 4r 150 kg Bench press 1s 3r 185 kg Dumbbell flies 4s 8r 22.5 kg	Snatch high pull 3s 8r 90 kg Squat 2s 8r 190 kg 2s 6r 200 kg Behind neck press 3s 3r 100 kg	Snatch high pull 3s 8r 90 kg Squat 1s 4r 210 kg Bench press 1s 6r 165 kg Dumbbell bench pr. 4s 10r 45 kg
4	Power clean 1s 5r 140 kg 1s 5r 150 kg 1s 5r 155 kg Good morning ex 4s 10r 90 kg Bench press 1s 10r 145 kg 1s 4r 155 kg	Snatch high pull 1s 4r 125 kg 1s 4r 140 kg 1s 4r 160 kg Squat 3s 10r 155 kg Behind neck press 3s 5r 85 kg	Bench press 6s 6r 190 kg Squat 1s 6r 145 kg Good morning ex 4s 10r 100 kg Dumbbell bench pr. 4s 10r 50 kg

s = sets, r = repetitions.

training sessions are characterized by low-intensity exercise. A typical load is 60–70% of maximum, moderate volume (4 to 6 sets), and high repetition (8 to 10 repetitions). Approximately every three weeks the athlete should employ a slightly higher intensity to maintain strength. This period provides both a physical and mental break from periods of heavy training, while ensuring that no significant deconditioning occurs.

Progressive resistance training programs using cycles are not appropriate for all types of athletes. To develop an appropriate, specific program, the

	TABLE 20-4		

Four-Week Peak Cycle of a World-Class Discus Thrower

Week	Tuesday	Friday	Sunday
1	Power snatch 1s 2r 100 kg 1s 1r 110 kg 1s 1r 120 kg 1s 1r 125 kg 1s 1r 130 kg Squat 1s 5r 160 kg 1s 5r 170 kg 1s 5r 180 kg 1s 3r 190 kg 1s 3r 200 kg Bench Press 1s 5r 365 kg Flies 1s 10r 22.5 kg	Power snatch 1s 1r 90 kg Dumbbell flies 3s 10r 12.5 kg Biceps curls 3s 10r 57.5 kg	Competition
Week	**Monday**	**Wednesday**	**Friday-Saturday**
2	Power snatch 1s 3r 90 kg Incline dumbbell 4s 6r 45 kg Bench press 1s 3r 175 kg 1s 3r 190 kg	Power snatch 1s 1r 130 kg 1s 1r 140 kg 1s 1r 142.5 kg 1s 1r 145.5 kg Squat 1s 2r 200 kg 1s 2r 220 kg 1s 2r 240 kg 1s 2r 255 kg	Competition
3	Squat 1s 5r 170 kg Bench press 1s 3r 185 kg 1s 2r 197.5 kg 1s 2r 212.5 kg 1s 2r 220 kg	Power snatch 1s 1r 90 kg Dumbbell flies 3s 10r 12.5 kg Biceps curls 3s 10r 57.5 kg	National Championship
4	Squat 1s 2r 200 kg 1s 2r 225 kg 1s 2r 250 kg	Bench press 1s 2r 175 kg 1s 2r 190 kg 1s 1r 202.5 kg 1s 1r 215 kg	No competition
5	United States Olympic Trials		

s = sets, r = repetitions.

strength requirements of each sport must be assessed. In general, though, sports requiring muscle endurance employ progressive resistance training schedules involving a higher number of repetitions, while those requiring strength use fewer repetitions. Progressive resistance training exercises should be chosen to develop the muscles used in the sport.

Physiological Basis for Periodization of Training

Periodization of training has been practiced around the world for at least 20 years, but there is surprisingly little evidence to demonstrate its effectiveness.

The purpose of periodization of training is to prevent overtraining and provide the muscles with the best biological environment for muscle growth and increased strength. Theoretically, periodization of training provides enough rest to prevent excessive release of corticosteroids or immunosuppression while allowing periodic intense training sessions. Several research groups have attempted to establish the existence of relative anabolic-catabolic states by measuring the ratio of testosterone to cortisol in blood. In general, during periods of heavy training in males, testosterone tends to decrease and cortisol increase. These hormone changes have been associated with overtraining. Other symptoms of overtraining include amenorrhea, muscle and liver glycogen depletion, fatigue, lethargy, immunosuppression, and psychological depression. These symptoms typically diminish with rest and altered training. Unfortunately, these hormone assays have not proved to be reliable indicators of performance.

■ Training for Flexibility

Flexibility—the ability of a joint to move through its full range of motion—is extremely important for general fitness and wellness. The smooth and easy performance of everyday and recreational activities is impossible if flexibility is poor. Fortunately, flexibility is a highly adaptable fitness component. It increases in response to a regular program of stretching exercises and decreases with inactivity. Flexibility is also specific. Good flexibility in one joint does not necessarily mean good flexibility in another. However, flexibility can be increased through stretching exercises for all major joints.

There are two basic types of flexibility: static and dynamic. *Static flexibility* refers to the ability to reach a point in a joint's range of motion. It is what most people mean by the term flexibility. *Dynamic flexibility,* unlike static flexibility, involves movement. It is the ability to move a joint quickly through its range of motion with little resistance. For example, static shoulder flexibility would determine how far you could extend your arm across the front of your body or out to the side. Dynamic flexibility would affect your ability to pitch a softball, swing a golf club, or swim the crawl stroke. For a gymnast to perform a split on a balance beam, she must have good static flexibility in her legs and hips; in order to perform a split leap, she must have good dynamic flexibility.

Static flexibility depends on many factors, including the structure of a joint and the tightness of muscles, tendons, and ligaments that are attached to it. Dynamic flexibility can be important for both daily activities and sports. However, because static flexibility is easier to measure and better researched, most assessment tests and stretching programs target static flexibility.

Benefits of Flexibility and Stretching Exercises

Good flexibility provides benefits for the entire musculoskeletal system. It may also prevent injuries and soreness and improve performance in sports and other activities.

Joint Health Good flexibility is essential to good joint health. When the muscles and other tissues that support a joint are tight, the joint is subject to abnormal stresses that can cause joint deterioration. For example, tight quadriceps muscles cause excessive patello-femoral compression, leading to conditions such as chondromalacia patella, which causes knee pain. Tight shoulder muscles can result in restricted shoulder mobility that can irritate and compress sensitive soft tissues, leading to pain and disability in the joint. Poor joint flexibility can also cause abnormalities in joint lubrication, leading to deterioration of sensitive chondrocytes (cartilage cells) lining the joint. Pain and further joint injury can result from this deterioration.

Improved flexibility can greatly improve quality of life, particularly in older people. Aging decreases the natural elasticity of muscles, tendons, and joints, resulting in stiffness. The problem is compounded in people with arthritis. Flexibility improves tissue elasticity, which facilitates movement.

Low-Back Pain and Injuries Low-back pain is sometimes related to poor spinal alignment, which puts pressure on the spinal nerve roots. Good strength and flexibility in the back, pelvis, and legs may help prevent this type of back pain. Unfortunately, research studies have not yet clearly defined the relationship between back pain and lack of flexibility or strength. For example, a study by Jackson and coworkers (1998) found no relationship between performance on the popular sit-and-reach or sit-up tests and the incidence of back pain.

Poor flexibility may increase the risk of injury. A general stretching program has been shown to be effective in reducing the frequency of injuries as well as their severity. When injuries do occur, flexibility exercises can be used in treatment. They reduce symptoms and help restore normal range of motion in affected joints. Unfortunately, not all studies have found a relationship between the risk of injury and poor flexibility. For example, Twellaar and coworkers (1997) found no influence of flexibility on the total number of injuries or the number of several specific injuries (ankle sprain, muscle rupture, dislocation, shinsplints, backache) in physical education students followed for four years.

Overstretching—stretching muscles to extreme ranges of motion—may actually decrease the stability of a joint. While some activities, such as gymnastics and ballet, require extreme joint movements, such flexibility is not recommended for the average person. In fact, due to decreased joint stability, extreme flexibility may increase the risk of injury in

activities such as skiing, basketball, and volleyball. As with other types of exercise, moderation is the key to safe training.

Additional Potential Benefits

- *Reduction of post-exercise muscle soreness.* Delayed-onset muscle soreness (DOMS; discussed in this chapter) is thought to be caused by damage to muscle fibers and supporting connective soft tissue. Some studies (though not all) have shown that stretching after exercise decreases the degree of muscle soreness.

- *Relief of aches and pains.* Flexibility exercises may help relieve pain and stiffness that develops from stress and prolonged sitting. Sitting in one place for a long time can cause muscle tension. Stretching relieves tension, so you can go back to work refreshed and effective.

- *Improved body position and strength for sports and life.* Good flexibility lets a person assume more efficient and biomechanically effective body positions and exert force through a greater range of motion. For example, swimmers with more flexible shoulders have stronger strokes because they can pull their arms through the water in the optimal position. Flexible joints and muscles allow more fluid movement without constraint. Some studies also suggest that flexibility training enhances strength development. However, others, such as a study by Klinge and coworkers (1997), found that flexibility had no influence on the capacity to gain strength.

In some sports, increasing flexibility may be contraindicated. Craib and coworkers (1996) found that runners who were less flexible in ankle dorsiflexion and standing hip rotation exhibited better running economy (i.e., lower oxygen consumption at a specific running velocity). They speculated that inflexibility in certain areas of the musculoskeletal system may enhance running economy by increasing power via the stretch-shortening cycle and minimizing the need for muscle-stabilizing activity.

• *Maintenance of good posture.* Good flexibility also contributes to body symmetry and good posture. Bad postural habits, muscle weakness, and poor flexibility can gradually change joint alignments. For example, sitting in a slumped position can lead to tightness in the anterior chest muscles and overstretching and looseness in the upper spine, causing a rounding of the upper back. This condition, called kyphosis, is common in older people. It may be prevented by doing stretching and strengthening exercises.

Factors Determining Flexibility

The flexibility of a joint is affected by its structure, by muscle elasticity and length, and by nervous system activity. Some factors—joint structure, for example—cannot be changed. Other factors, such as the length of resting muscle fibers, can be changed through exercise; these factors should be the focus of a program to develop flexibility.

Joint Structure   The amount of flexibility in a joint is determined in part by the nature and structure of the joint. Hinge joints, such as those in the fingers and knees, allow only limited forward and backward movement; and they lock when fully extended. Ball-and-socket joints, like those in the hips, enable movement in many different directions and have a greater range of motion. Knees, hips, and other major joints are surrounded by joint capsules, semi-elastic structures that give joints strength and stability but limit movement. Heredity also plays a role in joint structure and flexibility. For example, although everyone has a broad range of motion in the hip joint, not everyone can do "the splits."

Muscle Elasticity and Length   Soft tissues, including skin, muscles, tendons, and ligaments, also limit the flexibility of a joint. Muscle tissue is the key to developing flexibility because it can be lengthened if it is regularly stretched. Regular stretching adds sarcomeres (basic muscle contractile units) to the muscle fibers, which facilities a wider range of motion in the joint.

The most important component of muscle tissue related to flexibility is the connective tissue that surrounds and envelops every part of muscle tissue, from individual muscle fibers to entire muscles. Connective tissue provides structure, elasticity, and bulk and makes up about 30% of muscle mass. Two principal types of connective tissue are *collagen,* white fibers that provide structure and support, and *elastin,* yellow fibers that are elastic and flexible. Muscles contain both collagen and elastin, closely intertwined, so muscle tissue exhibits the properties of both types of fibers. A recently discovered structural protein in muscles called *titin* also has elastic properties and contributes to flexibility.

When a muscle is stretched, the wavelike elastin fibers straighten; when the stretch is relieved, they snap back to the resting position. If gently and regularly stretched, connective tissues will lengthen and flexibility will improve. Without regular stretching, the process reverses: these tissues shorten, resulting in decreased flexibility.

The stretch characteristics of connective tissue in muscle are important considerations for a stretching program. The amount of stretch a muscle will tolerate is limited, and as the limits of its flexibility are reached, connective tissue becomes more brittle and may rupture if overstretched (Figure 20-14). A safe and effective program stretches muscles enough to slightly elongate the tissues but not so much that they are damaged. Research has shown that flexibility is improved best by stretching when muscles are warm (following exercise or the application of heat) and the stretch is applied gradually and conservatively. Sudden, high-stress stretching is less effective and can lead to muscle damage.

Nervous System Activity   Muscle spindles (stretch receptors) also control muscle length. Sudden stretching of a muscle establishes a reflex loop between the muscle spindle, spinal cord, and muscle, causing a muscle contraction. These reflexes occur frequently in active muscles and are important for proprioception and controlling muscle length.

Small movements that only slightly stimulate these receptors cause small reflex actions. Rapid, powerful, and sudden movements that strongly

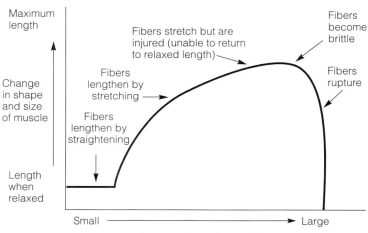

Figure 20-14 Stress–strain curve for skeletal muscle. As the muscle is first stressed, "looseness" in muscle structure creates little deformation. As muscle is stressed further, muscle deforms but quickly reverts to its normal length when the stretch is removed. Further stretch results in tissue deformation (i.e., the muscle does not revert to its normal length when the stretch is removed). Further stretch leads to rupture (failure) of the muscle.

stimulate the receptors cause large, powerful reflex muscle contractions. Stretches that involve rapid, bouncy movements are considered dangerous because they may stimulate a reflex muscle contraction during a stretch. A muscle that contracts at the same time it is being stretched can be easily injured, so slow gradual stretches are always safest.

During slow static stretching, muscle spindle activity increases for approximately 10 sec but then greatly diminishes. For this reason, stretches longer than 10 seconds are often recommended for effective development of flexibility. For the same reason, doing each stretching exercise several times in a row can "preset" the sensitivity of muscle stretch receptors. Stretching a muscle, relaxing, and then stretching it again cause the muscle spindles to become slightly less sensitive, thereby enabling the muscle to stretch farther. It is not known if stretch receptor sensitivity continues to change following prolonged flexibility training, but it is likely that neural changes do occur to help increase flexibility.

The decrease in muscle spindle sensitivity helps explain why stretching is often an effective way to alleviate muscle cramps. During a muscle cramp, muscle spindle activity greatly increases. Stretching the cramped muscle initially increases the muscle spindle. However, after about 10 seconds, muscle spindle activity decreases greatly, which causes a relaxation of the cramping muscle.

Strong muscle contractions produce a reflex of the opposite type—one that causes muscles to relax and keeps them from contracting too hard. This inverse stretch reflex has recently been introduced as an aid to improving flexibility. Contracting a muscle prior to stretching it causes it to relax, allowing it to stretch farther. The contraction-stretch technique for developing flexibility is called *proprioceptive neuromuscular facilitation* (PNF). PNF was discussed earlier in this chapter as a way of increasing strength. More research needs to be done to determine precisely the degree to which PNF techniques cause muscle relaxation and help develop flexibility.

Studies by Magnusson and associates (1996) and by McHugh and associates (1998) questioned the role of neural control in determining flexibility. They concluded that neural factors do not limit the range of movement during slow stretches and that the increased flexibility achieved from training is caused by increased stretch tolerance in the subject or changes in the mechanical properties of the muscle.

Developing Flexibility

Stretching techniques vary from simply stretching the muscles during the course of normal activities to sophisticated methods based on patterns of muscles reflexes. Some principles of training for

---

TABLE 20-5

Principles of Training for Flexibility

- Do stretching exercises statically. Stretch and hold the position for 10–30 seconds, rest 30–60 seconds, and repeat. Never "bounce" while stretching because it increases the risk of injury.
- Practice stretching exercises regularly and develop flexibility gradually over time. Improved flexibility, like other types of fitness, takes many months to develop.
- Do gentle warm-up exercises, such as easy jogging or calisthenics, before doing your pre-exercise stretching routine.
- You should feel a mild stretch rather than pain while performing these exercises. Emphasize relaxation.
- Do all flexibility exercises on both sides of the body.
- Avoid positions that increase the risk of low back injury. For example, if you are performing straight leg toe-touching exercises, bend your knees slightly when returning to a standing position.
- There are large individual differences in joint flexibility. Do not feel you have to compete with others during stretching workouts.

SOURCE: Fahey, Insel, and Roth, 1999.

flexibility are in Table 20-5. Three common stretching techniques for developing flexibility are *static stretches, ballistic stretches,* and *proprioceptive neuromuscular facilitation.*

Static Stretching   In static stretching, each muscle is gradually stretched, and the stretch is held for 10–30 seconds. Holding the stretch longer than 30 seconds will not further improve flexibility (Bandy et al., 1997), while stretching for less than 10 seconds will provide little benefit. A slow stretch prompts less reaction from stretch receptors, and the muscles can safely stretch farther than usual. The key to this technique is that a person must stretch the muscles and joints to the point where a pull is felt, but not to the point of pain.

Ballistic Stretching   The technique of ballistic stretching involves dynamic muscle action whereby the muscles are stretched suddenly in a bouncing movement. For example, toe touching in rapid succession is a ballistic stretch for the hamstrings. The

problem with this technique is that the heightened activity of muscle spindles caused by the rapid stretches can continue for some time, possibly causing injuries during physical activities that follow. For this reason, ballistic stretching is not recommended.

Proprioceptive Neuromuscular Facilitation (PNF)   As mentioned earlier, PNF techniques use reflexes initiated by muscle and joint receptors to cause greater training effects. The most popular PNF stretching technique is the contract-relax stretching method, in which a muscle is contracted before it is stretched (e.g., hamstring muscles are contracted before they are stretched). PNF appears to allow for more stretching, but it tends to cause more muscle stiffness and soreness than static stretching. Furthermore, it usually requires a partner and takes more time.

Passive and Active Stretching   Stretches can be done either passively or actively.

Passive Stretching   In passive stretching, an outside force or resistance provided by yourself, a partner, gravity, or a weight helps your joints move through their range of motion. You can achieve a greater range of motion using passive stretching. However, because the stretch is not controlled by the muscles themselves, there is a greater risk of injury.

Active Stretching   In active stretching, a muscle is stretched by a contraction of the opposing muscle. For example, an active seated stretch of the calf muscles occurs when a person actively contracts the muscles on the top of the shin. Contraction of this opposing muscle produces a reflex that relaxes the muscles to be stretched. The muscle can then be stretched farther with a low risk of injury.

The principal disadvantage of active stretching is that a person may not be able to produce enough stretch to increase flexibility using only the contraction of opposing muscle groups. The safest and most convenient technique is *active static stretching,* with an occasional passive assist. For example, a person might stretch the calves both by contracting the muscles on the top of the shin and by pulling the feet toward the body. This way, the advantages of active stretching—safety and the relaxation reflex—are combined with those of passive stretching—greater range of motion.

Intensity and Duration   In general, flexibility training should be of mild to moderate intensity. The muscles should slowly be stretched to the point of slight tension or mild discomfort, and then the stretch should be held for a duration of 10–30 seconds. Beyond the point of slight tension, a person risks injury. As the stretch is held, the feeling of tension generally subsides. At that point, a person can try stretching further. Rests of about 30–60 seconds between each stretch allow the muscles to recover. A minimum of four repetitions of each stretch is also required to improve flexibility.

Frequency   The American College of Sports Medicine recommends that stretching exercises be performed a minimum of 2–3 days per week. Many people do flexibility training more often—3–5 days per week—for even greater benefits. It is best to stretch when muscles are warm, so people should do these exercises after cardiorespiratory exercises or weight training. Stretching can also be a part of warm-up, but it is best to increase muscle temperature first by doing the active part of the warm-up (e.g., walking or slow jogging).

■ Summary

Strength-building exercises have become an important component of the training programs of many athletes. Strength exercises include isometrics, isotonic-variable resistance, isotonic-constant resistance, plyometrics, eccentrics, speed loading, and isokinetics. The ideal strength-building technique and protocol has yet to be determined.

A variety of factors must be considered when designing a program to build strength, including overload, specificity, reversibility, individual differences, and injury. The programs of proficient strength-trained athletes increasingly emphasize maximum loads but provide enough rest to prevent overtraining.

Three components of strength that should be addressed in the training program are neural-motor, contractile, and elastic. Neural-motor adaptation increases motor unit activation and the activation frequency. Contractile adaptation mainly involves increasing the cross-sectional area of the muscle. Elastic loading of a muscle as part of a stretch-shortening cycle increases the force exerted by the muscle.

Progressive resistance training programs of athletes have become very sophisticated. They include weight training, plyometrics, speed training, stretching, and development of specific skills. The program must be specific and intense enough to increase strength and power, but the volume of work should not be so great as to cause overtraining.

Flexibility—the ability of a joint to move through its full range of motion—is extremely important for general fitness and wellness. Good flexibility provides benefits for the entire musculoskeletal system. It may also prevent injuries and soreness and improve performance in sports and other activities. The flexibility of a joint is affected by its structure, by muscle elasticity and length, and by nervous system activity.

Three common stretching techniques are static stretches, ballistic stretches, and PNF. In static stretching (the most common stretching method), each muscle is gradually stretched, and the stretch is held for 10–30 seconds. The American College of Sports Medicine recommends that stretching exercises be performed a minimum of 2–3 days per week. Many people do flexibility training more often—3–5 days per week—for even greater benefits. It is best to stretch when muscles are warm, so people should do these exercises after cardiorespiratory exercises or weight training.

■ Selected Readings

Aagaard, P., and J. L. Andersen. Correlation between contractile strength and myosin heavy chain isoform composition in human skeletal muscle. *Med. Sci. Sport Exerc.* 30: 1217–1222, 1998.

American College of Sports Medicine. Position stand on the recommended quantity and quality of exercise for developing and maintaining cardiorespiratory and muscular fitness and flexibility in healthy adults. *Med. Sci. Sports Exerc.* 30: 975–991, 1998.

Andersen, J. L., H. Klitgaard, J. Bangsbo, and B. Saltin. Myosin heavy chain isoforms in single fibres from m. vastus lateralis of soccer players: effects of strength-training, *Acta Physiol. Scand.* 150: 21–26, 1994.

Armstrong, R. B., G. L. Warren, and J. A. Warren. Mechanisms of exercise-induced muscle fibre injury. *Sports Med.* 12: 184–207, 1991.

Avela, J., and P. V. Komi. Reduced stretch reflex sensitivity and muscle stiffness after long-lasting stretch-shortening cycle exercise in humans. *Eur. J. Appl. Physiol.* 78: 403–410, 1998.

Baker, D., G. Wilson, and B. Carlyon. Generality versus specificity: a comparison of dynamic and isometric measures of strength and speed-strength. *Eur. J. Appl. Physiol.* 68: 350–355, 1994.

Bandy, W. D., J. M. Irion, and M. Briggler. The effect of time and frequency of static stretching on flexibility of the hamstring muscles. *Physical Therapy* 77: 1090–1096, 1997.

Berger, R. Optimum repetitions for the development of strength. *Res. Quart. Am. Alliance Health Phys. Educ. Recrea.* 33: 334–338, 1962.

Berger, R. Comparative effects of three weight training programs. *Res. Quart. Am. Alliance Health Phys. Educ. Recrea.* 34: 396–398, 1963.

Blackburn, J. R., and M. C. Morrissey. The relationship between open and closed kinetic chain strength of the lower limb and jumping performance. *J. Orthop. Sports. Phys. Ther.* 27: 430–435, 1998.

Bobbert, M. F. Drop jumping as a training method for jumping ability. *Sports Med.* 9: 7–22, 1990.

Booth, F. W., and D. B. Thomason. Molecular and cellular adaptation of muscle in response to exercise: perspectives of various models. *Physiol. Rev.* 71: 541–585, 1991.

Caiozzo V. J., F. Haddad, M. J. Baker, and K. M. Baldwin. Influence of mechanical loading on myosin heavy-chain protein and mRNA isoform expression. *J. Appl. Physiol.* 80: 1503–1512, 1996.

Clarkson, P. M., and D. J. Newham. Associations between muscle soreness, damage, and fatigue. *Adv. Exp. Med. Biol.* 384: 457–469, 1995.

Costill, D. L., E. F. Coyle, W. F. Fink, G. R. Lesmes, and F. A. Witzmann. Adaptations in skeletal muscle following strength training. *J. Appl. Physiol.* 46: 96–99, 1979.

Coyle, E. F., S. Bell, D. L. Costill, and W. J. Fink. Skeletal muscle fiber characteristics of world class shot-putters. *Res. Quart.* 49: 278–284, 1978.

Craib, M. W., V. A. Mitchell, K. B. Fields T. R. Cooper, R. Hopewell, and D. W. Morgan. The association between flexibility and running economy in sub-elite male distance runners. *Med. Sci. Sports Exerc.* 28: 737–743, 1996.

Croisier, J. L., G. Camus, G. Deby-Dupont, F. Bertrand, C. Lhermerout, J. M. Crielaard, A. Juchmes-Ferir, C. Deby, A. Albert, and M. Lamy. Myocellular enzyme leakage, polymorphonuclear neutrophil activation and delayed onset muscle soreness induced by isokinetic eccentric exercise. *Arch. Physiol. Biochem.* 104: 322–329, 1996.

Delecluse, C. Influence of strength training on sprint running performance. *Sports Med.* 24: 147–156, 1997.

Delecluse, C., H. Van Coppenolle, E. Willems, et al. Influence of high-resistance and high-velocity training on sprint performance. *Med. Sci. Sports Exerc.* 27: 1203–1209, 1995.

Enoka, R. M. Morphological features and activation patterns of motor units. *J. Clin. Neurophysiol.* 12: 538–559, 1995.

Fahey, T. D. Basic Weight Training for Men and Women. Mountain View, Ca.: Mayfield, 1997.

Fahey, T. D., and G. Hutchinson. Weight Training for Women. Mountain View, Ca.: Mayfield, 1992.

Fahey, T. D., P. M. Insel, and W. T. Roth. Fit & Well: Core Concepts and Labs in Physical Fitness and Wellness. Mountain View, Ca.: Mayfield, 1998.

Fahey, T. D., and M. S. Pearl. The hormonal and perceptive effects of phosphatidylserine administration during two weeks of resistive exercise-induced overtraining. *Biol. Sport.* 15: 135–144, 1998.

Fleck, S. J., and W. J. Kraemer. Designing Resistance Training Programs. Champaign, Ill.: Human Kinetics, 1997.

Garfinkle, S., and E. Cafarelli. Relative changes in maximal force, EMG, and muscle cross-sectional area after isometric training. *Med. Sci. Sports Exerc.* 24: 1220–1227, 1992.

Garhammer, J. A comparison of maximal power outputs between elite male and female weightlifters in competition. *Int. J. Sport Biomech.* 7: 3–11, 1991.

Giddings, C. J., and W. J. Gonyea. Morphological observations supporting muscle fiber hyperplasia following weight-lifting exercise in cats. *Anat. Rec.* 233: 178–195, 1992.

Goldberg, A. L. Mechanisms of growth and atrophy of skeletal muscle. In Muscle Biology, R. G. Cassens (Ed.). New York: Marcel Dekker, 1972.

Gollnick, P. D., R. B. Armstrong, C. W. Saubertt, K. Piehl, and B. Saltin. Enzyme activity and fiber composition in skeletal muscle of trained and untrained men. *J. Appl. Physiol.* 33: 312–319, 1972.

Ground, M. D. Towards understanding skeletal muscle regeneration. *Path. Res. Pract.* 187: 1–22, 1991.

Harridge, S. D., R. Bottinelli, M. Canepari, M. Pellegrino, C. Reggiani, M. Esbjornsson, P. D. Balsom, and B. Saltin. Sprint training, *in vitro* and *in vivo* muscle function, and myosin heavy chain expression. *J. Appl. Physiol.* 84: 442–449, 1998.

Hettinger, T. L., and E. A. Muller. Muskelleistung und muskeltraining. *Int. Z. Angew. Physiol.* 15: 111, 1953.

Hickson, R. C. Interference of strength development by simultaneously training for strength and endurance. *Eur. J. Appl. Physiol.* 45: 255–263, 1980.

Hickson, R. C., K. Hidaka, and C. Foster. Skeletal muscle fiber type, resistance training, and strength-related performance. *Med. Sci. Sports Exerc.* 26: 593–598, 1994.

Higbie, E. J., K. J. Cureton, G. L. Warren III, and B. M. Prior. Effects of concentric and eccentric training on muscle strength, cross-sectional area, and neural activation. *J. Appl. Physiol.* 81: 2173–2181, 1996.

Hill, A. V. First and Last Experiments in Muscle Mechanics. Cambridge: Cambridge University Press, 1970.

Hortobagyi, T., J. P. Hill, J. A. Houmard, D. D. Fraser, N. J. Lambert, and R. G. Israel. Adaptive responses to muscle lengthening and shortening in humans. *J. Appl. Physiol.* 80: 765–772, 1996.

Jackson, A. W., J. R. Morrow Jr., P. A. Brill, H. W. Kohl III, N. F. Gordon, and S. N. Blair. Relations of sit-up and sit-and-reach tests to low back pain in adults. *J. Orthop. Sports Phys. Ther.* 27: 22–26, 1998.

Kawakami, Y., T. Abe, and T. Fukunaga. Muscle-fiber pennation angles are greater in hypertrophied than in normal muscles. *J. Appl. Physiol.* 74: 2740–2744, 1993.

Kelley, G. Mechanical overload and skeletal muscle fiber hyperplasia: a meta-analysis. *J. Appl. Physiol.* 81: 1584–1588, 1996.

Klinge, K., S. P. Magnusson, E. B. Simonsen, P. Aagaard, K. Klausen, and M. Kjaer. The effect of strength and flexibility training on skeletal muscle electromyographic activity, stiffness, and viscoelastic stress relaxation response. *Am. J. Sports Med.* 25: 710–716, 1997.

Komi, P. V. (Ed.). Strength and Power in Sport. London: Blackwell Scientific Publications, 1992.

Kraemer, W. J. Endocrine responses to resistance exercise. *Med. Sci. Sport Exerc.* 20 (Suppl.): S152–S157, 1988.

Kramer, J. F., A. Morrow, and A. Leger. Changes in rowing ergometer, weight lifting, vertical jump and isokinetic performance in response to standard and standard plus plyometric training programs. *Int. J. Sports Med.* 14(8): 449–454, 1993.

Larsson, L., and P. A. Tesch. Motor unit fiber density in extremely hypertrophied skeletal muscles in man. *Eur. J. Appl. Physiol.* 55: 130–136, 1986.

Lecomte, J. M., V. J. Lacroix, and D. L. Montgomery. A randomized controlled trial of the effect of naproxen on delayed onset muscle soreness and muscle strength. *Clin. J. Sports Med.* 8: 82–87, 1998.

Lesmes, G. R., D. Costill, E. F. Coyle, and W. J. Fink. Muscle strength and power changes during maximal isokinetic training. *Med. Sci. Sports* 10: 266–269, 1978.

MacDougall, J. D., D. G. Sale, S. E. Alway, and J. R. Sutton. Muscle fiber number in biceps brachii in body builders and control subjects. *J. Appl. Physiol.* 57: 1399–1403, 1984.

MacDougall, J. D., G. R. Ward, D. G. Sale, and J. R. Sutton. Biochemical adaptation of human skeletal muscle to heavy resistance training and immobilization. *J. Appl. Physiol.* 43: 700–703, 1977.

MacIntyre, D. L., W. D. Reid, and D. C. McKenzie. Delayed muscle soreness: The inflammatory response to muscle injury and its clinical implications. *Sports Med.* 20: 24–40, 1995.

Magnusson, S. P., E. B. Simonsen, P. Aagaard, H. Sorensen, and M. Kjaer. A mechanism for altered flexibility in human skeletal muscle. *J. Physiol.* (London) 497 (Pt. 1): 291–298, 1996.

Manning, R. J., J. E. Graves, D. M. Carpenter, S. H. Leggett, and M. L. Pollock. Constant vs. variable resistance knee extension training. *Med. Sci. Sports Exerc.* 22: 397–401, 1990.

Mastropaolo, J. A. A test of the maximum-power theory for strength. *Eur. J. Appl. Physiol.* 65: 415–420, 1992.

Matheson, G. O., R. J. Dunlop, D. C. McKenzie, C. F. Smith, and P. S. Allen. Force output and energy metabolism during neuromuscular electrical stimulation: a 31P-NMR study. *Scand. J. Rehabil. Med.* 29: 175–180, 1997.

McCall, G. E., W. C. Byrnes, A. Dickinson, P. M. Pattany, and S. J. Fleck. Muscle fiber hypertrophy, hyperplasia, and capillary density in college men after resistance training. *J. Appl. Physiol.* 81: 2004–2012, 1996.

McCarthy, J. P., J. C. Agre, B. K. Graf, M. A. Pozniak, and A. C. Vailas. Compatibility of adaptive responses with combining strength and endurance training. *Med. Sci. Sports Exerc.* 27: 429–436, 1995.

McHugh, M. P., I. J. Kremenic, M. B. Fox, and G. W. Gleim. The role of mechanical and neural restraints to a joint range of motion during passive stretch. *Med. Sci. Sports Exerc.* 30: 928–932, 1998.

Moritani, T., and H. A. deVries. Potential for gross muscle hypertrophy in older men. *J. Gerontol.* 35: 672–682, 1980.

Nardone, A., C. Romanò, and M. Schieppatgi. Selective recruitment of high-threshold human motor units during voluntary isotonic lengthening of active muscles. *J. Physiol.* 409: 451–471, 1989.

O'Shea, P. Effects of selected weight training programs on the development of strength and muscle hypertrophy. *Res. Quart.* 37: 95–102, 1964.

Osternig, L. R., R. H. Robertson, R. K. Troxel, and P. Hansen. Differential responses to proprioceptive neuromuscular facilitation (PNF) stretch techniques. *Med. Sci. Sports Exerc.* 22: 106–111, 1990.

Ozmun, J. C., A. E. Mikesky, and P. R. Surburg. Neuromuscular adaptations following prepubescent strength training. *Med. Sci. Sports Exerc.* 26: 514, 1994.

Phillips, C. A. Functional Electrical Rehabilitation. New York: Springer-Verlag, 1991.

Sale, D. G. Neural adaptation to resistance training. *Med. Sci. Sports Exerc.* 20 (Suppl.): S135–S145, 1988.

Saltin, B., and P. D. Gollnick. Skeletal muscle adaptability: significance for metabolism and performance. In Handbook of Physiology. Skeletal Muscle. Bethesda, Md.: Am. Physiol. Soc., 1983, pp. 555–631.

Soest, A. J. van, and M. F. Bobbert. The role of muscle properties in control of explosive movements. *Biol. Cybern.* 69: 195–204, 1993.

Staron, R. S., F. C. Hagerman, and R. S. Hikida. The effects of detraining on an elite power lifter. *J. Neuro. Sci.* 51: 247–257, 1981.

Stone, J., T. Brannon, F. Haddad, A. Qin, and K. M. Baldwin. Adaptive responses of hypertrophying skeletal muscle to endurance training. *J. Appl. Physiol.* 81: 665–672, 1996.

Sugden, P. H., and S. J. Fuller. Regulation of protein turnover in skeletal and cardiac muscle. *Biochem. J.* 273: 21–37, 1991.

Tamaki, T., T. Sekine, A. Akatsuka, S. Uchiyama, and S. Nakano. Detection of neuromuscular junctions on isolated branched muscle fibers: application of nitric acid fiber digestion method for scanning electron microscopy. *J. Electron Microsc.* 41: 76–81, 1992.

Tamaki, T., S. Uchiyama, and S. Nakano. A weight-lifting exercise model for inducing hypertrophy in the hindlimb muscles of rats. *Med. Sci. Sports Exerc.* 24: 881–886, 1992.

Tesch, P. A., and L. Larsson. Muscle hypertrophy in bodybuilders. *Eur. J. Appl. Physiol.* 49: 301–306, 1982.

Thomis, M. A. I., G. P. Beunen, H. H. Maes, C. J. Blimkie, M. Van Leemputte, A. L. Claessens, G. Marchal, E. Willems, and R. F. Vlietinck. Strength Training: Importance of genetic factors. *Med. Sci. Sports Exerc.* 30: 724–731, 1998.

Thorstensson, A. Muscle strength, fibre types and enzyme activities in man. *Acta Physiol. Scand.* (Suppl.) 443: 1–45, 1976.

Twellaar, M., F. T. Verstappen, A. Huson, and W. van Mechelen. Physical characteristics as risk factors for sports injuries: a four-year prospective study. *Int. J. Sports Med.* 18: 66–71, 1997

Van Etten, L. M., F. T. Verstappen, and K. R. Westerterp. Effect of body building on weight-training-induced adaptations in body composition and muscular strength. *Med. Sci. Sports Exerc.* 26: 515–521, 1994.

Verkhoshansky, U. How to set up a training program in speed-strength events. (Part 1). *Legkaya Atletika* 8: 8–10, 1979. Translated in *Sov. Sports Rev.* 16: 53–57, 1981.

Verkhoshansky, U. How to set up a training program in speed-strength events. (Part 2). *Legkaya Atletika* 8: 8–10, 1979. Translated in *Sov. Sports Rev.* 16: 123–126, 1981.

Wilson, G. J., A. J. Murphy, and A. Walshe. The specificity of strength training: the effect of posture. *Eur. J. Appl. Physiol.* 73: 346–352, 1996.

CHAPTER 21

CONDITIONING FOR ENDURANCE ATHLETIC EVENTS

It may seem that studying the physiology of various training methods to identify those that produce the greatest results should be simpler than, for instance, studying the biochemistry of fat metabolism in muscle. However, such is not the case. It has proved extremely difficult to manipulate, in a scientific way, the training regimens of quality athletes, who, understandably, do not want their training regimens "tampered" with. Therefore, studies of various training regimens frequently involve only a few trained athletes or a large number of untrained volunteers who train intensively over a period of only 8 to 15 weeks. Such results, however, are not directly applicable to the training of talented individuals who engage for many years in intense training. Moreover, the large number of training

World and Olympic heptathlon champion Jackie Joyner-Kersee performs in Olympic competition. Improvements in training techniques, intensity, and volume have resulted in vastly improved performances in most classes of athletes. PHOTO: © Reuters/Corbis.

regimens and all their possible combinations make it very difficult to include in a single study enough subjects to enable researchers to analyze the benefits of the different training protocols.

The sports of running and swimming epitomize those rhythmical activities where speed and endurance depend on physiological power. In general, running is the most universal sport, not only because of the worldwide interest in it, but also because skill in running is essential to success in many other sports. It is of interest to describe conditioning for running because internationally recognized (Olympic) competition includes events that span the range of pure sprint to pure endurance. In general, many of the same training principles that apply to running also apply to conditioning for swimming. As we will see, however, swimming differs in that Olympic competition does not include pure sprint events. Furthermore, in swimming, stroke technique is exceedingly important, perhaps more important than physiological power.

This chapter is based on three sources of information: (1) the limited studies available on training methods, (2) the application of basic physiological principles to athletic training, and (3) the practical experiences of athletes and coaches.

Training for Athletic Competition

Note that the following discussion refers to training for athletic competition and not the training of middle-aged, older, or recreational asymptomatic individuals. The training stimulus necessary to maintain or improve cardiovascular function in the population at large, as described by the American College of Sports Medicine (ACSM), is less in terms of intensity and volume than is the training of competitive athletes described here. The training stimulus recommended to maintain cardiovascular fitness (Chapters 24 and 27) will not develop the exceptional performance levels required for success in competitive athletes, but the ACSM training guidelines are probably no less effective for developing fitness for daily living and for promoting longevity than is training for competitive athletics.

Overload, Stimulus, and Response

The principle of overload is a rephrasing of the well-known general adaptation syndrome (Chapter 1), wherein physiological adaptations occur in response to appropriate stimuli. The amount of overload to a system can be varied by manipulating two basic factors:

1. Training intensity.
2. Training volume, which is made up of training frequency and training duration.

In general, the greater the overload, the greater the resulting adaptation and increase in functional capacity. Because it takes *time* for physiological responses to occur following application of a training stimulus, the progressive application of a training stimulus must be accomplished within particular constraints. The application of the stimulus (i.e., the increase in training volume and intensity) must be gradual and progressive. That is, rather than a rapid and continuous increase in training to achieve an increase in athletic performance (Figure 21-1), the training progression should be gradual and discontinuous in nature. Periods of heightened training should be interspersed with recovery periods involving decreased training intensity and volume. Adequate rest each day of training is also impor-

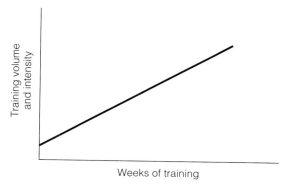

Figure 21-1 Over the course of a training season, the training volume and intensity should be progressively increased to achieve optimal performance. The linear increase in training illustrated here is probably not the best protocol for increasing training intensity (see Figure 21-2).

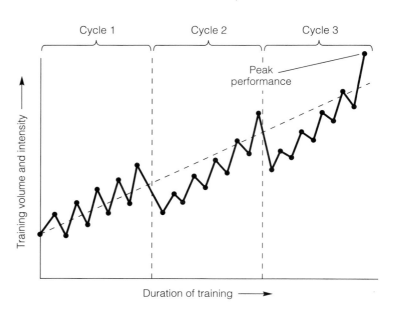

Figure 21-2   Over the course of a training season, training volume and intensity should increase in a cyclic, discontinuous manner. Days and weeks of heightened training should be followed by recovery days and weeks of lesser intensity. Similarly, each training cycle should begin at an intensity less than that of the previous cycle. In this system, the stimulus of progressive training is accompanied by periods of recovery and adaptation.

tant. In addition, hard-training days should be interspersed with easier days. Training should be scheduled in cycles, wherein a period of lesser training intensity and volume follows the peak training of the previous cycle. This is the concept of periodization of training, which was discussed in Chapter 20. Therefore, the training schedule of a serious athlete should look more like the broken zigzag shown in Figure 21-2 than the smooth line in Figure 21-1. Recuperation periods at the beginning of a training cycle are an essential part of the training regimen and should be followed scrupulously, for it is during recuperation periods that adaptation occurs.

■ Specificity, Skill Acquisition, and Developing Metabolic Machinery

Increasing performance as the result of training requires that training be appropriate for the event. Two factors are involved in the specificity of training response: learning the event of interest and developing the metabolic machinery to support the event. In sports such as running and, especially, swimming, where skill level can tremendously af-

fect performance outcome, long hours of training develop the neuromuscular control mechanisms needed to optimize results by minimizing the effort necessary to achieve a given result, or by maximizing the work output for a given metabolic power output.

The principle of specificity requires that the training regimen overload the metabolic system that supports the activity. In evaluating the relative contributions of immediate, glycolytic, and oxidative energy sources in supporting an activity, the duration of the activity is of primary importance. (See, for example, Figures 3-1 and 3-2.) Activities lasting a few seconds (e.g., the shot put) depend mainly on immediate energy sources (ATP and CP). Activities lasting from a few seconds to a minute (e.g., 200- to 400-m track race, 50-m swimming event) depend mostly on glycogenolysis and the glycolytic formation of lactic acid. Events lasting longer than a minute become increasingly dependent on oxidative metabolism. Because most athletic events last longer than a minute, training the cardiovascular mechanism of O_2 transport and the muscular mechanisms of O_2 utilization are very important. For endurance-type activities, training the systems of O_2 delivery and utilization are so important that their emphasis can allow athletes to be successful, even if they over-look and neglect training the other energy systems. At the conclusion of endurance events, athletes with the strongest finishing kicks are not necessarily the best sprinters: They are the freshest individuals with the best capabilities of prolonged oxidative metabolism.

Recovery from Sprint Activities

Being aerobically fit as the result of endurance train-ing can benefit speed and power athletes as well as endurance athletes. This is because many sports (e.g., soccer) require repeated bursts of activity with rapid recovery during the competition. Recovery is essentially an "aerobic" activity, so athletes en-gaged in repeated bursts of activity benefit from be-ing aerobically fit.

Over-Distance Training

The usually stated objectives of over-distance train-ing (sometimes also called long, slow distance train-ing) are twofold: (1) to increase $\dot{V}_{O_2max}$ and (2) to in-crease tissue respiratory (mitochondrial) capacity. Of the two, increasing respiratory capacity seems to be the more important. Although endurance ath-letes such as marathon runners display very high values of $\dot{V}_{O_2max}$, $\dot{V}_{O_2max}$ actually correlates poorly with performance in events such as the marathon.

Middle-distance runners sometimes record higher values of $\dot{V}_{O_2max}$ than do long-distance runners. Ob-viously, however, top marathon runners are faster in their event than are top mile or 1500-m runners. Some marathon runners can apparently sustain a pace eliciting over 90% of $\dot{V}_{O_2max}$ for several hours. The basis for this phenomenal endurance ability lies in the muscles and other tissues as well as in the capacity of the O_2 transport apparatus.

As shown by Gohil and associates (1983; Table 21-1), over-distance training, consisting of running or swimming mile after mile at a speed much less than the competitive pace, causes a proliferation of mitochondrial protein in muscle. In detailed animal studies, Davies and associates (1971) have shown that endurance capacity correlates better with mi-tochondrial density than with $\dot{V}_{O_2max}$. The ability to use fats as fuels and the ability to protect mitochon-dria against damage during prolonged work are qualities developed through over-distance training.

Each training regimen has its advantages and disadvantages. The advantage of over-distance training is that it develops tissue respiratory capac-ity; the disadvantages come from the fact that the training regimen overlooks the basic principle of specificity. All athletic activities, even the most ba-sic, such as running, require practice in technique. Therefore, an endurance athlete must prepare for competition by training at or near the race pace. A mile runner practicing to break the 4-min mile by

TABLE 21-1

Fiber-Type Composition and Activities of Mitochondrial Enzyme in Marathon Runners and Sedentary Controls

	% Type I, Slow-Twitch Muscle Fibers	Pyruvate-Cytochrome C Reductase[a]	Succinate-Cytochrome C Reductase	Cytochrome C Oxidase
Marathon Runners	82.1	2.4	2.0	21.0
Controls	40.6	0.3	1.3	6.7
% Difference	102	700	54	213
Average Difference (%)		270		

[a] Units are: μmol cyt. C $\cdot$ min^{-1} $\cdot$ g^{-1} wet wt

Source: From Gohil et al., 1983.

running at an 8-min · mi^{-1} pace makes about as much sense as a gymnast attempting to practice a double somersault in slow motion. Athletes must learn pace and improve skills by performing at high rates in practice. Over-distance training can build tissue respiratory capacity but does not develop sense of pace, skill, or the capacity to achieve a high $\dot{V}_{O_2max}$. However, athletes who attempt to practice at or near race pace cannot possibly sustain much of a training volume. One approach that provides specific skill and pace training along with volume training is the interval training regimen.

Interval Training

The interval training regimen is one in which periods of intensity during a workout are interspersed with periods of relief or rest. For instance, an interval workout for a runner practicing to dip under a 4-min mile might consist of 10 repeated 60-sec quarter-mile runs with a 2-min relief or rest interval between each run.

One advantage of interval training is that the athlete learns pace; he or she practices the specific competitive skill, and the cardiovascular training stimulus intensity is greater than it is in over-distance training. Another advantage of interval training is that the athlete can perform more total exercise and more high-intensity exercise in training. The interval training stimulus may maximize the improvement in $\dot{V}_{O_2max}$, as well as result in significant improvements in mitochondrial density.

In addition to these aerobic training benefits, high-intensity interval training stresses the glycolytic system in muscle and results in significant lactate accumulation. Because the accumulation of lactate is distressing to a competitor, he or she must learn to tolerate its presence by repeated exposure to it. More importantly, isotope studies of lactate metabolism indicate that training can improve the pathways of lactate removal (Chapter 10). The sites of lactate removal are heart and red skeletal muscle, which oxidize lactate, and gluconeogenic tissues (liver and kidney), which participate in the Cori cycle. Because the rate of lactate removal depends directly on its concentration (i.e., the greater the concentration, the greater the removal), interval training that increases blood lactate levels will stimulate improvement in its removal.

Training Variation and Peaking

With just two basic aerobic training regimens (over-distance and interval), the varying permutations and combinations of distance, speed, duration of rest interval, and so forth result in almost an infinite number of training possibilities. Coaches, who must often work to sustain interest and desire in athletes over months if not years of training, can benefit from this large number of choices. Varying the training regimen is one way to maintain interest in athletes.

As the training year progresses and the competitive season approaches, the training regimen should be adjusted to achieve peak performance. Let us illustrate by continuing to describe the training of a miler trying to break a time of 4 minutes. Early in the training year (i.e., fall and winter), the athlete should concentrate on over-distance training, with interval training sessions only once or twice a week. The athlete should alternate hard and easy days with an interval-training day substituted for the hard over-distance day. The over-distance day might consist of a 10-mi cross-country run at a pace of 6 min (30 sec · mi^{-1}). An easy over-distance day would consist of a 5-mi run at the same pace. Interval training on the track might focus on developing the 4-min mile rhythm. Therefore, an interval day might consist of either ten quarter-mile (400-m) runs at a 60-sec pace, each with a 5-min rest interval in between, or twenty 30-sec, 220-yd (200-m) runs, each with 2-min rests in between, or five 880-yd (800-m or ½-mi) runs in 2 minutes with 5-min rest intervals. During rest intervals, the athlete should walk or jog to stimulate oxidative recovery. By following such a training regimen, the athlete will be laying a solid base for the competitive season.

As the competitive season approaches, the athlete should gradually convert from a basic over-distance program to an interval program more specifically designed around the competitive event of interest. Instead of one or two interval sessions and five or six over-distance sessions a week, these training protocols should be reversed. Again, hard days should be alternated with easy days. If minor competitions begin to appear on the schedule, such as on every other Saturday, then the heavy volume of training should be centered early in the week (Sunday, Monday, Tuesday), and the training intensity and volume should taper at the end of the week. On hard interval-training days, the 4-min mile race pace should be the cornerstone of the interval training. For variety, the rest interval between runs may be shortened, distances lengthened, pace quickened, or number of interval repeats increased. If the distances and numbers of repeats are increased, care should be taken not to let the pace fall much below the intended race pace. In practice, if an athlete can manage 10 quarter-mile runs at 60 seconds with only a 2-min interval between, then he or she might expect to break 4 minutes.

Sprint Training

True sprints are events that last from a few seconds to approximately 30 seconds. Sprinting requires an extreme degree of skill, coordination, and metabolic power. This power comes from immediate and non-oxidative energy sources present in muscle before the activity begins. Although proficiency in sprinting certainly can be improved through training, the nature of the activity is such that genetic endowment determines in a major way the success an individual can achieve.

Whereas endurance athletes require daily training to develop the cardiovascular and muscle respiratory capacities necessary to be competitive, sprint training requires more intense but less frequent training.

An adequate training frequency for sprinters might be three to five days per week. The highly specialized nature of sprinting requires that training develop the specific skills used in the sport. Some of these skills might include starting, accelerating, relaxing while sprinting, and finishing. Running and sprinting drills should be all out, but the

distances should be kept at less than the competitive distances so that repeated bouts at maximal intensity can be practiced.

Analysis of a particular sprinter's performance during these high-intensity intervals at maximal speed may reveal a need for exercises to improve other specific aspects of his or her performance. For instance, analysis may point out the need for exercises to develop hip flexion (knee lift), which can be particularly important. Weight lifting and plyometric exercises can be used to develop both quadriceps and hamstring strength. High-speed filming may reveal inefficiencies in form that can be corrected in practice.

Although over-distance training is not strictly necessary for sprinting, under particular circumstances the training volume required by sprinters should be improved. For example, over-distance training early in the training session may be used to effect a weight loss in over-fat athletes. Long sprints (e.g., 400-m track running) require a significant aerobic component; therefore, interval training should be employed. Over-distance training can also be successful if the competitive situation requires repeated bursts of activity. For example, a particular competition may involve trials, quarter- and semifinals, as well as final heats; or a competition may involve participation in several events plus a relay. Even in soccer and American football, repeated sprinting is required. The ability to recover rapidly, therefore, is essential. Recovery is an aerobic process that can be improved through over-distance, interval training. Over-distance conditioning of sprinters may also reduce the incidence of injury.

■ Volume vs. Intensity of Training

A perpetual question in athletic training is whether increasing the volume (frequency and duration) of training is more beneficial than increasing the intensity. There appears to be no simple answer to this question, but consideration should be given to both the type of event involved and the phase of training in relation to the competitive season. In general, the more intense (sprint) types of activities require higher intensities of training. This is because of the principle of specificity, which dictates that attention be given to developing the metabolic apparatus and skill levels necessary to compete in the event. Of necessity, more intense training requires reduced training volume.

In general, during the early training season, athletes should focus on increasing the training volume. As the competitive season approaches, training intensity should be emphasized and volume diminished. In preparing for a major competition, both training volume and intensity should be tapered (i.e., reduced).

■ Central vs. Peripheral Adaptations

Even in untrained but otherwise healthy young adults, the oxidative capacity of mitochondria in the total muscle mass greatly exceeds the capacity for arterial oxygen transport (T_{O_2}, Chapter 11). However, as we have seen (Tables 6-1, 6-2, and 21-1), the ability to increase the density and amount of muscle mitochondria and increase the organization of the muscle mitochondrial reticulum (i.e., from about

50 to 300%) greatly exceeds the ability to increase $\dot{V}_{O_2max}$ in response to training (which is about 10 to 25%). We have interpreted these results to mean that at low to moderate exercise intensities, the increased muscle mitochondrial mass due to training enhances the ability to clear lactate produced from glycogenolysis and glycolysis and increases the sensitivity of mitochondrial respiratory control, thus better matching glycogenolysis to TCA cycle activity (Chapters 5 and 6).

Relative capacities in the abilities of the human cardiovascular system and muscle mitochondrial reticulum to expand and decline in response to training and detraining were demonstrated in a classic 1977 report by Henriksson and Reitman (Figure 21-3). In response to eight weeks of training, $\dot{V}_{O_2max}$ increased less than 20%, whereas the markers for mitochondrial capacity (SDH and cytochrome oxidase) increased over 30%. However, over six weeks of detraining, $\dot{V}_{O_2max}$ declined little, whereas the mitochondrial markers fell to pretraining levels, or below. Thus, mitochondrial capacity is far more labile than the total capacity to transport and use oxygen. Henriksson and Reitman's results form the basis for our approach to training and tapering for major athletic competitions.

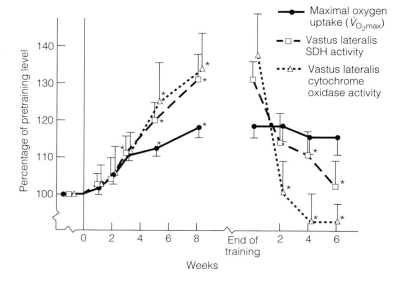

Figure 21-3 Change in $\dot{V}_{O_2max}$ and vastus lateralis (a quadriceps muscle) oxidative enzyme activities during training and detraining periods. Asterisks indicate significant differences from pre-training, or end-of-training values. Note that oxidative enzymes adapt more rapidly and to a relatively greater extent to both training and detraining. Hence, muscle mitochondrial capacity is more labile than whole-body $\dot{V}_{O_2max}$. Adapted from Henriksson and Reitman, 1977.

■ The Taper for Competition

The period prior to major competition, when athletes rest by decreasing training volume and, sometimes, intensity so that peak performances can be achieved, is termed the *taper period.* The taper can be understood generally in terms of the general adaptation syndrome. It is during the taper that athletes recover from the hard training and adaptive responses peak. The taper period used varies from sport to sport. In track, the most frequently used taper period ranges from one to two weeks; in swimming, the taper is frequently twice that used in track. At present, unfortunately, insufficient research has been done to allow us to calculate the duration or exercise regimen of the optimal taper period. A taper of two to three days should be sufficient to result in maximum *glycogen supercompensation,* a heightened level of muscle glycogen achieved by proper training, rest, and nutrition (Chapter 28). Within a week following intense training, minor injuries should have healed, enzymes should have adapted maximally, soreness should have disappeared, and nitrogen balance should have returned to zero from positive levels. In other words, the response to training overload should have peaked. On physiological bases, then, a taper period of around a week to 10 days would seem to be ideal.

In a report from McMaster University, Shepley, MacDougall, Cipriano, Sutton, Tarnopolsky and Coates (1992) provided needed information on the effects of a 1-week taper on 1500-m runners. Subjects tapered in three ways: by resting, by low intensity (60% $\dot{V}_{O_2max}$) running that averaged 6 (from 2 to 10) km, or by high-intensity running, starting with five, 500-m intense repeats on day 1, and decreasing one repeat each day. The investigators measured a variety of performance and physiological parameters, some of which are displayed in Figure 21-4.

When simple resting was compared with high-intensity (500-m repeat) tapering, the high-intensity taper provided the most consistent beneficial results. Subjects increased in total blood volume and red blood cell mass, muscle glycogen content, muscle mitochondrial activity, knee strength (MVC,

maximal voluntary contraction), and time to exhaustion in running at their best 1500-m race pace. The low-intensity taper produced inconsistent results, and they are not shown.

From these results, a pattern of tapering for competition emerges in which subjects greatly reduce volume of training, rest, and observe sound nutritional practices while at the same time maintaining training intensity. It appears, then, that tapering for the purposes of competition should involve preservation of training intensity but reduction of training volume.

■ High Altitude Training

Since the great East African runners burst on the scene in the 1970s, it has been widely speculated that the lifelong residency of these athletes at moderate altitudes has given them a competitive advantage in O_2 transport capacity over athletes who have lived and trained at sea level. Consequently, there has been considerable interest in high altitude training in the hope of enhancing sea-level performance, and several nations have developed high altitude training facilities. However, as described in Chapter 23, there are potentially deleterious effects of high altitude residency, including altitude sickness, loss of appetite, dehydration, loss of body weight and physical deconditioning. Consequently, while it is necessary to be adapted to high altitude for high altitude competition, it is not surprising that most attempts at using high altitude training to enhance sea-level performance have been unsuccessful. Not until very recently have the dedicated efforts of Ben Levine and associates (including James Stray-Gundersen and Robert Chapman) shown any significant benefits of altitude exposure to training regimens. Their formula of living moderately high to build blood volume and training low to maintain training volume and intensity has shown some variable but positive effects.

Levine and associates (1997) divided high caliber athletes into three groups: athletes who continued living and training at sea level (low–low), athletes who lived and trained at 2500-m (high–

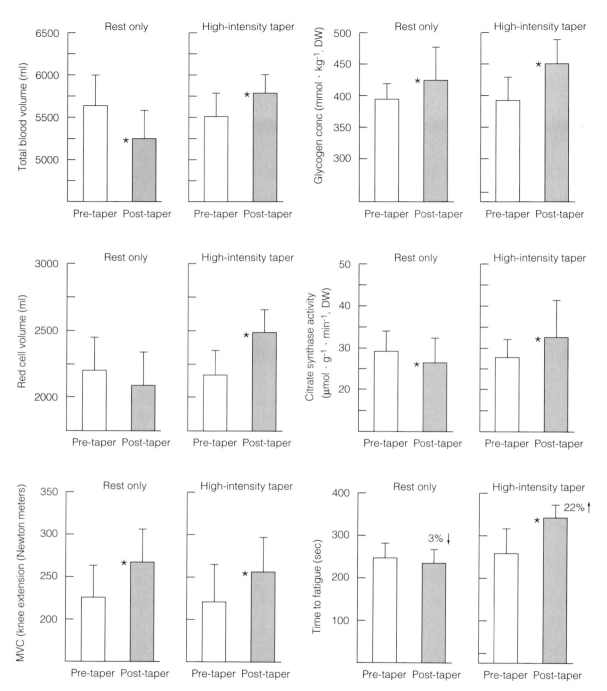

Figure 21-4 Changes in physiological and performance parameters in response to a week taper period by 1500-m runners. Results obtained with complete rest and intense 500-m running are compared. With complete rest, some physiological parameters deteriorate, as does running time to fatigue. However, with the high-intensity taper, blood volume, mitochondrial activity, muscle glycogen content, strength, and running performance all improve. *: significantly different from pre-taper. Modified from Shepley et al., 1992.

high), and those who lived at 2500-m but trained at 1250-m (high–low). Training was for four weeks. Because a sufficient number of athletes were studied, a pattern of the benefits of high–low living and training emerged. It also became evident that 2500-m altitude is not high, but quite moderate; that

some athletes do not respond to high–low living and training; and that such a training approach is very difficult and time consuming, especially when large numbers of athletes are involved.

Figure 21-5a from their study shows that altitude training did increase $\dot{V}_{O_2max}$. However, only high–

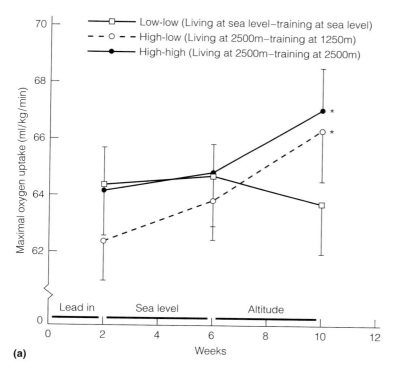

Figure 21-5 $\dot{V}_{O_2max}$ at baseline, after sea-level training and after altitude or continued sea-level training (a), and the ability to sustain high power output under the same conditions (b). Moderate altitude living and low altitude training (high-low) gave the best result. Adapted from Levine et al., 1997. Used with permission.

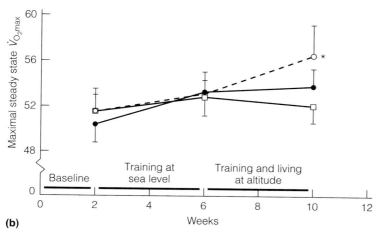

low trained athletes increased running endurance (Figure 21-5b). In a subsequent retrospective analysis of data from the same investigation, Chapman and associates (1998) were able to show that only about half (17 of 39) of the subjects responded positively to altitude exposure. The responders were characterized by having significantly greater levels of erythropoietin (EPO) in the body than nonresponders. Responders were also able to maintain training intensity, whereas non-responders could not.

On the basis of the experiments of Levine and colleagues, previous data on high altitude physiology research (Chapter 23) and athletic experience are more clearly understood. Properly conducted, residence at a moderately high altitude and lower altitude training can help some, but not all, athletes. In some cases, the cost of such training in terms of time, effort, and money may be justified. The reasons that some Kenyans are outstanding distance runners likely has more to do with geographic isolation and genetic predisposition for endurance exercise over generations than with altitude residency (Hochachka, 1998).

■ Three Components of a Training Session

Each training session should consist of three components: warm-up, training, and cool-down. There are several objectives to preliminary exercise (warm-up). The athlete attempts to increase the temperature of the tissues to take advantage of the Q_{10} effect (Chapter 2) on raising metabolic rate and the speed of muscle contraction. Preliminary exercise also raises the cardiac output and dilates capillary beds in muscle. In this way, circulation of blood and O_2 is raised before hard exercise starts. In addition, during a warm-up, an athlete attempts to stretch out the active tissues by stretching exercises and warming of the tissues. Through exercise and wearing heavy clothing, prior stretching and warm-up is thought to minimize the possibility of injury. Preliminary exercise also provides a "last-minute" practice session. Motor skills are fine-tuned and adjusted for the prevailing conditions. The warm-up procedure

also provides the athlete a time when she or he can prepare psychologically for the practice or competition. This neural aspect of preparing for exercise may exceed in importance the other benefits of preliminary exercise.

The nature of the preliminary exercise (warm-up bout) depends on the specificity activity to be performed. In general, however, several considerations apply. The exercise performed should utilize the major muscles to be involved in the training or competition. It should be the same or very similar to the activity to be engaged in, and it should progress from mild to hard intensity. Because about 10 minutes of exercise at a particular work intensity is required to reach a steady muscle temperature, a warm-up should be at least that long. Also, because athletes should avoid fatiguing exercise prior to practice or competition, and because tissues cool down more slowly than they heat up, there should be frequent rest periods. Then, once an athlete has warmed up, there should be a brief recovery period before the training or competition begins. Ideally, this period should be 5 to 10 minutes, although a well-clothed athlete may remain warmed up for 20 to 30 minutes.

By habit or tradition, most athletes include time in a training session to warm up, but they usually overlook the cool-down period following training. A cool-down period is very much like the reverse of a warm-up. The exercise intensity is gradually decreased and is followed by a period of passive stretching, wherein heavily used muscles are held in elongated positions for two 1-min intervals. Runners usually stretch the hamstrings and gastrocnemius muscles after training. In older or less fit individuals, particularly, a cool-down period may effectively minimize soreness and stiffness during the days following hard training or competition.

■ Methods of Evaluating Training Intensity

Is a workout schedule too hard, too easy, or just right? At various times, both coaches and athletes want some objective evaluation of an athlete's training regimen. To answer this question, several approaches can be taken, all of which involve evaluation of both performance and physiological criteria.

Of primary importance is an evaluation of the athlete's performance. If the athlete is meeting or exceeding performance criteria during workouts, time trials, and competitions, then the training regimen is obviously having good results. If the athlete feels good an hour or so after training, and if the athlete feels good the morning and day after training, then the training regimen is probably appropriate.

Heart Rate

Application of the above criteria may be able to answer the question whether the training schedule is too easy or too hard, but it cannot identify an optimal-training regimen or provide an objective assessment of training intensity. For this evaluation, determination of exercise heart rate can be useful. During interval training, the exercise intensity should be sufficient to stimulate heart rate to maximum. Maximal heart rate can be determined during an exercise stress test (Chapter 27) or immediately after a time trial by ECG or palpation (counting the pulse). The recovery interval should end and the next training interval should begin when heart rate falls to two-thirds of maximum (i.e., about 120 beats · min^{-1}). Such a regimen would constitute a very hard interval-training program. Heart rate during submaximal, over-distance training should stabilize at 85–90% of maximum and progress to maximum as the training session is completed.

In addition to measuring the intensity of exercise and recovery interval, heart rate can be used to assess adequacy of recovery. As part of training practice, athletes should learn to take their heart rate in bed before rising in the morning. If, after a hard day of training, the morning heart rate is elevated, then an easy day of training and more rest should be prescribed.

Blood Lactate

Determination of blood lactic acid level has been purported to be another "objective" means of evalu-

ating intensity of the training stimulus. Lactate level can easily be measured in a tiny drop of blood taken from the earlobe or fingertip by a pinprick. The former East German swim team reportedly identified the work intensity that elicits a 4-mM blood lactic acid level as the optimal aerobic (over-distance) training intensity. Unfortunately, no data have been put forward to justify this level as the ideal, and there is no theoretical basis for it. Certainly, for interval training, a 4-mM blood lactate level would lack sufficient intensity. Probably also for over-distance training, a 4-mM blood lactate level would be too easy. For instance, this author (who is now neither young nor trained) has been able to maintain for an hour exercise loads that elicit blood lactic acid levels from 6 to 8 mM. Because the ability to remove lactic acid is related to its concentration, an athlete should occasionally experience high circulating lactate levels (10 mM) in order to develop the mechanisms of lactate removal (Chapters 10 and 33).

Rather than identifying any particular value for blood lactate during a progressive exercise test (e.g., see Figure 10-1), it is probably more appropriate to identify the exercise power output at which blood lactate begins a continuous rise. This approach involves the concept of a *critical lactate clearance point* (maximal lactate steady state) and is illustrated in Figure 21-6. The rationale for using this approach to identify the power output at which lactate clearance mechanisms fail to keep pace with production is as follows.

From Figures 10-4 and 10-8, we realize that blood lactate rises as the result of beginning a submaximal exercise task. Referring to curves A–C in Figure 21-6, we see that after peaking 5–10 minutes after the start of exercise, blood lactate declines with continued effort. Therefore, even though the start of exercise results in lactate production and accumulation, compensatory mechanisms allow lactate disposal to match production (curve C) or exceed production (curves A and B).

Curve C represents exercise power output in which blood lactate initially rises above 4 mM, but then stabilizes at a lower level. Although blood lactate is elevated above resting, the subject can still

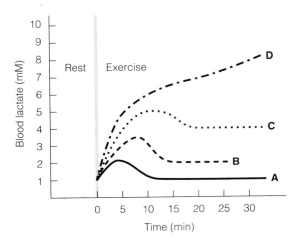

Figure 21-6 Illustration of the concept of a *critical lactate clearance point.* The relationship between lactate concentration and time is depicted during four continuous exercises of graded intensities, with (*a*) being the easiest and (*d*) the hardest. In this figure, exercise intensity (*c*) is the maximum that can be tolerated without evoking a continuous rise in blood lactate. During exercise (*c*), the capacity to clear lactate is sufficient to allow a maximal lactate concentration steady state. Though lactate clearance capacity may not be maximal at this point and clearance can increase if lactate rises (as in exercise (*d*), during exercise (*c*) a critical lactate clearance point has been achieved beyond which the dynamic steady state cannot be maintained. Note that in this diagram the critical lactate clearance point has been achieved at a concentration of 4 mM. A 4-mM concentration frequently occurs at the critical lactate clearance point, but the association is not causal.

cope metabolically with the exercise power output. In contrast, the power output represented in curve *D* elicits a continuous rise in blood lactate. We know, therefore, that this power output is beyond the subject's capability for lactate clearance (i.e., beyond the critical lactate clearance point), and he or she should not attempt to run a marathon at this pace. Blood lactate measurements made in continuous exercise protocols can help predict power outputs that can be sustained.

The exercise protocol we have described for identifying the lactate inflection point involves several bouts of prolonged continuous exercise. In contrast, some investigators have recommended exer-

cise tests in which the power output is increased every few minutes. In these so-called ramp tests (e.g., Figures 10-1 and 10-17), however, there is insufficient time for clearance mechanisms to compensate for lactate production and entry in blood. The lactate inflection point, therefore, frequently occurs at a low blood concentration (e.g., 2 mM), and at this level, the exercise intensity is too low for training improvements to occur (Figure 10-2).

The ventilatory threshold (T_{vent}) (Figure 21-7) is that point at which ventilation begins to increase nonlinearly in response to increments in work rate. The T_{vent} is sometimes called the "anaerobic threshold" and is usually associated with an increase in the blood lactic acid level (Chapter 10). However, several stress factors can elevate ventilation during exercise, and as such, determination of the ventilatory inflection point (T_{vent}) faces the same problems of interpretation as does determination of a blood lactate inflection point (T_{lact}) or the blood lactate level of 4 mM. Therefore, use of T_{vent} as a training or competition tool should involve continuous, as opposed to progressive (ramp), testing protocols. Another problem with using T_{vent} as a training guide is that its measurement requires a special apparatus that is not ordinarily available to the athlete and

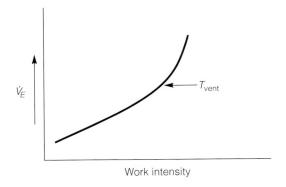

Figure 21-7 Pulmonary minute ventilation ($\dot{V}_E$) is a function of work intensity. As work rate increases, $\dot{V}_E$ increases linearly up to a particular work load, after which $\dot{V}_E$ increases disproportionately. This break in the ventilation versus work rate curve is termed the "ventilatory threshold" (T_{vent}). Blood lactate level (Figure 10-1b) and other factors affect the T_{vent}; please refer to Chapter 10.

coach. If a laboratory is available to a runner, the treadmill pace that elicits the T_{vent} can be determined with precision and then used in the field. However, repeated laboratory tests are required as the athlete improves with training. Some athletes learn to recognize the sensation of heavy breathing associated with lactic acidosis, which frequently makes possible the "talk test"; when lactic acidosis produces a hyperpnea sufficient that athletes cannot talk, they are at or above the T_{lact}.

Heart Rate Field Test for the Maximal Blood Lactate Steady State

Because of the obvious importance of the blood lactate response in assessing relative exercise intensity, there has been interest in finding a noninvasive means of identifying the maximal exercise intensity at which a stable circulating lactate concentration can be maintained. One proposal has been to use the heart rate response; in the past, however, application of this approach has been hampered by the lack of a theoretical basis as well as by the lack of systematic investigation linking heart rate and blood

lactate concentration responses. Fortunately, progress is being made in finding a solution to the latter problem.

Use of heart rate to identify maximal exercise intensity at which a stable circulating lactate concentration can be maintained has been studied in detail by Hofmann, Gaisl, and associates (Hofmann et al., 1994). Using computerized evaluations of heart rate response during graded, progressive leg cycle ergometer tests (Figure 21-8), they showed that in healthy young women the steep rise in blood lactate coincides with the failure of heart rate to increment linearly. On average, the so-called heart rate threshold (HRT) occurred at 90% of maximal heart rate and 70% of $\dot{V}_{O_2max}$. The usefulness of the HRT was demonstrated by these investigators because their subjects could sustain for 20 minutes a power output that was 10% less than that which elicited the HRT, but they could not sustain 20 minutes of exercise at a power output 10% greater than that which elicited the HRT.

In interpreting their results, Hofmann and associates were careful to note the influences of day-to-day variability among test results. We may add

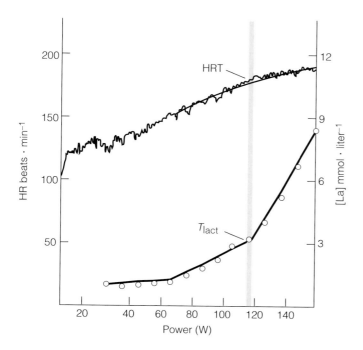

Figure 21-8   Relationship between the lactate threshold (T_{lact}) and the heart rate threshold (HRT) in young women during a progressive leg cycle ergometer test. Note that the downturn in heart rate response coincides with the T_{lact}. This approach requires precisely graded exercise and computer-aided analysis of heart rate response. Modified from Hofmann et al., 1994. Used with permission.

also that only healthy young women were studied and not elite athletes of either gender, and, further, that the cycle ergometer has the advantage of precisely controlling power output, which is not always the case in other forms of exercise. Still, the careful work done provides additional support for using exercise heart rate response as a means of assessing relative exercise intensity in athletes as well as others. This approach of using heart rate response to gauge exercise intensity is useful as long as its interpretation is not confused with representing an "anaerobic threshold."

In summary, all of the objective means available to most coaches to gauge training intensity and adequacy of recovery, perhaps the best approach is the monitoring of exercise and recovery heart rates. This approach conveniently allows training to be individualized to the athlete and event. Further, the method takes advantage of all the neural and humoral feedback mechanisms that regulate heart rate. Taking advantage of an individual's CNS cardiovascular control centers to evaluate adequacy of training stimuli and recovery is a viable approach to objective evaluation.

■ Planning a Training Schedule

In preparing for competition, many athletes expend a great deal of effort but are unsuccessful for some simple reasons. Athletic training is, in some cases and at some times, a very important endeavor. Therefore, a system or training strategy should be constructed and written out. A particular training schedule should be prescribed, and a log of training adherence and performance should be kept. Perhaps the most frequent failing of some training regimens is that insufficient overload is applied. Either training volume or training intensity is too little to result in training adaptations that will distinguish a particular competitor.

Some training regimens are often too shortsighted in their approach. Many athletes expect rapid improvements and become discouraged even as the foundation of real adaptation is being achieved. The progression of an athlete in training

and competition should realistically be planned to extend over several seasons. Particularly in American track and field competition, where rapid progress is frequently desired, hard interval training has been used to optimize performance in the short run. However, because no real basis has been laid down to support consistent improvement from season to season, athletes become discouraged when they fail to improve consistently. The temptation to develop racers rather than physiologically superior athletes must be suppressed.

Sometimes athletes fail to reach their level of competitive aspiration because they lack an individualized training program. In a group of 10 cross-country athletes running interval quarter miles together, perhaps 5 will be running very hard to keep up. Several others will not be training very hard, and at least 1 or 2 will be receiving a minimal training stimulus. Such a predicament happens when there are more athletes to train than can interact with the coach.

A log of other information can also be helpful to the athlete and coach. Such information can include but is not limited to: dietary assessment (Chapter 28), body weight and composition (Chapter 25), heart rate on arising, blood lactate threshold (T_{lact}), heart rate threshold (HRT), and ventilatory threshold (T_{vent}). Obviously, the availability of technology to perform such assessments will vary from location to location, but availability of any or all technologies is secondary in importance to having a plan organized around the basic principles of overload and specificity and recognizing the importance of rest and recovery in the periodization of training.

■ Summary

As with strength training (Chapter 20), training for rhythmical events involves applying of the basic training principles of overload, specificity, reversibility, and individuality. According to the principle of overload, applying an appropriate stimulus will result in adaptation; the greater the stimulus, the greater the adaptation. According to the principle of specificity, the adaptations will be specific to the type of stimulus provided and will occur only in the tissues and organs stressed. In other words, preparing for particular events involves very specific training regimens. The principle of reversibility reminds us that the adaptations due to overload are not permanent and that withdrawal of training will result in regression toward the untrained state. Finally, whereas according to the principles of overload and specificity, two individuals of equal initial ability will respond in the same direction as a result of a particular training regimen, the principle of individuality dictates that the degree of response will likely be different because each person is unique.

■ Selected Readings

American College of Sports Medicine. The recommended quality and quantity of exercise for developing and maintaining fitness in healthy adults. *Med. Sci. Sports* 10: vii–x, 1978.

Anderson, P., and J. Henrickson. Capillary supply of the quadriceps femoris muscle of man: adaptive response to exercise. *J. Physiol.* (London) 270: 677–690, 1977.

Bevegard, S., A. Holmgren, and B. Jonsson. Circulatory studies in well trained athletes at rest and during heavy exercise, with special reference to stroke volume and the influence of body position. *Acta Physiol. Scand.* 57: 26–50, 1963.

Bouchard, C., P. Godbout, and C. Leblanc. Specificity of maximal aerobic power. *Eur. J. Appl. Physiol.* 40: 85–93, 1979.

Brodal, P., F. Inger, and L. Hermansen. Capillary supply of skeletal muscle fibers in untrained and endurance-trained men. *Am. J. Physiol.* 232: H705–H712, 1977.

Brooks, G. A., "Anaerobic threshold:" an evolving concept. *Med. Sci. Sport Exer.* 17: 22–31, 1985.

Brooks, G. A. Lactate: glycolytic end product and oxidative substrate during sustained exercise in mammals—the "lactate shuttle." In Circulation, Respiration, and Metabolism: Current Comparative Approaches, vol. A, Respiration—Metabolism—Circulation, R. Gilles (Ed.). Berlin: Springer-Verlag, 1985, pp. 208–218.

Brooks, G. A. Current concepts in lactate exchange. *Med. Sci. Sports Exer.* 23: 895–906, 1991.

Brooks, G. A., and J. Mercier. The balance of carbohydrate and lipid utilization during exercise: The "crossover" concept. *J. Appl. Physiol.* 76: 2253–2261, 1994.

Brooks, G. A. Forty Years of Progress in Basic Exercise Physiology. American College of Sports Medicine, Indianapolis, June 1994.

Brynteson, P., and W. Sinning. The effects of training frequencies on the retention of cardiovascular fitness. *Med. Sci. Sports* 5: 29–33, 1973.

Chapman, R. F., J. Stray-Gundersen, and B. D. Levine. Individual variation in response to altitude training. *J. Appl. Physiol.* 85: 1448–1456, 1998.

Clausen, J. P. Effect of physical training on cardiovascular adjustments to exercise in man. *Physiol. Rev.* 57: 779–815, 1977.

Connett, R. J., C. R. Honig, T. E. J. Gayeski, and G. A. Brooks. Defining hypoxia: a systems view of $\dot{V}_{O_2}$, glycolysis, energetics and intracellular PO_2. *J. Appl. Physiol.* 68: 833–842, 1990.

Costill, D., J. Daniels, W. Evans, W. Fink, G. Krahenbuhl, and B. Saltin. Skeletal muscle enzymes and fiber composition in male and female track athletes. *J. Appl. Physiol.* 40: 149–154, 1976.

Costill, D., H. Thomason, and E. Roberts. Fractional utilization of the aerobic capacity during distance running. *Med. Sci. Sports* 5: 248–252, 1973.

Costill, D. L., D. S. King, R. Thomas, and M. Hargreaves. Effects of reduced training on muscular power in swimmers. *Physician Sports Med.* 13: 91–101, 1985.

Coyle, E. F., W. H. Martin III, S. A. Bloomfield, O. H. Lowry, and J. O. Holloszy. Effects of detraining on responses to submaximal exercise. *J. Appl. Physiol.* 59: 853–859, 1985.

Daniels, J. T., R. A. Yarbrough, and C. Foster. Changes in max $\dot{V}_{O_2}$ and running performance with training. *Eur. J. Appl. Physiol.* 39: 249–254, 1978.

Davies, C. T. M., and A. U. Knibbs. The training stimulus: the effects of intensity, duration and frequency of effort on maximum aerobic power output. *Int. Z. Angew. Physiol.* 29: 299–305, 1971.

Ekblom, B. Effect of physical training on the oxygen transport system in man. *Acta Physiol. Scand.* 328 (Suppl.): 11–45, 1969.

Ekblom, B., P. Astrand, B. Saltin, J. Stenberg, and B. Wallstrom. Effect of training on circulatory response to exercise. *J. Appl. Physiol.* 24: 518–528, 1968.

Ekblom, B., and L. Hermansen. Cardiac output in athletes. *J. Appl. Physiol.* 24: 619–625, 1968.

Faria, I. Cardiovascular response to exercise as influenced by training of various intensities. *Res. Quart.* 41: 44–50, 1970.

Fox, E. L. Difference in metabolic alteration with sprint versus endurance interval training programs. In Metabolic Adaptation to Prolonged Physical Education, H.

Howald and J. Poortmans (Eds.). Basel: Birkhauser Verlag, 1975, pp. 119–126.

Fox, E., R. Bartels, C. Billings, D. Mathews, R. Bason, and W. Webb. Intensity and distance of interval training programs and changes in aerobic power. *J. Appl. Physiol.* 38: 481–484, 1975.

Fox, E. L., R. L. Bartels, J. Klinzing, and K. Ragg. Metabolic responses to interval training programs of high and low power output. *Med. Sci. Sports* 9: 191–196, 1977.

Frick, M., A. Konttinen, and S. Sarajas. Effects of physical training on circulation at rest and during exercise. *Am. J. Cardiol.* 12: 142–147, 1963.

Fringer, M. N., and G. A. Stall. Changes in cardiorespiratory parameters during period of training and detraining in young adult females. *Med. Sci. Sports* 6: 20–25, 1974.

Gettman, L. R., M. L. Pollock, J. L. Durstine, A. Ward, J. Ayers, and A. C. Linnerud. Physiological responses of men to 1, 3, 5 day per week training programs. *Res. Quart.* 47: 638–646, 1976.

Gohil, K., D. A. Jones, G. G. Corbucci, S. Krywawych, G. McPhail, J. M. Round, G. Montanari, and R. H. T. Edwards. Mitochondrial substrate oxidations, muscle composition and plasma metabolite levels in marathon runners. In Biochemistry of Exercise, H. G. Knuttgen, J. A. Vogel, and J. Poortmans (Eds.). Champaign, Ill.: Human Kinetics, 1983, pp. 286–290.

Gollnick, P., R. Armstrong, B. Saltin, C. Saubert, W. Sembrowich, and R. Shepherd. Effect of training on enzyme activity and fiber composition of human skeletal muscle. *J. Appl. Physiol.* 34: 107–111, 1973.

Hagberg, J. M., R. C. Hickson, A. A. Ehsani, and J. O. Holloszy. Faster adjustment to and recovery from submaximal exercise in the trained state. *J. Appl. Physiol.* 48: 218–224, 1980.

Henriksson, J., and J. S. Reitman. Time course of changes in human skeletal muscle succinate dehydrogenase and cytochrome oxidase activities and maximal oxygen uptake with physical activity and inactivity. *Acta Physiol. Scand.* 99: 91–97, 1977.

Hickson, R. C., H. A. Bomze, and J. O. Holloszy. Faster adjustment of O_2 uptake to the energy requirement of exercise in the trained state. *J. Appl. Physiol.* 44: 877–881, 1978.

Hickson, R. C., C. Foster, M. L. Pollock, T. M. Galassi, and S. Rich. Reduced training intensities and loss of aerobic power, endurance, and cardiac growth. *J. Appl. Physiol.* 58: 492–499, 1985.

Hickson, R. C., J. M. Hagberg, A. A. Ehsani, and J. O. Holloszy. Time course of the adaptive responses of aerobic power and heart rate to training. *Med. Sci. Sports Exercise* 13: 17–29, 1981.

Hickson, R. C., C. Kanakis, Jr., J. R. Davis, A. M. Moore, and S. Rich. Reduced training duration effects on aerobic power, endurance, and cardiac growth. *J. Appl. Physiol.* 53: 225–229, 1982.

Hickson, R. C., and M. A. Rosenkoetter. Reduced trained frequencies and maintenance of increased aerobic power. *Med. Sci. Sports Exercise* 13: 13–16, 1981.

Hochachka, P. W. Mechanism and evolution of hypoxia-tolerance in humans. *J. Exp. Biol.* 201: 1243–1254, 1998.

Hochachka, P. W., H.-C. Gunga, and K. Kirsch. Our ancestral physiological phenotype: an adaptation for hypoxia tolerance and for endurance performance? *Proc. Nat. Acad. Sci. USA.* 95: 1915–1920, 1998.

Hofmann, P., V. Bunc, H. Leitner, R. Pokan, and G. Gaisl. Heart rate threshold related to lactate turn point and steady-state exercise on a cycle ergometer. *Eur. J. Appl. Physiol.* 69: 132–139, 1994.

Hoppeler, H., P. Lüthi, H. Classen, E. R. Weibel, and H. Howard. The ultrastructure of the normal human skeletal muscle. A morphometric analysis of untrained men, women, and well-trained orienteers. *Pflügers Arch.* 344: 217–232, 1973.

Inger, F. Capillary supply and mitochondrial content of different skeletal muscle fiber types on untrained and endurance trained men: a histochemical and ultrastructural study. *Eur. J. Appl. Physiol.* 40: 197–209, 1979.

Jackson, J., B. Sharkey, and L. Johnston. Cardiorespiratory adaptations to training at specified frequencies. *Res. Quart.* 39: 295–300, 1968.

Jansson, E., B. Sjokin, and P. Tesch. Changes in muscle fiber type distribution in men after physical training. *Acta Physiol. Scand.* 104: 235–237, 1978.

Karlsson, J., P. V. Komi, and J. H. T. Vitasalo. Muscle strength and muscle characteristics of monozygous and dizygous twins. *Acta Physiol. Scand.* 106: 319–325, 1979.

Karlsson, J., L.-O. Nordesjö, and B. Saltin. Muscle glycogen utilization during exercise after physical training. *Acta Physiol. Scand.* 90: 210–217, 1974.

Kiessling, K., K. Piehland, and C. Lundquist. Effect of physical training on ultrastructural features of human skeletal muscle. In Muscle Metabolism During Exercise, B. Pernow and B. Saltin (Eds.). New York: Plenum Press, 1971, pp. 97–101.

Klissouras, V. Genetic limit of functional adaptability. *Int. Z. Angew. Physiol.* 30: 85–94, 1972.

Klissouras, V. Heritability of adaptive variation. *J. Appl. Physiol.* 31: 338–344, 1981.

Klissouras, V., F. Pirnay, and J. Petit. Adaptation to maximal effort: genetics and age. *J. Appl. Physiol.* 35: 288–293, 1973.

Knuttgen, H., L. Nordesjö, B. Ollander, and B. Saltin. Physical conditioning through interval training with young male adults. *Med. Sci. Sports* 5: 206–226, 1973.

Komi, P. V., J. H. T. Viitasalo, M. Havy, A. Thorstensson, B. Sjödin, and T. Karlsson. Skeletal muscle fibers and muscle enzyme activities in monozygous and dizygous twins of both sexes. *Acta Physiol. Scand.* 100: 385–392, 1977.

Levine, B. D., and J. Stray-Gundersen. Living high-training low: effect of moderate-altitude acclimatization with low-altitude training on performance. *J. Appl. Physiol.* 83: 102–112, 1997.

Longhurst, J. C., A. R. Kelly, W. J. Gonyea, and J. H. Mitchell. Echocardiographic left ventricular masses in distance runners and weight lifters. *J. Appl. Physiol.* 210: 154–162, 1980.

Morganroth, J., B. Maron, W. Henry, and S. Epstein. Comparative left ventricular dimensions in trained athletes. *Ann. Intern. Med.* 82: 521–524, 1975.

Pedersen, P., and K. Jørgensen. Maximal oxygen uptake in young women with training, inactivity, and retraining. *Med. Sci. Sports* 10: 233–237, 1978.

Reeves, J. T., E. E. Wolfel, H. J. Green, R. S. Mazzeo, J. Young, J. R. Sutton, and G. A. Brooks. Oxygen transport during exercise at high altitude and the lactate paradox: lessons from Operation Everest II and Pikes Peak. *Exercise and Sport Sciences Reviews.* Vol. 20, Williams and Wilkins, 1992, pp. 275–296.

Roeske, W. R., R. A. O'Rourke, A. Klein, G. Leopold, and J. S. Karlinger. Noninvasive evaluation of ventricular hypertrophy in professional athletes. *Circulation* 53: 286–292, 1975.

Rowell, L. B. Human cardiovascular adjustments to exercise and thermal stress. *Physiol. Rev.* 51: 75–159, 1974.

Saltin, B., K. Nazar, D. L. Costill, E. Stein, E. Jansson, B. Essen, and P. D. Gollnick. The nature of the training response: peripheral and central adaptations to one-legged exercise. *Acta Physiol. Scand.* 96: 289–305, 1976.

Shephard, R. Intensity, duration and frequency of exercise as determinants of the response to a training regiment. *Int. Z. Agnew. Physiol.* 26: 272–278, 1968.

Shepley, B., J. D. MacDougall, N. Cipriano, J. R. Sutton, M. A. Tarnopolsky, and G. Coates. Physiological effects of tapering in highly trained athletes. *J. Appl. Physiol.* 72: 706–711, 1992.

Stainsby, W. N., and G. A. Brooks. Control of lactic acid metabolism in contracting muscles and during exercise. *Exercise and Sport Science Reviews,* Vol. 18, K. B. Pandolf and J. O. Holloszy (Eds.). Williams and Wilkins, 1990, pp. 29–63.

Stromme, S. B., F. Ingjer, and H. D. Meen. Assessment of maximal aerobic power in specifically trained athletes. *J. Appl. Physiol.* 42: 833–837, 1977.

Wilt, F. Training for competitive running. In Exercise Physiology, H. Fall (Ed.). New York: Academic Press, 1968.

Zeldis, S. M., J. Morganroth, and S. Rubler. Cardiac hypertrophy in response to dynamic coordinating in female athletes. *J. Appl. Physiol.* 44: 849–852, 1978.

CHAPTER 22

EXERCISE IN THE HEAT AND COLD

People have a remarkable ability to exercise in very hot and very cold environments. It is not unusual to see hundreds of diehard skiers schussing down the slopes at temperatures well below 0°C. Likewise, endurance competitions are sometimes run in desert climates at environmental temperatures exceeding 37°C (normal body temperature).

People often live and work in extreme temperatures. The mean annual temperature of much of Russia and Canada is lower than 0°C. In the winter, temperatures are often so cold that exposed skin will freeze within one minute. An even greater number of people are exposed to extreme heat. Summer temperatures regularly surpass 43°–49°C in places like Australia, the American Southwest, the Middle East, and India (Figure 22-1). In some desert areas, the inhabitants must endure extremely hot temperatures in the daytime and extremely cold temperatures at night.

We can tolerate these hot and cold climates because of a well-developed ability to regulate core body temperature—first, by behavioral means; then, if these do not work effectively, by physiological means. In some instances, behavioral means may also be concurrent with physiological means. When the environment is cold, we maintain our body core temperature by increasing the body's heat production, reducing the body's rate of heat loss by putting on more clothes, or turning on the heat. When it is

Gabrielle Andersen-Scheiss of Switzerland experienced extreme thermal distress during the marathon in the 1984 Los Angeles Olympics. The stresses imposed while exercising in hot and cold environments can compromise health and performance. © Corbis/Bettman-UPI

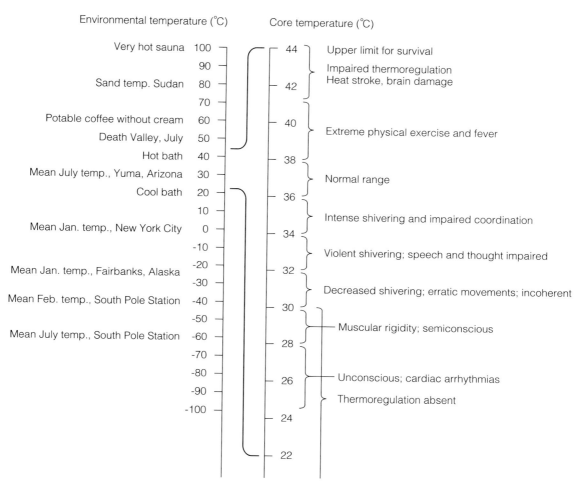

Figure 22-1   Environmental temperature extremes and the effects of alterations in core body temperature.

hot, we increase heat dissipation by sweating, increasing blood flow to the skin, removing clothes, or turning on the air conditioning.

■ Humans as Homeotherms

Animal body temperatures either remain constant (homeothermic and heterothermic) or vary with the environment (poikilothermic). *Poikilotherms,* such as lizards and insects, are at the mercy of the elements. When the climate is cold, their body temperatures can become so low that their metabolic rates drop to a level that forces inactivity. Likewise, in the heat, they must seek shelter or perish. Certain animals are *heterotherms* ("part-time" homeotherms). To escape extreme temperatures, a few homeotherms become heterothermic, allowing their core temperatures to fall near ambient temperature. Examples of such heterothermic behavior include the daily torpor (inactivity characterized by extreme sluggishness) of hummingbirds, bats, and pocket mice, and the hibernation of woodchucks, chipmunks, and black bears.

Advanced animals such as humans, monkeys, dogs, bears, and birds are *homeotherms*. They are able to function relatively independently of the environment because of their ability to maintain constant body core temperatures. Processes such as oxygen transport, cellular metabolism, and muscle contraction remain unimpaired in hot and cold environments as long as the internal temperature is maintained. Human mastery on the planet Earth certainly would have been impossible had we not been homeotherms.

In homeotherms, various physiological processes, such as neural function, depend on a normal body temperature to function properly. Abnormal increases and decreases in body temperature are catastrophic to the organism. At temperatures above 41°C, the interior of many cells begins to deteriorate. Heat stroke and permanent brain damage can occur if body temperature is not quickly brought down. At temperatures below 34°C, cellular metabolism slows greatly, leading to unconsciousness and cardiac arrhythmias (see Figure 22-1).

Poikilothermia syndrome, characterized by extreme difficulty in maintaining a constant body temperature, is a rare cause of thermoregulatory failure in humans. Compared to normal people, those with this syndrome begin sweating at a higher core temperature when exposed to the heat and their bodies vasoconstrict skin blood vessels and shiver at a lower core temperature when exposed to the cold. Like snakes and lizards, they must seek thermally neutral environments to survive.

Normal Body Temperature

Humans experience a range of normal resting body core temperatures that typically lie between 36.5°C and 37.5°C. However, in the early morning the temperature can fall to lower than 36°C and during exercise, body temperature can exceed 40°C with no ill effects. There is considerable temperature variation throughout the body. The internal temperature of the core remains relatively constant, while the skin temperature is closer to that of the environment. Body temperature is typically expressed in terms of the core temperature.

Core temperature is usually defined as the tem-

perature of the hypothalamus, the temperature regulatory center of the body. Oral temperature is the most common method of measuring core temperature. However, this method has severe limitations, particularly during exercise, because increased breathing cools the thermometer and produces an inaccurate measurement. In research, core temperature is most often measured rectally. Rectal temperature is typically 0.6°C higher than oral temperature. Although rectal temperature is more accurate as an estimate of hypothalamic temperature, it also has limitations. For example, vigorous exercise of local muscle groups produces a higher regional temperature that can lead to spurious results. In addition, there are temperature variations in the rectum itself. To ensure an accurate and reproducible measurement of rectal temperature, a thermistor should be inserted to a depth of at least 8 centimeters.

Researchers also estimate core temperatures by taking measurements in the auditory canal (tympanic temperature), esophagus, and stomach. The advantage of taking the tympanic temperature is the close proximity of the canal to the hypothalamus. However, its measurement can cause discomfort to the subject, and tympanic temperature is also affected by head skin temperature, which is lower than brain temperature. Stomach temperatures are obtained by telemetry. Subjects swallow a small radio transmitter that signals temperatures to the researcher. Unfortunately, stomach temperatures can differ considerably from the temperature of the hypothalamus. Changes in the ambient temperature or the digestion of food can result in variations as great as 3°C.

The assessment of mean body temperature (MBT) takes into consideration skin and core temperatures. Mean body temperature is typically found by adding together the rectal temperature and the average of a series of skin temperatures at various places on the body. It is expressed by Equation 22-1. Measuring skin and core temperature helps quantify temperature gradient and the rate of heat loss or gain.

$$\text{MBT} = (0.33 \cdot \text{Skin temperature}) \quad \quad (22\text{-}1)$$
$$+ (0.67 \cdot \text{Rectal temperature})$$

Thermal gradients determine the rate and direction of heat transfer. Thermal gradients are temperature differences from one point to another that lead to the movement of heat (e.g., core temperature is higher than skin temperature, so heat tends to move from the core to the skin). Heat transfer is always from higher to lower temperatures.

■ Heat Transfer

The temperature of an object is a measure of the kinetic activity of its molecules. If an object is hot, then its molecules are moving very rapidly; as it cools, its molecules move more slowly. The temperature of the body is directly proportional to the amount of heat it stores. When heat storage increases, such as in fever or during exercise, mean body temperature rises. When heat storage decreases, such as in hypothermia, mean body temperature falls.

Body core temperature is regulated by controlling the rate of heat gain and heat loss. When the rate of heat gain is exactly equal to the rate of heat loss, the body is in *thermal balance*. The body either gains or loses heat when it is out of thermal balance. The physiological means for controlling rates of heat gain or loss are directed by the hypothalamus with feedback from peripheral heat and cold receptors in the skin and from thermal receptors in the hypothalamus itself. The hypothalamus works like a thermostat by increasing the rate of heat production when body temperature falls and increasing the rate of heat dissipation when it rises. The energy balance equation is as follows:

$$S = M \pm C_v \pm C_d \pm R - E \qquad (22\text{-}2)$$

where:

S = heat storage (kcal/min or watts)
M = metabolic heat production (kcal/min or watts)
C_v = convective heat loss or gain (kcal/min or watts)
C_d = conductive heat loss or gain (kcal/min or watts)
R = radiant heat loss or gain (kcal/min or watts)
E = evaporative heat loss (kcal/min or watts)

Note that all of these factors function at a certain rate.

The rates of heat gain or loss can be divided into physical and chemical processes. Physical heat transfers occur principally by changing the resistance to heat flow, while chemical heat transfers occur by increasing metabolic rate (Figure 22-2).

Body Temperature, Environment, and Exercise Intensity

Within the broad range of 4°C to 30°C, body core temperature is independent of environmental temperature. Under such conditions, core temperature rises in direct proportion to relative exercise intensity—the greater the relative intensity, the more

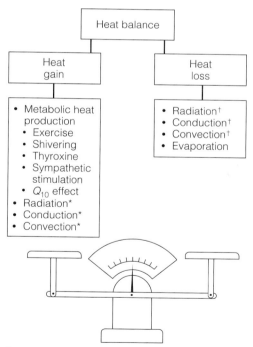

* When ambient temperature is greater than skin temperature: uncommon;
† When skin temperature is greater than ambient temperature: common.

Figure 22-2 Factors determining thermal balance.

core temperature will rise (see Figure 22-8b). Under these same conditions, the temperature of peripheral tissues (e.g., skin and muscle) will reflect the environmental temperature.

The temperatures of skeletal muscles are determined by muscle blood flow and metabolic rate. The temperature of contracting muscle rises during exercise because energy reactions involved in muscle contractions release heat. Muscle temperature tends to be higher on a warm day for a given exercise intensity than on a cool day.

Skin temperature is a function of environmental temperature, metabolic rate, clothing, and the state of hydration. For instance, on a warm day skin temperature at rest might be higher than on a cool day during exercise. However, heavy sweating during exercise on a warm, dry day may result in a skin temperature that is lower due to evaporation of sweat than it is during exercise on a cooler but more humid day, when evaporation of sweat is more difficult. During exercise in the heat with inadequate fluid intake, skin temperature tends to rise as sweat capacity decreases due to dehydration and because of increased movement (convection) of warm blood from the core to the surface. Clothing and air movement across the skin also affect skin temperature by influencing the capacity for heat loss.

The flow of heat between the body and the environment is determined by physical principles. Physiological processes, such as sweating, blood flow, shivering, and ventilation, can modify but not reverse the primary direction of heat flow. Heat always flows from hot to cold. Heat flux is considered positive if the body gains heat and negative if it loses heat.

Heat Production

Heat is a by-product of all biochemical reactions because these reactions are not 100% efficient (Chapter 3). Metabolism (M in Eq. 22-2) is the body's source of internal heat production. Metabolic heat production is defined as the rate of transformation of chemical energy into heat and is usually expressed per square meter of body surface area. Heat is produced even during deep sleep. During exer-

cise, heat production can be elevated 10–15 times above that during rest.

Metabolism Heat is generated naturally during normal metabolic reactions. At rest, all metabolically generated energy appears eventually as heat. Metabolic reactions during exercise lose approximately 75% of energy as heat. At basal metabolic rate (BMR) in the average person, heat loss is approximately 100 kcal · hr^{-1} (50 kcal · hr^{-1} · m^2). BMR is approximately proportional to the ¾ power of body weight (called the *surface rule*).

Shivering The main mechanism for increasing heat production during negative heat balance is shivering. Shivering is an involuntary contraction of muscle. Maximal shivering can increase the body's heat production by up to five times. Preshivering tone (an increase in muscle tone due to reduced muscle temperature) can increase heat production by 50 to 100%. Shivering is an effective way of increasing body temperature because no work is done by the shivering muscles and all of the expended energy appears as heat. Shivering is also beneficial because it increases cardiac output, mainly by increasing stroke volume via increased venous return (by stimulating the muscle pump). At the same time, however, shivering adds to heat loss to a small extent by increasing the thermal gradient between the environment and the body. Shivering capacity may be limited by glycogen depletion, hypoglycemia, fatigue, exercise, hypoxia, and drugs such as alcohol and barbiturates.

Nonshivering thermogenesis Increased thyroxin secretion from the thyroid and catecholamine secretion from the adrenals also increase metabolic rate. Thyroxin increases the metabolic rate of all the cells in the body. The catecholamines, principally norepinephrine, cause the release of fatty acids, which also increase metabolic heat production. Stimulation of brown adipose tissue (BAT), a mechanism very important in animals such as rodents and possibly in human infants, may also be a mechanism for increasing heat production, though researchers believe it is probably not very important in adult humans. In BAT, substrate oxidation is un-coupled from the production of adenosine triphosphate (ATP), so that energy appears mainly as heat and little is trapped in the formation of ATP. Uncoupling in BAT depends mainly on a protein called thermogenin, or uncoupling protein (UCP), located in the inner membrane of the mitochondria.

Leptin, a hormone produced by adipose tissue (see Chapter 25), is another source of nonshivering thermogenesis. It increases thermogenesis and energy expenditure by activating the sympathetic nervous system; namely, neurons in the retrochiasmatic area and lateral arcuate nucleus of the thoracic spinal cord. Leptin appears to be involved with thermogenesis that occurs in response to cold exposure.

Q_{10} Effect* Increased body temperature can be self-perpetuating and dangerous because metabolic rate increases with rising temperature resulting from the Q_{10} effect. At high temperatures, the hypothalamus begins to lose its ability to cool the body. Unfortunately, the rate of temperature increase is faster at these higher temperatures. At core temperatures above 41.5°C, sometimes the only recourse to preventing thermal damage is external cooling, because the hypothalamus may no longer be functional.

Heat Loss

When skin temperature (T_s) is greater than ambient temperature (T_a), the body loses heat by radiation, conduction, convection, and evaporation. Most heat is lost by outward heat flow caused by the negative thermal gradient. Thermal gradients are very important in determining thermal balance because heat moves from hot to cold. In the heat, or during heavy exercise, evaporation becomes the most important means of heat dissipation.

Radiation The loss (or gain) of heat in the form of electromagnetic waves is called *radiation* (R in Eq. 22-2). The parts of the electromagnetic

*Q_{10} is the ratio of the rate of a physiological process at a particular temperature to the rate at a temperature 10°C lower, when the logarithm of the rate is an approximately linear function of temperature (Precht, 1973, p. 17).

spectrum of significance in temperature regulation include the ultraviolet, visible, infrared, and microwave (0.25μm—100mm) sections. Infrared radiation is the main wavelength for this type of heat loss. At rest in a comfortable environment, radiation accounts for 60% of total heat loss.

The sun is the greatest source of radiant energy. Fortunately, the skin is never entirely exposed to the sun's rays. Radiant heat loss or gain varies with body position and clothing. For example, heat gain is greater when a lot of the body surface area is exposed to the elements (e.g., lying fully exposed in the sun with arms and legs spread apart) than when the surface area is protected (e.g., lying in the fetal position). Sunlight also strikes the skin at an angle, which modifies the full radiant heat load from the sun.

Any substance not at absolute zero (0° Kelvin) emits radiant heat waves. The body radiates and receives radiant heat at the same time. If skin temperature is greater than that of the surrounding environment, then more heat radiates from the body than to it. If the temperatures of solid bodies in the environment are greater, then the net flow of radiation is inward.

The color and texture of an object affects its ability to absorb radiant heat rays. Light-colored, shiny objects absorb radiant heat less easily than black rough objects. The human body has been called a perfect black body radiator. Human skin, regardless of color, absorbs about 97% of radiant energy that strikes it. So, a person exercising in the hot sun is better off wearing a light white cotton shirt than going bare-skinned. The environmental radiant heat load is measured with a black globe thermometer.

Conduction The transfer of heat from the body to an object with which it is in direct contact (or vice versa) or heat transfer within the organism down a thermal gradient is called *conduction* (C_d in Eq. 22-2). About 3% of total heat loss at room temperature occurs by this mechanism. A good example of conduction is the transfer of heat to a chair while a person is sitting on it. Heat loss in the urine and feces is also a form of conduction.

Convection   The conduction of heat to or from air or water is called *convection* (*Cv* in Eq. 22-2). Convection accounts for about 12% of the heat loss at normal room temperature. In convection, heat conducted to air or water moves so that other molecules can also be heated. As heat is transmitted to the surrounding air, it rises, allowing additional heat transfer to the surrounding air. Heat loss by conduction and convection occurs much more rapidly in water than in air because the layer of air that insulates the body is absent. Heat flux depends directly on the temperature gradient between the body and the water.

Heat loss by convection is greater in the wind because warmed air is quickly replaced by colder air, thus lowering the effective temperature. For example, heat loss at 10°C in a 2.2 mph wind is the same as that at −10°C in still air (Figure 22-3). The effect of wind on temperature is called the *windchill factor* and is expressed in kcal · hr^{-1} per m^2 exposed skin surface.

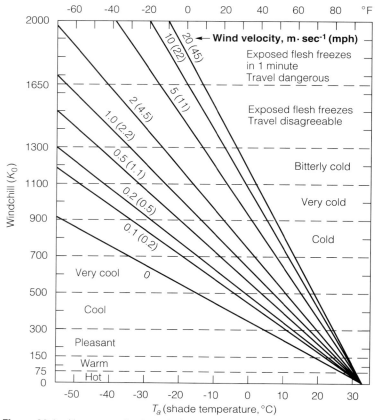

Figure 22-3   Nomogram for the calculation of windchill.

$$K_0 = \sqrt{(100v + 10.45 - v)}(33 - T_a) \quad (22\text{-}3)$$

where

K_0 = windchill as kcal $\cdot$ hr^{-1} $\cdot$ m^2 exposed skin surface in shade, ignoring evaporation.

V = Wind velocity in m $\cdot$ sec^{-1}

T_a = Ambient air temperature in °C

Wind speeds above 40 mph have no additional effect on heat loss because heat transfer to the skin does not occur rapidly enough.

Convection of heat from one site in the body to another also occurs via the circulatory system by a process called *circulatory convection*. Heat moves with the blood from the core to the periphery. As skin temperature increases, heat loss by convection and conduction into the environment also increases. In the skin, the countercurrent arrangement of arteries and veins facilitates circulatory convection (see Chapter 15). The rate of this process is determined by blood flow and the temperature gradient between the core and periphery.

Evaporation At rest in a comfortable environment, about 25% of heat loss is due to *evaporation* (E in Eq. 22-2). The quantity of heat absorbed by sweat as it evaporates is called the *latent heat of vaporization*. When 1 gram of sweat changes from water to vapor, the latent heat of vaporization is 2411.3 J or 0.58 kcal. In other words, the body loses 0.58 kcal of heat for each gram of water that evaporates.

Evaporation is the only means of cooling at high environmental temperature and is critically important during exercise. During maximal exercise, heat production in active muscle may be 100 times that of inactive muscle. When the environmental temperature is greater than skin temperature, the body gains heat by radiation and conduction. Sweating is critical—if the body cannot lose heat by evaporation, body temperature increases rapidly. During moderate exercise, if a person cannot dissipate heat, body temperature increases approximately 0.2°C per minute, which will cause thermal injury within 15–20 minutes.

Sweat is only effective for cooling if it evapo-

rates. If the humidity is high, the rate of evaporation may be greatly reduced or totally prevented so that the sweat remains in a fluid state. Effective evaporation is hampered by lack of air movement because the air surrounding the body becomes saturated with water vapor. This explains why fans are desirable on a hot day.

When males and females are matched for fitness, body surface area, body mass, and degree of heat acclimatization, few sex differences in temperature regulatory capacity are found.

Evaporation occurs as a result of sweating and insensible water loss. Insensible water loss is water loss through ventilation and diffusion through the skin; it does not include water excreted in sweat, urine, and feces. Water evaporates insensibly from the skin and lungs at a rate of about 600 ml $\cdot$ day^{-1}. This amounts to a continual heat loss of about 12 to 18 kcal $\cdot$ hr^{-1}. Insensible water loss cannot be controlled and it occurs regardless of body temperature. Insensible water loss is particularly significant in cold, dry air and may play an important role in water loss when exercising in the cold.

Except for insensible water loss, sweat rates are essentially zero when skin temperature is low. In hot weather, an unacclimatized individual (person not used to the heat) has a maximum sweat rate of about 1.5 liters $\cdot$ hr^{-1}, while an acclimatized person can sweat up to 4 liters $\cdot$ hr^{-1}. Evaporative sweat loss is proportional to the rate of sweating, which in contrast to insensible water loss is under physiological control.

The two kinds of sweat glands are called apocrine and eccrine. The apocrine sweat glands exist primarily under the arms and in the genital region. Their secretions contain a lipid material that produces a characteristic odor when acted on by bacteria. Apocrine sweat glands do not typically become fully operational until adolescence.

Eccrine sweat glands are tubular structures consisting of a deep coiled portion that secretes sweat and a duct portion that passes outward through the dermis of the skin. These glands cover most of the body and secrete a clear, essentially odorless sweat that accounts for most of the evaporative heat loss in the body (Figure 22-4). In addition to water,

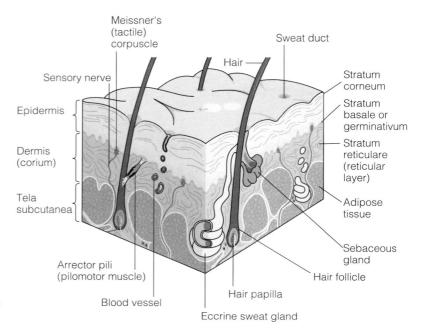

Figure 22-4 Structure of human skin and eccrine sweat gland

sodium chloride, urea, lactic acid, and potassium chloride are lost in sweat.

Eccrine sweat glands located on the hands and feet can be stimulated by sympathetic adrenergic fibers. The feeling of sweaty palms before competition is due to sympathetic nervous system discharge rather than temperature regulation. These sweat glands have the practical effect of providing traction for the feet and improved grip for the hands.

Temperature Regulation

The temperature regulatory center is located in the hypothalamus (Figure 22-5). The hypothalamus functions as the body's thermostat by keeping the core temperature within normal range. Thermoreceptors are the vehicles through which the hypothalamus responds to increases and decreases in temperature. Thermoreceptors are located in different parts of the body. There is a small proportion of thermoreceptors in the nervous system, blood vessels, and the abdominal cavity. A greater density of thermoreceptors lies in the skin and hypothalamus.

Of the receptors in the skin (peripheral receptors), there are more cold than heat receptors. The density of thermoreceptors in the skin also varies considerably—from one receptor per cm² on the back to one receptor per mm² on the lips. The thermoreceptors in the hypothalamus are by far the most important receptors in the body for regulating the core temperature.

The thermoreceptors work by transmitting nerve impulses to the spinal cord, which conducts the impulses to the hypothalamus. The hypothalamus then initiates the appropriate response. Heating the preoptic area of the hypothalamus stimulates heat loss mechanisms. The anterior hypothalamus stimulates the sweat glands, resulting in evaporative heat loss from the body. In addition, the cardiovascular control center (i.e., vasomotor center) is inhibited (see Chapter 15). This removes the normal vasoconstrictor tone to the skin vessels, allowing increased loss of heat through the skin. Nitric oxide (see Chapter 15), secreted by endothelial cells in the lining of the coronary arteries, is also thought to be involved in skin vasodilation in response to exercise in the heat. In animals such as dogs, the panting

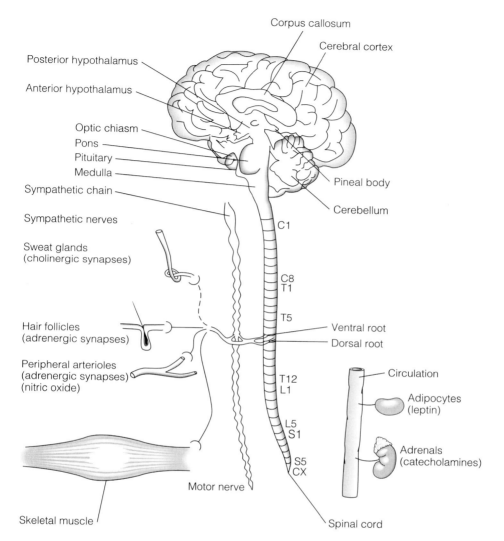

Figure 22-5 Temperature regulation schema for humans.

center is stimulated, which also increases evaporative heat loss.

When cold receptors in the skin and hypothalamus are stimulated, various processes increase heat production and insulation. Cooling causes inhibition of warm-sensitive neurons in the hypothalamus. This tends to prevent heat loss by interfering with processes, such as skin vasodilation, that favor loss of heat from the core of the body to the periph-

ery. The cardiovascular control center is stimulated, which results in vasoconstriction of blood vessels in the skin.

Intermittent vasodilation called the "hunting reflex" occurs in the hands and feet that maintains the health of peripheral tissues by sending warm blood to them. Hand and foot comfort in the cold are related to torso temperature. Applying heat to the torso during intense cold exposure can maintain finger and toe comfort for an extended period of time. A warmer torso prevents peripheral vasoconstriction and allows greater blood flow to the extremities.

Vasoconstriction can increase the effective insulation of the core by 1 to 2 inches. Stimulation of the shivering center causes shivering. Stimulation of the pilomotor center causes piloerection (i.e., goose bumps). In animals, piloerection results in involuntary bristling of hairs or ruffling of feathers. This traps warmer air next to the skin, which increases insulation. The posterior hypothalamus also initiates the release of norepinephrine, which results in the mobilization of fatty acids and an increase in metabolic heat production. It also indirectly increases thyroxin production by secreting thyrotropin-releasing factor, which, in turn, stimulates the secretion of thyrotropin by the pituitary gland.

When the core temperature goes above or below its set-point, the hypothalamus initiates processes to increase heat production or heat loss. Normally, sweating begins at almost precisely 37°C and heat production mechanisms begin below this point (Figure 22-6). The set-point can change temporarily in

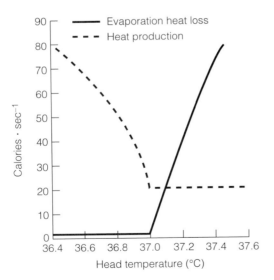

Figure 22-6 The effect of hypothalamic temperature on heat production and heat loss. The body's temperature regulatory system works very much like a thermostat in a house.

response to dehydration, starvation, or fever. In fever, the set-point is raised in response to a pathogen. The increased temperature is part of an immune response that helps the body get rid of the foreign organism by "overheating" it. Acclimatization to heat or cold can alter the hypothalamic set-point.

Behavior

Human behavior is an important component in temperature regulation. When the anterior hypothalamus is overheated, a person has the sensation of being hot and will do something about it, such as drink a cold glass of water, remove some clothing, or turn on the air conditioning. Likewise, a skier who is cold will put on more clothes or move indoors.

■ Exercise in the Cold

It is rare, except when survival is at stake, for people to exercise with low body temperatures. The combination of the increased metabolism of exercise and extra clothing minimize the chances of hypothermia during exercise or sports performance. However, cold exposure can be a factor in physical performance when environmental conditions present the potential for heat loss.

Movement in the Cold

The two biggest handicaps of exercising in the cold are the numbing of exposed flesh and the awkwardness and extra weight of protective clothing. Manipulative motor skills requiring finger dexterity, such as catching and throwing, are tremendously impaired in the cold because the cold effectively anesthetizes sensory receptors in the hands. Also, exposed flesh, particularly on the face, is susceptible to frostbite, which can become a serious medical problem.

Shivering or an increase in preshivering muscle tone may also decrease muscle efficiency. Both conditions increase the metabolic cost of exercise, which increases the perception of effort. Also, because shivering occurs in agonists and antagonists, normal movement patterns can be impaired. To compensate, more motor units may be recruited, which would also increase the difficulty of performing precise movements.

Clothing Clothing is an important consideration during physical activity in the cold. The insulation value of the clothing must be balanced with the increased metabolic heat production of exercise. If too much clothing is worn, the individual risks becoming a "tropical person" in a cold environment (i.e., even though it is cold, the person gets overheated from wearing too many clothes). Heat illnesses have occurred in overclothed persons exercising in extremely cold climates.

Clothes protect against the cold because they increase the body's insulation. Clothing traps warm air next to the skin and decreases heat loss by conduction and convection. The *clo unit* is a measure of the thermal insulation provided by clothing. One clo provides enough insulation to keep a resting person comfortable at 20°C, < 50% relative humidity, and wind velocity of 6 m · min⁻¹.

The best clothing for exercise in the cold allows for the evaporation of sweat while providing added protection from the cold. Clothing should be worn in layers so that it may be removed as the metabolic heat production increases during exercise. Select clothing that provides protection from the cold during rather than before exercise. Additional clothing should be available after exercise to prevent hypothermia. After exercise, metabolic rate decreases but the rate of heat loss remains high. Clothing manufacturers have made tremendous progress in recent years in developing lightweight clothing that provides sufficient insulation and freedom of movement during exercise.

Cardiopulmonary Responses to the Cold

Oxygen Consumption Maximal oxygen uptake is unaffected in the cold. However, submaximal $\dot{V}_{O_2}$ increases in the cold, particularly at lower intensities of exercise (Table 22-1). Differences between submaximal exercise $\dot{V}_{O_2}$ in cold and neutral thermal environments disappear as exercise approaches maximal intensity. Submaximal exercise $\dot{V}_{O_2}$ is higher in the cold because heat loss is greater than it is in a comfortable environment. Increases in skin and muscle blood flow, which occur during exercise in any environment, increase the thermal gradient between the surface of the body and the cold environment. Consequently, more heat is lost through convection and conduction. For the same reason, exercising with wet clothing, particularly

TABLE 22-1	
Physiological Response to Exercise in the Cold	
Physiological Response	**Physiological Effect or Mechanism**
Increased submaximal exercise $\dot{V}_{O_2}$	Greater heat loss
Decreased exercise capacity in water	Greater heat loss
Increased ventilation during submaximal exercise	Increased sympathetic stimulation
Reduced skin blood flow	Peripheral vasoconstriction
Lower lipid mobilization	Reduced blood flow to adipocytes
Increased glycogen use	Increased carbohydrate metabolism
Increased lactate concentration	Increased CHO metabolism and decreased lactate clearance
Increased central blood volume	Peripheral vasoconstriction
Decreased heart rate during submaximal exercise	Increased central blood volume
Decreased muscle strength*	Decreased muscle enzyme activity
Release of leptin from adipose tissue	Increased sympathetic stimulation

*Some studies have found increased muscle strength.

in the wind, increases $\dot{V}_{O_2}$ 15 to 20% more than the same exercise practiced in a comfortable environment.

Shivering can persist during exercise, which can increase $\dot{V}_{O_2}$. Some researchers have hypothesized that the higher $\dot{V}_{O_2}$ during submaximal exercise in the cold may also be due to increased nonshivering thermogenesis. This could be caused, in turn, by increased secretion of catecholamines or leptin.

Swimming in cold water can cause marked deterioration in exercise capacity and $\dot{V}_{O_2}$ peak (in this case, $\dot{V}_{O_2}$ peak is the maximal oxygen consumption measured during swimming; Figure 22-7). Heat conductance is about 25 times greater in water than in air. Body fat is an important factor in determining heat loss in cold water. Greater amounts of body fat increase insulation and slow the transfer of heat to the water. Long-distance channel swimmers typically have high fat percentages; this slows the rate of heat loss during their prolonged swims. It is unclear if this characteristic represents an adaptation to the cold or is simply a conscious effort on the part of the swimmer to increase insulation by increasing body fat.

Exercise can partially or totally replace the heat production of shivering during exposure to a cold environment. The peripheral blood vessels vasodilate during physical activity, which effectively decreases the body's insulation to the cold. Heat production from exercise and shivering must be adequate to maintain heat balance, or hypothermia will result.

Exercise followed by cold exposure raises postexercise thresholds for vasoconstriction and shivering. In other words, people who exercise and then are exposed to the cold shiver and vasoconstrict at a higher core temperature than if they had not exercised.

Ventilation Ventilation increases in the cold, particularly when cold exposure is sudden (e.g., falling from a boat into cold water). Abrupt exposure causes a gasping reflex. In the water, this may be responsible for the sudden disappearance syndrome—drowning caused by taking a gasping breath in cold water during sudden exposure. This "inspiratory gasp" during exposure to cold water is accompanied by hyperventilation, tachycardia, peripheral vasoconstriction, and hypertension. Increased ventilation also decreases CO_2 in the blood, which stimulates vasoconstriction in brain blood vessels and can cause confusion and unconsciousness. Taken together, these physiological responses are very dangerous and account for many cases of drowning each year.

The difference between ventilatory response to

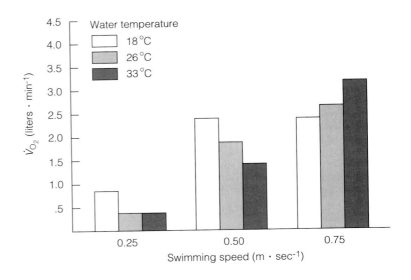

Figure 22-7 The effects of swimming speed and water temperature (T_w) on O_2 consumption for three different swimmers. Adapted from the data of Nadel, 1980.

Water temperature
☐ 18 °C
▨ 26 °C
▧ 33 °C

$\dot{V}_{O_2}$ (liters · min^{-1})

Swimming speed (m · sec^{-1})

cold and to normal temperature progressively narrows with increased exercise intensity. Also, elevated ventilatory response, at rest and during exercise, decreases with repeated exposure to cold. This may represent an habituation or acclimatization (cause unknown) to cold exposure.

Heart   Cold exposure causes peripheral vasoconstriction, which increases central blood volume. Consequently, blood pressure increases due to an increased afterload of the heart and, sometimes, an increased preload. Some, but not all, studies have found decreased heart rate and increased stroke volume. Stroke volume changes may be accompanied by increased ventricular end-diastolic and systolic volumes. Changes in cardiac performance with cold exposure are more common in males than females. This is probably because of higher subcutaneous fat in women, which provides better peripheral insulation against the cold.

The incidence of arrhythmias increases in the cold. Cold increases afferent impulses to the hypothalamus and cardiovascular control center, and increases adrenal secretion of epinephrine, all of which can increase the incidence of cardiac arrhythmias. Scuba diving in cold water may increase the incidence of fatal arrhythmias. Ventricular fibrillation is a leading cause of death in people with hypothermia.

Muscle Strength

Muscle strength and peak power output decrease as muscle temperature decreases. The enzyme activity necessary for energy transformations and muscle contractions are most efficient at temperatures slightly above those seen in normal resting muscle. Muscle strength and power decrease with the decreased muscle enzyme activity that occurs with cooling. However, some researchers have found that cooling the muscle actually increases strength.

Increased motor unit recruitment may be necessary to compensate for lower capacity. Muscle blood flow is less at rest and during submaximal exercise in cooled muscle. Increased motor unit recruitment and decreased blood flow may increase lactate production and decrease lactate clearance, which could speed the rate of fatigue. Other factors impairing muscle performance in the cold include increased muscle viscosity and decreased ATP metabolism and muscle contractile velocity.

Metabolic Changes

Cold exposure increases the use of carbohydrates as substrate. During light exercise, muscle glycogen decreases faster in the cold than it does in warmer air. During maximal exercise, glycogen depletion is independent of air temperature. Prolonged exposure to cold (e.g., being lost in the wilderness) often leads to hypoglycemia, which suppresses shivering and causes core temperature to drop.

Increased carbohydrate metabolism augments the rate of lactate formation. At a given exercise intensity, venous lactate is higher in the cold. A possible explanation for this is that cold causes decreased muscle blood flow during submaximal exercise, which decreases the rate of lactate clearance. Also, cold exposure increases catecholamine secretion, which stimulates lactate production.

Fat metabolism is depressed in the cold even though catecholamine concentrations increase (these hormones stimulate lipolysis in adipose tissue and speed intravascular breakdown of triglycerides). The probable reason for depressed fat metabolism in the cold is that blood flow to subcutaneous fat cells decreases. Some researchers have also found a decrease in the respiratory exchange ratio ($\dot{V}_{CO_2}/\dot{V}_{O_2}$, an indirect measure of the predominance of fat or carbohydrate metabolism), which would suggest an increase in fat metabolism. However, most evidence points to decreased fat metabolism during cold exposure.

Cold exposure also increases protein degradation, as measured by increased urea nitrogen excretion. Urea nitrogen, excreted principally in the urine, is a good measure of protein breakdown.

Problems with cold exposure may be compounded with fatigue, sleep loss, and underfeeding—conditions common in wartime or cold weather survival (e.g., poor elderly people living in marginal conditions). These conditions impair the body's ability to regulate its temperature in the cold and increase the risk of hypothermia. Likewise, repeated cold exposures decrease the body's ability to maintain its normal temperature because fatigue blunts the metabolic heat production capacity. It may be that the shivering response decreases with repeated cold exposures.

Acclimatization and Habituation to Cold

People exposed to environmental stresses such as heat, cold, and altitude usually make physiological and psychological adjustments (i.e., acclimatization and habituation) to improve their comfort. *Acclimatization* is physiological compensation to environmental stress occurring over a period of time. *Acclimation* is a related term meaning the adaptive changes occurring within the organism in response to experimentally induced changes in the environment (i.e., laboratory-induced changes). *Habituation* is the lessening of the sensation associated with a particular environmental stressor. Simply stated, in acclimatization and acclimation, definite physical alterations improve physiological function. In habituation, the person learns to live with the stressor.

It is difficult to demonstrate acclimatization in people who are not chronically exposed to cold. Although scientists have studied people in primitive societies who were chronically exposed to cold, studying cold exposure in more "civilized" societies is more difficult and provides less useful information about cold acclimatization. People who live in very cold areas, such as northern Canada, northern Scandinavia, or Siberia, protect themselves from low temperatures with winter clothing and shelter. Physiologically, they do not get the same chance for cold acclimatization as they would if they were less protected.

There are three basic tests of acclimatization to cold in humans: shivering threshold, hand and feet temperature, and the capacity to sleep. The first test is the threshold skin temperature that results in shivering. Cold-acclimatized people maintain heat production with less shivering. Shivering occurs later in subjects exposed to several weeks of cold temperatures. These subjects increase the secretion of thyroid hormones and their tissues become more sensitive to norepinephrine. This results in uncoupled oxidative phosphorylation—heat is released without the production of ATP. Leptin release from adipose tissue may also increase, which stimulates the sympathetic nervous system. Another mechanism affecting shivering threshold may be thickness of subcutaneous fat. Chronic exposure to cold may increase skinfold thicknesses. The mechanism for this change is not known.

The second test of acclimatization is the capacity to prevent large decreases in temperature in the hands and feet. In unacclimatized people, hand and foot temperatures drop progressively during cold exposure. Acclimatized people, however, are able to maintain almost normal hand and foot temperatures. Acclimatization results in improved intermittent peripheral vasodilation to make the hands and feet more comfortable. Habituation also seems to play a part. Some individuals seem to lose or learn to tolerate the pain sensations associated with cold feet and hands, even when there is little improvement in circulation or temperature.

The third test is the ability to sleep in the cold. Unacclimatized humans will shiver so much that it is impossible to sleep. Some studies show that it is possible to acclimatize enough to sleep, but these findings have not been consistently replicated. The ability to sleep in the cold seems to depend on the extent of nonshivering thermogenesis induced by increased secretion of norepinephrine. Some peoples, such as the Aborigines of Australia, are exceptions to this. They are capable of sleeping in the cold with little or no clothing without an increase in metabolism. They have a superior capacity to vasoconstrict peripheral blood vessels, which allows skin temperature to decrease without shivering.

Physical conditioning seems to be beneficial in acclimatization to the cold, just as it is in the heat and at altitude. Physical conditioning results in a higher body temperature in sleep tests in the cold. Consequently, the individual can sleep better and is more comfortable. The role of training in cold tolerance is not completely understood and remains controversial.

Cold Injury

Hypothermia   As noted in Figure 22-1, the hypothalamus ceases to control body temperature at extremely low core temperatures. Hypothermia depresses the central nervous system, which results in an inability to shiver, sleepiness, and eventually, coma. The lower temperature also results in a lower cellular metabolic rate, which further decreases temperature.

Hypothermia has profound effects on the cardiovascular system: Central blood volume decreases, while peripheral resistance and blood viscosity increase. Central blood volume decreases due to plasma sequestration, inadequate fluid intake, and cold diuresis. The heart rate decreases, and the heart is much more susceptible to life-threatening arrhythmias, such as ventricular fibrillation. Hypothermia can result from exposure to cold water, lack of protective clothing in the cold, leanness, high wind chill, use of alcohol or drugs in the cold, and the use of snow to relieve thirst.

Hypothermia is also possible during endurance exercise competitions (i.e., marathons) conducted in the cold. A person can become hypothermic if the rate of heat production during exercise is exceeded by the rate of heat loss. Glycogen depletion increases the risk of hypothermia because it leads to hypoglycemia and reduced central nervous system function.

Frostbite   Frostbite is caused by ice crystal formation within tissues and typically occurs to exposed body parts such as the earlobes, fingers, and toes. The risk of frostbite increases when the temperature drops below −6°C. It can cause permanent circulatory damage, and sometimes the frostbitten part is lost due to gangrene.

■ Exercise in the Heat

A naked human in still air can maintain a constant body temperature at an ambient temperature of 54–60°C. However, exercise in the heat sets the stage for a positive thermal balance (i.e., heat gain). The relative intensity of exercise is the most important factor increasing core temperature, which increases proportionally with increasing intensities of exercise. Although there is considerable variability in core temperature at any absolute work load, there is very little variation when the load is expressed as a percentage of maximum capacity (Figure 22-8). High environmental temperature adds to the metabolic heat stress of exercise.

Cardiovascular Effects

The extent of the effect of environmental temperature on exercise capacity depends on the body's ability to dissipate heat and maintain blood flow to the active muscles. During intense exercise in the heat, the combined circulatory demands of muscle and skin can effectively impair oxygen transport capacity.

Plasma volume decreases during exercise (due to increased blood pressure, which increases the rate of plasma filtration from the vascular space); this becomes increasingly acute as the intensity of the effort increases. The decrease in plasma volume is made worse by the loss of body fluids through sweating, which can become a particularly acute problem. Plasma volume decreases proportionately more than other body fluids during dehydration. Thus, during exercise in the heat, there may not be enough blood to adequately perfuse all tissues. Central blood volume may decrease and cause a decrease in cardiac filling pressure (i.e., preload). The result is increased heart rate in an attempt to compensate for the fall in stroke volume. However, at higher exercise intensities, increased heart rate cannot offset the deficit, so perceived exertion increases and the person fatigues.

During submaximal exercise, heart rate increases. This mechanism is less effective at higher intensities of exercise because of the approach of maximum heart rate. At maximal levels of exercise, skin vessels vasoconstrict rather than vasodilate (as during the early phases of exercise in the heat), which helps maintain blood pressure and cardiac output. Unfortunately, though, this response has a negative effect on heat transfer. In this instance, cir-

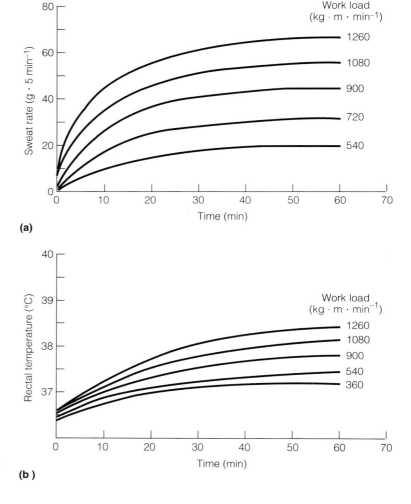

Figure 22-8 The effects of work load during prolonged exercise on (a) sweat rate and (b) rectal temperature. SOURCE: Neilson, 1938.

culatory regulation takes precedence over temperature regulation. These circumstances help to emphasize the danger of heat injury during exercise in hot climates.

$\dot{V}_{O_2max}$ is not impaired in the heat unless the subject is experiencing thermal imbalance when exercise begins. Athletes rarely have the luxury of going from the comforts of normal room temperature into the heat of an environmental chamber for athletic contests. They usually have to sleep, eat, and train in the adverse environment before they have to perform. Dehydration, lack of sleep, and anxiety can combine to cause a degree of physiological and psychological stress. Most studies show that $\dot{V}_{O_2max}$ is decreased by 6 to 8% in preheated subjects.

Sweating Response

Sweating is the primary means of heat dissipation during exercise, regardless of environmental temperature. In the heat, sweating is even more important because the body tends to gain rather than lose heat by radiation, conduction, and convection. Maximum sweat output of an unacclimatized person is approximately $30 \text{ g} \cdot \text{min}^{-1}$ (1.8 liters per hour). During exercise, sweat rate is related more to

TABLE 22-2

Physiological Changes Occurring with Acclimatization to Heat (heat exposure + exercise)

Physiological Change	Effect
Lower sweating threshold	Prevention of early increase in core temperature
Increased sweat rate	Greater evaporative heat loss, lower core temperature during exercise
Increased plasma volume	Better control of blood pressure, lower heart rate, maintenance of stroke volume, increased $\dot{V}_{O_2max}$
Decreased skin blood flow	Increased heat loss by radiation and convection

exercise intensity than to environmental temperature—that is, more to internal temperature than to skin temperature. As will be discussed, training and heat acclimatization increases sweat rate and causes an earlier onset of sweating.

The sweating rate differs in various parts of the body during intense exercise. For example, the head sweats more and earlier than the forearm. The increase in sweat rate during exercise depends first on the density of sweat glands in a skin area and then on sweat output per gland.

Light affects the onset of sweating. The onset of vasodilation of skin blood vessels and sweating occur at a lower core temperature if the athlete is exposed to bright light before exercise. This may have practical significance—taking a nap in a dark room before a competition may result in suboptimal temperature regulation during competition.

An interesting notion proposed by several research groups is pre-cooling the body prior to endurance performance in the heat. Booth and coworkers (1997) found increases in performance when exercise was preceded by cold-water immersion. Pre-cooling decreased the pre-exercise rectal and mean skin temperature by 0.7°C and 5.9°C, respectively, and had a marked effect on endurance performance.

Acclimatization to Heat

The first two weeks of heat exposure results in lower heart rate, core temperature, perceived exertion, and skin temperature at rest and during sub-

maximal exercise (see Table 22–2). Also, there is an increased stability in blood pressure during prolonged exercise. The primary physiological adjustments of acclimatization to heat include increased plasma volume and sweating capacity, a decreased core temperature at the onset of sweating (i.e., earlier sweating), and improved distribution of sweat over the skin. Sodium chloride losses in sweat and urine also decrease due to reduced electrolyte concentration in the sweat. In addition, erythrocyte magnesium increases during the first five days of acclimatization.

Individual Differences in the Rate of Acclimatization Two-thirds to three-quarters of the physiological adjustments and improvements in performance due to heat acclimatization are seen in 4 to 6 days. Aerobically fit people are partially but not fully acclimated to the heat. Acclimatization to heat is not complete unless the exposure is accompanied by exercise training. Intense exercise in the heat is more effective than over-distance exercise for inducing heat acclimatization. Optimal acclimatization seems to require elevated core temperature and an hourly sweat rate of at least 400–600 ml at temperatures greater than 30°C for at least five days. Acclimatization is also humidity specific—exposure to hot dry climates provides only partial acclimatization to hot humid environments.

Acclimatization rates differ among physiological systems. Significant changes in plasma volume, heart rate, and perceived exertion occur during the first 6 days of exposure to heat. Aerobically fit people acclimatize in these areas even faster, usually within about 4 days. Decreases in renal and sweat concentrations of Na+ and Cl- take about 10 days. Increases in sweat rate are not fully evident until 14 days after acclimatization begins.

Many older studies suggest that women and older people of both sexes do not acclimatize to heat as well as young men. Although the data are difficult to interpret, individual differences in physical fitness are probably more likely to account for most of the apparent age and sex differences in this rate of acclimatization.

Cardiovascular Adaptations Acclimatization to heat induces a 3–27% increase in plasma volume, provided that heat exposure is accompanied by exercise training. The increased plasma volume helps maintain stroke volume, central blood volume, and sweating capacity. In addition, it enables the body to hold more heat. The plasma volume increase is due mainly to an increase in plasma proteins and alterations in the secretion of the hormones vasopressin and renin. Fifteen grams of water are added to the plasma for every 1-gram increase in plasma protein. During the early days of heat acclimatization, there are increases in vasopressin (i.e., antidiuretic hormone), renin, and aldosterone, which also help increase plasma volume. As plasma volume increases, the levels of these circulating hormones decrease.

Blood flow to the skin decreases with acclimatization. This adjustment helps to restore central blood volume, which is vitally important for maintaining stroke volume, blood pressure, and muscle blood flow during exercise. Core temperature is lower during exercise in acclimatized people. The decrease in skin blood flow is accompanied by a large increase in sweating and evaporative cooling capacity, resulting in greater peripheral conductance of heat.

Sweating Response Acclimatization increases sweating capacity almost threefold, from about 1.5 liters per hour to 4 liters per hour. This is accompanied by a more complete and even distribution of sweating, which is particularly advantageous during exercise in humid heat. Sweat losses of sodium chloride decrease because of increased secretion of aldosterone. The increase in sweat rate and the onset of sweating is stimulated by repeated exposures to high core temperature (internal temperature). The increased sensitivity of the sweat glands after acclimatization may be due to an increase in receptor density for neural and humoral stimuli (hormone binding) and an increase in the size or number of active sweat glands.

The fall in the sweating threshold is important for keeping core temperature from increasing rapidly during the early stages of exercise. As discussed, increased temperature tends to cause further increases in temperature because of the Q_{10} effect—the early onset of sweating serves to partially negate this.

Exercise training is essential for acclimatization to the heat. However, training by itself does not provide a full measure of heat adaptation. A person from a temperate environment who must compete in the heat can become acclimatized for the effort by exercising (endurance training) in a hot room or wearing extra clothing. People should take care, however, not to completely stifle heat dissipation when using these artificial heat acclimatization techniques or hyperthermia could result.

Loss of Acclimatization to Heat

The absence of heat exposure and decrease in physical fitness will result in a gradual loss of heat acclimatization. Retention of heat acclimatization varies with the person and the environment. For example, physically fit people retain the benefits of heat acclimatization longer than unfit people. Physiological adaptations of heat acclimatization also remain longer for dry heat compared to humid heat.

Thermal Distress

With the popularity of distance running and competitive sports for the weekend athlete, thermal distress and heat injury are becoming increasingly common. Fortunately, many of the severe effects of heat stress in athletics can be avoided if the necessary precautions are taken. Thermal distress includes dehydration (loss of body fluid), heat cramps (involuntary cramping of skeletal muscle), heat exhaustion (hypotension and weakness caused by an inability of the circulation to compensate for vasodilation of skin blood vessels), heat syncope (fainting in the heat), and heat stroke (the failure of the temperature regulatory function of the hypothalamus). Heat disorders are typically points along a continuum, with the symptoms of one disorder blurring into the next.

Hyperthermia (high body temperature) is caused by an imbalance between heat gain and heat loss. It may result from a high rate of heat production, such as commonly occurs during exercise, that exceeds the capacity of the body's heat loss mechanism. It can also occur in high environmental temperatures, where an increase in skin temperature hampers thermal exchange.

Certain drugs can increase the risk of hyperthermia by increasing muscle heat production (e.g., amphetamines, cocaine), increasing metabolic rate (e.g., salicylates, thyroid hormone), impairing heat dissipation (e.g., antihistamines, atropine), and impair cardiovascular compensation (e.g., diuretics and beta-adrenergic blocking agents). Obesity, alcoholism, and pre-existing disease conditions may also predispose people to heat illnesses.

Dehydration

Dehydration is the loss of fluid from the body. Excessive dehydration can decrease sweat rate, plasma volume, cardiac output, maximal oxygen uptake, work capacity, muscle strength, and liver glycogen. Although it is a common condition during exercise in the heat, it can occur even in thermally neutral environments.

A water deficit of 700 ml (approximately 1% of body weight) increases plasma osmolality, which causes thirst. At a fluid deficit of 5% of body weight, symptoms include discomfort and alternating states of lethargy and nervousness. Irritability, fatigue,

and loss of appetite are also characteristic of this level of dehydration. Five percent dehydration is extremely common in sports such as football, tennis, and distance running. Dehydration levels greater than 7% are extremely dangerous. At these levels, salivating and swallowing food become difficult. At fluid losses above 10%, the ability to walk is impaired and is accompanied by discoordination and spasticity. As 15% dehydration approaches, the person experiences delirium and shriveled skin, along with decreased urine volume, loss of the ability to swallow food, and difficulty swallowing water. Above 20%, the skin bleeds and cracks. This is the upper limit of tolerance to dehydration before death ensues.

Other risks associated with dehydration include an increase in heat storage and a reduction in a person's ability to tolerate heat strain. These risks are linked to the fact that heat dissipation responses during exercise are more related to blood volume, which is critically affected by hydration status, than to aerobic power. Even moderate levels of dehydration (2% of body weight) impair cardiovascular and temperature regulation and decrease performance. In addition, dehydration and hyperthermia together cause significant reductions in cardiac output, muscle blood flow, skin blood flow, and blood pressure. Such a decrease in blood flow to muscles during exercise results from a lowering in perfusion pressure and systemic blood flow.

To avoid dehydration, athletes and coaches should follow an aggressive fluid replacement regimen during exercise in the heat. Athletes and non-athletes alike should drink fluids in direct proportion to sweat loss, even during exercise lasting only one hour. Researchers advocate consuming cold drinks containing moderate amounts of carbohydrate (i.e., 7% carbohydrate solution), which reduces the risk of heat illness and improves exercise performance by preventing dehydration and hypoglycemia.

Adequate hydration during exercise in the heat helps maintain blood pressure and cardiac output, which are essential to adequate skin blood flow and sweat rates. As discussed in Chapters 15 and 16, maintenance of blood pressure and the prevention of hypoxemia are central to cardiovascular regulation during exercise. Preventing dehydration is a practical way for athletes to maximize performance and delay fatigue; thus, maintaining adequate hydration must become part of an exercise regimen, and athletes must train themselves to drink large volumes of fluid frequently.

Osmoreceptors in the hypothalamus stimulate the drive to drink fluids. Unfortunately, though, thirst does not keep up with fluid requirements, so athletes can easily experience fluid deficits of 2 to 4% of body weight. It is very important for dehydrating athletes to have regular fluid breaks rather than relying only on their thirst for fluid replacement. Rehydration is facilitated if the fluid replacement beverage contains electrolytes and, possibly, carbohydrates.

Physical fitness helps prevent dehydration. Training, for example, increases blood volume (mainly plasma volume), which helps maintain central blood volume during dehydration. Training also causes more dilute sweat by reducing its electrolyte content.

The inadequacy of the thirst mechanism can be compounded by the type of fluid replacement. The ideal fluid replacement beverage should taste good—so people will drink it—and not induce gastrointestinal distress. The drink should contain some carbohydrate ($6–8$ g · 100 ml^{-1}), which helps maintain blood glucose. Carbohydrates also increase fluid osmolality, which helps maintain body water. Glucose and lactate polymers are excellent fuel sources in these drinks because they provide a relatively concentrated fuel source without excessive sweetness. The drink should also contain sodium. Sodium helps maintain extracellular fluid volume without affecting thirst. Most sports drinks contain $10–20$ mEq · liter, which is adequate for electrolyte restoration during dehydration. During prolonged exercise, people can develop hyponatremia (i.e., water intoxication), which is caused by excessive sodium loss in the sweat without adequate sodium replenishment. This can be prevented by including a small amount of sodium in the fluid replacement beverage.

Heat Cramps

Heat cramps are characterized by involuntary cramping and spasm in the muscle groups used during exercise. The term *heat cramps* is a misnomer because the condition is not caused by elevated body temperature. Until recently, heat cramps were thought to be caused by fluid or electrolyte abnormalities related to heat exposure, but there is little evidence for this. Muscle cramps can occur during or after exercise performed in heat or cold. Instead, cramps appear to be caused by a spinal neural mechanism linked to fatigue that is unrelated to biochemical changes in blood or in the cramping muscles.

Heat Exhaustion

Heat exhaustion is characterized by a rapid, weak pulse, hypotension, faintness, profuse sweating, and psychological disorientation (Figure 22-9). It results from an acute plasma volume loss and the inability of the circulation to compensate for the concurrent vasodilation in the skin and active skeletal muscles. Although core temperature may be elevated somewhat (usually $<39.5°C$), it does not reach the extremely high level seen in heat stroke ($>41°C$). Core temperature may even be normal or below normal. Paradoxically, studies have failed to show that exercise-related heat exhaustion is necessarily caused by specific fluid or electrolyte abnormalities.

The treatment for heat exhaustion includes having the person lie down in a cool area and administering fluids. Intravenous fluid administration may be appropriate in some instances. The person should not participate in any further activity for the rest of the day and should be encouraged to drink plenty of fluids for the next 24 hours.

Heat Syncope

Heat syncope is related to heat exhaustion but can occur even without major sweat loss. It typically occurs after exercise—the person stops moving, blood pools in the lower extremities, and the person faints.

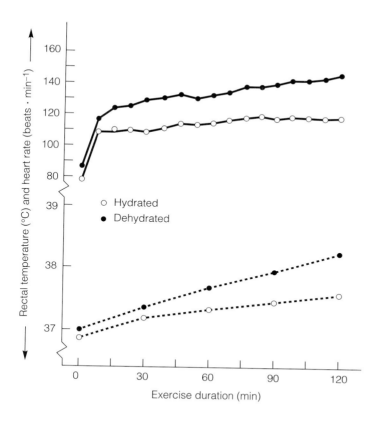

Figure 22-9 Effects of dehydration on heart rate and rectal temperature during prolonged cycling. Modified from Costill, 1977.

It tends to occur in unacclimatized people at the beginning of summer. Heat syncope can occur secondarily to heat exhaustion or can occur independently.

Heat Stroke

Heat stroke is the failure of the hypothalamic temperature regulatory center and represents a major medical emergency (Figure 22-10). The mortality rate for heat stroke may be over 20%, with death due to heart failure or cerebral edema. Although it occurs mainly in infants and the elderly, it sometimes also occurs in unacclimatized fit people. It is second only to head injuries in exercise-related deaths.

Heat stroke is principally caused by a failure of the temperature regulatory center in the hypothalamus, which, in turn, causes failure of the body's mechanisms to dissipate heat, resulting in an explo-

sive rise in body temperature. It is characterized by a high core temperature ($>41°C$); hot, dry skin; and extreme central nervous system dysfunction, such as extreme confusion, delirium, convulsions, and unconsciousness. In exercise-related heat stroke, approximately 50% of victims are still sweating. The cardiovascular effects are variable, with some people experiencing hypotension and others experiencing a full bounding pulse and high blood pressure. Heat stroke is associated with increases in plasma norepinephrine, epinephrine, endotoxin (lipopolysaccharide) levels that have been associated with vascular collapse.

A heat stroke victim should be cooled using tepid water, fanning, and ice packs placed on the neck, axillae, and groin. In the hospital, treatment includes cooling with a steady stream or spray of tepid water combined with an air stream directed at the patient. Rectal temperature is monitored to

Heat stroke: hypothalamic failure, peripheral
effector failure, massive heat storage

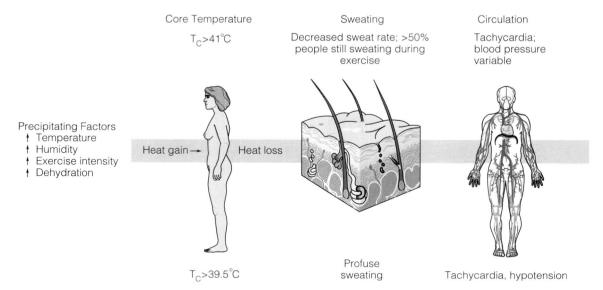

Core Temperature	Sweating	Circulation
$T_C > 41°C$	Decreased sweat rate; >50% people still sweating during exercise	Tachycardia; blood pressure variable

Precipitating Factors
↑ Temperature
↑ Humidity
↑ Exercise intensity
↑ Dehydration

Heat gain → Heat loss

$T_C > 39.5°C$

Profuse sweating

Tachycardia, hypotension

Heat exhaustion: decreased central blood
volume, fluid loss in extracellular space

Figure 22-10 Pathophysiology of heat stroke and heat exhaustion in the athlete.

guard against rebound hypothermia. When core temperature reaches 38.5°C, cooling procedures are discontinued. Fluids are administered to counteract hypotension. More drastic measures are sometimes required in the event of continued shock or renal failure.

Although heat stroke can occur in almost any environmental conditions, the risk is greatest during high temperatures, humidity, and solar radiation (hottest time of the day), and on days with little wind. People particularly susceptible are those who are obese, unfit, dehydrated, unacclimatized, ill, or have a history of heat stroke. As with heat exhaustion, there is little evidence that fluid and electrolyte abnormalities are critical for causing heat stroke during exercise. Damage to sweat glands is common in heat stroke victims, which may explain why these people are more susceptible to recurrent episodes of the problem.

Preventing Thermal Distress

The problems of thermal distress can be minimized by following a few simple principles:

- Ensure that athletes are in good physical condition. There should be a gradual increase in intensity and duration of training in the heat until the athletes are fully acclimatized. Optimal acclimatization to heat is accomplished by training in the heat at intensities above 50% of $\dot{V}_{O_2max}$.

- Ensure that athletes avoid becoming overheated before exercising in a hot environment (i.e., that they avoid pre-exercise heat storage).

- Be aware of the early symptoms of heat stress, such as thirst, fatigue, lethargy, and visual disturbances.

- Clinical signs of hyperthermia are typically more meaningful than temperature measurements. Core temperature is often underestimated because common sites of temperature measurement (i.e., mouth or axillary sites) are too superficial.

- Athletes should not run faster than their normal training intensity.

- Athletes should not compete if they have any illness accompanied by fever or had a fever shortly before the event. Other conditions that decrease heat tolerance include sleep loss, glycogen depletion or hypoglycemia, or recent heavy alcohol consumption (i.e., "hangover").

- Schedule practice sessions and games during the cooler times of the day.

- Modify or cancel sessions when the wet bulb globe temperature is 25.5°C or greater. *Wet bulb temperature* is used to determine humidity and *black globe temperature* is an indication of radiant heat.

- Plan for regular fluid breaks. Drink fluid in direct proportion to sweat loss (or close to it). Maintaining hydration will optimize physiological function and prevent deterioration of exercise performance, even during exercise lasting only one hour.

- Supply a drink that is cold (8–13°C) and contains some carbohydrate (6–8 g $\cdot$ 100 ml^{-1}), with a small amount of electrolytes (20 mEq $\cdot$ liter^{-1} sodium and 10mEq $\cdot$ liter^{-1} potassium).

- "Tank-up" before practice or games by drinking 600 ml ($\approx$2.5 cups) of fluid two hours before the activity and an additional 400 ml (1.5 cups) 15 minutes before.

- Fluid replacement should be particularly encouraged during the early stages of practice and competition. As exercise progresses, splanchnic blood flow decreases, which diminishes water absorption from the gut.

- Athletes should be weighed every day before practice. Any athlete showing a decrease of 2–3% should consume extra fluid. Athletes with weight losses of 4–6% should decrease training intensity, and those with weight losses greater than 7% should consult a physician. People who lose a lot of weight in the heat should be identified and closely monitored.

- Salt tablets are prohibited. However, athletes should be encouraged to consume ample amounts of salt at mealtime.

■ Summary

Humans can tolerate extremely hot and cold climates because of a well-developed ability to control body temperature. Body temperature is regulated by controlling the rate of heat production and heat loss. Mechanisms of heat production include basal metabolic rate, shivering, exercise, thermogenic hormone secretion, and the Q_{10} effect. The body loses heat by radiation, conduction, convection, evaporation, and excretion. The hypothalamus is the main center for temperature regulation.

Clothing is the most important limiting factor when exercising in the cold. However, swimming in the cold, or exercising with wet clothing, can result in a greatly diminished performance due to an accelerated heat loss. Acclimatization to cold is difficult to demonstrate but is known to occur.

The ability to exercise in high ambient temperatures depends on the ability to dissipate heat and maintain blood flow to active muscles. Sweating is the primary means of cooling in the heat. An increased sweat rate is the most important adaptation that occurs with heat acclimatization. Exercise training is necessary to maximize this process.

Thermal injuries are a serious problem in both competitive and recreational sports. Thermal injuries include heat cramps, heat exhaustion, and heat stroke. These injuries are typically preceded by dehydration. Preventive measures include adequate physical conditioning, exercising during cooler periods of the day, planning regular water breaks, and limiting workout sessions until adequate acclimatization is achieved.

■ Selected Readings

Adolf, E. F. (Ed.). Physiology of Man in the Desert. New York: Interscience, 1947.

Aizawa, S., and H. Tokura. Exposure to bright light for several hours during the daytime lowers tympanic temperature. *Int. J. Biometeorol.* 41: 90–93, 1997.

Aizawa, S., and H. Tokura. Influence of bright light exposure for several hours during the daytime on cutaneous vasodilatation and local sweating induced by an exercise heat load. *Eur. J. Appl. Physiol.* 78: 303–307, 1998.

Allison, T. G., and W. E. Reger. Comparison of responses of men to immersion in circulating water at 40.0 and 41.5 degrees C. *Aviat. Space Environ. Med.* 69: 845–850, 1998.

Aoki, K., N. Kondo, M. Shibasaki, S. Takano, and T. Katsuura. Circadian variation in skin blood flow responses to passive heat stress. *Physiol. Behav.* 63: 1–5, 1997.

Aoyagi, Y., T. M. McLellan, and R. J. Shephard. Interactions of physical training and heat acclimation. The thermophysiology of exercising in a hot climate. *Sports Med.* 23: 173–210, 1997.

Aoyagi, Y., T. M. McLellan, and R. J. Shephard. Effects of endurance training and heat acclimation on psychological strain in exercising men wearing protective clothing. *Ergonomics* 41: 328–357, 1998.

Armstrong, L. E., J. P. De Luca, and R. W. Hubbard. Time course of recovery and heat acclimation ability of prior exertional heat stroke patients. *Med. Sci. Sports Exer.* 22: 36–48, 1990.

Armstrong, L. E., R. W. Hubbard, E. W. Askew, J. P. De Luca. C. O'Brien, A. Pasqualicchio, and R. P. Francesconi. Responses to moderate and low sodium diets during exercise-heat acclimation. *Int. J. Sport Nutr.* 3: 207–221, 1993.

Armstrong, L. E., and C. M. Maresh. The induction and decay of heat acclimatization in trained athletes. *Sports Med.* 12: 302–312, 1991.

Armstrong, L. E., and C. M. Maresh. Effects of training, environment, and host factors on the sweating response to exercise. *Int. J. Sports Med.* 19 (Suppl. 2): S103–S105, 1998.

Bar-Or, O. Effects of age and gender on sweating pattern during exercise. *Int. J. Sports Med.* 19 (Suppl. 2): S106–S107, 1998.

Barros, R. C., and L. G. Branco. Effect of nitric oxide synthase inhibition on hypercapnia-induced hypothermia and hyperventilation. *J. Appl. Physiol.* 85: 967–972, 1998.

Basta, D., B. Tzschentke, and M. Nichelmann. Temperature guardian neurons in the preoptic area of the hypothalamus. *Brain Res.* 767: 361–362, 1997.

Booth, J., F. Marino, and J. J. Ward. Improved running performance in hot humid conditions following whole body precooling. *Med. Sci. Sports Exerc.* 29: 943–949, 1997.

Boulant, J. A. Hypothalamic neurons. Mechanisms of sensitivity to temperature. *Ann. NY. Acad. Sci.* 856: 108–115, 1998.

Brajkovic, D., M. B. Ducharme, and J. Frim. Influence of localized auxiliary heating on hand comfort during cold exposure. *J. Appl. Physiol.* 85: 2054–2065, 1998.

Bridgman, S. A. Peripheral cold acclimatization in Antarctic scuba divers. *Aviat. Space Environ. Med.* 62: 733–738, 1991.

Buono, M. J., J. H. Heaney, and K. M. Canine. Acclimation to humid heat lowers resting core temperature. *Am. J. Physiol.* 274: R1295–R1299, 1998.

Burke, L. M., and J. A. Hawley. Fluid balance in team sports. Guidelines for optimal practices. *Sports Med.* 24: 38–54, 1997.

Cannon, B., J. Houstek, and J. Nedergaard. Brown adipose tissue. More than an effector of thermogenesis? *Ann. NY. Acad. Sci.* 856: 171–187, 1998.

Commission for Thermal Physiology of the International Union of Physiological Sciences (IUPS Thermal Commission). Glossary of terms for thermal physiology. *Pflügers Arch.* 410: 567–587, 1987.

Convertino, V., J. Greenleaf, and E. Bernauer. Role of thermal and exercise factors in the mechanism of hypervolemia. *J. Appl. Physiol.* 48: 657–664, 1980.

Costill, D. The New Runner's Diet. Mountain View, Ca.: World, 1977.

Coyle, E. F., A. R. Coggan, M. K. Hemmert, and J. L. Ivy. Muscle glycogen utilization during prolonged strenuous exercise when fed carbohydrate. *J. Appl. Physiol.* 61: 165–172, 1986.

Dill, D. B. Life, Heat and Altitude. Cambridge: Harvard University Press, 1938.

Dill, D. B., L. F. Soholt, D. C. McLean, T. F. Drost, and M. T. Loughran. Capacity of young males and females for running in desert heat. *Med. Sci. Sports* 9: 137–142, 1977.

Dolny, D. G., and P. W. R. Lemon. Effect of ambient temperature on protein breakdown during prolonged exercise. *J. Appl. Physiol.* 64: 550–555, 1988.

Doubt, T. J., and P. A. Deuster. Fluid ingestion during exercise in 25 degrees C water at the surface and 5.5 ATA. *Med. Sci. Sports Exer.* 26: 75–80, 1994.

Doubt, T. J., and S. Hsieh. Additive effects of caffeine and cold water during submaximal leg exercise. *Med. Sci. Sports Exer.* 23: 435–442, 1991.

Elias, C. F., C. Lee, J. Kelly, C. Aschkenasi, R. S. Ahima, P. R. Couceyro, M. J. Kuhar, C. B. Saper, and J. K. Elmquist. Leptin activates hypothalamic CART neurons projecting to the spinal cord. *Neuron* 21: 1375–1385, 1998.

Epstein, Y. Heat intolerance: predisposing factor or residual injury? *Med. Sci. Sports Exer.* 22: 29–35, 1990.

Falk, B., O. Bar-Or, J. Smolander, and G. Frost. Response to rest and exercise in the cold: effects of age and aerobic fitness. *J. Appl. Physiol.* 76: 72–78, 1994.

Fortney, S. M., V. Mikhaylov, S. M. Lee, Y. Kobzev, R. R. Gonzalez, and J. E. Greenleaf. Body temperature and thermoregulation during submaximal exercise after 115-day spaceflight. *Aviat. Space Environ. Med.* 69: 137–141, 1998.

Frappell, P. Hypothermia and physiological control: the respiratory system. *Clin. Exp. Pharmacol. Physiol.* 25: 159–164, 1998.

Gabaree, C. L., B. S. Mair, M. A. Kolka, and L. A. Stephenson. Effects of topical skin protectant on heat exchange in humans. *Aviat. Space Environ. Med.* 68: 1019–1024, 1997.

Galloway, S. D., and R. J. Maughan. The effects of substrate and fluid provision on thermoregulatory, cardiorespiratory and metabolic responses to prolonged exercise in a cold environment in man. *Exp. Physiol.* 83: 419–430, 1998.

Geor, R. J., and L. J. McCutcheon. Thermoregulatory adaptations associated with training and heat acclimation. *Vet. Clin. North Am. Equine Pract.* 14: 97–120, 1998.

Gisolfi, C., and S. Robinson. Relations between physical training, acclimatization and heat tolerance. *J. Appl. Physiol.* 26: 530–534, 1969.

Gonzalez-Alonso, J. Separate and combined influences of dehydration and hyperthermia on cardiovascular responses to exercise. *Int. J. Sports Med.* 19 (Suppl. 2): S111–S114, 1998.

Gonzalez-Alonso, J., J. A. Calbet, and B. Nielsen. Muscle blood flow is reduced with dehydration during prolonged exercise in humans. *J. Physiol.* (London). 513: 895–905, 1998.

Greenleaf, J. E., R. Looft-Wilson, J. L. Wisherd, C. G. Jackson, P. P. Fung, A. C. Ertl, P. R. Barnes, C. D. Jensen, and J. H. Whittam. Hypervolemia in men from fluid ingestion at rest and during exercise. *Aviat. Space Environ. Med.* 69: 374–386, 1998.

Greenleaf, J., and F. Sargent, II. Voluntary dehydration in man. *J. Appl. Physiol.* 20: 719–724, 1965.

Hargreaves, M., and M. Febbraio. Limits to exercise performance in the heat. *Int. J. Sports Med.* 19 (Suppl. 2): S115–S116, 1998.

Havenith, G., J. M. Coenen, L. Kistemaker, and W. L. Kenney. Relevance of individual characteristics for human heat stress response is dependent on exercise intensity and climate type. *Eur. J. Appl. Physiol.* 77: 231–241, 1998.

Hokkanen, J. E. Thermal role of a blood vessel running through a temperature gradient. *Ann. NY. Acad. Sci.* 813: 56–62, 1997.

Hong, S. K., D. W. Rennie, and Y. S. Park. Cold acclimatization and deacclimatization of Korean women divers. *Exer. Sport Sci. Rev.* 14: 231–268, 1986.

Horswill, C. A. Effective fluid replacement. *Int. J. Sport Nutr.* 8: 175–195, 1998.

Horvath, S. M. Exercise in a cold environment. *Exer. Sport Sci. Rev.* 9: 221–263, 1981.

Hubbard, R. W. An introduction: the role of exercise in the etiology of exertional heatstroke. *Med. Sci. Sports Exer.* 22: 2–5, 1990.

Johnson, J. M. Physical training and the control of skin blood flow. *Med. Sci. Sports Exerc.* 30: 382–386, 1998.

Kandjov, I. M. Thermal resistance parameters of the air environment at various altitudes. *Int. J. Biometeorol.* 40: 91–94, 1997.

Kenney, W. L. Heat flux and storage in hot environments. *Int. J. Sports Med.* 19 (Suppl. 2): S92–S95, 1998.

Kenny, G. P., A. A. Chen, B. A. Nurbakhsh, P. M. Denis, C. E. Proulx, and G. G. Giesbrecht. Moderate exercise increases postexercise thresholds for vasoconstriction and shivering. *J. Appl. Physiol.* 85: 1357–1361, 1998.

Kerr, C. G., T. A. Trappe, R. D. Starling, and S. W. Trappe. Hyperthermia during Olympic triathlon: influence of body heat storage during the swimming stage. *Med. Sci. Sports Exerc.* 30: 99–104, 1998.

Kleiber, M. The Fire of Life. Huntington, N.Y.: R. E. Krieger, 1976.

Kondo, N., S. Takano, K. Aoki, M. Shibasaki, H. Tominaga, and Y. Inoue. Regional differences in the effect of exercise intensity on thermoregulatory sweating and cutaneous vasodilation. *Acta Physiol. Scand.* 164: 71–78, 1998.

Kurz, A., D. I. Sessler, F. Tayefeh, and R. Goldberger. Poikilothermia syndrome. *J. Intern. Med.* 244: 431–436, 1998.

Lee, D. T., M. M. Toner, W. D. McArdle, I. S. Vrabas, and K. B. Pandolf. Thermal and metabolic responses to cold-water immersion at knee, hip, and shoulder levels. *J. Appl. Physiol.* 82: 1523–1530, 1997.

Luetkemeier, M. J., M. G. Coles, and E. W. Askew. Dietary sodium and plasma volume levels with exercise. *Sports Med.* 23: 279–286, 1997.

Mack, G. W., H. Nose, A. Takamata, T. Okuno, and T. Morimoto. Influence of exercise intensity and plasma volume on active cutaneous vasodilation in humans. *Med. Sci. Sports Exer.* 26: 209–216, 1994.

Marino, F., and J. Booth. Whole body cooling by immersion in water at moderate temperatures. *J. Sci. Med. Sport* 1: 73–82, 1998.

Martens. W. J. Climate change, thermal stress and mortality changes. *Soc. Sci. Med.* 46: 331–344, 1998.

Maughan, R. J., and S. M. Shirreffs. Dehydration, rehydration and exercise in the heat: concluding remarks. *Int. J. Sports Med.* 19 (Suppl. 2): S167–S168, 1998.

McCann, D. J., and W. C. Adams. Wet bulb globe temperature index and performance in competitive distance runners. *Med. Sci. Sports Exerc.* 29: 955–961, 1997.

Mills, P. C., D. J. Marlin, C. M. Scott, and N. C. Smith. Nitric oxide and thermoregulation during exercise in the horse. *J. Appl. Physiol.* 82: 1035–1039, 1997.

Morimoto, T., and T. Itoh. Thermoregulation and body fluid osmolality. *J. Basic Clin. Physiol. Pharmacol.* 9: 51–72, 1998.

Murray, R. Rehydration strategies—balancing substrate, fluid, and electrolyte provision. *Int. J. Sports Med.* 19 (Suppl. 2): S133–S135, 1998.

Nadel, E. Circulatory and thermal regulations during exercise. *Fed. Proc.* 39: 1491–1497, 1980.

Nadel, E., I. Holmer, U. Bergh, P.-O. Åstrand, and J. Stolwijk. Energy exchanges of swimming man. *J. Appl. Physiol.* 36: 465–471, 1974.

Nadel, E. R., S. M. Fortney, and C. B. Wagner. Effect of hypohydration state on circulatory and thermal regulations. *J. Appl. Physiol.* 49: 715–721, 1980.

Neilson, B. Die regulation der korpertemperatur ber muskelarbeit. *Skand. Arch. Physiol.* 79: 193–230, 1938.

Nielsen, B. Heat acclimation—mechanisms of adaptation to exercise in the heat. *Int. J. Sports Med.* 19 (Suppl. 2): S154–S156, 1998.

Nielsen, B. Heat stress and acclimation. *Ergonomics* 37: 49–58, 1994.

Nielsen, B., J. R. Hales, S. Strange, N. J. Christensen, J. Warberg, and B. Saltin. Human circulatory and thermoregulatory adaptations with heat acclimation and exercise in a hot, dry environment. *J. Physiol.* (London) 460: 467–485, 1993.

Nielsen, B., G. Savard, E. A. Richter, M. Hargreaves, and B. Saltin. Muscle blood flow and muscle metabolism during exercise and heat stress. *J. Appl. Physiol.* 69: 1040–1046, 1990.

Nielsen, B., S. Strange, N. J. Christensen, J. Warberg, and B. Saltin. Acute and adaptive responses in humans to exercise in a warm, humid environment. *Pflügers Arch.* 434: 49–56, 1997.

Noakes, T. D. Fluid and electrolyte disturbances in heat illness. *Int. J. Sports Med.* 19 (Suppl. 2): S146–S149, 1998.

Nunneley, S. A. Heat stress in protective clothing. *Scand. J. Work, Environ., Health* 15 (Suppl.): 52–57, 1989.

O'Brien, C., R. W. Hoyt, M. J. Buller, J. W. Castellani, and A. J. Young. Telemetry pill measurement of core temperature in humans during active heating and cooling. *Med. Sci. Sports Exerc.* 30: 468–472, 1998.

O'Brien, C., A. J. Young, and M. N. Sawka. Hypohydration and thermoregulation in cold air. *J. Appl. Physiol.* 84: 185–189, 1998.

Oya, A., H. Asakura, T. Koshino, and T. Araki. Thermographic demonstration of nonshivering thermogenesis in human newborns after birth: its relation to umbilical gases. *J. Perinat. Med.* 25: 447–454, 1997.

Pandolf, K. B. Relation of thermoregulatory responses to aerobic power and blood volume. *Clin. J. Sport Med.* 8: 69, 1998.

Pandolf, K. B. Time course of heat acclimation and its decay. *Int. J. Sports Med.* (Suppl. 2): S157–S160, 1998.

Pirnay, F., R. Deroanne, and J. Petit. Maximal oxygen consumption in a hot environment. *J. Appl. Physiol.* 28: 642–645, 1970.

Precht, H. Temperature and Life. Berlin: Springer, 1973.

Purkayastha, S. S., G. Ilavazhagan, U. S. Ray, W. Selvamurthy. Responses of Arctic and tropical men to a standard cold test and peripheral vascular responses to local cold stress in the Arctic. *Aviat. Space Environ. Med.* 64: 1113–1119, 1993.

Rowell, L. B. Human cardiovascular adjustments to exercise and thermal stress. *Physiol. Rev.* 54: 75–159, 1974.

Rowell, L. B., H. J. Marx, R. A. Bruce, R. D. Conn, and F. Kusumi. Reductions in cardiac output, central blood volume and stroke volume with thermal stress in normal men during exercise. *J. Clin. Invest.* 45: 1801–1816, 1966.

Royburt, M., Y. Epstein, Z. Solomon, and J. Shemer. Long-term psychological and physiological effects of heat stroke. *Physiol. Behav.* 54: 265–267, 1993.

Savourey, G., and J. Bittel. Thermoregulatory changes in the cold induced by physical training in humans. *Eur. J. Appl. Physiol.* 78: 379–384, 1998.

Sawka, M. N., W. A. Latzka, R. P. Matott, and S. J. Montain. Hydration effects on temperature regulation. *Int. J. Sports Med.* 19 (Suppl. 2): S108–S110, 1998.

Sawka, M. N., A. J. Young, R. P. Francesconi, S. R. Muza, and K. B. Pandolf. Thermoregulatory and blood responses during exercise at graded hypohydration levels. *J. Appl. Physiol.* 59: 1394–1401, 1985.

Schmid, H. A., W. Riedel, and E. Simon. Role of nitric oxide in temperature regulation. *Prog. Brain Res.* 115: 87–110, 1998.

Scholander, P. F., H. T. Hammel, J. S. Hart, D. H. Le Messurier, and J. Steen. Cold adaptation in Australian aborigines. *J. Appl. Physiol.* 13: 201–210, 1958.

Shapiro, Y., D. Moran, and Y. Epstein. Acclimatization strategies—preparing for exercise in the heat. *Int. J. Sports Med.* 19 (Suppl. 2): S161–S163, 1998.

Sheffield-Moore, M., K. R. Short, C. G. Kerr, A. C. Parcell, D. R. Bolster, and D. L. Costill. Thermoregulatory responses to cycling with and without a helmet. *Med. Sci. Sports Exerc.* 29: 755–761, 1997.

Shimizu, T., M. Kosaka, and K. Fujishima. Human thermoregulatory responses during prolonged walking in water at 25, 30 and 35 degrees C. *Eur. J. Appl. Physiol.* 78: 473–478, 1998.

Shirreffs, S. M., and R. J. Maughan. Urine osmolality and conductivity as indices of hydration status in athletes in the heat. *Med. Sci. Sports Exerc.* 30: 1598–1602, 1998.

Shitzer, A. On the thermal efficiency of cold-stressed fingers. *Ann. NY. Acad. Sci.* 858: 74–87, 1998.

Simon, E. Nitric oxide as a peripheral and central mediator in temperature regulation. *Amino Acids* 14: 87–93, 1998.

Sink, K. R., T. R. Thomas, J. Araujo, and S. F. Hill. Fat energy use and plasma lipid changes associated with exercise intensity and temperature. *Eur. J. Appl. Physiol.* 58: 508–513, 1989.

Steiner, A. A., E. C. Carnio, J. Antunes-Rodrigues, and L. G. Branco. Role of nitric oxide in systemic vasopressin-induced hypothermia. *Am. J. Physiol.* 275: R937–R941, 1998.

Stendig-Lindberg, G., D. Moran, and Y. Shapiro. How significant is magnesium in thermoregulation? *J. Basic Clin. Physiol. Pharmacol.* 9: 73–85, 1998.

Stephenson, L. A., and M. A. Kolka. Thermoregulation in women. *Exer. Sport Sci. Rev.* 21: 231–262, 1993.

Sutton, J. R. Exercise and the environment. In Exercise, Fitness, and Health, C. Bouchard, R. J. Shephard, T. Stephens, J. R. Sutton, and B. D. McPherson (Eds.). Champaign, Ill.: Human Kinetics, 1990, pp. 165–178.

Szelenyi, Z. Neuroglia: possible role in thermogenesis and body temperature control. *Med. Hypotheses* 50: 191–197, 1998.

Takamata, A., K. Nagashima, H. Nose, and T. Morimoto. Role of plasma osmolality in the delayed onset of thermal cutaneous vasodilation during exercise in humans. *Am. J. Physiol.* 275: R286–R290, 1998.

Tek, D., and J. S. Olshaker. Heat illness. *Emerg. Med. Clin. North Am.* 10: 299–310, 1992.

Timmons, B. A., J. Araujo, and T. R. Thomas. Fat utilization enhanced by exercise in a cold environment. *Med. Sci. Sports Exer.* 17: 673–678, 1985.

Tipton, M. J., D. A. Stubbs, and D. H. Elliott. Human initial responses to immersion in cold water at three temperatures and after hyperventilation. *J. Appl. Physiol.* 70: 317–322, 1991.

Tuominen, J. A., P. Ebeling, M. L. Heiman, T. Stephens, and V. A. Koivisto. Leptin and thermogenesis in humans. *Acta Physiol. Scand.* 160: 83–87, 1997.

Webb, P. Heat storage and body temperature during cooling and rewarming. *Eur. J. Appl. Physiol.* 66: 18–24, 1993.

Wimer, G. S., D. R. Lamb, W. M. Sherman, and S. C. Swanson. Temperature of ingested water and thermoregulation during moderate-intensity exercise. *Can. J. Appl. Physiol.* 22: 479–493, 1997.

Yamauchi, M., T. Matsumoto, N. Ohwatari, and M. Kosaka. Sweating economy by graded control in well-trained athletes. *Pflügers Arch.* 433: 675–678, 1997.

Yoshida, T., K. Nagashima, H. Nose, T. Kawabata, S. Nakai, A. Yorimoto, and T. Morimoto. Relationship between aerobic power, blood volume, and thermoregulatory responses to exercise-heat stress. *Med. Sci. Sports Exerc.* 29: 867–873, 1997.

CHAPTER 23

EXERCISE, ATMOSPHERIC PRESSURE, AIR POLLUTION, AND TRAVEL

The environment often increases exercise stress. Low oxygen pressure at high altitude can stress the oxygen transport systems of even the fittest athletes. Cold temperatures can numb the flesh and sometimes suppress cellular metabolism to dangerous levels. Hot temperatures can result in a competition for blood between the skin and muscles that can severely limit physical performance and subject the body to thermal injury. Smoggy air can leave the athlete gasping for air during exercise, while trav-

eling across time zones can fatigue athletes and make them less able to compete.

Thousands ski and hike every year at altitudes over 3000 meters (9840 feet), subjecting themselves to both cold and altitude stresses. With the popularity of adventure travel, even novice mountaineers climb peaks in the rarefied air of the Andes, Himalayas, Alps, and Rockies. Environments at low pressures (*hypobaria*) place extraordinary stresses on the human body. Exposure to moderate altitudes—

Sojourners on Mt. Blanc (Chaminox, France) in the early nineteen hundreds. More leisure time and travel opportunities increased interest in environmental exercise physiology.

above 1524 meters (5000 feet)—results in decreased maximal oxygen consumption. Prolonged exposure to extremely high altitudes—over 6000 meters (19,685 feet)—leads to progressive deterioration that can eventually cause death unless the person is moved to a lower altitude.

At high pressures (*hyperbaria*), the driving force of gases is increased while their volumes are decreased. Prolonged exposure to depths greater than 10 meters (33 feet) leads to an increased retention of nitrogen, which can have a disastrous effect if the person fails to decompress. A rapid ascent from watery depths without expiration of air can actually explode the lungs. Also, gases that are either essential or benign at sea level can become toxic or narcotic at higher pressures.

Smoggy air can have negative effects on physical performance, health, and well-being. The active person should have an understanding of the nature of polluted air and its effect on exercise.

Traveling across time zones can sometimes have devastating effects on athletic performance. Jet lag, cramped muscles resulting from prolonged sitting, and dehydration caused by the dry air of pressurized aircraft can sometimes override the effects of even the most carefully planned physical conditioning program.

In this chapter, we pay particular attention to the effects of these conditions on acute and long-term exercise responses and to acclimatization. We examine the effects of exercise in low and high pressure, air pollution, and different time zones.

■ Altitude

The study of altitude physiology was begun late in the last century and during the early part of this century by physically active physiologists, such as Bert, Douglas, Haldane, and Barcroft, who made their investigations in conjunction with their alpine recreational pursuits. Further research was stimulated by aviation problems encountered during World War II, by mountain climbing expeditions, and by the 1968 Mexico City Olympics. An understanding of altitude physiology has become important because of the many athletic contests held at moderate altitudes and the tremendous popularity of mountain sports.

Altitude research has continued unabated in recent years. Studies include the Italian and American Medical Research Expeditions to Everest, the Everest II study (a simulated climb of Mt. Everest in an altitude chamber), and the Pikes Peak studies. These studies included extremely sophisticated measurements never before attempted at altitude. Ongoing altitude research is conducted in the Alps (e.g., at Mount Blanc), in the Andes, and in the Himalayas.

Exercising at altitude is stressful because less oxygen is available (P_{O_2}) than at sea level. Barometric pressure decreases with increasing altitude, resulting in less oxygen per volume of air (Figure 23-1). However, the percentage of oxygen in the air ($F_{I_{O_2}}$) at altitude is the same as at sea level. Although there are several mechanisms (i.e., increased ventilation and heart rate) for adjusting to *hypoxia*

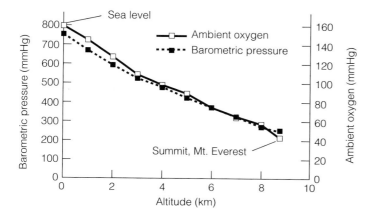

Figure 23-1 Changes in inspired oxygen partial pressure with decreasing barometric pressure.

TABLE 23-1	
Effects of Acute Exposure to Altitude	
Change	**Effect**
Increased resting and submaximal heart rate	Increased O_2 transport to tissue
Increased resting and submaximal ventilation	Increased alveolar P_{O_2}
	Decreased CO_2 and H^+ in CSF and blood
	Predominance of hypoxic ventilatory drive
	Left shift of oxyhemoglobin dissociation curve
	Acute mountain sickness
Increased blood pressure	Increased vascular resistance
Increased catecholamine secretion	Increased lactate production and increased vascular resistance
Decreased $\dot{V}_{O_2max}$	Decreased exercise capacity
Few acute changes in blood, muscle, or liver	

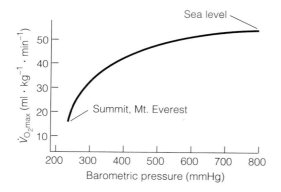

Figure 23-2 Effects of altitude on $\dot{V}_{O_2max}$.

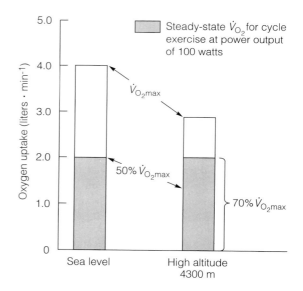

Figure 23-3 The effects of altitude on maximal and submaximal oxygen consumption. SOURCE: Young and Young, 1988.

(low levels of oxygen), oxygen transport capacity decreases steadily at higher and higher altitudes (Table 23-1).

Maximal oxygen consumption ($\dot{V}_{O_2max}$) begins to decrease in most people at about 1524 meters (5000 feet). Initially, $\dot{V}_{O_2max}$ decreases about 3% for each increase of 300 meters (1000 feet) of elevation. However, the rate of decrease is more severe at higher altitudes (Figure 23-2). In well-trained endurance athletes, $\dot{V}_{O_2max}$ may be impaired at an altitude as low as 580 meters.

The oxygen cost of work is similar at altitude and sea level, but the perception of effort is much greater as altitude increases. As altitude increases,

$\dot{V}_{O_2max}$ decreases. Thus, a given level of work represents a higher percentage of maximum at altitude than at sea level, and exercise becomes more difficult (Figure 23-3). $\dot{V}_{O_2max}$ tends to decrease in direct proportion to the decrease in the oxygen content of arterial blood (C_{aO_2}), which occurs with increasing altitude. There is very little difference among individuals in $\dot{V}_{O_2max}$ at very high altitudes (>6000 m). At the summit of Mt. Everest, for example, the driving pressure of oxygen is so low that people with high or low $\dot{V}_{O_2max}$ at sea level will be the same.

An interesting paradox is that after several weeks of altitude exposure, work capacity greatly increases, yet $\dot{V}_{O_2max}$ does not change. Biochemical changes no doubt account for improved exercise performance in the face of no change in oxygen transport capacity.

Human Responses to Altitude

The dangers of acute exposure to severe hypoxia have been known for over 100 years. In the late 1800s, French physiologist Paul Bert fainted during a balloon ascent to 6000 meters. Bert hypothesized that it was impossible for humans to survive at such high altitudes. However, as more people ascended to higher altitudes, it became apparent that humans could tolerate high altitudes if given time to adjust

(i.e., acclimatize). With proper acclimatization, humans have climbed Mount Everest without using supplemental oxygen. Acute exposure to such an altitude, though, would quickly lead to unconsciousness. Fortunately, humans have a remarkable ability to acclimatize (physiologically adapt) to the stresses of hypoxia (Table 23-2).

Mountain Illnesses Slow ascent to about 5486 meters (18,000 feet) can be accomplished with few adverse symptoms other than diminished exercise capacity, shortness of breath, elevated heart rate, and Cheyne–Stokes breathing at night. (Cheyne–Stokes is an irregular breathing pattern.) If ascent is rapid, however, as when going to a high altitude in a car or plane, acute mountain sickness (AMS) will often appear within two hours. Symptoms include headache, insomnia, irritability, weakness, poor appetite, vomiting, tachycardia, and disturbance of breathing. Above 3000 meters, AMS is common in a significant number of people. Forms of altitude illnesses include AMS, high-altitude

---

TABLE 23-2

Acclimatization to Altitude

Change	Effect
Decreased bicarbonate in CSF and excretion of bicarbonate by kidneys	Increases CO_2–H^+ control of ventilation Shifts oxyhemoglobin dissociation curve to the right
Increased RBC 2,3-DPG	Shifts HbO_2 curve to right
Decreased plasma volume; increased hemoglobin, RBC, and hematocrit	Improved oxygen-carrying capacity of blood
Reduction in resting and submaximal heart rate (below increases of early altitude exposure)	Restoration of more normal circulatory homeostasis
Increased blood pressure	Improved tissue perfusion
Increased pulmonary BP	Improved pulmonary perfusion
Increased pulmonary vascularity	Improved pulmonary perfusion
Increased size and number of mitochondria and in quantity of oxidative enzyme	Improved muscle biochemistry
Increased skeletal muscle vascularity	Improved O_2 transport
Increased tissue myoglobin	Improved cellular O_2 transport
Decreased catecholamine secretion (compared to acute exposure)	Lower lactate production

pulmonary edema (HAPE), high-altitude cerebral edema (HACE), and chronic mountain sickness. Retinal hemorrhage is another high-altitude disorder common above 5000-6000 meters.

Individual differences in the onset of and susceptibility to AMS seem to be caused by an inadequate ventilatory response to high altitude or abnormalities of gas exchange. People with blunted breathing responses to altitude (those whose ventilation fails to increase enough with altitude) are more susceptible to AMS. *Hypoxemia* (low levels of oxygen in the blood) may also disrupt the cellular sodium pump, which may contribute to AMS as well. AMS can be prevented to a certain extent by a slow rate of ascent and the use of acetazolamide (Diamox, 250 mg two times per day or 500 mg slow-release type once daily) or dexamethoasone (4 mg four times per day—should not be used more than 2–3 days). Experimental techniques include administration of nitric oxide and progesterone.

Stresses at Altitude   The effect of altitude on performance and the onset and severity of altitude illnesses are determined by the altitude, speed of ascent, and length of stay. However, other factors as well have severe effects on performance at altitude. Noted altitude researcher Charles Houston has summarized the stress of altitude as the "Four H's": *hypoxia, hypothermia, hypoglycemia* (low blood sugar), and *hypohydration.* Other factors include suppression of the immune system, emotional state, and seasonal and geographical variations in barometric pressure.* Responses to altitude are also affected by degree of exposure—they differ among visitors (sojourners), residents (people who live at altitude but were not born there), and natives.

Altitude stress is much more severe at higher altitudes. The degree of altitude is classified as moderate if less than 12,000 feet (alpine altitudes), high between 12,000 and 18,000 feet, and very high if greater than 20,000 feet. Speed of ascent is consid-

ered rapid if it occurs within a few hours (e.g., if travel is by car or ski tram), fast if it takes place over one or two days, and slow if it takes place over a longer period. During climbs to very high altitudes, adequate acclimatization requires many weeks.

Humans have a limited ability to adapt to high altitudes. The highest permanent altitude dwellers live at 5950 meters at a copper mine on Mt. Aucanquilcha in Chile. Climbers who have attempted to stay at such altitudes for prolonged periods deteriorate quickly—they lose weight rapidly and suffer long-lasting central nervous system impairment. Nevertheless, human capacity for acclimatization to altitude is impressive.

Systemic Responses to Altitude

Altitude exposure causes wide-ranging physiological responses. Changes occur in ventilation, heart rate, acid–base regulation, body composition, and substrate metabolism.

Pulmonary Function   Ventilation increases at rest and during submaximal exercise as altitude increases (Figure 23-4). Ventilation increases further

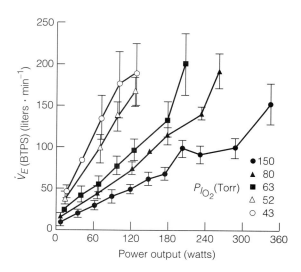

Figure 23-4   Ventilation during exercise at increasing altitude. The oxygen cascade—changes in P_{O_2} with altitude. SOURCE: Sutton et al., 1988.

---

*Barometric pressure is highest near the equator. It tends to decrease during the winter months. Exercise performance at very high altitudes is greatly affected by even small changes in barometric pressure—climbs of peaks such as Mt. Everest are more difficult or impossible during seasonal lows in barometric pressure.

during the first two weeks of exposure to a particular altitude. Hypoxia, by its effect on the aortic and carotid bodies, is the driving force for this increase in ventilation. Bicarbonate is excreted in the urine and there is increased central and peripheral chemosensitivity. At sea level, in addition to central nervous system drive, the most important factors regulating ventilation are P_{CO_2} (see Chapter 12) and H^+, through their effect on the central chemoreceptors located in the medulla. These controls are diminished at altitude because hyperventilation results in a decrease in P_{CO_2}. The practical effect of these changes is that ventilatory stimulation occurs at a lower level of carbon dioxide. Some of the adaptive changes that occur with exposure to altitude are aimed at increasing respiratory responsiveness to P_{CO_2}.

Changes in ventilation in response to hypoxia (i.e., hypoxic ventilatory response, HVR) is one of the most important elements determining successful adaptation to high altitude. Elevated ventilation helps preserve alveolar and arterial oxygen pressure and maximize arterial oxygen content ($C_{a_{O_2}}$). $C_{a_{O_2}}$ (determined by P_{O_2} and hemoglobin content) is the most important factor determining maximal oxygen consumption at altitude. People with a blunted HVR are more susceptible to high-altitude illnesses and poor physical performance at altitude. Fit endurance athletes often exhibit a blunted ventilatory response to exercise at sea level. Although this response seems to be an advantage at sea level (i.e., lower work of breathing), it is a disadvantage at altitude.

Heavy exercise at altitude stresses lung capacity to the limit. Arterial oxygen partial pressure ($P_{a_{O_2}}$) decreases during exercise at altitude. At sea level, maximum exercise $P_{a_{O_2}}$ remains at resting levels, except in some elite endurance athletes. Also, pulmonary gas exchange may be affected by incomplete diffusion or a mismatch between alveolar ventilation and cardiac output due to pulmonary or vascular shunting. Shunting could be caused by bronchoconstriction due to low levels of CO_2, vasoconstriction in the lung circulation, or pulmonary edema.

Diffusion Limitation at Altitude Diffusion across the pulmonary and tissue capillaries can become limiting at altitude. The rate of diffusion largely depends on the movement of gases from a high concentration to a lower one. The partial pressure of oxygen determines the driving force of the gas moving from one place to another. Physiologists call the oxygen driving force from the ambient air to lungs, blood, and cells the *oxygen cascade*. At altitude, this driving force is diminished (the slope of the oxygen cascade becomes flatter; see Figure 23-5) and the rate of diffusion slows.

Gases are composed of individual molecules that can be compressed or expanded. The low pressure of altitude allows the gases to expand, resulting in fewer molecules per unit volume. As discussed in Chapters 11 and 12, the ability to move oxygen from the ambient environment to the bloodstream depends on ventilation (movement of air into the lungs), diffusion (movement of oxygen from the alveoli to the pulmonary capillaries), and pulmonary perfusion (blood flow). With increasing altitude, ventilation is increased greatly to compensate for a decrease in ambient oxygen. However, ventilation cannot keep up with the declining P_{O_2} that occurs with increasing altitude. The inability of ventilation to compensate for decreased inspired P_{O_2} results in a slower rate of diffusion between the alveoli and the pulmonary circulation.

At sea level, the resting P_{O_2} of blood entering the pulmonary capillaries ($P_{v_{O_2}}$) is about 40 mmHg. At rest, blood takes approximately 0.75 second to travel the length of a capillary. Oxygen in the capillaries equilibrates with the oxygen in the alveoli within 0.25 second to reach a new $P_{a_{O_2}}$ of approximately 98–100 mmHg, illustrating the tremendous diffusion reserve of the lungs (Figure 23-6). During heavy exercise, blood travels through the capillaries more rapidly (0.25 second), but still reaches equilibrium with the alveoli in most people (see Chapter 13 for exceptions). Diffusion, then, is not normally a limiting factor of endurance performance in the normal lung at sea level.

Diffusion is severely affected during exercise at high altitudes, however. Although the transit time in the pulmonary capillaries remains at 0.25 second, the driving force for diffusion is much less than at sea level. The diffusion driving force is equal to the

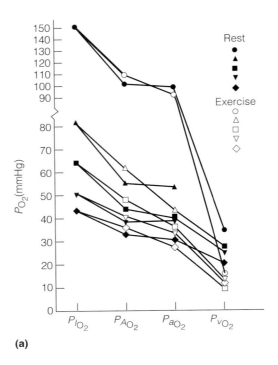

(a)

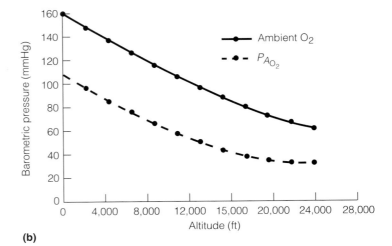

(b)

Figure 23-5   (a) Oxygen tension at rest and with exercise at crucial steps in the oxygen transport cascade from inspired air to mixed venous blood. Closed circles = sea level ($P_{I_{O_2}}$ = 149 torr); closed deltas = $P_{I_{O_2}}$ 80 torr; closed squares = $P_{I_{O_2}}$ = 63 torr; closed inverted deltas = $P_{I_{O_2}}$ 49 torr; closed diamonds = 43 torr. (b) Resting and exercise $P_{a_{O_2}}$ at sea level and high altitude (4300 m). SOURCE: Sutton et al., 1988, p. 130.

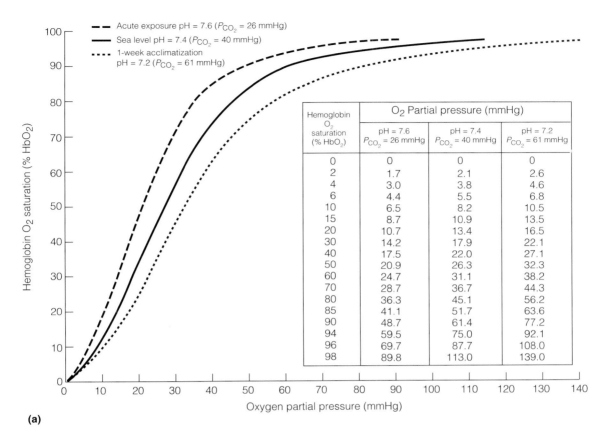

(a)

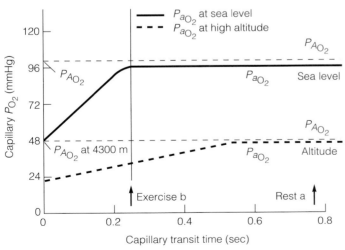

(b)

Figure 23-6   (a) The oxyhemoglobin dissociation curve during acute exposure to altitude and with acclimatization. (b) At rest and exercise pulmonary $P_{a_{O_2}}$ at sea level and high altitude. At rest at sea level and at 4300 m, blood is in the capillary long enough to achieve approximate equilibrium with alveolar O_2. However, during exercise at high altitude, the transit time of blood through the capillaries is too rapid to allow equilibrium, resulting in a diminished $P_{a_{O_2}}$. Thus, pulmonary diffusion is a major factor limiting oxygen transport during endurance exercise at high altitude.

difference between the alveolar ($P_{A_{O_2}}$) and $P_{v_{O_2}}$. As shown in Figure 23-6(b), the diffusion driving force is 48 mmHg at sea level [calculated by subtracting the $P_{v_{O_2}}$ (40 mmHg) from the $P_{A_{O_2}}$ (106 mmHg)] and only 25 mmHg at 4300 meters. This results in a drastic reduction in hemoglobin saturation during exercise that produces a decrease in maximal oxygen consumption.

Cardiovascular Adjustments With exposure to altitude, submaximal heart rate increases and stroke volume stays the same. With prolonged exposure to altitude, stroke volume decreases due to decreased filling pressure (i.e., blood returning to the heart), which declines because of a decrease in plasma volume and an increased hematocrit. After one to two weeks of altitude exposure, exercise cardiac output declines by 20–25%. Oxygen consumption at a particular power output stays the same because of greater tissue extraction (i.e., increased arteriovenous oxygen difference). Cardiovascular changes with acute and chronic exposure to altitude are shown in Figure 23-7.

Mean arterial blood pressure (MAP) gradually increases with exposure to altitude (in spite of decreased cardiac output) due to increased systemic vascular resistance. Compared to that at sea level, vascular resistance in active muscles at altitude acclimatization also increases.* It increases because of increased blood viscosity (due to increased hematocrit) and catecholamine secretion.

Blood viscosity increases with altitude exposure due to an increased number of red blood cells and decreased plasma volume. However, a threshold altitude of approximately 3000 meters seems to be necessary before marked changes are seen in hemoglobin and hematocrit. Increased red cell volume and hemoglobin are triggered by erythropoietin, which is increased quickly with exposure to altitude. However, full acclimatization of blood takes about 12 weeks. At that time, the sojourner's red cell volume and hemoglobin are similar to those of high altitude natives. The time for this adaptation can be

shortened considerably if sojourners receive a diet adequate in energy, protein, and iron.

The heart functions very well, even at very high altitudes. Maximal heart rate decreases at high altitude mainly because the maximum work rate is also less. Cardiac contractility is not affected by altitude. As discussed, stroke volume decreases at altitude because cardiac filling pressure is less. However, in sedentary adults the rate pressure product (heart rate × systolic blood pressure—an indirect measure of myocardial oxygen consumption) has been shown to increase by nearly 100% compared to sea level after exercise at 3000 m and above. Thus, in some travelers and workers exercise at altitude could pose an ischemic challenge to the heart.

Lungs *Pulmonary arterial pressure* (PAP) also increases during altitude exposure. In addition to sympathetic stimulation, PAP increases due to the increased size of the smooth muscle in the pulmonary arterioles, which narrows the arterioles, thus increasing vascular resistance and PAP. Increased PAP has been implicated as a cause of high-altitude pulmonary edema, a high-altitude illness that can be life-threatening.

Brain Blood flow to the brain is usually well maintained at altitude. However, altitude does provide challenges to cerebral circulation that can cause problems. Hypoxemia tends to stimulate vasodilation in brain blood vessels. The tendency toward vasodilation is balanced by *hypocapnia* (low CO_2), which is induced by hyperventilation, which tends to stimulate vasoconstriction. Reflexes from the sympathetic nervous system also dampen vasodilation induced by hypoxemia.

Hypoxemia increases the permeability of cerebral blood vessels during hypoxia. This may contribute to the development of high-altitude cerebral edema, another life-threatening high-altitude illness. Prolonged exposure to very high altitudes can also have long-lasting central nervous system effects. Several measures of CNS function, such as short-term memory, were impaired in climbers during the American Medical Research Expedition to Everest. Some impairment persisted one year after the climb. During the Operation Everest II study (a simulated climb of Mt. Everest in an altitude chamber), all the subjects developed edema and hemor-

---

*Vascular resistance in active muscles decreases when going from rest to exercise at sea level and at altitude. Vascular resistance decreases less during exercise after acclimatization than it does at sea level.

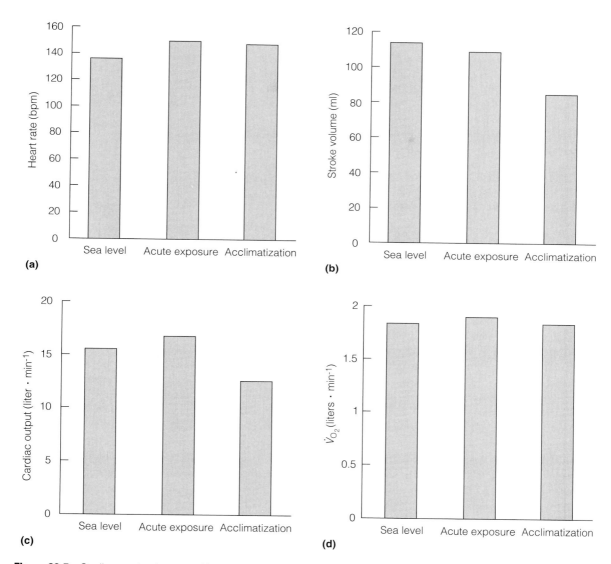

Figure 23-7 Cardiovascular changes with acute and chronic exposure to altitude. (a) Heart rate; (b) stroke volume; (c) cardiac output; and (d) $\dot{V}_{O_2}$ during submaximal exercise. Data from Wolfel et al., 1994.

rhages of the retina and diminished fine motor control, mental acuity, and judgment. The cerebrum appeared to be particularly susceptible to impairment from hypoxemia.

Muscle   Muscle blood flow during exercise is unchanged with acute exposure to altitude. With acclimatization, submaximal exercise muscle blood flow decreases by approximately 20–25%. Oxygen delivery to the muscles is unchanged because of an increased blood oxygen content. The reason for the decreased muscle blood flow during exercise with acclimatization seems to be an increase in norepinephrine secretion as well as decreased cardiac output.

Acclimatization causes no increase in the number of capillaries surrounding muscle fibers. However, capillary density increases because fibers tend to atrophy with chronic altitude exposure—capillaries get closer together. Although increased capillary density improves tissue diffusion, the reason for the increase cannot be considered a positive adaptation. Acclimatization also increases muscle myoglobin, muscle buffering capacity, and to a small extent, aerobic enzymes such as citrate synthetase. This enhances tissue oxygenation and acid-base balance.

Muscle Oxidative Capacity at Altitude

In the past, it has been assumed that altitude exposure results in the same sort of increases in muscle aerobic capacity that accompany endurance training at sea level. However, it is now clear that this is not the case. The first observations of muscle biopsies taken on Sherpas living at altitude in the Himalayas by Kayser and associates (1991) showed significantly less muscle mitochondrial density than in European climbers who were highly trained. Subsequently, the same group (Kayser, 1996) showed that low muscle mitochondrial capacity was a characteristic of Tibetans living at low altitude. In agreement with the Kayser group are Green and associates (1992), who showed that California sea-level residents did not increase muscle mitochondrial content after three weeks of residency at 4000 m altitude on Pikes Peak. Working in the Andes, Desplanches and associates (1996) also showed low muscle mitochondrial densities in a high-altitude native population. Further, the Desplanches group showed that the muscle mitochondrial density of Andean natives could increase with training.

The results of these biopsy studies on men exposed to high altitude have led to a revision in the understanding of muscle metabolism at high altitude. Even untrained muscle has sufficient mitochondria to reach pulmonary $\dot{V}_{O_2max}$ at high altitude. And, though stressful and exhausting, activity at high altitude is performed at low power outputs and metabolic rates. Because activity at altitude is most often limited by breathing and arterial oxygen content, there is no stimulus for the muscle mitochondrial mass to increase.

Exercise Capacity   Maximum strength (maximal voluntary torque) is unaffected by altitude exposure. However, the capacity for repeated contractions becomes progressively impaired at increasing altitude. Fatigue during repeated muscle contractions seems to be caused by decreasing motor drive and biochemical disruption in the interior of the muscle fiber.

Endurance capacity increases with acclimatization—this in spite of no change in $\dot{V}_{O_2max}$. Cellular adaptations are probably responsible for improved exercise capacity.

Acid–Base Balance   Among the most important initial adaptations to altitude are changes in the acid–base balance. These changes help increase ventilation control and facilitate the unloading of oxygen from hemoglobin. Some of these changes are a mixed blessing—although they improve ventilatory control and increase $C_{a_{O_2}}$, they impair blood buffering capacity, which is important during exercise.

During the first week of exposure to altitude, bicarbonate levels decrease in the cerebrospinal fluid and blood. Bicarbonate levels are reduced in the central nervous system by active transport and reduced in blood through excretion by the kidneys. Higher ventilation resulting from altitude exposure decreases P_{CO_2}, which makes the blood more alkaline. Bicarbonate excretion, then, helps normalize the pH of the blood and cerebrospinal fluid, which improves respiratory control at altitude. Bicarbonate excretion has only a limited capacity to control blood pH, however. At increasing altitudes, blood tends to become more alkaline because of hyperventilation and hypocapnia.

The preservation of blood pH affects the binding of hemoglobin and oxygen. Hyperventilation during acute altitude exposure has two conflicting effects: (1) a relative increase in blood P_{O_2} and (2) a shift in the oxyhemoglobin dissociation curve (ODC) to the left (see Figure 23-6). The leftward shift of the curve causes hemoglobin to bind more tightly to oxygen so that a lower P_{O_2} is necessary to release oxygen to the tissues. An added disadvantage is that this lower P_{O_2} decreases the capillary–

tissue diffusion gradient, which slows the movement of oxygen. The bicarbonate excretion that occurs with acclimatization shifts the curve back to the right, restoring the normal binding relationship between oxygen and hemoglobin and increasing the diffusion gradient. This improves oxygenation because more oxygen can be delivered at the same or higher oxygen tension.

Also occurring during the first few days of altitude exposure is a gradual increase in the concentration of 2,3-diphosphoglycerate (2,3-DPG) in the erythrocytes. This compound decreases the affinity of hemoglobin for oxygen and has the effect of further shifting the oxyhemoglobin dissociation curve to the right. The rightward shift of the oxyhemoglobin dissociation curve is much more pronounced in trained than in untrained people (at sea level and altitude). Trained people also exhibit a steeper ODC, which also helps in O_2 unloading. The rightward shift of the ODC is advantageous only to an altitude of approximately 5000 meters. Above this altitude, a rightward shift impairs the ability of hemoglobin to pick up oxygen in the lung and compromises $C_{a_{O_2}}$.

Body Composition and Nutrition

Weight loss and muscle atrophy are common at altitude. Most studies have shown weight losses of approximately $100-200$ g $\cdot$ day^{-1}. Weight losses are due to dehydration, which is due, in turn, to hyperventilation, energy deficit (energy expenditure greater than energy intake), increased activity level, and increased basal metabolic rate (BMR).

Increased ventilation is a leading cause of dehydration at altitude. Expired air is saturated with water vapor. Breathing more causes more water loss. Low relative humidity contributes to insensible water loss and evaporation of sweat (see Chapter 22). Inadequate fluid intake is another important reason for dehydration, particularly at very high altitudes, where water must often be obtained by melting snow.

Energy intake is often depressed at altitude. Appetite is less due to acute mountain sickness and the general malaise common at altitude. In the Pikes Peak study (Butterfield et al., 1992), subjects lost very little body weight during three weeks at 4300 meters because energy intake was carefully matched to energy expenditure. This was a well-controlled study, however, conducted in a high-altitude laboratory. In practice, it has been difficult to match energy intake and expenditure in the field and mountaineers have been unable to adopt lessons of laboratory research. For example, studies of mountaineers climbing above 6000 m found that energy intake was only 70% of energy expenditure. It is no wonder, therefore, that the physical condition of mountaineers deteriorates at high altitude. Nevertheless, the Pikes Peak data suggest that substantial weight loss at altitude is not inevitable.

Physical activity levels typically increase when people go to high altitude. During recreational pursuits, such as skiing, caloric expenditure often exceeds that of a person's normal exercise patterns. In climbers, activity levels are often so difficult that they would be considered extremely high activity levels even if conducted at sea level.

Fecal losses are greater at altitude than at sea level because absorption in the gut is less. Fecal losses may be minimized by ensuring energy intake and keeping the carbohydrate component of the diet high (>60% of caloric intake).

Basal metabolism rate increases during the first few days of altitude exposure. After $4-10$ days, however, BMR decreases but remains higher than at sea level. High-altitude natives typically have a higher BMR than lowlanders.

Exercise Energetics

Lactate Paradox Blood lactate concentration is higher at a given power output during acute exposure to altitude (Figure 23-8). With acclimatization, blood lactate decreases—yet, there is no change in maximal oxygen consumption. In addition, oxygen delivery to active muscles during the same exercise load is the same before and after acclimatization. The phenomenon of decreased lactate during exercise after acclimatization is called the *lactate paradox* because it contradicts the traditional view that hypoxemia results in lactate formation.

Increased lactate production (i.e., lactate appearance) is responsible for increased blood lactate concentration rather than for a decrease in lactate clearance. On acute exposure to altitude, epinephrine secretion increases, probably in response to a decrease in blood oxygen content ($C_{a_{O_2}}$). Epinephrine is a potent stimulator of glycogenolysis (breakdown of glycogen) and results in greatly increased lactate production. Both epinephrine and lactate production increase in concert with acute exposure to altitude (Figure 23-9). With acclimatization, $C_{a_{O_2}}$ increases (due to an increase in red blood cells and decreased plasma volume), causing a fall in epinephrine secretion and lactate production. Norepinephrine levels increase during altitude exposure, peaking after 4–6 days of exposure and then remaining elevated. Several cardiovascular and metabolic adaptations associated with high-altitude exposure, such as elevated heart rate, blood pressure,

and glucose utilization, are related to elevations of norepinephrine and to the increased sympathetic nerve activity it represents.

Measurements of the use of lactate and other fuels used by working muscles and the rest of the body, as well as measurements of circulating catecholamines in men on Pikes Peak, show that part of the responses at altitude that caused Edwards (1936), Hochachka (1989), Reeves and associates (1992), and Kayser (1996) to describe a "Lactate Paradox" at high altitude are due to changes in endocrine responses as well as to changes in the roles played by different tissues over time at high altitude. Initially at altitude, when contractions start, working muscles produce and release large amounts of lactate into the circulation. Additionally, epinephrine stimulation causes glycolysis and lactate production in other, nonworking tissue beds. Then, as exercise continues on acute exposure, working muscles switch to lactate consumption and oxidation as the other tissues produce and release lactate to fuel the working muscles.

After acclimatization to high altitude, the circulating epinephrine concentration is diminished, and nonworking tissues receive less stimulus to produce lactate as a fuel for consumption by the working muscles. In this situation, working muscles initially release less lactate but retain and more readily oxidize it, and they increase their dependence on blood glucose as a fuel.

Carbohydrate Metabolism At high altitude, carbohydrates are thought to be the preferred fuel because they have a high yield of ATP per mole of

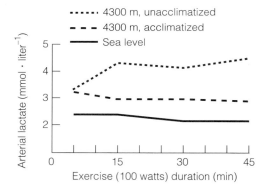

Figure 23-8 Blood lactate (100 W) at sea level, acute exposure, and acclimatization. SOURCE: Reeves et al., 1992. Used with permission.

Figure 23-9 The relationship between blood lactate and epinephrine. SOURCE: Reeves et al., 1992. Used with permission.

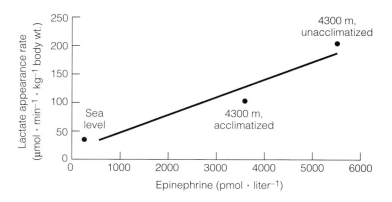

oxygen. However, regardless of altitude, relative exercise intensity is the major determinant of metabolic fuel selection. A basic problem with carbohydrates is that the oxygen-saving advantage of using them as fuel at altitude is counterbalanced by limited carbohydrate stores.

Hypoglycemia and decreased liver glycogen content are common findings at altitude. This appears to be partially due to inadequate energy intake, and possibly impaired intestinal absorption of glucose, increased glycogen breakdown, and increased rate of cellular glucose utilization. In the Pikes Peak study, glycogen breakdown decreased and physical well-being of the subjects improved with increased carbohydrates in the diet. High carbohydrate intake is essential at altitude because altitude exposure increases glucose use during rest and submaximal exercise (carbohydrates are the main fuel during maximal exercise at sea level or altitude). Nevertheless, with acclimatization, blood glucose decreases during prolonged exercise—glucose clearance increases, probably because of increased levels of norepinephrine.

Insulin metabolism seems to be unaffected at altitude. Glucose tolerance (blood glucose response to an oral glucose load) is less at altitude, probably because of impaired glucose absorption in the gut rather than because of an impairment in insulin metabolism.

Fat and Protein Metabolism Fat catabolism increases at altitude if diet is inadequate. Gluconeogenesis increases when loss of muscle mass and negative nitrogen balance (net loss of body protein) are significant.

By means of arterial-venous difference measurements of glucose, glycerol, and free fatty acids (FFA) in the legs of men at 4300 m altitude on Pikes Peak, Robert and associates (1996, 1996) clearly showed that working muscle prefers carbohydrate-derived fuel sources to the exclusion of fat at high altitude. After acclimatization, not only is FFA uptake by working muscle diminished, but so is glycerol release. This latter result means that neither blood-borne FFA nor intramuscular triglycerides are fuels for acclimatized working muscles. The work of Roberts and Butterfield and associates clearly shows the difficulties encountered by moun-

taineers. During hard exercise and at high altitude, there is a crossover to carbohydrate-derived energy sources in muscle. However, if dietary carbohydrates and total energy supply is diminished at altitude, the body must resort to using its own lean and adipose tissue reserves. For example, amino acids from protein breakdown are used to maintain blood glucose via gluconeogenesis. These are the least preferred fuels for muscle, and their use makes for poor exercise performance and growing weakness over time.

Changes in Muscle Energetics with Acclimatization Other than an increased $C_{a_{O_2}}$, there are few significant measurable changes in muscle energetics with acclimatization. Levels of adenosine triphosphate, creatine phosphate, muscle, and liver glycogen, oxidative and glycolytic enzymes remain largely unchanged with acclimatization, provided subjects consume adequate calories.

Competitive Athletics at Altitude

Altitude causes marked improvements in events of short duration and high intensity (sprints and throwing events) and deterioration in events of long duration and lower intensity (endurance events). The effect of gravity decreases by 0.3 cm · sec² for every 1000 meters of altitude, and wind resistance decreases with decreased air density. Both factors are beneficial at altitude for athletes involved in short-term maximal events. Table 23-3 compares the Olympic performances of the top three athletes in selected sprint events at the Mexico City Olympics (1968) with their previous personal bests. High-altitude performances were much better in almost every instance. In the long jump and triple jump (not shown on table), for example, the world records were surpassed by margins of 21¾ and 14¼ inches, respectively. The long jump record set at altitude in Mexico City was not broken for more than 20 years at sea level. In sprint events, some of the energy cost stems from overcoming air resistance, which is less at altitude.

In throwing events, performances can be helped or hindered by altitude. In events such as the discus or javelin, air mass provides lift to the implements, so performance tends to be hampered. In the shot

TABLE 23-3

Comparison of Personal Best and Mexico City Olympic Performances of Selected Sprint Athletes

Event, Placement, and Athlete's Name	Mexico City Olympics Time (sec)	Previous Personal Best Time (sec)
100 m—Men		
1st J. Hines (U.S.A.)	9.9	9.9
2nd L. Miller (Jamaica)	10.0	10.0
3rd C. Green (U.S.A.)	10.0	9.9
100 m—Women		
1st W. Tyus (U.S.A.)	11.0	11.1
2nd B. Farrell (U.S.A.)	11.1	11.2
3rd I. Szewinska (Poland)	11.1	11.1
200 m—Men		
1st T. Smith (U.S.A.)	19.8	19.9
2nd P. Norman (Australia)	20.0	20.5
3rd J. Carlos (U.S.A.)	20.0	19.7[a]
200 m—Women		
1st I. Szewinska (Poland)	22.5	22.7
2nd R. Boyle (Australia)	22.7	23.4
3rd J. Lamy (Australia)	22.8	23.1
400 m—Men		
1st L. Evans (U.S.A.)	43.8	44.0
2nd L. James (U.S.A.)	43.9	44.1
3rd R. Freeman (U.S.A.)	44.4	44.6
400 m—Women		
1st G. Besson (France)	52.0	53.8
2nd L. Board (Great Britain)	52.1	52.8
3rd N. Burda (U.S.S.R.)	52.2	53.1
400 m Hurdles—Men		
1st D. Hemery (Great Britain)	48.1	49.6
2nd G. Hennige (Federal Republic of Germany)	49.0	50.0
3rd J. Sherwood (Great Britain)	49.0	50.2
110 m Hurdles—Men		
1st W. Davenport (U.S.A.)	13.3	13.3[a]
2nd E. Hall (U.S.A.)	13.4	13.4[a]
3rd E. Ottoz (Italy)	13.4	13.5

[a] Previous personal best set at altitude

put and hammer throw, where the implements have minimal aerodynamic characteristics, performances improve marginally because of the decreased air resistance and effect of gravity.

At altitude, there is a decline in performance in running events greater than 800 meters. At the Mexico City Games, many athletes who dominated distance running at sea level were soundly defeated by athletes native to high altitude. High-altitude natives placed first or second in the 5000-meter run, 3000-meter steeplechase, 10,000-meter run, and the marathon.

Athletes who must compete at altitude benefit from a period of acclimatization of from one to twelve weeks. Athletes involved in activities of short duration, such as sprints, jumps, and throws need only acclimatize long enough to get over the effects of mountain sickness. Although adjustments in the acid–base balance take less than a week, changes in oxygen-carrying capacity can take many months. Athletes in championship form may risk losing their peak condition by too much acclimatization because they are unable to train as hard at altitude as at sea level.

Controversy exists over the effects of altitude training on subsequent performance at sea level. Most studies show no improvement in maximal oxygen consumption or maximal work capacity when returning from altitude. In the studies that show an improvement, the subjects may not have been in good condition to start with. At altitude, they improved their exercise capacity, but the improvement was no greater than they would have achieved by training at sea level.

The physiological adaptations to altitude are not necessarily beneficial at sea level. Although the increase in hemoglobin is probably helpful, the decreases in plasma volume and alkaline reserve (bicarbonate, HCO_3^-) are decided disadvantages. During high-intensity exercise, the decrease in HCO_3^- may result in a decreased lactate efflux from muscle to blood, leading to a decreased pH in muscle and, perhaps, an earlier onset of fatigue. Also, the decreased blood volume and increased hematocrit increase blood viscosity, which may have negative effects on oxygen transport capacity. Finally, the increased ventilatory response at altitude is counterproductive at sea level because it increases the work of breathing and could cause fatigue.

Training and intensity and duration are the most important factors in improving exercise performance—and athletes can't train as hard at high altitude. Even though they can reach the same relative percentage of maximum, their maximum is less. Nevertheless, some of the world's premier endurance coaches and athletes strongly believe in altitude training for competition at sea level. Also, high-altitude natives from countries such as Kenya have been remarkably successful at sea level. So, the possible benefits of high-altitude training cannot be ruled out.

Several studies have suggested that athletes "live high but train low" (LHTL)—that is, they should live at high altitude but train as close to sea level as possible. Theoretically, athletes who live at high altitude would increase oxygen delivery and respiratory extraction while maintaining the training intensity of sea level. A recent study of triathletes by Liu and coworkers (1998) found that LHTL improved left ventricular contractility. They hypothesized that this adaptation might be associated with increased β-adrenergic receptor density or an improved myocardial energy utilization.

Many coaches are adamant about the benefits of high-altitude training (training and living at altitude) for improved sea-level performance. While there may be an unknown physiological mechanism supporting their position, the benefit may be due to the increased focus athletes might get from training in a remote location.

If the LHTL technique is effective, the ideal training altitude appears to be between 2200–3500 m because red cell mass increases at that altitude. However, the elevation is not so high as to cause acute mountain sickness in most athletes. The problems and benefits of high altitude training are addressed more in Chapter 21.

■ Humans in High-Pressure Environments

Exposure to high-pressure environments, called *hyperbaria*, ranges from underwater breath holding to industrial and experimental saturation diving. Human exposure to high-pressure environments is mainly restricted to underwater diving. The scuba system (SCUBA—self-contained underwater breathing apparatus), invented by Jacques Cousteau and Emile Gagnon in 1943, has allowed millions of people to experience the beauties and challenges of the underwater environment.

The popularity of sports diving and ignorance of hyperbaric physiology have resulted in many accidents. High-pressure environments offer some extreme physiological challenges because gases can become toxic or narcotic, and rapid changes in pressure can sometimes be extremely traumatic or fatal. These dangers can be minimized with a consideration of the physics of gases and their physiological effects under increased pressure.

Divers encounter increasing pressure as the depth of a dive increases. Pressure increases by 1 atmosphere (760 mmHg) for every 10 meters (33 feet) of depth (Figure 23-10). Although the pressure subjects divers to increased air density and partial pressures, they have the added difficulty of exercising in an aqueous environment. There, they must contend with cold water, poor visibility, impaired communications, currents, restrictive equipment, and dangerous marine animals.

Physiological Effects of Exposure to High Pressure

The principal dangers of the hyperbaric environment stem from changes in gas volumes within enclosed spaces and the increased solubility of gases. Medical problems include barotrauma, gas toxicity, and decompression sickness (Table 23-4). Medical contraindications to diving are listed in Table 23-5.

Barotrauma   *Barotrauma* is tissue injury caused by changing pressure. The human body has a limited ability to distend and compress, and trauma results when these limits are exceeded. The volume of a gas decreases or increases as a diver descends or ascends (Boyle's law). On descent, volume decreases in enclosed spaces such as the middle ear, lungs, sinuses, mask, or teeth, creating a negative pressure. If the pressure is not allowed to equalize

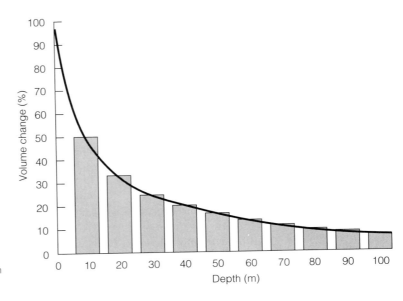

Figure 23-10 Percentage change in volume with increasing depth.

TABLE 23-4
Medical Problems in Scuba Diving

I. **Barotrauma**
  A. Barotrauma of descent
      Ears
      Sinuses
      Teeth
      Lungs
      Equipment (mask and suit)

  B. Barotrauma of ascent
      Ears
      Sinuses
      Teeth
      Lungs
      Suit

II. **Gas toxicity**
  A. O_2 toxicity

  B. Inert gas narcosis
      Nitrogen
      Helium

  C. Compressed air impurities

  D. CO_2 intoxication

  E. CO poisoning

III. **Decompression sickness**
  A. Joints, muscle, bone

  B. Respiratory system

  C. CNS

TABLE 23-5
Contraindications to Diving

Body System	Contraindications
Eyes, ears, nose, and throat	Vision <20/200 Perforated eardrum Sinusitis or severe allergy Residual or prior ear, nose, or throat surgery
Cardiovascular	Recent myocardial infarction Syncope- or arrhythmia-prone Hypertensive on therapy Cardiac murmurs
Pulmonary	Obstructive airway disease (asthma, COPD, bullae) Prior thoracotomy History of pneumothorax Recurrent pneumonia
Gastroesophageal	Severe gastroesophageal reflux
Musculoskeletal	Chronic disk disease
Central nervous system	Seizure disorder or syncope
Metabolic	Unstable diabetes mellitus
Other	Drug use or heavy smoking Pregnancy Emotional instability or proneness to accidents

SOURCE: Spivack, 1980.

with air from outside the space, tissue injury can result. On ascent, the volume in these areas increases, which can also cause injury if gas is not allowed to escape.

Ordinarily, descent barotrauma of the ears, teeth, or sinuses is prevented by pain, which is relieved by ascending. However, if descent occurs too rapidly, the tympanic membrane or round window in the ear can rupture, sinuses can fill with blood, or teeth may collapse. Preventive measures, such as slow descent, can minimize the risk of these problems. Patency of the eustachian tube, which is essential for equalizing pressure in the inner ear, is improved by increased muscular control or by decongestant medication (that does not cause drowsiness), such as Sudafed. Mask squeeze (intense pressure of the mask against the face) can be easily remedied by clearing the mask as negative pressure builds up. Individuals with serious nasal obstructions or colds should probably not dive because of the increased difficulty of equalizing pressure.

Pulmonary descent barotrauma (pulmonary squeeze) occurs when the lung volume decreases to the point where negative intrathoracic pressure is created, which increases pulmonary blood volume to the point of pulmonary capillary hemorrhage. This phenomenon can occur in breath-hold diving at extreme depths or in standard diving when the gas supply does not keep pace with the rate of descent. Pulmonary squeeze places a definite limit on the depths that can be achieved in breath-hold diving.

The leading cause of death in scuba diving is pulmonary ascent barotrauma. This occurs when the diver holds his or her breath while ascending, which results in overdistension and rupture of the lungs as the gas volume expands. The condition can occur at a surprisingly shallow depth: A diver whose lungs are filled to capacity at a depth of only 4 feet will experience lung rupture if he or she fails to expire while ascending to the surface. Problems associated with this serious condition include pulmonary tissue damage, surgical emphysema (gas escaping into interstitial pulmonary tissues), pneumothorax (gas entering the pleural cavity), and air embolism (gas entering the pulmonary veins and systemic circulation).

Gas Toxicity   Gases such as carbon monoxide, oxygen, carbon dioxide, nitrogen, and helium can be toxic or narcotic to a diver under certain circumstances. Sport divers depend on the compressed air they carry on their backs, and a contaminated air supply can lead to impaired performance and can be life-threatening.

Carbon Monoxide   Compressed air may contain CO if the air intake of the compressor was contaminated with an impure air source such as auto exhaust. The danger of CO lies in its great affinity for hemoglobin, which is 200 times greater than that of oxygen. CO poisoning leads to hypoxia, which results in progressive impairment of judgment and psychomotor performance, headache, nausea, weakness, dizziness, impaired vision, syncope, and coma. This problem is compounded if the individual has chronically high levels of CO from cigarette smoking or exposure to air pollution. Other contaminants in the compressed air include dust, oil vapor, and carbon dioxide. By removing oxygen from compressed air, rust may also be a source of contamination.

Oxygen Toxicity   Oxygen at high pressure is toxic to all life forms, with the degree of toxicity dependent on its concentration and the length of exposure. Physical exercise speeds its development. The principal sites of oxygen toxicity are the lungs and central nervous system. Pulmonary symptoms include substernal distress with soreness in the chest, usually accompanied by airway resistance on inspiration, histological changes in the alveoli, pulmonary edema (resembling pneumonia), flushing of the face, and a dry cough that eventually leads to a wet cough. Central nervous system (CNS) symptoms include nausea, contraction in the field of vision, convulsions, muscle twitching and spasm, lack of sphincter control, unconsciousness, and death. Oxygen toxicity can also cause cardiac arrhythmias. Factors increasing the risk of oxygen toxicity include increased exposure time to compressed air, deeper dives (>30m), the use of oxygen-enriched compressed air (i.e., nitrox), exercise, CO_2 retention, cold stress, and diseases that increase the metabolic rate, such as thyroid diseases.

The mechanism for oxygen toxicity is unknown. Hyperbaric oxygen may interfere with CO_2 transport. At high pressure, more oxygen is dissolved in blood, so the hemoglobin doesn't desaturate and become available for CO_2 transport. This may cause acidosis and increase the P_{O_2} in the brain because of the vasodilation effect of increased CO_2 on cerebral blood vessels. Also, high oxygen concentrations are known to interfere with several metabolic enzymes; such interference would disrupt cell function. High oxygen level also inhibits the action of g-amino gluteric acid, which is important for neural transmission in the CNS. The popularity of nitrox diving— that is, diving with oxygen-enriched compressed air—has increased the risk of oxygen toxicity in recreational divers.

Carbon Dioxide Toxicity CO_2 toxicity is most common in closed-circuit scuba systems and hose-supplied diving helmets, but can also occur in conventional scuba. Hypercapnia occurs when there is an inadequate respiratory exchange, and it has been shown to occur at depth during heavy exercise. Divers often try to suppress their ventilation in an attempt to conserve air, which can lead to a buildup of CO_2 that can have severe consequences. Symptoms of CO_2 toxicity include increased respiratory stimulation, dyspnea (uncomfortable breathing), headache, mental deterioration, violent respiratory distress, unconsciousness, and convulsions. Usually dyspnea will prevent the more serious consequences of CO_2 buildup. However, this may not be the case in an emergency, when a diver may not have the opportunity to surface or rest.

Nitrogen Narcosis Several gases also exert a narcotic or anesthetic effect at high pressure. The anesthetic effect of a gas depends on its partial pressure and its solubility in the body's tissues and fluids. Nitrogen can cause a condition called *nitrogen narcosis* beginning at about 30 meters (100 feet). Progressive symptoms of nitrogen narcosis include euphoria, impaired performance, weakness, drowsiness, and unconsciousness. It is caused by interference of the transfer of signals across neural synapses. This condition limits the use of compressed air to about 50 meters (165 feet).

Helium, because it is less anesthetic, usually replaces nitrogen on deep dives. However, helium breathing below 150 meters can cause a neuromuscular disorder called *high-pressure nervous syndrome* (HPNS), which is characterized by tremors, vertigo, and nausea. Researchers have attempted to prevent HPNS by adding nitrogen to the helium–oxygen mixture. Slowing down the rate of compression during the dive can also help prevent HPNS.

Nitrogen narcosis is almost impossible to avoid in commercial divers and is a limiting factor during deep dives. Nitrogen narcosis slows down information processing in the brain without distorting perception. Divers can improve performance and safety by slowing down their activities. Divers who do so commit fewer errors and improve efficiency.

Decompression Sickness (The Bends) Decompression sickness is caused by nitrogen bubble formation in the tissues as a result of a rapid ascent. It can occur in divers coming from the depths to the surface or in aviators in unpressurized aircraft going to great heights. Symptoms include itching of the skin; fatigue; pain in the muscles, joints, and bones; perspiration; nausea; and, less frequently, respiratory distress, ataxia, vascular obstruction, paralysis, unconsciousness, and death. When the bends affects the lungs, it is sometimes called "the chokes," and when it affects the central nervous system, it is sometimes called "the staggers." Although the symptoms usually appear about one hour after coming to the surface, they can occur either immediately or as long as 12 hours later.

A gas is absorbed into the tissues in proportion to its partial pressure and solubility (Figure 23-11). The uptake is exponential, with some tissues, such as blood, absorbing the N_2 rapidly and other tissues, such as fat, absorbing the N_2 more slowly (Figure 23-12). On decompression (going to the surface or to a shallower depth), N_2 leaves the tissue in proportion to the decrease in barometric pressure. It takes longer for N_2 to leave the slower tissues such as fat. If decompression occurs too rapidly, N_2 returns to the gaseous state and bubbles form in the blood and tissues.

Decompression tables, or dive tables, help divers safely reestablish surface nitrogen absorption levels (Figure 23-13). In general, decompression time increases with the length and depth of the dive.

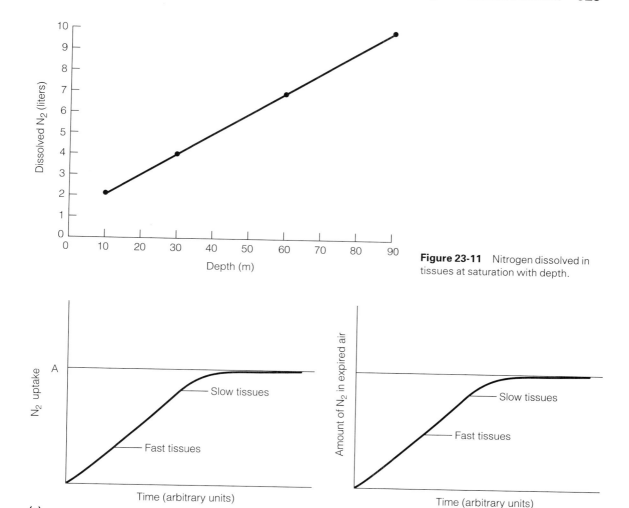

Figure 23-11 Nitrogen dissolved in tissues at saturation with depth.

(a)

(b)

Figure 23-12 The uptake and elimination of nitrogen follows a two-component curve.

Although these tables provide a margin of safety, factors such as percent body fat, age, physical condition, gas mixture, and altitude of the dive also affect decompression time. Repetitive diving or flying (airplane, hang glider, etc.) soon after a dive also affects the accuracy of the table.

The advent of dive computers has made the use of dive-table data practical and has increased the safety of scuba diving. In contrast to dive tables, which assume an entire dive was spent at maximum depth, dive computers can calculate decompression times during multilevel dives. They accomplish this by calculating nitrogen uptake and elimination at different depths and during repetitive dives. In spite of the advantages of dive computers, however, dive tables remain in use. Computer manufacturers recommend that divers bring printed tables on a dive as a backup in case of computer loss or failure.

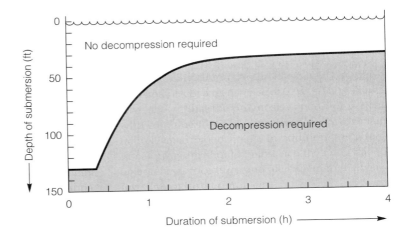

Figure 23-13 Decompression requirements in diving.

Nitrox Nitrox is an oxygen-enriched gas mixture that is used to prevent nitrogen narcosis or prolong diving times. Nitrox diving requires special training because use of the gas can lead to oxygen toxicity, which can cause a convulsion. During a convulsion, the regulator and mouthpiece may fall out, thus increasing the risk of drowning. Another problem with nitrox diving is that there are potential dangers associated with mixing gases, such as possible explosions and gas contamination. Many of these risks are minimized among commercial divers who use sophisticated gas mixers and diving helmets or band-masks with open-circuit demand regulators that stay on the diver in the event of convulsions.

Hyperbaric Exercise

Hyperbaric exercise studies are typically conducted in a hyperbaric chamber or underwater. The hyperbaric chamber provides the opportunity to isolate variables such as oxygen partial pressure, temperature, and gas mixtures, while the underwater experiments provide a more realistic environment. The hyperbaric chamber sometimes contains wet-pots (small pools about 3 to 4 feet deep) that are used to simulate ocean dives. It is important to understand the differences between these environments when making generalizations about hyperbaric exercise. Subjects in a hyperbaric chamber are

exposed to high pressure but not to the problems of buoyancy, fluid viscosity, temperature, airway obstruction, and awkwardness common underwater.

The cycle ergometer is the most common exercise device used in hyperbaric chambers and underwater experimental sessions. It is useful because results can be compared with the many bicycle ergometer studies done at sea level. However, because cycling is not a common method of transportation underwater, more specific exercise tests have been devised. For example, a diving ergometer has been developed that provides a reproducible resistance to the swimming diver (Figure 23-14). A standard exercise protocol can be administered by regulating swimming speed and resistance. Other researchers make physiological measurements with divers swimming at various speeds while tethered and connected to a strain gauge.

Biological measurements are exceedingly difficult in the hyperbaric environment. Measurements in hyperbaric chambers require expensive equipment and facilities and are technically exacting. Open-water measurements are complex because of the restraints imposed by the aqueous environment. Early open-water studies used pulse rate to study cardiovascular function and trapped air bubbles to study diving energetics. Metabolic systems have been developed to accurately measure oxygen consumption at different depths and exercise intensities during the same dive (Figure 23-15). These sys-

Figure 23-14   A diving ergometer developed by Pilmanis during the 1970s. Courtesy J. Dwyer.

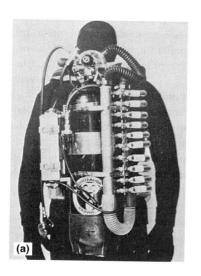
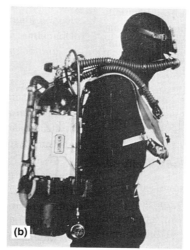

Figure 23-15   A device for monitoring $\dot{V}_{O_2}$ in a free-water diver. It was used in classic scuba energetics studies. Courtesy J. Dwyer.

tems are used with a recording system that allows simultaneous measurement of the electrocardiogram, ventilation, breathing frequency, and tidal volume during an open-water dive. One simple technique for estimating oxygen consumption and energy cost during scuba diving is to measure the difference in tank pressure at the beginning and end of a dive. A device for measuring oxygen consumption during shallow underwater swimming is a surface metabolic system connected to the diver who swims at a paced velocity.

Many factors add to the difficulty of exercising underwater—increased air density, cold, decreased efficiency, CO_2 retention, and inert gas narcosis.

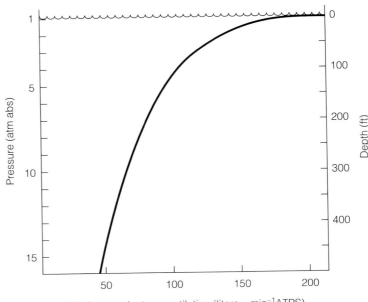

Figure 23-16 Maximum voluntary ventilation at depth. SOURCE: Modified from Miller et al., 1971.

Ventilation may be a limiting factor of physical performance in diving. Maximal voluntary ventilation decreases with depth, which results in a progressively smaller difference between exercise ventilation and maximal breathing capacity during heavy exercise (Figure 23-16). The higher density of air increases the flow resistance in the scuba equipment and airways, resulting in hypoventilation. This leads to increased retention of CO_2, increased work in breathing, and dyspnea. The ability to increase expiratory flow rate is limited. After reaching maximum flow rate, further effort only results in partial airway collapse. Maximum expiratory flow rate seems to be independent of effort at depth (Figure 23-17).

Although some studies have shown that the dyspnea accompanying heavy physical activity does not always decrease exercise capacity, it can be extremely dangerous during heavy exercise in an emergency. Heavy exercise in the presence of CO_2 retention increases the risk of CO_2 intoxication, which is compounded in the presence of nitrogen narcosis. Intense exercise while using nitrox might also increase the risk of oxygen toxicity. At depths

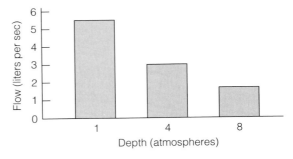

Figure 23-17 Respiratory effort and flow rate at depth. At increasing depth, increased air density will increase the work of breathing.

greater than 180 feet, exercise is made possible by using helium–oxygen mixtures, which have a lower density and lower anesthetic quality.

The energy cost of underwater swimming increases with the velocity of the swim, and at a given speed, it is determined by the ratio of active drag and overall net mechanical efficiency of the swimmer. The oxygen cost of swimming underwater is less for women than men (due to greater buoyancy).

Added equipment, such as double tanks, cameras, and dry suits, increases oxygen consumption, particularly at faster swimming speeds. The energy cost of swimming per kilometer actually decreases at faster speeds. Pendergast and associates (1996) found that for swimming longer distances underwater, one could conserve energy by swimming at 40 or 50 m · min^{-1} compared to 30 m · min^{-1}.

Most studies show that oxygen consumption increases in submaximal work with increasing depth (Figure 23-18). Depth affects $\dot{V}_{O_2}$ because of the increased energy costs of breathing, of maintaining body temperature in colder water, and of moving in higher hydrostatic pressures. Water becomes cooler with depth. Because wet suit material is compressed with greater pressure, thus decreasing insulation, oxygen consumption may be increased (due to shivering or increased muscle tone) to maintain body temperature. The increased hydrostatic pressure may, in turn, hamper mobility by its effects on wet suit compression and by effectively increasing the viscosity of the surrounding water.

Experienced divers can work at as high as 91% of their land-measured maximal oxygen consumption. However, the effective work that a diver can perform is much less because of the greatly reduced efficiency. Probably the most important factors dictating maximum capacity are tolerance to high levels of carbon dioxide and percentage of maximal oxygen consumption attained before reaching critical P_{CO_2}.

Swimming angle and the drag produced by the scuba equipment greatly affect the energy cost of swimming underwater. For example, swimming at a moderate speed of 30 m · min^{-1} in a partial feet-down position increases the oxygen cost by 30%. This can easily occur if the buoyancy compensator (BC vest) or weight belt is improperly adjusted. Individual differences in swimming efficiency can also affect energy costs.

The type of swim fin worn by the diver can affect propulsive force and oxygen consumption. The greater the surface area of the fin, the lower the oxygen consumption during swimming. Flexible fins allow divers to swim faster (lower oxygen cost), but propulsive force is greatest with more rigid fins. Women swim fastest with medium size, rigid fins, while men swim fastest with larger, rigid, non-vented fins.

The land-measured relationship between oxygen consumption and heart rate cannot be used in diving, and doing so can be potentially dangerous for the sports diver. Heart rate decreases as water temperature decreases and pressure increases, a phenomenon called *diving bradycardia*. Individual differences in swimming efficiency also affect heart rate. Heart rate can be used to estimate the energy cost in diving only when the heart rate–$\dot{V}_{O_2}$ relationship is known for a particular diver, at a particular depth.

Thermal Effects of Diving and Muscle Performance Cold water and heat loss pose significant challenges to divers. Heat transfer is about 200 times greater in water than air; so, even diving in tropical waters can cause significant heat loss. While exercise can prevent heat loss, intermittent exercise can actually increase heat loss by transiently widening the thermal gradient. Body temperature increases during exercise which increases the temperature difference between the diver's body and the water (i.e., the thermal gradient). Consequently, heat loss increases when the diver rests. By the same mechanism, exercising prior to diving will increase heat loss.

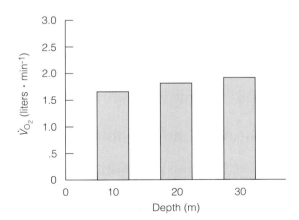

Figure 23-18 The effects of depth on oxygen uptake during a swim at 30 m · min^{-1}. From the data of J. Dwyer.

Strength decreases when muscle temperature lowers below 25°C. Biomechanically, physical tasks are more difficult underwater. Whole-body rotational movements are impaired between 20% and 30%. Pushing capacity decreases by 50%, while pulling capacity is unchanged. Handgrip strength is maintained when using gloves, but manual dexterity is greatly impaired.

Cold water dives can cause dyspnea and cardiac arrhythmias and are associated with a higher incidence of cardiac-related sudden deaths than are dives in warm water. The hands and feet are also at risk of thermal injury because they have very little heat generating capacity due to their small muscle mass.

Commercial Diving   Commercial divers are involved in a myriad of tasks that are vital to the industry. These tasks include underwater photography, salvage, search and rescue, inspection and repair, construction, maintenance, and scientific support. *Scuba* is the most familiar diving method but is used much less than other methods in commercial diving. *Surface demand diving* is commonly used at depths greater than 50 meters. The diver is connected to reinforced gas hoses and air is supplied from the surface.

Saturation diving involves prolonged exposure (up to 28 days) to hyperbaria and uses helium–oxygen gas mixtures. It is the most commonly used method during deep dives, below 50 meters. Divers become totally saturated with inert gases after 24–36 hours. After that, further exposure does not require any additional decompression time. When not working, divers live in a decompression chamber.

Atmospheric diving systems are becoming increasingly popular with commercial divers. Divers work in self-contained "submarines" that maintain normal atmospheric pressure. They use robot arms to manipulate objects outside the system.

Breath-Hold Diving   *Breath-hold diving* was the earliest human exposure to the hyperbaric environment. It is of interest to people involved in athletics and exercise physiology because of the popularity of snorkeling, the many associated deaths that occur each year, and the relatively recent attempts to set breath-hold diving performance records.

Breath-hold diving is dangerous if the dive is preceded by vigorous hyperventilation, which results in a decrease in carbon dioxide, thus depressing the urge to breathe. The diver may faint before the buildup of carbon dioxide is again high enough to force him or her to the surface to take a breath. A diver may suffer shallow water blackout on ascent when P_{O_2} in the alveoli and blood fall. Hyperventilation has little effect on hemoglobin saturation. It only serves to suppress the CO_2 ventilatory drive. Hyperventilation followed by exercise during the breath-hold dive may compound the dangers because of the more rapid rate of oxygen consumption—the combination results in depressed ventilatory drive and more rapid depletion of oxygen in the blood.

An added danger during breath-hold diving is an increased incidence of cardiac arrhythmias, which increases if the water temperature is low. This is related, in part, to the previously mentioned diver's bradycardia. Its effect on the death rate in breath-hold diving has not been determined.

The depth limits of the breath-hold diver are determined by the ratio of total lung capacity to residual lung volume. During descent, lung volume decreases until reaching residual lung volume. After that point, negative pressure is created in the thoracic cavity, drawing blood into the intrathoracic circulation until the capillaries rupture. Champion breath-hold divers, such as Robert Croft and Jacques Mayol, have extremely large total lung capacities with low residual lung volumes, which partially explains their exceptional abilities.

■ Exercise and Air Pollution

Air pollution is common in urban centers throughout the world. Smog can have a serious effect on physical performance, particularly in people suffering from pulmonary disorders, such as asthma and chronic obstructive pulmonary disease. The effects of polluted air are often additive to other environmental stressors, such as heat (Los Angeles or New York) and altitude (Denver or Mexico City).

Air pollutants directly emitted from automobiles, energy plants, and so on are classified as primary, while pollutants derived from the interaction of primary pollutants and ultraviolet rays or other agents found in the environment are called secondary. Examples of primary pollutants include carbon monoxide, nitrogen oxide, sulfur dioxide, and particulate matter. Secondary pollutants include ozone, peroxyacetyl nitrate, aldehydes, and nitrogen dioxide.

Studies of smog and exercise have tended to examine the effects of one substance at a time, including ozone, peroxyacetyl nitrate, nitrogen dioxide, sulfur dioxide, particulates, and carbon monoxide. Few studies have examined the effects of combinations of these substances. At the same time, few data are available on the effects of smog on elite athletes, although generalizations may be possible by extrapolating from studies of smog effects on young people.

Smog may negatively affect exercise performance because it causes tightening in the chest, difficulty in taking a deep breath, eye irritation, pharyngitis, headache, lassitude, malaise, nausea, and dryness of the throat. The effects of air pollution are exaggerated during physical activity (ventilation is greater, so more smog enters the pulmonary system) and become increasingly severe as the level of air pollution increases. Although most of these factors have little effect on measurable physiological factors, such as $\dot{V}_{O_2max}$, they certainly have a negative psychological impact. Studies of athletes in competition show marked decrements in performance when the air is significantly polluted. Increased airway resistance in the lungs may lead to discomfort, which limits the athletes' motivation to perform.

Smoking compounds the negative effects of smog. Carbon monoxide (CO), which is found in smoke and polluted air, competes with oxygen for hemoglobin in the blood, resulting in a decrease in oxygen transport capacity. CO also increases resting and submaximal exercise heart rates and reduces functional capacity.

Particulate air pollutants, such as dust and pollen, can have serious effects on the exercise performance of allergy-prone athletes. Many great athletes have suffered from the effects of airborne particulate matter to such an extent that their athletic careers were seriously compromised. Antihistamines (drugs that combat allergic reactions) can cause marked drowsiness and can hurt performance. Some people benefit from desensitization procedures available from allergy specialists.

During peak periods of air pollution concentration, practices and competitions should be limited. Air pollution levels tend to be higher during the commute hours (6–9 A.M. and 4–7 P.M. in most places) and during the middle of the day, when the temperature is highest. Air pollution is a fact of life in many areas, and there is no evidence that the human body can acclimatize to it. For this reason, attempts should be made to limit exposure, particularly in people with respiratory problems.

■ Biological Rhythms and Travel Across Time Zones

Many physiological functions follow a rhythm that varies approximately every 24 hours (circadian), weekly (circaseptian), monthly (circalunar), and yearly (circa-annual). Functions that exhibit a regular rhythm include sleep, body temperature, heart rate, blood pressure, metabolic rate, gonadotropin hormone secretion (menstrual cycle), and performance characteristics such as strength, power, reaction time, perceived exertion, and pattern recognition. Most studies indicate that physical performance is maximized in the afternoon. This time is characterized by the highest values in body temperature, strength, reaction time, pattern recognition, and heart rate, and a reduced level of perceived exertion and respiratory response to exercise. Unfortunately, studies of biological rhythms are invariably conducted in laboratories—and the increased stimulation of competition may supersede any arousing effect of biological rhythm.

Athletic contests are often scheduled at great distances from the home area, requiring the athlete to endure long airplane, bus, or automobile rides. These long trips may hamper performance by disturbing normal living habits and altering biological rhythms.

Jet lag is characterized by fatigue, malaise, sluggishness, decreased reaction time, and disorientation. This condition is caused by factors such as loss of sleep due to the excitement of travel, irregular and unfamiliar meals, dehydration, and disturbance of the biological clock. Some athletes may experience muscle stiffness and constipation caused by sitting for prolonged periods.

Eastbound travel seems to cause the most problems because it has the greatest effect on sleep. Proper travel scheduling can help alleviate the symptoms of jet lag. If possible, time the journey for arrival in the evening so athletes can get a full night's sleep. In addition, several days before the trip is scheduled, athletes should attempt to gradually shift the hours of eating and sleeping toward the time schedule of the destination. They should be well rested before beginning the journey.

Traveling in an airplane can lead to dehydration. The cabins of pressurized airplanes have a relative humidity of approximately 20%, resulting in an increased loss of body water through evaporation. To prevent this problem, athletes should be encouraged to consume more water than normal both before and during the trip. In addition, constipation, which occurs in some people following prolonged sitting, can often be prevented by ensuring adequate fluid intake.

A summary of the guidelines for preventing jet lag include:

- Make eastbound flights during daylight hours, leaving as early as possible; the longer the distance, the earlier the start.
- Make westbound flights late in the day, arriving as close to the athletes' retiring hour as possible.
- Drink plenty of water during the trip.
- Encourage light meals and discourage fatty foods.
- At regular intervals, get up from the seat and stretch or walk around.

Minimize the negative effects of travel on performance by planning the trip well and adhering to the athletes' normal schedule as much as possible.

■ Summary

Increasing numbers of people are exercising at moderate and high altitudes, which increases the stress on their physiological systems. Maximal oxygen consumption begins to deteriorate at an altitude of approximately 5000 feet, with pulmonary diffusion being the major limiting factor. As the partial pressure of oxygen decreases with altitude, the driving force of oxygen from the environment to the pulmonary circulation is impaired.

Acclimatization to altitude begins within the first few days of exposure. Early exposure to altitude diminishes the role of CO_2 in controlling ventilation. The early effects of acclimatization restore these CO_2 ventilatory control mechanisms by reducing bicarbonate levels in the blood and cerebrospinal fluid, and by increasing the ventilatory sensitivity to CO_2. Bicarbonate excretion and an increased concentration of 2,3-diphosphoglycerate shift the oxyhemoglobin dissociation curve to the right, which decreases the affinity of hemoglobin for oxygen. Epinephrine secretion increases during acute exposure to hypoxia. As epinephrine levels decrease, the lactate concentration in blood during exercise also decreases. Long-term changes include increased hemoglobin and hematocrit, which increase oxygen-carrying capacity.

The principal human exposure to hyperbaric environments is sports diving (skin or scuba diving). The sea provides many potential hazards, including gases that can be toxic or narcotic, currents that can tax physical stamina, and rapid changes in pressure that can be traumatic or fatal. Medical problems stemming from exposure to high pressure include barotrauma, gas toxicity, and decompression sickness.

Underwater research has focused on a number of factors that pose difficulty during exercise in this environment: increased air density, cold, decreased efficiency, CO_2 retention, and inert gas narcosis. The principles of exercise physiology determined on land cannot necessarily be extrapolated to underwater work. For example, pressure and temperature affect the relationship between oxygen consumption and heart rate, and heart rate–oxygen consumption relationships established on land can be dangerous if applied underwater.

Air pollution can have both psychological and physiological effects that can limit performance. Smog can produce tightening in the chest, difficulty taking a deep breath, eye irritation, pharyngitis, headache, lassitude, malaise, nausea, and dryness in the throat. During periods of extremely poor air quality, practices and competitions should be restricted, particularly for people with pulmonary diseases such as asthma.

Many physiological factors exhibit rhythms that span a day, week, month, or year. Available evidence suggests that several aspects of exercise performance are maximized in the late afternoon. Jet lag, which can cause fatigue and sluggishness, is caused by the effects of crossing time zones and by the disrupting effects of travel. Problems can be minimized by scheduling eastbound flights in the daylight hours and westbound flights late in the day.

■ Selected Readings

Anand, I. S., B. A. Prasad, S. S. Chugh, K. R. Rao, D. N. Cornfield, C. E. Milla, N. Singh, S. Singh, and W. Selvamurthy. Effects of inhaled nitric oxide and oxygen in high-altitude pulmonary edema. *Circulation* 98: 2441–2445, 1998.

Asano, M., K. Kaneoka, T. Nomura, K. Asano, H. Sone, K. Tsurumaru, K. Yamashita, K. Matsuo, H. Suzuki, and Y. Okuda. Increase in serum vascular endothelial growth factor levels during altitude training. *Acta Physiol. Scand.* 162: 455–459, 1998.

Bailey, D. M. and B. Davies. Physiological implications of altitude training for endurance performance at sea level: a review. *Br. J. Sports Med.* 31: 183–190, 1997.

Balke, B. Variations in altitude and its effects on exercise performance. In Exercise Physiology, H. Falls (Ed.). New York: Academic Press, 1968, pp. 240–265.

Beidleman, B. A., S. R. Muza, P. B. Rock, C. S. Fulco, T. P. Lyons, R. W. Hoyt, and A. Cymerman. Exercise re-

sponses after altitude acclimatization are retained during reintroduction to altitude. *Med. Sci. Sports Exerc.* 29: 1588–1595, 1997.

Boning, D. Altitude and hypoxia training—a short review. *Int. J. Sports Med.* 18: 565–570, 1997.

Bove, A. A. Medical aspects of sport diving. *Med. Sci. Sports Exerc.* 28: 591–595, 1996.

Bove, A. A., and J. C. Davis. (Eds.). Diving Medicine. Philadelphia: W. B. Saunders, 1990.

Brooks, G. A., G. E. Butterfield, R. R. Wolfe, B. M. Groves, R. S. Mazzeo, J. R. Sutton, E. E. Wolfel, and J. T. Reeves. Decreased reliance on lactate during exercise after acclimatization to 4300 m. *J. Appl. Physiol.* 71: 333–341, 1991.

Brooks, G. A., G. E. Butterfield, R. R. Wolfe, B. M. Groves, R. S. Mazzeo, J. R. Sutton, E. E. Wolfel, and J. T. Reeves. Increased dependence on blood glucose after acclimatization to 4300 m. *J. Appl. Physiol.* 70: 919–927, 1991.

Brooks, G. A., E. E. Wolfel, G. E. Butterfield, A. Cymerman, A. C. Roberts, R. S. Mazzeo, and J. T. Reeves. Poor relationship between arterial [lactate] and leg net release during steady-rate exercise at 4,300 m altitude. *Am. J. Physiol.* 275 (Regulatory Integrative and Comparative Physiol. 44): R1192–R1201, 1998.

Brooks, G. A., E. E. Wolfel, B. M. Groves, P. R. Bender, G. E. Butterfield, A. Cymerman, R. S. Mazzeo, J. R. Sutton, R. R. Wolfe, and J. T. Reeves. Muscle accounts for glucose disposal but not blood lactate appearance during exercise after acclimatization to 4300 m. *J. Appl. Physiol.* 72: 2435–2445, 1992.

Butterfield, G. E., J. Gates, S. Fleming, G. A. Brooks, J. R. Sutton, and J. T. Reeves. Increased energy intake minimizes weight loss in men at high altitude. *J. Appl. Physiol.* 72: 1741–1748, 1992.

Camporesi, E. M. Diving and pregnancy. *Semin. Perinatol.* 20: 292–302, 1996.

Cerretelli, P. Gas exchange at high altitude. In Pulmonary Gas Exchange, vol. 2, Organism and Environment, J. B. West (Ed.). New York: Academic Press, 1980.

Cogo, A., D. Legnani, and L. Allegra. Respiratory function at different altitudes. *Respiration* 64: 416–421, 1997.

Coote, J. H. Medicine and mechanisms in altitude sickness. Recommendations. *Sports Med.* 20: 148–159, 1995.

Derion, T., W. G. Reddan, and E. H. Lanphier. Static lung load and posture effects on pulmonary mechanics and comfort in underwater exercise. *Undersea Biomed. Res.* 19: 85–96, 1992.

Desplanches, D., H. Hoppeler, L. Tuscher, M. H. Mayet, H. Spielvogel, G. Ferretti, B. Kayser, M. Leuenberger, A. Grunenfelder, and R. Favier. Muscle tissue adaptations of high-altitude natives to training in chronic hypoxia or acute normoxia. *J. Appl. Physiol.* 81: 1946–1951, 1996.

Doubt, T. J. Cardiovascular and thermal responses to SCUBA diving. *Med. Sci. Sports Exerc.* 28: 581–586, 1996.

Duffner, G. J. Scuba diving. *Ciba Clinical Symposia* 10: 99, 1971.

Dwyer, J. Energetics of scuba diving and undersea work. *Proced. Intern. Sympos. on Man in the Sea:* 32–46, 1975.

Dwyer, J. Measurement of oxygen consumption in scuba divers working in open water. *Ergonomics* 4: 377–388, 1977.

Dwyer, J. Physiological studies of divers working at depths to 99 fsw in the open sea. In Underwater Physiology, VI, C. Shilling and M. Beckett (Eds.). Bethesda, Md.: FASEB, 1978, pp. 167–178.

Dwyer, J., H. Saltzman, and R. O'Bryan. Maximal physical-work capacity of man at 43.4 ATA. *Undersea Biomed. Res.* 4: 359–372, 1977.

Edwards, H. T. Lactic acid at rest and work at high altitudes. *Am. J. Physiol.* 116: 367–375, 1936.

Fagraeus, L., C. Hesser, and D. Linnarsson. Cardiorespiratory responses to graded exercise at increased ambient air pressure. *Acta Physiol. Scand.* 91: 259–274, 1974.

Faulkner, J. A., J. A. Opiteck, and S. V. Brooks. Injury to skeletal muscle during altitude training: induction and prevention. *Int. J. Sports Med.* 13 (Suppl. 1): S160–S162, 1992.

Ferretti, G. On maximal oxygen consumption in hypoxic humans. *Experientia* 46: 1188–1194, 1990.

Forte, V. A., Jr., D. E. Leith, S. R. Muza, C. S. Fulco, and A. Cymerman. Ventilatory capacities at sea level and high altitude. *Aviat. Space Environ. Med.* 68: 488–493, 1997.

Fowler, B. Nitrogen narcosis and diver performance. *Prog. Underwater Sci.* 12: 125–135, 1987.

Fulco, C. S., P. B. Rock, and A. Cymerman. Maximal and submaximal exercise performance at altitude. *Aviat. Space Environ. Med.* 69: 793–801, 1998.

Fusch, C., W. Gfrorer, H. H. Dickhuth, and H. Moeller. Physical fitness influences water turnover and body water changes during trekking. *Med. Sci. Sports Exerc.* 30: 704–708, 1998.

Garner, S. H., J. R. Sutton, R. L. Burse, A. J. McComas, A. Cymerman, and C. S. Houston. Operation Everest II: neuromuscular performance under conditions of extreme simulated altitude. *J. Appl. Physiol.* 68: 1167–1172, 1990.

Garrido, E., G. Rodas, C. Javierre, R. Segura, A. Estruch, and J. L. Ventura. Cardiorespiratory response to exercise in elite Sherpa climbers transferred to sea level. *Med. Sci. Sports Exerc.* 29: 937–942, 1997.

Gong, H. Effects of outdoor air pollutants on exercise performance. *Clin. Rev. Allergy.* 6: 361–383, 1988.

Gonzalez, N. C., M. Zamagni, and R. L. Clancy. Effects of alkalosis on maximum oxygen uptake in rates acclimated to simulated altitude. *J. Appl. Physiol.* 71: 1050–1056, 1991.

Gore, C., N. Craig, A. Hahn, A. Rice, P. Bourdon, S. Lawrence, C. Walsh, T. Stanef, P. Barnes, R. Parisotto, D. Martin, and D. Pyne. Altitude training at 2690 m does not increase total hemoglobin mass or sea level V_{O_2max} in world champion track cyclists. *J. Sci. Med. Sport* 1: 156–170, 1998.

Gore, C. J., A. G. Hahn, G. C. Scroop, D. B. Watson, K. I. Norton, R. J. Wood, D. P. Campbell, and D. L. Emonson. Increased arterial desaturation in trained cyclists during maximal exercise at 580 m altitude. *J. Appl. Physiol.* 80: 2204–2210, 1996.

Gore, C. J., S. C. Little, A. G. Hahn, G. C. Scroop, K. I. Norton, P. C. Bourdon, S. M. Woolford, J. D. Buckley, T. Stanef, D. P. Campbell, D. B. Watson, and D. L. Emonson. Reduced performance of male and female athletes at 580 m altitude. *Eur. J. Appl. Physiol.* 75: 136–143, 1997.

Gorman, D. F. Decompression sickness and arterial gas embolism in sports scuba divers. *Sports Med.* 8: 32–42, 1989.

Green, H. J., J. R. Sutton, E. E. Wolfel, J. T. Reeves, G. E. Butterfield, and G. A. Brooks. Altitude acclimatization and energy metabolic adaptations in skeletal muscle during exercise. *J. Appl. Physiol.* 73: 2701–2708, 1992.

Green, H. J., J. Sutton, P. Young, A. Cymerman, and C. S. Houston. Operation Everest II: muscle energetics during maximal exhaustive exercise. *J. Appl. Physiol.* 66: 142–150, 1989.

Hackett, P. H., R. C. Roach, R. A. Wood, R. G. Foutch, R. T. Meehan, D. Rennie, and W. J. Mills. Dexamethasone for prevention and treatment of acute mountain sickness. *Aviat. Space Environ. Med.* 59: 950–954, 1988.

Hanson, M. A. Role of chemoreceptors in effects of chronic hypoxia. *Comp. Biochem. Physiol. A. Mol. Integr. Physiol.* 119: 695–703, 1998.

Haykowsky, M. J., D. J. Smith, L. Malley, S. R. Norris, and E. R. Smith. Effects of short-term altitude training and tapering on left ventricular morphology in elite swimmers. *Can. J. Cardiol.* 14: 678–681, 1998.

Hobson, R. S. Airway efficiency during the use of SCUBA diving mouthpieces. *Br. J. Sports Med.* 30: 145–147, 1996.

Hochachka, P. W. The lactate paradox: analysis of underlying mechanisms. *Ann. Sports Med.* 4: 184–188, 1989.

Houston, C. S. Going Higher—The Story of Man and Altitude. Boston: Little, Brown, 1987.

Hoyt, R. W., T. E. Jones, C. J. Baker-Fulco, D. A. Schoeller, R. B. Schoene, R. S. Schwartz, E. W. Askew, and A. Cymerman. Doubly labeled water measurement of human energy expenditure during exercise at high altitude. *Am. J. Physiol.* 266: R966–R971, 1994.

Hultgren, H. N., B. Honigman, K. Theis, and D. Nicholas. High-altitude pulmonary edema at a ski resort. *West. J. Med.* 164: 222–227, 1996.

Jensen, J. B., B. Sperling, J. W. Severinghaus, and N. A. Lassen. Augmented hypoxic cerebral vasodilation in men during 5 days at 3,810 m altitude. *J. Appl. Physiol.* 80: 1214–1218, 1996.

Jiang, Z. L., J. He, H. Miyamoto, H. Tanaka, H. Yamaguchi, and Y. Kinouchi. Flow velocity in carotid artery in humans during immersions and underwater swimming. *Undersea Hyperb. Med.* 21: 159–167, 1994.

Kayser, B. Nutrition and energetics of exercise at altitude. Theory and possible practical implications. *Sports. Med.* 17: 309–323, 1994.

Kayser, B. Lactate during exercise at high altitude. *Eur. J. Appl. Physiol.* 74: 195–205, 1996.

Kayser, B., H. Hoppeler, H. Claassen, and P. Cerretelli. Muscle structure and performance capacity of Himalayan Sherpas. *J. Appl. Physiol.* 70: 1938–1942, 1991.

Kayser, B., H. Hoppeler, D. Desplanches, C. Marconi, B. Broers, and P. Cerretelli. Muscle ultrastructure and biochemistry of lowland Tibetans. *J. Appl. Physiol.* 8: 419–422, 1996.

Kayser, B., M. Narici, T. Binzoni, B. Grassi, and P. Cerretelli. Fatigue and exhaustion in chronic hypobaric hypoxia: influence of exercising muscle mass. *J. Appl. Physiol.* 76: 634–640, 1994.

Kerem, D., Y. I. Daskalovic, R. Arieli, and A. Shupak. CO_2 retention during hyperbaric exercise while breathing 40/60 nitrox. *Undersea Hyperb. Med.* 22: 339–346, 1995.

Krasney, J. A. A neurogenic basis for acute altitude illness. *Med. Sci. Sports Exer.* 26: 195–208, 1994.

Larsen, J. J., J. M. Hansen, N. V. Olsen, H. Galbo, and F. Dela. The effect of altitude hypoxia on glucose homeostasis in men. *J. Physiol.* (London) 504: 241–249, 1997.

Leach, J. W., and P. E. Morris. Commercial diving and diver performance. *Int. Rev. Ergonomics* 2: 105–122, 1988.

Levine, B. D., and J. Stray-Gundersen. "Living high-training low": effect of moderate-altitude acclimatization with low-altitude training on performance. *J. Appl. Physiol.* 83: 102–112, 1997.

Levine, B. D., J. H. Zuckerman, and C. R. deFilippi. Effect of high-altitude exposure in the elderly: the Tenth Mountain Division study. *Circulation* 96: 1224–1232, 1997.

Liu, Y., J. M. Steinacker, C. Dehnert, E. Menold, S. Baur, W. Lormes, and M. Lehmann. Effect of "living high–training low" on the cardiac functions at sea level. *Int. J. Sports Med.* 19: 380–384, 1998.

Luks, A. M., H. van Melick, R. R. Batarse, F. L. Powell, I. Grant, and J. B. West. Room oxygen enrichment improves sleep and subsequent day-time performance at high altitude. *Respir. Physiol.* 113: 247–258, 1998.

Lynch, P. R. Historical and basic perspectives of SCUBA diving. *Med. Sci. Sports Exerc.* 28: 570–572, 1996.

Maggiorini, M., A. Muller, D. Hofstetter, P. Bartsch, and O. Oelz. Assessment of acute mountain sickness by different score protocols in the Swiss Alps. *Aviat. Space Environ. Med.* 69: 1186–1192, 1998.

Mairbaurl, H. Red blood cell function in hypoxia at altitude and exercise. *Int. J. Sports Med.* 15: 51–63, 1994.

Malconian, M. K., P. B. Rock, J. T. Reeves, A. Cymerman, and C. S. Houston. Operation Everest II: gas tensions in expired air and arterial blood at extreme altitude. *Aviat. Space Environ. Med.* 64: 37–42, 1993.

Matheson, G. O., P. S. Allen, D. C. Ellinger, C. C. Hanstock, D. Gheorchiu, D. C. McKenzie, C. Stanley, W. S. Parkhouse, and P. W. Hochachka. Skeletal muscle metabolism and work capacity: a ^{31}P-NMR study of Andean natives and lowlanders. *J. Appl. Physiol.* 70: 1963–1976, 1991.

Mazzeo, R. S., P. R. Bender, G. A. Brooks, G. E. Butterfield, B. M. Groves, J. R. Sutton, E. E. Wolfel, and J. T. Reeves. Arterial catecholamine responses during exercise with acute and chronic high altitude exposure. *Am. J. Physiol.* 261: E419–E424, 1991.

Mazzeo, R. S., G. A. Brooks, G. E. Butterfield, A. Cymerman, A. C. Roberts, M. Selland, E. E. Wolfel, and J. T. Reeves. Beta-adrenergic blockade does not prevent the lactate response to exercise after acclimatization to high altitude. *J. Appl. Physiol.* 76: 610–615, 1994.

Mazzeo, R. S., A. Child, G. E. Butterfield, J. T. Mawson, S. Zamudio, and L. G. Moore. Catecholamine response during 12 days of high-altitude exposure (4300 m) in women. *J. Appl. Physiol.* 84: 1151–1157, 1998.

McClelland, G. B., P. W. Hochachka, and J. M. Weber. Carbohydrate utilization during exercise after high-altitude acclimation: a new perspective. *Proc. Natl. Acad. Sci. U.S.A.* 95: 10288–10293, 1998.

Melissa, L., J. D. MacDougall, M. A. Tarnopolsky, N. Cipriano, and H. J. Green. Skeletal muscle adaptations to training under normobaric hypoxic versus normoxic conditions. *Med. Sci. Sports Exerc.* 29: 238–243, 1997.

Miller, J. N., E. H. Lanphier, and O. D. Wagensteen. Respiratory limitations to work in diving and their significance to the design of underwater breathing apparatus. *Med. Dello Sport* 24: 231–237, 1971.

Mizuno, M., C. Juel, T. Bro-Rasmussen, E. Mygind, G. Schibye, B. Rasmussen, and B. Saltin. Limb skeletal muscle adaptation in athletes after training at altitude. *J. Appl. Physiol.* 68: 496–502, 1990.

Moore, L. G., S. Niermeyer, and S. Zamudio. Human adaptation to high altitude: regional and life-cycle perspectives. *Am. J. Phys. Anthropol.* 27: 25–64, 1998.

Muller, F. L. A field study of the ventilatory response to ambient temperature and pressure in sport diving. *Br. J. Sports Med.* 29: 185–190, 1995.

Nicholas, R., M. Yaron, and J. Reeves. Oxygen saturation in children living at moderate altitude. *J. Am. Board Fam. Pract.* 6: 452–456, 1993.

Nishihara, F., H. Shimada, and S. Saito. Rate pressure product and oxygen saturation in tourists at approximately 3000 m above sea level. *Int. Arch. Occup. Environ. Health* 71: 520–524, 1998.

Noakes, T. D. 1996 J. B. Wolffe Memorial Lecture. Challenging beliefs: ex Africa semper aliquid novi. *Med. Sci. Sports Exerc.* 29: 571–590, 1997.

Olszanski, R., Z. Sicko, Z. Baj, E. Czestochowska, M. Konarski, J. Kot, P. Radziwon, A. Raszeja-Specht, and A. Winnicka. Effect of saturated air and nitrox diving on selected parameters of haemostasis. *Bull. Inst. Marit. Trop. Med. Gdynia* 48: 75–82, 1997.

Pagani, M., G. Ravagnan, and D. Salmaso. Effect of acclimatisation to altitude on learning. *Cortex* 34: 243–251, 1998.

Patajan, J. H. The effects of high altitude on the nervous system and athletic performance. *Seminars in Neurology* 1: 253–261, 1981.

Peacock, A. J. and P. L. Jones. Gas exchange at extreme altitude: results from the British 40th Anniversary Everest Expedition. *Eur. Respir. J.* 10: 1439–1444, 1997.

Pendergast, D. R., M. Tedesco, D. M. Nawrocki, and N. M. Fisher. Energetics of underwater swimming with SCUBA. *Med. Sci. Sports Exerc.* 28: 573–580, 1996.

Pilmanis, A. A., J. Henriksen, and H. J. Dwyer. An underwater ergometer for diver work performance studies in the ocean. *Ergonomics* 20: 51–55, 1977.

Pulfrey, S. M. and P. J. Jones. Energy expenditure and requirement while climbing above 6,000 m. *J. Appl. Physiol.* 81: 1306–1311, 1996.

Reeves, J. T., R. S. Mazzeo, E. E. Wolfel, and A. J. Young. Increased arterial pressure after acclimatization to 4300 m: possible role of norepinephrine. *Int. J. Sports Med.* 13 (Suppl. 1): S18–S21, 1992.

Reeves, J. T., E. E. Wolfel, H. J. Green, R. S. Mazzeo, A. J. Young, J. R. Sutton, and G. A. Brooks. Oxygen transport during exercise at altitude and the lactate para-

dox: lessons from Operation Everest II and Pikes Peak. *Exer. Sport Sci. Rev.* 20: 275–296, 1992.

Reinertsen, R. E., V. Flook, S. Koteng, and A. O. Brubakk. Effect of oxygen tension and rate of pressure reduction during decompression on central gas bubbles. *J. Appl. Physiol.* 84: 351–356, 1998.

Roach, R. C., E. R. Greene, R. B. Schoene, and P. H. Hackett. Arterial oxygen saturation for prediction of acute mountain sickness. *Aviat. Space Environ. Med.* 69: 1182–1185, 1998.

Robach, P., D. Biou, J. P. Herry, D. Deberne, M. Letournel, J. Vaysse, and J. P. Richalet. Recovery processes after repeated supramaximal exercise at the altitude of 4,350 m. *J. Appl. Physiol.* 82: 1897–1904, 1997.

Roberts, A. C., G. E. Butterfield, J. T. Reeves, E. E. Wolfel, and G. A. Brooks. Acclimatization to 4,300 m altitude decreases reliance on fat as substrate. *J. Appl. Physiol.* 81: 1762–1771, 1996.

Roberts, A. C., J. T. Reeves, G. E. Butterfield, R. S. Mazzeo, J. R. Sutton, E. E. Wolfel, and G. A. Brooks. Altitude and β-blockade augment glucose utilization during submaximal exercise. *J. Appl. Physiol.* 80: 605–615, 1996.

Rodas, G., C. Javierre, E. Garrido, R. Segura, and J. L. Ventura. Normoxic ventilatory response in lowlander and Sherpa elite climbers. *Respir. Physiol.* 113: 57–64, 1998.

Samaja, M. Blood gas transport at high altitude. *Respiration* 64: 422–428, 1997.

Savourey, G., A. Guinet, Y. Besnard, N. Garcia, A. M. Hanniquet, and J. Bittel. Evaluation of the Lake Louise acute mountain sickness scoring system in a hypobaric chamber. *Aviat. Space Environ. Med.* 66: 963–967, 1995.

Savourey, G., N. Garcia, J. P. Caravel, C. Gharib, N. Pouzeratte, S. Martin, and J. Bittel. Pre-adaptation, adaptation and de-adaptation to high altitude in humans: hormonal and biochemical changes at sea level. *Eur. J. Appl. Physiol.* 77: 37–43, 1998.

Sawka, M. N., A. J. Young, P. B. Rock, T. P. Lyons, R. Boushel, B. J. Freund, S. R. Muza, A. Cymerman, R. C. Dennis, K. B. Pandolf, C. R. Valeri. Altitude acclimatization and blood volume: effects of exogenous erythrocyte volume expansion. *J. Appl. Physiol.* 81: 636–642, 1996.

Scherrer, U., L. Vollenweider, A. Delabays, M. Savcic, U. Eichenberger, G. R. Klege, A. Fikrle, P. E. Ballmer, P. Nicod, P. Bartsch. Inhaled nitric oxide for high-altitude pulmonary edema. *N. Engl. J. Med.* 334: 624–629, 1996.

Schmidt, W., H. W. Dahners, R. Correa, R. Ramirez, J. Rojas, and D. Böning. Blood gas transport properties in endurance-trained athletes living at different altitudes. *Int. J. Sports Med.* 11: 15–21, 1990.

Schoene, R. B. Control of breathing at high altitude. *Respiration* 64: 407–415, 1997.

Schoene, R. B., R. C. Roach, P. H. Hackett, J. R. Sutton, A. Cymerman, and C. S. Houston. Operation Everest II: ventilatory adaptation during gradual decompression to extreme altitude. *Med. Sci. Sports Exer.* 22: 804–810, 1990.

Selland, M. A., T. J. Stelzner, T. Stevens, R. S. Mazzeo, R. E. McCullough, and J. T. Reeves. Pulmonary function and hypoxic ventilatory response in subjects susceptible to high-altitude pulmonary edema. *Chest* 103: 111–116, 1993.

Smith, D. J. Diagnosis and management of diving accidents. *Med. Sci. Sports Exerc.* 28: 587–590, 1996.

Strauss, R. Diving Medicine. New York: Grune & Stratton, 1976.

Sutton, J. R., C. S. Houston, and G. Coates (Eds.). Hypoxia: The Tolerable Limits. Indianapolis: Benchmark, 1988.

Sutton, J. R., J. T. Reeves, B. M. Groves, P. D. Wagner, J. K. Alexander, H. N. Hultgren, A. Cymerman, and C. S. Houston. Oxygen transport and cardiovascular function at extreme altitude: lessons from Operation Everest II. *Int. J. Sports Med.* 3 (Suppl. 1): S13–S18, 1992.

Svedenhag, J., K. Piehl-Aulin, C. Skog, and B. Saltin. Increased left ventricular muscle mass after long-term altitude training in athletes. *Acta Physiol. Scand.* 161: 63–70, 1997.

Sykes, J. J. Medical aspects of scuba diving. *Br. Med. J.* 308: 1483–1488, 1994.

Tetzlaff, K., B. Neubauer, C. Buslaps, B. Rummel, and E. Bettinghausen. Respiratory responses to exercise in divers at 0.4 MPa ambient air pressure. *Int. Arch. Occup. Environ. Health* 71: 472–478, 1998.

Tetzlaff, K., L. Friege, M. Reuter, J. Haber, T. Mutzbauer, and B. Neubauer. Expiratory flow limitation in compressed air divers and oxygen divers. *Eur. Respir. J.* 12: 895–899, 1998.

Thalmann, E., D. Sponholtz, and C. Lundgren. Effects of immersion and static lung loading on submerged exercise at depth. *Undersea Biomed. Res.* 6: 259–289, 1979.

Thomas, R. G., P. C. LaStayo, H. Hoppeler, R. Favier, G. Ferretti, B. Kayser, D. Desplanches, H. Spielvogel, and S. L. Lindstedt. Exercise training in chronic hypoxia has no effect on ventilatory muscle function in humans. *Respir. Physiol.* 112: 195–202, 1998.

Tripathi, H. L., N. W. Eastman, K. G. Olson, D. A. Brase, and W. L. Dewey. Effects of hyperbaric simulation of scuba diving pressure on plasma β-endorphin. *Pharmacol. Biochem. Behav.* 38: 219–221, 1991.

Tschop, M., C. J. Strasburger, G. Hartmann, J. Biollaz, P. Bartsch. Raised leptin concentrations at high altitude associated with loss of appetite. *Lancet* 352: 1119–1120, 1998.

Wagner, P. D. Insensitivity of $\dot{V}_{O_2max}$ to hemoglobin-P50 at sea level and altitude. *Respir. Physiol.* 107: 205–212, 1997.

Webb, J. T., and A. A. Pilmanis. Breathing 100% oxygen compared with 50% oxygen: 50% nitrogen reduces altitude-induced venous gas emboli. *Aviat. Space Environ. Med.* 64: 808–812, 1993.

Wells, J. M. Hyperthermia in divers and diver support personnel. *Undersea Biomed. Res.* 18: 225–227, 1991.

Welsh, C. H., P. D. Wagner, J. T. Reeves, D. Lynch, T. M. Cink, J. Armstrong, M. K. Malconian, P. B. Rock, and C. S. Houston. Operation Everest II: spirometric and radiographic changes in acclimatized humans at simulated high altitudes. *Am. Rev. Respir. Dis.* 147: 1239–1244, 1993.

West, J. B. Limiting factors for exercise at extreme altitudes. *Clin. Physiol.* 10: 265–272, 1990.

West, J. B., and S. Lahiri. High Altitude and Man. Bethesda, Md.: American Physiological Society, 1984.

Wilks, J. Kitting up: an equipment profile of Queensland divers. *S. Pacific Underwater Med. Soc. J.* 20: 200–205, 1990.

Wilks, J. Diving dropouts: the Australian experience. *Austr. J. Sci. Med. Sport.* 21: 17–20, 1991.

Wilks, J., and V. O'Hagen. Queensland scuba divers and their tables. *S. Pacific Underwater Med. Soc. J.* 21: 11–14, 1991.

Williams, A. M., B. Clarke, and C. W. Edmonds. The influence of diving variables on perceptual and cognitive functions in professional shallow-water (abalone) divers. *Environ. Res.* 50: 93–102, 1989.

Wolfel, E. E., M. A. Selland, R. S. Mazzeo, and J. T. Reeves. Systemic hypertension at 4300 m is related to sympathoadrenal activity. *J. Appl. Physiol.* 76: 1643–1650, 1994.

Wolski, L. A., D. C. McKenzie, and H. A. Wenger. Altitude training for improvements in sea level performance. Is the scientific evidence of benefit? *Sports Med.* 22: 251–263, 1996.

Wood, R. J., A. G. Hahn, and M. E. McBride. The effect of 610 m altitude on rowing ergometer performance. *Aust. J. Sci. Med. Sport* 27: 98–102, 1995.

Young, A. J. Energy substrate utilization during exercise in extreme environments. *Exer. Sport Sci. Rev.* 18: 65–117, 1990.

Young, A. J., M. N. Sawka, S. R. Muza, R. Boushel, T. Lyons, P. B. Rock, B. J. Freund, R. Waters, A. Cymerman, K. B. Pandolf, and C. R. Valeri. Effects of erythrocyte infusion on $\dot{V}_{O_2max}$ at high altitude. *J. Appl. Physiol.* 81: 252–259, 1996.

Young, P. M., J. R. Sutton, H. J. Green, J. T. Reeves, P. B. Rock, C. S. Houston, and A. Cymerman. Operation Everest II: metabolic and hormonal responses to incremental exercise to exhaustion. *J. Appl. Physiol.* 73: 2574–2579, 1992.

Zaccaria, M., S. Rocco, D. Noventa, M. Varnier, and G. Opocher. Sodium regulating hormones at high altitude: basal and post-exercise levels. *J. Clin. Endocrinol. Metab.* 83: 570–574, 1998.

Zhuang, J., T. Droma, S. Sun, C. Janes, R. E. McCullough, R. G. McCullough, A. Cymerman, S. Y. Huang, J. T. Reeves, and L. G. Moore. Hypoxic ventilatory responsiveness in Tibetan compared with Han residents of 3658 m. *J. Appl. Physiol.* 74: 303–311, 1993.

CHAPTER 24

CARDIOVASCULAR DISEASES AND EXERCISE

Cardiovascular diseases (CVD) include coronary heart disease, valvular heart disease, chronic heart failure, cardiomyopathy, congenital heart defects, stroke, hypertension, and peripheral vascular disease. Cardiovascular diseases are the leading causes of death in Western countries. They have staggering economic and human costs.

Over 58 million people in the United States have one or more types of cardiovascular disease. This includes 50 million with hypertension, 14 million with coronary heart disease, 4 million with stroke, and 1.8 million with rheumatic heart disease. One in five males and females have some form of cardiovascular disease. The annual cost of CVD exceeds $260 billion, including the cost of hospitalization, nursing home costs, medical professional services, and lost productivity.

Even worse is the toll in human suffering. Scores of relatively young men and women die prematurely from CVD, resulting in irretrievable loss to their families and wasted human potential.

Current research suggests that many deaths from cardiovascular diseases are preventable, and increased awareness of risk factors have probably contributed to their decline in some countries. During the last 20 years, death rates from cardiovascular disease fell by 50% in Australia, Canada, France, and the United States and by more than 60% in Japan. In other countries, such as the Scandina-vian countries, Ireland, Portugal, and Spain, the decline has been only 20–25%. All of these countries have had significant heart-disease risk reduction programs.

The trend is just the opposite in former communist countries in eastern Europe. CVD death rates have risen 80% in Bulgaria, 40% in Hungary and the former Czechoslovakia, and 60% in Poland. By a large margin, Russia has the highest death rate in the world from cardiovascular diseases. Contributing factors to these high death rates may include emotional stress from social upheaval, cigarette smoking, alcoholism (possibly contributing to cardiomyopathies and insulin resistance), high-fat diet, and physical inactivity.

Cardiovascular diseases remain a severe problem worldwide. Each year, they kill over 12 million in the world and nearly 1 million people in the United States. Since 1900, these diseases have been the leading causes of death in the United States in every year except 1918—that year was the end of World War I and the year of a worldwide flu epidemic that killed over 20 million people. CVD continues to claim more lives each year than the next seven leading causes of death combined. Even though the death rate from CVD has been declining, the actual number of deaths due to these diseases has declined less than three percent.

Central among cardiovascular disease is coro-

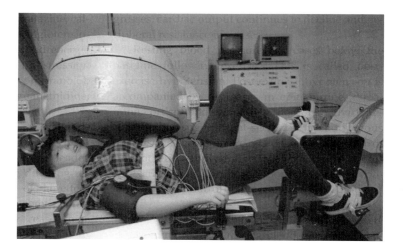

Recognition that appropriate exercise training can be used to prevent, diagnose, and treat several degenerative diseases, including heart disease, has led to increased participation by individuals of all ages in recreational and competitive activities. PHOTO © David Madison 1991.

nary artery disease (CAD). It involves a steady buildup of atherosclerotic plaques in the coronary arteries, leading to reduced myocardial blood flow. The effect on the heart depends on the degree of ischemia (inadequate blood flow). If blood flow is not totally blocked, but is insufficient to satisfy the myocardial oxygen demand, then angina (heart pain) or cardiac arrhythmias (abnormal impulse conduction in the heart) can develop. If blood flow is cut off (usually due to a thrombosis or blood clot), or reaches a critical level of ischemia, damage is irreparable. Hypoxemia does not just hamper cardiac function, it destroys it. The death of a portion of the myocardium is commonly called a "heart attack." Survivability, or the degree of impairment, depends on the amount of tissue destroyed.

Arteriosclerosis is a general term that describes age-related arterial changes to blood vessels, such as arterial thickening, loss of connective tissue, and increased diameter of the blood vessels. *Atherosclerosis* is the development of fibrotic, lipid-filled plaques in the walls of the larger arteries, such as the coronary, carotid, iliac, and femoral arteries. Its deadly effects are often centered in the heart, although it can impair blood flow to all vital organs, including the brain, kidneys, and liver. Atherosclerosis can result in extensive circulatory damage that can be ultimately impervious to medical interven-

tion. The most logical approach is to prevent the disease before it becomes life-threatening.

Current research suggests that CAD can be prevented and, to a certain extent, reversed. The incidence of CAD can be decreased by systematically reducing risk factors associated with the disease. This chapter discusses the physiology of CAD, risk factors for developing the disease, exercise and CAD, and cardiac rehabilitation. Cardiovascular diseases, such as stroke, congestive heart failure, hypertension, and peripheral vascular disease are also discussed.

■ Sudden Cardiac Death (SCD) and Exercise

Exercise and sports are associated with good health and wellness. Thus, it is extremely disconcerting when athletes die suddenly during or shortly after participating in physical activity. Often such deaths are caused by sudden cardiac death (SCD). SCD is death occurring immediately or within 48 hrs of a cardiac event, such as a myocardial infarction or cardiac arrest. Fortunately, SCD is extremely rare, occurring at an annual rate of 0.75 and 0.13 per 100,000 in young male and female athletes (respectively) and a rate of 6 per 100,000 in middle-aged men.

The risk of exercise is higher in cardiac patients. For these patients, cardiac arrest and heart attack occur at a rate of 1 for every 100,000–300,000 hours of exercise. Cardiac arrest is more common than myocardial infarction. The death rate is approximately 1 death for every 800,000 hours of exercise. The period during exercise is more dangerous for the cardiac patient than are less active times of the day. However, provided that precautions are taken, the long-term benefits of exercise for health and quality of life make it worthwhile.

Interest in SCD has a long history. In 490 B.C., 10,000 Athenians defeated a superior force of 15,000 Persians on the plain of Marathon in what is now Greece. A messenger named Pheidippides ran 25 miles from Marathon to Athens to deliver the news of victory to the anxiously waiting Athenians. Legend has it that he died from exhaustion after delivering the message. The Greek historian Herodotus reported that Pheidippides also ran 125 miles the day before his famous run trying to recruit the Spartans to join the battle. Thus, Pheidippides may be the first reported case of overtraining.

Throughout history, medical experts have vacillated over the relative risks and benefits of sports for the heart. Beginning with the ancient Greek physician Galen, medical experts have periodically warned against the health risks of competitive athletics. As recently as the 1970s, some medical experts questioned whether the benefits of exercise compensated for the increased risk of sudden death.

Conversely, others have rated endurance sports as an elixir of life that conferred immunity from coronary heart disease. For example, the Bassler hypothesis, formulated in the 1970s, suggested that marathon running provides immunity against coronary artery disease and fatal myocardial infarction. The hypothesis was based on the observation, which proved to be false, that marathon runners do not get coronary artery disease.

The truth is somewhere in between these medical opinions. Virtually all cases of SCD involve people with underlying cardiac disease. Further, it has been shown that exercise training reduces the risk of sudden death in people with latent coronary artery disease. However, it is also true that physically active people are not immune from heart disease—people with some kinds of heart disease are at increased risk of sudden death when they exercise.

The causes of sudden cardiac death during exercise vary with age (Table 24-1). In people under 35, it is most commonly associated with hypertrophic cardiomyopathy (HCM). HCM is usually a congenital condition characterized by unusual patterns of left

---

TABLE 24-1

Causes of Sudden Death During Exercise in Athletes

Conditions Causing Myocardial Ischemia
- Atherosclerotic coronary artery disease
- Coronary artery spasm
- *De novo* coronary artery thrombus
- Intramyocardial bridging
- Hypoplastic coronary artery
- Anomalous coronary arteries
- Coronary artery dissection

Structural Abnormalities
- Hypertrophic cardiomyopathy
- Mitral valve prolapse
- Valvular heart disease
- Lipomatous infiltration of the right ventricle
- Marfan's syndrome
- Right ventricular dysplasia
- Aortic stenosis
- Sickle cell trait

Arrhythmias
- Wolfe-Parkinson-White syndrome
- Lown-Ganong-Levine syndrome
- Long QT syndrome
- Ventricular arrhythmias
- Medial hyperplasia and intimal proliferation of the main sinus node artery

Miscellaneous
- Myocarditis
- Anabolic steroid use
- Cocaine abuse
- Sarcoidosis
- Nonpenetrating chest trauma (*commotio cordis*)

SOURCE: From Noakes, 1998.

ventricular and atrial hypertrophy, reduced size of the left ventricular chamber, and abnormal electrocardiogram and left ventricular filling patterns. Other causes of sudden cardiac death in young people include coronary artery anomalies, myocarditis, ruptured aortic aneurysms associated with Marfan's syndrome, aortic stenosis, various types of arrhythmias, myocarditis, and blunt trauma to the chest (*commotio cordis*). Coronary artery disease is the major cause of SCD death in people over 35. Coronary disease can also cause sudden death in young people, especially those with familial hypercholesterolemia (inherited high level of blood cholesterol).

Mechanisms of Sudden Death During Exercise

The mechanisms of SCD vary with the underlying cardiac pathology. However, there are several common characteristics. Prior to death, cardiac output is reduced drastically due to ischemia or arrhythmia. This leads to reduced brain bloodflow and unconsciousness. Coronary ischemia during exercise may be due to plaque rupture or thrombosis or exercise-induced coronary compression or coronary spasm.

In people with coronary artery disease, exercise-induced myocardial infarction might be due to increased blood pressure, which damages vulnerable atherosclerotic plaque. This condition might further induce thrombus formation and occlusion of the artery. Exercise may also trigger ventricular fibrillation and cardiac arrest by creating an imbalance between myocardial oxygen demand and supply. Coronary-artery spasms, occurring in areas of the artery with established disease, can also cause myocardial ischemia and trigger fatal arrhythmias during exercise.

Preventing Sudden Death During Exercise

The incidence of sudden death during exercise is extremely low. A person is more likely to die in an automobile accident on the way to the running track than to die suddenly during exercise. Predicting SCD has proved elusive, however. Only about 20%

of people with precipitating heart diseases can be identified. Even when disease is detected, it is not always possible to distinguish between people who will die suddenly during exercise from those with the same condition who are not at risk. Thus, routine evaluation of athletes for susceptibility to SCD is not practical or cost effective. However, when athletes show clinical signs or significant risk of heart disease, they should be evaluated thoroughly. Simple markers, such as a family history of sudden death due to SCD or recent viral illness, possibly suggesting viral myocarditis, are suggestive of increased risk. After medical evaluation, the nature of the exercise program and appropriate level of supervision can then be determined.

■ The Development of Atherosclerosis

Atherosclerosis is characterized by the formation of lipid-filled fibrous plaques that collect on the inner surface of the coronary arteries (Figure 24-1). The mechanism of atherosclerosis is not entirely clear—it cannot be explained by a single comprehensive theory. Incidents in the disease process include fatty streaking of the intima (inner lining of the coronary arteries), injury to the intima of the coronary artery, migration and aggregation of white blood cells and platelets to the injury site, smooth muscle proliferation, lipid accumulation, and plaque maturation. This process is summarized in Figure 24-1, 2.

Injury Atherosclerosis is thought to be initiated by small injuries to the tunica intima, the inner lining of the coronary arteries. The lining is composed of endothelial cells and an elastic membrane. The injury may be due to mechanical stress from the beating heart and from blood pressure, carbon monoxide, inflammation and infection, vasoconstrictor substances, immune responses, trauma, lipid impaction, or toxic chemicals.

Fatty Streak This is the earliest sign of the disease and is characterized by lipid deposition on the inner lining of the coronary arteries. In this process, monocytes that migrate to the area for re-

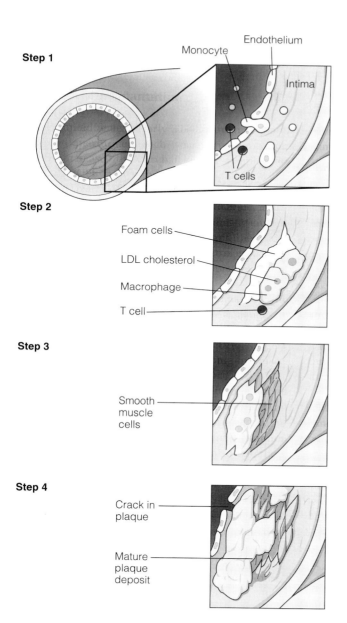

Step 1

Endothelium

Monocyte

Intima

T cells

Step 2

Foam cells

LDL cholesterol

Macrophage

T cell

Step 3

Smooth
muscle
cells

Step 4

Crack in
plaque

Mature
plaque
deposit

Figure 24-1 How atherosclerosis develops. *Step 1*—Injury to inner lining of intima. Monocytes and T cells penetrate into intima. *Step 2*—Monocytes are converted to macrophages. They then scavenge oxidized LDL, turning into foam cells. *Step 3*—Smooth-muscle cells migrate into intima and divide. Macrophages and smooth-muscle cells release collagen and other proteins. *Step 4*—The resulting mature plaque is a complex collection of foam cells, proteins, smooth-muscle cells, and cholesterol debris. The plaque can harden, crack, and then cause blood clots to form. SOURCE: The Johns Hopkins White Papers. Baltimore, MD: Johns Hopkins Medical Institutions, 1996.

pair of injured endothelial cells are converted to macrophages, which scavenge oxidized LDL, forming foam cells. These foam cells become part of fatty streaks in the lining of the coronary arteries. Some of these fatty streaks can progress to fibrous plaques. *Fibrous plaques* are elevated lesions on the lining of the arteries that are composed of fibrous connective tissue impacted with lipids (free and esterified cholesterol), necrotic (dead) cells, and smooth muscle that has migrated from the media (middle layer) of the artery. Fatty streaks have been observed in the coronary arteries of boys autopsied after accidental death, which suggests that the onset of coronary artery disease occurs at an early age in males.

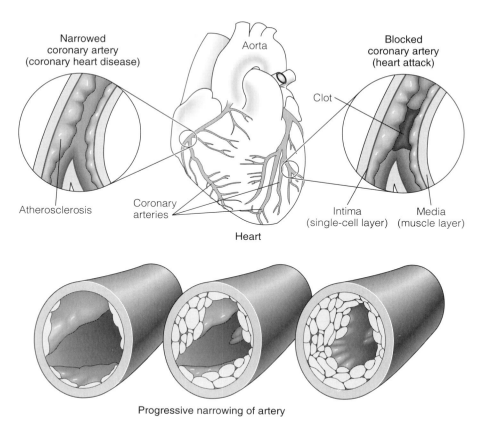

Progressive narrowing of artery

Figure 24-2 Atherosclerosis can form in the coronary arteries, resulting in a progressive narrowing of the lumen (artery passage). If a clot forms, blood flow through the coronary artery can be blocked, resulting in a heart attack.

Migration and Aggregation of White Blood Cells and Platelets, and Smooth Muscle Proliferation In response to injury to the intima, white blood cell monocytes, macrophages (derived from monocytes), and blood platelets migrate and collect in the area. This migration causes cellular responses that can lead to the formation of fibrous plaques. Hypertension and a diet high in cholesterol and fat stimulate monocytes to bind to the inner lining of the coronary arteries (i.e., to endothelial cells), and then to differentiate into macrophages. Macrophages can promote atherosclerosis in several ways:

- They become impacted with lipids to form foam cells. These foam cells are found in abundance in the interior of fibrous plaques.

- The injured endothelial cells, platelets, and the macrophages and the foam cells formed by them secrete growth factors that stimulate the growth of smooth muscle cells in fibrous plaque.

- Macrophages participate in phagocytosis, which is a scavenging process used by tissues to rid the body of unwanted material. During phagocytosis, reactive oxygen metabolites, such as superoxide anion, are produced, which causes low-density lipoprotein (LDL) to accumulate in the wall of the coronary artery.

- Also, macrophages produce enzymes such as elastase and collagenase that cause cell death in the interior of the fibrous plaque.

The critical event in the formation of fibrous plaque is the accumulation of smooth muscle cells in the intima (inner lining) of the artery. Stimulants of smooth muscle growth include polypeptide growth factor and platelet-derived growth factor. These factors also stimulate the formation of connective tissue (which make up the fibrous element of the plaque) and attract smooth muscle cells and monocytes to the fibrous plaque.

These growth factors, although necessary for normal physiological growth, are secreted in response to injury in the arterial wall and can lead to the pathological growth of the fibrous atherosclerotic plaque. There is some evidence that the growth of smooth muscle cells in the fibrous plaque resembles tumor growth. Viruses or substances in tobacco smoke may result in mutagenic stimulation of smooth muscle cells, leading to the formation of the fibrous plaque.

Lipid Accumulation   Ordinarily, endothelial cells act as a barrier to cholesterol and other lipids. However, the foam cells created by macrophages within the endothelial cells are filled with lipids, which accumulate and form part of the core of the fibrous plaque. Foam cells are produced largely by the macrophages taking up large amounts of LDL. Their formation is accelerated by hyperlipidemia and hypertension.

Maturation of the Lesion   Eventually, the plaque calcifies and connective tissue forms, producing a narrowed, rigid blood vessel. The mature plaque is a matrix of lipid-impacted smooth muscle cells, cellular debris, fibrous and collagenous connective tissue, and calcium complexes. Once the vessel has been narrowed, blood flow may be impaired, or the vessel can be blocked by a clot.

The maturation of the fibrous plaque is characterized by a loss of cells through necrosis (cell death) and a weakening of the arterial lining. This disease process hampers the function of the endothelial cells that line the artery. The cells' capacity to produce nitric oxide decreases, which impairs the vessels' ability to vasodilate. Decreased nitric oxide

production can cause coronary vasoconstriction or vasospasm that can result in coronary ischemia during exercise or emotional stress. Impaired endothelial function also increases permeability to lipids and platelet adhesiveness in the endothelial cells, both of which also speed the progression of the disease. As the disease progresses, the weak fibrous plaque can rupture or ulcerate, which leads to *thrombosis*—the formation of a clot. Over 90% of major myocardial infarctions (heart attacks) are associated with clot formation where a plaque has ruptured (see Figure 24-2). The clot that forms in response to a plaque rupture can itself be incorporated into the fibrous plaque, which can result in a rapid progression of the disease in a short time. Transient clot formation and clot absorption can cause temporary ischemia and angina. An elevated plasma fibrinogen (an important factor in clot formation) level is associated with an increased risk of coronary artery disease.

Overt symptoms of the disease, including angina, ST segment depression (abnormal electrocardiogram that suggests ischemia), and myocardial infarction typically begin after the artery is 60% occluded. The atherosclerotic lesion interferes with blood flow, decreases blood vessel elasticity, and increases the tendency to form clots, thrombi, and emboli.

In addition to coronary artery disease, a variety of other factors can cause coronary artery obstruction. These factors include coronary artery vasospasm, thrombus formation, and congenital abnormalities of the coronary arteries. Coronary vasospasm—coronary narrowing due to increased smooth muscle tone—can occur in normal coronaries, but it most often occurs in diseased arteries. While the causes of coronary artery spasm are not completely understood, it can be triggered by endothelial cell dysfunction, as mentioned above, and exaggerated sympathetic nervous stimulation (e.g., in response to stress, exercise, or cold). Thrombus formation can also occur in normal coronary arteries but is most likely to occur in the face of atherosclerosis. Structural abnormalities of the coronary arteries can result in coronary obstruction, particularly during exercise (see previous section on sudden cardiac death).

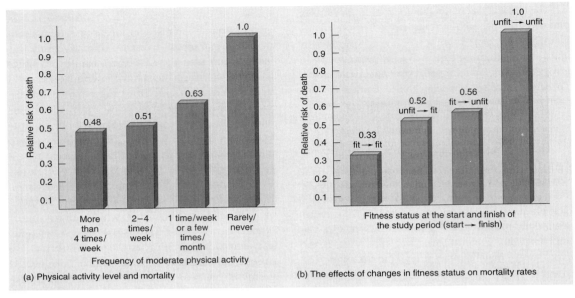

Figure 24-3 Physical activity, physical fitness, and mortality. Many long-term research studies have clearly shown the benefits of regular physical activity. (a) A study that followed more than 40,000 postmenopausal women for 7 years found a strong relatonship between physical activity and relative risk of death: the more frequent the activity, the lower the relative risk of death. (b) A study of more than 9500 men that tracked changes in fitness level for periods of 1–18 years found that fitness level was closely tied to risk of death: The lowest death rate was seen in men who were physically fit at the start and end of the study period; the highest death rate was in men who were unfit throughout the study. Men who improved from unfit to fit during the study experienced a 44% reduction in mortality risk relative to men who remained unfit. SOURCES: Kushi, L. H., et. al. 1997. Physical activity and mortality in post-menopausal women. *JAMA* 277(16): 1287–1292. Blair, S. N., et al. 1995. Changes in physical fitness and all-cause mortality: A prospective study of healthy and unhealthy men. *JAMA* 273(14): 1093–1098.

■ Risk Factors in the Development of Coronary Artery Disease

Heart disease is the number one killer in countries like the United States, Canada, and Great Britain—but in over half the world, the incidence of heart disease is low. In most Western countries, CAD begins to develop during childhood. By age 20, 75% of males have heart disease to a significant degree. Although females are less likely to suffer from coronary heart disease, the incidence is increasing.

The reasons for the different rates of heart disease between countries with high incidence (such as the United States) and those with low incidence (such as Japan) are difficult to explain. Several factors increase the risk of developing coronary heart disease (Table 24-2). The six most important risk factors are dyslipidemia, hypertension, physical inactivity, obesity, diabetes and smoking. Each of these increases the chances of developing CAD by 300%.

Other risk factors, such as family history, sex, age, emotional stress, hyperinsulinemia, and insulin resistance also increase the risk of heart disease, particularly when they are combined with other factors. Obesity and inactivity are considered particularly important. They directly affect other risk fac-

TABLE 24-2

Risk Factors for Heart Disease

Major Risk Factors
- Hypertension
- Dyslipidemia (e.g., elevated cholesterol, LDL, VLDL, lipoprotein (a), triglycerides; reduced HDL, elevated cholesterol/HDL ratio, cholesterol/LDL ratio)
- Smoking (includes second-hand smoke)
- Physical inactivity
- Obesity
- Diabetes/insulin resistance

Other Risk Factors
- Male gender
- Heredity
- Age
- Individual response to stress
- Abnormal ECG
- Chronic illness

tors, as well as exerting an independent effect of their own.

The incidence of heart disease has decreased in the United States in recent years. The reasons for this decrease are unclear although increased awareness of CAD risk factors and increased interest in preventive medicine and fitness may have played roles (see Figure 24-3).

Dyslipidemia

Dyslipidemia is a condition characterized by abnormal blood fats. High cholesterol and triglycerides, low levels of high-density lipoprotein (HDL), and a high cholesterol/HDL ratio increase the risk of developing CAD. Dyslipidemia leads to atherosclerosis by promoting lipid impaction in the lesion and precipitating the initial injury to the intima of the coronary arteries.

The American Heart Association estimated that about more than 50% of U.S. adults have blood cholesterol concentrations of more than 200 mg · dl^{-1}, and 20% have levels of 240 mg · dl^{-1} or higher. Further, more than 36% of U.S. youth age 19 and under have cholesterol levels of 170 mg · dl^{-1} or higher. A cholesterol level of 170 mg · dl^{-1} in young people is comparable to a level of 200 mg · dl^{-1} in adults. Par-

ticularly alarming are cholesterol levels in postmenopausal women. For ages 35–44, the mean total blood cholesterol level of women is 195 mg · dl^{-1}, but between ages 45 and 64, the average rises from 217 to 235 mg · dl^{-1}, putting women in this age group at equal risk for heart attack as men in the same age group.

Our knowledge of the relationship between blood lipids and coronary heart disease has increased greatly during the 1980s and 1990s. Numerous large-population studies (i.e., epidemiological studies) have shown a relationship between elevated cholesterol and CAD. For example, increasing cholesterol from 180 mg · dl^{-1} to 250 mg · dl^{-1} increases the risk of developing CAD by 350%. However, many people with proven CAD also have "normal" cholesterol levels. In these people, CAD may be associated with other lipid abnormalities in high-density lipoproteins (HDL), triglycerides, and HDL and LDL subfractions.

Lipoproteins Because lipids are not water soluble, they are transported as components of lipoproteins. The regulation of cholesterol and lipoproteins is shown in Figure 24-4. Lipoproteins are composed of cholesterol, protein, triglycerides, and phospholipids. There are five types of lipoproteins, classified by density, composition, and size. In order of increasing density, the five types are very-low-density lipoproteins (VLDL), low-density lipoproteins (LDL), intermediate-density lipoproteins (IDL), high-density lipoproteins (HDL), and chylomicrons. HDL are further subdivided into HDL$_1$, HDL$_2$, and HDL$_3$ (subfractions of HDL). Total cholesterol is the sum total of cholesterol carried by VLDL, LDL, HDL, and chylomicrons.

VLDL (55% triglycerides, 17% cholesterol, 20% phospholipids, and 8% protein*) are made in the liver. They carry triglycerides and cholesterol away from the liver. IDL are formed in the catabolism of VLDL and precede the development of LDL.

LDL (50% cholesterol, 20% phospholipids, 20% protein, and 10% triglycerides) are formed in the catabolism of IDL and are the principal cholesterol

---

*Percentages for this and other lipoproteins shown in parentheses are approximate.

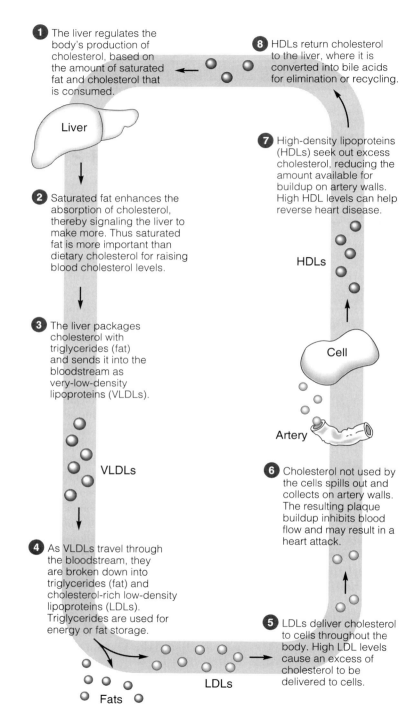

1 The liver regulates the body's production of cholesterol, based on the amount of saturated fat and cholesterol that is consumed.

Liver

2 Saturated fat enhances the absorption of cholesterol, thereby signaling the liver to make more. Thus saturated fat is more important than dietary cholesterol for raising blood cholesterol levels.

3 The liver packages cholesterol with triglycerides (fat) and sends it into the bloodstream as very-low-density lipoproteins (VLDLs).

VLDLs

4 As VLDLs travel through the bloodstream, they are broken down into triglycerides (fat) and cholesterol-rich low-density lipoproteins (LDLs). Triglycerides are used for energy or fat storage.

8 HDLs return cholesterol to the liver, where it is converted into bile acids for elimination or recycling.

7 High-density lipoproteins (HDLs) seek out excess cholesterol, reducing the amount available for buildup on artery walls. High HDL levels can help reverse heart disease.

HDLs

Cell

Artery

6 Cholesterol not used by the cells spills out and collects on artery walls. The resulting plaque buildup inhibits blood flow and may result in a heart attack.

5 LDLs deliver cholesterol to cells throughout the body. High LDL levels cause an excess of cholesterol to be delivered to cells.

LDLs

Fats

Figure 24-4 Regulation of cholesterol. SOURCE: Fahey et al., 1999.

carriers. LDL metabolism is regulated by the activity of LDL receptors in the liver. LDL receptor activity is, in turn, controlled by the liver's cholesterol requirements. Subfractions of LDL have also been identified. The small, dense LDL_3 subfraction is elevated in many people with CAD and is considered a significant risk factor for the disease. Exercise and weight reduction have been shown to decrease LDL_3. Lipoprotein (a), composed of LDL linked to apolipoprotein (a) (the protein constituent of lipoprotein), has many similar characteristics to LDL_3. It also increases the risk of CAD and is highly heritable.

Chylomicrons are formed in the intestinal wall from cholesterol and triglycerides in the diet. They leave the intestine through the lymphatic system and enter the circulation system at the left subclavian vein. Chylomicrons transport triglycerides to fat cells for storage and muscle for energy. These reactions are catalyzed by the action of lipoprotein lipase (LPL), which is found on the surface of the endothelial cells of the capillaries of fat and muscle. Liver LDL receptors bind with the cholesterol from the chylomicrons and remove it from the circulation.

HDL (50% protein, 5% triglycerides, 25% phospholipids, and 20% cholesterol) are produced in the liver, muscle, and intestines. They act as cholesterol scavengers via the enzyme lipoprotein lipase, which acts on the surface components of chylomicrons and VLDL. Approximately 17 different apolipoproteins, the protein constituents of lipoproteins, have been identified. Apolipoproteins play an important role of lipoprotein metabolism by acting as recognition sites for cell membrane receptors and as cofactors in enzymatic reactions. A low apolipoprotein A-I level is associated with an increased risk of CAD. Some researchers feel it is a better predictor of the disease than HDL or LDL levels. Other apolipoproteins that may have some value as predictors of CAD include apolipoproteins B-100, a, and E.

HDL retards the development of atherosclerosis by acting as a reverse cholesterol transfer system. HDL contains an enzyme called lecithin cholesterol acid transferase (LCAT). LCAT catalyzes a reaction that gathers free cholesterol and returns it to the liver. There it is released as cholesterol in the bile or is converted to bile salts.

Most people in the United States eat approximately 100 to 120 grams of fat a day. About two-thirds of this is transported to the liver, where it forms VLDL and LDL. The liver forms cholesterol even if the dietary intake of cholesterol itself is low. The rate of cholesterol formation is greater if fats are saturated. The level of LDL, particularly small-dense LDL, is far more important than total cholesterol in the development of coronary heart disease because it is the primary substance involved in the development of the atherosclerotic plaque.

The ratio of cholesterol to HDL is an extremely important factor in the development of heart disease. The average ratio in heart attack victims in the Framingham, Massachusetts, heart study (a long-term epidemiological study) was 5.4. The recommended ratio is less than 4.0. People with total cholesterols of 180 mg $\cdot$ dl^{-1}, considered favorable, may still be at risk if the level of LDL is high and the level of HDL is low. Populations with high levels of LDL, such as Eskimos, may have a low risk of heart disease because they eat fish that contains significant amounts of omega-3 fatty acids, which increases HDL.

Low HDL levels, regardless of total cholesterol, are an extremely important risk factor of CAD. HDL_2 levels are higher in women until menopause (approximately three times higher), at which time HDL_2 levels decrease and become similar to those of men. After menopause, heart disease rates follow a similar course in men and women. Low levels of HDL are also seen in hypertriglyceridemia, obesity, diabetes, and in cigarette smokers. Exercise increases HDL. Other factors affecting HDL levels include diet and alcohol consumption. Moderate alcohol has been shown to reduce mortality from CAD. However, because of the possibility of alcohol abuse, it is not recommended that nondrinkers begin consuming alcohol. Alcohol consumption increases the risk of hypertension and sudden death.

High levels of testosterone depress HDL levels. Athletes who take anabolic steroids typically have extremely low levels of HDL. Levels as low as 5 mg

$\cdot \, dl^{-1}$ have been recorded (20- to 40-year-old males average 40 mg $\cdot$ 100 ml^{-1}).

The research is equivocal about the significance of triglyceride levels in predicting CAD. Part of the problem stems from the complex interaction in the metabolism of HDL, LDL, and triglycerides. The link between CAD and triglycerides may be that the more dangerous, small-dense LDL particles (LDL$_3$) are associated with increases in plasma triglycerides.

Endurance training increases LDL receptor sensitivity (i.e., increases receptor activity), HDL, LCAT, and lipoprotein lipase and decreases LDL and triglycerides. The former group enhances the elimination of cholesterol in the bile and may be the principal mechanism through which exercise reduces the risk of CAD. A lipid management program should reduce LDL and increase HDL. Strategies include proper diet and exercise, cessation of smoking, and perhaps moderate alcohol consumption. (*Note:* High levels of alcohol consumption can be toxic to the heart and liver.) Researchers have speculated that CAD could be reduced 30–50% if the population could achieve cholesterol levels between 130 and 190 mg $\cdot$ dl^{-1}.

Cholesterol Reduction Therapy and CAD

Considerable evidence suggests that the progression of CAD can be delayed or prevented with cholesterol reduction therapy involving lifestyle modification or drugs. Aggressive dietary modification, smoking cessation, and drug treatment with agents such as clofibrate, probucol, cholestyramine, colestipol, and niacin have been shown to cause regression of atherosclerosis and prevention of myocardial infarction.

Hypertension

Hypertension (high blood pressure) is an arterial systolic blood pressure of 140 mmHg or greater or a diastolic pressure of 90 mmHg or greater. High blood pressure is sometimes called the "silent killer" because it usually has no apparent symptoms. In the United States, hypertension accounts for over 1.5 million deaths a year and affects more than 24% of Americans. In people between the ages

of 40 and 60 years, blood pressure above 150/90 mmHg increases the risk of heart disease by three times in men and by six times in women.

Hypertension affects the development of atherosclerosis, increases myocardial oxygen consumption at rest and exercise, and causes cellular changes in the heart and blood vessels. It has two main effects on the development of atherosclerosis: (1) It damages vascular endothelial cells by increasing shear force, torsion, and lateral wall pressure; and (2) it increases filtration of lipids into the atherosclerotic lesion.

Hypertension increases myocardial oxygen consumption by increasing the afterload of the heart, causing the heart to pump harder to deliver the same quantity of blood. This process usually results in cardiac hypertrophy, which, by itself, increases myocardial oxygen consumption. Myocardial hypertrophy with atherosclerosis increases the likelihood of the development of coronary ischemia.

The chronic afterload stress imposed by hypertension affects the ultrastructure of the myocardial cells. A chronically overloaded heart exhibits myofibril disorganization, disintegration and broadening of Z-bands, swelling and aggregation of mitochondria, intracellular edema, and decreased regional myocardial blood flow to the subendocardium during exercise. These changes can reduce the functional capacity of the heart and impair exercise capacity. Hypertension and exercise are discussed later in this chapter.

Smoking

Smoking and heart attack are related in the United States. In the Framingham study, 60% of the men and 40% of the women who had heart attacks were smokers. It has also been found that the risk of developing CAD in the United States is 70% greater in smokers than in nonsmokers. No such relationship, however, has been demonstrated in Finland, the Netherlands, Italy, Greece, the former Yugoslavia, Japan, or Puerto Rico.

The American Heart Association estimated that nearly 450,000 Americans die each year of smoking-related illnesses. Nearly one-fifth of deaths from cardiovascular diseases are attributable to smoking.

They also estimated that anywhere from 37,000–40,000 nonsmokers exposed to environmental tobacco smoke (ETS) die from cardiovascular diseases each year. Current estimates for the United States are that 25.9 million men (27.8%) and 23.5 million women (23.3%) are smokers, putting them at increased risk of heart attack. In addition, an estimated 4.4 million teenagers ages 12–17 years are smokers.

The role of smoking in atherosclerosis has not been clearly established. Smoking may stimulate CAD by increasing blood pressure, heart rate, platelet adhesiveness, and fatty acid mobilization, and decreasing high density lipoproteins. It may also be involved in fatal cardiac arrhythmias by lowering the ventricular fibrillation threshold and causing coronary artery vasoconstriction and vasospasm. Combining smoking with other primary risk factors increases geometrically the overall risk of CAD. Although the reason for accelerated atherosclerosis in smokers has not been clearly established, the statistical reality of a high number of heart attacks among smokers makes it a serious risk factor.

Family History, Gender, and Age

Family history, gender, and age are complex risk factors with genetic and sociological implications. Although the factors themselves cannot usually be altered, the relative risk of any individual can probably be reduced by a vigorous risk factor management program.

A person who had a close male relative (father or grandfather) die from a heart attack before age 60 has a three to six times greater risk of contracting coronary heart disease. The chances are even greater in people with familial hypercholesterolemia, as 50% of these die before age 60.

Although some people have a genetic tendency toward hypertension, diabetes, and obesity, many familial risk factor patterns are learned. It is likely that some of the dietary and psychological characteristics of families may be amenable to change under the right circumstances. Such people would undoubtedly be worthy candidates for high-risk factor management programs.

Premenopausal females have a lower risk of heart attack than men. The reason may lie in their higher levels of HDL (60 mg · 100 ml^{-1} in women versus 40 mg · 100 ml^{-1} in men) which appears to be supported by estrogen. As discussed, the higher levels of testosterone in males may be a factor in lower levels of HDL. Heart disease rates in women increase substantially with the onset of menopause. As mentioned, postmenopausal women experience a "catch-up" in deaths from all causes of cardiovascular disease. In 1995, 455,152 men (47.4% of total) died from CVD and 505,440 postmenopausal women (52.6% of total) died from cardiovascular disease. Death rates were similar for coronary artery disease: 244,819 deaths for men (50.9% of total) and 236,468 deaths for postmenopausal women (49.1% of total). The death rate for women from CVD disease even exceeds women's rate of death for cancer and osteoporosis, conditions that receive more publicity and research funding than heart disease.

Heart disease is a long-term degenerative disease. With far fewer people dying from infectious diseases because of antibiotics and modern medical techniques, it is not surprising that more people die from degenerative diseases, which progress in severity with age. Nevertheless, although it is extremely common in Western countries, atherosclerosis is not inevitable, as demonstrated by the low rate of heart disease among many peoples of the world. A risk factor reduction program will not allow people to live forever, but it may prevent them from dying prematurely from heart disease.

Obesity

Obesity is a major risk factor for CAD, and it is becoming a more serious problem in Western countries. Over 55% of U.S. adults are overweight and 22% are obese, an increase of over 40% since 1960 (Table 24-3). The problem is particularly acute in African American and Mexican American women and in Hawaiian natives of both sexes. Obesity is increasing among adolescents in the West as well. Among adolescents (ages 12–17), 12% of boys and 11% of girls are overweight, an increase of 7% and 5%, respectively, since 1980. It also negatively affects other risk factors, such as hypertension, dyslipidemia, insulin resistance, and hyperinsulinemia.

TABLE 24-3

The Prevalence of Overweight Populations of Special Concern

Group	Overweight and Obesity (BMI ≥25)	Obesity (BMI ≥30)
Adults (age ≥20)	54.9%	22.3%
Men	59.4	19.5
Women	50.7	25.0
Black women	66.0	36.7
Mexican American women	65.9	33.3
Men (age 50–59)	73.0	28.9
Women (age 50–59)	64.4	35.6

SOURCE: National Heart, Lung, and Blood Institute. 1998. *Clinical Guidelines on the Identification, Evaluation, and Treatment of Overweight and Obesity in Adults.* Bethesda, Md. National Institutes of Health.

CAD risk increases with total weight, body fat, and abdominal fat deposition (i.e., high waist-to-hip ratio). The health risks of obesity are discussed extensively in Chapter 25.

Emotional Stress

For many years, clinicians have identified the "heart attack personality" as aggressive, restless, impatient, and hard driving. Friedman and Rosenman (1974) coined the term "Type A personality" to describe this behavior pattern and defined it as "an action-emotional behavior complex exhibited by an individual in a lifelong struggle to obtain too much from the environment in too short a time, against the opposing efforts of objects or people in the environment." The less cardiac-prone behavioral type is the Type B personality. These people are less competitive and more relaxed. Extensive subsequent research has drawn a definite relationship between the Type A personality and increased risk of heart disease. For example, in the Framingham study, almost 22% of Type A men aged 55–64 had CAD compared to slightly over 10% of Type B men.

Although there is a statistical relationship between Type A personality and heart disease, there is no evidence directly relating this personality type to the process of coronary artery disease. It is possible that the adrenergic drive produced by emotional stress may induce mechanical trauma to the intima of the coronary arteries, but this has not been demonstrated. There is some evidence, however, to suggest that the Type A person releases more norepinephrine in the face of stress. These people have been called "hot reactors" because they exhibit increases in heart rate and blood pressure in response to an emotional stressor. This may enhance platelet aggregation, blood pressure, and lipid mobilization, which could accelerate development of the disease. There is no relationship, though, between the degree of Type A behavior and the degree of coronary occlusion. Also, no studies have shown that stress management programs reduce the incidence of mortality or morbidity from CAD. However, exercise training lessens depression and reduces anxiety, so it is an important part in a stress management program.

Other psychosocial factors, besides Type A behavior, also increase the risk of CAD. These include hostility, lower social class, neurotic behavior (i.e., anxiety and depression), sleep disturbances, and overwork. These factors promote atherosclerosis rather than cause sudden death by cardiac arrest.

Physical Activity

In 1987, the Centers for Disease Control and Prevention (CDC) in the United States classified physical inactivity as a major cardiovascular risk factor. Inactivity presented a risk similar to cigarette smoking, hypertension, obesity, and dyslipidemia. The bulk of evidence suggests that regular physical exercise is extremely important in reducing the risk of the disease. Fit people are less likely to get CAD, and when they do get it, it is less severe and happens at a later age than it does in sedentary people. Exercise training may result in regression of atheroma and reduction of the magnitude of other risk factors such as hypertension, obesity, and hyperlipidemia.

Based on epidemiological and experimental research, organizations responsible for making health and fitness recommendations to the public, such as the U.S. Centers for Disease Control and Prevention, the American College of Sports Medicine (ACSM), the American Heart Association, and the National Institutes of Health, have made a distinction between exercise that promotes health and exercise that promotes fitness. These studies suggest that chronic low-level physical activity (accumulation of 30 minutes a day) may improve metabolic fitness and reduce the risk of degenerative diseases, such as CAD, insulin resistance, some cancers, and osteoporosis, yet not significantly contribute to improved physical fitness.

The Surgeon General's report (1996) recommends that all Americans include a moderate amount of physical activity on most, preferably all, days of the week. The report suggests a goal of 150 calories per day, or about 1000 calories per week, expended in physical activity. Because energy expenditure is a function of both the intensity and duration of the activity, the same amount of activity can be obtained in longer sessions of more strenuous activities. Thus, 30 minutes of brisk walking or raking leaves is equivalent to 15 minutes of running or shoveling snow.

The position stand of ACSM (1998) views health and fitness as points along a continuum. People get benefits from exercise varying in quantity from ap-

proximately 700–2000 or more kilocalories of effort per week. People get significant health benefits going from a sedentary to a low level of physical activity (i.e., 700 kilocalories per week). They get additional benefits (i.e., improved fitness) by doing more exercise.

Another change from past recommendations is the increased emphasis on resistive exercise as an integral part of exercise requirements. The recommendations recognize the importance of resistance exercise for maintaining fat-free mass (FFM). FFM is critical for long-term control of body weight, maintenance of bone density, and movement capacity with aging. Resistance exercise also has some positive effects on CAD risk factors.

Since the early 1950s, epidemiological studies have demonstrated a relationship between physical activity and a reduced risk of heart disease. A number of studies compared active and sedentary populations. Subject populations included London bus drivers (sedentary) and bus conductors (active), postal clerks (sedentary) and postal deliverers (active), and active and sedentary longshoremen (Figure 24-5). In many studies, the relative risk of CAD death for people in sedentary jobs is consistently higher than for people in active jobs. Leisure-time

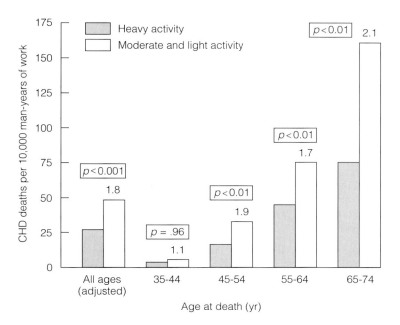

Figure 24-5 Deaths from CHD in longshoremen according to physical activity of work (range in kcal · min⁻¹) and age at death. Shaded bars = heavy activity (5.2–7.5 kcal · min⁻¹); unshaded bars = moderate and light activity (1.5–5.0 kcal · min⁻¹). The relative risk of developing CHD for moderate and light exercise groups compared to heavy exercise groups given above bars. Adapted from Paffenbarger and Hale, 1975. Used with permission.

physical activity is also an important marker of risk in developing CAD. When exercise consumes at least 700 kcal per week and is practiced three nonconsecutive days a week, it has a positive effect on protecting against CAD.

Several studies have demonstrated an inverse relationship between the level of physical fitness and the risk of heart disease (Figure 24-6). Extremely fit people tended to be less fat, with lower blood pressure, cholesterol, triglycerides, uric acid, and glucose.

Studies using animals provide direct experimental evidence that moderate endurance exercise can result in a reduction of coronary atherosclerosis. D. M. Kramsch and associates demonstrated that exercising monkeys who were fed an atherogenic diet had substantially reduced overall atherosclerotic involvement, lesion size, and collagen accumulation than did sedentary controls. Further, the trained monkeys had larger hearts and wider coronary arteries, which further reduced the degree of luminal narrowing (Figure 24-7). In contrast, the control animals, who were fed the same diet, suffered significant narrowing of the coronary arteries and, in one case, sudden death.

Autopsies on marathon runners who died from causes other than coronary heart disease consistently find enlarged and widely patent (open) coronary arteries. However, this does not actually prove that endurance training provides a measure of protection against the disease. CAD-related deaths are not uncommon in people involved in exercise—although most victims are usually relatively untrained and possess predisposing risk factors such as cigarette smoking, hypertension, or hyperlipidemia.

A summary of the mechanisms by which physical activity may reduce the occurrence or severity of coronary heart disease appears in Table 24-4. Endurance training helps reduce the risk of CAD by maintaining or increasing oxygen supply to the heart, decreasing work and oxygen demand by the heart, increasing function of the heart muscle (myocardium), and increasing the electrical stability of the heart.

Myocardial Oxygen Supply Myocardial oxygen is supplied by blood flow through the coronary arteries. Blood flow and myocardial oxygen supply decrease when the flow is restricted by coronary atherosclerosis. Exercise plays an important role in slowing the process of atherosclerosis. A technique developed at Stanford University allows precise measurement of the inner diameter of coronary arteries. Limited observations suggest it is possible to slow, and perhaps reverse, the process of atherosclerosis through lifestyle modification. However, current evidence does not suggest that exercise induces coronary collateral circulation (growth of the coronary circulation), except in cases of severe coronary ischemia.

Exercise training modifies several factors directly implicated in CAD. Endurance training increases HDL and the HDL/LDL ratio. These changes have been clearly demonstrated in males but are less pronounced in women. Exercise also increases the HDL_2 subfraction of HDL, which is thought to have a protective effect against CAD. A

Figure 24-6 ▶ Incidence and relative risks of coronary artery disease (CAD) per 10,000 man-years of observation, 1962–1972, among 16,936 Harvard alumni by cross-tabulations of alumnus physical activity and (a) college student sports play, (b) alumnus cigarette smoking, (c) alumnus body mass index, (d) alumnus and blood pressure status, (e) history of parental CHD, and (f) history of parental hypertension. Heights of the columns correspond to age-adjusted incidence rates of CHD (numbers atop columns). Relative risks are based on the incidence rate for the paired level presumed to induce the highest risk and placed as the back corner column. Numbers of cases appear in corresponding columns in the diamond key. Significance probabilities (*P*) are based on trends of incidence rate for one characteristic adjusted for age and paired characteristic. SOURCE: Paffenbarger et al., 1986.

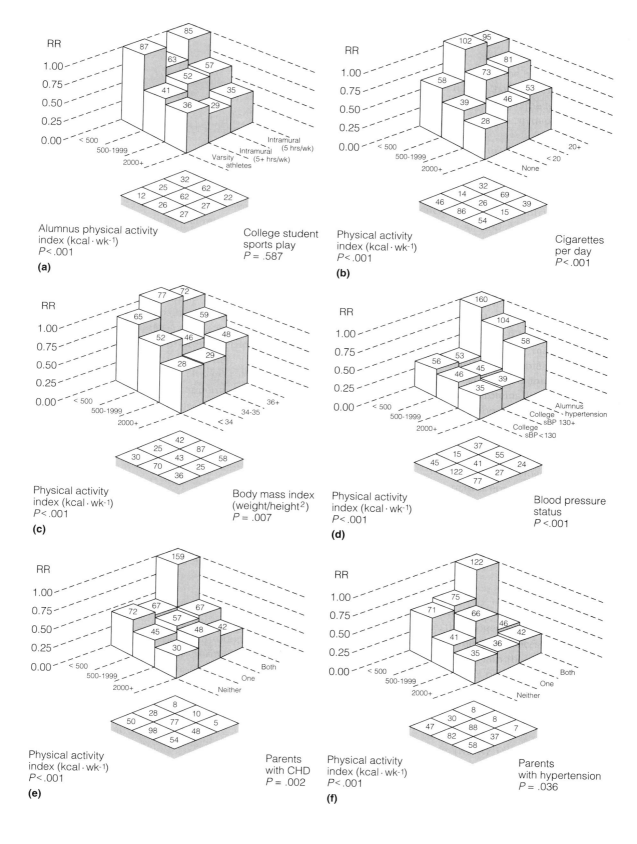

(a)
Alumnus physical activity index (kcal · wk⁻¹)
$P < .001$

College student sports play
$P = .587$

(b)
Physical activity index (kcal · wk⁻¹)
$P < .001$

Cigarettes per day
$P < .001$

(c)
Physical activity index (kcal · wk⁻¹)
$P < .001$

Body mass index (weight/height²)
$P = .007$

(d)
Physical activity index (kcal · wk⁻¹)
$P < .001$

Blood pressure status
$P < .001$

(e)
Physical activity index (kcal · wk⁻¹)
$P < .001$

Parents with CHD
$P = .002$

(f)
Physical activity index (kcal · wk⁻¹)
$P < .001$

Parents with hypertension
$P = .036$

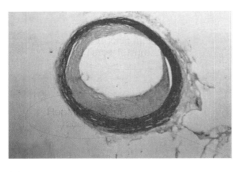

(a)   **(b)**

Figure 24-7 Micrographs of comparable sections of the left main coronary artery in sedentary (a) and active (b) monkeys on an atherogenic diet. The sedentary monkeys had significant development of atherosclerosis, while the coronary arteries of the active monkeys were healthy and had little evidence of the disease. SOURCE: Kramsch et al., 1981.

TABLE 24-4

Biologic Mechanisms by Which Exercise May Contribute to the Primary or Secondary Prevention of Coronary Heart Disease[a]

Maintain or increase myocardial oxygen supply
  Delay progression of coronary atherosclerosis (possible)
    Improve lipoprotein profile (increase HDL-C / LDL-C ratio) (probable)
    Improve carbohydrate metabolism (increase insulin sensitivity) (probable)
    Decrease platelet aggregation and increase fibrinolysis (probable)
    Decrease adiposity (usually)
  Increase coronary collateral vascularization (unlikely)
  Increase epicardial artery diameter (possible)
  Increase coronary blood flow (myocardial perfusion) or distribution (possible)
Decrease myocardial work and oxygen demand
  Decrease heart rate at rest and submaximal exercise (usually)
  Decrease systolic and mean systemic arterial pressure during submaximal exercise (usually) and at rest (possible)
  Decrease cardiac output during submaximal exercise (probable)

Decrease myocardial work and oxygen demand (*continued*)
  Decrease circulating plasma catecholamine levels (decrease sympathetic tone) at rest (probable) and at submaximal exercise (usually)
Increase myocardial function
  Increase stroke volume at rest and in submaximal and maximal exercise (likely)
  Increase ejection fraction at rest and during exercise (likely)
  Increase intrinsic myocardial contractility (possible)
  Increase myocardial function resulting from decreased "afterload" (probable)
  Increase myocardial hypertrophy (probable); but this may not reduce CAD risk
Increase electrical stability of myocardium
  Decrease regional ischemia or at submaximal exercise (possible)
  Decrease catecholamines in myocardium at rest (possible) and at submaximal exercise (probable)
  Increase ventricular fibrillation threshold due to reduction of cyclic AMP (possible)

[a] Expression of likelihood that effect will occur for an individual participating in endurance-type training program for 16 weeks or longer at 65% to 80% of functional capacity for 25 minutes or longer per session (300 kcal) for three or more sessions per week ranges from unlikely, possible, likely, probable, to usually. HDL-C = high-density lipoprotein cholesterol; LDL-C = low-density lipoprotein cholesterol; CAD = coronary artery disease; AMP = adenosine monophosphate.

SOURCE: American Heart Association, 1991.

minimum of 10 miles per week of walking or running seems to be necessary to increase HDL. Training also increases lipoprotein lipase (LPL) activity. LPL decreases HDL_2 breakdown and increases the use of triglycerides. In addition, exercise lowers triglycerides by increasing insulin receptor activity. Exercise training does not consistently reduce LDL, but it does tend to reduce the small-dense LDL subfraction, increase the large-dense LDL subfraction, and up-regulate LDL receptors. As discussed, small-dense LDL subfractions are associated with a more rapid progression of atherosclerosis.

Hyperinsulinemia stimulates the sympathetic nervous system, which raises blood pressure. The exercise-influenced increase in insulin receptor sensitivity and corresponding decrease in circulating insulin, therefore, may have important effects in reducing the progression of CAD. A single bout of exercise increases the rate of glucose uptake into the contracting skeletal muscles, a process that is regulated by the translocation of GLUT-4 glucose transporters to the plasma membrane and transverse tubules. Exercise training increases GLUT-4 expression in muscle, which contributes to a greater uptake of glucose when insulin binds to its receptor in the cell. Exercise also decreases body fat, which affects insulin metabolism, blood pressure, and blood lipids. Sedentary people with normal blood pressure have a 20-30% increased risk of developing hypertension. Thus, exercise directly influences most of the major factors associated with CAD.

Total body fat and abdominal fat deposition increase the risk of CAD. High body fat increases blood lipids and blood pressure and depresses insulin metabolism. Abdominal fat deposition patterns are particularly dangerous. Exercise reduces body fat and helps people maintain lost body weight. It does this by maintaining lean body weight and resting metabolic rate during weight reduction.

Exercise has an important influence on blood coagulation and fibrinolysis as well. Increases in platelet "stickiness" have been implicated in accelerated CAD, particularly in diabetics. Exercise decreases platelet aggregation and increases fibrinolysis. These changes are dramatic in diabetic subjects. Drugs such as aspirin, which also affect coagulation and fibrinolysis, have been shown to reduce the risk of sudden cardiac death. Exercise may present some of the same benefits.

Myocardial Work and Oxygen Demand
Myocardial work and oxygen demand improve the quality of life and decrease the symptoms of CAD. Moderate exercise reduces blood pressure at rest and during submaximal exercise, which directly affects the disease process. At the same time, training decreases resting and submaximal exercise heart rate and increases stroke volume, which improves functional capacity (i.e., fitness) and cardiac efficiency.

Myocardial Function
Atherosclerosis is accompanied by several abnormalities in myocardial cells. Exercise improves myocardial cellular capacity. This results in increased cardiac contractility and ejection fraction (percentage of end-diastolic volume ejected from the heart). Circulatory changes that decrease afterload make it easier for the heart to contract.

Electrical Stability of the Heart
Arrhythmias during exercise are not considered an independent risk factor of CAD. However, cardiac arrest is a leading mechanism of CAD-related morbidity and mortality. Exercise may decrease myocardial ischemia, which directly affects cardiac rhythm. Exercise training decreases catecholamine activity, including catecholamine response to emotional stress, at rest and during activity. Catecholamines greatly influence electrical stability at rest and during exercise and recovery. Training also reduces myocardial cyclic AMP (cAMP) concentration, which increases the ventricular fibrillation threshold.

■ Cardiac Rehabilitation

The use of exercise therapy is a radical change in the philosophy and treatment of the cardiac and cardiac-prone patient. Exercise therapy has gained widespread acceptance as an important tool in returning these people to physiological and psychological competence. Patients in cardiac rehabilitation programs have an improved sense of well-being, reduced anxiety and depression, improved exercise capacity, reduced ST segment depression on the electrocardiogram, reduced blood pressure, reduced heart rate at rest, reduced levels of serum cholesterol and triglycerides and elevated levels of HDL, and perhaps a lower mortality rate.

Candidates for the program include people with documented myocardial infarction or stable angina pectoris; postoperative cardiovascular surgical patients following procedures such as myocardial revascularization, peripheral arterial obstructive surgery, valve replacement, and repair of congenital heart defects; and those who are at high risk to develop heart disease because of significant predisposing factors. Patients with unstable conditions are generally excluded from the program.

The heart patient suffers both physical and emotional distress. After a myocardial infarction or heart surgery, there is a loss of cardiac function due to injury, and a loss of working capacity (fitness) because of deconditioning. In addition, the patient is faced with psychological difficulties stemming from fear of death or disability, and the perception of a changing lifestyle that may affect family, friends, economic well-being, and employment.

A primary goal of the cardiac rehabilitation program is to help the patient return to the quality of life he or she had prior to the onset of the event and to develop optimal physiological function. Initially, this consists of simple tasks such as dressing and showering. Ultimately, it may include a resumption of physically demanding occupations and recreational activities. Ideally, rehabilitation will return the individual to an acceptable level physiologically, mentally, socially, vocationally, and economically.

Exercise training has been shown to play an important role in the rehabilitation of most coronary patients. Physical activity should be integrated with a total program that includes cessation of cigarette smoking, weight control, proper diet, and a return to work and a normal social life. Interestingly, some heart patients actually improve their fitness over what it was prior to their heart attack. The cardiac event is potent motivation to become more active and increase awareness of coronary risk factors.

Education is an important part of the rehabilitation program. It should prepare patients and family for healthy alternatives in lifestyle that might reduce the risk of a recurrence of CAD—most studies have shown that the prognosis is poor for patients who continue to remain inactive, smoke, or maintain elevated blood lipids. In addition, the educational program should help teach the family about emergency procedures and alert them to the warning signs of further problems.

Cardiac rehabilitation reduces the cost of health care through shortened treatment time and prevention of disability. In addition, it may help to reduce occupational losses caused by cardiovascular disease.

The ideal cardiac rehabilitation program consists of three phases: the inpatient program, the outpatient therapeutic program, and the exercise maintenance program. During cardiac rehabilitation, the patient deals with a team that may include physicians, coronary care nurses, exercise physiologists, physical therapists, occupational therapists, dietitians, social workers, clinical psychologists, exercise specialists, and exercise technologists.

The Inpatient Program

The inpatient program is conducted in the hospital during the early period following the myocardial infarction or cardiovascular surgery. Early activity has been shown to reduce anxiety and depression, and to prevent some of the adverse effects of bedrest, such as muscle atrophy, thromboembolic complications, negative nitrogen balance, orthostatic intolerance, decreased aerobic capacity, tachycardia, and pulmonary atelectasis. In addition, the program helps to provide the physiological and psychological climate necessary for the resumption of normal activities.

The inpatient program includes appropriate medical treatment, patient and family education, and graded physical activity. By the end of this phase the patient should be able to meet low-level demands of daily activity, such as dressing, showering, going to the toilet, and walking up a flight of stairs.

The Outpatient Therapeutic Program

The second phase, the outpatient therapeutic program, is also usually conducted in the hospital and includes supervised exercise, vocational rehabilitation, and behavioral counseling. The exercise program is typically held at least three days per week and is closely monitored by a physician. Occupational task simulation is conducted to ensure that the individual is physiologically prepared to resume work.

In the past, endurance exercise training has been delayed for at least six weeks after myocardial infarction or cardiac surgery because of the fear that working above the 50% of maximum capacity required for the training effect might be dangerous. However, clinical trials indicate that active training can begin as early as three weeks after the event, provided that patients are adequately screened to rule out serious ventricular arrhythmias and evidence of heart failure.

Symptom-limited treadmill testing is important in the early exercise program and is typically administered prior to discharge from the hospital. Patients should be retested on the treadmill periodically to detect late-occurring ischemia and arrhythmias and to reassess the exercise prescription.

A principal goal of the outpatient program is to develop a functional capacity of 8 METs (an MET is equal to oxygen consumption at rest) or a maximal oxygen uptake of 28 ml $\cdot$ kg^{-1} $\cdot$ min^{-1}. This produces a fitness that will serve the needs of a sedentary lifestyle and act as a minimum criterion for discharge from the outpatient program. A person typically remains in most outpatient programs for about three months after a myocardial infarction.

The outpatient program allows the physician to ensure satisfactory medical status by reviewing resting and exercise ECG, blood pressure response to various stimuli, and symptoms. In addition, patients can obtain the education they need about the nature of heart disease and the signs and symptoms associated with their problem.

The Exercise Maintenance Program

The exercise maintenance program is ideally conducted in an environment especially organized for cardiac- and risk-prone individuals. Many fine programs have been established through hospitals, the YMCA, university physical education departments, medical schools, and private clinics. The purpose of this phase of the cardiac rehabilitation program is to prevent recurrence and to improve physical working capacity. This program can be used to evaluate the status and effectiveness of treatment regimens in the patient, help patients maintain compliance with lifestyle changes, and provide a safer environment for the exercise program. Exercises may include endurance exercises such as walking, jogging, cycling, or swimming; resistive exercise; and arm exercise. The design of the exercise program should reflect the physical requirements of the patient's job.

Because of the dangers of reinfarction, it's particularly important that attention be directed at resuscitation and safety procedures. Personnel, patients, and patients' families should be trained in cardiopulmonary resuscitation, and key program personnel should be certified in advanced cardiac life support. Proper equipment and drugs should be available for evaluating and correcting life-threatening cardiac disrhythmias.

Cardiac Rehabilitation and Reinfarction

In studies of dropouts versus adherents of the training program, the majority of recurrences seem to happen to dropouts. Their mortality rate has been reported to be 50% greater than that of patients who are faithful to their rehabilitation program. The reason for this isn't clear. It may be due to the actual effects of the program or to the personality or physiological characteristics of the dropouts. Factors that increase the risk of reinfarction are described in Table 24-5.

Cardiac Drugs and Exercise

A variety of drugs are used in cardiovascular disease that alter such things as the rate, contractility,

TABLE 24-5

Factors That Increase the Risk of Reinfarction

Nonadherence to exercise program
Prior history of angina or hypertension
Poor left ventricular function
Large arteriovenous O_2 difference
Ventricular arrhythmias—multifocal
Depression
Left atrial enlargement
Cardiac enlargement
Smoking
ST segment depression > 2 mV
Use of digitalis or diuretic agents
Cholesterol > 270 mg · dl^{-1}
Age, after age 55
Low systolic pressure during exercise
Increased ST segment depression with exercise
Diastolic pressure rise of 15 mmHg or more during
    exercise

conduction, preload, and afterload of the heart. These drugs can alter the electrocardiogram and the response to exercise. They can also have far-reaching effects on other physiological functions. Exercise response can help gauge the effectiveness of a drug, so it's important for people involved in prescribed exercise and training to understand the interactions between physical activity and drugs. A summary of the drugs used in heart disease, their actions, and their effects on exercise and the electrocardiogram appears in Table 24-6.

Digitalis drugs are used to treat or prevent heart failure and to suppress supraventricular arrhythmias. They can increase exercise capacity by improving the contractility and blood flow of the heart and normalizing fluid and electrolyte imbalances, all of which increase cardiac output. Digitalis frequently induces ST segment depression in the electrocardiogram, which can lead to false positive treadmill tests or mask ischemia if not taken into consideration (a false positive test incorrectly indicates ischemia). Digitalis can also be toxic, which can lead to heart block and to almost any arrhythmia, weakness, or depression, or to anorexia.

Diuretics are used primarily in the treatment of hypertension, fluid retention states, pulmonary edema, and heart failure. They are also sometimes misused by people seeking to lose weight. Diuretics that cause hypokalemia (low potassium) can render false positives on the electrocardiogram and an increased incidence of arrhythmias. There is also the danger of hypotension, particularly when more potent diuretics are used or when diuretics are combined with vasodilators. Work capacity may be impaired because of reduced plasma volume. Particular care should be taken with diuretics when exercising in the heat, as severe fluid diuresis will impair the capacity to sweat. Carbohydrate metabolism is further impaired in diabetics who are taking diuretics, so exercise leaders should be extremely careful with these individuals.

Vasodilators, such as nitroglycerin, have come to play an important part in the exercise program of cardiac patients. By reducing the afterload of the heart, they lower myocardial oxygen consumption and allow the patient to do more work before the onset of angina. Vasodilators have no effect on heart rate, but they tend to reduce blood pressure, systemic vascular resistance, and pulmonary wedge pressure. They also increase cardiac output in patients with acute myocardial infarction complicated by left ventricular failure.

Nitroglycerine is used to both relieve an angina attack and prevent one from occurring. This substance can, however, induce hypotension, which increases the risk of fainting during exercise. Nitroglycerine has a variable effect on the electrocardiogram. In many cases, it reduces ST segment depression and the incidence of arrhythmias. However, these changes are poorly correlated with changes in pain threshold.

Angiotensin-converting enzyme (ACE) inhibitors, such as enalapril, block the formation of angiotensin, a powerful vasoconstrictor. They are used to treat hypertension, heart failure, and kidney failure. These drugs have little effect on exercise capacity; however, they improve exercise capacity in patients with chronic heart failure. These are valuable drugs for people with diabetes because they protect the myocardium, retina, and kidneys. Potassium should be checked if the patient is on an ACE inhibitor and a potassium-retaining diuretic.

TABLE 24-6

Medications Relative to Exercise Training and Testing

Part I. GENERIC AND BRAND NAMES OF COMMON DRUGS BY CLASS

Generic Name	Brand Name
Beta Blockers	
Acebutolol	Sectral
Atenolol	Tenormin
Metoprolol	Lopressor, Toprol
Nadolol	Corgard
Pindolol	Visken
Propranolol	Inderal
Timolol	Blocadren
Carteolol	Cartrol
Betaxolal	Kerlone
Bisoprolol	Zebeta
Penbutolol	Levatol
Alpha₁ Blockers	
Prazosin	Minipress
Terazosin	Hytrin
Doxazosin	Cardura
Alpha and Beta Blocker	
Labetalol	Trandate, Normodyne
Antiadrenergic Agents Without Selective Receptor Blockade	
Clonidine	Catapres
Guanabenz	Wyntensin
Guanethidine	Ismelin
Guanfacine	Tenex
Methyldopa	Aldomet
Reserpine	Serapasil
Guanadrel	Hylorel
Nitrates and Nitroglycerin	
Isosorbide dinitrate	Isordil, Diltrate
Nitroglycerin	Nitrostat, Nitrolingual spray
Nitroglycerin ointment	Nitrol ointment
Nitroglycerin patches	Transderm Nitro, Nitro-Dur II, Nitrodisc
Isosorbide Mononitrate	Ismo, Monoket
Pentaerythritol tetranitrate	Cardilate

(continued)

TABLE 24-6 (continued)

Medications Relative to Exercise Training and Testing

Part I. GENERIC AND BRAND NAMES OF COMMON DRUGS BY CLASS

Generic Name	Brand Name
Calcium Channel Blockers	
Diltiazem	Cardizem
Nifedipine	Procardia, Adalat
Verapamil	Calan, Isoptin
Nicardipine	Cardene
Amlodipine	Norvasc
Felodipine	Plendil
Isradipine	DynaCirc
Nimodipine	Nimotop
Bepridil	Vascor
Digitalis	
Digoxin	Lanoxin
Diuretics	
Thiazides	
Hydrochlorothiazide (HCTZ)	Esidrix
"Loop"	
Furosemide	Lasix
Bumetanide	Bumex
Ethacrynic acid	Edecrin
Potassium-Sparing	
Spironolactone	Aldactone
Triamterene	Dyrenium
Amiloride	Midamor
Combinations	
Triamterene and hydrochlorothiazide	Dyazide, Maxzide
Amiloride and hydrochlorothiazide	Moduretic
Others	
Metolazone	Zaroxolyn
Peripheral Vasodilators (Nonadrenergic)	
Hydralazine	Apresoline
Minoxidil	Loniten
Angiotensin-Converting Enzyme (ACE) Inhibitors	
Captopril	Capoten
Enalapril	Vasotec
Lisinopril	Prinivil, Zestril
Ramipril	Altace
Benazepril	Lotensin
Fosinopril	Monopril
Quinapril	Accupril

TABLE 24-6 (continued)

Medications Relative to Exercise Training and Testing

Part I. GENERIC AND BRAND NAMES OF COMMON DRUGS BY CLASS

Generic Name	Brand Name
Antiarrhythmic Agents	
Class I	
IA	
Quinidine	Quinidex, Quinaglute
Procainamide	Pronestyl, Procan SR
Disopyramide	Norpace
IB	
Tocainide	Tonocard
Mexiletine	Mexitil
Lidocaine	Xylocaine, Xylocard
IC	
Encainide	Enkaid
Flecainide	Tambocor
Multiclass	
Ethmozine	Moricizine
Class II	
β-Blockers	
Class III	
Amiodarone	Cordarone
Bretylium	Bretylol
Sotalol	Betapace
Class IV	
Calcium channel blockers	
Sympathomimetic Agents	
Ephedrine	Adrenalin
Epinephrine	Alupent
Metaproterenol	Proventil, Ventolin
Albuterol	Bronkosol
Isoetharine	Brethine
Cromolyn sodium	Intal
Antihyperlipidemic Agents	
Cholestyramine	Questran
Colestipol	Colestid
Gemfibrozil	Lopid
Lovastatin	Mevacor
Nicotinic acid (niacin)	Nicobid, Nicolar, Slo-Niacin
Probucol	Lorelco
Pravastatin	Pravachol
Simvastatin	Zocor
Fluvastatin	Lescol
Other	
Dipyridamole	Persantine
Warfarin	Coumadin
Pentoxifylline	Trental

(continued)

TABLE 24-6 (continued)

Medications Relative to Exercise Training and Testing

Part II. EFFECTS OF MEDICATIONS ON HEART RATE, BLOOD PRESSURE, THE ELECTROCARDIOGRAM (ECG), AND EXERCISE CAPACITY

Medications	Heart Rate	Blood Pressure	ECG	Exercise Capacity
I. Beta blockers (including labetalol)	↓*(R and E)	↓ (R and E)	↓ HR*(R) ↓ ischemia† (E)	↑ in patients with angina; ↓ or ↔ in patients without angina
II. Nitrates	↑ (R) ↑ or ↔ (E)	↓ (R) ↓ or ↔ (E)	↑ HR(R) ↑ or ↔ HR (E) ↓ ischemia† (E)	↑ in patients with angina; ↔ in patients without angina; ↑ or ↔ in patients with congestive heart failure (CHF)
III. Calcium channel blockers Felodipine Isradipine Nicardipine Nifedipine	↑ or ↔ (R and E)	↓ (R and E)	↑ or ↔ HR(R and E) ↓ ischemia†(E)	↑ in patients with angina; ↔ in patients without angina
Bepridil Diltiazem Verapamil	↓ (R and E)		↓ HR (R and E) ↓ ischemia† (E)	
IV. Digitalis	↓ in patients w/ atrial fibrillation and possibly CHF Not significantly altered in patients w/sinus rhythm	↔	May produce nonspecific ST-T wave changes (R) May produce ST segment depression (E)	Improved only in patients with atrial fibrillation or in patients with CHF
V. Diuretics	↔	↔ or ↓ (R and E)	↔ (R) May cause PVCs and "false positive" test results if hypokalemia occurs. May cause PVCs if hypomagnesemia occurs (E)	↔, except possibly in patients with CHF
VI. Vasodilators, nonadrenergic	↑ or ↔ (R and E)	↓ (R and E)	↑ or ↔ HR (R and E)	↔, except ↑ or ↔ in patients with CHF

TABLE 24-6 (continued)

Medications Relative to Exercise Training and Testing

Part II. EFFECTS OF MEDICATIONS ON HEART RATE, BLOOD PRESSURE, THE ELECTROCARDIOGRAM (ECG), AND EXERCISE CAPACITY (continued)

Medications	Heart Rate	Blood Pressure	ECG	Exercise Capacity
ACE inhibitors	↔	↓ (R and E)	↔	↔, except ↑ or ↔ in patients with CHF
Alpha-adrenergic blockers	↔	↓ (R and E)	↔	↔
Antiadrenergic agents without selective blockade of peripheral receptors	↓ or ↔ (R and E)	↓ (R and E)	↓ or ↔ HR (R and E)	↔
VII. Antiarrhythmic agents		All antiarrhythmic agents may cause new or worsened arrhythmias (proarrhythmic effect)		
Class I Quinidine Disopyramide	↑ or ↔ (R and E)	↓ or ↔ (R) ↔ (E)	↑ or ↔ HR (R) May prolong QRS and QT intervals (R) Quinidine may result in "false negative" test results (E)	↔
Procainamide	↔	↔	May prolong QRS and QT intervals (R) May result in "false positive" test results (E)	↔
Phenytoin Tocainide Mexiletine	↔	↔	↔	↔
Flecainide Moricizine	↔	↔	May prolong QRS and QT intervals (R) ↔ (E)	↔
Propafenone	↓ (R) ↓ or ↔ (E)	↔	↓ HR (R) ↓ or ↔ HR (E)	↔

(continued)

TABLE 24-6 (continued)

Medications Relative to Exercise Training and Testing

Part II. EFFECTS OF MEDICATIONS ON HEART RATE, BLOOD PRESSURE, THE ELECTROCARDIOGRAM (ECG), AND EXERCISE CAPACITY (continued)

Medications	Heart Rate	Blood Pressure	ECG	Exercise Capacity
Class II Beta blockers (see I.)				
Class III Amiodarone	↓ (R and E)	↔	↓ HR (R) ↔ (E)	↔
Class IV Calcium channel blockers (see III.)				
VIII. Bronchodilators	↔	↔	↔	Bronchodilators ↑ exercise capacity in patients limited by bronchospasm
Anticholinergic agents Methylxanthines	↑ or ↔ (R and E)	↔	↑ or ↔ HR May produce PVCs (R and E)	
Sympathomimetic agents	↑ or ↔ (R and E)	↑, ↔, or ↓ (R and E)	↑ or ↔ HR (R and E)	↔
Cromolyn sodium	↔	↔	↔	↔
Corticosteroids	↔	↔	↔	↔
IX. Hyperlipidemic agents	Clofibrate may provoke arrhythmias, angina in patients with prior myocardial infarction Dextrothyroxine may ↑ HR and BP at rest and during exercise, provoke arrhythmias, and worsen myocardial ischemia and angina Nicotinic acid may ↓ BP Probucol may cause QT interval prolongation All other hyperlipidemic agents have no effect on HR, BP, and ECG			↔
X. Psychotropic medications Minor tranquilizers	May ↓ HR and BP by controlling anxiety. No other effects.			

TABLE 24-6 (continued)

Medications Relative to Exercise Training and Testing

Part II. EFFECTS OF MEDICATIONS ON HEART RATE, BLOOD PRESSURE, THE ELECTROCARDIOGRAM (ECG), AND EXERCISE CAPACITY (continued)

Medications	Heart Rate	Blood Pressure	ECG	Exercise Capacity
Antidepressants	↑ or ↔ (R and E)	↓ or ↔	Variable (R) May result in "false positive" test results (E)	
Major tranquilizers	↑ or ↔ (R and E)	↓ or ↔	Variable (R) May result in "false positive" or "false negative" test results (E)	
Lithium	↔	↔	May result in T wave changes and arrhythmias (R and E)	
XI. Nicotine	↑ or ↔ (R and E)	↑ (R and E)	↑ or ↔ HR, May provoke ischemia, arrhythmias (R and E)	↔, except ↓ or ↔ in patients with angina
XII. Antihistamines	↔	↔	↔	↔
XIII. Cold medications with sympathomimetic agents	Effects similar to those described in sympathomimetic agents, although magnitude of effects is usually smaller			↔
XIV. Thyroid medications Only levothyroxine	↑ (R and E)	↑ (R and E)	↑ HR May provoke arrhythmias ↑ ischemia (R and E)	↔, unless angina worsened
XV. Alcohol	↔	Chronic use may have role in ↑ BP (R and E)	May provoke arrhythmias (R and E)	↔
XVI. Hypoglycemic agents Insulin and oral agents	↔	↔	↔	↔

(continued)

TABLE 24-6 (continued)

Medications Relative to Exercise Training and Testing

Part II. EFFECTS OF MEDICATIONS ON HEART RATE, BLOOD PRESSURE, THE ELECTROCARDIOGRAM (ECG), AND EXERCISE CAPACITY (continued)

Medications	Heart Rate	Blood Pressure	ECG	Exercise Capacity
XVII. Dipyridamole	↔	↔	↔	↔
XVIII. Anticoagulants	↔	↔	↔	↔
XIX. Anti-gout medications	↔	↔	↔	↔
XX. Antiplatelet medications	↔	↔	↔	↔
XXI. Pentoxifylline	↔	↔	↔	↑ or ↔ in patients limited by intermittent claudication
XXII. Caffeine	Variable effects depending upon previous use / Variable effects on exercise capacity / May provoke arrhythmias			
XXIII. Diet pills	↑ or ↔	↑ or ↔	↑ or ↔ HR	

Key: ↑ = increase, ↔ = no effect, ↓ = decrease.
*Beta-blockers with ISA lower resting HR only slightly.
†May prevent or delay myocardial ischemia (see text).
R = rest; E = exercise.

SOURCE: American College of Sports Medicine, 1998. Used with permission.

Calcium channel blockers, such as verapamil, prevent the uptake of calcium in myocardial and vascular smooth muscle cells. They decrease cardiac afterload and contractility and reduce coronary artery ischemia. They are used for patients with hypertension, CAD, and angina. They increase exercise capacity in patients with angina but have no effect on work capacity in others.

Although catecholamines are used primarily in emergencies such as cardiac arrest and shock, they are also administered as an aerosol (epinephrine or isoproterenol) in chronic obstructive pulmonary disease. Because they can cause arrhythmias and increase blood pressure, adequate supervision is important for the heart patient using them. Ephed-

rine, another sympathomimetic drug that is used in asthma, can also increase heart rate and blood pressure.

β-adrenergic blocking drugs, such as propranolol, are used to decrease myocardial oxygen consumption and blood pressure. They do this by depressing heart rate and blood pressure. Because of the chronotropic effect of the β-blockers, the exercise prescribed must be based on an exercise stress test in which the drug was present in the typical therapeutic dosage.

Antiarrhythmic drugs such as procainamide and quinidine are used to suppress ventricular arrhythmias that can be particularly dangerous during exercise. These drugs can decrease myocardial contractility and cause a slight muscle weakness. In some people, they can cause nausea and vomiting, which may affect the desire to exercise. Because of their effects on suppressing arrhythmias, they can cause false negatives on the electrocardiogram during the exercise stress test.

■ Hypertension

As discussed, hypertension is high arterial blood pressure. It is a major health hazard, is related to 12% of all deaths in the United States, and affects 50 million Americans. It is a causal agent in stroke, congestive heart failure, kidney failure, and heart attack. Although there is presently no cure, it can usually be controlled by medication, diet, weight control, and exercise. Hypertension is classified as primary or secondary and categorized by severity. The causes of primary hypertension, which accounts for 95% of cases, are unknown. The causes of secondary hypertension are endocrine or structural disorders.

There is no absolute value of blood pressure below which the mortality rate is unaffected and above which it is increased. Symptoms of hypertension include headache, heart failure, renal disorders, neurological problems, claudication (leg pain caused by arterial occlusion), and chest pain. However, during the early stages of hypertension there are usually no symptoms, which is why hypertension is sometimes called the "silent killer."

Systolic Blood Pressure

Although much emphasis has been placed on the importance of diastolic pressure and health, epidemiological studies have shown that high systolic blood pressure is a significant risk factor for many diseases. Systolic pressures above 140 to 160 mmHg are considered hypertensive.

Because blood flow meets considerable resistance in the small capillaries and arterioles, blood must be pumped at high pressure to meet tissue demands. Blood pressure increases even more during exercise because tissue demands for blood are

higher. Normally, the body can easily tolerate increasing blood pressure during exercise. However, blood pressure can reach dangerous levels in the hypertensive individual.

In laboratory animals, systolic blood pressures of 300 mmHg have been shown to tear the intimal layer of the arterial wall. Presumably, in a vessel affected by arteriosclerosis, which would be less distensible than normal, damage could occur with even less pressure. Normally, maximal systolic pressure during exercise ranges between 150 to 220 mmHg. Pressure may be considerably higher during intense weight training. However, it is not uncommon for hypertensive people to exceed these levels at rest and to go considerably above them during exercise.

Diastolic Blood Pressure

Diastolic blood pressure is particularly important because of its effect on coronary blood flow. Diastolic blood pressures above 90 mmHg are considered hypertensive. During systole, when the heart muscle is contracting, intramuscular vessels constrict, which leads to a reduction in blood flow. This means that the most significant amount of coronary blood flow occurs during diastole. During exercise, myocardial oxygen demand may increase five or six times above that at rest. If, in the hypertensive individual, diastolic blood pressure increases during exercise, then coronary blood flow could be compromised at a time when more blood is needed. This would create a relative ischemia—more blood is needed than is delivered.

Increases in diastolic blood pressure during exercise of 15 mmHg above rest have been found to be highly predictive of coronary artery disease, and to indicate greater severity of disease and more frequent left ventricular contraction abnormalities.

Mechanisms of Hypertension

Blood pressure is the force exerted by the blood against any unit area of the vessel wall. As discussed in Chapter 15, Poiseuille's law accounts for the factors determining blood pressure and flow. Flow is directly proportional to pressure and the radius to the fourth power, and inversely proportional to viscosity and vessel length. Physiologically, the two most important factors regulating blood pressure are cardiac output (i.e., flow) and peripheral resistance (i.e., vessel diameter). Mechanisms that regulate blood pressure must directly or indirectly affect these critical biophysical measures.

As with other physiological functions, the body has many mechanisms for maintaining adequate blood flow to the tissues. Blood pressure is affected by increased higher central nervous system command, smooth muscle contraction, baroreceptor reflexes, autonomic discharge from the medulla and hypothalamus, arteriolar autoregulation, muscle mechanoreceptors, blood volume, and neurogenic and humeral influences. Although presently unknown, the cause of primary hypertension undoubtedly lies among these control mechanisms.

Factors affecting the risk of hypertension include family history, insulin resistance, diabetes, obesity, salt intake, age, inactivity, and alcohol consumption. Predictors of hypertension include ventricular mass, plasma hormonal and catecholamine concentrations, mental and emotional stressors, and the response to acute exercise.

Neural and Hormonal Mechanisms

As discussed in Chapter 15, the goal of central and sympathetic nervous mechanisms at rest and during exercise is to regulate blood pressure. Abnormalities in adrenal catecholamine secretion may be a cause of hypertension. Plasma catecholamine concentration and norepinephrine spillover rate tend to be higher in some people with hypertension.

Abnormalities in insulin metabolism may be related to hypertension. Inactive and obese people often have high blood insulin levels (hyperinsulinemia). Hyperinsulinemia increases sympathetic nervous system activity, which, in turn, increases blood pressure by increasing peripheral resistance and cardiac output. Increased insulin also enhances reabsorption of sodium in the kidneys.

Hyperinsulinemia and insulin resistance may be important reasons why hypertension is more common in overweight people. Obesity and abdominal fat deposition is associated with higher serum insulin levels and insensitivity of insulin receptors ("Syndrome X" [See Chapter 25]). Also, high fat diets, common in obese people, increase insulin and norepinephrine levels in the blood. These diets also increase blood pressure, elevate aldosterone, and enhance sodium retention. Exercise increases insulin receptor sensitivity. This may be an important reason for the decreased blood pressure seen in moderate-intensity endurance exercise programs.

Several other neural factors may be involved in hypertension. Recent findings suggest that endogenous opioids (i.e., endorphins) and serotonin may influence blood pressure control. These hormones may be particularly important in lowering resting blood pressure after exercise. More research is needed to determine their roles.

Baroreceptors may also influence blood pressure control. As discussed in Chapter 15, baroreceptors quickly monitor changes in blood pressure. They stimulate the cardiovascular control centers to maintain blood pressure by vasoconstricting blood vessels and increasing heart rate. Baroreceptors work by establishing a blood pressure set-point. If pressure increases above the set-point, the baroreceptors send impulses to the cardiovascular control center (CVC), which reduces the level of sympathetic activity and decreases blood pressure. When pressure again reaches the set-point, baroreceptor activity decreases. The reverse is true when blood pressure is below the set-point. Impulses from the baroreceptors to the CVC decrease, which causes blood pressure to increase. Abnormalities in set-point control may be a factor in hypertension. Altered barorector function in hypertension may be triggered by Syndrome X. People with Syndrome X (hypertension, hyperlipidemia, abdominal obesity, insulin resistance, hyperinsulinemia) have an altered autonomic control of the cardiovascular system characterized by impaired baroreceptor sensitivity and reduced heart rate variability.

The heart secretes a hormone-like substance called arterial natriuretic factor (ANF). ANF increases sodium loss in the kidney, reduces blood pressure and blood volume, and inhibits the sympathetic nervous system. ANF may be involved in blood pressure changes occurring with exercise training.

Vascular Structure Changing the diameters of the vessels influences peripheral resistance and blood pressure. Hypertension induces changes in endothelial function and vessel wall thickness and lumen diameter that increase peripheral resistance and make hypertension more intractable. Endurance exercise training tends to increase vascular lumen diameters without affecting wall thicknesses. This change may help to prevent the progressive vascular changes common in hypertension.

Long-term exposure to sports that induce significant pressure loads on the heart, such as weight lifting, may increase the risk of developing established hypertension (blood pressure that does not return to normal during relaxation). The chronic, periodic, explosive increases in blood pressure that occur during these activities may ultimately lead to structural changes in the arterial vasculature that could result in a permanent increase in peripheral vascular resistance and established hypertension.

Electrolytes and Nutrients Sodium increases vascular reactivity and plasma volume, both of which can increase blood pressure. Sodium restriction in hypertensives reduces plasma volume; aldosterone; renin; and systolic, mean, and diastolic blood pressures. The kidneys play a central role in secondary hypertension. They are related to the development of renal hypertension, pheochromocytma, corticoid hypersection, and primary aldosteronism.

Nutritional deficiencies have been found in many hypertensive patients. Low intakes of calcium, potassium, vitamin A, and vitamin C are common.

Labile Hypertension Blood pressure increases naturally in circumstances where physiological homeostasis is threatened. Examples of such situations include exercise; emotional responses such as anger, anxiety, or fear; and hostile environ-

ments such as extreme cold. Blood pressure typically increases before treadmill tests, athletic competition, and job interviews.

Increased catecholamine secretion due to psychologically induced excessive sympathetic stimulation may produce a labile hypertension that subsides with relaxation. Labile hypertension results from an increase in cardiac output rather than an increase in peripheral vascular resistance. It is most common in young people in the early stages of hypertension. Its significance is unknown.

Blood pressure response in labile hypertension is usually normal during exercise, with blood pressure often dropping to normal values during recovery. In true labile hypertension, no modification of the exercise program is required. Endurance exercise training reduces the pressor response to emotional stress.

Treatment of Hypertension

Drugs are currently the front line of defense against hypertension. The introduction of this mode of treatment in the 1950s has led to a 40% decrease in mortality from the disease. In general, the aim of hypertension management is to bring the standing blood pressure below 140/90, provided the side effects can be tolerated. In addition to medication, treatment for hypertension includes limiting sodium intake, weight reduction, regular exercise, cessation of smoking, and relaxation.

Hypertensives are often put on a low sodium diet. When body sodium levels exceed the kidneys' ability to excrete it, blood pressure increases. Hypernatremia (high sodium) tends to increase blood volume and adversely affect kidney function.

Three types of drugs are commonly employed in the treatment of hypertension, either singly or in combination: diuretics, vasodilators, and adrenergic inhibitors (β-blockers). Diuretics work by decreasing body sodium, potassium, and fluid volume, and eventually peripheral vascular resistance. Examples of diuretic drugs include hydrocholorothiazide, chlorothiazide, bendoroflumethiazide, and metolazone. Vasodilators decrease peripheral vascular resistance. Examples include hydralazine, calcium channel blockers, α-blockers, and angiotensin-converting enzyme (ACE) inhibitors. Adrenergic inhibitors interfere with sympathetic transmission to the heart. They include such drugs as propranolol, methyldopa, reserpine, metaprolol, guanethidine, and clonidine.

Control of hypertension is often a trial-and-error procedure that must balance the effects of drugs with their side effects. The exercise leader can help in this process by observing differing responses to physical activity. Because of the effect of exercise in lowering blood pressure, some people with hypertension may unilaterally cease their medication. The exercise leader should be aware that this may cause a rebound phenomenon, which results in intense vasoconstriction, with more serious hypertension that may be difficult to reverse.

Exercise and Hypertension

Acute Exercise During endurance exercise, systolic blood pressure increases to a maximum of 160–220 mmHg and diastolic blood pressure to 50–90 mmHg. Direct arterial measurements of diastolic blood pressure tend to be higher than indirect measurement. Exercise blood pressure is excessively high when systolic pressure is more than 225 mmHg or diastolic pressure is greater than 90 mmHg. Exercise and exercise testing should stop if systolic pressure is greater than 250 mmHg or if diastolic pressure is greater than 120 mmHg [American College of Sports Medicine (ACSM) guidelines].

In general, patients with high resting blood pressure have the highest systolic and diastolic blood pressures during exercise. However, exercise blood pressure is a poor predictor of hypertension. Many subjects who have an exaggerated blood pressure response to exercise will not develop hypertension (i.e., high resting blood pressure).

As hypertension progresses, maximal oxygen consumption is impaired. Hypertension eventually reduces stroke volume by diminishing ejection fraction. The cause is progressive cellular deterioration that affects cardiac contractility. As the disease pro-

gresses, cardiac output continues to decline and peripheral resistance increases.

Resting blood pressure decreases below the baseline following exercise. This is called post-exercise hypotension. Post-exercise hypotension is accompanied by decreases in epinephrine, dopamine, cortisol, and sympathetic nervous activity. Recent studies suggest that secretion of endogenous opioids and serotonin are related to the inhibition of sympathetic activity and reduced blood pressure after exercise. The significance of post-exercise hypotension in the control of hypertension is unknown.

Exercise Training In general, endurance exercise training reduces resting blood pressure. However, in most cases, exercise alone will not reduce high blood pressure to normal levels. Athletes and former athletes have a much lower incidence of hypertension than do sedentary people. African Americans have a higher incidence of hypertension than other races, but active black males have a lower incidence than their sedentary counterparts. Even chronic static exercise may decrease the chances of developing hypertension. For example, people who worked in jobs requiring them to perform static muscle contractions (e.g., laborers) had a lower incidence of the disease.

Training may reduce blood pressure by decreasing the activity of the sympathetic nervous system and by altering baroreflex sensitivity. A contributing mechanism for this may be increased insulin receptor sensitivity and reduction of hyperinsulinemia. As discussed, exercise may increase the secretion of arterial natriuretic factor by the heart, which also has an attenuating effect on the sympathetic nervous system. Endorphin and serotonin secretion may play a role as well in exercise-induced reduction of blood pressure. Long-term changes in blood vessel diameter induced by exercise training may also play a role.

Weight reduction is usually essential for blood pressure reduction. The majority of studies suggest that for every 1 kg decrease in body weight systolic pressure will fall by 1.6 mmHg and diastolic pressure will fall by 1.3 mmHg. Exercise, in combination with diet, can help decrease abdominal fat deposition, which is particularly important for reducing blood pressure. Exercise alone can decrease both systolic and diastolic blood pressure by 10 mmHg. Combining exercise and diet may reduce blood pressure enough so that medication can be reduced or eliminated. It can also have a marked effect on other risk factors of CAD, including reducing blood lipids, glucose and insulin metabolism, and body fat.

The level of intensity of exercise training correlates to the level of reduction in resting blood pressure. Moderate-intensity exercise, below 70% of maximum heart rate, typically causes a decrease in systolic blood pressure of approximately 20 mmHg and in diastolic blood pressure of 16 mmHg. These changes do not occur when exercise intensity is intense. Overtraining also increases resting blood pressure.

Deconditioning generally causes training-induced changes in resting blood pressure to reverse themselves. Resting blood pressure may increase 15–20 mmHg (diastolic and systolic pressures) with 3–4 weeks of inactivity.

Not all patients will respond to endurance exercise with a decrease in resting blood pressure. Nonresponders include patients with exaggerated exercise blood pressure responses, decreased cardiac compliance, and significant cardiac hypertrophy. Nonresponders also do not show decreased catecholamine levels with training.

Weight training decreases resting blood pressure in hypertensive patients. These results have been found for free-weight training and circuit training. However, weight training can also result in explosive increases in blood pressure, so caution is advised. Hypertensive patients should follow low-intensity, high-repetition training programs and should avoid performing Valsalva's maneuver (i.e., expiring against a closed glottis; holding their breath and straining) during the exercises.

Exercise Testing and Prescription Heart failure can occur due to an increased load on the heart resulting from hypertension and from the effects of coronary disease. During exercise stress tests, it is particularly important to take frequent measurements so that falling systolic blood pressure can be detected immediately.

During exercise stress testing, care should be taken to prevent patients from tightly grasping the supports of the treadmill or bicycle ergometer. Tightly grasping the bars involves isometric muscle contractions that may increase systolic and diastolic blood pressure. It also decreases the relative intensity and $\dot{V}_{O_2}$ of the exercise.

Unaccustomed static exercise, such as shoveling snow, water skiing, or carrying heavy suitcases, are particularly dangerous for people with hypertension. The combination of isotonic and isometric exercise, especially in the cold, greatly increases blood pressure. It may produce an imbalance between the oxygen supply and oxygen demand in the heart.

Care must be taken when prescribing exercise and administering exercise stress tests to patients with hypertension. Even though exercise is sometimes used as an adjunct treatment of the disease, the utmost caution should be used in its application. Whereas exercise training tends to lower resting blood pressure, acute exercise raises it. During or after exercise, hypertensive patients may be more susceptible to heart failure, coronary ischemia, angina, claudication, and possibly stroke.

Warm-up is very important for hypertensive people. Sudden, high-intensity exercise can result in explosive increases in blood pressure that compromise coronary artery blood flow. Hypertensives should increase the intensity of exercise very gradually. They should stress low-intensity, long-duration endurance activities rather than short-term, high-intensity exercise. Walking, jogging, or swimming is more appropriate than interval sprint training or start-and-stop sports such as racquetball.

Drug side effects should be considered in exercise prescription, and the patient warned of possible symptoms. Diuretics can cause dehydration, hypovolemia (low plasma volume), and muscle cramps, and could conceivably precipitate or augment heat injury during heavy exercise in hot environments. Certainly, if diuretics cause significant dehydration, they could hamper work capacity and increase the incidence of arrhythmias.

Adrenergic inhibitors, depending on the drug,

produce side effects such as orthostatic intolerance, severe post-exercise hypotension, bradycardia, asthma, fatigue, and drowsiness. Atenolol (a commonly prescribed β-blocker), for example, reduces heart rate at rest and during exercise. Its effects should be taken into consideration when prescribing exercise.

Vasodilators, again depending on the drug, can produce hypotension, tachycardia, and postural weakness. With all hypertension medication, particular attention should be directed toward the warm-down, or the continued low level of exercise after the workout, because of the dangers of hypotension, which can produce dizziness or fainting.

■ Chronic Heart Failure (CHF)

Chronic heart failure is the inability of the heart to meet circulatory demands. Common characteristics of the condition include impaired left ventricular systolic and diastolic function, baroreflex desensitization and sympathetic activation, abnormal neuroendocrine system activation, impaired circulation, and skeletal muscle abnormalities.

Myocardial infarction, causing loss of cardiac muscle, or ischemia can reduce cardiac contractility and cause depressed systolic function in the heart. Patients who have suffered from myocardial infarction often have a decreased end-diastolic volume (i.e., ventricular filling), which hampers their ability to increase stroke volume and cardiac output by the Frank–Starling mechanism (see Chapter 14).

The baroreceptors are important for controlling changes in blood pressure. In CHF, the baroreceptors are desensitized, which causes exaggerated activity of the sympathetic nervous system and catecholamine secretion. This reaction causes vasoconstriction of many tissues, particularly skeletal muscle, and contributes to decreased exercise capacity. In a manner similar to the lactate paradox at high altitude (see Chapter 23), the increased catecholamine secretion increases carbohydrate metabolism and lactate production. Lactate clearance is impaired because of diminished muscle blood flow.

Consequently, patients experience metabolic acidosis and reduced oxidative phosphorylation during exercise. Chronic underperfusion of muscle tissue and acidosis can lead to muscle fiber alteration (loss of type I fibers), decreased mitochondrial mass, and rhambdomyolysis (destruction of muscle tissue).

Neuroendocrine system activation causes increased secretion of renin, angiotensin II, aldosterone, arginine vasopressin, and atrial natriuretic peptide, all of which cause vasoconstriction in vascular beds throughout the body and increased plasma volume. The resulting increase in blood volume places a further load on the failing heart and vasoconstriction impairs cellular metabolism. Chronic vasoconstriction is thought to be responsible for endothelial cell abnormalities in blood vessels, which impair the cells' ability to secrete nitric oxide (an important vasodilator).

Circulatory vasoconstriction causes an increase in pulmonary blood pressure that results in a ventilation-perfusion mismatch in the lungs. This causes shortness of breath during exercise, a common symptom in CHF patients. Exertional dyspnea impairs functional capacity and motivation to exercise.

Exercise Training in CHF

Most studies show marked increases in maximal oxygen consumption with endurance training in CHF but little or no change in cardiac output, stroke volume, pulmonary wedge pressure (innate pulmonary pressure), and left ventricular ejection fraction (% of end-diastolic volume ejected). Thus, improvements in endurance capacity are due mainly to peripheral adaptations. Other changes from training include improved baroreflex sensitivity, decreased neuroendocrine system secretion and sympathetic activity, and increased parasympathetic activity, tissue vasodilation capacity, and muscle endurance.

Prior to the 1980s, CHF was rated as a major contraindication for exercise testing and training. However, with proper management, such as identifying and treating the underlying mechanisms of the disorder and controlling potentially fatal arrhythmias, exercise is an important treatment for most CHF patients.

■ Heart Transplantation

In the United States, nearly 3,000 heart transplants (HT) are performed each year. This procedure greatly increases the survival rate of people with end-stage CHF and improves their quality of life. However, exercise intolerance is common in these patients, due to such factors as the inability to recovery from the effects of CHF, reactions to corticosteroids used to prevent rejection of the transplanted heart, and loss of cardiac control associated with loss of sympathetic and parasympathetic nerves to the transplanted heart.

As discussed, cellular changes occurring with CHF combined with inactivity causes changes in skeletal muscle that decrease the exercise capacity in HT patients. Continued weakness from surgery combined with previous deconditioning in muscle to make exercise training very difficult.

Corticosteroid treatment to prevent rejection of the transplanted heart has major catabolic effects on bone and muscle. Osteoporosis, resulting in an increased incidence of fracture, is extremely common in HT patients. Muscle atrophy is also common, resulting in decreased muscle strength. Muscle strength is highly related to maximal oxygen consumption capacity in HT patients. Weight training helps prevent atrophy and prevents bone loss in these patients and is an important component of rehabilitation.

HT patients lack sympathetic and parasympathetic nerves to their transplanted hearts. Consequently, they have higher resting heart rates (100–120 bpm). During the early phases of exercise, they are dependent upon the Frank–Starling mechanism to increase cardiac output. However, as adrenal catecholamine secretion increases during exercise, heart rate can increase by humoral means. The HT patient experiences a sluggish increase in heart rate during exercise and slow decrease in heart rate during recovery. The recovery heart rate is slower be-cause the heart has no stimulus from the vagus nerve (parasympathetic) and is subject to elevated catecholamines that were secreted during exercise.

Exercise and the HT Patient

Warm-up and cool-down are particularly important for HT patients. Warm-up maximizes the action of the muscle pump and stimulates catecholamine secretion, which helps boost exercise cardiac output. Cool-down helps metabolize catecholamines gradually, which reduces the tendency to develop potentially dangerous arrhythmias.

Most HT patients should do endurance (e.g., walking) and resistance exercises. Because capacity is usually low, low intensity interval training, practiced several times a day for short intervals, is often highly effective. Resistance exercise, such as weight training, is important for increasing muscle strength and preventing loss of bone mass that normally accompanies corticosteroid treatment. Increasing muscle strength is very important for restoring functional capacity in these patients.

HT patients cannot increase cardiac output sympathetically (except through circulating epinephrine), so they may experience dizziness and fainting when lifting weights. They must be careful not to strain (i.e., not to perform Valsalva's maneuver) when doing exercises because this will decrease venous return of blood to the heart. The normal heart would compensate by increasing its heart rate, but this is not possible in the transplanted heart.

■ Summary

Over 58 million people in the United States have one or more types of cardiovascular disease. This includes 50 million with hypertension, 14 million with coronary heart disease, 4 million with stroke, and 1.8 million with rheumatic heart disease. One in five men and women have some form of cardiovascular disease.

Although rare, sudden cardiac death (SCD) sometimes occurs during exercise. In people under

35, it is most commonly associated with hypertrophic cardiomyopathy. Other causes of sudden cardiac death in young people include coronary artery anomalies, myocarditis, ruptured aortic aneurysms associated with Marfan's syndrome, aortic stenosis, various types of arrhythmias, myocarditis, and blunt trauma to the chest. Coronary artery disease is the major cause of SCD death in people over 35. Coronary disease can also cause sudden death in young people, especially those with familial hypercholesterolemia.

Coronary artery disease (CAD) is the leading cause of death in Western countries. Heart disease rates have decreased for over 20 years, probably due to changes in lifestyle caused by increased awareness of CAD risk factors. Coronary artery disease involves a steady buildup of atherosclerotic plaques in the coronary arteries, leading to reduced myocardial blood flow. Current research suggests that CAD can be prevented and, to a certain extent, reversed.

Atherosclerosis is characterized by the formation of lipid-filled fibrous plaques that collect on the inner surface of the coronary arteries. The critical event in the formation of the fibrous plaque is the accumulation of smooth muscle cells in the intima (inner lining) of the artery. Cell growth factors stimulate the growth of smooth muscles and the formation of connective tissues—that makes up the fibrous element of the plaque—and attract smooth muscle cell and monocytes to the fibrous plaque. Macrophages and smooth muscle cells create foam cells, which are filled with lipids, that accumulate and form part of the core of the fibrous plaque. Eventually the plaque calcifies and connective tissue forms, producing a narrowed, rigid blood vessel. As the disease progresses, the weak fibrous plaque can rupture or ulcerate, which leads to thrombosis.

The most important risk factors of CAD are hyperlipidemia, hypertension, lack of exercise, and cigarette smoking. High cholesterol and triglycerides, low high-density lipoprotein (LDL), and a high cholesterol/HDL ratio increase the risk of developing CAD. Hypertension (high blood pressure) is an arterial systolic blood pressure above 140 mmHg or a diastolic pressure greater than 90 mmHg. Hypertension affects the development of atherosclerosis, increases myocardial oxygen consumption at rest and exercise, and causes cellular changes in the heart and blood vessels. Smoking may stimulate CAD by decreasing blood pressure, heart rate, platelet adhesiveness, and fatty acid mobilization, and decreasing HDL. Exercise training may result in a regression of atheroma and a reduction in the magnitude of other risk factors such as hypertension, obesity, and dyslipidemia.

Using of exercise therapy is a radical change in the philosophy and treatment of the cardiac- and cardiac-prone patient. Patients in cardiac rehabilitation programs have an improved sense of well-being, reduced anxiety and depression, improved exercise capacity, reduced ST segment depression on the electrocardiogram, reduced blood pressure, reduced heart rate at rest, reduced levels of serum cholesterol and triglycerides and elevated levels of HDL, and perhaps a lower mortality rate. The ideal cardiac rehabilitation program consists of three phases: the inpatient program, the outpatient therapeutic program, and the exercise maintenance program.

■ Selected Readings

Adamopoulos, S., G. M. Rosano, P. Ponikowski, E. Cerquetani, M. Piepoli, F. Panagiota, P. Collins, P. Poole-Wilson, D. Kremastinos, and A. J. Coats. Impaired baroreflex sensitivity and sympathovagal balance in Syndrome X. *Am. J. Cardiol.* 82: 862–868, 1998.

American College of Sports Medicine. Guidelines for Exercise Testing and Prescription. 5th ed. Baltimore: Williams and Wilkins, 1995.

American College of Sports Medicine. ACSM's exercise management for persons with chronic diseases and disabilities. Champaign, IL: Human Kinetics, 1997.

American College of Sports Medicine. ACSM position stand on the recommended quantity and quality of exercise for developing and maintaining cardiorespiratory, muscular fitness, and flexibility in adults. *Med. Sci. Sports Exerc.* 30: 975–991, 1998.

American Heart Association. Exercise Standards: A Statement for Health Professionals. Dallas: American Heart Association, 1991.

Berlin, J. A., and G. A. Colditz. A meta-analysis of physical activity in the prevention of coronary heart disease. *Am. J. Epidemiol.* 132: 612–628, 1990.

Bijnen, F. C., C. J. Caspersen, and W. L. Mosterd. Physical inactivity as a risk factor for coronary heart disease: a WHO and International Society and Federation of Cardiology position statement. *Bull. World Health Organ.* 72(1): 1–4, 1994.

Blair, S. N., N. N. Goodyear, L. W. Gibbons, and K. H. Cooper. Physical fitness and incidence of hypertension in healthy normotensive men and women. *J.A.M.A.* 25: 487–490, 1984.

Blair, S. N., H. W. Kohl, R. S. Paffenbarger, D. G. Clark, K. H. Cooper, and L. W. Gibbons. Physical fitness and all-cause mortality: a prospective study of healthy men and women. *J.A.M.A.* 262: 2395–2401, 1989.

Braden, D. S., and W. B. Strong. Preparticipation screening for sudden cardiac death in high school and college athletes. *Physician Sportsmed.* 16(10): 128–140, 1988.

Braith, R. W. Exercise training in patients with CHF and heart transplant recipients. *Med. Sci. Sports Exerc.* 30(Suppl.): S367–S378, 1998.

Braunwald, E., J. Ross, and E. H. Sonnenblick. Mechanisms of contraction of the normal and failing heart. *N. Eng. J. Med.* 277: 794–799, 853–862, 910–919, 962– 971, 1012–1022, 1967.

Cannon, R. O., III. Role of nitric oxide in cardiovascular disease: focus on the endothelium. *Clin. Chem.* 44: 1809–1819, 1998.

Defronzo, R. A., and E. Ferrannini. Insulin resistance. A multifaceted syndrome responsible for NIDDM, obesity, hypertension, dyslipidemia, and atherosclerotic cardiovascular disease. *Diabetes Care* 14: 173–194, 1991.

Department of Health and Human Services. Physical Activity and Health: A Report of the Surgeon General. Atlanta: U.S. Department of Health and Human Services, Centers for Disease Control and Prevention, National Center for Chronic Disease Prevention and Health Promotion, 1996.

Dwyer, T., and L. E. Gibbons. The Australian Schools Health and Fitness Survey. Physical fitness related to blood pressure but not lipoproteins. *Circulation* 89: 1539–1544, 1994.

Eriksson, K. F., and F. Lindgarde. No excess 12-year mortality in men with impaired glucose tolerance who participated in the Malmo Preventive Trial with diet and exercise. *Diabetologia* 41: 1010–1016, 1998.

Fahey, T. D., P. Insel, and W. Roth. Fit and Well. 3d Ed. Mountain View, Ca.: Mayfield Publishing Company, 1999.

Farrell, S. W., J. B. Kampert, H. W. Kohl III, C. E. Barlow, C. A. Macera, R. S. Paffenbarger, Jr., L. W. Gibbons, and S. N. Blair. Influences of cardiorespiratory fitness levels and other predictors on cardiovascular disease mortality in men. *Med. Sci. Sports Exerc.* 30: 899–905, 1998.

Fontbonne, A. M., and E. M. Eschwege. Insulin and cardiovascular disease. Paris prospective study. *Diabetes Care* 14: 173–194, 1991.

Franklin, B. A., G. F. Fletcher, N. F. Gordon, T. D. Noakes, P. A. Ades, and G. J. Balady. Cardiovascular evaluation of the athlete. Issues regarding performance screening and sudden cardiac death. *Sports Med.* 24: 97–119, 1997.

Friedman, M., and R. H. Rosenman. Type A Behavior and Your Heart. New York: Knopf, 1974.

Friedman, M., R. H. Rosenman, R. Straus, M. Wurm, and R. Kositchek. The relationship of behavior pattern A to the state of the coronary vasculature. *Am. J. Med.* 44: 525–537, 1968.

Friman, G., E. Larsson, and C. Rolf. Interaction between infection and exercise with special reference to myocarditis and the increased frequency of sudden deaths among young Swedish orienteers 1997–92. *Scand. J. Infect. Dis.* 104 (Suppl.): 41–49, 1997.

Futterman, L. G., and R. Myerburg. Sudden Death in Athletes. *Sports Med.* 26: 335–350, 1998.

Goodyear, L. J., and B. B. Kahn. Exercise, glucose transport, and insulin sensitivity. *Annu. Rev. Med.* 49: 235–261, 1998.

Gordon, T., W. P. Castelli, M. C. Hjortland, W. B. Kannel, and T. R. Dawber. High density lipoprotein as a protective factor against coronary heart disease: the Framingham Study. *Am. J. Med.* 6: 707–714, 1977.

Guyton, J. R., and J. W. Heinecke. The pathogenesis of arteriosclerosis. In Training Program for the Clinical Management of Lipid Disorders. Dallas: American Heart Association, 1992.

Haskell, W. L., E. L. Alderman, J. M. Fair, D. J. Maron, S. F. Mackey, H. R. Superko, P. T. Williams, I. M. Johnstone, M. A. Champagne, and R. M. Krauss. Effects of intensive multiple risk factor reduction on coronary atherosclerosis and clinical cardiac events in men and women with coronary artery disease. The Stanford Coronary Risk Intervention Project (SCRIP). *Circulation* 89: 975–990, 1994.

Haskell, W. L., A. S. Leon, C. J. Caspersen, V. F. Froelicher, J. M. Hagberg, W. Harlan, J. O. Holloszy, J. G. Regensteiner, P. D. Thompson, R. A. Washburn, and P. W. F. Wilson. Cardiovascular benefits and assessment of physical activity and physical fitness in adults. *Med. Sci. Sports Exer.* 24: S201–S220, 1992.

Hiatt, W. R., J. G. Regensteiner, M. E. Hargarten, E. E. Wolfel, and E. P. Brass. Benefit of exercise conditioning for patients with peripheral arterial disease. *Circulation* 81: 602–609, 1990.

Hopkins, P. N., S. C. Hunt, P. J. Schreiner, J. H. Eckfeldt, I. B. Borecki, C. R. Ellison, R. R. Williams, and K. D. Siegmund. Lipoprotein(a) interactions with lipid and non-lipid risk factors in patients with early onset coronary artery disease: results from the NHLBI Family Heart Study. *Atherosclerosis* 141: 333–345, 1998.

Israel, R. G., M. J. Sullivan, R. H. Marks, R. S. Cayton, and T. C. Chenier. Relationship between cardiorespiratory fitness and lipoprotein(a) in men and women. *Med. Sci. Sports Exer.* 26: 425–431, 1994.

Kannel, W. B., P. W. F. Wilson, and S. N. Blair. Epidemiological assessment of the role of physical activity and fitness in the development of cardiovascular disease. *Am. Heart J.* 109: 855–876, 1985.

Kramsch, D. M., A. J. Aspen, B. M. Abramowitz, T. Kreimendahl, and W. B. Hood. Reduction of coronary atherosclerosis by moderate conditioning exercise in monkeys on an atherogenic diet. *New Eng. J. Med.* 305: 1483–1489, 1981.

Leon, A. S. Effects of exercise conditioning on physiologic precursors of coronary heart disease. *J. Cardiopulmon. Rehabil.* 11: 46–57, 1991.

McCartney, N. Role of resistance training in heart disease. *Med. Sci. Sports Exer.* 30(Suppl.): S396–S402, 1998.

McGinnis, J. M. The public health burden of a sedentary lifestyle. *Med. Sci. Sports Exer.* 24: S196–S200, 1992.

McMurray, R. G., B. E. Ainsworth, J. S. Harrell, T. R. Griggs, and O. D. Williams. Is physical activity or aerobic power more influential on reducing cardiovascular disease risk factors? *Med. Sci. Sports Exer.* 30: 1521–1529, 1998.

Morris, J. N., D. G. Clayton, M. G. Everitt, A. M. Semmence, and E. H. Burgess. Exercise in leisure time: coronary attack and death rates. *Br. Heart J.* 63: 325–334, 1990.

Morris, J. N., M. G. Everitt, R. Pollard, S. P. W. Chave, and A. M. Semmence. Vigorous exercise in leisure-time: protection against coronary heart-disease. *Lancet* 2: 1207–1210, 1980.

Noakes, T. D. Sudden death and exercise. In Encyclopedia of Sports Medicine and Science, T. D. Fahey (Ed.). Internet Society for Sport Science: http://sportsci.org, 8 Nov 1998.

Oldridge, N. B., G. H. Guyait, M. E. Fischer, and A. A. Rimm. Cardiac rehabilitation with exercise after myocardial infarction. *J.A.M.A.* 260: 945–950, 1988.

Paffenbarger, R. S. Contributions of epidemiology to exercise science and cardiovascular health. *Med. Sci. Sports Exer.* 20: 426–438, 1988.

Paffenbarger, R. S., and W. E. Hale. Work activity and coronary heart disease. *New Eng. J. Med.* 292: 545–550, 1975.

Paffenbarger, R. S., R. T. Hyde, A. L. Wing, and C-C. Hsieh. Physical activity, all-cause mortality, longevity of college alumni. *New Eng. J. Med.* 314: 605–614, 1986.

Paffenbarger, R. S., R. T. Hyde, A. L. Wing, and C-C. Hsieh. Physical activity and hypertension: an epidemiological view. *Ann. Med.* 23: 319–327, 1991.

Pini, O., and D. T. Lowenthal. Evaluation and treatment of hypertension in active individuals. *Med. Sci. Sports Exer.* 30 (Suppl.): S354–S366, 1998.

Pronk, N. P. Short term effects of exercise on plasma lipids and lipoproteins in humans. *Sports Med.* 16: 431–448, 1993.

Reaven, P. D., E. Barrett-Connor, and S. Edelstein. Relation between leisure-time physical activity and blood pressure in older women. *Circulation* 83: 559–565, 1991.

Rodriguez, B. L., J. D. Curb, C. M. Burchfiel, R. D. Abbott, H. Petrovitch, K. Masaki, and D. Chiu. Physical activity and 23-year incidence of coronary heart disease morbidity and mortality among middle-aged men. The Honolulu Heart Program. *Circulation* 89: 2540–2544, 1994.

Ross, R. The pathogenesis of atherosclerosis—an update. *N. Eng. J. Med.* 314: 488–500, 1986.

Ross, R., and J. A. Glomset. Pathogenesis of atherosclerosis. *N. Engl. J. Med.* 295: 369–377, 420–425, 1976.

Slattery, M. L., and D. R. Jacobs. Physical fitness and cardiovascular disease mortality: the U.S. Railroad Study. *Am. J. Epidemiol.* 127: 571–580, 1987.

Squires, R. W., G. T. Gau, T. D. Miller, T. G. Allison, and C. J. Lavie. Cardiovascular rehabilitation: status 1990. *Mayo Clin. Proc.* 65: 731–755, 1990.

Superko, H. R. Exercise training, serum lipids, and lipoprotein particles: is there a change threshold? *Med. Sci. Sports Exer.* 23(6): 677–685, 1991.

Superko, H. R. Prevention and regression of atherosclerosis with drug therapy. *Clin. Cardiol.* 14 (Suppl. 1): 140–147, 1991.

Superko, H. R. Did grandma give you heart disease? The new battle against coronary artery disease. *Am. J. Cardiol.* 82: 34Q–46Q, 1998.

Superko, H. R., and W. L. Haskell. The role of exercise training in the therapy of hyperlipoproteinemia. *Clin. Cardiol.* 5: 285–310, 1987.

Suzuki, I., H. Yamada, T. Sugiura, N. Kawakami, and H. Shimizu. Cardiovascular fitness, physical activity and selected coronary heart disease risk factors in adults. *J. Sports Med. Phys. Fitness* 38: 149–157, 1998.

Tanaka, H., C. M. Clevenger, P. P. Jones, D. R. Seals, and C. A. DeSouza. Influence of body fatness on the coronary risk profile of physically active postmenopausal women. *Metabolism* 47: 1112–1120, 1998.

Thompson, P. D. The cardiovascular complications of vigorous physical activity. *Arch. Int. Med.* 156: 2297–2302, 1996.

Thompson, P. D., F. J. Erik, R. A. Carelton, and W. Q. Sturner. Incidence of death during jogging in Rhode Island from 1975–1980. *J.A.M.A.* 247: 2535–2538, 1982.

Tipton, C. M. Exercise, training and hypertension: an update. *Exer. Sport Sci. Rev.* 19: 447–505, 1991.

Toobert, D. J., L. A. Strycker, and R. E. Glasgow. Lifestyle change in women with coronary heart disease: what do we know? *J. Women's Health* 7: 685–699, 1998.

Tuomi, K. Characteristics of work and life predicting coronary heart disease. Finnish research project on aging workers. *Soc. Sci. Med.* 38: 1509–1519, 1994.

Virmani, R., A. P. Burke, A. Farb, and J. A. Kark. Causes of sudden death in young and middle-aged competitive athletes. *Cardiol. Clin.* 15: 439–466, 1997.

CHAPTER 25

OBESITY, BODY COMPOSITION, AND EXERCISE

Obesity is excess body fat. It results from consuming more energy than is expended. It affects nearly 30% of the adults in Western countries and is a serious health problem (Figure 25-1). The causes of obesity are complex, and the problem is resistant to cure. Overfatness runs in families. Its roots are genetic and environmental (i.e., eating and physical activity patterns), and the disorder is often passed from one generation to the next

Obesity is a major health problem. High levels of body fat increase the risk of coronary heart disease, stroke, hypertension, hyperlipidemia, diabetes, osteoarthritis, gallstones, gallbladder disease, renal disease, hepatic cirrhosis, accident proneness, surgical complications, and back pain. The economic and emotional consequences of obesity are staggering.

Obesity impairs physical performance. Because greater mass requires more energy, overweight people must work harder at most tasks. Unfortunately, they are usually physically unfit, which compounds their handicap. The obese are also usually chronically inactive, which deprives them of an important means of weight loss.

Obesity carries a social stigma. Society often ridicules and subtly discriminates against overweight people. Social rejection sometimes causes emotional problems. Consequently, many people

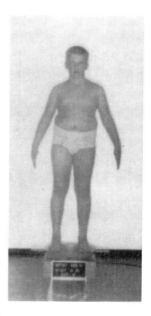

Obesity predisposes people to a variety of degenerative diseases such as diabetes and coronary artery disease.

battle constantly to lose weight in an attempt to escape obesity. Often, unrealistic expectations and pressure from society to look lean and fit compound the problem.

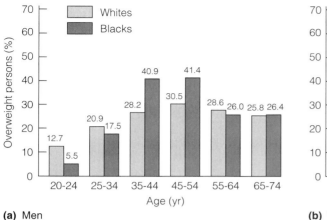
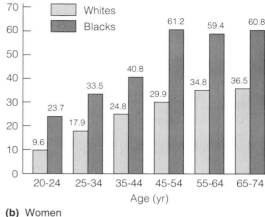

(a) Men **(b)** Women

Figure 25-1   Percentage of overweight men (a) and women (b) by age and race.

A diet industry has developed to cater to the national obsession with weight control. People spend millions of dollars on diet books, low-calorie foods, and appetite suppressants in the quest for leanness. These remedies usually provide only a short-term solution to the problem. In addition, chronic participation in quick-weight-loss programs causes large fluctuations in weight that can be unhealthy.

Habitual dieters usually emphasize weight loss rather than fat loss. Too often they see the loss of water or lean mass desirable. They concentrate on losing weight quickly rather than on changing the habits that result in obesity. Because food provides nutrients as well as energy, an adequate caloric intake is necessary for good health. The long-term solution lies in achieving a balance between energy intake and energy expenditure.

Body composition is an important consideration in health, aesthetics, and physical performance. In this chapter, we define body composition as the relative proportion of fat and fat-free weight. To a large extent, body composition is a product of caloric intake from the diet and the caloric expenditure from physical activity. In most people, obesity results from eating too much and not getting enough exercise.

Management of body composition is important in physical performance. In sports such as distance running, body composition can determine success or failure. In football, the drive toward a lean body must be balanced against the advantage of a large body mass. Each sport has unique requirements of body composition that must be considered to attain maximum performance.

This chapter explores the causes, consequences, and remedies of obesity. Topics also include body composition measurement techniques and the influence of body composition on health, aesthetics, and performance.

■ Obesity and Health

Obesity— the accumulation of body fat that is more than 25% of total body weight for men and 33% of total body weight for women—is associated with a wide variety of health problems (Table 25-1). Obese people have an overall mortality rate almost twice that of nonobese people. It increases health risks if body fat is stored disproportionately in the abdomen. In June 1998, the American Heart Association classified obesity as one of the major risk factors for

TABLE 25-1

Health Consequences of Obesity

Obesity increases the risk for the following:
- Premature death
- Death from CVD, including sudden death
- Hypertension
- Diabetes and insulin resistance (i.e., Syndrome X)
- Gallbladder disease
- Kidney disease
- Liver disease
- Cancer of the colon, prostate, esophagus, gallbladder, ovary, endometrium, breast, and cervix
- Arthritis
- Orthopedic disorders
- Fatal respiratory disease
- Stroke
- Gout
- Back pain
- Reproductive disorders in women
    - Menstrual disorders
    - Infertility
    - Miscarriage
    - Poor pregnancy outcome
    - Impaired fetal well-being
    - Diabetes mellitus during pregnancy
- Dyspnea
- Sleep apnea and snoring

Obesity is also associated with the following:
- Increased LDL
- Increased triglycerides
- Increased fasting glucose
- Increased fibrinogen levels
- Decreased HDL concentration
- Decreased GLUT-4 transporters
- Impaired cardiac function during exercise
- Impaired immune function
- Depression
- Chronic pain
- Impaired mobility

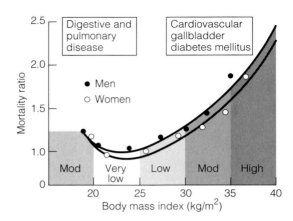

Figure 25-2 Relationship of body mass index and the risk of death from all causes. SOURCE: Bray and Gray, 1988.

heart disease. It is estimated that if all Americans had a healthy body composition, the incidence of coronary heart disease would drop by 25% and average life expectancy would increase by three years.

Total Body Fat The risk of premature death by all causes increases with the degree of obesity. Figure 25-2 shows the relationship of body mass index (BMI) to the death rate from all causes. BMI is weight/height^2, and it is fairly closely related to amount of body fat ($r \approx 0.80$). Although BMI is not as accurate as laboratory methods of determining body composition, it is widely used in epidemiological research because height and weight measurements are readily available. The relationship between BMI and mortality takes on a U-shaped curve: Both low and high BMI increase mortality risk. Extremely low BMI increases the risk of lung and gastrointestinal disease; high BMI increases the risk of arthritis, diabetes, hypertension, cardiovascular disease, gallbladder disease, and other disorders. The increased risk of diabetes at even fairly low values of BMI is of particular concern.

Optimal BMI depends on age and other health factors, but the BMI ranges for good health presented in Table 25-2 provide useful general guidelines. The relative risk for several health problems associated with different values of BMI is shown in Figure 25-3.

Body Fat Distribution People tend to store fat in the abdomen (abdominal, android, or male-type fat) or in the lower body (gynoid or female-type fat). Those with abdominal fat patterns have larger and perhaps a greater number of intra-abdominal fat cells. Such people are at greater risk for a variety of

TABLE 25-2

Classification of Overweight and Obesity by Body Mass Index (BMI)

Classification	BMI (kg/m²)	Obesity class	Disease risk relative to normal weight and waist circumference[a]	
			Men ≤ 40 in. (102 cm) Women ≤ 35 in. (88 cm)	>40 in. (102 cm) >35 in. (88 cm)
Underweight[b]	less than 18.5		—	—
Normal[c]	18.5–24.9		—	—
Overweight	25.0–29.9		Increased	High
Obesity	30.0–34.9	I	High	Very high
	35.0–39.9	II	Very high	Very high
Extreme obesity	40.0 and higher	III	Extremely high	Extremely high

[a]Disease risk for type 2 diabetes, hypertension, and CVD. The waist circumference cutoff points for increased risk are 40 inches (102 cm) for men and 35 inches (88 cm) for women.
[b]Research suggests that a low BMI can be healthy in some cases, as long as it is not the result of smoking, an eating disorder, or an underlying disease process.
[c]Increased waist circumference can also be a marker for increased risk even in persons of normal weight.

SOURCE: Adapted from National Heart, Lung, and Blood Institute, 1998.

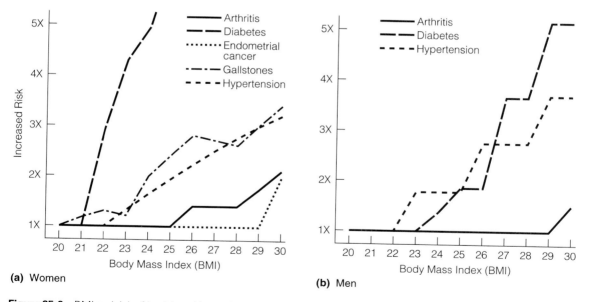

Figure 25-3 BMI and risk of health problems. SOURCE: How's your weight? 1997. *Nutrition Action Healthletter,* December. Copyright © 1997 CSPI. Used with permission.

health problems. Fat deposition patterns are commonly assessed by waist circumference and waist–hip ratio. A high ratio means that fat tends to be stored in the abdomen, while a low ratio means that more fat is stored in the lower body. Abdominal fat patterning is a greater health risk than high total body fat because abdominal fat cells release fat directly into the portal circulation and negatively affect liver metabolism. Abdominal fat deposition is associated with an increased risk of insulin resistance, diabetes, hypertension, and cardiovascular disease. Abdominal obesity is higher in men than women and increases with both age and BMI.

Obesity, Insulin Resistance, and the Metabolic Syndrome

In the late 1980s, several researchers (Reaven, 1988; Kaplan, 1989; and Foster, 1989) described a metabolic syndrome that greatly increased the risk of coronary artery disease. It became known as the metabolic syndrome, or Syndrome X. This syndrome is characterized by hyperinsulinemia (high insulin levels) and insulin resistance. Associated with this condition are hyperglycemia (high blood sugar), increased very-low-density lipoprotein concentrations, decreased high-density lipoprotein cholesterol, and hypertension. In 1990, Bjorntorp proposed that abdominal adipose tissue contributed to the metabolic syndrome because it released free fatty acids into the portal circulation, which went directly to the liver. This resulted in increased VLDL cholesterol and decreased the ability of the liver to clear insulin, causing hyperinsulinemia.

The role of obesity and abdominal fat deposition in the metabolic syndrome is unclear and controversial. Barnard and Wen (1994) speculated that physical inactivity and a high-fat, high-refined-sugar diet were the cause of the metabolic syndrome. As evidence, they cited studies that showed that feeding animals high-fat or high-refined-sugar diets caused insulin resistance within a few weeks. Studies also showed that changing exercise and dietary habits caused improvements in most factors associated with metabolic syndrome (i.e., levels of blood pressure, insulin, triglycerides), even when there is little or no loss of weight.

Insulin resistance and hyperinsulinemia may be a cause rather than a consequence of obesity. In skeletal muscle, lipoprotein lipase decreases when there is insulin resistance in the cells. In fat cells, on the other hand, high insulin levels stimulate lipoprotein lipase and inhibit hormone-sensitive lipase. These changes in muscle and fat may cause a decrease in fat metabolism in muscle and an increased storage of fat in the adipocytes (fat cells) (Barnard and Wen, 1994).

Exercise training appears to be vital in combating metabolic syndrome. Exercise improves insulin sensitivity by increasing glucose transporters (i.e., GLUT-4), oxidative enzymes, and muscle blood flow, and by reducing abdominal fat. Wallace and associates (1997) found that adding resistance exercise to an endurance training program reduces insulin resistance and reduce more body fat in people with hyperinsulinemia more than endurance training alone.

Obesity and Hypertension (cubation ie liver)

Hypertension is twice as common in overweight people as in normal-weight people and three times as common as in those who are underweight (Figure 25-4). This relationship is most striking in white

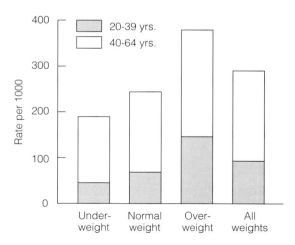

Figure 25-4 The frequency of hypertension in various weight categories. Data from Berchtold et al., 1981.

women and blacks of both sexes. It exists in all age groups, even in children as young as 5 years old. The incidence of hypertension increases with waist–hip ratio. In severely obese people with a ratio greater than 0.81, almost 70% are hypertensive. In Western countries, the incidence of obesity and hypertension increases with age. However, hypertension does not increase with age in populations that do not experience an age-related increase in body fat, such as the Masai and New Guinea tribal natives.

The causes of hypertension in the obese may include high insulin levels, increased insulin resistance related to abdominal fat, increased arterial peripheral resistance, or increased dietary salt intake. Qiao and associates (1998) found that a high body mass index (BMI) greatly increases the risk of developing hypertension. They also found that people with a high BMI have higher fasting insulin levels. Such elevated plasma insulin increases sodium retention in the kidneys, which, in turn, raises blood pressure (Reaven, 1997). Peripheral resistance may rise in obesity because of the expanded tissue volume and relatively smaller capillary density. High dietary sodium intake causes increased blood volume due to expansion of erythrocyte mass and greater plasma volume.

Weight loss in the obese almost always results in a reduction in blood pressure, even without restricted salt intake. This is probably due to the high sodium content of most diets. Salt intake decreases proportionately as people eat fewer calories, and systolic blood pressure decreases almost 7% with a 10% decrease in weight.

Other Hormones Obesity is associated with abnormal responses in prostaglandins, growth hormone, triiodothyronine (T_3), total thyroxin (T_4), serotonin, cortisol, and testosterone. These changes are largely reversed with weight loss. In women, obesity is related to a variety of reproductive problems. These problems include menstrual disorders, infertility, miscarriage, poor pregnancy outcome, impaired fetal well-being, and diabetes mellitus during pregnancy. Weight loss has marked effects on improving the menstrual cycle and promoting spontaneous ovulation and fertility (Norman et al., 1998).

Obesity and Hyperlipidemia

Hyperlipidemia (high blood fats) is common in obesity, with over 31% of obese persons exhibiting this disorder. In obese people with hyperlipidemia, three-quarters have high triglycerides, one-quarter have high cholesterol, and half have a combination of both. There is also a strong negative correlation between plasma high-density lipoproteins (HDL) and relative weight (proximity to ideal weight). The reason for this negative correlation may be that abdominal obesity is associated with a decrease in lipoprotein lipase activity in the liver, which impairs the breakdown of triglycerides. This impaired breakdown of triglycerides stimulates the production of small, dense lipoproteins that are highly atherogenic (i.e., greatly stimulate the development of atherosclerosis). This process also causes a more rapid breakdown of HDL apolipoproteins A-I and A-II, which reduces the levels of HDL. As discussed, HDL has a protective effect against heart disease. Exercise reduces triglycerides and increases HDL, which reinforces the importance of physical activity in weight reduction. A 10% decrease in weight typically decreases cholesterol by 11 mg · 100 ml^{-1}.

Gallbladder disease is a direct result of hyperlipidemia in the obese. Approximately one-third of obese women have gallbladder disease by age 60. Gallbladder disease is a consequence of higher cholesterol density in bile secretions.

Obesity and Musculoskeletal Injury

Obesity increases the risk of many musculoskeletal disorders, particularly osteoarthritis and backache. Increased body mass places chronic stress on joints, eventually leading to arthritic changes. A sagging abdomen, especially prominent in obese men, increases the lumbar lordosis (low back curve). An increased low back curve increases the risk of back pain. Generally, weak and inflexible abdominal, spinal, and leg muscles compound the problem.

■ Energy Balance: The Role of Exercise and Diet

Energy balance determines whether body fat increases, decreases, or remains the same. Energy balance occurs when energy intake equals energy expenditure (Figure 25-5). Excess energy is stored as fat when the balance becomes positive (i.e., more energy is consumed than expended). Once gained, the increased fat remains until there is a period of negative energy balance. Superficially, successful weight control involves not eating too much and exercising. However, people differ in metabolic rate, metabolic efficiency, and the drive to eat. The cause of the positive energy balance in obese people is not totally understood although the answer probably lies in the complex physiological and psychological factors that control food intake and energy expenditure.

Control of Food Intake

The control center for food intake is located in the hypothalamus. The hunger center lies in the lateral regions of the ventral hypothalamus; the satiety center is located primarily in the ventromedial area. The satiety center usually dominates, except when nutrient status declines. Then, the activity of the satiety center decreases, while that of the hunger center increases. Diseases of the hunger-satiety center are involved in some types of obesity, but this cause of obesity is rare.

A number of stimuli may trigger the hunger center, including glucose, stored triglycerides, plasma amino acid levels, and hypothalamic temperature. The *glucostatic theory* of hunger regulation proposes that glucose controls hunger and satiety with glucose sensors located in the hypothalamus. These sensors are sensitive to insulin and evoke satiety

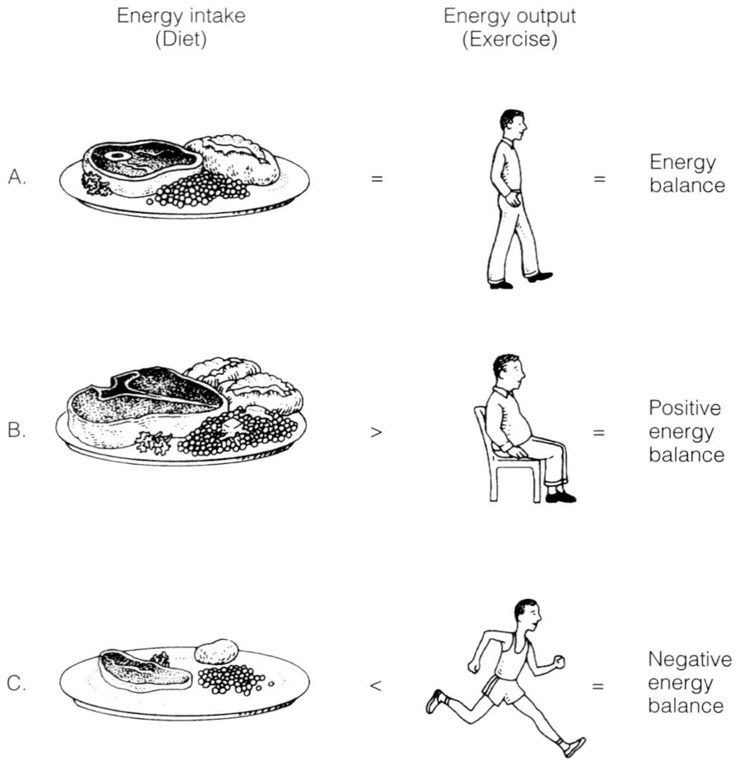

Figure 25-5 Energy balance.

Energy intake (Diet)

Energy output (Exercise)

A. = = Energy balance

B. > = Positive energy balance

C. < = Negative energy balance

signals as the utilization of glucose increases within the sensor cells.

Fat cells may also control long-term appetite control. The *lipostatic theory* proposes that stored triglycerides control the long-term regulation of appetite. This theory has been the focus of considerable research in recent years. The basis of the lipostatic theory is that genes, such as the ob gene, regulate appetite and metabolic rate based on fat cell content. A hormone-like substance called *leptin* is released from adipose tissue in amounts that reflect total body fat content. Leptin acts in the central nervous system (CNS) by binding with leptin receptors to decrease appetite and increase energy expenditure. Leptin controls body fat by acting as an afferent control agent between fat cells and the CNS. It also affects metabolic rate by increasing norepinephrine turnover and, in some animal species, sympathetic nerve activity to thermogenic brown adipose tissue (Haynes et al., 1997). Mice with a defective or missing ob gene gained weight. However, they experienced lost weight, decreased appetite, and increased metabolic rate when given injections of leptin.

In humans, obesity may be much more complicated than a single gene abnormality. Obese people often have normal or above normal levels of leptin, suggesting problems with the leptin receptors in the hypothalamus. Other substances have also been identified as important in appetite control. These substances include orexin, cortisol, glucagon-like peptide-1 (GLP-1), corticotropin-releasing factor, cholecystokinin, somatostatin, neuropeptide Y (NPY), and peptide YY. Researchers have been particularly interested in using leptin and neuropeptide Y analogues as potential antiobesity agents.

Other researchers have proposed an *aminostatic theory*, whereby hunger is controlled by plasma amino acids with an aminostat as a regulator. This theory, too, requires more experimental support.

Hypothalamic temperature may be another factor that controls appetite. This theory is called the *thermostatic theory*. It proposes that the specific heat of food (i.e., the heat given off during digestion) stimulates the satiety center, causing an increased energy expenditure that may be almost 15% of the caloric value of the meal. Increased appetite in the

cold and decreased appetite in the heat may be related to such changes in hypothalamic temperature. Increased temperature may also be the mechanism by which exercise exerts a short-term depressant effect on appetite.

Other factors involved in the hunger-satiety mechanism include gastrointestinal (GI) stretch receptors; nutrient levels; osmotic changes; GI hormones, such as cholecystokinin; hepatic nutrient receptors; hormones sensitive to nutrient status, such as insulin, glucagon, and growth hormone; and plasma levels of nutrients.

Psychological factors can affect food intake as well. This factor strongly influences the onset of obesity and is perhaps the most difficult to alter. Psychosocial factors such as food-centered social gatherings, structured mealtimes, and anxiety-stimulated eating binges are often independent of physiological hunger stimulatory mechanisms. Also, the availability of highly palatable foods may have an independent effect on food intake. Compared to carbohydrates and proteins, high-fat foods have a weak effect on satiation, which encourages passive overconsumption. This can lead to obesity because many high-fat foods taste good and have a high energy density (are calorie dense). Unfortunately, a high fat intake is not physiologically compensated for by an increased energy expenditure. There does not seem to be any link between fat oxidation and fat intake. A summary of hunger-satiety mechanisms appears in Figure 25-6.

■ Energy Expenditure

The components of energy expenditure include resting metabolic rate, thermogenesis, and physical activity. There are individual differences in each of these factors, which may explain the considerable variation in body fat among people on the same diets.

Resting Metabolic Rate

Resting metabolic rate (RMR) is the energy requirement of an awake, resting person measured

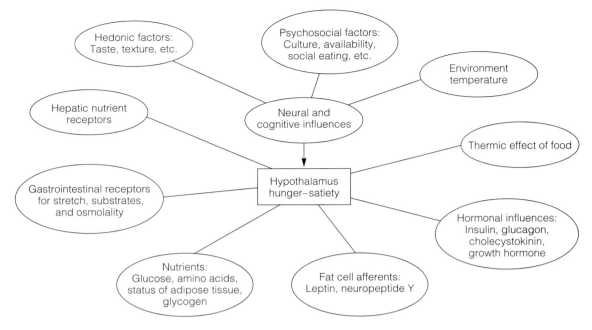

Figure 25-6 Factors controlling hunger and satiety.

8–12 hours after a meal or exercise. It is determined in a comfortable, stress-free environment. RMR accounts for over 70% of daily energy expenditure, varying less than 10% among subjects.

RMR decreases with age as percentage of fat increases and lean body mass decreases. The decline in RMR and caloric expenditure may become substantial over many years, contributing to the increased body fat that occurs with age. RMR is higher in men than women, and there is no difference in RMR between fat and lean people.

Thermogenesis

Thermogenesis is heat production not accounted for by resting, metabolic rate, or physical activity. Factors stimulating thermogenesis include food intake, thermogenic substances (i.e., catecholamines, drugs, and some types of food), cold exposure, and psychological stress. Thermogenesis accounts for approximately 15% of daily energy expenditure.

Abnormalities in thermogenesis may contribute to obesity (Figure 25-7).

Thermic Effect of Food Metabolic rate increases during the digestion, absorption, processing, and storing of food, continuing for several hours after a meal. The metabolic effects of food intake account for 5–10% of total energy expenditure. These effects are less for obese than for lean subjects. The energy cost of meals is highest for protein, lower for carbohydrates, and lowest for fats. Exercising before a meal increases the thermic effect of food in obese and lean subjects. Exercising after a meal increases the metabolic cost of food only in lean subjects.

The significance of diet-induced thermogenesis is controversial. Severe dieting reduces metabolic rate, partially because the thermic effect of food is less. Conversely, eating three to six meals a day during a weight-loss program may help dieters take advantage of the increased metabolic rate that ac-

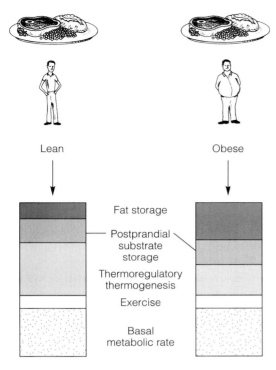

Lean

Obese

Fat storage

Postprandial substrate storage

Thermoregulatory thermogenesis

Exercise

Basal metabolic rate

Fate of energy from food

Figure 25-7 Thermoregulatory thermogenesis from tissues such as brown fat may be lower in the obese and may contribute to their overweight problem.

companies eating. Some researchers, however, feel that increases in metabolic rate are trivial and that people should emphasize caloric restriction instead.

Facultative Thermogenesis *Facultative thermogenesis* is heat production that does not result in mechanical work or net synthesis. Examples include increased activity of brown fat, sodium–potassium pump activity, and futile metabolic pathways. Futile metabolic pathways use ATP without contributing usable energy or synthesis. These processes decrease metabolic efficiency.

Brown adipose tissue (BAT) is an important center for thermogenesis in several animal species, including human infants, and may be an important center for facultative thermogenesis in adult humans. However, adults have only small amounts of the tissue. Small deposits of BAT exist in the interscapular, axillary, and peritoneal regions and around the great vessels in the thorax. Heleniak and Aston (1989) have speculated that 40–50 grams of active BAT could increase metabolic rate 20–25%.

The thermogenic property of BAT stems from the ability of its mitochondria to perform controlled uncoupling of oxidative phosphorylation (Figure 25-8). When the uncoupling pathway is activated, the use of substrate results in the production of heat rather than ATP. BAT is activated by the sympathetic nervous system via stimulation of β-adrenergic receptors.

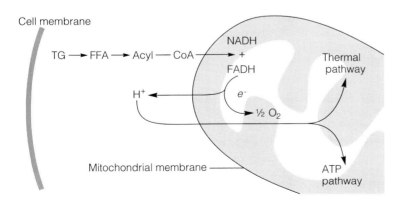

Figure 25-8 Thermogenesis in brown adipose tissue.

Cell membrane

TG $\longrightarrow$ FFA $\longrightarrow$ Acyl $\longrightarrow$ CoA

NADH + FADH

e^-

$\frac{1}{2}$ O$_2$

H$^+$

Thermal pathway

ATP pathway

Mitochondrial membrane

The small amount of BAT in adult humans has, until recently, excluded it from serious consideration as an important contributor to thermogenesis. Studies using thermography, a technique used to determine thermal gradients in the body, and direct thermistor readings following norepinephrine or ephedrine infusion, now suggest, however, that BAT may have a more important role in thermogenesis than previously suspected. The obese show a much lower thermogenic response to infusion of norepinephrine, which may indicate depressed sensitivity of the sympathetic nervous system or an abnormality in the thermogenic pathway in BAT.

Cold exposure, diet, starvation, and certain drugs influence BAT activity. Cold exposure, for example, causes brown fat hypertrophy. This has been demonstrated in Korean pearl divers, Finnish workers, and military recruits. Although controversial, some data also suggest that exercising in the cold may speed fat loss. Possible dietary stimulants of BAT include a high carbohydrate diet, supplements of essential fatty acids (i.e., cis-linoleic acid or cis-gamma-linoleic acid), and chromium. Ephedrine, a drug found naturally in some plants, may also stimulate BAT activity.

At present, the importance of BAT in human obesity is unknown. Data suggest, however, that this substance may account for some individual differences in body composition.

Physical Activity

Amount of physical activity is probably the most significant and variable factor of energy expenditure. Basal metabolic rate is less than 1000 kcal per day (depending on body size), but the additional caloric cost of physical activity can vary tremendously. Whereas a sedentary person typically expends approximately 1800 kcal per day, the endurance athlete may use as much as 6000 kcal. Although many health experts place little emphasis on the importance of exercise in caloric expenditure, this threefold difference between sedentary and extremely active people must be considered seriously. Even small differences in caloric expenditure can become substantial when accumulated over many months.

While most studies show that exercise alone is insufficient for long-term weight loss, exercise and dietary restriction combined appears to improve both early and longer-term outcomes of the weight-control program. Exercise is important because it helps preserve lean body mass, increases total energy output, and affects substrate utilization. It increases energy output by increasing resting metabolic rate and the thermic response to food, mainly by increasing fat-free weight in the body. Weight loss through dietary restriction alone typically causes loss of fat-free mass and improved fuel efficiency, but this makes continued weight loss more difficult and makes it easier to regain weight. For controlling body fat, the overall energy expenditure, mainly determined by the frequency and duration of exercise, is more important than exercise intensity. Most research studies suggest that a good exercise program for losing body fat should include both endurance and resistance exercises, practiced at least three times per week with an energy expenditure of approximately 1500 kilocalories.

During exercise, approximately 5 kcal are used for every liter of oxygen consumed. However, the increase in metabolic rate after exercise, due to the Q_{10} effect (heat increases metabolic rate) and the anabolic process elicited by the exercise stimulus, can increase the caloric cost of the exercise by 15–50%. These postexercise effects add to the potency of physical activity in caloric expenditure.

■ Hypercellularity of Adipose

Obesity can occur due to an increase in fat cell size and number, or both (Figure 25-9). Severely obese people have been shown to have three times as many fat cells as lean people and that these can be 40% larger than those in lean people. In adults, weight loss mainly results in a decrease in the size of the fat cells but little or no decrease in their number, so prevention of the initial development of fat cells during the growth period is important. Obesity

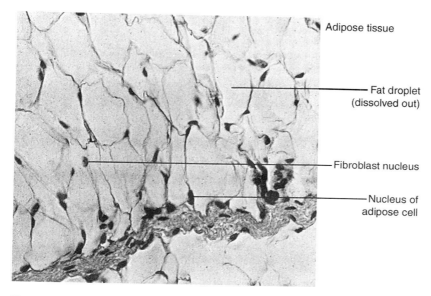

Adipose tissue

Fat droplet
(dissolved out)

Fibroblast nucleus

Nucleus of
adipose cell

Figure 25-9 Micrograph of adipose tissue. SOURCE: Crouch and McClintic, 1971, p. 112. Used with permission.

that begins in childhood is much more difficult to manage than obesity that develops in later life. Fat cells can be added or deleted during adulthood if there is a steady increase or decrease in body fat deposition.

Fat cells increase in both size and number during childhood and, to a certain extent, during adolescence. Adipocyte number can also increase in adults. However, in adults, fat cells increase in size before increasing in number. Fat mass is dependent on both the average volume and the number of fat cells. In adults, significant increases in body mass enhance adipocyte volume and number. Increases in fat cell number occur via replication and differentiation of preadipocytes, a process that occurs throughout life. Preadipocytes have been difficult to detect. In the past, this difficulty led researchers to erroneously conclude that the fat cell number is fixed in adults.

Regulation of fat cell number and size involves the endocrine, paracrine, and possibly autocrine systems. These regulators include insulin, retinoids, corticosteroids, and tumor necrosis factor-alpha.

The precise means by which the hypothalamus signals to adipose tissue to grow and proliferate is not fully understood.

Severe triglyceride depletion of the adipocytes during weight reduction seems to upset their homeostasis, which may result in stimulation of the appetite center in the hypothalamus to reinstitute the cellular *status quo*. This is the basis of the lipostatic theory of hunger and satiety discussed earlier. This process makes it extremely difficult for people with increased fat cell number to maintain a lower body weight after a period of weight reduction. People with hyperplastic obesity (more cells) gain lost weight back faster than people with hypertrophic obesity (larger fat cells).

Animal studies show that the development of fat cells can be reduced by endurance exercise and diet during growth. In addition, suppression of fat cell growth during childhood makes the development of obesity during adulthood much less of a problem, even if physical activity is discontinued. It is unfortunate that so many of our medical resources are devoted to fighting obesity in adults rather than to preventing the problem in children, where the effort might do some good.

■ The Treatment of Obesity

The positive energy balance of obesity may be caused by a number of factors, including genetic tendency, hypothalamic hunger-satiety defects, endocrine abnormalities, lipid or carbohydrate metabolism aberrations, inappropriate social attitudes toward feeding, and lack of exercise. A variety of treatments and remedies are used to deal with the problem (Table 25-3).

Diet: Caloric Restriction

Caloric restriction is the most common treatment of obesity and is an essential part of any weight-control program. Most quick-loss fad diets, however, stress weight loss rather than fat loss and fail to provide a regimen that can be followed for life. The goal of a dietary program should be to lose fat and then maintain the loss. Unfortunately, severe caloric restriction and the composition of fad diets are so unpleasant and unnatural that rebound weight gains are typical.

Many quick-loss diets promote low carbohydrate intake, which results in dehydration as muscle and liver glycogen stores are depleted. Low-carbohydrate diets lead to sodium diuresis and loss of intracellular and extracellular fluids. Although the weight loss appears impressive, most of it is in the form of water and lean mass rather than fat. Also, glycogen depletion greatly diminishes exercise capacity, which almost eliminates physical activity as a source of caloric expenditure (Figure 25-10).

The success or failure of any diet depends on its effect on energy balance. Studies of patients in metabolic wards, where diet and exercise are closely controlled and measured, have shown that people lose weight at a predictable rate when put on a low-calorie diet. Weight loss is independent of dietary composition. Fatter people tend to lose weight faster than people who are less fat. Men lose weight faster than women because of a higher lean body mass.

Although caloric content of the diet is most important for losing weight, dietary composition has an indirect influence on the quality of the weight loss. Protein (lean mass) loss is less in people consuming a high-carbohydrate diet. A high-carbohydrate diet also influences the sympathetic nervous system and triiodothyronine concentration, which, in turn, affect lean body mass and fat mass.

TABLE 25-3
Treatment of Obesity

Diet
Exercise
Anorectic and thermogenic drugs
Behavior modification
Jejunoileal bypass
Lipectomy
Hypnosis
Acupuncture
Miscellaneous agents: thyroid hormones, drugs that uncouple oxidative phosphorylation (e.g., dinitrophenol), human chorionic gonadotrophin, growth hormone, glucagon, progesterone, and biguanides

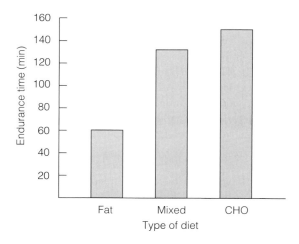

Figure 25-10 Endurance time on bicycle ergometer exercising at 75% of $\dot{V}_{O_2max}$ in subjects consuming high-fat, mixed, and high-carbohydrate (CHO) diets. Adapted from E. Hultman and L. H. Nilsson.

Studies in metabolic wards have also shown that anyone consuming 1200 kcal per day will lose weight. They also show that water and protein loss are common during the early stages of the diet. These losses are most pronounced in more obese subjects. Exercise can slow the loss of lean mass in obese people but cannot completely eliminate it.

Extremely low calorie diets (<200 to 800 kcal per day) can be dangerous. Many people have died from cardiac arrest while on these diets. Extremely low calorie diets have been associated with serious cardiac arrhythmias, loss of lean body mass, loss of hair, thinning of the skin, coldness in the extremities, gallbladder stones, gout, and elevated cholesterol. Such diets seldom result in the behavior changes necessary for permanent weight loss.

Dehydration

Diuretics and impermeable clothing have been advocated from time to time for losing weight. These, however, cause water loss rather than fat loss, and the resulting dehydration can be extremely dangerous. Although diuretics are often used in the treatment of hypertension in obese people, they have no place in the treatment of obesity itself.

Sauna belts and rubber suits are often used by the uninitiated in their quest to lose weight and inches. The resulting dehydration of the cells under the belts does cause a temporary loss of circumference. However, the change is short-lived, as normal fluid balance is quickly restored.

Medical Procedures

Medical intervention in the hospital is sometimes employed in extreme cases of obesity. Hospital procedures include: prolonged fasting, jejunoileal bypass, and lipectomy. These techniques, although often effective, are sometimes accompanied by undesirable side effects.

Starvation   Prolonged starvation has resulted in weight losses of over 100 pounds. However, this procedure causes loss of lean body mass and can have serious side effects, such as gout, anemia,

hypotension, and various metabolic disturbances. Also, starvation does little to modify eating habits, which will help maintain lost weight.

Invasive Treatments   Invasive procedures for treating obesity include gastric bubble, jaw wiring, gastric bypass, gastroplasty, jejunoileal bypass, and liposuction. The long-term effectiveness of temporary measures, such as gastric bubble and jaw wiring, is low—as with starvation, the resting metabolic rate is depressed and healthy eating habits are seldom substituted, so weight is usually regained.

Bypass Operations   Bypass operations, such as gastric and jejunoileal bypasses, are much more effective than starvation and jaw wiring. These procedures are very expensive, however, and much riskier than less invasive methods. Gastric operations reduce the gastric pouch size, which limits food intake. Clinical reports suggest that these procedures cause substantial weight loss. Although failure rates are as high as 50%, the procedures are satisfying to a large proportion of patients. Surgical risk is present but low. Other risks include vomiting, diarrhea, and peptic ulceration.

Jejunoileal bypass operations decrease the absorption of nutrients. These procedures cause weight losses similar to those from gastric operations. Side effects are much more severe, however, so the procedure is much less popular than it once was. Side effects include surgical complications, diarrhea, electrolyte disturbances, bacterial infection, and liver failure.

Suction Lipectomy   Liposuction has become the most popular type of elective surgery in the United States. The procedure involves removing limited amounts of fat from specific areas. Typically, no more than 2.5 kg of adipose is removed at a time. The procedure is usually successful if excess fat is limited and skin elasticity is good. The procedure is most effective, however, if integrated into a program of dietary restriction and exercise. Side effects include infection, dimpling, and wavy skin contours.

Drugs   Drugs used in weight control include those that suppress appetite, thermogenic drugs, and drugs affecting the gastrointestinal tract. Appetite-suppressing drugs work by acting on catecholamine neurotransmitters or serotonin neurotransmitters, or by blocking opioid receptors. Thermogenic drugs affect metabolic rate. Gastrointestinal drugs attempt to affect nutrient absorption. Other agents, with questionable effectiveness and desirability, include human chorionic gonadotropin, growth hormone, glucagon, progesterone, and biguanides.

Appetite suppressants include amphetamine, dexfenfluramine, sibutramine, diethylpropion, fenfluramine, and phenylpropanolamine. Although these drugs depress appetite, may have thermogenic effects, and may cause weight loss, some have serious side effects. Amphetamine, for example, is highly addictive and can cause cardiac arrhythmias and impair temperature regulation. Schedule III and IV drugs have a low potential for abuse and are preferred over Schedule II drugs, such as amphetamines. ("Scheduled" drugs are available only by prescription.)

Thermogenic drugs include thyroid hormone, ephedrine, and dinitrophenol. Thyroid hormones were commonly used in the 1970s. However, they reduced lean body mass and led to an increased incidence of cardiac arrhythmias. Combinations of ephedrine, caffeine, and aspirin have been used extensively as weight-loss agents by body builders. While effective, the combination of these drugs can cause cardiac arrhythmias, insomnia, irritability, and tremors. They are not recommended as weight-loss aids. As discussed, ephedrine may also be effective in weight control due to its ability to activate brown fat. Uncoupling agents, such as dinitrophenol, are toxic at effective dosages and are associated with neuropathy and cataracts.

Drugs affecting the gastrointestinal tract include dietary fiber and sucrose polyester. Dietary fiber causes gastrointestinal distension and may restrict energy intake. Limited clinical trials suggest that including fiber supplements in the diet may aid weight loss. Sucrose polyester (Olestra) is a new diet ingredient of some promise. This substance cannot be digested and can be substituted for fats in the diet. The limited number of studies done so far

on this substance show mixed results as to its effectiveness as a weight-loss inducing agent.

Other drugs are being developed to treat obesity. They include drugs that affect leptin and its receptors, various obesity genes, and drugs that induce satiety (i.e., neuropeptide Y, enterostatin, cholecystokinin, bombesin, and amylin).

Acupuncture Acupuncture, a quasi-medical technique involving the puncture of the body with small needles, has been suggested as a technique in the management of obesity. Little objective experimental evidence is available, however, and its effectiveness remains unestablished.

Behavior Modification

Behavior modification programs focus on the elimination of behavior associated with poor eating and exercise habits. The success rate of this technique has been mixed. Although some studies have demonstrated weight losses of 9 to 18 kg in over 80% of their subjects, others have shown considerably less success. Self-help groups, such as Weight Watchers and Take Off Pounds Sensibly (TOPS), sometimes achieve impressive results. However, membership attrition is a significant problem in these self-help groups.

The long-term success rate in behavior modification programs is less encouraging. Follow-up studies have found that 70% of subjects regained their lost weight after a year. However, 30% of subjects in behavior modification programs maintain weight loss after a year and continue to lose even more weight. The five-year success rate in such programs is approximately 50%, with an average weight loss of 6.0 kg in men and 4.0 kg in women. Behavior modification is definitely helpful for some people.

At present, little is known of the relationship between such factors as personality type, onset of obesity, age, or weight at entry into the program, and predictability of success in behavior modification programs. Unfortunately, the biggest failure, as a group, has been older obese persons with numerous health problems who have been referred by a physician to a behavior modification program—these are the people who need to reduce the most.

Hypnosis   Hypnosis is a type of behavior modification technique that is widely used. Unfortunately, no clinical research clearly substantiates its effectiveness. Most evidence is restricted to clinical observations and one-subject anecdotal reports. Techniques and hypnotic suggestions are not standardized and studies show little regard for scientific method. Hypnosis may be a valuable tool in the treatment of obesity, but its usefulness has not been established.

Exercise

Exercise is an important part of a successful long-term weight-control program. It increases resting metabolic rate, maintains lean body mass, and increases energy expenditure. It also allows the consumption of enough calories to supply the body with adequate nutrients as well as energy. Caloric restriction alone can lead to malnourishment because the low-calorie diet may not contain sufficient vitamins and minerals, and chronic caloric restriction may eventually have serious health consequences.

A single session of exercise causes little fat loss. However, regular training can make a substantial difference in the weight-control program. The expenditure of 300 calories during exercise, three or four times a week, can result in a loss of 13 to 23 pounds of fat a year, provided the caloric intake remains the same. That may not seem like much weight to a crash dieter. However, the weight loss consists largely of fat and is not a combination of water, lean tissue, and fat, which is what is commonly lost on most fad diets. Although dieting is a drudgery, exercise is an enjoyable way to expend calories.

As fitness improves, exercise has a more potent effect on caloric utilization. A change in maximal oxygen consumption from 3 liters per minute to 3.5 liters per minute increases the ability to use calories by almost 20%. Exercise for weight control should center on long-term endurance activity for a minimum of 20 minutes.

Exercise and Weight Loss   Short-term studies have shown that endurance exercise without caloric restriction causes weight loss. However, exercise by itself is less effective than caloric restriction (i.e., dieting) or caloric restriction plus exercise. Most studies have shown that exercise plus diet is more effective than dieting alone in maintaining weight loss (six months to three years after the beginning of the program).

Dieting by itself reduces lean body mass and decreases resting metabolic rate, sometimes by as much as 30%. During the initial stages of a starvation diet, 40% of the weight loss may be from lean mass. Most studies have shown that after one year, most weight lost through dieting alone is regained, and the dieter may become involved in a futile cycle of dieting and regained weight.

To be successful, long-term weight-control programs should include exercise. Exercise spares lean mass and increases resting metabolic rate during caloric restriction. Exercise also increases the thermic response to food (i.e., digesting and processing food increases metabolic rate). In addition, exercise exerts an independent effect on reducing the risk of coronary artery disease.

Exercise intensity may be important for increasing metabolic rate. Several studies have shown that intense exercise increases metabolic rate in obese subjects, while moderate exercise does not. The metabolic effects of exercise are temporary, however. In dieters, metabolic rate decreases when training is suspended for as little as three days.

The literature is unclear on the effects of exercise on food intake. Studies have found decreased, increased, and no change in food intake in people involved in moderate exercise. Likewise, it is not known if exercise influences dietary composition (i.e., the percentage of fats and carbohydrates in the diet).

Weight training has been suggested as an important component in a long-term weight-management program. Programs stressing caloric restriction cause decreases in lean body mass, negative nitrogen balance (i.e., body loses protein), and diminished muscle strength. Including weight training in the weight-reduction program helps spare lean body mass and maintain nitrogen balance. Also, improvements in strength between 17% and 22% have been reported in subjects who weight-trained during caloric restriction. Lean mass is the most important determinant of resting metabolic rate. Weight training increases or maintains lean mass in people on low-calorie diets.

Like endurance exercise, weight training has no effect on regional fat deposition (i.e., spot reducing is ineffective). Although the improved muscle tone that results from training usually makes a particular area of the body look better, the subcutaneous adipose layer that lies over the muscles is unaffected (except as it is affected by any negative caloric balance).

■ Body Composition

Body composition can be divided into fat-free mass (FFM) and body fat. Fat-free mass is fat-free weight. It includes the skeleton, water, muscle, connective tissue, organ tissues, and teeth. Body fat includes essential and nonessential fat stores. Essential fat includes lipid incorporated into organs and tissues such as nerves, brain, heart, lungs, liver, and mammary glands. Nonessential fat exists primarily within adipose tissue.

Ideal Body Composition

Perhaps the three most important considerations determining ideal body composition are health, aesthetics, and performance. The average person tends to be most concerned with health and aesthetics, while the athlete is concerned about all three.

Although the ideal healthy fat percentage has not been clearly established, it is generally agreed to be between 16% and 25% for women and less than 20% for men. Figures 25-11a and 25–11b show the 10th, 50th, and 90th percentiles of percent fat for men and women in different age groups. The aver-

age man and woman fall above the recommended fat percentage in almost all age categories.

The ideal aesthetic body composition is even more difficult to establish. In our society, the lean, athletic look is prized, while the more corpulent look of the turn of the century is disdained. Unfortunately, the quest for the "lean look" often leads to unhealthy dietary habits. This is especially disturbing among athletes, who usually have very high caloric requirements because of heavy training. Young teenage, female athletes seem to be especially prone to overzealous caloric restriction, which can be dangerous.

Body composition is an extremely important consideration in many sports. In sports with weight categories, such as wrestling, boxing, and weight lifting, serious abuses have taken place in an attempt to lose weight. Weight-loss practices have sometimes even compromised the health of the athletes. Successful athletes in various sports usually possess a characteristic body composition (Figure 25-12). Variability in body fat depends on the metabolic requirement of the activity and the relative disadvantage of carrying an extra load. For example, successful male distance runners have almost always less than 9% fat. For these athletes, excess fat is a decided disadvantage. The tremendous caloric cost of running long distances makes it difficult for runners to gain much fat.

Football linemen, however, have almost always greater than 15% fat. This may be advantageous because of the added mass and padding provided by the subcutaneous fat and the increase in lean mass that accompanies excess weight (muscle mass accompanies gains in fat to support the extra weight). Unfortunately, many younger athletes gain too much fat in an attempt to attain the high body weights of the professional football player.

Measuring Body Composition

The morning weighing ritual on the bathroom scale is really an attempt at estimating body composition. Using this method, it is impossible to determine accurately if a fluctuation in weight is due to a change

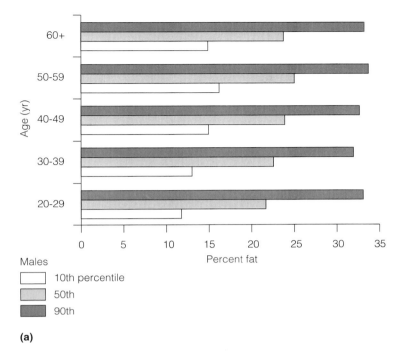

(a)

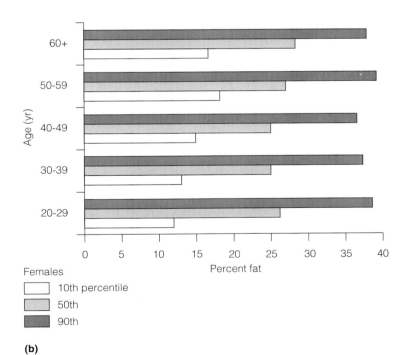

(b)

Figure 25-11a, b   Fat percentages in males and females as a function of age.

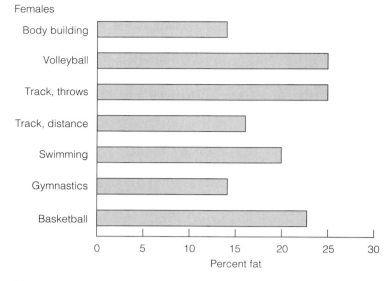

(a)

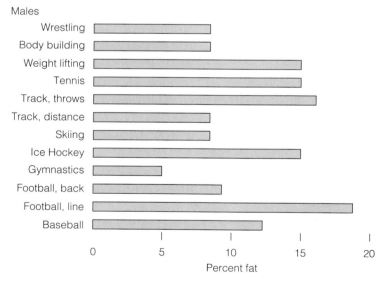

Figure 25-12a, b Body fat percentage of typical well-trained female and male athletes.

(b)

in muscle, body water, or fat. The method also cannot distinguish between overweight and overfat. A 260-pound muscular football player may be overweight according to population height–weight standards, yet actually have much less body fat than average. A 40-year-old woman may weigh exactly the same as when she was in high school, yet have a considerably different body composition. Despite these limitations, the U.S. Food and Drug Administration (FDA) and the Department of Health and Human Services have published new height and weight recommendations.

Several more precise methods have been developed that can provide an accurate estimation of body composition. Laboratory methods include dual-energy X-ray absorptiometry (DXA), densitometry (underwater weighing), magnetic resonance imaging (MRI), radiography, neutron activation analysis, and potassium-40 (^{40}K) analysis. Field test methods include ultrasound, anthropometry, skinfold measurement, and electrical impedance. Field tests are usually validated with standard laboratory techniques.

Basic Assumptions of Body Composition Methods   Dissection of the human body is the most accurate way to measure body composition. This is obviously not very practical, so indirect methods have been developed. Indirect methods such as densitometry, ^{40}K, and total body water make certain assumptions about body composition based on cadaver analysis. Densitometry, for example, assumes the density of fat-free tissue is $1.100 \text{ g} \cdot \text{ml}^{-1}$ and the density of fat is $0.901 \text{ g} \cdot \text{ml}^{-1}$. The total body water method assumes the water content of fat-free mass is 73.2%. The ^{40}K body composition method assumes the potassium content of fat-free mass is $68.1 \text{ mmol} \cdot \text{kg}^{-1}$.

The basic assumptions used in body composition measurement techniques were developed in the 1940s based on the dissection of eight cadavers. None of the cadavers were analyzed by indirect methods such as underwater weighing, so none of our present methods have been validated directly. Also, the assumed density value of $1.100 \text{ g} \cdot \text{ml}^{-1}$ was extrapolated to humans from research on eviscerated guinea pigs.

Clarys and coworkers (1984) reported data on dissections performed on cadavers. They found that bone accounted for approximately 16–26% of fat-free weight, while muscle accounted for 41–60%. This sample did not include athletes, children, or diverse racial groups.

Individual subjects may vary considerably from the basic assumption that fat-free mass has a density of $1.100 \text{ g} \cdot \text{ml}^{-1}$. For example, a person with a higher lean body mass density than expected would appear less fat than he or she actually was. Fat-free mass density may vary with age, gender, body fat, fitness, and ethnicity. Large and unacceptable measurement errors can result from errors in basic assumptions.

Individual differences in bone density account for most of the deviation from basic assumptions. Small deviations in bone density from the assumed value would be particularly important in lean athletes. Negative fat percentages have been reported for some athletes, undoubtedly because their tissue density varied from the basic assumption. The DXA method, because it can accurately measure bone density, holds great promise for increasing the accuracy of body composition measurement.

Methods such as skinfold measurement, electrical impedance, and anthropometry are doubly indirect because they employ underwater weighing and are thus two measurement techniques removed from dissection. Consequently, doubly indirect methods are subject to the errors inherent in densitometry (i.e., underwater weighing) as well as to errors in the procedures themselves.

Until body composition methods are validated directly, they remain useful—but sometimes unreliable—procedures for estimating body composition. The researcher, physician, or exercise leader should be aware of the limitations of these methods.

Hydrostatic Weighing   Until the development of DXA, hydrostatic or underwater weighing was considered the most accurate indirect means of measuring body composition. It served as a stan-

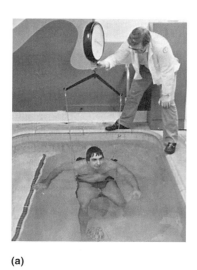

(a)

(b)

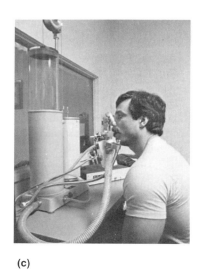

(c)

Figure 25-13a–c The underwater weighing technique. Subjects are two Olympic discus throwers. PHOTO: Wayne Glusker.

dard for other indirect techniques such as skinfold measurement (Figure 25-13) and remains an important tool. This procedure was popularized by researchers such as Behnke in the early 1940s and has become an important tool in exercise physiology and medicine. As discussed, the equations used in this method stem from assumptions made from the dissection of human cadavers.

Density is equal to mass divided by volume. Unfortunately, the irregular shape of the human body makes a simple geometric estimation of its volume impossible. However, the volume of the body can be measured by using Archimedes' principle of water displacement, which states that "a body immersed in water is buoyed up with a force equal to the weight of the water displaced." The volume of the body can be measured by determining the weight lost by complete immersion in water. The density of the body, and thus its percent of fat, can be calculated by dividing the body weight (scale weight) by the body volume (calculated by underwater weighing).

In this procedure, the subject is submerged and weighed underwater. Because muscle has a higher density and fat a lower density than water (approximately 1.100 g · ml^{-1} for muscle, 0.901 g · ml^{-1} for fat, and 1.00 g · ml^{-1} for water), fat people tend

to float and to weigh less underwater, while lean people tend to sink and weigh more underwater. At a given body weight, a fat person has a larger volume than a thin one, and thus a smaller density.

Many errors are possible, even in this relatively precise laboratory procedure (in addition to violations of basic assumptions already discussed). Failure to consider residual lung volume, intestinal gas, and water density decrease underwater weight and result in an overestimation of volume. During the measurement, the subject is weighed completely submerged in water after expelling as much air as possible from the lungs. A small but variable amount of air called the residual lung volume remains, however, and must be taken into account in the equation. Although residual lung volume can be estimated, accurate assessment of body composition requires that it be measured directly (Figure 25-13c). Intestinal gas also increases buoyancy and confounds results. Finally, water has a density of 1.00 g · ml^{-1} only at a temperature of 39.2°F. The calculation of volume must be corrected for a difference in water density, therefore, if the water temperature is other than 39.2°F. However, the method is accurate if the majority of the assumptions supporting the technique are met.

The two most widely used equations for calculating lean body mass and percent fat were derived by Brozek and associates (1963) and Siri (1956). Their slight variance stems from different estimations of the density of fat and muscle. These equations, together with sample calculations of density, percent fat, and lean mass appear in Table 25-4. Lohman has suggested a modification of these equations for use with children (Eq. 25-1; Lohman and Going, 1993):

$$F = \frac{530}{D - 489} \qquad (25\text{-}1)$$

where F = % fat and D = whole-body density. Alternate equations have also been developed that take into consideration ethnic and gender differences (Table 25-5).

Dual-energy X-ray Absorptiometry (DXA)

Clinically, dual-energy X-ray absorptiometry (DXA) is widely used to assess the risk and progression of osteoporosis. But researchers quickly found that the technique could also be used to assess the composition of soft tissue. DXA works by aiming X-rays at two different energies at the body. Differences in absorption of the X-ray beam at these two energies is used to calculate the bone mineral content and soft tissue composition in the scanned region. DXA scanning uses a very small dose of radiation, so the method is simple and safe for subjects ranging from children to the elderly.

DXA has been suggested as a replacement to densitometry (hydrostatic weighing) as the gold standard for the measurement of body composition. It relies on fewer assumptions about human tissue characteristics, and most importantly, comparisons of the method with autopsy in animals have demonstrated the validity of the technique.

The accuracy of DXA varies with different parts of the body and with the software used to make the

---

TABLE 25-4

Equations for the Calculation of Body Density, Percentage Fat, Body Fat, and Lean Body Mass

1. Body density = $\dfrac{BW}{\dfrac{BW - UWW}{D_{H_2O}} - RLV}$

   where BW = body weight (kg), UWW = underwater weight, D_{H_2O} = density of water (at submersion temperature), and RLV = residual lung volume.

2. Percentage fat = $\left(\dfrac{4.950}{D_b} - 4.50\right) \times 100$
   (Siri equation)
   where D_b = body density

3. Percentage fat = $\left(\dfrac{4.570}{D_b} - 4.142\right) \times 100$
   (Brozek equation)

4. Total body fat (kg) = Body weight (kg) × % $\dfrac{Fat}{100}$

5. Lean body weight = Body weight − Fat weight

TABLE 25-5

Equations for converting body density to percent for various ethnic groups

Ethnicity	Age (years)	Gender	% Body Fat	Fat-free Body density (g/cc)
American Indian	18–60	Female	(4.81/Db)−4.34	1.108
Black	18–32	Male	(4.37/Db)−3.93	1.113
	24–79	Female	(4.85/Db)−4.39	1.106
Hispanic	20–40	Female	(4.87/Db)−4.41	1.105
White	17–19	Male	(4.99/Db)−4.55	1.098
	17–19	Female	(5.05/Db)−4.62	1.095
	20–80	Male	(4.95/Db)−4.50	1.100
	20–80	Female	(5.01/Db)−4.57	1.097

Source: Adapted from Heywood, 1996. Used with permission.

analyses. The method works best in young healthy subjects but is less accurate in osteoporotic and obese subjects. Thus, because of limitations in the hydrostatic weighing (i.e., problems with underlying assumptions) and DXA (i.e., problems with standardization of technique and software), there is currently no gold standard for the measurement of body composition.

Biochemical Techniques Biochemical techniques of measuring body composition are based on the biological constants observed during direct chemical analysis of the body. These methods include potassium-40 (^{40}K) analysis, total body water, and inert gas absorption.

Lean body mass contains a relatively constant proportion of potassium, part of which is in the form of ^{40}K, a naturally occurring isotope. The gamma rays emitted by ^{40}K can be measured with a whole-body scintillation counter, which allows the prediction of total body potassium and lean body mass. The results of this method are very similar to those of hydrostatic weighing. Because of the expense and limited availability of whole-body scintillation counters, however, this method is almost completely restricted to research.

Diffusion techniques rely on the property of specific substances to diffuse into tissues or compartments. Tracers such as deuterium and tritium oxide (heavy water), antipyrine, and ethanol, for example, have been used to estimate total body water. Lean body mass can thus be estimated because body water represents an almost constant 73.2% of body weight, and the water is contained almost entirely within the lean body mass. Fat-free mass (FFM) can be estimated using the equation of Pace and Rathbun (Eq. 25-2; Pace and Rathbun, 1945):

$$FFM = \frac{TBW(L)}{0.732} \qquad (25\text{-}2)$$

where TBW = the total body water and L = body length. Fat-soluble inert gases such as krypton and cyclopropane have also been used to estimate percent fat by measuring their rate of absorption into the body.

Anthropometric and Skinfold Techniques

Anthropometric assessment of body composition uses various superficial measurements such as height, weight, and anatomical circumferences. Of these, height–weight is by far the most popular. Height–weight tables, periodically produced by insurance companies, are inadequate, however, because they are subject to individual interpretation—they require people to decide if they are of small, medium, or large frame. In addition, they do not take into consideration individual differences in lean body mass and relative fat.

Body Mass Index (BMI) The body mass index is weight/height^2, and it is moderately correlated to percent fat ($r = 0.80$). Because data are easy to collect, this index is widely used and reported in large epidemiological studies. A BMI of 25–30 for men and 27–30 for women is considered moderate obesity. A BMI of 30–40 is considered massive obesity, and a BMI greater than 40 is classified as morbid obesity.

Waist Circumference and Waist–Hip Ratio As discussed, abdominal fat deposition is a significant risk factor in a variety of diseases. A high waist circumference and waist–hip ratio (see Figure 25-14) increases the risk of myocardial infarction, angina pectoris, diabetes, gallbladder disease, and stroke.

Somatotype Sheldon devised a rating system for assessment of body composition based on three components, rated on a seven-point scale (Sheldon et al., 1954). The three components were endomorphy (the relative predominance of corpulence and roundness), mesomorphy (the relative predominance of muscularity), and ectomorphy (the relative predominance of linearity and fragility of body build). This method is very subjective and requires a test administrator who is well trained in the photographic protocol and rating system.

Anthropometric Method Behnke (1961) developed an anthropometric technique for estimating body composition that compares various circumferences of a subject with those of a reference man or woman. Like other field test methods, however, accuracy requires experienced test administrators.

Skinfolds Skinfold measurements are probably the most popular "scientific" means of assess-

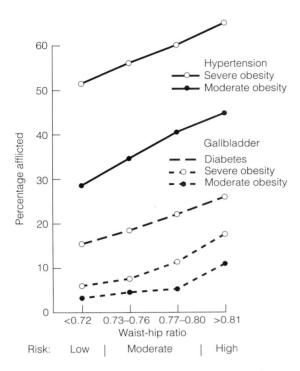

Figure 25-14 The risk of heart disease and waist–hip ratio in males and females.

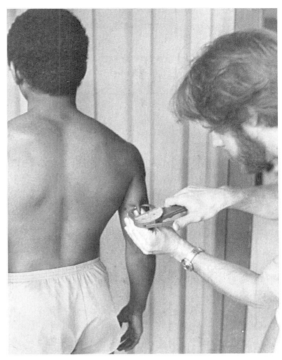

Figure 25-15 The skinfold technique. Photo: Wayne Glusker.

ing body composition (Figure 25-15). This method is inexpensive, rapid, and takes little time to learn. Skinfold equations are derived using a statistical technique called multiple regression that predicts the results of the hydrostatic weighing procedure from the measurement of skinfolds taken at various sites—few equations have been developed using the DXA method, which may replace underwater weighing. Subjects of skinfold measurements should be measured using an equation derived from a similar population. For example, it would be inappropriate to use a skinfold equation that was derived from 18-year-old college females to estimate the body fat of a 40-year-old man. Several equations have been developed for the general population (Table 25-6).

Several models of skinfold calipers are available. The ideal caliper should have parallel jaw surfaces and a constant spring tension, regardless of the de-

gree of opening. Many inexpensive, plastic calipers have become available recently, some of which provide excellent reliability.

The skinfold method is subject to severe limitations. It should be used only in field studies or when only gross estimations of body composition are required. The technique is subject to considerable measurement error, even among experienced observers. Dehydration, for example, decreases a skinfold thickness by as much as 15%, which, in turn, causes variability in measurements taken in the morning and evening. Also, skinfold measurements are relatively insensitive in predicting changes in body composition following weight loss. The accuracy of this method can be improved through multiple measurements by a single, experienced observer.

Ultrasound, Computerized Tomography (CT), and Electrical Impedance
Ultrasound is widely

TABLE 25-6

Skinfold Equations for Various Populations

Skinfold sites	Population subgroup	Gender	Age	Equation	Reference
Σ7SKF (chest + abdomen + thigh + triceps + subscapular + suprailiac + midaxilla)	Black or Hispanic	Women	18–55 yrs	Db (g/cc) = 1.0970 − 0.00046971 (Σ 7SKF) + 0.00000056 (Σ 7SKF)^2 − 0.00012828 (Age)	Jackson et al., 1980
	Black or athletes	Men	18–61 yrs	Db (g/cc) = 1.1120 − 0.00043499 (Σ 7SKF) + 0.00000055 (Σ 7SKF)^2 − 0.00028826 (Age)	Jackson and Pollock, 1978
Σ 4SKF (triceps + anterior suprailiac + abdomen + thigh)	Athletes	Women	18–29 yrs	Db (g/cc) = 1.096095 − 0.0006952 (Σ 4SKF) − 0.0000011 (Σ 4SKF)^2 − 0.0000714 (Age)	Jackson et al., 1980
Σ 3SKF (triceps + suprailiac + thigh)	White or anorexic	Women	18–55 yrs	Db (g/cc) = 1.0994921 − 0.0009929 (Σ 3SKF) + 0.0000023 (Σ 3SKF)^2 − 0.0001392 (Age)	Jackson et al., 1980
(chest + abdomen + thigh)	White	Men	18–61 yrs	Db (g/cc) = 1.109380 − 0.0008267 (Σ 3SKF) + 0.0000016 (Σ 3SKF)^2 − 0.0002574 (Age)	Jackson and Pollock, 1978
Σ 2SKF (triceps + calf)	Black or White	Boys	6–17 yrs	% BF = 0.735 (Σ SKF) + 1.0	Slaughter et al., 1988
	Black or White	Girls	6–17 yrs	% BF = 0.610 (Σ SKF) + 5.1	Slaughter et al., 1988

Σ SKF = sum of skinfolds (mm)
Convert body density (Db) to % fat using population specific equations shown in Table 25-5.

SOURCE: Heywood, 1998, p. 155. Used with permission.

used in cardiology to examine heart walls and valves noninvasively and in obstetrics to observe the developing fetus. Several investigators have also used ultrasound to estimate the thickness of layers of fat and muscle. The instrument emits sound waves directed at specific parts of the body, and an echo occurs as it strikes tissues of differing thicknesses and density. The echo is then quantified by the instrument.

Computerized tomography (CT) is an X-ray technique that provides a three-dimensional view of the body. CT scans allow accurate assessment of the volumes and thicknesses of various organs and tissue spaces and should add considerably to our knowledge of body composition. It is unlikely this technique will become widely used to measure body composition outside of research applications,

however, because it is expensive and involves radiation exposure.

Measurement of *total body electrical conductivity* (TOBEC) and *bioelectrical impedance analysis* (BIA) are new and popular methods of measuring body composition. TOBEC works by surrounding the body with an electromagnetic field and measuring the rate that the body conducts electrical energy. The operating principle is that lean tissue and body water conduct electricity more rapidly than fat. Although the TOBEC technique is an excellent predictor of body composition measured by underwater weighing, the instrument currently costs between $50,000 and $100,000. In BIA, an electric current is applied to an extremity (e.g., arm) and the impedance to the electrical flow is measured at an opposite extremity (e.g., leg). Impedance is greater in people with more fat and less in people with more lean tissue. Most studies have shown that BIA is approximately as accurate as the skinfold technique in estimating body composition, except in very lean and very obese people, where it is less accurate.

Application of Body Composition Measurements

Physicians, coaches, and trainers should use great care when applying body composition data in sports or wellness management. Under certain circumstances, the standard ways of measuring of body composition—hydrostatic weighing and DXA—can render false results. Doubly indirect measures—such as skinfolds and anthropometric techniques—which are based mainly on underwater weighing techniques are subject to even greater errors. One of the problems with misuse of body composition data is that it can exacerbate such common problems in young people as eating disorders and severe dehydration to make weight in weight-class sports.

For proper application of body composition measurements, people advising athletes must use common sense when using the data. Visual assessment of body fat by a "trained eye" has been shown to be just as accurate as using laboratory methods to assess body composition. When designing diet and exercise programs to modify body composition, physicians and coaches should consider body composition analysis but should also include social, psychological, and experiential factors in their recommendations.

■ Summary

Obesity is a major health problem in Western countries, affecting nearly 30% of adults. Obesity increases the risk of a variety of diseases, including heart disease and stroke. Total weight and fat deposition patterns are related to increased mortality. People who store fat in the abdomen are at greater risk than those who store it in the lower body.

Energy balance determines whether body fat increases, decreases, or remains the same. The cause of the positive energy balance in obese people is not totally understood although the answer probably lies in the complex physiological and psychological factors that control food intake and energy expenditure. The control center for food intake is located in the hypothalamus. Appetite and satiety may be regulated by glucose, stored triglycerides, plasma amino acid levels, and hypothalamic temperature. Substances such as leptin also help control appetite and metabolic rate.

The components of energy expenditure include resting metabolic rate (RMR), thermogenesis, and physical activity. RMR usually decreases during weight loss. Exercise increases metabolic rate and spares lean body weight during weight loss. Thermogenesis is stimulated by eating and processing meals. In some species, brown fat may be an important center of thermogenesis although its importance in adult humans is not known. Exercise contributes to weight loss and is important for maintaining weight loss.

Fat cells increase in size and number during growth. In adults, most increases in body fat occur due to fat cell hypertrophy. Recent evidence suggests that fat cells can be gained or lost in adults.

Common treatments for obesity include diet, exercise, drugs, surgery, liposuction, and hypnosis. Most weight-control programs are not successful. Successful weight-loss programs require caloric restriction and basic changes in lifestyle (behavior modification). Exercise enhances the chances of successful long-term weight loss.

The three most important considerations determining the ideal body composition are health, aesthetics, and performance. Exercise helps each of these by preserving lean body mass and decreasing fat mass.

Body composition is assessed by indirect and doubly-indirect methods. Indirect methods include hydrostatic weighing and dual-energy X-ray absorptiometry (DXA). Doubly indirect methods include skinfolds, bioelectrical impedance, and anthropometric techniques. The accuracy of any technique varies with the subject population, validity of basic assumptions, and experience of the test administrator.

■ Selected Readings

Abernathy, R. P., and D. R. Black. Is adipose tissue over-sold as a health risk? *J. Am. Diet Assoc.* 94: 641–644, 1994.

Bahr, R. Excess postexercise oxygen consumption—magnitude, mechanisms and practical implications. *Acta Physiol. Scand.* 605 (Suppl.): 1–70, 1992.

Ballor, D. L., and E. T. Poehlman. Exercise-training enhances fat-free mass preservation during diet-induced weight loss: a meta-analytical finding. *Int. J. Obes. Relat. Metab. Disord.* 18: 35–40, 1994.

Barnard, R. J., and S. J. Wen. Exercise and diet in the prevention and control of the metabolic syndrome. *Sports Med.* 18: 218–228, 1994.

Barr, S. I., L. J. McCargar, and S. M. Crawford. Practical use of body composition analysis in sport. *Sports Med.* 17: 277–282, 1994.

Behnke, A. R. Anthropometric fractionation of body weight. *J. Appl. Physiol.* 16: 949–954, 1961.

Behnke, A. R., and J. H. Wilmore. Evaluation and Regulation of Body Build and Composition. Englewood Cliffs, N.J.: Prentice-Hall, 1974.

Bjorntorp, P. 'Portal' adipose tissue as a generator of risk factors for cardiovascular disease and diabetes. *Atherosclerosis* 10: 495–496, 1990.

Blundell, J. E., and J. I. MacDiarmid. Fat as a risk factor for overconsumption: satiation, satiety, and patterns of eating. *J. Am. Diet. Assoc.* 97(Suppl.): S63–S69, 1997.

Bouchard, C., A. Tremblay, A. Nadeau, J. Dussault, J.-P. Despres, G. Theriault, P. J. Lupien, O. Serresse, M. R. Boulay, and G. Fournier. Long-term exercise training with constant energy intake. 1: Effect on body composition and selected metabolic variables. *Int. J. Obesity* 14: 57–73, 1990.

Bray, G. A. Exercise and obesity. In Exercise, Fitness, and Health, C. Bouchard et al. (Eds.). Champaign, Ill.: Human Kinetics, 1990, pp. 497–509.

Bray, G. A., J. A. Glennon, L. B. Salans, E. S. Horton, E. Danforth, and E. A. Sims. Spontaneous and experimental human obesity: effects of diet and adipose cell size on lipolysis and lipogenesis. *Metabolism* 26: 739–747, 1977.

Bray, G. A., and D. S. Gray. Obesity Part I—Pathogenesis. *West. J. Med.* 149: 429–441, 1988.

Bray, G. A., and D. S. Gray. Obesity Part II—Treatment. *West. J. Med.* 149: 555–571, 1988.

Brodie, D. A. Techniques of measurement of body composition, Part I. *Sports Med.* 5: 11–40, 1988.

Brodie, D. A. Techniques of measurement of body composition, Part II. *Sports Med.* 5: 74–98, 1988.

Brown, C. H., and J. H. Wilmore. The effects of maximal resistance training on the strength and body composition of women athletes. *Med. Sci. Sports* 6: 174–177, 1974.

Brozek, J., F. Grande, J. T. Anderson, and A. Keys. Densitometric analysis of body composition: review of some quantitative assumptions. *Ann. N.Y. Acad. Sci.* 110: 113–140, 1963.

Brozek, J., and A. Henschel (Eds.). Techniques for Measuring Body Composition. Washington, D.C.: National Academy of Sciences, National Research Council, 1961.

Chowdhury, B., L. Sjostrom, M. Alpsten, J. Kostanty, H. Kvist, and R. Lofgren. A multicompartment body composition technique based on computerized tomography. *Int. J. Obes. Relat. Metab. Disord.* 18: 219–234, 1994.

Clarys, J. P., A. D. Martin, and D. T. Drinkwater. Gross tissue weights in human body by cadaver dissection. *Human Biol.* 56: 459–473, 1984.

Council on Scientific Affairs. Treatment of obesity in adults. *J. Am. Med. Assoc.* 260: 2547–2551, 1988.

Crouch, J. E., and J. R. McClintic. Human Anatomy and Physiology. New York: John Wiley and Sons, 1971, p. 112.

Dengel, D. R., J. M. Hagberg, P. J. Coon, D. T. Drinkwater, and A. P. Goldberg. Effects of weight loss by diet alone or combined with aerobic exercise on body composition in older obese men. *Metabolism* 43: 867–871, 1994.

Ellis, K. J. Body composition of a young, multiethnic, male population. *Am. J. Clin. Nutr.* 66: 1323–1331, 1997.

Epstein, L. H., and R. R. Wing. Aerobic exercise and weight. *Addictive Behaviors* 5: 371–388, 1980.

Fahey, T. D., L. Akka, and R. Rolph. Body composition and $\dot{V}_{O_2max}$ of exceptional weight-trained athletes. *J. Appl. Physiol.* 39: 559–561, 1975.

Foster, D. W. Insulin resistance—a secret killer? *N. Engl. J. Med.* 320: 733–734, 1989.

Goran, M. I., E. T. Poehlman, E. Danforth, Jr., and K. S. Nair. Comparison of body composition methods in obese individuals. *Basic Life Sci.* 60: 85–86, 1993.

Greenwood, M. R. C., and P. R. Johnson. Genetic differences in adipose tissue metabolism and regulation. *Ann. N.Y. Acad. Sci.* 676: 253–269, 1993.

Hackman, R. M., B. K. Ellis, and R. L. Brown. Phosphorus magnetic resonance spectra and changes in body composition during weight loss. *J. Am. Coll. Nutr.* 13: 243–250, 1994.

Haus, G., S. L. Hoerr, B. Mavis, and J. Robison. Key modifiable factors in weight maintenance: fat intake, exercise, and weight cycling. *J. Am. Diet. Assoc.* 94: 409–413, 1994.

Haynes, W. G., W. I. Sivitz, D. A. Morgan, S. A. Walsh, and A. L. Mark. Sympathetic and cardiorenal actions of leptin. *Hypertension.* 30: 619–623, 1997.

Heleniak, E. P., and B. Aston. Prostaglandins, brown fat and weight loss. *Med. Hypoth.* 28: 13–33, 1989.

Heymsfield, S. B., and D. Matthews. Body composition: research and clinical advances—1993 A.S.P.E.N. research workshop. *J. Parenter Enteral. Nutr.* 18: 91–103, 1994.

Heywood, V. H. Evaluation of body composition. *Sports Med.* 22: 146–156, 1996.

Heywood, V. H. *Advanced Fitness Assessment & Exercise Prescription.* Champaign, IL.: Human Kinetics, 1998.

Jackson, A. S., and M. L. Pollock. Generalized equations for predicting body density in men. *Brit. J. Nutrition.* 40: 497–504, 1978.

Jackson, A. S., M. L. Pollock, and A. Ward. Generalized equations for predicting body density in women. *Med. Sci. Sports Exerc.* 12: 175–182, 1980.

James, W. P. T., and P. Trayhurn. Thermogenesis and obesity. *Br. Med. Bull.* 37: 43–48, 1981.

Jebb, S. A., and M. Elia. Techniques for the measurement of body composition: a practical guide. *Int. J. Obes. Relat. Metab. Disord.* 17: 611–621, 1993.

Kaplan, N. M. The deadly quartet: upper-body obesity, glucose intolerance, hypertriglyceridema, and hypertension. *Arch. Internal Med.* 149: 1514–1520, 1989.

King, A. C., B. Frey-Hewitt, D. Dreon, and P. Wood. Diet versus exercise in weight maintenance: the effects of minimal intervention strategies on long term outcomes in men. *Arch. Internal Med.* 149: 2741–2746, 1989.

King, A. C., and D. L. Tribble. The role of exercise in weight regulation in nonathletes. *Sports Med.* 11: 331–349, 1991.

Kleiber, M. The Fire of Life. New York: Wiley, 1961, pp. 41–59.

Lohman, T. G., and S. B. Going. Multicomponent models in body composition research: opportunities and pitfalls. *Basic Life Sci.* 60: 53–58, 1993.

Makan, S., H. S. Bayley, and C. E. Webber. Precision and accuracy of total body bone mass and body composition measurements in the rat using x-ray-based dual photon absorptiometry. *Can. J. Physiol. Pharmacol.* 75: 1257–1261, 1997.

Malcolm, R., P. M. O'Neil, A. A. Hirsch, H. S. Currey, and G. Moskowitz. Taste hedonics and thresholds in obesity. *Int. J. Obesity* 4: 203–212, 1980.

Martin, A. D., and D. T. Drinkwater. Variability in the measures of body fat: Assumptions or technique? *Sports Med.* 11: 277–288, 1991.

Molé, P. A. Impact of energy intake and exercise on resting metabolic rate. *Sports Med.* 10: 72–87, 1990.

Molé, P. A., J. S. Stern, C. L. Schultz, E. M. Bernauer, and B. J. Holcomb. Exercise reverses depressed metabolic rate produced by severe caloric restriction. *Med. Sci. Sports Exer.* 21: 29–33, 1989.

Mott, T., and J. Roberts. Obesity and hypnosis: a review of the literature. *Am. J. Clin. Hypnosis.* 22: 3–7, 1979.

National Heart, Lung, and Blood Institute. Clinical Guidelines on the Identification, Evaluation, and Treatment of Overweight and Obesity in Adults: The Evidence Report. Bethesda, Md.: National Institutes of Health, 1998.

Newham, D. J., R. A. Harrison, A. M. Tomkins, and C. G. Clark. The strength contractile properties and radiological density of skeletal muscle before and 1 year after gastroplasty. *Clinical Sci.* 74: 79–83, 1988.

Norman, R. J., and A. M. Clark. Obesity and reproductive disorders: a review. *Reprod. Fertil. Dev.* 10: 55–63, 1998.

Oscai, L. B., S. P. Babirak, J. A. McGarr, and C. N. Spirakis. Effect of exercise on adipose tissue cellularity. *Fed. Proc.* 33: 1956–1958, 1974.

Pace, N., and E. N. Rathbun. Studies in body composition III. The body water and chemically combined nitrigen content in relation to fat content. *J. Biol. Chem.* 158: 685–691, 1945.

Pollock, M. L., E. E. Laughridge, B. Coleman, A. C. Linnerud, and A. Jackson. Prediction of body density in young and middle-aged women. *J. Appl. Physiol.* 38: 745–749, 1975.

Prins, J. B., and S. O'Rahilly. Regulation of adipose cell number in man. *Clin. Sci.* (Colch) 92: 3–11, 1997.

Qiao, Q., U. Rajala, and S. Keinanen-Kiukaanniemi. Hypertension, hyperinsulinaemia and obesity in middle-aged Finns with impaired glucose tolerance. *J. Hum. Hypertens.* 12: 265–269, 1998.

Rathbun, E. N., and N. Pace. Studies on body composition: I. The determination of total body fat by means of the body specific gravity. *J. Biol. Chem.* 158: 667–676, 1945.

Reaven, G. M. Role of insulin resistance in human disease. *Diabetes* 37: 1595–1607, 1998.

Reaven, G. M. The kidney: an unwilling accomplice in syndrome X. *Am. J. Kidney Dis.* 30: 928–931, 1997.

Sheldon, W. H., C. W. Dupertuis, and E. McDermott. Atlas of Men. New York: Harper Brothers, 1954.

Shephard, R. J., and C. Bouchard. Principal components of fitness: relationship to physical activity and lifestyle. *Can. J. Appl. Physiol.* 19: 200–214, 1994.

Siri, W. E. Gross composition of the body. In Advances in Biological and Medical Physics, IV, J. H. Lawrence and C. A. Tobias (Eds.). New York: Academic Press, 1956.

Slaughter, M. H., T. G. Lohman, R. A. Boileau, C. A. Horswill, R. J. Stillman, M. D. VanLoan, and D. A. Bemben. Skinfold equations for estimation of body fatness in children and youth. *Hum. Biol.* 62: 709–723, 1988.

Spotti, D., M. C. Librenti, M. Melandri, G. Slaviero, R. Quartagno, P. Vedani, V. Tagliabue, and G. Pozza. Bioelectrical impedance in the evaluation of the nutritional status of hemodialyzed diabetic patients. *Clin. Nephrol.* 39 (3): 172–174, 1993.

St. Jeor, S. T. The role of weight management in the health of women. *J. Amer. Dietetic Assoc.* 93: 1007–1012, 1993.

Stunkard, A. J., J. R. Harris, N. L. Pederson, and G. E. McClearn. The body-mass index of twins who have been reared apart. *N. Engl. J. Med.* 322: 1483–1487, 1990.

VanLoan, M. D., N. L. Keim, K. Berg, and P. L. Mayclin. Evaluation of body composition by dual energy x-ray absorptiometry and two different software packages. *Med. Sci. Sports Exerc.* 27: 587–591, 1995.

Wabitsch, M., H. Hauner, E. Heinze, R. Muche, A. Bockmann, W. Parthon, H. Mayer, and W. Teller. Body-fat distribution and changes in the atherogenic risk-factor profile in obese adolescent girls during weight reduction. *Am. J. Clin. Nutr.* 60: 54–60, 1994.

Wadden, T. A., and T. B. Van Itallie. Treatment of the Seriously Obese Patient. New York: Guilford Press, 1992.

Walberg, J. L. Aerobic exercise and resistance weight-training during weight reduction. *Sports Med.* 47: 343–356, 1989.

Wallace, M. B., B. D. Mills, and C. L. Browning. Effects of cross-training on markers of insulin resistance/hyperinsulinemia. *Med. Sci. Sports Exerc.* 29: 1170–1175, 1997.

Weiser, M., W. H. Frishman, M. D. Michaelson, and M. A. Abdeen. The pharmacologic approach to the treatment of obesity. *J. Clin. Pharmacol.* 37: 453–473, 1997.

Weltman, A., S. Matter, and B. A. Stamford. Caloric restriction and/or mild exercise: effects on serum lipids and body composition. *Am. J. Clin. Nutr.* 33: 1002–1009, 1980.

Widdowson, E. M., R. A. McCance, and C. M. Spray. The chemical composition of the human body. *Clin. Sci.* 10: 113–115, 1951.

Wilmore, J. H. Body composition in sport and exercise: directions for future research. *Med. Sci. Sports Exer.* 15: 21–23, 1983.

Wilmore, J. H., C. H. Brown, and J. A. Davis. Body physique and composition of the female distance runner. *Ann. N.Y. Acad. Sci.* 301: 764–776, 1977.

Wilmore, J. H., R. N. Girandola, and D. L. Moody. Validity of skinfold and girth assessment for prediction alterations in body composition. *J. Appl. Physiol.* 29: 313–317, 1970.

Wilmore, J. H., and W. L. Haskell. Body composition and endurance capacity of professional football players. *J. Appl. Physiol.* 33: 564–567, 1972.

Wood, P. D., M. L. Stefanick, P. T. Williams, and W. L. Haskell. The effects on plasma lipoproteins, blood pressure and body composition of a calorie-reduced prudent diet, with and without exercise, in overweight men and women. *New Engl. J. Med.* 325: 461–466, 1991.

CHAPTER 26

EXERCISE, DISEASE, AND DISABILITY

The human body has a high tolerance for exercise. Heart rate can increase by 150% and stroke volume by 30%. The lungs can saturate the blood almost fully with oxygen, even during maximal exercise. Metabolism can increase three to four times above the resting rate in a debilitated person; in an elite endurance athlete, it can sometimes increase 25 times. In disease, however, the body is often under intolerable stress. At these times, the additional stress of exercise can lead to further deterioration—in some cases it can even cause death.

Exercise is used as an adjunct treatment for many diseases. Training allows the body to meet higher metabolic demands that otherwise would lead to further deterioration. The stress of exercise, however, must be applied cautiously and conservatively in disease states. In some cases, exercise is not appropriate. The decision to exercise a patient must be based on the nature and severity of the disease.

Research in this area has increased during the past ten years. We have new understanding, for example, of the influence of exercise on immunity. For people with diabetes and pulmonary diseases, exercise has become a conventional treatment technique. For the disabled, participation in sports has become extremely common, and there are disabled championships in most sports.

This chapter presents an overview of the physiological responses to exercise in selected diseases and disability. It is such a broad area that treatment of specific topics is brief, and the diseases included are those most often encountered by exercise

Despite disabilities due to disease and trauma, people use exercise to overcome limitations imposed by a variety of life circumstances. PHOTO: © Claus Anderson.

607

leaders or exercise physiologists. General factors affecting the body's response to exercise, including immunity and bedrest, are also explored. Heart disease and obesity are considered separately in Chapters 24 and 25, respectively.

■ Chronic Diseases

Cancer

Cancer is the abnormal, uncontrolled growth of cells, which, left untreated, can ultimately cause death. Cancers are classified as benign tumors or malignant. Benign tumors consist of cells similar to the surrounding cells. They are surrounded by normal cells and are enclosed in a membrane that prevents them from penetrating adjacent tissues. A malignant tumor can invade other tissues, including tissues of the lymphatic, circulatory, and nervous systems, by a process called metastasis. Consequently, they can form additional tumors at sites far from their origin.

In the United States 1.2 million people are diagnosed with cancer every year and more than 500,000 die of the disease. In women, breast cancer is the most common form of cancer, while prostate cancer is most common in men (Figure 26-1). Lung cancer, probably related to cigarette smoking, is the second most common type of cancer in both genders.

Cancer is the second leading cause of death in the United States. Unlike heart disease, the cancer rate has been increasing steadily since the 1950s. The causes of various cancers are not known. However, environmental factors are thought to play a strong role. Studies of immigrants have shown that these people assume the cancer risk of the people in their new homelands. Although exercise has been shown to decrease significantly the risk of coronary heart disease (see Chapter 24), the effects of exercise on cancer risk are much less clear.

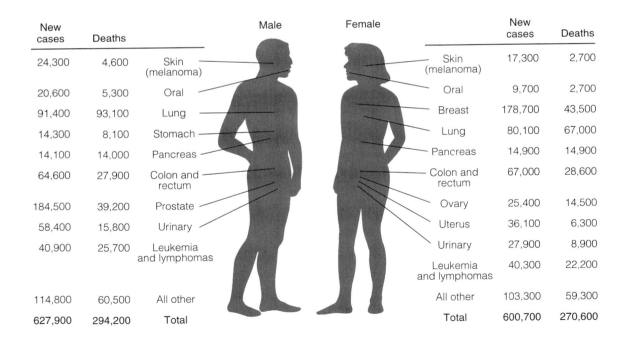

New cases	Deaths	Male		Female		New cases	Deaths
24,300	4,600	Skin (melanoma)		Skin (melanoma)		17,300	2,700
20,600	5,300	Oral		Oral		9,700	2,700
91,400	93,100	Lung		Breast		178,700	43,500
14,300	8,100	Stomach		Lung		80,100	67,000
14,100	14,000	Pancreas		Pancreas		14,900	14,900
64,600	27,900	Colon and rectum		Colon and rectum		67,000	28,600
184,500	39,200	Prostate		Ovary		25,400	14,500
58,400	15,800	Urinary		Uterus		36,100	6,300
40,900	25,700	Leukemia and lymphomas		Urinary		27,900	8,900
				Leukemia and lymphomas		40,300	22,200
114,800	60,500	All other		All other		103,300	59,300
627,900	294,200	Total		Total		600,700	270,600

Figure 26-1 Estimated new cancer cases and deaths by site and sex, United States, 1998. SOURCE: American Cancer Society.

The relationship between physical activity and cancers is not completely understood. Several studies have linked increased physical activity with reduced risk of breast, colon, and prostate cancers. While some data suggests that exercise reduces the risk of lung and liver cancers, the relationship between exercise and other types of cancer is unclear. Exercise may reduce the risk of breast and colon cancer by modifying other health factors (e.g., diet, smoking, alcohol, and medications). It may also work through metabolic effects (reduced obesity), hormonal and reproductive effects, mechanical effects, and enhancement of the immune system.

Exercise and Colon Cancer Many studies show a link between increased exercise and a reduced risk of colon cancer. Exercise may decrease the risk of colon cancer by improving gastrointestinal function and affecting lipid chemistry. Colon cancer may be related to transit time through the gastrointestinal (GI) tract because slow transit time increases the exposure of the GI tract to cancer-causing agents. Researchers believe that bile resins and acids may become carcinogenic in the large intestine.

Exercise may reduce the risk of colon cancer by speeding GI transit time. Peristalsis in the GI tract is under the control of the vagus nerve, and vagus nerve tone increases with exercise training. Exercise also decreases GI transit time by increasing prostaglandins F_{2a} and F_{1a} and increasing the secretion of several hormones important in gastrointestinal physiology, including motilin, vasoactive intestinal polypeptide, and pancreatic polypeptide.

Exercise, Breast, and Reproductive Organ Cancers in Women Several studies have shown a lower risk of breast and reproductive cancers in women who exercise or were athletically active in college. Former athletes showed a lower incidence of cancers of the breasts, cervix, ovaries, uterus, and vagina. A notable exception to these findings was the Nurses' Health Study II, a well-controlled prospective study of women aged 25–42 years conducted at Harvard Medical School and Brigham and Women's Hospital (Rockhill et al., 1998). The study did not support a link between physical activity and decreased breast cancer risk.

Prostate Cancer Oliveria and associates (1996) found an inverse relationship between cardiorespiratory fitness and the risk of prostate cancer. People who expended 2000–3000 kcal per week in exercise had the lowest risk of prostate cancer. Increases in insulin and estrogen may be the underlying mechanism in lowering prostate cancer risk. However, elevation of either of these hormones is associated with increased risk of neoplastic development (cancer cells).

Researchers have suggested that there is a possible link between an important risk factor of prostate cancer—hyperinsulinemia—and risk factors of coronary artery disease. Hyperinsulinemia is part of Syndrome X (see Chapters 24 and 25), which is characterized by insulin resistance, abdominal obesity, hyperlipidemia, and hypertension—all risk factors of coronary artery disease. Syndrome X may be at least partially caused by lack of exercise and a high-fat, high-refined-sugar diet.

Exercise and the Risk of Cancer Does exercise reduce the risk of cancer? The answer to this question is not known. Exercise and sports participation may decrease the risk of breast and other reproductive cancers by reducing exposure to estrogens, altering menstrual cycle patterns, delaying the age of menarche, increasing energy expenditure, reducing body fat, altering growth hormone and insulin-like growth factor, and enhancing natural immune mechanisms. Exercise may also be beneficial through mood elevation, decreased loss of lean tissue, and increased quality of life. The apparent decrease in cancer risk with exercise could also be due to factors such as diet, self-selection, genetics, alcohol consumption, and smoking. Diet may be different in the athletically active. Perhaps they smoke less or drink less alcohol. It is also possible that the same genetic factors that contribute to athletic success provide protection against cancer.

Exercise training may attenuate noncancerous prostate problems and lower urinary tract symptoms. Benign prostatic hyperplasia (BPH), leading to prostate enlargement and lower urinary tract symptoms, is highly prevalent among older men. Sympathetic nervous system activity, which is decreased by exercise training, is associated with increased prostatic smooth-muscle tone and prostatic symptoms. Platz and coworkers (1998) found that physical activity was inversely related with BPH. Their study showed that walking distance per week was inversely related to BPH risk: men who walked 2 to 3 hrs. · wk⁻¹ had a 25% lower risk of BPH.

Finally, healthier, less cancer prone people may select active occupations and lead a more active lifestyle. The link between reduced cancer risk and physical activity remains unknown.

Exercise and the Cancer Patient Exercise programming for cancer patients is in its infancy. Cancer treatment involves various combinations of surgery, radiation, and chemotherapy in an attempt to achieve cancer remission (i.e., eliminate cancer cells). It can result in a number of serious physical effects, including amputation, sensory and motor nerve damage, loss of joint range of motion, cardiac and lung scarring, cardiomyopathy, and anemia. These conditions, even when the cancer is in remission, leave the patient debilitated and deconditioned.

In active cancer patients, physical therapy and occupational therapy may prevent some deterioration that inevitably accompanies cancer treatment. For patients in remission, an exercise program must be tailored to the individual disability and exercise capacity. In such a program, care must be taken to prevent further damage to traumatized tissues. The goal of the program is to restore former physical and psychological quality of life.

Diabetes

Diabetes mellitus is characterized by a relative lack of or insensitivity to hormone insulin. Insulin is produced by the β-cells of the pancreas. The two major types of the disease are *insulin-dependent diabetes mellitus* (IDDM, Type I or juvenile onset diabetes) and *non–insulin-dependent diabetes mellitus* (NIDDM, Type II or adult onset diabetes). Two other

types of diabetes—gestational and secondary—also occur under certain circumstances. Gestational diabetes occurs during pregnancy and typically resolves itself after delivery. Secondary diabetes can result from diseases such as pancreatitis or Cushings syndrome, drugs such as prednisone, and genetic disorders such as Laurence-Moon-Bardet-Biedl syndrome.

IDDM requires injections of insulin, administered subcutaneously. Varieties of injected insulin are effective for long, intermediate, or short periods of time. Some diabetics use an insulin pump, which infuses insulin at a constant rate. The pump is helpful for controlling blood glucose. However, it is awkward and vulnerable to trauma during exercise. This will become less of a problem as pumps become smaller and more technically sophisticated.

NIDDM diabetes consists of excessive hepatic glucose production, peripheral insulin resistance (insensitivity), and defective β-cell secretory function. Insulin concentrations may be higher, lower, or equal to normal values. The main risk factors for NIDDM include family history, lack of exercise, obesity, ethnic background, age, and a history of gestational diabetes. Insulin insensitivity is particularly evident during exercise (Figure 26-2). Insulin

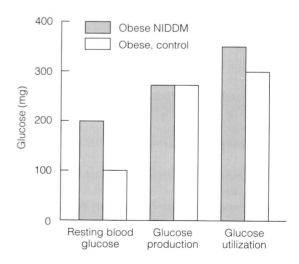

Figure 26-2 The effects of exercise on peak glucose production and use.

resistance is a common characteristic of insulin-dependent and non–insulin-dependent diabetes, particularly in skeletal muscle. The rate-limiting step for glucose utilization in muscle is glucose transport, which is mediated by glucose transporter membranes. Muscle has two types (isoforms) of glucose transporter proteins, called GLUT-1 and GLUT-4. Insulin causes an increase in glucose transport into the muscle cell by causing the GLUT-4 transporter to translocate from the interior to the outer membrane of the muscle cell. In diabetes, there seems to be a disruption in cellular communication that hampers the translocation of GLUT-4. Exercise training is thought to be beneficial in this process because it increases the number and activity of GLUT-4 glucose transporters (Chapter 5) and enzymes responsible for the phosphorylation, storage, and oxidation of glucose.

NIDDM can usually be controlled by diet alone although sometimes it requires a combination of diet and oral antidiabetic drugs to adequately control blood glucose. Oral antidiabetic agents include oral hypoglycemic agents (e.g., sulfonylureas), which stimulate insulin secretion; metformin, which inhibits excessive hepatic glucose production; acarbose, which delays the absorption of carbohydrates in the gut; and troglitazone, which reduces insulin resistance, primarily in skeletal muscle. Patients with NIDDM are also sometimes given injections of insulin.

Control of all types of diabetes involves maintaining normal or near normal blood glucose levels through the appropriate therapy. The appropriate therapy depends on the type of diabetes being treated and may include insulin, oral hypoglycemic agents, diet, and exercise.

There a number of benefits of exercise for the diabetic, Table 26-1 but exercise is not recommended unless the diabetes is under control—exercise in uncontrolled diabetes makes the condition worse. Blood glucose should be predictably regulated. Physical activity presents control problems that must be solved on a trial-and-error basis. The most immediate problem is hypoglycemia (low blood glucose). Glucose use is much greater during exercise than at rest. The diabetic who is taking insulin is involved in a daily and often hourly juggling act trying to control blood glucose. Control involves balancing energy intake (food), energy output (exercise and resting metabolism), and insulin (glucose regulator). A variation in any factor requires adjustment in the others.

Diet is very important for the diabetic. Food intake must be enough to satisfy the needs of growth (in children) and metabolism but should not produce obesity. The American Diabetes Association recommends that diabetics eat a balanced diet, with about 50% complex carbohydrates, <35% fat, and 15% proteins. Dietary cholesterol and fats should be curtailed to reduce the risk of heart disease.

Consequences of Diabetes Diseases of large and small blood vessels are common in diabetics. In fact, diabetes is a major risk factor of cardiovascular disease. Diabetes is typically accompanied by arteriosclerosis, which increases the risk of heart attack and stroke. Capillary abnormalities are also common in diabetes. The disease makes blood platelets more adhesive or stickier, allowing for a greater possibility of intravascular thrombosis (clotting). Diabetes also often causes nerve damage (neuropathy) through the deterioration of myelin, an important covering of the axons of myelinated nerve fibers. The specific effects of the disease include heart attack, gangrene, kidney failure, blindness, cataracts, muscle weakness, and ulceration of the skin.

Insulin resistance is associated with a host of cardiovascular risk factors that includes abdominal obesity, hypertension, hyperlipidemia, glucose intolerance, and hyperinsulinemia. This clustering of coronary-artery-disease risk factors is called Syndrome X or the insulin resistance syndrome. Syndrome X was discussed in Chapters 24 and 25.

Benefits of Exercise for the Diabetic There is no convincing evidence that exercise helps people with IDDM improve blood glucose control. However, exercise is very effective in improving insulin sensitivity in NIDDM.

Exercise is effective in reducing the side effects of the disease. In both IDDM and NIDDM, for instance, endurance exercise reduces platelet adhesiveness. Blood platelets, as noted, are more adhesive in diabetics. After exercises such as running, swimming, or cycling, normal adhesiveness can be achieved for about 24 hours. Regular endurance exercise, therefore, reduces the risk of coronary artery disease.

Exercise training is safe and improves health and quality of life in the diabetic if the diabetes is under control. Increased physical activity causes a loss of abdominal fat (a leading risk factor of insulin resistance), decreased blood pressure, and an improved lipid profile (i.e., decreased cholesterol, LDL, and triglycerides), including a preferential increase in the HDL3-C subfraction.

Weight training and circuit training decrease resting glucose and insulin levels in NIDDM. They also lower insulin response to a glucose challenge, as in patients who are given concentrated glucose orally. Important risk factors with exercise in diabetics include hypertension, obesity, blood lipids, and serum uric acid.

Exercise and IDDM   Insulin plays a pivotal role in fuel homeostasis. Many tissues require insulin for the uptake of glucose. In the metabolically normal person, more glucose is sometimes available than is needed. When this occurs, insulin is secreted, which acts to store glucose as glycogen in muscle tissue and the liver. Later, when blood level supplies are low, glycogen is broken down in the liver to provide glucose. Glycogen breakdown becomes very important during exercise, when considerable glucose is needed for optimum performance.

Blood glucose control in IDDM differs from that in the normal situation. Changes in blood glucose during exercise are often similar in both diabetics and nondiabetics. However, the diabetic relies more on gluconeogenesis than on liver glycogenolysis to maintain blood glucose. Glucose production is higher in the diabetic and glucose clearance is less. Diabetics oxidize glucose less completely due to the decreased activity of pyruvate dehydrogenase. People with IDDM compensate with an increased breakdown of fats and protein to meet energy needs. In inadequately controlled IDDM, there is less intramuscular glycogen and more intramuscular fat. In IDDM, if insulin is adequate, then metabolic response is similar to that in a metabolically normal person.

Preventing Hypoglycemia During Exercise Protecting against hypoglycemia is very important—particularly for the athletically active diabetic. Hypoglycemia can result in a severe impairment of judgment and loss of coordination, which could lead to injury. A hypoglycemic person can easily overestimate his or her capacity—and the results could be tragic. Extreme hypoglycemia due to too much insulin is called "insulin shock."

An injection of insulin is not the same physiologically as the natural secretion of insulin by the pancreas. Normally, insulin is released both continuously (basal secretion) and in response to blood-borne substrates. For example, after a meal, blood sugar rises and insulin is secreted to use the fuel effectively. In IDDM, insulin is administered all at once. Many of the side effects of the disease occur when there is a chronic imbalance between insulin levels and glucose. Thus, diabetes becomes very dangerous when it is not controlled.

Rapid absorption of insulin during exercise should be avoided. Diabetics who take insulin have a choice of where to inject it—usually beneath the skin covering the thigh, upper arms, abdomen, or buttocks. The site of injection is not terribly important unless exercise is to follow, and then the choice of site becomes critically important. If insulin is injected subcutaneously in an area adjacent to an exercising muscle, insulin will enter the bloodstream much more rapidly and hypoglycemia may follow. For example, in a runner who injects insulin into a leg before running, the insulin begins to act more quickly and more powerfully than normal. The runner can avoid the problem by injecting the insulin instead in the abdomen or arm. Still, the diabetic should reduce insulin dose before exercise as skin bloodflow increases all over the body during exercise, leading to an increased rate of insulin absorption.

In the metabolically normal person, insulin levels decrease during exercise. Insulin action is very important during exercise, so this decrease is paradoxical. Fortunately, though, insulin sensitivity increases during exercise, and this compensates for the decrease. Typically, the person with IDDM begins exercise with insulin levels appropriate for maintaining glucose levels at rest, but insulin levels that are appropriate at rest may lead to hypoglycemia during exercise. Increased sensitivity for exercise only adds to the problem of glucose control. Before exercise, then, it is a good idea to decrease insulin dosage or alter the type of insulin injected. Because glucose control in the exercising diabetic is a matter of trial and error, diabetics should monitor glucose regularly before, during, and after exercise.

Many active diabetics choose to counterbalance the increased glucose requirements of exercise by consuming extra food just before exercise. The food will often be a high-sugar type, such as candy, orange juice, or graham crackers. This approach works particularly well if the time or amount of exercise is irregular. If exercise habits are more predictable, then a change in the amount of insulin injected each day might help prevent hypoglycemia. Exercise at a given intensity for a given amount of time has a predictable fuel requirement, so regular, predictable exercise habits make diabetes control much easier. Diabetics should eat a meal within three hours of exercise. High-carbohydrate foods should be available during and after exercise. Also, it is a good idea for diabetics who are going running or cycling to carry a food high in carbohydrates.

If blood glucose is below 100 mg · 100 ml^{-1}, exercise should be delayed. Further, diabetics should be aware of changes in blood glucose levels. A fall in glucose (for example, from 130 to 100 mg · 100 ml^{-1}) may be a warning sign of potential hypoglycemia during exercise. Two measurements of 100 mg · 100 ml^{-1} would be more indicative of stability. If glucose levels fall below 100 mg · 100 ml^{-1}, additional carbohydrates should be eaten.

Exercise and Hyperglycemia If insulin is not adequate, elevated blood glucose and ketosis may result. Excessive secretion of glucagon, growth hormone, catecholamines, and cortisol also contribute to hyperglycemia and ketosis.

Ketosis is characterized by the accumulation of large amounts of acetoacetic acid, β-hydroxybutyric acid, and acetone in the blood. Ketosis occurs when the intracellular availability of carbohydrates is se-

verely diminished and fats must be used as the predominant fuel. The blood concentration of ketones can rise as much as 30 times above normal, leading to extreme acidosis. If left untreated, ketosis can lead to diabetic coma. Exercise is not recommended for a person with ketosis or hyperglycemia. Under certain conditions, exercise can make the problem worse by increasing the blood glucose level. Thus, diabetes should be controlled before an exercise program is begun. Exercise should be delayed if blood glucose is greater than 250 mg per 100 ml of blood. It should also be delayed if significant levels of ketones are present in the urine.

Excessive growth hormone release is also common when an uncontrolled diabetic exercises. Growth hormone is important in the regulation of blood lipids and in protein synthesis, particularly during growth. Excessive secretion of growth hormone in the uncontrolled diabetic, however, may complicate blood vessel disease. Again, the solution is to get diabetes well under control before beginning an exercise program.

Exercise and NIDDM Patients with NIDDM can expect excellent results from an exercise program. If they are taking insulin or hypoglycemic agents, they should take precautions to prevent hypoglycemia because even one exercise session increases insulin sensitivity (see Figure 26-2). Patients with vascular complications should heed the general exercise advice for diabetics given in the next section.

Precautions for Diabetics on an Exercise Program Table 26-1 lists general recommendations for the exercising diabetic. Because of the increased risk of heart disease, diabetics of any age should be tested on the treadmill before they start an exercise program. Heart disease appears in the general population beginning about 35–45 years of age. It appears sooner in diabetics. Diabetics should be thoroughly evaluated so they can safely participate in an exercise program. Also, specific complications should be considered when prescribing exercise. For example, patients with peripheral neuropathy should avoid running or basketball because of the danger of developing cuts and blisters.

TABLE 26-1

Recommendations for Physically Active Diabetics

Medical Evaluation and Exercise Prescription
- Get a medical evaluation before beginning an exercise program.
- Have an exercise stress test before beginning a program because of the increased risk of heart disease.
- Consider diabetic complications when choosing type of exercise.
- Estimate energy expenditure to help balance energy intake, energy output, and insulin requirement.

Metabolic Control
- Do not exercise if blood glucose exceeds 250 mg · 100 ml^{-1} and urinary ketones test positively.
- Eat within three hours before exercise.
- Consume additional carbohydrate if glucose levels are below 100 mg · 100 ml^{-1}.
- Carry high-carbohydrate foods during exercise.
- Be sure food is available during and after exercise.
- Consume adequate fluids before, during, and after exercise.

Blood Glucose Monitoring Before and After Exercise
- Control blood glucose systematically. Learn to balance energy intake, output, and insulin dose.
- Learn to identify physical responses to hypoglycemia.

Insulin Administration
- Avoid insulin injections one hour before exercise.
- Reduce dose when exercise is anticipated.
- Do not inject over an active limb.

General Precautions
- Protect the feet by wearing good-fitting shoes and cotton socks. Avoid activities that cause blisters.
- Carry medical identification.
- Have an emergency plan.

SOURCE: Modified from Wasserman and Abramrad, 1989. Used with permission.

■ Aging-Related Disorders

Arthritis

Arthritis is an inflammatory joint disease. More than 70% of people over 65 years of age in the United States have this disorder. The economic impact exceeds $13 billion a year. The most common forms of the disease include osteoarthritis, rheumatoid arthritis, juvenile rheumatoid arthritis, ankylosing spondylitis, systemic lupus erythematosus, and gout. Arthritis and related disorders are chronic, but they may go into remission periodically.

Osteoarthritis Osteoarthritis is the most common form of arthritis. It is caused by deterioration of the joints, a common effect of the aging process. Individual differences in the presence of arthritis can be accounted for by age, previous injury or trauma, heredity, previous joint disease, and metabolic diseases. This type of arthritis is specific to individual joints and will not necessarily spread to other joints. Pain related to osteoarthritis can vary from none to debilitating.

Osteoarthritis damages the articular cartilage of joints. This is often accompanied by bone spurs and adhesions in the membranes and ligaments of the joint. The pain that accompanies this usually results in decreased range of motion and disuse atrophy brought on by the avoidance of any motion that is uncomfortable. This activity limitation results in the formation of additional adhesions, and, in turn, further limits motion.

Exercise prescription for people with osteoarthritis includes range of motion and strength exercises. Activities that minimize weight bearing, such as swimming, are also recommended. It is often also helpful to reduce the amount of body fat, which relieves stress on the joints. Analgesics often help relieve the pain and allow the necessary exercises to be performed.

Does vigorous exercise over a lifetime predispose people to osteoarthritis? The risk of developing osteoarthritis appears to increase when people play sports involving high levels of impact or torsional loading of the joints. People who continue playing sports after joint injuries are also at higher risk. Felson and associates (1997), as part of the Framingham study (a famous epidemiological study that began in 1948), found that people at increased risk of developing osteoarthritis of the knee included those who were obese, nonsmokers, and physically active. Conversely, Fries and coworkers (1996) found in a longitudinal study of runners that vigorous running activity over many years was not associated with an increase in musculoskeletal pain with age, and that there may be a moderate decrease in pain, particularly in women. Otterness and coworkers (1998), in a study of the effects of exercise on hamster cartilage, found that a sedentary lifestyle in the hamster leads to a lower proteoglycan content in the cartilage and a lower synovial fluid volume. These changes are associated with cartilage fibrillation, pitting, and fissuring. Conversely, daily exercise prevented early cartilage degeneration and maintained normal articular cartilage. Thus, it appears that moderate exercise may enhance joint cartilage health while high-stress exercise may cause damage and deterioration.

Rheumatoid Diseases Arthritis is a common manifestation of rheumatoid diseases. Examples of these diseases include rheumatoid arthritis, lupus erythematosus, and polyarteritis nodosa. These are autoimmune diseases. Aberrant immunoglobulins combine to form anti-immunoglobulin antibodies called *rheumatoid factor.* The rheumatoid factor binds with antigens and immune complexes in tissue and produces an inflammatory response. With repeated bouts of inflammation, scar tissue eventually forms. Inflammation is a potent precipitator of tissue damage.

Until recently, exercise was contraindicated because it was thought to increase the severity of inflammation and tissue damage associated with rheumatoid diseases. Excessive bed rest and deconditioning can also lead to deterioration. However, over 30 studies have examined the effects of exercise on rheumatoid arthritis. In general, these studies have concluded that dynamic exercise training is effective in increasing aerobic capacity and muscle strength and that it has no detrimental effects on disease progression or pain. In fact, Van den Ende and associates (1996) showed that intense exercise (cycle ergometer at 60% or more of capacity) was superior to range-of-motion or isometric training for increasing aerobic capacity, joint mobility, and muscle strength. Other studies have shown that high-intensity strength training is feasible and safe if the rheumatoid arthritis is well controlled. In addition, weight training leads to significant improve-

ments in strength, pain, and fatigue without increasing the progression of the disease or joint pain.

Range of motion exercises and minimal weight-bearing endurance exercises such as swimming and walking are recommended. Exercise can be practiced as tolerated. Drugs, including aspirin, fenoprofen, indomethacin, tolmetin, corticosteroids, and phyenylbutazone, may be used to relieve pain and facilitate mobilization.

Another potentially serious problem associated with rheumatoid disease is heart and lung disease. Sixty percent of patients with systemic lupus erythematosis will have pleuritis, pericarditis, or both. Inflammation of small- or medium-sized blood vessels is relatively common in all rheumatoid diseases. Caution should therefore be used, as excessive exercise could be destructive to the heart.

Osteoporosis

Osteoporosis is a condition in which bone mass is so low that even a minor trauma can cause a fracture, most commonly in the hip, spine, and wrist. Because osteoporosis is a symptomless disease, it can progress undetected for decades. It is a major public health problem that affects more than 25 million Americans, 80% of whom are women. It is linked with more than 1.5 million fractures a year, and as the population ages in the United States, the incidence of osteoporotic fractures will increase.

People are at increased risk for osteoporosis if they are older, female, Caucasian or Asian, smoke cigarettes, have low body weight, drink alcohol excessively, are sedentary, consume inadequate calcium, or have a genetic predisposition to the disease. Women who have had normal or early menopause or prolonged premenopausal amenorrhea are also at increased risk. Prevention includes maximizing bone density during childhood and young adulthood, taking estrogen supplements at menopause, taking dietary calcium and perhaps vitamin D supplements, and practicing regular weight-bearing exercises.

Bone Formation The organic components of bone include a matrix of fibers, blood vessels, and lymphatics. The three types of bone cells are osteoblasts, osteocytes, and osteoclasts. Osteoblasts build bone tissue by producing the organic matrix where the crystallized salts are deposited. The matrix is called *osteoid;* the process of its formation is called *ossification*. In a process called *calcification,* calcium is deposited in the matrix after the matrix has been formed. The terms *ossification* and *calcification* are often used synonymously, but this is incorrect. An osteocyte is an osteoblast surrounded by the osteoid matrix.

Osteoclasts are cells that respond to the body's need for calcium by initiating a process called *osteoclasis*. Osteoclasis is the simultaneous removal of osteoid and salts (i.e., deossification and decalcification).

The inorganic components of bone are the crystalline salts, consisting mainly of calcium and phosphorous. Calcium is a very dynamic substance within bone. It is in equilibrium with the calcium in the body fluids and tissue. These calcium sources form a pool of exchangeable calcium that is available to satisfy calcium requirements.

Normally, osteoblastic activity (bone buildup) and osteoclastic activity (bone breakdown) balance each other, so there are no overt changes in bone. Peak bone mass occurs in women between 15 and 20 years of age. After that, osteoblast and osteoclast activity remains in balance until menopause. Then, osteoclastic activity accelerates, resulting in a gradual loss of bone-mineral mass.

Bone responds to stress and disuse by reducing tissue in areas where it is no longer needed and increasing it in areas that are subjected to stress. Increased bone density is called *sclerosis;* decreased bone density is called *rarefaction*. Sclerosis results from increased bone deposition with normal resorption and may occur in exercise-induced hypertrophy of bone, degenerative osteoarthritis, infection, neoplasms, and osteopetrosis.

Rarefaction eventually results in *osteopenia,* in which bone density decreases. Osteoporosis is a condition of extreme osteopenia, in which the bone is particularly susceptible to fracture. In osteoporosis, a relative decrease in osteoblastic activity reduces some bone, while the remaining bone is essentially normal in organic and inorganic components. Osteoporosis can be painlessly detected using dual-energy X-ray absorptiometry (DEXA).

Preventing Osteoporosis One of the best strategies for preventing post-menopausal osteoporosis is to maximize bone mass during childhood and premenopausal adulthood. This can be accomplished by practicing regular exercise and insuring adequate lifetime calcium intake. Habitual physical activity, particularly high-impact activities such as running and gymnastics, are associated with higher lifetime peak bone density and lower fracture risk. Recommended calcium intake to maximize bone mass is 1300 mg/day for children 9–18 years of age, 1000 mg for men and women 19–50 years of age, and 1200 mg for people over 50.

The American College of Obstetrics and Gynecology recommends that post-menopausal women should be encouraged to consider estrogen replacement therapy, and those with risk factors who decline estrogen replacement therapy should be recommended for bone density measurements. Women with low bone density or established osteoporosis should be offered estrogen replacement, alendronate, or calcitonin therapy.

Other factors to consider while developing osteoporosis-prevention strategies include the following (ACSM, 1995):

- **Bone adaptations are exercise specific**. Lower body exercise will strengthen leg bones but will not strengthen bones in the upper body.

- **Bone growth occurs when they are overloaded**. Nonweight bearing exercise, such as swimming, has no effect on bone mass.

- **Gains in bone mass are reversible**. Changes in bone mass are lost with subsequent deconditioning.

- **People with the lowest bone mass will improve the most from an exercise program.**

- **People have a genetic ceiling that limits their capacity to increase bone mass.** Gains in bone mass plateau as this ceiling is approached.

Exercise and Osteoporosis Prevention is the best method for dealing with osteoporosis. As discussed, a valid lifetime strategy is to maximize bone mass before menopause through diet and exercise and to maintain bone mass after menopause with estrogen, diet, and exercise. Bone mass at menopause is an important predictor of long-term prognosis.

People with established osteoporosis are typically unfit and may have significant orthopedic limitations. Exercise programs for such people should include endurance, resistance, and range-of-motion exercises. The principles of exercise training are the same as those described in Chapter 27. Walking and stationary cycling are excellent activities. Forward flexion (as in sit-ups) should be avoided, particularly in people with kyphosis, because it causes significant spinal loading that could lead to fracture. People should start conservatively and increase loads and intensity very gradually.

■ Pulmonary Disorders

Pulmonary diseases can impair the three processes of pulmonary physiology—ventilation, diffusion, and perfusion (lung blood flow). Ventilatory impairments may include increased airway resistance, reduced compliance (chest wall elasticity), increased work of breathing, ventilatory muscle weakness, ventilatory inefficiency, and ventilatory fatigue or failure. Diffusion can be compromised by destruction of the alveolar-capillary membranes and ventilation-perfusion inequality. Perfusion may decrease due to cardiovascular deconditioning, reduced pulmonary vascular conductance, or right heart failure. Patients with pulmonary disorders can also be affected by peripheral muscle disadaptations, dyspnea (distressed breathing), anxiety, and depression.

Chronic Obstructive Lung Disease

Chronic obstructive pulmonary disease (COPD) is progressive. It is characterized by destruction of alveoli in the airways, retention of mucous secretions, narrowed airways, and respiratory muscle weakness. Categories of this disease include emphysema, asthma, and bronchitis. Because a COPD patient typically has all or most of these conditions, it is difficult to subdivide patients.

Characteristics of the Disease Emphysema is characterized by a loss of alveoli and their related vasculature. In addition to the loss of functional lung tissue, the hypoxemia created by the reduction of blood supply results in pulmonary vasoconstriction, which tends to further reduce the surface area of the lung for gas exchange. The disease increases pulmonary artery pressure, first during exercise and then at rest. Chronic pressure overload leads to structural hypertrophy and hyperplasia of the smooth muscle pulmonary vascular bed. Fibrosis and atherosclerosis follow. Ultimately, pulmonary hypertension leads to right heart failure.

Emphysema patients often have chest deformities because they use accessory muscles for ventilation. Their diaphragm, for example, is fixed in an inspiratory position, their chest is overexpanded, and breathing muscles are weakened. In addition, adenosine triphosphate (ATP) and phosphocreatine (CP) levels decrease. These decreases are probably related to increased airway obstruction. Patients show a higher than normal residual lung volume and have difficulty expiring rapidly. Patients also exhibit reduced oxidative capacity in peripheral muscles (Serres, 1998).

Chronic bronchitis is chronic inflammation of the lower respiratory tract. Patients exhibit a persistent cough and often have episodes of infected sputum. Shortness of breath (dyspnea) is common. Bronchitis patients typically show blood gas abnormalities as well. These include reduced partial pressures of arterial oxygen and venous carbon dioxide. The cause is abnormal ventilation (airflow) and perfusion (blood flow) of the lungs. Extensive peripheral edema (swelling) is common during the later stages of the disease, caused by chronic pulmonary hypertension and right heart failure. Because the heart cannot keep up with circulatory demands, venous pressure increases, fluid is forced into the tissue spaces, and the tissues swell.

Exercise Response in COPD Ventilation and diffusion of respiratory gases are limiting in COPD. Weak respiratory muscles have difficulty ventilating the lung adequately. Airways are obstructed and narrowed, and alveoli are destroyed or compromised. Physical work capacity is usually low, and patients often complain of dyspnea at rest. (See Table 26-2.)

Maximal exercise ventilation often reaches or exceeds maximal voluntary ventilation (MVV). MVV is a pulmonary test done in the lab to measure maximum breathing capacity. In the test, a subject breathes as rapidly as possible for 10–15 seconds. At maximum exercise, people with normal lung function normally ventilate at 60–75% of MVV. Lack of breathing capacity in COPD patients causes dyspnea and fatigue at low exercise intensities.

Ventilation during submaximal exercise will usually be greater than normal in the COPD patient. These patients have an increased dead space and must breathe more to achieve the same level of alveolar ventilation. As discussed in Chapter 11, alveolar ventilation is the air that actually reaches the alveoli. As dead space increases, tidal volume must increase to compensate. COPD patients have reduced expiratory flow rates at all levels of ventilation. During exercise, they also have impaired inspiratory flow rates. COPD patients compensate for flow difficulties during exercise by increasing breathing frequency and decreasing tidal volume compared to normal subjects.

The increased work of breathing during an exercise test affects typical physiological responses. Maximal heart rate is less than normal, for example, and the maximal respiratory exchange ratio ($\dot{V}_{CO_2} / \dot{V}_{O_2}$) is usually less than 1.0. The latter indicates that the patient has not been metabolically exhausted during the test. Maximum blood lactates are usually less than 8 mM. These responses show that it is breathing capacity rather than metabolic capacity that is limiting. These people are often unable to push themselves during exercise. Anxiety produces shortness of breath.

During an exercise test, about 50% of COPD patients stop exercise because of leg pain or weakness. This is typically a sign of metabolic fatigue. Even though these patients have lung limitations, deconditioning is probably the primary mechanism of fatigue. In fact, they typically have less severe disease than those who stop exercise because of shortness of breath.

Arterial oxygen partial pressure ($P_{a_{O_2}}$) is decreased at rest, and usually decreases more with exercise. Hemoglobin saturation also decreases during exercise. Pulmonary limitations are due to inadequate diffusion and hypoventilation, problems that lead to dyspnea. During exercise, it is difficult to determine the true cause of fatigue. Whether COPD patients stop exercise because of breathing discomfort or because they have reached the limits of their pulmonary system is still a matter of controversy.

Exercise Prescription Unfortunately, in the COPD patient, lung function does not tend to improve with training although endurance training may delay the deterioration of pulmonary function and maximal oxygen consumption and work ca-

---

TABLE 26-2

Principal Changes in Oxygen Transport Capacity in Chronic Pulmonary Disease Compared to Normal Subjects

Factor	Change
Maximal heart rate	↓
Maximal oxygen consumption	↓
Maximal cardiac output	↓
Ventilation during submaximal exercise	↑
Exercise breathing frequency	↑
Exercise tidal volume	↓
Maximal $\dot{V}_E$/MVV	↑
Ventilatory reserve[a]	↓
Alveolar oxygen partial pressure	↑
Arterial oxygen partial pressure	↓
Arterial hemoglobin saturation	↓
Pulmonary artery blood pressure	↑

[a] Difference between maximum exercise ventilation and results of maximal voluntary ventilation test (MVV).

pacity may improve. However, the patient must be motivated and exercise intensity and duration should be moderate.

Exercise prescription for these patients should include progressive endurance exercises—walking and stationary cycling are good choices. Breathing exercises are also important. Breathing exercises help increase airflow to obstructed and restricted airways and improve respiratory muscle endurance. There is a close relationship between the degree of distressed breathing and breathlessness. Improving the strength of the respiratory muscles improves the quality of life for these patients: The strength of the inspiratory muscles improves and dyspnea decreases. Patients may also benefit from training in a hyperbaric environment (Kramer et al., 1998).

Other benefits of exercise include facilitation of mucus removal and maintenance of chest mobility. Ideally, hemoglobin saturation (a measure of pulmonary function) should be monitored with an oximeter (a device that specifically measures such saturation).

Asthma

Asthma is technically an obstructive lung disease, but because it is frequently experienced without emphysema or bronchitis, particularly in the young, it is discussed independently in this chapter. Asthma is a lung disorder characterized by edema in the walls of the small bronchioles. Secretions of thick mucus from the bronchial lumen are common, and spasm of the smooth muscle in the pulmonary trunk is typical. Symptoms include choking, shortness of breath, wheezing, tightness in the chest, increased mucus production, and fatigue. When an asthma attack occurs, breathing becomes labored. This is particularly true during expiration because of the decrease in bronchiolar diameter. An attack can be caused by emotional upset, dust, pollen, cold and damp weather, smoke, and exercise.

Asthma is becoming a more serious problem among elite athletes. Weiler and associates (1998) found that asthma was more prevalent in athletes who participated in the 1996 Summer Olympic Games (Atlanta) than in the general population or in those who participated in the 1984 Summer Olympic Games (Los Angeles). The degree of respiratory distress during exercise and exposure to environmental pollutants may be precipitating factors—asthma was extremely common in cyclists and mountain bikers (45%) but nonexistent in weight lifters.

Exercise-Induced Asthma   Although exercise can be beneficial, it is not without risk. Overexertion is a prominent precipitator of asthma. It is particularly important, therefore, that people with asthma increase the intensity of their programs gradually. Exercise-induced asthma (EIA) is thought to be caused by a reaction of the airways to changes in humidity and temperature. During exercise, increased ventilation leads to increased heat loss by convection and evaporation, and these changes can stimulate an asthma attack.

Exercise-induced asthma usually develops slowly during exercise (Figure 26-3). Peak symptoms are reached in six to eight minutes. Often, the attack will subside as exercise continues, but symptoms will often return or intensify during recovery from the exercise.

Laboratory performance measures typically change drastically during an asthmatic episode. Forced vital capacity (FVC) decreases at least 20%. Forced expiratory volume in one second and maximum voluntary ventilation (MVV) may decrease by 40%. These measures show the difficulty asthmatics have expiring air forcefully.

Residual lung volume (RLV) is a measure of respiratory dead space (see Chapter 12). Exercise-induced asthma can increase RLV by over 100%. As discussed in the section on COPD, increases in dead space must be matched by increased tidal volume. Without it, alveolar ventilation decreases.

People with asthma should avoid adverse environmental conditions. Air pollution and extreme cold can trigger an asthma attack, for example. In addition, exercise should be modified or curtailed if the asthmatic is too tired. Emotional stress can also increase the chances of an attack.

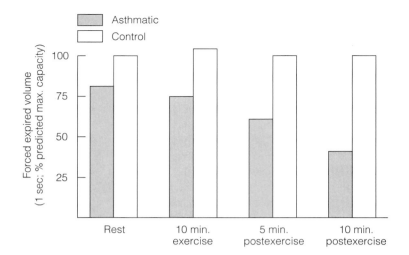

Figure 26-3 Exercise-induced asthma and forced expiratory volume in 1 second (FEV_1).

Under poor environmental conditions, exercise bouts should be kept short. Typically, they should be less than three minutes at a time, or as tolerated. Asthmatics should avoid swimming in excessively cold water because the cold water can induce an asthma attack. Conversely, swimming in water with a comfortable temperature is ideal exercise. The increased pressure on the chest wall makes it easier to exhale, and the warm, saturated air is also beneficial.

Exercise-induced asthma can be prevented by avoiding sudden changes in airway temperature. Preventive measures include warming up before and cooling down after vigorous exercise, covering the mouth and nose when exercising in cold weather, and training in warm, humidified environments. People with good fitness and well-controlled bronchial reactivity (through medication) have fewer problems with exercise-induced asthma.

Allergy Control Asthma is often associated with allergies, so some degree of allergic control is desirable before beginning a vigorous endurance program. Medical treatment may include desensitization against allergens or administration of appropriate medications. At the very least, the asthmatic can prevent allergic reactions by avoiding dusty areas, pollen, and smoke.

Exercise Prescription Exercise should increase in intensity gradually over time. The asthmatic should be thoroughly instructed about the disease, and basic symptoms and strategies for dealing with EIA should be discussed. Basic practices of warm-up and cool-down should be observed. These help ensure that temperature changes to the airways occur gradually.

Medical Screening The asthmatic should take a treadmill test before beginning an exercise program. Ideally, this test should include blood gas analysis. A measure of hemoglobin saturation, done with an oximeter, may also be helpful.

The lungs of people with normal lung function have a high capacity and are capable of almost completely oxygenating the blood, even during the heaviest levels of exercise. Asthmatics, however, may not be able to supply enough oxygen to the blood at the heavier levels of exertion. Ventilation and diffusion become the limiting factors. The treadmill and blood gas measurements, thus, make it possible to select a safe upper limit of exercise.

Type, Intensity, and Duration of Exercise People with asthma should participate in self-paced activities. They should avoid overly competitive situations when they are beginning their exercise programs. Endurance interval training—for example, 30-sec to 3-min bouts of exercise, followed by short rests—is good. Low-intensity interval training is well tolerated by most asthmatics and is recommended as a conditioning method during the early stages of physical training. Beginners should ideally exercise in a supervised environment.

After preliminary conditioning, exercise sessions can follow the American College of Sports Medicine guidelines for exercise in healthy adults. They recommend that people practice endurance exercise 20–60 minutes per session, 3–5 times per week, at 60–85% of maximum capacity. Walking and swimming in warm water are excellent exercises with which to begin the program.

Precautions During Exercise An inhaler and 100% oxygen should be available during exercise. Most asthmatics benefit from using an aerosol inhaler before exercise. β-2 agonists or cromolyn sodium are good choices for preexercise medication.

Someone should be available to assist the asthmatic if difficulty develops. In addition, the asthmatic should be instructed in relaxation positions and in breathing exercises, which are particularly helpful if there is shortness of breath or respiratory distress.

Asthmatics should also be encouraged to drink a lot of water when doing endurance exercise. Ingesting water reduces the thickness of lung secretions and makes breathing easier.

Medication Modern medications have enabled asthmatics to participate in even the most vigorous sports. These have included endurance events in the Olympic Games and in professional sports. In the 1972 Olympic Games, an asthmatic won a gold medal in swimming, only to be disqualified for taking his asthma medication. Today, several drugs are available for asthmatics that are effective and are not on the banned substance list. Inhaled β-agonists are the most commonly prescribed medications for preventing exercise-induced asthma. Inhaled sodium cromoglycate (cromolyn sodium) or nedocromil may also be used. Agents that may be added if inhaled β-agonists or sodium cromoglycate are not adequate include anticholinergic agents (such as ipratropium bromide), theophylline, calcium channel blockers, α-agonists, antihistamines and oral β-agonists. Newer agents include antileukotriene

agents, inhaled heparin, and inhaled furosemide (frusemide) (Tan and Spector, 1998).

Asthmatic drugs can prevent EIA. If exercise brings on an attack, the asthmatic should learn to take medication before a workout. Inhaling a bronchodilating drug before exercise can modify or eliminate symptoms and can also reduce the effects once they have occurred.

■ Exercise, Immunity, and Infection

The relationship between exercise and immunity has been a part of folklore and the subject of home-remedy medicine for centuries. It is popularly believed, for example, that regular exercise makes people more resistant to disease and that getting chilled makes people more prone to the common cold. What evidence do we have for these and other popular beliefs? In recent years, researchers have increased our understanding of the extremely complex human immune system. The system, working together with the neuroendocrine system, attempts to maintain homeostasis. In so doing, hormones such as catecholamines, insulin, and corticosteroids are secreted that influence the immune function—and exercise, as we saw in Chapter 9, has a profound effect on the secretion of these hormones.

Basic Structure of the Immune System

The immune system recognizes and works to rid the body of foreign material. The system can be subdivided into nonspecific and acquired immune mechanisms.

Nonspecific immune mechanisms include (Keast et al., 1988):

- Skin: resists penetration by foreign organisms and material
- Respiratory tract: filters particles entering from the air
- Acidic digestive secretions in the stomach: kill ingested organisms
- Capacity to lower iron levels in blood and intestinal fluid: affects growth of pathogens

- Phagocytosis of bacteria and viruses
- Inflammatory response to infection and injury: directs cells (macrophages) and causes release of chemicals (e.g., complements, histamine, bradykinin) in the area to rid body of foreign material or injured tissue
- Secretion of antibacterial substances (e.g., α-interferons)

Acquired immune mechanisms include those that are antibody-based and those that are cell-based. Antibody-based immunity involves B and T lymphocytes, which produce antibodies when exposed to a foreign substance, called an *antigen.* Classes of antibodies include immunoglobulin M (symbolized as IgM), IgA, IgG, IgE, and IgD. In antigen–antibody reactions, the antibodies destroy and remove the invading antigens (microorganisms).

Cell-based immunity is centered around the T lymphocyte. When exposed to an antigen, T lymphocytes may join with B lymphocytes to produce antibodies. They can also attach themselves to antigens directly and destroy them. Direct antigen destruction involves several subtypes of T lymphocytes, including cytotoxic T cells, helper T cells, and suppressor T cells. Refer to recent reviews of the immune function by Neiman (1997) and Neiman and Pedersen (1999) for a more in-depth coverage of this topic.

Exercise Training and Immunity

There is not enough evidence to state conclusively that exercise training either improves or impairs immune function, but several studies suggest that heavy training depresses the immune system while moderate exercise may improve immunity.

Moderate Exercise In animal studies, moderate exercise was found to increase antibody levels and longevity. Although similar data on humans are scarce, many studies do show that components of the immune system are affected by exercise. Several studies show that moderate exercise (walking) resulted in people getting half the colds of sedentary people. Moderate levels of physical activity may also decrease the incidence of colon cancer, which is thought to be linked to immune function.

Chronic Intense Exercise Several studies have reported an increased incidence of upper respiratory infection with exercise. Nieman and coworkers (1990), for example, found an increased incidence of upper respiratory infections in runners who ran a marathon. At the same time, several measures of immune function were found to be depressed in heavily training athletes. Immunoglobulin A, for example, important in protecting against upper respiratory infections, was found to be depressed in cross-country skiers, who otherwise reported no increased incidence of illness. In another study, however, athletes given antitetanus vaccine immediately after running a marathon had a normal antibody reaction (Eskola et al., 1978).

Untrained subjects show a greater increase in lymphocytes in reaction to intense physical exercise than do trained subjects, who typically have lower responses. Trained subjects have been found to have transient reductions in helper T cells, which could suppress immune function. Several studies have also shown that after vigorous exercise the ability of lymphocytes to react to specific antigens is reduced. These changes are temporary. The long-term effects on immune function are not known.

Overtraining is often associated with increases in corticosteroid hormones, such as cortisone. Elevated corticosteroids have a depressive effect on the immune system. For example, cortisone decreases the reaction of lymphocytes to an antigen. Although the effects of exercise on the immune system are probably small and transitory, if the exercise program leads to overtraining, the immune system may be suppressed. This is a critical consideration in the training of athletes. A sick athlete is in the same category as one who is injured or not making satisfactory progress.

Nieman (1999) made several recommendations for minimizing the effects of exercise and sport on the immune system:

- Keep other life stresses to a minimum. Mental stress in and of itself has been linked to an increased risk of upper respiratory infection.

- Eat a well-balanced diet to keep vitamin and mineral pools in the body at optimal levels. Although there is insufficient evidence to recommend nutrient supplements, ultramarathon runners may benefit by taking vitamin C supplements before ultramarathon races.

- Avoid overtraining and chronic fatigue.

- Obtain adequate sleep on a regular schedule. Sleep disruption has been linked to suppressed immunity.

- Avoid rapid weight loss, which has also been linked to negative immune changes.

- Avoid putting hands to the eyes and nose (primary routes of introducing viruses into the body). Before important athletic events, avoid sick people and large crowds when possible.

- Get flu shots when competing during the winter months.

- Use carbohydrate beverages before, during, and after marathon type race events or unusually heavy training bouts. This may lower the impact of stress hormones on the immune system.

- Do not resume intensive exercise training until a few days after the resolution of common cold syptoms (e.g., runny nose and sore throat without fever or general body aches and pains). Mild to moderate exercise (e.g., walking) when sick with the common cold does not appear to be harmful.

- With symptoms of fever, extreme tiredness, muscle aches, and swollen lymph glands, 2–4 weeks should probably be allowed before resumption of intensive training.

Viral Infections

Viral illnesses are extremely common, affecting the average person one to six times a year. They typically affect the upper respiratory tract, but sometimes they have systemic effects. Viral illnesses can have far-reaching effects on organs and tissues, and can impair skeletal muscle and cardiac function. The severity of viral infection can range from sub-

clinical to death. The most common viral groups are rhinovirus, Coxsackie A and B, echovirus, adenovirus, and influenza.

Viral Infections and Exercise In people with systemic viral infections, the risk of cardiac-related sudden death increases during exercise. The Coxsackie virus has been shown to have a tendency to invade the heart muscle, so exercising with a systemic Coxsackie infection may increase the risk of cardiac arrhythmias and sudden death. Several clinical studies have reported that exercising with a viral illness may have contributed to bacterial meningitis and acute rhabdomyolysis (muscle destruction). One study reported worsened symptoms with exercise in an asthma victim. In racehorses, upper respiratory infections have been associated with poor performance.

Viral illnesses decrease physical performance and affect muscle structure. Isometric strength decreases in patients with active viral infections, and recovery may take as long as one month. Reduced levels of several muscle substrates and enzymes have been found in patients with active viral illnesses. Muscle samples, observed with an electron microscope, have also revealed cellular abnormalities.

People experiencing a sudden loss in performance without any symptoms of illness may have a subclinical viral infection. Resumption of training after a viral illness depends on the symptoms. If symptoms are limited to the upper respiratory tract, then training can resume in a few days. However, if symptoms are general and more severe, then more rest is needed. Patients should rest at least one day for every day of illness.

Infectious Mononucleosis Infectious mononucleosis occurs often among young adults and teenagers. It can have a devastating effect on athletic performance. The disease is caused by the Epstein–Barr virus, a member of the herpes group. It is characterized by sore throat, fever, enlarged lymph nodes, malaise, and fatigue. Approximately 95% of college-age students are exposed to the virus. The acute phase of the disease lasts 5–14 days, and complete recovery takes 6–8 weeks. Highly trained athletes may not achieve pre-illness levels of fitness for up to three months.

Mononucleosis is self-limiting and rarely fatal. Most deaths have been due to complications such as bacterial sepsis, ruptured spleen, or asphyxia from airway obstruction. The disease is often accompanied by an enlarged spleen, which can be ruptured in contact sports or vigorous exercise. However, in approximately 50% of reported cases of ruptured spleen, there was no precipitating trauma or vigorous exercise. Ruptured spleens from mononucleosis occur most often in white males; they rarely occur in blacks or females. Vigorous exercise and contact sports should be avoided for at least one month after the illness has ended. Resumption of training should be delayed further if the spleen continues to be enlarged.

Acquired Immune Deficiency Syndrome (AIDS)

Over 12 million people worldwide are estimated to be infected with the human immunodeficiency virus (HIV). HIV causes a deterioration in immune function. It attacks CD4 cells, which are important in immune function and helper T cells, a subtype of T lymphocyte cells. Patients with AIDS are chronically ill. Because their immune systems cannot fight off common viruses and bacteria, they develop a wide variety of diseases. AIDS also results in deterioration of nerve and muscle tissue. People infected with the virus may have a milder illness called AIDS-related complex (ARC). Symptoms of ARC can be mild and flulike to physically debilitating. Many people are HIV-positive but have no symptoms. Many of these people will eventually develop AIDS or ARC.

AIDS is largely a sexually transmitted disease. In the Western industrialized countries, it affects mainly homosexual or bisexual males. Other groups include intravenous drug users, patients given contaminated blood, and heterosexuals (the incidence of AIDS among heterosexuals in the United States is increasing). In some African countries, heterosexual sex is the principal route of transmission. AIDS can also be transmitted through contact with infected body fluids. In several isolated incidents,

health care workers have contracted the disease. In sports such as boxing, wrestling, and football, there is a slight possibility that the disease will be transmitted through infected blood. For this reason, basic public health precautions should be taken, including the following:

- In boxing, (1) excessively bloody fights should be stopped, (2) referees and people working in the corners should wear rubber gloves, and (3) the boxers should wear head gear.

- People with open or oozing wounds should be excluded from participation in contact sports.

- Sharing water bottles and towels should be discouraged.

- Blood should be cleaned up from athletic surfaces as soon as possible.

- When blood samples are taken for exercise testing, proper procedures for blood collection and the disposal of biological wastes should be observed. Precautions include wearing gloves, lab coat, and protective eyewear.

Exercise and the AIDS Patient

Muscle atrophy, loss of lean body mass, and general metabolic dysfunction are characteristics of AIDS. AIDS patients are affected with neurological deterioration that contributes to skeletal muscle wasting.

Exercise training appears to be an important adjunct treatment of the disease, particularly in its early stages. During the early stages of AIDS, when the person is HIV positive but without symptoms, exercise increases CD4 cells, enhances fitness and fat-free weight, and possibly delays symptoms of the disease. During the second phase of the disease (ARC), CD4 cells also increase but to a lesser extent than during the early disease phase. The effects of exercise when the patient has AIDS are not well understood. It is not known if training improves CD4 concentration or AIDS symptoms.

As discussed earlier, low-intensity exercise training tends to improve immune function while intense exercise tends to cause immunosuppression. Therefore, people with HIV infections should avoid intense exercise training.

■ Mental and Physical Disability

Exercise and Mental Health

The relationship between mental health and exercise has only recently been explored. Many of the studies on the effects of exercise on mental disorders, such as schizophrenia and depression, have been case studies or have used poor research designs. These studies have also been hampered by the difficulty in adequately defining specific mental disorders.

However, although research in this area has been equivocal, exercise appears to be a promising treatment technique for many mental disorders because of its positive effects on the symptoms of anxiety and depression. Cross-sectional epidemiological studies have shown an inverse relationship between physical activity levels and the incidence of depressive symptoms. Inactivity in nondepressed men and women is a good predictor of future depression. Exercise has been shown to have both acute (state) and long-term (trait) effects on indices of mental health, such as trait anxiety and self-esteem.

Brown (1990) has summarized the mechanisms that have been proposed for the possible beneficial effects of exercise on mental health:

- Exercise may act as a temporary diversion to daily stresses.
- Exercise may provide an opportunity for social interaction that might otherwise be lacking in a person's life.
- Exercise provides an opportunity for self-mastery. Increasing fitness or improving body composition may thus improve self-esteem.
- Increased core body temperature during exercise may lead to reduced muscle tension or to alterations in brain neurotransmitters.
- Mood improvement may occur due to increased secretion of endogenous opiates.
- Psychological changes may occur due to alterations in norepinephrine, dopamine, or serotonin.

Intense exercise training may impair mental health. Overtraining in athletes such as swimmers and runners has been shown to cause mood disturbances and depression. There are also reports in the literature of people with "negative" addictions to exercise. These people exercise even when injured and even when training interferes with family or job; they experience withdrawal symptoms if they cannot train. The psychological characteristics of people with a negative exercise addiction has been described as similar to those of people with anorexia nervosa (an eating disorder).

Exercise for People Confined to Wheelchairs

People confined to wheelchairs include those with spinal cord injuries, cerebral palsy, stroke, anterior poliomyelitis, muscular dystrophy, and multiple sclerosis. Sports for the disabled began shortly after World War II. Today, thousands of disabled athletes compete or participate in almost every sport. Much has been learned about the physiology of the disabled. Although many of the acute and chronic effects of training are the same in both the disabled and the able-bodied, there are important differences.

Benefits of Exercise Exercise has physical and emotional benefits for the disabled. Psychologically, exercise training has been shown to improve self-esteem and depressive mood states. Quality of life is greatly improved. In addition to improved mental outlook, ambulation is enhanced. Improved fitness also increases career choices for the disabled.

Fitness reduces the risk of both cardiovascular disease and respiratory infection. At the beginning of the century, the mortality for spinal cord injured patients was approximately 75% per year. Antibiotics and physical conditioning have greatly affected the long-term prognosis. However, degenerative diseases, such as coronary heart disease, are still of real concern to people with spinal cord injuries.

Fitness is greatly increased in wheelchair athletes. $\dot{V}_{O_2max}$ has been shown to be 40% greater in these athletes than in sedentary wheelchair-bound controls. There are also large differences in peak power output, stroke volume, and respiratory minute volume. Lean body mass and strength are much better in the trained athlete.

Classifications in Wheelchair Sports The goal in sports competitions is to give competitors an equal chance to exhibit their skills. Because there is a large variance between type and site of injury among the disabled, a classification system has been established for disabled sports (Table 26-3). This classification system was designed by the International Stoke Mandeville Games Federation. Classifications consider level and completeness of the spinal cord lesion and muscle function.

Exercise for People with Spinal Cord Injuries Physical fitness in the spinal cord injured varies

from extreme debilitation to reasonably good fitness. Upper body muscle mass in well-trained athletes, such as weight lifters, is much better than in sedentary able-bodied subjects. However, muscle strength and muscle mass among athletes in sports such as basketball and track are not much better than those in sedentary able-bodied subjects.

Aerobic capacity in trained athletes is similar to that in sedentary able-bodied people (Figure 26-4). Power outputs of about 120 watts and $\dot{V}_{O_2max}$ of 35–38 ml · kg^{-1} · min^{-1} have been reported for Class V athletes. Peak power output during a 30-sec maximal test is approximately 225 watts. The endurance of elite wheelchair athletes is impressive. The world record for the marathon in a wheelchair is more than 20 minutes faster than that in running. Large

---

TABLE 26-3

Anatomical/Functional Classification for Spinal Injuries Designed by the International Stoke Mandeville Games Federation

Class	Cord Level	Functional Characteristics	Class	Cord Level	Functional Characteristics
IA	C4-6	Triceps 0–3 on MRC scale. Severe weakness of trunk and lower extremities, interfering with sitting balance and ability to walk.			costal muscles. No useful sitting balance.
			III	T6-10	Good upper abdominal muscles. No useful lower abdominal or lower trunk extensor muscles. Poor sitting balance.
IB	C4-7	Triceps 4–5. Wrist flexion and extension may be present. Generalized weakness of trunk and lower extremities, interfering significantly with sitting balance and ability to walk.			
			IV	T11-L3	Good abdominal and spinal extensor muscles. Some hip flexors and adductors. Weak or nonexistent quadriceps strength, limited gluteal control (0–2). Point 1–20, traumatic, 1–15 polio.
IC	C4-8	Triceps 4–5. Wrist flexion and extension present. Finger flexion and extension 4–5 permits grasping and release. No useful hand intrinsic muscles. Generalized weakness of trunk and lower extremities interfering significantly with sitting balance and ability to walk.	V	L4-S2	Good or fair quadriceps control. Points 21–40, traumatic, 16–35 polio[a].
			VI	L4-S2	Points 41–60, traumatic, 36–50 polio[a] (Class VI is a subdivision of V, applied only in swimming competitions).
II	T1-5	No useful abdominal muscles (0–2). No functional lower inter-			

[a]Each of the following muscle groups is awarded up to 5 points (5 = normal strength) per side: hip flexors, adductors, abductors, and extensors; knee flexors and extensors; ankle plantor and dorsiflexors. Potential score, 40 points per side.

Source: Shephard, 1988. Used with permission.

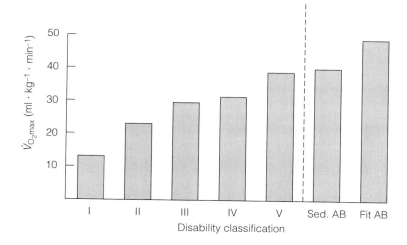

Figure 26-4 Aerobic capacity of college-aged wheelchair athletes.

increases in aerobic capacity, maximum power output, and muscle strength are possible with training.

Many wheelchair athletes have depressed cardiac contractile capacity during exercise. For a given power output, stroke volume is lower than in able-bodied subjects. Also, arteriovenous oxygen difference, a measure of oxygen extraction, tends to be higher. Paralysis has a profound effect on lower body circulation during exercise. Paralysis results in suppressed venomotor tone and muscle pump activity in the lower body. As a result, venous return of blood to the heart during exercise is inhibited.

Wheelchair athletes are subject to a greater variety of problems during exercise than are able-bodied subjects. Hypertension is more common, for example. The cause may be high levels of norepinephrine, increased β-adrenergic receptor activity, and reaction to spinal shock. Osteoporosis is also common. Fractures may occur due to muscle spasms or from the stress of muscle contractions or stretching exercises. Temperature regulation is impaired as well due to poor circulatory regulation and depressed activity of the sweat glands. Other common problems include soft tissue injuries, pressure sores, arthritis, and carpal tunnel syndrome. Carpal tunnel syndrome results from pressure on the median nerve and leads to numbness and pain in the hand and wrist.

Deconditioning and Bed Rest

Training results in improvement in exercise capacity. Inactivity leads to deterioration. Until quite recently, treatment for cardiac and postoperative patients involved prolonged bed rest. This forced inactivity, however, resulted in additional difficulties that were sometimes worse than the original problem. The negative effects of bed rest were aptly summed up by R. A. J. Asher (1947):

> Bed rest is anatomically and physiologically unsound. Look at a patient lying long in bed. What a pathetic picture he makes! The blood clotting in his veins, the lime draining from his bones, the scybala stacking up in his colon, the flesh rotting from his seat, the urine leaking from his distended bladder, and the spirit evaporating from his soul. Teach us to live that we may dread unnecessary time in bed. Get people up and we may save our patients from an early grave.

As shown in Tables 26-4 and 26-5, bed rest has far-reaching effects on most aspects of physiological function. Studies overwhelmingly show that bed rest leads to changes in the body's functional and psychological state. Some of these changes occur during the first few days. In many instances, disease can be made worse because of bed rest, and sometimes irreversible problems can develop. It is little wonder that current treatment for postoperative

TABLE 26-4

Effects of Prolonged Bed Rest on Physiological Function

Decreases	Increases	No Effect
Maximum stroke volume	Maximum heart rate	Mean corpuscular hemoglobin concentration
Orthostatic tolerance	Diastolic blood pressure	Forced vital capacity
Arterial vasomotor tone	Resting heart rate	Resting or exercise arteriovenous O_2 difference
Systolic time interval	Extra vascular and intravascular IgG[a]	
Coronary blood flow		Vital capacity
Maximal O_2 consumption	Submaximal exercise heart rate	Maximum voluntary ventilation
Pulmonary capillary blood volume	Submaximal exercise cardiac output	Total lung capacity
Plasma volume	Sleep disturbances	Proprioceptive reflexes
Skin blood flow	Diuresis	
Total diffusing capacity	Incidence of urinary infection	
Cerebrovascular tone	Incidence of deep vein thrombosis	
Sweating threshold temperature	Urinary excretion of calcium and phosphorus	
Vasomotor heat loss capacity	Nitrogen excretion	
Red blood cell production	Serum corticosteroids	
Red cell mass	Cultured staphylococci in nasal mucosa	
Hemoglobin		
Serum proteins	Extracellular fluid	
Serum albumin	Tendency to faint	
Intracellular fluid volume	Incidence of constipation	
Serum electrolytes	Cholesterol	
Coagulating capacity of blood	Low-density lipoproteins	
Bone calcium	Growth hormone	
Bone density	Electrocardiographic ST-segment depression	
Insulin sensitivity	Renal diurnal rhythms	
Acceleration tolerance		
Blood flow to extremities		
Catecholamines		
Serum androgens in males		
Muscular strength and mass		
Muscle tone		
Leukocyte phagocytic function		
Visual acuity		
Resistance to infection		
Systolic blood pressure		
Balance		

[a] Immunoglobulin G.

patients and for most diseases calls for early ambulation and physical activity.

There is a large body of research on deconditioning and bed rest, stemming largely from the space programs of the United States and the former Soviet Union. Well-controlled and well-funded studies have examined the effects of extreme deconditioning on most aspects of physiological function. Some of these studies were conducted for six months and involved almost complete restriction of movement.

The negative effects of bed rest are due primarily to the decrease in hydrostatic pressure within the cardiovascular system, the lower energy expenditure due to inactivity, decreased loading of the muscle and skeleton, changes in diet, and psychological stress.

	TABLE 26-5		
	Changes in Physiological Function During 15 Days of Bed Rest		
0–3 Days	**4–7 Days**	**8–14 Days**	**Over 15 days**
Increases in: Urine volume Urine Na$^+$, Cl$^-$, Ca^{2+}, and osmol excretion Plasma osmolality Hematocrit Venous compliance **Decreases in:** Total fluid intake Plasma volume Interstitial volume Intracellular volume Calf blood flow Resting heart rate Secretion of gastric juices Glucose tolerance Head-to-foot acceleration tolerance	**Increases in:** Urine creatinine, hydroxyproline, phosphate, nitrogen, and potassium excretion Plasma globulin, phosphate and glucose concentrations Blood fibrinogen Fibrinolytic activity and clotting time Focal point Hyperemia of eye conjuctiva and dilation of retinal arteries and veins Auditory threshold **Decreases in:** Near point of visual acuity Orthostatic tolerance Nitrogen balance	**Increases in:** Urine pyrophosphate Sweating sensitivity Exercise hyperthermia Exercise maximal heart rate **Decreases in:** Red blood cell mass Leucocyte phagocytosis Tissue heat conductance Lean body mass Body fat content	**Increases in:** Peak hypercalciuria Sensitivity to thermal stimuli Auditory threshold (secondary) **Decreases in:** Bone density

SOURCE: Greenleaf, 1988. Used with permission.

Bed Rest and the Cardiovascular System

The most profound changes from bed rest occur in the cardiovascular system. Impairments include diminished capacity of the heart, reduced plasma and blood volumes, and impaired control of blood vessels (Figure 26-5).

Maximal O$_2$ uptake ($\dot{V}_{O_2max}$) and exercise capacity decrease from as little as 1% to as much as 26%, depending on the type and duration of confinement. Exercise capacity is affected more in the upright than in the supine posture because of the added effects of orthostatic intolerance developed during bed rest. Interestingly, static exercise has been found to be effective in preventing some of the decrease in $\dot{V}_{O_2max}$ during bed rest, perhaps due to the positive effects of this exercise on muscle mass and strength, which influence $\dot{V}_{O_2max}$.

The changes in O$_2$ transport capacity are the result of reduced function in many parts of the system. Stroke volume and cardiac output decrease in upright and supine exercise due to impaired venous return of blood to the heart and decreased myocardial contractility. Tissue oxidative enzymes decrease, which affects submaximal exercise capacity. Diffusing capacity in the lungs decreases as well, because of reduced pulmonary blood volume.

Bed rest leads to reduced size and contractility of the heart. Some changes occur in the electrocardiogram: Heart rate increases at rest, conductivity through the atrioventricular node and bundle of His is slowed, ST-segment depression develops, and there is an increased incidence of sinarrhythmia. These changes seem to be caused by alterations in fluid and electrolyte metabolism that impair the K$^+$ and Na$^+$ cellular gradient and thus the electrocardiogram.

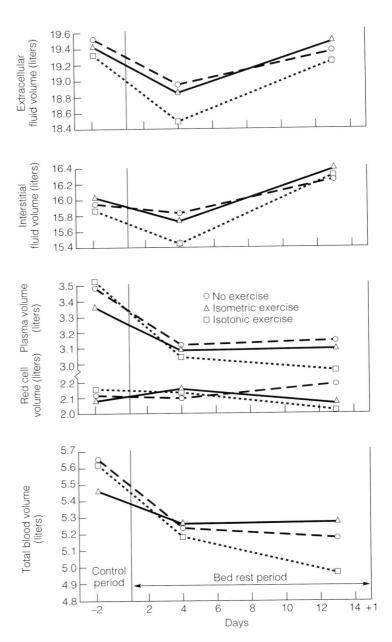

Figure 26-5 Fluid and blood changes during bed rest in subjects who practiced isometric exercise, isotonic exercise, no exercise. Modified from Greenleaf and Kozlowski, 1982. Used with permission.

Orthostatic Intolerance Bed rest results in *orthostatic intolerance* (OI), which is the inability of the circulation to adjust to the upright posture. The reduction in hydrostatic pressure is the primary stimulus for OI. When a bedridden patient assumes an upright posture, there is a sudden decrease in venous return of blood to the heart (Figure 26-6), caused by a reduction in vasomotor tone, blood volume, and muscle tone. The heart rate increases rapidly in an attempt to increase cardiac output. However, blood pressure falls and cerebral blood

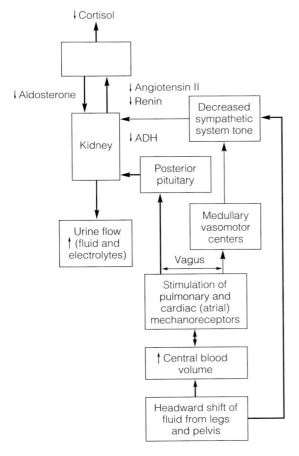

Figure 26-6 Circulatory events accompanying a change in posture (standing to lying) and possibly weightlessness. Modified from Sandler, 1980.

flow is impaired, resulting in dizziness, ataxia, and fainting.

Orthostatic tolerance is improved in direct proportion to the time spent in the upright posture. Exercise in the horizontal position during bed rest has no effect on OI although it does help prevent the deterioration in $\dot{V}_{O_2max}$ and exercise capacity. Sitting for eight hours a day helps maintain OI but has no effect on fitness.

Plasma Volume Plasma volume (PV) and blood volume decrease about 15% after a week of bed rest. Prolonged bed rest (100–200 days) results in 30% decreases in PV. The loss of red cell mass is due to a decrease in erythropoiesis. Plasma volume changes are due to fluid shifts. These changes occur early during bed rest and continue, to a certain extent, during the duration of the confinement.

During the first few days of bed rest, there are fluid shifts away from the legs and toward the head and torso. These fluid shifts cause discomfort. Symptoms include headache, dizziness, nasal congestion, and edema around the face and skull. Fluid shifts toward the heart increase central venous pressure, which tends to drive fluid out of the blood and into the interstitial tissue spaces. Thus, plasma volume decreases and upper body tissues swell. The increase in hydrostatic pressure in blood is accompanied by moderate hormone changes, including decreases in aldosterone, renin activity, and vasopressin. The hormone changes contribute to the decrease in plasma volume.

An increase in extravascular protein distribution (EPD) may help reduce plasma volume. EPD changes the extracellular osmolarity, which tends to draw fluid from the blood. The increased extravascular distribution of proteins is caused by an impaired capacity of lymph to transport proteins. Lymph transport depends on a massaging of the muscle, which is largely reduced during bed rest. Greenleaf (1988) has shown that heavy isotonic training during bed rest can maintain plasma volume at control levels for one month.

Circulatory Control Impairment of circulatory control is due primarily to reduced hydrostatic pressure. Bed rest appears to have a direct effect on the vasomotor center, as well as effects on local smooth muscle reflexes. Ordinarily, the smooth muscle contracts in response to the stretch from hydrostatic pressure. However, because this pressure is much less during bed rest, the smooth muscle of the precapillary sphincters loses some of its tone.

Effects on Substrate Metabolism Bed rest has far-reaching effects on metabolism. Insulin receptor sensitivity, for example, decreases during the first three days of bed rest. The extent of the effect of bed rest on carbohydrate metabolism depends on the length and degree of inactivity. The level of free fatty acids decreases and that of triglycerides increases. Cholesterol and low-density lipoproteins increase, which has possible implications in heart disease.

Effects on Muscle Mass There is a marked decrease in muscle tone due to the loss of intracellular fluid and contractile protein. This is particularly true in the lower body. Disuse atrophy is accompanied by a loss of muscular strength and endurance. Most studies show that bedridden patients go into negative nitrogen balance very early during convalescence. Losses in lean body mass and muscle strength can be prevented, however. The best methods are adequate caloric intake ($\approx$ 3000 kcal $\cdot$ day^{-1}) and intense isometric and isotonic exercise.

Bone and Calcium Metabolism Bone demineralization and loss of total body calcium occur at a rapid rate during bed rest. This is a result of reduced longitudinal stress on the bones rather than inactivity. The dissolution occurs at different rates in the various parts of the skeleton. The weight-bearing bones are particularly vulnerable. Calcium is lost at a rate of about 6 mg $\cdot$ day^{-1} during the first month of bed rest. In bed rest over 140 days, calcium loss is reduced to approximately 1 mg $\cdot$ day^{-1}. Standing for several hours a day significantly reduces calcium loss.

Psychological Effects Psychologists have described bed rest as a form of sensory deprivation. Bed rest produces a greater incidence of intellectual inefficiency, bizarre thoughts, and exaggerated emotional reactions. It also leads to time distortions, changes in body image, unusual body sensations, and an array of physical discomforts.

Reconditioning After Bed Rest The effects of prolonged bed rest are usually reversible with adequate ambulation. The physiological effects, if not too advanced, can be reversed with an appropriate training program. One problem is that the negative effects of bed rest lead to a vicious cycle, particularly in the elderly: Disuse leads to deterioration, which leads to the desire to continue to stay in bed and remain inactive.

The negative adaptability resulting from inactivity has implications for people interested in physical fitness. The atrophy in muscles resulting from a cast or prolonged bed rest are extreme examples of physical deterioration. The essence of the situation is similar to highly conditioned people who suddenly decondition. The body adapts to the lower stresses with reduced functional capacity.

■ Summary

Exercise is used as an adjunct treatment in many diseases. However, the stress of exercise must be applied cautiously and conservatively in disease states. The decision to exercise a patient must be based on the nature and severity of the disease. Sometimes, exercise is not appropriate.

Few studies have examined the relationship between physical activity and cancers. Although some studies have shown a relationship between exercise and a reduction of risk of all types of cancers, other studies have found that exercise may actually increase the risk of some cancers. Good evidence exists that exercise may reduce the risk of colon cancer and perhaps reproductive system cancers in women.

Exercise is a critical factor in the treatment of diabetes. Diabetic control involves balancing diet, insulin, and exercise. Therapy may include insulin, oral hypoglycemic agents, diet, and exercise.

Osteoporosis is a condition in which bone mass is so low that even minor trauma can cause fracture, most commonly in the hip, spine, and wrist. One of the best strategies for preventing post-menopausal osteoporosis is to maximize bone mass during childhood and premenopausal adulthood. This can be accomplished by practicing regular exercise and insuring adequate lifetime calcium intake. People with established osteoporosis are typically unfit and may have significant orthopedic limitations. Exercise programming should include endurance, resistance, and range-of-motion exercises.

Osteo- and rheumatoid arthritis can be helped by regular, measured physical activity. Exercise prescription for people with osteoarthritis includes range of motion and strength exercises and activities that minimize weight bearing, such as swimming. Exercise is contraindicated during an inflammatory period of rheumatoid arthritis because it increases the severity of inflammation and tissue damage. However, active people with the disease exhibit fewer pronounced degenerative changes, improved exercise tolerance, and fewer sick days from work.

Likewise, exercise is an important treatment in chronic obstructive pulmonary disease (COPD). Unfortunately, lung function does not improve with training in the COPD patient, but endurance training may delay the deterioration of pulmonary function. Exercise prescription for these patients should include progressive endurance exercise, such as walking and stationary cycling, and breathing exercises. Overexertion is a prominent precipitator of asthma (a category of COPD), so exercise intensity should increase gradually over time.

The relationship between exercise and viral illness is an emerging area in exercise physiology. However, more research is needed on the relationship between the development of viral diseases and exercise training, and the risks and benefits of exercise in people with these diseases need to be explored.

Inactivity and bed rest lead to physiological deterioration. Impairments include diminished capacity of the heart, reduced plasma and blood volumes, impaired blood vessel automaticity, decreased maximal O_2 consumption, muscle atrophy, orthostatic intolerance, and bone demineralization. Decreases in maximal O_2 consumption and work capacity range from 1 to 26%, depending on the type and duration of confinement.

The immune system can be weakened or strengthened by exercise. If the exercise program leads to overtraining, suppression of the immune system may result. In animal studies, moderate exercise increased antibody levels and increased longevity, but very few data are available on humans.

■ Selected Readings

Adami, S., D. Gatti, V. Braga, D. Bianchini, and M. Rossini. Site-specific effects of strength training on bone structure and geometry of ultradistal radius in postmenopausal women. *J. Bone Miner. Res.* 14: 120–124, 1999.

American College of Obstetricians and Gynecologists. ACOG educational bulletin: Osteoporosis. Number 246. April 1998. *Int. J. Gynaecol. Obstet.* 62: 193–201, 1998.

American College of Sports Medicine. Position stand on exercise and physical activity for older adults. *Med. Sci. Sports Exerc.* 30: 992–1008, 1998.

American College of Sports Medicine. ACSM's Exercise Management for Persons with Chronic Diseases and Disabilities. J. L. Durstine (Ed.). Champaign, Ill.: Human Kinetics, 1997.

American College of Sports Medicine and American Diabetes Association. Joint position statement: Diabetes mellitus and exercise. *Med. Sci. Sports Exerc.* 29: i–vi, 1997.

American College of Sports Medicine. Position stand on exercise and osteoporosis. *Med. Sci. Sports Exerc.* 27: i–vii, 1995.

Andrews, W. C. What's new in preventing and treating osteoporosis? *Postgrad. Med.* 104: 89–92; 95–97, 1998.

Arnaud, S., P. Berry, M. Cohen, J. Danellis, C. DeRoshia, J. Greenleaf, B. Harris, L. Kiel, E. Bernauer, M. Bond, S. Ellis, P. Lee, R. Selzer, and C. Wade. Exercise Countermeasures for Bed Rest Deconditioning. Washington, D.C.: NASA Space Life Sciences Symp., 1987, pp. 59–60.

Asher, R. A. J. The dangers of going to bed. *Brit. Med. J.* 4: 967–968, 1947.

Bernstein, L., B. E. Henderson, R. Hanisch, J. Sullivan-Halley, and R. K. Ross. Physical exercise and reduced risk of breast cancer in young women. *Natl. Cancer Inst.* 86: 1403–1408, 1994.

Briner, W. W., and A. L. Sheffer. Clinical symposium: exercise and allergy: physical allergy, exercise-induced anaphylaxis, and exercise-induced bronchospasm. *Med. Sci. Sports Exerc.* 24: 843–859, 1992.

Brown, D. R. Exercise, fitness, and mental health. In *Exercise, Fitness and Health: A Consensus of Current Knowledge*, C. Bouchard, R. J. Shephard, T. Stephens, J. R. Sutton, and B. D. McPherson (Eds.). Champaign, Ill.: Human Kinetics, 1990, pp. 607–626.

Camacho, T. C., R. E. Roberts, N. B. Lazarus, G. A. Kaplan, and R. D. Cohen. Physical activity and depression: evidence from the Alameda County study. *Am. J. Epidemiol.* 134: 220–231, 1991.

Cambach, W., R. C. Wagenaar, T. W. Koelman, A. R. van Keimpema, and H. C. Kemper. The long-term effects of pulmonary rehabilitation in patients with asthma and chronic obstructive pulmonary disease: a research synthesis. *Arch. Phys. Med. Rehabil.* 80: 103–111, 1999.

Cooper, R. A. An exploratory study of racing wheelchair propulsion dynamics. *Adapted Phys. Act. Quart.* 7: 74–85, 1990.

Cooper, R. A. High tech wheelchairs gain the competitive edge. *IEEE Eng. in Med. and Biol. Mag.* 10: 49–55, 1991.

Cooper, R. A., and J. F. Bedi. Function of class and disability groups in the results of elite wheelchair road racers. *Palaestra.* 8: 36–41, 1992.

Cooper, R. A., S. M. Horvath, J. F. Bedi, D. M. Drechsler-Parks, and R. E. Williams. Maximal exercise responses of paraplegic wheelchair road racers. *Paraplegia* 30: 573–581, 1992.

Davis, G. M. Exercise capacity of individuals with paraplegia. *Med. Sci. Sports Exerc.* 25: 423–432, 1993.

Dela, F., T. Ploug, A. Handberg, L. N. Petersen, J. J. Larsen, K. J. Mikines, and H. Galbo. Physical training increases muscle GLUT4 protein and mRNA in patients with NIDDM. *Diabetes* 43: 862–865, 1994.

Dorgan, J. F., C. Brown, M. Barrett, G. L. Splansky, B. E. Kreger, R. B. D'Agostino, D. Albanes, and A. Schatzkin. Physical activity and risk of breast cancer in the Framingham Heart Study. *Am. J. Epidemiol.* 139: 662–669, 1994.

Duncan, K., S. Harris, and C. M. Ardies. Running exercise may reduce risk for lung and liver cancer by inducing activity of antioxidant and phase II enzymes. *Cancer Lett.* 116: 151–158, 1997.

Dunstan, D. W., I. B. Puddey, L. J. Beilin, V. Burke, A. R. Morton, and K. G. Stanton. Effects of a short-term circuit weight training program on glycaemic control in NIDDM. *Diabetes Res. Clin. Pract.* 40: 53–61, 1998.

Edelman, S. V. Type II diabetes mellitus. *Adv. Intern. Med.* 43: 449–500, 1998.

Erikssen, G., K. Liestol, J. Bjornholt, E. Thaulow, L. Sandvik, and J. Erikssen. Changes in physical fitness and changes in mortality. *Lancet* 352: 759–762, 1998.

Fahey, T. D., P. Insel, and W. Roth. *Fit and Well.* 3d. Ed. Mountain View, Ca.: Mayfield Publishing Co., 1999.

Felson, D. T., Y. Zhang, M. T. Hannan, A. Naimark, B. Weissman, P. Aliabadi, and D. Levy. Risk factors for incident radiographic knee osteoarthritis in the elderly: the Framingham Study. *Arthritis Rheum.* 40: 728–733, 1997.

Feskens, E. J., J. G. Loeber, and D. Kromhout. Diet and physical activity as determinants of hyperinsulinemia: the Zutphen Elderly Study. *Am. J. Epidemiol.* 140: 350–360, 1994.

Franco, M. J., E. M. Olmstead, A. N. Tosteson, T. Lentine, J. Ward, and D. A. Mahler. Comparison of dyspnea ratings during submaximal constant work exercise with incremental testing. *Med. Sci. Sports Exerc.* 30: 479–482, 1998.

Fries, J. F., G. Singh, D. Morfeld, P. O'Driscoll, and H. Hubert. Relationship of running to musculoskeletal pain with age: a six-year longitudinal study. *Arthritis Rheum.* 39: 64–72, 1996.

Frisch, R. E., G. Wyshak, N. L. Albright, T. E. Albright, I. Schiff, K. P. Jones, J. Witschi, E. Shiang, E. Koff, and M. Marguglio. Lower prevalence of breast cancer and cancers of the reproductive system among former college athletes compared to non-athletes. *Br. J. Cancer* 52: 885–891, 1985.

Gallagher, C. G. Exercise and chronic obstructive pulmonary disease. *Med. Clin. N. Amer.* 74: 619–641, 1990.

Garvey, C. COPD and exercise. *Lippincotts Prim. Care Pract.* 2: 589–598, 1998.

Goodyear, L. J., and B. B. Kahn. Exercise, glucose transport, and insulin sensitivity. *Ann. Rev. Med.* 49: 235–261, 1998.

Granberry, M. C., and V. A. Fonseca. Insulin resistance syndrome: options for treatment. *South. Med. J.* 92: 2–15, 1999.

Greendale, G. A., E. Barrett-Connor, S. Edelstein, S. Ingles, and R. Haile. Lifetime leisure exercise and osteoporosis: the Rancho Bernardo study. *Am. J. Epidemiol.* 141: 951–959, 1995.

Greenleaf, J. E. Physiology of prolonged bed rest. *NASA Technical Memorandum 101010*, August 1988, pp. 1–8.

Greenleaf, J. E., and S. Kozlowski. Physiological consequences of reduced physical activity during bed rest. *Exer. Sci. Sport Sci. Rev.* 10: 84–119, 1982.

Hakim, A. A., H. Petrovitch, C. M. Burchfiel, G. W. Ross, B. L. Rodriguez, L. R White, K. Yano, J. D. Curb, and R. D. Abbott. Effects of walking on mortality among nonsmoking retired men. *N. Engl. J. Med.* 338: 94–99, 1998.

Hayashi, T., J. F. Wojtaszewski, and L. J. Goodyear. Exercise regulation of glucose transport in skeletal muscle. *Am. J. Physiol.* 273: E1039–E1051, 1997.

Heath, G. W., C. A. Macera, and D. C. Nieman. Exercise and upper respiratory tract infection: is there a relationship? *Sports Med.* 14: 353–365, 1992.

Heinonen, A., P. Kannus, H. Sievanen, M. Pasanen, P. Oja, and I. Vuori. Good maintenance of high-impact activity-induced bone gain by voluntary, unsupervised exercises: an 8-month follow-up of a randomized controlled trial. *J. Bone Miner. Res.* 14: 125–128, 1999.

Hoffman-Goetz, L. Exercise, natural immunity, and tumor metastasis. *Med. Sci. Sports Exerc.* 26: 157–163, 1994.

Hoffman-Goetz, L., D. Apter, W. Demark-Wahnefried, M. I. Goran, A. McTiernan, and M. E. Reichman. Possible mechanisms mediating an association between physical activity and breast cancer. *Cancer* 83(Suppl.): 621–628, 1998.

Hopman, M. T. E., B. Oeseburg, and R. A. Binkhorst. Cardiovascular responses in paraplegic subjects during arm exercise. *Eur. J. Appl. Physiol.* 65: 73–78, 1992.

Hopman, M. T. E., B. Oeseburg, and R. A. Binkhorst. Cardiovascular responses in paraplegics to prolonged arm exercise and thermal stress. *Med. Sci. Sports Exer.* 25: 577–583, 1993.

Ilarde, A., and M. Tuck. Treatment of non–insulin-dependent diabetes mellitus and its complications: a state of the art review. *Drugs Aging* 4: 470–491, 1994.

Ishi, T., T. Yamakita, T. Sato, S. Tanaka, and S. Fuji. Resistance training improves insulin sensitivity in NIDDM subjects without altering maximal oxygen uptake. *Diabetes Care* 21: 1353–1355, 1998.

Ivy, J. L. Role of exercise training in the prevention and treatment of insulin resistance and non-insulin-dependent diabetes mellitus. *Sports Med.* 24: 321–336, 1997.

Jobin, J., F. Maltais, J. F. Doyon, P. LeBlanc, P. M. Simard, A. A. Simard, and C. Simard. Chronic obstructive pulmonary disease: capillarity and fiber-type characteristics of skeletal muscle. *J. Cardiopulm. Rehabil.* 18: 432–437, 1998.

Jones, N. L., and K. J. Killian. Exercise in chronic airway obstruction. In Exercise, Fitness, and Health, C. Bouchard, R. J. Shephard, T. Stephens, J. R. Sutton, and B. D. McPherson (Eds.). Champaign, Ill.: Human Kinetics Books, 1990, pp. 547–559.

Kano, K. Relationship between exercise and bone mineral density among over 5,000 women aged 40 years and above. *J. Epidemiol.* 8: 28–32, 1998.

Keast, D., K. Cameron, and A. R. Morton. Exercise and the immune response. *Sports Med.* 5: 248–267, 1988.

King, A. C., C. B. Taylor, and W. L. Haskell. Effects of differing intensities and formats of 12 months of exercise training on psychological outcomes in older adults. *Health Psychol.* 12: 292–300, 1993.

Kirsten, D. K., C. Taube, B. Lehnigk, R. A. Jorres, and H. Magnussen. Exercise training improves recovery in patients with COPD after an acute exacerbation. *Respir. Med.* 92: 1191–1198, 1998.

Kohrt, W. M., A. A. Ehsani, and S. J. Birge, Jr. Effects of exercise involving predominantly either joint-reaction or ground-reaction forces on bone mineral density in older women. *J. Bone Miner. Res.* 12: 1253–1261, 1997.

Kramer, M. R., C. Springer, N. Berkman, M. Glazer, M. Bublil, E. Bar-Yishay, and S. Godfrey. Rehabilitation of hypoxemic patients with COPD at low altitude at the Dead Sea, the lowest place on earth. *Chest* 113: 571–575, 1998.

Kronhed, A. C., and M. Moller. Effects of physical exercise on bone mass, balance skill and aerobic capacity in women and men with low bone mineral density, after one year of training—a prospective study. *Scand J. Med. Sci. Sports* 8: 290–298, 1998.

Lane, N. E., and J. M. Thompson. Management of osteoarthritis in the primary-care setting: an evidence-based approach to treatment. *Am. J. Med.* 103: S25–S30, 1997.

Lane, N. E. Physical activity at leisure and risk of osteoarthritis. *Ann. Rheum. Dis.* 55: 682–684, 1996.

Layne, J. E., and M. E. Nelson. The effects of progressive resistance training on bone density: a review. *Med. Sci. Sports Exerc.* 31: 25–30, 1999.

Lee, I. M., J. E. Manson, U. Ajani, R. S. Paffenbarger, Jr., C. H. Hennekens, and J. E. Buring. Physical activity and risk of colon cancer: the Physicians' Health Study (United States). *Cancer Causes Control* 8: 568–574, 1997.

Lee, I. M., R. S. Paffenbarger, Jr., and C. C. Hsieh. Physical activity and risk of prostatic cancer among college alumni. *Am. J. Epidemiol.* 135: 169–179, 1992.

Lees, A., and S. Arthur. An investigation into anaerobic performance of wheelchair athletes. *Ergonomics* 31: 1529–1537, 1988.

Lehmann, R., V. Kaplan, R. Bingisser, K. E. Bloch, and G. A. Spinas. Impact of physical activity on cardiovascular risk factors in IDDM. *Diabetes Care* 20: 1603–1611, 1997.

Lemanske, R. F., Jr., and W. W. Busse. Asthma. *J. Am. Med. Assoc.* 278: 1855–1873, 1997.

Le Marchand, L., L. N. Kolonel, and C. N. Yoshizawa. Lifetime occupational physical activity and prostate cancer risk. *Amer. J. Epidemiol.* 133: 103–111, 1991.

Lipkin, E. New strategies for the treatment of type 2 diabetes. *J. Am. Diet. Assoc.* 99: 329–334, 1999.

Lyngberg, K. K., B. U. Ramsing, A. Nawrocki, M. Harreby, and B. Danneskiold-Samsoe. Safe and effective isokinetic knee extension training in rheumatoid arthritis. *Arthritis Rheum.* 37: 623–628, 1994.

Mackinnon, L. T. Current challenges and future expectations in exercise immunology: back to the future. *Med. Sci. Sports Exerc.* 26: 191–194, 1994.

MacNeil, B., and L. Hoffmann-Goetz. Effect of exercise on natural cytotoxicity and pulmonary tumor metastases in mice. *Med. Sci. Sports Exerc.* 25: 922–928, 1993.

Madsen, K. L., W. C. Adams, and M. D. Van Loan. Effects of physical activity, body weight and composition, and muscular strength on bone density in young women. *Med. Sci. Sports Exerc.* 30: 114–120, 1998.

McTiernan, A., C. Ulrich, S. Slate, and J. Potter. Physical activity and cancer etiology: associations and mechanisms. *Cancer Causes Control* 9: 487–509, 1998.

Moses, J., A. Steptoe, A. Mathews, and S. Edwards. The effects of exercise training on mental well-being in the normal population: a controlled trial. *J. Psychosom. Res.* 3: 47–61, 1989.

Neufer, P. D., and G. L. Dohm. Exercise induces a transient increase in transcription of the GLUT-4 gene in skeletal muscle. *Am. J. Physiol.* 265: C1597–C1603, 1993.

Nieman, D. C. Exercise immunology: practical applications. *Int. J. Sports Med.* 18 (Suppl. 1): S91–S100, 1997.

Nieman, D. C. Immune response to heavy exertion. *J. Appl. Physiol.* 82: 1385–1394, 1997.

Nieman, D. C. Exercise Testing and Prescription. Mountain View, CA: Mayfield Publishing Co., 1999.

Nieman, D. C., D. A. Henson, G. Gusewitch, B. J. Warren, R. C. Dotson, D. E. Butterworth, and S. L. Nehlsen-Cannarella. Physical activity and immune function in elderly women. *Med. Sci. Sports Exerc.* 25: 823–831, 1993.

Nieman, D. C., L. M. Johanssen, J. W. Lee, and K. Arabatzis. Infectious episodes in runners before and after the Los Angeles Marathon. *J. Sports Med. Phys. Fitness* 30: 316–328, 1990.

Nieman, D. C., and B. K. Pedersen. Exercise and immune function. Recent developments. *Sports Med.* 27: 73–80, 1999.

Nordemar, R., B. Ekblom, L. Zachrisson, and K. Lundqvist. Physical training in rheumatoid arthritis: a controlled long-term study (I and II). *Scand. J. Rheumatol.* 10: 17–30, 1981.

Oliveria, S. A., H. W. Kohl, III, D. Trichopoulos, and S. N. Blair. The association between cardiorespiratory fitness and prostate cancer. *Med. Sci. Sports Exerc.* 28: 97–104, 1996.

Oremek, G. M., and U. B. Seiffert. Physical activity releases prostate-specific antigen (PSA) from the prostate gland into blood and increases serum PSA concentrations. *Clin. Chem.* 42: 691–695, 1996.

Ossip-Klein, D. J., E. J. Doyne, E. D. Bowman, K. M. Osborn, I. B. McDougall-Wilson, and R. A. Neimeyer. Effects of running or weight lifting on self-concept in clinically depressed women. *J. Consult. Clin. Psychol.* 57: 158–161, 1989.

Otterness, I. G., J. D. Eskra, M. L. Bliven, A. K. Shay, J. P. Pelletier, and A. J. Milici. Exercise protects against articular cartilage degeneration in the hamster. *Arthritis Rheum.* 41: 2068–2076, 1998.

Paffenbarger, R. S., W. E. Hale, and A. L. Wing. Physical activity and incidence of cancer in diverse populations: a preliminary report. *Amer. J. Clin. Nutr.* 45: 312–317, 1987.

Panush, R. S., and H. A. Holtz. Is exercise good or bad for arthritis in the elderly? *South Med. J.* 87: S74–S78, 1994.

Persky, V., A. R. Dyer, J. Leonas, J. Stamler, D. M. Berkson, H. A. Lindberg, O. Paul, R. B. Shekelle, M. H. Lepper, and J. A. Schoenberger. Heart rate: a risk factor for cancer? *Am. J. Epidemiol.* 114: 477–487, 1981.

Platz, E. A., I. Kawachi, E. B. Rimm, G. A. Colditz, M. J. Stampfer, W. C. Willett, and E. Giovannucci. Physical activity and benign prostatic hyperplasia. *Arch. Intern. Med.* 158: 2349–2356, 1998.

Polednak, A. P. College athletics, body size, and cancer mortality. *Cancer* 38: 382–387, 1976.

Polednak, A. P. Epidemiology of breast cancer in Connecticut women. *Conn. Med.* 63: 7–16, 1999.

Rall, L. C., S. N. Meydani, J. J. Kehayias, B. Dawson-Hughes, and R. Roubenoff. The effect of progressive resistance training in rheumatoid arthritis: increased strength without changes in energy balance or body composition. *Arthritis Rheum.* 39: 415–426, 1996.

Richter, E. A., K. J. Mikines, H. Galbo, and B. Kiens. Effects of exercise on insulin action in human skeletal muscle. *J. Appl. Physiol.* 66: 876–885, 1989.

Roberts, J. A. Viral illnesses and sports performance. *Sports Med.* 3: 296–303, 1986.

Rockhill, B., W. C. Willett, D. J. Hunter, J. E. Manson, S. E. Hankinson, D. Spiegelman, and G. A. Colditz. Physical activity and breast cancer risk in a cohort of young women. *J. Natl. Cancer Inst.* 90: 1155–1160, 1998.

Rook, A. An investigation into the longevity of Cambridge sportsmen. *Brit. Med. J.* 1: 773–777, 1954.

Rosholt, M. N., P. A. King, and E. S. Horton. High-fat diet reduces glucose transporter responses to both insulin and exercise. *Am. J. Physiol.* 266: R95–R101, 1994.

Sandler, H. Effects of bed rest and weightlessness on the heart. *Hearts and Heart-like Organs* 2: 435–524, 1980.

Serres, I., M. Hayot, C. Prefaut, and J. Mercier. Skeletal muscle abnormalities in patients with COPD: contribution to exercise intolerance. *Med. Sci. Sports Exerc.* 30: 1019–1027, 1998.

Sharkey, A. M., A. B. Carey, C. T. Heise, and G. Barber. Cardiac rehabilitation after cancer therapy in children and young adults. *Am. J. Cardiol.* 71: 1488–1490, 1993.

Shephard, R. J. Sports medicine and the wheelchair athlete. *Sports Med.* 4: 226–247, 1988.

Shephard, R. J. Exercise, immune function and HIV infection. *J. Sports Med. Phys. Fitness* 38: 101–110, 1998.

Shepard, R. J., and P. N. Shek. Associations between physical activity and susceptibility to cancer: possible mechanisms. *Sports Med.* 26: 293–315, 1998.

Siegel, A. J. Medical conditions arising during sports. In Women and Exercise: Physiology and Sports Medicine, M. Shangold and G. Merkin (Eds.). Philadelphia: F. A. Davis, 1988, pp. 220–238.

Smith, W. G. Adult heart disease due to Coxsackie virus group B. *Brit. Heart J.* 28: 204–208, 1966.

Spence, D. W., M. L. A. Galantino, K. A. Mossberg, and S. O. Zimmerman. Progressive resistance exercise: effect on muscle function and anthropometry of a select AIDS population. *Arch. Phys. Med. Rehabil.* 71: 644–648, 1990.

St. Pierre, B. A., C. E. Kasper, and A. M. Lindsey. Fatigue mechanisms in patients with cancer: effects of tumor necrosis factor and exercise on skeletal muscle. *Oncol. Nurs. Forum* 19: 419–425, 1992.

Sternfeld, B. Cancer and the protective effect of physical activity: the epidemiological evidence. *Med. Sci. Sports Exerc.* 24: 1195–1209, 1992.

Taaffe, D. R., T. L. Robinson, C. M. Snow, and R. Marcus. High-impact exercise promotes bone gain in well-trained female athletes. *J. Bone Miner. Res.* 12: 255–260, 1997.

Tan, R. A., and S. L. Spector. Exercise-induced asthma. *Sports Med.* 25: 1–6, 1998.

Thomasi, T. B., F. B. Trudeau, D. Czerwinski, and S. Erredge. Immune parameters in athletes before and after strenuous exercise. *J. Clin. Immun.* 2: 173–178, 1982.

Van den Ende, C. H., J. M. Hazes, S. le Cessie, W. J. Mulder, D. G. Belfor, F. C. Breedveld, and B. A. Dijkmans. Comparison of high and low intensity training in well controlled rheumatoid arthritis: results of a randomised clinical trial. *Ann. Rheum. Dis.* 55: 798–805, 1996.

Van den Ende, C. H., T. P. Vliet Vlieland, M. Munneke, and J. M. Hazes. Dynamic exercise therapy in rheumatoid arthritis: a systematic review. *Br. J. Rheumatol.* 37: 677–687, 1998.

Van Loan, M. D., S. McCluer, J. M. Loftin, and R. A. Boileau. Comparison of physiological responses to maximal arm exercise among able-bodied, paraplegics and quadriplegics. *Paraplegia* 25: 397–405, 1987.

Vico, L., D. Chappard, C. Alexandre, S. Palle, P. Minaire, G. Riffat, B. Morukov, and S. Rakhmanov. Effects of a 120 day period of bed rest on bone mass and bone cell activities in man: attempts at countermeasures. *Bone & Mineral* 2: 383–394, 1987.

Wasserman, D. H., and N. N. Abumrad. Physiological basis for the treatment of the physically active individual with diabetes. *Sports Med.* 7: 376–392, 1989.

Wasserman, D. H., R. J. Greer, D. E. Rice, D. Bracy, P. J. Flakoll, L. L. Brown, J. O. Hill, and N. N. Abrumrad. Interaction of exercise and insulin action in humans. *Amer. J. Physiol.* 260 (*Endocrinol. Metab.* 23): E37–E45, 1991.

Watanabe, K. T., R. A. Cooper, A. J. Vosse, F. D. Baldini, and R. N. Robertson. Training practices of athletes who participated in the National Wheelchair Athletic Association training camps. *Adapted Phys. Act. Quart.* 9: 249–260, 1992.

Watchie, J., C. N. Coleman, T. A. Raffin, R. S. Cox, A. A. Raubitschek, T. Fahey, R. T. Hoppe, and A. VanKessel. Minimal long-term cardiopulmonary dysfunction following treatment for Hodgkin's disease. *Int. J. Radiation Oncology Biol. Phys.* 13: 517–524, 1987.

Watson, R. P., S. Moriguchi, J. C. Jackson, L. Werner, J. H. Wilmore, and B. J. Freund. Modification of cellular immune functions in humans by endurance exercise training during β-adrenergic blockade with atenolol on propranolol. *Med. Sci. Sports Exerc.* 18: 95–100, 1986.

Wei, M., L. W. Gibbons, T. L. Mitchell, J. B. Kampert, C. D. Lee, and S. N. Blair. The association between cardiorespiratory fitness and impaired fasting glucose and type 2 diabetes mellitus in men. *Ann. Intern. Med.* 130: 89–96, 1999.

Weiler, J. M., T. Layton, and M. Hunt. Asthma in United States Olympic athletes who participated in the 1996 Summer Games. *J. Allergy Clin. Immunol.* 102: 722–726, 1998.

West, R. V. The female athlete: the triad of disordered eating, amenorrhoea and osteoporosis. *Sports Med.* 26: 63–71, 1998.

Weyer, S. Physical inactivity and depression in the community: evidence from the Upper Bavarian Field Study. *Int. J. Sports Med.* 13: 492–496, 1992.

Wicks, J., K. Lymburner, S. Dinsdale, and N. Jones. The use of multistage exercise testing with wheelchair ergometry and arm cranking in subjects with spinal cord lesions. *Paraplegia* 15: 252–261, 1978.

Wicks, J., N. B. Oldridge, N. B. Cameron, and N. L. Jones. Arm-cranking and wheelchair ergometry in elite spinal cord injured athletes. *Med. Sci. Sports Exerc.* 15: 224–231, 1983.

Woods, J. A. Exercise and resistance to neoplasia. *Can. J. Physiol. Pharmacol.* 76: 581–588, 1998.

Woods, J. A., and J. M. Davis. Exercise, monocyte/macrophage function, and cancer. *Med. Sci. Sports Exerc.* 26: 147–156, 1994.

Woods, J. A., J. M. Davis, J. A. Smith, and D. C. Nieman. Exercise and cellular innate immune function. *Med. Sci. Sports Exerc.* 31: 57–66, 1999.

CHAPTER 27

EXERCISE TESTING AND PRESCRIPTION

The process of exercise testing and prescription is exercise physiology at its most applied level. Accurate exercise prescription requires an accurate profile of functional capacity. The results of exercise tests are often used as this basis for exercise prescription. Testing and prescription require knowledge of the physiological response to exercise.

The purpose of exercise training is to stress the body's physiological systems so they adapt and increase their capacity. As physically fit people continually adapt to a series of exercise stresses, they improve their ability to meet the demands of physical effort. Physical improvements—including increased cardiac output, strength, flexibility, power, muscle capillary density, and muscle oxidative capacity—depend on the nature of the exercise training program.

Exercise testing uses the stimulus-response method of inquiry. A standard exercise stimulus is applied and the response is measured against recognized standards that are themselves based on typical reactions to the exercise stress. Changes in heart rate, breathing, the electrocardiogram, and subjective sensations during the exercise test are much more meaningful if the test is standardized.

The exercise training program should stress the body at an appropriate level, or an injury may occur. Conversely, the body will fail to adapt if the exercise stresses are insufficient. The ideal exercise prescription stimulates the body to adapt at a rapid rate but with a low risk of injury.

Based on considerable research, several organizations—including the American College of Sports Medicine (ACSM), the Centers for Disease Control and Prevention, and the American Heart Association—and the Surgeon General have made a distinction between physical fitness and physical activity as it relates to health. Low levels of exercise practiced almost every day appear to reduce the risk of certain degenerative diseases and enhance metabolic health; yet they do not contribute to improved physical fitness.

The ACSM (1998) views the benefits of increasing amounts and intensity of exercise as points along a continuum. People get significant benefits from doing low levels of physical activity. They receive additional benefits when they exercise more intensely or more frequently.

Fitness improves most rapidly when intensity reaches a threshold percentage of maximum—not when training is done at specific exercise intensity. For example, running a mile in 10 minutes (a specific intensity) would be too intense for a person who recently had a heart attack, but would be extremely easy for an elite endurance runner. The ideal intensity for each person should be based on a

639

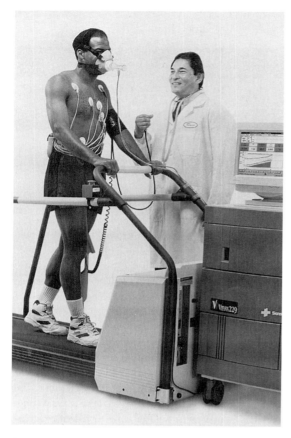

Treadmills are frequently used in the laboratory and clinical settings to apply exercise stress and record physiological responses or relatively stationary subjects and patients during exercise. Photo of Vmax 229 Cardiorespiratory Exercise System, courtesy of Sensor Medic Corporation.

percentage of capacity. The best intensity for improving fitness is between 40% and 85% of maximal oxygen consumption ($\dot{V}_{O_2max}$), or 55 to 90% of maximal heart rate. This intensity range results in the fastest rate of development in endurance capacity, while posing an acceptable risk of injury.

Effective exercise prescription (program design and recommendations) depends on knowledge of the maximum capacity. Many laboratory and field tests have been devised to measure physical fitness. They may be conducted at places such as a running track, a treadmill, bicycle ergometer, step-

ping bench, arm ergometer, or swimming flume. Fitness tests vary in their safety, accuracy, and specificity.

This chapter discusses the principles of measuring cardiovascular fitness and prescribing exercise programs. While this chapter stresses endurance-type exercise, strength and flexibility exercises are also important for health and well-being. Other topics that are discussed include contraindications to exercise, principles of ergometry, field tests of physical fitness, and exercise stress testing.

■ Medical Screening Prior to Beginning an Exercise Program

Most people can exercise safely if they are in good health and follow basic training principles. However, exercise may pose a risk to health and well-being if there are pre-existing medical conditions, such as heart disease. People sometimes die suddenly from heart attacks and cardiac arrest—some of them during exercise. These people are usually over 35–45 years of age and have coronary artery disease (CAD) or risk factors of CAD, such as high blood pressure or cigarette smoking, which predisposes them to the disease. Medical screening can help identify people who should not exercise or who should exercise only on a modified program. For most people, however, it is safer to exercise than to remain sedentary. To paraphrase the famous exercise scientist Per-Olaf Åstrand: If you don't want to exercise, you should see a physician to determine if you can withstand the physical deterioration that occurs with the sedentary lifestyle.

Medical Examination

The ACSM (1995) recommends that males 40 years and older and females 50 years and older, or any person with significant health problems, should get a medical examination and a maximal treadmill test before beginning a vigorous exercise program. People with significant health problems include those with two or more of the major risk factors of

TABLE 27-1a

Major Symptoms or Signs Suggestive of Cardiopulmonary or Metabolic Disease

Pain or discomfort in your chest (pain may radiate to surrounding areas, such as the neck or arms). Pain may appear during physical exertion or emotional stress.
Unaccustomed shortness of breath or shortness of breath with mild exertion
Dizziness or fainting
Distressed breathing while lying in bed at night
Swelling in the ankles
Skipped or racing heartbeats
Leg pains that increase in severity with increased exercise intensity
Known heart murmur

SOURCE: Adapted from American College of Sports Medicine, 1995, p. 17. Used with permission.

coronary heart disease (see Chapter 24). Table 27-1a lists symptoms suggestive of cardio-pulmonary or metabolic diseases that could make exercise dangerous. People with these symptoms should seek medical evaluation. The examination should include a medical history and appropriate procedures (blood chemistry, blood pressure, resting electrocardiogram, and so on) to identify such conditions. People of any age who experience any unusual symptoms, such as chest pain, viral infections, irregular heartbeats, shortness of breath, or severe pain in muscles, joints, or skeleton, should consult a physician before continuing with an exercise program.

The American Heart Association recommends that anyone 40 yrs. and older planning to exercise more vigorously than walking should get a physical examination and take a maximum exercise test. Table 27-1b and Table 27-1c list health status criteria for exercise testing and prescription.

The British Columbia Ministry of Health has developed a physical activity readiness questionnaire, or PAR-Q, to determine if a person can safely participate in an exercise program (see Table 27-2). People who answer yes to any question should see their physician before beginning such a program. PAR-Q is an excellent tool for helping determine if medical clearance is necessary.

Contraindications to Exercise Training and Testing

For most people, exercise training will be safe and enjoyable. However, for those with medical conditions, exercise may be dangerous. Anyone testing or prescribing exercise to these people may need to take special precautions. For this group, the exercise program should be modified, medically supervised, or not undertaken at all. Table 27-3 lists absolute and relative contraindications to exercise testing. An absolute contraindication means that exercise testing or exercise training should not be done. Relative contraindications mean that special precautions (e.g., medical supervision) should be taken before exercising and before exercise testing (Table 27-4).

Medical conditions that significantly limit the body's ability to adapt to physiological challenges make exercise dangerous. For example, in severe left ventricular dysfunction (i.e., congestive heart failure) or active unstable angina (heart-related chest pain that occurs at unpredictable times), the heart is under significant stress at rest. The additional stress induced by exercise can only lead to a deterioration in the medical condition. A serious noncardiac disorder, such as uncontrolled diabetes, can also be made worse by exercise. In these patients, exercise may lead to ketoacidosis or diabetic coma. Clearly, although exercise training is beneficial for most people, it is not for everyone.

■ Functional Capacity

Functional capacity is the extent to which a person can increase exercise intensities and maintain those increased levels. It is determined by the person's maximum ability to convert chemical energy into mechanical energy. Cardiovascular fitness is an important component of functional capacity.

Maximal oxygen consumption ($\dot{V}_{O_2max}$) is considered the best measure of cardiovascular capacity. It is often used synonymously with functional capacity. The limitations of using maximal oxygen consumption as a measure of fitness were discussed

TABLE 27-1b

Classification of Individuals by Health Status Prior to Exercise Testing or Exercise Prescription

Category	Description
Apparently healthy	Those who are asymptomatic* and apparently healthy, with no more than one major coronary risk factor**
Increased risk	Those who have signs or symptoms suggestive of possible cardiopulmonary or metabolic disease (see next category) and/or two or more major coronary risk factors
Known disease	Those with known cardiac, pulmonary, or metabolic disease (diabetes, thyroid disorders, renal disease, liver disease)

*Major symptoms or signs suggestive of cardiopulmonary or metabolic disease include (1) pain or discomfort in the chest or surrounding areas that appear to be ischemic (deficiency of blood supply due to obstruction of the circulation) in nature; (2) unaccustomed shortness of breath at rest or with mild exertion; (3) dizziness or fainting; (4) difficulty in breathing when standing or sudden breathing problems during the night; (5) ankle edema; (6) rapid throbbing or fluttering of the heart; (7) severe pain in leg muscles during walking; (8) known heart murmur; (9) unusual fatigue or shortness of breath with usual activities.
**If an individual has two or more of the following coronary risk factors, the classification is "increased risk":

Positive Risk Factors	Defining Criteria
1. Age	Men >45 years; women >55 or premature menopause without estrogen replacement therapy
2. Family history	MI or sudden death before 55 years of age in father or other male first-degree relative, or before 65 years of age in mother or other female first-degree relative
3. Current cigarette smoking	
4. Hypertension	Blood pressure ≥140/90 mm Hg, confirmed by measurements on at least two separate occasions; or on antihypertensive medication
5. Hypercholesterolemia	Total serum cholesterol >200 mg/dl (if lipoprotein profile is unavailable) or HDL <35mg/dl. (Most other organizations use >240 mg/dl.)
6. Diabetes mellitus	Persons with insulin-dependent diabetes mellitus (type 1) who are >30 years of age, or have had type 1 for >15 years, and persons with non-insulin-dependent diabetes mellitus (type 2) who are >35 years of age should be classified as patients with disease
7. Sedentary lifestyle	Persons composing the least active 25% of the population, as defined by the combination of sedentary jobs involving sitting for a large part of the day and no regular exercise or active recreational pursuits

Negative Risk Factor	Comments
1. High serum HDL cholesterol	>60 mg/dl

Notes: (1) It is common to sum risk factors in making clinical judgments. If HDL is high, subtract one risk factor from the sum of positive risk factors because high HDL decreases CAD risk; (2) obesity is not listed as an independent positive risk factor because its effects are exerted through other risk factors (e.g., hypertension, hyperlipidemia, diabetes). Obesity should be considered as an independent target for intervention, however.

SOURCE: American College of Sports Medicine, *ACSM's Guidelines for Exercise Testing and Prescription* (5th ed.). Baltimore: Williams & Wilkins, 1995. Used with permission.

TABLE 27-1c

American Heart Association Risk-Stratification Criteria

After the medical clearance stratification is complete, subjects can be classified by risk, based on their characteristics. The following classifications are recommended, and subsequent electrocardiogram (ECG) monitoring is advised, based on this classification.

Class A: Apparently Healthy

There is no evidence of increased cardiovascular risk for exercise. This classification includes (1) individuals under age 40 who have no symptoms or known presence of heart disease or major coronary risk factors, and (2) individuals of any age without known heart disease or major risk factors and who have a normal exercise test.

Activity guidelines: No restrictions other than basic guidelines

ECG and blood pressure monitoring: Not required

Supervision required: None

Class B: Presence of Known, Stable Cardiovascular Disease with Low Risk for Vigorous Exercise but Slightly Greater Than for Apparently Healthy Individuals

Moderate activity is not believed to be associated with increased risk in this group. This classification includes individuals with (1) coronary artery disease (CAD) (myocardial infarction, coronary artery bypass surgery, per-

cutaneous transluminal coronary angioplasty, angina pectoris, abnormal exercise test, and abnormal coronary angiograms) whose condition is stable and who have the following clinical characteristics; (2) valvular heart disease; (3) congenital heart disease; (4) cardiomyopathy; and (5) exercise test abnormalities that do not meet the criteria outlined in class C.

Clinical characteristics—(1) New York Heart Association (NYHA) class 1 or 2, (2) exercise capacity over 6 METs, (3) no evidence of heart failure, (4) free of ischemia or angina at rest or on the exercise test at or below 6 METs, (5) appropriate rise in systolic blood pressure during exercise, (6) no sequential ectopic ventricular contractions, and (7) ability to satisfactorily self-monitor intensity of activity

Activity guidelines—individualized activity with exercise prescription by qualified personnel trained in basic CPR or with electronic monitoring at home

ECG and blood pressure monitoring—only during the early prescription phase of training, usually 6–12 sessions

Supervision required—medical supervision during prescription sessions and nonmedical supervision for other exercise sessions until the individual understands how to monitor his or her activity

(continued)

Class C: Those at Moderate-to-High Risk for Cardiac Complications during Exercise and/or Unable to Self-Regulate Activity or to Understand Recommended Activity Level

This classification includes individuals with (1) CAD with the following clinical characteristics, (2) cardiomyopathy, (3) valvular heart disease, (4) exercise test abnormalities not directly related to ischemia, (5) previous episode of ventricular fibrillation or cardiac arrest that did not occur in the presence of an acute ischemic event or cardiac procedure, (6) complex ventricular arrhythmias that are uncontrolled at mild-to-moderate work intensities with medication, (7) three-vessel disease or left main disease, and (8) low-ejection fractions (less than 30%).

Clinical characteristics—(1) Two or more myocardial infarctions, (2) NYHA class 3 or greater, (3) exercise capacity less than 6 METs, (4) ischemic horizontal or downsloping ST depression of 4 mm or more or angina during exercise, (5) fall in systolic blood pressure with exercise, (6) a medical problem that the physician believes may be life-threatening, (7) previous episode of primary cardiac arrest, and (8) ventricular tachycardia at a workload of less than 6 METs

Activity guidelines—individualized activity with exercise prescription by qualified personnel

ECG and blood pressure monitoring—continuous during exercise sessions until safety is established, usually in 6–12 sessions or more

Supervision—medical supervision during all exercise sessions until safety is established

Class D: Unstable Disease with Activity Restriction

This classification includes individuals with (1) unstable ischemia, (2) heart failure that is not compensated, (3) uncontrolled arrhythmias, (4) severe and symptomatic aortic stenosis, and (5) other conditions that could be aggravated by exercise.

Activity guidelines—no activity recommended for conditioning purposes. Attention should be directed to treating the subject and restoring him or her to class C or higher. Daily activities must be prescribed based on individual assessment by the subject's personal physician.

The foregoing classifications are presented as a means of beginning exercise with the lowest possible risk. They do not consider accompanying morbidities (e.g., insulin-dependent diabetes mellitus, morbid obesity, severe pulmonary disease, or debilitating neurological or orthopedic conditions) that may necessitate closer supervision during training sessions. As the individual gains experience, the decision may be made to place the subject in another category. In most cases, as the safety of exercise and improvement in working capacity are demonstrated, graduation to classes nearer A and B is appropriate.

TABLE 27-2

Safety of Exercise Participation: PAR-Q

PAR-Q & YOU
(A Questionnaire for People Aged 15 to 69)

Regular physical activity is fun and healthy, and increasingly more people are starting to become more active every day. Being more active is very safe for most people. However, some people should check with their doctor before they start becoming much more physically active.

If you are planning to become much more physically active than you are now, start by answering the seven questions in the box below. If you are between the ages of 15 and 69, the PAR-Q will tell you if you should check with your doctor before you start. If you are over 69 years of age, and you are not used to being very active, check with your doctor.

Common sense is your best guide when you answer these questions. Please read the questions carefully and answer each one honestly: check YES or NO.

1. Has your doctor ever said that you have a heart condition *and* that you should only do physical activity recommended by a doctor?
2. Do you feel pain in your chest when you do physical activity?
3. In the past month, have you had chest pain when you were not doing physical activity?
4. Do you lose your balance because of dizziness or do you ever lose consciousness?
5. Do you have a bone or joint problem that could be made worse by a change in your physical activity?
6. Is your doctor currently prescribing drugs (for example, water pills) for your blood pressure or heart condition?
7. Do you know of *any other reason* why you should not do physical activity?

If you answered YES to one or more questions
Talk with your doctor by phone or in person BEFORE you start becoming much more physically active or BEFORE you have a fitness appraisal. Tell your doctor about the PAR-Q and which questions you answered YES.
- You may be able to do any activity you want—as long as you start slowly and build up gradually. Or, you may need to restrict your activities to those which are safe for you. Talk with your doctor about the kinds of activities you wish to participate in and follow his/her advice.
- Find out which community programs are safe and helpful for you.

If you answered NO to all questions
If you answered NO honestly to *all* PAR-Q questions, you can be reasonably sure that you can:
- start becoming much more physically active—begin slowly and build up gradually. This is the safest and easiest way to go.
- take part in a fitness appraisal—this is an excellent way to determine your basic fitness so that you can plan the best way for you to live actively.

DELAY BECOMING MUCH MORE ACTIVE:
- if you are not feeling well because of a temporary illness such as a cold or a fever—wait until you feel better; or
- if you are or may be pregnant—talk to your doctor before you start becoming more active

Please note:
If your health changes so that you then answer YES to any of the above questions, tell your fitness or health professional. Ask whether you should change your physical activity plan.

SOURCE: 1994 rev. version of the Physical Activity Readiness Questionnaire. Copyright © 1994 Canadian Society for Exercise Physiology. Used with permission.

TABLE 27-3

Absolute and Relative Contraindications to Exercise Testing

Absolute Contraindications
   Acute myocardial infarction (within 3 to 5 days)
   Unstable angina
   Uncontrolled cardiac arrhythmias causing symptoms
      or hemodynamic compromise
   Active endocarditis
   Symptomatic severe aortic stenosis
   Uncontrolled symptomatic heart failure
   Acute pulmonary embolus or pulmonary infarction
   Acute noncardiac disorder that may affect exercise per-
      formance or be aggravated by exercise (e.g., infec-
      tion, renal failure, thyrotoxicosis)
   Acute myocarditis or pericarditis
   Physical disability that would preclude safe and ade-
      quate test performance
   Thrombosis of lower extremity

Relative Contraindications
   Left main coronary stenosis or its equivalent
   Moderate stenotic valvular heart disease
   Electrolyte abnormalities
   Significant arterial or pulmonary hypertension
   Tachyarrhythmias or bradyarrhythmias
   Hypertrophic cardiomyopathy
   Mental impairment leading to inability to cooperate
   High-degree atrioventricular block

*Relative contraindications can be superseded if benefits outweigh risks of exercise.

SOURCE: American Heart Association, 1995.

in Chapter 16. To better understand the role of functional capacity in exercise testing and prescription, readers should review the basic ideas presented in Chapters 14–16.

Functional capacity is typically expressed as MET_{max} (Eq. 27-1), $\dot{V}_{O_2max}$ (ml · kg^{-1} · min^{-1}), or $\dot{V}_{O_2max}$ (liters · min^{-1}). One MET is equivalent to:

$$MET = \dot{V}_{O_2max} \text{ (ml · kg}^{-1} \text{ · min}^{-1})/3.5 \quad (27\text{-}1)$$

or

$$MET \approx \dot{V}_{O_2max} \text{ (ml · min}^{-1})/300$$

or

$$MET = 1 \text{ kcal · kg}^{-1} \text{ · hr}^{-1}$$

$\dot{V}_{O_2max}$ (ml · kg^{-1} · min^{-1}) is *relative maximal oxygen consumption*. It is derived by dividing oxygen consumption (ml · min^{-1}) by body mass. $\dot{V}_{O_2max}$ (liters · min^{-1}) is the *absolute oxygen consumption*. All of these measures are largely determined by the ability to maintain blood pressure, power output capacity of the muscles, and the capacity of the cardiovascular system.

METs

A MET (metabolic equivalent) is the resting metabolic rate, the metabolic rate while sitting quietly in a chair. One MET is approximately equal to an oxygen consumption of 3.5 ml · kg^{-1} · min^{-1}. When using METs, exercise intensity is expressed in multiples of the resting metabolic rate (see Table 27-5). For example, an exercise intensity of 2 METs means the exercise intensity is two times the resting metabolic rate (i.e., two times the resting oxygen consumption).

METs are used to describe exercise intensity for occupational activities and exercise prescription. Exercise intensities of less than 3–4 METs are considered low. Household chores and most industrial jobs fall into this category. Although exercise at these intensities does not improve fitness for most people, it does improve fitness for people with low physical capacities (i.e., $MET_{max} < 6$). Activities that increase metabolism to 6–8 METs are classified as moderate-intensity exercise. This is a suitable training intensity for most people beginning an exercise program. Vigorous exercise increases metabolic rate by more than 10 METs. Fast running or cycling, as well as intense play in sports like racquetball, can place people in this category.

METs are only intended to be an approximation of exercise intensity. Skill, body weight, body fat, and environment affect the accuracy of METs. As a practical matter, however, these limitations are disregarded. METs are a good way to express exercise

<table>
<tr><td colspan="6" align="center">**TABLE 27-4**</td></tr>
</table>

ACSM Recommendations for (A) Medical Examination and Exercise Testing Prior to Participation and (B) Physician Supervision of Exercise Tests

A. Medical examination and clinical exercise test recommended prior to:

	Apparently Healthy		Increased Risk*		
	Younger‡	Older	No Symptoms	Symptoms	Known Disease†
Moderate exercise§	No‖	No	No	Yes	Yes
Vigorous exercise¶	No	Yes#	Yes	Yes	Yes

B. Physician supervision recommended during exercise test:

	Apparently Healthy		Increased Risk*		
	Younger‡	Older	No Symptoms	Symptoms	Known Disease†
Submaximal testing	No‖	No	No	Yes	Yes
Maximal testing	No	Yes#	Yes	Yes	Yes

*Persons with two or more risk factors or one or more signs or symptoms.
†Persons with known cardiac, pulmonary, or metabolic disease.
‡Younger implies ≤ 40 years for men, ≤ 50 years for women.
§Moderate exercise as defined by an intensity of 40% to 60% $\dot{V}_{O_2max}$; if intensity is uncertain, moderate exercise may alternately be defined as an intensity well within the individual's current capacity, one which can be comfortably sustained for a prolonged period of time—that is, 60 minutes, with a gradual initiation and progression—and is generally noncompetitive.
‖A "No" response means that an item is deemed "not necessary." The "No" response does **not** mean that the item should not be done.
¶Vigorous exercise is defined by an exercise intensity > 60% $\dot{V}_{O_2max}$; if intensity is uncertain, moderate exercise may alternately be defined as exercise intense enough to represent a substantial cardiorespiratory challenge or if it results in fatigue within 20 minutes.
#A "Yes" response means that an item is recommended. For physician supervision, this suggests that a physician is in close proximity and readily available should there be an emergent need.

SOURCE: American College of Sports Medicine, 1995. Used with permission.

intensity because the system is easy for people to remember and understand.

$\dot{V}_{O_2max}$ Expressed per Kilogram Body Mass (ml · kg⁻¹ · min⁻¹)

Most sports scientists use oxygen consumption expressed in ml · kg⁻¹ · min⁻¹ to describe functional capacity. METs are calculated from $\dot{V}_{O_2max}$ (ml · kg⁻¹ · min⁻¹), so the terms are synonymous (see Table 27-5). The use of METs and $\dot{V}_{O_2max}$ per unit body mass is useful because it facilitates the comparison of people with different body masses. Larger people tend to consume more oxygen because they have more tissue. Also, it is possible for a large, sedentary person to have a considerably larger absolute $\dot{V}_{O_2max}$ (liters · min⁻¹) than a more physically fit, but smaller person. The use of $\dot{V}_{O_2max}$ expressed per unit body mass factors out the effects of body mass and it makes it possible to compare people based on fitness capacity.

Using maximal oxygen consumption divided by body mass presents several problems, however. It assumes that body mass accounts for all of the dif-

TABLE 27-5

Energy Cost of Various Activities [a]

Activity	Calories [b] (cal · min⁻¹)	METs [c]	Oxygen Cost (ml · kg⁻¹ · min⁻¹)
Archery	3.7–5	3–4	10.5–14
Backpacking	6–13.5	5–11	17.5–38.5
Badminton	5–11	4–9	14–31.5
Basketball			
Nongame	3.7–11	3–9	10.5–31.5
Game	8.5–15	7–12	24.5–42
Bed exercise (arm movement, supine or sitting)	1.1–2.5	1–2	3.5–7
Bicycling (pleasure or to work)	3.7–10	3–8	10.5–28
Bowling	2.5–5	2–4	7–14
Canoeing (rowing and kayaking)	3.7–10	3–8	10.5–28
Calisthenics	3.7–10	3–8	10.5–28
Dancing (social and square)	3.7–8.5	3–7	10.5–24.5
Fencing	7.5–12	6–10	21–35
Fishing			
(bank, boat, or ice)	2.5–5	2–4	7–14
(stream, wading)	6–7.5	5–6	17.5–21
Football (touch)	7.5–12	6–10	21–35
Golf			
(using power cart)	2.5–3.7	2–3	7–10.5
(walking, carrying bag, or pulling cart)	5–8.5	4–7	14–24.5
Handball	10–15	8–12	28–42
Hiking (cross-country)	3.7–8.5	3–7	10.5–24.5
Horseback riding	3.7–10	3–8	10.5–28
Horseshoe pitching	2.5–3.7	2–3	7–10.5
Hunting, walking			
Small game	3.7–8.5	3–7	10.5–24.5
Big game	3.7–17	3–14	10.5–49
Jogging	10–15	8–12	28–42
Mountain climbing	6–12	5–10	17.5–35
Paddleball (racquet)	10–15	8–12	28–42
Sailing	2.5–6	2–5	7–17.5
Scuba diving	6–12	5–10	17.5–35
Shuffleboard	2.5–3.7	2–3	7–10.5
Skating (ice or roller)	6–10	5–8	17.5–28
Skiing (snow)			
Downhill	6–10	5–8	17.5–28
Cross-country	7.5–15	6–12	21–42
Skiing (water)	6–8.5	5–7	17.5–24.5
Snow shoeing	8.5–17	7–14	24.5–49
Squash	10–15	8–12	28–42
Soccer	6–15	5–12	17.5–42
Softball	3.7–7.5	3–6	10.5–21
Stair climbing	5–10	4–8	14–28
Swimming	5–10	4–8	14–28
Table tennis	3.7–6	3–5	10.5–17.5
Tennis	5–11	4–9	14–31.5
Volleyball	3.7–7.5	3–6	10.5–21
Walking	2.5–8.5	2–7	7–24.5

[a] Energy cost values based on an individual of 154 lb of body weight (70 kg).
[b] Calorie = a unit of measure based on heat production. One calorie equals approximately 200 ml of oxygen consumed.
[c] MET = basal oxygen requirement of the body sitting quietly. One MET equals 3.5 ml of oxygen per kilogram of body weight per minute.

SOURCE: Pollock, Wilmore, and Fox, 1978, p. 124. Used with permission.

ferences in $\dot{V}_{O_2max}$ observed among people. However, only about 50% of $\dot{V}_{O_2max}$ is determined by body mass. Other components, such as training and genetics, account for the remainder. This problem is of more than theoretical interest. $\dot{V}_{O_2max}$ expressed per unit body mass can occasionally create confusion when attempting to assess physical fitness. For example, rapid weight loss will result in an apparent increase in functional capacity expressed as MET_{max}, or $\dot{V}_{O_2max}$ (ml · kg^{-1} · min^{-1}), but there will be no actual improvement in exercise capacity. In this circumstance, functional capacity would be better expressed in liters · min^{-1}. In most instances, though, expressing functional capacity per kilogram of body mass is useful and practical.

■ Measuring Maximal Oxygen Consumption

Oxygen consumption is measured in most laboratories by indirect, open-circuit calorimetry. Many manufacturers have developed systems to measure $\dot{V}_{O_2}$. Although these systems are sophisticated and expensive, they all determine $\dot{V}_{O_2}$ by taking some basic measurements, including ventilation and the fractional concentration of oxygen and carbon dioxide. Required laboratory instruments include a flow meter or spirometer to measure expiratory volume, gas analyzers to measure oxygen and carbon dioxide, a thermistor to measure the temperature of the expired gases, and a barometer to measure barometric pressure.

Ventilation

Ventilation is the volume of air breathed. It is measured breath by breath or per minute.* The most common ventilatory measurement instruments are the pneumotachograph, flow meter, gasometer, and Tissot spirometer (Figure 27-1). Ventilation is measured directly by one of these instruments (expired

---

*Expiratory volume can also be determined by measuring inspired volume.

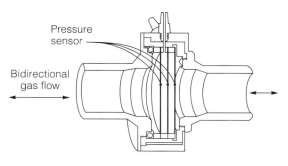

(a)

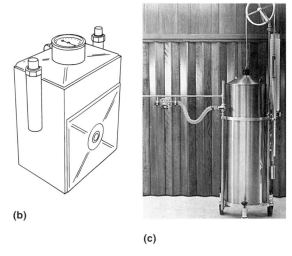

(b)

(c)

Figure 27-1   Three devices for measuring pulmonary minute volume: (a) pneumotachometer, (b) Parkinson-Cowan gasometer, and (c) chain-compensated (Tissot) spirometer. Source: Tissot spirometer courtesy of W. E. Collins, Inc., Braintree, Mass.

gases flow from the subject's mouth directly into the ventilation meter) or indirectly by collecting the gases in large bags or balloons and then measuring the ventilatory volumes at the end of the test.

Gas Temperature and Barometric Pressure

Temperature and barometric pressure are measured so that gas volumes can be expressed at standard conditions (i.e., BTPS and STPD). Lung measurements, such as minute volume, tidal volume, and maximal voluntary ventilation, are typically corrected to BTPS: body temperature, ambient pressure, saturated with water vapor. When minute ventilation volume (the quantity of air expired per minute) is used to calculate oxygen consumption, it is corrected to STPD: standard temperature ($0°C$), standard pressure (760 mmHg), dry (no water vapor). The temperature of expired gases must be measured continuously because it changes rapidly during progressive intensity exercise tests. Barometric pressure needs to be measured only at the beginning of a test because it remains relatively stable.

Composition of Expired Air

Sample Collection The composition of expired gases is measured in a number of ways: continuously, using a mixing chamber, using aliquots (small rubber bags containing a portion of the expired air), using a large collection bag, or using computers. The technique selected depends on available equipment, budget, and expertise.

Continuous measurements are made when breath-by-breath oxygen consumption measurements are required. High-speed gas analyzers, computers, and sophisticated computer programs are necessary for continuous gas analysis because the composition of each expired breath changes so quickly.

Mixing chambers are used in most laboratories for gas sampling during oxygen consumption measurements (Figure 27-2). The mixing chamber is a plastic box with baffles that help to mix the expired breaths. The mixing chamber allows for continuous gas sampling without the technological problems presented by continuous breath-by-breath sampling. This method works best when measurements are made during "steady state" (when the metabolism has reached a relative equilibrium). Although some error is introduced when the composition of expired gases is changing rapidly, which occurs at the end of a maximal oxygen consumption test, for most applications (i.e., those that do not require extreme accuracy) this is of minor significance.

Figure 27-2 Example of a mixing chamber. This device mixes expired air during the open-circuit measurement of oxygen consumption. SOURCE: Courtesy of Vacumed, Inc., Ventura, Calif.

Aliquot sampling techniques have many of the same benefits and problems of mixing chambers. This technique involves continuous sampling of a portion of expired air that has been collected in a small rubber bag (aliquot). Three bags are usually attached to a three-way valve: gas is being collected in one bag, being analyzed in another bag, and being emptied from a third bag by a vacuum pump (Figure 27-3).

Large-bag collection techniques involve collecting and analyzing all of the expired gases during a particular time interval. These techniques typically use either Douglas bags, weather balloons, or a large Tissot spirometer. Although previously widely used, the techniques have been largely replaced by more automated techniques. Large-bag techniques are awkward and not very adaptable to rapid and repeated measurements. They were necessary when gas analysis was done by manual chemical analyzers, such as the Scholander and Haldane apparatuses. However, they are accurate and do serve a purpose when other systems are unavailable or extreme accuracy is needed (e.g., with 100% O_2 breathing experiments).

Figure 27-3 Wilmore–Costill valve. This valve is used as part of a semiautomated procedure to measure oxygen consumption. SOURCE: Courtesy of Jack Wilmore.

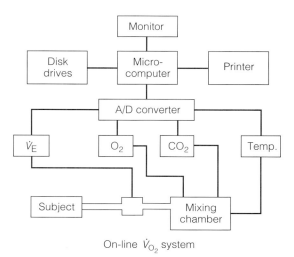

On-line $\dot{V}_{O_2}$ system

(a)

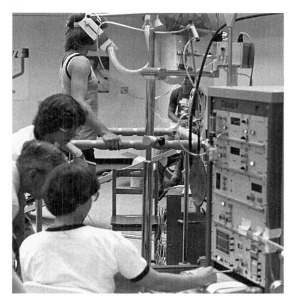

(b)

Figure 27-4 (a) A schematic diagram of an on-line oxygen consumption system using a low-cost microcomputer. The output from gas analyzers and other physiological equipment is automatically translated to the computer through an analog-to-digital converter. Oxygen consumption is calculated, displayed, and stored. (b) A commercially constructed system for measuring $\dot{V}_{O_2}$. $\dot{V}_E$ is ventilation—exhaled air volume.

Low-cost microcomputers and instrument interfaces have enabled the widespread use of on-line oxygen consumption systems, in both research laboratories and clinical settings (Figure 27-4). On-line systems automatically measure gas volume and composition, calculate oxygen consumption, display information, and store data. These systems, which include a computer, analog-to-digital converter (translates information from the analyzers to the computer), ventilation meter, oxygen analyzer, carbon dioxide analyzer, mixing chamber, respiratory valve (e.g., Daniels, Hans-Rudolf, Triple J), tubing, thermistor, and nose clip, can be constructed from commercially available components. These systems can be expanded to include computer disk drives (floppy and hard disks), a printer, a plotter, and even a human voice simulator. Commercially constructed systems are also available and are in common use.

Gas Analysis Gas composition is determined by gas analyzers that use various chemical methods. Devices such as Scholander and Haldane analyzers determine gas composition by analyzing the reactions among oxygen, carbon dioxide, and specific chemical reagents. These devices are highly accurate but very time-consuming to use. High-speed gas analyzers use infrared and paramagnetic methods to measure the oxygen and carbon dioxide content of expired air. These are the most common types of analyzers in exercise laboratories. The most sophisticated gas analyzers are mass spectrometers, which work by measuring the molecular weight of gases. These are capable of analyzing the composition of almost any gas in a sample. However, they are very expensive and difficult to maintain.

■ Tests of Functional Capacity

Functional capacity tests can be subdivided into laboratory tests and field tests. Laboratory tests are more accurate and reliable than field tests, but they tend to be considerably more expensive and technologically difficult to perform. In laboratory tests, maximal oxygen consumption is measured while subjects exercise on devices such as a treadmill, bicycle ergometer, arm ergometer, or swimming flume. Field tests estimate maximal oxygen consumption from established formulas based on performance in running, bench stepping, cycling, or swimming.

Maximal Versus Submaximal Tests

Tests can be either maximal or submaximal. The advantages of submaximal tests are that they are physically less demanding of the subject, take less time, and may be safer (this point is controversial). Maximal tests are more accurate, provide a better physiological profile of the subject, and can produce diagnostic changes in the electrocardiogram that may be missed during submaximal tests.

Medical Supervision and Safety Procedures
Testing should follow safety procedures established by professional organizations such as the American College of Sports Medicine (ACSM, 1995) and the American Heart Association (1995). Safety consid-

erations include physician supervision, ECG monitoring, and availability of defibrillator and emergency medications.

Although unmonitored submaximal tests are safe in young subjects, they could be dangerous in older or symptomatic people, who often have extremely low levels of functional capacity or may develop problems at submaximal-level exercise or during recovery. Submaximal tests may unwittingly become supramaximal tests in some subjects. In addition, problems that develop during unmonitored submaximal tests may develop into medical emergencies if left to progress.

In apparently healthy people, physician supervision is required during maximal tests in males 40 years older and in females 50 years and older (see Table 27-4). A healthy person of any age can take a submaximal test without physician supervision, provided the test is conducted by people who are well-trained. People of any age who have major risk factors of coronary artery disease (i.e., hypertension, hyperlipidemia, smoking, diabetes, or family history) or symptoms of cardiorespiratory or metabolic diseases should have physician supervision during maximal exercise testing. These people may take a submaximal exercise test (i.e., <75% of maximum) without physician supervision, provided they are not having active symptoms. The ACSM guidelines recommend that persons of any age with symptoms suggestive of coronary, pulmonary, or metabolic disease have a medical examination and a physician-supervised maximal exercise test prior to beginning an exercise program.

Exercise Test Measurements

Many physiological measurements are possible during an exercise tolerance test. Some of these measurements—including the electrocardiogram (ECG), heart rate, blood pressure, and rate–pressure product (heart rate · systolic blood pressure)—examine the performance of the heart during exercise.

The ECG measures the electrical activity of the heart. It is valuable for helping assess cardiac metabolism and blood flow. Heart rate is usually obtained from an ECG, but it can also be estimated by counting the pulse rate. Blood pressure, an extremely important measure of cardiovascular function, measures the force of blood flow in the arteries. Rate–pressure product is an indirect measure of the oxygen consumption of the heart.

More sophisticated measures of heart function include radionuclear perfusion scintigraphy ([201]Thallium), tests of left ventricular performance (radionuclide cineangiography, blood pool labeling studies), cardiac output, and echocardiography. Measures of general metabolic response include ventilation (volume and rate), oxygen consumption, perceived exertion (Figure 27-5), core and skin temperature, and substrates (i.e., glucose, lactate).

Clinically, the most common stress test measurements are the ECG, blood pressure, oxygen consumption, and test duration. Test duration (treadmill stage, power, etc.) is an adequate predictor of maximal oxygen consumption in untrained subjects

(but not heart patients). The nomogram in Figure 27-6 provides a simple way of predicting $\dot{V}_{O_2}$ from exercise duration during common treadmill tests.

All laboratories should follow established guidelines (such as those of the American College of Sports Medicine and the American Heart Association) for conducting graded exercise tests. As noted earlier, these guidelines specify requirements for physician supervision and safety equipment (defibrillator, emergency medication, and so on), and criteria for stopping a test (fatigue, critical levels of ST segment depression, arrhythmias, angina, and so on).

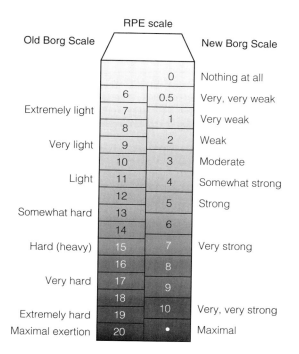

Figure 27-5 **(left)** Ratings of perceived exertion (RPE) (Old Borg Scale). **(right)** Ratings of perceived exertion (RPE) (New Borg Scale).

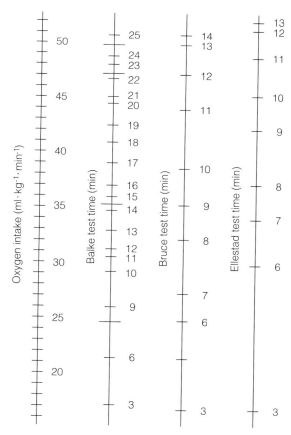

Figure 27-6 A nomogram to predict oxygen consumption from exercise duration during the Balke, Bruce, and Ellestad maximal treadmill stress test protocols. Modified from Pollock, 1976.

TABLE 27-6

Indications for Stopping an Exercise Test

1. Progressive angina (stop at 3+ level or earlier on a scale of 1+ to 4+)
2. Ventricular tachycardia
3. Any significant drop (20 mmHg) of systolic blood pressure or a failure of the systolic blood pressure to rise with an increase in exercise load
4. Lightheadedness, confusion, ataxia, pallor, cyanosis, nausea, or signs of severe peripheral circulatory insufficiency
5. Early onset deep (>4 mm) horizontal or downsloping ST depression or elevation
6. Onset of second- or third-degree a–v block
7. Increasing ventricular ectopy, multiform PVCs, or R on T PVCs
8. Excessive rise in blood pressure: systolic pressure >250 mmHg; diastolic pressure >120 mmHg
9. Chronotropic impairment: increase in heart rate that is <25 beats · min^{-1} below age-predicted normal value (in the absence of beta blockade)
10. Sustained supraventricular tachycardia
11. Exercise-induced left bundle branch block
12. Subject requests to stop
13. Failure of the monitoring system

SOURCE: Adapted from American Heart Association, 1991.

The test administrator should be especially aware of criteria for stopping a test. Physical signs suggesting that a subject is at increased risk during a test (Table 27-6) include indications of a decrease in central blood volume, ischemia (the heart muscle is not getting enough blood), and arrhythmia (aberrant ECG rhythm). Indications of decreased central blood volume include a falling systolic blood pressure during exercise, central nervous system distress (weaving or dizziness), and diaphoresis (cold sweat). Ischemia is indicated by ST segment depression, arrhythmias, and angina (chest pain; see Figures 27-7 and 27-8). Arrhythmias are abnormal electrocardiogram rhythms that can result in decreases or total interruption in cardiac output (see Figure 27-9). Each laboratory should have its own written criteria for stopping functional capacity tests, based on sound guidelines, such as those in *Guidelines for Graded Exercise Testing and Exercise Prescription* (ACSM, 1995).

Precise methods are necessary to ensure accuracy and subject safety. All instruments must be regularly calibrated and serviced (including the bicycle ergometer and treadmill). Skin electrode preparation must be meticulous in order to obtain artifact-free ECGs. Tests should be stopped if a good ECG tracing cannot be obtained while the subject is exercising.

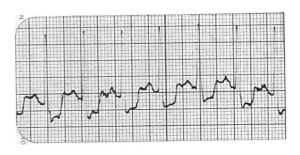

(a)

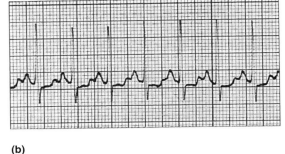

(b)

Figure 27-7 ST segment depression before (a) and after coronary artery bypass surgery (b). Coronary artery disease may result in coronary ischemia (inadequate blood flow) that is often detected by ST segment depression on the electrocardiogram. In this patient, bypassing the obstructed coronary artery resulted in relief of ischemia and a marked reduction in ST segment depression.

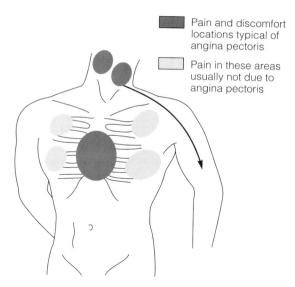

■ Pain and discomfort locations typical of angina pectoris

☐ Pain in these areas usually not due to angina pectoris

Figure 27-8 Angina pectoris is pain caused by coronary artery ischemia. Pain in the center of the chest, neck, or radiating down the left arm may be angina pectoris (*dark ovals*). Pain in the more lateral aspects of the chest (*light ovals*) is more typical of a "stitch" (exercise or gastrointestinally related side pain) and is not usually due to coronary ischemia.

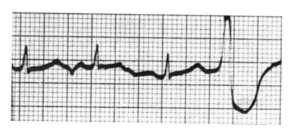

Figure 27-9 Example of a premature ventricular contraction (PVC), a type of arrhythmia. PVCs arise from ventricular ectopic foci (emergency pacemakers).

Physical Preparation for Exercise Tolerance Testing

A number of physical factors must be considered to ensure test standardization and consistency, including pretest diet and exercise, clothing, room temperature, and medication. Failure to consider these factors can decrease the reliability of the measurements. They may even place the subject at risk of injury. In addition, to protect the examiner and reduce anxiety in the subject, factors such as informed consent, human rights, and documentation should be given serious thought.

Meals and Exercise Subjects should report to the laboratory in a postabsorptive state. However, a light carbohydrate meal three to four hours prior to the test may be permitted. Cigarettes and caffeine products should be avoided because they could affect the results of the test by increasing heart rate, blood pressure, and the incidence of arrhythmias. Subjects should refrain from intense exercise for 12 hours before the test.

Clothing Proper clothing is important for comfort, safety, and artifact-free electrocardiograms. Subjects should wear running shoes and shorts. Women should also wear a tight-fitting bra that provides support but does not interfere with the ECG electrodes.

Environment Tests should be conducted in a comfortable environmental temperature 22°C (72°F) or less, and at a relative humidity of 60% or lower. For accuracy, they should be conducted as close to sea-level elevation as possible. $\dot{V}_{O_2max}$ is decreased by approximately 3% at 5500 feet altitude (see Chapter 23).

Pretest Medical Considerations When a physician is present, pretest preparations should include an evaluation of medical history and a brief physical examination. These are conducted to detect important clinical signs and conditions that could make exercise testing dangerous. These procedures are also valuable when evaluating the results of the exercise test.

When a physician is not present, the technician should look for potentially dangerous conditions, and the medical history should be carefully evaluated for potentially disqualifying conditions. Subjects should be questioned about their health and well-being. The test should not be conducted if there is any question about the safety of the test.

Prior to the test, a resting ECG should be examined. This pre-exercise ECG should be obtained with subjects seated, standing, and hyperventilating. These measures may be valuable in detecting false-positive tests. A false-positive test is incorrectly finding an abnormality—the data suggest the existence of heart disease when none exists.

Medication   Use of medication during an exercise test depends on the reason for the test. Some medications interfere with exercise performance and test interpretation. If the test is used for exercise prescription, however, then normal medications should be continued. Decisions regarding the use of medication during the test fall within the domain of the physician.

ECG Electrodes—Skin Preparation and Electrode Placement   Skin preparation is vital to obtaining clear, artifact-free ECG recordings. The best recordings are obtained when the electrical resistance between the electrode and the skin is low ($<5000 \ \Omega$). First, the skin should be shaved to remove body hair. Next, the outer layer of skin and surface oil and dirt should be removed using gauze saturated with alcohol. Use good-quality

electrodes. Cables should be in good repair to get a good signal.

Bipolar leads are commonly used for young, healthy subjects when routine ECG monitoring is desired. The CM-5 lead configuration is favored by most laboratories (Figure 27-10a). In this configuration, the positive reference is the V_5 position (the fifth intercostal space at the midclavicular line). The negative reference M is placed at the top of the manubrium. The indifferent reference is placed at the fifth intercostal level and the anterior axillary line.

Multiple leads (12 leads) are used when diagnostic testing is desired (Figure 27-10b). Standard electrode placement is not possible during exercise. Arm electrodes are placed below the sternum, close to the shoulders. Limb electrodes are placed anterior to the ilium, below the umbilicus, and the precordial electrodes are placed in their standard positions.

Legal Considerations   Informed consent and human rights are important considerations during exercise stress testing. Subjects must always be treated courteously and with respect. Noise and distractions must be minimized and all attention should be focused on the subjects and the mea-

Figure 27-10   (a) The CM-5 bipolar ECG configuration. (b) The 10-electrode placement for obtaining a 12-lead ECG during exercise. Source: ACSM, 1991, p. 67. Used with permission.

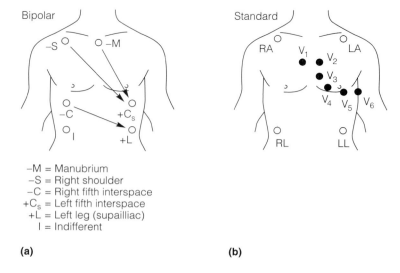

−M = Manubrium
−S = Right shoulder
−C = Right fifth interspace
+C$_s$ = Left fifth interspace
+L = Left leg (supailliac)
 I = Indifferent

(a)   **(b)**

surement instruments. Subjects should understand what is expected of them and be fully aware of the possible risks of taking a maximal exercise test. An informed consent declaration should be signed and witnessed (also by a parent or guardian if the subject is a minor). Ideally, subjects should have the opportunity to practice on the exercise device prior to the test. This may involve as little as walking on the treadmill briefly during a preliminary visit to the laboratory.

All aspects of an exercise tolerance test should be documented. Any abnormal symptoms or physiological responses should be recorded on a standardized form. This precaution is important for both the subject and test administrator.

The Treadmill

The treadmill is the most common apparatus for determining functional capacity. This device allows precise exercise intensities using exercise familiar to the vast majority of people (walking and running). The treadmill is preferred by most researchers and clinicians because it renders a higher peak oxygen consumption than can be obtained with other techniques (Table 27-7). Disadvantages include high cost, size, inability to measure power output directly, subject anxiety, and difficulty making measurements such as blood pressure (and, to a certain extent, electrocardiograms). The treadmill cannot be used for people who are unable to walk (i.e., par-

aplegics, those with severe fractures or advanced neurological disease, and so on).

A variety of treadmill test protocols are used to measure functional capacity. The most common include the Bruce, Balke, Naughton, and Ellestad, and various modifications of these tests (Table 27-8). Although these protocols vary in their use of speed and slope, they have several factors in common: They are all continuous, standardized, include a warm-up, and involve a minimum to moderate metabolic transition between stages. Tests should be selected on the basis of subject population and the measurements desired.

Warm-up is important during the test to prevent abnormalities in the electrocardiogram, to avoid musculoskeletal injury, and to facilitate the redirection of blood to the working muscles. The warm-up stage also gives the subject time to become familiar and comfortable with the exercise and allows the metabolic instruments time to achieve a degree of equilibrium before the more important maximal measurements are taken. A monitored cool-down is important as well, to prevent venous pooling and minimize stress on the heart.

In continuous treadmill tests exercise intensities are increased without rest periods between the stages. In discontinuous tests a rest interval is included between stages. A *stage* is an individual segment of the exercise test characterized by a specific speed and grade. Continuous tests are the method of choice because they are less time-consuming and produce the same results as discontinuous tests.

Standardization is an extremely important consideration in exercise testing, as in all experimental research. The treadmill test is an example of the common stimulus–response technique of experimental inquiry. A stimulus is introduced (i.e., a precise, standardized exercise intensity) and the response measured (heart rate, ECG, $\dot{V}_{O_2max}$, blood pressure, perceived exertion, and so on). Lack of standardization leads to increased variability in the metabolic requirement of the test. Treadmill programmers that automatically change the speed and elevation of the treadmill improve standardization.

All subjects should take the test in the same way. Subjects must walk and run at the appointed time

---

TABLE 27-7

Comparison of Functional Capacity Achieved on a Treadmill and Other Devices (Treadmill = 100%)

	Males (%)	Females (%)
Bicycle ergometer (seated)	93	91
Bicycle ergometer (reclining)	90	88
Arm ergometer	88	85
Step test	96	98

TABLE 27-8

Principal Treadmill Protocols

Min	Bruce Test Speed (mph)	Bruce Test % Grade	Modified Bruce Test Speed (mph)	Modified Bruce Test % Grade	Balke Test Speed (mph)	Balke Test % Grade	Naughton Test Speed (mph)	Naughton Test % Grade	Ellestad Test Speed (mph)	Ellestad Test % Grade	Modified Åstrand Test Speed (mph)	Modified Åstrand Test % Grade
1	1.7 →	10 →	1.5 →	5 →	3.3 →	0	1.2 →	0 →	1.7 →	10 →	5–8.5 →	0 →
2						2						
3						3						
4	2.5 →	12 →	1.7 →	10 →		4	2.0 →		3.0 →			2.5 →
5						5	2.0 →	3.5 →				
6						6			4.0 →			5.0 →
7	3.4 →	14 →	2.5 →	12 →		7	2.0 →	7.0 →				
8						8			5.0 →			7.5 →
9						9	2.0 →	10.5 →				
10	4.2 →	16 →	3.4 →	14 →		10			6.0 →	15 →		10.0 →
11						11	2.0 →	14.0 →				
12						12			7.0 →			
13	5.0 →	18 →	4.2 →	16 →		13	2.0 →	17.5 →				
14						14			8.0 →			
15						15	3.0 →	12.5 →				
16	5.5 →	20 →	5.0 →	18 →		16						
17						17	3.0 →	15.0 →				
18						18						
19	6.0 →	22 →				19	3.0 →	17.5 →				
20						20						
21						21	3.0 →	20.0 →				
22						22						
23						23						
24						24						
25						25						

Figure 27-11 The effects of holding the railing during treadmill running on oxygen consumption and heart rate. It is important that treadmill tests be standardized, particularly when oxygen consumption is predicted on the basis of test duration. Adapted from Åstrand, 1984. Used with permission.

Oxygen uptake (liters · min^{-1})	3.0	2.5
Heart rate (beats · min^{-1})	165	155

during the test. For example, it is inappropriate for a subject to run during stage 3 of a Bruce test because the test protocol requires subjects to walk during that stage. Running would increase the O_2 cost of the stage. Also, subjects must not be allowed to hold onto the railing of the treadmill during the test because it reduces the test's metabolic cost (Figure 27-11).

The metabolic increment between stages (the increase in speed and grade) should not exceed 3 METs in healthy subjects. Stage increments should not exceed 0.5–1.0 MET in cardiac patients. Most exercise tests use 2–3 min stages. The test should be chosen on the basis of the subject's capacity. For example, a modified Bruce test with a lower exercise intensity during the warm-up stage may be more appropriate than the more strenuous standard Bruce test for a recent myocardial infarction patient. Walking tests, such as the Balke, render lower values for $\dot{V}_{O_2max}$ and maximal heart rate, and are less reliable for young, fit individuals. The Balke might be the method of choice if blood pressure measurements are particularly critical (e.g., in subjects with hypertension).

Predicting Oxygen Consumption During Walking and Running Equations and nomograms have been established to predict maximal oxygen uptake from exercise time or speed on a treadmill. (See Figure 27-6, Eqs. 27-2 and 27-3, and Table 27-9a, b.) These enable test administrators to predict $\dot{V}_{O_2}$ without using elaborate metabolic measurement equipment. They are useful for exercise testing and prescription. The equations are

from *Guidelines for Graded Exercise Testing and Exercise Prescription* (ACSM, 1995, pp. 269–287).

These equations, nomograms, and tables are not as accurate as direct measurements because subjects will vary in efficiency and in the oxygen cost of their running or walking style. The equations are equally accurate for men and women, but tend to be less accurate for children than adults. Despite these disadvantages, the equations are useful when direct measurements are not possible.

The Bicycle Ergometer

The bicycle ergometer is the preferred instrument for stress testing in some laboratories, particularly in Europe. This device has the advantage of being relatively inexpensive, portable, and less intimidating than a treadmill. Also, physiological measurements are easier to obtain because the upper body is more stationary than it is on a treadmill or bench step. In addition, power output can be measured directly, which makes it attractive for energetic studies. Bicycle ergometry can be conducted in either upright or supine positions.

There are many disadvantages to its use as well, particularly in testing women, elderly subjects, and people unfamiliar with cycling. Cycling places a considerable load on the quadriceps, which can lead to local muscle fatigue and can limit performance before the capacity of the cardiovascular system is maximized (recall that the purpose of a functional capacity test is to test the capacity of the cardiovascular system and not the endurance capacity of particular muscle groups). Functional capacity

Energy cost of walking at speeds between 50 to 100 m · min⁻¹ (1.9 to 3.7 mph):

$$\dot{V}_{O_2} \text{ (ml · kg}^{-1}\text{ · min}^{-1}\text{)} = \textbf{Horizontal component} + \textbf{Vertical component} \tag{27-2}$$

Horizontal component:

$$\dot{V}_{O_2} \text{ (ml · kg}^{-1}\text{ · min}^{-1}\text{)} = (\text{Speed} \cdot 0.1) + 3.5$$

where Speed is expressed in m · min⁻¹ (1 m · min⁻¹ = 0.0373 mph); 0.1 (ml · kg⁻¹ · min⁻¹) is the oxygen cost of walking at 1 m · min⁻¹; and 3.5 (ml · kg⁻¹ · min⁻¹) is resting oxygen consumption, or 1 MET.

Vertical component:

$$\dot{V}_{O_2} \text{ (ml · kg}^{-1}\text{ · min}^{-1}\text{)} = \% \text{ Grade} \cdot \text{Speed} \cdot 1.8$$

where % Grade is the slope of the treadmill expressed as a fraction (vertical distance climbed /belt speed); Speed is expressed in m · min⁻¹; and 1.8 is the oxygen cost of a power output of 1 kgm (kilogram-meter, the force required to move 1 kilogram 1 meter).

Energy cost of running at speeds greater than 134 m · min⁻¹ (5 mph):

$$\dot{V}_{O_2} \text{ (ml · kg}^{-1}\text{ · min}^{-1}\text{)} = \textbf{Horizontal component} + \textbf{Vertical component} \tag{27-3}$$

Horizontal component:

$$\dot{V}_{O_2} \text{ (ml · kg}^{-1}\text{ · min}^{-1}\text{)} = (\text{Speed} \cdot 0.2) + 3.5$$

where Speed is expressed in m · min⁻¹; 0.2 (ml · kg⁻¹ · min⁻¹) is the oxygen cost of running in m · m⁻¹; and 3.5 (ml · kg⁻¹ · min⁻¹) is resting oxygen consumption, or 1 MET. (*Note:* This equation can be used for speeds between 80 and 134 m · min⁻¹ if the person is jogging rather than walking.)

Vertical component:

$$\dot{V}_{O_2} \text{ (ml · kg}^{-1}\text{ · min}^{-1}\text{)} = \% \text{ Grade} \cdot \text{Speed} \cdot 1.8 \cdot 0.5$$

where % Grade is the slope of the treadmill expressed as a fraction; Speed is expressed in m · min⁻¹; 1.8 is the oxygen cost of a power output of 1 kgm; and 0.5 is the factor used to correct for the difference in O_2 cost of the vertical component of treadmill and outdoor running. (*Note:* The factor 0.5 should not be used if this equation is to be used for outdoor running.)

TABLE 27-9a

Approximate Oxygen Cost of Walking (ml · kg^{-1} · min^{-1})

Speed		Grade					
m · min^{-1}	mph	0%	5%	10%	15%	20%	25%
50	1.87	8.5	13.0	17.5	22.0	26.5	31.0
55	2.05	9.0	14.0	18.9	23.9	28.8	33.8
60	2.24	9.5	14.9	20.3	25.7	31.1	36.5
65	2.43	10.0	15.9	21.7	27.6	33.4	39.3
70	2.61	10.5	16.8	23.1	29.4	35.7	42.0
75	2.80	11.0	17.8	24.5	31.3	38.0	44.8
80	2.99	11.5	18.7	25.9	33.1	40.3	47.5
85	3.17	12.0	19.7	27.3	35.0	42.6	50.3
90	3.36	12.5	20.6	28.7	36.8	44.9	53.0
95	3.54	13.0	21.6	30.1	38.7	47.2	55.8
100	3.73	13.5	22.5	31.5	40.5	49.5	58.5

TABLE 27-9b

Approximate Oxygen Cost of Running (ml · kg^{-1} · min^{-1}) at a 0% Grade

Speed				Speed			
m · min^{-1}	mph	$\dot{V}_{O_2}$	METs	m · min^{-1}	mph	$\dot{V}_{O_2}$	METs
140	5.2	31.5	9.0	195	7.3	42.5	12.1
145	5.4	32.5	9.3	200	7.5	43.5	12.4
150	5.6	33.5	9.6	205	7.6	44.5	12.7
155	5.8	34.5	9.9	210	7.8	45.5	13.0
160	6.0	35.5	10.1	215	8.0	46.5	13.3
165	6.2	36.5	10.4	220	8.2	47.5	13.6
170	6.3	37.5	10.7	225	8.4	48.5	13.9
175	6.5	38.5	11.0	230	8.6	49.5	14.1
180	6.7	39.5	11.3	235	8.8	50.5	14.4
185	6.9	40.5	11.6	240	9.0	51.5	14.7
190	7.1	41.5	11.9				

measured on a bicycle ergometer is typically 8 to 15% lower than on a treadmill. Bicycle ergometer performance is largely dependent on the subject's motivation. If the pedal revolutions decrease, then oxygen uptake decreases. Fatiguing subjects can volitionally and prematurely slow down when they begin to fatigue, resulting in an underestimation of their functional capacity.

The three basic types of bicycle ergometers are electrically braked, mechanically braked, and isoki-

netic. Electrically and mechanically braked ergometers place a known load on a flywheel (rotating mass analogous to the front wheel on a bicycle). These ergometers depend on the subject to pedal the ergometer at a set rate in order to obtain a given power output. In the isokinetic dynamometer, the pedal revolution is constant but the resistance is variable. Braked ergometers are by far the most commonly used type for functional capacity testing.

Power is the term used to define the intensity of

exercise performed on a bicycle ergometer. Power is work per unit of time (Eq. 27-4):

$$\text{Power} = \text{Work} \cdot \text{Time}^{-1} \qquad (27\text{-}4)$$

where Work = Force × Distance (Force = Frictional resistance to pedaling, Distance = Revolutions per minute (rpm) and distance that the flywheel travels per revolution.

Watts is the accepted unit of measurement for power output (in SI units). Previously, the most common unit of measurement was the kilopond-meter per minute (kpm · min^{-1}; also called kilogram-meter per minute, kgm · min^{-1}). A kilopond-meter is the force required to move a mass of 1 kilogram a distance of 1 meter at the normal acceleration of gravity. One watt is equal to 6.12 kpm · min^{-1}.

A bicycle ergometer that is to be used for testing should include indicators of frictional resistance and rpm. Revolutions per minute must be measured exactly with a device such as a microswitch and counter rather than estimated from a speedometer because deviations in specified rpm affect total power output. The ergometer should be regularly calibrated to ensure accuracy. This is simple to do in a friction-braked ergometer, but electrically braked ergometers typically require calibration by the manufacturer.

No specific test protocol is used by a majority of laboratories. However, a few basic principles should be followed when designing a bicycle test. Basic considerations include starting power output, power increments, subject population, and standardization.

Exercise loads are administered in fixed increments or determined on the basis of heart rate response (Figure 27-12). In healthy subjects, starting power is 50 to 100 watts (typically, 50 watts for women and 100 watts for men) and increased by 25 to 50 watts every 2–3 minutes. Patient populations typically begin at 25 to 50 watts with power increments of 5–25 watts per stage.

Bicycle ergometer tests are usually conducted at 50–70 rpm in both healthy and patient populations. Subjects should be prompted to maintain the specified cadence with a metronome set at twice the rpm

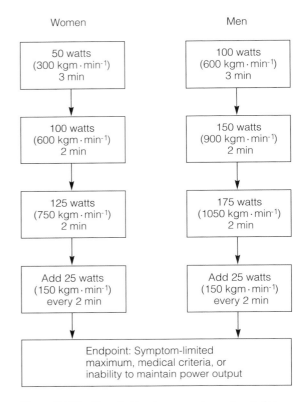

Figure 27-12 Sample bicycle ergometer protocols for men and women.

(e.g., if the cadence is 60 rpm, then the metronome should be set for 120 beats per minute). Trained cyclists are often tested at 90 rpm to simulate the pedaling cadence typically used in the sport. In addition, when testing cyclists, the seat, pedals, and handlebars may be modified to help the athletes reach a higher functional capacity.

Predicting Oxygen Consumption Oxygen consumption can be measured directly during bicycle ergometry, or estimated from a predictive equation (ACSM, 1995, pp. 269–287) for power outputs between 300–1200 kgm · min^{-1} (50–200 watts) (Eq. 27-5, Table 27-10) where 2.0 is the oxygen cost (ml · min^{-1}) of a power output of 1 kgm · min^{-1}, and 300(ml) is resting oxygen consumption.

The equation predicts oxygen consumption expressed in liters per minute. These values must be

TABLE 27-10

Approximate $\dot{V}_{O_2}$ and METs on Bicycle Ergometer for Subjects with Different Body Weights

Power Output		$\dot{V}_{O_2}$ (liters · min^{-1})	METs[b]					
W[a]	kgm · min^{-1}		50 kg	60 kg	70 kg	80 kg	90 kg	100 kg
25	150	0.6	3.4	2.8	2.4	2.1	1.9	1.7
50	300	0.9	5.1	4.2	3.6	3.2	2.8	2.5
75	450	1.2	6.8	5.7	4.8	4.2	3.8	3.4
100	600	1.5	8.5	7.1	6.1	5.3	4.7	4.2
125	750	1.8	10.2	8.5	7.3	6.4	5.7	5.1
150	900	2.1	12	10	8.5	7.5	6.6	6
175	1050	2.4	13.7	11.4	9.7	8.5	7.6	6.8
200	1200	2.7	15.4	12.8	11	9.6	8.5	7.7
225	1350	3	17.1	14.2	12.2	10.7	9.5	8.5
250	1500	3.3	18.8	15.7	13.4	11.7	10.4	9.4
275	1650	3.6	20.5	17.1	14.6	12.8	11.4	10.2
300	1800	3.9	22.2	18.5	15.9	13.9	12.3	11.1
325	1950	4.2	24	20	17.1	15	13.3	12
350	2100	4.5	25.7	21.4	18.3	16	14.2	12.8

[a] The equivalency of watts and kilogram-meter per minute is approximate.
[b] 1 MET is approximately equal to 3.5 ml · kg^{-1} · min^{-1}.

divided by body weight to calculate oxygen consumption in ml · kg^{-1} · min^{-1}, or METs. Standardization is critical when predicting $\dot{V}_{O_2}$. Factors that must be considered include seat height, rpm, static exercise performed while gripping the handlebars, and ambient environment.

The Arm Ergometer

Arm ergometry is an important testing method for disabled people who do not have functional use of their legs; for athletes, such as canoeists and kayakers, who use their upper bodies in their sport. During this type of exercise, maximal heart rate is less and submaximal heart rate (at the same power output) is higher than during leg exercise.

Predicting Oxygen Consumption In Equation 27-6 (ACSM, 1995, pp. 269–287), note that the oxygen cost per kgm · min^{-1} is higher during arm ergometry (3 ml · kg^{-1} · min^{-1}) than during leg ergometry (2 ml · kg^{-1} · min^{-1}). This is due to the increased energy cost of stabilizing the upper body during this type of exercise.

Bench Steps

Step tests, done with bench steps, are among the oldest techniques for measuring functional capacity. Popular procedures, such as the Harvard step test, involve completion of an exercise protocol and the measurement of recovery heart rate. Recovery heart rate is used in these tests because it is technically difficult to measure exercise pulse rates, and accurate equations to predict oxygen consumption during bench stepping are unavailable. However, recovery pulse rate is only moderately related to maximal oxygen consumption, so these types of tests are less appropriate than current procedures for measuring physical fitness.

$$\dot{V}_{O_2} \text{ (ml · min}^{-1}) = [\text{Power output (kgm · min}^{-1}) \cdot 2.0] + 300 \qquad (27\text{-}5)$$

$$\dot{V}_{O_2} \text{ (ml · min}^{-1}) = [\text{Power output (kgm · min}^{-1}) \cdot 3.0] + 300 \qquad (27\text{-}6)$$

Step tests have the advantage of being inexpensive, portable, and technically simple to perform. They can be adapted to field situations such as physical education classes and can be administered to large numbers of people in a relatively short time. The disadvantages make this test inappropriate for the laboratory. Taking physiological measurements during step tests is difficult, and there is a danger of tripping. Safety problems also make step tests inappropriate for the elderly and certain types of medically impaired subjects (particularly the orthopedically and neurologically impaired). In addition, large step heights must be used when testing young, healthy subjects. However, this can lead to local muscle fatigue and an underestimation of functional capacity. Functional capacity measured on a step test is typically 5–7% less than that measured on a treadmill. To ensure accuracy, subjects must stand upright and not lean over as they step up.

Bench stepping and bicycle ergometer test protocols are similar. Step tests require a variable height bench and a metronome to maintain cadence. The most common procedure is to begin with a step height of 2 centimeters and a stepping cadence of 30 repetitions per minute. One repetition consists of the subject standing in front of the bench step, placing one foot and then the other on the bench step, then returning to the starting position, one step at a time (the metronome should be adjusted to make four sounds per repetition). The step height is increased 2–4 centimeters every 1–2 minutes. Another variation is to use a bench with a fixed height, but increase the number of repetitions per exercise stage. This method is less desirable than the adjustable bench step technique because it requires greater cooperation from the subject and puts the subjects at a greater risk of injury (subjects have to move faster when fatigued).

Predicting Oxygen Consumption In step tests, oxygen consumption is usually estimated from an equation (ACSM, 1995, pp. 269–287) rather than directly measured. Predicting oxygen con-

Oxygen Costs of Bench Stepping

$$\dot{V}_{O_2}(ml \cdot kg^{-1} \cdot min^{-1}) = \textbf{O}^2 \textbf{ cost of positive and negative work} + \textbf{O}^2 \textbf{ cost of horizontal movement} \qquad (27\text{-}7)$$

The oxygen cost of the positive and negative work of bench stepping:

$$\dot{V}_{O_2} (ml \cdot kg^{-1} \cdot min^{-1}) = \text{Bench height (m)} \times \text{Step rate} \cdot min^{-1} \times 1.8 \times 1.33$$

where 1.8 is the oxygen cost (ml) of a power output of 1 kgm and 1.33 is the factor used to calculate the oxygen cost of the total work (positive work + negative work, where the oxygen cost of negative work is one-third that of positive work).

The oxygen cost of horizontal movements (moving back and forth while stepping on and off the step):

1. Calculate METs from the O_2 cost of positive and negative work ($\dot{V}_{O_2}$ /3.5).
2. Divide the step rate by 10 to estimate the MET cost of the horizontal movements.
3. Multiply the MET cost of the horizontal movements by 3.5 to get the oxygen cost of these horizontal movements.
4. Add the oxygen cost of the horizontal movements to the oxygen cost of the positive and negative work to get the total oxygen cost of the bench-stepping exercise.

TABLE 27-11

Oxygen Cost of Bench Stepping (ml · kg^{-1} · min^{-1})

Step Height		Step Frequency (steps · min^{-1})				
cm	in.	15	20	25	30	35
4	1.6	6.7	8.9	11.1	13.4	15.6
6	2.4	7.4	9.9	12.3	14.8	17.3
8	3.1	8.1	10.8	13.5	16.2	19
10	3.9	8.8	11.8	14.7	17.7	20.6
12	4.7	9.6	12.7	15.9	19.1	22.3
14	5.5	10.3	13.7	17.1	20.6	24
16	6.3	11	14.7	18.3	22	25.7
18	7.1	11.7	15.6	19.5	23.4	27.3
20	7.9	12.4	16.6	20.7	24.9	29
22	8.7	13.2	17.5	21.9	26.3	30.7
24	9.4	13.9	18.5	23.1	27.7	32.4
26	10.2	14.6	19.4	24.3	29.2	34
28	11	15.3	20.4	25.5	30.6	35.7
30	11.8	16	21.4	26.7	32	37.4
32	12.6	16.7	22.3	27.9	33.5	39.1
34	13.4	17.5	23.3	29.1	34.9	40.7
36	14.2	18.2	24.2	30.3	36.4	42.4
38	15	18.9	25.2	31.5	37.8	44.1
40	15.7	19.6	26.2	32.7	39.2	45.8
42	16.5	20.3	27.1	33.9	40.7	47.4

sumption during bench stepping requires a knowledge of the bench height and the number of lifts per minute. Oxygen consumption is predicted in ml · kg^{-1} · min^{-1} (Eq. 27-7; Table 27-11).

■ Exercise-Specific Tests

As discussed in Chapter 16, the determination of maximal oxygen consumption requires that the test endpoint be determined by the cardiovascular system rather than by local muscle fatigue. Using this criterion, tests involving arm ergometry, swimming, and wheelchair ergometry will not render true measurements of functional capacity (Figure 27-13a,b).

Field Tests of Maximal Oxygen Consumption

Running Tests The Cooper 12-min run, the Balke 15-min run, and the 1½-mile run are among the most popular exercise-specific tests (Table 27-12). These tests have a fairly good relationship to $\dot{V}_{O_2max}$ in healthy college-age subjects. The relationship is not as good in other populations, perhaps because of problems with motivation and lack of familiarity with exercise training.

Studies comparing $\dot{V}_{O_2max}$ in running tests with its measurement in the laboratory have been plagued by small sample sizes and questionable motivation by the subjects. These considerations are important because running tests, such as the 12-min run are widely used in schools to measure cardiovascular fitness. However, they may be dangerous for some people because the electrocardiogram is not monitored during the test. Before the test is taken, a low-level conditioning program is recommended.

Walking Tests The 1-mile walk test provides a reasonable measure of $\dot{V}_{O_2max}$. This test was developed by Kline and colleagues (1987). It has been

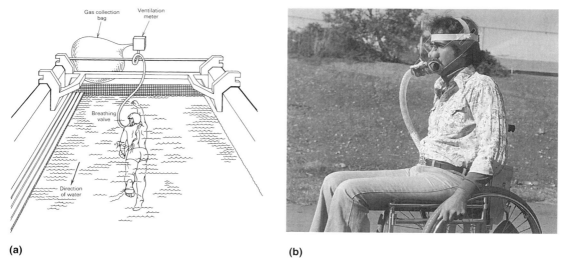

(a)   **(b)**

Figure 27-13 (a) Swimming flume for measuring exercise capacities during swimming. (b) Field measurement of $\dot{V}_{O_2}$ during wheelchair propulsion. [(a) Adapted from Holmer, 1974.]

TABLE 27-12

The 1.5-Mi Run Test

Equipment

1. A running track or course that is flat and provides exact measurements of up to 1.5 miles
2. A stopwatch, clock, or watch with a second hand

Preparation

You may want to practice pacing yourself prior to taking the test to avoid going too fast at the start and becoming prematurely fatigued. Allow yourself a day or two to recover from your practice run before taking the test.

Instructions

1. Warm up before taking the test. Do some walking, easy jogging, and stretching exercises.
2. Try to cover the distance as fast as possible, at a pace that is comfortable for you. If possible, monitor your own time or have someone call out your time at various intervals of the test to determine whether your pace is correct.

3. Record the amount of time, in minutes and seconds, it takes for you to complete the 1.5-mi distance.
4. Cool down after the test by walking or jogging slowly for about 5 minutes.

Determining Maximal Oxygen Consumption

1. Convert your running time from minutes and seconds to a decimal figure. For example, a time of 14 minutes and 25 seconds would be $14 + (25/60)$ or 14.4 minutes.
2. Insert your running time in the equation below, where

T = running time (in minutes)

$$\dot{V}_{O_2max} = (483 \div T) + 3.5$$

For example, a person who completes 1.5 miles in 14.4 minutes would calculate maximal oxygen consumption as follows:

$$\dot{V}_{O_2max} = (483 \div 14.4) + 3.5 = 37 \text{ ml} \cdot \text{kg}^{-1} \cdot \text{min}^{-1}$$

validated in subjects ranging in age from 20 to 80 years. The test predicts $\dot{V}_{O_2max}$ based on time to walk 1 mile, heart rate during the last 2 minutes of exercise, age, weight, and sex. Equation 27-8 gives the regression equation for predicting $\dot{V}_{O_2max}$ from a 1-mile walk where W = Body weight in pounds, A = Age in years, G = Gender (1 for male, 2 for female), T = Time to complete 1-mile course (min), H = average heart rate for last 2 minutes of exercise (immediate postexercise heart rate acceptable).

$$\dot{V}_{O_2max} = 132.853 - (0.0769 \times W) - (0.3877 \\ \times A) + (6.315 \times G) - (3.2649 \qquad (27\text{-}8) \\ \times T) - (0.1565 \times H)$$

Swimming Swimming tests performed on a heterogeneous population (i.e., trained and untrained swimmers) measure swimming skill rather than functional capacity. The energy cost in swimming is determined by the metabolic requirements of forward propulsion and body drag (resistance provided by the body in water). Propulsion can be increased and drag decreased by improving technique. Swimming tests are valuable for developing an exercise prescription for swimming and for measuring functional capacity in well-trained swimmers.

Upper-Body Endurance Tests Wheelchairs and arm ergometers stress the relatively small upper-body muscles. Maximal exercise on these devices is limited by local muscle fatigue rather than by cardiovascular capacity. Decrements of 25% or more between these upper-body tests and treadmill tests are not uncommon. However, for subjects such as paraplegics, who are unable to use their legs, these tests are the only ones available for assessing functional capacity.

Franklin and coworkers (1990) have developed a field test to estimate $\dot{V}_{O_2max}$ in wheelchair users (Eq. 27-9). In this test, $\dot{V}_{O_2max}$ is estimated based on the distance a person can travel in a wheelchair in 12 minutes. This test was validated using Class II-V paraplegics, postpolio patients, and Class V amputees. (See Chapter 26 for a discussion of disability classifications.) Table 27-13 shows norms for peak

TABLE 27-13

Levels of Aerobic Fitness Based on 12-Mi Wheelchair Performance Test and $\dot{V}_{O_2max}$ (ml · kg^{-1} · min^{-1})

Distance (miles)	$\dot{V}_{O_2max}$ (ml · kg^{-1} · min^{-1})	Fitness Level
<0.63	<7.7	Poor
0.6–0.86	7.7–14.5	Below average
0.87–1.35	14.6–29.1	Fair
1.36–1.59	29.2–36.2	Good
≥1.60	≥36.3	Excellent

SOURCE: Franklin et al., 1990.

oxygen uptake and 12-min wheelchair propulsion capacity.

$$\dot{V}_{O_2max} \ (ml \cdot kg^{-1} \cdot min^{-1}) \\ = -10.916 + 29.623 \times \text{Distance in miles} \qquad (27\text{-}9)$$

Submaximal Tests

Submaximal exercise tests attempt to predict functional capacity from the heart rate response during a submaximal bout of exercise. Procedures such as the Åstrand bicycle ergometer test rely on the nearly linear relationship between oxygen consumption and heart rate and an assumed maximal heart rate (i.e., $HR_{max} = 220 - $ age).

A simple procedure that requires one or two 5-min exercise bouts is shown in Figure 27-14. This technique involves the prediction of maximal power output derived from this assumed linear relationship between heart rate and power output. $\dot{V}_{O_2max}$ is predicted using Equation 27-5. The accuracy of the technique is based on a series of assumptions (age-predicted maximum heart rate, linear relationship between HR and power output,* and the accuracy of the $\dot{V}_{O_2}$-bicycle ergometry equation). Small errors in any one of these assumed values can lead to considerable error.

The pulse working capacity 170 (PWC$_{170}$) is a popular submaximal test of cardiovascular capacity

---

*Heart rate and power output are linear only within certain ranges. Above 90–95% of maximal heart rate, heart rate begins to level off with increasing power output.

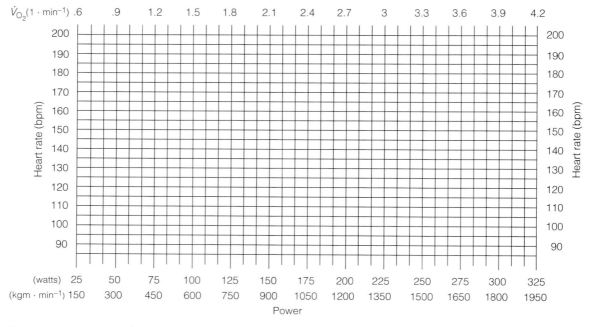

Figure 27-14 Predicting $\dot{V}_{O_2max}$ from submaximal power output and heart rate. Note that this technique may be subject to considerable error and should be used with caution. Instructions: (1) Estimate maximal heart rate (subtract age from 220). (2) Plot heart rate response during two to three different 5-min exercise bouts on a bicycle ergometer. The power output for the first exercise bout should be approximately 50 W for men and 25 W for women. Subsequent power outputs should not raise the heart rate above 75% of predicted maximum heart rate. (3) Extend power–heart rate line to predicted maximum heart rate. (4) Read predicted $\dot{V}_{O_2max}$ from the top of the chart.

that is similar to the procedure just mentioned. PWC$_{170}$ involves plotting the relationship between heart rate and power output based on two to five repetitions of 2-3-min exercise bouts of increasing intensity. The power output at a heart rate of 170 beats per minute is predicted from the power-heart rate relationship.

As discussed, unmonitored submaximal exercise tests should not be assumed acceptable for testing older or symptomatic patients. Problems can and do occur at any level of exercise. In addition, these deceptively simple unmonitored tests can unwittingly become supramaximal exercise tests for older or deconditioned subjects.

Tests of High-Intensity Exercise Capacity

High-intensity exercise tests are often called *anaerobic power tests*. Although intense exercise relies heavily on glycolysis (an anaerobic metabolic pathway), the use of glycolysis during intense exercise is due to motor unit selection rather than to hypoxemia. Therefore, it is more correct to refer to these tests as *high-intensity exercise tests*. This term avoids linking this type of exercise with a dubious physiological concept. The three most common high-intensity tests are the vertical jump, the Margaria-Kalaman test, and the Wingate test.

Problems common to high-intensity exercise tests include:

- Motivational problems in the subjects
- Using suboptimal resistance when calculating power output
- Measuring mean power rather than peak power
- Including submaximum power measurement in estimation of maximum power
- Limitations in the exercise device

For these reasons, more than one test should probably be used when assessing high-intensity exercise capacity. Such tests should be used only as an aid in prescribing exercise or assessing fitness—they should not be considered the final answer. In addition, choose a power test that uses a movement close to that used in an athlete's sport.

Vertical Jump   The vertical jump is widely used as a test of maximal leg power. It is highly related to the maximum leg force that can be generated while jumping from a force platform (a device that measures force as you push against it). It is important that the technique used during the test be the standardized version. The test must be performed exactly as it is described in the test instructions. For example, because swinging the arms and hands improves performance, they must not be moved during the test.

Leg power can be calculated from vertical jump using equation 27-10:

$$\text{Leg power (kgm/second)} = 2.21 \times \text{Weight (kg)} \times \sqrt{\text{Vertical jump (m)}} \qquad (27\text{-}10)$$

Norms for vertical jump by age and gender are shown in Table 27-14.

TABLE 27-14

Age Group and Gender Classifications for Leg Power (kgm/second) from Vertical Jump

Age	Excellent	Very Good	Good	Fair	Needs Improvement
15–19					
Male	≥ 104	88–103	73–87	61–72	≤ 60
Female	≥ 74	67–73	58–66	51–57	≤ 50
20–29					
Male	≥ 121	102–120	89–101	74–88	≤ 73
Female	≥ 78	65–77	56–64	52–55	≤ 51
30–39					
Male	≥ 120	102–119	87–101	70–86	≤ 69
Female	≥ 74	64–73	56–63	51–55	≤ 50
40–49					
Male	≥ 113	96–112	81–95	73–80	≤ 72
Female	≥ 72	60–71	56–59	52–55	≤ 51
50–59					
Male	≥ 105	93–104	76–92	68–75	≤ 67
Female	≥ 71	63–70	57–62	54–56	≤ 53
60–69					
Male	≥ 98	84–97	75–83	67–74	≤ 66
Female	≥ 64	56–63	53–55	49–52	≤ 48

Source: Canadian Society for Exercise Physiology. *The Canadian Physical Activity, Fitness & Lifestyle Appraisal*, 1996. Used with permission.

	TABLE 27-15			
	Young Adult (Ages 18–25) Norms for the Wingate Anaerobic Test			
	Males		**Females**	
Classification	Peak Power (watts/kg)	Mean Power (watts/kg)	Peak Power (watts/kg)	Mean Power (watts/kg)
Very poor	5.4–6.8	5.1–6.0	6.3–7.3	4.3–4.9
Poor	6.8–7.5	6.0–6.4	7.3–7.8	4.9–5.2
Below average	7.5–8.2	6.4–6.9	7.8–8.3	5.2–5.5
Average	8.2–8.8	6.9–7.3	8.3–8.8	5.5–5.8
Good	8.8–9.5	7.3–7.7	8.8–9.3	5.8–6.1
Very good	9.5–10.2	7.7–8.2	9.3–9.8	6.1–6.4
Excellent	10.2–11.6	8.2–9.0	9.8–10.8	6.4–7.0
Elite sprinters/jumpers	11.0–12.2	8.5–9.5		
Elite rowers	11.2–12.2	9.9–10.9		

SOURCE: Inbar O., Bar-Or O., Skinner J. S. *The Wingate Anaerobic Test.* Champaign, IL: Human Kinetics, 1996. Used with permission.

Stair-Climbing Tests (Margaria-Kalaman Test) The Margaria-Kalaman test is a common method of measuring leg power. It measures maximum power output during stair climbing. Subjects run up a flight of stairs as quickly as possible, three stairs at a time. Timing devices are placed on the third and ninth stairs. Maximum power output is determined using Equation 27-11 where Distance is the distance between the third and ninth steps; 9.81 is the acceleration of gravity in meters · sec²; and Time is the time it takes to go from the third to the ninth steps.

$$\text{Power (W)} = [\text{Body weight (kg)} \times 9.81] \times [\text{Distance (m)/Time(s)}] \quad (27\text{-}11)$$

Wingate Test The Wingate test can be done on a bicycle or arm ergometer. Subjects exercise maximally for 30 seconds against a constant resistance. Resistance is set at 0.075 kp · kg⁻¹ (kp = kilopond, a measure of force). Pedal revolutions are counted during the test, and power output is calculated every three–five seconds.

Measurements during the Wingate test include peak power output, mean power for 30 seconds, and fatigue rate. Norms for adults ages 18–25 are shown in Table 27-15.

The test can be done using only a bicycle ergometer and stopwatch, but the most accurate assessment of power output is obtained when the ergometer is interfaced to a counter and computer.

Performance on the Wingate test is highly related to performance on high-intensity exercises, such as the 50-m dash, 25-m swim, and 91.5-m speed skate. It is less well related to the vertical jump, 500-m speed skate, and percent fast-twitch muscle fiber area.

■ **Exercise Prescription for Health and Fitness**

In July 1996, the U.S. Surgeon General's office issued the landmark *Report on Physical Activity and Health.* It stated that regular, moderate activity dramatically reduces the risk of many diseases and health problems.

The report highlighted the large body of research on exercise, disease, and longevity that has been published during the past 30 years. It summarized two findings that are particularly significant to the average person.

1. **Regular physical activity** may be the single most important health practice for promoting wellness.

2. **Intense exercise** is not necessary to get most of the benefits of exercise.

Exercise and Health

The Surgeon General's report cited several major health benefits of regular exercise:

- **Lower risk of premature death.** People who exercise experience a lower death rate from all causes. The lower all-cause of death risk includes the leading killers—coronary artery disease, some types of cancer, and accidents.

- **Lower risk of coronary artery disease:** People who exercise have a lower risk of developing and dying from heart disease (see Chapter 24). Active people also have a lower risk of developing hypertension (high blood pressure), an important risk factor of heart disease and stroke. Exercise also affects other risk factors of heart disease. Regular physical activity lowers cholesterol and low density lipoproteins, raises high density lipoproteins, reduces the risk of developing non–insulin-dependent diabetes (NIDDM), makes blood platelets less sticky, and helps reduce body fat. Exercise also reduces the risk of heart disease for people who smoke cigarettes.

- **Reduced risk of some cancers.** Physical activity during work or leisure lowers the risk of colon cancer, possibly by speeding up the transit time of food through the gastrointestinal tract. Although less clear, several studies have shown a link between physical activity and a reduced risk of breast cancer and perhaps other reproductive cancers. Physical activity during high school and college years may be particularly important for preventing breast cancer during adulthood.

- **Mental health.** Exercise helps relieve symptoms of depression and anxiety and improves mood and sense of well-being. Increasingly, mental health professionals use exercise as an inexpensive way to treat patients.

- **Osteoporosis and muscle mass.** Osteoporosis is characterized by loss of bone mass. Exercise during growth and early adulthood help increase bone mass, which may be very important in postmenopausal bone loss. Physical activity may also help postmenopausal women maintain bone mass. In older people, exercise helps preserve muscle mass and movement skills which help prevent accidents and life-threatening fractures.

- **Arthritis.** Physical activity is important for maintaining joint mobility in people who have osteo- and rheumatoid arthritis. If the program is not too severe, exercise can reduce symptoms of the disease.

For more detailed discussions on the effects of exercise on selected conditions, see Chapters 24–26, 30, and 32.

Activity Recommended 30–45 Minutes a Day Recommendations for exercise and fitness by professional organizations such as ACSM and the American Heart Association have changed during the past 40 years. Participation in sports or intense exercise was once thought necessary for optimum health and fitness. Until recently, people were advised to exercise intensely (above 60 percent of maximal oxygen consumption) in activities such as jogging and cycling. Golfing, gardening, or walking were thought to provide few health benefits.

Recent studies by researchers such as Steven Blair of the Cooper Aerobics Institute in Dallas and Ralph Paffenbarger of Stanford University showed that almost any physical activity practiced 30–45 minutes a day is beneficial to health (while providing little or no contribution to fitness). Activities such as waxing the car, mowing the lawn, walking a couple of extra blocks to work, and gardening benefit health. Significantly, low-intensity exercises are beneficial, even if done in 10-minute segments. For example, a person could go for a short walk in the morning, walk up several flights of stairs at work, and go dancing at night. The goal is to be more physically active during the day. As stated in the introduction to this chapter, people receive additional benefits when they exercise more intensely or more frequently than the minimum recommendations.

Regular Exercise in the United States In spite of current knowledge of the benefits of exercise, only 10–20 percent of people in the United States do regular vigorous exercise. Over 60% of adults participate in no formal exercise program and 25% do not do any exercise at all. Half of young people ages 12–21 do not exercise vigorously on a regular basis. Among high school students, participation in daily physical education classes decreased from 42% in 1991 to 25% in 1995. The problem is most severe among young women and gets worse in both genders throughout the teen years. By college age, young men and women approach the exercise levels of the adult population.

In 1990, the federal government issued its health goals for the turn of the century called "Healthy People 2000." This report set three major goals:

1. Increase the span of healthy life for Americans.

2. Reduce health disparities among Americans.

3. Achieve access to preventive services for all Americans.

To help meet these goals, the report made 300 individual and collective recommendations. Two suggestions were to increase the activity level of the average American and to increase the proportion of people who exercise vigorously. Unfortunately, the nation is not making much progress in achieving these goals. It is hoped that the Surgeon General's report, which promotes more habitual levels of low-intensity activity, will encourage people to become more active and to reap the health benefits of physical activity.

How Much Exercise is Necessary? An ideal fitness program combines a physically active lifestyle with a systematic exercise program to develop and maintain physical fitness. An ideal overall program is shown in Figure 27-15, which uses a pyra-

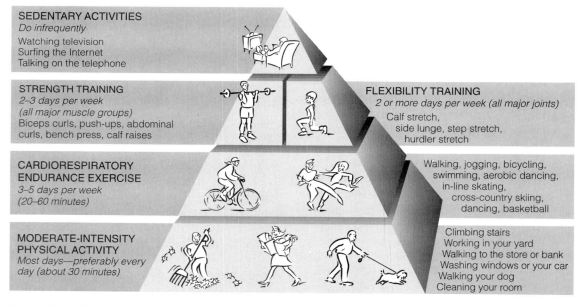

Figure 27-15 Physical activity pyramid.   Similar to the Food Guide Pyramid, this physical activity pyramid is designed to help people become more active. If you are currently sedentary, begin at the bottom of the pyramid and gradually increase the amount of moderate-intensity physical activity in your life. If you are already moderately active, begin a formal exercise program that includes cardiorespiratory endurance exercise, flexibility training, and strength training to help you develop all the health-related components of fitness. SOURCE: Fahey et al., 1999, p. 28

mid to illustrate ideal physical activity patterns. People should try to do more moderate-intensity physical activities—such as walking, gardening, stair climbing—which are described in the base of the pyramid. They should also include endurance, strength, and flexibility exercises that contribute to fitness and optimal body composition. In addition, they should cut down on sedentary activities—such as watching television and playing video games—located at the top of the pyramid.

While moderate-intensity physical activity promotes health, ideally people should include regular vigorous exercise as part of their lifestyle. The goal of the fitness-oriented program should be to increase functional capacity to at least the "good" level of physical fitness (relative to age and gender), shown in Table 27-16. Essential factors in achieving a good level of physical fitness include the type, frequency, duration, and intensity of the exercise (Table 27-17).

Type A fitness-oriented exercise program should emphasize rhythmic, continuous endurance activities such as walking, running, cycling, swimming, and cross-country skiing. "Stop-and-start" activities, such as tennis and racquetball, are acceptable if the skill level is high enough for the activity to be continuous and the intensity is vigorous. However, the personality of the person should be considered when prescribing competitive sports as exercise. Some health experts have suggested that excessive competition by a Type A personality may actually negate the positive effects of physical activity and increase the risk of coronary artery disease.

The choice of an activity depends on the individual's functional capacity and factors such as body fat percentage, previous musculoskeletal injuries, and individual preference. Running, for example, is inappropriate if physical fitness is low because the stress of running will probably result in injury. A person who dislikes swimming will be unlikely to continue the program if that type of exercise is prescribed.

Planning is important in making progressive improvements in fitness. People should have sev-

---

TABLE 27-16

Cardiorespiratory Fitness Classification

Age (years)	Maximal Oxygen Consumption (ml · kg^{-1} · min^{-1})				
	Very Low	Low	Moderate	High	Very High
Women					
Under 29	Below 24	24–30	31–37	38–48	Above 48
30–39	Below 20	20–27	28–33	34–44	Above 44
40–49	Below 17	17–23	24–30	31–41	Above 41
50–59	Below 15	15–20	21–27	28–37	Above 37
60–69	Below 13	13–17	18–23	24–34	Above 34
Men					
Under 29	Below 25	25–33	34–42	43–52	Above 52
30–39	Below 23	23–30	31–38	39–48	Above 48
40–49	Below 20	20–26	27–35	36–44	Above 44
50–59	Below 18	18–24	25–33	34–42	Above 42
60–69	Below 16	16–22	23–30	31–40	Above 40

SOURCE: Preventive Medicine Center, Palo Alto, Calif., and a survey of published sources.

TABLE 27-17

Fitness Recommendations for Healthy Adults

Cardiorespiratory Fitness and Body Composition
1. **Frequency of training:** 3–5 days per week.
2. **Intensity of training:** 55/65%—90% of maximum heart rate (HR_{max}), or 40/50%—85% of maximum oxygen uptake reserve ($\dot{V}_{O_2}R$) or HR_{max} reserve (HRR).[1]
3. **Duration of training:** 20–60 min of continuous or intermittent (minimum of 10-min bouts accumulated throughout the day) aerobic activity. Moderate-intensity activity of longer duration is recommended for adults not training for athletic competition.
4. **Mode of activity:** Continuous, rhythmical aerobic activities that use large muscle groups, such as walking, hiking, running, jogging, cycling, cross-country skiing, group exercise, rope skipping, rowing, stair climbing, swimming, skating, and endurance games.

[1] Maximum heart rate reserve (HRR) and maximum oxygen uptake reserve ($\dot{V}_{O_2}R$) are calculated from the difference between resting and maximum heart rate and resting and maximum $\dot{V}_{O_2}$, respectively. To estimate training intensity, a percentage of this value is added to the resting heart rate and/or resting $\dot{V}_{O_2}$ and is expressed as a percentage of HRR or $\dot{V}_{O_2}R$.

Muscular Strength and Endurance, Body Composition, and Flexibility
1. **Resistance training:** Practice 8–12 repetitions of at least one set of 8–10 progressive resistance exercises, 2–3 days per week. Exercises should work the major muscle groups and be of sufficient intensity to enhance strength and muscular endurance and to maintain fat-free mass.
2. **Flexibility training:** Practice flexibility exercises 2–3 days per week that are sufficient to develop and maintain range of motion in the major joints of the body.

SOURCE: Adapted from American College of Sports Medicine, 1998, pp. 975–991. Used with permission.

eral activity options available to them. For example, if a person is injured while running, that person should switch to swimming or cycling. A "fair-weather" walker should have an alternative activity for rainy days.

Frequency For most people, the ideal exercise frequency varies from three to five days per week. However, in sedentary people or heart patients who are just beginning a program, more frequent, shorter exercise training sessions are often appropriate. In these people, daily or twice daily exercise sessions of five minutes duration may be easily tolerated and lead to the most rapid rate of physiological adaptation. As fitness improves, the person can extend the time of each session until a longer and more beneficial exercise training session can be tolerated. However, excessive exercise frequency should be avoided by most people beginning a program because they will rarely adhere to a high-frequency–short-duration exercise training program (they tend to overtrain and consequently develop overuse injuries). In general, deconditioned persons should be advised to begin with three or four sessions of moderate activity per week and should progress to more frequent sessions only as fitness improves. Research has shown that people sustain more musculoskeletal injuries as the frequency of exercise increases. People must realize that their bodies will not adapt at a faster rate than they are capable. Overtraining usually results in injury rather than accelerated training gains.

Also, inflexible training schedules are inadvisable. People should avoid exercising when ill, in extreme heat or cold, or when overly tired, even if they have not completed the number of workout sessions they had planned for the week.

Duration The ideal duration of a training session for a healthy adult is between 20 and 60 minutes (continuous, or minimum of 10-min bouts accumulated during the day). Low-intensity exercise, such as walking, and "stop-and-start" sports, such as tennis and racquetball, should be practiced at least 45 minutes. As with frequency, no exercise duration is appropriate for everyone. People should be instructed to stop when they are overly tired or experience symptoms such as leg cramps, chest pains, dyspnea, or irregular heart rhythms.

Intensity Intensity is perhaps the most critical aspect of exercise prescription for developing physical fitness. If intensity is insufficient, there will be little improvement in fitness. However, if intensity is excessive, then injury may occur. For example, in a study of people who sustained a nonfatal heart attack, excessive exercise intensity was cited as a precipitating factor. More frequently, excessive exercise intensity leads to a variety of musculoskeletal injuries that necessitates a drastic reduction or cessation of training activities.

As discussed, the ideal exercise intensity for healthy adults is between 40/50%–85% of $\dot{V}_{O_2max}$ reserve, or approximately 55/65%–90% of maximum heart rate. Exercise intensity can be estimated from heart rate. Figure 27-16 shows that a given percentage of maximum heart rate does not correspond to the same percentage of $\dot{V}_{O_2max}$ (e.g., 60% HR_{max} ≈ 43% $\dot{V}_{O_2max}$). An accurate assessment of proper training intensity can be obtained using the heart rate reserve ($HR_{max} - HR_{rest}$) and the Karvonen formula for calculating target heart rate (THR):

$$THR = 0.65 \text{ to } 0.85 \cdot (HR_{max} - HR_{rest}) + HR_{rest} \quad (27\text{-}12)$$

Where 0.65 to 0.85 is a range of percentages of HR reserve (choose one for each calculation), HR_{max} is the measured or predicted maximum heart rate, and HR_{rest} is a typical resting value for resting heart rate.

Maximum heart rate can be measured or estimated. Measurement during a monitored exercise stress test is preferred because it is safer. However, in young, fit, asymptomatic people, HR_{max} can be measured after a maximum field test such as a 300–400-m sprint. This is a dangerous procedure in older, sedentary, or symptomatic populations. HR_{max} can also be estimated by subtracting a person's age from 220.

Developing Muscular Strength and Endurance   Muscular strength and endurance can be developed through resistance training. For people interested in general fitness, a recommended program includes one set of 8–12 repetitions of 8–10 exercises that condition the major muscle groups, performed 2–3 days per week. People may get more benefits from doing more than one set of each exercise. Principles of strength training were described in Chapter 20.

Flexibility   Flexibility is a type of fitness that is developed by regularly stretching the major muscle groups. People should do stretching exercises at least 2–3 days per week. Principles of flexibility training are also discussed in Chapter 20.

Healthy Body Composition   A person can promote healthy body composition by combining a sensible diet and regular exercise. Endurance exercise is best for reducing body fat; and resistance training builds muscle mass, which helps maintain metabolic rate.

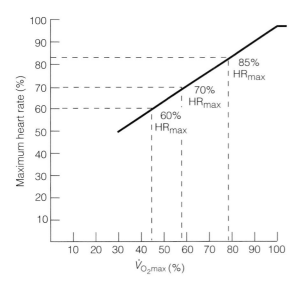

Figure 27-16   The relationship between $\dot{V}_{O_2}$ and heart rate. Notice that a given percentage of maximal heart rate is not equal to the same percentage of $\dot{V}_{O_2max}$. In addition, note that the relationship is curvilinear during heavy exercise, which introduces error when one tries to estimate metabolic rate and functional capacity from heart rate. Adapted from Pollock, Wilmore, and Fox, 1978.

Fine-Tuning the Program   Exercise prescription principles should be used only as rough guidelines because people will vary in the way they feel on a particular day. If they feel ill, they should not exercise; if they feel particularly good, then they can sometimes exceed the exercise prescription with few adverse side effects (provided they are healthy and reasonably fit).

Also, people should be told about other indicators of intensity and fatigue, such as sweating, ventilation, and perceived exertion. The basic exercise prescription should cause people to sweat during the training session, and they should understand that this is normal and desirable. The "talk test" is a rough indicator of proper exercise intensity. People should exercise at the fastest rate at which they can carry on a conversation. This places them at an intensity of approximately 60–70% of $\dot{V}_{O_2max}$. Finally, people should learn to associate a particular level of perceived exertion with the target heart rate, which will enable them to exercise more consistently at the proper intensity.

■ Exercise Prescription for the Cardiac Patient

Exercise training may be valuable for a wide variety of patients with cardiovascular disease. Patients may include those with (American Heart Association, 1995):

- Abnormal exercise test, indicating ischemia
- Angina pectoris
- Myocardial infarction
- Coronary artery bypass surgery (CABS)
- Percutaneous transluminal coronary angioplasty (PTCA)
- Heart transplant
- Cardiomyopathies
- Valvular heart disease
- Hypertension
- Pacemakers

Benefits Exercise training in people with coronary artery disease improves exercise capacity, $\dot{V}_{O_2max}$, and maximum cardiac output, and decreases myocardial oxygen uptake during submaximal exercise. There does not seem to be any formation of coronary collateral circulation (except in cases of significant ischemia during exercise), improved myocardial function, or enhanced blood flow to the myocardium (i.e., [201]thallium perfusion). Cardiac rehabilitation reduces the risk of sudden cardiac death, results in earlier discharge from the hospital, and increases the chances of returning to work. However, it does not seem to lead to a reduction in the incidence of reinfarction (i.e., additional heart attacks).

Risks Exercise is beneficial for the cardiac patient. However, the program must be well structured and prudent guidelines must be followed. The incidence of cardiac arrest during exercise in patients with coronary artery disease is approximately 1 in 80,000 to 160,000 exercise hours. The incidence is highest when the intensity of exercise is high, such as during uncontrolled jogging. Controlled, low-intensity activities have a much lower incidence of problems.

Exercise Testing and Prescription The exercise training program should be based on an exercise test that includes measurement of heart rate, blood pressure, rate-pressure product, and, ideally, oxygen consumption (during graded exercise). Patients should be evaluated for ischemia, arrhythmias, and adequate blood flow to the extremities. The last is important because, for example, claudication (i.e., intense leg pains) in these patients is sometimes caused by vascular disease in the lower extremities.

Medications such as β-blocking agents, digitalis, diuretics, vasodilators, and antiarrhythmics should be considered in the exercise prescription. Exercise tests designed to aid in the exercise prescription should be conducted while the patient is taking his or her normal medications.

Exercise testing can begin as early as one or two days prior to discharge from the hospital. The exercise prescription should be evaluated regularly. For example, the test should be repeated six weeks after the initial testing to reevaluate the prescription. The test should then be repeated yearly.

A training diary, ideally kept by the exercise leader (the person administering the exercise program), should be used to record the nature, intensity, and duration of each exercise session. Clinical studies have shown that consideration of exercise intensity is especially vital to improved capacity. In the absence of ischemia or significant arrhythmias, training should be 40–85% of $\dot{V}_{O_2max}$. Training at these intensities improves work capacity and lowers resting and submaximal exercise heart rates. Lower-intensity recreational activities generally do not appreciably improve fitness, while training at intensities greater than 75% of $\dot{V}_{O_2max}$ increases the risk of reinfarction and should be done only in a controlled environment with great care.

Myocardial oxygen consumption ($M\dot{V}_{O_2}$) is an important consideration when prescribing calisthenics and resistive exercises to cardiac patients. As discussed in Chapter 16, an acceptable indirect index of myocardial oxygen consumption is heart rate times systolic blood pressure (i.e., rate–pressure product, RPP). Exercises of the same metabolic cost can vary considerably in $\dot{V}_{O_2}$. Exercises that require use of the upper extremities, or those that induce the Valsalva maneuver should be administered with care.

Exercise should be prescribed with discretion in the presence of angina, ST segment depression of 2 mm or more, or a fall in systolic blood pressure of 20 mmHG or more. If appropriate, patients with these abnormalities should exercise at a heart rate 10 beats per minute less than the heart rate associated with the symptom. Of course, the exercise leader must appreciate individual differences. An exercise that might be easily tolerated by one patient may produce an ischemic response in another.

The amount of improvement in cardiovascular capacity in the cardiac rehabilitation program depends on the patient's adherence to the program over months and years. Adherence depends on program design, attitude of the spouse, nature of the activities, and accessibility of facilities and personnel. Reinforcers such as awareness of a reduction in percent fat or body weight, reduced heart rate and blood pressure at rest and during exercise, decreased angina, and improved working capacity

are potent motivators. Realistic short-term goals can also be very effective in maximizing adherence to the program.

There has been considerable interest in marathon training for the postcardiac patient. It has even been suggested that such training may offer immunity from coronary disease. However, because the vast majority of postcardiacs simply do not have the aerobic capacity to run a marathon safely, marathon training has definite limitations. In addition, most occupations require training in the upper extremities rather than the legs, and crossover benefits between lower and upper body exercise are minimal. Muscle specificity, therefore, must be considered in prescribing exercise for the cardiac patient who will return to vocational and avocational activities.

Cardiac patients involved in heavy exercise should take precautions and avoid unnecessary risks—over 60% of coronary patients who die do so before they get to the hospital. Care should be taken to avoid competition and exercising in adverse environments. Patients should run with a group of people trained in CPR and, ideally, in a supervised, monitored environment. A summary of basic requirements for medically supervised cardiac rehabilitation programs appears in Table 27-18.

■ Summary

An exercise training program should stress the body at a level appropriate for the individual, or injury may occur. Exercise training is not appropriate for everyone. Some medical conditions make exercise dangerous. People with significant disease risk factors should engage in an exercise program only after seeking medical advice.

Maximal oxygen consumption ($\dot{V}_{O_2max}$) is considered the best measure of cardiovascular capacity. It is measured in a laboratory but can be estimated with various field tests. Oxygen consumption at various exercise intensities during walking, running, cycling, and bench stepping can also be estimated using established equations.

The U.S. Surgeon General recommends that everyone exercise at least 30 min on most days of the

TABLE 27-18

Basic Requirements for Medically Supervised Programs for Moderate- to High-Risk Patients

Adequately ventilated and temperature-controlled space

Capability to assess patients with blood pressure and electrocardiographic analysis

ECG monitoring during initial sessions to ascertain desirable exercise levels

Supervision by either a nurse or physician in the exercise room. If a physician is not present, he or she must be immediately available (in the facility) for consultation.

Medically qualified staff (completion of an AHA-sponsored advanced cardiac life support course [or the equivalent] and a minimum of two staff members present who are trained in cardiopulmonary resuscitation)

Appropriate drugs and equipment (emergency medications [as outlined in the AHA's *Textbook of Advanced Cardiac Life Support*] and cardioverter/defibrillator)

Standard orders for the nurse if physician is not immediately available

Written procedures for the following:

Identification of conditions needed to conduct session

Management of problems that do not require hospitalization, such as acute, well-tolerated arrhythmias and neuromuscular injuries

Ruling out myocardial infarction and management of problems requiring hospitalization, including post-resuscitation problems

Management of cardiac arrest, including procedure for immediate treatment and transportation to hospital

SOURCE: American Heart Association, 1991.

week. Moderate levels of habitual activity (e.g., walking from your car to the office or mowing the lawn) promotes health but contributes little to physical fitness. The minimum exercise program for fitness promotion has been determined to be participation in an endurance-type exercise (walking, running, cycling, swimming, etc.), three to five days per week, for 20–60 minutes, at 40–85% of $\dot{V}_{O_2max}$, or 55 to 90% of maximum heart rate. Heart rate can be used to help select proper exercise intensity.

■ Selected Readings

Ahmaidi, S., K. Collomp, C. Caillaud, and C. Prefaut. Maximal and functional aerobic capacity as assessed by two graduated field methods in comparison to laboratory exercise testing in moderately trained subjects. *Int. J. Sports Med.* 13: 243–248, 1992.

Alexander, K. P., L. J. Shaw, E. R. Delong, D. B. Mark, and E. D. Peterson. Value of exercise treadmill testing in women. *J. Am. Coll. Cardiol.* 32: 1657–1664, 1998.

American College of Sports Medicine. Guidelines for Graded Exercise Testing and Exercise Prescription. Baltimore: Williams and Wilkins, 1995.

American College of Sports Medicine. Position stand on the recommended quantity and quality of exercise for developing and maintaining cardiorespiratory and muscular fitness in healthy adults. *Med. Sci. Sports Exerc.* 30: 975–991, 1998.

American College of Sports Medicine. ACSM's Exercise Management for Persons with Chronic Diseases and Disabilities, J. L. Durstine (Ed.). Champaign, Ill: Human Kinetics, 1997.

American College of Sports Medicine. ACSM's Guidelines for Exercise Testing and Prescription, W. L. Kenny (Ed.). Baltimore: Williams and Wilkins, 1998.

American College of Sports Medicine. ACSM's Resource Manual for Guidelines for Exercise Testing and Prescription, J. L. Roitman (Ed.). Baltimore: Williams and Wilkins, 1998.

American Heart Association. Exercise Standards: A Statement for Health Professionals. Dallas: American Heart Association, 1995.

Åstrand, P.-O. Principles in ergometry and their implications in sports practice. *Sports Med.* 1: 1–5, 1984.

Barnard, R. J., G. W. Gardner, N. V. Diaco, R. N. McAlpin, and A. A. Katus. Cardiovascular responses to sudden strenuous exercise: heart rate, blood pressure, and ECG. *J. Appl. Physiol.* 34: 833–837, 1973.

Bar-Or, O. The Wingate anaerobic test: an update on methodology, reliability and validity. *Sports Med.* 4: 381–394, 1987.

Barnekow-Bergkvist, M., G. Hedberg, U. Janlert, and E. Jansson. Prediction of physical fitness and physical activity level in adulthood by physical performance and physical activity in adolescence—an 18-year follow-up study. *Scand. J. Med. Sci. Sports* 8: 299–308, 1998.

Berthon, P., N. Fellmann, M. Bedu, B. Beaune, M. Dabonneville, J. Coudert, and A. Chamoux. A 5-minute running field test as a measurement of maximal aerobic velocity. *Eur. J. Appl. Physiol.* 75: 233–238, 1997.

Billat, V., M. Faina, F. Sardella, C. Marini, F. Fanton, S. Lupo, P. Faccini, M. de Angelis, J. P. Koralsztein, and A. Dalmonte. A comparison of time to exhaustion at $\dot{V}_{O_2max}$ in elite cyclists, kayak paddlers, swimmers and runners. *Ergonomics* 39: 267–277, 1996.

Bilodeau, B., B. Roy, and M. R. Boulay. Upper-body testing of cross-country skiers. *Med. Sci. Sports Exerc.* 27: 1557–1562, 1995.

Blackie, S. P., M. S. Fairbarn, N. G. McElvaney, P. G. Wilcox, N. J. Morrison, and R. L. Pardy. Normal values and ranges for ventilation and breathing pattern at maximal exercise. *Chest* 100: 136–142, 1991.

Blair, S. N., H. W. Kohl, N. F. Gordon, and R. S. Paffenbarger. How much physical activity is good for health? *Annu. Rev. Public Health* 13: 99–126, 1992.

Blair, S. N., P. Painter, R. R. Pate, L. K. Smith, and C. B. Taylor (Eds.). Resource Manual for Guidelines for Exercise Testing and Prescription, American College of Sports Medicine. Philadelphia: Lea & Febiger, 1988.

Borg, G. A. Psychological bases of perceived exertion. *Med. Sci. Sports Exerc.* 14: 377–381, 1982.

Bruce, R. A., and T. A. DeRouen. Exercise testing as a predictor of heart disease and sudden death. *Hosp. Pract.* 14: 69–75, 1978.

Bulbulian, R., J. W. Jeong, and M. Murphy. Comparison of anaerobic components of the Wingate and Critical Power tests in males and females. *Med. Sci. Sports Exerc.* 28: 1336–1341, 1996.

Calbet, J. A., J. Chavarren, and C. Dorado. Fractional use of anaerobic capacity during a 30- and a 45-s Wingate test. *Eur. J. Appl. Physiol.* 76: 308–313, 1997.

Coleman, S. G., and T. Hale. The effect of different calculation methods of flywheel parameters on the Wingate Anaerobic Test. *Can. J. Appl. Physiol.* 23: 409–417, 1998.

Consolazio, C. F., R. E. Johnson, and L. J. Pecora. Physiological Measurements of Metabolic Functions in Man. New York: McGraw-Hill, 1963.

Cox, M. H. Exercise training programs and cardiorespiratory adaptation. *Clin. Sports Med.* 10 (1): 19–32, 1991.

Cox, M. H., D. S. Miles, and J. P. Bomze. Reporting aerobic power values using breath-by-breath technology. *Can. J. Appl. Spt. Sci.* 14: 107, 1989.

Das, S. K., and A. Dutta. Relation of speed of a mile run, maximum energy cost of running, and maximum oxygen consumption: a field study. *Br. J. Sports Med.* 29: 271–272, 1995.

DiCarlo, L. J., P. B. Sparling, M. L. Millard-Stafford, and J. C. Rupp. Peak heart rates during maximal running and swimming: implications for exercise prescription. *Int. J. Sports Med.* 12:309–312, 1991.

Donovan, C. M., and G. A. Brooks. Muscular efficiency during steady-rate exercise II: effects of walking speed on work rate. *J. Appl. Physiol.* 43: 431–439, 1977.

Dunbar, C. C., E. L. Glickman-Weiss, D. A. Bursztyn, M. Kurtich, A. Quiroz, and P. Conley. A submaximal treadmill test for developing target ratings of perceived exertion for outpatient cardiac rehabilitation. *Percept. Mot. Skills* 87: 755–759, 1998.

Fahey, T. D. Basic Weight Training for Men and Women. 4th Ed. Mountain View, Calf: Mayfield Publishing Co., 2000.

Fahey, T. D., P. M. Insel, W. T. Roth. Fit & Well. 3d Ed. Mountain View, Calf; Mayfield Publishing Co., 1999.

Falk, B., Y. Weinstein, R. Dotan, D. A. Abramson, D. Mann-Segal, J. R. Hoffman. A treadmill test of sprint running. *Scand. J. Med. Sci. Sports* 6: 259–264, 1996.

Fearon, W. F., L. Voodi, J. E. Atwood, and V. Froelicher. Should only the squeaky wheel get the grease? The prognostic significance of silent ischemia detected by exercise treadmill testing. *Am. Heart J.* 136: 759–761, 1998.

Fletcher, G. F., G. Balady, V. F. Froelicher, L. H. Hartley, W. L. Haskell, M. L. Pollock. Exercise standards: A statement for healthcare professionals from the American Heart Association. *Circulation* 91: 580–615, 1995.

Franklin, B. A., K. I. Swantek, S. L. Grais, K. S. Johnstone, S. Gordon, and G. C. Timmis. Field test estimation of maximal oxygen consumption in wheelchair users. *Arch. Phys. Med. Rehab.* 71: 574–578, 1990.

Fredriksen, P. M., F. Ingjer, W. Nystad, and E. Thaulow. Aerobic endurance testing of children and adolescents—a comparison of two treadmill-protocols. *Scand. J. Med. Sci. Sports* 8: 203–207, 1998.

Gaesser, G. A., and G. A. Brooks. Muscular efficiency during steady-rate exercise: effects of speed and work rate. *J. Appl. Physiol.* 38: 1132–1139, 1975.

George, J. D. Alternative approach to maximal exercise testing and $\dot{V}_{O_2max}$ prediction in college students. *Res. Q. Exerc. Sport* 67: 452–457, 1996.

George, J. D., G. W. Fellingham, and A. G. Fisher. A modified version of the Rockport Fitness Walking Test for college men and women. *Res. Q. Exerc. Sport* 69: 205–209, 1998.

Glass, S. C., R. G. Knowlton, and M. D. Becque. Accuracy of RPE from graded exercise to establish exercise training intensity. *Med. Sci. Sports Exerc.* 24: 1303–1307, 1992.

Glass, S. C., M. H. Whaley, and M. S. Wegner. A comparison between ratings of perceived exertion among standard test protocols and steady state running. *Int. J. Sports Med.* 12: 77–82, 1991.

Golding, L. A., C. R. Myers, and W. E. Sinning. The Y's Way to Physical Fitness. 2d Ed. Champaign, Ill.: Human Kinetics, 1989.

Granier, P., B. Mercier, J. Mercier, F. Anselme, and C. Prefaut. Aerobic and anaerobic contribution to Wingate test performance in sprint and middle-distance runners. *Eur. J. Appl. Physiol.* 70: 58–65, 1995.

Green, J. J., and A. E. Patla. Maximal aerobic power: neuromuscular and metabolic considerations. *Med. Sci. Sports Exerc.* 24: 38–46, 1992.

Green, S. Measurement of anaerobic work capacities in humans. *Sports Med.* 19: 32–42, 1995.

Hauber, C., R. L. Sharp, and W. D. Franke. Heart rate response to submaximal and maximal workloads during running and swimming. *Int. J. Sports Med.* 18: 347–353, 1997.

Holland, G. J., J. J. Hoffmann, W. Vincent, M. Mayers, and A. Caston. Treadmill vs steptreadmill ergometry. *Physician Sportsmed.* 18: 79–86, 1990.

Holmer, I. Physiology of swimming man. *Acta Physiol. Scand.* (Suppl. 407), 1974.

Honig, C. R., R. J. Connett, and T. E. J. Gayeski. O_2 transport and its interaction with metabolism: a systems view of aerobic capacity. *Med. Sci. Sports Exerc.* 24: 47–53, 1992.

Howley, E. T., D. L. Colacino, and T. C. Swensen. Factors affecting the oxygen cost of stepping on an electronic stepping ergometer. *Med. Sci. Sports Exerc.* 24: 1055–1058, 1992.

Hubert, H. B., and J. F. Fries. Predictors of physical disability after age 50. Six-year longitudinal study in a runners club and a university population. *Ann. Epidemiol.* 4: 285–294, 1994.

Huskey, T., J. L. Mayhew, T. E. Ball, and M. D. Arnold. Factors affecting anaerobic power output in the Margaria-Kalaman test. *Ergonomics* 32: 959–965, 1989.

Huonker, M., A. Schmid, S. Sorichter, A. Schmidt-Trucksab, P. Mrosek, and J. Keul. Cardiovascular differences between sedentary and wheelchair-trained subjects with paraplegia. *Med. Sci. Sports Exerc.* 30: 609–613, 1998.

Iyriboz, Y., S. Powers, J. Morrow, D. Ayers, and G. Landry. Accuracy of pulse oximeters in estimating heart rate at rest and during exercise. *Br. J. Sports Med.* 25: 162–164, 1991.

Jetté, M., K. Sidney, and G. Blümchen. Metabolic equivalents (METs) in exercise testing, exercise prescription, and evaluation of functional capacity. *Clin. Cardiol.* 13: 555–565, 1990.

Judge, J. O. Exercise programs for older persons: writing an exercise prescription. *Conn. Med.* 57: 269–275, 1993.

Kaminsky, L. A., and M. H. Whaley. Evaluation of a new standardized ramp protocol: the BSU/Bruce Ramp protocol. *J. Cardiopulm. Rehabil.* 18: 438–444, 1998.

Katch, F. I., F. N. Girandola, and V. L. Katch. The relationship of body weight on maximum oxygen uptake and heavy work endurance capacity on the bicycle ergometer. *Med. Sci. Sports* 3: 101–106, 1971.

Kligman, E. W., and E. Pepin. Prescribing physical activity for older patients. *Geriatrics* 47: 33–34, 37–44, 47, 1992.

Kline, G. M., J. P. Porcari, R. Hintermeister, P. S. Freedson, A. Ward, R. F. McCarron, J. Ross, and J. M. Rippe. Estimation of $\dot{V}_{O_2 max}$ from a one-mile track walk, gender, age, and body weight. *Med. Sci. Sports Exerc.* 19: 253–259, 1987.

Koyanagi, A., K. Yamamoto, K. Nishijima, K. Kurahara, Y. Kuroki, and W. J. Koyama. Recommendation for an exercise prescription to prevent coronary heart disease. *Med. Syst.* 17: 213–217, 1993.

Lakomy, H. K., and T. Williams. The responses of an able bodied person to wheelchair training: a case study. *Br. J. Sports Med.* 30: 236–237, 1996.

Lauer, M. S., G. S. Francis, P. M. Okin, F. J. Pashkow, C. E. Snader, and T. H. Marwick. Impaired chronotropic response to exercise stress testing as a predictor of mortality. *J. Am. Med. Assoc.* 281: 524–529, 1999.

Lavoie, J.-M., and R. R. Montpetit. Applied physiology of swimming. *Sports Med.* 3: 165–189, 1986.

Lee, C. D., S. N. Blair, and A. S. Jackson. Cardiorespiratory fitness, body composition, and all-cause and cardiovascular disease mortality in men. *Am. J. Clin. Nutr.* 69: 373–380, 1999.

Levine, G. N., and G. J. Balady. The benefits and risks of exercise training: the exercise prescription. *Adv. Intern. Med.* 38: 57–79, 1993.

Lightfoot, J. T. Can blood pressure be measured during exercise?: a review. *Sports Med.* 12: 290–301, 1991.

Lightfoot, J. T., C. Tankersley, S. A. Rowe, A. N. Freed, and S. M. Fortney. Automated blood pressure measurements during exercise. *Med. Sci. Sports Exerc.* 21: 698–707, 1989.

Loudon, J. K., P. E. Cagle, S. F. Figoni, K. L. Nau, and R. M. Klein. A submaximal all-extremity exercise test to predict maximal oxygen consumption. *Med. Sci. Sports Exerc.* 30: 1299–1303, 1998.

Mahon, A. D., P. Del Corral, C. A. Howe, G. E. Duncan, and M. L. Ray. Physiological correlates of 3-kilometer running performance in male children. *Int. J. Sports Med.* 17: 580–584, 1996.

McClenaghan, B. A., and W. Literowich. Fundamentals of computerized data acquisition in human performance laboratory. *Sports Med.* 4: 425–445, 1987.

McConell, T. R. Practical considerations in the testing of $\dot{V}_{O_2max}$ in runners. *Sports Med.* 5: 57–68, 1988.

McInnis, K. J., D. S. Bader, G. L. Pierce, and G. J. Balady. Comparison of cardiopulonary responses in obese women using ramp versus step treadmill protocols. *Am. J. Cardiol.* 83: 289–291, 1999.

McNaughton, L., P. Hall, and D. Cooley. Validation of several methods of estimating maximal oxygen uptake in young men. *Percept. Mot. Skills* 87: 575–584, 1998.

Mellerowicz, H., and V. N. Smodlaka. Ergometry: Basics of Medical Exercise Testing. Baltimore: Urban & Schwarzenberg, 1981.

Miura, H., K. Kitagawa, and T. Ishiko. Economy during a simulated laboratory test triathlon is highly related to Olympic distance triathlon. *Int. J. Sports Med.* 18: 276–280, 1997.

Mundal, R., S. E. Kjeldsen, L. Sandvik, G. Erikssen, E. Thaulow, and J. Eriksson. Exercise blood pressure predicts cardiovascular mortality in middle-aged men. *Hypertension* 24: 56–62, 1994.

Myers, J., D. Walsh, M. Sullivan, and V. Froelicher. Effect of sampling on variability and plateau in oxygen uptake. *J. Appl. Physiol.* 68: 404–410, 1990.

Nakagaichi, M., and K. Tanaka. Development of a 12-min treadmill walk test at a self-selected pace for the evaluation of cardiorespiratory fitness in adult men. *Appl. Human Sci.* 17: 281–288, 1998.

Noakes, T. D. Implications of exercise testing for prediction of athletic performance: a contemporary perspective. *Med. Sci. Sports Exerc.* 20: 319–330, 1988.

Noakes, T. D. Lore of Running. 3d. Ed. Champaign, Ill.: Leisure Press, 1991.

Okazaki, Y., K. Kodama, H. Sato, M. Kitakaze, A. Hirayama, M. Mishima, M. Hori, and M. Inoue. Attenuation of increased regional myocardial oxygen consumption during exercise as a major cause of warm-up phenomenon. *J. Am. Coll. Cardiol.* 21: 1597–1604, 1993.

Peel, C., and C. Utsey. Oxygen consumption using the K2 telemetry system and a metabolic cart. *Med. Sci. Sports Exerc.* 25: 396–400, 1993.

Pfitzinger, P. and P. S. Freedson. The reliability of lactate measurements during exercise. *Int. J. Sports Med.* 19: 349–357, 1998.

Pilote, L., F. Pashkow, J. D. Thomas, C. E. Snader, S. A. Harvey, T. H. Marwick, M. S. Lauer. Clinical yield and cost of exercise treadmill testing to screen for coronary artery disease in asymptomatic adults. *Am. J. Cardiol.* 81: 219–224, 1998.

Pollock, M. L. A comparison of four protocols for maximal treadmill stress testing. *Am. Heart J.* 92: 39–46, 1976.

Porcari, J. P., C. B. Ebbeling, A. Ward, P. S. Freedson, and J. M. Rippe. Walking for exercise testing and training. *Sports Med.* 8: 189–200, 1989.

Rady, M. Y., J. D. Edwards, E. P. Rivers, and M. Alexander. Measurement of oxygen consumption after uncomplicated acute myocardial infarction. *Chest* 104: 930–934, 1993.

Rintala, P., J. A. McCubbin, S. B. Downs, S. D. Fox. Cross validation of the 1-mile walking test for men with mental retardation. *Med. Sci. Sports Exerc.* 29: 133–137, 1997.

Saltin, B., and P.-O. Åstrand. Maximal oxygen uptake in athletes. *J. Appl. Physiol.* 23: 353–358, 1967.

Saltin, B., and S. Strange. Maximal oxygen uptake: "old" and "new" arguments for a cardiovascular limitation. *Med. Sci. Sports Exerc.* 24: 30–37, 1992.

Sheldahl, L. M., F. E. Tristani, J. E. Hastings, R. B. Wenzler, and S. G. Levandoski. Comparison of adaptations and compliance to exercise training between middle-aged and older men. *J. Am. Geriatr. Soc.* 41: 795–801, 1993.

Shephard, R. J., S. Thomas, and I. Weller. The Canadian Home Fitness Test: 1991 Update. *Sports Med.* 11: 358–366, 1991.

Smithies, M. N., B. Royston, K. Makita, K. Konieczko, and J. F. Nunn. Comparison of oxygen consumption measurements: indirect calorimetry versus the reversed Fick method. *Crit. Care Med.* 19: 1401–1406, 1991.

Stachenfeld, N. S., M. Eskenazi, G. W. Gleim, N. L. Coplan, and J. A. Nicholas. Predictive accuracy of criteria used to assess maximal oxygen consumption. *Am. Heart J.* 123: 922–925, 1992.

St. Clair Gibson, A., S. Broomhead, M. I. Lambert, and J. A. Hawley. Prediction of maximal oxygen uptake from a 20-m shuttle run as measured directly in runners and squash players. *J. Sports Sci.* 16: 331–335, 1998.

Stein, P. K., and S. H. Boutcher. The effect of participation in an exercise training program on cardiovascular reactivity in sedentary middle-aged males. *Int. J. Psychophysiol.* 13: 215–223, 1992.

Stevens, N., and K. Sykes. Aerobic fitness testing: an update. *Occup. Health* (London) 48: 436–438, 1996.

Stewart, M. J., and P. L. Padfield. Blood pressure measurement: an epitaph for the mercury sphygmomanometer? *Clin. Science* 83: 1–12, 1992.

Swain, D. P., K. S. Abernathy, C. S. Smith, S. J. Lee, and S. A. Bunn. Target heart rates for the development of cardiorespiratory fitness. *Med. Sci. Sports Exerc.* 26: 112–116, 1994.

Swain, D. P., and B. C. Leutholtz. Metabolic Calculations Simplified. Baltimore: Williams and Wilkins, 1997.

Taylor, H. L., E. R. Buskirk, and A. Henschel. Maximal oxygen intake as an objective measure of cardiorespiratory performance. *J. Appl. Physiol.* 8: 73–80, 1955.

Tropp, H., K. Samuelsson, and L. Jorfeldt. Power output for wheelchair driving on a treadmill compared with arm crank ergometry. *Br. J. Sports Med.* 31: 41–44, 1997.

U.S. Department of Health and Human Services. *Physical Activity and Health: A Report of the Surgeon General.* Department of Health and Human Services, 1996.

Weltman, A., R. L. Seip, D. Snead, J. Y. Weltman, E. M. Haskvitz, W. S. Evans, J. D. Veldhuis, and A. D. Rogol. Exercise training at and above the lactate threshold in previously untrained women. *Int. J. Sports Med.* 13: 257–263, 1992.

Williford, H. N., M. Scharff-Olson, and D. L. Blessing. Exercise prescription for women. Special considerations. *Sports Med.* 15: 299–311, 1993.

Wilmore, J. H., and D. L. Costill. Semi-automated systems approach to the assessment of oxygen uptake during exercise. *J. Appl. Physiol.* 36: 618–620, 1974.

Wilmore, J. H., J. A. Davis, and A. C. Norton. An automated system for assessing metabolic and respiratory function during exercise. *J. Appl Physiol.* 40: 619–624, 1976.

Wisloff, U., and J. Helgerud. Methods for evaluating peak oxygen uptake and anaerobic threshold in upper body of cross-country skiers. *Med. Sci. Sports Exerc.* 30: 963–970, 1998.

Wyndham, C. H., N. B. Strydom, W. P. Leary, and C. G. Williams. A comparison of methods of assessing the maximum oxygen intake. *Int. Z. Angew. Physiol. Arbets.* 22: 285–295, 1966.

Zhang, Y. Y., M. C. Johnson, II, N. Chow, and K. Wasserman. Effect of exercise testing protocol on parameters of aerobic function. *Med. Sci. Sports Exerc.* 23: 625–630, 1991.

Zhou, S., S. J. Robson, M. J. King, and A. J. Davie. Correlations between short-course triathlon performance and physiological variables determined in laboratory cycle and treadmill tests. *J. Sports Med. Phys. Fitness* 37: 122–130, 1997.

CHAPTER 28

NUTRITION AND ATHLETIC PERFORMANCE

Because diet affects metabolism during both rest and exercise, and because dietary manipulation has been used to enhance performance in several types of athletic endeavors, we now address the dietary requirements of people who are required to perform at high metabolic rates for extended periods in athletic training and competition. In recent years there has been a convergence of opinion by nutritionists and physiologists: The normal balanced diet, which is based on the consumption of complex carbohydrate foods, will enable an individual to perform at his or her optimum in both practice and competition. Further, because dietary practices are the same for athletes and those in the general population, habits learned by young athletes will serve them well throughout life. The extreme dietary manipulations performed by some athletes immediately before a competition are more likely to affect a performance negatively than positively.

■ Nutritional Practice In Athletics

Perhaps no greater mythology exists in sports than that built around the subject of nutrition. To gain an immediate competitive edge, athletes have engaged in all sorts of odd dietary habits—and if the athlete is successful, he or she will continue those habits, even if the success is really due, for example, to genetic endowment. What really allows an athlete to perform up to his or her potential in practice and competition is sound nutritional practices of what to eat, how much to eat, and when to eat. Sound nutrition combined with habitual overload training can enable athletes to achieve incredible feats. At present, there is no known foodstuff that will allow a mediocre, moderately conditioned athlete to become an Olympic champion. Fortunately, nutritionists and registered dietitians have begun to specialize in "sports nutrition." As a result, it is now possible to obtain reliable information and personal counseling on the nutrition problems of active people. In the near future, it is hoped, the American College of Sports Medicine and the American Dietetic Association will join in a certification system to formalize this effort.

To perform at its maximum, the body requires specific nutritional elements. However, those elements are available in a wide variety of foods. The body even has some flexibility in terms of interconversions of materials, such as carbohydrates substituting for amino acid energy. With regard to nutrition and sports, it is *not* true that you are what you eat, and it is *not* true that you metabolize exactly what you consume. The approach we advocate is that a balanced diet is appropriate for athletes as well as for those in the general population. A bal-

Performances such as those of Michelle Kwan require proper training and sound nutrition. PHOTO: AP/Wide World Photos.

anced diet should involve at least three meals per day, in which the daily protein content is 10 to 15% *of total energy,* carbohydrate content is 55 to 65%, and fat consumption is 30% or less. The energy should be consumed in the form of breads and whole-grain products or potatoes, dark green or yellow fresh vegetables, citrus and other fruits, as well as milk and lean meats (including fish, poultry, eggs, and cheese). The consumption of fats and refined sugar products should be held to a minimum. For a review of substrate utilization during exercise and recovery, see Chapters 5 through 8.

■ Muscle Glycogen and Dietary Carbohydrate

The diets of athletes and other physically active people should be designed to supply them with the nutrients they need. Figure 28-1 illustrates the rela-

tionship between the respiratory gas exchange ratio ($R = \dot{V}_{CO_2}/\dot{V}_{O_2}$) and relative work load. From the R, a crude estimate can be made of the relative contribution to energy yield of carbohydrates (glycogen, hexoses, and lactate) and fats. Once exercise starts, carbohydrates contribute most of the energy released to support the exercise. The relative contribution provided by carbohydrates increases as relative work load increases. Endurance training has the effect of shifting the curve slightly down and to the right, but *carbohydrate is almost always the most important fuel* for muscle exercise, especially at high intensities.

The introduction of the muscle biopsy technique in the late 1960s by Bergström, Hultman, and other Scandinavian scientists, significantly increased our understanding of exercise metabolism. In the biopsy procedure, a hollow needle is inserted into a muscle through a small incision in the skin and fascia. A stylet is then inserted into the biopsy needle and a small piece of muscle is excised within the needle. The needle is then removed and the muscle specimen is quick-frozen for biochemical assays or for microscopy. Information contained in the now-famous one-legged experiments (Figure 28-2) summarizes much of what was learned by the biopsy technique. In these experiments, two subjects pedaled the same bicycle ergometer; each subject used one leg to pedal, and the exercise continued until exhaustion. During the days following the exercise, the subjects rested and ate a high-carbohydrate diet. These one-legged and subsequent studies indicated the following:

1. Prolonged exercise of submaximal intensity can result in glycogen depletion of the active muscle.
2. Following exhausting exercise and a high-carbohydrate diet, glycogen level in the exercised muscle is restored to a higher level than before the exercise.
3. Glycogen level in the inactive muscle is little affected by the exercise and diet procedure unless blood glucose falls and epinephrine rises (Chapters 9 and 10).

The relationship between muscle glycogen content and endurance during hard but submaximal

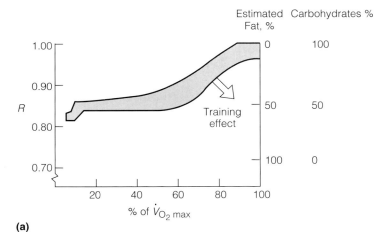

(a)

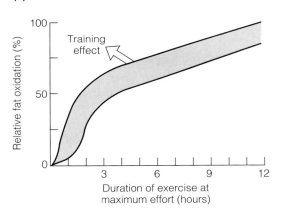

(b)

Figure 28-1   (a) Relationship between the ventilatory respiratory exchange ratio ($R = \dot{V}_{CO_2}/\dot{V}_{O_2}$) and relative exercise intensity in well-nourished individuals. Utilization of fat and carbohydrate can be estimated from the R. As relative exercise intensity increases, the proportional use of carbohydrate increases. Training shifts the curve to the right. (b) Relationship between the relative utilization of fat and the duration of exercise. The utilization of fat was estimated from the R. As exercise duration increases, the relative utilization of fat increases. Training shifts the curve to the left. [Source: (a) Based on tables of Zuntz and Schumberg (1901) and other sources.]

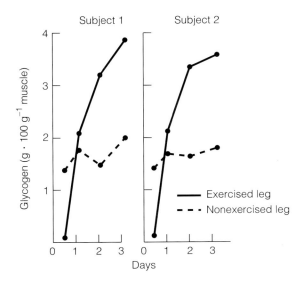

Figure 28-2   The effects of exhausting exercise and diet on quadriceps muscle glycogen content. Exhausting exercise depletes glycogen content in active muscle. Rest and a high-carbohydrate diet results in a glycogen overshoot (supercompensation) in the exercised muscle only. Source: Based on data of Bergström and Hultman, 1967.

leg cycling exercise (at about 75% $\dot{V}_{O_2max}$) is illustrated in Figure 28-3. In these experiments, muscle glycogen content was manipulated by diet and previous exercise. Subjects then attempted to maintain a given work load on a bicycle ergometer for as long as possible. Based on these studies, it is apparent that for exercises such as cycling, the amount of glycogen present when exercise begins can determine the endurance time.

Respiratory gas exchange, muscle biopsy, and other procedures have demonstrated the overwhelming importance of glycogen as the fuel for exercise. Because glycogen reserves can be influenced by diet, it is now thought that one of the major ways diet can influence training and competitive performances is through its effects on glycogen reserves. Therefore, the athlete's diet should supply the carbohydrate energy needed for hard training and competition.

The High-Carbohydrate Diet as the Norm

The process by which glycogen concentration is raised to levels two or three times greater than normal is called *glycogen supercompensation.* Glycogen supercompensation results from a program of exercise (submaximal exercise to exhaustion) followed by a high-carbohydrate diet. In the past, this procedure was called "carbohydrate (carbo) loading" and was performed as follows. Three or four days prior to competition, the athlete exercised to near exhaustion. For some sports, the procedure also included some very hard intervals (sprints) of exercise during and after a brief recovery from the prolonged exercise. The purpose of the sprints was to deplete the fast-glycolytic (Type IIx) and fast-oxidative-glycolytic (Type IIa) fibers of glycogen. The athlete then rested and tapered off by reducing training volume (see Chapter 21) and eating a high-carbohydrate (60–70%) diet during the remaining days until competition. The athlete was then considered well prepared to compete in activities requiring hard continuous exercise (e.g., cycling, distance running) or activities wherein the activity is not continuous but contains intervals of very hard exercise (e.g., soccer, American football).

Over the last 15 years, we have come to appreciate the contributions of sports nutritionists and physiologists, and so the high-carbohydrate diet of the past has now become the norm recommended for active people. Currently, questions remain over the advisability of depleting and then trying to replete glycogen to higher than normal levels. However, complex carbohydrates have nonetheless been recognized as the basis for good nutrition for athletes and the general public.

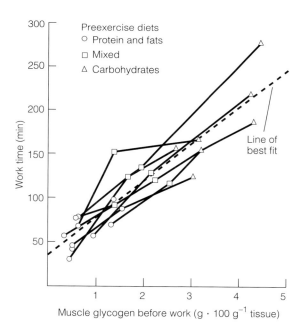

Figure 28-3 Work time on a bicycle ergometer at a set work rate for subjects on three different diets. Endurance (work time) depends in part on the muscle glycogen content before exercise. The diet several days prior to exercise affects the glycogen content of muscle, and therefore exercise endurance. SOURCE: Based on data from Bergström et al., 1967.

Problems with Carbohydrate Loading

Although the high-carbohydrate diet has come to be that recommended for athletes and the general population alike, in the more than 20 years of experience we've had with carbohydrate loading (i.e., glycogen depletion followed by repletion), several potential problems with the practice have been recognized. For instance, from the work of Sherman and associates (1981, 1982), we know that carbohydrate loading does not provide the same advantage for distance running as it does for cycling. Moreover, unschooled athletes are not always familiar with components of a "high-carbohydrate" diet and need to be educated about foods and nutrition. Athletes need to learn that a high-carbohydrate diet is one rich in whole-grain cereals, pasta, potatoes, and starchy vegetables. Conversely, refined sugar products (cakes, pies, candies) and alcohol are not complex carbohydrates and are not recommended. Further, in manipulating an athlete's diet immediately prior to competition, care should be taken to include necessary nutrients as well as energy. Too little protein or energy input will result in lean tissue wasting, and too much carbohydrate and total energy will result in fat gain. Moreover, a preliminary trial is necessary to ensure that the diet is agreeable and palatable to the athlete.

Another potential problem or asset in glycogen supercompensation, depending on the activity, is that glycogen storage requires H_2O retention. Because storage of each gram of glycogen requires almost 3 grams of H_2O, each gram in effect adds 4 grams of body weight. In some sports (e.g., sprinting, gymnastics, or wrestling), normal glycogen reserves are adequate and added weight would be a liability. For endurance exercise, however, the added H_2O is useful for preventing dehydration.

In the past, it was sometimes advocated that, during the interim between exercise to exhaustion and carbohydrate loading, a 2-day period of carbohydrate starvation be imposed as part of the carbo-loading regimen. Supposedly, this period when proteins and fats were allowed, but no sugars or starches were consumed, resulted in a heightened glycogen supercompensation effect—a "super-supercompensation." However, it has never been convincingly demonstrated that adding such a carbohydrate starvation period is beneficial. To the contrary, carbohydrate starvation after exhausting exercise is a form of malnutrition and has often been reported to produce serious side effects, including behavioral changes, ketosis, and negative nitrogen balance. Moreover, Sherman and associates (1981, 1982) have also shown that the carbohydrate starvation period is unlikely to improve either muscle glycogen storage or running performance. Consequently, inducing a period of muscle wasting, depression, and lethargy immediately prior to a competition is no longer thought to be the best way to prepare an athlete psychologically or physiologically.

Liver Glycogen and Blood Glucose

The release of glucose from the liver during exercise as the result of glycogenolysis and gluconeogenesis is very important during prolonged exercise. However, muscle glycogen is relatively more important as a carbohydrate fuel than is bloodborne glucose. On the basis of isotope tracer studies and arteriovenous difference measurements of blood glucose across working limbs and muscle biopsy studies, it has been estimated that muscle glycogen contributes three to five times as much fuel as does blood glucose during prolonged submaximal exercise. Nevertheless, although muscle glycogen is more important as a fuel than blood glucose, maintaining the fuel supply to the brain and nerve tissues is critically important, and these rely almost exclusively on glucose delivered by the blood. This means that the morning and precompetition meal should be designed to replete liver glycogen so that during subsequent activity the liver will be able to maintain blood glucose concentration, leaving muscle glycogen to fuel the exercise.

■ Amino Acid Participation in Exercise

Initial attempts to evaluate the roles of amino acids and proteins in supporting exercise were based on urinary nitrogen excretion determinations that did not take into account the substantial nitrogen loss in sweat or the complete balance of nitrogen input versus loss resulting from exercise. Hence, it was widely believed that only carbohydrates and fats were oxidized during exercise. We now know, however, that amino aids (e.g., leucine) are important for the preservation of homeostasis during exercise.

On the basis of isotope tracer studies performed on experimental animals and, more recently, on humans (Chapter 8), it now appears that the oxidation of particular amino acids, including essential amino acids such as leucine, is increased in proportion to the increase in $\dot{V}_{O_2}$ during prolonged submaximal exercise (Figures 8-11 through 8-13). The contribution of amino acids in supplying fuels is not great (approximately 5 to 10%), and may seem insignificant. However, keep in mind that small metabolic differences can produce major differences in performance rankings because the performances by winners of athletic competitions are seldom superior to those of "also rans" by a margin as large as 5%.

Considering amino acids only as substrates (i.e., their deamination and oxidation) would be a mistake. Amino acids participate in a number of transamination reactions, not only among themselves but also with glycolytic and TCA-cycle intermediates (Figure 8-10). Consequences of these transfer reactions are illustrated in Figure 28-4, which shows two-dimensional radiochromatograms of rat muscle following injection of [^{14}C]lactate and -glucose after exhausting exercise. Obvious in the chromatograms is an exchange of carbon between a variety of metabolites. The amino acid alanine (a participant in the glucose–alanine cycle) plays an important role in maintaining blood glucose homeostasis (Figure 8-9). The energy returning to muscle during exercise as the result of glucose–alanine cycle activity is not great, but the effect on maintaining blood glucose level is most important.

As the result of muscular contraction, skeletal muscle produces glutamine (see Figure 28-4) in significant amounts. The carbon source for this amino acid is α-ketoglutarate (α-KG), a TCA cycle product, and the ammonia source is from the purine nucleotide cycle (Figure 8-16). In order for the TCA cycle to continue functioning [citrate cannot be formed unless oxaloacetic (OAA) is present], material must be added to the TCA cycle to compensate for the

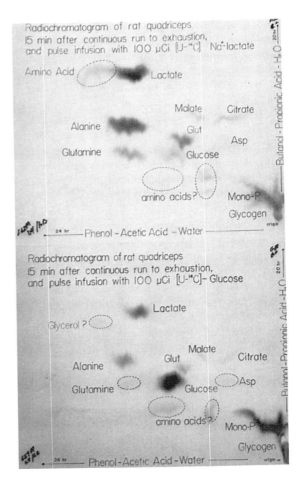

Figure 28-4 Two-dimensional radiochromatograms of rat skeletal muscle 15 minutes after exercise to exhaustion and injection with [U-^{14}C]lactate (top), and -glucose (bottom). Note the spots indicating incorporation of label from injected tracer into a variety of metabolic intermediates including amino acids (alanine, glutamate, and glutamine). These results indicate some of the complex interactions among metabolites during exercise. SOURCE: Gaesser and Brooks, 1984. Used with permission.

loss of α-KG. When material is lost from the TCA cycle, it is said to be cataplerotic; when material is added to ("fills up") the TCA cycle to sustain its function, the material is said to be anaplerotic. The anaplerotic addition of glycolytic and amino acid derivatives is probably one reason why fat cannot be utilized as the single substrate in humans and most other mammals. Further, glutamine has recently been shown to be important for preserving intestinal function during exercise and recovery. Thus, glutamine plays a role both during exercise and in postexercise nutrition.

To summarize this section, amino acids and proteins are important in the diets of active people. Amino acids are necessary to maintain lean tissue mass; they also supply some energy to sustain exercise. Moreover, amino acids fulfill a variety of other roles that are equally, or more, important than just as an energy substrate supply that can actually be accommodated by other fuel sources.

■ Lean Tissue Maintenance and Accretion (Bulking Up)

For most individuals, including those who are moderately active, a dietary protein content from 0.8 to 1.0 g of protein $\cdot$ kg^{-1} body weight $\cdot$ day^{-1} is recommended to ensure nitrogen balance (see Figure 8-4). In a normal balanced North American or European diet, this amount of protein is easily met, especially in active individuals who tend to eat more. Therefore, at present we are fairly certain that a normal balanced diet provides adequate protein for maintaining lean tissue in active individuals. However, we are uncertain of the energy and protein requirements necessary in those attempting to increase lean body mass—that is, "bulk up."

It is extremely difficult to evaluate nitrogen balance in athletes during training. Determination of nitrogen loss involves collecting and analyzing all nitrogen-containing material from the body (including sweat, feces, urine, whiskers, semen, menses, and phlegm). Despite these difficulties, however, progress has been made, especially in the area of nitrogen balance and lean tissue maintenance. For instance, studies by Gontzea and colleagues (1975) indicate that individuals on a normal recommended diet for protein content (1 g $\cdot$ kg^{-1} $\cdot$ day^{-1}) go into negative nitrogen balance when they initiate a program of exercise training, even if the diet contains adequate energy to cover the increased need at the

start of training (Figure 28-5). Therefore, when starting to train, or when increasing training intensity, it is advisable to increase protein intake transiently, being sure also to include sufficient energy to cover the increased need. However, even if extra protein is not included in the diet, the condition of negative nitrogen balance would resolve itself anyway in two weeks as the body adapts to the demands of exercise.

On the basis of experimentally determined protein requirements for heavyweight individuals engaged in resistance training, it is thought that 1 to 1.5 g $\cdot$ kg^{-1} $\cdot$ day^{-1} is adequate for maintaining lean body mass. Unfortunately, however, information on the dietary protein content required for lean tissue accretion, or "bulking up," is inadequate. Increasing lean body mass is often a requirement for success in some track and field events, American football, wrestling, and weight lifting. However, available information shows that endurance athletes need as much or more protein in their diet than resistance athletes. The testimonials of athletes who believe that their success depends on consumption of large amounts of protein and energy, and the examples of Japanese sumo wrestlers and Olympic weight lifters, suggest that further laboratory investigation is necessary before the question of protein need in bulking up is settled. For the present, it is our judgment that the interactions among energy and protein intake and nitrogen balance hold for resistance as well as endurance athletes. For example, Butterfield and associates (1987) have shown that runners given 2 g $\cdot$ kg^{-1} $\cdot$ day^{-1} were in negative nitrogen balance if energy was inadequate to meet need. Thus, high-protein intake alone is not sufficient to maintain or increase lean body mass; energy intake must also be adequate. Nevertheless, the possibility exists that consumption of high-protein, high-calorie diets coupled with extremely high resistance training can positively affect lean, if not total, body mass. In Chapter 24, the negative side effects of such practices are described. For the present, the example of sumo wrestlers, in whom overnutrition results in fat as well as lean tissue accretion, is sufficient to keep us wary of claims about the need for great amounts of protein.

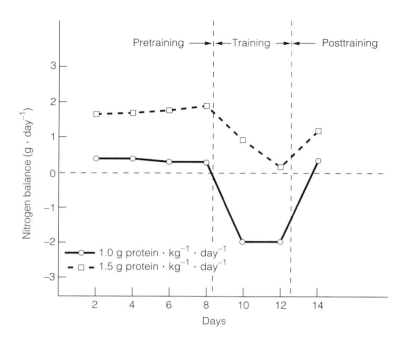

Figure 28-5 The initiation of endurance training causes individuals on a diet containing "normal" amounts of dietary protein (1.0 g protein · kg^{-1} body weight · day^{-1}) to go into negative nitrogen balance because energy was not added to the diet to cover the added need. The negative nitrogen balance condition, which occurs when training intensity increases, may be prevented by increasing energy and dietary protein content, but the condition of negative nitrogen balance resolves itself anyway in 12–14 days without increasing protein intake. SOURCE: Based on data from Gontzea et al., 1975.

■ Fat Utilization During Exercise

From Figure 28-1, as well as from similar results in Chapter 7 that describe the "crossover concept," we know that carbohydrate-derived energy sources predominate during exercise. However, bloodborne fatty acids and intramuscular triglycerides are energy sources during prolonged exercise. We also know now that intramuscular triglycerides are used in the post-exercise period (Kiens and Richter, 1998). Moreover, we know that fats are necessary in the diet both to provide energy and to absorb fat-soluble vitamins. The question then is, Should athletes consume extra fat in the diet? The answer to this question is a resounding "NO!" The recommended diet, with an emphasis on complex carbohydrate and protein food sources, necessarily contains sufficient fat. In fact, active individuals with large energy requirements find it difficult to regularly consume a diet with less than 30% fat by energy content.

Whereas it is possible to deplete glycogen reserves in prolonged difficult exercise, for practical purposes during athletics, the stores of lipid in adipose tissue are essentially inexhaustible. In addition to adipose tissue, significant amounts of lipid are also deposited in muscle and liver. On the basis of dry weight, more than twice as much energy is stored in fatty acids (9 kcal · g^{-1}) than in glycogen (4 kcal · g^{-1}). Because the storage of lipid entails far less additional water weight than glycogen does, the kilocalories stored per gram of lipid plus associated water may exceed that in glycogen by six to eight times. Consequently, storing energy in adipose tissue rather than in glycogen is far more efficient in terms of weight and space. Adipose tissue should be considered a great fuel reserve. As described in Chapter 7, however, the mobilization of energy from adipose tissue, its delivery to muscle, as well as its uptake and utilization in muscle, is a complicated and lengthy process. Furthermore, untrained subjects seldom have the circulatory capacity or the muscle enzyme activity to utilize fat as the predominant fuel during endurance exercise. Consequently, glycogen becomes the preferred fuel (Figure 28-1a). Through training, however, both circulatory capacity (Chapter 16) and muscle oxidative capacity (Chapter 6) improve, so that the curve shown in Figure 28-1a shifts slightly down and to the right.

So great has been the emphasis on a high complex carbohydrate diet for athletes in the last 25 years, as just discussed, that the notion that athletes might consider consuming a high-fat diet to improve endurance was considered untenable. However, recognizing the previous work of Phinney and associates on highly trained cyclists (1983), Muoio and Pendergast and their associates (1994) showed that runners could adapt to a high (38%) fat diet and perform well. Only a small number of athletes were studied, and respiratory gas exchange (R) data indicated that carbohydrate was the predominant fuel for exercise. Moreover, in neither study were subjects challenged to perform bursts of activity that require muscle glycogen. Therefore, to summarize thinking in the field: A well-balanced diet is important for athletic training and performance that will involve prolonged, sustained exercise as well as exercise involving brief bursts of high power. For the present, it is our opinion that the high complex carbohydrate diet appears optimal for athletes and the general population alike. However, because the human body shows great adaptability, athletes who consume high protein (20%) and fat (40%) diets can still perform well.

■ The Precompetition Meal

The timing, size, and composition of the precompetition meal are all important considerations for optimal performance. Ideally, the athlete will enter competition with neither feelings of hunger or weakness (from having eaten too little or not at all) nor feelings of fullness (from having eaten too much or too recently). Ideally also, blood glucose level will be in the high–normal range (100 mg · dl^{-1}, or 5.5 mM) and will not be falling. Blood insulin levels should be constant or falling slightly, but not rising.

It is important that athletes eat a moderate-sized meal 2.5 to 3 hours prior to competition. The nervous athlete who cannot eat, or the athlete with an unfavorable competition schedule, could be at a severe disadvantage in terms of maintaining blood glucose homeostasis. Even in well-nourished athletes, during the morning, after 8 hours of sleep, the liver will be essentially empty of glycogen, and the blood glucose level will be falling. The hepatic glucose output will be the result of protein catabolism and gluconeogenesis. Because a nutritious breakfast is necessary to stop the fall in blood glucose and begin replenishing liver glycogen content, the sleep schedule of an athlete may need to be modified so that if the athlete trains early in the morning, he or she can arise a half hour early to ensure fueling before the training begins.

The practice of carbohydrate loading has unfortunately encouraged the habit, among some athletes, of consuming large amounts of high-glycemic index foods, such as simple starches, syrups, and sugars immediately prior to competition. The glycemic index of a food is determined from the integrated rise in blood glucose following consumption of a standard amount (e.g., 50–100 g) of a particular food or metabolite. The reference base is the curve showing the rise in blood glucose following the consumption of glucose. The concept of a glycemic index was conceived by David Jenkins of Toronto, and the index provides a measure of the rapidity with which sugar enters into the blood following food consumption. Glycemic indices for particular foods are given in Table 28-1. Based on the blood glucose (glycemic) response to particular foods and based on our experience, we recommend the consumption of low-glycemic index foods before training and competition, and high-glycemic index foods immediately after.

With regard to our suggestion that low-glycemic index foods be eaten before competition, we were appalled to see Frank Shorter on television prior to the 1976 Olympic marathon in Montreal eating a heap of pancakes topped with a chocolate bar and immersed in syrup. If Frank really ate all of that and washed it down with soda pop as he pre-

TABLE 28-1

Glycemic Index of Different Foods [a]

Food	
Glucose	100
Fructose	20
Sucrose	59
Honey	87
Grain products	
Bread, white	69
Bread, wholemeal	72
Rice, white	72
Rice, brown	66
Spaghetti, white	50
Spaghetti, wholemeal	42
Cereals	
All-bran	51
Corn flakes	80
Oatmeal	49
Shredded wheat	67
Vegetables	
Beets	64
Carrots	92
Corn	59
Peas	51
Potatoes, new white	70
Potatoes, sweet	48
Legumes	
Beans, butter	36
Beans, kidney	29
Beans, navy	31
Lentils	29
Fruit	
Apples	39
Bananas	62
Oranges	40
Raisins	64
Miscellaneous	
Fish sticks	38
Peanuts	13
Potato chips	51
Sausages	28

[a] Values expressed as percentage of values observed after ingesting 50 g of glucose

SOURCE: Data from Jenkins et al., 1981.

tended, his failure to repeat his 1972 Olympic victory may be explained. Simple sugars and refined carbohydrates are digested rather rapidly, and the rush of glucose into the portal bloodstream results in a tremendous release of insulin. High portal glucose and insulin promote the desired synthesis of liver glycogen, but too much glucose escapes into the systemic circulation and causes a persistent elevation of insulin. Over time, insulin acts to clear the glucose from the circulation, but the hormone lingers, causing a continuous fall in blood glucose and an increased utilization of muscle glycogen. This is no way to be entering competition!

Instead of consuming exclusively high-glycemic index foods prior to competition, the athlete should be encouraged to continue with a normal balanced diet, determined by complex carbohydrates and proteins and resulting in lesser, but prolonged, elevations in blood glucose. The beneficial effects of a low-glycemic index meal taken prior to exercise was recently shown by DeMarco, Butterfield, and associates (1999). They compared blood glucose levels in people after they consumed low-glycemic index (LGI) meals, high-glycemic index (HGI) meals, and no meal (CON). They found better maintenance of blood glucose over time after LGI meals during leg cycling at 70% of $\dot{V}_{O_2max}$ (Figure 28-6a). Further, exercise stress—as measured by a rating of perceived exertion (RPE)—was less during exercise following LGI meals than under HGI and CON conditions (Figure 28-6c). The higher blood glucose and lower RPE following LGI meals allowed subjects to sprint 60% longer at $\dot{V}_{O_2max}$ after two hours of exercise. Notably, the insulin response to the LGI meal was less than the insulin response to the HGI meal (Figure 28-6b). During exercise, circulating insulin fell in all cases, but was similar in LGI and CON conditions.

Prior to the advent of carbohydrate loading, athletes were frequently advised to consume a meal of steak or eggs and toast or potatoes. With perhaps one exception, this precompetition meal is today considered to contain too much fat and to be digested only slowly. The exception to substituting a high-energy, high-fat, and high-protein meal for a balanced meal prior to competition involves the

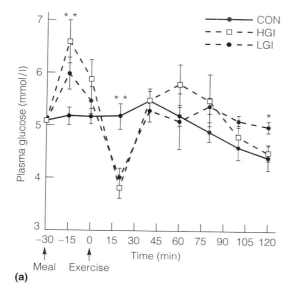

(a)

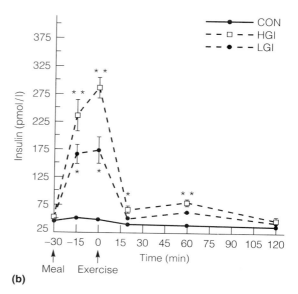

(b)

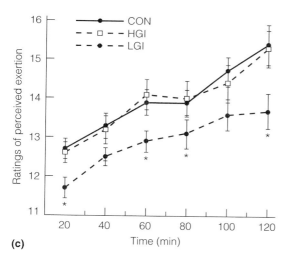

(c)

Figure 28-6 Mean ± SEM plasma glucose (a) and insulin levels (b) before and during exercise at 70% of V_{O_2max} following isoenergetic low (LGI) and high (HGI) glycemic index meals as well as control (CON) meals taken 30 minutes prior to exercise. A corresponding rating of perceived exertion (RPE) is shown in (c). * Indicates LGI significantly different from HGI and CON; ** indicates LGI significantly different from HGI and CON ($p < 0.05$). Over time, the low-glycemic index meal produced the best results in terms of supporting blood glucose and lowering perceived levels of stress. Modified from DeMarco et al., 1999. Used with permission.

necessity of timing and the availability of alternatives. If travel and other requirements of the competitive schedule require that a single meal be consumed hours before competition and, further, that the meal "hold on" for most of a day, then an energy-dense, high-protein–high-fat meal could be taken. Again, though, the admonition would be to emphasize protein energy sources that present minimal problems with digestibility.

■ Water in the Diet

During heavy exercise in the heat, athletes can easily lose several liters of sweat per hour. Added to sweat losses in training are water losses in urine, feces, and breathing (i.e., the so-called insensible water losses). Therefore, the diet should contain approximately 4 liters of water and other fluids per day.

Because thirst is *not* always an adequate mechanism to ensure proper rehydration, athletes need to form the habit of weighing themselves daily prior to training. Athletes can easily lose 1–2% of body weight from the previous day's training, which can result in decreased performance and the increased possibility of heat injury. Therefore, each athlete needs to record body weight before training, and coaches need to be *aware* of decreases in weight that are attributable to dehydration. Athletes who are down 1–2% in body weight from the previous day can be assumed to be dehydrated. They should be rehydrated and should train lightly until body weight is restored. In this regard, Nadel and associates (Takamata et al., 1994) have shown that the consumption of dilute aqueous solutions containing

4–8% glucose and 20 mEq · liter^{-1} of Na$^+$, such as in some commercially available sports drinks, may promote rehydration more effectively than water alone by maintaining the thirst mechanism longer. Further, such drinks may be more palatable than water alone. The overriding concern is to drink adequate volumes of fluid.

■ Fluid, Energy, and Electrolyte Ingestion During and After Exercise

The consumption of dilute aqueous solutions containing electrolytes, simple sugars, and carbohydrates during exercise may enhance performance (Figure 28-7). When dilute aqueous solutions are

Figure 28-7 Ingestion of aqueous glucose solutions during prolonged exercise can result in greater work output and endurance. SOURCE: Based on data from Ivy et al., 1979.

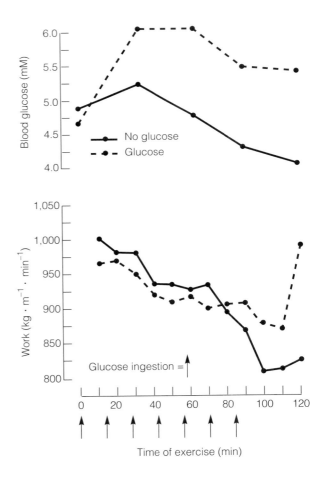

taken during exercise, the fuel can enter the circulation without eliciting a large insulin response. Thus, the glucose fuel becomes available to working muscle, and "normal" blood glucose levels are maintained to benefit other glucose-requiring tissues such as the brain. As discussed in Chapter 22, glucose should be taken in dilute aqueous solution to maximize its rate of entry into the blood. Today, some commercial sports drinks also contain small amounts of amino acids such as glutamine and arginine to carry other metabolites (e.g., lactate) to enhance intestinal function.

Immediately after prolonged training, athletes can be dehydrated as well as mildly hypoglycemic, especially if they have delayed or skipped a meal in order to train. Even though muscle glucose uptake and glycogen synthesis is high after exercise, in the glycogen-depleted, high-cortisol post-exercise state, lean tissue is degraded to supply precursors for gluconeogenesis. Therefore, as soon as possible after training, an athlete should take steps to reverse this catabolic condition. Fluid-electrolyte-energy drinks such as CYTOMAX$_R$ are ideal for this purpose.

CYTOMAX$_R$ and other fluid-electrolyte-energy (sports) drinks such as PERFORM$_R$ are blended to hasten the entry of fluids, energy-containing metabolites, and electrolytes into the blood. Sports drinks should be consumed on an empty stomach during and immediately after exercise in order to initiate the restoration of blood glucose and liver and muscle glycogen. For instance, a swimmer completing training at 7:45 A.M. and then heading for class or work should have a 500-ml bottle on the pool deck and start consuming the sports drink as he or she leaves the deck for the shower. Then, as the stomach empties, 30 minutes later, the athlete would benefit from consuming some solid foods with higher-energy densities than those in the fluid-electrolyte-energy drinks.

■ The Normal Balanced Diet

It is very difficult to make specific dietary recommendations for the general population because individual dietary needs can be quite unique. Periods of growth and physical activity, for example, require extra energy and protein intake; for older individuals, the dietary essentials should include less energy. During unique physiological situations such as pregnancy, proper nutrition becomes especially important. Dietary recommendations are difficult to make also because our knowledge of essential dietary constituents, as well as their proportions, is inadequate. Therefore, dietary recommendations today generalize that individuals should eat a wide variety of foods to increase the probability that all the essentials will be included.

To help the general population select from available foods to form a daily, nutritionally adequate and balanced diet, the U.S. Department of Agriculture (USDA) in 1958 classified foods into four basic groups and made recommendations as to how many items from the four groups should be consumed daily. These four food groups became known as the "Basic Four." They included: (1) milk and calcium equivalent foods, (2) meat and protein equivalent foods, (3) fruits and vegetables, and (4) grains. Today the Basic Four are considered to contain too much fat and cholesterol and not to provide the carbohydrate energy necessary for active people such as athletes.

Recognizing the need to decrease fat and increase carbohydrate consumption, nutrition scientists have recommended more fruits, vegetables, whole-grain cereals, and other complex carbohydrate food sources in the diet. The effect of these recommendations has been to "lighten up" our diet by diluting the input from meats and milk and meat products containing cholesterol and saturated fats. Fats are included as a separate group to ensure assimilation of fat-soluble vitamins. In the contemporary balanced diet, contributions toward the caloric intake (not the contribution by weight) should be approximately as follows:

Protein: 10 to 15%
Carbohydrate: 55 to 65%
Fat: 30% maximum

Meal Planning

Providing practical information to athletes and others in the general population is always a challenge,

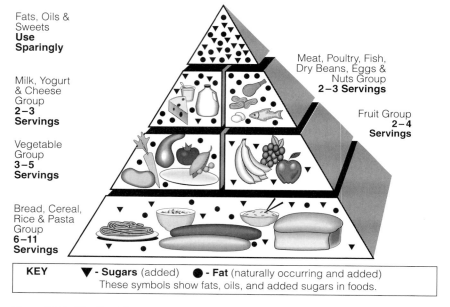

Fats, Oils &
Sweets
**Use
Sparingly**

Meat, Poultry, Fish,
Dry Beans, Eggs &
Nuts Group
2–3 Servings

Milk, Yogurt
& Cheese
Group
**2–3
Servings**

Fruit Group
**2–4
Servings**

Vegetable
Group
**3–5
Servings**

Bread, Cereal,
Rice & Pasta
Group
**6–11
Servings**

KEY	▼ - **Sugars** (added) ● - **Fat** (naturally occurring and added)
	These symbols show fats, oils, and added sugars in foods.

Figure 28-8 The Food Guide Pyramid recommended by the USDA has, as its basis, the consumption of bread, cereals, pasta, fruits, and vegetables. Milk and protein foods have lower priority, and fats and sweets are to be consumed only sparingly. SOURCE: Food Guide Pyramid, Consumer Information Center, USDA, Pueblo, Colo.

but in 1992, heeding the advice of the nutrition scientists, the U.S. Department of Agriculture developed a new model of proper nutrition, termed the "Food Guide Pyramid" (Figure 28-8). The Food Guide Pyramid illustrates the need to base dietary practices on the consumption of complex carbohydrate foods; fruits and vegetables are the next level of priority. Milk and protein foods then constitute the third level of priority, with fats and sweets at the top and no recommended serving allocated. At a practical level, adopting such a high-carbohydrate–low-fat diet will require alterations in the cooking and eating styles of many. However, in California, Peter Wood (1983) and, more recently, in Australia, Helen O'Connor and Donna Hay (1993) have shown that a fitness diet can be both relatively easy to prepare as well as extremely satisfying.

To help educate people on the composition of foods and how to select foods for a nutritious diet, various nutrition advisory boards, such as the U.S. National Dairy Council, publish tables (e.g., Table 28-2), which are very helpful in identifying the nutrients present in various foods and the physiological functions served by those foods.

■ The Athlete's Diet and Trace Elements

In eating, the athlete should be regulated by appetite once the competitive weight is achieved. Athletes should eat when hungry, but eat from the bottom of the Food Pyramid. At present, we simply do not know if active people require more of the trace dietary elements than less active people. Vitamin and mineral pills contain only what the manufacturers put into them, not what the body's unique needs are. The best way to meet calorie, protein, and other dietary requirements is to eat a broad diet to ensure purposefully that all the essentials are accidentally consumed.

TABLE 28-2

Food Sources and Physiological Functions of Nutrients

Nutrient	Important Sources of Nutrient	Some Major Physiological Functions		
		Provide Energy	Build and Maintain Body Cells	Regulate Body Processes
Protein	Meat, poultry, fish Dried beans and peas Egg Cheese Milk	Supplies 4 kcal · g^{-1}	Constitutes part of the structure of every cell, such as muscles, blood, and bone; supports growth and maintains healthy body cells	Constitutes part of enzymes, some hormones and body fluids, and antibodies that increase resistance to infection
Carbohydrate	Cereal Potatoes Dried beans Corn Bread Sugar	Supplies 4 kcal · g^{-1} Major source of energy for central nervous system	Supplies energy so protein can be used for growth and maintenance of body cells	Unrefined products supply fiber—complex carbohydrates in fruits, vegetables, and whole grains—for regular elimination Assists in fat utilization
Fat	Shortening, oil Butter, margarine Salad dressing Sausages	Supplies 9 kcal · g^{-1}	Constitutes part of the structure of every cell Supplies essential fatty acids	Provides and carries fat-soluble vitamins (A, D, E, and K)
Vitamin A (retinol)	Liver Carrots Sweet potatoes Greens Butter, margarine		Assists formation and maintenance of skin and mucous membranes that line body cavities and tracts, such as nasal passages and intestinal tract, thus increasing resistance to infection	Functions in visual processes and forms visual purple, thus promoting healthy eye tissues and eye adaptation in dim light
Vitamin C (ascorbic acid)	Broccoli Orange Grapefruit Papaya Mango Strawberries		Forms cementing substances, such as collagen, that hold body cells together, thus strengthening blood vessels, hastening healing of wounds and bones, and increasing resistance to infection	Aids absorption and utilization of iron
Thiamin (B$_1$)	Lean pork Nuts Fortified cereal products	Aids in utilization of energy		Functions as part of a coenzyme to promote the utilization of carbohydrate

(continued)

TABLE 28-2 (*continued*)

Food Sources and Physiological Functions of Nutrients

Nutrient	Important Sources of Nutrient	Some Major Physiological Functions		
		Provide Energy	Build and Maintain Body Cells	Regulate Body Processes
Thiamin (B_1) (*continued*)				Promotes normal appetite Contributes to normal functioning of nervous system
Riboflavin (B_2)	Liver Milk Yogurt Cottage cheese	Aids in utilization of energy		Functions as part of a coenzyme in the production of energy within body cells Promotes healthy skin, eyes, and clear vision
Niacin	Liver Meat, poultry, fish Peanuts Fortified cereal products	Aids in utilization of energy		Functions as part of a coenzyme in fat synthesis, tissue respiration, and utilization of carbohydrates Promotes healthy skin, nerves, and digestive tract Aids digestion and fosters normal appetite
Calcium	Milk, yogurt Cheese Sardines and salmon with bones Collard, kale, mustard, and turnip greens		Combines with other minerals within a protein framework to give structure and strength to bones and teeth Minimizes risk of osteoporosis later in life	Assists in blood clotting Functions in normal muscle contraction and relaxation, and normal nerve transmission
Iron	Enriched farina Prune juice Liver Dried beans and peas Red meat	Aids in utilization of energy	Combines with protein to form hemoglobin, the red substance in blood that carries oxygen to and carbon dioxide from the cells Prevents nutritional anemia and its accompanying fatigue Increases resistance to infection	Functions as part of enzymes involved in tissue respiration

SOURCE: National Dairy Council, 1980. Used with permission.

The number of meals consumed by an athlete is influenced not only by appetite and caloric requirements but also by the training schedule. For instance, highly competitive collegiate swimmers often train twice a day for two hours at a time (e.g., from 6 to 8 A.M. and 3 to 5 P.M.). Consequently, a light breakfast before the first workout should be followed by fluid-electrolyte-energy replacement immediately after training and a more substantial breakfast after the workout. Snacks in mid-morning and afternoon as well as late evening may even be appropriate. With a training schedule of this type, malnutrition can easily result if breakfast is consistently missed.

In Chapter 25, we saw that in active individuals, the mechanisms of appetite result in a stable body weight. Young, active athletes sometimes discover that they can eat "all they want, whenever they want" and not get fat. Unfortunately, instead of eating from the bottom of the Food Guide Pyramid, athletes often resort, for convenience, to consuming "junk foods." Junk foods are usually highly processed foods—composed primarily of sugar, salt, starch, and fat—that supply energy without providing other essential nutrients. Two sets of problems can result from consuming too much junk food. First, meeting caloric requirements with junk foods lessens the probability that essential trace elements will be provided; second, consuming junk food upsets the balance of nutrients in the diet. This may or may not be a problem to an athlete in heavy training who is consuming many other nutritional foods. Over the long term, however, the persistence of poor nutritional habits in the retired athlete could work to his or her disadvantage.

■ Cutting Weight and Special Needs of the Female Athlete

Some athletes are under the mistaken impression that body weight should be minimized for successful performance in many sports. However, the evidence that cutting weight leads to success is mostly anecdotal and lacks a scientific basis. Nevertheless, many female endurance athletes, and some males also, are concerned with body weight. They frequently approach the problem incorrectly, though.

We have already emphasized in this chapter and in Chapter 8 that the body can substitute one energy source for another. An example is the interplay between carbohydrate and protein intake. Because some female distance runners have been found to consume surprisingly little energy when training and competing, they may have the highest relative protein requirements (1.5 to 2.0 g $\cdot$ kg^{-1} $\cdot$ day^{-1}) of any class of athletes. Therefore, female athletes need to be counseled to meet energy and protein needs in the diet.

Rather than be chronically undernourished and try to compensate with a high-protein diet, active women and others concerned about body weight should be concerned with proper nutrition during the competitive season. *Dieting to lose weight is best accomplished in the off-season,* when energy demands associated with training and competition are less.

If weight loss is desirable and is accomplished in the off-season, and the diet is adjusted to provide adequate energy before and during the competitive season, the athlete avoids a number of problems. With adequate energy, training volume and intensity will be greater and recovery faster than if the diet were restricted. With more and better training, performance improvements will occur more rapidly. Moreover, other side effects of poor nutrition will be minimized. For instance, a low-energy but high-protein diet has been associated with calcium loss in both sexes. Particularly for women in hard training, there is no need to risk bone demineralization and stress fractures. Important also is the necessity for athletes to eat well to feel well—and this aspect of training, competing, and living a full life cannot be overemphasized.

Although in this chapter we have deemphasized the need for dietary supplementation in preference for the consumption of real foods, iron supplementation may be necessary for female athletes. There are several reasons for this: Women have higher iron needs than men, the iron density in the food supply is low (i.e., 6 mg $\cdot$ 1000 kcal^{-1}), and some women athletes have low energy intakes ($1800-2200$ kcal · day^{-1}). Low food consumption by women athletes can translate to dietary iron deficiency. As a consequence, heme compounds needed for O_2 transport and utilization can be in short supply, and performance in training and competition negatively affected. Moreover, it is essential that girls and young women take in 1200 mg of calcium per day. Such calcium consumption is necessary to develop sufficient bone density to minimize the possibility of osteoporosis later in life (Chapter 30).

Summary

Whether for supporting daily life or for athletic training and competition, the normal balanced diet has not been surpassed. Eating a wide variety of foods ensures that necessary energy, proteins, carbohydrates, fats, vitamins, minerals, and trace elements are included. Further, consumption of such a diet develops sound nutritional habits necessary to sustain the individual throughout his or her life.

Extreme manipulation of the diet immediately prior to competition can more likely upset or handicap the athlete than trigger outstanding performance. Two exceptions to this rule bear mention, however. First, in heavyweight athletes concerned with increasing body mass and lean body mass, the total energy and dietary protein content may have to be emphasized to increase the deposition of lean and total tissue. Second, water and fluid intake are as important as any other part of athletic nutrition. Athletes need to be weighed daily before training to ensure sufficient rehydration. Also important, fluid-energy-electrolyte drinks need to be consumed during as well as immediately after training to begin rehydration and repletion of blood glucose and tissue glycogen stores. Fluid-energy-electrolyte drinks can also be used to rehydrate chronically dehydrated individuals. The precompetition meal following carbohydrate loading should be a balanced meal, timed so that the stomach has an opportunity to empty before competition begins.

■ Selected Readings

Ahlborg, G., and P. Felig. Influence of glucose ingestion on fuel-hormone response during prolonged exercise. *J. Appl. Physiol.* 41: 683–688, 1976.

American Association for Health, Physical Education, and Recreation. Nutrition for the Athlete. Washington, D.C.: AAHPER, 1971.

Bergström, J., L. Hermansen, E. Hultman, and B. Saltin. Diet, muscle glycogen and physical performance. *Acta Physiol. Scand.* 71: 140–150, 1967.

Bergström, J., and E. Hultman. Muscle glycogen synthesis after exercise: an enhancing factor localized to the muscle cells in man. *Nature* 210: 309–310, 1966.

Bergström, J., and E. Hultman. A study of the glycogen metabolism during exercise in man. *Scand. J. Clin. Lab. Invest.* 19: 218–228, 1967.

Bergström, J., and E. Hultman. Synthesis of muscle glycogen in man after glucose and fructose infusion. *Acta Med. Scand.* 182: 93–107, 1967.

Bergström, J., and E. Hultman. Nutrition for maximal sports performance. *J. Am. Med. Assoc.* 221: 999–1006, 1972.

Bergström, J., E. Hultman, L. Jorfeldt, B. Pernow, and J. Wahren. Effect of nicotinic acid on physical working capacity and on metabolism of muscle glycogen in man. *J. Appl. Physiol.* 26: 170–177, 1969.

Buskirk, E. R. Diet and athletic performance. *Postgrad. Med.* 61: 229–236, 1977.

Buskirk, E. R. Some nutritional considerations in the conditioning of athletes. *Ann. Rev. Nutr.* 1: 319–350, 1981.

Butterfield, G. E. Whole body protein utilization in humans. *Med. Sci. Sports Exerc.* 19: 157–171, 1987.

Butterfield, G., and D. Calloway. Physical activity improves protein utilization in young men. *Brit. J. Nutr.* 51: 171–184, 1984.

Cathcart, E. P., and W. A. Burnett. Influence of muscle work on metabolism in varying conditions of diet. *Proc. Roy. Soc. (Biol.)* 99: 405, 1926.

Celejowa, I., and M. Homa. Food intake, nitrogen and energy balance in Polish weight lifters during a training camp. *Nutr. and Metab.* 12: 259–274, 1970.

Christensen, E. H., and O. Hansen. Arbeitsfähigkeit und Ehrnährung. *Sand. Arch. Physiol.* 81: 160–163, 1939.

Clark, K. L. Working with college athletes, coaches, and trainers at a major university. *Int. J. Sport Nutr.* 4: 135–141, 1994.

Cole, K. J., P. W. Grandjean, R. J. Sobszak, and J. B. Mitchell. Effect of carbohydrate composition on fluid balance, gastric emptying, and exercise performance. *Int. J. Sport Nutr.* 3: 408–417, 1993.

Conlee, R. K. Muscle glycogen and endurance exercise: a twenty-year perspective. *Exer. Sport Sci. Rev.* 15: 1–28, 1987.

Consolazio, C. F., H. L. Johnson, R. A. Dramise, and J. A. Skata. Protein metabolism during intensive physical training in the young adult. *Am. J. Clin. Nutr.* 28: 29–35, 1975.

Costill, D. L., A. Bennett, G. Brahnam, and D. Eddy. Glucose ingestion at rest and during prolonged severe exercise. *J. Appl. Physiol.* 34: 764–769, 1973.

Costill, D. L., E. Coyle, G. Dalsky, W. Evans, W. Fink, and D. Hopes. Effects of elevated plasma FFA and insulin on muscle glycogen usage during exercise. *J. Appl. Physiol.: Respirat. Environ. Exercise Physiol.* 43: 695–699, 1977.

Costill, D. L., and J. M. Miller. Nutrition for endurance sport: carbohydrate and fluid balance. *Int. J. Sports Med.* 1: 2–14, 1980.

Costill, D. L., and B. Saltin. Factors limiting gastric emptying during rest and exercise. *J. Appl. Physiol.* 37: 679–683, 1974.

Darling, R. C., R. E. Johnson, G. Pitts, R. C. Consolazio, and R. F. Robinson. Effects of variations in dietary protein on the physical well-being of men doing manual work. *J. Nutr.* 28: 273–281, 1955.

DeMarco, H. M., K. P. Sucher, C. J. Cisar, and G. E. Butterfield. Pre-exercise carbohydrate meals: application of glycemic index. *Med. Sci. Sports Exerc.* 1: 164–170, 1999.

Dengel, D. R., P. G. Weyand, D. M. Black, and K. J. Cureton. Effects of varying levels of hypohydration on ratings of perceived exertion. *Int. J. Sport Nutr.* 3: 376–386, 1993.

Dubois, E. F. A graphic representation of the respiratory quotient and the percentage of calories from protein, fat, and carbohydrate. *J. Biol. Chem.* 59: 43–49, 1924.

Durnin, J. V. G. A. Protein requirements and physical activity. In Nutrition, Physical Fitness and Health, J. Parizkova and V. A. Rogozkin (Eds.). Baltimore: University Press, 1978, pp. 53–60.

Essen, B. Intramuscular substrate utilization. *N.Y. Acad. Sci.* 301: 30–44, 1977.

The Food Guide Pyramid. Leaflet No. 572. U.S. Department of Agriculture, Human Information Service, Consumer Information Center, Department 159-Y, Pueblo, Colo. 81009, 1992.

Foster, C., D. L. Costill, and W. J. Fink. Effects of preexercise feedings on endurance performance. *Med. Sci. Sports* 11: 1–5, 1979.

Fox, E. L. Sports Physiology. Philadelphia: W. B. Saunders, 1979, pp. 242–281.

Gaesser, G. A., and G. A. Brooks. Metabolic bases of excess post-exercise oxygen consumption: a review. *Med. Sci. Sports Exerc.* 16: 29–43, 1984.

Gollnick, P. D., K. Piehl, I. V. Saubert, C. W. Armstrong, and B. Saltin. Diet, exercise, and glycogen changes in human muscle fibers. *J. Appl. Physiol.* 33: 421–425, 1972.

Gontzea, I., R. Sutzescu, and S. Dumitrache. The influence of adaptation to physical effort on nitrogen balance in man. *Nutr. Rep. Int.* 11: 231–236, 1975.

Grandjean, A. C. Practices and recommendations of sports nutritionists. *Int. J. Sport Nutr.* 3: 232–242, 1993.

Henderson, S. A., A. L. Black, and G. A. Brooks. Leucine turnover and oxidation in trained and untrained rats during rest and exercise. *Med. Sci. Sports Exerc.* 15: 98, 1983.

Hickson, R. C., M. J. Rennie, W. W. Winder, and J. O. Holloszy. Effects of increased plasma fatty acids on glycogen utilization and endurance. *J. Appl. Physiol.* 43: 829–833, 1977.

Hultman, E. Muscle glycogen in man determined in needle biopsy specimens. Method and normal values. *Scand. J. Clin. Lab. Invest.* 19: 209–217, 1967.

Ivy, J. L., D. L. Costill, W. J. Fink, and R. W. Lower. Influence of caffeine and carbohydrate feedings on endurance performance. *Med. Sci. Sports* 11: 6–11, 1979.

Jansson, E. Diet and muscle metabolism in man. *Acta Physiol. Scand.*, 487 (Suppl.): 1–24, 1980.

Jansson, E., and L. Kaijser. Effect of diet on the utilization of blood-borne and intramuscular substrate during exercise in man. *Acta Physiol. Scand.* 115: 19–30, 1982.

Jenkins, D. J. A., T. M. S. Wolever, A. L. Jenkins, R. G. Josse, and G. S. Wong. The glycaemic response to carbohydrate foods. *Lancet* 2(8399): 388–391, 18 August 1984.

Jenkins, D. J. A., T. M. S. Wolever, R. H. Taylor, H. Barker, H. Fielden, J. M. Baldwin, A. C. Bowling, H. C. Newman, A. L. Jenkins, and D. V. Goff. Glycemic index of foods: a physiological basis for carbohydrate exchange. *Am. J. Clin. Nutr.* 34: 362–366, 1981.

Jette, M., O. Pelletier, L. Parker, and J. Thoden. The nutritional and metabolic effects of a carbohydrate-rich diet in a glycogen supracompensation training regimen. *Am. J. Clin. Nutr.* 31: 2140–2148, 1978.

Karlsson, J., and B. Saltin. Diet, muscle glycogen and endurance performance. *J. Appl. Physiol.* 31: 203–206, 1971.

Kiens, B., and E. A. Richter. Utilization of skeletal muscle triacylglycerol during postexercise recovery in humans. *Am. J. Physiol.* 275: E332–E337, 1998.

Koivisto, V. A., S. L. Kavonen, and E. A. Nikkila. Carbohydrate ingestion before exercise: comparison of glucose, fructose, and sweet placebo. *J. Appl. Physiol.: Respirat. Environ. Exercise Physiol.* 51: 783–787, 1981.

Krause, M. V., and L. K. Mahan. Food, Nutrition, and Diet Therapy. Philadelphia: W. B. Saunders, 1979.

Lemon, P. W. R., and J. P. Mullin. Effect of initial muscle glycogen levels on protein catabolism during exercise. *J. Appl. Physiol.: Respirat. Environ. Exercise Physiol.* 48: 624–629, 1980.

Lindinger, M. I., L. L. Spriet, E. Hultman, T. Putman, R. S. McKelvie, L. C. Lands, N. L. Jones, and G. J. Heigenhauser. Plasma volume and ion regulation during exercise after low- and high-carbohydrate diets. *Am. J. Physiol.* 266: R1896–R1906, 1994.

Lusk, G. Analysis of the oxidation of mixtures of carbohydrate and fat. *J. Biol. Chem.* 59: 41–42, 1924.

Mayer, J., and B. Bullin. Nutrition, weight control and exercise. In Science and Medicine of Exercise and Sport, W. Johnson and E. R. Buskirk (Eds.). New York: Harper & Row, 1974.

Merkin, G. Carbohydrate loading: a dangerous practice. *J. Am. Med. Assoc.* 223: 1511–1512, 1973.

Meydani, M., W. J. Evans, G. Handelman, L. Biddle, R. A. Fielding, S. N. Meydani, J. Burrill, M. A. Fiatarone, J. B. Blumberg, and J. G. Cannon. Protective effect of vitamin E on exercise-induced oxidative damage in young and older adults. *Am. J. Physiol.* 264: R992–R998, 1993.

Morgan, W. (Ed.). Ergogenic Aids and Muscular Performance. New York: Academic Press, 1972.

Mulligan, K., and G. E. Butterfield. Discrepancies between energy intake and expenditure in physically active women. *Brit. J. Nutr.* 64: 23–36, 1990.

Muoio, D. M., J. J. Leddy, P. J. Horvath, A. B. Awad, and D. R. Pendergast. Effect of dietary fat on metabolic adjustments to maximal VO_2 and endurance in runners. *Med. Sci. Sports Exerc.* 26: 81–88, 1994.

National Dairy Council. Nutrition and human performance. *Dairy Counc. Dig.* 51: 13–17, 1980.

Nutrition for physical fitness and athletic performance for adults: position of the American Dietetic Association and the Canadian Dietetic Association. *J. Am. Diet Assoc.* 6: 691–696, 1993.

O'Connor, H., and D. Hay. A Taste of Fitness. Rushcutters Bay, NSW, Australia: J. B. Fairfax Press, 1993.

Olsson, K., and B. Saltin. Diet and fluids in training and competition. *Scand. J. Rehabil. Med.* 3: 31–38, 1971.

Page, L., and E. Phippard. Essentials of an adequate diet. Home Economics Research Report No. 3. Washington, D.C.: U.S. Department of Agriculture, 1957.

Pernow, B., and B. Saltin (Eds.). Muscle Metabolism During Exercise. New York: Plenum Press, 1971.

Phinney, S. D., B. R. Bistrian, W. J. Evans, E. Gervino, and G. L. Blackman. The human metabolic response to ketosis without caloric restriction: preservation of submaximal exercise capacity with reduced carbohydrate oxidation. *Metabolism* 32: 769–776, 1983.

Pirnay, F., M. Lacroix, F. Mosora, A. Luyckx, and P. Lefebvre. Glucose oxidation during prolonged exercise evaluated with naturally labeled [^{13}C] glucose. *J. Appl. Physiol.* 43: 258–261, 1977.

Rennie, M. J., R. H. T. Edwards, D. Halliday, C. T. M. Davies, E. E. Mathews, and D. J. Millward. Protein metabolism during exercise. In Nitrogen Metabolism in Man, J. C. Warterlow and J. M. L. Stephensen (Eds.). London: Applied Science Publishers, 1981, pp. 509–523.

Sherman, W. M., D. L. Costill, W. J. Fink, and J. M. Miller. Effect of exercise-diet manipulation on muscle glycogen and its subsequent utilization during performance. *Int. J. Sports Med.* 2: 114–118, 1981.

Sherman, W. M., M. J. Plyley, R. L. Sharp, P. J. Van Handle, R. M. McAllister, W. J. Fink, and D. L. Costill. Muscle glycogen storage and its relationship to water. *Int. J. Sports Med.* 3: 22–24, 1982.

Takamata, A., G. W. Mack, C. M. Gillen, and E. R. Nadel. Sodium appetite, thirst, and body fluid regulation in humans during rehydration without sodium replacement. *Am. J. Physiol.* R1493–R1502, 1994.

Wahren, J. Glucose turnover during exercise in man. *Ann. N.Y. Acad. Sci.* 301: 45–55, 1977.

White, T. P., and G. A. Brooks. [U-^{14}C]glucose, -alanine, and -leucine oxidation in rats at rest and two intensities of running. *Am. J. Physiol. (Endocrinol. Metab.* 3): E155–E165, 1981.

Wolever, T. M. S., and D. J. A. Jenkins. The use of the glycemic index in predicting the blood glucose response to mixed meals. *Am. J. Clin. Nutr.* 43: 167–172, 1986.

Wolever, T. M. S., D. J. A. Jenkins, A. L. Jenkins, and R. G. Josse. The glycemic index: methodology and clinical implications. *Am. J. Clin. Nutr.* 54: 846–854, 1991.

Wood, P. D. California Diet and Exercise Program. Mountain View, Calif.: Anderson World Press, 1983.

Zuntz, N. Betrachtungen über drei Beziehungen zwischen Währstoffen und Leistungen des Körpers. *Oppenheimers Handbuch der Biochemie* 4: 826, 1911.

Zuntz, N., and W. A. E. F. Schumburg. Studien zu einer Physiologie des Marches. In Sammlung von Werken aus dem Bereiche der medizinischen Wissenschaften, 6, O. Schjerning (Ed.). Berlin: Bibliotek v. Coler, 1901.

CHAPTER 29

ERGOGENIC AIDS

*E*rgogenic aids are substances or techniques used to improve athletic performance. They are employed to improve exercise capacity, enhance physiological processes, depress psychological inhibition, or provide a mechanical advantage in the performance of sports skills. The use of ergogenic aids has a long history. In ancient times, athletes and soldiers preparing for battle consumed specific animal parts to confer agility, speed, or strength associated with that animal. Today, the use of ergogenic aids is extremely widespread in athletics. Athletes take a variety of substances, such as creatine monohydrate, to improve performance. The widespread use of ergogenic aids should not be surprising. The extreme competitiveness within the sports world has forced athletes to use any substance or technique that might provide them with an edge over the competition.

The vast majority of ergogenic aids are simply placebos. A *placebo* causes an improvement through the power of suggestion—if a person believes it works, it will. The importance of placebos and the placebo effect has long been recognized in medicine and athletics. There are anecdotes, for example, of coaches administering "super pills" to athletes to improve performance, and although the pills contained only sugar, the athletes' strong belief in the substance did improve their performance.

The popularity of specific ergogenic aids usually stems from their use by top athletes. The successful Japanese swim teams of the 1930s used supplemental oxygen during competition, so the practice spread. Many champion weight-trained athletes use anabolic steroids, thus encouraging their use by younger athletes. Testimonials for vitamin and protein supplements by Mr. America or Mr. Universe in bodybuilding magazines perpetuate the use of sometimes dubious products, even when prestigious research journals present evidence that shows these products to be useless or worse.

Research on ergogenic aids is often filled with contradiction and uncertainty stemming from poor experimental controls and the use of few subjects. Often, the only information available on a specific agent comes from studies using animal models or untrained subjects, whose response to training varies widely. These studies make it difficult to extrapolate the results to athletes.

The most popular ergogenic aids are listed in Table 29-1. Although most are ineffective, their mystique continues to captivate almost everyone associated with sport. This chapter explores the most popular ergogenic aids and discusses their possible side effects.

■ Banned Substances

The International Olympic Committee (IOC) has banned the use of substances taken for the purpose

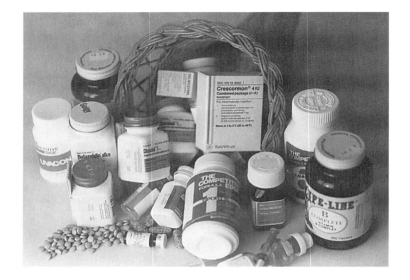

The ergogenic aids in one athlete's medicine chest more than fill a cornucopia. Athletic performances attributed to many substances and inadequate policing by sports authorities leads athletes to jeopardize their immediate and long-term health through the consumption of banned and illegal substances.

TABLE 29-1

Some Substances or Techniques Used as Ergogenic Aids

Alcohol	Inosine
Alkalies	Insulin
Amino acids	Insulin-like ˉrowth factor
Amphetamines	(IGF-1)
Anabolic steroids	Large-head tennis rackets
Aspartates	and golf clubs
Bee pollen	Low wind resistance
β-blockers	clothing
Blood doping	Marijuana
Caffeine	Massage
Camphor	Metal and plastic skis
Cocaine	Mineral supplements
Cold	Negatively ionized air
Creatine monohydrate	Nicotine
Digitalis	Nitroglycerine
Dopamine	Norepinephrine
Electrical stimulation	Organ extracts
Ephedrine	Oxygen
Epinephrine	Periactin
Erythropoietin (EPO)	Protein supplements
Fiberglass vaulting pole	Pyridoxine α-ketoglutarate
Gelatin	(PAK)
Ginseng	Strychnine
Growth hormone	Succinates
Heat	Sulfa drugs
Human chorionic	Vitamin supplements
gonadotropin	Wheat germ oil
Hypnosis	Yeast

of unfairly and artificially improving performance in competition (Table 29-2). The banned list appears to be based on an attempt to discourage any pretext of unfair competition even though the beneficial effects of many of these substances have not been demonstrated. Banned substance classifications include stimulants, anabolic steroids, narcotic analgesics (pain killers), β-blockers, restricted β-2 agonists, diuretics, and peptide and glycoprotein hormones and their analogues. Restrictions are also placed on the use of alcohol, marijuana, local anesthetics, and corticosteroids. Artificial methods of improving physiological capacity, such as blood doping, are also not permitted.

Attempts at detecting the use of banned substances in international sport have escalated from the relatively modest efforts made at the 1960 Olympics in Rome, to the multimillion dollar efforts made today at international competitions. The use of random drug testing has both curtailed the use of many drugs and forced athletes to become more sophisticated in their drug use to avoid detection.

Testing procedures involve collecting urine specimens from medal winners or randomly selected athletes and determining if any of the banned substances are present in the samples (Figure 29-1). A strict protocol for sample collection ensures fairness for all competitors. Samples are analyzed

TABLE 29-2

Categories of Substances (with examples)[a] Banned by the International Olympic Committee

I. **Doping Classes**
   A. Stimulants
      Amfepramone (Apisate, Tenuate, Tepanil)
      Amfetaminil (AN-1 [Germany])
      Amineptine (Survector [Europe])
      Amiphenazole (Dapti, Daptizole, Amphisol)
      Amphetamine (Delcobese, Obetrol, Benzedrine)
      Bemegride (Megimide)
      Benzphetamine (Didrex)
      Bromantin
      Caffeine, 12 MCG/ML
      Cathine (Norpseudoephedrine, Adiposetten N [Germany])
      Chlorphentermine (Pre Sate, Lucofen)
      Clobenzorex (Dinintel [France])
      Clorprenaline (Vortel; Asthone [Japan])
      Cocaine (Methyl-Benzoylecgonine)
      Cropropamide (Component of "Micoren")
      Crothetamide (Component of "Micoren")
      Desoxyephedrine (Vicks Inhaler)
      Diethylpropion (HCL Tenuate, Tepanil)
      Dimetamfetamine (Amphetamine)
      Ephedrine (Tedral, Bronkotabs, Rynatuss, Primatene)
      Etafedrine (Mercodal, Decapryn, Nethaprin)
      Ethamivan (Emivan, Vandid)
      Etilamfetamine (Apetinil [Netherlands])
      Fencamfamine (Envitrol, Altimine, Phencamine)
      Fenetylline (Captagon [Germany])
      Fenproporex (Antiobes Retard [Spain], Appeti-zugler [Germany])
      Furfenorex (Frugal [Argentina], Frugalan [Spain])
      Isoetharine—HCL (Bronkosol, Bronkometer, Nu-motac, Dilabron)
      Isoproterenol (Isuprel, Norisodrine, Metihaler–ISO)
      Meclofenoxate (Lucidril, Brenal)
      Mefenorex (Doracil [Argentina], Pondinil [Switzerland], Rondimen [Germany])
      Mesocarbe (Mesocarb, Mesocarbi, Sydnocarb [Europe])
      Metaproterenol (Alupent, Metaprel)
      Methamphetamine (Desoxyn, Met–Ampi)
      Methoxyphenamine (Orthoxicol Cough Syrup)
      Methyl-ephedrine (Tzbraine, Methep [Germany, United Kingdom])
      Methylphenidate HCL (Ritalin)
      Morazone (Rosimon-Neu [Germany])

      Nikethamide (Coramine)
      Pemoline (Cylert, Deltamine, Stimul)
      Pentetrazol/Pentylenetetrazol (Leptazol)
      Phendimetrazine (Phenzine, Bontril, Plegine)
      Phenmetrazine (Preludin)
      Phentermine HCL (Apidex-P, Fastin, Ionamin)
      Picrotoxine (Cocculin)
      Pipradol (Meratran, Constituent of Alertonic)
      Prolintane (Villescon, Promotil, Katovit)
      Propylhexedrine (Benzedrex Inhaler)
      Pyrovalerone (Centroton, Thymergix)
      Selegiline (Eldepryl, Plurimen [Spain], Deprenyl)
      Strychnine and Related Substances (Movellan [Germany])

   B. Restricted β-2 Agonists[b]
      Salbutamol (Albuterol, Ventolin Inhaler, Proventil Inhaler)
      Salmeterol (Serevent)
      Terbutaline (Brethaire)
      Salbutamol/Ipratropium (Combivent)

   C. Narcotic Analgesics or Pain Killers
      Alphaprodine (Nisentil)
      Anileridine (Leritine, Apodol)
      Buprenorphine (Buprenex)
      Dextromoramide (Palflum, Jetrium, D-Moramid, Dimorlin)
      Diamorphine (Heroin)
      Dipipanone (Pipadone, Diconal, Wellconal)
      Ethoheptazine (Panalgin [Italy])
      Levorphanol (Levo-Dromoran)
      Meperidine (Demorol, Mepergan)
      Methadone HCL (Dolophine, Amidon)
      Morphine (Cyclimorph 10, Duromorph, MST-Continus)
      Nalbuphine (Nubain)
      Pentazocine (Talwin)
      Pethidine (Demerol, Centralgin, Dolantin, Dolosal, Pethold)
      Phenazocine (Narphen)
      And Related Compounds
            Hydrocodone (Hycodan, Tussionex, Vicodin)
            Oxycodone (Percodan, Tylox)
            Oxymorphone (Numorphan)
            Hydromorphone (Dilaudid)
            Tincture Opium (Paragoric)

(continued)

TABLE 29-2 (continued)

Categories of Substances (with examples)[a] Banned by the International Olympic Committee

D. Anabolic Steroids
  Androstendione (Androsten)
  Bolasterone (Vebonol)
  Boldenone (Equipoise)
  Clostebol (Sternanobol)
  Dehydrochlormethyl testosterone (Turnibol)
  Dihydrotestosterone (Stanolone)
  Fluoxymesterone (Android F, Halotestin, Ora-Testryl, Ultandren)
  Mesterolone (Androviron, Proviron)
  Metandienone (Danabol, Dianabol)
  Metenolone (Primobolan, Primonabol-Depot)
  Methandrostenolone (Dianabol)
  Methyltestosterone (Android, Estratest, Methandren, Oreton, Testred)
  Nandrolone (Durabolin, Deca-Durabolin, Kabolin, Nandrobolic)
  Norethandrolone (Nilevar)
  Oxandrolone (Anavar)
  Oxymesterone (Oranabol, Theranabol)
  Oxymetholone (Anadrol, Nilevar, Anapolon 50, Adroyd)
  Stanozolol (Winstrol, Stromba)
  Testosterone (Malogen, Malogex, Delatestryl, Oreton)
  And Related Compounds:
    Danazol
    Danocrine
    DHEA
    Zeranol
    Clenbuterol
    Growth Hormone
    Human Chorionic Gonadotrophin

E. Diuretics
  Acetazolamide (Diamox, AK-ZOL, Dazamide)
  Amiloride (Midamor)
  Bendroflumethiazide (Naturetin)
  Benzthiazide (Aquatag, Exna, Hyrex, Marazide, Proaqua)
  Bumetanide (Bumex)
  Canrenone (Aldadiene, Aldactone [Germany], Phanurane [France])
  Soldactone (Switzerland)

Chlormerodrin (Orimercur [Spain])
Chlortalidone (Hygroton, Hylidone, Thalitone)
Diclofenamide (Daranide, Oratrol, Fenamide)
Ethacrynic Acid (Edecrin)
Furosemide (Lasix)
Hydrochlorothiazide (Esidrix, Hydro-Diuril, Oretic, Thiuretic)
Mannitol (IV Only, Osmitol)
Mersalyl (Salyrgan)
Spironolactone (Alatone, Aldactone)
Torsemide (Demadex)
Triamterene and related substances (Dyrenium, Dyazide)

F. Peptide and Glycoprotein Hormones and Analogues
  Chorionic Gonadotrophin (HCG-Human Chorionic Gonadotrophin)
  Corticotrophin (ACTH)
  Growth Hormone (hGH, Somatotrophin)
  Erythropoietin (EPO)

G. β Blockers
  Acebutol
  Alprenolol
  Atenolol
  Labetalol
  Metoprolol
  Nadolol
  Oxprenolol
  Propranolol
  Sotalol
  Timolol

II. Doping Methods
Blood Doping
Pharmocological, Chemical, and Physical Manipulation of the Urine
Epitestosterone ($>200\mu g \cdot ml^{-1}$)

III. Classes of Drugs Subject to Certain Restrictions
Alcohol
Local Anesthetics
Corticosteroids
Marijuana

[a]Origin of brand name used outside United States given in brackets.
[b]Permissible with prescription from a physician and written notification to U.S. Olympic Committee

Figure 29-1 Urine samples from successful athletes, such as these competitors in the Olympic 5,000-m, are scrutinized for several classes of banned substances. Federations in several countries require unannounced testing of athletes intermittently during the year. PHOTO: © Claus Andersen.

at officially sanctioned laboratories designated by the IOC.

Many sports have begun drug testing, even at local competitions. Drug tests are now common in track and field, swimming, weight lifting (power and Olympic lifting), bodybuilding, and highland games. Strict drug testing programs have been started by the National Football League, National Collegiate Athletic Association, and the National Basketball Association. There is also a trend toward "drug-free" contests in bodybuilding and weight lifting. Many smaller competitions use polygraphs (i.e., lie detector tests) rather than blood or urine tests to screen for steroid users.

Unfortunately, many banned substances are found in a variety of over-the-counter and prescribed medication. For example, banned substances are contained in some nonprescription medications, such as decongestants, throat lozenges, topical nasal decongestants, and eye drops. This is a tremendous problem for sports coaches and administrators. In one instance, accidental consumption of substances found on the list resulted in the disqualification of an Olympic gold medal winner. Consequently, many of the larger countries involved in international sports competition have engaged pharmacology consultants to help them sift through the complications involved in the consumption and detection of drugs. Guidelines for avoiding problems with doping control can be found on Internet sites of Olympic committees in various countries.

■ Anabolic–Androgenic Steroids

Anabolic steroids are drugs that resemble androgenic hormones (sometimes called male hormones), such as testosterone (Figure 29-2). Athletes consume them in the hope of gaining weight, strength, power, speed, endurance, and aggressiveness. They are widely used by athletes involved in such sports as track and field (mostly in the throwing events), weight lifting, and American football. Recent studies have shown these drugs effective in preventing muscle atrophy in AIDS patients. However, in spite of their tremendous popularity, their effectiveness is controversial. The research literature is divided on whether anabolic steroids enhance physical performance. Yet, almost all athletes who consume these substances acclaim their beneficial effects. Many athletes feel they would not have been as successful without them.

There are several possible reasons for the large differences between experimental findings and empirical observations. For example, an incredible mystique has arisen around these substances, providing fertile ground for the placebo effect. In addition, the use of anabolic steroids in the "real world" is considerably different from that in rigidly controlled, double-blind experiments (in a double-blind study, neither the subject nor the experimenter knows who is actually taking the drug and who is actually getting a placebo). Most studies, for instance, did not use the same drug dosage used by athletes because institutional safeguards prohibit administration of high dosages of possibly dangerous substances to human subjects. Also,

Generic Name (Drug name) — **Structure**

Testosterone

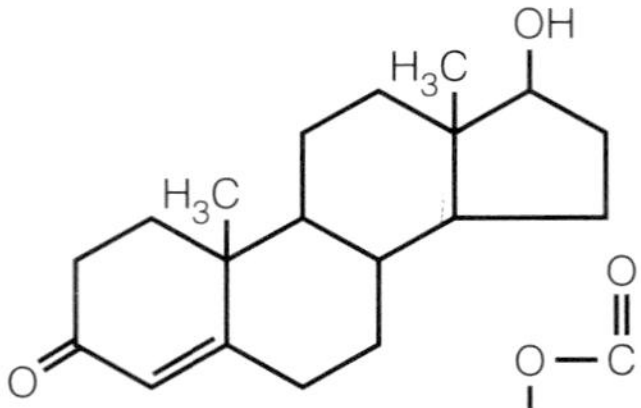

Nandrolone decanoate
(Deca-Durabolin)

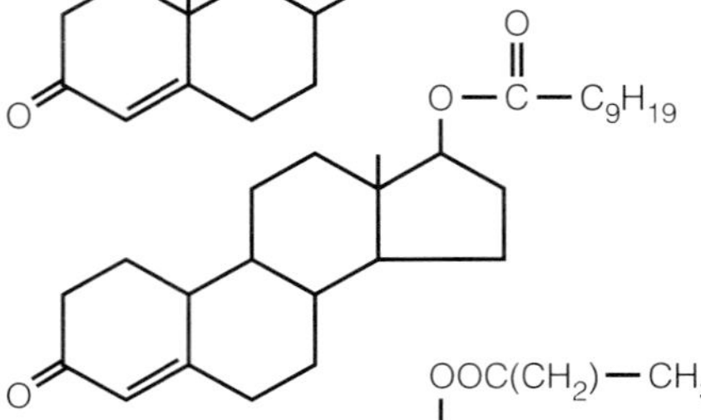

Testosterone enanthate
(Delatestryl)

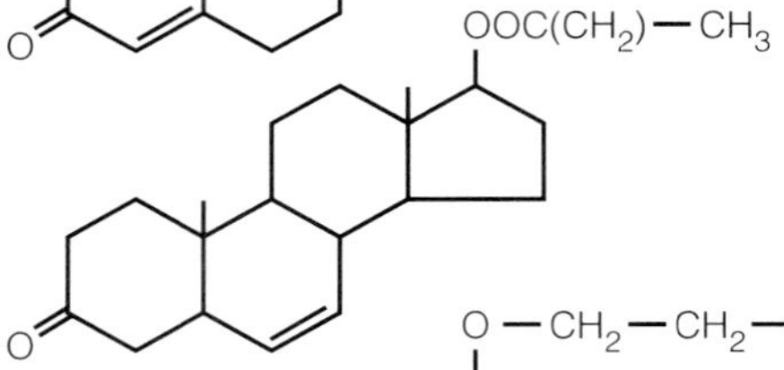

Testosterone cypionate
(Depo-Testosterone)

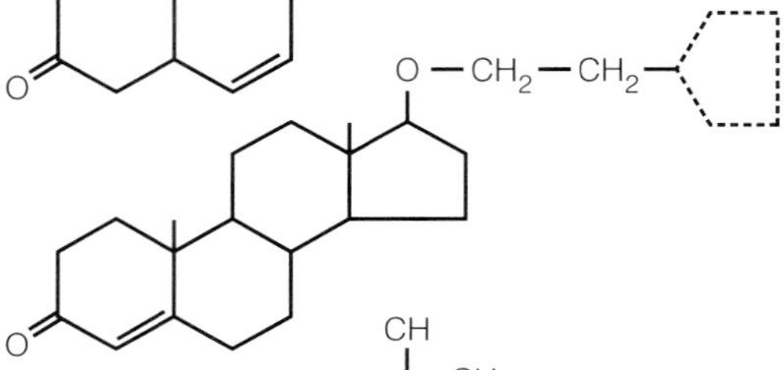

Methandrostenolone
(Dianabol)

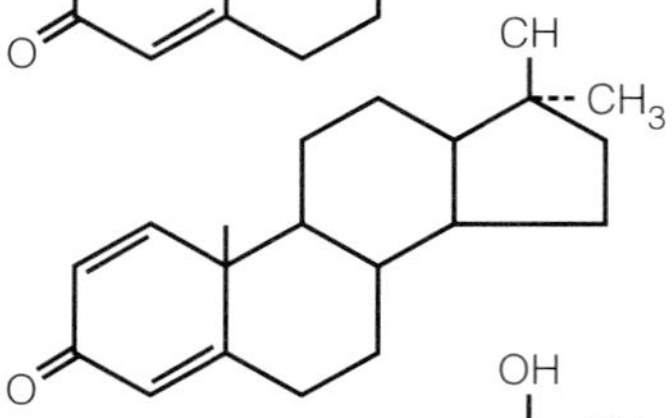

Oxandrolone
(Anavar)

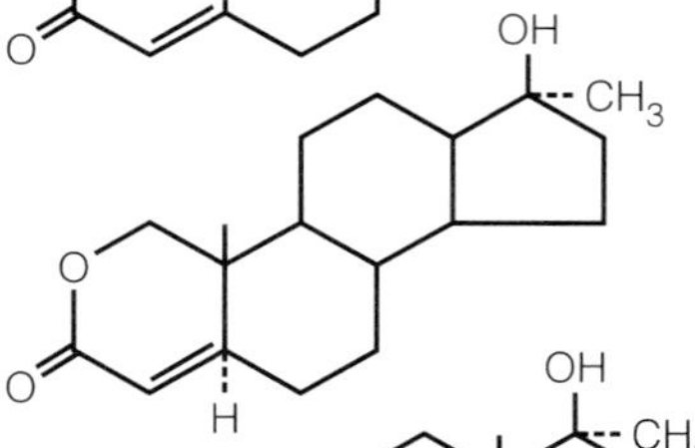

Stanozolol
(Winstrol)

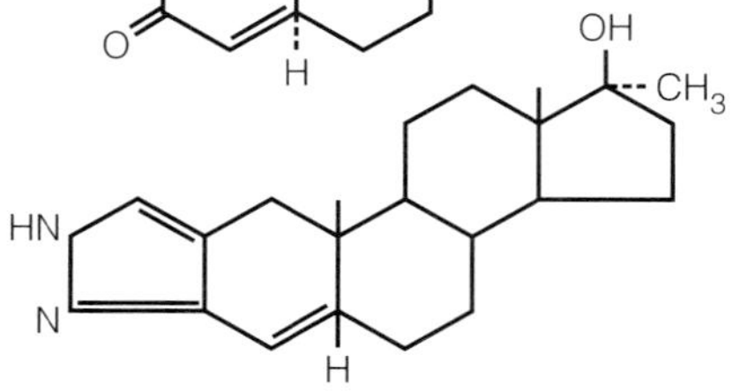

Figure 29-2 Structure of testosterone and the principal anabolic steroids used by athletes.

subjects in research experiments seldom resemble accomplished weight-trained athletes. Under these conditions, we must assess the results of research studies, as well as clinical and empirical field observations, in order to obtain a realistic profile of the use, effects on performance, and side effects of these substances.

How Anabolic Steroids Work

Male hormones, principally testosterone, are partially responsible for the tremendous developmental changes that occur during puberty and adolescence. Male hormones have androgenic and anabolic effects. *Androgenic effects* are changes in primary and secondary sexual characteristics. These include enlargement of the penis and testes; voice changes; hair growth on the face, axilla, and genital areas; and increased aggressiveness. The *anabolic effects* of androgens include accelerated growth of muscle, bone, and red blood cells, and enhanced neural conduction.

Anabolic steroids have been manufactured to enhance the anabolic properties (tissue building) of the androgens and minimize the androgenic (sex-linked) properties. However, no steroid has actually eliminated the androgenic effects because these so-called androgenic effects are really anabolic effects in sex-linked tissues. The effects of male hormones on accessory sex glands, genital hair growth, and oiliness of the skin are anabolic processes in those tissues. The steroids with the most potent anabolic effect are also those with the greatest androgenic effect.

Steroid receptors Anabolic steroids work by binding with intracellular androgen receptors (Figure 29-3). The resulting hormone-receptor complex is translocated (transferred) to the nucleus where it attaches to the nuclear chromatin. This results in the transcription of genes and results in the synthesis of specific messenger RNA molecules. These molecules then form cell proteins, such as those involved in the ribonucleic acid-polymerase system, which promotes a cellular anabolic effect.

Heavy resistance training seems to be necessary for anabolic steroids to exert any beneficial effect on physical performance. Most research studies that have demonstrated improved performance with anabolic steroids have used experienced weight lifters who were capable of training with heavier weights and producing relatively greater muscle tension during exercise than novice subjects. The effectiveness of anabolic steroids depends on unbound receptor sites in muscle, and intense strength training may increase the number of unbound recep-

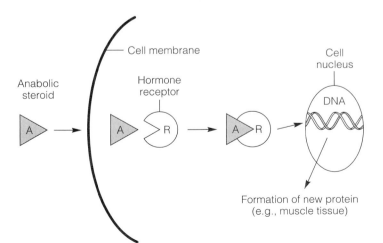

Figure 29-3 How anabolic steroids cause muscle growth.

tor sites, thus increasing the effectiveness of anabolic steroids.

Anticatabolic Effects of Anabolic Steroids

Many athletes have said that anabolic steroids help them train harder and recover faster. They have also said that they have difficulty making progress (or even holding onto the gains) when they are off the drugs. Anabolic steroids may, thus, have an anti-catabolic effect, which means that the drugs may prevent the muscle catabolism (breakdown) that of-ten accompanies intense exercise training (see Chapter 20). At present, this hypothesis has not been fully proven.

The anticatabolic effect of anabolic steroids may manifest itself by cross-binding with glucocorticoid receptors, interfering with post-exercise glucocorti-coid catabolic activity. Glucocorticoids, such as cor-tisol, are released following intense exercise and other stresses. Blocking their effects may allow ath-letes to recover faster and train harder. However, several recent studies have shown that anabolic ste-roids also have an anabolic effect in sedentary pa-tients. Therefore, the anticatabolic effect of these drugs is unlikely to be their primary mechanism of action.

Cortisol causes protein breakdown and is se-creted during exercise to enhance the use of pro-teins for fuel and to suppress the inflammation that accompanies tissue injury. Cortisol and related hor-mones, secreted by the adrenal cortex, have recep-tor sites within skeletal muscle cells. As mentioned, anabolic steroids may block the binding of cortisol to its receptor sites, which would both prevent this muscle breakdown and enhance recovery. Al-though this is beneficial while the athlete is taking the drug, the effect backfires when he or she stops taking it. Hormonal adaptations occur in response to the abnormal amount of male hormone present in the athlete's body. Although the effects are blocked, cortisol receptor sites and cortisol secretion from the adrenal cortex increase.

Also, anabolic steroid use decreases testoster-one secretion. People who stop taking steroids are also hampered with less male hormone than usual during the "off" periods. The catabolic effects of cortisol are enhanced when the athlete stops taking

the drugs, and strength and muscle size are lost at a rapid rate.

The rebound effect of cortisol and its receptors presents people who use anabolic steroids with sev-eral serious problems. (1) Because they tend to lose strength and size rapidly when off steroids, athletes may want to take the drugs for long periods of time to prevent deconditioning and falling behind. They thus become dependent on the drugs, and psycho-logical addiction is more probable. (2) Long-term administration increases the chance of serious side effects. (3) Cortisol suppresses the immune system. This makes steroid users more prone to diseases such as colds and flu during the period immediately following steroid withdrawal.

Psychological Effects

Some researchers have speculated that the real effect of anabolic steroids is to create a "psychosomatic state" characterized by sensations of well-being, euphoria, increased ag-gressiveness, and tolerance to stress that allows the athlete to train harder. Such a psychosomatic state would be more beneficial to experienced weight lift-ers, who have developed the motor skills to exert maximal force during strength training. Diets high in protein and calories may also be important in maximizing the effectiveness of anabolic steroids.

Anabolic Steroids and Performance

The effects of anabolic steroids on physical perfor-mance are unclear. Well-controlled, double-blind studies have shown conflicting results. In studies showing beneficial effects, body weight increased by an average of about 4 pounds, lean body weight by about 6 pounds (fat loss accounts for the discrep-ancy between gains in lean mass and body weight), bench press increased by about 15 pounds, and squats by about 30 pounds (these values represent the average gains for all studies showing a bene-ficial effect; Table 29-3). Almost all studies have failed to demonstrate a beneficial effect on maximal oxygen consumption or endurance capacity. Ana-bolic steroid studies have typically lasted six to eight weeks and have usually involved relatively untrained subjects.

Most changes in strength during the early part of

TABLE 29-3

Summary of the Actions of Anabolic Steroids and Their Effects on Athletic Performance

- Anabolic steroids are synthetic male hormones resembling testosterone that have been manufactured to enhance the anabolic properties of male hormones while minimizing their androgenic properties.
- Anabolic steroids work by increasing the rate of protein synthesis in target cells, slowing protein breakdown during and after exercise, and increasing aggressiveness and feelings of well-being. These effects help some athletes improve faster and recover more rapidly from intense workouts.
- Anabolic steroids are effective for increasing strength, power, muscle size, and speed in some athletes if the athlete is involved in an intense weight-training routine.
- Anabolic steroids do not improve maximal oxygen consumption but may improve endurance capacity by allowing athletes to exercise at higher percentages of maximal oxygen consumption.
- Most studies have not shown any effects of the drugs on body fat, although testosterone is known to affect fat metabolism. Food intake and physical activity in athletes may be the cause of this confusion.
- There are major differences between published research findings and popular conceptions among athletes about the effectiveness of anabolic steroids in improving athletic performance. These differences may be due to (1) the placebo effect, (2) the unrealistic use of the drugs in research studies, (3) the use of untrained test subjects, and (4) the failure of the studies to use peaking techniques.

training are neural; that is, increased strength is mainly due to an improved ability to recruit motor units (see Chapter 20). However, anabolic steroids affect processes associated with protein synthesis in muscle. Studies lasting six weeks (typical study length) would largely reflect these neural changes and could easily miss the cellular effects of the drugs.

The gains made by athletes in uncontrolled observations have been much more impressive. Weight gains of 30 or 40 pounds, coupled with 30% increases in strength, are not unusual. Although such case studies lack credibility because of the absence of scientific controls, it would be foolish to completely disregard these observations because the "subjects" have been highly trained and motivated athletes.

Side Effects of Anabolic Steroids

The principal side effects of anabolic steroids can be subdivided into those attributable to (1) the normal physiological actions of male hormones that are inappropriate in the recipient, and (2) toxic effects caused by the chemical structure of the drug (principally "C-17 alpha alkylated" oral anabolic steroids; Table 29-4).

TABLE 29-4

Major Side Effects of Anabolic Steroids

Liver toxicity
    Elevated levels of critical liver enzymes
    Increased bromsulphalein (BSP) retention
    Hepatocellular carcinoma
    Peliosis hepatis
    Cholestasis
Elevated CK and LDH
Elevated blood pressure
Edema
Alterations in clotting factors
Elevated cholesterol and triglycerides
Decreased HDL
Elevated blood glucose
Increased nervous tension
Psychosis
Altered electrolyte balance
Depressed spermatogenesis
Lowered testosterone production
Reduced gonadotropin hormone production (LH, FSH)
Increased urine volume
Premature closure of epiphyses in children
Masculinization
Increased or decreased libido
Sore nipples
Acne
Lowered voice in women and children
Clitoral enlargement in women
Increased aggressiveness
Nosebleeds
Muscle cramps and spasms
GI distress
Dizziness
Disturbed thyroid function
Wilms' tumor
Postatic hypertrophy
Prostic cancer
Increased activity of apocrine sweat glands
Increased risk of AIDS if needles shared
Depressed immune function

Effects on Endocrine Control Physiological side effects include reduced production of endogenous testosterone, pituitary gonadotropin hormones, and hypothalamic-releasing factors (all of which control testicular function and sperm cell production). Libido may be increased or decreased. In addition, the structural similarity of anabolic steroids to aldosterone causes them to increase fluid retention. Steroid use by women and immature children may cause masculine effects such as hair growth on the face and body, deepening of the voice, oily skin, increased activity of the apocrine sweat glands, acne, and baldness. In women, some of these masculine changes are irreversible. Children will initially experience accelerated maturation, followed by premature closure of the epiphyseal growth centers in the long bones. Women may also experience clitoral enlargement and menstrual irregularity.

Anabolic steroids affect the regulation of the hypothalamic-gonadotropin-testicular axis, which controls normal reproductive processes (Figure 29-4). They also suppress luteinizing hormone (LH) and follicle-stimulating hormone (FSH), resulting in decreased production of testosterone by the testes. This process can result in testicular atrophy and decreased sperm production. Although these changes reverse themselves after withdrawal from the substance, their prolonged use may permanently disturb this delicate regulatory system. During periods when they are not taking anabolic steroids, some athletes use drugs (i.e., human chorionic gonadotropin, HCG) that stimulate the natural production of testosterone.

Liver Toxicity Oral anabolic steroids, such as methandrostenolone (Dianabol), present the greatest risk of biological toxicity, particularly to the liver, because their structure is altered to make them more biologically active (Table 29-5). This increased activity causes the steroid to become concentrated in the liver much earlier and in greater quantity than the injectable varieties. Athletes using anabolic steroids often have elevated serum levels of liver enzymes such as glutamic-oxalacetic transaminase (SGOT), glutamic-pyruvic transaminase (SGPT), and alkaline phosphatase, which sometimes

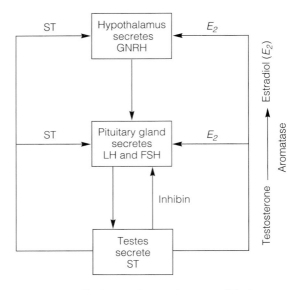

Figure 29-4 The feedback control system of the hypothalamic-pituitary-gonadal axis. Testosterone (ST) exerts independent control over the hypothalamus, which secretes gonadotrophin-releasing hormone (GNRH), and the pituitary, which secretes luteinizing hormone (LH) and follicle-stimulating hormone (FSH). The aromatization of testosterone to estradiol provides additional regulation of the hypothalamus and pituitary. Inhibin, which is produced in the testes, also controls the production of FSH.

TABLE 29-5
The Effects of 17-Alpha-Alkylated (Oral) Anabolic Steroids on Liver Function

Anabolic steroids may cause:

- Intrahepatic cholestasis (impairment of bile flow)
- Ultrastructural changes in liver canaliculi, microvilli, and mitochondria
- Biochemical changes resulting in reduced hepatic excretory function
- Vascular complications: peliosis hepatis (blood-filled cysts in the liver)
- Benign and malignant liver tumors
- Focal nodal hyperplasia

indicates liver toxicity. Elevated levels of blood glucose, creatine kinase, and bilirubin have also been noted. These changes are usually reversible upon withdrawal from the drug. Prolonged administration in some groups of patients, however, has been linked to severe liver disorders, such as peliosis hepatis, hepatocellular carcinoma, and cholestasis.

Coronary Heart Disease Anabolic steroid use may increase the risk of coronary heart disease. Anabolic steroids increase cholesterol, triglycerides, and blood pressure, and decrease levels of high-density lipoproteins (HDL; Figure 29-5). Cholesterol and triglyceride levels above 300 mg % and HDL levels of less than 10 mg % have been noted in athletes taking large doses of these steroids (respective ideal levels for cholesterol, triglycerides, and HDL are 180, 100, and 60 mg % in males). Anabolic steroid use in athletes has also been shown to increase the concentration of hepatic triglyceride lipase (an enzyme that catabolizes HDL) by 300%. High cholesterol and triglyceride levels may be par-

tially related to diet because many weight-trained athletes consume high-fat, high-cholesterol diets. However, the low HDL level seems to be directly related to the steroids. Many weight-trained athletes compete from 10 to more than 20 years, subjecting themselves to the real possibility of premature death from atherosclerosis due to steroid use. Hypertension is also characteristic, probably due to the fluid retention properties of these drugs.

In addition, severe cardiovascular side effects of anabolic steroid use have been reported in athletes, including myocardial infarction and ventricular tachycardia. Anabolic steroid use may cause myocardial infarction by inducing thrombosis and increasing vascular reactivity. An increased risk of thrombosis may occur due to altered platelet function with anabolic steroid use. Anabolic steroids have been shown to depress the nitric oxide dilator system in blood vessels. As discussed in Chapter 14, nitric oxide is an important vasodilator. As a result, anabolic steroid use could cause coronary ischemia and vasospasm that could contribute to infarction.

A number of studies have suggested that anabolic steroids may promote cardiac hypertrophy in weight-trained athletes. Testosterone is the preferred ligand of the human androgen receptor in the myocardium and directly modulates transcription, translation, and enzyme function, which promotes cardiac hypertrophy. However, strength training tends to cause cardiac hypertrophy even in the absence of anabolic steroid use. It is possible that the increased strength induced by anabolic steroid use may create a greater pressure load on the heart with a greater tendency toward cardiac hypertrophy.

Psychiatric Effects Anabolic steroids have significant psychiatric effects. They can produce dependence similar to that seen in opioid-type drugs. Some people who stop taking the drugs experience depression, fatigue, and craving for them. This effect is not well studied, however.

Many case studies of the psychiatric side effects of anabolic steroid use have appeared in the literature on the subject. Reported side effects of their use include depression, paranoia, irritability, hyperactivity, hallucinations, and grandiose delusions.

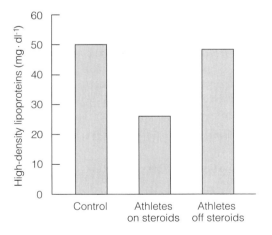

Figure 29-5 The effects of anabolic steroids on HDL-cholesterol. When athletes were on steroids, HDL was depressed. It rebounded toward control values when they went off the drugs. HDL-cholesterol is thought to provide protection against hardening of the arteries. Anabolic steroids greatly decrease the amount of HDL-cholesterol in the blood, which may increase the risk of getting heart disease. Data from Peterson and Fahey, 1984.

Anabolic steroid use has been used many times in the last 10 years as a defense in criminal cases.

Effects on the Immune System Prolonged anabolic steroids may also weaken the immune system. As discussed, these drugs are thought to block the action of corticosteroid hormones involved with the breakdown and repair process that follows a heavy workout. In reaction to this, the body increases the production of corticosteroids and their receptors. Corticosteroids are known to suppress the immune system (the system fights off diseases). Athletes going off steroids often get colds or flu because of the increase in corticosteroids. However, short-term use in AIDS patients has been shown to boost immune function.

Effects of Steroid Use on Female Athletes
Anabolic steroid use has been reported in women athletes involved in swimming, track and field, and bodybuilding. Anabolic steroids have been detected in female athletes during drug tests done during international competitions. The incredible level of muscle development in some female athletes who train with weights (i.e., bodybuilders, throwers, and weight lifters) has increased suspicions about drug use in athletes in these sports.

Women can expect greater gains than men from steroid use. Levels of circulating androgens are low and the relative proportions of unbound androgen receptor sites are greater. However, the side effects are more severe. Beside the masculine changes already mentioned, typical side effects include acne, changes in skin texture, severe fluid retention, unnatural increases in muscle mass, and radically changed cholesterol metabolism. As mentioned, females may also experience clitoral enlargement and menstrual irregularity. Many of these changes are irreversible. The effects on future fertility are unknown.

Effects of Steroid Use on Children Children taking anabolic steroids first experience accelerated maturation, followed by premature closure of the epiphyseal growth centers in the long bones. If anabolic steroid use begins early in adolescence, the ultimate height of the athlete may be less than it otherwise would be. It is suspected that some young female gymnasts have taken anabolic steroids specifically to stunt growth (small stature is thought to be an advantage in women's gymnastics).

Anabolic steroid use by adolescent athletes may predispose them to an increased risk of musculoskeletal injury. Studies in rats have shown that soft tissue tensile strength diminishes in animals given anabolic steroids, perhaps because of the rebound increase in corticosteroid hormones that happens after a period of drug use. Corticosteroids, as discussed, have a catabolic effect that results in tissue breakdown. This catabolic rebound effect may make young athletes' tendons and ligaments more prone to injury. Anabolic steroids may also have this effect in adults.

Acquired Immune Deficiency Syndrome (AIDS) Several cases of AIDS by athletes using shared needles have been reported in the literature. This is a particularly alarming development.

Miscellaneous Side Effects Miscellaneous side effects include muscle cramps, gastrointestinal distress, headache, dizziness, sore nipples, and abnormal thyroid function. Some of these side effects have occurred in people who took low doses for short periods of time. To summarize, anabolic steroids are probably effective for increasing strength and lean body mass in trained athletes but present possible grave health risks that could be life-threatening.

Use of Anabolic Steroids by Athletes

Anabolic steroids have become a recreational drug, even among high school students. Buckley and co-workers (1988) studied a nationwide sample of 3400 high school seniors. They found that over 6% of students use anabolic steroids. Although the majority of these were athletes, 35% were nonathletes. Thirty percent of steroid users cited improved appearance as the main reason for taking the drugs. Of particular concern was the source of the drugs—20% of steroid users obtained the drugs from a health care professional.

The findings of Buckley and coworkers have been replicated in a large epidemiological study conducted by the U.S. Centers for Disease Control and Prevention (DuRant et al., 1995). Faigenbaum and colleagues (1998) found that anabolic steroid use has even filtered down to the elementary-school level. They found that in a sample of nearly 1000 subjects, 2.7% of all middle-school students reported using steroids, with use patterns similar in girls and boys.

In some sports, anabolic steroids are used by the majority of accomplished athletes. Unfortunately, administration of these dangerous substances is done without medical supervision. Table 29-6 presents the most popular oral and injectable anabolic steroids, listed in order of androgenicity. Athletes tend to take less of androgenic steroids during building periods (off-season conditioning) and more during competitive periods. They also often take combinations of steroids (a practice called stacking) to obtain the most beneficial effects. Table 29-7 shows the anabolic steroid therapy program of a world-class athlete during a transition period between conditioning and competition.

Physicians, coaches, and educators should learn the signs of anabolic steroid use in young, athletic people (Table 29-8). Unusual changes in appearance or performance are often indicative of steroid use. Look for abrupt changes in behavior, such as increased aggressiveness, hostility, or rebelliousness. An excellent marker is an abnormally low level of high-density lipoproteins in the blood. An HDL of less than 30–35 mg · dl^{-1} almost always indicates anabolic steroid use in a young, fit person.

Counterfeit drugs have become a serious problem for anabolic steroid users. The risk of legal sanctions against physicians and other health care workers has forced athletes to purchase anabolic steroids on the black market. A recent report by van der Kuy and coworkers (1997) detailed an account of counterfeit Dianabol (methandrostenolone, a popular anabolic steroid). The tablets contained methyltestosterone and clenbuterol, also anabolic agents, instead of methandrostenolone. Two patients were admitted to the hospital for clenbuterol poisoning.

Human Chorionic Gonadotropin Prolonged use of anabolic steroids suppresses serum testosterone and spermatogenesis (Lamb, 1987) and training gains are lost rapidly. To boost natural testosterone production and prevent muscle atrophy that is common during withdrawal from anabolic steroids, athletes sometimes take human chorionic gonadotropin (HCG).

Martikainen and associates (1986) examined testosterone concentration after a single dose of HCG in six power athletes who had used high doses of testosterone and anabolic steroids for three months. After HCG injection, serum testosterone and 5 α-dihydrotestosterone concentrations increased without affecting estradiol and 17-hydroxyprogesterone concentrations. They also found that high doses of HCG were no more effective than lower doses in increasing serum testosterone—administration of 1500, 3000, or 4500 IU of HCG on successive days did not increase serum testosterone above the level elevated by the 1500 IU dosage.

TABLE 29-6

Anabolic–Androgenic Steroids Used by Athletes Listed in Order of Androgenicity[a]

Orals	Injectables
Maxibolin (ethyloestrenol)	Deca-Durabolin (nandrolone decanoate)
Anavar (oxandrolone)	Durabolin 50 (nandrolone phenpropionate)
Winstrol (stanozolol)	Delatestryl (testosterone enanthate)
Dianabol (methandrostenolone)	Testosterone propionate
Anadrol (oxymetholone)	Depo-Testosterone (testosterone cypionate)
	Aqueous testosterone

[a] Trade names are given with generic names in parentheses, except for testosterone propionate and aqueous testosterone, which are generic names.

TABLE 29-7

Anabolic Steroid Therapy Program of a World-Class, Weight-Trained Athlete[a]

Week	Injectable Steroid	Daily Oral Steroid	Training Program
1–2	100 mg Deca-Durabolin every 5 days	8 mg Winstrol	Heavy weight training—high volume, medium intensity
3–4	100 mg Deca-Durabolin every 5 days	10 mg Winstrol 10 mg Dianabol	Heavy weight training—high volume, medium intensity
5	100 mg Deca-Durabolin every 5 days	10 mg Winstrol 20 mg Dianabol	Heavy weight training—high volume, medium intensity
6	100 mg Deca-Durabolin every 5 days	Decreasing dosage	Low volume, low intensity— rest week
7–8	No injectables	No orals	Low volume, high intensity
9	No injectables	10 mg Dianabol	Low volume, high intensity Lift heavy once during week
10	200 mg Delatestryl every 5 days	15 mg Dianabol	Low volume, high intensity Lift heavy once during week
11	200 mg Delatestryl every 5 days	20 mg Dianabol	Low-volume, low-intensity weight training
12	200 mg Delatestryl every 5 days 100 mg testosterone propionate day before meet	25 mg Dianabol	Competition, no lifting

[a] This table should not be considered an endorsement for use of anabolic steroids. It is merely intended to demonstrate use patterns in strength-trained athletes.

TABLE 29-8

Recognizing Anabolic Steroid Users

- Obvious abnormal gains in size and strength
- Uncharacteristic changes in behavior: aggression, rebelliousness, hostility, extreme anxiety, insomnia, increased libido
- Recent onset of extreme muscle cramping, nosebleeds, acne
- Abnormally low HDL (<35 mg %)

■ Growth Hormone

Growth hormone (GH) is a polypeptide hormone produced by the anterior pituitary gland. Some power lifters, bodybuilders, and throwers use it to increase muscle mass and strength. Reports in the news media suggest that, like anabolic steroids, its use has filtered down to nonathlete high school students. GH may be valuable in rehabilitating injured muscle and has been used extensively as an anabolic agent in geriatric patients.

To date, there has been no large-scale research of its effect on athletic performance. Although animal studies suggest that growth hormone stimulates muscle hypertrophy, its effects on the strength of normal muscle are unclear. Table 29-9 summarizes the measurements of a world-class power lifter who was on a 6-wk, self-administered growth hormone regimen. This athlete experienced a large increase in lean body mass and a decrease in fat. This observation is consistent with the testimonials of athletes who have claimed weight gains of 30 to 40 pounds in ten weeks, accompanied by equally remarkable gains in strength. Nitrogen bal-

TABLE 29-9

Effects of 6-Week Administration of Human Growth Hormone to a World-Class Power Lifter on Body Composition and Left Ventricular (LV) Wall Thicknesses of the Heart[a]

Week	Weight (lb)	% Fat	Lean Body Mass (lb)	LV Septal Thickness (mm)	Posterior Wall Thickness (mm)
Pre	272.1	20.0	217.7	11.0	12.0
4	279.6	17.4	230.9	——	——
6	280.3	16.5	234.1	13.0	12.0
8	277.9	15.7	234.1	——	——

[a] Body composition measured by underwater weighing and left ventricular wall thicknesses determined by M-mode echocardiography. Human growth hormone was administered during weeks 1–6.

ance becomes highly positive when GH is administered to adult humans, which lends some credence to these claims. However, uncontrolled observations and testimonials from athletes are *not* acceptable substitutes for well-structured scientific studies. Judgment must be reserved until evaluations on the effects of this substance have been made on performance or body composition.

Growth hormone is a potent anabolic agent. It facilitates the transport of amino acids into cells and their incorporation into protein. An increase rate of amino acid transport into muscle cells is associated with muscle hypertrophy. Growth hormone is also involved in the formation of connective tissue. It exerts a stimulatory effect on insulin-like growth factor (IGF-1), which is also a potent anabolic agent. Growth hormone affects carbohydrate and fat metabolism, as well, stimulating glucose uptake in muscle and fat and mobilizing free fatty acids from adipose tissue.

The use of growth hormone by athletes can have severe consequences. Growth hormone can cause diabetes, which could become permanent if large doses are administered. Because of its general anabolic effect, growth hormone can also cause cardiomegally (i.e., enlarged heart, see Table 29-9), which could result in an increased myocardial oxygen demand. Combining growth hormone and anabolic steroids can increase the risk of heart disease, with growth hormone increasing heart size

and anabolic steroids causing atherosclerosis. GH can also cause acromegaly, which is characterized by enlarged bones in the head, face, and hands; myopathy; peripheral neuropathy; osteoporosis; arthritis; and heart disease. The skeletal changes are irreversible. Administration of growth hormone may also cause antibody formation to the hormone, which could affect normal growth hormone metabolism.

In the past, growth hormone has been very expensive and in short supply because it was obtained only from cadavers or rhesus monkeys. However, recent advances in genetic engineering have made this substance more widely available. Even with its severe side effects, the possibility that it may allow dramatic increases in performance in a short time make its use irresistible to some athletes.

Some athletes take drugs that increase exogenous growth hormone secretion. These drugs include propanolol, vasopressin, clonidine, and levodopa. There is no evidence that these practices enhance muscle hypertrophy.

IGF-1

Insulin-like growth factor (IGF-1) is an anabolic agent. The use of recombinant IGF-1 has recently become a popular drug among weight-trained athletes. IGF-1, also called somotomedin C, is one of several hormones known as somatomedins.

IGF-1 production is stimulated by growth hormone. It is released mainly by the liver but may also be secreted by bone, fat cells, the testes, and the heart. IGF-1 is an extremely anabolic hormone. It facilitates amino-acid and glucose transport and glycogen synthesis, and it has anabolic effects in bone and cartilage. IGF-1 also causes a positive nitrogen balance. While IGF-1 is a powerful anabolic agent, its effects in athletes are unknown. Nevertheless, it is a popular topic in bodybuilding magazines.

The side effects of IGF-1 administration are thought to be similar to those of growth hormone. It has been suggested that long-term increases in IGF-1 levels may promote cancer growth, but there is no direct evidence. People with acromegaly, a risk associated with prolonged IGF-1 use, have a higher cancer risk so the relationship is inferred.

Dehydroepiandrosterone (DHEA) and Androstenedione

DHEA and androstenedione are relatively weak adrenal androgens. They began to receive increased notice by bodybuilders in 1996, when their sale as food supplements was permitted in the United States. Bodybuilders take these drugs to stimulate muscle hypertrophy and aid in weight control.

DHEA administration in middle-aged and older adults (Morales et al., 1994) improved "energy levels," increased muscle mass, mental acuity, and immune function. However, these subjects had blood levels of the hormone at least 20% below the average level of a 20-year-old. Androstenedione is thought to have similar effects as DHEA.

The effects of DHEA in young, healthy athletes are unknown. Young men given 1600 mg/d supplements of DHEA experienced a 30% decline in fat mass (Nestler et al., 1988), but these results have not been replicated by others (Casson and Buster, 1995). Its value as an ergogenic aid may lie in its ability to increase serum testosterone concentrations. It may act as an anti-obesity agent by increasing futile cycling (energy lost as heat; not captured as ATP) in mitochondria.

The side effects of DHEA in bodybuilders are also unknown. High doses (1600 mg/d) of the drug in young men led to a reduction in LDL and HDL cholesterol. However, in elderly subjects, DHEA reduced LDL and had no effect on HDL (Morales et al., 1994). Doses of 150–300 mg/d have a marked effect on testosterone concentration. This could lead to masculinization in female bodybuilders and interfere with the hypothalamic-pituitary-gonadal axis in both sexes. The HPG axis controls estrogen release in women and testosterone release in men.

Insulin

Bodybuilders sometimes inject insulin subcutaneously or elevate levels during exercise by dietary stimulation (Fahey et al., 1993) in the hope of benefitting from the hormone's anabolic effects. Exogenous insulin injection is not, fortunately, a widespread practice among bodybuilders.

Insulin, aside from its influence on glucose and fat metabolism, has wide-ranging effects on protein metabolism. These effects include enhanced amino acid transport into cells, increased rate of incorporation of amino acids into protein, and suppression of protein catabolism. The effectiveness of insulin supplementation or elevation in stimulating muscle hypertrophy is not known. The most serious side effect of exogenous insulin administration in exercising athletes is insulin shock.

Clenbuterol

One of the newest substances used by athletes as an anabolic agent is clenbuterol, a β-2 adrenergic agonist. Bodybuilders take the drug to prevent muscle atrophy, increase lean body mass, and decrease body fat. Reports citing serious medical complications in bodybuilders listed clenbuterol as one of their ergogenic aids.

The proposed benefit from this substance emanates from its capacity to stimulate the central nervous system. Adrenergic agonists are widely used as bronchodilators to prevent and treat symptoms of exercise-induced asthma. Clenbuterol may promote muscle hypertrophy by stimulating protein metabolism in the cell via increased calcium transport, increased cyclic adenosine monophosphate (cAMP) levels, and an activation of protein kinase.

Most studies on the influence of this substance on body composition and exercise capacity have been done in the rat. These studies have shown that the drug increases muscle mass, decreases body fat, and exerts metabolic effects that augment the endurance training effect. Carter and Lynch (1994) found that albutamol (anther β-2 adrenergic agonist) and clenbuterol increased carcass protein content 20% and 30% in young rats, with 12% and 21% increases in old rats. Torgan and associates (1993), using obese Zucker rats, found that the drug assisted endurance-exercise-induced increases in citrate synthase activity and skeletal muscle insulin resistance. The drugs quickly lose their capacity to promote muscle hypertrophy. Rothwell and coworkers (1987) showed that muscle-β-receptor density decreased 50% after 18 days of clenbuterol administration in rats. For this reason, bodybuilders often "cycle" clenbuterol, taking it on and off in 2-day cycles. This cycle is generally continued for 8 to 10 weeks, followed by 10 to 12 weeks without the drug.

Side effects include insomnia, arrhythmias, anxiety, anorexia, and nausea. More serious side effects include cardiomegally and myocardial infarction.

Creatine Monohydrate

Currently, creatine monohydrate is among the most popular and widely used supplements. Athletes use it to enhance recovery, power output, and muscle hypertrophy.

While creatine feeding was attempted early this century, interest in this technique as an ergogenic aid stems from the work of Harris and associates (1992). They found that creatine feeding (in the form of creatine monohydrate) increased the creatine phosphate content of the muscle by 20%. The optimal dosage for achieving maximum values of muscle creatine phosphate appear to be 20 g/day for 5 days followed by daily doses of approximately 3–5 g/day. Increasing the intake to 20 g/day resulted in no further increase in total muscle creatine. Creatine supplementation has been shown in some studies to improve performance in short-term, high-intensity, repetitive exercise, which would make it a valuable supplement for strength-speed athletes and bodybuilders. Creatine supplementation may improve performance by enhancing the availability of creatine phosphate and possibly regulating the rate of muscle glycolysis. It may also enhance muscle hypertrophy by allowing athletes to train harder.

The health risks of long-term creatine monohydrate administration are unknown. Anecdotal evidence suggests that athletes taking creatine monohydrate are more susceptible to muscle cramps during intense exercise in the heat. However, these observations have not been investigated systematically.

Phosphatidylserine (PS)

Phosphatidylserine (PS) is a naturally occurring phospholipid derived from soybeans or bovine cerebral cortex. It is used in the treatment of Alzheimer's disease and attention deficit disorder. Several studies have shown that PS attenuates cortisol during endurance and weight-training exercise, which may help speed recovery from intense training. Fahey and Pearl (1998), using overtrained weight lifters, found that PS administration decreased post-exercise cortisol levels, decreased muscle soreness, and improved perception of well-being. Little is known of the long-term efficacy or side effects of PS. However, selected patient populations have taken this substance without incident.

β-hydroxy-beta-methylbutyrate (HMB)

HMB, a metabolite of the amino acid leucine, is a popular supplement used to increase muscle mass and strength. While several animal studies have shown that the substance has promise as an ergogenic aid, only a few studies have been done in humans. Nissen and colleagues (1996) found that subjects taking 1.5 or 3 g HMB/day increased strength and experienced less muscle proteolysis than control subjects. However, little is known about long-term risks and benefits of this substance.

■ Amphetamines

Amphetamines are among the most abused drugs in sports. They are particularly popular in football, basketball, track and field, and cycling. Athletes use them to prevent fatigue, and to increase confidence, cardiovascular endurance, muscle endurance, speed, power, and reaction time. Generic examples of amphetamines include benzedrine, dexedrine, dexamyl, and methedrine.

These drugs act as both central nervous system and sympathomimetic stimulants. Sympathomimetic drugs bind with α- and β-adrenergic receptors and have the same effects as catecholamines. Amphetamines stimulate the central nervous system by directly affecting the reticular-activating system and postganglionic nerves. Central effects include increased arousal, wakefulness, confidence, and the feeling of an enhanced capability to make decisions. Sympathomimetic effects include increased blood pressure, heart rate, oxygen consumption in the brain, and glycolysis in muscle and liver, vasoconstriction in the arterioles of the skin and spleen, and vasodilation in muscle arterioles.

Amphetamines and Athletic Performance

The effects of amphetamines on athletic performance are controversial. Many studies have been poorly controlled, have used low dosages, and did not allow enough time for absorption of the drug. Although administration of approximately 15–50 mg of d-amphetamine is a common dose in athletics, for example, some studies have used a dosage as low as 5 mg. Also, these drugs are readily absorbed orally but take 1½ to 2 hours to reach peak levels in the body (peak levels are reached in 30 minutes if the amphetamine is injected). Yet, some studies began performance testing within a half hour of administration. In addition, amphetamines create a euphoric sensation that is easily identifiable, thus making a true double-blind study almost impossible.

Amphetamines became popular during World War II with soldiers who used them to ward off fatigue. Studies have generally supported the effectiveness of amphetamines as psychotropic drugs that mask fatigue, but have been equivocal on their ability to improve endurance performance. Many studies have demonstrated enhanced feelings of well-being and improved exercise capacity in fatigued subjects. Fatigued animal and human subjects, for example, improved endurance in marching, cycling, swimming, and treadmill exercise, and improved simple reaction time. Amphetamines have no effect on reaction time in rested subjects.

Most studies have failed to demonstrate an effect on cardiovascular function although exercise time to exhaustion and peak blood lactate concentration are often increased. Maximal values for oxygen consumption, heart rate, minute volume, respiratory exchange ratio, respiratory rate, oxygen pulse, CO_2 production, and ventilatory equivalent were unaffected by amphetamines.

The effects of amphetamines on strength and power seem to depend on the size of the muscle studied. Substantial increases in knee extension strength have been demonstrated after administration of amphetamines. The drugs had no effect on elbow flexion strength in the same subjects. In sprinting, acceleration, but not top speed, is enhanced by the drug. This points to an increased excitability of the muscles but not to an increase in their maximal capacity. Most studies have shown increases in static strength but mixed results in muscle endurance.

The effects of amphetamines on sports performance are also unclear. They appear to help power-oriented movement skills in activities that employ constant motor patterns, such as shot putting and hammer throwing. They are probably less effective in sports requiring the execution of motor skills in an unpredictable order, such as football, basketball, and tennis. In these sports, amphetamines may be deleterious because they may interfere with the body's fatigue alarm system, cause confusion, impair judgment, and, in high concentrations, cause neuromuscular blockade, and loss of effective motor control.

Side Effects of Amphetamines

Amphetamines can cause a variety of severe side effects (Table 29-10). They are very addictive. They can have severe neural and psychological effects,

TABLE 29-10

Side Effects of Amphetamines

Mild	Severe	Chronic
Restlessness	Confusion	Addiction
Dizziness	Assaultiveness	Weight loss
Tremor	Delirium	Psychosis
Irritability	Paranoia	Paranoid delusions
Insomnia	Hallucinations	Dyskinesias
Euphoria	Convulsions	Compulsive/stereotypic/repetitive behavior
Uncontrolled movements	Cerebral hemorrhage	Vasculitis
Headache	Angina/myocardial infarction	Neuropathy
Palpitations	Hypertension	
Anorexia	Circulatory collapse	
Nausea		
Vomiting		

SOURCE: Wadler and Hainline, 1989, p. 84. Used with permission.

including aggressiveness, paranoia, hallucinations, compulsive behavior, restlessness, and irritability. They can cause arrhythmias, hypertension, and angina. They also increase the risk of hyperthermia because of their vasoconstriction effect on the arterioles of the skin. Numerous deaths in endurance sports have been reported in athletes competing in the heat while under the influence of these drugs. Amphetamine use can also result in cerebral hemorrhage.

■ Cocaine

Cocaine, a central nervous system stimulant, is an alkaloid derived from the leaves of the coca plant. Its use is illegal but it is used widely as a "recreational" drug. Cocaine has become popular with some well-known athletes, and reports have increased in the news media recently of cocaine use by prominent professional football and basketball players. Several players have died as a result of cocaine use.

The use of cocaine as a stimulant has a long history, beginning with the Incas in Peru. Its use was advocated by Sigmund Freud, and it was an ingredient in Coca-Cola during the early days of the product. It has been a controlled substance since 1906. It is administered orally, nasally, or intrave-nously. Cocaine is available as cocaine hydrochloride, freebase cocaine, and crack.

Cocaine works by inhibiting the reuptake of dopamine and norepinephrine in sympathetic neurons. It has a direct sympathomimetic effect (i.e., increased heart rate and increased systolic and diastolic blood pressures). It is also thought to affect the mesocortical brain, sometimes called the pleasure center. This effect may account for the extreme compulsive-addictive behavior common in cocaine users.

Drug tolerance, caused by dopamine depletion and an increased number of dopamine receptors in postsynaptic neurons, develops with prolonged use. Tolerance is a consequence of cocaine addiction, so there is a tendency to take higher doses.

Cocaine produces a feeling of exhilaration and an enhanced sense of well-being, and depresses fatigue. It increased work capacity in several poorly controlled studies but had no effect on swim time to exhaustion in rats. In low doses, cocaine probably has beneficial effects on performance, similar to amphetamines. However, its effects are short-lived, which makes it impractical as an ergogenic aid. In higher doses, the drug appears to have negative effects on performance. Empirical evidence from professional athletes suggests that prolonged cocaine use causes deterioration in performance. Common

TABLE 29-11

Side Effects of Cocaine Use

Cardiac
  Ventricular arrhythmia
  Sudden death
  Angina pectoris
  Myocardial infarction
  Myocarditis

Neuropsychiatric
  Cerebrovascular
    Cerebral infarction
    Cerebral hemorrhage
    Subarachnoid hemorrhage
    Cerebral vasculitis
    Transient ischemic attacks
  Addiction
  Seizures
  Tourette's exacerbation
  Headache
  Visual scotoma/blindness
  Optical neuropathy
  Behavioral
    Insomnia
    Euphoria/dysphoria
    Confusion
    Assaultiveness
    Delirium
    Paranoia
    Hallucinations
    Psychosis
    Repetitive behavior
    Anorexia

Ob/Gyn
  Abruptio placentae
  Spontaneous abortion
  Congenital malformations of fetus
  Placental transfer to infant and secondary acute toxicity
  Breast milk transfer to infant and secondary acute toxicity

Other Complications
  Sexual dysfunction
  Liver toxicity
  Osteolytic sinusitis
  Pneumomediastinum
  Aortic dissection
  Gastrointestinal ischemia
  Necrosis/perforation of nasal septum
  Loss of smell
  Hyperthermia/tachycardia

problems include impairments in visual acuity, judgment, hand–eye coordination, concentration, sense of time, and social behavior.

Cocaine use can have severe health consequences (see Table 29-11). Side effects include serious arrhythmias, myocardial infarction, coronary artery vasospasm, dilated cardiomyopathy, and cerebral hemorrhage. Several well-publicized incidents of sudden death in well-known, healthy young athletes has heightened awareness of the danger of cocaine.

■ Caffeine

Caffeine is a xanthine and is found naturally in numerous plant species. It stimulates the central nervous system by causing the adrenal medulla to release epinephrine. It stimulates the heart (increasing the rate and contractility of the heart at rest), causes peripheral vasodilation, and acts as a diuretic by blocking renal tubular reabsorption of sodium. Caffeine increases calcium transport by increasing calcium permeability in the sarcoplasmic reticulum. It increases cellular cyclic AMP. In subjects who have not taken caffeine regularly, this stimulates glycogenolysis, resulting in increased blood glucose. The increase in cyclic AMP also causes increased fatty acid mobilization from fat cells. Caffeine may also block adenosine receptors. Adenosine has a calming effect. Blocking adenosine may partially explain the stimulating effects of the drug.

Caffeine and Performance

In athletics, caffeine is used as a stimulant and as a fatty acid mobilizer. It is found in a variety of food products such as coffee, tea, and chocolate. Although there is some evidence that caffeine may improve endurance, the drug does not appear to enhance short-term maximal exercise capacity.

Caffeine is a much weaker stimulant than amphetamines, yet it is widely used by weight lifters and throwers (discus, shot, javelin, and hammer) to enhance strength and power. These athletes take the caffeine in the form of strong coffee or over-the-counter medications such as Vivarin or No-doze. Although some older, poorly controlled studies found improvements in strength and power from this substance, these findings have not been replicated by well-controlled studies.

Caffeine appears to enhance performance in prolonged endurance exercise by mobilizing free fatty acids and sparing muscle glycogen. However, the drug is most effective in well-trained, caffeine-naive subjects. The most effective dose is approximately 200–300 mg. Many studies have shown that caffeine ingestion prior to endurance exercise increases time to exhaustion or decreases race times. Metabolic changes have been less consistent. Most studies show that free fatty acids and glucose in blood increase. However, the respiratory exchange ratio (R) has not decreased consistently in these studies. A decrease in R would suggest increased fat oxidation, which is the proposed benefit of caffeine as an ergogenic aid in endurance exercise. Several studies using untrained subjects have shown that caffeine increases glycogen depletion during endurance exercise. This was true even when blood free fatty acids were increased. Several well-controlled studies have shown no beneficial effects of caffeine ingestion on endurance exercise.

The use of caffeine as an ergogenic aid is not without danger. The diuretic and cardiac stimulatory properties of this substance can combine to increase the risk of arrhythmias, such as ventricular ectopic beats and paroxysmal atrial tachycardia. This is particularly alarming for older, less-conditioned individuals. Caffeine can also cause insomnia and is addicting. Bodybuilders often take caffeine, ephedrine, and aspirin in combination as a weight-loss agent. While effective, unsupervised use of this "weight-loss cocktail" has resulted in dangerous cardiac arrhythmias.

■ Other Stimulants

Other agents used by bodybuilders to enhance training intensity include ephedrine and ginseng. Ephedrine, a weak stimulant, is widely used by athletes during workouts. Sidney and Lefcoe (1977) found that despite a slight stimulating effect on blood pressure and on exercise and recovery heart rates, ephedrine had no effect on physical work capacity. Their study included measurements of muscle strength, endurance and power, lung function, reaction time, hand-eye coordination, anaerobic capacity and speed, cardiorespiratory endurance, and responses to maximal and submaximal effort, including maximum oxygen intake, ratings of perceived exertion and speed of recovery from effort. Ginseng is also very popular with athletes. However, a review by Bahrke and Morgan (1994) concluded that there is an absence of compelling research evidence to support its use as an ergogenic aid.

■ Nutritional Supplements

Athletes spend a fortune on an endless variety of dietary supplements, such as proteins, vitamins, and weight-gain products. However, an overwhelming body of literature has demonstrated that as long as an athlete is receiving a balanced diet, most dietary supplements have no effect on performance. Of course, if the diet is deficient in any essential nutrient, then supplementation may very well be beneficial. Diet and performance are discussed in Chapter 28, so this section focuses on those nutritional supplements that are specifically used as ergogenic aids.

Carbohydrate Feeding

Carbohydrate (CHO) feeding, often in the form of glucose, dextrose, or honey, has long been used as an ergogenic aid to increase strength, speed, and endurance. CHO feeding has no effect on strength, power, or high-intensity, short-term exercise.

Consuming liquid meals before and during exercise has been shown to be beneficial in endurance exercise. Compared to no feeding, carbohydrate feeding during exercise increases blood glucose concentration. During prolonged exercise, carbohydrate intake increases endurance and the ability to exercise intensely late in exercise. The longer exercise continues, the more important ingested carbohydrate becomes as a fuel source. Preexercise carbohydrate feedings have improved endurance performance, even when serum insulin levels are increased at the start of exercise. Carbohydrate feeding is most effective in well-trained athletes when exercise duration exceeds three hours. The effective exercise duration decreases with the training status of the subject. In moderately trained people, carbohydrate feeding may be effective in events lasting as little as 1½ hours. Ivy (1998) also found that the addition of protein to a carbohydrate supplement may also increase the rate of glycogen storage due to the ability of protein and carbohydrate to act synergistically on insulin secretion.

Lactate has recently been suggested as a component in an energy-enhanced fluid replacement. Polylactate (polymerized lactate bound to arginine) was able to maintain blood glucose as well as glucose polymers during a 3-hr ride on a cycle ergometer. This substance was also found superior to a nonnutritive placebo in allowing recovery from intense interval training. Polylactate by itself causes gastrointestinal distress in some subjects, so it should be used with other carbohydrate sources when incorporated into a liquid meal.

Vitamins

Vitamins are substances needed in very small quantities. They are involved in numerous metabolic reactions and must be consumed in the diet. Vitamins are water-soluble or fat-soluble. Water-soluble vitamins, such as vitamin C, tend to act as coenzymes (work with enzymes to cause a chemical reaction). They are stored in small quantities, so regular intake is essential. Fat-soluble vitamins do not act as coenzymes and are stored in greater quantity than water-soluble vitamins. With a few exceptions, active people do not benefit from vitamin supplements. The best dietary strategy for adequate vitamin intake continues to be selecting a balanced diet from the basic food groups.

A variety of vitamins has been used as ergogenic aids. As discussed in Chapter 28, these substances, although essential, are required in extremely small

quantities. Because mitochondria increase due to endurance training, more vitamins may be needed to support the increased metabolic activity. However, vitamin deficiency in athletes has not been consistently demonstrated. In fact, most athletes take many times the minimum daily requirement for these substances. Only levels of vitamin C, thiamin, pyridoxine, and riboflavin decrease with exercise (due to increased metabolism). These vitamins are not lost in sweat.

B Vitamins

B vitamins include thiamin, riboflavin, niacin, pyridoxine, pantothenic acid, cobalamin, folacin, and biotin. Cobalamin (vitamin B_{12}) is one of the most popular ergogenic aids in sports.

Thiamin (Vitamin B_1) Thiamin levels in human blood and in rat liver, kidney, and blood decrease with intense exercise training. However, most well-controlled studies have shown no effect of thiamin supplementation on strength or endurance. Data on rats suggest that thiamin supplementation may improve endurance capacity in subjects consuming a high-carbohydrate diet.

Riboflavin (Vitamin B_2) Exercise decreases riboflavin in blood. These effects increase with exercise intensity and duration. Several studies have suggested that RDA for riboflavin (1.6 mg for men and 1 mg for women) is inadequate for people involved in intense exercise. Except for several early, poorly controlled studies, riboflavin supplementation above the RDA has not been shown to improve exercise performance.

Niacin (Vitamin B_3) Niacin, also known as nicotinic acid or nicotinamide, decreases endurance performance. It inhibits fatty acid mobilization and speeds glycogen depletion during endurance exercise. Using niacin as an ergogenic aid is contraindicated.

Pyridoxine (Vitamin B_6) Pyridoxine levels in blood decrease with exercise, particularly when subjects are on a low-carbohydrate diet. Its effectiveness as an ergogenic aid is unclear and needs further study. Pyridoxine infusion causes an increase in plasma growth hormone, which suggests a possible role in enhancing muscle hypertrophy in strength athletes. No well-controlled study has shown that pyridoxine supplementation by itself improves performance.

PAK, a combination of pyridoxine and α-ketoglutarate, an intermediate in the Krebs cycle, has been suggested as an ergogenic aid. PAK has been shown to reduce lactate levels in patients with hepatitis. In normal subjects, Marconi and coworkers (1982) found lower blood lactate during static and dynamic exercise and increased aerobic capacity. However, Linderman and Fahey (1991) found that PAK had no effect on short-term maximal exercise capacity. Also, these results were not supported by other studies (Trappe et al., 1994, Decombaz et al., 1993).

Pantothenic Acid Pantothenic acid may improve performance. Its administration increased work output in isolated frog muscle, and rats fed diets high in pantothenic acid improved swim times to exhaustion. In humans, pantothenic acid administration improved running and cycling performance. However, not all studies have had positive results.

Cobalamin (Vitamin B$_{12}$) As noted, vitamin B$_{12}$ is one of the most popular ergogenic aids in athletics. The vast majority of studies have shown, however, that when dietary intake of cobalamin is normal, no extra benefit is gained from taking it in high doses.

Pangamic Acid ("Vitamin B$_{15}$") Pangamic acid, sometimes called vitamin B$_{15}$, has been hailed as a miracle substance that improves endurance and fights off fatigue. However, the product has come under increasing criticism from the Food and Drug Administration because it is not an identifiable substance (not a vitamin or provitamin) and it has no established medical or nutritional usefulness. The active ingredient in this product is apparently N,

N-dimethylglycine (DMG), which is said to increase oxygen utilization. Although some studies support the claims of improved endurance, the most stringently controlled investigations generally have not. The use of this substance may be dangerous because of its potential for mutagenesis (change in DNA structure).

Vitamin C

Vitamin C is an antioxidant and is involved in the synthesis of catecholamines, collagen, carnitine, and serotonin. Vitamin C has been used to improve both cardiovascular and muscle endurance. The research is divided on the beneficial effects of the vitamin on performance. Discrepancies in the literature may be due to poor experimental controls in some studies and low initial levels of the vitamins in some of the subjects. Intense exercise training decreases vitamin C levels in a variety of tissues.

Although vitamin C has been shown to speed up the process of acclimatization to heat, most studies have not demonstrated any effect on factors important to endurance performance. Exercise increases the generation of free radicals and lipid peroxides that damage cell structures. The role of vitamin C as an antioxidant is receiving increasing interest. However, its role in performance and recovery from exercise are poorly understood. Vitamin C supplementation is considered safe although megadoses may cause diarrhea and nausea.

Vitamin E and Wheat Germ Oil

Vitamin E and wheat germ oil have been widely used as ergogenic aids for many years. The beneficial constituents of wheat germ oil are purported to be vitamin E and octacosanol. Vitamin E includes a number of substances from the tocopherol group (α-, β-, γ-, and δ-tocopherol). It has received a great deal of interest because of its antioxidant properties. Vitamin E prevents lipid peroxidation in cellular membranes and is thought to regulate the immune function. Because of the widespread availability of tocopherols (i.e., vitamin E) in foods, vitamin E deficiency is very rare in humans.

Wheat germ oil has been highly touted as an ergogenic aid that increases endurance. The proposed mechanism of its beneficial effect is its reduction of the oxygen requirement of the tissues and its improvement of coronary collateral circulation. As discussed in Chapter 14, however, coronary collateral development in healthy exercising humans is unlikely. Numerous studies by Cureton have shown a beneficial effect of wheat germ oil on endurance performance; however, these findings have generally not been replicated by others. Although Cureton's work generally appears to be well controlled, it has often been criticized for its use of unorthodox statistical methods.

Exercise increases the need for vitamin E although these increased requirements in humans have not yet been determined. Prolonged exercise increases membrane breakdown in cells by a process called lipid peroxidation (LPO). A marker of LPO is the pentane content in expired air. The administration of 1200 IU of vitamin E for two weeks eliminated pentane from the expired air in subjects who rode a cycle ergometer for one hour.

Also, as discussed in Chapter 20, delayed onset muscle soreness may be partially caused by mechanical disruption of the sarcoplasmic reticulum, which leads to calcium leakage and necrosis of muscle tissue. To the extent that lipid peroxidation is involved in this process, vitamin E may provide some protection. This hypothesis has not been tested, however.

Neither wheat germ nor vitamin E has been shown to increase maximal oxygen consumption. However, vitamin E may be an effective ergogenic aid in the face of environmental hypoxic stress. Several studies have found that vitamin E increased $\dot{V}_{O_2max}$ at altitude. In one study, $\dot{V}_{O_2max}$ increased by 9% at 1524-m (5000-ft) and 14% at 4572-m (15,000-ft) altitude. Another study found improved aerobic capacity following exposure to ozone.

At sea level, there are no apparent benefits to consuming megadoses of vitamin E. Although vitamin E supplementation is generally safe, some people do experience side effects. Intakes of over 150 mg · day^{-1} have been associated with weakness, fatigue, headache, diarrhea, nausea, phlebitis, and high cholesterol.

Aspartates

Aspartates, which are the potassium and magnesium salts of aspartic acid, have been used to reduce fatigue. It has been hypothesized that they work by accelerating the resynthesis of ATP and CP in muscle, detoxifying ammonia, increasing Krebs cycle flux, and sparing glycogen. As with many so-called ergogenic aids, however, their beneficial effects have been demonstrated only in poorly controlled studies that used few subjects.

L-Carnitine

L-carnitine is an important substance in fat metabolism. It acts as a carrier of fatty acids within the mitochondria and as an acceptor of acyl groups from acyl-CoA. Lipid metabolism during exercise is affected by its concentration in muscle and blood and by the rate that fatty acids can enter and be used in the mitochondria. The latter factor is affected by intramitochondrial L-carnitine acyl-transferase and L-carnitine. L-carnitine has also been shown to be important in facilitating the use of branched-chain amino acids for energy.

Acute and chronic exercise affects L-carnitine metabolism, and L-carnitine levels fall during exercise. Trained people have higher L-carnitine levels than untrained people. Marconi and coworkers (1982) showed that subjects fed supplements of L-carnitine for two weeks improved their aerobic capacity by 6%. They speculated that these results were due to enhanced flow of fuels through the Krebs cycle. Gorostiaga and colleagues (1989) reported an almost twofold increase in endurance capacity after three weeks of L-carnitine administration. Although more research needs to be done, L-carnitine supplementation appears to be a valuable practice for athletes who wish to enhance fat metabolism.

D,L-carnitine is toxic and decreases performance. D,L-carnitine has been shown to cause muscle weakness, muscle cramps, myoglobin in the urine (this condition suggests muscle damage), and muscle weakness. One runner who took D,L-carnitine for two days before a marathon experienced these symptoms and did not fully recover for two months.

Succinates

Succinate, like α-ketoglutarate (see the section on pyridoxine), is a Krebs cycle intermediate. Several studies suggest that supplementing the diet with either potassium succinate or sodium succinate enhances maximal oxygen uptake and high-intensity exercise capacity. Although succinate supplementation is not well studied in athletes, cell and animal studies suggest that it may improve performance by enhancing the ability to use lactic acid as a fuel during exercise and speed up the activity of the Krebs cycle. Succinates stimulate calcium pump activity, so they may be valuable in lessening the severity of delayed onset muscle soreness.

Chromium

Chromium is the latest ergogenic aid popular with weight lifters and bodybuilders. Studies have shown that it may increase lean body mass, decrease body fat, and reduce the risk of heart disease by lowering total and low-density lipoprotein (LDL) cholesterol. Chromium may work by increasing the sensitivity of insulin receptors. Insulin is an important anabolic hormone, playing an important role in many aspects of protein synthesis in skeletal muscle.

Although studies show this substance has potential merit as an ergogenic aid, it has not been definitely shown to be effective. As for now, chromium supplementation remains an unproven aid to athletic performance.

"Adaptogens"—Herbal Stimulants

A variety of largely herbal substances are used in Eastern European countries and China as biostimulants. These include eleuthrococcus senticosis, ginseng, schizandra chinensis, pantocrine, rantarin, saparal, mumie, adaptozol, and tonedrin. Although they are widely used by athletes in these countries, their effectiveness is unknown. Numerous studies have been published in the Russian literature, but few credible reports have appeared in Western journals. More work needs to be done on these substances before they can be accepted as effective ergogenic aids.

Other Nutritional Ergogenic Aids

A variety of nutritional substances have been proposed as ergogenic aids, including bee pollen, gelatin, lecithin, phosphates, and organ extracts. Although little research is available concerning them, there is little reason to support their effectiveness. The case for organ extracts, for example, seems to almost be a throwback to the time when warriors ate the hearts or livers of brave adversaries to obtain some of their prowess in battle.

■ Sodium Bicarbonate

The administration of buffering substances such as sodium bicarbonate ($NaHCO_3^-$) has been suggested as an ergogenic aid in preventing fatigue during exercise lasting one to seven minutes. Studies on isolated muscle have shown that decreased pH slows the recovery rate of fatigued muscle, which is related to the time it takes to clear lactate from the muscle and reduce intracellular acidosis. Low intracellular pH contributes to fatigue by inhibiting glycolysis. Extracellular HCO_3^- concentration indirectly affects intracellular pH by maintaining a pH gradient between intra- and extracellular compartments. This enhances the movement of lactate and H^+. These observations suggest that increased extracellular buffer concentration may affect maximal exercise performance.

Administering sodium bicarbonate increases plasma pH, HCO_3^-, and lactate following short-term maximal exercise. Sodium bicarbonate administration affects intracellular acid–base balance indirectly. The extracellular effect of increased HCO_3^- in blood is to enhance the pH gradient between intracellular and extracellular compartments in favor of the lactate efflux (movement of lactate from muscle to blood).

Studies on the effects of sodium bicarbonate on exercise capacity have shown conflicting results. Of those evaluated in a review by Linderman and Fahey (1991), approximately 40% showed a beneficial effect. Discrepancies in the literature appear to be due to differences in exercise protocol and dosage. Sodium bicarbonate ingestion seems to be most effective during maximal exercise bouts between one and seven minutes, as noted. Most studies (but not all) showing a beneficial effect used a dosage of 300 mg · kg^{-1} body weight. Alternative exercise protocols and lower dosages have not consistently demonstrated improved exercise performance. Sodium bicarbonate ingestion causes extreme gastrointestinal distress in at least 50% of subjects, which greatly diminishes its usefulness as an ergogenic aid.

■ Blood Doping and Erythropoietin

Blood doping, or induced erythrocythemia, is an increase in blood volume by transfusion for the purpose of increasing oxygen transport capacity. Although the transfusion can utilize blood from a matched donor, it is typically accomplished by the removal, storage, and subsequent reinfusion of the subject's own blood. This procedure received considerable publicity with the publication of a paper by Ekblom and colleagues (1972), who reported increases of 23% in work capacity and 9% in $\dot{V}_{O_2max}$. It also caught the interest of the media. At every Olympic competition, there are rumors of endurance athletes using this procedure to improve performance.

The majority of studies show improvements in maximal oxygen consumption (ranging from a 1 to 26% increase) and endurance exercise capacity (ranging from a 2.5 to 37% increase) following induced erythrocythemia. The increase in blood volume induces a decreased heart rate and cardiac output. There is usually no change in stroke volume during submaximal exercise. During maximal exercise, however, stroke volume and cardiac output increase with no increase in maximal heart rate.

Training may be an important prerequisite to the beneficial effects of blood doping. In untrained rats, for example, maximal oxygen consumption increases with hematocrit until the hematocrit reaches 40%. After that, further increases in hemoconcentration do not result in increased $\dot{V}_{O_2max}$.

Blood doping appears to be a relatively safe procedure in normal, healthy individuals. Reinfusion of autologous blood caused no abnormal changes in the exercise electrocardiogram or in blood pressure. However, homologous transfusions carry with them the risk of hepatitis, bacterial communication, and blood type incompatibility. Blood doping has been banned since the 1984 Olympics.

Erythropoietin Erythropoietin is a hormone produced by the kidneys that is a growth factor for erythrocytes. It is produced by DNA recombinant techniques, so it is readily available. But its use is banned by the International Olympic Committee.

Erythropoietin administration increases red blood cell production. Hematocrit was shown to increase 35% in patients treated for kidney failure. The effects are dose-dependent: Higher doses have a greater effect on hematocrit and hemoglobin. Erythropoietin administration has been shown to increase $\dot{V}_{O_2max}$ and exercise performance.

It can have severe side effects, however. Use of erythropoietin has been linked with hypertension, hypertensive encephalopathy, seizures, congestive heart failure, and stroke. Numerous unexplained deaths among elite European cyclists may be due to erythropoietin use.

Oxygen

Supplemental oxygen has been used before and during exercise in an attempt to enhance performance, and after exercise to hasten recovery. The first use of this technique was by Japanese swimmers in the 1932 Olympic Games held in Los Angeles. Because they were successful, the practice became institutionalized in several sports. The sight of a gasping football player on the sidelines with an oxygen mask over his face has become common.

Oxygen breathing immediately prior to exercise increases the oxygen content of the blood by 80–100 ml. In exercise lasting less than two minutes, supplemental oxygen improves maximal work capacity by approximately 1% and decreases submaximal exercise heart rate. Most studies (but not all) have shown that oxygen can be beneficial in short-term maximal efforts in swimming and running. It is not a very practical ergogenic aid, however, because of the time lag between the administration of oxygen and the start of performance—runners and swimmers must take the time to prepare themselves in the starting blocks, and football players must align themselves in formation. During this time, the increase in the oxygen content of blood has dissipated.

Although supplemental oxygen can be of theoretical benefit in recovery from repeated maximal efforts, several studies have been unable to demonstrate an accelerated decrease in ventilation, heart rate, or blood pressure following oxygen breathing. In fact, arterial oxygen partial pressure increases by 20% during the first few seconds of recovery from maximal exercise (Figure 29-6).

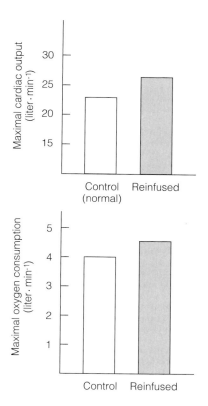

Figure 29-6 The effects of "blood doping" on cardiac output and maximal oxygen consumption. Blood doping increases cardiac output by increasing blood volume, which enhances venous return of blood to the heart. Arterial oxygen content is enhanced because more red blood cells are available to transport oxygen. Maximal oxygen consumption increases because of enhanced arterial oxygen content and a higher cardiac output. Adapted from data of Thomson et al., 1982.

Oxygen breathing during exercise improves work capacity, increases maximal oxygen consumption, and decreases submaximal heart rate. Supplemental oxygen effectively increases oxygen transport capacity by increasing the oxygen dissolved in the blood and perhaps by increasing maximal myocardial oxygen consumption. With the exception of mountain climbing, though, supplemental oxygen breathing during exercise is of little practical significance, as few sports allow an athlete to carry an oxygen tank on the field.

■ Summary

Ergogenic aids are substances or techniques that improve performance. The vast majority of ergogenic aids provide no benefit other than the placebo effect and often pose serious health hazards. Research studies in this area are often poorly controlled and fail to examine the ergogenic aids in the same manner as they are used in athletics.

The governing bodies of sports (e.g., International Olympic Committee, National Collegiate Athletic Association, U.S. Olympic Committee) have established stringent doping control regulations. Major classifications of banned substances include stimulants, narcotics, anabolic steroids, β blockers, diuretics, and peptide hormones and analogues. Urine samples are analyzed for these substances following major competitions.

Substances and techniques used as ergogenic aids include anabolic steroids, amphetamines, cocaine, growth hormone, caffeine, wheat germ oil, vitamins, blood doping, and oxygen breathing.

■ Selected Readings

Alen, M., K. Hakkinen, and P. V. Komi. Changes in neuromuscular performance and muscle fiber characteristics of elite power athletes self administering androgenic and anabolic steroids. *Acta Physiol. Scand.* 122: 535–544, 1984.

Alen, M., and P. Rahkila. Reduced high-density lipoprotein cholesterol in power athletes: use of male sex hormone derivates, an atherogenic factor. *Int. J. Sports Med.* 5: 341–342, 1984.

American Medical Association, Council on Scientific Affairs. Medical and non-medical uses of anabolic-androgenic steroids. *J. Amer. Med. Assoc.* 264: 2923–2927, 1990.

Appleby, M., M. Fisher, and M. Martin. Myocardial infarction, hyperkalaemia and ventricular tachycardia in a young male body-builder. *Int. J. Cardiol.* 44: 171–174, 1994.

Applegate, E. A., and L. E. Grivetti. Search for the competitive edge: a history of dietary fads and supplements. *J. Nutr.* 127(5 Suppl.): S869–S873, 1997.

Ariel, G., and W. Saville. Anabolic steroids: the physiological effects of placebo. *Med. Sci. Sports* 4: 124–126, 1972.

Bahrke, M. S., C. E. Yesalosk, and J. E. Wright. Psychological and behavioural effects of endogenous testosterone levels and anabolic-androgenic steroids among males: a review. *Sports Med.* 10: 303–337, 1990.

Bahrke, M. S., and W. P. Morgan. Evaluation of the ergogenic properties of ginseng. *Sports Med.* 18: 229–248, 1994.

Balsom, P. D., B. Ekblom, and K. Söderlund et al. Creatine supplementation and dynamic high-intensity intermittent exercise. *Scand. J. Med. Sci. Sports.* 3: 143–149, 1993.

Balsom, P. D., K. Söderlund, and B. Ekblom. Creatine in humans with special reference to creatine supplementation. *Sports Med.* 18: 268–280, 1994.

Beltz, S. D., and P. L. Doering. Efficacy of nutritional supplements used by athletes. *Clin. Pharm.* 12: 900–908, 1993.

Berdanier, C. D., J. A. Parente, Jr., and M. K. McIntosh. Is dehydroepiandrosterone an anti-obesity agent? *FASEB J.* 7: 414–419, 1993.

Bergström, J., E. Hultman, L. Jorfeldt, B. Pernow, and J. Wahren. Effect of nicotinic acid on physical working capacity and on metabolism of muscle glycogen in man. *J. Appl. Physiol.* 26: 170–176, 1969.

Buckley, W. E., C. E. Yasalis, K. E. Friedl, W. A. Anderson, A. L. Streit, and J. E. Wright. Estimated prevalance of anabolic steroid use among male high school seniors. *J. Amer. Med. Assoc.* 260: 3441–3445, 1988.

Burkett, L. N., and A. K. Bernstein. Strength testing after jaw repositioning with a mandibular orthopedic appliance. *Physician Sportsmed.* 10: 101–107, 1982.

Buskirk, E. R. Some nutritional considerations in the conditioning of athletes. *Ann. Rev. Nutr.* 1: 319–350, 1981.

Cartana, J., T. Segues, M. Yebras, N. J. Rothwell, and M. J. Stock. Anabolic effects of clenbuterol after long-term treatment and withdrawal in the rat. *Metabolism* 43: 1086–1092, 1994.

Carter, W. J., and M. E. Lynch. Comparison of the effects of salbutamol and clenbuterol on skeletal muscle mass and carcass composition in senescent rats. *Metabolism* 43: 1119–1125, 1994.

Casson, P. R., and J. E. Buster. DHEA administration to humans: panacea or palaver? *Sem. Reprod. Endocrinol.* 13: 247–256, 1995.

Castner, S., R. Early, and B. R. Carlton. Anabolic steroid effects on body composition in normal young men. *J. Sports Med. Phys. Fitness* 11: 98–103, 1971.

Chandler, J. V., and S. N. Blair. The effect of amphetamines on selected physiological components related to athletic success. *Med. Sci. Sports Exercise* 12: 65–69, 1980.

Coyle, E. F., J. M. Hagberg, B. F. Hurley, W. H. Martin, A. A. Ehsnai, and J. O. Holloszy. Carbohydrate feeding during prolonged strenuous exercise can delay fatigue. *J. Appl. Physiol.: Environ. Exercise Physiol.* 55: 230–235, 1983.

Cureton, T. K. The Physiological Effects of Wheat Germ Oil on Humans in Exercise. Springfield, Ill.: Charles C. Thomas, 1972.

Currier, D., J. Lehman, and P. Lightfoot. Electrical stimulation in exercise of the quadriceps femoris muscle. *Phys. Ther.* 59: 1508–1512, 1979.

Decombaz, J., O. Deriaz, K. Acheson, B. Gmuender, and E. Jequier. Effect of L- carnitine on submaximal exercise metabolism after depletion of muscle glycogen. *Med. Sci. Sports Exerc.* 25: 733–740, 1993.

Deyssig, R., H. Frisch, W. F. Blum, and T. Waldhor. Effect of growth hormone treatment on hormonal parameters, body composition and strength in athletes. *Acta Endo.* 128: 313–318, 1993.

Dickerman, R. D., F. Schaller, and W. J. McConathy. Left ventricular wall thickening does occur in elite power athletes with or without anabolic steroid use. *Cardiology* 90: 145–148, 1998.

DuRant, R. H., L. G. Escobedo, and G. W. Heath. Anabolic-steroid use, strength training, and multiple drug use among adolescents in the United States. *Pediatrics* 96: 23–28, 1995.

Ekblom, B., A. N. Goldbarg, and B. Gullbring. Response to exercise after blood loss and reinfusion. *J. Appl. Physiol.* 33: 175–180, 1972.

Evans, G. W. The effect of chromium picolinate on insulin controlled parameters in humans. *Int. J. Biosocial Med. Res.* 11: 163–180, 1989.

Fahey, T. Pharmacology of bodybuilding. In The Clinical Pharmacology of Sport and Exercise, T. Reilly and M. Orme (Eds.). Amsterdam: Elsevier Science B.V., 1997.

Fahey, T. D. Anabolic-androgenic steroids: mechanism of action and effects on performance. In Encyclopedia of Sports Medicine and Science. T. D. Fahey (Ed.). Internet Society for Sport Science: http://sportsci.org., 7 March 1998.

Fahey, T. D., and C. H. Brown. The effects of an anabolic steroid on the strength, body composition, and endurance of college males when accompanied by a weight training program. *Med. Sci. Sports* 5: 272–276, 1973.

Fahey, T. D., Hoffman, K. W. Colvin, and G. Lauten. The effects of intermittent liquid meal feeding on selected hormones and substrates during intense weight training. *Int. J. Sports Nutrition* 3: 67–75, 1993.

Fahey, T. D., and M. S. Pearl. The hormonal and perceptive effects of phosphatidylserine administration during two weeks of resistive exercise-induced overtraining. *Biol. Sport.* 15: 135–144, 1998.

Faigenbaum, A. D., L. D. Zaichkowsky, D. E. Gardner, and L. J. Micheli. Anabolic steroid use by male and female middle school students. *Pediatrics* 101: E6, 1998.

Ferner, R. E. Drug-induced diabetes. *Baillieres Clin. Endocrinol. Metab.* 6: 849–866, 1992.

Fields, L., W. R. Lange, N. A. Kreiter, and P. J. Fudala. A national survey of drug testing policies for college athletes. *Med. Sci. Sports Exerc.* 26: 682–686, 1994.

Fowler, W. M., G. W. Gardner, and G. H. Egstrom. Effect of an anabolic steroid on the physical performance of young men. *J. Appl. Physiol.* 20: 1038–1040, 1965.

Gledhill, N. Blood doping and related issues: a brief review. *Med. Sci. Sports Exerc.* 14: 183–189, 1982.

Gorostiaga, E. M., C. A. Maurer, and J. P. Eclache. Decrease in respiratory quotient during exercise following L-carnitine supplementation. *Int. J. Sports Med.* 10: 169–174, 1989.

Greenhaff, P. L., A. Casey, A. H. Short, R. Harris, K. Söderlund, and E. Hultman. Influence of oral creatine supplementation of muscle torque during repeated bouts of maximal voluntary exercise in man. *Clin. Sci.* 84: 565–571, 1993.

Harris, R. C., K. Soderlund, and E. Hultman. Elevation of creatine in resting and exercised muscle of normal subjects by creatine supplementation. *Clin. Sci.* 83: 367–374, 1992.

Harvey, G. R., A. V. Knibbs, L. Burkinshaw, D. B. Morgan, P. R. M. Jones, D. R. Chettle, and D. R. Vartsky. Effects of methandienone on the performance and body composition of men undergoing athletic training. *Clin. Sci.* 60: 457–461, 1981.

Huie, M. J. An acute myocardial infarction occurring in an anabolic steroid user. *Med. Sci. Sports Exerc.* 26: 408–413, 1994.

Ivy, J. L. Glycogen resynthesis after exercise: effect of carbohydrate intake. *Int. J. Sports Med.* 19 (Suppl. 2): S142–S145, 1998.

Ivy, J. L., D. L. Costill, W. J. Fink, and R. W. Lower. Influence of caffeine and carbohydrate feedings on endurance performance. *Med. Sci. Sports* 11: 6–11, 1979.

Johnson, C. C., M. H. Stone, R. J. Byrd, and S. A. Lopez. The response of serum lipids and plasma androgens to weight training exercise in sedentary males. *J. Sports Med.* 23: 39–44, 1983.

Kindermann, W., J. Keul, and G. Huber. Physical exercise after induced alkalosis (bicarbonate or tris-buffer). *Europ. J. Appl. Physiol.* 37: 197–204, 1977.

Kochakian, C. D. (Ed.). Anabolic-androgenic Steroids. Berlin: Springer-Verlag, 1976.

Lamb, D. R. Anabolic steroids in athletics: how well do they work and how dangerous are they? *Amer. J. Sports Med.* 12: 31–38, 1987.

Linderman, J., and T. D. Fahey. Sodium bicarbonate ingestion and exercise performance. *Sports Med.* 11: 71–77, 1991.

Linderman, J., T. D. Fahey, L. Kirk, J. Musselman, and B. Dolinar. The effects of sodium bicarbonate and pyridoxine-alpha ketoglutarate on short-term maximal exercise capacity. *J. Sports Sci.* 10: 243–253, 1992.

Litoff, D., H. Scherzer, and J. Harrison. Effects of panothenic acid supplementation on human exercise. *Med. Sci. Sports Exerc.* 17 (abstract): 87, 1985.

Lopes, J. M., M. Aubier, J. Jardim, J. V. Aranda, and P. T. Macklem. Effect of caffeine on skeletal muscle function before and after fatigue. *J. Appl. Physiol.* 54: 1303–1305, 1983.

Marconi, C., G. Sassi, and P. Cerretelli. The effects of alpha-ketoglutarate-pyridoxine complex on human maximal aerobic and anaerobic performance. *Eur. J. Appl. Physiol.* 49: 307–317, 1982.

Martikainen, H. M. Alen, P. Rahkila, and R. Vihko. Testicular responsiveness to human chorionic gonadotrophin during transient hypogonadotrophic hypogonadism induced by androgenic/anabolic steroids in power athletes. *J. Steroid Biochem.* 25: 109–112, 1986.

McKeever, K. H. Effects of sympathomimetic and sympatholytic drugs on exercise performance. *Vet. Clin. North Am. Equine Pract.* 9: 635–647, 1993.

Mohan, P. F., and M. P. Cleary. Effect of short term DHEA administration on liver metabolism of lean and obese rats. *Am. J. Physiol.* 255: E1-E8, 1988.

Morales, A. J., J. J. Nolan, J. C. Nelson, and S. S. Yen. Effects of replacement dose of dehydroepiandrosterone in men and women of advancing age. *J. Clin. Endocrinol. Metab.* 78: 1360–1367, 1994.

Nestler, J. E., C. O. Barlascin, J. N. Clore, and W. G. Blackard. Dehydroepiandrosterone reduces serum low density lipoprotein levels and body fat but does not alter insulin sensitivity in normal men. *J. Clin. Endocrinol. Metab.* 66: 57–61, 1988.

Nissen, S., R. Sharp, M. Ray, J. A. Rathmacher, D. Rice, J. C. Fuller, Jr., A. S. Connelly, and N. Abumrad. Effect of leucine metabolite beta-hydroxy-beta-methylbutyrate on muscle metabolism during resistance-exercise training. *J. Appl. Physiol.* 81: 2095–2104, 1996.

Peterson, G. E., and T. D. Fahey. HDL-C in five elite athletes using anabolic-androgenic steroids. *Physician Sportsmed.* 12(6): 120–130, 1984.

Philen, R. M., D. I. Ortiz, S. B. Auerbach, and H. Falk. Survey of advertising for nutritional supplements in health and bodybuilding magazines. *J.A.M.A.* 268: 1008–1011, 1992.

Phillips, B. 1998 Supplement Review. Denver: Mile High Publisher, 1998.

Pope, H. G., and D. L. Katz. Affective disorders and psychotic symptoms associated with anabolic steroids. *Amer. J. Psychiatry* 145: 487–490, 1988.

Pope, H. G., and D. L. Katz. Psychiatric and medical effects of anabolic-androgenic steroid use. A controlled study of 160 athletes. *Arch. Gen. Psychiatry* 51: 375–382, 1994.

Rechler, M. M., and S. P. Nissley. Insulin-like growth factors. In Peptide growth factors and their receptors. M. B. Sporn and A. B. Roberts (Eds.). New York: Springer-Verlag, 1990, pp. 263–367.

Reilly, T., and M. Orme. *The Clinical Pharmacology of Sport and Exercise.* Amsterdam: Elsevier Science B.V., 1997.

Repcekova, D., and L. Mikulaj. Plasma testosterone response to HCG in normal men without and after administration of anabolic drug. *Endokrinologie* 69: 115–118, 1977.

Reverter, J. L., C. Tural, A. Rosell, M. Dominguez, and A. Sanmarti. Self-induced insulin hypoglycemia in a bodybuilder. *Arch. Intern. Med.* 154: 225–226, 1994.

Rogozkin, V. Metabolism of Anabolic Androgenic Steroids. Leningrad: Hayka, 1988.

Rogozkin, V., and B. Feldkoren. The effect of retabolil and training on activity of RNA polymerase in skeletal muscles. *Med. Sci. Sports* 11: 345–347, 1979.

Rothwell, N. J., M. J. Stock, and D. K. Sudera. Changes in blood tissue flow and beta-receptor density of skeletal muscle in rats treated with the $\beta2$-adrenoceptor agonist clenbuterol. *Br. J. Pharmacol.* 90: 601–607, 1987.

Saartok, T., and T. Häggmark. Human growth hormone: friend or foe in sports medicine. *Bul. Hosp. Joint Dis. Orthop. Inst.* 48: 159–163, 1988.

Shephard, R. J., D. Killinger, and T. Fried. Response to sustained use of anabolic steroid. *Brit. J. Sports Med.* 11: 170–173, 1977.

Sidney, K. H. and N. M. Lefcoe. The effects of ephedrine on the physiological and psychological responses to submaximal and maximal exercise in man. *Med. Sci. Sports* 9: 95–99, 1977.

Smith, G. M., and H. K. Beecher. Amphetamine sulfate and athletic performance. *J.A.M.A.* 170: 542–557, 1959.

Smith, G. M., and H. K. Beecher. Amphetamine, secobarbital and athletic performance. *J.A.M.A.* 172: 1502–1514 and 1623–1629, 1960.

Souccar, C., A. J. Lapa, and R. B. doValle. The influence of testosterone on neuromuscular transmission in hormone sensitive mammalian skeletal muscles. *Muscle and Nerve* 5: 232–237, 1982.

Spivak, J. L. Erythropoietin: a brief review. *Nephron* 59: 289–294, 1989.

Strauss, R. H., J. E. Wright, G. A. M. Finerman, and D. H. Catlin. Side effects of anabolic steroids in weight-trained men. *Physician and Sports Med.* 11: 87–96, 1983.

Street, C., and J. Antonio. Effects of growth hormone administration on muscle and viscera in rats undergoing compensatory hypertrophy. *J. Strength and Cond. Res.* 10: 144–147, 1996.

Stromme, S. B., H. D. Meen, and A. Aakvaag. Effects of an androgenic-anabolic steroid on strength development and plasma testosterone levels in normal males. *Med. Sci. Sports* 6: 203–208, 1974.

Taylor, W. N. Anabolic Steroids and the Athlete. Jefferson, N.C.: McFarland, 1982.

Tennant, F., D. L. Black, and R. O. Voy. Anabolic steroid dependence with opioid-features. *New Eng. J. Med.* 319: 578, 1988.

Thein, L. A., J. M. Thein, G. L. Landry, and L. Gregory. Ergogenic aids. *Physical Therapy* 75: 426–439, 1995.

Thomson, J. M., J. A. Stone, A. D. Ginsburg, and P. Hamilton. O_2 transport during exercise following blood reinfusion. *J. Appl. Physiol.* 53: 1213–1219, 1982.

Trappe, S. W., D. L. Costill, B. Goodpaster, M. D. Vukovich, and W. J. Fink. The effects of L-carnitine supplementation on performance during interval swimming. *Int. J. Sports Med.* 15: 181–185, 1994.

Torgan, C. E., G. J. Etgen, J. T. Brozinick, R. E. Wilcox, and J. L. Ivy. Interaction of aerobic exercise training and clenbuterol: effects on insulin-resistant muscle. *J. Appl. Physiol.* 75: 1471–1476, 1993.

Tricker, R., M. R. O'Neill, and D. Cook. The incidence of anabolic steroid use among competitive bodybuilders. *J. Drug Educ.* 19: 313–325, 1989.

United States Olympic Committee. USOC Drug Education Handbook. Colorado Springs: USOC, 1989.

Van der Kuy, P. H., A. Stegeman, B. J. Looij, Jr., and P. M. Hooymans. Falsification of Thai Dianabol. *Pharm. World Sci.* 19: 208–209, 1997.

Wadler, G. I., and B. Hainline. Drugs and the Athlete. Philadelphia: F. A. Davis, 1989.

Ward, P. The effect of an anabolic steroid on strength and lean body mass. *Med. Sci. Sports* 5: 277–282, 1973.

Webb, O. L., P. M. Laskarzewski, and C. J. Glueck. Severe depression of high-density lipoprotein cholesterol levels in weight lifters and body builders by self-administered exogenous testosterone and anabolic-androgenic steroids. *Metabolism* 33: 971–975, 1984.

Williams, M. H. Drugs and Athletic Performance. Springfield, Ill.: Charles C. Thomas, 1974.

Williams, M. H. Nutritional Aspects of Human Physical and Athletic Performance. Springfield, Ill.: Charles C. Thomas, 1976.

Williams, M. H. The use of nutritional ergogenic aids in sports: is it an ethical issue? *Int. J. Sport Nutr.* 4: 120–131, 1994.

Williams, M. H. (Ed.). Ergogenic Aids in Sport. Champaign, Ill.: Human Kinetics, 1983.

Win-May, M., and M. Mya-Tu. The effect of anabolic steroids on physical fitness. *J. Sports Med. Phys. Fitness* 15: 266–271, 1975.

Wright, J. Anabolic Steroids and Sports. Natick, Mass.: Sports Science Consultants, 1978.

Wright, J. Anabolic steroids and athletics. *Exer. and Sports Sci. Rev.* 8: 149–202, 1980.

Wright, J. Anabolic Steroids and Sports, vol. 2. Natick, Mass.: Sports Science Consultants, 1982.

Wynn, V. Metabolic effects of anabolic steroids. *Br. J. Sports Med.* 9: 60–64, 1975.

Yang, Y. T., and M. A. McElligott. Multiple action of beta-adrenergic agonists on skeletal muscle and adipose tissue. *Biochem. J.* 261: 1–10. 1989.

Yarasheski, K. E., J. A. Cambell, K. Smith, M. J. Rennie, J. O. Holloszy, and D. M. Bier. Effect of growth hormone and resistance exercise on muscle growth in young men. *Am. J. Physiol.* 262 (Endo. Metab. 25): E261–E267, 1992.

Yesalis, C. E. and M. S. Bahrke. Anabolic-androgenic steroids. *Sports Med.* 19: 326–340, 1995.

CHAPTER 30

GENDER DIFFERENCES IN PHYSICAL PERFORMANCE

When prescribing exercise, it is important that gender be taken into consideration because the physiological characteristics of women and men are different. Menstruation and pregnancy affect the exercise response. Differences exist, too, in body composition, calcium and iron metabolism, and stature. These differences must be appreciated when assessing exercise responses in men and women. In addition, girls mature earlier than boys, which affects performance and patterns of physical activity in young people.

Mass participation in exercise and sport by women is relatively recent. Until 1958, the longest event in women's track and field in competitions hosted by the Amateur Athletic Union of the United States was 440 yards. In 1965, top female runners were threatened with banishment from international competition if they ran in a race longer than 1.5 miles. In 1984, the first Olympic marathon for women was held in Los Angeles. Now, it is common for women to compete in endurance events such as ultramarathons, triathlons, and long-distance swimming and cycling.

There are gender differences in physical performance. Males are larger, with more muscle mass. This gives them more strength and power. They also have larger hearts, which gives them a greater oxygen transport capacity. Aside from body and or-

Performances of Olympic and World speed skating champion Bonnie Blair exceed those of men only a generation ago. PHOTO: © Reuters/Corbis.

gan size, however, few gender differences affect responses to exercise training. Oxygen transport capacity is comparable when the effects of size are equalized. Responses to strength-training programs are relatively the same. At elite athlete levels, the performance gap between the sexes is closing (Figure 30-1). In most sports, the difference in performance between men and women is 6–13%. Regardless of differences in performance, sports and physical activity are good for both sexes.

■ Physiological Sex Differences

Sex Chromosomes

Gender is determined by the sex chromosomes. Human cells have 46 chromosomes, distributed in 23 pairs. In both sexes, 22 of the pairs are alike. The last pair, the sex chromosomes, differ in men and women: Women have two X chromosomes; men have one X chromosome and a smaller Y chromosome.

Sex chromosomes are responsible for the sex differences between men and women. The genetic material in the chromosomes acts as cellular blueprints, determining the sexual characteristics of genitals, muscle mass, heart size, and body fat. It is also involved in determining psychological traits such as aggression.

Sex chromosome tests have proved necessary because of scandals involving men posing as women athletes. Perhaps the most celebrated case involved Hermann Ratjen, a high jumper for Germany in the 1936 Olympics, who posed as a woman. He never would have been detected had he not publicly released his story in 1956. In 1968, a group of athletes,

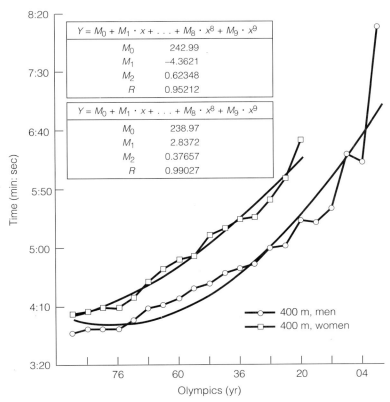

Figure 30-1 Olympic gold medal performances in the 400-m freestyle (swimming). The winning times for men and women were fitted to a two-factor polynomial regression equation. The trends show that performances are getting closer together and could theoretically match each other in the near future. However, mathematical trends can be misleading. Gender differences in the past could have been due to a myriad of factors, including differences in opportunity (training, facilities, coaching, etc.), talent pool (relative number of men and women active in the sport), as well as physiological sex differences (strength, muscle mass, size, etc.).

$Y = M_0 + M_1 \cdot x + \ldots + M_8 \cdot x^8 + M_9 \cdot x^9$

M_0	242.99
M_1	−4.3621
M_2	0.62348
R	0.95212

$Y = M_0 + M_1 \cdot x + \ldots + M_8 \cdot x^8 + M_9 \cdot x^9$

M_0	238.97
M_1	2.8372
M_2	0.37657
R	0.99027

—○— 400 m, men
—□— 400 m, women

Time (min: sec)

Olympics (yr)

including two former gold medal winners, withdrew from the Olympics rather than take the chromosome test. There are also several instances of Olympic medals won by women who later had surgery to change their sex.

The goal of chromosome tests is the same as that for drug tests: to give all athletes a fair chance in the competition. The first gender verification tests occurred in 1966 and involved physical examination. This was soon replaced by a buccal smear test for the Barr body (X chromatin). At the 1992 Olympics, the Barr test was replaced by the polymerase chain reaction (PCR) test, which detects the presence of the Y chromosome.

The tests are not without controversy. Since the onset of sex testing, no males with unusual chromosomes have been detected. In contrast, an estimated 1 in 500 women have been excluded due to unusual chromosomes. Genetic makeup is an important ingredient for Olympic success. Why should a woman with a genetic tendency toward superior endurance, for example, be awarded a gold medal, while a woman with a slight gender variant (e.g., single X, XY chromosomes) be disqualified? Women with XY chromosomes appear physically just like "normal" women. Their condition is called "androgen insensitivity syndrome." Although they produce more testosterone than the average woman, their cells are much less responsive to it than are those of men, so they derive no athletic advantage. The condition is part of normal human diversity.

Increased knowledge of human gene variations is changing the way governing bodies of sport think about gender. In track and field, gender determination tests have been eliminated. The feeling is that the risk of false positive tests, estimated at 1%, creates intolerable embarrassment to the participants. Also, supervised urine testing for doping control makes it nearly impossible for a man to masquerade as a woman. Unfortunately, this common sense approach has not been applied in all sports or adopted by the International Olympic Committee. Groups on record against requiring gender verification for athletic competitions include the American Academy of Pediatrics, the American College of Obstetricians and Gynecologists, the American College of Physicians, the American Medical Association, the American Society of Human Genetics, and the Endocrine Society.

Growth and Maturation

Growth rates in both sexes proceed in four major stages. The first stage is an accelerated growth during the prenatal and infant period. The second stage is a plateau in the rate of growth during childhood. The third stage is an acceleration in growth during puberty. The fourth stage is a plateau in growth rate during late adolescence.

Girls mature faster than boys. During childhood, it is not unusual for some girls to be bigger and more physically skilled than some boys. The pubertal growth spurt begins between 10 and 13 years in girls, and between 12 and 15 years in boys. Young adolescent girls are often taller than boys of the same age, but boys catch up and eventually surpass girls (see Chapter 31). Boys also grow for a longer time than girls, so males are, on average, about 10% taller and 17% heavier than females.

Body composition changes during puberty are caused by the increased secretion of gonadotropin hormones from the pituitary. This secretion is controlled by releasing factors in the hypothalamus. Hormone changes stimulate and increase estrogens in girls and androgens in boys. Estrogens increase adipose tissue, and have a slight retardant effect on lean body mass. Androgens, conversely, increase lean tissue and inhibit the development of body fat. Hormone changes during puberty partially cause the increase in sex differences in physical performance that occur during adolescence.

Boys have slightly more muscle mass than girls throughout growth, but at puberty, muscle mass in boys increases greatly. Girls' muscle mass also increases but much less than boys'. Gender differences in lean mass growth during puberty are due to higher androgen levels in boys. In females, the adrenal gland secretes low levels of androgens, principally androstenedione. Androgens are important hormones for protein synthesis. Testosterone in blood is similar in prepubertal boys and girls, ranging between 20 and 60 ng · 100 ml^{-1}. During adolescence, however, testosterone levels in males increase to adult levels of about 600 ng · 100 ml^{-1}, while those in females remain at prepubertal levels.

Anthropometric characteristics, with a few exceptions, are similar in boys and girls before puberty. Until about age 10, the sexes are similar in leg length, upper arm circumference, body weight, sitting height, and ponderal index. Ponderal index is height divided by the cubed root of weight. Throughout growth, boys are larger in the biepicondylar diameter of the femur and humerus, arm length, and chest and shoulder circumference. In relation to height, girls have larger hips throughout growth.

Anthropometric characteristics change considerably during puberty. Boys develop larger shoulders; girls develop larger hips. The smaller shoulder girdle makes it much more difficult for women to develop upper body strength, and the wider hips makes the angle of the femur much more pronounced than in males. Wider, larger hips give women a slightly lower center of gravity. Also,

women have relatively shorter legs. In women, the average leg length is 51.2% of height; in men, leg length averages 52% of height (Wells, 1985). Wider hips and shorter legs give women an advantage in activities requiring balance.

Body Composition

Figure 30-2 shows the body composition of the reference man and woman. The typical college-age female has more fat and less muscle and bone than the average college male. Active women have lower fat percentages than average (Table 30-1). Values below 22% are typical in recreational runners. Women in higher social classes are usually leaner. Ideal fat percentages are dictated as much by individual interpretation of fashion as by natural fat patterns.

Fat can be classified as essential and nonessential. Essential fat is stored in a wide variety of cells, including those in the central nervous system, muscle, heart, lungs, and bone marrow. Fat in these tissues is critical for metabolism, conduction of nerve impulses, cell structure, and protection from trauma. In women, sex-specific fat is also considered essential. Sex-specific essential fat in women

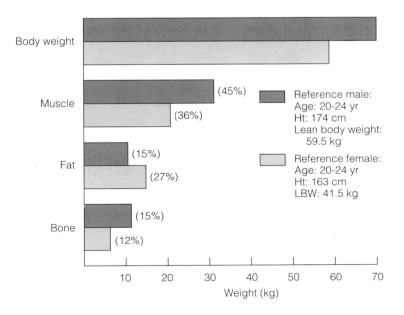

Figure 30-2 Gender differences in body composition of a reference man and woman. Source: Based on data from Behnke and Wilmore, 1974.

TABLE 30-1

Gender Differences in Fat Percentage
(50th percentile)

Age (yr)	Men	Women
20–29	21.6	25.0
30–39	22.4	24.8
40–49	23.4	26.1
50–59	24.1	29.3
Over 60	23.1	28.3

SOURCE: Data from Pollock, Wilmore, and Fox, 1978.

is stored in the breasts, subcutaneously (particularly the lower body), and genitals. In the reference woman of Figure 30-2, essential fat represents 40% of the total stored fat; the other 60% is storage fat. In the reference male, essential fat represents only 20% of total body fat.

The minimum percent fat for good health in women is unknown. In the reference woman, the minimum weight is approximately 49 kg. Extremely low fat percentages in women may have adverse health consequences. Health habits associated with low body fat include excessive exercise and eating disorders, such as anorexia nervosa and bulimia (see Chapter 25). Women must balance the need for low body fat with a body composition associated with good health.

The ideal lower limit of body fat for performance is not known. There is a 40 to 60% difference between men and women in $\dot{V}_{O_2max}$ (liters · min^{-1}). Gender differences are reduced to less than 10% when $\dot{V}_{O_2max}$ is expressed per kilogram lean body mass. So, low levels of body fat are desirable for peak endurance performance in women. However, world distance running records have been set by women with greater than 15% fat. All categories of female endurance athletes are leaner than sedentary women of the same age. Swimmers have more body fat than runners, cyclists, and cross-country skiers.

Higher body fat may be an advantage for women swimmers. When swimming at comparable velocities, women have lower body drag than men. This is due to more subcutaneous fat, which makes women more efficient at the sport. The ideal fat percentage of the female swimmer is also affected by fitness and stroke mechanics. Women could swim faster than men if they could develop comparable power. There is a 10% difference between the sexes in the world record in the 1500-m run. There is only a 6% difference between them in the 400-m swim. The lower drag among women swimmers may account for this reduced sex difference. Also, the availability of top-level coaching for women may also partially explain the improved performances. Top women swimmers today are swimming faster than did 1972 Olympic champion Mark Spitz.

Differences in body composition affect the exercise capacity and performance capacity of women. Women are at a distinct disadvantage in sports where they must lift or move their mass against gravity. In activities such as running, climbing, and jumping, they propel more body fat with less muscle mass than men. Greater body fat is also a disadvantage in that it slows the release of body heat during exercise.

Oxygen Transport and Endurance

Women's performance times are about 10% slower than men's in most track sports (Table 30-2). However, there is considerable variance in performance in specific events. In the 400-m swim, the difference between the men's and women's world record is slightly more than 8% (see Figure 30-1). Differences in world record times at distances between 800m and 100 km are consistently about 10%. However, in a study that used males and females with identical times in the marathon, females ran faster for 90 km (Speechly et al., 1996). They achieved this by maintaining a higher fraction of $\dot{V}_{O_2max}$. In cycling, men rode longer distances in the 1984 Olympic cycling road race competition (79.2 km for women and 190.2 km for men). Yet, the average velocity of the winning man was only 5% faster than that of the winning woman. There are slightly larger differences between the sexes in upper body endurance events, such as canoeing, because men have relatively more muscle in the upper body, which allows them to generate more power.

TABLE 30-2

Comparison of Performance Times Between Male and Female in Selected World Records in Athletics (Track), Swimming, and Cycling[a]

	Event	Males	Females	% Difference
Athletics	100	9.84	10.49	6.2
	200	19.32	21.34	9.5
	400	43.18	47.60	9.3
	800	1:41.73	1:53.28	10.2
	1500	3:26.00	3:50.46	10.6
	Mile	3:44.39	4:12.56	11.2
	3000	7:20.67	8:06.11	9.3
	5000	12:39.36	14.28.09	12.5
	10000	26:22.52	29:31.78	10.2
	Marathon	2:06.50	2:21.06	10.3
	100 km	9:16.41	7:00.47	10.5
Swimming	50 freestyle	21.81	24.51	11.0
	100 freestyle	48.21	54.01	10.7
	200 freestyle	1:46.69	1:56.78	8.6
	400 freestyle	3:43.80	4:03.85	8.2
	800 freestyle	7:46.00	8:16.22	6.1
	1500 freestyle	14:41.66	15:52.10	7.4
Cycling	Flying 200	9.865	10.831	9.0
	Flying 500	26.649	34.017	21.7
	1hr	56.375 km	48.195 km	14.5

[a] Distances are in meters unless otherwise noted

Determining gender differences has been difficult. In the past, some events, such as ultra-marathons, were not contested very often by women. This made it difficult to determine gender differences based on performance comparisons. Now, such races are common, which will facilitate gender comparisons. In the physiological responses to exercise, gender differences are also often unclear. In many studies, physically fit male subjects often were compared with sedentary female subjects.

Organ size and body mass are important in determining sex differences in endurance performance. Greater body size and stature provide a greater power output capacity, and men have more muscle mass, both in relative and absolute terms, while women have more fat. Greater fat-free weight is an asset in endurance performance, while more fat weight is a hindrance. Although muscle fiber composition is similar between the sexes, both fast- and slow-twitch muscle fibers are usually larger in men. Some studies have found males to have a slightly greater percentage of fast-twitch fibers.

Gender differences in endurance performance increase as sports level decreases. Thus, there are fewer sex differences between male and female elite athletes than between those of lesser standing. Strength and power differences are major reasons for sex differences in performance. Males and females make the same relative gains in strength when they are subjected to the same training stimuli. At elite levels, the training programs of men and women may be closer to each other in intensity than those of lower-level athletes. With years of training, men and women get closer to their absolute potential, and as they approach this point, it becomes possible to make more realistic comparisons of true sex differences.

Absolute maximal oxygen consumption (liters · min^{-1}) is typically more than 40% greater in men than women. This difference is reduced to approximately 20% when $\dot{V}_{O_2max}$ is expressed per kilogram body weight (Table 30-3). It decreases further, to less than 10%, when expressed per kilogram lean body weight. Although excess fat is a handicap to women endurance athletes, it does not appear to account for all sex differences in performance.

Adding extra weight to men in an attempt to equalize fat masses experimentally eliminates the differences between men and women in relative $\dot{V}_{O_2max}$. However, even with added weight, sex differences remain in the distance run in 12 minutes, maximum treadmill run time, and running efficiency. Fat percentage accounts for approximately 75% of the sex differences in running performance. The higher $\dot{V}_{O_2max}$ of men (ml · kg^{-1} lean body weight) accounts for 20%.

The average man has a larger heart size and heart volume than the average woman (in both absolute and relative terms). Studies have shown that in similarly trained endurance athletes, males have larger left ventricular heart mass, expressed per kilogram body weight, than women. The larger heart size results in a greater stroke volume during intense exercise and contributes to the sex differences in $\dot{V}_{O_2max}$. Even though women have a higher relative heart rate (higher % of max HR) during ex-

ercise, it is not enough to compensate for their lower stroke volume. The resultant smaller cardiac output of women contributes to their lower aerobic capacity. The amount and concentration of hemoglobin are also higher in men. This gives male blood greater oxygen-carrying capacity. Women average about 13.7 g Hb · 100 ml^{-1}. Men average 15.8 g Hb · 100 ml^{-1}. This difference is attributed to the stimulating effect of androgens on hemoglobin production and to the effects of menstrual blood loss and differences in dietary intake.

The literature is equivocal regarding the comparability in energy cost of running between men and women. Many studies have shown that males are more economical runners than females, largely because of lower body fat and greater lean body mass. However, it is extremely difficult to match male and female subjects on fitness level, training background, and running performance.

Some researchers suggest that the energy cost of running is related to the percentage of fast-twitch fibers. They have hypothesized that many women runners have a higher proportion of slow-twitch fibers than most men. From the standpoint of fiber distribution, women may have a predisposition for higher running economy. However, this fact is counterbalanced by their higher body fat. As discussed, other investigators have found no difference between men and women in the distribution of muscle fiber types. Saltin and colleagues (1977) found that men have greater extremes in fiber distribution. This may be significant in accounting for sex differences in athletes.

No gender differences are found in the ability to improve $\dot{V}_{O_2max}$ through training. As discussed in Chapter 21, $\dot{V}_{O_2max}$ can increase by up to 20% in both sexes. No sex differences are found either in improvement in endurance performance through interval and continuous exercise programs.

Resting and submaximal exercise heart rates are higher in women than men (Table 30-4). Because exercise oxygen requirements per kilogram body weight are the same for men and women, this means that the female heart must beat faster to make up for its lower pumping capacity. In similarly trained people, there is little difference in

TABLE 30-3

Gender Differences in Maximal Oxygen Consumption (ml $O_2 \cdot kg^{-1} \cdot min^{-1}$) (50th percentile)

Age (yr)	Men	Women
20–29	40.0	31.1
30–39	37.5	30.3
40–49	36.0	28.0
50–59	33.6	25.7
Over 60	30.0	22.9

SOURCE: Data from Pollock, Wilmore, and Fox, 1978.

TABLE 30-4

Gender Differences in Resting and Maximal Heart Rate (50th percentile)

| | Heart Rate (beats · min⁻¹) | | | |
| | Resting | | Maximal | |
Age (yr)	Men	Women	Men	Women
20–29	64	67	192	188
30–39	63	68	188	183
40–49	64	68	181	175
50–59	63	68	171	169
Over 60	63	65	159	151

SOURCE: Data from Pollock, Wilmore, and Fox, 1978.

maximal heart rate between the sexes. Because of larger maximal cardiac output, males have a higher systolic blood pressure during maximal exercise. Heart rate recovery is slower in women.

Physical Performance

Men have about a 10–20% greater capacity for endurance and short-duration, high-intensity exercise. This is largely due to factors already discussed, such as body and organ size. Gender differences in performance are greater in activities benefiting from strength and size, such as weight lifting and sprinting. Sex differences in swimming are reduced to 6–11% because of women's greater buoyancy.

Boys are slightly better than girls in sprinting, long jump, and high jump, and are much better in throwing skills. At puberty, sex differences become greater in most physical skills. Boys improve at a much greater rate; girls change very little or even regress. Gender differences in sports participation probably accounts for some of these changing trends.

Men and women respond similarly to all types of physical training. However, in the literature there are conflicting findings. When evaluating these studies, we must consider the relative fitness of the male and female subjects. Some studies show that women improve more than men, but this is usually because of a relatively higher initial fitness in males.

Males in these studies were usually closer to their maximum fitness potential. Many recent studies of gender differences have attempted to control for differences in fitness and training experience. The majority of these studies have shown equal training responses in men and women.

Men have a higher capacity than women for high intensity exercise. Men do much better on tests of maximal exercise capacity, such as the Wingate test (see Chapter 27), even when the results are expressed per kilogram lean body mass (fat-free weight). During short-term maximal exercise, men respond with higher epinephrine and lactate concentrations. This suggests either that men have a greater adrenergic response to intense exercise or that the studies compared fit men with less fit women.

Gender differences in temperature regulation is a controversial topic. Many early studies found that males better tolerated exercise in the heat. However, these studies usually used fit men and sedentary women as subjects. In studies using fit women, sweat rates during exercise were usually less. Acclimatization and control of body temperature was similar. In a study by Anderson and associates (1997), men and women were emersed in 28°C water, in which they exercised and then recovered. No gender differences were evident in the sweating, shivering threshold, or esophageal temperature (i.e., core temperature). Women as well as men may rely on circulatory mechanisms, such as altering vascular tone, to control body temperature. In general, relative fitness rather than sex differences are more important in determining heat tolerance and ability to acclimatize to heat.

Muscle Metabolism and Substrate Utilization

In equally trained male and female subjects, muscle glycogen, maximal exercise blood lactate, fat metabolism capacity, and muscle fiber composition are similar. Some aspects of female muscle biochemistry do not respond to training as well as those aspects in males. Examples include muscle glycogen synthesis and fat oxidation capacity. Muscle succinate dehydrogenase and carnitine palmitoyl transferase activities are higher in men. Muscle glycogen production is stimulated by testosterone, and testosterone concentrations are much higher in males.

Hormones such as estrogen (estradiol) and progesterone, which vary during the menstrual cycle in women, affect carbohydrate and fat metabolism. These hormones promote lipid oxidation and decrease glucose oxidation during exercise. In amenorrheic women (absence of menses), mean cycle estrogen and progesterone are ⅓ and ⅕, respectively, that of women with normal menses (eumenorrheic women). Consequently, women with amenorrhea experience decreased lipid mobilization and increased carbohydrate metabolism, while women with normal cycles experience the opposite effect. In some studies using sedentary subjects, gender differences appear in the relative selection and metabolism of fats and carbohydrates during exercise. These differences disappear as women become better trained. Highly trained female endurance athletes, who have lower levels of estrogens and progesterone, appear to metabolize such substrates as fats and carbohydrates similarly to trained males.

Hellstrom and coworkers (1996) found that during short-term submaximal exercise circulating lipids increase more in women than in men. This appears, at least in part, to be due to a sex difference in the adrenergic regulation of lipid mobilization during exercise. In their experiment, they selectively blocked α and β receptors with drugs. They found that exercise activates β- as well as α-adrenergic receptors in the adipose tissue of men, whereas only β receptors were activated in the fat cells of women.

Men appear to be better able to benefit from a high carbohydrate diet during exercise than women. Tarnopolsky and coworkers (1995) increased the dietary carbohydrate percentage from approximately 55 to 75% of the total caloric intake. The men increased muscle glycogen concentration by 41% and made a small improvement in performance in response to the diet, while muscle glycogen and performance in women did not increase at all. The women oxidized significantly more fat and less carbohydrate and protein compared with the men during moderately high-intensity exercise (75% $\dot{V}_{O_2max}$).

They concluded that women do not increase muscle glycogen in response to carbohydrate loading and that women exhibit greater lipid and lower carbohydrate and protein oxidation than men during submaximal endurance exercise.

Strength

Boys are stronger than girls throughout childhood, and the gap widens during adolescence (Chapter 19). Body weight and lean body mass account for much of this difference. In childhood, body weight and muscle mass are similar in boys and girls although, even then, male muscle can develop slightly more tension per cross-sectional area. In the adult, men are 40–50% stronger than women in most muscle groups. Sex differences in the strength of the lower body muscles are slightly less. As discussed, males have a greater proportion of their muscle mass in the upper body, which gives them an advantage in upper body strength. In common fitness tests, the greatest gender differences in strength are found in pulling hand grip and vertical jump performance.

Most studies show that adult males and females develop similar force per cross-sectional area of muscle. Average force development for men and women is 6 kg · cm². Some studies have shown slightly higher values in men (Ryushi et al., 1988). Power output is approximately 33% greater in men than in women. Garhammer examined power output in male and female Olympic weight lifters during the pulling phase of the snatch. Relative power output was 22.5 W · kg^{-1} in females and 34.4 W · kg^{-1} in males.

Muscle fibers in women and men are similar in distribution and histochemical characteristics. This is true in athletes and nonathletes. In women, all muscle fiber types have smaller cross-sectional areas. However, in male and female elite body builders, muscle fiber cross-sectional areas are similar. Most studies have found equal numbers of muscle fibers in men and women. Men have greater muscle pinnation angles than women, but this appears to be due to gender differences in muscle size.

Higher androgen levels in men partly account

for the large strength differences between the sexes. Androgens are potent anabolic hormones responsible for much of the muscle hypertrophy seen in men during the adolescent growth spurt. One hypothesis is that due to low levels of androgens, women will not develop larger muscles from strength training. And, in fact, several studies have found that strength training causes no changes in muscle cross-sectional area in women (Bailey et al., 1987; Wilmore, 1974; Mayhew and Gross, 1974; and Capen et al., 1961). Other studies, however, do not support this hypothesis. When men and women are put on identical training programs, relative changes in strength are the same. Cureton and colleagues (1988) and Hakkinen and colleagues (1992) found similar changes between the sexes in muscle hypertrophy following 16 and 10 weeks, respectively, of weight training. Both studies used sophisticated techniques (e.g., computer tomography and ultrasound) to accurately measure changes in cross-sectional area.

In women, testosterone may be of great importance for muscular power and strength development during weight training and an important indicator of trainability. Hakkinen and coworkers (1990) found that serum levels of both total and free testosterone were related to strength gains during a resistive exercise-training program.

As discussed in Chapter 20, early changes in strength during a weight-training program are due to neural factors. Changes in muscle cell size take much longer to develop. It is not known if the equal changes from training observed in men and women would have continued in longer studies, so it is premature to state that the potential for muscle hypertrophy is the same in males and females. Judging by the extreme muscular development in elite female bodybuilders, considerable muscle hypertrophy is possible in women. However, some of these women may have taken anabolic steroids. This makes it impossible to use these observations to determine sex differences in the potential for muscle hypertrophy.

To summarize, based on present evidence, women do experience hypertrophy from weight training. During the first few months of training, increases in muscle size are relatively the same in men and women. Gender differences in maximum attainable muscle hypertrophy (without the aid of anabolic steroids), however, are not known.

Exercise and the Menstrual Cycle

Menarche

Sexual maturation in the female is marked by *menarche*, the first menstruation. Menarche typically occurs after the peak of the adolescent growth spurt. It follows the appearance of pubic hair, breast development, and mature patterns of fat deposition. Menarche may occur when enough gonadotropin hormones are released from the pituitary and hypothalamus.

Athletes, particularly in sports such as gymnastics, swimming and running, reach menarche later than nonathletes. Menarche occurs at approximately 12.9 years. Stager and Hatler showed that menarche occurred later in swimmers than in their sedentary sisters. The study also showed that menarche occurred later than normal in the sedentary sisters, which suggests that delayed menarche is influenced by genetics as well as training. Pathological eating behaviors, relatively common in young girls, also contribute to delayed menarche.

Ethnic differences in the onset of menarche also exist. Eastern European girls attain menarche later than Western girls. Latin American girls reach menarche earlier than Caucasian Americans. These differences may be due to genetics and diet. Low protein intake combined with rigorous training may delay menarche. The age of menarche was much later in Europe after both world wars; these were periods of protein shortages. However, the influence of diet on the onset of menarche is unknown.

The age at which menarche and altered estrogen levels begin in girls has public health implications. In Western countries, young girls often consume high-fat, low-fiber diets and get little exercise. This lifestyle is thought to advance the onset of puberty, which means an earlier onset of menarche, breast development, and adolescent growth spurt. Both earlier menarche and earlier adult height in girls are markers of increased risk of breast cancer. Girls who have experienced early menarche exhibit higher estrogen and progesterone levels and lower levels of sex hormone binding globulin. Such differences can persist in women 20 to 30 years of age. These higher hormone levels result in a higher degree of breast epithelial proliferation, which is associated with an increased cancer risk.

Breast cancer is the most common cancer in women, and it ranks second behind lung cancer as a cause of cancer-related death. In Western countries, it affects one in ten women. Physical activity, particularly during late childhood and adolescence, may reduce the risk of breast cancer. Frisch and colleagues (1985) found a decreased incidence of reproductive organ cancer in women who were athletes in college. These women attained menarche later than women in the nonathletic control group. Likewise, recreational physical activity has been associated with a reduced risk of breast cancer in Japanese women (Ueji et al., 1998). However, a study by Chen and coworkers (1997) showed that neither leisure-time physical activity in the adolescent years nor in adulthood had a protective effect on breast cancer in young women. If exercise is effective in reducing breast cancer risk, it may be because exercise can decrease exposure to estrogen and progesterone, alter menstrual cycle patterns, delay menarche, increase energy expenditure, reduce body weight, influence insulin-like and other growth factors, and enhance natural immune mechanisms.

Amenorrhea and Oligomenorrhea

Menstrual cycles may be characterized as eumenorrheic, oligomenorrheic, or amenorrheic. *Eumenorrhea* is a normal menstrual cycle. *Oligomenorrhea* is irregular menstruation. *Amenorrhea* is absence of menstruation. Amenorrhea may be primary or secondary. In primary amenorrhea, menstruation has never occurred. In secondary amenorrhea, menstruation has occurred in the past but is currently absent. Irregular or absent menstruation occurs more in athletes than in sedentary women.

Regulation of the Menstrual Cycle The normal menstrual cycle is 28 days, although normal

cycles vary between 22 and 36 days. The menstrual cycle consists of follicular and luteal phases. The follicular phase begins on the first day of menstruation. It ends with ovulation. The luteal phase begins at ovulation and ends with menstruation. The luteal phase lasts 10–16 days. In a given woman, luteal-phase length is very consistent. The follicular phase is much more variable.

Events in the menstrual cycle are regulated by a variety of hormones in delicate balance with each other (Figure 30-3). Estrogen (estradiol) is produced continuously during the cycle. It is very low during the early follicular phase, rising to a peak during the late follicular phase and triggering ovulation. Progesterone and estrogen increase during the luteal phase. During the luteal phase, this increased estrogen causes the endometrium, the inner lining of the uterus, to thicken and mature.

Progesterone helps to mature and stabilize the endometrium. It prevents further endometrial proliferation and mitosis. Progesterone also changes the endometrium to a secretory structure that is ready for implantation of the fertilized ovum. At the end of the luteal phase, concentrations of estrogen and progesterone decrease, which triggers menstruation. Menstruation occurs from the breakdown of the endometrium.

The release of estrogens and progesterone is controlled by hypothalamic and pituitary hormones. Gonadotropin-releasing hormone (GnRH) is secreted from the hypothalamus. This controls the secretion from the pituitary of follicle-stimulating hormone (FSH) and luteinizing hormone (LH). FSH stimulates the growth of the ovarian follicle and estrogen production. It is also involved in the production of estrogen from androgens. Androgens are produced in the ovaries and can be chemically converted to estrogens. LH is released when estrogen reaches a critical level. LH triggers ovulation and is involved in transforming the follicle into a corpus luteum. Estrogen and progesterone are produced by the corpus luteum.

Factors That Disrupt the Menstrual Cycle
Many factors can disrupt the menstrual cycle (Table 30-5), and identifying a single cause has been very

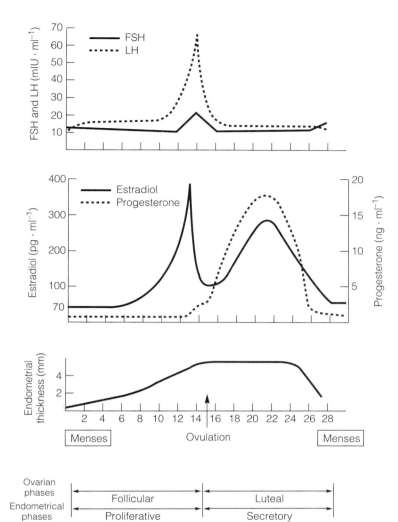

Figure 30-3   Hormonal events of the menstrual cycle, phases of the ovarian and endometrial cycles, and endometrial height throughout the menstrual cycle. mIU = micro international unit. SOURCE: Shangold, 1994. Used with permission.

difficult. A runner, for example, may be exercising heavily for her sport—but she may also be under stress and eating a poor diet. The present state of knowledge makes it difficult to determine accurately if exercise or a combination of factors alters menstrual function. Training may simply make pre-existing problems worse.

The incidence of menstrual irregularity is much higher in athletes than in sedentary women. Secondary amenorrhea in athletes is as high as 43%. The incidence in the general population is only 2–5%. Several studies have found that high training volume increases the incidence of amenorrhea (Figure 30-4). Increased training intensity is also a likely cause. Women involved in strength sports, such as bodybuilding or weight lifting, have an increased incidence of amenorrhea, as well, although the incidence is lower than that in endurance athletes.

Reproductive maturation may be important in establishing normal menstrual cycles. Women who have been pregnant before becoming involved in exercise training have a much lower incidence of

TABLE 30-5

Factors That May Disrupt the Menstrual Cycle

- Body composition
    - Weight loss
    - Low weight
    - Low body fat
- Poor nutrition
    - Low calorie intake
    - Inadequate protein intake
- Exercise
    - Prolonged
    - High intensity
    - Overtraining
    - Physical stress from job or activities
- Pregnancy

- Psychological stress
    - Training and competition
    - Personal
- Reproductive immaturity
- Acute and chronic endocrine alterations
    - Reduced luteinizing hormone (LH) pulse frequency and amplitude
    - Reduced gonadotropin-releasing hormone (GnRH)
    - Increases in stress hormones: endogenous opioids, cortisol, prolactin
    - Decreased estrogens
    - Increased testosterone
    - Increased catecholestrogen formation

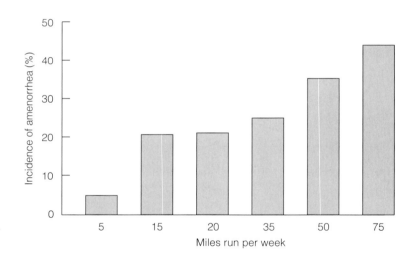

Figure 30-4 The effects of training mileage on the incidence of amenorrhea in runners.

amenorrhea than women who have not. Pregnancy may be a sign of maturation of reproductive function and may account for the more stable cycles. Women with previous menstrual irregularity are much more likely to have amenorrhea with training.

Diet may also play a role in amenorrhea. As discussed, delayed menarche and menstrual irregularity were reported in Europe following both world wars. These were periods of food shortages. Bullen and colleagues (1985) found that weight loss and intense exercise had a severe effect on the menstrual cycle. They studied women involved in heavy exercise training—with and without caloric restric-

tion. Of 14 subjects in an exercise–weight reduction group, only one had a normal cycle during the experiment. Heavy exercise, by itself, also disrupted most menstrual cycles. However, the effects were not as great as those when exercise and dietary restriction were combined.

Body composition may influence the menstrual cycle as well. Frisch and McArthur hypothesized that a minimum body fat percentage (i.e., 17%) was necessary to maintain normal menstrual function. Other studies have shown that amenorrheic runners were lighter, thinner, and lost more weight than runners with normal cycles. The basis of this theory is that fat is an important site for the production of estrogen. In adipose tissue, androstenedione is converted into the estrone by a process called aromatization. Androstenedione is an androgen and estrone, an estrogen. This theory is controversial, however. Studies have not found decreased estrone levels in amenorrheic women. Also, several studies have found no relationship between missed cycles and body composition (Abraham et al., 1982; Sinning et al., 1985).

The physiological basis for secondary amenorrhea in active women is controversial and not completely understood. The initial trigger may be a decrease in the secretion of GnRH from the hypothalamus and LH from the pituitary. Heavy training increases the secretion of hormones such as cortisol, endogenous opioids, and prolactin, which suppress the secretion of GnRH and LH. Constantini and Warren (1995) have suggested that mild hyperandrogenism in swimmers might contribute to menstrual irregularities. Exercise also increases catecholamine production, which results in an increased production of catecholestrogens. These substances also inhibit the production of GnRH. Decreased GnRH and LH reduce estrogen levels, resulting in the disruption of the menstrual cycle.

Consequences of Amenorrhea The consequences of amenorrhea may include infertility, failure to reach peak bone mass, premature bone loss, musculoskeletal injuries (mainly stress fractures), and scoliosis.

Fertility may be affected in overtrained women, regardless of whether they are menstruating normally. Often, the luteal phase of the cycle is affected. A study of marathon runners found that 8 of 24 women did not ovulate, even though they were menstruating normally. Disruptions in the cycle may be recognized only when attempts are made to conceive. The effects of exercise on fertility are temporary. Normal fertility and normal menstrual cycles are usually restored with rest and improved diet.

Amenorrhea (due to suppressed estrogen levels) can cause osteoporosis. Osteoporosis is the loss of bone minerals. It occurs in all postmenopausal women. Numerous studies have shown that young amenorrheic women athletes experience decreased bone density and fail to achieve their predicted peak bone density. The incidence of stress fractures in amenorrheic female athletes is more than 50% greater than in women with normal menses. Stress fractures in these athletes tend to occur in weight-bearing cortical bone. The reason for this is unknown.

Athletic women such as ballet dancers are more susceptible to scoliosis (a condition characterized by lateral curvature of the spine). Low estrogen and delayed puberty have been associated with scoliosis.

Exercise associated amenorrhea may also increase the risk of coronary heart disease. Estrogen increases the blood levels of high-density lipoproteins and apoprotein A-1 (thought to provide some protection against coronary heart disease) and decreases low-density lipoproteins. Intense training, leading to estrogen suppression and amenorrhea, may decrease the protective effects of estrogen.

Low levels of estrogen is the cause of bone loss in these women. Drinkwater and colleagues (1990) found that estradiol levels in amenorrheic runners was less than half that of normally menstruating runners. Trabecular bone, because of its greater turnover rate, is more affected than cortical bone.

There is no evidence that the bone loss can be reversed. Osteoporosis is a serious problem in older people. It predisposes people to fractures that can sometimes be life-threatening. Early bone loss in young athletic women may predispose them to fractures earlier in life. For that reason, early medical evaluation is very important. Estrogen therapy or oral contraceptives may be helpful in preventing bone loss in young amenorrheic athletic women. Such therapy should be initiated only after a thorough medical evaluation, however.

Irregular menses may be a sign of overtraining. Overtraining has many symptoms, including decreased performance, fatigue, irritability, depression, and suppression of the immune system. Biochemically, overtraining is accompanied by increased or decreased cortisol and increased endorphins. Exercise associated amenorrhea may not be a problem of women exercising but more of women overtraining.

Dysmenorrhea and Premenstrual Syndrome (PMS)

Dysmenorrhea is painful menstruation. In practical terms, it affects athletic performance more than amenorrhea. Many female athletes suffer from dysmenorrhea. Symptoms include PMS, abdominal cramps, backache, nausea, and vomiting. Dysmenorrhea detracts from the feeling of well-being. For that reason, it negatively affects performance. Some of the problem may be psychologically induced.

Exercise is considered beneficial to dysmenorrhea and is used by about 20% of women to treat its symptoms. However, the role of exercise in the treatment of dysmenorrhea is not completely understood. Presently, the first lines of defense against the condition are pharmacological agents, such as gonadotropin-releasing hormone (GnRH) agonists, benzodiazepines, and selective serotonin reuptake inhibitors (SSRIs).

Premenstrual syndrome (PMS) is characterized by mood change, bloating, cramping, headaches, weight gain, fatigue, and altered sex drive. It occurs during the luteal phase of the menstrual cycle. Symptoms typically begin at ovulation and are relieved with menstruation. The effects of PMS on exercise performance are small. However, the effects of menstrual cycle disturbances on exercise performance increased with the complexity and difficulty of the exercise. The risk of athletic injury is higher in women with PMS symptoms. The most important symptoms associated with athletic injuries are irritability, breast swelling, and abdominal congestion. Oral contraceptive use may decrease the risk of injuries in athletes with PMS.

The cause of PMS is unknown. Women with the syndrome do not show different levels from normal in the primary hormones controlling the menstrual cycle. Giannini and colleagues (1994) found that PMS may be related to altered β-endorphin metabolism in the central nervous system. Increased activity of norepinephrine may also be important. PMS symptoms such as backache, headache, cramps, and emotional distress may improve with exercise. Exercise may be beneficial because it increases β-endorphins in the central nervous system.

■ Pregnancy

Pregnancy has traditionally been treated almost as an illness. Pregnant women have been advised to rest and avoid unnecessary physical activity. However, moderate exercise is usually beneficial to the mother, and it does not place the fetus at increased risk—provided the intensity and duration of exercise are not too excessive (Figure 30-5).

Regular exercise during pregnancy counteracts the effects of deconditioning. Thus, it combats one of the causes of fatigue. Muscle strength is also better maintained, which may speed delivery and recovery from pregnancy. Faster delivery benefits both mother and baby. In addition, exercise may improve posture and decrease back pain. Other benefits include prevention of Type II diabetes and insulin resistance, improved emotional well-being, and prevention of excessive weight gain.

Exercise can be dangerous if excessive. The mother is susceptible to acute hypoglycemia, chronic fatigue, and increased risk of injury. In addition, excessive exercise may expose the fetus to hypoxemia, hyperthermia, and hypoglycemia. Heavy exercise can hypothetically induce premature labor and impair fetal development. Reduced birth weight is also theoretically possible if the mother does too much exercise.

Energy Cost of Pregnancy

Pregnancy increases the energy cost of daily activities. Likewise, pregnant women become deconditioned as they approach term. The result is often chronic fatigue and backache. Pregnancy has a higher energy cost because of the metabolic needs of the fetus. Metabolic rate increases—it takes more energy to move the larger body mass.

Considerable controversy surrounds the added caloric requirement of pregnancy. Some researchers estimate an added requirement of 80,000 kcal for the entire pregnancy. Increased energy is needed to build new tissues, and the higher energy cost of daily activities must be met. Other researchers believe that pregnant women usually decrease their

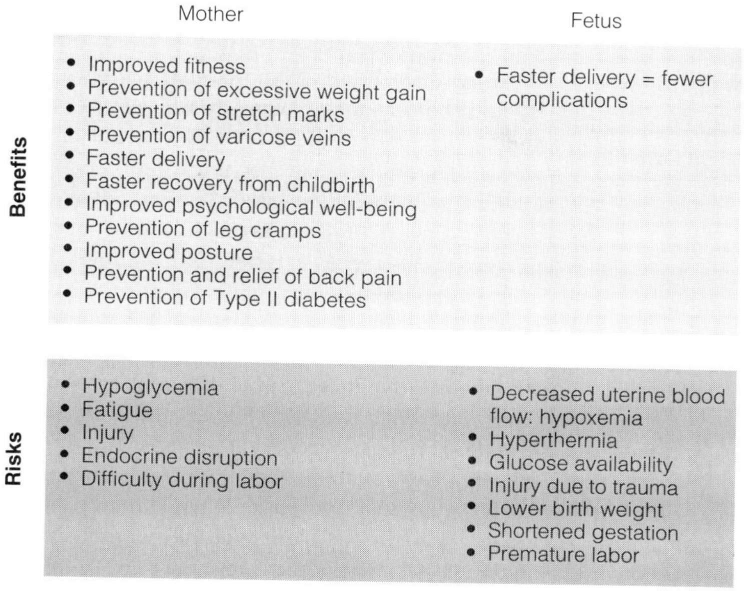

	Mother	Fetus

Benefits
- Improved fitness
- Prevention of excessive weight gain
- Prevention of stretch marks
- Prevention of varicose veins
- Faster delivery
- Faster recovery from childbirth
- Improved psychological well-being
- Prevention of leg cramps
- Improved posture
- Prevention and relief of back pain
- Prevention of Type II diabetes

- Faster delivery = fewer complications

Risks
- Hypoglycemia
- Fatigue
- Injury
- Endocrine disruption
- Difficulty during labor

- Decreased uterine blood flow: hypoxemia
- Hyperthermia
- Glucose availability
- Injury due to trauma
- Lower birth weight
- Shortened gestation
- Premature labor

Figure 30-5 Possible risks and benefits of exercise during pregnancy.

activity levels, due to the increased difficulty of movement. If true, then the added energy cost of pregnancy is minimal.

Energy balance is the ratio of energy consumed to energy expended. Management of caloric balance depends on the amount and intensity of daily activity. The caloric cost of activities is higher during pregnancy than the caloric cost of those same activities done in the same way when not pregnant. A sample of the caloric cost of common household activities during pregnancy is shown in Table 30-6.

Most obstetricians recommend an average weight gain of 11–16 kg (24–35 lb; Table 30-7). However, there are large individual differences in ideal weight gain. Also, experts do not agree on how much weight gain is desirable. Some caution against excessive weight gain, pointing to an increased risk of toxemia of pregnancy. In addition, extreme obesity can cause medical complications during pregnancy, and gained weight is often difficult to lose after delivery. Exercise may be helpful for avoiding excessive weight gain during pregnancy. However, few studies have shown effectively that exercise is responsible for lower maternal weight gains.

TABLE 30-6

Energy Consumed in Normal Activities During Pregnancy

Activity	Energy Consumed (kcal · min⁻¹)	
	Pregnant	Not Pregnant
Lying quietly	1.11	.95
Sitting	1.32	1.02
Sitting, combing hair	1.36	1.22
Sitting, knitting	1.55	1.47
Standing	1.41	1.12
Standing, washing dishes	1.63	1.33
Standing, cooking	1.66	1.41
Sweeping with broom	2.90	2.50
Bed making	2.98	2.66

Rigidly adhering to predetermined weight standards could lead to fetal and maternal malnutrition. Malnutrition during pregnancy may be an important factor in low birth weight. It may also increase the risk of toxemia. Toxemia, also called

TABLE 30-7

Components of Typical Weight Gain During Pregnancy

Component	Weight Gain (lb)
Baby	7
Amniotic fluid, placenta, and fetal membranes	4
Breasts	3–4
Uterus	2
Mother's tissue (protein 3 lb, fluid 3 lb, fat 2 lb)	8–18
TOTAL	24–35

eclampsia, is characterized by hypertension and proteinuria. Maternal malnutrition can affect the nutrition of the fetus, and in extreme cases, can lead to intrauterine growth retardation—the fetus stops growing. Many experts emphasize sound nutritional practices rather than caloric restriction during pregnancy.

Effects of Physical Fitness on Childbirth

Controversy surrounds the effects of fitness on ease of delivery. This probably stems from a lack of objective definition of fitness. Some studies consider a woman physically fit and active if she considers herself to be—even though most people are poor judges of their own physical condition. Studies of extremely fit women have shown that superior fitness speeds delivery and makes the experience less tiring.

Zaharieva (1972) studied Olympians who became pregnant. She found that the first stage of labor was the same as for nonathletes, which is understandable. This phase of childbirth involves involuntary uterine contractions. The second, or explosive, stage of labor was 50% shorter for athletes, probably due to the stronger abdominal muscles. No difference was found for the third phase, which involves expulsion of the placenta. Erdelyi (1962) found that athletes had less toxemia, threatened abortions, forceps deliveries, and cesarean sections.

Fitness was more important in ease of delivery than was innate physical superiority.

Several studies have shown that aerobic conditioning shortens labor. Wong and McKenzie (1987) and Clapp (1990) found that labor was shorter in fit women. Active women get the same benefits as Olympians in their deliveries. Physically fit mothers do not experience fatigue as much as unfit ones, and fit mothers also recover faster. Other studies, however, have shown no effect of exercise on the duration of labor.

Another consideration is that intense exercise training may have negative effects on the rate of delivery and birth rate. But Kardel and Kase (1998) found that intense or moderate exercise training did not negatively affect delivery rate and birth weight.

Fit mothers have fewer stretch marks (striae gravidarum) in the skin over the womb. In one study, stretch marks were found in 26.3% of athletes but in over 90% of sedentary women. Strong muscles distend less in response to the growing abdomen. Less stress is placed on the skin. Fit mothers also have less subcutaneous fat, which plays a role in the development of stretch marks.

Effects of Pregnancy on Fitness

Athletes Olympic medals have been won by women in the early stages of pregnancy. Nevertheless, in general, pregnant athletes are unable to maintain vigorous levels of training. The level of Olympic performance is so high that pregnancy and preparation for peak performance are incompatible.

Pregnancy probably has no effect on later athletic performance—although the increased time commitment of child care might. In 1972, 27% of Olympians who became pregnant participated in at least one more Olympic Games. Their performances usually improved. The most likely explanation for this improved performance is that only the most highly motivated athletes continue to participate in sports after childbirth. Their improvement merely represents the normal progression in their event. A more recent study of 30 elite Finnish endurance athletes found similar results (Penttinen and Erkkola, 1997). Eighteen of the thirty continued to compete within 2 to 24 months after delivery. Two of them (11%) achieved a better condition than before the pregnancy, eleven (61%) reached the same level and five (28%) did not achieve the same performance level.

Another hypothesis is that increased flexibility following pregnancy accounts for some of the increased success of these athletes. During pregnancy, a hormone called relaxin is released that facilitates the opening of the birth canal during delivery. Relaxin increases the flexibility of all the body's soft tissues, and this effect persists after pregnancy. It is extremely difficult to quantify accurately the effect of increased flexibility on performance, however. Consequently, the contribution of relaxin to later success in athletics is not known. Relaxin is withdrawn after delivery and probably has no long-term effects on flexibility. It is known, however, that increases in laxity of knee ligaments pose a small risk of injury to women practicing weight-bearing activities.

Aerobic Capacity   Pregnant women lose very little capacity for non–weight-bearing exercise such as cycling or swimming during the first two trimesters. However, the capacity for exercise against gravity (i.e., walking) decreases. For example, $\dot{V}_{O_2}$ increases by 13% during treadmill walking. However, there is no increase in the energy cost of submaximal cycling (a non-weight-bearing exercise). Throughout pregnancy, there are progressive increases in many measures of oxygen transport during weight-bearing submaximal exercise. These include O_2 consumption, cardiac output, stroke volume, heart rate, ventilation, ventilatory equivalent ($\dot{V}_e / \dot{V}_{O_2}$), and respiratory exchange ratio.

Cardiac output increases at rest and during exercise throughout pregnancy, and submaximal exercise heart rate increases progressively. Heart rate tends to increase in greater proportion than the increase in body weight. Heart rate at a given exercise intensity also continues to increase throughout pregnancy. Stroke volume contributes to the increase in cardiac output. Stroke volume increases largely due to an increase in maternal blood volume, which may increase by as much as 40% during

pregnancy. Stroke volume increases mostly during the first 20–24 weeks of gestation. Increased cardiac output makes it possible to supply skeletal muscle, while maintaining blood flow to the uterus. Cardiac output and stroke volume increase during pregnancy mainly due to a 35–45% increase in blood volume. Such a large increase in blood volume should increase blood pressure. Fortunately, increased blood volume is compensated for by an increase in venous capacitance (capacity to store more blood in the veins). Consequently, blood pressure does not increase and may actually decrease in the second trimester. The increased venous capacitance contributes to skin vasodilation. This promotes heat loss that may help to prevent hyperthermia during exercise.

Training and Aerobic Capacity   Exercise training during pregnancy results in improved aerobic capacity. In absolute terms ($\dot{V}_{O_2max}$ expressed in liters · min^{-1}), improvements from training are only slightly less than those seen in nonpregnant women. In relative terms ($\dot{V}_{O_2max}$ expressed per kilogram body weight), exercise maintains prepregnancy aerobic capacity.

Improvements in various cardiovascular indices have occurred from exercise programs involving walking, cycling, jogging, aerobics, cross-country skiing, and gymnastics. Physical training can increase endurance capacity in previously sedentary pregnant women and can minimize the effects of deconditioning in those who were previously trained. In spite of the benefits, however, even highly motivated women find it difficult to continue training during the first few months of pregnancy because of morning sickness.

In research studies, direct measurements of $\dot{V}_{O_2max}$ during pregnancy are rare. For safety reasons, $\dot{V}_{O_2max}$ is usually predicted from the relationship of heart rate and power output. As noted, during pregnancy, submaximal exercise heart rate progressively increases, while maximum heart rate tends to decrease.

Pulmonary Function   Ventilation increases during submaximal exercise in pregnancy as a result of changes in tidal volume (V_T) rather than breathing frequency (f). Elevated V_T increases alveolar ventilation. As a result, alveolar P_{CO_2} decreases. As pregnancy continues, respiratory sensitivity to P_{CO_2} increases. This helps to maintain respiratory responsiveness during pregnancy.

The work of breathing is higher during late gestation due to restriction of the diaphragm. Dyspnea (distressed breathing), a common complaint during late pregnancy, may contribute to the decreased activity patterns observed in some women.

The Effects of Exercise on the Fetus

Exercise may subject the fetus to trauma, inadequate oxygen and glucose availability, and hyperthermia. These risks can be significant if exercise is excessive or the trauma severe. Chronic fetal insults from excessive exercise can lead to fetal death or altered development.

Trauma The female anatomy provides good protection to the fetus during pregnancy. The pelvis, uterine wall, and amniotic fluid can absorb considerable shock. These tissues enable safe participation in most activities. During the embryonic phase (early pregnancy), amniotic fluid provides a buoyant medium for the delicate tissues. During the later stages of pregnancy, it allows the fetus to move freely and change body position.

Blows to the abdomen resulting from trauma can be hazardous, however. Sudden rushes of water through the vagina, such as might occur during water skiing, should also be avoided. Changes of pressure, such as occur during scuba diving, also place the fetus and mother at risk.

Fetal Circulation The fetus depends on placental circulation for gas exchange and nutrients. As discussed in Chapter 15, circulatory controls are aimed at maintaining blood pressure and central blood volume. During exercise in the pregnant woman, these circulatory priorities change very little. During intense or prolonged exercise, for example, a general circulatory vasoconstriction still helps maintain blood flow to the heart. However,

now, this process directs blood away from the placenta, which could have severe consequences for the fetus. Chronic fetal hypoxemia can result in fetal death or hampered development.

Pregnancy places an added load on the cardiovascular system of the mother. During exercise, the circulatory needs of the fetus must be added to those of the maternal physiology. There is increased competition for blood in the pregnant woman during physical activity. The needs of muscle, placenta, and central circulation all must be met—and this may not be possible during intense or prolonged exercise. Studies in animals and humans have shown that exercise reduces placental blood flow. In spite of this, the vast majority of studies show no differences in birth weight or gestational age between exercising and nonexercising pregnant women.

The fetus has several ways of protecting itself from hypoxemia. During maternal exercise, there is an increased hemoconcentration, and placental oxygen extraction increases. Thus, although placental blood flow decreases during exercise, uterine and umbilical oxygen uptakes are unchanged because of increased oxygen extraction. There is a redistribution of blood flow to the placenta after exercise. This is a reactive hyperemia that might represent a compensatory mechanism for the fetus.

Fetal circulation is well prepared to meet the stresses of hypoxemia created by the exercising mother. Fetal hemoglobin can carry 20 to 30% more oxygen than maternal hemoglobin. Fetal hemoglobin is also about 50% more concentrated. The fetus has a relatively small demand for oxygen: Its muscular activity is low and its body temperature is kept fairly constant by the fluid environment.

Fetal heart rate is a good measure of hypoxemia. Fetal hypoxemia is suggested if fetal heart rate differs from the normal range of 120–160 bpm. Exercise stress tests may be an early means of detecting inadequate gas exchange in the uterus and placenta. In the normal fetus, heart rate goes up during maternal exercise and, during recovery, it gradually decreases. In uteroplacental insufficiency, fetal heart rate response is irregular during maternal exercise, followed by bradycardia during recovery. The higher the degree of potential fetal distress from hypoxemia, the more likely is fetal bradycardia to occur during and after maternal exercise.

Fetal Development Most human studies have found no effects of moderate exercise on the outcome of pregnancy. Some animal studies have also shown no ill effects of exercising while pregnant. Animal studies have suggested, however, a variety of risks of excessive exercise during pregnancy. Negative effects, which occur when exercise duration or intensity are excessive, include increased fetal mortality, low birth weight, decreased diffusing capacity of the placenta, reduced kidney weight, and abnormal liver function.

Most studies, but not all, show that physically fit women tend to give birth to heavier babies. In contrast, animal studies using African pygmy goats, mice, sheep, and guinea pigs found that chronic exercise reduced fetal weight at term. In guinea pigs, birth weight, placental diffusion capacity, and placental weight decreased with increasing intensity and duration of exercise. Exercise training lasting more than 30 minutes per session caused the most serious effects. If the results apply to humans, excessive exercise may have negative effects on the quality of fetal and child development. Occupations that require prolonged standing or other forms of physical activity may increase the risk of low birth weight and pre-term delivery. In some, but not all, studies of low-income women, the number of hours spent standing during the day was related to the risk of premature childbirth. In Western countries, prolonged standing does not seem to be related to birth weight. However, in developing countries, occupational exercise is highly related to lower birth weight and adverse pregnancy outcome.

Fetal Hyperthermia Exercise-induced fetal hyperthermia is a risk, particularly during the first trimester of pregnancy. Rectal temperatures of 40°C are not unusual in recreational marathoners and road racers. A woman could easily mistake pregnancy for the irregular menses that is so common among distance runners.

Central nervous system defects have been reported in laboratory animals in response to hyper-thermia. The greatest danger is between 21 and 28 days following conception. This is a critical time of central nervous system development during embryonic growth. Central nervous system defects in fetuses exposed to maternal hyperthermia have been reported by many studies. The women in many of these studies suffered from febrile illnesses. It may be that the illness and not the fever caused the CNS defects. It is not known if the results apply to exercise.

Hyperthermia decreases placental blood flow. This effect is most serious during the later months of gestation. The problem is worse if the environmental temperature is high, the mother is dehydrated, or the exercise is prolonged or intense. Pregnant women or women who could possibly be pregnant should avoid core temperatures greater than 38.9°C.

Exercise Prescription in Pregnancy

Many clinical studies have been done on the effects of exercise on pregnancy in humans. In general, these studies have been positive. Findings have included easier delivery, no apparent fetal defects, and normal birth weight. Many factors are involved in producing a successful outcome of pregnancy, including diet, genetics, stress, alcohol and drug consumption, and smoking. The negative effects of excessive exercise during pregnancy, shown in animal studies, may be balanced by positive nutritional and health practices. Studies show that excessive exercise can be harmful. Prudence should be used and individual differences taken into consideration when prescribing endurance exercise. Moderate exercise is beneficial, but excessive exercise can compromise fetal well-being.

No pregnant woman should begin an exercise program without a medical evaluation. Several absolute and relative contraindications for exercise during pregnancy exist (Table 30-8). During the evaluation, the risks and benefits of exercise should be assessed. Several groups and authorities, including the American College of Obstetricians and Gynecologists (ACOG) and the American College of Sports Medicine, have established guidelines for ex-

---

TABLE 30-8

Absolute and Relative Contraindications for Aerobic Exercise During Pregnancy

Absolute Contraindications
- Valvular or ischemic heart disease
- Type I diabetes, peripheral vascular disease, uncontrolled hypertension, thyroid disease, or other serious systemic disorder (e.g., hepatitis, mononucleosis)
- Incompetent cervix
- History of two or more spontaneous abortions in previous pregnancies
- Bleeding or placenta praevia
- Ruptured membranes or premature labor in current pregnancy
- Toxemia or preeclampsia in current pregnancy
- Smoking or excessive alcohol intake (>2 drinks/day)
- Low body fat, history of anorexia nervosa
- Multiple pregnancy

Relative contraindications
- History of premature labor, intrauterine growth retardation, preeclampsia or toxemia
- Anemia or iron deficiency (Hb < 10 g $\cdot$ dl^{-1})
- Significant pulmonary disease (e.g., asthma, chronic obstructive pulmonary disease)
- Mild valvular heart disease, clinically significant cardiac arrhythmias
- Obesity or type II diabetes mellitus
- Very low physical fitness prior to pregnancy
- Medications that can alter cardiac output or blood flow distribution
- Breech presentation in third trimester
- Presence of twins (after 24 weeks of gestation)

SOURCE: Wolfe et al., 1989. Used with permission.

---

ercising during pregnancy. Because of the many unknowns and the possible consequences of over-exercising, these guidelines are conservative.

Exercise should consist of moderate-intensity physical activity. Appropriate activities include walking, cycling, swimming, aerobics, and jogging. Pregnant women should avoid contact sports and sports that pose unknown physiological effects to the fetus. These latter include scuba diving and mountaineering at high altitudes. High-intensity sports involving rapid direction changes should also be avoided, as these sports increase the risk of injury. Weight training is recommended by some

authorities and discouraged by others. If used, resistance should be low and performance of the valsalva maneuver should be minimized. Women should not learn new sports during pregnancy but may start an exercise program if monitored.

In 1994, the ACOG modified its 1985 recommendations for exercise during pregnancy to allow women to train more intensely. They suggest that pregnant women do not need to limit the intensity of exercise to any particular heart rate. Rather, they should instead monitor intensity with subjective perception of exertion. A summary of the ACOG's exercise recommendations (1994) during pregnancy is as follows:

- During pregnancy, women can continue their mild to moderate exercise routines. Regular exercise (at least three times per week) is preferable to intermittent activity.

- Pregnant women should not exercise in the supine position after the first trimester because it can decrease placental and fetal brain blood. Pregnant women should also avoid prolonged standing.

- Because of competition for blood from placental circulation and for temperature regulation, less oxygen is available in the mother for aerobic exercise during pregnancy. Consequently, exercise intensity should be modified accordingly. Women should stop exercising when fatigued and should not exercise to exhaustion. Pregnant women may be able to continue doing weight-bearing exercise at close to their usual intensity throughout pregnancy, but non–weight-bearing exercises such as swimming or cycling are easier to continue, and they carry less risk of injury.

- Pregnant women should not do exercises that could cause loss of balance, especially in the third trimester. They should avoid any exercise that risks even mild abdominal trauma.

- Pregnant women need an additional 300 calories a day during pregnancy. If exercising, women should be particularly careful to ensure they have an adequate diet.

- Pregnant women should be particularly careful to avoid thermal distress during the first trimester. They should drink enough water, wear cool clothing, and not exercise in excessively hot environments.

- After giving birth, pregnant women should resume their prepregnancy exercise routines gradually, based on physical capacity.

Breasts should be supported during exercise. During running, the movement of unsupported breasts against the chest creates between 50 to 100 pounds of force. Prolonged periods of training without adequate breast support can result in inflamed nipples and eventually lead to sagging breast tissue. Sports bras are currently available that adequately support the breasts during physical activity.

In general, pregnancy calls for moderate, sensible levels of exercise. High-intensity, competitive exercises are not in the interest of the fetus and may also increase the risk of injury to the mother.

■ Summary

When prescribing exercise, it is important that gender be taken into consideration. There are gender differences in physical performance. Men are larger, with more muscle mass. This gives them more strength and power. They also have larger hearts, which gives them a greater oxygen transport capacity. Aside from body and organ size, however, few gender differences affect responses to exercise training.

Sex chromosomes are responsible for the sex differences between men and women. For many years, gender verification was accomplished with a buccal smear test for the Barr body (X chromatin). This has recently been replaced by the polymerase chain reaction (PCR) test, which detects the presence of the Y chromosome. The tests are controversial, and many sports leaders are attempting to eliminate them from athletic competitions.

Girls mature faster than boys. Young adolescent girls are often taller than boys of the same age, but the boys catch up and eventually surpass girls in almost all aspects of physical performance.

When children stop growing, the performance times of young women are 6 to 15% slower than those of young men in most endurance sports. Organ size and body mass are important in determining sex differences in endurance performance. Greater size provides a greater power output capacity. Men have more muscle mass, both in relative and absolute terms, while women have more fat. The average man has a larger heart size and heart volume than the average woman (in both absolute and relative terms). This results in a greater stroke volume during intense exercising. It also contributes to the gender differences in $\dot{V}_{O_2max}$. Gender differences in swimming are reduced to 7 to 13% because of the greater buoyancy of women.

In equally trained men and women subjects, muscle glycogen, maximal exercise blood lactate, fat metabolism capacity, and muscle fiber composition are similar.

Males are stronger than females throughout childhood, and the gap widens during adolescence. Body weight and lean body mass account for most of the sex differences.

Sexual maturation in the female is marked by menarche, the first menstruation. Menstrual cycles may be characterized as eumenorrheic, oligomenorrheic, or amenorrheic. The incidence of menstrual irregularity is much higher in athletes than in sedentary women. The causes of menstrual irregularity are not completely understood. Factors such as low body fat, eating disorders, excessive exercise, or stress may play a role.

The mechanism for amenorrhea may be a decrease in secretion of gonadotropin-releasing hormone (GnRH) from the hypothalamus triggered by the secretion of hormones such as cortisol, endogenous opioids, and prolactin. These hormones suppress the secretion of GnRH and luteinizing hormone (LH), which reduce estrogen levels. A relatively common occurrence among some women athletes is the female triad: amenorrhea, eating disorders, and osteoporosis.

Moderate exercising during pregnancy is usually beneficial to the mother and does not place the fetus at increased risk. Regular exercise during pregnancy counteracts the effects of deconditioning and helps maintain muscle strength. This may speed delivery and recovery from pregnancy, which benefits both mother and baby. Exercise may improve posture and decrease back pain, prevent Type II diabetes, improve emotional well-being, and prevent excessive weight gain. However, exercise can be dangerous for mother and fetus if excessive.

■ Selected Readings

Abe, T., W. F. Brechue, S. Fujita, and J. B. Brown. Gender differences in FFM accumulation and architectural characteristics of muscle. *Med. Sci. Sports Exerc.* 30: 1066–1070, 1998.

Abraham, S., P. J. Beaumont, I. F. Fraser, and D. Llewellyn Jones. Body weight, exercise, and menstrual states among ballet dancers in training. *Brit. J. Obstet. Gynec.* 99: 507–510, 1982.

Agostini, R. (Ed.). Medical and Orthopedic Issues of Active and Athletic Women. Philadelphia: Hanley & Belfus, Inc., 1994.

American College of Obstetricians and Gynecologists. Pregnancy and the postnatal period. *ACOG Home Exercise Program,* 1985.

American College of Obstetricians and Gynecologists. Exercise during pregnancy and the postpartum period. ACOG Technical Bulletin Number 189—February 1994. *Int. J. Gynaecol. Obstet.* 45: 65–70, 1994.

American College of Sports Medicine. Position Stand on the Female Athletic Triad, 1995.

American College of Sports Medicine. Guidelines for Graded Exercise Testing and Prescription. 5th Ed. Baltimore: Williams and Wilkins, 1995.

Anderson, G. S., R. Ward, and I. B. Mekjavic. Gender differences in physiological reactions to thermal stress. *Eur. J. Appl. Physiol.* 71: 95–101, 1995.

Apter, D. Hormonal events during female puberty in relation to breast cancer risk. *Eur. J. Cancer Prev.* 5; 476–482, 1996.

Bailey, L. L., W. C. Byrnes, A. L. Dickinson, and V. L. Foster. Muscular hypertrophy in women following a concentrated resistance training program. *Med. Sci. Sports Exerc.* 19 (Suppl.): S16, 1987.

Batterham, A. M., and K. M. Birch. Allometry of anaerobic performance: a gender comparison. *Can. J. Appl. Physiol.* 21: 48–62, 1996.

Behnke, A. R., and J. H. Wilmore. Evaluation and Regulation of Body Building and Composition. Englewood Cliffs, N.J.: Prentice-Hall, 1974.

Bernstein, L., B. E. Henderson, R. Hanisch, J. Sullivan-Halley, and R. K. Ross. Physical exercise and reduced risk of breast cancer in young women. *Natl. Cancer Inst.* 86: 1403–1408, 1994.

Blackburn, M., and D. Calloway. Energy expenditure of pregnant adolescents. *J. Am. Dietetics* 65: 24–30, 1974.

Bourdin, M., J. Pastene, M. Germain, and J. R. Lacour. Influence of training, sex, age and body mass on the energy cost of running. *Eur. J. Appl. Physiol.* 66: 439–444, 1993.

Bullen, B. A., G. S. Skrinar, I. Z. Beitins, G. von Mering, B. A. Turnbull, and J. W. McArthur. Induction of menstrual disorders by strenuous exercise in untrained women. *N. Engl. J. Med.* 312: 1349–1353, 1985.

Bunc, V., and J. Heller. Energy cost of running in similarly trained men and women. *Eur. J. Appl. Physiol.* 59: 178–183, 1989.

Capen, E. K., J. A. Bright, and P. A. Line. The effects of weight training on strength, power, muscular endurance and anthropometric measurements on a select group of college women. *J. Assoc. Phys. Ment. Rehab.* 15: 169–173, 1961.

Chen, C. L., E. White, K. E. Malone, and J. R. Daling. Leisure-time physical activity in relation to breast cancer among young women (Washington, United States). *Cancer Causes Control* 8: 77–84, 1997.

Choi, P. Y., and P. Salmon. Symptom changes across the menstrual cycle in competitive sportswomen, exercisers and sedentary women. *Br. J. Clin. Psychol.* 34: 447–460, 1995.

Clapp, J. F. Acute exercise stress in the pregnant ewe. *Am. J. Obstet. Gynecol.* 136: 489–494, 1980.

Clapp, J. F., III. The course of labor after endurance exercise. *Am. J. Obstet. Gynecol.* 163: 1799–1805, 1990.

Clapp, J. F. A clinical approach to exercise during pregnancy. *Clin. Sports Med.* 13: 443–458, 1994.

Collings, C. A., L. B. Curet, and J. P. Mullin. Maternal and fetal responses to a maternal aerobic exercise program. *Amer. J. Obstet. Gynecol.* 146: 702–707, 1983.

Constantini, N. W. Clinical consequences of athletic amenorrhoea. *Sports Med.* 17: 213–223, 1994.

Constantini, N. W., and M. P. Warren. Special problems of the female athlete. *Baillieres Clin. Rheumatol.* 8: 199–219, 1994.

Constantini, N. W., and M. P. Warren. Menstrual dysfunction in swimmers: a distinct entity. *J. Clin. Endocrinol. Metab.* 80: 2740–2744, 1995.

Cureton, K. J., M. A. Collins, D. W. Hill, and F. M. McElhannon. Muscle hypertrophy in men and women. *Med. Sci. Sports Exerc.* 20: 338–344, 1988.

Cureton, K. J., and P. B. Sparling. Distance running performance and metabolic responses to running in men and women with excess weight experimentally equated. *Med. Sci. Sports Exerc.* 12: 288–294, 1980.

Dingeon, B., J. L. Simpson, A. Ljungqvist, A. Chapelle, M. Ferguson-Smith, E. Ferris, M. Genel, A. Ehrhardt, and A. Carlson. Gender verification in the next Olympic Games. *J. Am. Med. Assoc.* 269: 357–360, 1993.

Drinkwater, B. L. Physical exercise and bone health. *J. Am. Med. Wom. Assoc.* 45: 91–97, 1990.

Drinkwater, B. L. Exercise in the prevention of osteoporosis. *Osteoporos. Int.* 3 (Suppl. 1): 169–171, 1993.

Drinkwater, B. L., B. Bruemner, and C. H. Chesnut. Menstrual history as a determinant of current bone density in young athletes. *JAMA* 263: 545–548, 1990.

Drinkwater, B. L., and C. H. Chesnut. Bone density changes during pregnancy and lactation in active women: a longitudinal study. *Bone Miner* 14: 153–160, 1991.

Dugowson, C. E., B. L. Drinkwater, and J. M. Clark. Nontraumatic femur fracture in an oligomenorrheic athlete. *Med. Sci. Sports Exerc.* 23: 1323–1325, 1991.

Dumas, G. A., and J. G. Reid. Laxity of knee cruciate ligaments during pregnancy. *J. Orthop. Sports Phys. Ther.* 26: 2–6, 1997.

Erdelyi, G. J. Gynecological survey of female athletes. *J. Sports Med. Phys. Fitness* 2: 174–179, 1962.

Erkkola, R. The influence of physical training during pregnancy on physical working capacity and circulatory parameters. *Scand. J. Clin. Invest.* 36: 747–754, 1976.

Fahey, T. D. Endurance. In Women and Exercise: Physiology and Sports Medicine, M. Shangold and G. Merkin (Eds.). Philadelphia: F. A. Davis, 1994.

Fahey, T. D., S. I. Gates, W. Colvin, G. D. Swanson, and J. K. Linderman. The effects of prolonged, intense exercise on estradiol, progesterone, LH and FSH concentrations during mid-menstrual cycle. *Biol. Sport.* 14: 175–183, 1997.

Friedenreich, C. M., and T. E. Rohan. A review of physical activity and breast cancer. *Epidemiology* 6: 311–317, 1995.

Frisch, R. E., and J. W. McArthur. Menstrual cycles: fatness as a determinant of minimum weight for height necessary for their maintenance or onset. *Science* 185: 949–951, 1974.

Frisch, R. E., G. Wyshak, N. L. Albright, T. E. Albright, I. Schiff et al. Lower prevalence of breast cancer and cancers of the reproductive system among former college athletes compared to non-athletes. *Brit. J. Cancer* 52: 885–891, 1985.

Fuster, V., A. Jerez, and A. Ortega. Anthropometry and strength relationship: male-female differences. *Anthropol. Anz.* 56: 49–56, 1998.

The gender unbenders. *Economist* 322 (7744): 97–98, 1992.

George, K. P., L. A. Wolfe, and G. W. Burggraf. Electrocardiographic and echocardiographic characteristics of female athletes. *Med. Sci. Sports Exerc.* 27: 1362–1370, 1995.

Giannini, A. J., S. M. Melemis, D. M. Martin, and D. J. Folts. Symptoms of premenstrual syndrome as a function of beta-endorphin: two subtypes. *Prog. Neuropsychopharmacol. Biol. Psychiatry* 18: 321–327, 1994.

Glenmark, B., G. Hedberg, L. Kaijser, and E. Jansson. Muscle strength from adolescence to adulthood—relationship to muscle fibre types. *Eur. J. Appl. Physiol.* 68: 9–19, 1994.

Golomb, L. M., A. A. Solidum, and M. P. Warren. Primary dysmenorrhea and physical activity. *Med. Sci. Sports Exerc.* 30: 906–909, 1998.

Grumbt, W. H., S. E. Aalway, W. J. Gonyea, and J. Stray-Gunderson. Isometric strength and morphological characteristics in the arm flexors of elite male and female body builders. *Med. Sci. Sports Exerc.* 20 (Suppl.): S71, 1988.

Hakkinen, K., A. Pakarinen, and M. Kallinen. Neuromuscular adaptations and serum hormones in women during short-term intensive strength training. *Eur. J. Appl. Physiol.* 64: 106–111, 1992.

Hakkinen, K., A. Pakarinen, H. Kyrolainen, S. Cheng, D. H. Kim, and P. V. Komi. Neuromuscular adaptations and serum hormones in females during prolonged power training. *Int. J. Sports Med.* 11: 91–98, 1990.

Hall, D. C., and D. A. Kaufman. Effects of aerobic and strength conditioning on pregnancy outcome. *Amer. J. Obstet. Gynecol.* 157: 1199–1203, 1987.

Hellstrom, L., E. Blaak, and E. Hagstrom-Toft. Gender differences in adrenergic regulation of lipid mobilization during exercise. *Int. J. Sports Med.* 17: 439–447, 1996.

Holloway, J. B., and T. R. Baechle. Strength training for female athletes. *Sports Med.* 9: 216–228, 1990.

Kardel, K. R., and T. Kase. Training in pregnant women: effects on fetal development and birth. *Am. J. Obstet. Gynecol.* 178: 280–286, 1998.

Knuttgen, H. G., and K. Emerson. Physiological responses to pregnancy at rest and during exercise. *J. Appl. Physiol.* 36: 549–553, 1974.

Latikka, P., E. Pukkala, and V. Vihko. Relationship between the risk of breast cancer and physical activity. *Sports Med.* 26: 133–143, 1988.

Lieberman, L., M. Meana, and D. Stewart. Cardiac rehabilitation: gender differences in factors influencing participation. *J. Womens Health* 7: 717–723, 1998.

Lindholm, C., K. Hagenfeldt, and B. M. Ringertz. Pubertal development in elite juvenile gymnasts. Effects of physical training. *Acta Obstet. Gynecol. Scand.* 73: 269–273, 1994.

Lindholm, C., K. Hagenfeldt, and H. Ringertz. Bone mineral content of young female former gymnasts. *Acta Paediatr.* 84: 1109–1112, 1995.

MacDougal, J. D., D. G. Sale, S. E. Alway, and J. R. Sutton. Differences in muscle fibre number in biceps brachii between males and females. *Can. J. Applied Sports Sci.* 8: 221, 1983.

Marcus, R., B. Drinkwater, G. Dalsky, J. Dufek, D. Raab, C. Slemenda, and C. Snow-Harter. Osteoporosis and exercise in women. *Med. Sci. Sports Exerc.* 24 (Suppl. 6): S301–S307, 1992.

Marshall, L. A. Clinical evaluation of amenorrhea in active and athletic women. *Clin. Sports Med.* 13: 371–387, 1994.

Mayhew, J. L., and P. M. Gross. Body composition changes in young women with high resistance weight training. *Res. Q.* 45: 433–440, 1974.

McMurray, R. G., and V. L. Katz. Thermoregulation in pregnancy: implications for exercise. *Sports Med.* 10: 146–158, 1990.

Meyer, W. R., E. F. Pierce, and V. L. Katz. The effect of exercise on reproductive function and pregnancy. *Curr. Opin. Obstet. Gynecol.* 6: 293–299, 1994.

Möller-Nielsen, J., and M. Hammar. Women's soccer injuries in relation to the menstrual cycle and oral contraceptives. *Med. Sci. Sports Exerc.* 21: 126–129, 1989.

Morris, F. L., G. A. Naughton, J. L. Gibbs, J. S. Carlson, and J. D. Wark. Prospective ten-month exercise intervention in premenarcheal girls: positive effects on bone and lean mass. *J. Bone Miner. Res.* 12: 1453–1462, 1997.

Nielsen, J. M., and M. Hammar. Sports injuries and oral contraceptive use: is there a relationship? *Sports Med.* 12: 152–160, 1991.

Penttinen, J., and R. Erkkola. Pregnancy in endurance athletes. *Scand. J. Med. Sci. Sports* 7: 226–228, 1997.

Pollock, M. W., J. H. Wilmore, and S. M. Fox. Health and Fitness Through Physical Activity. New York: Wiley, 1978.

Prior, J. C., and Y. Vigna. Conditioning exercise and premenstrual symptoms. *J. Reproductive Med.* 32: 423–428, 1987.

Putukian, M. The female triad. Eating disorders, amenorrhea, and osteoporosis. *Med. Clin. North Am.* 78: 345–356, 1994.

Roger, V. L., S. J. Jacobsen, P. A. Pellikka, T. D. Miller, K. R. Bailey, and B. J. Gersh. Gender differences in use of stress testing and coronary heart disease mortality: a population-based study in Olmsted County, Minnesota. *J. Am. Coll. Cardiol.* 32: 345–352, 1998.

Ruby, B. C., and R. A. Roberts. Gender differences in substrate utilisation during exercise. *Sports Med.* 17: 393–410, 1994.

Ruffin, M. T., R. E. Hunter, and E. A. Arendt. Exercise and secondary amenorrhoea linked through endogenous opioids. *Sports Med.* 10: 65–71, 1990.

Ryushi, T., K. Hakkinen, H. Kauhanen, and P. V. Komi. Muscle fiber characteristics, muscle cross-sectional area, and force production in strength athletes, physically active males and females. *Scand. J. Sports Sci.* 10: 7–15, 1988.

Sady, S. P., M. W. Carpenter, P. D. Thompson, M. A. Sady, B. Haydon, and D. R. Coustan. Cardiovascular response to cycle exercise during and after pregnancy. *J. Appl. Physiol.* 66: 336–341, 1989.

Saltin, B., J. Hakkinen, H. Kauhanen, and P. V. Komi. Fiber types and metabolic potentials of skeletal muscles in sedentary man and endurance runners. *Ann. N.Y. Acad. Sci.* 301: 3–29, 1977.

Shangold, M. Menstruation. In Women and Exercise: Physiology and Sports Medicine, M. Shangold and G. Merkin (Eds.). Philadelphia: F. A. Davis, 1994.

Shangold, M. M., and H. S. Levine. The effect of marathon training upon menstrual function. *Amer. J. Obstet. Gynecol.* 143: 862–869, 1982.

Singh, B. B., B. M. Berman, R. L. Simpson, and A. Annechild. Incidence of premenstrual syndrome and remedy usage: a national probability sample study. *Altern. Ther. Health Med.* 4: 75–79, 1998.

Sinning, W. E., and K. D. Little. Body composition and menstrual function in athletes. *Sports Med.* 4: 34–45, 1987.

Sparling, P. B. A meta-analysis of studies comparing maximal oxygen uptake in men and women. *Res. Q.* 51: 542, 1980.

Speechly, D. P., S. R. Taylor, and G. G. Rogers. Differences in ultra-endurance exercise in performance-matched male and female runners. *Med. Sci. Sports Exerc.* 28: 359–365, 1996.

Stager, J. M., and L. K. Hatler. Menarche in athletes: the influence of genetics and training. *Med. Sci. Sports Exerc.* 20: 369–373, 1988.

Stephenson, J. Female Olympians' sex tests outmoded. *J. Amer. Med. Assoc.* 276: 177–178, 1996.

Sternfeld, B. Physical activity and pregnancy outcome: Review and recommendations. *Sports Med.* 23: 33–47, 1997.

Stoll, B. A. Western diet, early puberty, and breast cancer risk. *Breast Cancer Res. Treat.* 49: 187–193, 1998.

Tarnopolsky, M. A., S. A. Atkinson, S. M. Phillips, and J. D. MacDougall. Carbohydrate loading and metabolism during exercise in men and women. *J. Appl. Physiol.* 78: 1360–1368, 1995.

Ueji, M., E. Ueno, D. Osei-Hyiaman, H. Takahashi, and K. Kano. Physical activity and the risk of breast cancer: a case-control study of Japanese women. *J. Epidemiol.* 8: 116–122, 1998.

Veille, J. C., H. K. Hellerstein, B. Cherry, and A. E. Bacevice, Jr. Effects of advancing pregnancy on left ventricular function during bicycle exercise. *Am. J. Cardiol.* 73: 609–610, 1994.

Warren, M. P., J. Brooks-Gunn, L. H. Hamilton, L. F. Warren, and W. G. Hamilton. Scoliosis and fractures in young ballet dancers. Relation to delayed menarche and secondary amenorrhea. *N. Engl. J. Med.* 314: 1348–1353, 1986.

Webb, K. A., L. A. Wolfe, J. E. Tranmer, and M. J. McGrath. Pregnancy outcome following physical fitness training. *Can. J. Sports Sci.* 13: 93P–94P, 1988.

Wells, C. L. Women, Sport & Performance: A Physiological Perspective. Champaign, Ill.: Human Kinetics, 1985.

Wells, C. L., M. A. Boorman, and D. M. Riggs. Effect of age and menopausal status on cardiorespiratory fitness in masters women runners. *Med. Sci. Sports Exerc.* 24: 1147–1154, 1992.

Wilmore, J. Alterations in strength, body composition, and anthropometric measurements consequent to a 10-week weight training program. *Med. Sci. Sports* 6: 133–138, 1974.

Wolfe, L. A., and M. F. Mottola. Aerobic exercise in pregnancy: an update. *Can. J. Appl. Physiol.* 18: 119–147, 1993.

Wolfe, L. A., R. M. Walker, A. Bonen, and M. J. McGrath. Effects of pregnancy and chronic exercise on respiratory responses to graded exercises. *J. Appl. Physiol.* 76: 1928–1936, 1994.

Zaharieva, E. Olympic participation by women: effects on pregnancy and childbirth. *J. Am. Med. Assoc.* 221: 992–995, 1972.

CHAPTER 31

GROWTH AND DEVELOPMENT

Physical performance in children and adolescents must always be assessed in light of the growth process. Growth involves a series of developmental stages that are remarkably similar in all people. Individual differences in diet, exercise, and health may affect these stages to a certain extent, but the basic pattern remains the same.

Each growth stage has a profound influence on individual capability for physical performance. For example, the ability to throw a ball in the adult fashion of weight transfer and trunk rotation has a prerequisite neurological development. Likewise, prepubertal boys have a limited ability to increase muscle size because they do not produce male hormones in sufficient quantities to induce significant muscle hypertrophy. Attempts at hurrying the developmental process are futile and may lead to physical and emotional harm.

It is important to recognize the capabilities and limitations imposed by growth and maturation. For instance, a contact sport such as American football could pose a severe risk to an immature overweight prepubertal child but might be perfectly appropriate for a mature and muscular 16-year-old high school student. Likewise, the apparent excess body weight in an 8-year-old boy may not reflect extreme deconditioning, but simply be a common characteristic of prepubertal youngsters.

Proper exercise and sound nutrition allow individuals to reach their full potential.

This chapter examines the relationship among growth, exercise capacity, and training. Emphasis is placed on the importance of exercise during growth and the relative influence of environmental and genetic factors on development.

■ Nature of the Growth Process

Growth involves the transformation of nutrients into living tissue. It implies the development of the organism in an orderly fashion and represents a predominance of anabolic over catabolic processes. Growth is characterized by the progressive transformation of the organism into the adult form. Somatic changes are best represented by increases in height and weight. Musculoskeletal changes are characterized by increases in muscle mass and bone maturation. Sexual maturation involves the progressive development of primary and secondary sexual characteristics.

There are three types of growth: (1) statural or incremental growth, (2) hypertrophic growth, and (3) reparative growth. Body mass increases in size and length in structural growth. This process begins at conception and concludes some time during adolescence. In hypertrophic growth, body mass increases in size in response to functional demands. For example, weight training increases muscle size through hypertrophy. Reparative growth involves structural maintenance of tissues and repair of damaged tissues. An example of reparative growth is a wound healing after a cut, burn, or other soft tissue injury.

Growth Curves

Growth proceeds along a basic curve that is S-shaped (Figure 31-1). During the first two years of life, height and weight increase rapidly. This early increase is followed by a progressively declining growth rate during childhood. At puberty the trend reverses itself in an abrupt increase in growth rate called the *adolescent growth spurt*.

Different tissues and organs of the body have their own rates of growth. Humans grow from the head down, a process called *cephalocaudal development* (Figure 31-2). The head grows more rapidly than the chest and arms, which grow more rapidly than the legs, and the growth of skeletal muscle, heart, liver, and kidney is slower than the growth rate of the skeleton. Differences in these growth rates

affect physical performance at different stages of development.

Although all humans follow the basic growth curve, individual differences in environment can affect the rate of change in height, weight, and physiological development. For example, over the past 100 years, there has been a progressive increase in the growth rate of children living in Western industrial nations, which has been accompanied by earlier maturation. This trend, which is not evident in less developed countries, is probably due to improved public health and better nutrition. Of course, even in a country with a high standard of living, environmental conditions such as malnutrition or excessive training may adversely affect children's growth.

This change in growth patterns has been particularly evident in sports. Overall, performances in sports such as track and field, swimming, and American football have improved tremendously since the early days of competition. These improvements are most certainly due, in large part, to increases in height, weight, and muscle mass. Changes in training methods, facilities, and a societal emphasis on sport have also played a role.

Neuroendocrine Control of Growth

Statural and reparative growth depend on anabolic hormones such as growth hormone (GH) and an adequate supply of energy. During early childhood, growth rate is determined by the volume and rate of growth hormone secretion from the pituitary gland. The decrease in growth rate seen in later childhood (years preceding puberty) coincides with a decrease in growth-hormone concentration in blood. Thyroid hormone, secreted by the thyroid gland, works synergistically to enhance the action of growth hormone on growth. Insulin and gonadal hormones, such as testosterone and estradiol (estrogen), also influence growth.

Growth hormone affects the growth process by two primary mechanisms. First, it directly stimulates differentiation and proliferation of cells throughout the body. Second, it stimulates growth factors in the

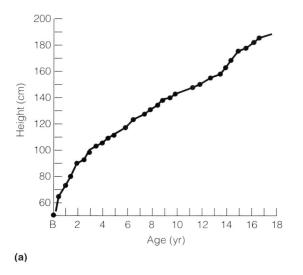

(a)

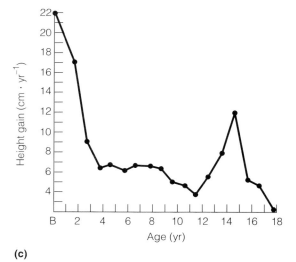

(b)

Figure 31-1 Growth in height and weight follows an S-shaped curve. The rate of growth is rapid during the first two years of life, levels off during childhood, then accelerates again during adolescence. (a and b) Absolute changes in growth of height and weight to age 18. (c) Gain in height during each year of the growth period.

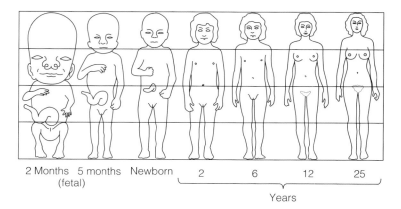

(c)

Figure 31-2 Changes in the human body during growth. The head becomes smaller in proportion to the rest of the body, while the legs become relatively longer.

cells, such as insulin-like growth factor 1 (IGF-I; also called somatomedin), which also stimulates cell differentiation and proliferation.

Optimal function of GH depends on up regulation (increased sensitivity) of its receptor and binding protein. GH function can be enhanced or impaired by exercise and nutrition. Moderate exercise and consumption of ample nutrients up regulates GH receptors and stimulates cell growth factor production, which stimulates statural and reparative growth. On the other hand, excessive exercise, particularly if accompanied by inadequate caloric intake, down regulates GH receptors and inhibits production of growth factors, which impairs growth.

Mechanical stress rather than growth hormone is mainly responsible for hypertrophic growth. Stress activates tissue growth factors, such as IGF-I and IGF-II, and promotes synthesis of structural proteins. While not essential for hypertrophy, this type of cell growth can be enhanced by testosterone and insulin.

Exercise and Growth

The influence of sports participation on growth and its neuroendocrine controllers is unclear. Exercise can impair growth when the energy demands of the sport compete with the energy demands of growth. Depressed stature is common among gymnasts and ballet dancers. Participation in these sports typically involves long hours of practice combined with caloric restriction. Conversely, young football, basketball, and soccer players are typically taller and bigger than their sedentary peers.

In sports in which athletes exhibit decreased or increased stature, it is difficult to differentiate between selection process and the physiological consequences of participation. In gymnastics, extremely lean and short girls have an advantage over taller heavier girls. Likewise, in football or basketball, larger and taller children have a natural advantage. It is uncertain whether these sports influence height and weight or whether children with certain characteristics gravitate toward sports in which they might experience greater success. For example, are small-statured gymnasts common in

the sport because of the stresses of training or because small stature is an advantage in gymnastics?

Growth in Infancy and Childhood

The first two years of life are called *infancy*. During this period tremendous changes occur in body proportions. In the neonate, the head represents about one-fourth the total height, with the trunk slightly longer than the legs. After about six months, the rate of growth in the skull slows, while that of the legs and trunk proceeds more rapidly. The fastest growth takes place in the legs, so by the time the child is two years old, the legs and trunk are about equal in length. (See Figure 31-2.)

Males tend to be taller and heavier than females although females have relatively longer legs, reflecting their relatively greater level of maturity. The growth rate is faster in males (although it occurs later), but relative changes are similar in both sexes.

During childhood, the period between about two years of age and the onset of puberty, there is a gradual but steady increase in height and weight, with height increasing relatively faster than weight. The legs continue to grow faster than the trunk, and there is a proportional increase in the growth of the pelvis and shoulders.

The growth rate of boys and girls throughout childhood is similar. Boys tend to be slightly larger, but the skeletal age of girls is more advanced. Between the ages of 6 and 10, girls gain in pelvic width faster than boys, while boys tend to have larger thoraxes and forearms. Despite these few differences, anthropometric characteristics are almost the same until puberty.

The gradual growth rate during childhood is conducive to learning motor skills. The relatively constant ratio between height and lean body mass provides a stable environment for developing coordination and neuromuscular skill. This is an important period for the introduction and development of gross motor activities such as running, jumping, hopping, and throwing. However, the limited muscle mass and fragile skeletal epiphyseal growth centers make vigorous strength training less appropriate.

Growth at Puberty and Adolescence

Adolescence is the final period in the growth process leading to maturity. It is a time of rapid increases in height and weight and is accompanied by puberty, the period when the sex organs become fully developed (Figure 31-3). Puberty begins in girls and boys when they reach critical levels of fat, body mass, stature, and skeletal maturity. Nutrient availability and energy expenditure have a profound effect on the onset of puberty. In sports such as girls' gymnastics, in which athletes train for long hours with inadequate nutrition, it is not unusual for menarche to occur at 16 years or later.

Adolescent growth characteristics differ between early and late maturers. The earlier the growth spurt occurs, the more intense is the rate of growth (Figure 31-4). Boys who mature earlier tend to be more muscular, with shorter legs and broader hips. Girls who mature earlier have shorter legs and narrower shoulders than those who mature later. Also, late maturers tend to become slightly taller adults. Although the growth spurt is less sudden in late maturers, these individuals have been growing over a longer period of time. The Amsterdam Growth and Health study (Kemper et al., 1997) found that late maturers had a higher energetic food intake and a slightly higher activity pattern than people who experienced early maturation during adolescence. This could have health consequences. This study (van Lenthe et al., 1996) also found that children who matured rapidly in adolescence were, in general, more obese than slowly maturing adolescents. Rapid maturation seems to have long-term consequences for obesity and should be considered a risk indicator for the development of obesity.

The characteristic sexual differences in anthropometric measures arise during puberty. Males develop greater height and weight with larger musculature and broader shoulders; females develop broader hips. The growth spurt is slightly over 7.5 cm · year^{-1} in girls and 10 cm · year^{-1} in boys.

In girls, evidence of sexual development can occur at a surprisingly young age. Herman-Giddens and coworkers (1997) studied 17,077 white and African-American girls between the ages of 3 and 12 years who were examined by pediatricians in the United States. At age 3, 3% of African-American

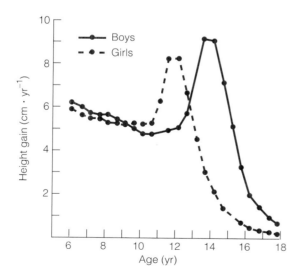

Figure 31-3   The adolescent growth spurt in height for girls and boys. SOURCE: Tanner, 1962.

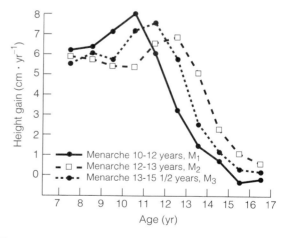

Figure 31-4   Peak velocities in height for early- (M_1), average- (M_2) and late-maturing (M_3) girls, based on average time of menarche for each group. SOURCE: Tanner, 1962.

girls and 1% of white girls showed breast or pubic hair development. By age 8, 48% of African-American girls and 15% of white girls had begun development. Menarche occurred at 12.2 years in African-American girls and 12.9 years of age in white girls. Their study showed that not only are girls maturing earlier than before, but there are significant racial differences.

The adolescent growth spurt has profound effects on physical performance. Males markedly improve in endurance, strength, speed, power, and various motor skills. In females, conversely, physical performance often levels off.

Assessing Maturation in Adolescents

Maturational assessment is important because most indices of physical performance are most highly related to tissue growth (i.e., increases in muscle mass, heart size, and so on). Chronological age is a poor measure of biological age during growth. Adolescent girls will typically be two years more mature than boys. Among 13–14 year olds, biological age may vary from 9 to 16 years. Maturation may be assessed by determining skeletal, sexual, or somatic maturation.

Skeletal Maturation Skeletal maturation is assessed by estimating bone age, which is determined by observing the maturational development of the epiphyseal growth plates. Usually, this is done by comparing X-rays of the child with reference X-rays and making a subjective evaluation of relative maturation. An obvious problem with this technique is the risk of exposing children to X rays. Magnetic resonance imaging (MRI) is also used to assess skeletal maturation, but this technique is expensive and is not likely to become widely available for routine assessment.

Sexual Maturation Degree of sexual maturation is the most common method of assessing biological age (Table 31-1). This technique, which was developed by J. M. Tanner in the 1950s, involves assessing development of secondary sexual characteristics: breast development in girls, genital development in boys, and pubic hair development in both sexes. Development occurs in five stages: Stage 1 is prepuberty; stage 5 is full sexual maturation. Although commonly used, this technique could be considered an invasion of personal privacy, which makes it problematic for children at a time when they are particularly concerned with their changing maturational characteristics.

In girls, sexual maturation begins when the breasts and pubic hair first appear. These indicators are followed by menarche, the first menstruation. Menarche provides a definite landmark for the assessment of maturation that is not available for boys. The closest indicators of maturity in boys, in addition to the appearance of pubic hair, are the appearance of facial hair and deepening of the voice. In boys and girls, body mass index (BMI = weight/height^2) is related to age, pubertal stage, and major body segment girths (e.g., hip and shoulder circumferences). It is more closely related to maturation stage than age.

Hormonal assessment of sexual maturation Puberty and adolescence are also characterized by marked changes in hormone secretions (testosterone in boys and estrogen in girls). In boys, testosterone levels (particularly free testosterone) are related to maturational development. With technological advances, testosterone can be measured easily in serum and saliva (Figure 31-5). Because of marked diurnal variations in testosterone, care must be taken to standardize specimen collection when studies are conducted or field evaluations made of sexual maturation.

Hormone assessment of maturation in girls is more difficult than in boys. Although estrogen is primarily responsible for the development of secondary sexual characteristics in girls, menstrual cycle variations in the hormone make its use extremely difficult and unreliable. In addition, secretion of the hormone is pulsatile; that is, the hormone is released in spurts rather than at a constant rate.

As discussed, critical fat accumulation has been suggested as a triggering mechanism for the onset of puberty. Leptin, a hormone secreted by adipose

TABLE 31-1

Pubertal Stage Ratings for Boys and Girls[a]

Pubertal Stage	Pubic Hair	Male Genital Development	Breast Development
1	None	Testes, scrotum about same size and proportions as in early childhood	Elevation of papilla only
2	Sparse growth of long, slightly pigmented downy hair, straight or only slightly curled, appearing chiefly at base of penis or along labia	Enlargement of scrotum and testes; skin of scrotum reddens and changes in texture; little or no enlargement of penis at this stage	Breast bud stage; elevation of breast and papilla as small mound; enlargement of areolar diameter
3	Considerably darker, coarser, and more curled; hair spreads sparsely over junction of pubes	Enlargement of penis, which occurs at first mainly in length; further growth of testes and scrotum	Further enlargement and elevation of breast and areola, with no separation of their contours
4	Hair now resembles adult in type, but area covered is still considerably smaller than in adult; no spread to medial surface of thighs	Increased size of penis with growth in breadth and development of glans; further enlargement of testes and scrotum; increased darkening of scrotal skin	Projection of areola and papilla to form a secondary mound above the level of the breast
5	Adult in quantity and type with distribution of horizontal (or classically "feminine" pattern) growth; spread to medial surface of thighs but not up the linea alba or elsewhere above the base of the inverse triangle. In about 80% of Caucasian men and 10% of women, pubic hair spreads further, but this takes some time to occur after stage 5 is reached. This may not be completed until the midtwenties or later.	Genitalia adult in size and shape; no further enlargement after stage 5 is reached	The mature stage; projection of papilla only, due to recession of areola to general contour of breast

[a]Pubertal stage is rated on a scale of 1 to 5. Stage 1 represents prepubertal development, whereas stage 5 represents the characteristics of the adult. Stages are determined by the degree of development of pubic hair in both sexes, genital development in boys, and breast development in girls.

SOURCE: Adapted from Larson, 1974.

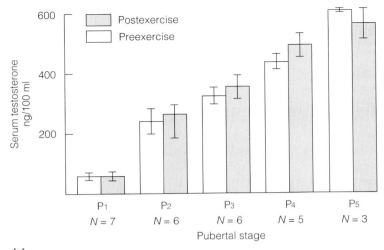

(a)

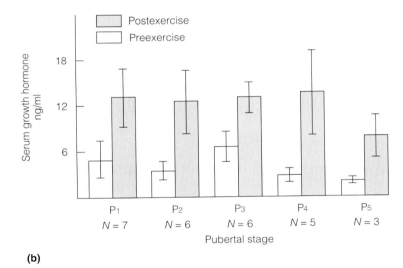

(b)

Figure 31-5 Testosterone (a) and growth hormone (b) levels in pubertal stages 1–5. Testosterone levels are closely related to sexual maturation. Growth hormone levels are highest during the early pubertal years (stages 1–3) and decrease somewhat during stages 4–5. These hormones show no pubertal stage differences in their responses to exercise. Care must be taken to standardize measurements because of diurnal variations in secretion of the hormone. SOURCE: Fahey et al., 1979.

tissue that sends signals to the brain regarding fat-cell status, increases by 50% just prior to the onset of puberty in girls and boys (Mantzoros et al., 1997). Leptin levels return to baseline levels after puberty begins. These data suggest that leptin is an important triggering mechanism of puberty. The stimulus for a surge in leptin levels just before the onset of puberty is unknown.

A new technique for assessing sexual maturation in girls is the measurement of the adrenal hormone dehydroepiandrosterone (DHEA). Increased secretion of this hormone occurs at age 7 or 8, followed by a sharp increase at age 13. The levels stabilize to adult values at approximately age 15.

Somatic Maturation Somatic maturation is determined by assessing the time when peak growth velocity is occurring—that is, when children are at the peak of their adolescent growth spurt. This method requires a minimum of three height measurements per year. The period of peak height velocity typically occurs between 11 and 12 years in girls and 13 and 14 years in boys.

Training and Maturation Most studies show no relationship between sports training and maturation. However, some studies suggest that heavy training in sports such as gymnastics may delay maturation, particularly in girls. Delayed menarche (first menstruation) may occur due to interactions among caloric restriction, intense training, and a genetic tendency toward irregular menses. Conversely, children with a disposition toward delayed maturation may self-select themselves into sports such as gymnastics, where small stature and body proportions are advantages.

■ Skeletal Changes During Growth

Increases in height occur because of skeletal growth, principally in the long bones. Bone growth occurs at the epiphyseal growth plates, which are located at both ends of long bones between the articular epiphysis and the central diaphysis (Figure 31-6). The process of bone formation is called *ossification*.

Bone Growth

Bone is a matrix composed of collagen fibers and ground substance. Embedded in the matrix are osteocytes, osteoblasts, and osteoclasts. Bone grows as a result of the secretion of collagen by the osteoblasts. During this process, some osteoblasts are trapped in the tissue (now called osteoid). These trapped osteoblasts, called osteocytes, begin to collect calcium and phosphorus. This process continues until the growth plates themselves are finally ossified.

The nature of the bone matrix gives it great compressional and tensile strength. Collagen fibers are arranged along vectors of the bone receiving stress. The bone salts form crystals that are arranged to provide compressional strength greater than that of concrete.

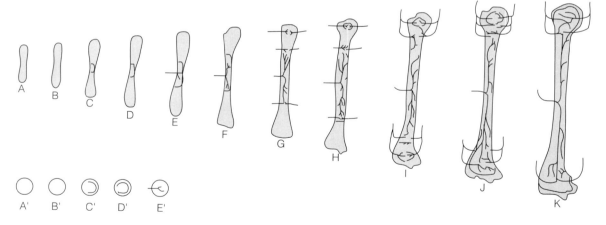

Figure 31-6 Diagram of the development of a typical long bone and its blood supply: (a) cartilage model; (b) development of the bone collar; (c) development of calcified cartilage in primary center; (d) extension of the bone collar; (e and f) invasion of the cartilage by vascular sprouts and mesenchyme, with formation of two areas of bone toward the bone ends; (g, h, i) secondary centers develop in the bone ends as the central area expands; (j and k) epiphyseal plates disappear and the blood vessels of the diaphysis and epiphyses intercommunicate. A′–E′ = growth of blood supply.

As in most processes in physiology, bone formation is extremely dynamic. Approximately 100% of the infant skeleton and 10–30% of the adult skeleton is replaced each year. This process of breakdown and buildup of bone is called *remodeling*. Bone is absorbed by the osteoclasts and deposited by the osteoblasts. The process allows precise control of calcium and phosphate metabolism and helps maintain healthy bone. It also enables bones to adapt to stress: Bone forms when subjected to increased loads and is absorbed when stresses decrease. As a result of bone remodeling, then, bone mass may increase, stay the same, or decrease. Bone mass tends to increase during growth and decrease in old age.

During growth, bone formation exceeds bone absorption by osteoclasts. Exercise, which stresses bones and stimulates bone growth, does not seem to enhance linear bone growth but does increase bone density and width. During the growth period, general exercise creates a skeleton composed of denser, stronger bone that is better able to withstand stress. Moderate exercise, combined with a calcium-rich diet, may maximize bone density in children.

Some research has shown that overtraining (imbalance between training and recovery) can lead to decreased bone density, particularly in women, by suppressing gonadal hormones (estrogen and testosterone) important for bone growth. Those particularly at risk include athletes such as gymnasts, figure skaters, and ballet dancers, who train heavily and diet to maintain a lean body composition. Fortunately, most studies show that exercise during childhood and adolescence increases bone mass. While overtraining may delay menarche in some prepubertal girls and cause secondary amenorrhea and bone loss in some adolescents, in the majority of girls sports participation increases peak bone mineral density (BMD). Increases in peak bone mineral density during growth may contribute to the prevention of osteoporosis. Since the years before puberty are relatively sex hormone independent, participating in sports during the prepubertal years appears to result in higher BMD. Bass and coworkers (1998) found that BMD in prepubertal gymnasts

was 30–85% greater than in prepubertal controls. Adult gymnasts, who were retired for 20 years, showed higher BMD than age-matched sedentary controls, even though the adult gymnasts were no longer as physically active.

Linear growth continues as long as the ossification centers are open. Approximately 45% of bone mass is accumulated during the adolescent growth spurt. Depending on the bone, peak bone mass occurs anywhere between the late teens to mid-thirties.

Epiphyseal Injury

Epiphyseal-diaphyseal union begins at puberty in some bones but is not completed until age 18 or later. Disease or trauma can injure the growth plates, which can, in turn, affect growth. It is extremely important, therefore, to avoid situations that could adversely damage these areas. For example, excessive baseball pitching by youngsters can cause epiphyseal damage. Related problems are focal growth disturbances (osteochondroses), such as Little League elbow, Sever's disease, Osgood-Schlatter disease, and osteochondritis dissecans. These are often seen in children and are thought to be related to (although not necessarily caused by) overuse. Late-maturing children are thought to be more prone to growth plate injuries because the growth plates are open longer.

■ Body Composition and Obesity

The ratio of fat to fat-free weight is particularly important during growth because overweight children tend to become overweight adults. Physical activity is vitally important in the maintenance of ideal body composition. Children and adolescents are increasingly inactive; these patterns tend to persist throughout life. However, programs of vigorous physical activity and proper diet can reverse the process of chronic obesity.

Body weight is regulated by the balance between caloric intake and energy expenditure. Increases in body fat occur because of increases in fat cell size (hypertrophy), cell number (hyperplasia), or both.

In humans, fat cell hyperplasia occurs mainly during the third trimester of pregnancy and just before and during puberty.

The prevalence of childhood and adolescent obesity is increasing. Almost one-fourth of U.S. children are obese, which is an increase of over 20% in the past decade. Adults who were obese as children have increased mortality independent of adult weight. Obesity is a major risk factor for insulin resistance and diabetes, hypertension, cancer, gallbladder disease, and atherosclerosis. Although genetics accounts for as much 50% of the body weight variation, much of the increased incidence of obesity is due to behavioral reasons such as lack of physical activity and overeating. Because obesity increases risk of many health problems, obesity prevention programs for children and adolescents will have long-term benefits. Unfortunately, young people are not very physically active (Figure 31-7a and Figure 31-7b).

The Surgeon General's Report on Exercise and Health (1996) summarized exercise habits in children and adolescents as follows:

- Only about one-half of U.S. young people (ages 12–21) regularly participate in vigorous physical activity. One-fourth report no vigorous physical activity.

- Approximately one-fourth of young people walk or bicycle (i.e., engage in light to moderate activity) nearly every day.

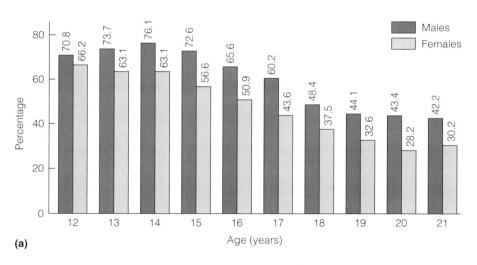

(a)

Figure 31-7 (a) Vigorous exercise among young people during three or more of the seven days preceding the survey. Female young people exercise vigorously less often than males, and the percentage for both males and females falls sharply with increasing age. (b) Percentage of high school students attending physical education classes daily. A year 2000 goal is that 50% of students have daily P.E. Overall, only 25.4% of high school students have daily P.E., with 40% not enrolled in any P.E. classes.

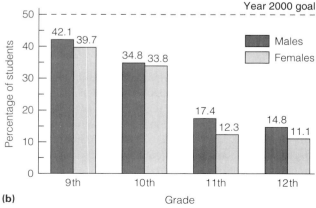

(b)

- About 14% of young people report no recent vigorous or light to moderate physical activity. This indicator of inactivity is higher among females than males and higher among black females than white females.

- Males are more likely than females to participate in vigorous physical activity, strengthening activities, and walking or bicycling.

- Participation in all types of physical activity declines strikingly with increase of age or grade in school.

- Among high school students, enrollment in physical education remained unchanged during the first half of the 1990s. However, daily attendance in physical education declined from approximately 42 to 25%.

- The percentage of high school students who were enrolled in physical education and who reported being physically active for at least 20 minutes in physical education classes declined from approximately 81 to 70% during the first half of this decade.

- Only 19% of all high school students report being physically active for 20 minutes or more in daily physical education classes.

Changes in Body Composition During Growth   During childhood, females have slightly more fat than males. Typical body fat percentages are about 16% for an 8-year-old boy and 18% for an 8-year-old girl. At puberty, marked changes take place in body composition. Boys make rapid increases in lean body weight and decreases in percentage of fat, which typically drops 3 to 5% between the ages of 12 and 17. Girls also increase in lean mass, although less than boys, and increase in body fat. An untrained 17-year-old girl typically has around 25% fat. However, an athlete often has less than 16 to 18% (Figure 31-8).

Physical activity will not change the essence of these growth-related stages of body composition. In females, for example, vigorous training will not prevent the increase in fat that occurs during adolescence. However, the increases are less than those seen in sedentary females. Studies comparing

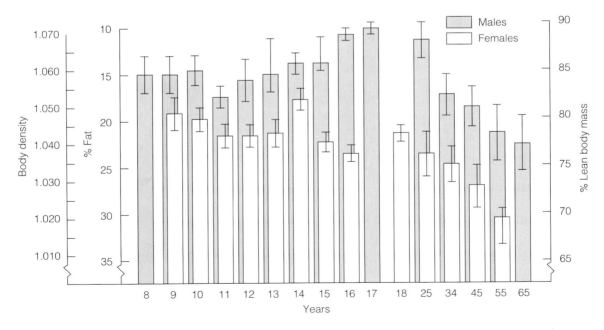

Figure 31-8   Body composition during growth and aging in males (shaded) and females (unshaded). Source: Parizkova, 1968.

active versus inactive youngsters show that training consistently results in higher lean body mass and lower body fat. These advantages do not persist into adulthood, though, unless the training is continued. Inactivity results in decreased muscle mass and increased fat.

Body composition is an important consideration in the development of motor skills. The relatively slow growth period is an important time for learning gross motor skills. In overweight children, however, these efforts may be compromised. Their relatively weak muscles could have considerable difficulty overcoming the burden of excess body fat.

■ Muscle

Muscle growth accounts for a considerable portion of weight gain during growth. Muscle tissue grows steadily during the first seven years of life, followed by a slowing trend in the years immediately preceding puberty. In a normal child, muscle and skeletal growth keep pace with one another. During the adolescent growth spurt following puberty, muscle grows at a rapid rate, particularly in boys. The increase in muscle tissue typically occurs slightly after the greatest increases in height. This explains the awkward gangliness in children that age.

In boys, increases in muscle size are related to improvements in strength. Rapid increases begin at about 14 years and continue throughout adolescence. However, such factors as maturational level, body build, and physical activity lead to numerous individual differences. Muscle strength and size are of considerable concern to adolescents because they are the most important predictors of athletic success.

The extent of muscle development and performance depends on the relative maturation of the nervous system. High levels of strength, power, and skill are impossible if the child has not reached neural maturity. Myelination of nerves is not complete until sexual maturity has been achieved, so an immature child cannot be expected to respond to training or reach the same level of skill as an adult.

Significant training-induced muscle hypertrophy does not occur until adolescence in boys due to the low levels of male hormones (principally testosterone) found in sexually immature children. Testosterone (also growth hormone) is an important regulator of protein synthesis. Adult males have about 10 times the amount of testosterone as prepubertal children and adult women (Figure 31-5). Maximal strength-gaining potential is not possible until adult levels of testosterone are achieved. However, many recent studies on the effect of resistance training in young boys (less than 13 years) show that boys can gain strength (although not significant muscle growth) from resistance training. These studies suggest that the risk of injury in supervised programs is extremely low.

■ Cardiorespiratory and Metabolic Function

Cardiac Performance

Heart Rate The resting heart rate declines progressively during childhood and adolescence. A typical HR_{rest} of 85 beats per minute (bpm) in a 5-year-old child decreases to 62 bpm by age 15. Heart rate tends to be the same in both sexes during childhood but is about three to four beats per minute higher in girls during and after adolescence. At puberty in boys, submaximal exercise and recovery heart rate tend to be lower, even in the absence of vigorous training. This decreased heart rate is accompanied, and perhaps caused, by greater resting and exercise stroke volume, which occurs because of increased heart size and blood volume.

Maximum heart rate in children varies between 195 and 205 bpm. It does not change during childhood, so estimating it with equations such as $HR_{max} = 220 -$ age is inappropriate. There are considerable individual differences in maximum heart rate (SD = ± 5–9 bpm). Because maximum heart rates stay the same and resting heart rates decrease, heart rate reserves ($HR_{max} - HR_{rest}$) increase progressively during childhood. This could contribute to the improved aerobic capacity seen during childhood.

Stroke Volume Stroke volume (SV) increases with age during childhood and adolescence (e.g., SV = 5 ml at birth, 25 ml at 5 years, and 85 ml at

15 years). It is closely related to body weight and body surface area. Cardiac contractility does not appear to improve with age in growing children, so the increase in stroke volume is due entirely to increased left ventricular size. There is no difference in ejection fraction (i.e., the percentage of end diastolic volume ejected from the left ventricle) between pre- and postpubertal children.

Cardiac Output Maximum cardiac output increases during growth, advancing from around 12 to 21 liters · min^{-1} in boys and 10.5 to 15.5 liters · min^{-1} in girls between the ages of 10 and 20. Increases in exercise cardiac output closely match increased size and body surface area, so there is little change in maximum cardiac output relative to body surface area.

Blood Pressure Resting systolic blood pressure is relatively low in the newborn (60 mmHg), but increases progressively through childhood toward adult values (Figure 31-9). Because of the relatively high tissue perfusion levels in children [i.e., $\dot{Q}$ (cardiac output)/body mass], the low blood pressures in children indicate low levels of total peripheral resistance (TPR). Although circulatory regulation in children during exercise has not been studied in detail, the pressure–resistance relationship in children suggests that increased muscle perfusion during exercise results more from increments in cardiac output than from the shunting mechanism (that is, from controlling total peripheral resistance by vasodilation and vasoconstriction in blood vessels). Thus, even though values of $\dot{V}_{O_2max}$ seen in children are relatively high, they are significantly less than in trained adults.

Blood pressure assumes adult values shortly after the adolescent growth spurt. The pubertal acceleration in blood pressure occurs in females shortly before it occurs in males, but blood pressure in males soon surpasses that in females.

Pulmonary Function

Minute Ventilation Minute ventilation ($\dot{V}_E$) increases with exercise intensity in children and adults due to increases in breathing frequency (f) and tidal volume (V_T). To achieve high minute volumes of exercise, children rely on a higher f and lower V_T than adults (Figures 31-10 and 31-11). Maximal V_T, which is highly related to total lung volume, increases with age until age 13 years in girls and 15 years in boys. After that, the increases are much more gradual and variable until adult values are assumed. There are no sex differences in maximal exercise f until late adolescence.

The smaller tidal volume and higher breathing frequency during exercise might be expected to lead

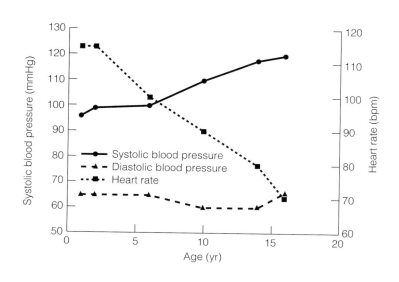

Figure 31-9 Normal average resting blood pressures and heart rates at various ages.

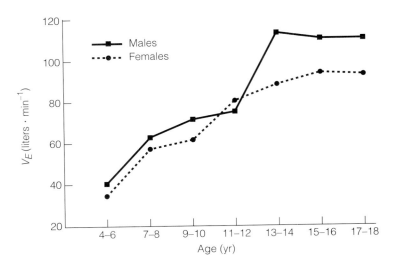

Figure 31-10 Changes in $\dot{V}_{Emax}$ during growth. $\dot{V}_{Emax}$ increases similarly in both sexes during childhood, but increases more abruptly in males than females during the adolescent growth spurt. These changes are probably due more to changes in weight and height than to innate sex differences.

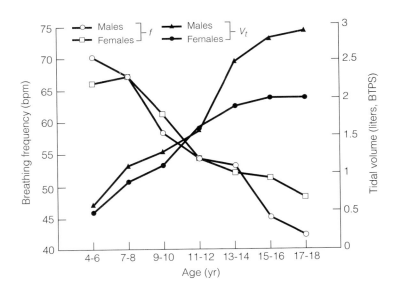

Figure 31-11 Changes in maximal exercise breathing frequency and tidal volume during growth. Children rely more on breathing frequency than adults to achieve maximum exercise minute volume.

to a higher dead space/tidal volume ratio (V_D/V_T) and thus a lower alveolar ventilation. However, this effect is reduced by a smaller anatomical dead space in children. The V_D/V_T of approximately 20% found in children during exercise is similar to that found in adults.

$\dot{V}_E$ accelerates faster in young children during maximal exercise, reaching 73% of $\dot{V}_{Emax}$ within one minute in 6 year olds and 50 to 62% in 11 to 17 year olds. $\dot{V}_{Emax}$ increases with age until young adulthood. However, it is approximately the same in growing children and adults when the values are expressed per unit body weight, height, or body surface area. The $\dot{V}_{Emax}$/body weight (in kilograms) is approximately 1.7 in children between 6 and 25 years.

Training and Pulmonary Function Exercise training produces a more "adultlike" pulmonary exercise response in children. Maximal exercise ventilation increases with training, while submaximal exercise ventilation decreases. Respiratory frequency decreases and tidal volume increases at any intensity of exercise. Exercise tidal volume in the trained child represents a higher percentage of vital capacity than in the untrained child. The changes in ventilatory characteristics may be due, in part, to improved endurance of the ventilatory muscles.

$\dot{V}_{E\max}$ is related to lung volume. Eriksson and coworkers (1971) found that girls involved in competitive swimming for three or more years showed substantially greater vital capacity, total lung capacity, and functional residual capacity than sedentary girls or girls who had trained for one year or less. Training resulted in an increase in vital capacity that was disproportional to linear growth and its attendant growth in lung capacity. Follow-up studies on the "girl swimmers" as adults showed that the increases in vital capacity, total lung capacity, and functional residual capacity remained higher than would be expected by growth alone. However, the higher values for $\dot{V}_{E\max}$, $\dot{V}_{O_2\max}$, blood volume, and hemoglobin demonstrated during the subjects' competitive period were no longer evident 10 years later. It is possible that large lung volumes, which have been evident in trained swimmers studied in several investigations, are a prerequisite to becoming a successful swimmer.

Oxygen Consumption Exercise capacity and $\dot{V}_{O_2\max}$ (liters $\cdot$ min^{-1}) increase gradually throughout childhood. During puberty, there is a dramatic increase in boys and a leveling off in girls. In absolute terms, boys have an aerobic capacity that is approximately 50% greater than girls by age 16. Gender differences are mainly due to differences in body composition. The improvements in oxygen transport capacity in boys during puberty are largely due to increased muscle mass and heart size.

At a given O_2 consumption level during running or cycling, cardiac output and stroke volume are lower and heart rate and total peripheral resistance are higher in children than adults. Children compensate for the lower cardiac output by increasing oxygen extraction (i.e., arterial-venous O_2 difference) so they can achieve the same or similar oxygen consumption. The stroke volume is lower because stroke volume at a given exercise intensity is closely related to left ventricular mass.

Ventilatory Inflection Point The ventilation inflection point (T_{vent}), the point where the relationship between ventilation and oxygen consumption becomes nonlinear, is highly correlated to height. However, T_{vent} occurs at approximately the same point in children and adults when it is expressed relative to weight. T_{vent} occurs at a lower percentage of $\dot{V}_{O_2\max}$ in girls than boys, perhaps due to differences in fitness.

Diffusion and Perfusion The diffusing capacity of the lungs (D_L) increases with age, height, weight, body surface area, and pulmonary capillary blood volume (Q_c), but is comparable to adults when expressed in relation to body weight. During mild treadmill exercise, D_L varies from 12 ml $\cdot$ min^{-1} $\cdot$ mmHg^{-1} at age 4 to 30 ml $\cdot$ min^{-1} $\cdot$ mmHg^{-1} at age 12. Diffusion of carbon monoxide ($D_{L_{CO}}$) in 11–13-year-old boys is approximately 24 ml $\cdot$ min^{-1} $\cdot$ mmHg^{-1}, which, when expressed in body surface area, is approximately the same as that seen in healthy adults.

Children show an increase in capillary blood volume with age until adolescence, when the changes are more variable. The ratio of capillary blood volume to alveolar ventilation (Q_c/V_A) is lower in children, probably due to the lower relative number of alveoli in children compared to adults. Yet, arterial oxygen pressure ($P_{a_{O_2}}$) and hemoglobin saturation (Hb_S) during maximal exercise are similar in children and adults. This suggests that the diffusion reserve (lung diffusion time − blood cell transit time in the pulmonary capillary required for hemoglobin saturation) is adequate in children and probably does not limit oxygen transport.

Increases in $\dot{V}_{O_2\max}$ during growth are mainly due to organ development, which greatly increases stroke volume. Maximal heart rate changes little during childhood, and arteriovenous O_2 difference is slightly lower in prepubertal than postpubertal children. So, according to the Fick equation ($\dot{V}_{O_2}$ = HR · SV · Arteriovenous oxygen difference) increases in $\dot{V}_{O_2max}$ must be due to an increased stroke volume. In boys, stroke volume increases 200% between ages 5 and 15.

$\dot{V}_{O_2max}$ does not increase relative to body size (i.e., ml $O_2 \cdot min^{-1} \cdot kg^{-1}$) during growth. Yet, during puberty, boys typically improve running performance by 100% over prepubertal values. This paradox reinforces the reservations expressed by Noakes (1998; see Chapter 16) of using $\dot{V}_{O_2max}$ as a benchmark of physical fitness. Their observations suggest that factors such as exertional dyspnea and processes involved in muscle contraction may limit performance as much as oxygen transport capacity. During puberty, children may improve endurance performance (in events such as the 1.5-mi run) by improving movement efficiency, substrate utilization, ventilatory efficiency, and coordination of motor unit recruitment and muscle elastic recoil.

Temperature Regulation

Prepubertal children have immature temperature regulatory systems, which limits their ability to sweat. However, there is no evidence that their limited ability to dissipate heat is of extraordinary concern. In a study comparing rectal temperature of children and adults entered in a 10K race, the children had consistently lower temperatures. It is possible that the children did not exercise at the same intensity as the adults. Muscle and liver glycogen levels are lower in children than in adults. This may prevent children from running for extended periods at high intensities, thus minimizing exercise-induced temperature increases.

Children have a large body surface area compared to their muscle mass. This gives them a very large area from which to dissipate heat. However, when they compete in environments with high temperature and humidity, the potential exists for thermal injury. In such environments, conduction and convection are useless for heat dissipation. Under these circumstances, great care should be taken to prevent heat injury.

Training

Prepubertal children do not seem to respond to endurance training as well as adolescents and adults. Some studies show no improvement in $\dot{V}_{O_2max}$ from training beyond that expected during normal growth. Many of these studies show that children failed to achieve the minimum training intensity and duration necessary to improve aerobic capacity (see Chapter 16). Even in studies that meet these requirements, prepubertal subjects improved $\dot{V}_{O_2max}$ by 10% versus the 10–25% increases experienced by adults from comparable levels of training.

There are several possible explanations for lesser training responsiveness in children: They have greater habitual levels of physical activity and they have a higher training threshold (i.e., children may have to train at higher heart rates than adults to get the same training effects). However, the former explanation is unlikely because children seldom achieve the prerequisite increases in $\dot{V}_{O_2}$ during habitual physical activity to increase $\dot{V}_{O_2max}$. Children typically have higher heart rates at the ventilatory threshold (T_{vent}) than adults.

The habitual exercise of children during the adolescent growth spurt has long been hypothesized as a means to optimize the physiological, biochemical, and anatomical changes occurring during that time. However, this hypothesis remains controversial. Although increased cardiovascular capacity resulting from endurance training can be observed during the growth spurt (see Figure 31-12), the increases in cardiorespiratory capacity appear to be proportional to the increase in lean body mass. In a study of identical twins (Weber et al., 1976), in which one twin trained but the other did not, the rate of increase in $\dot{V}_{O_2max}$ within twin pairs during the adolescent growth spurt was similar.

The enhancement of developmental processes by training during the adolescent growth spurt, if any, have also been hypothesized to be due to endocrine influences. However, although exercise may increase concentrations of anabolic hormones such as testosterone and growth hormone, exaggerated hormonal responses do not occur at any pubertal stage. (See Figure 31-5.)

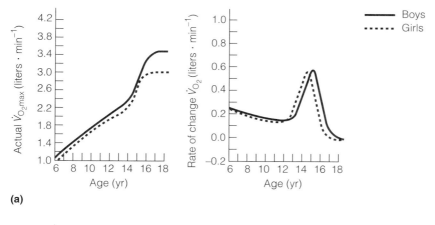

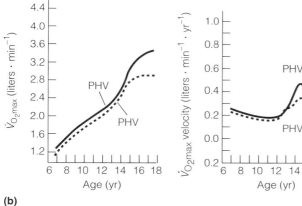

Figure 31-12 $\dot{V}_{O_2max}$ increases with age and is related to increased size of skeletal muscle and the heart. During the adolescent growth spurt (expressed as peak height velocity, PHV), $\dot{V}_{O_2max}$ increases at an accelerated rate. Training at this time also increases $\dot{V}_{O_2max}$ at an accelerated rate. (a) $\dot{V}_{O_2max}$ increases with age in active and inactive children. (b) $\dot{V}_{O_2max}$ when aligned on peak height velocity. Modified from Mirwald and Bailey, 1986.

■ "Anaerobic" Work Capacity

Cardiovascular responses of children to exercise mimic those of adults. However, during exercise, blood and muscle lactate levels in children are lower than in adults. Muscle biopsies of children taken during rest and exercise indicate that muscle adenosine triphosphate (ATP) and creatine phosphate (CP) levels during rest and exercise are similar to those in adults. Therefore, the lower lactate levels in children during exercise appear to be due to lower resting glycogen levels and to lower levels of phosphofructokinase (PFK). These differences indicate that the capacity for rapid glycogenolysis and glycolysis in immature skeletal muscle is not completely developed. Because the levels of muscle phosphagens are similar to those in adults and because muscle glycogenolytic capacity in children is not yet fully developed, the propensity for children to perform short, sprint-type exercise of 5- to 10-s duration is understood.

■ Genetics and Physical Performance

Endurance physical performance by genetic and nongenetic factors is described by equation 31-1:

$$V_p = V_g + V_e + V_{ge} + E \qquad (31\text{-}1)$$

where V_p is the total variation in endurance performance; V_g is the variation attributable to genotype; V_e represents nongenetic, environmental effects (such as training and nutrition); V_{ge} represents the interaction between genotype and environment (trainability); and E is the random error component.

The genetic effects on cardiovascular capacity, muscle fiber type and, hence, endurance, have been described in studies of identical and fraternal twins (Figure 31-13). The high correlation between the $\dot{V}_{O_2max}$ of fraternal twin siblings and the even higher correlation in identical twins leads to the conclusion that $\dot{V}_{O_2max}$ is greatly influenced by heredity.

There are strong family resemblances in maximal oxygen consumption. The Heritage Family Study (Bouchard et al., 1998) examined similarity in exercise capacity in 429 sedentary people (170 parents and 259 of their offspring) between 16 and 65 years of age. This study found that genetic and environmental factors contributed to the familial resemblance in maximal oxygen consumption. According to this study, approximately 50% of endurance performance is due to genetic factors. Further, the mother's genes, which determine mitochondrial mass, are the most important factors for determining endurance capacity, while the father's contribution is largely environmental.

There is also a strong genetic influence on muscle fiber type (Figure 31-14). Twin studies have shown a strong correlation between percent slow-twitch (high-oxidative) fiber types. Komi and Karlsson (1979) demonstrated a genetic link between $\dot{V}_{O_2max}$ and muscle fiber types. Muscle fiber type was correlated to $\dot{V}_{O_2max}$ in monozygous twins ($r = 0.47$) but not in dizygous twins.

Although the effects of heredity are important in determining the cardiovascular, pulmonary and muscular capacity, environmental factors such as training are also important. The effects of training are manifest in two ways: (1) the amount and quality of training and (2) the interaction between genetics and environment (training). Genetics studies suggest that there are high and low responders to endurance training and that sensitivity to training is genetically determined (hence the V_{ge} component in Eq. 31-1).

■ Summary

Growth involves a series of developmental stages that are similar in all people. Individual differences in diet, exercise, and health may affect these stages to a certain extent, but the basic pattern remains the same. Each growth stage has a profound influence on individual capability for physical performance.

Growth proceeds along an S-shaped curve. During the first two years of life height and weight increase rapidly. This early increase is followed by a progressively declining growth rate during childhood. At puberty the growth rate abruptly increases in the adolescent growth spurt.

Maturational assessment is important because most indices of physical performance are most highly related to tissue growth (i.e., increases in muscle mass, heart size, and so on). Maturation may be assessed by determining skeletal, sexual, or somatic maturation. Most studies show no relationship between sports training and maturation.

During growth, bone formation exceeds absorption. Exercise, which stresses bones and stimulates bone growth, does not seem to enhance linear bone growth but does increase bone density and width.

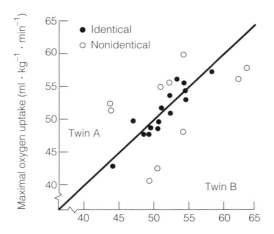

Figure 31-13 The correlation between maximal oxygen uptake ($\dot{V}_{O_2max}$) values in monozygous (identical) and dizygous (nonidentical) twins. The correlation is clearly higher in identical than in nonidentical twins.

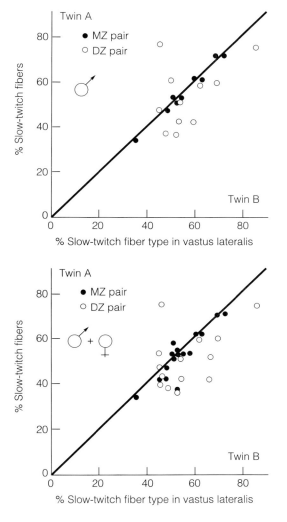

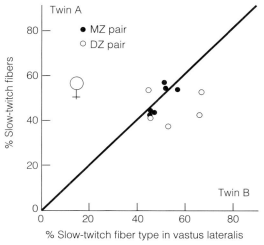

Figure 31-14 Interpair comparisons of slow-twitch fiber type distribution in medial vastus lateralis in mono- and dizygous twins. SOURCE: Komi and Karlsson, 1979.

Moderate exercise, combined with a calcium-rich diet, may maximize bone density in children. Over-training (imbalance between training and recovery) can lead to decreased bone density, particularly in women, by suppressing gonadal hormones important for bone growth.

Exercise capacity and $\dot{V}_{O_2max}$ (liters · min⁻¹) increase gradually throughout childhood. During puberty, these increase dramatically in boys and level off in girls. The improvements in oxygen transport capacity in boys during puberty are largely due to increased muscle mass and heart size. Cardiovas-

cular responses of children to exercise mimic those of adults.

The capacity for rapid glycogenolysis and glycolysis in immature skeletal muscle is not completely developed. Because the levels of muscle phosphagens are similar to those in adults and because muscle glycogenolytic capacity in children is not yet fully developed, the propensity for children to perform short, sprint-type exercise of 5- to 10-s duration is understood.

Genetics and training are important determinants of aerobic capacity. Genetic studies have shown that the differential sensitivity to training is genetically determined—there are low and high responders to training.

■ Selected Readings

Abe, D., K. Yanagawa, K. Yamanobe, and K. Tamura. Assessment of middle-distance running performance in sub-elite young runners using energy cost of running. *Eur. J. Appl. Physiol.* 77: 320–325, 1998.

American College of Sports Medicine and American Heart Association. Recommendations for cardiovascular screening, staffing, and emergency policies at health/fitness facilities. *Med. Sci. Sports Exerc.* 30: 1009–1018, 1998.

Armstrong, N., J. Williams, J. Balding, P. Gentle, and B. Kirby. The peak oxygen uptake of British children with reference to age, sex, and sexual maturity. *Eur. J. Appl. Physiol.* 62: 369–375, 1991.

Åstrand, P.-O. Experimental Studies of Physical Work Capacity in Relation to Sex and Age. Copenhagen: Munksgaard, 1952.

Bailey, D. A., W. D. Ross, R. L. Mirwald, and C. Weese. Size dissociation of maximal aerobic power during growth in boys. *Med. Sports* 11: 140–151, 1978.

Bass, S., G. Pearce, M. Bradney, E. Hendrich, P. D. Delmas, A. Harding, and E. Seeman. Exercise before puberty may confer residual benefits in bone density in adulthood: studies in active prepubertal and retired female gymnasts. *J. Bone Miner. Res.* 13: 500–507, 1998.

Blimkie, C. R. J. Resistance training during preadolescence: issues and controversies. *Sports Med.* 15: 389–509, 1993.

Bonjour, J., G. Theintz, B. Buchs, D. Slosman, and R. Rizzoli. Critical years and stages of puberty for spinal and femoral bone mass accumulation during adolescence. *J. Clin. Endocrinol. Metab.* 73: 555–563, 1991.

Borer, K. T. The effects of exercise on growth. *Sports Med.* 20: 375–397, 1995.

Bouchard, C., E. W. Daw, T. Rice, L. Perusse, J. Gagnon, M. A. Province, A. S. Leon, D. C. Rao, J. S. Skinner, and J. H. Wilmore. Familial resemblance for $\dot{V}_{O_2max}$ in the sedentary state: the Heritage Family Study. *Med. Sci. Sports Exerc.* 30: 252–258, 1998.

Caine, D. C. Growth plate injury and bone growth: an update. *Pediatr. Exer. Sci.* 2: 209–229, 1990.

Daniels, S. R., P. R. Khoury, and J. A. Morrison. The utility of body mass index as a measure of body fatness in children and adolescents: differences by race and gender. *Pediatrics* 99: 804–807, 1997.

DeLany, J. P. Role of energy expenditure in the development of pediatric obesity. *Am. J. Clin. Nutr.* 68: S950–S955, 1998.

DiLorenzo, T. M., R. C. Stucky-Ropp, J. S. Vander Wal, and H. J. Gotham. Determinants of exercise among children: a longitudinal analysis. *Prev. Med.* 27: 470–477, 1998.

Driscoll, D. J., B. A. Staats, and K. C. Beck. Measurement of cardiac output in children during exercise: a review. *Pediatr. Exerc. Science* 1: 102–115, 1989.

Ekblom, B. Effects of physical training in adolescent boys. *J. Appl. Physiol.* 27: 350–355, 1969.

Eliakim, A., J. A. Brasel, T. J. Barstow, S. Mohan, and D. M. Cooper. Peak oxygen uptake, muscle volume, and the growth hormone insulin-like growth factor-1 axis in adolescent males. *Med. Sci. Sports Exerc.* 30: 512–517, 1998.

Eliakim, A., J. A. Brasel, S. Mohan, W. L. T. Wong, and D. M. Cooper. Increased physical activity and the growth hormone IGF-1 axis in adolescent males. *Am. J. Physiol.* 275: R308–R314, 1998.

Ericksson, B. O., I. Engstrom, A. Karlberg, A. Lundin, B. Saltin, and C. Thoren. Long-term effect of previous swim training in girls: a 10-year follow-up study of the "girl swimmers." *Acta Paediatr. Scand.* 67: 285–292, 1978.

Ericksson, B. O., I. Engstrom, A. Karlberg, B. Saltin, and C. Thoren. Preliminary report on the development of lung volumes in young girl swimmers. *Acta Paediatr. Scand.* 217(Suppl.): 73–77, 1971.

Eriksson, B. O., and G. Koch. Effect of physical training on hemodynamic response during submaximal and maximal exercise in 11–13-year-old boys. *Acta Physiol. Scand.* 82: 27–39, 1973.

Fahey, T. D. The development of respiratory capacity during exercise in children and adolescents. In Handbook of Human Growth and Developmental Biology, vol. 3, part B, E. Meisami and P. S. Timiras (Eds.). Boca Raton, Fl.: CRC Press, 1988, pp. 209–218.

Fahey, T. D., A. Del Valle-Zuris, G. Oehlesen, M. Trieb, and J. Seymour. Pubertal stage differences in hormonal and hematological responses to maximal exercise in males. *J. Appl. Physiol.* 46: 823–827, 1979.

Fehling, P. C., R. J. Stillman, R. A. Boileau, M. H. Slaughter, J. L. Clasey, A. Alekel, and A. Rector. Total body bone mineral content and density in males and females aged 10–80 years. *Med. Sci. Sports Exerc.* 24: S11, 1992.

Gilliam, T. B., S. Sady, W. G. Thorland, and A. C. Weltman. Comparison of peak performance measures in children ages 6 to 8, 9 to 10, and 11 to 13 years. *Res. Quart.* 48: 695–702, 1977.

Gilsanz, V., D. T. Gibbens, T. F. Roe, M. Carlson, M. O. Senac, M. Biechat, H. K. Huang, E. E. Schulz, C. R. Libanati, and C. C. Cann. Vertebral bone density in children: effect of puberty. *Radiology* 166: 847–850, 1989.

Goran, M. I., and M. Sun. Total energy expenditure and physical activity in prepubertal children: recent advances based on the application of the doubly labeled water method. *Am. J. Clin. Nutr.* 68: S944–S949, 1998.

Herman-Giddens, M. E., E. J. Slora, R. C. Wasserman, C. J. Bourdony, M. V. Bhapkar, G. G. Koch, and C. M. Hasemeier. Secondary sexual characteristics and menses in young girls seen in office practice: a study from the Pediatric Research in Office Settings network. *Pediatrics* 99: 505–512, 1997.

Ilich, J. Z., M. Skugor, T. Hangartner, A. Baoshe, and V. Matkovic. Relation of nutrition, body composition and physical activity to skeletal development: a cross-sectional study in preadolescent females. *J. Am. Coll. Nutr.* 17: 136–147, 1998.

Kemper, H. C., G. B. Post, and J. W. Twisk. Rate of maturation during the teenage years: nutrient intake and physical activity between ages 12 and 22. *Int. J. Sport Nutr.* 7: 229–240, 1997.

Klissouras, V. Prediction of athletic performance: genetic considerations. *Can. J. Sport Sci.* 1: 195–200, 1976.

Koch, G. and B. O. Erickson. Anatomical right-to-left shunt at rest and ventilation, gas exchange and pulmonary diffusing capacity during exercise in 11 to 13 year old boys before physical training. In Pediatric Work Physiology, O. Bar-Or (Ed.). Tel-Aviv: Wingate Institute, 1973, p. 151.

Komi, P. V., and J. Karlsson. Physical performance, skeletal muscle enzyme activities and fiber types in monozygous and dizygous twins of both sexes. *Acta Physiol. Scand.* 462 (Suppl.): 1–43, 1979.

Krahenbuhl, G. S., D. W. Morgan, and R. P. Pangrazi. Longitudinal changes in distance running performance in young males. *Int. J. Sports Med.* 10: 92–96, 1989.

Krahenbuhl, G. S., J. S. Skinner, and W. M. Kort. Developmental aspects of maximal aerobic power in children. *Exer. Sports Sci. Rev.* 13: 503–538, 1985.

Larson, L. Fitness, Health, and Work Capacity. New York: Macmillan, 1974, pp. 516–517.

Malina, R. M., and C. Bouchard. Growth Maturation and Physical Activity. Champaign, Ill.: Human Kinetics, 1991.

Mantzoros, C. S., J. S. Flier, and A. D. Rogol. A longitudinal assessment of hormonal and physical alterations during normal puberty in boys. Rising leptin levels may signal the onset of puberty. *J. Clin. Endocrinol. Metab.* 82: 1066–1070, 1997.

Mirwald, R. L., and D. A. Bailey. Maximal Aerobic Power. London, Ont.: Sports Dynamics, 1986.

Nagano, Y., R. Baba, K. Kuraishi, T. Yasuda, M. Ikoma, K. Nishibata, M. Yokota, and M. Nagashima. Ventilatory control during exercise in normal children. *Pediatr. Res.* 43: 704–707, 1998.

Nevill, A. M., R. Ramsbottom, and C. Williams. Scaling physiological measurement for individuals of different size. *Eur. J. Appl. Physiol.* 65: 110–117, 1992.

Noakes, T. D. Implications of exercise testing for prediction of athletic performance: a contemporary perspective. *Med. Sci. Sports. Exerc.* 20: 319–330, 1988.

Noakes, T. D. Maximal oxygen uptake: "Classical" versus "contemporary" viewpoints: a rebuttal. *Med. Sci. Sports Exerc.* 30: 1381–1398, 1998.

Parizkova, J. Longitudinal study of the development of body composition and build in boys of various physical activity. *Hum. Biol.* 40: 212–223, 1968.

Page, A., and K. R. Fox. Is body composition important in young people's weight management decision-making? *Int. J. Obes. Relat. Metab. Disord.* 22: 786–792, 1998.

Pate, R. R., and D. S. Ward. Endurance exercise trainability in children and youth. *Adv. Sports Med. Fitness* 3: 37–55, 1990.

Rarick, G. L. Physical Activity: Human Growth and Development. New York: Academic Press, 1973.

Robinson, S. Experimental studies of physical fitness in relation to age. *Arbeitsphysiologie* 10: 18–23, 1938.

Rosenthal, M., and A. Bush. Haemodynamics in children during rest and exercise: methods and normal values. *Eur. Respir. J.* 11: 854–865, 1998.

Rossner, S. Childhood obesity and adulthood consequences. *Acta Paediatr.* 87: 1–5, 1998.

Rowland, T. W. Exercise and Children's Health. Champaign, Ill.: Human Kinetics, 1990.

Rowland, T. W. Trainability of the cardiorespiratory system during childhood. *Can. J. Sport Sci.* 17: 259–263, 1992.

Rowland, T. W. (Ed.). Pediatric Laboratory Exercise Testing. Clinical Guidelines. Champaign, Ill.: Human Kinetics, 1993.

Rowland, T., D. Goff, P. DeLuca, and B. Popowski. Cardiac effects of a competitive road race in trained child runners. *Pediatrics* 100: E2, 1997.

Rutenfranz, J., M. Macek, K. Lange Anderson, R. D. Bell, J. Vavra, J. Radvansky, F. Klimmer, and H. Kylian. The relationship between changing body height and growth related changes in maximal aerobic power. *Eur. J. Appl. Physiol.* 60: 282–287, 1990.

Schonfeld-Warden, N., and C. H. Warden. Pediatric obesity: an overview of etiology and treatment. *Pediatr. Clin. North Am.* 44: 339–361, 1997.

Shephard, R. J. Effectiveness of training programmes for prepubescent children. *Sports Med.* 13: 194–213, 1992.

Shephard, R. J. Physical Activity and Growth. Chicago: Year Book Publishers, 1982.

Sjodin, B., and J. Svedenhag. Oxygen uptake during running as related to body mass in circumpubertal boys: a longitudinal study. *Eur. J. Appl. Physiol.* 65: 150–157, 1992.

Stergioulas, A., A. Tripolitsioti, D. Messinis, A. Bouloukos, and C. Nounopoulos. The effects of endurance training on selected coronary risk factors in children. *Acta Paediatr.* 87: 401–404, 1998.

Stunkard, A. J., J. R. Harris, N. L. Pederson, and G. E. McClearn. The body-mass index of twins who have been reared apart. *N. Engl. J. Med.* 322: 1483–1487, 1990.

Tanner, J. M. Growth at Adolescence. 2d Ed. Oxford: Blackwell Scientific Publications, 1962.

Theintz, G. E., H. Howald, U. Weiss, and P. C. Slzoneko. Evidence for a reduction of growth potential in adolescent female gymnasts. *J. Pediatr.* 122: 306–313, 1993.

Timaris, P. S. Developmental Physiology and Aging. New York: Macmillan, 1972.

U.S. Department of Health and Human Services. Physical Activity and Health: A Report of the Surgeon General. Atlanta: U.S. Department of Health and Human Services, National Centers for Disease Control and Prevention, National Center for Chronic Disease Prevention and Health Promotion, 1996.

Van Lenthe, F. J., C. G. Kemper, and W. van Mechelen. Rapid maturation in adolescence results in greater obesity in adulthood: the Amsterdam Growth and Health Study. *Am. J. Clin. Nutr.* 64: 18–24, 1996.

Viru, A., L. Laaneots, K. Karelson, T. Smirnova, and M. Viru. Exercise-induced hormone responses in girls at different stages of sexual maturation. *Eur. J. Appl. Physiol.* 77: 401–408, 1998.

Washington, R. L., J. C. van Gundy, C. Cohen, H. M. Sondheimer, and R. R. Wolfe. Normal aerobic and anaerobic exercise data for North American school-age children. *J. Pediatr.* 112: 223–233, 1988.

Weber, G., W. Kartodihardjo, and V. Klissouras. Growth and physical training with reference to heredity. *J. Appl. Physiol.* 40: 211–215, 1976.

Williams, J. R., N. Armstrong, E. M. Winter, and N. Crichton. Changes in peak oxygen uptake with age and sexual maturation in boys: physiological fact or statistical anomaly? In Children and Exercise XVI, J. Coudert and E. Van Praagh (Eds.). Paris: Masson, 1992, pp. 35–37.

CHAPTER 32

AGING AND EXERCISE

It is extremely difficult to fully quantify the effects of aging on physiological function and physical performance. Lifestyle may exacerbate or delay the effects of aging. Some people deteriorate in their 20s and 30s due to lack of exercise, obesity, poor diet, smoking, and stress. Others are active and robust in their 50s, 60s, 70s, and beyond, and are capable of remarkable athletic performances.

Disease can further complicate our understanding of the aging process. Various systematic disorders, such as osteoarthritis and atherosclerosis, are so common in older people that they are sometimes considered a normal part of the aging process. So, although aging contributes to deteriorating physiological capacity, it is extremely difficult to separate its effect from those of deconditioning and disease. Physiological systems also vary in the extent to which they experience age-related changes. For example, respiratory and kidney functions deteriorate markedly with age, while nerve impulse conduction slows only slightly.

There has been considerable research in recent years on aging and physical activity. This topic is of particular concern to students of exercise physiology because the aging adult is steadily comprising a larger percentage of the population, and they are very likely to have contact with these individuals. The average age of the American population is 36 years, up from 33 years in 1990. By the year 2030, 70

In 1999, 82-year old Payton Jordon, the legendary coach and athlete, broke the world age-group record in th 100m dash with a time of 14.8 seconds. SOURCE: © Suzy Hess.

million people in the United States will be 65 years of age or older, and the group of people aged 85 and older will be the fastest growing segment of the population. By that time, out of financial necessity, scarce medical resources will require a greater emphasis on disease prevention than on medical crisis intervention.

Lifestyle (i.e., exercise, diet, etc.) may influence performance and health during aging, but it does not halt the aging process. The average life expectancy has increased from approximately 47 years in 1900 to approximately 76 years in 2000. Advances in public health, medicine, and standard of living increasingly protect people against premature death from disease and accidents. However, the maximum lifespan of slightly more than 100 years has not changed.

The increase in average life expectancy in the last 100 years has been due largely to improved medical and environmental practices. As the population ages, quality of life becomes increasingly important. Most Americans maintain a healthy life for only about 85% their lives. During the remaining 15%, most people are impaired significantly with degenerative diseases and decreased mobility.

The leading causes of death are largely related to lifestyle and are, to a large extent, preventable or modifiable (Table 32-1). Older people readily adapt and respond to endurance and strength training. Endurance training helps maintain cardiovascular function, enhances exercise capacity, and reduces risk factors associated with heart disease, diabetes, insulin resistance, and some cancers. Strength training helps prevent loss of muscle mass and strength normally associated with aging. Exercise training prevents bone loss and improves postural stability, which reduces the risk of fractures and falling. Mobility exercises improve flexibility and joint health.

TABLE 32-1

Leading Causes of Death in the United States

Rank	Cause of Death	Number of Deaths	Percent of Total Deaths	Female/Male Ratio*	Lifestyle Factors
1	Heart disease	733,834	31.6	50/50	D I S A
2	Cancer	544,278	23.4	48/52	D I S A
3	Stroke	160,431	6.9	61/39	D I S
4	Chronic obstructive lung diseases	106,146	4.6	48/52	S
5	Unintentional injuries	93,874	4.0	34/66	S A
	Motor vehicle-related	(43,449)	(1.9)	(33/67)	
	All others	(50,425)	(2.2)	(35/65)	
6	Pneumonia and influenza	82,579	3.6	54/46	S
7	Diabetes mellitus	61,559	2.7	56/44	D I
8	HIV infection	32,655	1.4	7/93	
9	Suicide	30,862	1.3	19/81	A
10	Chronic liver disease and cirrhosis	25,135	1.1	34/66	A
	All causes	2,322,421			

Key: D Cause of death in which diet plays a part
   I Cause of death in which an inactive lifestyle plays a part
   S Cause of death in which smoking plays a part
   A Cause of death in which excessive alcohol consumption plays a part

*Ratio of females to males who died of each cause. For example, an equal number of women and men died of heart disease, but only about half as many women as men died of motor vehicle–related injuries.

Source: Fahey et al., 1999.

Exercise also provides psychological benefits, such as preservation of cognitive function, reduced incidence of depression, and enhanced self-efficacy.

■ The Nature of the Aging Process

Aging involves a diminished capacity to regulate the internal environment. It impairs the ability to meet metabolic and external challenges and results in a reduced probability of survival. Simply stated, the physiological control mechanisms do not work as well in old people. Reaction time is slowed, resistance to disease is impaired, work capacity is diminished, recovery from effort is prolonged, and body structures are less capable and resilient.

Biological systems are noted for their large capacity. In the heart, for example, maintenance of cardiac output is possible even in the face of extreme vascular disease and drastically impaired cardiac function. Likewise, the pulmonary system has a large reserve capacity for ventilating the alveoli. Its function is limited only by extreme environmental conditions or disease. Unfortunately, however, the capacity of these systems diminishes with age (Table 32-2).

No matter how well people take care of themselves, their physiological processes eventually become prey to the ravages of old age. This fact is consistently observed in all animal species. Even cultured bacteria can reproduce only a limited number of times. Animals such as rats, dogs, horses, and aardvarks have life spans characteristic of their species. Of course, all animals of a species do not live to be precisely the same age—there are individual differences. Likewise, the quality of life can be considerably different. In humans, some can be fit and alert at 90 years, while others are invalids in their 60s.

Genetic factors have a profound influence on length of life, while a combination of environmental and genetic factors govern quality of life (i.e., robustness). Observations of identical and nonidentical twins showed that the life spans of twins are remarkably similar. The identical twins usually died within two to four years of each other, the nonidentical twins within seven to nine years.

Although little is known of the process, aging may be due to accumulated injury, immune responses to the body's own tissues, and problems associated with cell division. Aging is likely related in some way to abnormalities in the genetic functions of cells. With time, progressive genetic damage impairs the ability of the cells to reproduce and function normally. This process undoubtedly affects the cellular communications systems, which control metabolic processes such as protein synthesis. The result is the formation of tissues that do not function as well, such as stiff and brittle cartilage and inactive enzymes. This process also affects the immune system, which can cause the body to react to its own proteins. Possible causes for these "genetic errors" include damage from free radicals produced during metabolism, or external agents such as radiation and other environmental toxins.

The aging process is also associated with an accumulation of insults and wear-and-tear that lead to the gradual loss of the ability to adapt to stress. As we have seen, most physiological control mechanisms are highly adaptable. Joints, for example, adapt to mobilization by maximizing their range of motion to the extent of the stress. If the joint is injured repeatedly over a lifetime, however, the joint capsule thickens, resulting in a gradual loss of range of motion. Likewise, progressive arteriosclerosis eventually leads to impaired circulation and accompanying catastrophic cardiovascular events.

Although the maximum lifespan is finite, the quality of life is extremely variable. Exercise seems to be an important factor in maximizing physiological function throughout life. Physical activity has significant effects on many biological activities. In fact, for some processes, it almost appears to make "time stand still."

The human life span may be subdivided into childhood, adolescence (puberty–19 years), young adulthood (20–35 years), young middle age (35–45 years), later middle age (45–65 years), early old age (65–75 years), middle old age (75–85 years), and very old age (>85 years). By middle old age, most

TABLE 32-2	

Physiological Effects of Aging

Effect	Functional Significance
Cardiovascular	
Blood, plasma, and red cell volumes ↓	Decreased venous return and stroke volume
Capillary/fiber ratio ↓	Decreased muscle blood flow
Cardiac compliance ↓	Decreased early diastolic filling and increased contribution of atrial priming
Cardiac muscle and heart volume ↓	Decreased maximal stroke volume and cardiac output
Elasticity of blood vessels ↓	Increased peripheral resistance, blood pressure, and cardiac afterload
Myocardial myosin-ATPase ↓	Decreased myocardial contractility
Sympathetic stimulation of SA node ↓	Decreased maximum heart rate
Respiration	
Condition of elastic lung support structures ↓	Increased work of breathing
Elasticity of support structures ↓	Decreased lung elastic recoil
Size of alveoli ↑	Decreased diffusion capacity and increased dead space
Number of pulmonary capillaries ↓	Decreased ventilation/perfusion equality
Muscle, joints, and other soft tissues	
Accumulated mechanical stress in joints ↑	Stiffness, loss of flexibility, and osteoarthritis
Action potential threshold ↓	Loss of strength and power
Blood insulin ↑	Hypertension, coronary heart disease
$(Ca^{2+}$, Myosin)-ATPase ↓	
Insulin sensitivity ↓	Diabetes, coronary heart disease, obesity, hyperlipidemia, hypertension
Lactate dehydrogenase ↓	Slows glycolysis
Muscle mass ↓	
Number of type IIa and IIb fibers ↓	
Oxidative enzymes: SDH, cytochrome oxidase, and MDH ↓	Decreased muscle respiratory capacity
Size and number of mitochondria ↓	Decreased muscle respiratory capacity
Size of motor units ↓	
Stiffness of connective tissue in joints ↑	Decreased joint stability and mobility
Total protein and N_2 concentration ↓	
Water content in intervertebral cartilage ↓	Atrophy and increased chance of compression fractures in spine
Bone	
Bone minerals ↓	Osteoporosis—increased risk of fracture
Body composition and stature	
Abdominal fat deposition ↑	Coronary heart disease, insulin resistance (?), hyperlipidemia, back pain
Body fat ↑	Impaired mobility and increased risk of disease
Fat-free weight ↓	Decreased metabolic rate
Kyphosis ↑	Loss of height

people will have developed some degree of impairment. People who reach very old age are generally dependent on others for their care. However, graded physical exercise can make a significant contribution to the quality of life, even in people older than 90 years.

■ The Aging Process and the Effects of Exercise

Aging affects cellular function and systemic regulation (see Table 32-2). Peak physiological function, for the most part, occurs at about 30 years of age. After that, most factors decline at a rate of 0.75 to 1.0% a year.

Decline in physical capacity is characterized by a decrease in maximal oxygen consumption, maximal cardiac output, muscle strength and power, neural function, flexibility, and increased body fat. All of these factors can be positively affected by

training. In fact, remarkable levels of performance are possible, particularly if physical training has been maintained throughout life.

Exercise training does not retard the aging process; it just allows the individual to perform at a higher level. Comparisons of trained and sedentary individuals indicate a similar decrease in work capacity with age. Of course, the trained subjects achieved a higher level of performance at all ages. In some cases, the decrease in performance is greater in the trained than untrained individual, indicating the difficulty of maintaining a high physical capacity with advancing age (Figure 32-1).

Cardiovascular Capacity

Maximal Oxygen Consumption Maximal oxygen consumption ($\dot{V}_{O_2max}$) decreases approximately 30% between the ages of 20 and 65, with the rate of decline greatest after age 40. Decreases in $\dot{V}_{O_2max}$ with age are extremely variable. For example,

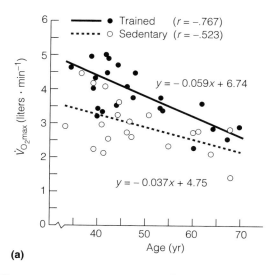

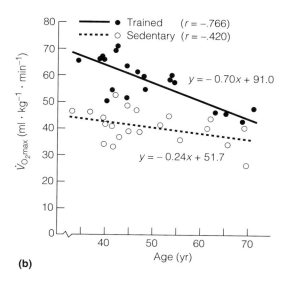

(a)

(b)

Figure 32-1 The effects of age on $\dot{V}_{O_2max}$ in trained and sedentary men. (a) $v = \dot{V}_{O_2max}$ (liters · min⁻¹). (b) $y = \dot{V}_{O_2max}$ (ml · kg⁻¹ · min⁻¹). Although training will improve $\dot{V}_{O_2max}$ and the quality of life in the elderly, it will not prevent indefinitely the decline in functional capacity. x = age (yr). Adapted from Suominen et al., 1980.

Wells and coworkers (1992) have reported that masters women runners at age 70 had $\dot{V}_{O_2max}$ values equivalent to those of sedentary 20-year-olds.

Decreased $\dot{V}_{O_2max}$ occurs because of decreases in maximal heart rate, stroke volume, power output capacity, fat-free mass, and arteriovenous oxygen difference. Relative is % of a person's maximum capacity. As discussed in Chapter 16, the relationship between $\dot{V}_{O_2}$ and exercise intensity is linear and probably related as much to muscle biochemical factors that limit power output as to limitations in oxygen transport capacity. Decreased muscle mass in the elderly contributes significantly to the fall in $\dot{V}_{O_2max}$ with age.

Heart Rate   During submaximal exercise, heart rate is lower with age at any relative exercise intensity but the same at any absolute intensity. Also, cardiovascular drift is greater in the aged. Cardiovascular drift is the tendency of physiological factors such as heart rate, core temperature, and ventilation to rise at a constant exercise intensity. There is also a longer recovery rate for submaximal and maximal exercise.

β-adrenergic responsiveness decreases with age, which tends to decrease maximal heart rate. However, the age-related decrease in maximal heart rate is at least partially due to deconditioning and loss of muscle mass. HR_{max} during maximal treadmill testing is typically above 170 beats · min^{-1} in fit people 60–70 years (Figure 32-2). Reduced HR_{max} is particularly apparent during exercise that is greatly affected by muscle mass (e.g., cycle ergometer) or when the person is affected by exertional dyspnea or angina. The effects of aging on functional capacity in males are summarized in Table 32-3.

Stroke Volume and Cardiac Output   Aging impairs the heart's capacity to pump blood. Thus, cardiac output and stroke volume is less in older adults at the same absolute and relative exercise intensities than it is in younger people. There is a gradual loss of contractile strength caused in part by a decrease in Ca^{2+}-myosin ATPase activity and often myocardial ischemia. The heart wall stiff-

ens, which delays ventricular filling. This delay decreases stroke volume. The impairments in maximal heart rate and stroke volume result in substantial decreases in cardiac output. Other changes affecting stroke volume and end-diastolic volume include poor peripheral venous tone, varicose veins, decreased plasma, red cell, and total blood volumes, and slow relaxation of the ventricular wall.

To increase stroke volume during exercise, older adults rely largely on the Frank–Starling mechanism rather than increased cardiac contractility because of loss of contractile strength. Older adults also rely more on the atria to facilitate end-diastolic volume (ventricular filling).

Numerous cellular changes occur in the cardiovascular system that help explain its diminished capacity to transport gases and substrates (see Table 32-2). Aging cardiovascular tissue typically experiences increased deposits of amyloid-, basophilic-, lipid-, and collagen-type substances. The elasticity of the major blood vessels and the heart decreases due to connective tissue changes. This is usually accompanied by narrowing of the blood vessels in the muscles, heart, and other organs. Heart mass also usually decreases, and there are sometimes fibrotic changes in the heart valves. The fiber/capillary ratio is reduced, as well, which impairs blood flow to the muscles. The venous valves also deteriorate.

These changes have several consequences for cardiovascular performance. Vascular stiffness increases the peripheral resistance to blood flow, which increases the afterload of the heart. This forces the heart to work harder to push blood into the circulation and increases myocardial oxygen consumption at a given intensity of effort. Cardiac hypoxemia can ensue because of atherosclerotic changes in the coronary arteries. The higher peripheral resistance also raises systolic blood pressure at rest and during maximal exercise. However, there is little or no increase in diastolic pressure.

Arteriovenous Oxygen Difference ($(a-v)O_2$)
Arteriovenous oxygen ($(a-v)O_2$) decreases with age, which contributes to diminished aerobic capacity in older adults. The decrease is from approximately 16

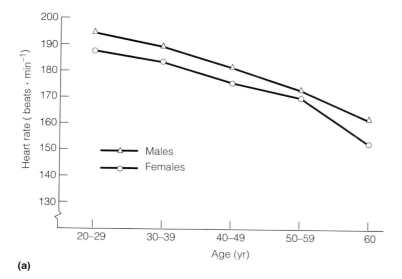

Figure 32-2 Average (a) maximal and (b) resting heart rate with age. Triangles = males; circles = females.

volumes % in a 20-year-old to 12–13 volumes % in a 65-year-old. The decrease is due to a reduction in the fiber/capillary ratio, total hemoglobin, the respiratory capacity of muscle, and the shunting of a larger portion of the cardiac output to areas with limited oxygen extraction, such as the skin and viscera. Muscle mitochondrial mass decreases, along with decreases in several oxidative enzymes. However, $(a-v)O_2$ is higher at any absolute exercise intensity.

Circulatory Regulation The capacity of the autonomic reflexes, which control blood flow, also seems to diminish. At rest, for example, circulation to the skin is often poor, which can make peripheral body parts uncomfortably cold. During physical activity, however, a disproportionate amount of blood is directed to the skin, which can further hamper oxygen extraction. In addition, the elderly are more subject to orthostatic intolerance—they sometimes

TABLE 32-3

Changes in Functional Capacity and Body Composition with Age in Males

Effect	Age (yr) 20	Age (yr) 60
$\dot{V}_{O_2max}$ (ml · kg^{-1} · min^{-1})	39	29
Maximal heart rate (beats · min^{-1})	194	162
Resting heart rate (beats · min^{-1})	63	62
Maximal stroke volume (ml)	115	100
Maximal (a–v) O_2 difference (ml · liter^{-1})	150	140
Maximal cardiac output (liters · min^{-1})	22	16
Resting systolic blood pressure (mmHg)	121	131
Resting diastolic blood pressure (mmHg)	80	81
Total lung capacity (liters)	6.7	6.5
Vital capacity (liters)	5.1	4.4
Residual lung volume (liters)	1.5	2.0
Percentage fat	20.1	22.3

have difficulty maintaining blood pressure when going from a horizontal to vertical posture.

Endurance Training Endurance training induces improvements in aerobic capacity in the aged that are similar to those in young people. Most studies indicate that gains of about 20% in $\dot{V}_{O_2max}$ can be expected in a 6-mo endurance exercise program. As in the young, individual differences can be accounted for by initial fitness, motivation, intensity and duration of the program, and genetic characteristics.

Endurance exercise results in a decrease in submaximal heart rate at a given work load, decreased resting and exercise systolic blood pressure, and faster recovery heart rate. Abnormal electrocardiographic findings, such as ST segment depression, have also been shown to improve with training. Although the reason for this is unclear, it could be due to reduced oxygen consumption in the heart, possibly due to regression of atherosclerosis.

Training can increase stroke volume and cardiac output in older subjects by over 3 liters · min^{-1}. The increased stroke volume may be due in part to peripheral adaptations, such as increased plasma and blood volume, which facilitate the venous return of blood to the heart and increased end-diastolic vol-

ume. Training improves resting and exercise diastolic filling and decreases reliance on atrial contraction for enhancing end-diastolic volume. It also reduces arterial stiffness, which may reduce cardiac afterload and enhance stroke volume. Older women do not seem to experience the same increases in stroke volume as men. While they improve $\dot{V}_{O_2max}$ the same percentage as men, the increase is largely due to a larger arteriovenous O_2 difference.

Older athletes are often capable of notable physical performances. For example, 69-year-old Lynn Edwards, who was the oldest person to complete the Hawaii Ironman Triathlon, swam 2.4 ocean miles in 2 hours, bicycled 112 miles in 8 hours, and ran the marathon in 5 hours 41 minutes. Payton Jordan, the legendary track coach from Stanford University, ran 100 meters in 12.91 seconds at the age of 74 and 14.8 sec at the age of 82. Throwers, such as Al Oerter, Carl Wallin, and Joe Keshmiri, performed at levels comparable to collegiate athletes well into their 50s. Although not considered old, Eamon Coughland was the first person over age 40 to break the 4-min mile.

Assessing the effects of age on athletic performance is difficult. The number of participants in masters athletics is relatively small, and participants are usually less serious about competition than are younger athletes. However, the number of participants is increasing, and competition is becoming more serious. This may help us to understand the ultimate limitations imposed by aging.

To live an independent lifestyle, an elderly person requires a minimum $\dot{V}_{O_2max}$ of approximately 20 ml O_2 · kg^{-1} · min^{-1}. A conservative, well-structured endurance training program can increase the fitness of an elderly person to this level within approximately three months. Although the program prevents premature disability, it does not seem to affect longevity. Typically, the fit older adult will live in good health until shortly before death.

■ Pulmonary Function

The normal lung enjoys a large reserve capacity that can meet ventilatory requirements even during maximal exercise. This reserve capacity begins to deteriorate gradually between 30 and 60 years of age, and accelerates after that. The process may be faster if the individual is a smoker or is chronically subjected to significant amounts of airborne contaminants.

The three most important changes that occur in this system with aging are a gradual increase in the size of the alveoli, the disintegration of the elastic support structure of the lungs, and a weakening of the respiratory muscle. These changes can interfere with the ventilation and perfusion of the lung, both of which can impair oxygen transport capacity.

Enlargement of the alveoli also occurs in chronic obstructive pulmonary disease. Thus, it is difficult to separate the effects of disease from those of aging. Alveolar enlargement is accompanied by a decrease in pulmonary vascularization. Both of these changes decrease the effective area available for diffusion.

The loss of pulmonary elasticity and the weakening of respiratory muscles can have a marked effect during exercise. Both changes make expiration more difficult and increase the work of breathing (oxygen cost). The loss of elastic recoil results in the premature closing of airways, which impairs ventilation in some of the alveoli. Because of these flow restrictions, there is an increased dependence on breathing frequency, rather than tidal volume, for ventilation during increasing intensities of exercise.

The deterioration of pulmonary function is similar in magnitude to that in the cardiovascular system. Thus, unless the decline in pulmonary function is accentuated by disease, ventilation remains adequate during exercise in the aged. Ventilation does not seem to limit endurance performance in the young or old—the most important limiting factor is probably cardiac output.

Training increases maximum ventilation, but the improvements parallel those of cardiac output. Although the breathing muscles can be strengthened through exercise, most changes are irreversible. Because of the large reserve capacity of the lungs, though, the changes can be tolerated quite well.

■ Skeletal System

Bone loss is a serious problem in older people, particularly in women. Women begin to lose bone mineral at about 30 and men at about 50 years of age. Bone loss, called osteoporosis, results in bone with less density and tensile strength. Osteoporosis increases the risk of fracture, which drastically increases the short-term mortality rate. The mortality rate within one year after hip fractures (usually due to falls) is approximately 50%.

Although the cause of bone loss in the aged is not completely understood, it seems to be related to a combination of factors including inactivity, diet, skeletal blood flow, and endocrine function. These factors may induce a negative calcium balance that steadily saps the bones of this important mineral. Calcium insufficiency during the pre- and post-pubertal years in girls seems related to the propensity to develop osteoporosis later in life. Estrogen deficiency in postmenopausal women is thought to be the major factor responsible for the accelerated trabecular bone loss in women 45–60 years old. Currently, estrogen replacement therapy and dietary calcium supplementation is recommended by many researchers for women in this age category to prevent bone mineral loss.

Exercise has been shown to be important in the prevention and treatment of osteoporosis. Bones, like other tissues, adapt to stresses placed on them. They become stronger when stressed and weaker when not stressed. Elderly male athletes have higher bone mineral content and density than nonathletes. In women, the results are less clear—some studies show a benefit of training for competitive sports in the elderly, while others do not. In both sexes, training seems to have a greater effect in trabecular bone than in cortical bone. Among nonathletic elderly women, factors that determine bone density include activity levels, body fat, sources of mechanical stress for weight-bearing bones, and muscle strength.

Previous exercise may also impact bone mineral content and density in elderly women. Although there is no relationship between the number of years of training and bone mineral content, previously active women tend to have a higher bone mineral content than those who have always been sedentary. At the same time, however, excessive training (i.e., chronic overtraining) in young people is associated with decreased trabecular bone mineral density in men and women. These results have also been found in elderly men and women.

In young women, chronic amenorrhea is related to decreased bone density—apparently overtraining depresses estrogen levels, which induces amenorrhea. The problem appears to be more severe in sedentary women with amenorrhea than in active women with this problem. Women who remain amenorrheic continue to lose bone mass, while women who resume normal menses increase bone mineral content. Fortunately, exercise enhances bone density in young women; and lifelong participation in physical activity may provide a buffer against bone loss during aging.

Bone mineral content can be improved by exercise, even in people over 80 years of age.

Joints

Joints become less stable and less mobile with age. Aging is often associated with degradation of collagen fibers, cross-linkage formation between adjacent fibrils of collagen, fibrous synovial membranes, joint surface deterioration, and decreased viscosity of synovial fluid. Joint stiffness and loss of flexibility are common in the elderly. In fact, some researchers feel that osteoarthritis is a natural result of the aging process.

It is difficult to separate aging from accumulated wear and tear, however. Trauma to joint cartilage results in the formation of scar tissue, characterized by the buildup of fibrous material that makes the connective tissue stiffer and less responsive to stress. This can result in a thickened joint capsule that often contains debris which impairs range of motion. Osteoarthritis in the elderly is principally located in areas receiving the most mechanical stress. So, it is unclear if the restricted range of motion found in joints is principally the result of aging or repeated trauma.

Range-of-motion exercises have been shown to dramatically increase flexibility in a relatively short time. Several months of flexibility training can result in improved range of motion of approximately 8% in the shoulder to 48% in the ankle. Flexibility training can be facilitated in the elderly by performing the exercises in warm water. However, exercise cannot significantly affect severely degenerated joints.

Skeletal Muscle

Loss of muscle mass and strength can have severe consequences in the quality of life of the elderly. Elderly people can become so weak that they are unable to perform basic essential tasks, such as carrying household objects or lifting the body weight from the tub, toilet, or bed. Weak, atrophied muscles contribute to obesity by decreasing metabolic rate. Poor muscular support can also lead to unstable joints, and increased incidence of falls and contribute to the development of osteoarthritis. Poor muscle strength can impair functional capacity to the point where exertional dyspnea creates anxiety, which further contributes to inactivity. Refer to Chapter 19 for underlying mechanisms of muscle adaptations to age.

Strength

Marked deterioration in muscle mass occurs with aging. This has important functional consequences because strength is highly related to the muscles' cross-sectional area. Muscle strength decreases approximately 8% a decade after age 45 years, with a total decrement of 30–40% between peak strength and strength at the time of death.

Strength losses are associated with decreased size and, to a lesser degree, loss of muscle fibers, selective loss or atrophy of type II fibers, decreases in the muscles' respiratory capacity, and increases in connective tissue and fat. These changes can have severe consequences in the elderly: Mobility is hampered, incidents of soft tissue pain are more frequent, and work capacity is impaired. Aging results in decreases in isometric and dynamic strength and speed of movement. However, lean body mass is maintained into the 70s in masters athletes who have remained consistently active.

Muscle Fiber Types and Motor Units

With age, there is a selective loss of type II fibers, which diminishes available strength and power. Type II fibers decrease more rapidly in the lower than upper body, which could reflect differences in aging rates between the two parts of the body or differences in disuse patterns. The mechanism for selective loss of type II fibers is unclear. One explanation is a loss of terminal sprouting (a normal repair mechanism), which causes denervation of the fibers. In elderly endurance athletes, there is a higher type I fiber area, coupled with loss of type II fibers, which may reflect selective hypertrophy from training.

A loss of fibers from individual motor units results in less available contractile force when a motor unit is recruited. Computer tomography studies have confirmed that decreased muscle cross-sectional area is related to loss of muscle fibers.

The mechanisms involved in muscle contraction are also impaired, which contributes to the loss of strength and power. Aging muscle is less excitable and has a greater refractory period. Thus, a greater stimulus is needed for contraction and a longer period of time is required before the muscle can respond to another stimulus. Contraction time and half relaxation time increase with age. Concentrations of ATP and CP are also reduced, particularly in fast-twitch muscle, which further impairs muscle function. Maximum contractile velocity decreases, probably reflecting the decreased proportion of type II fibers.

Muscle Biochemistry

The loss of the muscle's biochemical capacity is characterized by decreases in glycolytic enzymes, such as myokinase, lactate dehydrogenase, and triose phosphate dehydrogenase. There are either no changes or slight increases in oxidative enzymes with age. In very old age, there is evidence of the formation of incomplete or relatively inactive enzymes. In addition, some researchers have found decreases in mitochondrial mass. All of these changes affect ATP production and thus impair physical working capacity.

Relative strength increases from training are similar in the young and old, at least in short-term programs. Because of the difficulty conducting such research, available data are restricted to observations of a few months. Strength training renders beneficial effects even in subjects in their 90s.

■ Exercise and Cardiovascular Disease Risk Factors

The effects of exercise training on cardiovascular disease risk factors are the same as those discussed in Chapter 24. However, several training effects are particularly important in the elderly. Even low-intensity exercise has marked metabolic and systemic effects that influence cardiovascular disease. Training lowers fasting and glucose-stimulated blood insulin levels and improves glucose tolerance (if impaired) and insulin sensitivity. However, improvements in glucose metabolism are not as great as those that occur in younger people.

Older people experience reductions in systolic and diastolic blood pressure with training. The greatest improvements occur with low to moderate intensities of exercise, particularly in older hypertensive adults.

Other risk factors influenced by training in older adults include improvements in plasma lipoprotein lipid profiles and body composition. Training in the elderly increases HDL and HDL_2 levels and decreases plasma triglyceride and cholesterol levels. It also decreases body fat (1–4% in most studies) and reduces abdominal fat deposition.

■ Body Composition and Stature

Body composition and stature change markedly with age. Body weight increases steadily beginning in the 20s, continues to increase until about 55 or 60 years, and then begins to decline. Height decreases gradually. Weight gain is accompanied by increased body fat and decreased lean body mass. Males advance from an average of 15% fat at 17 years to about 28% at 60. Women change from about 25% fat at 17 years to about 39% at 60 years. These values are highly variable and tend to be much less in the trained individual.

Table 32-4 presents longitudinal body composition measurements made on Dr. Albert Behnke, who is widely regarded as the father of body composition. Although Dr. Behnke remained physically active all of his life, his lean body mass tended to decrease and his percent fat tended to increase with age. Of particular interest are the measurements made during the periods of weight loss (3/12/40 to 10/9/40, 12/20/55 to 10/16/56, and 1967 to 1971). Note that Dr. Behnke seemed to be able to lose fat and maintain lean body mass much more easily when he was younger. It is difficult to draw any conclusions from these data, but they certainly conform to cross-sectional studies of body composition in older adults.

The distribution of fat tends to change in aging. There is a tendency for a greater proportion of the total body fat to be located internally rather than subcutaneously. Therefore, if body composition is measured by a skinfold technique, an equation developed specifically for elderly subjects must be employed.

Increased body fat with age is of concern because of its possible relationship with disease and premature mortality. Exercise, particularly resis-

			TABLE 32-4		
			Serial Body Composition Measurements on Dr. A. R. Behnke		
Year	Age (yr)	Weight (kg)	Density (g · cm^{-3})	Lean Body Weight (kg)	Percent Fat
1940(3/12)	36	92.0	1.056(SG)	72.7	21.0
1940(7/1)	36	88.4	1.060(SG)	71.6	19.0
1940(8/13)	37	85.0	1.066(SG)	71.3	16.1
1940(10/9)	37	83.2	1.071(SG)	71.9	13.6
1944(5/2)	40	90.7	1.056(SG)	71.7	21.0
1949(—)	46	91.6	1.054(SG)	69.5	24.1
1953(—)	50	95.5	1.044(SG)	69.7	27.0
1953(12/9)	50	95.5	1.045	69.0	27.8
1955(12/20)	52	101.5	1.035	69.9	31.1
1956(3/30)	52	86.3	1.047	66.6	22.8
1956(6/29)	52	84.4	1.053	66.4	21.3
1956(10/16)	53	86.5	1.058	67.5	22.0
1956(11/12)[a]	53	83.0	1.058	64.6	22.2
1967(—)	64	95.6	1.032	67.2	29.7
1970(—)	67	87.9	1.043	66.3	24.6
1971(—)	68	84.3	1.049	65.8	21.9
1971(—)	68	81.8	1.047	65.9	19.4
1973(4/6)	69	87.9	1.040	65.0	26.1
1973(10/31)	70	89.2	1.041	66.5	25.4
1977(3/8)	73	83.2	1.042	62.4	25.0
1977(—)	74	86.2	1.038	63.7	26.1

[a] Measured after week on Pemmican, 1000 kcal · day^{-1} diet (45% fat, protein, 10% carbohydrate). (SG) = Specific gravity.

SOURCE: Data courtesy of Dr. A. R. Behnke.

tance exercise, is extremely important in managing body composition in the elderly. Metabolic rate decreases by about 10% from age 20 to 65 years, and another 10% during the later years of life. Decreased metabolic rate is mainly caused by a smaller lean body mass. This falling metabolic rate necessitates a low caloric diet in order to minimize fat gain. However, this diet is often low in necessary vitamins and minerals, particularly calcium, as well. Regular exercise enables the elderly to consume more calories, which allows them to satisfy their nutritional requirements.

Stature decreases with age due to an increased kyphosis (rounding of the back), compression of intravertebral discs, and deterioration of vertebrae. Height typically decreases by 6 centimeters between 17 and 60 years.

■ Neural Function

Many neurophysiological changes occur with aging. Again, it's often difficult to separate the changes resulting from aging from those caused by disease. Principal changes include decreased visual acuity, hearing loss, deterioration of short-term memory, inability to handle several pieces of information simultaneously, and decreased reaction time.

Physical training seems to have little effect on the deterioration of neural function. There is no difference in neurobiological factors between extremely fit elderly endurance athletes and sedentary men. The effects of endurance training in the elderly are largely limited to functions relevant to physical performance. In other words, exercise training will improve performance in the elderly, but it will not prevent aging processes.

■ Exercise Prescription for the Elderly

The principles of exercise prescription (Chapter 27) apply to people of all ages. However, because of the increased risk of exercise for the elderly, caution is required.

Electrocardiographic abnormalities, particularly during exercise, are found regularly in the elderly. Aberrations such as ST segment depression, heart blocks, and disrhythmias are common enough to warrant routine exercise stress testing for any elderly person wishing to undertake an exercise program.

Great care must be taken when determining the type and intensity of exercise. For a sedentary person, the exercise should be one that minimizes soft tissue injuries. Good choices are walking and swimming. More vigorous exercise, such as running and racquetball, should be attempted only when the person has achieved sufficient fitness.

Ideally, intensity of exercise should coincide with the target heart rate, which is calculated from the maximal heart rate, which itself is often calculated from the formula $220 - $ age. However, for the elderly, this calculation is extremely misleading and can result in problems with exercise prescription. Although it is fairly accurate for estimating maximal heart rate at the 50th percentile of the population, it does not take into consideration the tremendous variability observed in the older age group. Maximal heart rates in people over 60 can range from highs of over 200 to lows of 105 bpm. Thus, target heart rates predicted on the basis of age can often either underestimate or overestimate the ideal exercise intensity. Whenever possible, an accurate measurement of maximal heart rate should be used rather than a predicted one.

Often older people have misperceptions about physical conditioning. They sometimes overestimate their stamina and exercise at extremely low intensities. For some elderly persons, it's important to increase the frequency of exercise until sufficient fitness is developed to also increase the intensity.

Basic principles for prescribing exercise for the elderly include the following:

- Follow a careful progression in intensity and duration of exercise.

- Warm up slowly and carefully, particularly for activities that may suddenly overload local muscle groups (such as tennis and racquetball).

- Cool down slowly. Before showering, elderly persons should continue very light exercise, such as walking or unloaded pedaling, until heart rate is below 100.

- Static stretching after exercise is helpful in preventing the soft tissue pain that is often associated with exercise training in older persons.

◼ Summary

The effects of aging are difficult to separate from those of degenerative diseases and deconditioning. However, it is clear that aging takes its toll on almost every facet of physiological function. Although the life span of any species is identifiable within relatively narrow limits, quality of life can be extremely variable. Regular exercise training can improve quality of life by increasing, or at least slowing deterioration in, physical capacity.

After 30 years of age, most physiological functions decline at a rate of approximately 0.75 to 1% a year. This decline in physical capacity is characterized by a decrease in $\dot{V}_{O_2max}$, maximal cardiac output, muscle strength and power, neural function, and flexibility, and an increase in body fat.

Exercise training does not retard the aging process; it merely allows the individual to perform at a higher level. Although older individuals cannot expect to reach the same absolute capacity as the young, they can improve, relatively, about the same amount.

Care should be taken when prescribing exercise for older people because of possible complications from degenerative diseases such as atherosclerosis, arthritis, and osteoporosis. Physical capacity should be carefully measured and considered. The duration and intensity of training should be mild at the beginning of the program and then progress slowly.

◼ Selected Readings

Abbasi, A., E. H. Duthie, Jr., L. Sheldahl, C. Wilson, E. Sasse, I. Rudman, and D. E. Mattson. Association of dehydroepiandrosterone sulfate, body composition, and physical fitness in independent community-dwelling older men and women. *J. Am. Geriatr. Soc.* 46: 263–273, 1998.

Adams, G., and H. DeVries. Physiological effects of an exercise training regimen upon women aged 52–79. *J. Gerontol.* 28: 50–55, 1973.

Alessio, H. M., and E. R. Blasi. Physical activity as a natural antioxidant booster and its effect on a healthy life span. *Res. Q. Exerc. Sport* 68: 292–302, 1997.

American College of Sports Medicine. Position stand on exercise and physical activity for older adults. *Med. Sci. Sports Exerc.* 30: 992–1008, 1998.

Aoyagi, Y., and R. J. Shephard. Aging and muscle function. *Sports Med.* 14: 376–396, 1992.

Barnard, J., G. Grimditch, and J. Wilmore. Physiological characteristics of sprint and endurance runners. *Med. Sci. Sports* 11: 167–171, 1979.

Bar-Or, O. Effects of age and gender on sweating pattern during exercise. *Int. J. Sports Med.* 19 (Suppl. 2): S106–S107, 1998.

Bijnen, F. C., E. J. Feskens, C. J. Caspersen, W. L. Mosterd, and D. Kromhout. Age, period, and cohort effects on physical activity among elderly men during 10 years of follow-up: the Zutphen Elderly Study. *J. Gerontol. A. Biol. Sci. Med. Sci.* 53: M235–M241, 1998.

Brown, D. A., and W. C. Miller. Normative data for strength and flexibility of women throughout life. *Eur. J. Appl. Physiol.* 78: 77–82, 1998.

Buchner, D. M. Preserving mobility in older adults. *West. J. Med.* 167: 258–264, 1997.

Butler, R. N., R. Davis, C. B. Lewis, M. E. Nelson, and E. Strauss. Physical fitness: benefits of exercise for the older patient. Part 2 of a roundtable discussion. *Geriatrics* 53: 46, 49–52, 61–62, 1998.

Campbell, W. W., M. C. Crim, V. R. Young, and W. J. Evans. Increased energy requirements and changes in body composition with resistance training in older adults. *Am. J. Clin. Nutr.* 60: 167–175, 1994.

Carmeli, E., and A. Z. Reznick. The physiology and biochemistry of skeletal muscle atrophy as a function of age. *Proc. Soc. Exp. Biol. Med.* 206: 103–113, 1994.

Cartee, G. D. Influence of age on skeletal muscle glucose transport and glycogen metabolism. *Med. Sci. Sports Exerc.* 26: 577–585, 1994.

Castelo-Branco, C. Management of osteoporosis: an overview. *Drugs Aging* 12 (Suppl. 1): 25–32, 1998.

Cooper, C. S., D. R. Taaffe, D. Guido, E. Packer, L. Holloway, and R. Marcus. Relationship of chronic endurance exercise to the somatotropic and sex hormone status of older men. *Eur. J. Endocrinol.* 138: 517–523, 1998.

Croisant, P. Metabolic and heart rate response to chair exercise. *Med. Sci. Sports Exerc.* 24 (Suppl. 5): S22, 1992.

DeVries, H. Physiological effects of an exercise training program regimen upon men aged 52 to 88. *J. Gerontol.* 25: 325–336, 1970.

DiPietro, L., T. E. Seeman, N. S. Stachenfeld, L. D. Katz, and E. R. Nadel. Moderate-intensity aerobic training improves glucose tolerance in aging independent of abdominal adiposity. *J. Am. Geriatr. Soc.* 46: 875–879, 1998.

Drinkwater, B. L., S. M. Horvath, and C. L. Wells. Aerobic power of females, ages 10 to 68. *J. Gerontol.* 30: 385–394, 1975.

Drinkwater, B. L., K. Nilson, S. Ott, and C. H. Chesnut. Bone mineral density after resumption of menses in amenorrheic athletes. *J. Am. Med. Assoc.* 256: 380–382, 1986.

Ettinger, W. H., Jr. Physical activity, arthritis, and disability in older people. *Clin. Geriatr. Med.* 14: 633–640, 1998.

Fahey, T. D., P. Insel, and W. Roth. Fit and Well. 3d. Ed. Mountain View, Ca.: Mayfield Publishing Company, 1999.

Fiatarone, M. A., E. C. Marks, N. D. Ryan, C. N. Meredith, L. A. Lipsitz, and W. J. Evans. High intensity strength training in nonagenerians. Effects on skeletal muscle. *J. Am. Med. Assoc.* 263: 3029–3034, 1990.

Fitzgerald, M. D., H. Tanaka, Z. V. Tran, and D. R. Seals. Age-related declines in maximal aerobic capacity in regularly exercising vs. sedentary women: a meta-analysis. *J. Appl. Physiol.* 83: 160–165, 1997.

Frischknecht, R. Effect of training on muscle strength and motor function in the elderly. *Reprod. Nutr. Dev.* 38: 167–174, 1998.

Giada, F., E. Bertaglia, B. De Piccoli, M. Franceschi, F. Sartori, A. Raviele, and P. Pascotto. Cardiovascular adaptations to endurance training and detraining in young and older athletes. *Int. J. Cardiol.* 65: 149–155, 1998.

Graves, J. E., M. L. Pollock, and J. F. Carroll. Exercise, age, and skeletal muscle function. *South. Med. J.* 87: S17–S22, 1994.

Green, J. S., and S. F. Crouse. Endurance training, cardiovascular function and the aged. *Sports Med.* 16: 331–341, 1993.

Hakkinen, K., A. Pakarinen, R. U. Newton, and W. J. Kraemer. Acute hormone responses to heavy resistance lower and upper extremity exercise in young versus old men. *Eur. J. Appl. Physiol.* 77: 312–319, 1998.

Higuchi, M., K. Iwaoka, K. Ishii, S. Matsuo, S. Kobayashi, T. Tamai, H. Takai, and T. Nakai. Plasma lipid and lipoprotein profiles in pre- and post-menopausal middle-aged runners. *Clin. Physiol.* 10: 69–76, 1990.

Hiltunen, L., E. Laara, S. L. Kivela, and S. Keinanen-Kiukaanniemi. Glucose tolerance and mortality in an elderly Finnish population. *Diabetes Res. Clin. Pract.* 39: 75–81, 1998.

Hurley, B. F., and J. M. Hagberg. Optimizing health in older persons: aerobic or strength training? *Exer. Sport Sci. Rev.* 26: 61–89, 1998.

Inbar, O., A. Oren, M. Scheinowitz, A. Rotstein, R. Dlin, and R. Casaburi. Normal cardiopulmonary responses during incremental exercise in 20- to 70-yr-old men. *Med. Sci. Sports Exerc.* 26: 538–546, 1994.

Johnson, B. D., M. S. Badr, and J. A. Dempsey. Impact of the aging pulmonary system on the response to exercise. *Clin. Chest Med.* 15: 229–246, 1994.

Joyner, M. J. Physiological limiting factors and distance running: influence of gender and age on record performances. *Exer. Sport Sci. Rev.* 21: 103–133, 1993.

Judge, J. O., R. H. Whipple, and L. I. Wolfson. Effects of resistive and balance exercises on isokinetic strength in older persons. *J. Am. Geriatr. Soc.* 42: 937–946, 1994.

Kraemer, W. J., K. Hakkinen, R. U. Newton, M. McCormick, B. C. Nindl, J. S. Volek, L. A. Gotshalk, S. J. Fleck, W. W. Campbell, S. E. Gordon, P. A. Farrell, and W. J. Evans. Acute hormonal responses to heavy resistance exercise in younger and older men. *Eur. J. Appl. Physiol.* 77: 206–211, 1998.

Krishnan, R. K., J. M. Hernandez, D. L. Williamson, D. J. O'Gorman, J. P. Evans, and W. J. Kirwan. Age-related differences in the pancreatic beta-cell response to hyperglycemia after eccentric exercise. *Am. J. Physiol.* 275: E463–E470, 1998.

Leeuwenburgh, C., R. Fiebig, R. Chandwaney, L. L. Ji. Aging and exercise training in skeletal muscle: responses of glutathione and antioxidant enzyme systems. *Am. J. Physiol.* 267: R439–R445, 1994.

Limacher, M. C. Aging and cardiac function: influence of exercise. *South Med. J.* 87: S13–S16, 1994.

Mackinnon, L. T. Future directions in exercise and immunology: regulation and integration. *Int. J. Sports Med.* 19 (Suppl. 3): S205–S211, 1998.

Martin, F. C., A. L. Yeo, and P. H. Sonksen. Growth hormone secretion in the elderly: aging and the somatopause. *Baillieres Clin. Endocrinol. Metab.* 11: 223–250, 1997.

Mazzeo, R. S. The influence of exercise and aging on immune function. *Med. Sci. Sports Exerc.* 26: 586–592, 1994.

Mengelkoch, L. J., M. L. Pollock, M. C. Limacher, J. E. Graves, R. B. Shireman, W. J. Riley, D. T. Lowenthal, and A. S. Leon. Effects of age, physical training, and physical fitness on coronary heart disease risk factors in older track athletes at twenty-year follow-up. *J. Am. Geriatr. Soc.* 45: 1446–1453, 1997.

Navarro-Arevalo, A., and M. J. Sanchez-del-Pino. Age and exercise-related changes in lipid peroxidation and superoxide dismutase activity in liver and soleus muscle tissues of rats. *Mech. Aging Dev.* 104: 91–102, 1998.

Okada, M., K. Mitsunami, T. Inubushi, and M. Kinoshita. Influence of aging or left ventricular hypertrophy on the human heart: contents of phosphorus metabolites measured by 31P MRS. *Magn. Reson. Med.* 39: 772–782, 1998.

Paffenbarger, R. S., R. T. Hyde, A. L. Wing, and C. C. Hsieh. Physical activity, all-case mortality, and longevity in older adults. *N. Eng. J. Med.* 314: 605–613, 1986.

Preuss, H. G. Effects of glucose/insulin perturbations on aging and chronic disorders of aging: the evidence. *J. Am. Coll. Nutr.* 16: 397–403, 1997.

Proctor, D. N., K. C. Beck, P. H. Shen, T. J. Eickhoff, J. R. Halliwill, and M. J. Joyner. Influence of age and gender on cardiac output-$\dot{V}_{O_2}$ relationships during submaximal cycle ergometry. *J. Appl. Physiol.* 84: 599–605, 1998.

Proctor, D. N., P. H. Shen, N. M. Dietz, T. J. Eickhoff, L. A. Lawler, E. J. Ebersold, D. L. Leoffler, and M. J. Joyner. Reduced leg blood flow during dynamic exercise in older endurance-trained men. *J. Appl. Physiol.* 85: 68–75, 1998.

Pullinen, T., A. Mero, E. MacDonald, A. Pakarinen, and P. V. Komi. Plasma catecholamine and serum testosterone responses to four units of resistance exercise in young and adult male athletes. *Eur. J. Appl. Physiol.* 77: 413–420, 1998.

Rantanen, T., J. M. Guralnik, G. Izmirlian, J. D. Williamson, E. M. Simonsick, L. Ferrucci, and L. P. Fried. Association of muscle strength with maximum walking speed in disabled older women. *Am. J. Phys Med. Rehabil.* 77: 299–305, 1998.

Reimers, C. D., T. Harder, and H. Saxe. Age-related muscle atrophy does not affect all muscles and can partly be compensated by physical activity: an ultrasound study. *J. Neurol. Sci.* 159: 60–66, 1998.

Roberts, S. B., and G. E. Dallal. Effects of age on energy balance. *Am. J. Clin. Nutr.* 68: S975–S979, 1998.

Rosen, M. J., J. D. Sorkin, A. P. Goldberg, J. M. Hagberg, and L. I. Katzel. Predictors of age-associated decline in maximal aerobic capacity: a comparison of four statistical models. *J. Appl. Physiol.* 84: 2163–2170, 1998.

Saltin, B., and G. Grimby. Physiological analysis of middle-aged and former athletes. *Circulation* 38: 1104–1115, 1968.

Saltin, B., L. Hartley, A. Kilbom, and I. Åstrand. Physical training in sedentary middle-aged and older men. *Scand. J. Clin. Lab. Invest.* 24: 323–334, 1969.

Saltzman, J. R., and R. M. Russell. The aging gut: nutritional issues. *Gastroenterol. Clin. North Am.* 27: 309–324, 1998.

Schroll, M. Are cardiovascular risk factors still predictive in the elderly? The longitudinal study of epidemiology of aging in Copenhagen County, Denmark. *Aging* (Milano) 10: 167–168, 1998.

Schut, L. J. Motor system changes in the aging brain: what is normal and what is not. *Geriatrics* 53 (Suppl. 1): S16–S19, 1998.

Seals, D. R., J. A. Taylor, A. V. Ng, and M. D. Esler. Exercise and aging: autonomic control of the circulation. *Med. Sci. Sports Exerc.* 26: 568–576, 1994.

Seiler, K. S., W. W. Spirduso, and J. C. Martin. Gender differences in rowing performance and power with aging. *Med. Sci. Sports Exerc.* 30: 121–127, 1998.

Shephard, R. J. Physical Activity and Aging, 2d Ed. London: Croom Helm, 1987.

Shephard, R. J. Fitness and aging. In Aging into the Twenty-First Century, C. Blais (Ed.). Downsview, Ont.: Captus University, 1991, pp. 22–35.

Shephard, R. J., and W. Montelpare. Geriatric benefits of exercise as an adult. *J. Gerontol. (Med. Sci.)* 43: M86–M90, 1988.

Simonen, R. L., T. Videman, M. C. Battie, and L. E. Gibbons. Determinants of psychomotor speed among 61 pairs of adult male monozygotic twins. *J. Gerontol. A. Biol. Sci. Med. Sci.* 53: M228–M234, 1998.

Starling, R. D., M. J. Toth, D. E. Matthews, and E. T. Poehlman. Energy requirements and physical activity of older free-living African-Americans: a doubly labeled water study. *J. Clin. Endocrinol. Metab.* 83: 1529–1534, 1998.

Struder, H. K., W. Hollmann, P. Platen, R. Rost, H. Weicker, and K. Weber. Hypothalamic-pituitary-adrenal and -gonadal axis function after exercise in sedentary and endurance trained elderly males. *Eur. J. Appl. Physiol.* 77: 285–288, 1998.

Suominen, H. Bone mineral density and long term exercise. *Sports Med.* 16: 316–330, 1993.

Suominen, H., E. Heikkinen, T. Parkatti, S. Forsberg, and A. Kiiskinen. Effects of lifelong physical training on functional aging in men. *Scand. J. Soc. Med.* (Suppl.) 14: 225–240, 1980.

Tanaka, H., C. A. Desouza, P. P. Jones, E. T. Stevenson, K. P. Davy, and D. R. Seals. Greater rate of decline in maximal aerobic capacity with age in physically active vs. sedentary healthy women. *J. Appl. Physiol.* 83: 1947–1953, 1997.

Tate, C. A., M. F. Hyek, and G. E. Taffet. Mechanisms for the responses of cardiac muscle to physical activity in old age. *Med. Sci. Sports Exerc.* 26: 561–567, 1994.

Thomas, W. C., Jr. Exercise, age, and bones. *South. Med. J.* 87: S23–S25, 1994.

Thompson, J. L., G. E. Butterfield, U. K. Gylfadottir, J. Yesavage, R. Marcus, R. L. Hintz, A. Pearman, and A. R. Hoffman. Effects of human growth hormone, insulin-like growth factor 1, and diet and exercise on body composition of obese postmenopausal women. *J. Clin. Endocrinol. Metab.* 83: 1477–1484, 1998.

Thune, I., I. Njolstad, M. L. Lochen, and O. H. Forde. Physical activity improves the metabolic risk profiles in men and women: the Tromso Study. *Arch. Intern. Med.* 158: 1633–1640, 1998.

Timaris, P. S. Developmental Physiology and Aging. New York: Macmillan, 1972.

Tremblay, A., V. Drapeau, E. Doucet, N. Almeras, J. P. Despres, and C. Bouchard. Fat balance and aging: results from the Quebec Family Study. *Br. J. Nutr.* 79: 413–418, 1998.

Tulppo, M. P., T. H. Makikallio, T. Seppanen, R. T. Laukkanen, and H. V. Huikuri. Vagal modulation of heart rate during exercise: effects of age and physical fitness. *Am. J. Physiol.* 274: H424–H429, 1998.

Van Boxtel, M. P., F. G. Paas, P. J. Houx, J. J. Adam, J. C. Teeken, and J. Jolles. Aerobic capacity and cognitive performance in a cross-sectional aging study. *Med. Sci. Sports Exerc.* 29: 1357–1365, 1997.

Van Heuvelen, M. J., G. I. Kempen, J. Ormel, and P. Rispens. Physical fitness related to age and physical activity in older persons. *Med. Sci. Sports Exerc.* 30: 434–441, 1998.

Vance, M. L. The Gordon Wilson Lecture: growth hormone replacement in adults and other uses. *Trans. Am. Clin. Climatol. Assoc.* 109: 87–96, 1998.

Wells, C. L. Women, Sport, and Performance: A Physiological Perspective, 2d Ed. Champaign, Ill.: Human Kinetics, 1991.

Wells, C. L., M. A. Boorman, and D. M. Riggs. Effect of age and menopausal status on cardiorespiratory fitness in masters women runners. *Med. Sci. Sports Exerc.* 24: 1147–1154, 1992.

White, T. P., and S. T. Devor. Skeletal muscle regeneration and plasticity of grafts. In *Exercise and Sports Sciences Reviews,* J. O. Holloszy (Ed.). Vol. 21. 1993, pp. 263–296.

Willis, P. E., S. G. Chadan, V Baracos, and W. S. Parkhouse. Restoration of insulin-like growth factor 1 action in skeletal muscle of old mice. *Am J. Physiol.* 275: E525–E530, 1998.

Young, A. Aging and physiological functions. *Philos. Trans. R. Soc. Lond. B. Biol. Sci.* 352: 1837–1843, 1997.

CHAPTER 33

FATIGUE DURING MUSCULAR EXERCISE

Muscular fatigue is usually defined as the inability to maintain a given exercise intensity. As we will see, there is no one cause of fatigue. Fatigue is task-specific and its causes are multifocal and vary from occasion to occasion. Fatigue during muscular exercise is often due to impairment within the active muscles themselves, in which case the fatigue is peripheral to the CNS and is due to muscle fatigue. Muscular fatigue can also be due to more diffuse, or more central, factors. For example, for psychological reasons an athlete may be unable to bring his or her full muscle power to bear in performing an activity. Alternatively, environmental factors such as hot, humid conditions may precipitate a whole series of physiological responses that detract from performance. In such cases, the cause of the fatigue resides outside the muscles.

Not only does the cause of fatigue vary with the

Fatigue and elation are expressed simultaneously in these competitors in the New York Marathon. Rod Dixon of New Zealand raises his arms in triumph after crossing the finish line, while Britain's Geoff Smith, who finished second, lies on the ground. PHOTO: © UPI/Corbis.

nature of the activity, but the training and physiological status of the individual, as well as environmental conditions, affect the progress of fatigue during exercise. Fatigue can be due specifically to depletion of key metabolites in muscle or to the accumulation of other metabolites, which can affect the intracellular environment and also spill out into the circulation and affect the general homeostasis. The failure of one enzyme system, cell, or muscle group is likely to affect numerous other cells, organs, and tissues. Therefore, the causes of fatigue are interactive.

The study of fatigue in exercise has occupied the attention of many of the best biological scientists. Identifying a cause of fatigue is not simple, as it is often difficult to separate causality from concurrent appearance.

■ Identifying Fatigue

Before the cause and site of fatigue can be identified, we must define it. Usually, in athletic competition, fatigue means the inability to maintain a given exercise intensity. For instance, in distance running or swimming, when the competitor is unable to keep the pace, he or she is considered to have fatigued for the particular event. However, an athlete is rarely completely fatigued and can usually maintain a lesser power output for some time. In a few instances, such as during wilderness hiking, individuals will push themselves until they are completely unable to move. In such cases of complete exhaustion, death can result from exposure.

Sometimes the cause of a decrement in work performance (fatigue) can be identified as to its specific cause and site. For instance, the depletion of a particular metabolite in a particular fiber type within a specific muscle may be identified. At other times, as in dehydration, the causes of fatigue are diffuse and involve several factors that contribute to the disturbance in homeostasis.

It is often easier to identify a factor whose presence is correlated with the onset of fatigue than it is to determine that the presence (or absence) of the

factor and fatigue are causally related. For instance, heavy muscular exercise has long been observed to be associated with lactic acid accumulation. Naturally, lactic acid was assumed to be the cause of fatigue. However, whenever lactate is high, so are other metabolites (e.g., phosphate); yet still other metabolites (glycogen, ATP, CP) can be low. Is fatigue the result of lactic acidosis and CP depletion, or is some other factor causing lactate and CP levels to change? As we will see, many of the data relating lactic acid accumulation to onset of fatigue are often circumstantial.

Compartmentalization in physiological organization, involving the division of the body into various systems, organs, tissues, and cells, and the subdivisions of cells into various subfractions and organelles, also makes it difficult to identify the fatigue site. For instance, ATP may be depleted at a particular site within a cell (e.g., on the head of myosin crossbridges), but may be adequate elsewhere. In such a case, it would be extremely difficult to identify ATP depletion as the cause of fatigue, even if a muscle biopsy were performed or NMR technique used. Compartmentalization can mask the site of fatigue.

The effect of exercise at given absolute or relative values of $\dot{V}_{O_2max}$ can be more severe on untrained than on trained individuals. For instance, fatigue occurs sooner in an untrained person exercising at 75% of $\dot{V}_{O_2max}$ than in an endurance-trained individual exercising at the same work rate, or at a higher work rate that elicits 75% of the trained individual's $\dot{V}_{O_2max}$.

It is well known that environment can affect exercise endurance. For example, endurance is reduced during exercise in heat. This reduced endurance is due to the redistribution of cardiac output from contracting muscles and hepatic gluconeogenic areas to include greater cutaneous circulation as well. High muscle temperatures can also loosen the coupling between oxidation and phosphorylation in mitochondria. In this case, $\dot{V}_{O_2}$ is unchanged or increased, but ATP production is decreased. To the extent that the exercise is submaximal and there exists a reserve for cardiac output expansion, exercise under conditions of heat can be continued. The stress level, however, is greater. If the need to circulate to exercising muscles, cutaneous areas, and other essential areas exceeds cardiac output, then endurance is reduced.

During exercise in hot environments, the rate of sweat loss increases and body heat is gained over time. Severe sweating results in dehydration and shifts both fluid and electrolytes among body compartments. These shifts, as well as increased body temperature, represent direct irritants to the CNS that can additionally affect an individual's subjective perception of the exercise.

The physiological status of the individual can also easily affect exercise tolerance. For instance, if an individual exercises to fatigue in the heat on Monday, his or her ability to repeat that performance on Monday, Tuesday, or perhaps even Wednesday may be impaired. Similarly, individuals may have less endurance when they are glycogen-depleted than when glycogen levels are normal or supercompensated.

■ Metabolite Depletion

The Phosphagens (ATP and CP)

Adenosine triphosphate (ATP) is the high-energy intermediate that supports muscle contraction as well as most other cellular endergonic processes. The immediate source of ATP rephosphorylation is creatine phosphate (CP) (Chapter 3). Because the catalyzing enzyme, creatine kinase, functions so rapidly, the muscle concentration of ATP is little affected until the CP level is significantly depleted.

Endurance and its converse, fatigue, can be likened to two opposing and competing forces—those that utilize ATP and those that restore it. Because the quantities of ATP and CP on hand in a resting muscle are fairly small, any significant utilization must be immediately matched with an equivalent restoration. If the rates of ATP and CP restoration are even a little less than utilization, exercise cannot continue very long.

Biopsies of human quadriceps muscle during cycling exercise have demonstrated that the CP level

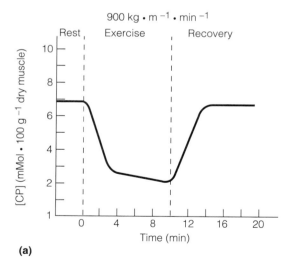

(a)

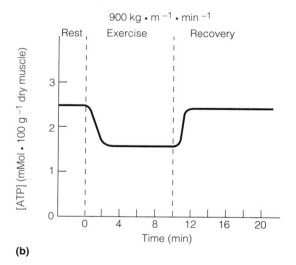

(b)

Figure 33-1 Creatine phosphate [CP] and adenosine triphosphate levels [ATP] during steady-rate exercise. When exercise starts, both [CP] (a) and [ATP] (b) decline. Although [CP] may continue to decline if exercise intensity is difficult for the subject, ATP levels will be well maintained until CP is exhausted. In recovery, both CP and ATP levels recover rapidly. Modified from Bergström, 1967.

in muscle declines in two phases (Figure 33-1a). When exercise starts, the CP level drops rapidly at first, and then slowly. Both the severity of the initial drop and the extent of the final drop (Figure 33-2) appear to be related to the relative work intensity for the subject. In general, the greater the relative work load, the greater will be the CP depletion. In subjects willing to push themselves to their limit, the point of fatigue in isometric exercise and in supermaximal cycling exercise coincides with CP depletion in muscle. Detailed studies on isolated muscles stimulated to contract indicate that tension development is related to CP level. Thus, it is clear that *CP depletion leads to muscle fatigue.*

Although ATP level declines somewhat when exercise starts (Figure 33-1b), ATP level is apparently well maintained (largely at the expense of CP) until the level of CP is greatly reduced. There-

fore, at the fatigue point, both CP and ATP become depleted.

Why it is that creatine phosphate—and not ATP depletion—is associated with muscle fatigue is puzzling to physiologists, and at least two views have emerged to explain the phenomenon of ATP levels being better maintained than CP levels. One explanation, that of compartmentalization, has already been mentioned. Again, because of the gross nature of biopsy and NMR techniques, we may simply, at present, be unable to measure ATP depletion from critical cell compartments and sites. Another explanation is that in order to protect itself from the potential devastating consequences of ATP depletion, muscle cells *down regulate* some discretionary functions (e.g., contraction) to preserve other essential functions (e.g., maintenance of ion gradients). The down regulation theory of muscle fatigue is extremely interesting, and work is currently under way to explain how such a mechanism might operate.

Free Energy of ATP and other ATP-Related Fatigue Effects

Before leaving the general topic of high energy phosphate depletion and muscle fatigue, it should be noted that muscle ATP concentration is well maintained up to maximum effort. This fact remains perplexing to physiologists, who have thus considered subtle effects of muscle contraction on ATP hydrolysis. One effect is that of the fall in muscle pH on the free energy of ATP hydrolysis. In Chapter 2 (Equation 2-17 and Figure 2-7), we discussed that the free energy available from ATP hydrolysis depends on several factors that can change in fatiguing muscle. The standard free energy of ATP hydrolysis ($\Delta G^{0\prime}$) is pH sensitive and can decrease 14% in the physiological pH range (Figure 2-7). Further, the actual free energy of ATP hydrolysis (i.e., ΔG) depends not only on the standard ($\Delta G^{0\prime}$), but also on the [ATP]/[ADP][Pi] ratio as it exists in the myofibrils. Thus, even if the concentration of ATP does not decrease, but concentrations of ADP and Pi rise, then the ΔG of ATP could fall. A consequence of this fall would be that less

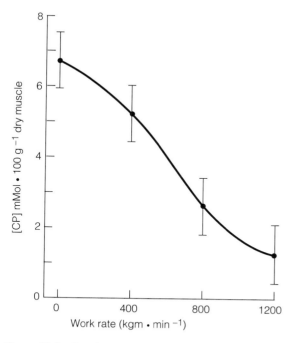

Figure 33-2 Creatine phosphate levels ([CP]) in quadriceps muscles after five minutes of cycling. A strong negative relationship is indicated between work rate and CP levels. Extremely hard exercise can lead to CP and ATP depletion, a situation that causes fatigue. Modified from Bergström, 1967.

energy is available to do work for a given $\dot{V}_{O_2}$ or glycolytic flux.

In addition to the possible role of changes in the free energy of ATP in fatigued muscle, physiologists also consider the possibility that the fatigue process somehow influences the binding of ATP in the cross-bridge cycle (see Chapter 17, Figure 17-14). For the present, we must concede that we are baffled by the inability to explain why muscle ATP concentration is well defended in the fatigue process.

Glycogen

Glycogen deletion in skeletal muscle is associated with fatigue during prolonged submaximal exercise to exhaustion (Chapter 28). In cycling, when the pedaling rate is moderate (i.e., 60 to 70 revolutions · min^{-1}), glycogen appears to be depleted uniformly from the different fiber types. However, according to the work of Gollnick and associates (1973), when subjects perform at a given work rate, rapid cycling at 100 revolutions · min^{-1} (low resistance) results in selective recruitment and depletion of glycogen from the low-force-producing, slow-twitch fibers (Chapter 17). Maintaining the same work rate at a slow high-force pedaling frequency (i.e., 50 revolutions · min^{-1}) results in recruitment of high-force, fast-twitch fibers. Thus, it is possible for an athlete to exercise to exhaustion and fatigue because of glycogen depletion from specific muscle fibers, while glycogen remains in adjacent fibers within the tissue. These glycogen reserves can be mobilized if epinephrine levels rise, stimulating glycogenolysis, glycolysis, lactate production and release, and energy (lactate) exchange via the lactate shuttle (Chapter 10).

Blood Glucose

During short, intense exercise bouts, blood glucose rises above preexercise levels because the autonomic nervous system stimulates hepatic glycogenolysis (Chapter 9). The ability of the liver to maintain a high rate of glucose release over time is limited by the amount of glycogen stored and by the activities of the hepatic glycogenolytic and gluconeogenic enzymes. During prolonged exercise, glucose production may be limited to gluconeogenesis because of hepatic glycogen depletion; thus, glucose production may fall below that required by working muscle and other essential tissues such as the brain. Also, in prolonged exercise leading to dehydration and hyperthermia shunting of blood flow away from the liver and kidneys occurs. Thus, the levels of gluconeogenic precursors (lactate, pyruvate, alanine) rise, and hepatic glucose production falls. In this case of falling blood glucose, the exercise becomes subjectively more difficult because of CNS starvation and difficulty in oxidizing fats in muscle due to the absence of anaplerotic substrates (Chapter 8).

■ Metabolite Accumulation

Not long ago, the story of muscle fatigue was easy for scientists, textbook authors, and students to explain. The explanation went like this: When exercise was too difficult, an athlete went into "O₂ debt." The athlete then built up lactic acid, which caused fatigue. During recovery, the "O₂ debt" was repaid, and lactate was reconverted to glycogen. Unfortunately, the lactic acid explanation is *not* now universally accepted as an explanation of either the "O₂ debt" (Chapter 10) or fatigue. Most of the data concerning lactic acid and fatigue reveal that the relationship is circumstantial at best. Certainly during prolonged exercise, glucose, glycogen, and lactate levels are low. Today, one popular sports drink (CYTOMAX_R) even contains organic and inorganic lactate salts as a major component. In addition, injected lactate actually enhances the performance of people with genetic defects in the glycolytic pathway. More likely than the lactate anion, it is the accumulation of the associated hydrogen ion that is detrimental to performance.

Lactic Acid Accumulation (Lactic Acidosis)

During short-term, high-intensity exercise, lactate accumulates as the result of lactic acid production being greater than its removal. At a physiological pH, lactic acid, a strong organic acid, dissociates a proton (H^+). It is the H^+ rather than the lactate ion that causes pH to decrease. Although lactate accumulation in blood is directly related to H^+ accumulation in blood because the muscle cell membrane exports into blood both lactate anions and protons, in muscle the cause of acidosis is different. All the glycolytic intermediates of glycolysis are weak organic acids and dissociate protons. Further, as pointed out by Gevers (1977), the degradation of ATP results in H^+ formation. Thus, lactate accumulation is associated with acidosis for more than one reason, but it is important to recognize that it is unbuffered protons (i.e., H^+) and not lactate anions that pose difficulties for the performer.

The H^+ accumulation resulting from glycolysis and ATP catabolism as the result of lactic acid production can have several negative effects. Within the muscle, the lower pH may inhibit phosphofructokinase (PFK) and slow glycolysis. In addition, H^+ may act to displace Ca^{2+} from troponin, thereby interfering with muscle contraction. Further, the low pH may stimulate pain receptors.

Hydrogen ion liberated into the blood and reacting in the brain causes severe side effects, including pain, nausea, and disorientation. Within the blood itself, H^+ inhibits the combination of O_2 with hemoglobin in the lung. Some species actually run themselves to the point where O_2 delivery is reduced by lactic acid formation and the blocking of oxyhemoglobin formation. High circulating H^+ levels also thwart the action of hormone-sensitive lipase activity in adipose tissue by stimulating phosphodiesterase and the reesterification of fatty acids to triglycerides. The net result is a limiting of the release of free fatty acids (FFA) into the circulation. Fat oxidation in muscle is directly dependent on circulating FFA levels.

As debilitating as high levels of H^+ from lactic acid dissociation may be in the muscles and blood, it is uncertain whether the pH decrement actually stops exercise. Because of a muscle's gross and microanatomy, muscle biopsies actually yield little information on the pH at critical sites of metabolism. Many active sites on enzymes are hydrophobic, and the environment pH has minimal effect. In theory, a lowered cytoplasmic pH should benefit mitochondrial function (Chapter 6). Recent studies utilizing nuclear magnetic resonance (NMR) technique to look within muscles noninvasively during exercise and recovery suggest that fatigue is due to CP depletion, as noted, rather than to lactic acid accumulation.

Muscle and blood lactate accumulation during exercise are symptomatic of more than muscle and blood acidosis. Lactate accumulation means that the mechanisms of lactate disposal and clearance have been exceeded (Chapter 10). Thus, the overall system is failing to cope with metabolic demands. Further, lactate accumulation is indicative of glycogen depletion, as noted (see Chapter 28).

Phosphate and Diprotenated Phosphates

Phosphagen depletion during exercise (see Figures 33-1 and 33-2) results in phosphate (P_i, or HPO_4^{2-}) accumulation. Studies on isolated muscles and enzyme systems indicate that phosphate behaves much like hydrogen ion in interfering with glycolysis (PFK) and excitation-contraction coupling (Ca^{2+} binding to troponin) in muscle. In fact, some investigators such as Nosek, Godt, and associates (1987) have obtained results that hydrogen ion and phosphate act to produce hydrogen phosphate, and that this diprotenated phosphate

$$H^+ + HPO_4^{2-} \leftrightarrow H_2PO_4^- \qquad (33\text{-}1)$$

is the most deleterious metabolite to accumulate in muscle, which is working hard and suffering both acidosis and phosphagen depletion (Figure 33-3).

At present, not all investigators agree that $H_2PO_4^-$ accumulation is more deleterious than HPO_4^{2-} accumulation, but it is apparent that when concentrations of H^+ and HPO_4^{2-} rise, the muscle is not working in a steady state: oxidative metabolism is unable to maintain high levels of CP and ATP and low levels of phosphate, and fatigue is imminent.

Calcium Ion

There are several reasons why the calcium ion (Ca^{2+}) may be involved in muscle fatigue. First, Ca^{2+}

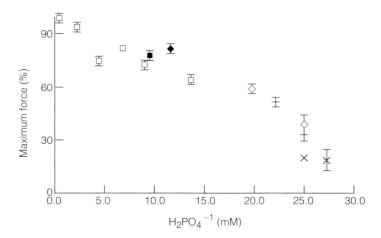

Figure 33-3 Correlation of phosphate-induced depression of maximal Ca^{2+}-activated force with concentration of $H_2PO_4^-$ in skinned skeletal muscle fibers. Concentration of $H_2PO_4^-$ calculated from total $[P_i]$ with pK of 6.79. Data in this figure are the means of at least five fibers. Vertical bars show standard error of the mean. SOURCE: Nosek et al., 1987.

loss from the sarcoplasmic reticulum (SR) during excitation-contraction (EC) coupling may be taken up by mitochondria, ultimately interfering with the efficiency of mitochondrial function. Second, the ability of SR to release Ca^{2+} during twitches may be reduced; thus, the intracellular signal for contraction may be reduced, resulting in a less forceful contraction. Third, sensitivity of the actin-myosin contractile apparatus to Ca^{2+} is reduced. This latter effect is probably largely due to interference by hydrogen ion (H^+) in Ca^{2+}-troponin binding. And fourth, Ca^{2+} re-uptake by the SR may be slowed, thereby prolonging contractions and slowing relaxation of the muscle.

Mitochondrial Coupling Efficiency

The accumulation of Ca^{2+} within muscle mitochondria during prolonged exercise may be more debilitating than the decrease in cytoplasmic pH from lactic acid formation. Some of the Ca^{2+} liberated in muscle from the sarcoplasmic reticulum during excitation-contraction coupling may be sequestered by mitochondria. Some increase in mitochondrial Ca^{2+} is probably beneficial, as it stimulates the dehydrogenases of the TCA cycle (Chapter 6). However, Ca^{2+} excretion from mitochondria is energy-linked, and so too much Ca^{2+} sequestration results in O_2 consumption and saps the mitochondrial energy potential for phosphorylating ADP to ATP. When muscle mitochondria are allowed to respire in a test tube in the presence of Ca^{2+}, they take up so much Ca^{2+} that oxidative phosphorylation is eventually uncoupled. The extent to which this occurs in the performing athlete remains to be determined. However, such a Ca^{2+} effect could account for the situation in which athletes can repeatedly reach the same $\dot{V}_{O_2max}$ during interval exercise but not match the initial work output.

The problem of Ca^{2+} sequestration by muscle mitochondria in heavy exercise is exacerbated by the finding of depressed Ca^{2+} sequestration by the sarcoplasmic reticulum obtained from fatigued muscle. Thus, both processes of excitation-contraction coupling and mitochondrial oxidative phosphorylation are likely to be affected by fatiguing exercise.

Ryanodine Receptor Fatigue

Figure 33-4, which has been modified from Westerblad and Allen (1993), will be used to illustrate changes in Ca^{2+} flux and signaling in fatigued muscle. When isolated muscles are stimulated electrically, a symptom of fatigue is a decrease in force generated in response to single or to tetanizing stimulations. This decrease has been related to the SR Ca^{2+} channels' (i.e., the ryanodine receptors) re-

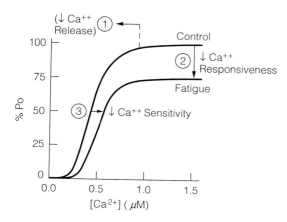

Figure 33-4 High intensity exercise results in decreased muscle force output (fatigue). The responses of isolated muscles to tetanizing electrical stimulations depend on the amount and effectiveness of Ca^{2+}. In fatigued muscle, less Ca^{2+} is released in response to stimulation. Arrow number one shows the effect of decreased Ca^{2+} release from sarcoplasmic reticulum Ca^{2+} channels (ryanodine receptors). Acidosis, if present, decreases the response to available Ca^{2+} because H^+ interferes with Ca^{2+} binding to troponin; this effect is shown by arrow number two. And finally, fatigued muscle displays reduced sensitivity to a given $[Ca^{2+}]$ (arrow number three). Po is maximal (optimal) isometric force. Modified from Westerblad and Allen (1993).

lease of Ca^{2+} (see Chapter 17). In this case, the addition of large amounts of caffeine to the bathing solutions (caffeine concentrations are far in excess of those possible through oral consumption) the tension developed by isolated muscles in response to electric stimulation is increased. Because caffeine is known to increase Ca^{2+} release from the SR, a component of fatigue is attributable to depressed function of the ryanodine receptors. In Figure 33-4, the right to left shift (arrow number one) is caused by decreased levels of free Ca^{2+}. The reason for the depression in Ca^{2+} release in fatigued muscle is unknown, but Hogan and associates (1995 and personal communication) suspect that lactate anion (which is released during fatigue) may interfere with ryanodine receptors in muscle. More recently, Favero and associates (1997) showed that high (10–20 mm) concentrations of the lactate anion

$(C_3H_5O_3-)$ inhibit Ca^{2+} from the ryanodine receptor in isolated SR vesicles. Thus, lactic acid formation not compensated by intramuscular oxidation or efflux via the sarcolemmal lactate transport mechanism could have multiple effects on muscle contraction because of the combined effects of H^+ and $C_3H_5O_3-$.

H^+ Effects on Ca^{2+}-Troponin Interactions (Responsiveness) A consequence of high-intensity exercise is that rapid glycolysis results in lactic acid accumulation and decreased muscle pH. In Figure 33-4, the downward shift (arrow number two) is attributable to the effect of H^+ interfering with the effect of a given concentration of Ca^{2+}. This decrease in the maximal effect of Ca^{2+} on muscle force is termed *decreased responsiveness*. Thus, as indicated in Figure 33-4, decreases in Ca^{2+} release and responsiveness have major effects on muscle force.

Ca^{2+}-Troponin Interactions (Sensitivity) And finally, in Figure 33-4 a small left to right curve shift (arrow number three) is indicated in fatigued muscle. In fatigued muscle, a given free concentration of Ca^{2+} elicits a lesser force than in fresh muscle. In other words, Ca^{2+} sensitivity is reduced. Decreased Ca^{2+} sensitivity in fatigued muscle is of lesser consequence than decreased Ca^{2+} release or responsiveness.

SR Ca^{2+}–ATPase Fatiguing exercise affects intramuscular calcium flux in still another way. Not only is the sarcoplasmic reticulum in fatigued muscle less capable of releasing Ca^{2+}, but the SR is also less capable of taking up calcium ion and, therefore, less able to allow the muscle to relax. This latter effect is attributable to depressed Ca^{2+}–ATPase activity in the sarcoplasmic reticulum.

■ **O₂ Depletion and Muscle Mitochondrial Density**

The depletion of muscle O₂ stores, or rather the inadequacy of circulating O₂ delivery to muscle, can result in fatigue. Those with impaired circulatory or ventilatory function, those engaged in exercise at high altitudes, or those engaged in strenuous exercise at sea level can fall short in the balance between muscle respiratory requirement and the actual O_2 supply. Because most of the ATP required to perform any activity lasting 90 seconds or more is from cell respiration, adequate O_2 supply is essential to support maximal aerobic work. Maximal aerobic work is defined as the situation in which the ability to work is directly dependent on $\dot{V}_{O_2max}$, or in which the required work rate elicits close to 100% of $\dot{V}_{O_2max}$.

The effects of inadequate O_2 supply or utilization can be represented by increased lactate production or decreased CP levels, or both. Thus, inadequate oxygenation of contracting muscle can result in at least two fatigue-causing effects.

Skeletal muscles, even untrained muscles, contain a greater mitochondrial respiratory capacity than can be supplied by the circulation. Therefore, the maximal rate of muscle oxidation delivery is ultimately limited by cardiac output. For activities that require very high metabolic rates for several minutes, the capacities for a high $\dot{V}_{O_2max}$ and cardiac output are important (Chapter 13).

The doubling of muscle mitochondrial activity in response to endurance training benefits endurance capacity by a means other than increasing the $\dot{V}_{O_2max}$. Twice as much mitochondrial content increases respiratory control and effectively doubles the capacity to oxidize fatty acids as fuel. This results in a glycogen-sparing effect (Chapter 7). In addition, it now appears that more than the minimal number of mitochondria needed to achieve a circulatory-limited $\dot{V}_{O_2max}$ are necessary to minimize the effect of mitochondrial damage during exercise. The utilization of O_2 in mitochondria is associated with the liberation of free radicals, which present a real threat to mitochondria. Indeed, K. J. A. Davies and colleagues (1982) at the University of California have supplied evidence of mitochondrial damage due to free radical accumulation at the point of fatigue. In this sense, having more mitochondria or a larger mitochondrial reticulum to sustain prolonged submaximal exercise bouts is analogous to having more pawns with which to play the game

of chess. More pawns (mitochondria) can lessen the effect of an opponent's attack (free radicals).

■ Disturbances to Homeostasis

In Chapter 22, we illustrated the fatiguing effects of thermal dehydration. It should be emphasized that the continuance of exercise depends on an integrated functioning of many systems, each containing many elements. Any factor or set of factors that upsets this integrated functioning can cause fatigue. Levels of K^+, Na^+, and Ca^{2+}, compartmentalization of these ions within a cell, levels of blood glucose and FFA, plasma volume, pH and osmolality, body temperature, and hormone levels are a few of the factors involved in maintenance of homeostasis.

■ Central and Neuromuscular Fatigue

In the linkage between afferent input and the performance of an appropriate motor task, several sites require adequate function. A performance decrement at any of these sites results in a decrease in motor performance. Thus, it is possible to have muscular fatigue when the muscle itself is not impaired. Specifically, the proper functioning of receptors, CNS integrating centers, sensory cortex, spinal cord, α-motoneurons, γ-loops, motor end plates, muscle cells, and the cerebellum are frequently required to perform motor tasks. As discussed in Chapter 18, it has not been possible to obtain many direct data on CNS function during exercise. Therefore, researchers have tended to focus on the musculature itself as a point of fatigue. In many, if not most, cases, the cause of fatigue may well lie in the musculature, but the interrelationships between central and peripheral functions should not be overlooked. At the very least, painful afferent inputs from muscle and joints might negatively affect an athlete's willingness to continue competition. Physiological signals can lead to psychological inhibition.

In a now classic series of experiments, Merton (1954) determined that the site of fatigue in the adductor pollicis muscle was peripheral to the neuro-

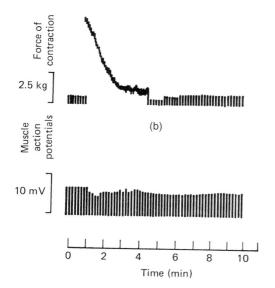

Figure 33-5 Maximum voluntary isometric contraction force of thumb adductor muscles (a) and corresponding muscle action potentials (b). Superimposed are electrical shocks to the controlling motor nerve. Results indicate that voluntary effort produces the maximal contraction possible in the muscle, and that the motor nerve and motor end plate are not fatigued when muscle performance declines. Fatigue is then peripheral to the nervous system in the muscle. Modified from Merton, 1954.

muscular junction (i.e., it was within the muscle itself and not at the neuromuscular junction), or at some site more centrally located. When the voluntary tension developed by the adductor pollicis decreased to indicate significant fatigue, Merton applied electrical shocks to the ulnar nerve to increase stimulation of the muscle (Figure 33-5). As indicated in the figure, muscle action potentials (part b) indicated clear conduction of the signal from nerve to muscle, but there was no tension response in the fatigued muscle (part a).

When the electrical potentials of muscle are recorded with electromyography (EMG), there frequently occurs a distinct change in the point at which fatigue approaches during both dynamic and static muscle exercise. The so-called fatigue EMG can have one or perhaps two characteristics. The first is a large increase in the integrated EMG signal, wherein all action potentials are weighted to represent a single voltage (Figure 33-6). Failure of the muscle to respond to increased stimulation indicates that fatigue is peripheral. The second characteristic is a shift to the left in the EMG power frequency spectrum (PFS; Figure 33-7). The PFS shows the relative electrical activity contributed by slow (on the left) and fast (on the right) motor units. The leftward shift indicates increased stimulation of the smaller, slow, fatigue-resistant motor units.

Perhaps the best evidence for a central basis of

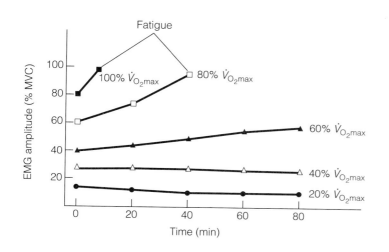

Figure 33-6 Integrated EMG signal in a rectus femoris muscle during cycle exercise of graded increasing intensity. The integrated EMG represents the sum total (frequency and amplitude) of all muscle action potentials. Values are given as a percentage of that recorded during a maximal voluntary contraction (MVC). At higher work rates, the EMG signal increases disproportionally, reaching a maximum as fatigue approaches. The so-called fatigue EMG represents the attempt to recruit additional motor units as the work output of fatiguing motor units declines. Modified from Petrofsky, 1979.

Figure 33-7 An EMG power frequency spectrum (PFS) represents a plot of muscle action potential amplitude versus frequency. Slow, type I motor units fire at lower frequency than do fast, type II motor units. In the experiment illustrated, subjects were asked to hold 50% of their maximum voluntary contraction force by means of visual feedback. The nonfatigued EMG is bimodal in its distribution, indicating participation of fast (high frequency) and slow (low frequency) muscle fibers. The EMG power frequency spectrum shifts to the left and has only one peak after fatigue as slow-twitch fibers persist and fast fibers drop out. The shift probably indicates recruitment of slow motor units in preference to fast units. Modified from Naeije and Zorn, 1982.

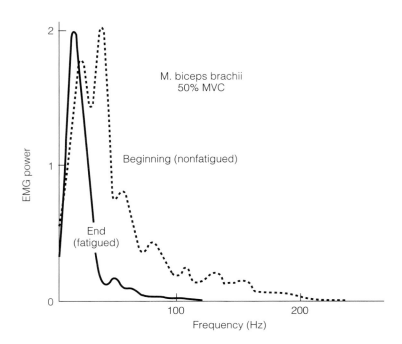

Figure 33-8 An illustration of the Setchenov phenomenon. Work output on a Musso, single-arm ergometer is recorded after two minutes of maximal work (bout 1). Thereafter, subjects rested for two minutes prior to each subsequent work bout. Between exercise bouts, subjects rested by means of either an active pause, a.p. (consisting of mental activity or light physical activity), or a passive pause, p.p. (consisting of minimal activity). Work output appears to be significantly greater after an active pause than after a passive pause. Modified from Asmussen and Mazin, 1978.

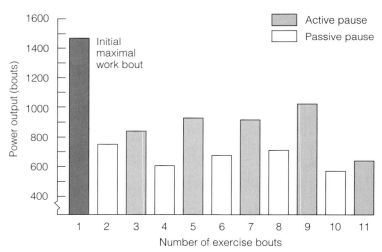

fatigue follows from the work of the Russian scientist Setchenov. He observed that the exhausted muscles of one limb recovered faster if the opposite limb was exercised moderately during a recovery period. The so-called Setchenov phenomenon, wherein an exhausted muscle recovers more rapidly if a diverting activity (either physical or mental) is performed, has been repeated several times, notably by Asmussen and Mazin. In Figure 33-8, the work performed following "active pauses" (gray bars) is clearly greater than that after "passive pauses" (white bars). This greater recovery following a diverting activity is apparently not associated with a greater recovery in muscle blood flow.

Rather, the Setchenov phenomenon is attributed to afferent input having a facilitatory effect on the brain's reticular formation and motor centers.

Psychological Fatigue

The Setchenov phenomenon is one example of the overlapping of the sciences of physiology and psychology. Eventually our understanding of the brain will advance to the point where its underlying physiology and biochemistry are understood and the effects of afferent input can be predicted. For the present, however, we can really only begin to address the question of how afferent input during competition (pain, breathlessness, nausea, audience response) can influence the physiology of the CNS. For now, a behavioral (psychological) approach to understanding these questions may be beneficial. Through training or intrinsic mechanisms, some athletes learn to minimize the influences of distressing afferent input and therefore approach performance limits set in the musculature. At times, such as at high altitudes, athletes slacken their pace to reduce discomforting inputs to a tolerable level. Consequently, work output decreases, but not because of a muscle limitation. On the opposite end of the spectrum, we frequently see examples of inexperienced or foolhardy athletes who set a blistering pace for most of a race and then experience real muscle fatigue before the end. In track events, these are the competitors who, in the home stretch, look like someone reading the newspaper. In order to perform optimally in an activity, the experiences of prior training and competition are often necessary for an athlete to evaluate afferent inputs properly and to utilize them in determining the maximal rate at which physiological power can be meted out during competition.

The Heart as a Site of Fatigue

In healthy individuals, there is no direct evidence that exercise, even prolonged endurance exercise, is limited by fatigue of the heart muscle. Because arterial P_{O_2} is well maintained during exercise, and because the heart in effect gets first choice at cardiac output, the heart is well oxygenated and nourished, even at maximal heart rate. In addition, because the heart is "omnivorous" in its appetite for fuels, it can be sustained by either lactic acid (which rises in short-term work) or fatty acids (which rise in long-term work). In individuals free of heart disease, the ECG does not reveal signs of ischemia (inadequate blood flow) during exercise. If ischemic symptoms are observed (Chapter 27), then this is, in fact, evidence of heart disease.

During prolonged work that leads to severe dehydration and major fluid and electrolyte shifts, or other situations in which exercise is performed after thermal dehydration or diarrhea, changes in plasma, Na^+, K^+, or Ca^{2+} can affect excitation-contraction coupling of the heart. In these cases, cardiac arrhythmias are possible, and exercise is not advised.

$\dot{V}_{O_2max}$ and Endurance

Given the obvious importance of maximal O_2 consumption ($\dot{V}_{O_2max}$) in setting the upper limit for aerobic metabolism, many scientists have been interested in studying the factors that limit $\dot{V}_{O_2max}$ and in how $\dot{V}_{O_2max}$ may affect exercise performance. As far as $\dot{V}_{O_2max}$ is concerned, researchers have considered two possibilities. One is that $\dot{V}_{O_2max}$ is limited by O_2 transport from the lungs to contracting muscles (i.e., limited by cardiac output and arterial O_2 content). The other possibility is that $\dot{V}_{O_2max}$ is limited by the respiratory capacity of contracting muscles. As far as endurance is concerned, some researchers believe that endurance is directly related to $\dot{V}_{O_2max}$ (i.e., the greater the $\dot{V}_{O_2max}$, the greater the endurance); other researchers support the possibility that endurance is determined by peripheral factors, such as the capacity of muscle mitochondria to sustain high rates of respiration. For the present, we must conclude that $\dot{V}_{O_2max}$ is a parameter set by maximal oxygen transport, but that endurance is determined also by muscle respiratory (mitochondrial) capacity. Several important lines of evidence lead to this conclusion.

Muscle Mass

Although it is true that the mass of contracting muscle can influence $\dot{V}_{O_2}$ during exercise, once a critical mass of muscle is utilized, $\dot{V}_{O_2\text{max}}$ is independent of muscle mass. The influence of muscle mass is revealed in the observations that $\dot{V}_{O_2\text{max}}$ during arm cranking or bicycling with one leg is less than that when cycling with two legs. Furthermore, $\dot{V}_{O_2\text{max}}$ during running is greater than that when bicycling or swimming, probably because of the greater muscle mass involved in running. However, $\dot{V}_{O_2\text{max}}$ while simultaneously arm cranking and leg cycling, or cross-country skiing, is not greater than that during running. In addition, twice the $\dot{V}_{O_2\text{max}}$ of one leg is much greater than the $\dot{V}_{O_2\text{max}}$ measured for both legs. Thus, in healthy individuals $\dot{V}_{O_2\text{max}}$ apparently increases as active muscle mass increases to a point beyond which O_2 delivered through the circulation is inadequate to supply the active muscle mass (Chapter 15).

Muscle Mitochondria

Several groups of investigators, including Booth and Narahara (1974) and Ivy and associates (1980), have observed the correlation between $\dot{V}_{O_2\text{max}}$ and leg muscle mitochondrial activity to approximate 0.8 (a perfect correlation is 1.0; no correlation is 0.0). Although these results suggest a direct relationship between $\dot{V}_{O_2\text{max}}$ and muscle mitochondrial content, the results do not directly address the issue of whether muscle mitochondrial respiratory capacity limits $\dot{V}_{O_2\text{max}}$. To address this question, Henriksson and Reitman (1977; see Chapter 21) followed the course of changes in $\dot{V}_{O_2\text{max}}$ and leg muscle mitochondrial activity during training and detraining. The results of these investigators indicate differential responses during training and detraining. Compared to $\dot{V}_{O_2\text{max}}$, muscle mitochondrial capacity increased relatively more during training (30% versus 19%), but improvements in $\dot{V}_{O_2\text{max}}$ persisted longer during detraining than does muscle respiratory capacity. The results demonstrate an independence of $\dot{V}_{O_2\text{max}}$ and muscle respiratory capacity.

In their work at the University of California, K. J. A. Davies and associates (1981, 1982, 1984)

approached issues related to $\dot{V}_{O_2\text{max}}$, O_2 transport, and muscle respiratory capacity in several ways. In endurance-trained and untrained control animals, $\dot{V}_{O_2\text{max}}$ correlated only moderately well (0.74) with endurance capacity, but muscle respiratory capacity correlated very well (0.92) with running endurance (see Table 6-3). Endurance training elicited a 100% increase in muscle mitochondrial content and a 400% increase in running endurance. $\dot{V}_{O_2\text{max}}$ increased only 15% in response to endurance training. However, sprint training, which also increased $\dot{V}_{O_2\text{max}}$ 15%, did not improve muscle mitochondrial content or running endurance.

Other approaches by Davies and coworkers to assess whether $\dot{V}_{O_2\text{max}}$ or muscle mitochondrial content better predicted exercise endurance involved dietary iron deficiency. In one study, animals were raised on an iron-deficient diet so that blood hemoglobin and hematocrit and muscle mitochondrial content were depressed. When iron was restored in the diet (Figure 33-9), hematocrit and $\dot{V}_{O_2\text{max}}$ re-

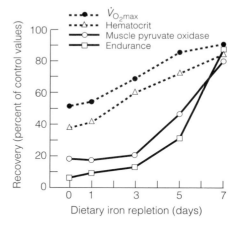

Figure 33-9 Recovery of $\dot{V}_{O_2\text{max}}$, hematocrit, leg muscle pyruvate oxidase activity, and running endurance in rats restored to a diet containing normal levels of iron following a zero iron-containing diet. Results indicate that hematocrit and $\dot{V}_{O_2\text{max}}$ respond in parallel, and that pyruvate oxidase (a mitochondrial marker) and endurance respond in parallel. The results suggest that $\dot{V}_{O_2\text{max}}$ is a function of circulatory, O_2 transport capacity, whereas running endurance is related to muscle mitochondrial content. Source: K. J. A. Davies et al., 1982.

sponded rapidly and in parallel. Muscle mitochondrial capacity (pyruvate oxidase) and running endurance also improved in parallel, but more slowly. In still another series of experiments, iron-deficient animals received cross-transfusions of packed red blood cells in exchange for equal volumes of anemic blood. This procedure raised blood hemoglobin and hematocrit immediately and restored $\dot{V}_{O_2max}$ to within 90% of normal. However, despite the fact that hematocrit, hemoglobin, and $\dot{V}_{O_2max}$ were improved in iron-deficient animals by cross-transfusion of red blood cells, running endurance was unimproved.

Taken together, these results strongly suggest that $\dot{V}_{O_2max}$ is a function of O_2 transport (a cardiovascular parameter), whereas endurance is more dependent on muscle mitochondrial capacity. The maximal ability of mitochondria in muscle to consume O_2 is apparently several times greater than the ability of the circulation to supply O_2. Hence, $\dot{V}_{O_2max}$ is probably limited by arterial O_2 transport.

Arterial O_2 Transport

Researchers have utilized several other techniques to study the relationship of arterial O_2 transport and $\dot{V}_{O_2max}$. Arterial O_2 transport ($T_{a_{O_2}}$) is equal to the product of cardiac output ($\dot{Q}$) and arterial O_2 content ($C_{a_{O_2}}$):

$$T_{a_{O_2}} = \dot{Q}(C_{a_{O_2}}) \tag{33-2}$$

In general, attempts to raise arterial O_2 content by having subjects breathe gases high in O_2 content, or by inducing an increase in red blood cell mass (blood doping), raise $\dot{V}_{O_2max}$. Conversely, $\dot{V}_{O_2max}$ is depressed when subjects breathe low O_2 content gas mixtures, have depressed cardiac function, are anemic, or are exposed to carbon monoxide (CO), a gas that binds to hemoglobin and interferes with arterial O_2 binding and transport.

Muscle Oxygenation at $\dot{V}_{O_2max}$

Despite the large volume of evidence that $\dot{V}_{O_2max}$ is limited by O_2 transport, scientists have reported that

under some circumstances $\dot{V}_{O_2max}$ is *not* limited by O_2 transport. In dog gastrocnemius muscle preparations stimulated to contract to $\dot{V}_{O_2max}$, Jöbsis and Stainsby (1968) made spectrophotometric measurements on mitochondrial cytochromes and found them to be oxidized. Thus, mitochondrial O_2 content was adequate. Furthermore, Barclay and Stainsby (1975) raised $\dot{V}_{O_2}$ and the contractile tension of dog muscle preparations by assisting arterial blood flow with a pump. In humans during bicycling, Pirnay and colleagues (1972) noted a significant O_2 content of femoral venous blood draining from active muscles. As noted in Chapter 6, through the work of Richardson and associates (1998), who used nuclear magnetic resonance (NMR) to determine the saturation of muscle myoglobin during exercise (Figure 6-16) indicates that muscle PO_2 is well above the critical mitochondrial O_2 tension. The critical O_2 tension for mitochondria (i.e., the O_2 tension below which the maximal rate of mitochondrial O_2 consumption declines) is very low. Based on studies of isolated mitochondria, it is probable that mitochondria are capable of achieving the maximal rates of respiration when the PO_2 is as low as 1 mmHg (Figure 6-15). Thus, when small muscle masses are made to work maximally, $\dot{V}_{O_2max}$ may be limited by factors within muscle and not by O_2 transport.

Arterial O_2 Transport, $\dot{V}_{O_2max}$, and Exercise Endurance

From Figure 13-1 and the preceding discussion, it should be clear that maximal oxygen consumption ($\dot{V}_{O_2max}$) is limited by capacity for arterial oxygen transport ($T_{a_{O_2}}$), which is, in turn, limited by maximal cardiac output ($\dot{Q}_{max}$). Therefore, over a broad range of cardiac outputs and values of $\dot{V}_{O_2max}$ in a population including fit and unfit, young and old, healthy and ill individuals, $\dot{V}_{O_2max}$ appears to correlate well with endurance and physical performance. This is because on a broad scale, $\dot{V}_{O_2max}$ and the parameters that determine it set the boundaries for physical performance. Further, in addition to oxygen transport, other key physiological functions, such as delivery of hormones and substrates, removal of wastes, acid–base balance and temperature regulation all depend on the circulation. Thus, we can say that the maximal capacities of $\dot{Q}_{max}$, $T_{a_{O_2}max}$, $\dot{V}_{O_2max}$, and physical performance are all interrelated.

Despite the common correlates among parameters of oxygen transport and use, $\dot{V}_{O_2max}$, among athletes is a poor predictor of performance. This is because of the importance of peripheral (i.e., intramuscular) as opposed to central (i.e., cardiovascular) factors in determining endurance.

■ Catastrophe Theory

In the development and writing of this text, we have repeatedly emphasized that physiological processes are highly controlled and often redundant in function. Consequently, it comes as no surprise that successes and failures in integrated functions involve multiple cells, tissues, organs, and systems. The complexity of physiological control of function in the development of muscle fatigue has best been described by Edwards (1983), who has articulated a "catastrophe theory." According to this theory, the failure of one enzyme system, cell, tissue organ or system places an undue burden on related systems, such that several may begin to fail simultaneously. To quote from Edwards, "An important prediction from a catastrophe theory of fatigue is that it may prove very difficult if not impossible to recognize the limiting factor, e.g., at the end of a marathon." In such circumstances, multiple organs and organ systems may be unable to sustain high metabolic rates.

■ The Future of Fatigue

Through the wonders of contemporary science and technology, a whole series of devices and techniques are becoming available to researchers. Many of these devices will be of assistance in understanding muscle fatigue, and three deserve special mention: nuclear magnetic resonance (NMR), positron emission tomography (PET), and near infrared spectroscopy (NIRS).

NMR (MRS and MRI)

The technique of nuclear magnetic resonance (NMR) spectroscopy is simple in theory, but very difficult to perform in actuality. The technique takes advantage of the fact that nuclei with odd numbers of nucleons (neutrons + protons) are slightly magnetic. Thus, atoms containing naturally occurring, nonradioactive elemental isotopes such as ^{31}P, ^{13}C, and ^{1}H can be aligned in a uniform and powerful magnetic field. If a brief pulse of oscillating, radio frequency magnetic field is applied at a right angle to the fixed field, the influenceable nuclei resonate between the directions of the two fields. This resonance persists briefly after the radio frequency field is switched off, and the nuclei themselves emit a radio frequency signal. Because the radio frequency signal emitted by a particular atomic species is influenced by the nature of the compound in which that species exists, the resulting signal varies and produces a particular NMR spectrum (Figure 33-10). Thus, it is possible to identify the various compounds in which the susceptible nuclei exist, and the area under each peak (i.e., the integral) is proportional to the concentration of each compound.

Through the technique of NMR spectroscopy, it is now possible to determine concentrations of ATP, CP, P_i, H^+ (pH), and concentrations of water, fat, and particular metabolites, continuously and painlessly, without breaking the skin. In some instances, a cross-sectional image through a body part (i.e., arm, leg, or brain) can be obtained through magnetic resonance imaging (MRI). The technique of MRI takes advantage of naturally occurring isotopes in the body; or, in some cases, nonradioactive isotopes such as ^{13}C can be infused or injected to enhance the NMR signal.

Figure 33-10a displays the phosphorus NMR spectrum from a flexor forearm muscle in a resting individual. In this procedure, called topical nuclear magnetic resonance (TNMR), a radio frequency (RF) sensor coil is applied to the skin directly over the muscle. Initial attempts at utilizing NMR to study muscle metabolism have focused on ^{31}P-containing compounds. Table 33-1 presents a comparison of phosphate compound results in rest-

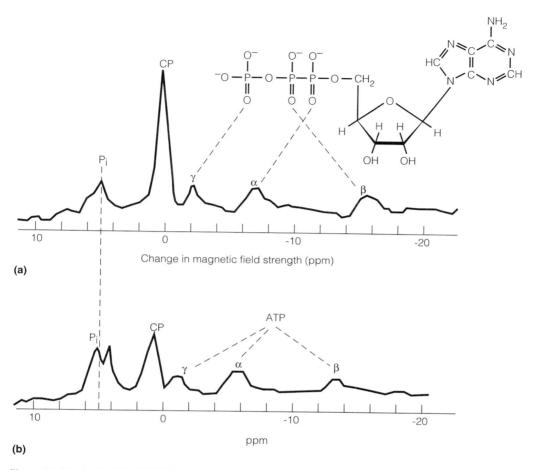

Figure 33-10   Topical NMR (TNMR) phosphate spectrum from a human forearm before (a) and after (b) fatiguing isometric exercise. The area under each
peak is proportional to the concentration of each metabolite. After fatigue, the
concentrations of free phosphate increase and the levels of CP and ATP decrease. In addition, the phosphate peak splits and shifts to the right, indicating
a decrease in pH. Source: Brooks, Nunnally, and Budinger, 1983.

ing human muscle obtained by means of muscle biopsy and NMR spectroscopy. Although ATP levels
are similar, creatine phosphate (CP) levels are significantly higher in NMR. Thus, the CP to ATP ratio
in NMR is approximately 5 to 6, while that from the
biopsy is only 3 to 4. The difference is likely due to
CP degradation during the biopsy procedure.

In part b of Figure 33-10, recorded after intermittent isometric contraction to fatigue, the TNMR
spectrum has changed dramatically. Both ATP and
CP peaks have decreased, P_i has increased, and the
P_i peak has split and shifted right toward the CP
peak, indicating a decrease in pH. Experiments such
as these will probably make the muscle biopsy procedure obsolete for measuring muscle phosphates
and pH during rest, exercise, and in the clinical
laboratory. The NMR signal represents a larger,
more representative sample, and the result prob-

TABLE 33-1

Comparison of Phosphate Compounds in Healthy, Resting Human Muscle Obtained by Muscle Biopsy and NMR Spectroscopy

P Metabolite	Needle Biopsy Mean ± SE mmol · kg⁻¹ Wet	^{31}P NMR Mean ± SE mmol · kg⁻¹ Wet
ATP	5.5±0.07	5.11±0.12
PC	17.4±0.19	28.52±0.43
P_i	10	4.27±0.17
CP+Cr	28.6±0.28	—
CP+P_i	31.6±3.27	32.79±0.41
Phosphodiester		0.83±0.17
NAD⁺		0.88±0
pH		
Homogenates	7.08±0.008	—
Intracellular	7.00±0.02	7.13±0.02

Source: Edwards et al., 1982, p. 3.

ably reflects more accurately the situation in vivo. Of course, the biopsy procedure will be retained for particular situations, such as when there is a need to do histochemistry. NMR imaging allows the degree of atherosclerosis and the presence of tumors to be evaluated noninvasively.

PET

The technique of *positron emission tomography* (PET) offers great potential for studying regional blood flow and metabolism. Like NMR in principle, PET is conceptually not difficult to understand, but its successful execution requires enormous theoretical and technical competence. Nuclei of particular short half-life isotopes such as ^{11}C and ^{82}Rb emit positrons when they decay. These positrons (β^+) have the same mass as electrons but have an opposite or positive charge. When an emitted β^+ (a particle of antimatter) strikes an electron (matter), the two particles annihilate each other. The result is two photons, roughly equivalent to gamma rays, which fly off in opposite directions. Thus, if an atom decays within a ring of crystal and photomultiplier sensors, two crystals (on opposite sides of the ring) may be struck simultaneously as the result of nu-

clear decay. By means of some extremely complex electronics and computer programming, the point within the ring where the nucleus has decayed can be pinpointed.

Figure 33-11 illustrates the University of California, Donner Research Medicine PET. The device is large enough to insert the head or thorax of an adult. When a positron-emitting isotope is injected into the circulation supplying a tissue, such as the heart, the isotope flows where the blood flows through the myocardium, and pinpoints areas that receive little or no flow.

Compared to the NMR technique, the PET technique has the disadvantage of requiring a radioactive isotope. These B⁺-emitting isotopes, however, have short half-lives (on the order of minutes) so that the radioactivity decays rapidly. Thus, repeated measurements can be made in a relatively short period of time. Furthermore, PET is useful not only for imaging tissues of interest, but also for obtaining quantitative data on tissue uptake, metabolism, and washout. The short half-life of B⁺ emitters makes it possible to follow the injection of one isotope with another. For instance, in the heart, regional blood flow can be determined with an injection of ^{82}Rb. This can then be followed with an injection of ^{11}C-labeled metabolite (e.g., lactate, pyruvate, fatty acid), and regional myocardial metabolism can be studied. In this way, detailed studies of myocardial flow and metabolism can be performed. Positron emission tomography can be used to study blood flow and metabolism in any tissue that can fit in a tomograph. To date, most studies have focused on the heart and brain, but it is also possible to study exercising limb skeletal muscles.

NIRS

An emerging technique termed *near infrared spectroscopy* (NIRS) is being developed. This technology offers the possibility of noninvasively and continuously monitoring the state of oxygenation of iron-containing compounds such as myoglobin and mitochondrial cytochrome oxidase in working muscle. Though extremely sophisticated in design, in practice the approach will be relatively simple and allow placement of skin probes at several sites during

Figure 33-11 The University of California, Donner Research Medicine (Lawrence Berkeley Laboratory) positron emission tomograph. When sensors on opposite sides of the ring are simultaneously struck by photons, the locus of the disintegrating nucleus can be pinpointed. See the text for a description of how PET can be used to image heart, brain, and other parts, and to study regional blood flow and metabolism. SOURCE: Courtesy of T. F. Budinger, Donner Research Medicine Group, Lawrence Berkeley Laboratory.

most forms of exercise. At present, preliminary data obtained using NIRS indicate complete, or near complete, muscle deoxygenation at power outputs that elicit mild to moderate intensity exercises. Because it is known from previous work that venous drainage of muscle working at such intensities contains substantial oxygen, and because recent studies by MacDonald and associates (1999) show NIRS muscle O_2 and femoral venous O_2 values to be poorly related, it is likely that present NIRs technology grossly underestimates the state of muscle oxygenation. Hence, there is at present no firm basis on which to use information from NIRS. Nonetheless, given the rate of progress in NIRS technology, NIRS may soon provide us with information we have long desired on the state of muscle oxygenation during exercise.

In the future, devices using NMR, PET, NIRS, as well as other devices may become available to provide us with a better view of the metabolic changes in muscle associated with fatigue processes.

■ Summary

The reasons why people are unable to continue muscular work and why they experience fatigue depend on several general factors, including the nature of the activity, the training and physiological status of the individual, and environmental conditions. The continuance of exercise sometimes depends on the availability of some key metabolite in a particular tissue. Sometimes also the continuance of exercise depends on maintenance of the functional integrity of several individual systems, as well as their proper interaction. Because physiological systems are often matched in their capacities, during exercise stress it is sometimes difficult to identify a single factor that failed in an attempt to maintain exercise intensity. In fact, fatigue sites may be multiple and interrelated.

Fatigue may result from the depletion of key metabolites such as muscle creatine phosphate (CP) and ATP, muscle glycogen, liver glycogen, blood glucose, and arterial as well as muscle O_2. Without adequate levels of each of these, high rates of work by muscles cannot continue.

The accumulation of particular metabolites can also hinder the continuance of exercise. In the past, the accumulation of lactic acid (which dissociates a proton and decreases pH) was thought to be the main fatigue-causing agent. It is unlikely, however, that lactic acid accumulation always causes the muscle fatigue, pain, or soreness usually attributed to it. Other potential fatigue factors include the loss of Ca^{2+} responsiveness and sensitivity as well as the accumulation of Ca^{2+} in mitochondria and in cytosol. In mitochondria, Ca^{2+} accumulation affects oxidative phosphorylation, and in cytosol it affects glycolysis and excitation-contraction coupling. Phosphate (HPO_4^{2-}) and diprotenated phosphate ($H_2PO_4^-$) affect a number of steps in the overall process of excitation-contraction coupling. When phosphate and hydrogen ions start to accumulate in muscle cells, performance soon diminishes. Ca^{2+} is essential for triggering excitation-contraction coupling, and changes in the ability to release and re-sequester Ca^{2+} as well as loss of responsiveness to Ca^{2+} release, are detrimental to the force of muscle contraction.

Overheating can lead to a number of general as well as local effects, including blood flow redistribution away from muscle and liver, dehydration, fluid and electrolyte redistribution, CNS dysfunction, and mitochondrial uncoupling of oxidative phosphorylation. In the case of overheating, we have an example of how a single specific factor (overheating) can result in a generalized disturbance to homeostasis.

During contraction in athletic competitions as we know them, muscular fatigue usually appears to be a peripheral phenomenon and due to fatigue of the muscles. The central nervous system and the associated motor neurons and neuromuscular junctions appear to be far superior to skeletal muscle in maintaining function. In some instances, however, CNS factors may operate to limit the performance of muscles. In healthy individuals, the heart is not thought to fatigue and cause reduced exercise power output.

■ Selected Readings

Allen, D. G., L. Lännergren, and H. Westerblad. Muscle cell function during prolonged activity: cellular mechanisms of fatigue. *Experimental Biology* 80: 497–527, 1995.

Alpert, N. Lactate production and removal and the regulation of metabolism. *Ann. N. Y. Acad. Sci.* 119: 995–1001, 1965.

Anderson, P., and J. Henriksson. Capillary supply to the quadriceps femoris muscle of man: adaptive response to exercise. *J. Physiol.* (London) 270: 677–690, 1977.

Asmussen, E. Muscle Fatigue. *Med. Sci. Sports* 11: 313–321, 1979.

Asmussen, E., and B. Mazin. Recuperation after muscular fatigue by "diverting activities." *Europ. J. Appl. Physiol.* 38: 1–8, 1978.

Asmussen, E., and B. Mazin. A central nervous component in local muscular fatigue. *Europ. J. Appl. Physiol.* 38: 9–15, 1978.

Åstrand, P.-O., and K. Rodahl. Textbook of Work Physiology. New York: McGraw-Hill, 1970, pp. 154–178 and 187–254.

Åstrand, P.-O., and B. Saltin. Maximal oxygen uptake and heart rate in various types of muscular activity. *J. Appl. Physiol.* 16: 977–981, 1961.

Bang, O. The lactate content of blood during and after exercise in man. *Skand. Arch. Physiol.* 74 (Suppl.) 10: 51–82, 1936.

Bannister, R. G., and C. J. C. Cunningham. The effects on the respiration and performance during exercise of adding oxygen to the inspired air. *J. Physiol.* (London) 125: 118–121, 1954.

Barclay, J. K., and W. N. Stainsby. The role of blood flow in limiting maximal metabolic rate in muscle. *Med. Sci. Sports* 7: 116–119, 1975.

Barcroft, H., and J. L. E. Millen. The blood flow through muscle during sustained contraction. *J. Physiol.* 97: 17–31, 1939.

Basmajian, J. V. Muscles Alive. Baltimore: Williams and Wilkins, 1962.

Bergh, U., I.-L. Kaustrap, and B. Ekblom. Maximal oxygen uptake during exercise with various combinations of arm and leg work. *J. Appl. Physiol.* 41: 191–196, 1976.

Bergström, J. Local changes of ATP and phosphorylcreatine in human muscle tissue in connection with exercise. In Physiology of Muscular Exercise, C. B. Chapman (Ed.). New York: American Heart Association, 1967, pp. 191–196. (Monograph No. 15).

Bergström, J., and E. Hultman. The effects of exercise on muscle glycogen and electrolytes in normals. *Scand. J. Clin. Lab. Invest.* 18: 16–20, 1966.

Bigland-Ritchie, B., E. Cafarelli, and N. K. Vøllestad. Fatigue of submaximal static contractions. *Acta Physiol. Scand.* 128 (Suppl. 556): 137–148, 1986.

Bigland-Ritchie, B., F. Furbush, and J. J. Woods. Fatigue of intermittent submaximal voluntary contractions: central and peripheral factors. *J. Appl. Physiol.* 61: 421–429, 1986.

Bigland-Ritchie, B., D. A. Jones, and J. J. Woods. Excitation frequency and muscular fatigue: electrical responses during human voluntary and stimulated contractions. *Exp. Neurol.* 64: 414–427, 1979.

Booth, F. W., and K. A. Narahara. Vastus lateralis cytochrome oxidase activity and its relationship to maximal oxygen consumption in man. *Pflügers Archiv.* 349: 319–324, 1974.

Bowie, W., and G. R. Cumming. Sustained handgrip reproducibility: effects of hypoxia. *Med. Sci. Sports* 3: 48–52, 1971.

Brasil-Neto, J. P., L. G. Cohen, and M. Hallett. Central fatigue as revealed by postexercise decrement of motor evoked potentials. *Muscle and Nerve* 17: 713–719, 1994.

Brooks, G. A., K. E. Brauner, and R. G. Cassens. Glycogen synthesis and metabolism of lactic acid after exercise. *Am. J. Physiol.* 220: 1053–1059, 1971.

Brooks, G. A., K. J. Hittelman, J. A. Faulkner, and R. E. Beyer. Temperature, liver mitochondrial respiratory functions, and oxygen debt. *Med. Sci. Sports* 2: 71–74, 1971.

Brooks, G. A., K. J. Hittelman, J. A. Faulkner, and R. E. Beyer. Tissue temperatures and whole-animal oxygen consumption after exercise. *Am. J. Physiol.* 221: 427–431, 1971.

Brooks, G. A., R. Nunnally, and T. F. Budinger. Human forearm muscular fatigue investigated by phosphorus nuclear magnetic resonance. Unpublished manuscript, 1983.

Buick, F. J., N. Gledhill, A. B. Frosese, L. Spriet, and E. C. Meyers. Effect of induced erythrocythemia on aerobic work capacity. *J. Appl. Physiol: Respirat. Environ. Exer. Physiol.* 48: 636–642, 1980.

Capelli, C., G. Antonutto, P. Zamparo, M. Girardis, and P. E. di Prampero. Effects of prolonged cycle ergometer exercise on maximal muscle power and oxygen uptake in humans. *Eur. J. Appl. Physiol.* 66: 189–195, 1993.

Chance, B., S. Eleff, J. S. Leigh, D. Sokolow, and A. Sapega. Mitochondrial regulation of phosphocreatine/phosphate ratios in exercised human limbs: gated ^{31}P NMR study. *Proc. Natl. Acad. Sci. USA* 78: 6714–6718, 1981.

Chapman, C. B. (Ed.). Physiology of Muscular Exercise. New York: American Heart Association, 1967. (Monograph No. 15).

Christensen, E. H. Fatigue of the working individual. In Structure and Function of Muscle II: Biochemistry and physiology, G. H. Bourne (Ed.). New York: Academic Press, 1960, pp. 455–465.

Ciba Foundation. 82nd Symposium on Human Muscle Fatigue: Physiological Mechanisms. London: Pitman Medical, 1981.

Costill, D. L., R. Bowers, G. Branam, and K. Sparks. Muscle glycogen utilization during prolonged exercise on successive days. *J. Appl. Physiol.* 31: 834–838, 1971.

Davies, C. T. M., and A. J. Sargent. Physiological response to one- and two-leg exercise breathing air and 45% oxygen. *J. Appl. Physiol.* 267: 703–735, 1977.

Davies, K. J. A., C. M. Donovan, C. J. Refino, G. A. Brooks, L. Packer, and P. R. Dallman. Distinguishing effects of anemia and muscle iron deficiency on exercise bioenergetics in the rat. *Am. J. Physiol.* 246 (*Endocrinol. Metab.* 9): E535-E543, 1984.

Davies, K. J. A., J. L. Maguire, G. A. Brooks, P. R. Dallman, and L. Packer. Muscle mitochondrial bioenergetics, oxygen supply, and work capacity during dietary iron deficiency and repletion. *Am. J. Physiol.* 242 (*Endocrinol. Metab.*): E418-E427, 1982.

Davies K. J. A., L. Packer, and G. A. Brooks. Biochemical adaptation of mitochondria, muscle, and whole-animal respiration to endurance training. *Arch. Biochem. Biophs.* 209: 539–554, 1981.

Davies, K. J. A., L. Packer, and G. A. Brooks. Exercise bioenergetics following sprint training. *Arch. Biochem. Biophs.* 215: 260–265, 1982.

Dawson, M. J., D. G. Gadian, and D. R. Wilkie. Contraction and recovery of living muscles studied by ^{31}P nuclear magnetic resonance. *J. Physiol.* (London) 267: 703–735, 1977.

Dawson, M. J., D. G. Gadian, and D. R. Wilkie. Muscular fatigue investigated by phosphorus nuclear magnetic resonance. *Nature* 274: 861-866, 1978.

Dennis, S. C., W. Gevers, and L. H. Opie. Protons in ischemia: where do they come from; where do they go to? *J. Mol. Cell Cardiol.* 23: 1077–1086, 1991.

Derman, W. E., F. Dunbar, M. Haus, M. Lambert, and T. D. Noakes. Chronic beta-blockade does not influence muscle power output during high-intensity exercise of short-duration. *Eur. J. Appl. Physiol.* 67: 415–419, 1993.

Dill, D. B. Fatigue and physical fitness. In Structural and Physiological Aspects of Exercise and Sport, W. R. Johnson and E. R. Buskirk (Eds.). Princeton, N.J.: Princeton Book, 1980.

Dobson, G. P., E. Yamamoto, and P. W. Hochachka. Phosphofructokinase control in muscle: nature and reversal of pH-dependent ATP inhibition. *Am. J. Physiol.* 250 (*Reg. Integ. Comp. Physiol.* 19): R71–R76, 1986.

Duty, S., and D. G. Allen. The distribution of intracellular calcium concentration in isolated single fibres of mouse skeletal muscle during fatiguing stimulation. *Pflügers Arch.* 427: 102–109, 1994.

Edwards, R. H. T. Biochemical bases of fatigue in exercise performance: catastrophe theory in muscular fatigue. In Biochemistry of Exercise, H. G. Knuttgen, J. A. Vogel, and J. Poortmans (Eds.). Champaign, Ill.: Human Kinetics, 1983, pp. 1–28.

Edwards, R. H. T. Muscle fatigue and pain. *Acta Med. Scand.* (Suppl. 711): 179–188, 1986.

Edwards, R. H. T., D. K. Hill, and P. A. Merton. Fatigue of long duration in human skeletal muscle after exercise. *J. Physiol.* (London) 272: 769–778, 1977.

Edwards, R. H. T., D. R. Wilkie, M. J. Dawson, R. E. Gordon, and D. Shaw. Clinical use of nuclear magnetic resonance in the investigation of myopathy. *Lancet* (March): 725–731, 1982.

Ekblom, B., A. N. Goldbarg, and B. Gullbring. Response to exercise after blood loss and reinfusion. *J. Appl. Physiol.* 33: 175–189, 1972.

Ekblom, B., R. Hout, E. Stein, and A. Thorstensson. Effect of changes in arterial oxygen content on circulation and physical performance. *J. Appl. Physiol.* 39: 71–75, 1975.

Ekblom, B., G. Wilson, and P.-O. Åstrand. Central circulation during exercise after venesection and reinfusion of red blood cells. *J. Appl. Physiol.* 40: 379–383, 1976.

Ekelund, L. G., and A. Holmgren. Circulatory and respiratory adaptation during long-term, non-steady state exercise in the sitting position. *Acta Physiol. Scand.* 62: 240, 1964.

Ekelund, L. G., and A. Holmgren. Central hemodynamics during exercise. In Physiology of Muscular Exercise, C. B. Chapman (Ed.). New York: American Heart Association, 1967, pp. 133–143. (Monograph No. 15).

Fallowfield, J. L., and C. Williams. Carbohydrate intake and recovery from prolonged exercise. *Int. J. Sport Nutr.* 3: 150–164, 1993.

Faulkner, J. A., D. E. Roberts, R. L. Elk, and J. Conway. Cardiovascular responses to submaximum and maximum effort cycling and running. *J. Appl. Physiol.* 30: 457–461, 1971.

Favero, T. G., A. C. Zable, D. Colter, and J. J. Abramson. Lactate inhibits Ca^{2+}-activated Ca^{2+}-channel activity from skeletal muscle sarcoplasmic reticulum. *J. Appl. Physiol.* 82: 447–452, 1997.

Fitts, R. H. Cellular mechanisms of muscle fatigue. *Physiological Reviews* 74: 49–94, 1994.

Gadian, D. G., and G. K. Radda. NMR studies of tissue metabolism. *Ann. Rev. Biochem.* 50: 69–83, 1981.

Gardner, G. W., V. R. Edgerton, R. Senewirathe, R. J. Barnard, and Y. Ohira. Physical work capacity and metabolic stress in subjects with iron deficiency anaemia. *Am. J. Clin. Nutr.* 30: 910–917, 1977.

Gevers, W. Generation of protons by metabolic processes in heart cells. *J. Mol. Cell Cardiol.* 9: 867–874, 1977.

Gleser, M. A. Effect of hypoxia and physical training on hemodynamic adjustments to one-legged exercise. *J. Appl. Physiol.* 34: 655–659, 1973.

Gleser, M. A., D. H. Horstoman, and R. P. Mello. The effects on $\dot{V}_{O_2max}$ of adding arm work to maximal leg work. *Med. Sci. Sports* 6: 104–107, 1974.

Gollnick, P. D., R. B. Armstrong, C. W. Saubert, W. L. Sembrowich, R. E. Shepherd, and B. Saltin. Glycogen depletion patterns in human skeletal muscle fibers during prolonged work. *Pflügers Archiv.* 344: 1–12, 1973.

Gollnick, P. D., P. Körge, J. Karpakka, and B. Saltin. Elongation of skeletal muscle relaxation during exercise is linked to reduced calcium uptake by the sarcoplasmic reticulum in man. *Acta Physiol. Scand.* 142: 135–136, 1991.

Gordon, R. E., P. E. Hanley, and D. Shaw. Topical magnetic resonance. *Prog. NMR Spectr.* 15: 1–47, 1982.

Grimby, G., E. Haggendal, and B. Saltin. Local Xenon 133 clearance from the quadriceps muscle during exercise in man. *J. Appl. Physiol.* 22: 305–310, 1967.

Häkkinen, K. Neuromuscular fatigue in males and females during strenuous heavy resistance loading. *Electromyogr. Clin. Neurophysiol.* 34: 205–214, 1994.

Häkkinen, K., and P. V. Komi. Effects of fatigue and recovery on electromyographic and isometric force- and relaxation-time characteristics of human skeletal muscle. *Eur. J. Appl. Physiol.* 55: 588–596, 1986.

Hanson, P., A. Claremont, J. Dempsey, and W. Reddan. Determinants and consequences of ventilatory responses to competitive endurance running. *J. Appl. Physiol.: Respirat. Environ. Exercise Physiol.* 52: 615–623, 1982.

Henriksson, J., and J. S. Reitman. Time course of changes in human skeletal muscle succinate dehydrogenase and cytochrome exidas activities and maximal oxygen up-take with physical activity and inactivity. *Acta Physiol. Scand.* 99: 91–97, 1977.

Hogan, M. C., L. B. Gladden, S. S. Kurdak, and D. C. Poole. Increased [lactate] in working dog muscle reduces tension development independent of pH. *Med. Sci. Sports Exerc.* 27: 371–377, 1995.

Horstman, D. H., M. Gleser, and J. Delehunt. Effects of altering O_2 delivery on VO_2 of isolated working muscle. *Am. J. Physiol.* 230: 327–334, 1976.

Horstman, D. H., M. Gleser, D. Wolfe, T. Tyron, and J. Delehunt. Effects of hemoglobin reduction on $\dot{V}_{O_2max}$ and related hemodynamics in exercising dogs. *J. Appl. Physiol.* 37: 97–100, 1974.

Hughes, R. L., M. Clode, R. H. Edwards, T. J. Goodwin, and N. Jones. Effect of inspired O_2 on cardiopulmonary and metabolic responses to exercise in man. *J. Appl. Physiol.* 24: 336–347, 1968.

Ivy, J. L., D. L. Costill, and B. D. Maxwell. Skeletal muscle determinants of maximal aerobic power in man. *Eur. J. Appl. Physiol.* 44: 1–8, 1980.

Ivy, J. L., R. T. Withers, D. J. Van Handel, D. H. Elger, and D. L. Costill. Muscle respiratory capacity and fiber type as determinants of lactate threshold. *J. Appl. Physiol.: Respirat. Environ. Exercise Physiol.* 48: 523–527, 1980.

Jessup, G. T. Changes in forearm blood flow associated with sustained handgrip performance. *Med. Sci. Sports* 5: 258–261, 1973.

Ji, L. L., F. W. Stratman, and H. A. Lardy. Enzymatic down regulation with exercise in rat skeletal muscle. *Arch. Biochem. Biophys.* 263: 137–149, 1988.

Jöbsis, F. F., and W. N. Stainsby. Oxidation of NADH during contractions of circulated mammalian skeletal muscle. *Resp. Physiol.* 4: 292–300, 1968.

Jones, D. A., B. Bigland-Ritchie, and R. H. T. Edwards. Excitation frequency and muscle fatigue: mechanical responses during voluntary and stimulated contractions. *Exp. Neurol.* 64: 401–413, 1979.

Kayser, B., M. Narici, T. Binzoni, B. Grassi, and P. Cerretelli. Fatigue and exhaustion in chronic hypobaric hypoxia: influence of exercising muscle mass. *J. Appl. Physiol.* 76: 634–640, 1994.

Kroll, W. Isometric fatigue curves under varied intertrial recuperation periods. *Res. Q.* 39: 106–115, 1968.

Leonard, C. T., J. Kane, J. Perdaems, C. Frank, D. G. Graetzer, and T. Moritani. Neural modulation of muscle contractile properties during fatigue: afferent feedback dependence. *Electroencephalogr. Clin. Neurophysiol.* 93: 209–217, 1993.

Lewis, S. F., and R. G. Haller. The pathophysiology of McArdle's disease: clues to regulation in exercise and fatigue. *J. Appl. Physiol.* 61: 391–401, 1986.

Lind, A. R. Muscle fatigue and recovery from fatigue induced by sustained contractions. *J. Physiol.* (London) 127: 162–171, 1959.

Lippold, D. C. J., J. W. T. Redfearn, and J. Vuco. The electromyography of fatigue. *Ergonomics* 3: 121–131, 1960.

MacDonald, M. J., M. A. Tarnopolsky, H. J. Green, and R. L. Hughson. Comparison of femoral blood gases and muscle near-infrared spectroscopy at exercise onset in humans. *J. Appl. Physiol.* 86: 687–693, 1999.

Madsen, K., P. K. Pedersen, M. S. Djurhuusand, and N. A. Klitgaard. Effects of detraining on endurance capacity and metabolic changes during prolonged exhaustive exercise. *J. Appl. Physiol.* 75: 1444–1451, 1993.

Margaria, R., H. T. Edwards, and D. B. Dill. The possible mechanisms of contracting and paying the oxygen debt and the role of lactic acid in muscular contraction. *Am. J. Physiol.* 106: 689–715, 1933.

Martin, B. J., and G. M. Gaddis. Exercise after sleep deprivation. *Med Sci. Sports Exerc.* 13: 220–223, 1981.

Mathews, D. E., D. M. Bier, M. J. Rennie, R. H. T. Edwards, D. Halliday, D. J. Millard, and G. Clugston. Regulation of leucine metabolism in man: a stable isotope study. *Science* 214: 1129–1131, 1981.

Merton, P. A. Voluntary strength and fatigue. *J. Physiol.* (London) 123: 553–564, 1954.

Metzger, J. M., and R. H. Fitts. Fatigue from high- and low-frequency muscle stimulation: role of sarcolemma action potentials. *Experimental Neurol.* 93: 320–333, 1986.

Meyers, S., and W. Sullivan. Effect of circulatory occlusion on time to muscular fatigue. *J. Appl. Physiol.* 24: 54–59, 1968.

Mitchell, J. H., B. J. Sproule, and C. B. Chapman. Physiological meaning of the maximal oxygen uptake test. *J. Clin. Invest.* 37: 538, 1958.

Moxham, J., R. H. T. Edwards, M. Aubrer, A. De Troyer, G. Farkas, P. T. Macklem, and C. Roussos. Changes in EMG power spectrum (high-to-low ratio) with force fatigue in humans. *J. Appl. Physiol.: Respirat. Environ. Exercise Physiol.* 53: 1094–1099, 1982.

Naeije, M., and H. Zorn. Relation between EMG power spectrum shifts and muscle fibre action potential conduction velocity changes during local muscular fatigue in man. *Eur. J. Appl. Physiol.* 50: 25–33, 1982.

Nakamura, K., D. A. Schoelher, F. J. Winkler, and H.-L. Schmidt. Geographical variations in the carbon isotope composition of hair in contemporary man. *Biomed. Mass Spectr.* 9: 390–394, 1982.

Nosek, T. M., K. Y. Fender, and R. E. Godt. It is dipronated inorganic phosphate that depresses force in skinned skeletal muscle fibers. *Science* 236: 191–193, 1987.

Parkhouse, W. S. The effects of ATP, inorganic phosphate, protons, and lactate on isolated myofibrillar ATPase activity. *Can. J. Physiol. Pharmacol.* 70: 1175–1181, 1991.

Petrofsky, J. S. Frequency and amplitude analysis of EMG during exercise on the bicycle ergometer. *Europ. J. Appl. Physiol.* 41: 1–15, 1979.

Pirnay, F., M. Lamy, J. Dujardin, R. Deroanne, and J. M. Petit. Analysis of femoral venous blood during maximum muscular exercise. *J. Appl. Physiol.* 33: 289–292, 1972.

Pirnay, F., R. Marechal, R. Radermechker, and J. M. Petit. Muscle blood flow during submaximum and maximum exercise on a bicycle ergometer. *J. Appl. Physiol.* 32: 210–212, 1972.

Reybrouck, T., G. F. Heigenhauser, and J. A. Faulkner. Limitation to maximum oxygen uptake in arm, leg, and combined arm-leg ergometry. *J. Appl. Physiol.* 38: 774–779, 1975.

Richardson, R. S., D. C. Poole, D. R. Knight, S. S. Kurdak, M. C. Hogan, B. Grassi, E. C. Johnson, K. F. Kendrick, B. K. Erickson, and P. D. Wagner. High muscle blood flow in man: is maximal O_2 extraction compromised? *J. Appl. Physiol.* 75: 1911–1916, 1993.

Richardson, R. S., E. A. Noyszewski, J. S. Leigh, and P. D. Wagner. Lactate efflux from exercising human skeletal muscle: role of intracellular PO_2. *J. Appl. Physiol.* 85: 627–634, 1998.

Rodbard, S., and M. Farbstein. Improved exercise tolerance during venous congestion. *J. Appl. Physiol.* 33: 704–710, 1972.

Rodbard, S., and E. B. Pragay. Contraction frequency, blood supply, and muscle pain. *J. Appl. Physiol.* 24: 142–145, 1968.

Saltin, B., R. F. Grover, C. G. Blomquist, L. H. Hartley, and R. J. Johnson. Maximal oxygen uptake and cardiac output after 2 weeks at 4300 m. *J. Appl. Physiol.* 25: 406–409, 1968.

Sargeant, A. J. Human power output and muscle fatigue. *Int. J. Sports Med.* 15: 116–121, 1994.

Secher, N. H., N. Ruberg-Larsen, R. Binkhorst, and F. Bonde-Peterson. Maximal oxygen uptake during arm cranking and combined arm plus leg exercise. *J. Appl. Physiol.* 36: 515–518, 1974.

Setchenov, I. M. Zur frage nach der einwirkung sensitiver reize auf die muskelarbeit des menschen. In Selected Works. Moscow: 1935, pp. 246–260.

Shulman, R. G., T. R. Brown, K. Ugurbil, S. Ogawa, S. M. Cohen, and J. A. den Hollander. Cellular applications of ^{31}P and ^{13}C nuclear magnetic resonance. *Science* 205: 160–165, 1979.

Simonson, E. (Ed.). Physiology of Work Capacity and Fatigue. Springfield, Ill.: Charles C. Thomas, 1971.

Sinoway, L. I., M. B. Smith, B. Enders, U. Leuenberger, T. Dzwonczyk, K. Gray, S. Whisler, and R. L. Moore. Role of diprotonated phosphate in evoking muscle reflex responses in cats and humans. *Am. J. Physiol.* 267: H770–778, 1994.

Spriet, L. L., K. Söderlund, M. Bergström, and E. Hultman. Skeletal muscle glycogenolysis, glycolysis, and pH during electrical stimulation in men. *J. Appl. Physiol.* 62: 616–621, 1987.

Stephens, J. A., and A. Taylor. Fatigue of maintained voluntary muscle contraction in man. *J. Physiol.* (London) 220: 1–18, 1972.

Stienen, G. J., I. A. van Graas, and G. Elzinga. Uptake and caffeine-induced release of calcium in fast muscle fibers of Xenopus laevis: effects of MgATP and Pi. *Am. J. Physiol.* 265: C650-C657, 1993.

Stull, C. A., and J. T. Kearney. Recovery of muscular endurance following submaximal, isometric exercise. *Med. Sci. Sports* 10: 109–112, 1978.

Sundberg, C. J., O. Eiken, A. Nygren, and L. Kaijser. Effects of ischaemic training on local aerobic muscle performance in man. *Acta Physiol. Scand.* 148: 13–19, 1993.

Taylor, H. E., E. R. Buskirk, and A. Henschel. Maximal oxygen uptake as an objective measure of cardiorespiratory performance. *J. Appl. Physiol.* 8: 73–84, 1955.

Thompson, L. V., and R. H. Fitts. Muscle fatigue in the frog semitendinosus: role of the high-energy phosphates and P$_i$. *Am. J. Physiol.* 263 (*Cell Physiol.* 32): C803-C809, 1992.

Waller, A. The sense of effort: an objective study. *Brain* 14: 179–249, 1881.

Westerblad, H., and D. G. Allen. Changes in myoplasmic calcium concentration during fatigue in single mouse muscle fibers. *J. Gen. Physiol.* 98: 615–635, 1991.

Westerblad, H., and D. G. Allen. The influence of intracellular pH on contraction, relaxation and [Ca^{++}]$_i$ in intact fibers of mouse muscle. *J. Physiol.* 466: 611–628, 1993.

Westerblad, H., and D. G. Allen. The contribution of [Ca^{++}]$_i$ to the slowing of relaxation in fatigued fibers from mouse skeletal muscle. *J. Physiol.* 468: 729–740, 1993.

Westerblad, H., S. Duty, and D. G. Allen. Intracellular calcium concentration during low-frequency fatigue in isolated single fibers of mouse skeletal muscle. *J. Appl. Physiol.* 75: 382–388, 1993.

Westerblad, H., J. A. Lee, J. Lännergren, and D. G. Allen. Cellular mechanisms of fatigue in skeletal muscle. *Am. J. Physiol.* 261 (*Cell Physiol.* 30): C195–C209, 1991.

Wickler, S. J., and T. T. Gleeson. Lactate and glucose metabolism in mouse (Mus musculus) and reptile (Anolis carolinensis) skeletal muscle. *Am. J. Physiol.* 264: R487–R491, 1993.

Wilkie, D. R. Muscular fatigue: effects of hydrogen ions and inorganic phosphate. *Federation Proc.* 45: 2921–2923, 1986.

LIST OF SYMBOLS AND ABBREVIATIONS

Note: a dash over any symbol indicates a mean value (e.g., $\bar{x}$); a dot above any symbol indicates time derivate (e.g., $\dot{V}$).

■ Respiratory and Hemodynamic Notations

V	gas volume
$\dot{V}$	gas volume · unit time^{-1} (usually liters · min^{-1}); pulmonary minute volume
R	ventilatory respiratory exchange ratio (volume CO_2/volume O_2 = $\dot{V}_{CO_2}/\dot{V}_{O_2}$)
RQ	cellular respiratory quotient ($\dot{V}_{CO_2}/\dot{V}_{O_2}$)
I	inspired gas
E	expired gas
A	alveolar gas
F	fractional concentration in dry gas phase
f	respiratory frequency (breath · unit time^{-1})

TLC	total lung capacity
IRV	inspiratory reserve volume
ERV	expiratory reserve volume
IC	inspiratory capacity
VC	vital capacity
FRC	functional residual capacity
RV	residual volume
T	tidal gas
D	dead space
FEV	forced expiratory volume
FEV$_{1.0}$	forced expiratory volume in 1 s
MET	multiple of the resting metabolic rate, approximately equal to 3.5 ml O_2 · kg body wt^{-1} · min^{-1}
MVV	maximal voluntary ventilation (also termed maximum breathing capacity, MBC)
D_L	diffusing capacity of the lungs (ml · min^{-1} · mmHg^{-1})
P	gas pressure
T	transport of O_2 in blood (ml · min^{-1})
B or Bar	barometric
STPD	standard temperature and pressure, dry: 0°C, 760 mmHg, dry

Based on *Federation Proceedings* 9:602–605, 1960.

823

BTPS — body temperature, ambient pressure, saturated with water vapor

ATPD — ambient temperature and pressure, dry

ATPS — ambient temperature and pressure, saturated with water vapor

Q — blood flow or volume or cardiac output (usually l)

$\dot{Q}$ — blood flow · unit time^{-1} (without other notation, cardiac output; usually liters · min^{-1})

SV or V_s — stroke volume (usually ml · beat^{-1})

HR or f_H — heart rate (usually beats · min^{-1})

BV — blood volume

Hb — hemoglobin concentration (g · 100 ml^{-1})

HbO$_2$ — oxyhemoglobin

Hct — hematocrit (usually %)

BP — blood pressure (usually in mmHg)

TPR — total peripheral resistance (usually in mmHg · ml^{-1} · min^{-1}, also known as PRU, peripheral resistance unit)

C — concentration in blood phase

S — percent saturation of Hb

a — arterial

c — capillary

v — venous

■ Temperature Notations

T — temperature

r — rectal

s — skin

e — esophageal

m — muscle

ty — tympanic

M — metabolic energy yield

C_d — conductive heat exchange

C_v — convective heat exchange

R — radiation heat exchange

E — evaporative heat loss

S — storage of body heat

°C — temperature in degrees Celsius

°F — temperature in degrees Fahrenheit

°K — temperature in Kelvin

■ Body Dimensions

W — weight

H — height

L — length

LBM — lean body mass

BSA — body surface area

BMI — body mass index (W · H^{-2})

■ Statistical Notations

$M, \bar{x}$ — arithmetic mean

SD or S.D. — standard deviation

SE or S.E. — standard error of the mean

n — number of observations

r — correlation coefficient

range — smallest and largest observed value

Σ — summation

P — probability

* — denotes a (probably) significant difference; (e.g., $P < 0.05$)

■ Examples

V_A — volume of alveolar gas

$\dot{V}_E$ — expiratory gas volume · minute^{-1}; pulmonary minute flow or ventilation

$\dot{V}_{O_2}$ — volume of oxygen · minute^{-1} (oxygen uptake · min^{-1})

$\dot{V}_{O_2max}$ — maximal volume of O$_2$ consumed (l · min^{-1})

V_T	tidal volume		m	meter
P_A	alveolar gas pressure		meq	milliequivalent
P_B	barometric pressure		mg %	milligrams per 100 ml (dl) of blood
$F_{I_{O_2}}$	fractional concentration of O_2 in inspired gas		min	minute

V_T — tidal volume
P_A — alveolar gas pressure
P_B — barometric pressure
$F_{I_{O_2}}$ — fractional concentration of O_2 in inspired gas
$P_{A_{O_2}}$ — alveolar oxygen pressure
$\dot{T}_{O_2}$ — oxygen transport capacity
pH_a — arterial pH
$C_{a_{O_2}}$ — oxygen content in arterial blood
$C_{a_{O_2}} - C_{\bar{v}_{O_2}}$ — difference in oxygen content between arterial and mixed venous blood, often written $(a - \bar{v})_{O_2}$
T_r — rectal temperature
$(a - \bar{v})_{O_2}$ — arteriovenous O_2 difference

m — meter
meq — milliequivalent
mg % — milligrams per 100 ml (dl) of blood
min — minute
ml — milliliter
mm — millimeter
mM — millimoles per liter ($mmol \cdot liter^{-1}$)
mmol — millimole
nM — nanomoles per liter ($nmol \cdot liter^{-1}$)
Na^+ — sodium ion
O_2 — oxygen
P — power
PC — phosphocreatine
P_i — inorganic phosphate
PP_i — pyrophosphate
s or sec — second

■ Chemical and Physical Notations

ADP — adenosine diphosphate
AMP — adenosine monophosphate
ATP — adenosine triphosphate
Ca^{2+} — calcium ion
Cl^- — chloride ion
CO_2 — carbon dioxide
CP — creatine phosphate
e^- — electron
FFA — free fatty acids
H — hydrogen atom
H^+ — hydrogen ion
$H\cdot$ — hydride ion
HCO_3^- — bicarbonate ion
Hg — mercury
H_2CO_3 — carbonic acid
H_2O — water
K^+ — potassium ion
kcal — kilocalorie
kcal/min — kilocalories per minute ($kcal \cdot min^{-1}$)
kg — kilogram

■ Notations of Blood Metabolite Concentration and Flux

[x] — the concentration of metabolite x, in terms of mass per unit volume, usually in mM, or mg %

R_a — rate of appearance of a metabolite in the blood, in terms of mass per unit time, usually in $mg \cdot min^{-1}$, or normalized to body mass, $mg \cdot kg^{-1} \cdot min^{-1}$

R_d — rate of disappearance (or disposal) of a metabolite from the blood, in terms of mass per unit time, usually in $mg \cdot min^{-1}$, or normalized to body mass, $mg \cdot kg^{-1} \cdot min^{-1}$

R_{ox} — rate of oxidation R_a of a metabolite in the body, in terms of mass per unit time, usually in $mg \cdot min^{-1}$, or normalized to body mass, $mg \cdot kg^{-1} \cdot min^{-1}$. For most metabolites oxidation is a major pathway of disposal.

R_t rate of turnover or renewal. R_t is defined for the metabolic steady state when [metabolite] is constant, and $R_a = R_d$.

NOD   nonoxidative disposal, the disposal of a metabolite not accounted for by oxidation ($= R_d - R_{ox}$). For glucose, NOD may estimate glycogen synthesis.

MCR   metabolic clearance rate gives the effective volume from which a metabolite is cleared (removed) per unit time. The MCR is also the R_d normalized to concentration ($= R_d/[x]$). The usual units are liters $\cdot$ min^{-1}.

UNITS AND MEASURES

Note: The basic unit of measurement adopted by the Système International d'Unités will be denoted as the basic SI unit.

▣ Density

Basic SI unit: kilogram per cubic meter ($kg \cdot m^{-3}$)
   1 microgram per cubic centimeter ($\mu g \cdot cm^{-3}$)
      $= 10^{-3} kg \cdot m^{-3}$
   1 milligram per cubic centimeter ($mg \cdot cm^{-3}$) =
      $1 kg \cdot m^{-3}$
   1 gram per cubic centimeter ($g \cdot cm^{-3}$) = $10^3 kg$
      $\cdot m^{-3}$

Common Anglo-Saxon unit of density:
   1 pound per cubic foot = $1.601 \times 10 \ kg \cdot m^{-3}$

▣ Force

Basic SI unit: newton (N)
   1 pond (p) = $9.80665 \times 10^{-3} N$
   1 kilopond (kp) = $9.80665 N$
   1 dyne (dyn) = $10^{-5} N$

Common Anglo-Saxon unit of force:
   1 foot-pound ($ft \ lb \cdot s^{-2}$) = 0.13825495437 N

▣ Frequency

Basic SI unit: hertz (Hz)
   1 kilohertz (kHz) = 1000 Hz

Common Anglo-Saxon unit of frequency:
   1 cycle per second ($c \cdot s^{-1}$) = 1 Hz

▣ Length

Basic SI unit: meter (m):
   1 angstrom (Å) = $10^{-10} m$
   1 nanometer (nm) = $10^{-9} m$
   1 micrometer (μm) = $10^{-6} m$
   1 millimeter (mm) = $10^{-3} m$
   1 centimeter (cm) = $10^{-2} m$
   1 decimeter (dm) = $10^{-1} m$
   1 decameter (dam) = 10 m
   1 hectometer (hm) = $10^2 m$
   1 kilometer (km) = $10^3 m$

Common Anglo-Saxon units of length:
   1 inch (in.) = $2.540 \times 10^{-2} m$
   1 foot (ft) = $3.048 \times 10^{-1} m$
   1 yard (yd) = $9.144 \times 10^{-1} m$
   1 fathom (fath) = 1.828 m
   1 furlong (fur) = $2.011 \times 10^2 m$
   1 statute mile (mi) = $1.609 \times 10^3 m$

■ Linear Velocity

Basic SI unit: meter per second (m · s^{-1})
  1 centimeter per sec (cm · s^{-1}) = 10^{-2} m · s^{-2}
  1 kilometer per hour (km · h^{-1}) = 2.8 × 10^{-1}
    m · s^{-1}

Common Anglo-Saxon units of linear velocity:
  1 foot per second (ft · s^{-1}) = 3.048 × 10^{-1} m ·
    s^{-1}
  1 mile per hour (mile · h^{-1}) = 4.470 × 10^{-1} m ·
    s^{-1}

■ Mass

Basic SI unit: kilogram (kg)
  1 picogram (pg) = 10^{-15} kg
  1 nanogram (ng) = 10^{-12} kg
  1 microgram (μg) = 10^{-9} kg
  1 milligram (mg) = 10^{-6} kg
  1 gram (g) = 10^{-3} kg
  1 metric ton (t) = 10^{3} kg

Common Anglo-Saxon units of mass:
  1 ounce (oz) = 2.834 × 10^{-2} kg
  1 pound (lb) = 4.535 × 10^{-1} kg
  1 short ton (sh tn) = 9.071 × 10^{2} kg

■ Power

Basic SI unit: watt (W)
  1 joule per second (J · s^{-1}) = 1 W
  1 erg per second (erg · s^{-1}) = 10^{7} W
  1 kilopond meter per minute (kpm · min^{-1}) =
    0.1635 W

Common Anglo-Saxon units of power:
  1 British thermal unit per hour (Btu · h^{-1}) =
    2.931 × 10^{-1} W
  1 horsepower (hp) = 7.457 × 10^{2} W

■ Pressure

Basic SI unit: newton per square meter (N · m^{-2})

Common laboratory unit: millimeters of mercury
  (mmHg)
  1 mmHg = 133.322 N · m^{-2}
  1 atmosphere (atm) = 101325 N · m^{-2}
  1 torr (torr) = 133.322368 N · m^{-2}

■ Temperature

Basic SI unit: Kelvin (K)

Common laboratory unit: degrees Celsius (°C).
  Note: The Celsius scale is subdivided into the
  same intervals as the Kelvin scale but has its
  zero point displaced by 273.15 K.
  1 degree Celsius (°C) = 1 K
  1 degree Celsius (°C) = 9/5 °Fahrenheit + 32

Common Anglo-Saxon units of temperature:
  1 degree Fahrenheit (°F) = 5/9 K
  1 degree Rankine (°R) = 5/9 K

■ Time

Basic SI unit: second (s)
  1 minute (min) = 6 × 10 s
  1 hour (hr) = 3.6 × 10^{3} s
  1 day (d) = 8.64 × 10^{4} s
  1 year (y, 365 d) = 3.1536 × 10^{7} s

■ Volume

Basic SI unit: cubic meter (m^{3})
Common laboratory unit: liter (l)
  1 cubic nanometer (nm^{3}) = 10^{-27} m^{3}
  1 cubic micrometer (μm^{3}) = 10^{-18} m^{3}
  1 cubic millimeter (mm^{3}) = 10^{-9} m^{3} =
    1 microliter (μl)
  1 cubic centimeter (cc^{3}) = 10^{-6} m^{3} =
    1 milliter (ml)

1 cubic decimeter (dm³) = 10^{-3} m³ =
1 liter (l)

1 cubic kilometer (km³) = 10^9 m³ =
1 hectoliter (hl)

Common Anglo-Saxon units of volume:
1 cubic inch (in.³) = 1.638×10^{-5} m³
1 cubic foot (ft³) = 2.831×10^{-2} m³
1 cubic yard (yd³) = 7.645×10^{-1} m³
1 fluid ounce (fl oz) = 2.957×10^{-2} l
1 liquid quart (liq qt) = 9.463×10^{-1} l
1 gallon (gal) = 3.785 l

■ Work and Energy

Basic SI unit: joule (J)
1 kilocalorie (kcal) = 4186 J
1 kilopond meter (kp · m) = 9.807 J
1 newton meter (Nm) = 1 J

Common Anglo-Saxon units of power:
1 British thermal unit (Btu) = 1.055×10^3 J
1 horsepower-hour (hph) = 2.685×10^6 J
1 foot pound-force (ft · lbf) = 1.356 J